ENVIRONMENTAL ARSENIC IN A CHANGING WORLD

Arsenic in the Environment – Proceedings

Series Editors

Jochen Bundschuh
UNSECO Chair on Groundwater Arsenic within 2030 Agenda for Sustainable Development & Faculty of Health, Engineering and Sciences, The University of Southern Queensland, Toowoomba, Queensland, Australia

Prosun Bhattacharya
KTH-International Groundwater Arsenic Research Group, Department of Sustainable Development, Environmental Science and Engineering, KTH Royal Institute of Technology, Stockholm, Sweden
School of Civil Engineering and Surveying and International Centre for Applied Climate Science, The University of Southern Queensland, Toowoomba, Queensland, Australia

ISSN: 2154-6568

7TH INTERNATIONAL CONGRESS AND EXHIBITION
ARSENIC IN THE ENVIRONMENT, BEIJING, P.R. CHINA, JULY 1–6, 2018

Environmental Arsenic in a Changing World

As2018

Editors

Yong-Guan Zhu
Key Laboratory of Urban Environment and Health, Institute of Urban Environment, Chinese Academy of Sciences, Xiamen, P.R. China
Research Centre for Eco-environmental Sciences, Chinese Academy of Sciences, Beijing, P.R. China

Huaming Guo
State Key Laboratory of Biogeology and Environmental Geology, China University of Geosciences (Beijing), Beijing, P.R. China
MOE Key Laboratory of Groundwater Circulation and Environment Evolution, School of Water Resources and Environment, China University of Geosciences (Beijing), Beijing, P.R. China

Prosun Bhattacharya
KTH-International Groundwater Arsenic Research Group, Department of Sustainable Development, Environmental Sciences and Engineering, KTH Royal Institute of Technology, Stockholm, Sweden
School of Civil Engineering and Surveying and International Centre for Applied Climate Science, The University of Southern Queensland, Toowoomba, Queensland, Australia

Jochen Bundschuh
UNSECO Chair on Groundwater Arsenic within 2030 Agenda for Sustainable Development & Faculty of Health, Engineering and Sciences, The University of Southern Queensland, Toowoomba, Australia

Arslan Ahmad
KWR Water Cycle Research Institute, Nieuwegein, The Netherlands
KTH-International Groundwater Arsenic Research Group, Department of Sustainable Development, Environmental Science and Engineering, KTH Royal Institute of Technology, Stockholm, Sweden
Department of Environmental Technology, Wageningen University and Research (WUR), Wageningen, The Netherlands

Ravi Naidu
Global Centre for Environmental Remediation (GCER), Faculty of Science & Information Technology, The University of Newcastle, Callaghan, NSW, Australia
Cooperative Research Centre for Contamination Assessment and Remediation of the Environment (CRC CARE), University of Newcastle, Newcastle, New South Wales, Australia

CRC Press
Taylor & Francis Group
Boca Raton London New York

CRC Press is an imprint of the
Taylor & Francis Group, an **informa** business
A BALKEMA BOOK

Cover photo

The cover photo shows the paddy field in mountainous Fujian province, Southeast China. Fujian (means "Happy Establishment")[1] is one of the country's smaller provinces, on the southeastern coast of China. The province is also known historically as Min, for the "seven Min tribes" that inhabited the area during the Zhou dynasty (1046–256 BCE). It was, however, during the Song dynasty (960–1279 CE) that the name Fujian was given to a superprefecture created in the area and the basic geographical boundaries of the province were established. Covering an area of approximately 123,100 square km, the Fujian province is bordered by the provinces of Zhejiang to the north, Jiangxi to the west, and Guangdong to the southwest; the East China Sea lies to the northeast, the Taiwan Strait (between the mainland and Taiwan) to the east, and the South China Sea to the southeast. The Fujian province is traversed by several ranges of moderate elevation that constitute a part of a system of ancient blocks of mountains trending from southwest to northeast, parallel to the coast although some narrow coastal plains are prevailing towards the soth-eastern part of the province. Rivers are of great importance in Fujian, and have been the only means of transport for centuries. Most of the rivers flow into estuaries that form natural harbours, and provide water supplies for domestic consumption and irrigation of the myriad rice fields in the alluvial plains. The area is also characterized by several mineral deposites which have been mined over centuries for mining of lead, zinc and copper deposits.

One of the most picturesque region in Asia, Fujian province is endowed with wooded hills and winding streams, orchards, tea gardens, and terraced rice fields on the gentler slopes. Its major crops are sugarcane, peanuts (groundnuts), citrus fruit, rice, and tea. Two crops of rice are harvested each year, the first in June, the second in September. After centuries of rice cultivation, soils in the valley plains have been greatly modified. Well-developed gray-brown forest soils are widely distributed in the forest areas of the interior mountains, whereas mature red soils are common in the low hills and on high terraces. Rice is the staple food in China, and it has been estimated that rice ingestion contributes to about 60% of arsenic exposure from food in China[2].

CRC Press
Taylor & Francis Group
6000 Broken Sound Parkway NW, Suite 300
Boca Raton, FL 33487-2742

First issued in paperback 2020

Typeset by MPS Limited, Chennai, India

ISBN-13: 978-1-138-48609-6 (hbk)
ISBN-13: 978-0-367-77921-4 (pbk)

[1] https://www.britannica.com/place/Fujian

[2] Li, G., Sun, G.X., Williams, P.N., Nunes, L. & Zhu, Y.G., 2011. Inorganic arsenic in Chinese food and its cancer risk. Environ. Int. 37(7): 1219–1225

Environmental Arsenic in a Changing World –
Zhu, Guo, Bhattacharya, Ahmad, Bundschuh & Naidu (Eds)
ISBN 978-1-138-48609-6

Table of contents

Section 1: Arsenic behaviour in changing environmental media

1.1 Sources, transport and fate of arsenic in changing groundwater systems

Section 2: Arsenic in a changing agricultural ecosystem and food chain effects

2.1 Processes and pathways of arsenic in agricultural ecosystems

2.2 Arsenic dynamics in rhizosphere

2.3 Microbial ecology of arsenic biotransformation in soils

Section 3: Health impacts of environmental arsenic

3.1 Exposure and epidemiology of arsenic impacts on human health

3.2 Genetic predisposition of chronic arsenic poisoning

Section 4: Technologies for arsenic immobilization and clean water blueprints

4.1 Adsorption and co-precipitation for arsenic removal

Section 5: Sustainable mitigation and management

5.1 Societal involvement for mitigations of long-term exposure

5.2 Policy instruments to regulate arsenic exposure

5.3 Risk assessments and remediation of contaminated land and water environments – Case studies

5.5 Drinking water regulations of water safety plan

5.6 Arsenic in drinking water and implementation plan for safe drinking water supply from sustainable development perspectives

ISBN 978-1-138-48609-6

About the book series

Although arsenic has been known as a 'silent toxin' since ancient times, and the contamination of drinking water resources by geogenic arsenic was described in different locations around the world long ago—e.g. in Argentina in 1914—it was only two decades ago that it received overwhelming worldwide public attention. As a consequence of the biggest arsenic calamity in the world, which was detected more than twenty years back in West Bengal, India and other parts of Southeast Asia, there has been an exponential rise in scientific interest that has triggered high quality research. Since then, arsenic contamination (predominantly of geogenic origin) of drinking water resources, soils, plants and air, the propagation of arsenic in the food chain, the chronic affects of arsenic ingestion by humans, and their toxicological and related public health consequences, have been described in many parts of the world, and every year, even more new countries or regions are discovered to have elevated levels of arsenic in environmental matrices.

Arsenic is found as a drinking water contaminant, in many regions all around the world, in both developing as well as industrialized countries. However, addressing the problem requires different approaches which take into account, the differential economic and social conditions in both country groups. It has been estimated that 200 million people worldwide are at risk from drinking water containing high concentrations of As, a number which is expected to further increase due to the recent lowering of the limits of arsenic concentration in drinking water to 10 μg/L, which has already been adopted by many countries, and some authorities are even considering decreasing this value further.

The book series "Arsenic in the Environment – Proceedings" is an inter- and multidisciplinary source of information, making an effort to link the occurrence of geogenic arsenic in different environments and the potential contamination of ground- and surface water, soil and air and their effect on the human society. The series fulfills the growing interest in the worldwide arsenic issue, which is being accompanied by stronger regulations on the permissible Maximum Contaminant Levels (MCL) of arsenic in drinking water and food, which are being adopted not only by the industrialized countries, but increasingly by developing countries.

Consequently, we see the book series *Arsenic in the Environment-Proceedings* with the outcomes of the International Congress Series – Arsenic in the Environment, which we organize biannually in different parts of the world, as a regular update on the latest developments of arsenic research. It is further a platform to present the results from other from international or regional congresses or other scientific events. This Proceedings series acts as an ideal complement to the books of the series *Arsenic in the Environment*, which includes authored or edited books from world-leading scientists on their specific field of arsenic research, giving a comprehensive information base. Supported by a strong multi-disciplinary editorial board, book proposals and manuscripts are peer reviewed and evaluated. Both of the two series will be open for any person, scientific association, society or scientific network, for the submission of new book projects.

We have an ambition to establish an international, multi- and interdisciplinary source of knowledge and a platform for arsenic research oriented to the direct solution of problems with considerable social impact and relevance rather than simply focusing on cutting edge and breakthrough research in physical, chemical, toxicological and medical sciences. It shall form a consolidated source of information on the worldwide occurrences of arsenic, which otherwise is dispersed and often hard to access. It will also have role in increasing the awareness and knowledge of the arsenic problem among administrators, policy makers and company executives and improving international and bilateral cooperation on arsenic contamination and its effects.

Both of the book series cover all fields of research concerning arsenic in the environment and aims to present an integrated approach from its occurrence in rocks and mobilization into the ground- and surface water, soil and air, its transport therein, and the pathways of arsenic introduction into the food chain including uptake by humans. Human arsenic exposure, arsenic bioavailability, metabolism and toxicology are treated together with related public health effects and risk assessments in order to better manage the contaminated land and aquatic environments and to reduce human arsenic exposure. Arsenic removal technologies and other methodologies to mitigate the arsenic problem are addressed not only from the technological perspective, but also from an economic and social point of view. Only such inter- and multidisciplinary approaches will allow a case-specific selection of optimal mitigation measures for each specific arsenic problem and provide the local population with arsenic-safe drinking water, food, and air

Jochen Bundschuh
Prosun Bhattacharya
(Series Editors)

*Environmental Arsenic in a Changing World –
Zhu, Guo, Bhattacharya, Ahmad, Bundschuh & Naidu (Eds)
ISBN 978-1-138-48609-6*

Dedication

Dipankar Chakraborti, Ph.D.
'Arsenic Legend of India'
Analytical Chemist who made legendary contributions to arsenic research in India and Bangladesh
Former Director (Research), School of Environmental Studies, Jadavpur University, Kolkata, India
* 29 October 1943 †28 February 2018

We dedicate the 7th International Congress on Arsenic in the Environment (As 2018) and the Volume of Proceedings of the International Congress of Arsenic in the Environment – *Environmental Arsenic in a Changing World (As 2018)* to the memory of Professor Dr. Dipankar Chakraborti (popularly known as Dip), who passed away on the 28th of February 2018 at the age of 74.

Dr. Chakraborti had established himself as a legendary scientist in the field of arsenic research across the globe for a period of more than three decades through his contributions towards raising a global awareness about the growing arsenic crisis in the Bengal delta. Born in Ujirpur in the district of Barisal in Bangladesh, Dip was raised in Madaripur in Faridpur district where he had spent his childhood before moving to West Bengal, India around 1949.

Following the Bachelor and Master of Science degrees in Chemistry, he received his Ph.D. in Analytical Chemistry from Jadavpur University, Kolkata, India in 1973. His carreer in academics commenced as early as in 1967, when he joined as an Assistant Professor (lecturer) at Jadavpur University. Later in 1977, he moved abroad, to join University of Prague of Czech Republic as UNESCO Fellow. In 1978, he moved to Universitaire Instelling Antwerpen, Wilrijk in Belgium, where he worked with Prof. Freddy Adams. He joined Texas A & M University at College Station, Texas, USA in 1981 and worked with Professor K. J. Irgolic until 1983. He was once again invited as a Visiting Scientist to work at Universitaire Instellig Antwerpen, Wilrijk, Belgium during the period between 1984 and 1986. After an illustrative carreer as a scientist abroad, he returned back to to India to join Jadavpur University in Kolkata in 1987. He became the Director of School of Environmental Studies (SOES) at the Jadavpur University and continued until his formal retirement.

It did not take a long time for Dr. Chakraborti to form the internationally recognized Arsenic Research Team at SOES and he started working on arsenic toxicity since 1988, in collaboration with Dr. K.C. Saha (School of Tropical Medicine, Kolkata), Dr. D.N. Guha Mazumder (Institute of Post Graduate Medical Education and Research, Kolkata) and Dr. Allan H. Smith (School of Public Health, University of California-Berkeley, USA) over an extended period of time to highlight the epidemiological impact of arsenic-laden groundwater in West Bengal, India. Since then, along with his team he had not only been engaged in research on groundwater arsenic in the Ganga-Meghna-Brahmaputra (GMB) Plain, but also worked on fluoride one of the most widespread geogenic contaminant in groundwaters of India. He and his co-workers have played a pivotal role in documenting the magnitude of the arsenic calamity in the Bengal delta, both in India and Bangladesh.

Chakraborti's work on arsenic contamination in the environment brought him international recognition. He was the key proponent of the hypothesis on the arsenic mobilization mechanism in groundwater of Bengal delta, known as the *pyrite oxidation theory*. He was the person who raised an alarm on a possible widespread groundwater arsenic contamination in Bangladesh and also discovered the contamination in Bihar state of India. Dissemination of science among a broader public was certainly one of Chakraborti's aims and his research outputs helped Governments of both Bangladesh and West Bengal to take the necessary steps to mitigate the well neglected arsenic calamity. To the end of his life, though he continued to fight for clean water he became increasingly tolerant to failures of drinking water supply mitigation schemes across both countries.

He was member of WHO working group for "Environmental Health Criteria 224 for Arsenic and Arsenic compounds (2nd edition)", and IARC Monographs on the "Evaluation of Carcinogenic Risks to Humans – Some Drinking Water Disinfectants and Contaminants, including Arsenic, Volume 58". His ground-breaking research on arsenic field testing kits turned attention of UNICEF and led to the discontinuation of the use of improper test kits for arsenic measurement in Bangladesh and India.He authored more than 200 publications in highly acclaimed international peer-reviewed journals of high impact which include Environmental Health Perspectives (IF 9.78), International Journal of Epidemiology (IF 7.73), Science of the Total Environment (IF 4.9) and co-authored 20 chapters in books/monographs. The scholastic achievements of Dipankar's publications are demonstrated through more than 17000 citations, with 131 publications cited more than 10 times (i10 score) and a h-index of 53 till date. His citations peaked both in 2015 and 2016 with ca 1100 citations, which definitely is one of the highest among the community of scientists in India. He had organized five international conferences on the groundwater arsenic problem including International Conference on Arsenic Pollution of Groundwater in India (Kolkata, India, 1995) and the International Conference on Arsenic Pollution of Groundwater in Bangladesh – Causes, Effects and Remedies (Dhaka, Bangladesh, 1998), where he highlighted the interdisciplinary aspects of arsenic pollution in groundwater in Bangladesh and West Bengal, India and kindled global attention of multidisciplinary group of scientists on the environmental health calamity caused by arsenic in drinking water from groundwater sources affecting health of millions of exposed population.

He owed a great deal to his birth place, as later in his life he did spend more than 400 days in the remote villages of Bangladesh fighting for the victims of arsenic poisoning. He was married to Dr. Reena Chakraborty and one daughter, who have remained extremely supportive to his scientific acheivements. He has left behind his legacy through a number of his students who also made signifigant progress in the field of arsenic research in India and abroad. His pronouncements carried great authority, and he might ask his students to follow his life style including routine physical activities, healthy diet and yoga, himself being an addict to yoga. He was an extraordinarily determined person. He would never ask for funding to support his research and SOES was a self-funded and self-sustained unit and as per his principle, he did never accept any foreign grant.

Those who knew him as Dip, as he was so fondly called, would clearly appreciate the real human being, with an enormous zest for life, and tremendous determination, yet with normal human weaknesses, as well as his more obvious strengths. As a true environmentalist he cared deeply for Mother Nature and was the pioneer of arsenic research in India and Bangladesh. He was extremely regarded for his contributions to the understanding of the contamination of drinking water and the subsequent consequences to human suffering.

We deeply mourn the death of Dr. Dipankar Chakraborti. We lost a beloved colleague, friend, the kindest and most generous soul and a great personality, who devoted his entire life to the victims of arsenic poisoning. The arsenic community will always remember his contributions in the field of arsenic research and related problems and will miss his supportive, hard working and optimistic company.

M.M. Rahman
Debapriya Mandal
Prosun Bhattacharya

"Jodi tor daak shune keu na ashe tobe ekla cholore"

ISBN 978-1-138-48609-6

Organizers

ORGANIZERS OF BIANNUAL CONGRESS AND EXHIBITION SERIES:
ARSENIC IN THE ENVIRONMENT

Jochen Bundschuh
University of Southern Queensland (USQ), Toowomba, QLD, Australia
International Society of Groundwater for Sustainable
Development (ISGSD) Stockholm, Sweden

Prosun Bhattacharya
KTH-International Groundwater Arsenic Research Group
Department of Sustainable Development, Environmental Sciences
and Engineering, KTH Royal Institute of Technology, Stockholm, Sweden
University of Southern Queensland (USQ), Toowomba, QLD, Australia
International Society of Groundwater
for Sustainable Development (ISGSD) Stockholm, Sweden

Arslan Ahmad
KWR Water Cycle Research Institute, Nieuwegein, The Netherlands

Ravi Naidu
Global Centre for Environmental Remediation, The University of Newcastle,
Callaghan, NSW, Australia
CRC CARE, University of Newcastle, Callaghan, NSWAustralia

Local Organizing Committee

Yong-Guan Zhu
Key Laboratory of Urban Environment and Health, Institute of
Urban Environment, Chinese Academy of Sciences, Xiamen, P.R. China
Research Centre for Eco-environmental Sciences,
Chinese Academy of Sciences, Beijing, P.R. China

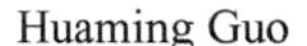

Huaming Guo
State Key Laboratory of Biogeology and Environmental Geology,
China University of Geosciences (Beijing), Beijing, P.R. China
MOE Key Laboratory of Groundwater Circulation and Environment
Evolution, School of Water Resources and Environment,
China University of Geosciences (Beijing), Beijing, P.R. China

ISBN 978-1-138-48609-6

Sponsors and Contributors

中國科協　中國國際科技交流中心
China Centre for International Science
and Technology Exchange(CISTE)
China Association for Science and Technology(CAST)

ISBN 978-1-138-48609-6

Scientific committee

B. Dousova: *ICT, Prague, Czech Republic*
Ö. Ekengren: *IVL, Swedish Environmental Reseasrch Institute, Stockholm, Sweden*
M. Ersoz: *Department of Chemistry, Selcuk University, Konya, Turkey*
M.L. Esparza: *CEPIS, Lima, Peru*
S. Farías: *National Atomic Energy Commission, Buenos Aires, Argentina*
J. Feldman: *University of Aberdeen, Aberdeen, Scotland, UK*
R. Fernández: *National University of Rosario, Rosario, Argentina*
A. Fernández-Cirelli: *University of Buenos Aires, Buenos Aires, Argentina*
A. Figoli: *Institute on Membrane Technology, ITM-CNR c/o University of Calabria, Rende (CS), Italy*
B. Figueiredo: *UNICAMP, Campinas, SP, Brazil*
R.B. Finkelman: *US Geological Survey, Menlo Park, CA, USA*
A. Fiúza: *University of Porto, Porto, Portugal*
A.E. Fryar: *University of Kentucky, Lexington, KY, USA*
S.E. Garrido Hoyos: *Mexican Institute of Water Technology, Jiutepec, Mor., Mexico*
M. Gasparon: *The University of Queensland, Brisbane, Australia*
A.K. Ghosh: *Mahavir Cancer Institute and Research Centre, Patna, India; Bihar State Pollution Control Board, Patna, India*
A.K. Giri: *CSIR-Indian Institute of Chemical Biology, Kolkata, India*
W. Goessler: *University of Graz, Austria*
D.N. Guha Mazumder: *DNGM Research Foundation, Kolkata, India*
L.R. Guimaraes Guilherme: *Federal University of Lavras, Lavras, M.G., Brazil*
H.M. Guo: *China University of Geosciences (Beijing), P.R. China*
X. Guo: *Peking University, Peking, PR China*
J.P. Gustafsson: *Swedish University of Agricultural Sciences, Uppsala, Sweden*
S. Hahn-Tomer: *UFZ, Leipzig, Germany*
B. Hendry: *Cape Peninsula University of Technology, Cape Town, South Africa*
M. Hernández: *National University of La Plata, La Plata, Argentina*
J. Hoinkis: *Karlsruhe University of Applied Sciences, Karlsruhe, Germany*
C. Hopenhayn: *University of Kentucky, Lexington, KY, USA.*
M.F. Hughes: *Environmental Protection Agency, Research Triangle Park, NC, USA*
A.M. Ingallinella: *Centro de Ingeniería Sanitaria (CIS), Facultad de Ciencias Exactas, Ingeniería y Agrimensura, Universidad Nacional de Rosario, Rosario, Prov. de Santa Fe, Argentina*
G. Jacks: *Department of Sustainable Development, Environmental Sciences and Engineering, KTH Royal Institute of Technology, Stockholm, Sweden*
J. Jarsjö: *Stockholm University, Stockholm, Sweden*
J.-S. Jean: *National Cheng Kung University, Tainan, R. O. China*
B. Johnson: *Dundee Precious Metals, Toronto, ON, Canada*
R. Johnston: *World Health Organization, Geneva, Switzerland*
N. Kabay: *Chemical Engineering Department, Engineering Faculty, Ege University, Izmir, Turkey*
I.B. Karadjova: *Faculty of Chemistry, University of Sofia, Sofia, Bulgaria*
A. Karczewska: *Institute of Soil Sciences and Environmental Protection, Wroclaw University of Environmental and Life Sciences, Poland*
G. Kassenga: *Ardhi University, Dar es Salaam, Tanzania*
D.B. Kent: *US Geological Survey, Menlo Park, CA, USA*
N.I. Khan: *The Australian National University, Canberra, Australia*
K.-W. Kim: *Department of Environmental Science and Engineering, Gwangju Institute of Science and Technology, Gwangju, South Korea*
W. Klimecki: *Department of Pharmacology and Toxicology, University of Arizona, Tucson, Arizona, USA*
J. Kumpiene: *Department of Civil, Environmental and Natural Resources Engineering, Luleå University of Technology, Luleå, Sweden*
M.I. Litter: *Comisión Nacional de Energía Atómica, and Universidad de Gral. San Martín, San Martín, Argentina*
D.L. López: *Ohio University, Athens, Ohio, USA*
L.Q. Ma: *Research Center for Soil Contamination and Environment Remediation, Southwest Forestry University, Kunming, P.R. China; Soil and Water Sciences Department, University of Florida, Gainesville, FL, USA*
M. Mallavarapu: *Faculty of Science and Information Technology, The University of Newcastle Callaghan, NSW, Australia*

N. Mañay: *De la República University, Montevideo, Uruguay*
R.R. Mato: *Ardhi University, Dar es Salaam, Tanzania*
J. Matschullat: *Interdisciplinary Environmental Research Centre (IÖZ), TU Bergakademie Freiberg, Freiberg, Germany*
F. Mtalo: *Department of Water Resources Engineering, University of Dar E Salaam, Dar es Salaam, Tanzania*
M. Mörth: *Stockholm University, Sweden*
A.A. Meharg: *Institute of Biological and Environmental Sciences, University of Aberdeen, Aberdeen, UK*
A. Mukherjee: *Department of Geology and Geophysics, Indian Institute of Technology (IIT) – Kharagpur, India*
R. Naidu: *CRC Care, University of Newcastle, NSW, Australia*
J. Ng: *National Research Centre for Environmental Toxicology, The University of Queensland, Brisbane, Australia*
H.B. Nicolli: *Instituto de Geoquímica (INGEOQUI), San Miguel, Prov. de Buenos Aires, Argentina and Consejo Nacional de Investigaciones Científicas y Técnicas (CONICET), Argentina*
B. Noller: *The University of Queensland, Australia*
D.K. Nordstrom: *U. S. Geological Survey, Menlo Park, CA, USA*
G. Owens: *Mawson Institute, University of South Australia, Australia*
P. Pastén González: *Pontificia Universidad Católica de Chile, Chile*
C.A. Pérez: *LNLS -Brazilian Synchrotron Light Source Laboratory, Campinas, SP, Brazil*
A. Pérez Carrera: *University of Buenos Aires, Buenos Aires, Argentina*
C. Pérez Coll: *National San Martín University, San Martín, Argentina*
B. Petrusevski: *UNESCO-IHE, Institute for Water Education, Delft, The Netherlands*
L. Pflüger: *Secretary of Environment, Buenos Aires, Argentina*
G.E. Pizarro Puccio: *Pontificia Universidad Católica de Chile, Chile*
D.A. Polya: *The University of Manchester, UK.*
I. Queralt: *Institute of Earth Sciences Jaume Almera – CSIC, Spain*
J. Quintanilla: *Institute of Chemical Research, Universidad Mayor de San Andrés, La Paz, Bolivia*
M.M. Rahman: *The University of Newcastle Callaghan, NSW, Australia*
AL. Ramanathan: *School of Environmental Science, Jawaharlal Nehru University, New Delhi, India*
B. Rosen: *Florida International University, Miami, USA*
J. Routh: *Linköping University, Linköping, Sweden*
D. Saha: *Central Ground Water Board, New Delhi, India*
A.M. Sancha: *Department of Civil Engineering, University of Chile, Santiago de Chile, Chile*
M. Schreiber: *Virginia Polytechnic Institute and State University, VA, USA*
C. Schulz: *La Pampa National University, Argentina*
O. Selinus: *Linneaus University, Kalmar, Sweden*
A. SenGupta: *Lehigh University, Bethlehem, PA, USA*
V.K. Sharma: *Florida Institute of Technology, Melbourne, FL, USA*
A. Shraim: *The University of Queensland, Brisbane, Australia*
M. Sillanpää: *Lappeenranta University of Technology, Finland*
P.L. Smedley: *British Geological Survey, Keyworth, UK*
A.H. Smith: *University of California, Berkeley, CA, USA*
M. Styblo: *University of North Carolina, Chapel Hill, USA*
C. Tsakiroglou: *Foundation for Research and Technology, Hellas, Greece*
M. Vahter: *Karolinska Institutet Stockholm, Sweden*
D. van Halem: *Delft University of Technology, Delft, The Netherlands*
J. W. Vargas de Mello: *Federal University of Viçosa, Viçosa, MG, Brazil*
D. Velez: *Institute of Agrochemistry and Food Technology, Valencia, Spain*
M. Vithanage: *Institute of Fundamental Studies, Hantana Road, Kandy, Sri Lanka*
Y.X. Wang: *China University of Geosciences (Wuhan), Wuhan, P.R. China*
F.J. Zhao: *Nanjing Agricultural University, Nanjing, P.R. China*
Y. Zheng: *School of Environmental Science and Technology, Southern University of Science and Technology, Shenzhen, China*
Y.G. Zhu: *Chinese Academy of Sciences, Research Centre for Eco-environmental Sciences, Beijing, China*

Environmental Arsenic in a Changing World –
Zhu, Guo, Bhattacharya, Ahmad, Bundschuh & Naidu (Eds)
ISBN 978-1-138-48609-6

Foreword (Director General, Institute of Urban Environment, Chinese Academy of Sciences)

Arsenic, as a global contaminant, is impacting the health of millions of people around the world, through water, food and potential also air pollution. It has recently been estimated that in China alone, there are about 19 million people may be drinking water above the World Health Organization guideline of 10 μg/L. Arsenic is also a cultural element, as it is a notorious poison. It has been suggested that a Qing Dynasty Emperor was poisoned by arsenic. With rapid industrialization and urbanization in modern China, arsenic pollution is a major environmental challenge. According to the recent China national soil pollution survey, about 3% of China's arable land exceeded the soil quality standards for arsenic.

The Institute of Urban Environment, Chinese Academy of Science (IUE-CAS) is very happy to host the 7th International Congress on Arsenic in the Environment together with China University of Geosciences (Beijing). The institute was established on 4 July 2006. It is located in the beautiful coastal city-Xiamen. IUE-CAS is a unique national research institute engaged in comprehensive studies on the world's urban environment and the impacts of urbanization on ecosystem and human health. IUE-CAS hosts over 200 staff scientists plus around 200 graduate students. There are number of scientists within the institute working on arsenic biogeochemistry, human health impacts. Over the last 10 years, IUE-CAS has published about 150 papers related to arsenic in international journals, covering topics ranging from environmental chemistry, ecotoxicology, risk assessment and microbial ecology and genomics etc. in terrestrial and aquatic environments.

We cordially invite delegates from all corners of the globe to participate this important congress, and forge new friendship and collaborative linkages.

Professor Yong-Guan Zhu
Director-General
Institute of Urban Environment
Chinese Academy of Sciences
Xiamen, P.R. China
May 2018

ISBN 978-1-138-48609-6

Foreword (Vice President, China University of Geosciences, Beijing)

As a toxicant and carcinogen for humans, environmental arsenic is one of the biggest issues in the world. Hundreds of millions of people are suffering from chronical arsenic poisoning worldwide, including Bangladesh, India, China, Pakistan, Nepal, Cambodia, and Vietnam. China is a typical country facing ecologic poisoning of environment arsenic, where there were more than 5 million residents being at risk of chronic arsenic poisoning in both inland basins experiencing an arid/semiarid continental climate, and river deltas experiencing a humid tropical climate. Around forty million people were exposed to drinking water with arsenic concentration $>10\,\mu g/L$ (WHO drinking water guideline value).

As a major media hosting environment arsenic, high arsenic groundwater is closely related to human health due to the pathways for arsenic from water to human via ingestion of drinking water and digestion of the groundwater-irrigated foods. Hydrogeological and biogeochemical studies showed that redox milieu, the source of dissolved organic carbon, microbial diversity, sedimentation sequences and groundwater hydraulics are the major contributors for spatial and temporal variation in arsenic concentrations of groundwater from aquifers which do not contain abnormal arsenic contents. In addition, irrigation with high arsenic groundwater not only affects arsenic contents of food products but also deteriorates soil quality. Soil pollution was correlated with arsenic concentration of irrigation water extracted from groundwater aquifers, which led to arsenic accumulation from soil to food chains. Via drinking arsenic-contaminated groundwater or the food chain, arsenic is entering and accumulating in the human body since only approximately one-third of the uptaken-arsenic can be excreted daily, which causes chronic poisoning arsenicosis (such as keratosis, hyperpigmentation, diarrhoe, respiratory disorders, hypertension and malignancy). Elimination of drinking groundwater arsenic is an effective method to alleviate arsenic exposure to human body via drinking water pathway. Many new materials and filter systems have recently been developed to fix arsenic from aqueous solutions, some of which are available for practical applications in both the house-hold unit and the water supply plant scale. Although geochemical, health and mitigation investigations on environmental arsenic have made promising advances, interdisciplinary scientific exchanges among physicians, chemists, biologists and geologists and among different countries are quite limited. The coming 7th International Congress on Arsenic in the Environment, with a theme of Environmental Arsenic in a Changing World, is therefore quite necessary to strengthen and highlight the scientific exchanges among scientists from different disciplines and from different countries.

China University of Geosciences (Beijing) co-hosts the 7th International Congress on Arsenic in the Environment with The Institute of Urban Environment, Chinese Academy of Science. China University of Geosciences (Beijing), being founded in 1952, has become one of the national key universities and the advanced education center for geoscience studies in China. The university has 14,000 full-time enrolled students, 1400 teaching and research staff members. Good moral, sound background, wide knowledge, and high profession are the mutual goals of our students and staffs. Aiming at the first-class international university in the field of geosciences, the university values international and interdisciplinary cooperation and exchange. There are several groups working on arsenic geochemistry, biogeochemistry, remediation techniques and mechanisms in our university.

I am proud to write this forward to the Proceeding Series Volume of Arsenic in the Environment, which contains over 240 extended abstracts to be presented in the coming 7th International Congress on Arsenic in the Environment. This volume would be the state of the art of contributions to the arsenic research society around the world, which is related to geological arsenic in aquifers, geochemical and biogeological processes for arsenic mobilization, microbe-mineral-plant interactions, arsenic toxication, and mitigation techniques for arsenic fixation.

I deeply thank the Local Organizersf rom China University of Geosciences (Beijing) and the Institute of Urban Environment, and the Intemational Organizers from KTH Royal Institute of Technology (Sweden), the University of Southem Queensland (Australia), the KWR Watercycle Research Institute (KWR) (The Netherlands), and the International Society of Groundwater for Sustainable Development (ISGSD) for their elaborate work on this volume, which, I hope, will greatly improve our understanding of arsenic cycling in the system of biosphere, hydrosphere, geosphere, and anthroposphere.

Wan Li

Professor Dr. Li Wan
Vice President
China University of Geosciences (Beijing)
Beijing, China
May 2018

ISBN 978-1-138-48609-6

Foreword (KTH Royal Institute of Technology)

Arsenic is a natural or anthropogenic contaminant in many areas around the globe, where human subsistence is at risk. It is considered as a class 1 carcinogen, and its presence in groundwater has emerged as a major environmental calamity in several parts of the world. It has been estimated that nearly 137 million people drink water contaminated with arsenic globally. The widespread discovery of arsenic in Asia has paved the way to the discovery of the presence of this element in different environmental compartments as a "silent" toxin, especially in countries such as Bangladesh, Cambodia, China, India, Nepal, Pakistan, Taiwan, Thailand and Vietnam, the situation of arsenic toxicity is alarming and severe health problems are reported amongst the inhabitants relying on groundwater as drinking water. It is important to note that approximately 250 000 people in Sweden rely on drinking water from private wells with arsenic concentrations above the drinking water guideline value of 10 μg/L. However recent investigations have also shown that the problem of arsenic in groundwater exists in many countries in Latin America, Europe, Africa and Australia. The use of arsenic contaminated groundwater in irrigation landscapes especially in the rice cultivating regions and its bioaccumulation in rice and several other food crops has emerged as an additional pathway for arsenic exposure to humans and livestock through the food chain. New areas with elevated arsenic occurrences are reported in groundwater exceeding the maximal contamination levels set by the WHO and other national and international regulatory organizations are identified each year. It therefore requires innovative solutions to ensure access to clean drinking water. The Netherlands is now focusing on reducing their arsenic levels to below 1 μg/L, as there is a healthy arsenic content, and therefore assumes that the requirements will be tightened.

Since 2000, we have witnessed a remarkable rise in interest on research in the field of arsenic. Many research councils and international donor organizations have provided significant support to local and international research teams to develop strategies to address the problem with an aim to minimize the risk of arsenic exposure among the population. As a consequence, there has been a radical increase in the number of scientific publications that give a holistic overview on the occurrence, fate and cycling of arsenic in natural environment, its impact on human health, and implications on the society.

The WHO/FAO Joint Expert Committee (JEC) review document on Food Additives, resulted in withdrawal of the provisional tolerable weekly intake (PTWI) since 2010. The other important gaps identified by the JEC is particularly related to the need for accurate quantification of arsenic in dietary and other exposure routes as well as the speciation of arsenic and bioavailability that account for the total daily intake. Long-term exposure to arsenic is related to non-specific pathological irreversible effects and has significant social and economic impacts. The presence of arsenic in rice and rice products available in the markets has raised a critical concern – and this includes rice cakes, breakfast cereals as well as plain rice. Daily intake of inorganic arsenic in small quantities in rice and all rice products leads to high levels of arsenic exposure—and especially to the group of population with rice as the staple diet. Children are vulnerable to arsenic exposure, where the risk of arsenic exposure is exceptionally high due to the consumption of rice cakes especially in the pre-schools. Thus, arsenic in environment is clearly a concern that needs an inter- and multi-disciplinary and cross-disciplinary platform of research including hydrogeology and hydrogeochemistry, environmental sciences, food and nutrition, toxicology, health and medical sciences, remediation technologies and social sciences.

The biennial International Congress Series on Arsenic in the Environment is providing a common platform for sharing knowledge and experience on multidisciplinary issues on arsenic occurrences in groundwater and other environmental compartments on a worldwide scale to identify, assess, develop and promote approaches for management of arsenic in the environment and health effects. Since the first International Congress on "Arsenic in the Environment" at the UNAM, Mexico City in 2006, there has been an overwhelming response from the scientific community engaged with multidisciplinary facets of arsenic research to participate and present their research findings on this platform. The conference has been taken a form of biennial congress series with rotating venues at different continents. The following three events namely the 2nd International Congress (As 2008), with the theme "Arsenic from Nature to Humans" (Valencia, Spain) and the 3rd International Congress (As 2010) with the theme "Arsenic in Geosphere and Human Diseases" (Tainan, Taiwan), the 4th International Congress on Arsenic in the Environment (As 2012) with a theme "Understanding the Geological and Medical Interface" (Cairns, Australia), the 5th International Congress on Arsenic in the Environment (As 2014) with a theme "One Century of the Discovery of Arsenicosis in Latin America (1914–2014)" (Buenos Aires, Argentina) and the 6th International Congress on Arsenic in the Environment (As 2016) is envisioned with a theme "Arsenic Research and Global Sustainability" (Stockholm, Sweden) have been successfully organized and participated by the leading scientific community across the globe. The upcoming 7th International Congress on Arsenic in the Environment (As 2018) is envisioned with a theme "Environmental Arsenic in a Changing World" to be organized in Beijing, Peoples Republic of China between 1st and 6th July, 2018, with an aim to provide another international, multi- and interdisciplinary discussion platform for the presentation of cutting edge scientific research involving arsenic in natural systems, food chain, health impacts, clean water technology and other related social issues linked with environmental arsenic by bringing together scientific, medical, engineering and regulatory professionals.

I feel proud to write this foreword to this Volume of Arsenic in the Environment-Proceedings Series, containing the extended abstracts of the presentations to be made during the forthcoming 7th International Congress & Exihibition on Arsenic in the Environment – As 2018. The present volume "Environmental Arsenic in a Changing World" being published as a new volume of the book series "Arsenic in the Environment-Proceedings under the auspices of the International Society of Groundwater for Sustainable Development (ISGSD), will be an important updated contribution, comprising a large number of over 240 extended abstracts submitted by various researchers, health workers, technologists, students, legislators, and decision makers around the world that would be discussed during the conference. Apart from exchanging ideas, and discovering common interests, the scientific community involved in this specialized field needs to carry out researches, which not only address academic interests but also contribute to the societal needs through prevention or reduction of exposure to arsenic and its toxic effects in millions of exposed people throughout the world.

I deeply appreciate the efforts of the International Organizers from KTH-International Groundwater Arsenic Research Group, Department of Sustainable Development, Environmental Science and Engineering, School of Architecture and Built Environment KTH Royal Institute of Technology and the University of Southern Queensland, Toowoomba, Australia, the KWR Watercycle Research Institute (KWR) and the International Society of Groundwater for Sustainable Development (ISGSD) together with the Local Organizers from Institute of Urban Environment, Xiamen and the China University of Geosciences (Beijing) and the entire editorial team for their untiring work with this volume. I hope that the book will reflect the update on the current state-of-the-art knowledge on the interdisciplinary facets of arsenic in the environment required for the management of arsenic in the environment for protecting human health.

Professor Dr. Sigbritt Karlsson
KTH Royal Institute of Technology
Stockholm, Sweden
April 2018

Environmental Arsenic in a Changing World –
Zhu, Guo, Bhattacharya, Ahmad, Bundschuh & Naidu (Eds)
ISBN 978-1-138-48609-6

Foreword (Deputy Vice Chancellor, University of Southern Queensland)

The University of Southern Queensland (USQ) has great pleasure in co-organising the 7th International Congress & Exhibition on Arsenic in the Environment (As2018) themed 'Environmental Arsenic in a Changing World' in July 2018 in Beijing, China.

Arsenic originating from geogenic sources is a global issue as over 200 million people, so far known from over 80 countries, is at risk due to ingestion of arsenic-contaminated food and drinking water. In food, arsenic is particularly accumulated as a result of irrigation with arsenic-rich water – the staple food rice is thereby especially affected. Despite the fact that the problem occurs equally in developing and industrialized countries, the problem is most severe in the first country group where the poor are those who are at the highest risk and suffer most. Hence, arsenic pollution is an increasing global problem that will require a global approach and world wide solutions. Thereby, transdisciplinary research into the occurrence, mobility and bioavailability of arsenic in different environments including aquifers, soils, sediments as well as the food chain, will all become increasingly important.

It gives me pleasure to congratulate the organisers for their success in bringing this Congress to China and acknowledge the collaborative and cooperative efforts of the KTH Royal Institute of Technology. I hope that these proceedings will serve as a lasting record in co-organising this international Congress.

Professor Mark Harvey
Deputy Vice Chancellor (Research & Innovation)
The University of Southern Queensland
Toowoomba, Australia
April 2018

ISBN 978-1-138-48609-6

Foreword (Director, KWR Watercycle Research Institute)

It is with great pleasure and expectations that I write this Foreword to the Proceedings of the 6th International Congress and Exhibition on Arsenic in the Environment (As2018), themed 'Environmental Arsenic in a Changing World' held in Beijing, Peoples Republic of China, July 1–6, 2018.

The International Congress on Arsenic in the Environment has been previously held five times: Mexico 2006, Spain 2008, R. O. China 2010 Australia, 2012, Argentina 2014 and Sweden 2016. The Congress series has evolved into a highly reputable platform for sharing and assessing global knowledge on various aspects of arsenic research. Arsenic in drinking water is a global problem affecting populations on all five continents. Despite historical recognition of arsenic toxicity, more than 200 million people around the world are still exposed to above acceptable arsenic levels. This situation is alarming. Arsenic contamination of drinking water can be caused both by natural and anthropogenic processes. For example, in Poland and Brazil, arsenic contamination of groundwater due to anthropogenic mining activities have been reported. On the other hand, in some parts of Turkey elevated arsenic in groundwater is attributed to natural geothermal factors, and in Bangladesh geogenic processes are the major cause of large scale arsenic contamination. Whatever the origin may be, once detected in drinking water sources, suitable arsenic remediation measures should be taken to ensure supply of safe drinking water – as this is the fundamental right of every human being.

In the Netherlands, drinking water companies have recently updated their policy on arsenic and they will present their rationale at As2016. KWR Watercycle Research Institute is collaborating with the water companies in various fundamental and applied research projects to support the realization of this policy. Recognizing the global significance of arsenic for safe water supply, KWR has gladly invested in realizing As 2018 via participation in the organizing committee and the scientific board of As2018, by our research scientist, Mr. Arslan Ahmad, from our Knowledge Group Water Systems and Technology.

I congratulate all the authors, reviewers and editors for providing excellent content and structure to this book. I hope that these proceedings will serve as a deep-rooted record of the state-of-the-arsenic-related-science in the year 2018 and serve as a reference base for future research and support water suppliers and policy makers all over the world in addressing the arsenic problem efficiently and effectively.

Prof. Dr. Wim van Vierssen
Director
KWR Watercycle Research Institute
Nieuwegein, The Netherlands
April, 2018

Environmental Arsenic in a Changing World –
Zhu, Guo, Bhattacharya, Ahmad, Bundschuh & Naidu (Eds)
ISBN 978-1-138-48609-6

Foreword (Vice-Chancellor and President, The University of Newcastle)

It is with deep satisfaction that I write this foreword to the 7th International Congress & Exhibition on Arsenic in the Environment (As 2018). 'Environmental Arsenic in a Changing World' will be held in July 2018 in Beijing, China. The University of Newcastle (UON) is very proud to be part of this international congress series as co-organizer.

The first arsenic workshop (Arsenic in the Asia-Pacific Region) was organized by Professor Ravi Naidu in Adelaide, South Australia in 2001 where the extent, severity and potential risks arising from exposure to arsenic, as well as the fate of arsenic in water, soil and food was discussed. The continuation of this as a global congress - "*Arsenic in the Environment*" - was then held in Mexico in 2006. Since then, the international congress series has been held every two years at various locations around the globe. Thus, this arsenic congress has received enormous attention and is a platform where scientists, government officials, policy makers and regulators share their knowledge on the recent developments in arsenic research.

Arsenic is a toxic element and is categorized as a Class I carcinogen, which is ubiquitous in the environment. Arsenic is present in our environment as a naturally occurring substance and because of anthropogenic activities. It is generally found in waters (both surface and sub-surface), soil, food and the air and can occur in both organic and inorganic forms. Arsenic occurrence in water in the Australian landscape is generally low but major pollution can occur due to mining activities, the use of arsenic based pesticides and herbicides, as well as from CCA treated wood. Arsenic concentrations in cattle dip and sheep dip soils and railway corridor soils in Australia are also at levels two to five times above the health screening levels. These are the major causes of arsenic contaminated sites in Australia.

The first arsenic contamination was reported in Germany in 1885 and arsenic related health effects (widely known as Bell Ville Disease) were first reported in 1917 from the province of Cordoba, Argentina. Later, gangrene was reported in the population of arsenic impacted villages in south-western Taiwan, generally known as Blackfoot disease. During the 1960s, high levels of arsenic were detected in the groundwater of the Lagunera region of Mexico where various arsenic related diseases were also reported at this time. The global epidemiological research based on the data from these studies played a crucial role in establishing arsenic related health effects.

UON is a leading research organisation, which contributes to both Australian and international social, economic, cultural and environmental well-being through its innovative research activities that supports research in identified areas of strength, to address national and international challenges. UON continues to build its global reputation for delivering world-class research and innovation, a reputation that has been built on high quality performance in a wide range of specialist research fields.

We are ranked in the top 1% of the world's universities, according to the QS World University Rankings 2017/18. Our Engineering – Mineral and Mining discipline - was ranked in the top 30 in the world for the second consecutive year in the 2018 QS World University Rankings by Subject list. The University also had six subjects ranked in the world's top 100 and 15 subjects ranked in the world's top 200. The University was 8th in Australia for research deemed to be 'well above world standard' in the 2015 Excellence in Research Australia (ERA) exercise.

UON researchers have been working in Bangladesh and India over many years and have made substantial contributions to various aspects of arsenic research including arsenic chemistry, toxicity and bioavailability, human health effects, and food quality and safety. By combining laboratory-based studies with field surveys, they have contributed significantly to the generation of new knowledge in this important research field. UON researchers have also made major contributions by developing new and novel analytical techniques for arsenic speciation in various environmental matrices, which has helped to understand the toxicity, bioavailability, and accurate estimation of human health risks. UON's current activities include: researching geographical variations and age related dietary exposure in rice along with cancer and non-cancer effects; inorganic arsenic levels in rice and rice based diets and the potential risk to babies and toddlers; lowering arsenic levels in rice by managing irrigation options with enhanced productivity; and arsenic bioavailability in various rice varieties using a swine model to understand the human health risk.

We sincerely hope that the congress proceedings will become an excellent and much-used resource for researchers and others who are working on arsenic and related research fields. We would like to thank the contributors and conference delegates for their active participation. We would also like to express our whole-hearted appreciation to all co-organizers and others who will be involved in the congress series and who will no doubt make this congress a great success.

Professor Caroline McMillen
Vice-Chancellor and President
The University of Newcastle (UON)
Newcastle, Australia
May 2018

ISBN 978-1-138-48609-6

Editors' foreword

Occurrence of elevated arsenic concentrations in ground water used for drinking purpose, and associated health risks, were reported at first international conference on environmental arsenic, which was held in Fort Lauderdale, USA, almost exactly 40 year ago; October, 1976. Over the past 2 to 3 decades arsenic in drinking water, and more recently, in plant based foods, especially rice, has been recognized as a major public health concern in many parts of the world. Latest surveys estimated that currently more than 200 million people around the world are exposed to unacceptably high arsenic levels. The geological, geomorphological and geochemical reasons for high arsenic concentrations in groundwater vary from place to place and require different mitigation policies and practices. Although, the high income countries may invest in research and development of suitable remediation techniques, arsenic in private water sources is not always tested. On the other hand, low to lower-middle income countries, such as many areas in South-East Asia, Africa and South America, where millions of people still use arsenic-contaminated drinking water, are still coping with stagnated mitigation efforts and slow progress towards safe drinking water. It is disturbing to enter almost any village of the Bengal basin today and find that groundwater drawn from untested shallow wells continues to be used routinely for drinking and cooking, given that the arsenic problem was already recognized in the mid-1980s in West Bengal and the mid-1990s in Bangladesh. Equally problematic is the fact that hundreds of millions of wells world-wide are not yet tested for arsenic. Moreover, many low and lower-middle income countries have yet not been able to revise their standards for arsenic in drinking water to 10 μg/L, the guideline value of the World Health Organization. We sincerely believe that sharing knowledge and experience on arsenic related science and practices on a world-wide scale and across varied disciplines can serve as an effective strategy to support global arsenic management and mitigation efforts.

The biannual International Congress Series on Arsenic in the Environment aims at providing a common platform for sharing knowledge and experience on multidisciplinary issues on arsenic occurrences in groundwater and other environmental compartments on a worldwide scale for identifying and promoting optimal approaches for the assessment and management of arsenic in the environment. The International Congress on Arsenic in the Environment has previously been held six times; Mexico 2006, Spain 2008, R. O. China 2010, Australia, 2012, Argentina 2014 and Sweden 2016. The seventh International Congress on Arsenic in the Environment (As2018) is being organized in Beijing, the Capital of the Peoples Republic of China, between 1 and 6 July, 2018 and with a theme "Environmental Arsenic in a Changing World". The UN Agenda 2030 for Sustainable Development adopted in September 2015, list 17 Sustainable Development Goals (SDGs) of raise the global profile of arsenic in order to achieve universal and equitable access to safe and affordable drinking water for all. This emphasizes holistic management of drinking water services and monitoring of drinking water quality and deployment of clean water technology in the across the world for protecting human health. We envision As2018 as a global interdisciplinary platform to exchange and disseminate research results to improve our understanding of the occurrence, mobility, bioavailability, toxicity and dose-response relationship with various health effects of environmental arsenic in the current epoch of a changing world.

We have received a large number of (over 250) extended abstracts which were submitted mainly from researchers, but also health workers, technologists, students, legislators, government officials. The topics to be covered during the Congress As 2018 have been grouped under the five general thematic areas:

Theme 1: Arsenic Behaviour in Changing Environmental Media
Theme 2: Arsenic in a Changing Agricultural Ecosystem
Theme 3: Health Impacts of Environmental Arsenic
Theme 4: Technologies for Arsenic Immobilization and Clean Water Blueprints
Theme 5: Sustainable Mitigation and Management.

We thank the international scientific committee members, for their efforts on reviewing the extended abstracts. Further, we thank the sponsors of the Congress from around the world: KTH Royal Institute of Technology (Sweden), University of Southern Queensland (Australia), KWR Watercycle Research Institute (The Netherlands), The University of Newcastle (Australia) and the CRC-CARE, at the University of South Australia and OPCW for their generous support – Thank you all sponsors for your support that contributed to the success of the congress As2018.

The International Organizers would like thank Instutute of Urban Environment, Chinese Academy of Sciences and the China University of Geosciences, Beijing, China Centre for International Science and Technology Exchange (CISTE), KWR Watercycle Research Institute, The Netherlands, the Global Centre for Environmental Remediation, The University of Newcastle and CRC CARE, University of Newcastle, NSW, Australia and the University of Southern Queensland, Australia for their support to organize the 7th International Congress and Exhibition on Arsenic in the Environment (As2018). We thank the KTH Royal Institute of Technology, especially the KTH School of Architecture and Built Environment for supporting the KTH-International Groundwater Arsenic Research Group at the Department of Sustainable Development, Environmental Sciences and Engineering, Stockholm as an International Organizer of this Congress. We would like to thank Dr. M. Mahmudur Rahman, Professor M. Alauddin, P. Kumarathilaka, Dr. G. Sun, Dr. J. Ye and M. Tahmidul Islam for their help with the preparation and formatting of the content of this volume. Lastly, the editors thank Janjaap Blom and Lukas Goosen of the CRC Press/Taylor and Francis (A.A. Balkema) Publishers, The Netherlands for their patience and skill for the final production of this volume.

Yong-Guan Zhu
Huaming Guo
Prosun Bhattacharya
Jochen Bundschuh
Arslan Ahmad
Ravi Naidu
(Editors)

ISBN 978-1-138-48609-6

List of contributors

Abdulmutalimova, T.O.: *Institute of Geology, Dagestan Center of Science, Russian Academy of Sciences, Makhachkala, Dagestan, Russia*
Abhinav, S.: *Mahavir Cancer Institute & Research Centre, Patna, Bihar, India*
Abiye, T.A.: *School of Geosciences, University of the Witwatersrand, Johannesburg, South Africa*
Afroz, H.: *Institute for Global Food Security, Queens University Belfast, Belfast, UK*
Ahmad, A.: *KWR Water Cycle Research Institute, Nieuwegein, The Netherlands; KTH-International Groundwater Arsenic Research Group, Department of Sustainable Development, Environmental Science and Engineering, KTH Royal Institute of Technology, Stockholm, Sweden; Department of Environmental Technology, Wageningen University and Research (WUR), Wageningen, The Netherlands*
Ahmad, S.A.: *Department of Occupational and Environmental Health, Bangladesh University of Health Sciences, Dhaka, Bangladesh*
Ahmed, F.: *UNICEF Bangladesh, Dhaka, Bangladesh*
Ahmed, K.M.: *Department of Geology, University of Dhaka, Dhaka, Bangladesh*
Ahmed, S.: *Department of Chemistry, Sripat Singh College, Murshidabad, India*
Ahmed, S.: *Environment and Population Research Centre (EPRC), Dhaka, Bangladesh*
Ahmed, S.: *Institute of Child and Mother Health, Sk Hospital, Dhaka, Bangladesh*
Ahsan, H.: *Departments of Health Studies, Medicine and Human Genetics and Cancer Research Center, The University of Chicago, Chicago, IL, USA; Institute for Population and Precision Health, Chicago Center for Health and Environment, The University of Chicago, Chicago, IL, USA*
Ahuja, S.: *Department of Civil Engineering, Indian Institute of Technology, Guwahati, India*
Akter, N.: *UNICEF Bangladesh, Dhaka, Bangladesh*
Alam, K.: *UNICEF Bangladesh, Dhaka, Bangladesh*
Alam, M.O.: *Department of Civil and Environmental Engineering, Birla Institute of Technology, Mesra, India*
Alarcón-Herrera, M.T.: *Centro de Investigación Materiales Avanzados (Sede Chihuahua-CIMAV), Chihuahua, Chih., Mexico*
Alauddin, M.: *Department of Chemistry, Wagner College, Staten Island, New York, USA*
Alauddin, S.: *Department of Chemistry, Wagner College, Staten Island, New York, USA*
Ali, M.: *Mahavir Cancer Sansthan & Research Centre, Patna, Bihar, India*
Allen, J.: *PS Analytical Ltd, Arthur House, Orpington, Kent, UK*
Alvarez Gonçalvez, C.V.: *Instituto de Investigaciones en Producción Animal UBA-CONICET (INPA), CONICET – Universidad de Buenos Aires, Buenos Aires, Argentina; Universidad de Buenos Aires. Centro de Estudios Transdisciplinarios del Agua (CETA), Universidad de Buenos Aires, Buenos Aires, Argentina; Universidad de Buenos Aires. Cátedra de Qumica Orgánica de Biomoléculas, Universidad de Buenos Aires, Buenos Aires, Argentina*
Álvarez Vargas, A.: *Departamento de Biología, División de Ciencias Naturales y Exactas, Universidad de Guanajuato, Campus Guanajuato, México*
Amin, R.: *UNICEF Bangladesh, Dhaka, Bangladesh*
Amirnia, S.: *Department of Environmental Science, Saitama University, Saitama, Japan*
Anantharaman, G.: *Department of Chemistry, Indian Institute of Technology, Kanpur, India*
Annaduzzaman, M.: *Sanitary Engineering Section, Faculty of Civil Engineering and Geoscience, Delft University of Technology, Delft, The Netherlands*
Antelo, J.: *Technological Research Institute, University of Santiago de Compostela, Santiago de Compostela, Spain*
Apollaro, C.: *Department of Biology, Ecology and Earth Sciences (DIBEST), University of Calabria, Rende (CS), Italy*

Arellano, F.E.: *Instituto de Investigaciones en Producción Animal UBA-CONICET (INPA), CONICET – Universidad de Buenos Aires, Buenos Aires, Argentina; Universidad de Buenos Aires. Centro de Estudios Transdisciplinarios del Agua (CETA), Universidad de Buenos Aires, Buenos Aires, Argentina; Universidad de Buenos Aires. Cátedra de Qumica Orgánica de Biomoléculas, Universidad de Buenos Aires, Buenos Aires, Argentina*
Asaeda, T.: *Department of Environmental Science, Saitama University, Saitama, Japan*
Ascott, M.: *Groundwater Science Directorate, British Geological Survey, Wallingford, UK*
Awasthi, S.: *CSIR – National Botanical Research Institute, Lucknow, Uttar Pradesh, India*
Ayora, C.: *Instituto de Diagnóstico Ambiental y Estudios del Agua (IDAEA-CSIC), Barcelona, Spain*
Baba, A.: *Izmir Institute of Technology, Geothermal Energy Research and Application Center, Izmir, Turkey*
Baeyens, W.: *Analytical, Environmental and Geochemical Department (AMGC), Vrije Universiteit Brussel, Brussels, Belgium*
Bagade, A.V.: *Department of Chemistry, Savitribai Phule Pune University, Pune, India*
Bahr, C.: *GEH Wasserchemie GmbH & Co. KG, Osnabrück, Germany*
Bai, L.Y.: *Institute of Environment and Sustainable Development in Agriculture, Chinese Academy of Agricultural Sciences/Key Laboratory of Agro-Environment, Ministry of Agriculture, Beijing, P.R. China*
Baig, Z.U.: *Environmental Geochemistry Laboratory, Department of Environmental Sciences, Faculty of Biological Sciences, Quaid-i-Azam University, Islamabad, Pakistan*
Baisch, P.: *Laboratório de Oceanografia Geológica, Instituto de Oceanografia, Universidade Federal do Rio Grande (FURG), Rio Grande, RS, Brazil*
Bakshi, K.: *Kalyani Institute for Study, Planning and Action for Rural Change (KINSPARC), Kalyani West Bengal, India*
Ballentine, C.J.: *Department of Earth Sciences, University of Oxford, Oxford, UK*
Ballinas-Casarrubias, M.L.: *Faculty of Chemical Sciences, Autonomous University of Chihuahua, Chihuahua, Mexico*
Balseiro, M.: *Department of Chemical Engineering, Centre for Research in Environmental Technologies (CRETUS), Universidade de Santiago de Compostela, Spain*
Bandyopadhyay, A.K.: *Health Point Multispecialty Hospital, Kolkata, India*
Banerjee, N.: *Molecular Genetics Division, CSIR-Indian Institute of Chemical Biology, Kolkata, India*
Barla, A.: *Earth and Environmental Science Research Laboratory, Department of Earth Sciences, Indian Institute of Science Education and Research Kolkata (IISER-K), Mohanpur, West Bengal, India*
Bassil, N.: *School of Earth and Environmental Sciences, The University of Manchester, UK*
Basu, A.: *Sripat Singh College, Jiaganj, Murshidabad, India*
Battaglia-Brunet, F.: *BRGM, Orléans, France*
Battistel, M.: *Department of Environmental Engineering, Technical University of Denmark, Kongens Lyngby, Denmark*
Bea, P.A.: *Instituto de Hidrologa de Llanuras "Dr. Eduardo J. Usunoff" (IHLLA), Azul, Buenos Aires, Argentina*
Bea, S.: *Instituto de Hidrologa de Llanuras "Dr. Eduardo J. Usunoff" (IHLLA), Azul, Buenos Aires, Argentina*
Beaulieu, M.: *BRGM, Orléans, France*
Bedogni, G.R.: *Laboratorio de Microbiologa General, Universidad Nacional del Chaco Austral, P.R. Sáenz Peña, Chaco, Argentina*
Behrends, T.: *Department of Earth Sciences-Geochemistry, Faculty of Geosciences, Utrecht University, Utrecht, The Netherlands*
Beldowski, J.: *Institute of Oceanology of the Polish Academy of Sciences, Sopot, Poland*
Benning, L.G.: *Helmholtz Centre Potsdam – GFZ German Research Centre for Geosciences, Potsdam, Germany; Institute of Geological Sciences, Department of Earth Sciences, Free University of Berlin, Berlin, Germany*
Berg, M.: *Eawag, Swiss Federal Institute of Aquatic Science and Technology, Dübendorf, Switzerland*
Berube, M.: *Department of Geology, Kansas State University, Manhattan, KS, USA*
Bhattacharya, P.: *KTH International Groundwater Arsenic Research Group, Department of Sustainable Development, Environmental Science and Engineering, KTH Royal Institute of Technology, Stockholm, Sweden*
Bhattacharya, T.: *Department of Civil and Environmental Engineering, Birla Institute of Technology, Mesra, India*
Bianucci, E.C.: *Departamento de Ciencias Naturales, Facultad de Ciencias Exactas, Fsico-Qumicas y Naturales, Universidad Nacional de Ro Cuarto, Córdoba, Argentina*

Bibi, I.: *Institute of Soil and Environmental Sciences, University of Agriculture Faisalabad, Faisalabad, Pakistan*
Biswas, M.: *Sripat Singh College, Jiaganj, Murshidabad, India*
Bolaños, D.: *Department of Earth and Construction Science, Universidad de las Fuerzas Armadas ESPE, Sangolqu, Ecuador*
Bolevic, L.: *Department of Chemistry, Wagner College, Staten Island, New York, USA*
Bolton, M.: *Golder Associates Ltd., Canada*
Bose, N.: *Department of Geography, A.N. College, Patna, India*
Bose, S.: *Earth and Environmental Science Research Laboratory, Department of Earth Sciences, Indian Institute of Science Education and Research Kolkata (IISER-K), Mohanpur, West Bengal, India*
Bostick, B.C.: *Lamont-Doherty Earth Observatory, Columbia University, New York, NY, USA*
Boyce, A.J.: *Scottish Universities Environmental Research Centre, East Kilbride, UK*
Bravo, J.C.: *EcoMetales Limited, Providencia, Región Metropolitana, Santiago de Chile, Chile*
Briseño, J.: *Universidad Politécnica del Estado de Morelos. Jiutepec, Mor., Mexico*
Brown, P.E.: *Giant Mine Oversight Board, Yellowknife, NT, Canada*
Bryant, C.: *NERC Radiocarbon Facility, East Kilbride, UK*
Bundschuh, J.: *UNESCO Chair on Groundwater Arsenic within the 2030 Agenda for Sustainable Development & Faculty of Health, Engineering and Sciences, University of Southern Queensland, Toowoomba, Australia*
Burchiel, S.W.: *College of Pharmacy, University of New Mexico, Albuquerque, NM, USA*
Burgers, F.: *Department of Applied Geoscience and Engineering, Delft University of Technology, Delft, The Netherlands*
Buzek, F.: *Czech Geological Survey, Prague, Czech Republic*
Böhlke, J.K.: *U.S. Geological Survey, Reston, VA, USA*
Bühl, V.: *Analytical Chemistry, Faculty of Chemistry, DEC, Universidad de la República (UdelaR), Montevideo, Uruguay*
Cacciabue, L.: *Instituto de Hidrologa de Llanuras "Dr. Eduardo J. Usunoff" (IHLLA), Azul, Buenos Aires, Argentina*
Cai, Y.: *Department of Chemistry and Biochemistry, Florida International University, Miami, FL, USA; Southeast Environmental Research Center, Florida International University, Miami, FL, USA*
Calderón, E.: *Facultad de Ciencias Veterinarias, Centro de Estudios Transdisciplinarios del Agua (CETA – UBA), Universidad de Buenos Aires,Buenos Aires, Argentina*
Cano Canchola, C.: *Departamento de Biología, División de Ciencias Naturales y Exactas, Universidad de Guanajuato, Campus Guanajuato, México*
Cañas Kurz, E.E.: *Department of Mechatronics and Sensor Systems Technology, Vietnamese-German University, Binh Duong Province, Vietnam; Center of Applied Research, Karlsruhe University of Applied Sciences, Karlsruhe, Germany*
Cao, Y.: *State Key Lab of Pollution Control and Resource Reuse, School of the Environment, Nanjing University, Jiangsu, P.R. China*
Cao, Y.W.: *State Key Laboratory of Biogeology and Environmental Geology & School of Environmental Studies, China University of Geosciences, Hubei, P.R. China*
Carabante, I.: *Waste Science and Technology, Luleå University of Technology, Luleå, Sweden; Department of Earth System Science, Stanford University, Stanford, CA, USA*
Carpio, E.: *UNI, Universidad Nacional de Ingeniera, Lima, Peru*
Castro, S.: *Departamento de Ciencias Naturales, Facultad de Ciencias Exactas, Fsico-Qumicas y Naturales, Universidad Nacional de Ro Cuarto, Córdoba, Argentina*
Cejkova, B.: *Czech Geological Survey, Prague, Czech Republic*
Cekovic, R.: *Department of Chemistry, Wagner College, Staten Island, New York, USA*
Centeno, J.A.: *Division of Biology, Chemistry and Materials Science, US Food and Drug Administration, Washington DC, USA*
Chakraborty, M.: *Department of Geology and Geophysics, Indian IIT Kharagpur, Kharagpur, West Bengal, India*
Chakraborty, S.: *Department of Civil and Environmental Engineering, Birla Institute of Technology, Mesra, India*
Chakradhari, S.: *School of Studies in Chemistry/Environmental Science, Pt. Ravishankar Shukla University, Raipur, India*

Chambers, L.: *Lancaster Environment Centre, Lancaster University, Lancaster, UK; Department of Civil and Environmental Engineering, University of Strathclyde, Glasgow, UK*

Chandrbhushan: *Innervoice Foundation, Varanasi, India*

Chang, H.C.: *Department of Environmental Science, Xi'an Jiaotong-Liverpool University, Suzhou, Jiangsu, P.R. China*

Charron, M.: *BRGM, Orléans, France*

Charron, R.: *Health Canada*

Chatterjee, D.: *Department of Chemistry, University of Kalyani, Kalyani, West Bengal, India*

Chatterjee, D.: *Molecular Genetics Division, CSIR-Indian Institute of Chemical Biology, Kolkata, India*

Chaudhary, H.J.: *Department of Plant Sciences, Faculty of Biological Sciences, Quaid-i-Azam University, Islamabad, Pakistan*

Chauhan, I.S.: *Department of Civil Engineering, Indian Institute of Technology (BHU) Varanasi, Varanasi, India*

Chauhan, R.: *CSIR – National Botanical Research Institute, Lucknow, Uttar Pradesh, India*

Chen, C.J.: *Genomics Research Centre, Academia Sinica, Taipei, Taiwan*

Chen, J.: *Department of Cellular Biology and Pharmacology, Herbert Wertheim College of Medicine, Florida International University, Miami, Florida, USA*

Chen, M.: *School of Environmental Studies, China University of Geosciences, Wuhan, Hubei, P.R. China*

Chen, P.: *State Key Laboratory of Urban and Regional Ecology, Research Center for Eco-Environmental Sciences, Chinese Academy of Sciences, Beijing, P.R. China*

Chen, X.M.: *State Key Laboratory of Biogeology and Environmental Geology and Department of Biological Science and Technology, School of Environmental Studies, China University of Geosciences, Wuhan, P.R. China*

Chen, Y.: *State Key Lab of Pollution Control and Resource Reuse, School of the Environment, Nanjing University, Jiangsu, P.R. China*

Chen, Y.S.: *Key Laboratory of Urban Environment and Health, Institute of Urban Environment, Chinese Academy of Sciences, Xiamen, P.R. China*

Chen, Y.S.: *State Key Lab of Pollution Control and Resource Reuse, School of the Environment, Nanjing University, Jiangsu, P.R. China*

Chen, Z.: *Department of Environmental Science, Xi'an Jiaotong-Liverpool University, Suzhou, P.R. China*

Cheng, S.C.: *College of Resources and Environment, Yunnan Agricultural University, Kunming, P.R. China*

Cheng, S.G.: *School of Environmental Studies, China University of Geosciences (Wuhan), Wuhan, Hubei, P.R. China*

Chi, Z.Y.: *School of Environmental Studies & State Key Laboratory of Biogeology and Environmental Geology, China University of Geosciences, Wuhan, China*

Chillrud, S.N.: *Barnard College and Lamont-Doherty Earth Observatory, Columbia University, New York, NY, USA*

Chiou, H.Y.: *Department of Public Health, Taipei Medical University, Taipei, Taiwan*

Choi, B.S.: *College of Medicine, Chung-Ang University, Seoul, South Korea*

Chorover, J.: *Department of Soil, Water and Environmental Science, University of Arizona, Tucson, AZ, USA*

Choudhury, R.: *Department of Civil Engineering, Indian Institute of Technology, Guwahati, Assam, India*

Ciminelli, V.S.T.: *Universidade Federal de Minas Gerais-UFMG, Department of Metallurgical and Materials Engineering, Belo Horizonte, Brazil; National Institute of Science and Technology on Mineral Resources, Water and Biodiversity, INCT-Acqua, Brazil*

Cirpka, O.: *Geomicrobiology and Hydrology, University of Tübingen, Tübingen, Germany*

Cook, P.A.: *School of Health Sciences, University of Salford, Salford, UK*

Corns, W.: *PS Analytical Ltd, Arthur House, Orpington, Kent, UK*

Corroto, C.E.: *Agua y Saneamientos Argentinos S.A. (AySA S.A.) and Centro de Estudios Transdisciplinarios del Agua, Universidad de Buenos Aires and Buenos Aires, Argentina*

Criscuoli, A.: *Institute on Membrane Technology (ITM-CNR), Rende (CS), Italy*

Cruz, G.: *Universidad Nacional de Tumbes, Campus Universitario, Tumbes, Peru*

Cui, J.L.: *Research Center for Eco-Environmental Sciences, Chinese Academy of Sciences, Beijing, P.R. China*

Cumbal, L.: *Department of Life Science and Agriculture, Universidad de las Fuerzas Armadas ESPE, Sangolqu, Ecuador; Center of Nanoscience and Nanotechnology (CENCINAT), Universidad de las Fuerzas Armadas ESPE, Sangolqu, Ecuador*

Datta, S.: *Department of Geology, Kansas State University, Manhattan, KS, USA*

De Lary De Latour, L.: *BRGM, Orléans, France*
de Meyer, C.M.C.: *Eawag, Swiss Federal Institute of Aquatic Science and Technology, Dübendorf, Switzerland*
de Ridder, D.J.: *Sanitary Engineering Section, Faculty of Civil Engineering and Geoscience, Delft University of Technology, Delft, The Netherlands*
De Rosa, R.: *Department of Biology, Ecology and Earth Sciences (DIBEST), University of Calabria, Rende (CS), Italy*
Delbem, I.D.: *National Institute of Science and Technology on Mineral Resources, Water and Biodiversity, INCT-Acqua, Brazil; Universidade Federal de Minas Gerais-UFMG, Center of Microscopy, Belo Horizonte, Brazil*
Deng, Y.: *Geological Survey, China University of Geosciences, Wuhan, China*
Deng, Y.X.: *Agro-Environmental Protection Institute, Ministry of Agriculture, Tianjin, P.R. China*
Desiderio, G.: *DeltaE, University of Calabria, Rende (CS), Italy*
Desmet, M.: *University of Tours, Tours, France*
Devau, N.: *BRGM, Orléans, France*
Dey, N.C.: *BRAC Research and Evaluation Division, BRAC Centre, Dhaka, Bangladesh*
Dheeman, D.S.: *Herbert Wertheim College of Medicine, Florida International University, Miami, FL, USA*
Dhotre, D.: *Division of Biochemical Sciences, CSIR-National Chemical Laboratory, Pune, India*
Diacomanolis, V.: *Queensland Alliance for Environmental Health Sciences (QAEHS), The University of Queensland, Brisbane, QLD, Australia*
Dideriksen, K.: *Nano-Science Center, Department of Chemistry, University of Copenhagen, Copenhagen, Denmark*
Dietrich, S.: *Instituto de Hidrologa de Llanuras "Dr. Eduardo J. Usunoff" (IHLLA), Azul, Buenos Aires, Argentina*
Ding, C.F.: *Key Laboratory of Soil Environment and Pollution Remediation, Institute of Soil Science, Chinese Academy of Sciences, Nanjing, P.R. China*
Ding, J.J.: *Key Laboratory of Dryland Agriculture, Ministry of Agriculture, Institute of Environment and Sustainable Development in Agriculture, Beijing, P.R. China*
Dong, B.B.: *College of Resources and Environmental Sciences, Jiangsu Provincial Key Laboratory of Marine Biology, Nanjing Agricultural University, Nanjing, P.R. China*
Dong, L.J.: *Sanitation & Environment Technology Institute, Soochow University, Jiangsu, P.R. China*
Dong, Y.H.: *School of Environmental Sciences, China University of Geosciences, Wuhan, P.R. China; School of Water Resources and Environmental Engineering, East China University of Technology, Nanchang, P.R. China*
Donselaar, M.E.: *Faculty of Civil Engineering and Geosciences, Delft University of Technology, Delft, The Netherlands; KU Leuven, Department of Earth and Environmental Sciences, Leuven-Heverlee, Belgium*
Dousova, B.: *University of Chemistry and Technology Prague, Prague, Czech Republic*
Du, H.H.: *College of Resource and Environment, Hunan Agricultural University, Changsha, P.R. China*
Du, Y.H.: *Guangdong Key Laboratory of Integrated Agro-environmental Pollution Control and Management, Guangdong Institute of Eco-Environmental Science & Technology, Guangzhou, P.R. China*
Duan, G.L: *Research Center for Eco-Environmental Sciences, Chinese Academy of Sciences, Beijing, P.R. China*
Duan, X.X.: *Department of Toxicology, School of Public Health, Shenyang Medical College, Shenyang, Liaoning, P.R. China*
Duan, Y.H.: *State Key Laboratory of Biogeology and Environmental Geology, China University of Geosciences, Wuhan, P.R. China; School of Environmental Studies, China University of Geosciences, Wuhan, P.R. China*
Duncan, E.: *Future Industries Institute, University of South Australia, Adelaide, SA, Australia*
Eiche, E.: *Institute of Applied Geosciences, KIT, Karlsruhe, Germany*
Elert, M.: *Kemakta Konsult AB, Stockholm, Sweden*
Ellis, T.: *Lamont-Doherty Earth Observatory of Columbia University, Palisades, NY, USA*
Ellwood, M.: *Research School of Earth Sciences, Australian National University, Canberra, ACT, Australia*
Eqani, S.A.M.A.S: *COMSATS Institute of Information Technology, Islamabad, Pakistan*
Espinoza Gonzales, R.: *Department of Chemical Engineering Biotechnology and Materials, Faculty of Physical and Mathematical Sciences, University of Chile, Santiago de Chile, Chile*
Eunus, M: *Columbia University Arsenic Research Project, Dhaka, Bangladesh*
Fakhreddine, S.: *Department of Civil and Environmental Engineering, Stanford University, Stanford, CA, USA*
Falnoga, I.: *Department of Environmental Sciences, Jožef Stefan Institute, Ljubljana, Slovenia*
Fan, C.: *Department of Chemistry and Biochemistry, Florida International University, Miami, FL, USA*

Fan, X.: *State Key Laboratory of Agricultural Microbiology, College of Life Science and Technology, Huazhong Agricultural University, Wuhan, P.R. China*
Fang, W.: *State Key Laboratory of Pollution Control and Resource Reuse, School of the Environment, Nanjing University, Jiangsu, P.R. China*
Farooqi, A.: *Environmental Geochemistry Laboratory, Department of Environmental Sciences, Faculty of Biological Sciences, Quaid-i-Azam University, Islamabad, Pakistan*
Fendorf, S.: *Department of Civil and Environmental Engineering, Stanford University, Stanford, CA, USA*
Fernandes, C.S.: *Environmental Engineering Graduating Program (ProAmb), Pharmacy Department (DEFAR), School of Pharmacy, Federal University of Ouro Preto, MG, Brazil*
Fernández-Cirelli, A.: *Instituto de Investigaciones en Producción Animal UBA-CONICET (INPA), CONICET – Universidad de Buenos Aires, Buenos Aires, Argentina Centro de Estudios Transdisciplinarios del Agua (CETA), Universidad de Buenos Aires, Buenos Aires, Argentina Cátedra de Qumica Orgánica de Biomoléculas, Universidad de Buenos Aires, Buenos Aires, Argentina*
Figoli, A.: *Institute on Membrane Technology (ITM-CNR), Rende (CS), Italy*
Filardi Vasques, I.C.: *Universidade Federal de Lavras, Lavras, MG, Brazil*
Fiol, S.: *Technological Research Institute, University of Santiago de Compostela, Santiago de Compostela, Spain*
Freeman, H.M.: *Helmholtz Centre Potsdam – GFZ German Research Centre for Geosciences, Potsdam, Germany*
Freitas, E.: *National Institute of Science and Technology on Mineral Resources, Water and Biodiversity, INCT-Acqua, Brazil; Universidade Federal de Minas Gerais-UFMG, Center of Microscopy, Belo Horizonte, Brazil*
Fryar, A.E.: *Department of Earth and Environmental Sciences, University of Kentucky, Lexington, KY, USA*
Fu, J.W.: *State Key Lab of Pollution Control and Resource Reuse, School of the Environment, Nanjing University, Jiangsu, P.R. China*
Fuoco, I.: *Department of Biology, Ecology and Earth Sciences (DIBEST), University of Calabria, Rende (CS), Italy*
Furlan, A.: *Departamento de Ciencias Naturales, Facultad de Ciencias Exactas, Fsico-Qumicas y Naturales, Universidad Nacional de Ro Cuarto, Córdoba, Argentina*
Gabriele, B.: *LISOC Group, Department of Chemistry and Chemical Technologies, University of Calabria, Rende (CS), Italy*
Gaikwad, S.: *Department of Chemistry, Savitribai Phule Pune University, Pune, India*
Gailer, J.: *Department of Chemistry, University of Calgary, Calgary, Canada*
Gan, Y.Q.: *State Key Laboratory of Biogeology and Environmental Geology, China University of Geosciences, Wuhan, P.R. China; School of Environmental Studies, China University of Geosciences, Wuhan, P.R. China*
Gao, B.: *School of Environmental Studies, China University of Geosciences, Wuhan, Hubei, P.R. China*
Gao, F.: *College of Resources and Environmental Sciences, Nanjing Agricultural University, Nanjing, P.R. China*
Gao, J.: *Geological Survey, China University of Geosciences, Wuhan, China*
Gao, X.B.: *School of Environmental Studies, China University of Geosciences, Wuhan, P.R. China*
Gao, Y.: *Analytical, Environmental and Geochemical Department (AMGC), Vrije Universiteit Brussel, Brussels, Belgium*
Garcia, K: *Instituto Mexicano de Tecnología del Agua, Subcoordinación de Posgrado, Jiutepec, Mor., Mexico*
Garcia, P.: *Universidad Nacional de Ingeniera, Lima, Peru*
Garcia-Chirino, J.: *Instituto de Ingeniera, Universidad Nacional Autónoma de México (UNAM), Mexico City, Mexico*
Garrido Hoyos, S.E.: *Instituto Mexicano de Tecnologa del Agua, Subcoordinación de Posgrado, Jiutepec, Mor., Mexico*
Gasparon, M.: *National Institute of Science and Technology on Mineral Resources, Water and Biodiversity, INCT-Acqua, Brazil; The University of Queensland, School of Earth and Environmental Sciences, St Lucia, Australia*
Gautret, P.: *BRGM, ISTO, Orléans, France*
Ge, Y.: *Demonstration Laboratory of Elements and Life Science Research, Laboratory Centre of Life Science, Nanjing Agricultural University, Nanjing, P.R. China*
George, G.N.: *Department of Geological Sciences, University of Saskatchewan, Saskatchewan, Canada*
Ghosal, P.S.: *Environmental Engineering Division, Department of Civil Engineering, Indian Institute of Technology, Kharagpur, India*
Ghosh, A.K.: *Mahavir Cancer Institute and Research Centre, Patna, India; Bihar State Pollution Control Board, Patna, India*

Ghosh, D.: *Centre for Earth Sciences, Indian Institute of Science, Bangalore, India*
Ghosh, P.: *Department of Chemistry, University of Kalyani, Kalyani, West Bengal, India*
Ghosh, S.K.: *Department of Public Health Engineering, Ministry of Local Government, Rural Development and Cooperatives, Dhaka, Bangladesh*
Ghosh, U.C.: *Department of Chemistry, Presidency University, Kolkata, India*
Gil, R.A.: *Laboratorio de Espectrometra de Masas – Instituto de Qumica de San Luis (CCT-San Luis), Área de Qumica Analtica, Universidad Nacional de San Luis, San Luis, Argentina*
Gimenez, M.C.: *Laboratorio de Microbiologa General, Universidad Nacional del Chaco Austral, P.R. Sáenz Peña, Chaco, Argentina*
Giménez-Forcada, E.: *Instituto Geológico y Minero de España – IGME, Salamanca, Spain*
Giri, A.: *Microbial Culture Collection, National Centre for Cell Science, Pune, India*
Giri, A.K.: *Molecular Genetics Division, CSIR-Indian Institute of Chemical Biology, Kolkata, India*
Glodowska, M.: *Geomicrobiology and Hydrology, University of Tübingen, Tübingen, Germany*
Gnanaprakasam, E.T.: *School of Earth and Environmental Sciences, The University of Manchester, UK*
Gomez, M.M.: *Universidad Nacional de Ingeniera, Lima, Peru*
Gong, P.L.: *School of Environmental Studies, China University of Geosciences, Wuhan, P.R. China*
Gonzalez, B.J.: *Faculty of Civil Engineering and Geosciences, Delft University of Technology, Delft, The Netherlands*
González-Chávez, J.L.: *Facultad de Qumica, Universidad Nacional Autónoma de México (UNAM), Mexico City, Mexico*
Gooddy, D.C.: *Groundwater Science Directorate, British Geological Survey, Wallingford, UK*
Gómez Samus, M.L.: *Lab. de Entrenamiento Multidisciplinario para la Investigación Tecnológica (LEMIT), La Plata, Argentina*
Gómez-Gómez, M.: *Department of Analytical Chemistry, Faculty of Chemistry, Universidad Complutense de Madrid, Madrid, Spain*
Gracia Caroca, F.: *Department of Chemical Engineering Biotechnology and Materials, Faculty of Physical and Mathematical Sciences, University of Chile, Santiago de Chile, Chile*
Graziano, J.H.: *Mailman School of Public Health, Columbia University, New York City, NY, USA; Department of Environmental Health, Columbia University New York, New York, NY, USA*
Grosbois, C.: *University of Tours, Tours, France*
Guan, X.: *State Key Laboratory of Pollution Control and Resources Reuse, Tongji University, Shanghai, P.R. China*
Guay, M.: *Health Canada*
Gude, J.C.J.: *Department Water Management, Section Sanitary Engineering, Delft University of Technology, The Netherlands*
Guilherme, L.R.G.: *Department of Soil Science, Federal University of Lavras, Minas Gerais, Brazil*
Guo, H.M.: *State Key Laboratory of Biogeology and Environmental Geology, China University of Geosciences, Beijing, P.R. China; School of Water Resources and Environment, China University of Geosciences (Beijing), Beijing, P.R. China*
Guo, J.: *School of Chemistry and Environment, South China Normal University, Guangzhou, P.R. China*
Guo, L.D.: *School of Freshwater Sciences, University of Wisconsin-Milwaukee, WI, USA*
Guo, Q.H.: *State Key Laboratory of Biogeology and Environmental Geology & School of Environmental Studies, China University of Geosciences, Hubei, P.R. China*
Guo, Y.Y.: *Environment and Non-Communicable Disease Research Center, Key Laboratory of Arsenic-related Biological Effects and Prevention and Treatment in Liaoning Province, School of Public Health, China Medical University, Shenyang, P.R. China*
Gupta, A.K.: *Environmental Engineering Division, Department of Civil Engineering, Indian Institute of Technology, Kharagpur, India*
Gupta, K.: *Department of Chemistry, Presidency University, Kolkata, India*
Gustave, W.: *Department of Environmental Science, Xi'an Jiaotong-Liverpool University, Suzhou, Jiangsu, P.R. China*
Haidari, A.H.: *Faculty of Civil Engineering and Geosciences, Delft University of Technology, Delft, The Netherlands*
Han, S.B.: *Center for Hydrogeology and Environmental Geology Survey, China Geological Survey, Baoding, China*

Han, Y.H.: *State Key Lab of Pollution Control and Resource Reuse, School of the Environment, Nanjing University, Jiangsu, P.R. China*
Harris, H.H.: *Cooperative Research Centre for Contamination Assessment and Remediation of the Environment (CRC CARE), Newcastle, NSW, Australia*
Hasibuzzaman, M.M.: *Department of Biochemistry and Molecular Biology, University of Rajshahi, Rajshahi, Bangladesh*
Hassan, M.: *Village Education and Resource Center, Dhaka, Bangladesh*
Hatakka, T.: *Geological Survey of Finland, Espoo, Finland*
He, M.C.: *State Key Laboratory of Water Environment Simulation, School of Environment, Beijing Normal University, Beijing, P.R. China*
Heijman, S.G.J.: *Faculty of Civil Engineering and Geosciences, Delft University of Technology, Delft, The Netherlands*
Hellal, J.: *BRGM, Orléans, France*
Hellriegel, U.: *Center of Applied Research, Karlsruhe University of Applied Sciences, Karlsruhe, Germany; Department of Mechatronics and Sensor Systems Technology, Vietnamese-German University, Binh Duong Province, Vietnam*
Herath, I.: *Faculty of Health, Engineering and Sciences, University of Southern Queensland, Toowoomba, Queensland, Australia*
Himeno, S.: *Laboratory of Molecular Nutrition and Toxicology, Faculty of Pharmaceutical Sciences, Tokushima Bunri University, Tokushima, Japan*
Hirai, N.: *Department of Biology and Geosciences, Osaka City University, Sumiyoshi-ku, Osaka, Japan*
Hocevar, B.: *Department of Environmental and Occupational Health, School of Public Health, Indiana University, Bloomington, Indiana; USA Department of Environmental and Occupational Health, Indiana University Bloomington, Bloomington, IN, USA*
Hofs, B.: *Department of Watertechnology, Evides Waterbedrijf, Rotterdam, The Netherlands*
Hoinkis, J.: *Center of Applied Research, Karlsruhe University of Applied Sciences, Karlsruhe, Germany; Department of Mechatronics and Sensor Systems Technology, Vietnamese-German University, Binh Duong Province, Vietnam*
Hong, Y.: *School of Chemistry and Environment, South China Normal University, Guangzhou, P.R. China*
Honma, T.: *Niigata Agricultural Research Institute, Nagaoka, Japan*
Hoque, B.A.: *Environment and Population Research Centre (EPRC), Dhaka, Bangladesh*
Hoque, M.M.: *Environment and Population Research Centre (EPRC), Dhaka, Bangladesh*
Horvat, M.: *Department of Environmental Sciences, Jožef Stefan Institute, Ljubljana, Slovenia; Jožef Stefan International Postgraduate School, Ljubljana, Slovenia*
Hosomi, M.: *Department of Chemical Engineering, Tokyo University of Agriculture and Technology, Tokyo, Japan*
Hossain, K.: *Department of Biochemistry and Molecular Biology, University of Rajshahi, Rajshahi, Bangladesh*
Hou, D.Y.: *School of Environment, Tsinghua University, Beijing, P.R. China*
Hou, L.: *Research Institute of Rural Sewage Treatment, Southwest Forestry University, Kunming, P.R. China; College of Ecology and Soil & Water Conservation, Southwest Forestry University, Kunming, P.R. China*
Hsu, K.H.: *Department of Health Care Management, Chang-Gung University, Taoyuan, Taiwan*
Hsu, L.I.: *Genomics Research Centre, Academia Sinica, Taipei, Taiwan*
Hua, C.Y.: *State Key Lab of Pollution Control and Resource Reuse, School of the Environment, Nanjing University, Jiangsu, P.R. China*
Huang, H.: *Research and Development Department, China State Science Dingshi Environmental Engineering Co., Ltd, Beijing, P.R. China*
Huang, K.: *College of Resources and Environmental Sciences, Nanjing Agricultural University, Nanjing, P.R. China*
Hube, D.: *BRGM, Orléans, France*
Huq, M.E.: *School of Environmental Studies, China University of Geosciences (Wuhan), Wuhan, Hubei, P.R. China*
Huque, S.: *Environment and Population Research Centre (EPRC), Dhaka, Bangladesh*
Hussain, I.: *Environmental Geochemistry Laboratory, Faculty of Biological Sciences, Department of Environmental Sciences, Quaid-i-Azam University, Islamabad, Pakistan*

Huynh, T.: *Centre for Mined Land Rehabilitation, Sustainable Minerals Institute, The University of Queensland, Brisbane, QLD, Australia*

Ijumulana, J.: *DAFWAT Research Group, Department of Water Resources Engineering, College of Engineering and Technology, University of Dar es Salaam, Dar es Salaam, Tanzania; KTH-International Groundwater Arsenic Research Group, Department of Sustainable Development, Environmental Science and Engineering, KTH Royal Institute of Technology, Stockholm, Sweden*

Iriel, A.: *Agua y Saneamientos Argentinos S.A. (AySA S.A.) and Centro de Estudios Transdisciplinarios del Agua, Universidad de Buenos Aires and BuenosAires, Argentina*

Irunde, R.: *DAFWAT Research Group, Department of Water Resources Engineering, College of Engineering and Technology, University of Dar es Salaam, Dar es Salaam, Tanzania; KTH-International Groundwater Arsenic Research Group, Department of Sustainable Development, Environmental Science and Engineering, KTH Royal Institute of Technology, Stockholm, Sweden*

Isela, M.F.: *Posgrado de Ingeniera, UNAM, Instituto Mexicano de Tecnologa del Agua, Jiutepec, Mor., Mexico*

Ishikawa, S.: *Institute for Agro-Environmental Sciences, NARO, Tsukuba, Ibaraki, Japan*

Islam, A.B.M.R.: *Department of Occupational and Environmental Health, Bangladesh University of Health Sciences, Dhaka, Bangladesh; Coordination of Environment and Life Line (CELL), Next International Co. Ltd, Sangenjaya, Tokyo, Japan*

Islam, M.S.: *Department of Biochemistry and Molecular Biology, University of Rajshahi, Rajshahi, Bangladesh; Department of Applied Nutrition and Food Technology, Islamic University, Kushtia, Bangladesh*

Islam, M.T.: *KTH-International Groundwater Arsenic Research Group, Department of Sustainable Development, Environmental Sciences and Engineering, KTH Royal Institute of Technology, Stockholm, Sweden*

Islam, S.: *Global Centre for Environmental Remediation (GCER), Faculty of Science, The University of Newcastle, Callaghan, NSW, Australia, Cooperative Research Centre for Contamination Assessment and Remediation of the Environment (CRC CARE), The University of Newcastle, Callaghan, NSW, Australia; Department of Soil Science, Bangladesh Agricultural University, Mymensingh, Bangladesh*

Islam, T.: *Columbia University Arsenic Research Project, Dhaka, Bangladesh*

Jackova, I.: *Czech Geological Survey, Prague, Czech Republic*

Jacks, G.: *Division of Water and Environmental Engineering, Department of Sustainable Development, Environmental Science and Engineering, KTH Royal Institute of Technology, Stockholm, Sweden*

Jakariya, M.: *Department of Environmental Science and Management, North South University, Dhaka, Bangladesh*

Jamieson, J.: *University of Western Australia and CSIRO Land and Water, Perth, Australia*

Javed, A.: *Department of Earth and Environmental Sciences, Bahria University, Islamabad, Pakistan; Environmental Geochemistry Laboratory, Department of Environmental Sciences, Faculty of Biological Sciences, Quaid-i-Azam University, Islamabad, Pakistan*

Jayawardhana, Y.: *Environmental Chemodynamics Project, National Institute of Fundamental Studies, Kandy, Sri Lanka*

Jeworrek, A.: *Department of Environmental Technology, Wageningen University and Research (WUR), Wageningen, The Netherlands*

Jia, M.R.: *State Key Lab of Pollution Control and Resource Reuse, School of the Environment, Nanjing University, Jiangsu, P.R. China*

Jia, X.Y.: *School of Environment, Tsinghua University, Beijing, P.R. China*

Jia, Y.F.: *State Key Laboratory of Environmental Criteria and Risk Assessment, Chinese Research Academy of Environmental Sciences, Beijing, P.R. China; State Environmental Protection Key Laboratory of Simulation and Control of Groundwater Pollution, Chinese Research Academy of Environmental Sciences, Beijing, P.R. China*

Jiang, F.: *School of Chemistry and Environment, South China Normal University, Guangzhou, P.R. China*

Jiang, H.C.: *State Key Laboratory of Biogeology and Environmental Geology, China University of Geosciences, Wuhan, P.R. China*

Jiang, Y.H.: *State Key Laboratory of Environmental Criteria and Risk Assessment, Chinese Research Academy of Environmental Sciences, Beijing, P.R. China; State Environmental Protection Key Laboratory of Simulation and Control of Groundwater Pollution, Chinese Research Academy of Environmental Sciences, Beijing, P.R. China*

Jiang, Z: *State Key Laboratory of Biogeology and Environmental Geology, China University of Geosciences, Wuhan, P.R. China*

Jing, C.Y.: *Research Center for Eco-Environmental Sciences, Chinese Academy of Sciences, Beijing, P.R. China*
Johnston, D.: *UNICEF Bangladesh, Dhaka, Bangladesh*
Jones-Johansson, C.: *Kemakta Konsult AB, Stockholm, Sweden*
Jordan, I.: *G.E.O.S. Ingenieurgesellschaft mbH, Halsbrücke, Sachsen, Germany*
Joshi, H.: *Department of Hydrology, Indian Institute of Technology, Roorkee, India*
Joulian, C.: *BRGM, Orléans, France*
Jovanović, D.D.: *Institute of Public Health of Serbia "Dr Milan Jovanović Batut", Belgrade, Serbia*
Juhasz, A.L.: *Centre for Environmental Risk Assessment and Remediation, University of South Australia, Mawson Lakes, SA, Australia*
Júnior, F.M.: *Laboratório de Ensaios Farmacológicos e Toxicológicos, Instituto de Ciências Biológicas, Universidade Federal do Rio Grande (FURG), Rio Grande, RS, Brazil*
Kaija, J.: *Geological Survey of Finland, Espoo, Finland*
Kamendulis, L.: *Department of Environmental and Occupational Health, School of Public Health, Indiana University, Bloomington, Indiana; USA Department of Environmental and Occupational Health, Indiana University Bloomington, Bloomington, IN, USA*
Kang, I.G.: *College of Medicine, Chung-Ang University, Seoul, South Korea*
Kappler, A.: *Geomicrobiology and Hydrology, University of Tübingen, Tübingen, Germany*
Kar, K.K.: *Department of Mechanical Engineering, IIT Kanpur, India*
Karana, E.: *Faculty of Industrial Design Engineering, Delft University of Technology, Delft, The Netherlands*
Karthikeyan, S.: *Health Canada*
Kashif Hayat, A.B.: *School of Agriculture and Biology, Key Laboratory of Urban Agriculture, Ministry of Agriculture, Bor S. Luh Food Safety Research Center, Shanghai Jiao Tong University, Shanghai, China; Department of Plant Sciences, Faculty of Biological Sciences, Quaid-i-Azam University, Islamabad, Pakistan*
Kasiuliene, A.: *Waste Science and Technology Research Group, Department of Civil, Environmental and Natural Resources Engineering, Lulea University of Technology, Lulea, Sweden*
Kassenga, G.R.: *Department of Environmental Science and Management, Ardhi University, Dar es Salaam, Tanzania*
Katou, H.: *Institute for Agro-Environmental Sciences, NARO, Tsukuba, Japan*
Ke, T.T.: *School of Water Resources and Environment, China University of Geosciences (Beijing), Beijing, P.R. China*
Kent, D.B.: *U.S. Geological Survey, Menlo Park, CA, USA*
Khan, E.: *Department of Public Health Engineering, Dhaka, Bangladesh*
Khan, K.M: *Department of Environmental and Occupational Health, School of Public Health, Indiana University, Bloomington, Indiana; USA Department of Environmental and Occupational Health, Indiana University Bloomington, Bloomington, IN, USA*
Khan, M.R.: *Department of Geology, University of Dhaka, Dhaka, Bangladesh*
Khanam, S.: *Environment and Population Research Centre (EPRC), Dhaka, Bangladesh*
Khattak, J.A.: *Environmental Geochemistry Laboratory, Faculty of Biological Sciences, Department of Environmental Sciences, Quaid-i-Azam University, Islamabad, Pakistan*
Khiadani, M.: *School of Engineering, Edith Cowan University, Joondalup, WA, Australia*
Kim, H.: *Chungbuk National University, Cheongju, South Korea*
Kipfer, R.: *Eawag, Swiss Federal Institute of Aquatic Science and Technology, Dübendorf, Switzerland*
Kitanidis, P.K.: *Department of Civil and Environmental Engineering, Stanford University, Stanford, CA, USA*
Kodam, K.: *Department of Chemistry, Savitribai Phule Pune University, Pune, India*
Kong, S.Q.: *School of Environmental Studies and State Key Laboratory of Biogeology and Environmental Geology, China University of Geosciences, Wuhan, P.R. China; Laboratory of Basin Hydrology and Wetland Eco-restoration, School of Environmental Studies, China University of Geosciences, Wuhan, P.R. China*
Kontny, A.: *Institute of Applied Geosciences, KIT, Karlsruhe, Germany*
Korevaar, M.W.: *KWR Watercycle Research Institute, Nieuwegein, The Netherlands*
Kraal, P.: *Department of Earth Sciences-Geochemistry, Faculty of Geosciences, Utrecht University, Utrecht, The Netherlands*
Krikowa, F.: *Institute for Applied Ecology, University of Canberra, Canberra, ACT, Australia*
Krohn, R.M.: *Faculty of Veterinary Medicine, University of Calgary, Calgary, Alberta, Canada*
Kshirsagar, S.: *Department of Chemistry, Savitribai Phule Pune University, Pune, India*
Kulkarni, H.: *Department of Geology, Kansas State University, Manhattan, KS, USA*

Kumar, A.: *Department of Chemistry, Indian Institute of Technology, Roorkee, India*
Kumar, A.: *Department of Hydrology, Indian Institute of Technology, Roorkee, India*
Kumar, A.: *Mahavir Cancer Institute and Research Centre, Patna, Bihar, India*
Kumar, D.: *Exceldot AB, Bromma, Sweden*
Kumar, M.: *Department of Environmental Science, School of Earth, Environment & Space Studies, Central University of Haryana, Jant Pali, Mahendergarh, India; School of Environmental Sciences, Jawaharlal Nehru University, New Delhi, India*
Kumar, R.: *Mahavir Cancer Institute & Research Centre, Patna, Bihar, India*
Kumar, S.: *Department of Applied Geoscience and Engineering, Delft University of Technology, Delft, The Netherlands*
Kumarathilaka, P.: *School of Civil Engineering and Surveying, Faculty of Health, Engineering and Sciences, University of Southern Queensland, Toowoomba, QLD, Australia*
Kumari, P.: *Mahavir Cancer Institute and Research Center, Patna, India*
Kumari, S.: *B.R.A. Bihar University, Muzaffarpur, Bihar, India; Mahavir Cancer Institute and Research Center, Patna, India*
Kumpiene, J.: *Waste Science and Technology Research Group, Department of Civil, Environmental and Natural Resources Engineering, Lulea University of Technology, Lulea, Sweden*
Kuramata, M.: *Institute for Agro-Environmental Sciences, NARO, Tsukuba, Ibaraki, Japan*
Kwon, H.J.: *Dankook University, Cheongju, South Korea*
Lal, V.: *Queensland Alliance for Environmental Health Sciences (QAEHS), The University of Queensland, Brisbane, QLD, Australia*
Lan, V.M.: *CETASD, Vietnam National University, Hanoi, Vietnam*
LaPorte, P.F.: *Division of Hematology-Oncology, University of California Los Angeles, Los Angeles, California, USA*
Lapworth, D.J.: *British Geological Survey, Wallingford, UK*
Larebeke, N.V.: *Analytical, Environmental and Geochemical Department (AMGC), Vrije Universiteit Brussel, Brussels, Belgium*
Lata, L.: *Department of Soil Science/Geology, Maria Curie-Sklodowska University, Lublin, Poland*
Lauer, F.T.: *College of Pharmacy, University of New Mexico, Albuquerque, NM, USA*
Launder, J.: *School of Earth and Environmental Sciences and Williamson Research Centre for Molecular Environmental Science, University of Manchester, Manchester, UK*
Le Forestier, L.: *BRGM, ISTO, Orléans, France*
Le Guédard, M.: *LEB Aquitaine Transfert, Villenave d'Ornon, France*
Lee, J.H.: *Department of Civil and Environmental Engineering, Stanford University, Stanford, CA, USA; Department of Environmental Engineering, University of Hawaii, Honolulu, HA, USA*
Lee, S.G.: *College of Medicine, Chung-Ang University, Seoul, South Korea*
Leermakers, M.: *Analytical, Environmental and Geochemical Department (AMGC), Vrije Universiteit Brussel, Brussels, Belgium*
Lei, M.: *College of Resource & Environment, Hunan Agricultural University, Changsha, P.R. China*
Lezama-Pacheco, J.: *Department of Earth System Science, Stanford University, Stanford, CA, USA*
Lhotka, M.: *University of Chemistry and Technology Prague, Prague, Czech Republic*
Li, B.: *Environment and Non-Communicable Disease Research Center, Key Laboratory of Arsenic-related Biological Effects and Prevention and Treatment in Liaoning Province, School of Public Health, China Medical University, Shenyang, P.R. China*
Li, B.: *College of Resource & Environment, Hunan Agricultural University, Changsha, P.R. China*
Li, B.Q.: *Guangdong Key Laboratory of Agricultural Environment Pollution Integrated Control, Guangdong Institute of Eco-Environmental Science & Technology, Guangzhou, P.R. China*
Li, C.C.: *School of Environmental Studies, China University of Geosciences, Wuhan, P.R. China*
Li, C.H.: *College of Resources and Environmental Sciences, Jiangsu Provincial Key Laboratory of Marine Biology, Nanjing Agricultural University, Nanjing, P.R. China*
Li, F.: *School of Environmental Studies, China University of Geosciences, Wuhan, Hubei, P.R. China*
Li, F.B.: *Guangdong Key Laboratory of Integrated Agro-environmental Pollution Control and Management, Guangdong Institute of Eco-Environmental Science & Technology, Guangzhou, P.R. China*
Li, H.B.: *State Key Laboratory of Pollution Control and Resource Reuse, School of the Environment, Nanjing University, Nanjing, P.R. China*

Li, J.: *Herbert Wertheim College of Medicine, Florida International University, Miami, FL, USA; College of Basic Medicine, Dali University, Dali, Yunnan, P.R. China*
Li, J.L.: *School of Environmental Sciences, China University of Geosciences, Wuhan, P.R. China*
Li, J.L.: *Department of Occupational and Environmental Health, Key Laboratory of Occupational Health and Safety for Coal Industry in Hebei Province, School of Public Health, North China University of Science and Technology, Tangshan, Hebei, P.R. China*
Li, J.N.: *Department of Chemical Engineering, Tokyo University of Agriculture and Technology, Tokyo, Japan; Research Fellow of Japan Society for the Promotion of Science, Tokyo, Japan*
Li, J.X.: *School of Environmental Studies & State Key Laboratory of Biogeology and Environmental Geology, China University of Geosciences, Wuhan, China*
Li, N.: *College of Resources and Environment, Yunnan Agricultural University, Kunming, P.R. China*
Li, P.: *PS Analytical Ltd, Arthur House, Orpington, Kent, UK*
Li, P.: *State Key Laboratory of Biogeology and Environmental Geology, China University of Geosciences, Wuhan, P.R. China*
Li, Q.: *Central South University, Changsha, Hunan, P.R. China*
Li, Q.: *College of Resources and Environment, Yunnan Agricultural University, Kunming, P.R. China*
Li, X.: *Central South University, Changsha, Hunan, P.R. China*
Li, X.: *Research Center of Environment and Non-Communicable Disease, School of Public Health, China Medical University, Shenyang, P.R. China*
Li, X.M.: *Guangdong Key Laboratory of Integrated Agro-environmental Pollution Control and Management, Guangdong Institute of Eco-Environmental Science & Technology, Guangzhou, P.R. China*
Li, Y.: *School of Environmental Studies, China University of Geosciences, Wuhan, P.R. China*
Li, Y.: *Research Center of Environment and Non-Communicable Disease, School of Public Health, China Medical University, Shenyang, P.R. China*
Li, Y.T.: *Agro-Environmental Protection Institute, Ministry of Agriculture, Tianjin, P.R. China*
Liamtsau, V: *Department of Chemistry and Biochemistry, Florida International University, Miami, FL, USA; Southeast Environmental Research Center, Florida International University, Miami, FL, USA*
Liang, J.H.: *Key Laboratory of Karst Ecosystem and Treatment of Rocky Desertification, Institute of Karst Geology, Chinese Academy of Geological Sciences, Guilin, P.R. China*
Ligate, F.J. *DAFWAT Research Group, Department of Water Resources Engineering, College of Engineering and Technology, University of Dar es Salaam, Dar es Salaam, Tanzania; KTH-International Groundwater Arsenic Research Group, Department of Sustainable Development, Environmental Science and Engineering, KTH Royal Institute of Technology, Stockholm, Sweden*
Lightfoot, A.: *Eawag, Swiss Federal Institute of Aquatic Science and Technology, Dübendorf, Switzerland*
Lin, C.Y.: *State Key Laboratory of Water Environment Simulation, School of Environment, Beijing Normal University, Beijing, P.R. China*
Lin, H.: *School of Freshwater Sciences, University of Wisconsin-Milwaukee, WI, USA*
Lin, L.: *Research Center of Environment and Non-Communicable Disease, School of Public Health, China Medical University, Shenyang, P.R. China*
Lions, J.: *BRGM, Orléans, France*
Litter, M.I.: *Comisión Nacional de Energa Atómica, Buenos Aires, Argentina*
Liu, C.P.: *Guangdong Key Laboratory of Integrated Agro-environmental Pollution Control and Management,Guangdong Institute of Eco-Environmental Science & Technology, Guangzhou, P.R. China*
Liu, F.: *School of Environmental Studies, China University of Geosciences, Wuhan, P.R. China*
Liu, G.: *Department of Chemistry and Biochemistry, Florida International University, Miami, FL, USA*
Liu, H.: *State Key Laboratory of Biogeology and Environmental Geology, China University of Geosciences, Wuhan, P.R. China*
Liu, R.: *School of Environmental Studies, China University of Geosciences (Wuhan), Wuhan, Hubei, P.R. China*
Liu, W.J.: *College of Resources and Environmental Science, Hebei Agricultural University, Baoding, P.R. China*
Liu, X.: *Mailman School of Public Health, Columbia University, New York City, NY, USA*
Liu, X.: *Research Center for Soil Contamination and Environment Remediation, Southwest Forestry University, Kunming, P.R. China*
Liu, X.: *State Key Lab of Pollution Control and Resource Reuse, School of the Environment, Nanjing University, Jiangsu, P.R. China*

Liu, Y.C.: *Research and Development Department, China State Science Dingshi Environmental Engineering Co., Ltd, Beijing, P.R. China*
Liu, Y.F.: *School of Environmental Studies, China University of Geosciences, Wuhan, P.R. China*
Liu, Y.G.: *Research Institute of Rural Sewage Treatment, Southwest Forestry University, Kunming, P.R. China; College of Ecology and Soil & Water Conservation, Southwest Forestry University, Kunming, P.R. China*
Lloyd, J.R.: *School of Earth and Environmental Sciences, The University of Manchester, UK*
Lopez, B.: *Universidad Politécnica del Estado de Morelos. Jiutepec, Mor., Mexico*
Loukola-Ruskeeniemi, K.: *Geological Survey of Finland, Espoo, Finland*
Lu, Z.J.: *Geological Survey, China University of Geosciences, Wuhan, P.R. China*
Luo, J.: *State Key Laboratory of Pollution Control and Resource Reuse, School of the Environment, Nanjing University, Jiangsu, P.R. China*
Luo, L.: *Hunan Agricultural University, Changsha, Hunan, P.R. China*
Luo, W.T.: *School of Environmental Studies, China University of Geosciences, Wuhan, P.R. China*
Luo, Z.X.: *Key Laboratory of Urban Environment and Health, Institute of Urban Environment, Chinese Academy of Sciences, Xiamen, P.R. China; College of Chemistry and Environment, Fujian Province Key Laboratory of Modern Analytical Science and Separation Technology, Minnan Normal University, Zhangzhou, P.R. China*
Luong, T.V.: *Vietnamese-German University, Ho Chi Minh City, Vietnam; UNESCO Chair on Groundwater Arsenic within the 2030 Agenda for Sustainable Development & Faculty of Health, Engineering and Sciences, University of Southern Queensland, Toowoomba, Australia*
Luong, V.T.: *Faculty of Health, Engineering and Sciences, University of Southern Queensland, Toowoomba, QLD, Australia*
Lv, Y.H.: *Guangdong Key Laboratory of Integrated Agro-environmental Pollution Control and Management, Guangdong Institute of Eco-Environmental Science & Technology, Guangzhou, P.R. China*
Ma, C.L.: *State Key Laboratory of Water Environment Simulation, School of Environment, Beijing Normal University, Beijing, P.R. China*
Ma, L.Q.: *Research Center for Soil Contamination and Environment Remediation, Southwest Forestry University, Kunming, P.R. China; Soil and Water Sciences Department, University of Florida, Gainesville, FL, USA*
Ma, M.: *College of Engineering, Peking University, Beijing, China; School of Environmental Science and Technology, Southern University of Science and Technology, Shenzhen, China*
Ma, R.: *School of Environmental Studies, China University of Geosciences, Wuhan, P.R. China; Laboratory of Basin Hydrology and Wetland Eco-restoration, China University of Geosciences, Wuhan, P.R. China*
Ma, T.: *School of Water Resources and Environmental Engineering, East China University of Technology, Nanchang, P.R. China*
Ma, Y.B.: *Institute of Agricultural Resources and Regional Planning, Chinese Academy of Agricultural Sciences, Beijing, P.R. China*
Machado, I.: *Analytical Chemistry, Faculty of Chemistry, DEC, Universidad de la República (UdelaR), Montevideo, Uruguay*
Machovic, V.: *University of Chemistry and Technology Prague, Prague, Czech Republic*
Magnone, D.: *School of Earth, Atmospheric and Environmental Sciences and Williamson Research Centre for Molecular Environmental Science, University of Manchester, Manchester, UK; School of Geography, University of Lincoln, Lincoln, UK*
Maguffin, S.C.: *Civil & Environmental Engineering, Cornell University, Ithaca, NY, USA; Dale Bumpers National Rice Research Center, Stuttgart, AR, USA*
Mahanta, C.: *Department of Civil Engineering, Indian Institute of Technology, Guwahati, Assam, India*
Mahdyarfar, M.: *Delta Niroo Gameron Co., Tehran, Iran*
Maher, W.: *Institute for Applied Ecology, University of Canberra, Canberra, ACT, Australia*
Mahmud, M.N.: *UNICEF Bangladesh, Dhaka, Bangladesh*
Mailloux, B.J.: *Barnard College and Lamont-Doherty Earth Observatory, Columbia University, New York, USA; Department of Environmental Sciences, Barnard College, New York, NY, USA*
Majumdar, A.: *Earth and Environmental Science Research Laboratory, Department of Earth Sciences, Indian Institute of Science Education and Research Kolkata (IISER-K), Mohanpur, West Bengal, India*
Makino, T.: *Institute for Agro-Environmental Sciences, NARO, Tsukuba, Japan*
Mallick, S.: *CSIR – National Botanical Research Institute, Lucknow, Uttar Pradesh, India*
Mancuso, R.: *LISOC Group, Department of Chemistry and Chemical Technologies, University of Calabria, Rende (CS), Italy*

Mandal, D.: *Sripat Singh College, Jiaganj, Murshidabad, India*
Mandal, U.: *Department of Chemistry, University of Kalyani, Kalyani, Nadia, West Bengal, India*
Manojlović, D.D.: *Institute of Chemistry, Technology and Metallurgy, Center of Chemistry, Belgrade, Serbia*
Mañay, N.: *Toxicology, Faculty of Chemistry, DEC, Universidad de la República (UdelaR), Montevideo, Uruguay*
Martin, H.P.: *Research School of Earth Sciences, Australian National University, Canberra, ACT, Australia*
Martin, M.: *Dipartimento di Scienze Agrarie, Forestali e Alimentari, University of Torino, Turin, Italy*
Martin-Domnguez, I.R.: *Centro de Investigación en Materiales Avanzados SC (CIMAV), Chihuahua, Chih., Mexico*
Martins, G.C.: *Department of Soil Science, Federal University of Lavras, Minas Gerais, Brazil*
Masuda, H.: *Department of Biology and Geosciences, Osaka City University, Sumiyoshi-ku, Osaka, Japan*
Mato, R.R.A.M.: *Department of Environmental Science and Management, Ardhi University, Dar es Salaam, Tanzania*
Matthews, H.: *School of Health Sciences, University of Salford, Salford, UK*
Mazej, D.: *Department of Environmental Sciences, Jožef Stefan Institute, Ljubljana, Slovenia*
McClung, A.: *Civil & Environmental Engineering, Cornell University, Ithaca, NY, USA; Dale Bumpers National Rice Research Center, Stuttgart, AR, USA*
Meharg, A.A.: *Institute for Global Food Security, Queens University Belfast, Belfast, UK*
Meharg, C.: *Institute for Global Food Security, Queens University Belfast, Belfast, UK*
Mei, X.Y.: *College of Resources and Environment, Yunnan Agricultural University, Kunming, P.R. China*
Mella, F.P.: *EcoMetales Limited, Providencia, Región Metropolitana, Santiago de Chile, Chile*
Menan, L.C.: *Department of Civil Engineering, Indian Institute of Technology, Guwahati, India*
Mendoza, R.R.: *Facultad de Qumica, Universidad Autónoma de Querétaro, Santiago de Querétaro, Qro, Mexico*
Meng, C.: *Hunan Agricultural University, Changsha, Hunan, P.R. China*
Mercado-Borrayo, B.M.: *Instituto de Ingeniera, Universidad Nacional Autónoma de México (UNAM),Mexico City, Mexico*
Mercedes, E.R.C.: *Posgrado de Ingeniera, UNAM, Instituto Mexicano de Tecnologa del Agua, Jiutepec, Mor., Mexico*
Milosh, H.: *Department of Soil Science/Geology, Maria Curie-Sklodowska University, Lublin, Poland*
Mink, A.: *Watermanagement Department, Faculty of Civil Engineering and Geosciences, Delft University of Technology, Delft, The Netherlands*
Mirlean, N.: *Laboratório de Oceanografia Geológica, Instituto de Oceanografia, Universidade Federal do Rio Grande (FURG), Rio Grande, RS, Brazil*
Mohan, D.: *School of Environmental Sciences, Jawaharlal Nehru University, New Delhi, India*
Mondal, D.: *School of Environment & Life Sciences, University of Salford, Salford, UK*
Monzon, E.H.: *Laboratorio de Microbiologa General, Universidad Nacional del Chaco Austral, P.R. Sáenz Peña, Chaco, Argentina*
Morais, M.: *Kinross Brasil Mineração, Brazil*
Moreno, R.V.: *Facultad de Qumica, Universidad Autónoma de Querétaro, Santiago de Querétaro, Qro, Mexico*
Mortensen, R.: *SDC University, Beijing, China*
Motelica, M.: *BRGM, ISTO, Orléans, France*
Mtalo, F.: *DAFWAT Research Group, Department of Water Resources Engineering, College of Engineering and Technology, University of Dar es Salaam, Dar es Salaam, Tanzania*
Mtamba, J.: *DAFWAT Research Group, Department of Water Resources Engineering, College of Engineering and Technology, University of Dar es Salaam, Dar es Salaam, Tanzania*
Mueller, I.: *Saxon State Office for Environment, Agriculture and Geology, Freiberg, Germany*
Mukherjee, A.: *Department of Geology and Geophysics, Indian IIT Kharagpur, Kharagpur, West Bengal, India; School of Environmental Science and Engineering, IIT Kharagpur, Kharagpur, West Bengal, India; Applied Policy Advisory to Hydrogeosciences Group, IIT Kharagpur, Kharagpur, West Bengal, India*
Mukhopadhyay, S.: *Kalyani Institute for Study, Planning and Action for Rural Change (KINSPARC), Kalyani West Bengal, India*
Müller, S.: *Department of Earth Sciences-Geochemistry, Faculty of Geosciences, Utrecht University, Utrecht, The Netherlands*
Mushtaq, N.: *Environmental Geochemistry Laboratory, Faculty of Biological Sciences, Department of Environmental Sciences, Quaid-i-Azam University, Islamabad, Pakistan*

Muthusamy, S.: *Queensland Alliance for Environmental Health Sciences (QAEHS), The University of Queensland, Brisbane, QLD, Australia*
Nadar, V.S.: *Department of Cellular Biology and Pharmacology, Herbert Wertheim College of Medicine, Florida International University, Miami, Florida, USA*
Naidu, R.: *Global Centre for Environmental Remediation (GCER), Faculty of Science, The University of Newcastle, Callaghan, NSW, Australia, Cooperative Research Centre for Contamination Assessment and Remediation of the Environment (CRC CARE), The University of Newcastle, Callaghan, NSW, Australia*
Nakamura, K.: *Institute for Agro-Environmental Sciences, NARO, Tsukuba, Japan*
Nandre, V.: *Department of Chemistry, Savitribai Phule Pune University, Pune, India*
Natal-da-Luz, T.: *Centre for Functional Ecology, Department of Life Sciences, University of Coimbra, Coimbra, Portugal*
Nath, B.: *Lamont-Doherty Earth Observatory of Columbia University, Palisades, NY, USA*
Navarro Solis, H.I.: *Escuela de Ingeniera Ambiental, Universidad Popular Autónoma del Estado de Puebla A.C., Puebla, Pue., Mexico*
Navin, S.: *Mahavir Cancer Institute & Research Centre, Patna, Bihar, India*
Neumann, T.: *Institute of Applied Geosciences, KIT, Karlsruhe, Germany*
Ng, J.C.: *Queensland Alliance for Environmental Health Sciences (QAEHS), The University of Queensland, Brisbane, QLD, Australia; Cooperative Research Centre for Contamination Assessment and Remediation of the Environment (CRC CARE), Newcastle, NSW, Australia*
Nghiem, A.A.: *Barnard College and Lamont-Doherty Earth Observatory, Columbia University, New York, USA*
Niazi, N.K.: *Institute of Soil and Environmental Sciences, University of Agriculture Faisalabad, Faisalabad, Pakistan*
Niraj, P.K.: *Mahavir Cancer Institute & Research Centre, Patna, Bihar, India*
Noller, B.N.: *Centre for Mined Land Rehabilitation, Sustainable Minerals Institute, The University of Queensland, Brisbane, QLD, Australia*
Norini, M.P.: *BRGM, ISTO, Orléans, France*
Ochoa, J.: *Instituto Nacional de Investigaciones Forestales, Agrcolas y Pecuarias, Campo Experimental La Laguna, Matamoros, Coah., Mexico*
Ochoa-Riveros, J.M.: *Instituto Nacional de Investigaciones Forestales, Agrcolas y Pecuarias (INIFAP)-CIRNOC-Experimental Station La Campana, Aldama, Chih., Mexico*
Ogata, R.: *Japan International Cooperation Agency (JICA), Japan; The University of Tokyo, Tokyo, Japan*
Ok, Y.S.: *Korea Biochar Research Center, Korea University, Republic of Korea*
Oliveira, C.: *Department of Soil Science, Federal University of Lavras, Minas Gerais, Brazil*
Olmos-Márquez, M.A.: *Universidad Autónoma de Chihuahua, Chihuahua, Chih., Mexico; Junta Municipal de Agua y Saneamiento, Chihuahua, Chih., Mexico*
Olopade, C.: *Departments of Health Studies, Medicine and Human Genetics and Cancer Research Center, The University of Chicago, Chicago, IL, USA*
Onabolu, B.: *UNICEF Bangladesh, Dhaka, Bangladesh*
Onses, F.: *Department of Environmental Engineering, Technical University of Denmark, Kongens Lyngby, Denmark*
Ormachea, M.: *Laboratorio de Hidroqumica, Instituto de Investigaciones Qumicas, Universidad Mayor de San Andrés, La Paz, Bolivia*
Packianathan, C.: *Herbert Wertheim College of Medicine, Florida International University, Miami, FL, USA*
Paing, J.: *O-pure, Beaumont-la-Ronce, France*
Palacios, M.A.: *Department of Analytical Chemistry, Faculty of Chemistry, Universidad Complutense de Madrid, Madrid, Spain*
Pande, V.: *Department of Biotechnology, Kumaun University, Bhimtal, Nainital, Uttarakhand, India*
Park, J.D.: *College of Medicine, Chung-Ang University, Seoul, South Korea*
Parvez, A.: *State Key Laboratory of Water Environment Simulation, School of Environment, Beijing Normal University, Beijing, P.R. China*
Parvez, F.: *Department of Environmental Health, Columbia University New York, NY, USA*
Patel, K.S.: *School of Studies in Chemistry/Environmental Science, Pt. Ravishankar Shukla University, Raipur, India*
Patzner, M.: *Geomicrobiology and Hydrology, University of Tübingen, Tübingen, Germany*
Paul, D.: *Division of Biochemical Sciences, CSIR-National Chemical Laboratory, Pune, India*

Paunović, K.: *Institute of Hygiene and Medical Ecology, Faculty of Medicine, University of Belgrade, Belgrade, Serbia*
Pawar, S.: *Division of Biochemical Sciences, CSIR-National Chemical Laboratory, Pune, India*
Paz, J.: *Department of Life Science and Agriculture, Universidad de las Fuerzas Armadas ESPE, Sangolqu, Ecuador*
Pellizzari, E.E.: *Laboratorio de Microbiologa General, Universidad Nacional del Chaco Austral, P.R. Sáenz Peña, Chaco, Argentina*
Peng, C.: *Queensland Alliance for Environmental Health Sciences (QAEHS), The University of Queensland, Brisbane, QLD, Australia*
Peng, L.: *College of Resource and Environment, Hunan Agricultural University, Changsha, P.R. China*
Peng, Q.: *ASEM Water Resources Research and Development Center, Changsha, Hunan, P.R. China*
Penido, E.S.: *Department of Chemistry, Federal University of Lavras, Minas Gerais, Brazil*
Penke, Y.K.: *Materials Science Programme, IIT Kanpur, India*
Peralta, J.M.: *Departamento de Ciencias Naturales, Facultad de Ciencias Exactas, Fsico-Qumicas y Naturales, Universidad Nacional de Ro Cuarto, Córdoba, Argentina*
Perez, J.P.H.: *Helmholtz Centre Potsdam – GFZ German Research Centre for Geosciences, Potsdam, Germany; Institute of Geological Sciences, Department of Earth Sciences, Free University of Berlin, Berlin, Germany*
Pérez Carrera, A.L.: *Instituto de Investigaciones en Producción Animal UBA-CONICET (INPA), CONICET – Universidad de Buenos Aires, Buenos Aires, Argentina; Universidad de Buenos Aires. Centro de Estudios Transdisciplinarios del Agua (CETA), Universidad de Buenos Aires, Buenos Aires, Argentina; Universidad de Buenos Aires. Cátedra de Qumica Orgánica de Biomoléculas, Universidad de Buenos Aires, Buenos Aires, Argentina*
Perugupalli, P.: *Spectralinslights Pvt Ltd., Bangalore, India*
Pi, K.F.: *School of Environmental Studies & State Key Laboratory of Biogeology and Environmental Geology, China University of Geosciences, Wuhan, China*
Pickering, I.J.: *Department of Geological Sciences, University of Saskatchewan, Canada*
Pierce, B.: *Institute for Population and Precision Health, Chicago Center for Health and Environment, The University of Chicago, Chicago, IL, USA*
Pincetti Zúñiga, G.P.: *School of Earth, Atmospheric and Environmental Sciences and Williamson Research Centre for Molecular Environmental Science, University of Manchester, Manchester, UK*
Pinedo-Alvarez, A.: *Universidad Autónoma de Chihuahua, Chihuahua, Chih., Mexico*
Pizarro, I.: *Facultad de Ciencias Básicas, Universidad de Antofagasta, Antofagasta, Chile*
Planer-Friedrich, B.: *Environmental Geochemistry, Bayreuth University, Bayreuth, Germany*
Podgorski, J.E.: *Eawag, Swiss Federal Institute of Aquatic Science and Technology, Dübendorf, Switzerland*
Polya, D.A.: *School of Earth and Environmental Sciences and Williamson Research Centre for Molecular Environmental Science, University of Manchester, Manchester, UK*
Ponomarenko, O.: *Department of Geological Sciences, University of Saskatchewan, Canada; Poonam Institute of Environment and Sustainable Development, Banaras Hindu University, Varanasi, India*
Poonam: *Institute of Environment and Sustainable Development, Banaras Hindu University, Varanasi, India*
Prommer, H.: *University of Western Australia and CSIRO Land and Water, Perth, Australia*
Qian, K.: *School of Environmental Studies & State Key Laboratory of Biogeology and Environmental Geology, China University of Geosciences, Wuhan, China*
Qiao, J.T.: *Guangdong Key Laboratory of Integrated Agro-environmental Pollution Control and Management, Guangdong Institute of Eco-Environmental Science & Technology, Guangzhou, P.R. China*
Quino, I.: *Laboratorio de Hidroqumica, Instituto de Investigaciones Qumicas, Universidad Mayor de San Andrés, La Paz, Bolivia; KTH-International Groundwater Arsenic Research Group, Department of Sustainable Development, Environmental Science and Engineering, KTH Royal Institute of Technology, Stockholm, Sweden*
Quintanilla, J.: *International Center for Applied Climate Science, University of Southern Queensland, Toowoomba, Queensland, Australia*
Rahman, A.: *Department of Biochemistry and Molecular Biology, University of Rajshahi, Rajshahi, Bangladesh*
Rahman, M.: *Research and Evaluation Division, BRAC, Mohakhali, Dhaka, Bangladesh*
Rahman, M.M.: *Global Centre for Environmental Remediation (GCER), Faculty of Science, The University of Newcastle, Callaghan, NSW, Australia, Cooperative Research Centre for Contamination Assessment and Remediation of the Environment (CRC CARE), The University of Newcastle, Callaghan, NSW, Australia*

Rahman, M.S.: *Department of Public Health Engineering, Dhaka, Bangladesh*
Rahman, M.S.: *Mahavir Cancer Institute & Research Centre, Patna, Bihar, India*
Rahman, S.: *Department of Public Health Engineering, Ministry of Local Government, Rural Development and Cooperatives, Dhaka, Bangladesh*
Ramrez-Zamora, R.M.: *Instituto de Ingeniera, Universidad Nacional Autónoma de México (UNAM),Mexico City, Mexico*
Ramanathan, AL.: *School of Environmental Sciences, Jawaharlal Nehru University, New Delhi, India*
Ramazanov, O.M.: *Institute of Geothermal Problems, Dagestan Center of Science, Russian Academy of Sciences, Makhachkala, Dagestan, Russia*
Ramkumar, J.: *Department of Mechanical Engineering, IIT Kanpur, India*
Ramos, O.: *Laboratorio de Hidroqumica, Instituto de Investigaciones Qumicas, Universidad Mayor de San Andrés, La Paz, Bolivia*
Ramírez-Zamora, R.M.: *Instituto de Ingeniería, Universidad Nacional Autónoma de México (UNAM), Mexico City, Mexico*
Ramteke, S.: *School of Studies in Chemistry/Environmental Science, Pt. Ravishankar Shukla University, Raipur, India*
Rane, N.: *Department of Chemistry, Savitribai Phule Pune University, Pune, India*
Raqib, R.: *Nutritional Biochemistry Laboratory, Dhaka, Bangladesh*
Rasic-Milutinović, Z.: *Department of Endocrinology, University Hospital Zemun, Belgrade, Serbia*
Rathi, B.: *Centre for Applied Geoscience, University Tübingen, Tübingen, Germany*
Reid, M.C.: *Civil & Environmental Engineering, Cornell University, Ithaca, NY, USA; Dale Bumpers National Rice Research Center, Stuttgart, AR, USA*
Ren, W.: *Research Institute of Rural Sewage Treatment and College of Ecology and Soil & Water Conservation, Southwest Forestry University, Kunming, P.R. China*
Rensing, C.: *Key Laboratory of Urban Environment and Health, Institute of Urban Environment, Chinese Academy of Sciences, Xiamen, P.R. China; Fujian Provincial Key Laboratory of Soil Environmental Health and Regulation, College of Resources and the Environment, Fujian Agriculture & Forestry University, Fuzhou, P.R. China*
Repert, D.A.: *U.S. Geological Survey, Boulder, CO, USA*
Revich, B.A.: *Institute for Forecasting of the Russian Academy of Science, Moscow, Russia*
Richards, L.A.: *School of Earth, Atmospheric and Environmental Sciences and Williamson Research Centre for Molecular Environmental Science, University of Manchester, Manchester, UK*
Rietveld, L.C.: *Department Water Management, Section Sanitary Engineering, Delft University of Technology, The Netherlands*
Riya, S.: *Department of Chemical Engineering, Tokyo University of Agriculture and Technology, Tokyo, Japan*
Rocío, G.M.: *Centro de Ciencias de la Atmósfera, UNAM, Ciudad Universitaria, Coyoacán, Mexico*
Rodrguez Castrejón, U.E.: *Departamento de Biología, División de Ciencias Naturales y Exactas, Universidad de Guanajuato, Campus Guanajuato, México*
Rodriguez, J.: *UNI, Universidad Nacional de Ingeniera, Lima, Peru*
Rohila, J.: *Civil & Environmental Engineering, Cornell University, Ithaca, NY, USA; Dale Bumpers National Rice Research Center, Stuttgart, AR, USA*
Rolle, M.: *Department of Environmental Engineering, Technical University of Denmark, Kongens Lyngby, Denmark; SDC University, Beijing, China*
Román, D.: *Department of Analytical Chemistry, Faculty of Chemistry, Universidad Complutense de Madrid, Madrid, Spain*
Romani, M.: *Ente Nazionale Risi, Milan, Italy*
Root, R.A.: *Department of Soil, Water and Environmental Science, University of Arizona, Tucson, AZ, USA*
Rosano Ortega, G.: *Escuela de Ingeniera Ambiental,Universidad Popular Autónoma del Estado de Puebla A.C., Puebla, Pue., Mexico*
Rosen, B.P.: *Department of Cellular Biology and Pharmacology, Herbert Wertheim College of Medicine, Florida International University, Miami, Florida, USA*
Routh, J.: *Department of Thematic Studies – Environmental Change, Linköping University, Linköping, Sweden*
Roy, S.: *Kalyani Institute for Study, Planning and Action for Rural Change (KINSPARC), Kalyani West Bengal, India*
Saha, I.: *Department of Chemistry, Sripat Singh College, Murshidabad, India*

Saha, R.: *BRAC Research and Evaluation Division, BRAC Centre, Dhaka, Bangladesh*
Saha, S.: *BIRDEM Hospital, Dhaka, Bangladesh*
Sáenz-Uribe, C.G.: *Facultad de Zootecnia y Ecología, Universidad Autónoma de Chihuahua, Chihuahua, Chih., Mexico*
Saiqa Menhas, C.D.: *Department of Plant Sciences, Faculty of Biological Sciences, Quaid-i-Azam University, Islamabad, Pakistan*
Sakamoto, M.: *The University of Tokyo, Tokyo, Japan*
Salaun, P.: *Department of Environmental Science, University of Liverpool, Brownlow Hill, Liverpool, UK*
Samarasinghe, GS.: *de S. Jayasinghe Central College, Dehiwala, Sri Lanka*
Sánchez, V.: *Department of Earth and Construction Science, Universidad de las Fuerzas Armadas ESPE, Sangolqu, Ecuador*
Sandaruwan, L.S.: *de S. Jayasinghe Central College, Dehiwala, Sri Lanka*
Santos, A.C.: *Environmental Engineering Graduating Program (ProAmb), Pharmacy Department (DEFAR), School of Pharmacy, Federal University of Ouro Preto, MG, Brazil*
Santos, G.R.: *Environmental Engineering Graduating Program (ProAmb), Pharmacy Department (DEFAR), School of Pharmacy, Federal University of Ouro Preto, MG, Brazil*
Sargsyan, H.: *UNICEF Bangladesh, Dhaka, Bangladesh*
Sathe, S.: *Department of Civil Engineering, Indian Institute of Technology, Guwahati, India*
Schaaf, B.: *Evides Water Company N.V. Rotterdam, The Netherlands*
Schneider, M.: *Institute of Applied Geosciences, KIT, Karlsruhe, Germany*
Schouwenaars, R.: *Facultad de Ingeniera, Universidad Nacional Autónoma de México (UNAM), Mexico City, Mexico*
Schuessler, J.A.: *Helmholtz Centre Potsdam – GFZ German Research Centre for Geosciences, Potsdam, Germany*
Sekar, R.: *Department of Environmental Science, Xi'an Jiaotong-Liverpool University, Suzhou, Jiangsu, P.R. China*
Selim Reza, A.H.M.: *Department of Geology and Mining, University of Rajshahi, Rajshahi, Bangladesh*
Seneweera, S.: *Center for Crop Health, Faculty of Health, Engineering and Sciences, University of Southern Queensland, Toowoomba, QLD, Australia*
Serafn Muñoz, A.H.: *Departamento de Biología, División de Ciencias Naturales y Exactas, Universidad de Guanajuato, Campus Guanajuato, México*
Shahid, M.: *COMSATS Institute of Information and Technology, Vehari, Pakistan*
Shankar, P.: *Mahavir Cancer Institute & Research Centre, Patna, Bihar, India*
Sharma, P.: *Department of Geography, A.N. College, Patna, India*
Sharma, P.K.: *Mahavir Cancer Institute and Research Center, Patna, Bihar, India*
Sharma, R.: *School of Studies in Chemistry/Environmental Science, Pt. Ravishankar Shukla University, Raipur, India*
Sharma, S.: *Exceldot AB, Bromma, Sweden*
Shen, Y.: *Key Laboratory of Urban Environment and Health, Institute of Urban Environment, Chinese Academy of Sciences, Xiamen, P.R. China*
Shi, K.X.: *State Key Laboratory of Agricultural Microbiology, College of Life Science and Technology, Huazhong Agricultural University, Wuhan, P.R. China*
Shi, W.X.: *State Key Laboratory of Biogeology and Environmental Geology and Department of Biological Science and Technology, School of Environmental Studies, China University of Geosciences, Wuhan, P.R. China*
Siade, A.: *University of Western Australia and CSIRO Land and Water, Perth, Australia*
Sierra, L.: *Instituto de Hidrologa de Llanuras "Dr. Eduardo J. Usunoff" (IHLLA), Azul, Buenos Aires, Argentina*
Singh, P.: *School of Environmental Sciences, Jawaharlal Nehru University, New Delhi, India*
Singh, S.: *Innervoice Foundation, Varanasi, India*
Singh, S.K.: *Department of Earth and Environmental Studies, Montclair State University, Montclair, NJ, USA*
Singh, S.K.: *Department of Environment and Water Management, A.N. College, Patna, India*
Singh, S.P.: *Department of Geological Sciences, University of Saskatchewan, Canada*
Slavkovich, V.: *Mailman School of Public Health, Columbia University, New York City, NY, USA*
Šlejkovec, Z.: *Department of Environmental Sciences, Jožef Stefan Institute, Ljubljana, Slovenia*
Smith, R.L.: *U.S. Geological Survey, Boulder, CO, USA*

Smits, J.E.G.: *Faculty of Veterinary Medicine, University of Calgary, Calgary, Alberta, Canada*
Snell, P.: *Department of Primary Industries, Yanco Agricultural Institute, Orange, NSW, Australia*
Solis, J.L.: *Universidad Nacional de Ingeniera, Lima, Peru*
Sosa, N.N.: *Centro de Investigaciones Geológicas (CONICET-UNLP), Universidad Nacional de La Plata, La Plata, Argentina*
Sousa, J.P.: *Centre for Functional Ecology, Department of Life Sciences, University of Coimbra, Coimbra, Portugal*
Sovann, C.: *Department of Environmental Science, Royal University of Phnom Penh, Cambodia*
Spallholz, J.E.: *Division of Nutritional Sciences, Texas Tech University, Lubbock, Texas, USA*
Srivastava, S.: *Institute of Environment & Sustainable Development, Banaras Hindu University, Varanasi, Uttar Pradesh, India*
Stajnko, A.: *Department of Environmental Sciences, Jožef Stefan Institute, Ljubljana, Slovenia; Jožef Stefan International Postgraduate School, Ljubljana, Slovenia*
St-Amand, A: *Health Canada*
Stipp, S.L.S.: *Nano-Science Center, Department of Chemistry, University of Copenhagen, Copenhagen, Denmark*
Stolze, L.: *Department of Environmental Engineering, Technical University of Denmark, Lyngby, Denmark*
Stopelli, E.: *Eawag, Swiss Federal Institute of Aquatic Science and Technology, Dübendorf, Switzerland*
Su, C.: *School of Environmental Studies, China University of Geosciences (Wuhan), Wuhan, Hubei, P.R. China*
Su, J.Q.: *Key Laboratory of Urban Environment and Health, Institute of Urban Environment, Chinese Academy of Sciences, Xiamen, P.R. China*
Su, S.M.: *Institute of Environment and Sustainable Development in Agriculture, Chinese Academy of Agricultural Sciences/Key Laboratory of Agro-Environment, Ministry of Agriculture, Beijing, P.R. China*
Sültenfuß, J.: *Institute of Environmental Physics, University of Bremen, Bremen, Germany*
Suman, S.: *Mahavir Cancer Institute and Research Center, Patna, India*
Sumi, D.: *Faculty of Pharmaceutical Sciences, Tokushima Bunri University, Tokushima, Japan*
Sun, G.: *School of Engineering, Edith Cowan University, Joondalup, WA, Australia*
Sun, G.F.: *Environment and Non-Communicable Disease Research Center, Key Laboratory of Arsenic-related Biological Effects and Prevention and Treatment in Liaoning Province, School of Public Health, China Medical University, Shenyang, P.R. China*
Sun, G.X.: *State Key Lab of Urban and Regional Ecology, Research Center for Eco-Environmental Sciences, Chinese Academy of Sciences, Beijing, P.R. China*
Sun, J.: *School of Earth Sciences, University of Western Australia, Perth, Western Australia, Australia*
Sun, W.: *Guangdong Key Laboratory of Agricultural Environment Pollution Integrated Control, Guangdong Institute of Eco-Environmental Science & Technology, Guangzhou, P.R. China*
Sun, W.J.: *Department of Civil and Environmental Engineering, Southern Methodist University, Dallas, Texas, USA*
Sun, Y.: *College of Resources and Environmental Science, Hebei Agricultural University, Baoding, P.R. China*
Sun, Y.: *State Key Laboratory of Pollution Control and Resources Reuse, Tongji University, Shanghai, P.R. China*
Sun, Y.Q.: *College of Engineering, Peking University, Beijing, China; School of Environmental Science and Technology, Southern University of Science and Technology, Shenzhen, China*
Szubska, M.: *Institute of Oceanology of the Polish Academy of Sciences, Sopot, Poland*
Taga, R.: *Centre for Mined Land Rehabilitation, Sustainable Minerals Institute, The University of Queensland, Brisbane, QLD, Australia; Queensland Alliance for Environmental Health Sciences (QAEHS), The University of Queensland, Brisbane, QLD, Australia*
Takeuchi, C.: *Department of Environmental Science, Saitama University, Saitama, Japan*
Tamayo Calderón, R.M: *Department of Chemical Engineering Biotechnology and Materials, Faculty of Physical and Mathematical Sciences, University of Chile, Santiago de Chile, Chile*
Tang, L.: *College of Resource and Environment, Hunan Agricultural University, Changsha, P.R. China*
Tang, X.J.: *Institute of Soil and Water Resources and Environmental Science, College of Environmental and Resource Sciences, Zhejiang Provincial Key Laboratory of Agricultural Resources and Environment, Zhejiang University, Hangzhou, P.R. China*
Tang, Z.: *College of Resources and Environmental Science, Nanjing Agricultural University, Nanjing, P.R. China*
Tarah, F.: *Delta Niroo Gameron Co., Tehran, Iran*
Tarvainen, T.: *Geological Survey of Finland, Espoo, Finland*

Taylor, R.W.: *Department of Earth and Environmental Studies, Montclair State University, Montclair, NJ, USA*
Tazaki, K.: *Department of Earth Sciences, Faculty of Science, Kanazawa University, Kanazawa, Ishikawa, Japan*
Teixeira, M.C.: *Environmental Engineering Graduating Program (ProAmb), Pharmacy Department (DEFAR), School of Pharmacy, Federal University of Ouro Preto, MG, Brazil*
Terada, A.: *Department of Chemical Engineering, Tokyo University of Agriculture and Technology, Tokyo, Japan*
Thouin, H.: *BRGM, ISTO, Orléans, France*
Tie, B.Q.: *College of Resource and Environment, Hunan Agricultural University, Changsha, P.R. China*
Timón-Sánchez, S.: *Instituto Geológico y Minero de España – IGME, Salamanca, Spain*
Touzé, S.: *BRGM, Orléans, France*
Tran, L.L.: *Center of Applied Research, Karlsruhe University of Applied Sciences, Karlsruhe, Germany*
Tran, M.: *CETASD, Vietnam National University, Hanoi, Vietnam*
Trang, P.T.K.: *CETASD, Vietnam National University, Hanoi, Vietnam*
Travaglia, C.N.: *Departamento de Ciencias Naturales, Facultad de Ciencias Exactas, Fsico-Qumicas y Naturales, Universidad Nacional de Ro Cuarto, Córdoba, Argentina*
Tripathi, R.D.: *CSIR – National Botanical Research Institute, Lucknow, Uttar Pradesh, India*
Tsai, S.-F.: *National Institute of Environmental Heath Sciences, National Health Research Institutes, Miaoli, Taiwan*
Uddin, S.: *Asia Arsenic Network, Dhaka, Bangladesh*
Upadhyay, M.K.: *Institute of Environment & Sustainable Development, Banaras Hindu University, Varanasi, India*
Valles-Aragón, M.C.: *Faculty of Agrotechnological Science, Autonomous University of Chihuahua, Chihuahua, Chih., Mexico*
van der Wal, A.: *Department of Environmental Technology, Wageningen University and Research (WUR), Wageningen, The Netherlands; Department of Water Technology, Evides Water Company N.V., Rotterdam, The Netherlands*
van der Wens, P.: *Brabant Water, Department of Drinking Water Production, Breda, The Netherlands*
van Dongen, B.E.: *School of Earth, Atmospheric and Environmental Sciences and Williamson Research Centre for Molecular Environmental Science, University of Manchester, Manchester, UK*
van Geen, A.: *Lamont Doherty Earth Observatory, Columbia University, Palisades, New York, USA*
van Genuchten, C.M.: *Department of Earth Sciences-Geochemistry, Faculty of Geosciences, Utrecht University, Utrecht, The Netherlands; Nano-Science Center, Department of Chemistry, University of Copenhagen, Copenhagen, Denmark*
van Halem, D.: *Department Water Management, Section Sanitary Engineering, Delft University of Technology, The Netherlands*
van Mook, J.: *Department of Watertechnology, Evides Waterbedrijf, Rotterdam, The Netherlands*
Vandenberg, A.: *Department of Plant Sciences, University of Saskatchewan, Saskatoon, Canada*
Vargas de Mello, J.W.: *Universidade Federal de Viçosa, Lavras, MG, Brazil*
Vega, M.: *Hydrologic Science and Engineering Program, Colorado School of Mines, Golden, CO, USA*
Vega-Alegre, M.: *Departamento de Qumica Analtica, Universidad de Valladolid – UVA, Valladolid, Spain*
Velázquez, C.: *Facultad de Qumica, Universidad Autónoma de Querétaro, Santiago de Querétaro, Qro, Mexico*
Velázquez, L.R.: *Facultad de Qumica, Universidad Autónoma de Querétaro, Santiago de Querétaro, Qro, Mexico*
Vezina, A.: *Health Canada*
Viet, N.: *CETASD, Vietnam National University, Hanoi, Vietnam*
Viet, P.H.: *CETASD, Vietnam National University, Hanoi, Vietnam*
Vijayakumar, S.: *Kemakta Konsult AB, Stockholm, Sweden; KTH-International Groundwater Arsenic Research Group, Department of Sustainable Development, Environmental Science and Engineering, KTH Royal Institute of Technology, Stockholm, Sweden*
Villalobos, J.S.: *EcoMetales Limited, Providencia, Región Metropolitana, Santiago de Chile, Chile*
Vipasha, M.: *Department of Civil Engineering, Indian Institute of Technology, Guwahati, India*
Vithanage, M.: *Ecosphere Resilience Research Center, Faculty of Applied Science, University of Sri Jayewardenepura, Nugegoda, Sri Lanka*
von Brömssen, M.: *Ramböll Sweden AB, Stockholm, Sweden*
Vries, D.: *KWR Watercycle Research Institute, Nieuwegein, The Netherlands*
Vu, D.: *CETASD, Vietnam National University, Hanoi, Vietnam*
Wahnfried, I.: *UFAM, Universidade Federal do Amazonas, Manaus, Brazil*

Wallis, I.: *School of Environment, Flinders University, Adelaide, Australia*
Wang, C.: *College of Resources and Environmental Science, Nanjing Agricultural University, Nanjing, P.R. China*
Wang, D.: *Research Center of Environment and Non-Communicable Disease, School of Public Health, China Medical University, Shenyang, P.R. China*
Wang, G.J.: *State Key Laboratory of Agricultural Microbiology, College of Life Science and Technology, Huazhong Agricultural University, Wuhan, P.R. China*
Wang, H.L.: *State Key Laboratory of Biogeology and Environmental Geology, China University of Geosciences, Wuhan, P.R. China*
Wang, H.Y.: *State Key Laboratory of Urban and Regional Ecology, Research Center for Eco-Environmental Sciences, Chinese Academy of Sciences, Beijing, P.R. China*
Wang, H.Y.: *Institute of Mineralogy and Geochemistry, Karlsruhe Institute of Technology, Karlsruhe, Germany*
Wang, J.: *Environmental Geochemistry, Bayreuth University, Bayreuth, Germany*
Wang, J.: *School of Chemistry and Environment, South China Normal University, Guangzhou, P.R. China*
Wang, J.Y.: *School of Environmental Studies, China University of Geosciences, Wuhan, P.R. China*
Wang, L.Y.: *Research Center for Eco-Environmental Sciences, Chinese Academy of Sciences, Beijing, P.R. China*
Wang, Q.: *State Key Laboratory of Agricultural Microbiology, College of Life Science and Technology, Huazhong Agricultural University, Wuhan, P.R. China*
Wang, S.: *School of Environmental Studies, China University of Geosciences, Wuhan, P.R. China*
Wang, S.F.: *Institute of Applied Ecology, Chinese Academy of Sciences, Shenyang, China*
Wang, S.J.: *Research and Development Department, China State Science Dingshi Environmental Engineering Co., Ltd, Beijing, P.R. China*
Wang, S.L.: *National Institute of Environmental Heath Sciences, National Health Research Institutes, Miaoli, Taiwan*
Wang, X.Q.: *Guangdong Key Laboratory of Integrated Agro-environmental Pollution Control and Management, Guangdong Institute of Eco-Environmental Science & Technology, Guangzhou, P.R. China*
Wang, X.X.: *Key Laboratory of Soil Environment and Pollution Remediation, Institute of Soil Science, Chinese Academy of Sciences, Nanjing, P.R. China*
Wang, Y.F.: *Research and Development Department, China State Science Dingshi Environmental Engineering Co., Ltd, Beijing, P.R. China*
Wang, Y.H.: *State Key Laboratory of Biogeology and Environmental Geology, China University of Geosciences, Wuhan, P.R. China*
Wang, Y.N.: *Institute of Environment and Sustainable Development in Agriculture, Chinese Academy of Agricultural Sciences/Key Laboratory of Agro-Environment, Ministry of Agriculture, Beijing, P.R. China*
Wang, Y.X.: *School of Environmental Studies, China University of Geosciences, Wuhan, China; State Key Laboratory of Biogeology and Environmental Geology, China University of Geosciences, Wuhan, China*
Wang, Z.H.: *College of Chemistry and Environment, Fujian Province Key Laboratory of Modern Analytical Science and Separation Technology, Minnan Normal University, Zhangzhou, China; Key Laboratory of Urban Environment and Health, Institute of Urban Environment, Chinese Academy of Sciences, Xiamen, China*
Weerasundara, L.: *Environmental Chemodynamics Project, National Institute of Fundamental Studies, Kandy, Sri Lanka*
Wei, C.Y.: *Institute of Geographic Sciences and Natural Resources Research, Chinese Academy of Sciences, Beijing, P.R. China; Key Laboratory of Land Surface Pattern and Simulation, Chinese Academy of Sciences, Beijing, P.R. China*
Wei, D.Z.: *State Key Laboratory of Biogeology and Environmental Geology, China University of Geosciences, Wuhan, P.R. China*
Wei, X.: *School of Environmental Studies, China University of Geosciences, Wuhan, Hubei, P.R. China*
Weinzettel, P.A.: *Instituto de Hidrologa de Llanuras "Dr. Eduardo J. Usunoff" (IHLLA), Azul, Buenos Aires, Argentina*
Welmer Veloso, R.: *Instituto Federal de Rondônia, Lavras, MG, Brazil*
Weng, L.P.: *Agro-Environmental Protection Institute, Ministry of Agriculture, Tianjin, P.R. China*
Werry, K.: *Health Canada*
Werth, N.P.: *EcoMetales Limited, Providencia, Santiago, Chile*
Williams, P.N.: *Institute for Global Food Security, Queen's University Belfast, Belfast, UK*
Williamson, G.: *Department of Environmental Science, University of Liverpool, Liverpool, UK*
Winkel, L.H.E.: *Eawag and Institute of Biogeochemistry and Pollutant Dynamics, ETH Zürich, Zürich, Switzerland*

Wu, C.X.: *Institute of Environment and Sustainable Development in Agriculture, Chinese Academy of Agricultural Sciences/Key Laboratory of Agro-Environment, Ministry of Agriculture, Beijing, P.R. China*
Wu, W.W.: *School of Environmental Studies, China University of Geosciences (Wuhan), Wuhan, Hubei, P.R. China*
Xia, Q.: *Queensland Alliance for Environmental Health Sciences (QAEHS), The University of Queensland, Brisbane, QLD, Australia*
Xiao, E.Z.: *Key Laboratory of Water Quality and Conservation in the Pearl River Delta, Ministry of Education, School of Environmental Science and Engineering, Guangzhou University, Guangzhou, P.R. China*
Xiao, T.F.: *Key Laboratory of Water Quality and Conservation in the Pearl River Delta, Ministry of Education, School of Environmental Science and Engineering, Guangzhou University, Guangzhou, P.R. China*
Xiao, Z.Y.: *State Key Laboratory of Biogeology and Environmental Geology & School of Environmental Studies, China University of Geosciences, Wuhan, P.R. China*
Xie, X.J.: *School of Environmental Studies & State Key Laboratory of Biogeology and Environmental Geology, China University of Geosciences, Wuhan, P.R. China*
Xie, Z.: *School of Environmental Studies, China University of Geosciences, Wuhan, Hubei, P.R. China*
Xiu, W.: *State Key Laboratory of Biogeology and Environmental Geology, China University of Geosciences, Beijing, P.R. China*
Xu, J.M.: *Institute of Soil and Water Resources and Environmental Science, College of Environmental and Resource Sciences, Zhejiang Provincial Key Laboratory of Agricultural Resources and Environment, Zhejiang University, Hangzhou, P.R. China*
Xu, L.: *School of Earth and Environmental Sciences and Williamson Research Centre for Molecular Environmental Science, University of Manchester, Manchester, UK*
Xu, Z.X.: *Guangdong Key Laboratory of Agricultural Environment Pollution Integrated Control, Guangdong Institute of Eco-Environmental Science & Technology, Guangzhou, P.R. China*
Xue, X.M.: *Key Lab of Urban Environment and Health, Institute of Urban Environment, Chinese Academy of Sciences, Xiamen, China*
Yadav, A.: *School of Studies in Chemistry/Environmental Science, Pt. Ravishankar Shukla University, Raipur, India*
Yadav, M.K.: *School of Environmental Science and Engineering, Indian Institute of Technology, Kharagpur, India*
Yan, C.Z.: *Key Laboratory of Urban Environment and Health, Institute of Urban Environment, Chinese Academy of Sciences, Xiamen, P.R. China*
Yan, L.: *School of Environmental Studies & State Key Laboratory of Biogeology and Environmental Geology, China University of Geosciences, Wuhan, P.R. China*
Yan, W.: *Research Institute of Rural Sewage Treatment and College of Ecology and Soil & Water Conservation, Southwest Forestry University, Kunming, P.R. China*
Yan, Y.: *Key Lab of Urban Environment and Health, Institute of Urban Environment, Chinese Academy of Sciences, Xiamen, China*
Yang, F.: *Institute of Geographic Sciences and Natural Resources Research, Chinese Academy of Sciences, Beijing, P.R. China; Key Laboratory of Land Surface Pattern and Simulation, Chinese Academy of Sciences, Beijing, P.R. China; University of Chinese Academy of Sciences, Beijing, P.R. China*
Yang, G.Y.: *Research Institute of Rural Sewage Treatment and College of Ecology and Soil & Water Conservation, Southwest Forestry University, Kunming, P.R. China*
Yang, S.: *Environment and Non-Communicable Disease Research Center, Key Laboratory of Arsenic-related Biological Effects and Prevention and Treatment in Liaoning Province, School of Public Health, China Medical University, Shenyang, P.R. China*
Yang, Y.: *State Key Laboratory of Biogeology and Environmental Geology and Department of Biological Science and Technology, School of Environmental Studies, China University of Geosciences, Wuhan, P.R. China*
Yang, Y.F.: *State Key Joint Laboratory of Environment Simulation and Pollution Control, School of Environment, Tsinghua University, Beijing, P.R. China*
Yang, Y.P.: *Research Center for Eco-Environmental Sciences, Chinese Academy of Sciences, Beijing, P.R. China*
Ye, J.: *Key Lab of Urban Environment and Health, Institute of Urban Environment, Chinese Academy of Sciences, Xiamen, China*
Ye, L.: *Research Center for Eco-Environmental Sciences, Chinese Academy of Sciences, Beijing, China*
Yi, Q.Q.: *School of Environmental Studies, China University of Geosciences, Wuhan, P.R. China*
Yi, X.Y.: *Research Center for Eco-Environmental Sciences, Chinese Academy of Sciences, Beijing, P.R. China*

Yin, D.: *State Key Laboratory of Pollution Control and Resource Reuse, School of the Environment, Nanjing University, Jiangsu, P.R. China*
Yoshinaga, M.: *Herbert Wertheim College of Medicine, Florida International University, Miami, FL, USA*
Yu, Q.N.: *College of Resources and Environmental Sciences, Jiangsu Provincial Key Laboratory of Marine Biology, Nanjing Agricultural University, Nanjing, P.R. China*
Yu, X.: *Geological Survey, China University of Geosciences, Wuhan, P.R. China*
Yu, X.N.: *School of Water Resources and Environment, China University of Geosciences (Beijing), Beijing, P.R. China*
Yuan, W.J.: *School of Water Resources and Environment, China University of Geosciences (Beijing), Beijing, P.R. China*
Yuan, X.F.: *Geological Survey, China University of Geosciences, Wuhan, P.R. China*
Yuan, Z.F.: *Department of Environmental Science, Xi'an Jiaotong-Liverpool University, Suzhou, Jiangsu, P.R. China*
Yunus, M.: *icddrb, Dhaka, Bangladesh*
Zahid, M.A.: *Environment and Population Research Centre (EPRC), Dhaka, Bangladesh*
Zeng, X.B.: *Institute of Environment and Sustainable Development in Agriculture, Chinese Academy of Agricultural Sciences/Key Laboratory of Agro-Environment, Ministry of Agriculture, Beijing, P.R. China*
Zeng, X.C.: *State Key Laboratory of Biogeology and Environmental Geology and Department of Biological Science and Technology, School of Environmental Studies, China University of Geosciences, Wuhan, P.R. China*
Zhai, W.W.: *Institute of Soil and Water Resources and Environmental Science, College of Environmental and Resource Sciences, Zhejiang Provincial Key Laboratory of Agricultural Resources and Environment, Zhejiang University, Hangzhou, P.R. China*
Zhang, B.: *Research and Development Department, China State Science Dingshi Environmental Engineering Co., Ltd, Beijing, P.R. China*
Zhang, C.H.: *Demonstration Laboratory of Elements and Life Science Research, Laboratory Centre of Life Science, Nanjing Agricultural University, Nanjing, P.R. China*
Zhang, D.: *School of Water Resources and Environment, China University of Geosciences, Beijing, P.R. China*
Zhang, F.: *Hunan Agricultural University, Changsha, Hunan, P.R. China*
Zhang, J.Y.: *College of Resources and Environmental Sciences, Jiangsu Provincial Key Laboratory of Marine Biology, Nanjing Agricultural University, Nanjing, P.R. China*
Zhang, S.Y.: *State Key Lab of Urban and Regional Ecology, Research Center for Eco-Environmental Sciences, Chinese Academy of Sciences, Beijing, P.R. China*
Zhang, X.: *Key Lab of Urban Environment and Health, Institute of Urban Environment, Chinese Academy of Sciences, Xiamen, China*
Zhao, D.: *State Key Laboratory of Pollution Control and Resource Reuse, School of the Environment, Nanjing University, Nanjing, P.R. China*
Zhao, F.J.: *College of Resources and Environmental Science, Nanjing Agricultural University, Nanjing, P.R. China*
Zhao, L.: *Environment and Non-Communicable Disease Research Center, Key Laboratory of Arsenic-related Biological Effects and Prevention and Treatment in Liaoning Province, School of Public Health, China Medical University, Shenyang, P.R. China*
Zhao, Q.L.: *College of Resources and Environmental Science, Hebei Agricultural University, Baoding, P.R. China*
Zhao, R.: *Research Institute of Rural Sewage Treatment, Southwest Forestry University, Kunming, P.R. China; College of Ecology and Soil & Water Conservation, Southwest Forestry University, Kunming, P.R. China*
Zhao, Y.: *Key Laboratory of Urban Environment and Health, Institute of Urban Environment, Chinese Academy of Sciences, Xiamen, P.R. China*
Zhao, Z.X.: *College of Chemistry and Chemical Engineering, Xinjiang Normal University, Urumqi, China*
Zheng, J.: *Centre for Mined Land Rehabilitation, Sustainable Minerals Institute, The University of Queensland, Brisbane, QLD, Australia*
Zheng, Q.M.: *Sanitation & Environment Technology Institute, Soochow University, Jiangsu, P.R. China*
Zheng, R.L.: *Beijing Research & Development Center for Grasses and Environment, Beijing Academy of Agriculture and Forestry Sciences, Beijing, P.R. China*
Zheng, T.I.: *Geological Survey, China University of Geosciences, Wuhan, China*
Zheng, T.L.: *Geological Survey, China University of Geosciences, Wuhan, P.R. China*
Zheng, Y.: *School of Environmental Science and Technology, Southern University of Science and Technology, Shenzhen, China*

Zhong, Q.Y.: *State Key Laboratory of Water Environment Simulation, School of Environment, Beijing Normal University, Beijing, P.R. China*
Zhou, P.: *School of Agriculture and Biology, Key Laboratory of Urban Agriculture, Ministry of Agriculture, Bor S. Luh Food Safety Research Center, Shanghai Jiao Tong University, Shanghai, China*
Zhou, S.: *College of Resource & Environment, Hunan Agricultural University, Changsha, P.R. China*
Zhu, G.B.: *Key Laboratory of Drinking Water Science and Technology, Research Center for Eco-Environmental Sciences, Chinese Academy of Sciences, Beijing, P.R. China*
Zhu, M.: *Department of Ecosystem Science and Management, University of Wyoming, Laramie, WY, USA*
Zhu, X.B.: *State Key Laboratory of Biogeology and Environmental Geology and Department of Biological Science and Technology, School of Environmental Studies, China University of Geosciences, Wuhan, P.R. China*
Zhu, Y.G.: *State Key Lab of Urban and Regional Ecology, Research Center for Eco-Environmental Sciences, Chinese Academy of Sciences, Beijing, P.R. China; Key Laboratory of Urban Environment and Health, Institute of Urban Environment, Chinese Academy of Sciences, Xiamen, P.R. China*
Zoeller, T.: *Department of Biology, University of Massachusetts Amherst, Amherst, MA, USA*
Zu, Y.Q.: *College of Resources and Environment, Yunnan Agricultural University, Kunming, P.R. China*

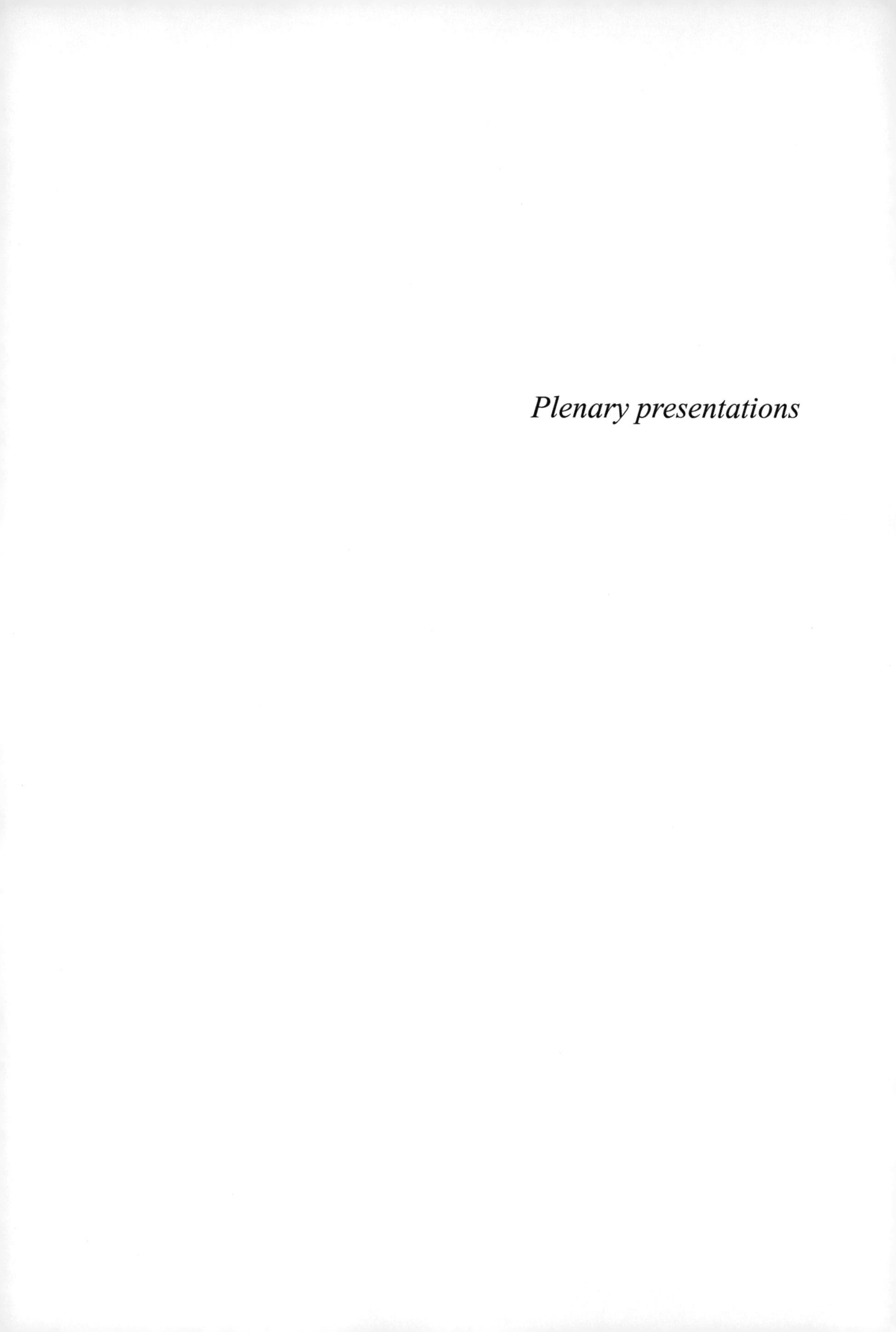

Plenary presentations

Environmental Arsenic in a Changing World –
Zhu, Guo, Bhattacharya, Ahmad, Bundschuh & Naidu (Eds)
ISBN 978-1-138-48609-6

Sedimentological and hydro-biogeochemical processes controlling arsenic behavior in the Holocene and upper Pleistocene aquifers of the central Yangtze River Basin

Y.X. Wang[1,2], Y.Q. Gan[1,2], Y. Deng[3], Y.H. Duan[1,2], T.L. Zheng[3] & S. Fendorf[4]
[1] *State Key Laboratory of Biogeology and Environmental Geology, China University of Geosciences, Wuhan, P.R. China*
[2] *School of Environmental Studies, China University of Geosciences, Wuhan, P.R. China*
[3] *Geological Survey, China University of Geosciences, Wuhan, P.R. China*
[4] *Department of Earth System Science, Stanford University, Stanford, CA, USA*

ABSTRACT: Understanding the mechanism of geogenic arsenic mobilization from sediments to groundwater is important for safe and sustainable drinking water supply in the central Yangtze River Basin. Unlike the preponderance of observations within the deltas of South and Southeast Asia, groundwater As concentrations in the Holocene and upper Pleistocene aquifers at Yangtze vary by up to an order of magnitude seasonally. Bulk sediment geochemistry, arsenic associated mineralogical analysis and high-resolution OSL dating were applied to decipher the sedimentological controls on the formation of high arsenic aquifers. Sedimentological processes and paleoclimatic optima after the Last Glacial Maximum (LGM) have created favorable conditions for the formation of high-As aquifer systems. Intensive chemical weathering leading to sulfur depletion after the LGM could facilitate As enrichment in the Holocene and upper Pleistocene aquifers. Arsenic release depends on the iron mineralogy and microbial community in the aquifer sediments. Results of batch experiments and reactive transport modeling indicated that seasonal changes in surface water and groundwater levels drive changes in redox conditions. Seasonal variation of biogeochemical iron redox cycle may essentially control the mobilization/immobilization of As, while bacterial sulfate reduction could cause temporal attenuation of groundwater As concentration.

1 INTRODUCTION

Intake of geogenic arsenic (As) contaminated groundwater in sedimentary aquifers has caused serious health problems among tens of millions of people worldwide. Within river systems draining the Himalaya, tectonic movement, sedimentological processes and paleoclimatic optima after the Last Glacial Maximum (LGM) have created favorable conditions for the formation of high-As aquifer systems mainly in the late Pleistocene–Holocene deposits (McArthur *et al.*, 2011; Wang *et al.*, 2017). The Jianghan Plain in central Yangtze River Basins was documented with severe As contamination in the shallow aquifers (Deng *et al.*, 2018), and As concentrations exhibited appreciable seasonal variations, with lower concentrations corresponding to lower water levels during the dry season and higher concentrations with increased water levels during the rainy season (Duan *et al.*, 2015; Schaefer *et al.*, 2016). However, the underlying mechanism responsible for the formation of high arsenic aquifers and the cause of seasonal release and retention of As between sediments and groundwater remain poorly understood. The major objective of this study was to understand the causes of the seasonal As concentration variation and the effects of Quaternary sediment deposition responding to sea level change since LGM on the arsenic enrichment in groundwater at Jianghan.

2 METHODS

2.1 *Study area*

Jianghan alluvial plain is located between the Yangtze River and its largest tributary, the Han River (Fig. 1).

2.2 *Sampling and analysis*

Sediment samples from six boreholes with depths up to 200 m in typical arsenic-affected area at Jianghan were collected for bulk geochemistry analysis, As and Fe speciation analysis, optical stimulated luminescence (OSL) dating, grain size analysis and mineralogy SEM-EDS analysis.

To simulate the arsenic concentration changes observed in the field, batch sediment incubation experiments were conducted under varying redox conditions by purging N_2 and O_2 gas respectively. Two sets of

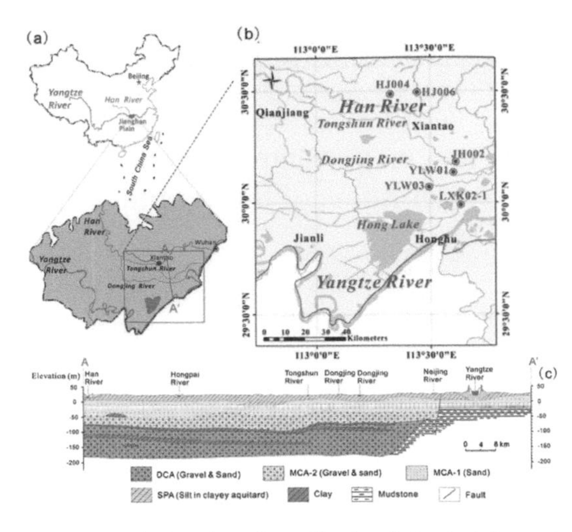

Figure 1. The location of sampling boreholes (red dots and numbers) and a typical hydrogeological section (A–A') across the study area in the Jianghan Plain. SPA, MCA and DCA are respectively shallow phreatic aquifer, middle and deep confined aquifers.

incubations were respectively subjected to two different patterns of redox cycles by switching the order of oxic and anoxic conditions: Anoxic-oxic-anoxic (AOA) and oxic-anoxic-oxic (OAO) incubations.

3 RESULTS AND DISCUSSION

3.1 *Sedimentological controls on the enrichment of arsenic in shallow aquifers since LGM*

Sea level in LGM was 120 m lower than today, a quick response of the Yangtze River 1000 km away from the estuary, which led to depositional break 40–20 ka ago. As was correlated with sulfur in the pre-LGM sediments and the pyrite was observed to be the main sink of As in the deep aquifer (deeper than 60 m); by contrast, arsenic was closely correlated with iron in the post-LGM sediments. Results of sequential extraction of Fe indicated that Fe mainly occurs as iron-oxides in Holocene and upper Pleistocene sediments, and as siderite in middle–lower Pleistocene sediments.

The intense chemical weathering after LGM could be a major factor controlling As enrichment in the Holocene and upper Pleistocene aquifer. As the paleoclimate became warm and humid after LGM, intense chemical weathering led to sulfur depletion, secondary As-bearing Fe/Mn oxides were formed and buried jointly with biodegradable NOM in the sediments.

3.2 *Redox fluctuation controls the seasonal variation of arsenic in groundwater*

Aqueous As and iron concentrations increase during anoxic periods and decrease during oxic periods in both AOA and OAO incubations. Dominant geochemical reactions deciphered from the results of our batch incubations explain the observed seasonal release and retention of As between sediments and groundwater at Jianghan. During groundwater recharging, oxygen and nitrate delivered to groundwater result in oxic condition in the aquifer. Fe(II) and As(III) are oxidized to form Fe(III) and As(V). Formation of fresh Fe(III) oxides provides additional sorption sites for As(III) and As(V), causing decrease in aqueous As concentration. During groundwater discharging, anoxic condition in the aquifer prevails due to limited input of oxygen. Fe(III) reductive dissolution and As(V) reduction cause release of As (mostly As(III)) into groundwater.

4 CONCLUSIONS

Sedimentological processes and paleoclimatic optima after LGM have created favorable conditions for the formation of high-As aquifer systems at Jianghan. Oscillation of redox conditions driven by temporal changes in surface and groundwater levels is the predominant factor affecting the observed seasonal As concentration variation in aquifers. Arsenic release depends on the iron mineralogy and microbial community in the aquifer sediments. Biogeochemical Fe-S redox cycle may essentially control the mobilization/immobilization of As.

ACKNOWLEDGEMENTS

The research was financially supported by the National Natural Science Foundation of China (Nos. 41521001 & 41572226), China Geological Survey (Nos. 121201001000150121) and the 111 Program (State Administration of Foreign Experts Affairs & the Ministry of Education of China, grant B18049).

REFERENCES

Deng, Y., Zheng, T., Wang, Y., Liu, L., Jiang, H. & Ma, T. 2018. Effect of microbially mediated iron mineral transformation on temporal variation of arsenic in the Pleistocene aquifers of the central Yangtze River basin. *Sci. Total Environ.* 619–620: 1247–1258.

Duan, Y., Gan, Y., Wang, Y., Deng, Y., Guo, X. & Dong, C. 2015. Temporal variation of groundwater level and arsenic concentration at Jianghan Plain, central China. *J. Geochem. Explor.* 149: 106–119.

McArthur, J.M., Nath, B., Banejee, D.M., Purohit, R. & Grassineau, N. 2011. Palaeosol control on groundwater flow and pollutant distribution: the example of arsenic. *Environ. Sci. Technol.* 45(4): 1376–1383.

Schaefer, M.V., Ying, S.C., Benner, S.G., Duan, Y., Wang, Y. & Fendorf, S. 2016. Aquifer arsenic cycling induced by seasonal hydrologic changes within the Yangtze River Basin. *Environ. Sci. Technol.* 50(7): 3521–3529.

Wang, Y.X., Pi, K.F., Fendorf, S., Deng, Y.M. & Xie, X.J. 2017. Sedimentogenesis and hydrobiogeochemistry of high arsenic Late Pleistocene–Holocene aquifer systems. *Earth Sci. Rev.* DOI:10.1016/j.earscirev.2017.10.007.

Environmental Arsenic in a Changing World –
Zhu, Guo, Bhattacharya, Ahmad, Bundschuh & Naidu (Eds)
ISBN 978-1-138-48609-6

Groundwater Assessment Platform (GAP): A new GIS tool for risk forecasting and mitigation of geogenic groundwater contamination

M. Berg & J.E. Podgorski
Eawag, Swiss Federal Institute of Aquatic Science and Technology, Dübendorf, Switzerland

ABSTRACT: Over 400 million people worldwide use groundwater contaminated with arsenic and/or fluoride as a source of drinking water. The Swiss Federal Institute of Aquatic Science and Technology (Eawag) has developed a method to estimate the risk of contamination in a given area using geological, topographical and other environmental data without having to test samples from every single well. The research group's knowledge is now being made available free of charge on the interactive Groundwater Assessment Platform (GAP, www.gapmaps.org). GAP is an online GIS platform for risk forecasting and mitigation of geogenic groundwater contamination. GAP enables researchers, authorities, NGOs and other professionals to visualize their own data and generate hazard risk maps for their areas of interest.

1 INTRODUCTION

A third of the world's population uses groundwater for drinking and cooking. Groundwater is generally a safe alternative to untreated, microbially contaminated surface water. However, about 10% of wells are contaminated with arsenic and/or fluoride.

Since health symptoms often only become visible after long-term exposure, it is important to identify safe and unsafe groundwater sources as early on as possible. In recent years, significant progress has been made in predictive risk modeling of arsenic contamination. Several studies identified a relatively small number of geological and hydrogeochemical parameters as significant spatial proxies that can be used to predict the regional distribution of high and low arsenic concentrations across entire regions, even in areas without survey data (Amini *et al.*, 2008; Ayotte *et al.*, 2017; Bretzler *et al.*, 2017; Podgorski *et al.*, 2017; Rodriguez-Lado *et al.*, 2013; Winkel *et al.*, 2008, 2011).

This approach has tremendous potential and provided the basis for Eawag to develop GAP for the mapping, sharing and predictive risk modeling of groundwater data.

2 METHODS

2.1 *Internet platform*

GAP was developed as an online open-source data and information-sharing portal for groundwater-related questions, with a special focus on geogenic contaminants (Fig. 1). It offers two main sections:

(1) GAP Maps where users can view existing data and risk maps, but most importantly use their own data to create risk maps (Fig. 2);

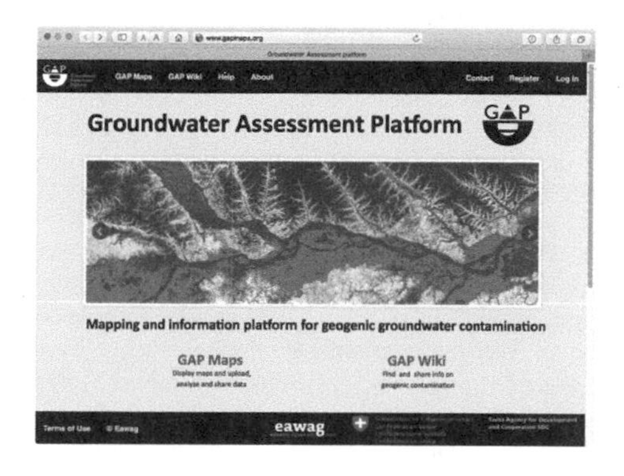

Figure 1. Entry page of the Groundwater Assessment Platform (www.gapmaps.org).

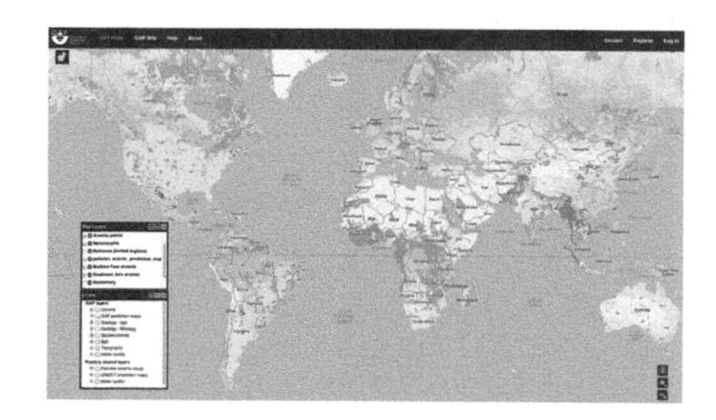

Figure 2. Existing arsenic data in GAP as of November 2017.

(2) GAP Wiki where users can view and contribute information on mitigation and issues surrounding geogenic contamination.

The high-level goals of GAP are aligned with the UN Sustainable Development Goal (SDG) 6, in particular 6.1 – Access to safe drinking water,

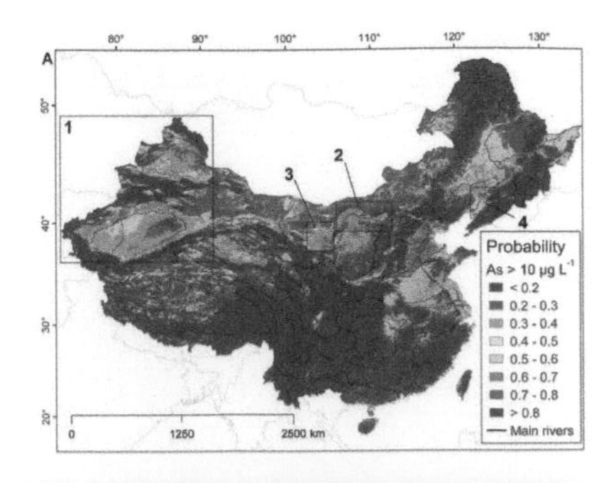

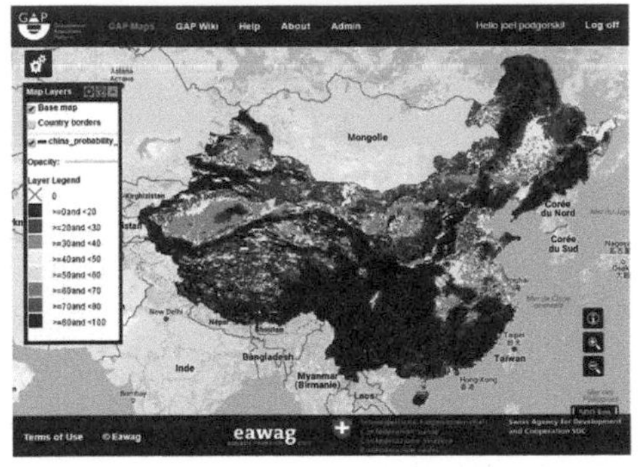

Figure 3. Arsenic prediction map for China modeled offline (Rodriguez-Lado *et al.*, 2013) and the analog created in GAP (lower panel). The results between modeling via manual coding and modeling with GAP are obviously very similar.

6.2 – Pollution and hazardous chemicals and 6.B – Community involvement in water management. Hence, GAP's mission is "*To assist communities, national and international institutions, civil society and research organizations in having access to maps, data and relevant information to enable all people and the environment to have an equitable right to safe groundwater.*"

3 RESULTS

3.1 *Simple and fast creation of hazard prediction maps*

To showcase its modeling functionality, GAP was used to reproduce the arsenic hazard map of China that had been previously created (Rodríguez-Lado *et al.*, 2013). The question was if the relatively basic features offered by GAP could achieve similar results and reproduce this published map, which was generated by a complex statistical modeling scheme that involved taking a weighted average of the best of 100 logistic regression iterations that used different random combinations of training and testing data. GAP indeed produced a very similar map as can be seen in Figure 3. Although some alternative datasets and a somewhat lower degree of sophistication (e.g. only one iteration and no random sampling) were necessary with GAP, the result below shows that a comparable model can be produced in GAP in a fraction of the time.

4 OUTLOOK

GAP enables users to visualize their own data with relatively little effort and to produce hazard maps with them. This makes it simple to identify wells that should be investigated as a matter of priority, such that available funds and resources can be deployed in a targeted manner.

This new platform facilitates rapid localization of geogenic contamination, which itself is a milestone in the protection of public health. But just as important as early detection is the development of practical methods for removing these toxic substances from water.

In spite of the progress made, a lot of research still needs to be done, especially on the development of locally adapted treatment technologies. Therefore, GAP was also designed with a wiki forum for sharing information. The sooner existing knowledge can be effectively disseminated, the better.

ACKNOWLEDGEMENTS

GAP is co-funded by the Swiss Agency for Development and Cooperation (SDC) and Eawag. We thank Manouchehr Amini, Ruth Arnheiter, Anja Bretzler, Jay Matta, Annette Johnson, Fabian Suter, Tobias Siegfried, Andreas Steiner, Jakob Steiner, and Chris Zurbrügg for their various contributions to the development of the GAP tool.

REFERENCES

Amini, M., Abbaspour, K.C., Berg, M., Winkel, L., Hug, S.J., Hoehn, E., Yang, H. & Johnson, C.A. 2008. Statistical Modeling of global geogenic arsenic contamination in groundwater. *Environ. Sci. Technol.* 42: 3669–3675.

Ayotte, J.D., Medalie, L., Qi, S.L., Backer, L.C. & Nolan, B.T. 2017. Estimating the high-arsenic domestic-well population in the conterminous United States. *Environ. Sci. Technol.* 51: 12443–12454.

Bretzler, A., Lalanne, F., Nikiema, J., Podgorski, J., Pfenninger, N., Berg, M. & Schirmer, M. 2017. Groundwater arsenic contamination in Burkina Faso, West Africa: predicting and verifying regions at risk. *Sci. Total Environ.* 584–585, 958–970.

Podgorski, J.E., Eqani, S.A.M.A.S., Khanam, T., Ullah, R., Shen, H. & Berg, M. 2017. Extensive arsenic contamination in high-pH unconfined aquifers in the Indus Valley. *Science Advances* 3, e1700935.

Rodriguez-Lado, L., Sun, G., Berg, M., Zhang, Q., Xue, H., Zheng, Q. & Johnson, C.A. 2013. Groundwater arsenic contamination throughout China. *Science* 341: 866–868.

Winkel, L., Berg, M., Amini, M., Hug, S.J. & Johnson, C.A. 2008. Predicting groundwater arsenic contamination in Southeast Asia from surface parameters. *Nature Geosci.* 1: 536–542.

Winkel, L.H.E., Trang, P.T.K., Lan, V.M., Stengel, C., Amini, M., Ha, N.T., Viet, P.H. & Berg, M. 2011. Arsenic pollution of groundwater in Vietnam exacerbated by deep aquifer exploitation for more than a century. *Proc. Natl. Acad. Sci. U.S.A.* 108: 1246–1251.

Environmental Arsenic in a Changing World –
Zhu, Guo, Bhattacharya, Ahmad, Bundschuh & Naidu (Eds)
ISBN 978-1-138-48609-6

Arsenic biogeochemistry from paddy soil to rice

F.-J. Zhao
College of Resources and Environmental Sciences, Nanjing Agricultural University, Nanjing, P.R. China

ABSTRACT: Rice is a major dietary source of inorganic arsenic for the population who consumes rice as the staple food. Paddy rice accumulates much more arsenic than other cereal crops because the anaerobic conditions in submerged paddy soil is conducive to the mobilization of arsenite, and arsenite hitchhikes on the silicon uptake pathway which is highly expressed in rice roots. Although arsenite is the predominant arsenic species in submerged paddy soil, arsenate can still account for a significant portion of the soluble arsenic species. A number of denitrifying bacteria from paddy soils are able to mediate anaerobic arsenite oxidation by coupling with denitrification. Additions of nitrate enhanced anaerobic arsenite oxidation and decreased arsenic mobility in paddy soil. Anaerobic conditions also enhance arsenic methylation, whereas methylated arsenic species can be demethylated by some microbial populations. The dynamics of arsenic methylation and demethylation control the accumulation of dimethylarsenate, which is highly phytotoxic. Different arsenic species are taken up by rice via different transporters. Recent studies have identified a new class of enzymes that can reduce arsenate to arsenite, allowing arsenite to be extruded from rice roots to the external medium for detoxification.

1 INTRODUCTION

Rice is the staple food for about half of the world population and is also a major dietary source of inorganic arsenic (As), a class-one carcinogen for humans. Rice accumulates As much more efficiently than other cereal crops. This results from a combination of an elevated As bioavailability in anaerobic paddy soil and efficient uptake of As by rice roots (Zhao *et al.*, 2010). Paddy soils in many areas in Asia are contaminated with As due to mining activities and irrigation of As-laden groundwater, leading to phytotoxicity in rice crop and substantial yield losses (Panaullah *et al.*, 2009). The straight-head disease, a physiological disorder in rice prevalent in the USA and China, is thought to be related to As. Understanding the As biogeochemistry in paddy environments and the mechanisms of As uptake and transport in rice plants is important for both food security and safety.

2 RESULTS AND DISCUSSION

2.1 *Arsenic biogeochemistry in paddy environments*

Arsenic is a redox sensitive metalloid. Arsenate [As(V)] is the predominant As species under aerobic conditions. In flooded paddy soil, As(V) is readily reduced to arsenite [As(III)], which is less strongly adsorbed by iron oxyhydroxides and therefore more mobile than As(V). Arsenic bioavailability in paddy soils is greatly enhanced under anaerobic conditions. As(V) reduction is mediated by soil microbes via either the detoxification mechanism, in which As(V) is reduced to As(III) by microbial As(V) reductases (e.g. ArsC) and extruded out of the cell, or the dissimilatory pathway, in which As(V) serves as the terminal electron acceptor for microbial respiration. Paradoxically, As(V) can still account for considerable proportions (10–30%) of the As in the solution of anaerobic paddy soils, even though thermodynamic calculations predict negligible presence of As(V). Our recent studies have shown that some denitrifiers in paddy soils are able to couple denitrification with anaerobic As(III) oxidation (Zhang *et al.*, 2015; Zhang *et al.*, 2017). Additions of nitrate to anaerobic paddy soils enhanced the population of denitrifiers, the abundance of the genes encoding As(III) oxidases and the oxidation of As(III) to As(V). These effects attenuated As bioavailability in flooded paddy soil.

Arsenic can be methylated into mono-, di- and tri-methylated As species by soil microbes. Some of the methylated As species, methylarsines, are volatile. Arsenic methylation and volatilization are enhanced under anaerobic conditions in paddy soils. Arsenic methylation in anaerobic paddy soils appears to be driven primarily by sulphate reducing bacteria. In contrast, dimethylarsenate (DMA) can be demethylated by methanogens. Arsenic methylation is catalyzed by arsenite S-adenosylmethionine methyltransferase (ArsM), which is present in some soil microbes (Huang *et al.*, 2016; Qin *et al.*, 2006). Higher plants do not appear to possess the ability to methylate As, but are able to take up methylated As species from the soil (Lomax *et al.*, 2012). Transgenic rice expressing microbial ArsM was able to convert most inorganic As into DMA and TMAO, which caused phytotoxicity similar to the straight-head disease in rice.

2.2 *Arsenic uptake and detoxification in rice*

Arsenite is taken up mainly by the silicon transporters in rice (Ma *et al.*, 2008), whilst As(V) is taken up by phosphate transporters (e.g. OsPT8) (Wang *et al.*, 2016). As(V) reduction followed by As(III) efflux is a major mechanism of As detoxification in microorganisms. This has been found to be true in plants as well. Recent studies have identified a new class of As(V) reductases, named HAC, in Arabidopsis thaliana (Chao *et al.*, 2014). Our recent studies have identified OsHAC1;1, OsHAC1;2 and OsHAC4 as new As(V) reductases in rice that play a key role in As(V) tolerance and As accumulation in rice plants (Shi *et al.*, 2016; Xu *et al.*, 2017). These enzymes reduce As(V) to As(III); the latter is then effluxed to the external medium to avoid excessive build-up of As in the cells. Mutation in the OsHAC genes resulted in greatly increased accumulation of As in the above-ground tissues of rice, whereas overexpression of OsHAC genes decreased As accumulation.

DMA is taken up by rice roots partly via the silicon transporter Lsi1 (Li *et al.*, 2009) and transported to rice grain via the peptide transporter OsPTR7 (Tang *et al.*, 2017). DMA is more toxic to plants than inorganic As (Tang *et al.*, 2016). This is because DMA cannot be detoxified by complexation with phytochelatins and, as a result, is highly mobile during the long-distance translocation in rice plants. The high mobility and high toxicity of DMA explains its high potency in causing spikelet infertility in rice panicle, the key symptom of the straight-head disease.

ACKNOWLEDGEMENTS

The study was supported by the National Natural Science Foundation of China (grant No. 41330853 and 21661132001).

REFERENCES

Chao, D.Y., Chen, Y., Chen, J.G., Shi, S.L., Chen, Z.R., Wang, C.C., Danku, J.M. Zhao, F.J. & Salt, D.E. 2014. Genome-wide association mapping identifies a new arsenate reductase enzyme critical for limiting arsenic accumulation in plants. *PLoS Biol.* 12(2): e1002009.

Huang, K., Chen, C., Zhang, J., Tang, Z., Shen, Q., Rosen, B.P. & Zhao, F.J. 2016. Efficient arsenic methylation and volatilization mediated by a novel bacterium from an arsenic-contaminated paddy soil. *Environ. Sci. Technol.* 50(12): 6389–6396.

Li, R.Y., Ago, Y., Liu, W.J., Mitani, N., Feldmann, J., McGrath, S.P., Ma, J.F. & Zhao, F.J. 2009. The rice aquaporin Lsi1 mediates uptake of methylated arsenic species. *Plant Physiol.* 150(4): 2071–2080.

Lomax, C., Liu, W.J., Wu, L., Xue, K., Xiong, J., Zhou, J., McGrath, S.P., Meharg, A.A., Miller, A.J. & Zhao, F.J. 2012. Methylated arsenic species in plants originate from soil microorganisms. *New Phytol.* 193(3): 665–672.

Ma, J.F., Yamaji, N., Mitani, N., Xu, X.Y., Su, Y.H., McGrath, S.P. & Zhao, F.J. 2008. Transporters of arsenite in rice and their role in arsenic accumulation in rice grain. *Proc. Nat. Acad. Sci. U.S.A.* 105(29): 9931–9935.

Panaullah, G.M., Alam, T., Hossain, M.B., Loeppert, R.H., Lauren, J.G., Meisner, C.A., Ahmed, Z.U. & Duxbury, J.M. 2009. Arsenic toxicity to rice (*Oryza sativa* L.) in Bangladesh. *Plant Soil* 317(1–2): 31–39.

Qin, J., Rosen, B.P., Zhang, Y., Wang, G.J., Franke, S. & Rensing, C. 2006. Arsenic detoxification and evolution of trimethylarsine gas by a microbial arsenite S-adenosylmethionine methyltransferase. *Proc. Nat. Acad. Sci. U.S.A.* 103(7): 2075–2080.

Shi, S., Wang, T., Chen, Z., Tang, Z., Wu, Z.C., Salt, D.E., Chao, D.Y. & Zhao, F.J. 2016. OsHAC1;1 and OsHAC1;2 function as arsenate reductases and regulate arsenic accumulation. *Plant Physiol.* 172: 1708–1719.

Tang, Z., Kang, Y.Y., Wang, P.T. & Zhao, F.J. 2016. Phytotoxicity and detoxification mechanism differ among inorganic and methylated arsenic species in *Arabidopsis thaliana*. *Plant Soil* 401(1–2): 243–257.

Tang, Z., Chen, Y., Chen, F., Ji, Y.C. & Zhao, F.J. 2017. OsPTR7 (OsNPF8.1), a putative peptide transporter in rice, is involved in dimethylarsenate accumulation in rice grain. *Plant Cell Physiol.* 58(5): 904–913.

Wang, P.T., Zhang, W.W., Mao, C.Z., Xu, G.H. & Zhao, F.J. 2016. The role of OsPT8 in arsenate uptake and varietal difference in arsenate tolerance in rice. *J. Exp. Bot.* 67(21): 6051–6059.

Xu, J.M., Shi, S.L., Wang, L., Tang, Z., lv, T.T., Zhu, X.L., Ding, X.M., Wang, Y.F., Zhao, F.J. & Wu, Z.C. 2017. OsHAC4 is critical for arsenate tolerance and regulates arsenic accumulation in rice. *New Phytol.* 215(3): 1090–1101.

Zhang, J., Zhou, W., Liu, B., He, J., Shen, Q. & Zhao, F.-J. 2015. Anaerobic arsenite oxidation by an autotrophic arsenite-oxidizing bacterium from an arsenic-contaminated paddy soil. *Environ. Sci. Technol.* 49(10): 5956–5964.

Zhang, J., Zhao, S.C., Xu, Y., Zhou, W.X., Huang, K., Tang, Z. & Zhao, F.-J. 2017. Nitrate stimulates anaerobic microbial arsenite oxidation in paddy soils. *Environ. Sci. Technol.* 51(8): 4377–4386.

Zhao, F.J., McGrath, S.P. & Meharg, A.A. 2010. Arsenic as a food-chain contaminant: mechanisms of plant uptake and metabolism and mitigation strategies. *Ann. Rev. Plant Biol.* 61: 535–559.

Environmental Arsenic in a Changing World –
Zhu, Guo, Bhattacharya, Ahmad, Bundschuh & Naidu (Eds)
ISBN 978-1-138-48609-6

Mechanism of As(III) S-adenosylmethionine methyltransferases and the consequences of human polymorphisms in hAS3MT

B.P. Rosen[1], C. Packianathan[1] & J. Li[1,2]
[1]*Herbert Wertheim College of Medicine, Florida International University, Miami, FL, USA*
[2]*College of Basic Medicine, Dali University, Dali, Yunnan, P. R. China*

ABSTRACT: Humans detoxify As(III) by methylation. Rapid methylation results in rapid clearance of arsenic from the body, leading to lower risk of arsenic-related diseases. Paradoxically, the methylated products, MAs(III) and DMAs(III) are more toxic and potentially more carcinogenic than inorganic arsenic. When methylation is slow, body clearance is slower, and the risk of arsenic-related diseases increases. We demonstrate that human single nucleotide polymorphisms in the AS3MT coding region lead to less active enzymes that slow the rate of methylation. We predict that these SNPs lead to increased risk of arsenic-related diseases. To understand the relationship between slower methylation and AS3MT activity, we conducted a structure-function analysis of an AS3MT ortholog and identified the rate limiting step in the catalytic cycle as slow reorientation of the methyl group of enzyme-bound MAs(III).

1 INTRODUCTION

Arsenic methylation, the primary biotransformation in the human body, is catalyzed by the enzyme As(III) S-adenosylmethionine (SAM) methyltransferases (hAS3MT). This process is thought to be protective from acute high-level arsenic exposure. However, with long term low-level exposure, hAS3MT produces intracellular methylarsenite (MAs(III)) and dimethylarsenite (DMAs(III)), which are considerably more toxic than inorganic As(III) and may contribute to arsenic-related diseases. Three previously identified exonic single nucleotide polymorphisms (SNPs) (R173W, M287T and T306I) may be deleterious. We identified five additional intragenic variants in hAS3MT (H51R, C61W, I136T, W203C and R251H) (Li *et al.*, 2017). We purified the eight polymorphic hAS3MT proteins and characterized their enzymatic properties. Each enzyme had low methylation activity through decreased affinity for substrate, lower overall rates of catalysis and/or lower stability. We propose that amino acid substitutions in hAS3MT with decreased catalytic activity lead to detrimental responses to environmental arsenic and may increase the risk of arsenic-related diseases.

Methylarsenite (MAs(III)) is both the product of the first methylation step and the substrate of the second methylation step. When the rate of the overall methylation reaction was determined with As(III) as substrate, the first methylation step was rapid, while the second methylation step was slow (Dheeman *et al.*, 2014). In contrast, when MAs(III) was used as substrate, the second methylation step was as fast as the first methylation step when As(III) was used substrate. These results indicate that there is a slow conformational change between the first and second methylation steps. The structure of the AS3MT ortholog CmArsM was determined with bound MAs(III) at 2.27 Å resolution (Packianathan & Rosen, unpublished). The methyl group is facing solvent, as would be expected when MAs(III) is bound as substrate rather than facing the SAM binding site, as would be expected for MAs(III) as a product. We propose that the rate-limiting step in arsenic methylation is slow reorientation of the methyl group from the SAM binding site to solvent.

2 METHODS

2.1 *Construction of hAS3MT variant*

The eight mutations in the synthetic hAS3MT gene (Dheeman *et al.*, 2014) were introduced by site-directed mutagenesis. The variants were purified and assayed by previously reported methods (Dheeman *et al.*, 2014).

2.2 *Crystallization of CmArsM with bound MAs(III)*

Co-crystallization of CmArsM with MAs(III) was performed in hanging drops (Ajees *et al.*, 2012). Data sets were collected at the Southeast Regional Collaborative Access Team (SER-CAT) facility at Advanced Photon Source (APS), Argonne National Laboratory. The structure was by molecular replacement with the unliganded crystal structure of CmArsM (PDB ID 4FS8) as a template.

3 RESULTS AND DISCUSSION

3.1 *Effect of SNPs on hAS3MT activity*

In this study we searched repositories of human genomic data and identified eight nonsynonymous missense variants for further analysis. We introduced each mutation into a synthetic hAS3MT gene (Li *et al.*, 2017) and purified the resulting enzymes. We compared their enzymatic properties and stability with the most common form of hAS3MT using either As(III) or MAs(III) as substrate. The location of each substitution in the structure of hAS3MT was identified using a homology structural model of hAS3MT (Dheeman *et al.*, 2014), allowing correlation of structure with enzymatic properties. For example, a I136T substitution could be placed in the N-terminal domain that contains the SAM binding domain, and this single amino acid substitution resulted in lower affinity for SAM. A C61W substitution eliminates one of the four conserved cysteine residues (Cys32, Cys61, Cys156 and Cys206) involved in substrate binding and specificity (Fig. 1). As predicted, the C61W enzyme methylated MAs(III) but was unable to methylate As(III). Each of the other hAS3MT polymorphic enzymes has lower methylation capacity, with reduction in the Vmax, Kcat/Km and affinity for As(III), resulting in lower DMA:MAs ratios. Compared to wild type hAS3MT, those SNPs also show lower thermal stability. Thus, individuals with any of the eight variants would be predicted to have a longer total arsenic retention time in the body, leading to elimination of more iAs and MAs and less DMAs with a lower urinary DMAs:MAs ratio. These individuals could be at greater risk for arsenic-related diseases.

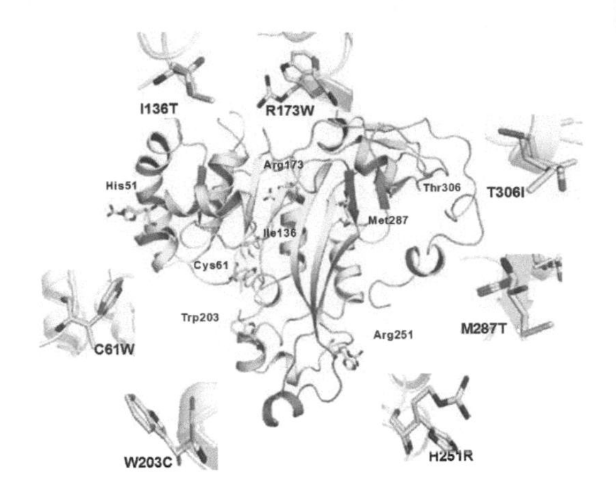

Figure 1. Homology model of hAS3MT and polymorphisms. The structural model of the human AS3MT consists of three domain, the N-terminal SAM binding domain (light blue), the central As(III) binding domain (grey) and the C-terminal domain (green). The location of the eight residues altered by the SNPs are shown in stick form, and the predicted structure of the amino acid substitutions are shown on the boarder superimposed on the original residues. Four (H51R, C61W, I136T and R173W) are located in the large N-terminal domain that includes the SAM binding domain, two (W203C and R251H) are located in the As(III) binding domain, and two (M287T and T306I) are in the C-terminal domain.

3.2 *Reorientation of the methyl group of enzyme-bound MAs(III) is the rate-limiting step in the arsenic methylation pathway*

We determined the rates of the first two steps of methylation, As(III) to MAs(III) and MAs(III) to DMAs. When As(III) was used as substrate, the first methylation step, (As(III)→MAs(III)), was rapid, while the second methylation step, MAs(III)→DMAs, was considerably slower. In contrast, when the second methylation step was assayed using MAs(III) as substrate, it was slightly faster than the first methylation step. Thus, the rate of the second methylation step, MAs(III)→DMAs, depends on whether the initial substrate is As(III) or MAs(III). Immediately following methylation of As(III) to MAs(III), the orientation of the methyl group is inferred to be facing the S-adenosylhomocysteine (SAH) product in the SAM binding site. In order to carry out the next methylation reaction, the methyl group must leave the SAM site to allow a new molecule of SAM to bind. We propose that, when MAs(III) is the product of As(III) methylation, the orientation of the methyl group adjacent to the SAM binding site prevents the SAH product from leaving and entrance of a second SAM molecule, creating a kinetic block between the first and second methylation steps. To examine the structural basis for this slow step, we obtained a new crystal structure of MAs(III)-bound CmArsM. In this structure the orientation of the methyl group is facing toward solvent, the predicted conformation when MAs(III) is substrate. In this conformation an aqueous channel to the active site is open, allowing exchange of the product SAH for another SAM substrate. We propose that slow reorientation of the methyl group from the SAM binding site to solvent is rate-limiting in methylation of inorganic arsenic (Fig. 2).

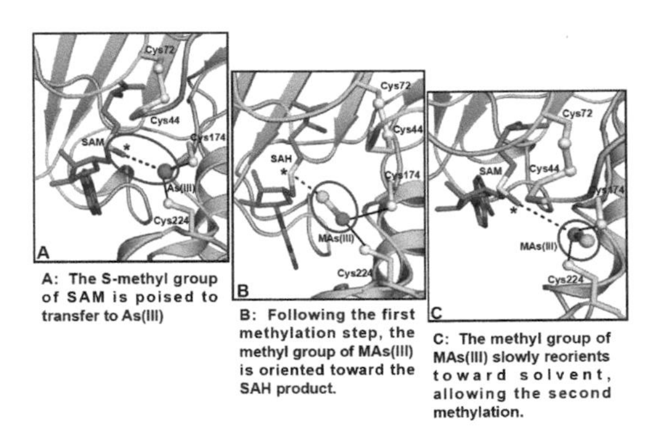

Figure 2. Reorientation of the methyl group is the rate-limiting step in As(III) methylation.

4 CONCLUSIONS

Eight nonsynonymous missense variants in hAS3MT all had reduced catalytic activity, leading to high MAs:DMAs ratios. We propose that individuals with these SNPs may be at increased risk of arsenic-related diseases. To understand how the enzyme controls formation of MAs and DMAs, we identified the rate-limiting step in the overall reaction as a slow reorientation of the methyl group of enzyme-bound MAs(III).

ACKNOWLEDGEMENTS

This work was supported by NIH grants R01 ES023779 and R01 GM055425, and a pilot project grant to CP from Herbert Wertheim College of Medicine.

REFERENCES

Ajees, A.A., Marapakala, K., Packianathan, C., Sankaran, B. & Rosen, B.P. 2012. Structure of an As(III) S-adenosylmethionine methyltransferase: insights into the mechanism of arsenic biotransformation. *Biochemistry* 51(27):5476–5485.

Dheeman, D.S., Packianathan C., Pillai, J.K. & Rosen, B.P. 2014. Pathway of human AS3MT arsenic methylation. *Chem. Res. Toxicol.* 27(11):1979–1989.

Li, J., Packianathan, C., Rossman, T.G. & Rosen, B.P. 2017. Nonsynonymous polymorphisms in the human AS3MT arsenic methylation gene: implications for arsenic toxicity. *Chem. Res. Toxicol.* 30(7):1481–1491.

Environmental Arsenic in a Changing World –
Zhu, Guo, Bhattacharya, Ahmad, Bundschuh & Naidu (Eds)
ISBN 978-1-138-48609-6

Genetic susceptibility and alterations in relation to arsenic exposure, metabolism and toxicity

H. Ahsan & B. Pierce
Institute for Population and Precision Health, Chicago Center for Health and Environment, The University of Chicago, Chicago, IL, USA

ABSTRACT: Chronic arsenic exposure increases risks for mortality and morbidity in humans. In addition to dose, duration of exposure and other host factors, the risks vary based on genetic status of individuals. We undertook some of the largest and most comprehensive genome-wide investigations of genetic susceptibility to metabolism and health effects of arsenic. Two major genetic loci have been identified to play significant roles: arsenic methyl-transferase gene on chromosome 10 and formiminotransferase cyclodeaminase gene on chromosome 21. Candidate gene studies found genes in the oxidative stress, DNA repair and immune pathways also to play roles. Our genome-wide investigations of gene expression and epigenetic variations identified novel genomic alterations induced by arsenic exposure in human. The presentation at the conference will integrate these findings, both published and unpublished, along with other recent key genomic investigations of arsenic metabolism and toxicity in different populations in the world.

1 INTRODUCTION

Arsenic contamination of drinking water is a major public health problem affecting >100 million people in many countries, increasing their risk for a wide array of diseases and mortality. Specific diseases liked with arsenic exposure include skin lesions, cancer, diabetes, cardiovascular disease, non-malignant lung disease, and overall mortality. There is inter-individual variation in arsenic metabolism efficiency and susceptibility to arsenic toxicity; however, the basis of this variation is not well understood. Once absorbed into the blood, most inorganic arsenic (iAs) is converted to mono-methylated (MMA) and then di-methylated (DMA) forms, facilitating excretion in urine. Arsenic metabolism is influenced by lifestyle and demographic factors, as well as inherited genetic variation. Here we focus on the genetic susceptibility to arsenic metabolism and toxicity and associated mechanisms and molecular alterations.

2 METHODS

We conducted the largest and most comprehensive genomic investigations of arsenic metabolism and toxicity to date, through a series of genome-wide and epigenome-wide association studies and molecular genomic studies involving gene expression and methylation. GWAS studies of common variants as well as whole-exome study involving nonsynonymous, protein-coding variations were conducted in relation to arsenic metabolism efficiency and toxicity phenotypes. These studies were conducted within two large cohorts: Health Effects of Arsenic Longitudinal Study (HEALS) and Bangladesh vitamin E and Selenium Trial (BEST). Because variants influencing arsenic metabolism may alter susceptibility to arsenic toxicity, we also investigated the roles of metabolite-associated SNPs in arsenic-induced premalignant skin lesions, the hallmark of chronic arsenic toxicity.

Genomic DNA and RNA from individuals participating in HEALS and BEST were utilized for SNP array, exome array, gene expression array and methylation array to generate data on common variants, exomic rare variants, mRNA expression and CpG methylation variations, respectively. The SNP variants were imputed to 1KG dataset to expand genomic coverage. Arsenic metabolites in urine were separated using high-performance liquid chromatography and detected using inductively coupled plasma-mass spectrometry, as previously described (Ahsan *et al.*, 2007). Percentages of iAs, MMA and DMA were calculated after subtracting asenobetaine and arsenocholine from total arsenic.

For association analyses, GEMMA (Genome-wide Efficient Mixed Model Association) was used to account for cryptic relatedness, as a substantial number of our participants have a close relative pair in the study. Allele frequencies and linkage disequilibrium (LD) patterns were examined using LDlink and the Geography of Genetic Variants browser. Gene expression and methylation analyses used Illumina software for calling and a number of different specialized software for specific statistical analyses.

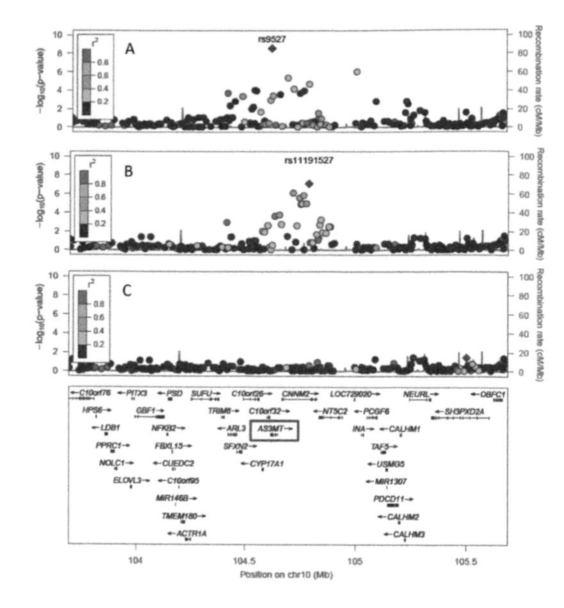

Figure 1. 10q24.32 region variants showing independent associations with DMA%. A: Overall association; B: Adjusted for rs9527; C: Adjusted for rs9527 and rs11191527.

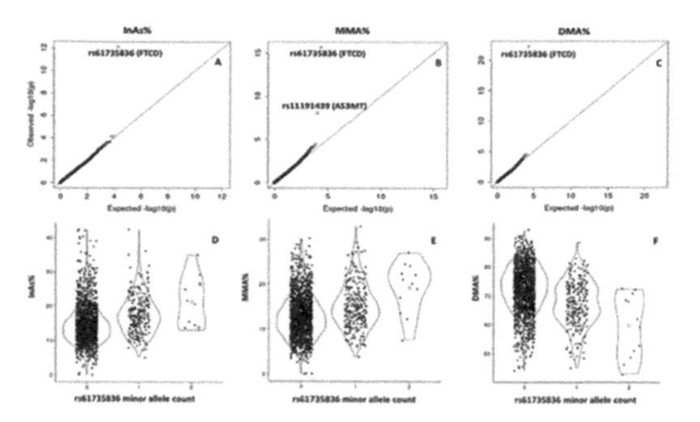

Figure 2. Principal component (PC) scores and correlations between each PC and As species.

3 RESULTS AND DISCUSSION

In the first genome-wide association study of arsenic-related metabolism and toxicity phenotypes we previously identified genetic variation near *AS3MT* (arsenite methyltransferase; 10q24.32; $P < 5 \times 10^{-8}$) to modify arsenic metabolism efficiency and toxicity (Pierce *et al.*, 2012). We reported two *AS3MT* genetic variants showing independent associations with arsenic metabolism (Fig. 1) and one of these five variants (rs9527) was also associated with arsenical skin lesion risk ($P = 0.0005$). The rs9527 variant was also found to interact with arsenic to influence incident arsenical skin lesion risk ($P = 0.01$). Expression quantitative trait locus (eQTL) analyses of genome-wide expression data from based on lymphocyte RNA suggested that several of our lead variants represent cis-eQTLs for AS3MT ($P = 10^{-12}$) and neighboring gene C10orf32 ($P = 10^{-44}$), which are involved in C10orf32-AS3MT read-through transcription. For our five lead SNPs, we found the rs9527 allele associated with decreased DMA% and increased skin lesion risk ($P = 0.0005$), consistent with the hypothesis that DMA is less toxic than MMA.

In a follow-up investigation (Fig. 2), we used a complementary statistical approach to document that two arsenic-methylated species (MMA% and DMA%) involve associations with distinct 10q24.32 variants (Jansen *et al.*, CEBP 2016).

To identify additional arsenic metabolism loci, we measured protein-coding variants across the human exome for nearly 5,000 HEALS and BEST individuals. Among 19,992 coding variants genome-wide, the minor allele (A) of rs61735836 was associated with increased urinary iAs% ($P = 2 \times 10^{-12}$) and MMA% ($P = 7 \times 10^{-16}$) and decreased DMA% ($P = 8 \times 10^{-22}$) (Fig. 3). Among 2,401 individuals with arsenic-induced skin lesions (an indicator of toxicity and cancer risk) and 2,472 controls, carrying the low-efficiency A allele (frequency = 7%) was associated with increased skin lesion risk (odds ratio = 1.35; $P = 1 \times 10^{-5}$). rs61735836 (p.Val101Met) resides in exon 3 of *FTCD* (formiminotransferase cyclodeaminase) and is in weak linkage disequilibrium with all nearby variants (Pierce *et al.*, submitted, 2018). In exome-wide analyses of all 19,992 post-QC variants, rs61735836 (chr21:47572637 based on hg19) showed a clear association with all three metabolite percentages (Fig. 3 A–C). P-values for this association were $P = 8 \times 10^{-13}$ for iAs%, $P = 2 \times 10^{-16}$ for MMA%, and $P = 6 \times 10^{-23}$ for DMA%. The minor allele (A) was associated with decreased DMA% and increased MMA% and iAs% (Fig. 3 D–F), consistent with the associations previously observed for SNPs in the AS3MT region (Pierce *et al.*, submitted 2018). Together, *FTCD* and *AS3MT* SNPs explain ~10% of the variation in DMA% and support a causal effect of arsenic metabolism efficiency on arsenic toxicity in Bangladesh population. Variants in both loci also show suggestive gene-arsenic interactions Using genomewide SNP variations data among related individuals in our study population, we have previously shown that as high as 61% of variations in DMA% could be explained by genetic variations (Gao *et al.*, 2015). We have proposed novel integrated approaches to evaluate subtle, but more pervasive, gene-arsenic interactions beyond these two major loci by using other omics (gene expression and epigenomic) data (Argos *et al.*, 2018).

To identify alterations in genome that are induced by arsenic exposure as an attempt to elucidate potential biological mechanism underlying health effects of arsenic we also evaluated genome-level variations in gene expression and DNA methylation. We identified a number of novel gene expression and DNA methylation alterations that are significantly associated with arsenic exposure (Argos *et al.* 2015; submitted 2018; Gao *et al.*, 2015). In an independent integrated genomics study, we have recently shown that gene expression and DNA methylation

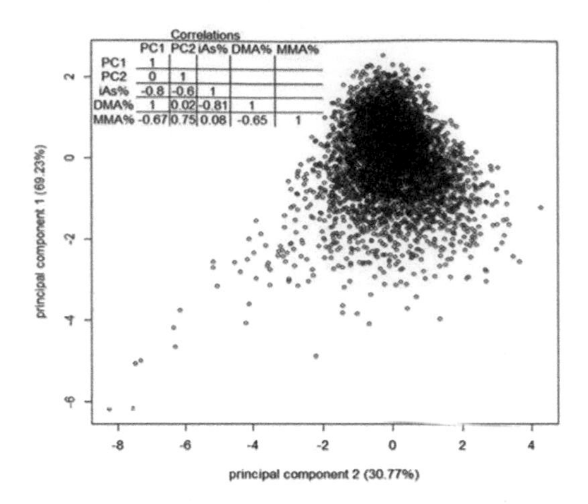

Figure 3. *FTCD* protein-coding SNP rs61735836 is associated with the all three arsenic species measured in urine.

alterations are co-regulated in the genome (Pierce *et al.*, 2018). Future studies will investigate whether arsenic-induced gene expression and DNA methylation alterations are co-regulated in a manner that can be amenable to interventions.

4 CONCLUSIONS

In summary, our body of work implicates two genetic loci, a common variant locus in the *AS3MT*/C10orf32 locus at 10q24.32 and a protein-altering variant in *FTCD* locus at chr21:47572637 influencing arsenic metabolism efficiency and risk for arsenic-induced skin lesions, the most common sign of arsenic toxicity. We also identified novel genomic alterations potentially induced by arsenic exposure and novel approaches to investigate them. Future studies can use these results, to study the effects of arsenic exposure and metabolism efficiency on health outcomes believed to be affected by arsenic (e.g., cancer and cardiovascular disease), even in the absence of data on arsenic exposure. Knowledge of variation in these regions and associated biological processes could be used to develop intervention and pharmacological strategies aimed at preventing large numbers of arsenic-related deaths in arsenic-exposed populations worldwide.

REFERENCES

Ahsan, H., Chen, Y., Kibriya, M.G., Slavkovich, V., Parvez, F., Jasmine, F., Gamble, M.V. & Graziano, J.H. 2007. Arsenic metabolism, genetic susceptibility, and risk of premalignant skin lesions in Bangladesh. *Cancer Epidemiol. Biomarkers Prev.* 16(6): 1270–1278.

Argos, M., Chen, L., Jasmine, F., Tong, L., Pierce, B.L., Roy, S., Paul-Brutus, R., Gamble, M.V., Harper, K.N., Parvez, F., Rahman, M., Rakibuz-Zaman, M., Slavkovich, V., Baron, J.A., Graziano, J.H., Kibriya, M.G. & Ahsan, H. 2015. Gene-specific differential DNA methylation and chronic arsenic exposure in an epigenome-wide association study of adults in Bangladesh. *Environ. Health Perspect.* 123(1): 64–71.

Argos, M., Tong, L., Roy, S., Sabarinathan, M., Ahmed, A., Islam, M.T., Islam, T., Rakibuz-Zaman, M., Sarwar, G., Shahriar, H., Rahman, M., Yunus, M., Graziano, J.H., Jasmine, F., Kibriya, M.G., Zhou, X., Ahsan, H. & Pierce, B.L. 2018. Screening for gene-environment (G×E) interaction using omics data from exposed individuals: an application to gene-arsenic interaction. *Mamm. Genome* 29(1–2): 101–111.

Farzan, S.F., Karagas, M.R., Jiang, J., Wu, F., Liu, M., Newman, J.D., Jasmine, F., Kibriya, M.G., Paul-Brutus, R., Parvez, F., Argos, M., Scannell Bryan, M., Eunus, M., Ahmed, A., Islam, T., Rakibuz-Zaman, M., Hasan, R., Sarwar, G., Slavkovich, V., Graziano, J., Ahsan, H. & Chen Y. 2015. Gene-arsenic interaction in longitudinal changes of blood pressure: Findings from the Health Effects of Arsenic Longitudinal Study (HEALS) in Bangladesh. *Toxicol. Appl. Pharmacol.* 288(1): 95–105.

Gao, J., Tong, L., Argos, M., Scannell Bryan, M., Ahmed, A., Rakibuz-Zaman, M., Kibriya, M.G., Jasmine, F., Slavkovich, V., Graziano, J.H., Ahsan, H. & Pierce, B.L. 2015. The genetic architecture of arsenic metabolism efficiency: a SNP-based heritability study of Bangladeshi adults. *Environ. Health Perspect.* 123(10): 985–992.

Gao, J., Roy, S., Tong, L., Argos, M., Jasmine, F., Rahaman, R., Rakibuz-Zaman, M., Parvez, F., Ahmed, A., Hore, S.K., Sarwar, G., Slavkovich, V., Yunus, M., Rahman, M., Baron, J.A., Graziano, J.H., Ahsan, H. & Pierce, B.L. 2015. Arsenic exposure, telomere length, and expression of telomere-related genes among Bangladeshi individuals. *Environ. Res.* 136: 462–469.

Jansen, R.J., Argos, M., Tong, L., Li, J., Rakibuz-Zaman, M., Islam, M.T., Slavkovich, V., Ahmed, A., Navas-Acien, A., Parvez, F., Chen, Y., Gamble, M.V., Graziano, J.H., Pierce, B.L. & Ahsan, H. 2016. Determinants and consequences of arsenic metabolism efficiency among 4,794 individuals: demographics, lifestyle, genetics, and toxicity. *Cancer Epidemiol. Biomarkers Prev.* 25(2): 381–390.

Pierce, B.L., Kibriya, M.G., Tong, L., Jasmine, F., Argos, M., Roy, S., Paul-Brutus, R., Rahaman, R., Rakibuz-Zaman, M., Parvez, F., Ahmed, A., Quasem, I., Hore, S.K., Alam, S., Islam, T., Slavkovich, V., Gamble, M.V., Yunus, M., Rahman, M., Baron, J.A., Graziano, J.H. & Ahsan, H. 2012. Genome-wide association study identifies chromosome 10q24.32 variants associated with arsenic metabolism and toxicity phenotypes in Bangladesh. *PLoS Genet.* 8(2): e1002522.

Environmental Arsenic in a Changing World –
Zhu, Guo, Bhattacharya, Ahmad, Bundschuh & Naidu (Eds)
ISBN 978-1-138-48609-6

Arsenic oral bioavailability in soils, housedust, and food: implications for human health

L.Q. Ma[1,2], H.B. Li[3], D. Zhao[3] & A.L. Juhasz[4]
[1]*Research Center for Soil Contamination and Environment Remediation, Southwest Forestry University, Kunming, P.R. China*
[2]*Soil and Water Sciences Department, University of Florida, Gainesville, FL, USA*
[3]*State Key Laboratory of Pollution Control and Resource Reuse, School of the Environment, Nanjing University, Nanjing, P.R. China*
[4]*Centre for Environmental Risk Assessment and Remediation, University of South Australia, Mawson Lakes, SA, Australia*

ABSTRACT: Incidental ingestion of As-contaminated soils and housedust is an important non-dietary As exposure contributor for children living nearby contaminated sites, while consumption of rice has been recognized as the most important dietary contributor. However, to accurately assess the health risk associated with soil and dust ingestion and rice consumption, determination of total concentration and oral bioavailability of As are both important. However, compared to soils, assessment of As relative bioavailability (RBA) in housedust and rice is limited. In addition, the suitability of in vitro bioaccessibility assays to predict As bioavailability has not been compared between different exposure scenarios. Recently, by combining in vivo mouse bioassay and in vitro bioaccessibility assays, we measured As-RBA (relative to the absorption of sodium arsenate) in samples of contaminated soils, slightly-contaminated housedust, co-contaminated housedust, and rice. Results showed that As-RBA in these media showed significant variation among individual samples, suggesting the need to incorporate bioavailability to accurately assess the associated health risk. In addition, by establishing in vivo-in vitro correlations, the most suitable in vitro assay to predict As-RBA varied with the target media, suggesting the need of developing specific methodologies to predict As-RBA in different environmental media.

1 INTRODUCTION

Arsenic (As) is a ubiquitous contaminant present in food and the environment. Chronic exposure to As causes various adverse health effects including cancers, skin disorders, vascular disease, and diabetes mellitus (Hinwood *et al.*, 2003). Humans are exposed to As via both dietary and non-dietary pathways, with incidental ingestion of contaminated soils and housedust being important non-dietary contributor, while rice consumption is recognized as an important dietary contributor to inorganic As intake (Juhasz, *et al.*, 2006; Li *et al.*, 2011, 2014, 2017; Ruby *et al.*, 1999; Williams *et al.*, 2009). Reliable assessment of human health risk from both the dietary and non-dietary pathways depends not only on total As concentration in soils, housedust, and rice, but also its bioavailability (Li *et al.*, 2016a; Liao *et al.*, 2005; Ruby & Lowney, 2012; Zhu *et al.*, 2008; Zhu *et al.*, 2013).

While studies have utilized in vivo animal bioassays to determine As relative bioavailability (RBA, relative to the absorption of sodium arsenate) in contaminated soils, limited researches have focused on housedust and rice (Ollson *et al.*, 2017; Torres-Escribano *et al.*, 2008; Williams *et al.*, 2009). Compared to animal models, bioaccessibility methods offer a simple, inexpensive way to estimate As-RBA. However, it is important to establish the relationship between As bioaccessibility and As relative bioavailability (RBA) so the assay is valid to predict As-RBA in soils, housedust, and food (Li *et al.*, 2016b).

2 METHODS/EXPERIMENTAL

2.1 *Arsenic RBA and bioaccessibility in contaminated soils*

Twelve As-contaminated soils (22.2 to 4172 mg As kg^{-1}) were collected from typical mining/smelting contaminated sites across China. As-RBA was assessed using in vivo mouse blood AUC (area under the blood As time curve over 48-h period) bioassay following a single dust gavage dose, while As bioaccessibility was measured using five in vitro assays (UBM, SBRC, IVG, DIN, and PBET). Linear correlation between As-RBA and As bioaccessibility was established to investigate the As-RBA prediction ability of in vitro assays.

2.2 *Arsenic RBA and bioaccessibility in housedust*

As-RBA in 12 housedust samples across China slightly contaminated with As contamination (7–38 mg kg^{-1}) was measured using an in vivo mouse blood AUC model and compared to As bioaccessibility determined using 4 in vitro bioaccessibility assays (SBRC, IVG, DIN, and PBET).

In addition, 15 housedust samples co-contaminated with As (43–2380 mg kg^{-1}), Cd (8.46–329 mg kg^{-1}), and Pb (127–15486 mg kg^{-1}) collected from Pb-Zn ore mining/smelting sites at Hunan, China were assessed for As-, Cd-, and Pb-RBA using an in vivo mouse bioassay with metal(loid) accumulation in liver and kidneys as the biomarker of dust-amended diet consumption over a 10-d period. Arsenic bioaccessibility was assessed using 3 bioaccessibility assays (SBRC, PBET, and UBM).

2.3 *Arsenic RBA in rice*

A mouse model based on steady state As urinary excretion (SSUE) was developed to determine As-RBA in rice to refine inorganic As exposure in humans (Li *et al.*, 2011). Fifty-five rice samples from 15 provinces in China were analyzed for total As, with 11 cooked for As speciation and bioavailability assessment using A mouse bioassay.

3 RESULTS AND DISCUSSION

3.1 *Correlation between As-RBA and bioaccessibility for contaminated soils*

Arsenic RBA in the 12 soils ranged from 6.38 ± 2.80% to 73.1 ± 17.7% with soils containing higher free Fe contents showing lower values (Li *et al.*, 2016). Arsenic bioaccessibility varied within and between in vitro assays. All in vitro assays show strong linear correlation with As-RBA data ($R^2 = 0.50$–0.83), with the IVG assay providing the best estimate of As-RBA ($R^2 = 0.81$–0.83) although the IVIVC slope did not meet the criteria of 0.8–1.2 (Fig. 1). This suggests that the IVG assay has the potential to predict As-RBA in contaminated soils from China.

3.2 *Correlation between As-RBA and bioaccessibility for housedust*

For the 12 slightly contaminated housedust samples, As-RBA ranged from 21.8 ± 1.6 to 85.6 ± 7.2% with samples containing low Fe and high total organic carbon content having higher As-RBA (Li *et al.*, 2014). Strong in vivo–in vitro correlations (IVIVC) were observed between As-RBA and As bioaccessibility for SBRC and DIN assays ($R^2 = 0.63$–0.85), but weaker ones for IVG and PBET assays ($R^2 = 0.29$–0.55).

For the 15 As, Cd, and Pb co-contaminated housedust samples, As-, Cd-, and Pb-RBA was 7.64–37.0, 20.3–94.3, 7.21–52.1%, averaging 21, 52, and 30%, respectively. Analyses of relationship between

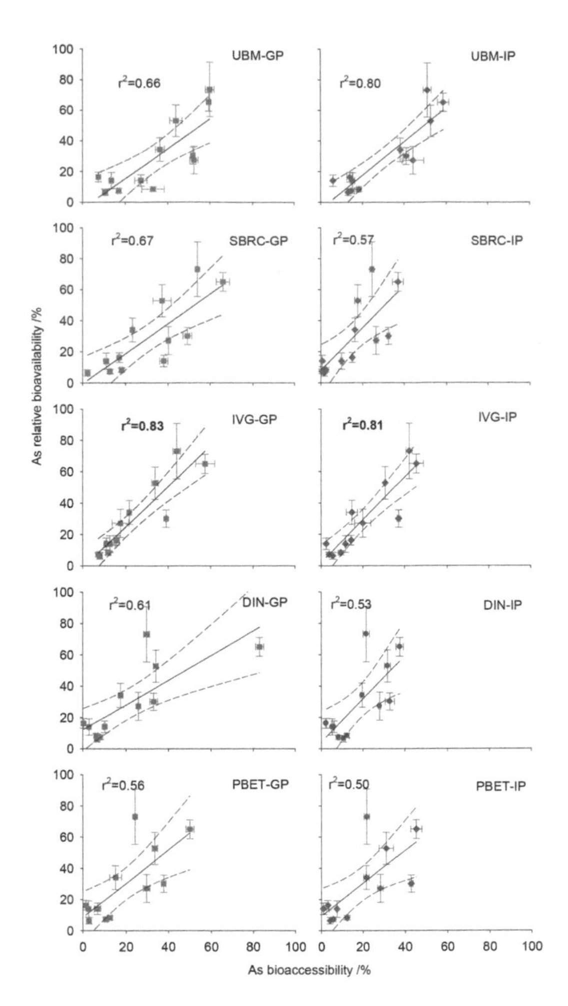

Figure 1. In vivo–in vitro correlations between As relative bioavailability (RBA) in 12 As-contaminated soils from China using a mouse model and As bioaccessibility determined using gastric (GP) and intestinal phases (IP) of five in vitro assays (UBM, SBRC, IVG, DIN, and PBET).

metal(loid) RBA and concentrations showed that Pb-RBA decreased with increasing As concentration, while As-RBA decreased with increasing Pb and Cd concentration ($R^2 = 0.62$–0.72), suggesting that co-presence of As and Pb decreased each other's RBA, while Cd reduced As-RBA.

By correlating metal(loid) RBA data with bioaccessibility data using the SBRC, PBET, and UBM assays, the strongest As in vivo-in vitro correlations were observed for the PBET assay ($R^2 = 0.69$).

3.3 *As-RBA in rice*

Following oral doses of individual As species to mice at low levels of As exposure (2.5–15 µg As per mouse) over a 7-d period, strong linear correlations ($R^2 = 0.99$) were observed between As urinary excretion and cumulative As intake, suggesting the suitability and sensitivity of the mouse bioassay to measure

As-RBA in rice. RBA in cooked rice ranged from 26.2 ± 7.0% to 49.5 ± 4.7% (averaging 39.9 ± 8.3%) for inorganic As. Calculation of inorganic As intake based on inorganic As concentration in rice led to an overestimate of As exposure by 2.02–3.67-fold compared to that calculated based on bioavailable inorganic As.

4 CONCLUSIONS

The results suggested that in vitro assays suitable to predict As-RBA varied with the target media. For accurate assessment of the health risk associated with incidental ingestion of soil and housedust and consumption of rice, it is important to consider As bioavailability.

ACKNOWLEDGEMENTS

This work was supported by National Natural Science Foundation of China (21507057; 41673101; 21637002; BK20150573).

REFERENCES

Hinwood, A.L., Sim, M.R., Jolley, D., de Klerk, N., Bastone, E.B., Gerostamoulos, J. & Drummer, O.H. 2003. Hair and toenail arsenic concentrations of residents living in areas with high environmental arsenic concentrations. *Environ. Health Perspect.* 111(2): 187–193.

Juhasz, A.L., Smith, E., Weber, J., Rees, M., Rofe, A., Kuchel, T., Sansom, L. & Naidu, R. 2006. In vivo assessment of arsenic bioavailability in rice and its significance for human health risk assessment. *Environ. Health Perspect.* 114(12): 1826–1831.

Li, G., Sun, G.X., Williams, P.N., Nunes, L. & Zhu, Y.G. 2011. Inorganic arsenic in Chinese food and its cancer risk. *Environ. Int.* 37(7): 1219–1225.

Li, H.B., Li, J., Juhasz, A.L. & Ma, L.Q. 2014. Correlation of in vivo relative bioavailability to in vitro bioaccessibility for arsenic in household dust from China and its implication for human exposure assessment. *Environ. Sci. Technol.* 48(23): 13652–13659.

Li, H.B., Zhao, D., Li, J., Li, S.W., Wang, N., Juhasz, A.L., Zhu, Y.G. & Ma, L.Q. 2016a. Using the SBRC assay to predict lead relative bioavailability in urban soils: contaminant source and correlation model. *Environ. Sci. Technol.* 50(10), 4989–4996.

Li, H.B., Li, J., Zhao, D., Li, C., Wang, X.J., Sun, H.J., Juhasz, A.L. & Ma, L.Q. 2017. Arsenic relative bioavailability in rice using a mouse arsenic urinary excretion bioassay and its application to assess human health risk. *Environ. Sci. Technol.* 51(8): 4689–4696.

Li, J., Li, C., Sun, H.J., Juhasz, A.L., Luo, J., Li, H.B. & Ma, L.Q. 2016b. Arsenic relative bioavailability in contaminated soils: comparison of animal models, dosing schemes, and biological end points. *Environ. Sci. Technol.* 50(1): 453–461.

Liao, X.Y., Chen, T.B., Xie, H. & Liu, Y.R. 2005. Soil As contamination and its risk assessment in areas near the industrial districts of Chenzhou City, Southern China. *Environ. Int.* 31(6): 791–798.

Ollson, C.J., Smith, E., Herde, P. & Juhasz, A.L. 2017. Influence of sample matrix on the bioavailability of arsenic, cadmium and lead during co-contaminant exposure. *Sci. Total Environ.* 595: 660–665.

Ruby, M.V. & Lowney, Y.W. 2012. Selective soil particle adherence to hands: implications for understanding oral exposure to soil contaminants. *Environ. Sci. Technol.* 46(23): 12759–12771.

Ruby, M.V., Schoof, R., Brattin, W., Goldade, M., Post, G., Harnois, M., Mosby, D.E., Casteel, S.W., Berti, W., Carpenter, M., Edwards, D., Cragin, D. & Chappell, W. 1999. Advances in evaluating the oral bioavailability of inorganics in soil for use in human health risk assessment. *Environ. Sci. Technol.* 33(21): 3697–3705.

Torres-Escribano, S., Leal, M., Vélez, D. & Montoro, R. 2008. Total and inorganic arsenic concentrations in rice sold in Spain, effect of cooking, and risk assessments. *Environ. Sci. Technol.* 42(10): 3867–3872.

Williams, P.N., Lei, M., Sun, G., Huang, Q., Lu, Y., Deacon, C., Meharg, A.A. & Zhu, Y.G. 2009. Occurrence and partitioning of cadmium, arsenic and lead in mine impacted paddy rice: Hunan, China. *Environ. Sci. Technol.* 43(3): 637–642.

Zhu, Y.G., Sun, G.X., Lei, M., Teng, M., Liu, Y.X., Chen, N.C., Wang, L.H., Carey, A.M., Deacon, C., Raab, A., Meharg, A.A. & Williams, P.N. 2008. High percentage inorganic arsenic content of mining impacted and non-impacted Chinese rice. *Environ. Sci. Technol.* 42(13): 5008–5013.

Zhu, Z., Sun, G., Bi, X., Li, Z. & Yu, G. 2013. Identification of trace metal pollution in urban dust from kindergartens using magnetic, geochemical and lead isotopic analyses. *Atmos. Environ.* 77: 9–15.

Environmental Arsenic in a Changing World –
Zhu, Guo, Bhattacharya, Ahmad, Bundschuh & Naidu (Eds)
ISBN 978-1-138-48609-6

Arsenic removal by iron-based nanomaterials

M.I. Litter
Comisión Nacional de Energía Atómica, Buenos Aires, Argentina

ABSTRACT: Processes based on the use of iron-based nanoparticles (FeNPs) have proved to be a promising technology for the removal of a wide range of pollutants including metals and metalloids in water, being simple and low-cost procedures for removal of As(III)/As(V). FeNPs such as zerovalent iron nanoparticles (nZVI), nanoparticles of iron oxides (FeONPs) as hematite or magnetite, and iron-based nanoparticles prepared from natural extracts of plants or seeds (as *yerba mate* and green tea) have a larger surface area/volume ratio and exceptional adsorption properties compared with micro- or macrosized materials, with the advantages of an effective removal of pollutants at very low concentrations and a low generation of sludge. Laboratory experiments have been performed to evaluate the efficiency of some iron-based nanoparticles for removal of As(III)/(V), studying the influence of the initial concentration of the pollutant, the NP mass, the pollutant:FeNP molar ratio, pH, the presence of O_2, among other variables. In this presentation, a brief review on the use of FeNPs for As removal from water and some results of our laboratory will be presented.

1 INTRODUCTION

Processes based on the use of iron nanoparticles (FeNPs) can be simple and low-cost procedures for removal of As(III)/As(V) in the pH range typical of groundwaters (Lata & Samadder, 2016). FeNPs such as zerovalent iron nanoparticles (nZVI), nanoparticles of iron oxides (FeONPs) as hematite or magnetite, and iron-based nanoparticles prepared from natural extracts of plants or seeds have a larger surface area/volume ratio and exceptional adsorption properties compared with micro- or macrosized materials, with the advantages of an effective removal of pollutants at very low concentrations and a low generation of sludge. The use of FeNPs has proved to be a promising technology for the removal of a wide range of pollutants including metals and metalloids in water (Crane & Scott, 2012; Quici *et al.*, 2018). These nanomaterials can be used to develop low-cost nanotechnologies for *in situ*, on site and *ex situ* water treatments, including drinking water provision, residual effluents and remediation of contaminated sites.

2 MECHANISMS OF REMOVAL OF METALS AND METALLOIDS WITH IRON NANOPARTICLES

Figure 1 shows the possible mechanisms for removal of metals and metalloids by FeNPs, including surface adsorption, complexation, redox processes, precipitation and coprecipitation.

When different FeNPs are compared, As removal capacity is higher for the samples of higher Fe content, i.e., usually the most reduced materials: nZVI > Fe_3O_4 > Fe_2O_3 > FeOOH (Crane & Scott, 2012).

Mixed metal oxide nanoparticles, with Fe combined with other metals, have been also tested for As removal (Lata & Samadder, 2016).

As(V) adsorption is higher than As(III) adsorption at acid and neutral pH values because As(III) species are neutral up to pH 9, while As(V) species are charged. As(V) adsorption is always higher under acidic conditions, as the anionic species are dominant even at pH 2.

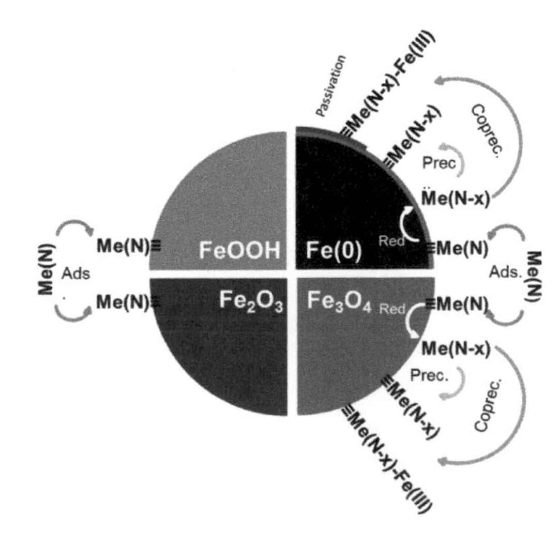

Figure 1. Proposed removal mechanisms of metals and metalloids taking place on nFe surfaces (based on Simeonidis *et al.*, 2013). M: metal or metalloid, N: oxidation state.

Some studies report that the optimal As(III) adsorption on magnetite and maghemite nanoparticles takes place at pH 2 (Quici *et al.*, 2018).

In the presence of dissolved oxygen, arsenite adsorption over magnetite is followed by As(III) partial oxidation to As(V) by the action of reactive oxygen species formed after Fe(II) oxidation by O_2; incomplete As(III) oxidation was ascribed to magnetite passivation (Ona-Nguema *et al.*, 2010).

One alternative low-cost, green method of synthesis of FeNPs is the reaction of iron salts with extracts of leaves or grains of natural plants, such as green tea (GT, *Camellia sinensis*), yerba mate (YM, *Ilex paraguariensis Saint Hilaire*), coffee and wine, where the active compounds are polyphenols and other reducing or complexing species (Oakes, 2013). The preparation of these green FeNPs has been reviewed (Genuino *et al.*, 2013; Herlekar *et al.*, 2014). Hoag *et al.* (2009) reported the first synthesis of FeNPs prepared from GT extracts and $Fe(NO_3)_3$ solutions. Later, similar products using a variety of natural materials and other Fe common salts were reported (Wang, 2015). However, a discrepancy exists on the composition and chemical structure of these materials, some authors claiming the presence of zerovalent iron (Fe(0)), others indicating the formation of amorphous iron complexes and some others presenting the formation of iron oxides. YM has been recently used for producing iron-based nanomaterials (Trotte *et al.*, 2016; Mercado *et al.*, 2017). There are few reports on As removal by green FeNPs (Martínez Cabanas *et al.*, 2016; Poguberović *et al.*, 2016), but the informed As removal capacity (up to 300 mg g^{-1}) makes these low-cost materials very interesting.

3 RESULTS OF OUR LABORATORY

3.1 *Introduction*

In our laboratory, the efficiency of FeNPs has been evaluated for removal of As(III)/(V), studying the influence of the nature and amount of the iron material, the initial concentration of the pollutant, the pollutant: FeNP molar ratio (MR), the pH, the presence of O_2, and other variables (Morgada *et al.*, 2009, Pabón *et al.*, 2018). Recently, FeNPs were rapidly synthesized using YM extracts and Fe(III) salts, and the applicability of the materials for the removal of As(III)/As(V) from aqueous solutions has been tested and compared with similar materials prepared from GT and with commercial nZVI (García *et al.*, 2018). The last advances of our laboratory will be briefly presented.

3.2 *Experimental*

3.2.1 *Materials and methods*

Ethanol (99.5%), NaOH (97%), $FeSO_4{\cdot}7H_2O$ (99%), $FeCl_3.6H_2O$ (99.6%), orthophosphoric acid (85%), and H_2SO_4 (98%) were analytical grade (Biopack). $K_2Cr_2O_7$ was Mallinkrodt. $Na_2HAsO_4{\cdot}7H_2O$, $NaAsO_2$, 1,5-diphenylcarbazide (UCB), acetone (99.8%, Anedra) and gallic acid monohydrate (J.T. Baker) were used. NanoFe® was synthetized by Nanotek S.A. (Santa Fe, Argentina) by a proprietary technology based on chemical reduction of ferric salts with sodium borohydride in a stabilized multiphase nanoemulsion (www.nanoteksa.com). nZVI (NANOFER 25, hereafter N25) was provided by NANO IRON s.r.o. (Czech Rep.) as a concentrated aqueous suspension. Commercial dried YM (Taragui *sin palo*) and GT (Hierbas del Oasis, S.R.L.) leaves were purchased from a local supermarket. In all experiments, Milli-Q water was used (resistivity = 18 MΩ cm). All other chemicals were of the highest purity.

3.2.2 *Synthesis of iron-based nanoparticles from GT and YM extracts*

Commercial GT leaves were processed for 30 s with a hand blender (Philips, HR 1368/00) to obtain a size similar to commercial YM (around 1–3 mm diameter). A 50 g L^{-1} suspension of each powder was extracted during 1 h at 79–85°C and filtered through a 0.22 μm pore size nylon membrane.

Fresh 0.1 M $FeCl_3.6H_2O$ solutions were prepared and vacuum filtered through a 0.22 μm pore size nylon membrane. To 2.5 mL of the Fe(III) solution, the corresponding volume of YM or GT extract, previously diluted to 25 mL, was added at a 2.8:1 extract:Fe MR. The color of the mixture with YM changed from yellow to greenish-black, and to black for GT, indicating in both cases the formation of nanoparticles. The corresponding iron suspensions will be named YM-Fe(III) and GT-Fe(III).

3.2.3 *Experiments of As removal with NanoFe®*

In a typical experiment, 150 mL of an As(V) solution (1 mg L^{-1}, pH 7.8), were introduced into a 250 mL glass Erlenmeyer flask (covered by an aluminum foil to avoid the entrance of ambient light) and bubbled with air (1 mL min^{-1}) for 30 min under orbital stirring in order to reach oxygen saturation. Then, air bubbling was stopped, and a volume of the NanoFe® suspension containing the corresponding amount of the material (to have final concentrations ranging 0.005–0.1 g L^{-1}) was added, under vigorous stirring. The flask was left open to the air all throughout the run. Temperature was ambient, never higher than 30°C. Samples (2.5 mL) were taken periodically, filtered through a 0.45 μm Millipore membrane and analyzed for total As in solution by ICP-OES, using a Perkin-Elmer Optima 3100 XL apparatus.

3.2.4 *Experiments of As removal with N25*

Experiments with As(III) (1 mg L^{-1}, pH 6) were performed in a recirculating batch system with a cylindrical Pyrex glass reactor, adding the corresponding amount of nZVI suspension in different As:Fe RMs, 1:10, 1:20, 1:30, 1:50 and 1:100. The system was stirred, open to air and the temperature was 25°C. Changes in As(V) and As(III) concentration in solution were measured using a modified arsenomolybdate technique at 868 nm (Levy *et al.*, 2012).

3.2.5 *Experiments of As removal with YM-Fe(III) and GT-Fe(III)*

The corresponding volume of YM-Fe(III) and GT-Fe(III) suspensions was added to an Erlenmeyer flask containing 25 mL of a solution of As(V) or As(III) (1 mg L^{-1}, pH 7, As:Fe MR = 1:30), under continuous agitation. Control experiments with Fe(III), Fe(II) at As:Fe MR = 1:30 were performed. Samples (0.5 mL) were taken at periodical times, filtered through a Millipore membrane (0.22 μm) and analyzed for remaining total As in the filtrate by TXRF using a Bruker S2 Picofox spectrometer.

3.3 *Results and discussion*

3.3.1 *Experiments of As(V) removal with NanoFe®*

Experiments of As(V) (1 mg L^{-1}, pH 7.8) removal using different NanoFe® masses are shown in Figure 2. As removal was rapid and increased with NanoFe® dosage in the range 0.005–0.1 g L^{-1}. Removal attained more than 90% after 150 min of contact time at the highest NanoFe® concentrations; 0.05–0.1 g L^{-1} were enough to reach an excellent and rapid efficiency.

3.3.2 *As(III) removal with N25*

Results of As(III) removal (1 mg L^{-1}) are presented in Figure 3. The removal increased with the increase of the amount of nZVI, being almost complete at 1:50 MR. As(V) was detected in the final solution in all cases but at a low concentrations compared with the initial As(III) concentration (lower than 0.4 mg L^{-1}), indicating that part of the material was immobilized on the remaining solid.

3.3.3 *As removal with YM-Fe(III)-5 and GT-Fe(III)-5*

In Figure 4, results of total As removal with YM-Fe(III)-5 and GT-Fe(III)-5 starting from As(III) (1 mg L^{-1}, pH 7) are shown. With both Fe(III)-5 and GT-Fe(III)-5, low removals, 22 and 32% respectively, were obtained; however, at this As:Fe MR = 1:30, N25 was not very efficient either (44%) and the removal was similar to that performed adding soluble Fe(II). Control with Fe(III) gave negligible removal.

Figure 5 shows the results of total As removal with YM-Fe(III)-5 and GT-Fe(III)-5 starting from As(V).

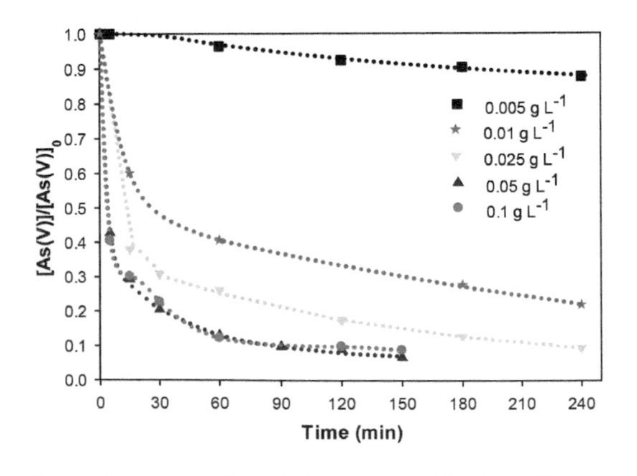

Figure 2. Temporal evolution of normalized As(V) concentration during treatment with NanoFe® in different iron masses. $[As(V)]_0 = 1$ mg L^1 pH 7.8, $T = 30°C$.

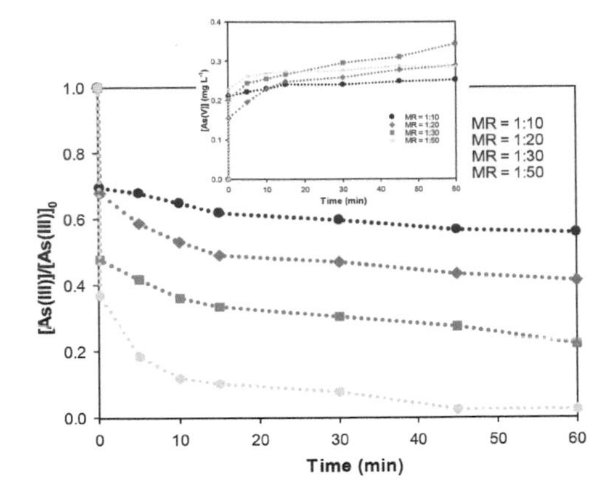

Figure 3. Temporal evolution of normalized As(III) concentration during treatment with N25. Conditions: $[As(III)]_0 = 1$ mg L^{-1}, MR = 1:10, 1:20, 1:30, 1:50, pH 6-7, open to air, $T = 25°C$. Inset: temporal evolution of As(V) concentration.

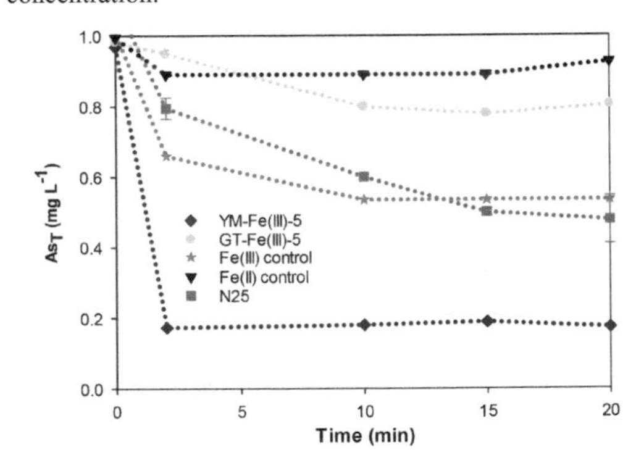

Figure 4. Temporal evolution of normalized total As concentration during treatment with YM-Fe(III)-5 and GT-Fe(III)-5. Conditions: $[As(III)]_0 = 1$ mg L^{-1}, pH 7, As:Fe MR = 1:30, $T = 25°C$.

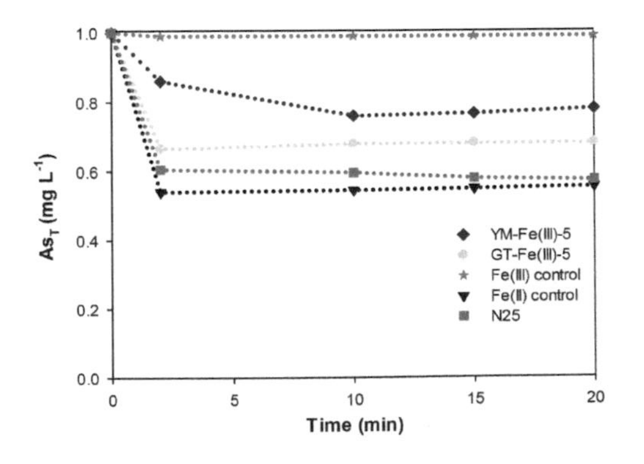

Figure 5. Temporal evolution of total As concentration during treatment with YM-Fe(III)-5 and GT-Fe(III)-5. Conditions: $[As(V)]_0 = 1$ mg L^{-1}, pH 7, As:Fe MR = 1:30, $T = 25°C$.

In contrast with the results of As(III), YM-Fe(III)-5 gave a very good total As removal (83%), better than that obtained with N25 (52%), while GT-Fe(III)-5 gave only 20%. Controls with Fe(III) and Fe(II) gave removals of 28 and 7%, respectively.

4 CONCLUSIONS

Use of iron-based nanoparticles is a very efficient and low-cost method for arsenic removal from water. These nanomaterials, including those prepared from natural extracts and cheap iron salts, can be used to develop low-cost nanotechnologies for *in situ*, *on site* and *ex situ* water treatments, including drinking water provision, residual effluents and remediation of contaminated sites. Laboratory tests must be done to continue testing the efficiency of these materials and are underway.

ACKNOWLEDGEMENTS

To M.E. Morgada, I.K. Levya, V. Salomone, S.S. Farías, G. López, C. Pabón, F. García, A. Senn, N. Quici, and M. Meichtry for collaborating in the experiments. This work was performed as part of Agencia Nacional de Promoción Científica y Tecnológica of Argentina PICT2011-0463, 2015-208 and PICT 3640 projects.

REFERENCES

Crane, R.A & Scott, T.B. 2012. Nanoscale zero-valent iron: future prospects for an emerging water treatment technology. *J. Hazard. Mater.* 211–212: 112–125.

García, F.E., Senn, A.M., Meichtry, J.M., Scott, T.B., Poulin, H., Leyva, G., Halac, E., Ramos, C., Sacanell, J., Mizrahi, M., Requejo, F. & Litter, M.I. 2018. Chromium and arsenic removal using iron based nanoparticles prepared from yerba mate and green tea extracts, submitted.

Genuino, H.C., Luo, Z., Mazrui, N., Seraji, M.I. & Hoag, G.E. 2013. Green synthesized iron nanomaterials for oxidative catalysis of organic environmental pollutants. New and Future Developments in Catalysis, First edition. *Catalysis for Remediation and Environmental Concerns*, pp. 41–61.

Herlekar, M., Barve, S. & Kumar R. 2014. Plant-mediated green synthesis of iron nanoparticles. *J. Nanopart.*, http://dx.doi.org/10.1155/2014/140614.

Hoag, G.E., Collins, J.B., Holcomb, J.L., Hoag, J.R., Nadagouda, M.N. & Varma, R.S. 2009. Degradation of bromothymol blue by 'greener' nano-scale zero-valent iron synthesized using tea polyphenols. *J. Mater. Chem.*, 19(45): 8671–8677.

Lata, S. & Samadder, S.R. 2016. Removal of arsenic from water using nano adsorbents and challenges: a review. *J. Environ. Manag.* 166: 387–406.

Levy, I.K., Mizrahi, M., Ruano, G., Requejo, F., Zampieri, G. & Litter, M.I. 2012. TiO_2-Photocatalytic reduction of pentavalent and trivalent arsenic: production of elemental arsenic and arsine. *Environ. Sci. Technol.* 46(4): 2299–2308.

Martínez-Cabanas, M., López-García, M., Barriada, J.L., Herrero, R. & Sastre de Vicente, M.E. 2016. Green synthesis of iron oxide nanoparticles. Development of magnetic hybrid materials for efficient As(V) removal. *Chem. Eng. J.* 301: 83–91.

Mercado, D.F., Caregnato, P., Villata L.S. & Gonzalez M.C. 2017. Ilex paraguariensis, extract-coated magnetite nanoparticles: a sustainable nano-adsorbent and antioxidant. *J. Inorg. Organomet. Polym.*, Mater. 28(2): 519–527.

Morgada, M.E., Levy, I.K., Salomone, V., Farías, S.S., López, G. & Litter, M.I. 2009. Arsenic (V) removal with nanoparticulate zerovalent iron: effect of UV light and humic acids. *Catal. Today* 143(3–4): 261–268.

Oakes, J.S. 2013. *Investigation of Iron Reduction by Green Tea Polyphenols for Application in Soil Remediation*. Master's Thesis, University of Connecticut Graduate School.

Ona-Nguema, G., Morin, G., Wang, Y.H., Foster, A.L., Juillot, F., Galas, G. & Brown, G.E. 2010. XANES evidence for rapid Arsenic(III) oxidation at magnetite and ferrihydrite surfaces by dissolved O_2 via Fe^{2+}-mediated reactions. *Environ. Sci. Technol.* 44(14): 5416–5422.

Poguberović, S.S., Krčmar, D.M., Maletić, S.P., Kónya, Z., Pilipović, D.D. T., Kerkez, D.V. & Rončević, S.D. 2016. Removal of As(III) and Cr(VI) from aqueous solutions using "green" zero-valent iron nanoparticles produced by oak, mulberry and cherry leaf extracts. *Ecol. Eng.* 90: 42–49.

Quici, N., Meichtry, M. & Montesinos, V.N. 2018. Use of Nanoparticulated Iron Materials for Chromium, Arsenic and Uranium Removal from Water. In: Marta I. Litter, Natalia Quici, Martín Meichtry (eds) *Iron Nanomaterials for Water and Soil Treatment*, Pan Stanford Publishing Pte. Ltd., Singapore, in press.

Simeonidis, K., Tziomaki, M., Angelakeris, M., Martinez-Boubeta, C., Balcells, L., Monty, C., Mitrakas, M., Vourlias, G. & Andritsos, N. 2013. Development of iron-based nanoparticles for Cr(VI) removal from drinking water. *EPJ Web Conf.*, 40: 8007.

Trotte, N.S.F., Aben-Athar M.T G. & Carvalho N.M.F. 2016. Yerba mate tea extract: a green approach for the synthesis of silica supported iron nanoparticles for dye degradation. *J. Braz. Chem. Soc.* 27: 2093–2104.

Wang, Z. 2015. *Iron-Polyphenol Complex Nanoparticles Synthesized by Plant Leaves and Their Application in Environmental Remediation*. PhD Thesis, University of South Australia. www.nanoteksa.com

Environmental Arsenic in a Changing World –
Zhu, Guo, Bhattacharya, Ahmad, Bundschuh & Naidu (Eds)
ISBN 978-1-138-48609-6

Distribution of arsenic hazard in public water supplies in the United Kingdom – methods, implications for health risks and recommendations

D.A. Polya[1], L. Xu[1], J. Launder[1], D.C. Gooddy[2] & M. Ascott[2]
[1]*School of Earth and Environmental Sciences and Williamson Research Centre for Molecular Environmental Science, University of Manchester, Manchester, UK*
[2]*Groundwater Science Directorate, British Geological Survey, Wallingford, UK*

ABSTRACT: Public water supplies in the United Kingdom are highly regulated and monitored and, in particular, have an outstanding compliance with regulatory standards, particularly with respect to the UK PCV (prescribed concentration value) for arsenic of 10 μg L^{-1}. Nevertheless, many UK public water supplies contain arsenic at concentrations within a factor of 10 of the PCV. Given increasing concerns over detrimental health outcomes arising from chronic exposure to drinking water containing arsenic at sub-regulatory concentrations in the $10^0 \mu$g L^{-1} range, quantifying the distribution of arsenic intake from consumers exposed to arsenic via drinking water in the UK is indicated. Using the limited secondary summary water quality data available in the public domain from the Drinking Water Inspectorate and assuming a log normal distribution, we calculate that, in 2015, on the order of 10^5 consumers in the UK were supplied with drinking water with arsenic concentrations at or above 5 μg L^{-1}; 10^6 at or above 2 μg L^{-1} and 10^7 at or above 1 μg L^{-1}. However, examination of much more detailed secondary data kindly supplied by individual UK water supply companies indicates that the overall distribution of arsenic hazard is not log normally distributed and results in an overestimate of the number of the consumers exposed to high As concentrations using that assumption. Our more detailed analysis shows that approximately 130,000 consumers in the UK are supplied with drinking water with arsenic concentrations at or above 5 μg L^{-1}; the equivalent figures for other concentrations being 1,080,000 for at or above 2 μg L^{-1} and 9,750,000 for at or above 1 μg L^{-1}. Epidemiological evidence seems currently insufficiently powerful to reliably quantify the detrimental health outcomes arising from such sub-regulatory exposures, but arsenic-attributable premature avoidable deaths in the UK on the order of 100 to 1000 per annum are plausibly estimated here from combined cancer and cardiovascular disease causes. There are considerable uncertainties in these estimates due to (i) model (e.g. linearity, threshold) and parameter uncertainties in the dose-response relationships at such low concentrations; (ii) partial reliance on ecological studies, which may be sensitive to the nature of adjustment for socio-economic and other potential confounders of risk and (iii) the lack of explicit consideration of the many other sequela for which arsenic is known, at higher concentrations in drinking water, to contribute. We note that, the estimates here, however, are broadly equivalent to the number of annual fatalities of car occupants in road traffic accidents in the UK.

1 INTRODUCTION

Although arsenic from (mostly) groundwater sourced drinking water supplies is a widely known major global health risk (Polya & Middleton, 2017; Ravenscroft *et al.*, 2009; Smith *et al.*, 2000) concerns in the UK over arsenic in drinking water supplies have been limited largely to private water supplies, which face less scrutiny than volumetrically more important public water supplies, and which sometimes contain arsenic at concentrations exceeding the WHO provisional guide value of 10 μg L^{-1} (Ander *et al.*, 2016; Crabbe *et al.*, 2017; Middleton *et al.*, 2016). This relatively limited concern has arisen, in part, because public water supplies in the UK are highly regulated and monitored and have outstanding compliance with respect to the UK PCV (prescribed concentration value) for arsenic of 10 μg L^{-1} (DWI, 2016).

Notwithstanding this, many public water supplies in the UK contain arsenic at concentrations within a factor of 10 of the PCV. Given that there are increasing concerns over detrimental health outcomes – particularly including premature death from cancers and cardiovascular diseases – arising from chronic exposure to drinking water containing arsenic at sub-regulatory concentrations in the low μg L^{-1} range (e.g. Garcia-Esquinas *et al.*, 2013; Medrano *et al.*, 2010; Moon *et al.*, 2017; Pompili *et al.*, 2017; Roh *et al.*, 2017), quantifying the distribution of arsenic intake by UK consumers exposed to arsenic via drinking water from public water supplies is indicated.

Here we report a study, based upon collated secondary data obtained from the DWI (Drinking Water Inspectorate) and other public websites or kindly provided by UK water supply companies, aimed at (i) calculating the distribution of arsenic hazard in UK public water supplies; (ii) estimating the resultant human uptake of arsenic; and (iii) obtaining a scoping estimate of plausible human health risks arising from this exposure. Data and model uncertainties are discussed and, in the light of these, research and public policy recommendations made. Finally, we speculate on the potential health implications of exposure in other countries to drinking water with (sub-regulatory) concentrations of arsenic within a factor of 10 of the WHO provisional guide value.

2 METHODS/EXPERIMENTAL

2.1 *As hazard distribution – scoping calculation*

DWI (2016) provide summary data, viz. 1st and 99th percentiles of arsenic concentrations, of arsenic concentrations in sampled tap waters supplied by each large water supply company in England and Wales for 2015. For scoping calculations, by assuming that the distribution of such arsenic concentrations for each water supply company was log normal, the mean, μ_i, and standard deviation, σ_i, of the modelled log normal arsenic concentration distribution for each water supply company, i, was determined by:

$$\mu_i = (P_{1,i} + P_{99,i})/2 \quad (1)$$

and then

$$\sigma_i = (P_{99,i} - \mu_i)/2.3263 \quad (2)$$

where $P_{1,i}$ and $P_{99,i}$ are the 1st and 99th percentile respectively of the log (arsenic concentration) distribution of the subscripted water supply company, i, and where 2.3263 corresponds to $t_{.99}$ for a normal distribution with an infinite number of degrees of freedom. As $t_{.99}$ for df = 100 is 2.364, the error introduced by ignoring the finite number of tests for each water supply company with over 100 sample tests was accordingly less than 2%. In 2015, water supply companies carrying out <100 sample tests in total supplied less than 0.2% of the total population supplied by water supply companies in the UK and overwhelmingly at arsenic concentrations lower than the overall mean supplied, so no substantive model errors arise from the use of the $t_{.99}$ figure *per se*.

In the absence of equivalent summary data for Scotland, for the scoping calculation, the contribution of public water supplies in Scotland to the overall UK distribution was calculated by correcting for the relative populations of Scotland and the whole UK.

2.2 *As hazard distribution – detailed method*

For a more robust calculation, more detailed data, including mean arsenic concentrations in sampled taps and population, of the approximately 1900 water supply zones in the UK was obtained directly from the 23 largest water supply companies in the UK excluding Scotland. Water quality data and water supply zone data for Scotland was obtained for 2016 from public websites as the required comprehensive data for water regulation zones for 2015 was not provided by Scottish Water. Across the UK, for the 10 individual water supply zones for which data was not available, the arsenic concentration was assumed to be the mean of the arsenic concentration supplied by the relevant water supply company. For the 5 smallest (in terms of population supplied) UK water supply companies, arsenic concentration and population supplied over the entire area of their water supply was used as published by DWI (2016).

2.3 *Water consumption rates*

Mean daily total liquid consumption for the population was taken to be 2.003 L day^{-1} (Accent, 2008). This accounts for both water consumed directly via taps (1.314 L day^{-1}) and that consumed indirectly via incorporation into food and is broadly similar to that determined for the USA (Burmaster, 1998).

2.4 *Dose-response relationships for arsenic-attributable sequela at low exposures*

Reflecting the paucity of relevant epidemiological studies and focusing on perceived likely first order effects, simplified scoping calculations of population-level detrimental health outcomes arising from chronic exposure to arsenic in drinking water at concentrations less than 10 µg L^{-1}, here, were restricted to (i) premature death end-points (i.e. not considering morbidity); (ii) the, arguably, most important two broad categories of sequela, cancers and cardiovascular disease, and (iii) without detailed consideration of age, gender, genetic, dietary and socio-economic confounders.

The model all-cancer annual risk, $R_{c,annual}$, of mortality attributable to tap water arsenic with concentration, C, was calculated from equation (3)

$$R_{c,annual} = (C / C_{ref})\, R_{ref} / E \quad (3)$$

where C_{ref} is the reference concentration of arsenic, taken here to be 10 µg L^{-1} and for which the lifetime all-cancer risk, R_{ref}, attributable to the arsenic is taken to be 2000 per 100,000 (NRC, 2001), and E is the estimated mean life expectancy of the population of UK calculated from ONS (2017). Total model arsenic attributable premature deaths were then calculated by numerically integrating these risks over the distribution of tap water arsenic concentration.

We explored the use of the log-linear models reported in the meta-analysis of Moon *et al.* (2017) of studies of CVD mortality arising from low level arsenic exposures, however, the nature of the model precludes its robust extrapolation to concentrations much below their reference concentration of 10 µg L^{-1}, even with the adoption of a reference arsenic concentration corrected to 1 µg L^{-1}, being a conservative (with respect to risk estimates; upper) estimate of the mean arsenic concentration supplied by UK water supply companies as well as an approximation for the volunteer-weighted mean of median references concentrations in the studies reviewed by Moon *et al.*

(2017). Rather, we based our estimates of excess CVD mortality for exposures >1 μg As L^{-1} by combining ONS (2017) reported overall UK CVD mortality rates with the 2.2 % (CI-0.9%, 5.5%) increase in fully adjusted CVD mortality rates for those drinking water with 1–10 μg As L^{-1} compared to a <1 μg As L^{-1} reference, as reported in the ecological study for Spain of Medrano *et al.* (2010).

3 RESULTS AND DISCUSSION

3.1 *As hazard distribution – scoping calculation*

Scoping calculations, assuming a log normal distribution of arsenic concentrations for each water supply company, indicated on the order of 10^5 persons supplied with tap water with a ≥5 μg As L^{-1}; 10^6 with ≥2 μg L^{-1}; and 10^7 with ≥1 μg L^{-1}. However, although such log normal distributions have been widely found to be functional models of the distribution of chemical constituents in groundwater (e.g. Zikovsky & Chah, 1990) substantial deviations, including from right-skewness and polymodality have been observed in many datasets (Shand *et al.*, 2007). In this study, comparison with the more robustly calculated distribution (Fig. 1) shows that large deviations from log normality arise, in part, also because of (i) data censoring at low concentrations near method detection limits and (ii) truncation of high concentrations near the UK PCV because of the regulatory compliance by water supply companies. Here, the magnitude of the model errors as a function of concentration over the range 1 to 7 μg As L^{-1} are on the order of 10s to 100s% – although high, indicating the utility of the method, if used with caution, for scoping calculations, in the absence of more detailed data, from the limited summary water supply quality data often available in the public domain.

3.2 *As hazard distribution – detailed method*

Our more detailed analysis shows that approximately 130,000 consumers in the UK are supplied with drinking water with ≥5 μg As L^{-1}; the equivalent figures for other concentrations being 1,080,000 for ≥2 μg L^{-1} and 9,750,000 for ≥1 μg L^{-1} (see Fig. 1).

3.3 *Estimated arsenic attributable health risks*

Calculated model health risks are shown in Figure 2 and are of the order 100 to 1000 per annum – this is similar in magnitude to UK deaths arising from car occupants in road traffic accidents (ONS, 2017). Most of the modelled excess mortality arises from arsenic concentrations less than 5 μg L^{-1} with broadly similar contributions from all-cancer and CVD.

There are considerable model and parameter uncertainties inherent in these estimates (cf. Monrad *et al.*, 2017), but nevertheless they suggest that there could be public health benefits in the UK by tightening the UK PCV for arsenic in drinking water from 10 μg L^{-1} to a lower value (cf. Moon *et al.* (2017) with respect to the USA). This is likely to be highly relevant to many, if not most, other countries across the globe and might, of itself, help to drive innovation to find more cost-effective arsenic remediation and mitigation technologies and other instruments relevant to these hitherto considered relatively low concentrations (Polya & Richards, 2017).

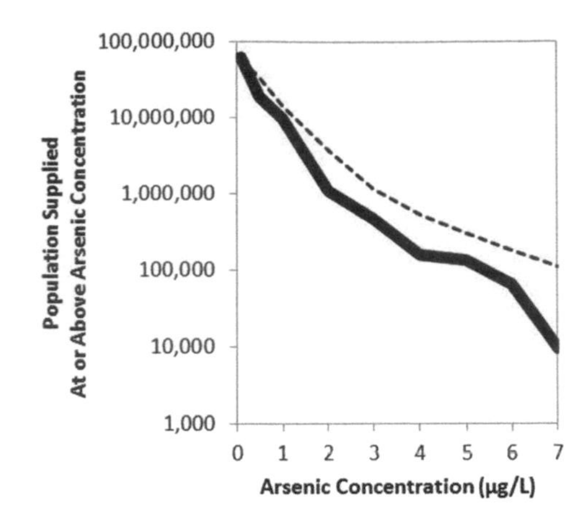

Figure 1. Cumulative distribution of arsenic hazard in UK tap water supplied by water supply companies in 2015 (includes Scotland data instead for 2016). The dashed line shows overestimated hazard distribution using limited public domain DWI (2016) summary data and calculated assuming a log normal model for waters supplied by each water supply company.

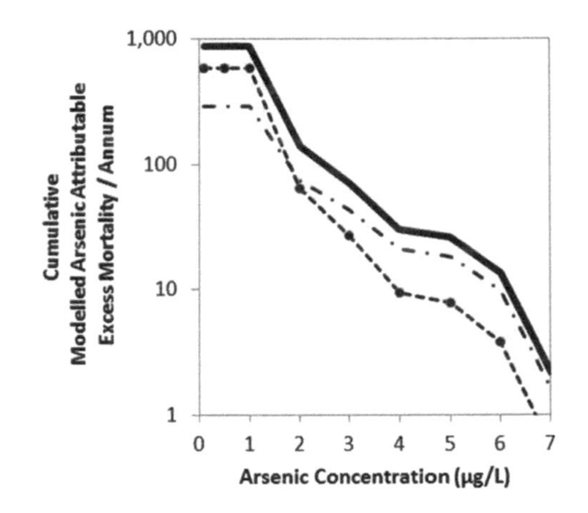

Figure 2. Modelled cumulative distribution of excess mortality per annum due to cardiovascular disease (dashes), all cancers (dot-dash) and combined (solid line). See text for limitations.

4 CONCLUSIONS

Although regulatory compliance of UK water supply companies with the UK PCV of 10 μg L^{-1} is outstanding, about 1,080,000 consumers in the UK are supplied by water supply companies with drinking water with As ≥2 μg L^{-1} giving rise to modelled arsenic attributable excess mortality on the order of 100 to 1000 per annum. The nature and magnitude of dose-response relationships for arsenic in drinking water at concentrations below 10 $\mu g\ L^{-1}$ is currently a controversial matter – indeed some argue that the excess health risks at these concentrations are negligible – further research into the mechanisms of action and epidemiology of arsenic-attributable detrimental health outcomes at concentrations around and below the WHO provision guide value are indicated.

ACKNOWLEDGEMENTS

We acknowledge with thanks the data provided by the 23 largest UK (excluding Scotland) water supply companies as well as their approvals to use their data for this research published in the public interest. Any opinions expressed in this abstract are those of the authors and do not necessarily reflect those of any of these companies. Data from the UK Drinking Water Inspectorate (DWI) for England and Wales are used here under the terms of a UK Open Government License for public sector information, conditions at: http://www.nationalarchives.gov.uk/doc/open-government-licence/version/3/. LX acknowledges a University of Manchester President's PhD Scholarship and JL a NERC CASE (Health and Safety Executive's Health and Safety Laboratory (HSE's HSL)) PhD Studentship through the Manchester-Liverpool NERC EAO DTP. We thank Nick Warren (HSE's HSL) for statistical advice. DCG and MJA publish with the permission of the Executive Director, British Geological Survey (NERC).

REFERENCES

Accent 2008. National Tap Water Consumption Study Drinking Water Inspectorate 70/2/217 Phase Two Final Report.

Ander, E.L., Watts, M.J., Smedley, P.L., Hamilton, E.M., Close, R., Crabbe, H., Fletcher, T., Rimell, A., Studden, M. & Leonardi, G. 2016. Variability in the chemistry of private drinking water supplies and the impact of domestic treatment systems on water quality. *Environ. Geochem. Hlth.* 38(6): 1313–1332.

Burmaster, D.E. 1998. Lognormal distributions for total water intake and tap water intake by pregnant and lactating women in the United States. *Risk Anal.* 18(2): 215–219.

Crabbe, H., Close, R., Rimell, A., Leonardi, G., Watts, M.J., Ander, E.L., Hamilton, E.M., Middleton, D.R.S., Smedley, P.L., Gregory, M., Robjohns, S., Sepai, O., Studden, M., Polya, D.A. & Fletcher, T. 2017. Estimating the population exposed to arsenic from groundwater-sourced private drinking water supplies in Cornwall, UK. In: Bhattacharya, P., Polya, D.A. & Jovanovic. D. (eds) *Best Practice Guide for the Control of Arsenic in Drinking Water*, IWA Publishing, London, UK; Chapter A3, pp. 161–170.

DWI (Drinking Water Inspectorate). 2016. Drinking water 2015 – chief inspectors' report. DWI. Available from: http://www.dwi.gov.uk/about/annualreport/2015/index.html Last updated 5 August 2016. Last accessed 2 October 2017.

García-Esquinas, E., Pollán, M., Umans, J.G., Francesconi, K.A., Goessler, W., Guallar, E., Howard, B., Farley, J., Best, L.G. & Navas-Acien, A. 2013. Arsenic exposure and cancer mortality in a US-based prospective cohort: the strong heart study. *Cancer Epidemiol. Biomarkers Prev.* 22(11): 1944–1953.

Medrano, M.J., Boix, B, Pastor-Barriuso, R., Palau, M., Damián, J., Ramis, R., del Barrio, J.L. & Navas-Acien, A. 2010. Arsenic in public water supplies and cardiovascular mortality in Spain. *Environ. Res.* 110(5): 448–454.

Middleton, D.R.S., Watts, M.J., Hamilton, E.M., Ander, E.L., Close, R.M., Exley, K.S., Crabbe, H., Leonardi, G.S., Fletcher, T. & Polya, D.A. 2016. Urinary arsenic profiles reveal exposures to inorganic arsenic from private drinking water supplies in Cornwall, UK. *Sci. Rep.* 6: Art. 25656.

Monrad, M., Ersboll, A.K., Soresnen, M., Baastrup, R., Hansen, B., Gammelmark, A., Tjonnneland, A., Overvad, K. & Tassschou-Nielsen, O. 2017. Low-level arsenic in drinking water and risk of indident myocardial infarction: a cohort study. *Environ. Res.* 154: 318–324.

Moon, K.A., Oberoi, S., Barchowsky, A., Chen, Y., Guallar, E., Nachman, K.E., Rahman, M., Sohel, N., D'Ippoliti, D., Wade, T.J., James, K.A., Farzan, S.F., Karagas, M.R., Ahasan, H. & Navas-Acien, A. 2017. A dose-response meta-analysis of chronic arsenic exposure and incident cardiovascular disease. *Int. J. Epidemiol.* 46(6): 1924–1939.

NRC (National Research Council). 2001. Arsenic in Drinking Water: Update. National Academy Press, Washington, DC.

Polya, D.A. & Middleton, D.R.S. 2017. Arsenic in drinking water: sources & human exposure routes. In: Bhattacharya, P., Polya, D.A., Jovanovic. D. (eds) *Best Practice Guide for the Control of Arsenic in Drinking Water*. IWA Publishing, London, UK; Chapter 1, pp. 1–23.

Polya, D.A. & Richards, L.A. 2017. Arsenic and the provision of safe and sustainable drinking water. Asia Pacific Tech Monitor, July–Sept 2017, 23–30, UNESCAP.

Pompili, M., Vichi, M., Dinelli, E., Erbuto, D., Pycha, R., Serafini, G., Giordano, G., Valera, P., Albanese, S., Lima, A., de Vivo, B., Cicchella, D., Rihmer, Z., Fiorillo, A., Amore, M., Girardi, P. & Baldessarini, R.J. 2017. Arsenic: association of regional concentrations in drinking water with suicide and natural causes of death in Italy. Psychiatry Res. 249: 311–317.

Ravenscroft, P., Brammer, H., Richards, K. 2009. *Arsenic Pollution*. Wiley-Blackwell: London, UK.

Roh, T., Lynch, C.F., Weyer, P., Wang, K., Kelly, K.M. & Ludewig, G. 2017. Low-level arsenic exposure from drinking water is associated with prostate cancer in Iowa. *Environ. Res.*, 159: 338–343.

Shand, P., Edmunds, W.M., Lawrence, A.R., Smedley, P.L. & Burke, S. The natural (baseline) quality of groundwater in England and Wales. 2007. British Geological Survey Research Report RR/07/06. Environment Agency Science Group: Air, Land and Water Technical Report NC/99/74/24.

Smith, A.H., Lingas, E.O. & Rahman, M. 2000. Contamination of drinking-water by arsenic in Bangladesh: a public health emergency. *Bull. World Health Organ.* 78(9): 1093–1103.

Zikovsky, L. & Chah, B. 1990. The log-normal distribution of radon concentration in ground-water. *Ground Water* 28(5): 673–676.

Section 1: Arsenic behaviour in changing environmental media

1.1 Sources, transport and fate of arsenic in changing groundwater systems

Environmental Arsenic in a Changing World –
Zhu, Guo, Bhattacharya, Ahmad, Bundschuh & Naidu (Eds)
ISBN 978-1-138-48609-6

AdvectAs challenge: multidisciplinary research on groundwater arsenic dissolution, transport, and retardation under advective flow conditions

E. Stopelli[1], A. Lightfoot[1], R. Kipfer[1], M. Berg[1], L.H.E. Winkel[2], M. Glodowska[3], M. Patzner[3], B. Rathi[3], O. Cirpka[3], A. Kappler[3], M. Schneider[4], E. Eiche[4], A. Kontny[4], T. Neumann[4], H. Prommer[5], I. Wallis[6], D. Vu[7], M. Tran[7], N. Viet[7], V.M. Lan[7], P.K. Trang[7] & P.H. Viet[7]
[1] *Eawag, Swiss Federal Institute of Aquatic Science and Technology, Dübendorf, Switzerland*
[2] *Eawag and Institute of Biogeochemistry and Pollutant Dynamics, ETH Zürich, Zürich, Switzerland*
[3] *Geomicrobiology and Hydrology, University of Tübingen, Tübingen, Germany*
[4] *Institute of Applied Geosciences, KIT, Karlsruhe, Germany*
[5] *University of Western Australia and CSIRO Land and Water, Perth, Australia*
[6] *School of Environment, Flinders University, Adelaide, Australia*
[7] *CETASD, Vietnam National University, Hanoi, Vietnam*

ABSTRACT: Constant exposure to groundwater contaminated by arsenic (As) constitutes a major health risk for millions of people worldwide. Therefore, the biogeochemical and physical processes responsible for the release, transport and retardation of As in groundwater need to be identified to optimize groundwater management strategies, both in rural and urban areas. River delta regions in Asia constitute a relevant spot to study the presence of As in drinking water, due to the interplay among the natural release of As from sediments, the onset of redox fronts in aquifers, which can sharply change the mobility of As, and the anthropogenic increased groundwater extraction for the supply of growing urban areas. The AdvectAs project involves several teams to promote transdisciplinary research on the environmental behaviour and spatial heterogeneity of As in groundwater under advective conditions associated with extensive pumping of pristine groundwater from the city of Hanoi, Red River delta region in Vietnam. In autumn 2017 a sampling campaign in the vicinity of Hanoi was carried out, integrating sediment analyses, geochemical, mineralogical and microbiological studies, hydrogeochemical characterization of groundwater, groundwater dating and transport modelling. The first outcomes of the hydrogeochemical analyses will be presented in this contribution.

1 INTRODUCTION

The health of an estimated 200 million people is threatened by high levels of arsenic (As) in drinking water. This is specifically the case for river plains and river delta regions of Asia, where rural inhabitants often rely on naturally contaminated groundwater as primary drinking water source. It has been shown that exposure to groundwater with elevated As levels can be exacerbated by large-scale pumping of groundwater from growing urban areas (Erban *et al.*, 2013; Winkel *et al.*, 2011), which can promote the horizontal and vertical advection of As from shallow aquifers to previously uncontaminated groundwater resources.

Several studies have indicated that the biogeochemical reduction of iron minerals coupled to oxidation of organic matter plays a key role on the release of As from Holocenic sediments (Eiche *et al.*, 2017; Fendorf *et al.*, 2010; Muehe *et al.*, 2014; Rodriguez-Lado *et al.*, 2013; Smedley *et al.*, 2002). Nevertheless, many aspects concerning As release, transport and (im)mobilisation remain largely unravelled, like the kinetics of As mobilisation processes, the role of bacteria in these processes, and the relative contribution of locally released or of advected organic matter and As to the contamination of groundwater bodies.

AdvectAs is aimed at understanding and predicting As mobility under advection induced by extensive pumping of groundwater by means of transdisciplinary research involving mineralogy, microbiology, hydrochemistry and modelling.

2 METHODS

2.1 *Site selection*

The site selected for the AdvectAs project is located on a peninsula of the Red River delta region, some 10 km southeast of the city of Hanoi, Vietnam. The growth of Hanoi has promoted a tremendous increase of groundwater abstraction from Pleistocene aquifers, causing an expansion of the drawdown of several kilometres (Winkel *et al.*, 2011). This has impacted the site of our investigation, where the advective front has reversed the natural direction of the groundwater flow towards Hanoi (Stahl *et al.*, 2016). The peninsula is also characterised by the patchy presence of dissolved As in shallow wells (Berg *et al.*, 2008; Eiche *et al.*, 2008). Part of this spatial heterogeneity is linked to the presence of a Fe-dominated redox transition zone between Holocene and Pleistocene aquifers which plays a relevant role in the mobility of As (Postma *et al.*, 2007; Rathi *et al.*, 2017). All these anthropogenic and biogeochemical features make the site ideal to provide deeper multidisciplinary understanding of the factors responsible for the spatial heterogeneity of dissolved As in wells and to extend the results to other delta regions in Asia impacted by As contamination and extensive groundwater pumping.

2.2 *Set of analyses*

An interdisciplinary sampling campaign was carried out in September and October 2017, aimed at investigating the interactions of biogeochemical processes responsible for As behaviour associated to redox transition zones. Sediments are being analysed for their geochemical and mineralogical composition, including magnetic susceptibility, and for the stability of mineralogical phases, particularly redox-sensitive Fe-bearing oxides and hydroxides. The microbial abundance, bacterial community characterization and metabolic potential are being described to understand the role of the microbial component on As release and stabilization. Analyses of the hydrogeochemical composition of groundwater and groundwater dating are being carried out to improve the understanding and description of the water flow, the movement of pollution and the frame of chemical and physical conditions where biogeochemical processes can take place. Finally, reactive transport modelling is combining the data in order to quantify not only the small-scale processes determining the presence of As in groundwater but also to make predictions of As behaviour at larger-scale.

3 RESULTS

3.1 *Hydrogeochemical analyses*

Focus of this conference contribution will be the presentation of the first outcomes from the hydrogeochemical analyses. An accent will be put on the implications for the factors responsible for small-scale spatial heterogeneity of As and the consequences of extensive pumping on the mobility of As, to promote a better long-term management of groundwater resources.

3.2 *Further results*

A brief overview of the outcomes from sediment analyses, microbial characterization, and groundwater dating and modelling will be also presented.

4 OUTLOOK

The interdisciplinary research of the AdvectAs project is aiming at promoting a better understanding of the natural and anthropogenic factors responsible for the spatial spread and heterogeneity of As contamination in groundwater in the Hanoi area. A further goal is to provide evidence for optimised management strategies of water resources, to ensure better quality of drinking water both to rural and urban areas characterised by extensive pumping.

ACKNOWLEDGEMENTS

We thank Caroline Stengel, Reto Britt, Dirk Radny for technical support; DFG and SNF for co-funding this project; the people at the field site for their friendly support during field work; and Dieke Postma, Rasmus Jakobsen and Alexander van Geen for providing important information on drilling techniques and drilling sites.

REFERENCES

Berg, M., Trang, P.T.K., Stengel, C., Buschmann, J., Viet, P. H., Dan, N.V., Giger, W. & Stüben, D. 2008. Hydrological and sedimentary controls leading to arsenic contamination of groundwater in the Hanoi area, Vietnam: The impact of iron-arsenic ratios, peat, river bank deposits, and excessive groundwater abstraction. *Chem. Geol.* 294: 91–112.

Eiche, E., Neumann, T., Berg, M., Weinman, B., van Geen, A., Norra, S., Berner, Z., Trang, P.T.K., Viet, P.H. & Stüben, D. 2008. Geochemical processes underlying a sharp contrast in groundwater arsenic concentrations in a village on the Red River Delta, Vietnam. *Appl. Geochem.* 23: 3143–3154.

Eiche, E., Berg, M., Hönig, S.-M., Neuman, T., Lan, V.M., Pham, T.K.T. & Pham, H.V. 2017. Origin and availability of organic matter leading to arsenic mobilisation in aquifers of the Red River Delta, Vietnam. *Appl. Geochem.* 77: 184–193.

Erban, L.E., Gorelick, S.M., Zebker, H. A. & Fendorf S. 2013. Release of arsenic to deep groundwater in the Mekong Delta, Vietnam, linked to pumping-induced land subsidence. *PNAS* 110: 13751–13756.

Fendorf, S., Michael, H.A. & van Geen, A. 2010. Spatial and Temporal Variations of Groundwater Arsenic in South and Southeast Asia. *Science* 328: 1123–1127.

Muehe, E. M. & Kappler, A. 2014. Arsenic mobility and toxicity in South and South-east Asia-a review on biogeochemistry, health and socio-economic effects, remediation and risk predictions. *Environ. Chem.* 11: 483–495.

Postma, D., Larsen, F., Hue, N.T.M., Duc, M.T., Viet, P.H., Nhan, P.Q. & Jessen, S. 2007. Arsenic in groundwater of the Red River floodplain, Vietnam: controlling geochemical processes and re-active transport modelling. *Geochim. Cosmochim. Acta* 71: 5054–5071.

Rathi, B., Neidhardt, H., Berg, M., Siade, A. & Prommer, H. 2017: Processes governing arsenic retardation on Pleistocene sediments: Adsorption experiments and model-based analysis. *Wat. Resour. Res.* 53: 4344–4360.

Rodriguez-Lado, L., Sun, G., Berg, M., Zhang, Q., Xue, H., Zheng, Q. & Johnson, C.A. 2013. Groundwater Arsenic Contamination throughout China. *Science* 341: 866–868.

Smedley, P. L. & Kinninburgh, D. G. 2002. A review of the source, behaviour and distribution of arsenic in natural waters. *Appl. Geochem.* 17: 517–568.
Stahl, M.O., Harvey, C.F., van Geen, A., Sun, J., Trang, P.T.K., Lan, V.M., Phuong, T.M., Viet, P.H. & Bostick, B.C. 2016. River bank geomorphology controls groundwater arsenic concentrations in aquifers adjacent to the Red River, Hanoi, Vietnam. *Water Resour. Res.* 52: 6321–6334.
Winkel, L. H., Pham, T. K., Vi, M. L., Stengel, C., Amini, M., Nguyen, T.H., Pham H.V. & Berg, M. 2011. Arsenic pollution of groundwater in vietnam exacerbated by deep aquifer exploitation for more than a century. *Proc. Natl. Acad. Sci. U.S.A.* 108(4), 1246–1251.

Environmental Arsenic in a Changing World –
Zhu, Guo, Bhattacharya, Ahmad, Bundschuh & Naidu (Eds)
ISBN 978-1-138-48609-6

Arsenic in the Baltic Sea sediments – past, present, and future

M. Szubska & J. Bełdowski
Institute of Oceanology of the Polish Academy of Sciences, Sopot, Poland

ABSTRACT: Baltic Sea is a very specific water body. Its geographical situation, geological development, hydrological features and physical drivers, make it very susceptible to pollutants, including arsenic contamination. Baltic Sea is surrounded by nine countries with different industrialization levels, affecting the riverine outflow and surface runoff into the sea. Some parts of the Baltic Sea are characterized with highly elevated arsenic concentrations resulting from the anthropogenic activity on land. There is also a significant inner source of arsenic in the Baltic – arsenic containing Chemical Warfare Agents (CWA) from chemical munitions dumped after the World War II and remaining on the sea bottom.

1 INTRODUCTION

Baltic Sea is one of the biggest brackish water bodies on earth. It is a non-tidal, epicontinental sea – partly cut off from the ocean. Its shape and location make it unique on a global scale. Geographical position, geological development, hydrographical features and physical drivers together create the Baltic Sea environment. The presence of particular separate basins plays role in the transport of water masses, chemical substances and sedimentary material. Therefore different areas of the Baltic Sea are characterized by diversified hydrological properties of water masses, variable bottom types and sediment composition (Snoeijs-Leijonmalm *et al.*, 2017; Szubska, 2018).

Baltic Sea is directly surrounded by nine countries, with five more within the drainage area (Fig. 1).

The drainage area is about 4 times bigger than the area of the Baltic Sea itself, heavily industrialized and relatively polluted. Large input of freshwater results in low salinity and characteristic salinity gradient due to mixing with saline waters entering via the straits of the Kattegat. This limited inflow of oceanic water masses results in a long residence (turnover) time, estimated on 30–40 years for the whole volume of Baltic waters (Snoeijs-Leijonmalm *et al.*, 2017). As a result, all compounds which entered the Baltic Sea are circulating for a very long time. And due to the increase of anoxic and hypoxic areas in the Baltic Sea, arsenic might be released from the sediments and enter the biogeochemical cycle in its most harmful form.

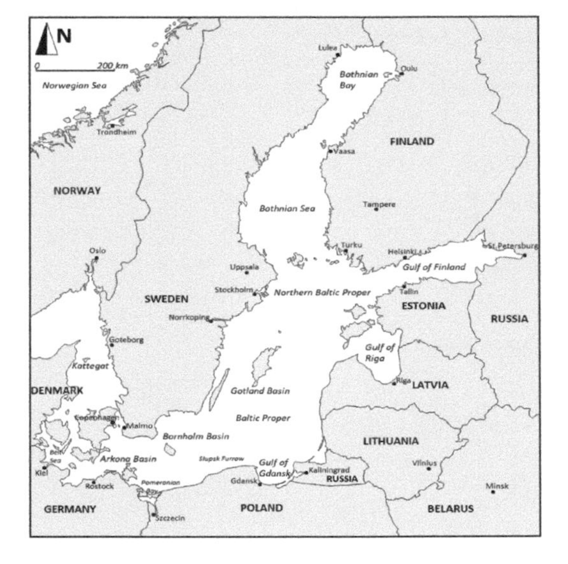

Figure 1. Location, subdivision and surrounding of the Baltic Sea.

2 METHODS/EXPERIMENTAL

Presented study is a summary of literature data on arsenic concentrations in Baltic Sea sediments and authors own research.

2.1 *Analytical methods*

For total arsenic concentrations measurements Atomic Absorption Spectrometry with Hydride Generation (HG-AAS) was used. Sediment samples underwent dry digestion in 550°C and were dissolved in HCl before analysis.

Arsenic speciation was analyzed with Ion-Exchange Chromatography column (CF-Kit-As35) combined with Inductively Coupled Plasma Mass Spectrometry (ICP-MS). Samples extraction with H_3PO_4 in microwave was conducted to obtain concentrations of As(III), As(V), MMA, DMA and arsenobetaine AsB.

3 RESULTS AND DISCUSSION

3.1 *Overall view*

In the area of most countries surrounding the Baltic Sea, levels of arsenic in top soil and stream sediment are very low. This part of northern Europe can be characterized by the lowest concentrations of arsenic in soils on the whole continent. Only the area of Northern Sweden stands out of this trend, exceeding 9 μg/g in subsoil, topsoil and reaching up to $36\,\mu g \cdot g^{-1}$ in stream sediments, as it stays under the strong influence of mining (Uścinowicz, 2011; Salminen, 2005). Total arsenic concentration levels in Baltic Sea bottom sediments are highly variable and depend on the distance from arsenic source and its type, as well as on the type of sediments covering the bottom and environmental conditions. They vary spatially over a very wide range from 0.3 to 277 μg/g (Uścinowicz, 2011).

In the oscillating oxygen conditions in eutrophic areas of the Baltic Sea, deposits of iron-manganese concretions forming a large storage of arsenic and phosphorus may be an inner source of arsenic returning to the biogeochemical cycle as the concretions may dissolve through microbial reactions in anoxic conditions. A preliminary research was made on arsenic speciation in sediments from southern Baltic Sea and until now only As(V) was noted in samples collected in areas with different oxygen levels.

3.2 *Arsenic in heavily contaminated Bothnian Bay*

Bothnian Bay is one of the most polluted areas of the Baltic Sea and arsenic is assumed to be the biggest threat of all pollutants in this area. Arsenic loads to the Bothnian Bay sediments origin from the Rönnskärsverket smelter in Sweden and may be harmful to marine organisms (Vallius, 2014). Arsenic concentrations were increasing since the middle 1970s and in the 1990s reached up to 100 μg/g. Even though the cessation of smelter activity arsenic levels declined only slightly after reaching this maximum and still remain elevated comparing to other areas.

3.3 *Arsenic in chemical ammunition dumpsites*

Nearly 30000 tons of chemical weapons, including 10000 tons of arsenic based Clark I and II, Lewisite and Adamsite, were dumped in the Baltic Se. Sediments in the dumping areas are characterized with significantly higher As concentrations in comparison with reference areas. Even though the concentrations are lower than in the area of Bothnian Sea, it is worth to point out a different behavior of arsenic in dumpsite areas in terms of correlation between As concentrations and such parameters as organic matter contents, sediment grain size, iron and manganese concentrations in samples with and without detected arsenic based CWA.

4 CONCLUSIONS

Even though the concentrations of arsenic are relatively low in most compartments of the Baltic Sea environment, it is still worth observing and understanding the processes occurring in this ecosystem. Specificity of the Baltic Sea (steep salinity and temperature gradients, oxygen deficiency in deep basins and different sediment coverage) makes it impossible to compare this sea with oceanic ecosystems and the results of arsenic research concerning other regions. Also the issue of arsenic speciation was already raised by other researchers, studying both marine and freshwater ecosystems, however so far there is no information on the behavior of arsenic species in marine waters with very low salinity – like the brackish waters of the Baltic Sea. Changing environment of the Baltic Sea may result in fluctuations of parameters controlling arsenic release from the sediments, including the most toxic arsenic species – As (III).

ACKNOWLEDGEMENTS

Presented results were compiled within a research grant “Arsenic speciation in the environment of southern Baltic Sea” nr 2016/21/N/ST10/03245 financed by the National Science Centre, Poland. Particular results were obtained within NATO SPS G4589 project “MODUM”, #069 CHEMSEA and #013 DAIMON partly financed by European Union European Regional Development Fund under the Interreg Baltic Sea Region Programme 2007–2013/2014–2020.

REFERENCES

Salminen, R. (ed.) 2005. Geochemical atlas of Europe. Available on-line at: http://weppi.gtk.fi/publ/foregsatlas/

Snoeijs-Leijonmalm, P., Schubert, H. & Radziejewska, T. (eds) 2017. *Biological Oceanography of the Baltic Sea.* Springer.

Szubska, M. 2018. Arsenic in the environment of the Baltic Sea – a review. In T. Zieliński, I. Sagan, W. Surosz (eds), Inter-disciplinary approaches for Sustainable Development Goals. Springer Series, GeoPlanet, *Earth Planet. Sci.* (in press).

Uścinowicz, S. (ed.) 2011. *Geochemistry of Baltic Sea surface sediments.* Polish Geological Institute – National Research institute. Warsaw, Poland.

Vallius, H. 2014. Heavy metals concentrations in sediment cores from the northern Baltic Sea: Declines in the last two decades. *Mar. Pollut. Bull.* 78: 359–364.

Environmental Arsenic in a Changing World –
Zhu, Guo, Bhattacharya, Ahmad, Bundschuh & Naidu (Eds)
ISBN 978-1-138-48609-6

The role of aquifer flushing on groundwater arsenic across a 35-km transect in the upper Brahmaputra River in Assam, India

R. Choudhury[1], C. Mahanta[1], M.R. Khan[2], B. Nath[3], T. Ellis[3] & A. van Geen[3]
[1]*Department of Civil Engineering, Indian Institute of Technology, Guwahati, Assam, India*
[2]*Department of Geology, University of Dhaka, Dhaka, Bangladesh*
[3]*Lamont-Doherty Earth Observatory of Columbia University, Palisades, NY, USA*

ABSTRACT: In contrast to high concentrations observed near the Brahmaputra River in Bangladesh, well testing in an upstream portion of floodplain in the Indian state of Assam has shown groundwater arsenic (As) concentrations increasing with distance from the river. To study the cause of this pattern, over 900 wells <60 m deep were tested for As and 9 sites were drilled manually over the same depth range along a 35-km transect perpendicular to the river. No relation was observed between groundwater As concentrations ranging from <0.04 to 660 $\mu g\ L^{-1}$ along the transect and the As <1–5 $mg\ kg^{-1}$ content of recovered sand cuttings. Drilling showed a marked increase in the thickness of a clay layer capping the aquifer starting from <5 m closest to the river to >60 m at the most distant site near the base of the Naga foothills. Radiocarbon ages of 18–46 kyr indicate pre-Holocene deposition of the underlying sands across the entire transect. With the exception of two drill sites closest to the river, the cuttings were consistently grey and indication of reduced Fe oxides. Radiocarbon ages of DIC of 0.2, 4.7, and 17.8 kyr were measured in groundwater from 3 monitoring wells installed to 30–60 m depth at distances of 10, 20, and 40 km from the river, respectively. A conceptual groundwater flow model consistent with monitored heads and groundwater ages suggests that thick clay layers capping the aquifer inhibited the flushing of As towards the foothills and, as a result, maintained higher levels of As in the groundwater.

1 INTRODUCTION

Elevated arsenic (As) concentrations in well water is a major public health concern for over 100 million people across South and Southeast Asia (Smedley & Kinniburgh, 2002). Reductive dissolution of Fe(III) oxyhydroxides minerals has emerged as the leading mechanism for As release to groundwater (Bhattacharya *et al.*, 1997). Yet, several key factors including the relative importance of sedimentary and advected sources of reduced carbon for dissolution of Fe(III) oxides remain poorly understood (Datta *et al.*, 2011). This is one a reason why As heterogeneity across different regions has been difficult to explain (Winkel *et al.*, 2011).

This study aims to understand the underlying processes contributing to spatial As variability along a 35 km transect in the upper Brahmaputra floodplains in India, where observed patterns demonstrates groundwater As concentrations increasing with distance from the river.

2 METHODS AND EXPERIMENTAL

2.1 *Groundwater sampling and analysis*

private and public wells <60 m deep straddling the districts of Golaghat and Jorhat in the upper Brahmaputra Valley bounded by the Brahmaputra River to the north, and Naga Patkai Hill ranges in the south were analyzed using a field kit between June and August 2015, and an additional 136 wells in November 2015.

2.2 *Sediment sampling and analysis*

Sediment cuttings were recovered from 9 boreholes drilled along a 35-km transect perpendicular to the Brahmaputra river. At 3 of the drill sites located 10, 20, and 40 km from the river, respectively, monitoring wells screened at 45, 32, and 57 m depth were installed. Field analysis included reflectance measurement using a CM700d diffuse reflectance spectrophotometer and bulk solid phase estimation using an InnovX Delta Premium X-ray fluorescence analyzer. Field data were complemented by radiocarbon dates for organic carbon in a set of clay cuttings and dissolved inorganic carbon (DIC) in groundwater samples collected from 3 monitoring wells installed along the same transect. The various observations are then linked with a simplified 2-D cross sectional groundwater flow model that extends from the Naga foothills to the center of the Brahmaputra river.

3 RESULTS AND DISCUSSION

3.1 *Groundwater and sediment data*

Field kit data for a total 913 wells <60 m deep indicate that 33% of wells in the area met the WHO guideline of 10 $\mu g\ L^{-1}$ for As in drinking water and are mostly concentrated within 10 km of the main course of the Brahmaputra. Another 21% contained >10 to 50 μg L^{-1}, and 46% samples exceeded 50 μg L^{-1} and concentrated within 10 km of the Naga foothills. Comparison of field and laboratory data for a subset of 288 wells confirms that the kit results are by and large consistent with the ICP-MS measurements, with 94% of wells containing <10 μg L^{-1} As and 83% of the wells with >50 μg L^{-1} correctly classified by the kit.

Borehole lithology indicates a marked increase in thickness of a surface clay layer starting from none to <5 m at the 3 sites closest to the river to 60 m of almost continuous clay at the most distant site near the base of the Naga foothills. The reflectance data indicate particularly reduced Fe oxides (ΔR <0.1%, Horneman *et al.*, 2004) prevail in the middle portion of the transect when compared to both the less reduced sands above the layer of brown sand at the site closest to the river and the thinner sand layers at two sites closer to the Naga Hills.

Bulk As concentrations measured by X-ray fluorescence in as many as 121 of total 147 analyzed sand cuttings were below the limit of detection of 1.4–1.8 mg kg^{-1} calculated by the manufacturer's software. Concentrations of As in the remaining samples range from 1.5 to 4.6 mg kg^{-1}, without any clear pattern distinguishing drill sites close to the Brahmaputra River and the Naga Hills.

Cuttings were also analyzed for exchangeable As with the field kit. Readings were limited to 0 and 10 μg L^{-1} As for all 49 samples of sand cuttings from the 5 sites closest to the river. The corresponding range in leachable As concentrations is 0–1 mg kg^{-1} without adjustment in reference chart. Out of the 9 samples of sand cuttings analyzed from the 4 sites closest to the Naga Hills instead, 3 samples gave readings of 25 and 50 μg L^{-1} corresponding to 2.5–5 mg kg^{-1} leachable As.

With the exception of one shallow sample at 0.3 m bgl, uncorrected radiocarbon age of organic matter in all clay cuttings ranged from 18 to 46 kyr, indicating pre-Holocene (>12 kyr) deposition along most of the transect. Uncorrected DIC ages for groundwater span a wide range from 0.2 kyr at site closest to river to 4.7 kyr near the middle of the transect and 17.8 kyr closest the Naga Hills.

Measured groundwater levels range from 2 to 10 m below the local land surface across the transect. When referred to a constant datum, shallow groundwater elevations decrease 95 m near the hills to 80 m at the river, corresponding to a lateral head gradient of 0.3×10^{-3}.

Observed spatial pattern along the 35 km transect can only partially be explained by difference in extent of reduced Fe oxide in aquifer sediment. Redox of the sediment alone cannot account for the trend in groundwater because the most reduced sands prevail in the middle portion of the transect instead of the region closest to the Naga Hills.

3.2 *Groundwater flow modeling*

Among the different groundwater flow models that were tested, the best fit model that replicated field observations was the one where a constant head boundary near the hill was assumed. The implication is that water recharges far from the transect and flows a considerable distance before reaching it. It is therefore already old when it reaches the hillside boundary of the transect and ages further while flowing towards the river. The model indicates that thickness of this flow-system depends on the horizontal hydraulic conductivity as well as the recharge rate.

4 CONCLUSIONS

Observed increase in As concentrations in shallow aquifers, with distance away from the Brahmaputra River documented in this study can be attributed to flushing of exchangeable As playing a dominant role. High surface permeability and therefore recharge and flushing of coarse sands closest to the Brahmaputra River has flushed the aquifers of their exchangeable As. In contrast, such flushing was inhibited by the thick clay capping aquifers near the Naga foothills.

ACKNOWLEDGEMENTS

This study was supported by a grant from CE/C/CM/57 to CM, a Fulbright fellowship to RC, US NIEHS grant P42 ES10349, and US NSF ICER 1414131. We are grateful to PHED Assam for their generous support. We also thank the students of IIT Guwahati, Dr. Chander Kumar Singh and Imtiaz Choudhury for their help in the field.

REFERENCES

Bhattacharya, P., Chatterjee, D. & Jacks, G. 1997. Occurrence of arsenic-contaminated groundwater in alluvial aquifers from Delta Plains, Eastern India: Options for Safe Drinking Water Supply. *Int. J. Water Res. Dev.* 13(1): 79–92.

Datta, S., Neal, A., Mohajerin, W., Ocheltree, T., Rosenheim, B.E., White, C.E. & Johannesson, K.H. 2011. Perennial ponds are not an important source of water or dissolved organic matter to groundwaters with high arsenic concentrations in West Bengal, India. *Geophys. Res. Lett.* 38: L20404. 8(20): 582-582.

Horneman, A., van Geen, A., Kent, D.V., Mathe, P.E., Zheng, Y., Dhar, Y.R., Oonnell, S., Hoque, M.A., Aziz, Z., Shamsudduha, M., Seddique, A.A. & Ahmed, K.M. 2004. Decoupling of As and Fe release to Bangladesh groundwater under reducing conditions. Part I: Evidence from sediment profiles. *Geochim. Cosmochim. Acta* 68(17): 3459–3473.

Smedley, P.L. & Kinniburgh, D.G. 2002. A review of the source, behavior and distribution of arsenic in natural waters. *Appl. Geochem.* 17(5): 517–568.

Winkel, L.H.E., Trang, P.T.K., Lan, V.M., Stengel, C., Amini, M., Ha, N.T., Viet, P.H. & Berg, M. 2011. Arsenic pollution of groundwater in Vietnam exacerbated by deep aquifer exploitation for more than a century. *Proc. Natl. Acad. Sci. U.S.A.* 108(4): 1246–1251.

Environmental Arsenic in a Changing World –
Zhu, Guo, Bhattacharya, Ahmad, Bundschuh & Naidu (Eds)
ISBN 978-1-138-48609-6

Geographical controls on arsenic variability in groundwater of Upper Indus Basin, Punjab, Pakistan

A. Farooqi[1], N. Mushtaq[1], J.A. Khattak[1], I. Hussain[1] & A. van Geen[2]
[1]*Environmental Geochemistry Laboratory, Faculty of Biological Sciences, Department of Environmental Sciences, Quaid-i-Azam University, Islamabad, Pakistan*
[2]*Lamont Doherty Earth Observatory, Columbia University, Palisades, New York, USA*

ABSTRACT: Blanket testing was carried out in 179 villages in Upper Indus Basin of Punjab, Pakistan. Out of total 19551 wells tested, 79% were found to have arsenic levels within WHO prescribed limit of 10 $\mu g\ L^{-1}$ while 11% had arsenic levels within National limit of 50 $\mu g\ L^{-1}$. Ravi flood plain was found to be comparatively more contaminated than the rest of the study area. The prevailing redox conditions for arsenic release in Punjab is found to be mixed in nature, with conditions reducing enough for Fe oxide reduction but typically not enough for complete reduction of SO_4.

1 INTRODUCTION

Naturally occurring high arsenic (As) concentrations in groundwater of flood plains of South East and South Asia are a threat to health of more than 100 million people living there (Postma et al., 2016). The source of As in groundwater is generally associated with sediments deposited by rivers originating in Himalayas. The general geochemical conditions or settings that lead to the mobilization of arsenic into groundwater include environments that are either reducing, arid or oxidizing with elevated pH, geothermal or sulfide mineralization (Smedley & Kinniburgh, 2002). It is a well-known fact that distribution of As in groundwater is highly variable spatially, and varies with the geology and soil properties of the area (Winkel *et al.*, 2011). Occurrence of high As in groundwater has also been linked to recently abandoned river channels deposits (Sahu & Saha, 2015), with concentrations changing as hydrogeology and water table also further influence mobilization of As (Postma *et al.*, 2016). The groundwater and surface water of Indus River system is a key source of drinking water in Pakistan. A survey carried out jointly by local government and rural development department of Pakistan, public health engineering department, and United Nations International Children's Emergency Fund (UNICEF) indicated hot spots of As abundance in Indus alluvial basin of Pakistan specifically in the central and southern region of the country. (Rabbani *et al.*, 2017) reported 13 million people to be at risk from drinking As contaminated groundwater from 27 districts of Sindh. Keeping these in mind, blanket testing was carried out in the Punjab region to identify areas contaminated with As and pinpoint the locations of safe wells using field kits. This transboundary research also focuses on the difference in geochemical settings in both India and Pakistan responsible for As release into groundwater.

2 METHODS/EXPERIMENTAL

2.1 *Geological background*

Blanket testing was carried out in Upper Indus Basin, Punjab region for the testing of As and selected redox parameters. The villages covered so far were selected using Google Earth imagery along the Pakistan part of the transects (Fig. 1). The aquifer system under doabs (area between two rivers) are underlain by unconsolidated alluvial sediments deposited by Indus river tributaries during Pleistocene-Recent age. The alluvial complex has an average thickness of more than 400 meters. The sediments consist of brownish-grey, fine to medium sand, silt and clay in varying proportions with quartz, muscovite, biotite, and chlorite making up the chief mineral constituents of the sediments. The shifting course of tributaries has imparted heterogeneous nature to alluvial complex resulting in little to no vertical or lateral continuity. But still the sediments on regional scale behave as homogeneous aquifer (Farooqi *et al.*, 2014).

2.2 *Field sampling and analysis*

Between 2015 and 2017, more than 19551 wells have been tested for As by Arsenic Econo-Quick (™) (EQ) kit in almost 14 districts of Punjab covering 175 villages across Punjab. pH/EC and redox sensitive parameters NO_3, SO_4, and Fe were also measured on field using HANNA field kits. Total of 18533 tests were performed for all 5 parameters. 10% of samples

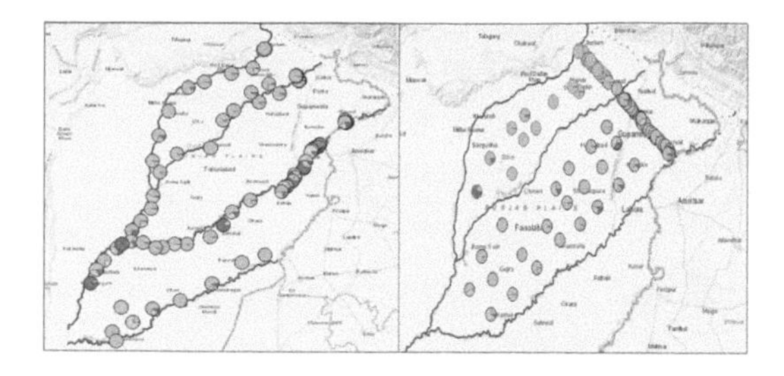

Figure 1. Distribution of arsenic in upper Indus Basin, Punjab, along Flood plains (*left*) and Doabs (*right*). The flood plain of Ravi is more affected by As contamination than Jhelum and Chenab, while northern part of the Rachna doab is comparatively more affected than Chaj doab.

were randomly collected for quality assurance for later laboratory analysis.

The sampling covered all the three-major flood plain areas of River Ravi, Chenab and Jhelum, and also covered both Rachna and Chaj Doabs.

3 RESULTS AND DISCUSSION

3.1 *Groundwater As distribution along Flood Plains*

Overall blanket testing results reveal that almost 79% of the wells tested had As concentration within WHO prescribed limit of $10\,\mu g\ L^{-1}$ while 11% had concentrations within the National Environmental Quality Standards (NEQS) of $50\,\mu g\ L^{-1}$. The testing covered major flood plains of the area and also the doabs (the area between two rivers). For better understanding of the arsenic distribution, the results are discussed separately for flood plains and doabs. 67% of the wells tested along flood plains of all three rivers (Ravi, Jhelum and Chenab) were found to be safe (As $<10\,\mu g\ L^{-1}$) while 16% wells had As levels within the prescribed National Limit.

The comparison of proportion of wells unsafe regarding As (>WHO limit) concentrations showed Ravi flood plain to be more contaminated as compared to the Jhelum and Chenab Flood plains. 49% of the wells had As levels higher than $10\,\mu g\ L^{-1}$. 29 out of 34 villages tested along River Ravi had As >50 ug L^{-1}, more than that of national limit.

Along Chenab and Jhelum flood plains, only 13% and 4% wells exceeded WHO limit, respectively.

The previously published data suggest the oxidative dissolution of As at higher pH to be the source of As release along Ravi river (Sultana *et al.*, 2014).

3.2 *Arsenic in Chaj and Rachna Doabs*

Rachna Doab had comparatively higher proportion of unsafe wells (As $>10\,\mu g\ L^{-1}$) than Chaj Doab (Fig. 1). 20% wells tested in Rachna and 2% wells in Chaj had As levels above the prescribed WHO limit. Out of total 73 villages tested in both doabs, 10 villages had As levels above $50\,\mu g\ L^{-1}$. As seen from the Figure 1, the spatial distribution of As is highly variable and changes with the varying geological and topographic settings.

Table 1. Classification of redox parameters (percentage wise) on the basis of arsenic concentrations

Arsenic Classes	*Fe > 1 (mg L^{-1})	*SO_4 > 20 (mg L^{-1})	*NO_3-N > 10 (mg L^{-1})
As $\leq 10\,\mu g\ L^{-1}$	24	81	91
As 11–50 $\mu g\ L^{-1}$	16	11	6
As $>50\,\mu g\ L^{-1}$	18	8	3

*Detectable levels by Kits in mg L^{-1}

3.3 *Redox indicators*

Among the 11352 wells meeting the WHO prescribed limit of $10\,\mu g\ L^{-1}$, only 24% samples had Fe > 1 mg L^{-1} and 76% samples having SO_4 >20 mg L^{-1} detectable by kits. NO_3-N>10 mg L^{-1} was detected in 1603 wells only using kits (Table 1). The overall redox parameter results indicate mixed conditions to be prevailing throughout study area with conditions often reducing enough for Fe oxide reduction but not enough for complete reduction of SO_4.

4 CONCLUSIONS

The testing reveals wide variability in arsenic levels in groundwater with most of the samples meeting the WHO limit for drinking water in the Punjab region, Pakistan. Mixed redox settings were found to be responsible were elevated levels of As.

ACKNOWLEDGEMENTS

This project is funded by HEC-USAID under the Pak-USAID project.

REFERENCES

Postma, D., Larsen, F., Jakobsen, R., Sø, H.U. Kazmierczak, J.,Trang P.T.K., Lan, V.M., Viet, P.H., Hoan, H., Trung, D. & Nhan, P.Q. 2016. On the spatial variation of arsenic in groundwater of the Red River floodplain, Vietnam. In: P. Bhattacharya, M. Vahter, J. Jarsjö, J. Kumpiene, A. Ahmad, C. Sparrenbom, G. Jacks, M.E. Donselaar, J. Bundschuh, & R. Naidu (eds.) "*Arsenic Research and Global Sustainability As 2016*". CRC Press/Taylor and Francis (ISBN 978-1-138-02941-5), pp. 3–4.

Rabbani, U., Mahar, G., Siddique, A., & Fatmi, Z. 2017. Risk assessment for arsenic-contaminated groundwater along River Indus in Pakistan.*Environ. Geochem. Hlth.* 39(1): 179–190.

Sahu, S. & Saha, D. 2015. Role of shallow alluvial stratigraphy and Holocene geomorphology on groundwater arsenic contamination in the Middle Ganga plain, India. *Environ. Earth. Sci.* 73: 3523–3536.

Smedley, P.L. & Kinniburgh, D.G. 2002. A review of the source, behaviour and distribution of arsenic in natural waters. *Appl. Geochem.* 17(5): 517–568.
Sultana, J., Farooqi, A., & Ali, U. 2014. Arsenic concentration variability, health risk assessment, and source identification using multivariate analysis in selected villages of public water system, Lahore, Pakistan. *Environ. Mon Ass. 186*(2): 1241–1251.
Winkel, L.H.E., Pham, T.K.T., Vi, M.L., Stengel, C., Amini, M., Nguyen, T.H., Viet, P.H. & Berg, M. 2011. Arsenic pollution of groundwater in Vietnam exacerbated by deep aquifer exploitation for more than a century. *Proc. Natl. Acad. Sci. U.S.A.* 108: 1246–1251.

Environmental Arsenic in a Changing World –
Zhu, Guo, Bhattacharya, Ahmad, Bundschuh & Naidu (Eds)
ISBN 978-1-138-48609-6

The influence of irrigation-induced water table fluctuation on iron redistribution and arsenic fate in unsaturated zone

Z.Y. Chi, X.J. Xie, K.F. Pi & Y.X. Wang
State Key Laboratory of Biogeology and Environmental Geology & School of Environmental Studies, China University of Geosciences, Wuhan, China

ABSTRACT: Using high arsenic (As) groundwater as the main source of irrigation water is a common practice in vast rural regions due to the water resource shortage of surface water. This study aims to describes the As and Fe redistribution during irrigation and consequent water table fluctuation through column experiment. Results clearly demonstrate the oxygen exchange between atmosphere and column pore during both water table decrease and recovery. The exchanged oxygen has resulted in the As and Fe redistribution. Moreover, the fixed As and Fe were accumulated mostly on the top of the column during percolation. However, topsoil is precisely the site where nutrients come from for crops. As a result, As uptake by crops from soil may continually occur. Thus, it is of great significance for food security to reduce the agricultural use of high-As groundwater.

1 INTRODUCTION

Recent studies have shown that consumption of As-contaminated foods became an alternative pathway for As exposed to human besides water drinking (Fransisca *et al.*, 2015). Actually, As compounds in crops come mainly from soil and irrigation water. The As content in soil can reach up to 83 $\mu g \cdot g^{-1}$ in some irrigated areas, which far exceeds the background level or safety level of agricultural soil (Williams *et al.*, 2006). This poses potentially considerable risk to public health. Given above reasons, it is imperative to comprehend the As and Fe redistribution behavior during irrigation. In order to achieve the goal, a column experiment was carried out in this study. Furthermore, water table fluctuation was set to the main artificially manageable process since it has been confirmed to be a most important factor influenced the redox reactions (Tong *et al.*, 2016).

2 METHODS/EXPERIMENTAL

The column experimental setup and procedure was shown in Figure 1. In this experiment, groundwater and soil was replaced by deoxidized As(III)-Fe(II) solution and acid washed quartz sand, respectively. External environment of this experiment is mainly maintained in anaerobic condition. The ventilation valve was open to atmosphere and the height of the splitter controlled the water table inside the column. The liquid samples were collected from T-value, and the flow cell was used for pH, ORP and DO measurements. The artificially manageable process is water table fluctuations, while the vertical infiltration recharge is uninterrupted. The sand inside the column was then sampled every 6 cm. The sampled sand was dried up under anaerobic atmosphere and then used for As and Fe sequential extraction.

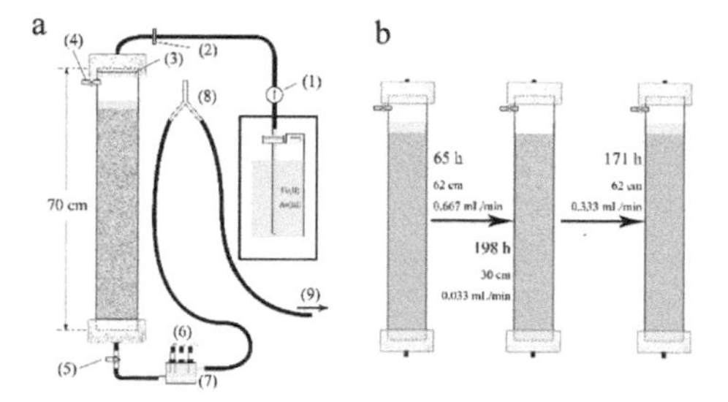

Figure 1. (a) The column experiment setup. (1) Peristaltic pump, (2) Syringe-driven filter, (3) Water distributor, (4) Ventilation valve, (5) T valve, (6) Detector probe, (7) Small flow cell, (8) 3-way splitter, (9) Waste. (b) The experimental procedure.

3 RESULTS AND DISCUSSION

3.1 *Water table fluctuation and oxygen transfer*

As shown in Figue 2a, during the equilibrium phase (0–65 h), oxygen inside the column was washed away, and thus effluent dissolved oxygen content reached the lowest level at 65 h. Then, effluent DO rose rapidly when valve was opened and water table lowered down to 30 cm. This indicates that the decrease of water table can result in the infiltration of air into column pore. Effluent DO then decreased due to consumption of dissolved oxygen. Notably, the experimental results showed that oxygen exchange between atmosphere and column pore also happened during the period of water table recovery. The effluent DO showed a sharp rise when water table ascended, and the DO remained at about 6.7 $mg \cdot L^{-1}$ for a long time after recovery. This demonstrates that atmospheric oxygen was trapped by solution before the injected solution flowed into sand pores.

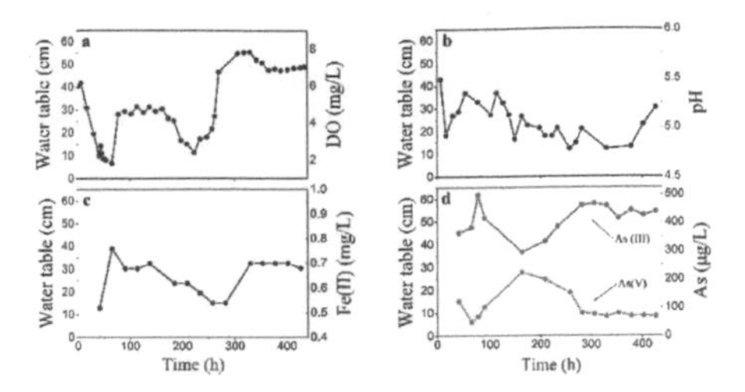

Figure 2. Effluent monitoring results.

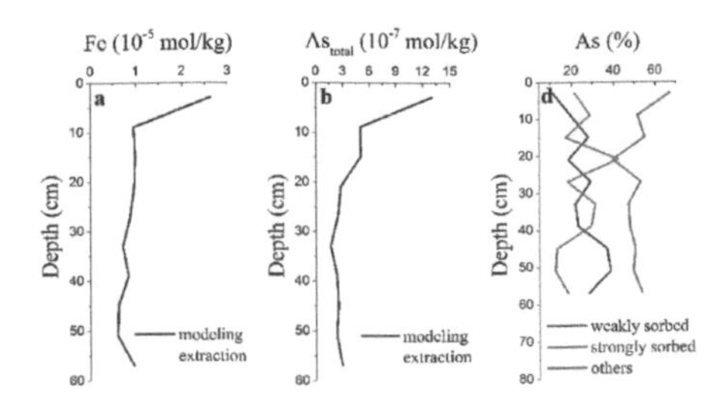

Figure 3. The results of As, Fe extraction.

3.2 *The As(III) and Fe(II) oxidation*

As shown in Figure 2c, it is a synchronous variation in DO and Fe(II) contents at the beginning of LWP, which illustrates that DO initiates Fe(II) oxidation. The effluent pH and Fe(II) content showed the same varying trend (Fig. 2b), which indicates the Fe(III) hydrolysis controlled the aqueous-phase pH in this experiment. However, the Fe(II) and DO content varying trends were not completely the opposite. Especially at the latter half of LWP, the effluent Fe(II) content and DO, solution pH consistently decreased. This may result from the self-catalyzed characteristic of Fe(II) oxidation in the presence of Fe(III) mineral.

Moreover, As(III) was observed to be partially oxidized during percolation (Fig. 2d). However, it has been confirmed that As(III) oxidation is quite slow in the absence of strong oxidant. The two processes that can accelerate As(III) oxidation under natural conditions are microbial mediation and hydroxyl radical-evolving function. Jia *et al.* has confirmed the production of hydroxyl radical when Fe(II) was oxidized by O_2 (Jia *et al.*, 2017). Thus, the mechanism promoting the As(III) oxidation was considered as the generation of hydroxyl radical when the Fe(II) was oxidized to Fe(III) during this column experiment.

3.3 *As and Fe redistribution*

The As and Fe extraction results are shown in Figure 3. It is obvious that both As and Fe contents were higher in the sands of top 9 cm, but decreased dramatically with depth increased. The As extraction results showed that the As in each sample mainly existed as the form of adsorbed As (especially strongly adsorbed As), accounting for 58.0%–88.0% with an average of 76.8% of total solid-phase As. Furthermore, the proportion of weakly adsorbed As was gradually increased with depth, while strongly adsorbed and non-adsorbed As decreased. This was likely due to the transformation from As-Fe surface complexation into As-Fe surface precipitation during the growth of Fe(III) mineral at the top of column.

3.4 *Environmental implications*

The As/Fe molar ratios ranged from 0.025 to 0.048 with an average of 0.035, which far exceeded the background level of soils. These fixed As were more likely to be released again due to the transformation of Fe(III) mineral or microbial actions after irrigation. This release could be very slow and last for a long time. Coincidentally, topsoil is precisely where nutrients come from for crops. Thus, irrigation and consequent water table fluctuation may pose a long-term potential risk to public health via dietary approach.

4 CONCLUSIONS

The column experiment results show that irrigation and consequent water table fluctuations (both decrease and increase) can result in the oxygen exchange between atmosphere and column pore. The exchanged oxygen was then initiated the Fe(II) and As(III) oxidation. As a results, As and Fe was accumulated on the sand column, especially top of column. These processes are highly likely to cause a long-term potential risk to public health via dietary approach. Thus, it is of great significance to reduce the agricultural use of high-As groundwater.

ACKNOWLEDGEMENTS

This study was jointly supported by the Natural Science Foundation of China (Nos: 41372254, 41772255 and 41521001) and Hubei Science and Technology Innovation Project (2016ACA167).

REFERENCES

Fransisca, Y., Small, D.M., Morrison, P.D., Spencer, M.J.S., Ball, A.S. & Jones, O.A.H. 2015. Assessment of arsenic in Australian grown and imported rice varieties on sale in Australia and potential links with irrigation practises and soil geochemistry. *Chemosphere* 138: 1008–1013.

Jia, M., Bian, X. & Yuan, S. 2017. Production of hydroxyl radicals from Fe(II) oxygenation induced by groundwater table fluctuations in a sand column. *Sci. Total Environ.* 584: 41–47.

Tong, M., Yuan, S., Ma, S., Jin, M., Liu, D., Cheng, D., Liu X., Gan, Y. & Wang, Y. 2016. Production of abundant hydroxyl radicals from oxygenation of subsurface sediments. *Environ. Sci. Technol.* 50: 214–221.

Williams, P.N., Islam, M.R., Adomako, E.E., Raab, A., Hossain, S.A., Zhu, Y.G., Feldmann, J. & Meharg, A.A. 2006. Increase in rice grain arsenic for regions of Bangladesh irrigating paddies with elevated arsenic in groundwaters. *Environ. Sci. Technol.* 40: 4903–4908.

Environmental Arsenic in a Changing World –
Zhu, Guo, Bhattacharya, Ahmad, Bundschuh & Naidu (Eds)
ISBN 978-1-138-48609-6

Towards imaging the spatial distribution of geochemical heterogeneities and arsenic sources

M. Rolle[1,2], M. Battistel[1], F. Onses[1], R. Mortensen[2], S. Fakhreddine[3], S. Fendorf[3], P.K. Kitanidis[3] & J.H. Lee[3,4]

[1]*Department of Environmental Engineering, Technical University of Denmark, Kongens Lyngby, Denmark*
[2]*SDC University, Beijing, China*
[3]*Department of Civil and Environmental Engineering, Stanford University, Stanford, CA, USA*
[4]*Department of Environmental Engineering, University of Hawaii, Honolulu, HA, USA*

ABSTRACT: We propose a methodology to image the spatial distribution of reactive minerals in the subsurface based on distributed sensor measurements of water quality parameters coupled with forward and inverse reactive transport modeling. We focus on kinetic oxidative dissolution of pyrite and As-bearing pyrite minerals and illustrate the potential of the methodology in synthetic modeling applications at different scales, as well as in laboratory flow-through experiments in 1-D column setups and 2-D flow-through chambers.

1 INTRODUCTION

The spatial distribution of physical and chemical heterogeneities is critical in many subsurface applications. For instance, the location of reactive minerals is a primary factor controlling the fate and transport of organic and inorganic pollutants in groundwater. The latter include geogenic contamination causing the release of heavy metals and metalloids such as arsenic. A number of studies have focused on using hydrologic measurements and inverse modeling techniques to image physical heterogeneity and the spatial distribution of hydraulic conductivity. However, the applications of such approaches to water quality and reactive transport problems are rare.

In this study we focus on oxidation of pyrite and As-bearing pyrite. This process is of key importance in many natural settings as well as in engineering applications such as managed aquifer recharge.

2 METHODS

2.1 *Flow-through experiments*

We studied the oxidative dissolution of pyrite in different experimental setups, including batch systems, 1-D column setups and 2-D flow-through chambers. Measurements of water quality parameters such as pH, dissolved oxygen, iron and sulfur were useful to formulate and constrain pyrite dissolution kinetics. In particular, spatially-distributed measurements of dissolved oxygen in the 1-D and 2-D setups were instrumental for imaging pyrite inclusions. Non-invasive optode sensors along the column setups and at different cross sections in the 2-D system allowed us to measure oxygen transport and consumption at high spatial resolution (2.5 mm spacing). A schematic illustration of the experimental setup with two pyrite inclusions is provided in Figure 1.

(a) True Pyrite Distribution

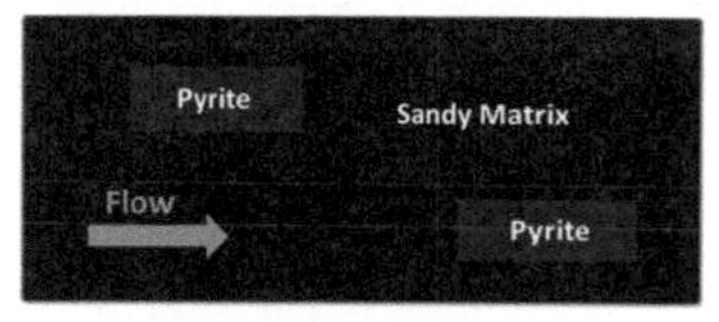

(b) Oxygen Concentration

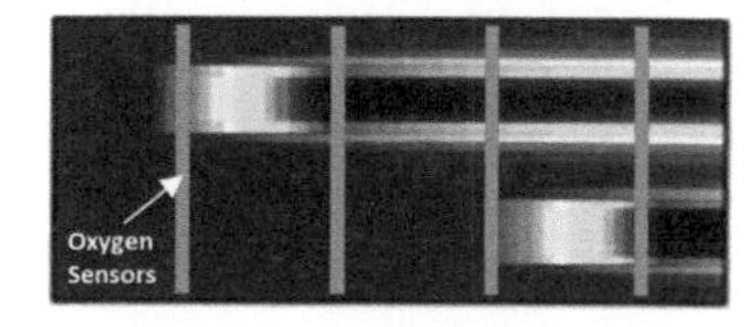

(c) Estimated Pyrite Distribution

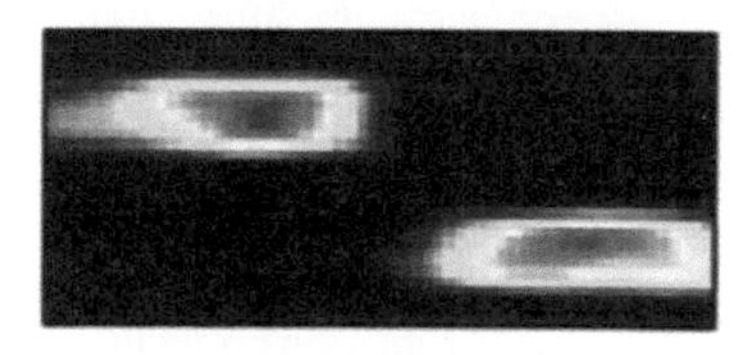

Figure 1. Schematic illustration of the 2-D experimental setup with two pyrite inclusions (a), location of distributed sensor measurements (b), and results of the inverse reactive transport simulations (c).

2.2 *Reactive transport modeling*

A reactive transport network, including the kinetics of pyrite and As-bearing pyrite oxidative dissolution,

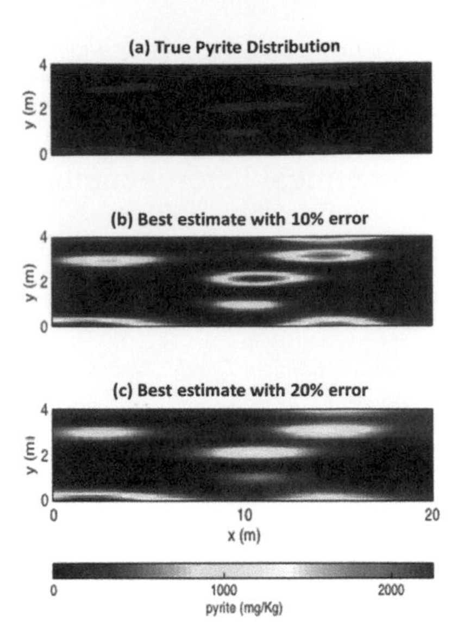

Figure 2. Maps of randomly distributed As-bearing pyrite: true distribution (a) and best estimate in case of 10% error (b) and 20% error (c) in DO measurement (modified from Fakhreddine *et al*., 2016).

was implemented in PHT3D that served as the forward reactive transport simulator. The forward model was applied for synthetic simulations at different scales as well as to quantitatively interpret the laboratory results.

As inverse model, we used the Principal Component Geostatistical Approach (PCGA, Kitanidis and Lee, 2014) to determine the spatial distribution of pyrite and As-bearing pyrite. The PCGA approach was used in combination with the developed forward model and with a limited number of dissolved oxygen observations.

3 RESULTS AND DISCUSSION

The results of the laboratory experiments performed in a number of columns with different size, concentration and number of pyrite inclusions, as well as the outcomes of the 2-D flow-through experiments are presented by Battistel *et al.* (2017). Here we show some of the results of the synthetic applications at the field scale. Figure 2a shows a 2-D cross section of a shallow aquifer with randomly distributed inclusions of As-bearing pyrite. The inversion was based on synthetic oxygen data, mimicking measurements in multilevel observation wells with 4 m spacing along the flow path. The results show the capability of the approach to correctly locate the As-bearing pyrite inclusions as well as their concentration. The outcomes show a good performance of the inversion also in presence of significant measurement errors (Fig. 2b and 2c).

Figure 3 shows the outcomes of simulations performed in the true geochemically heterogeneous aquifer as well as in the best estimate fields obtained with the proposed inverse method. The inverse results

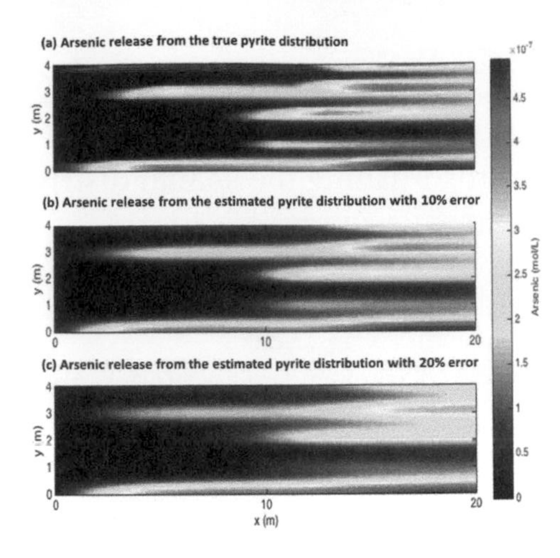

Figure 3. Arsenic plumes in the geochemically heterogeneous aquifers for the true As-bearing pyrite distribution (a), and for the best estimates (b and c) obtained with the proposed inverse modeling approach (reprinted from Fakhreddine *et al.*, 2016, with permission from Elsevier).

show the capability to reproduce the multiple plumes of dissolved arsenic in the heterogeneous domain. An increase in the oxygen measurement error still allows representing the main arsenic plumes, however, the small scale As-sources were more difficult to capture.

4 CONCLUSIONS

We proposed an inverse method for tomography of geochemical heterogeneity. The investigation was focused on imaging the spatial distribution of pyrite and As-bearing pyrite based on spatially distributed oxygen measurements. The methodology proposed is flexible and can be extended to different processes, reactive minerals, as well as distributed measurements of other water quality parameters.

ACKNOWLEDGEMENTS

This study was supported by a grant from the Villum Foundation and by the Sino-Danish Center.

REFERENCES

Battistel, M., Lee, J., Onses, F., Mortensen, R. & Rolle, M. 2017. Imaging the spatial distribution of pyrite in porous media: multidimensional flow-through experiments and inverse modeling (in preparation).

Fakhreddine, S., Lee, J., Kitanidis, P.K., Fendorf, S. & Rolle, M. 2016. Imaging geochemical heterogeneity using inverse reactive transport modeling: An example for characterizing arsenic mobility and distribution. *Adv. Water Res.* 88: 186–197.

Kitanidis, P.K & Lee, J. 2014. Principal component geostatistical approach for large-dimensional inverse problems. *Water Resour. Res.* 50, 5428–5443.

Environmental Arsenic in a Changing World –
Zhu, Guo, Bhattacharya, Ahmad, Bundschuh & Naidu (Eds)
ISBN 978-1-138-48609-6

Source of arsenic bearing detrital minerals in shallow aquifer of southeastern Bangladesh

A.H.M. Selim Reza[1] & H. Masuda[2]
[1]*Department of Geology and Mining, University of Rajshahi, Rajshahi, Bangladesh*
[2]*Department of Biology and Geosciences, Osaka City University, Sumiyoshi-ku, Japan*

ABSTRACT: Forty seven (47) sediment samples from two bore holes from different depths and two river bed sediments from Meghna River were collected from arsenic hot spot area of Bangladesh for geochemical analyses. Concentrations of major elements (SiO_2, CaO, MgO, K_2O, Na_2O, Fe_2O_3, Al_2O_3, Mno, TiO_2) and trace elements (V, Cr, Co, Ni, Cu, Zn, Rb, Sr, Y, Zr, Nb, Ba, La, Ce, Nd, Th and Pb) of sediment were analyzed by XRF (X-ray Fluorescence Spectrometer). Total As of bulk sediment was measured by ICP-MS. Total As content in sediment ranges from 1.91 mg kg^{-1} to 7.97 mg kg^{-1} (average 3.60 ± 2.04 mg kg^{-1}). Higher content of As (7.97 mg kg^{-1}) is observed in clayey sediment and lower content (1.91 mg kg^{-1}) is found in sandy sediments. The river bed sediment contains arsenic ranges from 1.65 mg kg^{-1} to 2.27 mg kg^{-1}. The sequential extraction of arsenic along with the strong positive correlation between As, Fe and Mn suggest that As is adsorbed by FeOOH/MnOOH and liberated from sediment into groundwater due to reductive dissolution of FeOOH and MnOOH. The bulk mineralogy of the sediments was determined by X-ray powder diffraction (XRD) method, which shows that quartz, feldspar, mica and chlorite are dominant mineral in sediment of the study area. The positive correlation between As and intensity of mica and chlorite mineral also reveals that the possible source of As in sediment are mica and clay minerals.

1 INTRODUCTION

Arsenic (As) contamination of groundwater of Bangladesh is one of the greatest environmental disasters in the world especially in the southeastern part of Bangladesh (Ahmed *et al.*, 2004; von Brömssen *et al.*, 2007). About 10 million people in Bangladesh are facing various diseases from drinking of arsenic contaminated groundwater. Ganges, Brahmaputra and Meghna carry huge amount of sediment with arsenic rich iron oxides mineral and deposited in the south eastern part of Bangladesh (Seddique *et al.*, 2008). There are several hypotheses regarding the release mechanism of arsenic in groundwater. One of the well established hypotheses is the reductive dissolution of iron oxyhydroxides or manganese oxyhydroxides in presence of organic matter by anaerobic bacteria release arsenic from sediment into groundwater under reducing condition. Few studies have been carried out regarding the source and mobilization of arsenic in the south eastern part of Bangladesh especially in Chandpur district.

2 METHODS/EXPERIMENTAL

2.1 *Samples collection in the field*

Forty seven (47) sediment samples from two bore holes using hand percussion method at different depths from Hajiganj areas of Chandpur district and two river bed sediments were collected from Meghna river of Chandpur districts in April 2015.

2.2 *Samples analyses in the laboratory*

The bulk mineralogy of the sediments was determined by X-ray powder diffraction (XRD) using a Rigaku Geigerflex instrument (RAD-IA system) with Ni-filtered Cu Ka radiation (30 kV, 10 mA) operating in step scan mode, over an angular range of 2–65° 2θ with 0.02° 2θ steps and 2-scount time on 200 mg unoriented side-packed powder mounts. Prior to analysis, 1 g samples, <63 lm particles passing through 230 mesh, from bulk dried sediments, were ground by hand with a mortar. Approximate relative abundance ratios of major minerals were estimated from the relative intensities of the most intense and specified peak of each mineral. The analytical and sampling-related error in determining mineral abundance was about ±5%.

The major element composition of bulk dried and powdered sediments were determined by X-ray fluorescence (VXQ-160S, Shimadzu) using glass bead samples, which were prepared by fusion of sediment samples with lithium borate (1:3 ratio). Calibration lines were obtained using glass beads made by the same procedure with reference sedimentary rock samples (sedimentary rock series, JSd-1, JSd-2, JSd-3, JSI-1 and JLk-1; Geological Survey of Japan). Analytical error for major elements was within ±3%, estimated using the duplicate standard samples.

Total As concentration of bulk sediment was determined by ICP-MS (ICP-MS; SPQ 9900) using calibration lines made with commercially distributing standard and spike solutions (WAKO Lmt., Japan). Analytical error is <7% for As elements. BCR method was applied for sequential extraction of As.

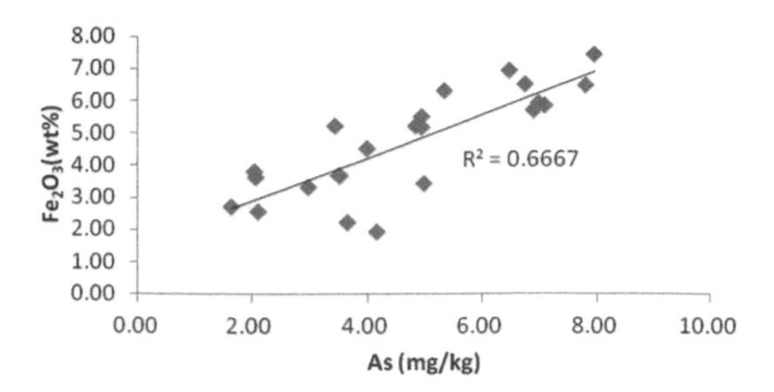

Figure 1. As vs Fe_2O_3 in the drilling site 1.

3 RESULTS AND DISCUSSION

3.1 *Chemical properties of the sediment samples*

Based on XRD analysis, abundant quartz, mica (muscovite and biotite), feldspar (both K-feldspar and plagioclase), and chlorite were observed in both sandy and silty clay sediments. In southeastern Bangladesh, sediments are rich in quartz, mica, feldspar and chlorite (Anawar *et al.*, 2003). The aquifer sediments become gray when the reduction of Fe (oxy) hydroxides is complete and the majority load of sorbed As is released to solution. The depth profile relationships of arsenic with iron, manganese and aluminum oxides and the correlations among these variables suggest that arsenic is strongly associated with iron and manganese oxides and noncrystalline aluminosilicate phases. The correlation analyses (Fig. 1) revealed that the concentrations of As in the sediments are well correlated with those of Fe, Mn and Al contents ($R^2 = 0.666$, $p < 0.05$ for Fe_2O_3 $R^2 = 0.66$, $p < 0.05$ for MnO; $R^2 = 0.666$, $p < 0.05$ for Al_2O_3) in the drilling site 1.

3.2 *Mineralogical and geochemical composition of sediments*

Positive correlation of As and Fe_2O_3 ($R^2 = 0.666$, $p < 0.05$) observed in the core sediments of Chandpur is consistent with the previous reports by Nickson *et al.* (2000) for the Ganges flood plain sediments, Bangladesh. Ahmed *et al.* (2004) and Hasan *et al.* (2007) also reported that Fe oxyhydroxides are the major hosts of As in the Ganges delta plain sediments.

3.3 *Hydrogeochemical factors controlling arsenic mobilization*

Sequential extraction method suggests that Fe_2O_3 or MnO phase act as a scavenger of As in sediment (Table 1). A strong positive correlation between As and total Fe (denoted by Fe_2O_3) is observed throughout the sediment column (Fig. 1). Although sediments of the clay unit contain the greatest amounts of Fe_2O_3 (6.96–7.45 wt.%), probably due to enrichment of clay-size chlorite, a correlation with As is observed (Fig. 1). The total As content shows a positive relationship with the XRD intensity of mica and chlorite in both core sediment samples (Seddique *et al.*, 2011).

Table 1. Sequentially extracted As from sediment using BCR method.

Sample	Acid soluble ppm	Reducible ppm	Oxidizable ppm	Insoluble ppm	Residue ppm
D1-10	0.33	0.22	0.62	3.82	0.49
D1-20	0.13	0.44	0.82	2.08	0.00
D1-30	0.16	0.25	0.84	1.77	0.00
D1-40	0.13	0.47	0.76	1.49	0.00
D1-50	0.28	1.30	0.89	5.41	0.59
D1-60	0.14	0.39	0.49	0.90	0.00
D1-70	0.31	1.13	0.93	2.55	0.29
D1-80	0.20	0.68	0.91	1.01	0.20
D1-90	0.20	0.30	0.42	0.52	0.15
D1-100	0.07	0.93	1.37	3.87	0.51
D1-110	0.34	1.08	0.84	4.08	0.51
D1-120	0.28	0.89	0.52	3.19	0.53

4 CONCLUSIONS

Positive correlations among As, MnO, and Fe_2O_3 suggest that reductive dissolution of MnOOH and FeOOH mediated by anaerobic bacteria represents an important mechanism for releasing arsenic into the groundwater. Fe and Mn oxides may be the possible host of As in sediment of Bengal basin. The positive correlation between As and intensity of mica and chlorite mineral also reveals that the possible source of As in sediment are mica and clay minerals.

ACKNOWLEDGEMENTS

The authors are grateful to JSPS for financial support.

REFERENCES

Ahmed, K.M., Bhattacharya, P., Hasan, M.A., Akhter, S.H., Alam, S.M.M., Bhuyian, M.A.H., Imam, M.B., Khan, A.A. & Sracek, O. 2004. Arsenic contamination in groundwater of alluvial aquifers in Bangladesh: An overview. *Appl. Geochem.* 19(2): 181–200.

Anawar, H.M., Akai, J., Komaki, K., Terao, H., Yoshioka, T., Ishizuka, T., Safiullah, S. & Kato, K. 2003. Geochemical occurrence of arsenic in groundwater of Bangladesh: sources and mobilization processes. *J. Geochem. Explor.* 77(2): 109–131.

Hasan, M. A., Ahmed, K. M., Sracek, O., Bhattacharya, P., von Brömssen, M., Broms, S., Fogelström, J., Mazumder, M.L. & Jacks, G. 2007. Arsenic in shallow groundwater of Bangladesh: investigations from three different physiographic settings. *Hydrogeol. Jour.* 15: 1507–1522.

Nickson, R.T., McArthur, J.M., Ravenscroft, P. Burgess, W.G. & Ahmed, K.M. 2000. Mechanism of arsenic release to groundwater, Bangladesh and West Bengal. *Appl. Geochem.* 15: 403–413.

Seddique, A.A., Masuda, H., Mitamura, M., Shinoda, K., Yamanaka, T., Itai, T., Maruoka, T., Uesugi, K., Ahmed, K.M. & Biswas, D.K. 2008. Arsenic release from biotite into a Holocene groundwater aquifer in Bangladesh. *Appl. Geochem.* 23(8): 2236–2248.

von Brömssen, M., Jakariya, Md., Bhattacharya, P., Ahmed, K. M., Hasan, M.A., Sracek, O., Jonsson, L., Lundell, L. & Jacks G. 2007. Targeting low-arsenic aquifers in groundwater of Matlab Upazila, Southeastern Bangladesh. *Sci. Total Environ.* 379: 121–132.

Environmental Arsenic in a Changing World –
Zhu, Guo, Bhattacharya, Ahmad, Bundschuh & Naidu (Eds)
ISBN 978-1-138-48609-6

Groundwater arsenic contamination in selected area of Bihar

S. Kumari[1], A.K. Ghosh[2], D. Mondal[3], S. Suman[4], P. Sharma[4], P. Kumari[2], N. Bose[4] & S.K. Singh[4]
[1] *B.R.A. Bihar University, Muzaffarpur, Bihar, India*
[2] *Mahavir Cancer Institute and Research Center, Patna, India*
[3] *School of Environment & Life Sciences, University of Salford, Salford, UK*
[4] *Department of Geography, A.N. College, Patna, India*

ABSTRACT: Arsenic (As) is more concentrated near river plain of Bihar, like Ganga and Gandak rivers. It is not reported in places where fluoride content is high in ground water. The current study observed high As contents in the alluvial aquifer of Ganga basin, but no As in hard rock area away from Ganga like Banka district of Bihar.

1 INTRODUCTION

Arsenic (As) is an element, which is highly toxic in nature. The occurrence of As in ground water was first reported in 1980 in West Bengal in India. Groundwater is the main source of drinking water and constitutes about more than 80% drinking source in rural Bihar. The groundwater As contamination in Bihar was first reported in Semaria Ojha Patti village of Shahpur, a block of Bhojpur district in 2002 (Chakraborti *et al.*, 2003). So far, 17 of 37 districts and a total of 87 of 532 community blocks have been investigated for groundwater As contamination. Three community blocks were As safe, as all the sources tested in these blocks had As levels below the detection limit of the measurement method used (Ghosh *et al.*, 2009). Currently, the groundwater As contamination has spread to 16 districts, threatening more than 10 million people in Bihar (Ghosh *et al.*, 2007). Out of a total 240,000 water supply hand pumps (public and private) in As affected blocks of Bihar, only 27% (66,623) of the sources were tested for elevated As levels (Nickson *et al.*, 2007). The continuous consumption of As through drinking water and food sources may lead to As poisoning popularly known as '*Arsenicosis*'. The United States National Research Council has reported that, one in 10 people who drink water containing 500 $\mu g\ L^{-1}$ of As may ultimately develop lung, bladder, and skin cancers. Arsenic health effects range from skin lesions to cancer at values from 10 $\mu g\ L^{-1}$ to up to 2000 $\mu g\ L^{-1}$ of As in drinking water.

2 METHODS

2.1 *Study area*

Samples (N = 42) were collected from volunteers residing in As contaminated areas from Bhagalpur (Rannuchak and Nandgola), Banka (Rajpura and Kakna), Saran (Sabalpur), Patna (Haldi Chapara, Maner) and Samastipur (hanssopur) districts of Bihar, India as a part of ongoing Nutri-SAM project.

2.2 *Data collection*

After taking informed consent, data on socio-economic, nutritional and health status were collected using questionnaire. Water sample were collected and As content was estimated by Atomic Absorption Spectrophotometer (PerkinElmer PinAAcle 900 T) using standard APHA protocol.

2.3 *Analysis method of water*

Sterilized bottles were used for collection of water sample in which 6–10 drops of 1% HNO_3 was added as preservative in each sample bottle. Water samples were collected after 10–15 minutes of flushing the hand pump. Sample bottle were filled up to neck. Samples were brought to lab and stored in refrigerator at 5°C. Arsenic content was analyzed by AAS.

Figure 1. Site visited during field study for water sample analysis.

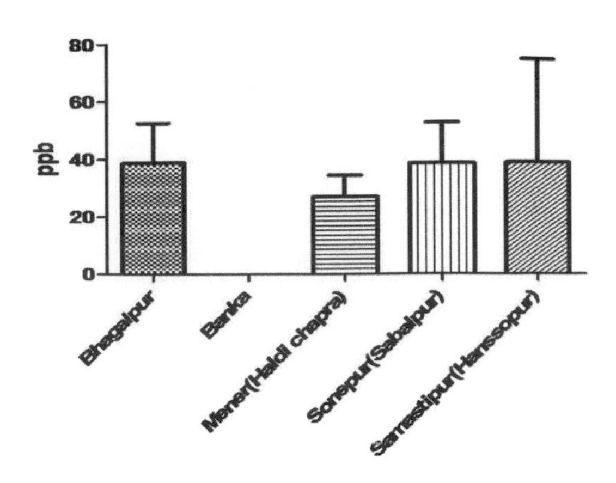

Figure 2. Arsenic level in water samples.

Table 1. Summary of arsenic concentrations ($\mu g\ L^{-1}$).

Area	Mean	SEM	Maximum
Bhagalpur	38.79	13.75	99.47
Banka	00	00	00
Mener (Haldi Chapra)	26.88	7.527	50.62
Sonepur (Sabalpur)	38.62	14.09	107.1
Samastipur (Hanssopur)	38.74	35.96	254.5

2.4 *Statistical analysis*

Descriptive data as mean ± standard deviation along with comparison tests, association and correlations represented using Graph Pad Prism was done.

3 RESULTS AND DISCUSSION

3.1 *Arsenic level in water samples*

There were significant differences between As level in water of different study site. Mean ± SEM of As level in Bhagalpur was $38.79 \pm 13.75\ \mu g\ L^{-1}$ (n = 11), mean ± SEM of As level in Maner (Haldi Chapra) was $26.88 \pm 7.53\ \mu g\ L^{-1}$ (n = 7), mean ± SEM of As level in Sonepur (Sabalpur) was $38.62 \pm 14.09\ \mu g\ L^{-1}$ (n = 8), mean ± SEM of As level in Samastipur (Hanssopur) was $38.74 \pm 35.96\ \mu g\ L^{-1}$ (n = 7).

High As concentration was observed at 50 to 140 ft depth, while groundwater samples below 250 ft had no As in this area. There was no significant difference in demography and nutrient intake between participants. Skin pigmentation and keratosis were reported in low socioeconomic people residing in these areas, while with high socio economic status have little or no symptoms, which indicates that nutritional supplement has a great role in detoxification of As toxicity.

4 CONCLUSIONS

It is concluded from study that As is more concentrated near river plain of Bihar, like Ganga and Gandak rivers. It is not reported in places where high fluoride content is reported for example Banka district but has no As.

ACKNOWLEDGEMENTS

This study is funded by the DST-UKIERI Thematic partnership project. We thank all the participants.

REFERENCES

Chakraborti, D., Mukherjee, S.C., Pati, S., Sengupta, M.K., Rahman, M.M. & Chowdhury, U.K. 2003. Arsenic groundwater contamination in middle Ganga plain, Bihar, India: A future danger. *Environ. Health Persp.* 111(9): 1194–201.

Ghosh, A.K., Singh, S.K., Bose, N., Roy, N.P., Singh, S.K., Upadhyay, A.K., Kumar, S. & Singh A. 2009. Arsenic hot spots detected in the state of Bihar (India): A serious health hazard for estimated human population of 5.5 lakhs. In Assessment of Ground Water Resources and Management, I.K. *International Publishing House Pvt. Ltd., New Delhi, India.*, 62–70.

Ghosh, A. K., Singh, S.K., Bose, N., Singh, S.K., Singh, A., Chaudhary, S. Mishra, R., Roy, N.P. & Upadhyaya, A. 2007. Study of arsenic contamination in ground water of Bihar (India) along the river Ganges. In: *International Workshop on Arsenic Sourcing and Mobilization in Holocene Deltas.* Department of Science and Technology, Government of India, pp. 83–87.

Nickson, R., Sengupta, C., Mitra, P., Dave, S.N., Banerjee, A. K., Bhattacharya, A., Basu, S., Kakoti, N., Moorthy, N. S., Wasuja, M., Kumar, M., Mishra, D.S., Ghosh, A., Vaish, D. P., Srivastava, A. K., Tripathi, R. M., Singh, S. N., Prasad, R., Bhattacharya, S. & Deverill, P. 2007. Current knowledge on the distribution of arsenic in groundwater in five states of India. *J. Environ. Sci. Heal. A,* 42(12): 1707–1718.

Environmental Arsenic in a Changing World –
Zhu, Guo, Bhattacharya, Ahmad, Bundschuh & Naidu (Eds)
ISBN 978-1-138-48609-6

Naturally occurring arsenic in geothermal systems in Turkey

A. Baba
Izmir Institute of Technology, Geothermal Energy Research and Application Center, Izmir, Turkey

ABSTRACT: Human beings have been benefiting from geothermal energy for different uses since the dawn of the civilization in many parts of the world. One of the earliest uses of geothermal energy was for heating and was used extensively by Romans in Turkey, where is an area of complex geology with active tectonics and high geothermal potential. The highest concentrations of naturally occurring aqueous arsenic (As) are found in certain types of geothermal waters, which are generally related to faults and alteration zone. The especially volcanic activity led to the delineation of wide-ranging areas of alteration within mineral assemblages, from advanced argillic type to silica type to prophylitic type at deep levels. The advanced argillic alteration zones are typified by enrichment of sulfur in volcanic rocks that have been dominant in the geological formation of Turkey and the primary mechanism for the presence of numerous trace elements in earth's crust, including but not limited to arsenic. Also, secondary epithermal gypsum has a high concentration of As in the form of realgar and orpiment along the fracture zones of metamorphic and carbonate aquifers. The temperature of geothermal fluid ranges from 40 to 295°C in Turkey. The high arsenic concentrations in geothermal resources have been detected in different part of Turkey from 1 to 6000 μg L^{-1} in geothermal fluids.

1 INTRODUCTION

Plate tectonics control the thermal conditions in the crust. These large-scale movements of plates produce geothermal systems in different part of the world. For example, geothermal systems in Turkey located in the active Alpine-Himalayan Fold and Thrust Belt where the collision of African and Eurasian plates and also the closure of the Tethys Ocean occurs today (Bozkurt, 2001). The graben systems and major faults accompanied by young volcanism form abundant geothermal areas in tectonically-active in most part of Turkey. Generally, geothermal field in Turkey roughly parallel the trends of the graben-bounding faults, young volcanism and hydrothermally altered areas (Mutlu & Gülec, 1998; Simsek *et al.*, 2002; Baba & Ármannsson, 2006; Baba & Sözbilir, 2012) (Fig. 1).

Turkey is favored by a large number of thermal springs known since classical and even prehistoric times. Most important geothermal exploration studies in Turkey began in 1962 by the General Directorate of Mineral Research and Exploration (MTA). There are a total of about 1,500 thermal and mineral water spring groups sped all over the country (Simsek *et al.*, 2002; MTA, 1980; Simsek, 2009). The highest (295°C) bottom hole temperatures have been measured in central Turkey. In Turkey, geothermal energy is also used in various applications such as power generation, greenhouse, district heating, industrial processes, and balneology. The installed capacity is 3322 MWt for direct use (heating) and 1053 MWe for power generation (Akkuş, 2017). The application of geothermal energy for power generation has increased exponential in Turkey (Fig. 2).

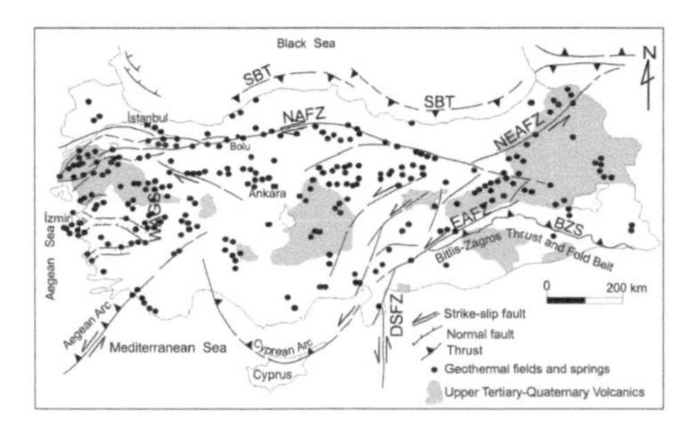

Figure 1. Simplified tectonic map of Turkey showing major neotectonic structures, volcanic province and geothermal spring area in Turkey (from Şimşek *et al.*, 2002; Yiğitbaş *et al.*, 2004).

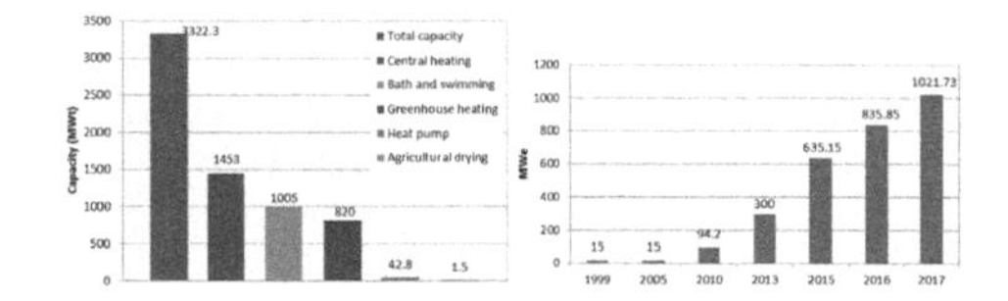

Figure 2. Application of geothermal system in Turkey a) direct use and b) power generation.

2 ARSENIC IN GEOTHERMAL FLUID

Based on the tectonic characteristics and the geological structure, many parts of Turkey are likely to have arsenic containing geological formations. Particularly, high arsenic levels have been naturally detected in along the fault system where volcanic, metamorphic and sedimentary formations out croup. The especially

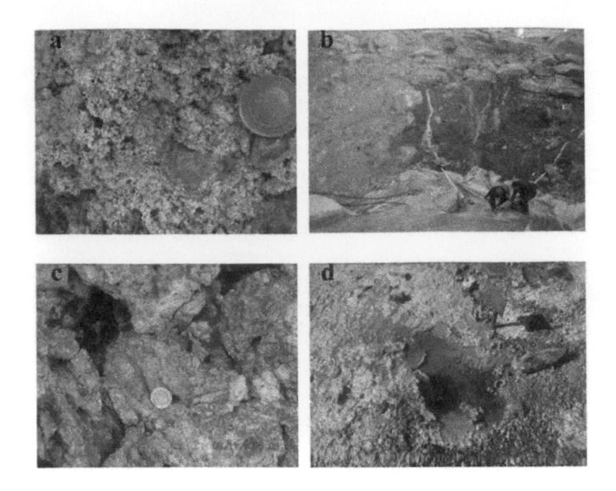

Figure 3. Alteration zone which is including arsenic (a: Lif (Siirt), b: Alaşehir (Manisa), c: Varto (Muş) and d: Tuzla (Çanakkale)).

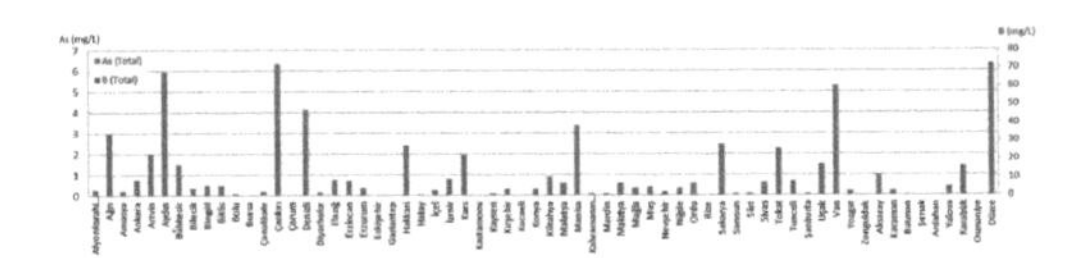

Figure 4. Concentration of arsenic in some geothermal field in Turkey (Some data taken from MTA, 2005).

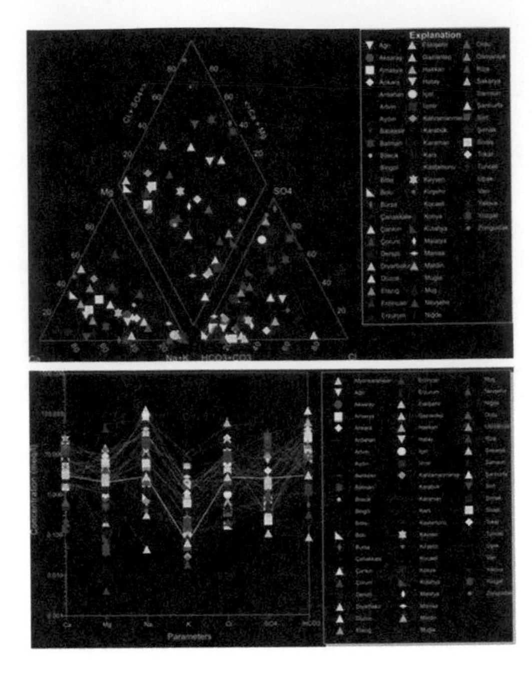

Figure 5. Chemical properties of geothermal fluid in Turkey a) Piper diagram and b) Schoeller diagram.

volcanic activity led to the delineation of wide-ranging areas of alteration within mineral assemblages, from advanced argillic type to silica type to prophylitic type at deep levels (Fig. 3). The advanced argillic alteration zones are typified by enrichment of sulfur in volcanic rocks that have been dominant in the geological formation of Turkey is the primary mechanism for the presence of numerous trace elements in earth's crust, including but not limited to arsenic (Baba & Gunduz, 2010; Baba & Sözbilir, 2012). The concentration of arsenic in geothermal fluid changes in each field because of geological properties. The concentration of arsenic is range from 10 to 6936 mg L^{-1}. The highest As concentration was found in the Hamamboğazı (Uşak) geothermal spring with values of 6936 mg L^{-1}. The concentration of arsenic in some geothermal field is given in Figure 4. Except for arsenic, boron values also are quite in the geothermal system of Turkey. The concentration of boron reaches about 70 mg L^{-1} in some geothermal field such as Aydın, Manisa and Düzce Region. Boron concentration is related to volcanic and sedimentary rocks, but may also be controlled by degassing of magma intrusive (Baba & Armmansson, 2006).

Figure 5 illustrates dominant hydrochemical features of geothermal fluid in Turkey. Each geothermal fluid has different compositions generally most of geothermal fluids which have a deep circulation are of Na-HCO_3^- type, whereas shallow fluids are mostly of the Ca-HCO_3^- type. Along the coastal region such as western Turkey, hot spring exhibited a Na-Cl type with high concentrations of Na^+ and Cl^-.

3 CONCLUSIONS

The study shows that geothermal fluid has a different water types in the different region and some geothermal fields have high concentrations of arsenic. Re-injection is one of the important processes to minimize environmental problem and sustainability of the system. It was observed that the deterioration of local shallow groundwater resources through arsenic contamination was due to the mixing of geothermal fluid or utilization of geothermal resources for energy regeneration. Therefore, proper management and control strategy must be adopted in order to ensure environmental safety to freshwater resources which is currently under threat from geothermal (Bundschuh *et al.*, 2013).

REFERENCES

Akkuş, İ. 2017. Importance of geothermal energy in Turkey. TMMOB, JMO, *Mavi. Gezegen.* 23: 25–39.

Baba, A. & Ármannsson, H. 2006. Environmental impact of the utilization of a geothermal area in Turkey. *Energ. Source* 1(3): 267–278.

Baba, A. & Gunduz, O. 2010. Effect of alteration zones on water quality: a case study from biga peninsula, turkey. *Arch. Environ. Con. Tox.* 58(3): 499–513.

Baba, A. & Sözbilir, H. 2012. Source of arsenic based on geological and hydrogeochemical properties of geothermal systems in Western Turkey. *Chem. Geol.* 334: 364–377.

Bozkurt, E. 2001. Neotectonics of Turkey – a synthesis. *Geodin. Acta* 14(1–3): 3–30.

Bundschuh, J., Maity J. P., Nath B., Baba A., Gunduz O., Kulp T.R., Jean J.S., Kar S., Tseng Y., Bhattacharya P. & Chen C.Y. 2013. Naturally occurring arsenic in terrestrial geothermal systems of western Anatolia, Turkey: potential role in contamination of freshwater resources. *J. Hazard. Mater.* 262: 951–959.

Mineral Research and Exploration (MTA). 1980. Hot and Mineral Water Inventory. *General Directorate of Mineral Research and Exploration (MTA),* MTA Rap., Ankara.

Mineral Research and Exploration (MTA). 2005. Geothermal Resources in Turkey. General *Directorate of Mineral Research and Exploration (MTA),* MTA Rap.21, Ankara.

Mutlu, H. & Güleç, N. 1998. Geochemical characteristics of thermal waters from Anatolia (Turkey). *J. Volcanol. Geoth. Res.* 85: 495–515.

Simsek, S, Yildirim, N., Simsek, Z.N. & Karakus, H. 2002. Changes in geothermal resources at earthquake regions and their importance. *Proceedings of Middle Anatolian Geothermal Energy and Environmental Symposium*, pp. 1–13.

Simsek, S. 2009. Geothermal energy development possibilities in Turkey. *NUMOW Conference on 'Geothermal Energy in Turkey'*, Potsdam-Germany, 1 October, 1–6.

Yigitbas, E., Elmas, A., Sefunc, A. & Ozer, N. 2004. Major neotectonic features of eastern Marmara region, Turkey: development of the Adapazari-Karasu corridor and its tectonic significance. *Geol. J.* 39(2): 179–198.

Environmental Arsenic in a Changing World –
Zhu, Guo, Bhattacharya, Ahmad, Bundschuh & Naidu (Eds)
ISBN 978-1-138-48609-6

Exploring arsenic and other geogenic groundwater contaminants in the vast and scarcely studied Amazon Basin

C.M.C. de Meyer[1], M. Berg[1], J. Rodriguez[2], E. Carpio[2], P. Garcia[2] & I. Wahnfried[3]
[1] *Eawag, Swiss Federal Institute of Aquatic Science and Technology, Dübendorf, Switzerland*
[2] *UNI, Universidad Nacional de Ingeniería, Lima, Peru*
[3] *UFAM, Universidade Federal do Amazonas, Manaus, Brazil*

ABSTRACT: We conducted groundwater surveys in the Amazon Basin to determine the occurrence and distribution of arsenic and other contaminants. Initial results indicate that in groundwater resources in the recent floodplains of the Amazon River and of other white-water rivers, arsenic concentrations are often above the WHO-guideline value of $10\,\mu g\,L^{-1}$ for drinking water. We compared the chemical analyses with geospatial data to identify vulnerable areas in the whole region. Understanding the regional geochemical mechanism(s) triggering the enrichment of arsenic in groundwater is important to raise awareness and implement mitigation where needed.

1 INTRODUCTION

To date, the groundwater quality in the Amazon region is very poorly known. This is of particular concern because people in both rural and urban areas rely on groundwater as their source of drinking water.

The Amazon Basin covers a vast region with heterogeneous geological and geographical features. In the floodplains of the Amazon river, and some of its main tributaries, high loads of Andean sediments get buried together with fresh organic matter. In those areas, reducing conditions in the subsurface are to be expected. The question remains if, where, and to what extend this may lead to reductive dissolution of arsenic and other redox sensitive elements in groundwater in concentrations harmful for human health. We seek answers to these questions by combining local scale groundwater surveys in selected areas of the Amazon Basin, with predictions on a regional scale using geospatial data.

2 METHODS

We sampled groundwater from domestic and community wells in the Peruvian Amazon (see Fig. 1). We chose the locations to be able to study groundwater from various depositional environments. In addition to

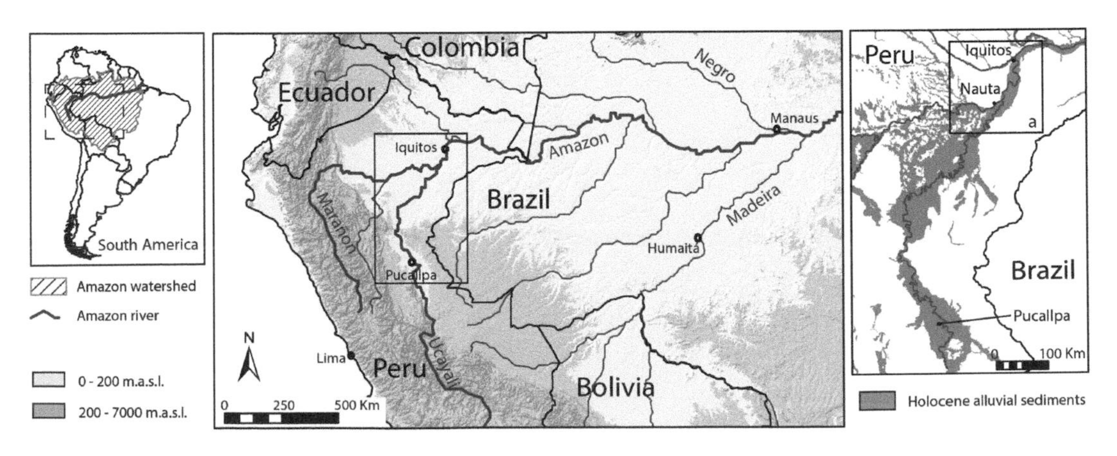

Figure 1. Location of the studied areas in the Western Amazon in Peru. Reducing subsurface conditions, and hence arsenic contaminated groundwater, is expected in the Holocene alluvial deposits along the Amazon River and some of its' main tributaries. Adapted from de Meyer *et al.* (2017).

arsenic we analyzed some 50 groundwater parameters including major and minor elements.

We linked the groundwater data to environmental factors such as geology, topography, river geomorphology and soil, with the aim to develop a predictive map on the distribution of arsenic contamination on the sub-continental scale (Bretzler *et al.*, 2017, Rodriguez-Lado *et al.*, 2013).

3 RESULTS AND DISCUSSION

3.1 *Groundwater survey*

Groundwater arsenic concentrations exceeded the WHO-limit of 10 μg L^{-1} (WHO, 2011) in a considerable amount of wells. Maximum values up to 700 μg L^{-1} were determined. High As-groundwater is of the Ca-Mg-HCO_3 type, has a negative redox potential, slightly acidic pH and elevated concentrations of Fe, Mn, dissolved organic carbon, phosphate and ammonium (de Meyer *et al.*, 2017). Based on these hydrochemical characteristics we concluded that arsenic is likely released by reductive dissolution of Fe-(hydr)oxides.

3.2 *Distribution of arsenic geogenic contamination*

de Meyer *et al.* (2017) observed that the distribution of arsenic in groundwater is linked to the depositional environment of the aquifer sediments. The contaminated wells tap groundwater from young alluvial deposits along the Amazon river and its main tributaries, as illustrated in Figure 2. These so-called white water rivers originate in the Andes and carry a high sediment load. We used these observations to delineate the areas at risk of arsenic contaminated groundwater.

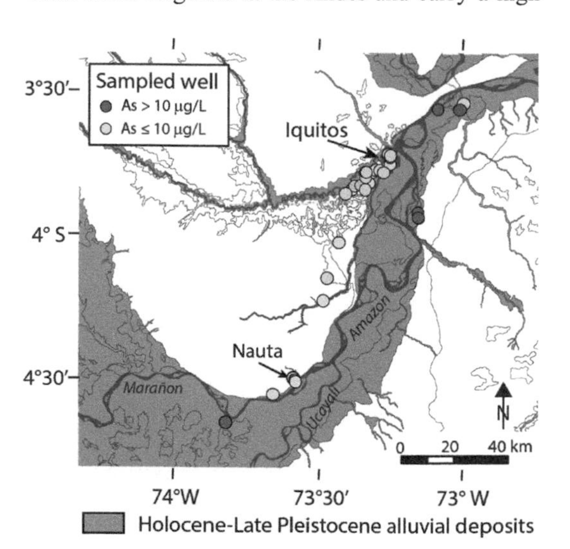

Figure 2. Indication of arsenic contaminated and non-contaminated wells around the city of Iquitos, in the Peruvian Western Amazon. Note that the wells where arsenic contamination was found are located in the floodplain of the Amazon river, i.e. on the recent alluvial deposits. (After de Meyer *et al.*, 2017).

The effectiveness of the use of a small number of geological and hydrogeochemical parameters as spatial proxies to predict the regional distribution of high and low risk to arsenic concentrations has been demonstrated for As-affected basins in e.g. South-East Asia (Winkel *et al.*, 2008). The Amazon Basin has hydrological and climatic conditions similar to South-east Asia. We therefore tested the relevance of the parameters used in the model of Winkel *et al.* (2008), and explored the significance of other parameters, to provide a risk map for the Amazon region.

4 CONCLUSION

Our study provides first insights on the presence and distribution of arsenic in groundwater resources of the Amazon Basin. The high concentrations of arsenic encountered in groundwater along the populated floodplains of the Amazon river require urgent investigation on possible health effects.

ACKNOWLEDGEMENTS

The authors would like to thank Caroline Stengel, Thomas Rüttimann, Numa Pfenninger and the AuA laboratory at Eawag for chemical analysis of the samples. We are grateful to the well owners for access to their wells, and permission for sampling. The ETRAS-team of PAHO in Lima and the DIRESA of Peru are sincerely thanked for their logistic support.

REFERENCES

Bretzler, A., Berg, M., Winkel, L., Amini, M., Rodriguez-Lado, L., Sovann, C., Polya, D.A. & Johnson, A. 2017. Geostatistical modelling of arsenic hazard in groundwater. In "Best Practice Guide on the Control of Arsenic in Drinking Water", Eds. Bhattacharya P., Polya D.A., Jovanovic D. IWA Publishing, London, UK. 153–160.

de Meyer, C.M.C., Rodríguez, J.M., Carpio, E.A., García, P.A., Stengel, C. & Berg, M. 2017. Arsenic, manganese and aluminum contamination in groundwater resources of Western Amazonia (Peru). *Sci. Total Environ.* 607–608: 1437–1450.

Rodriguez-Lado, L., Sun, G., Berg, M., Zhang, Q., Xue, H., Zheng, Q. & Johnson, C.A. 2013. Groundwater arsenic contamination throughout China. *Science* 341: 866–868.

WHO. 2011. Guidelines for Drinking-Water Quality. 4th ed. (Geneva).

Winkel, L., Berg, M., Amini, M., Hug, S.J. & Johnson, A.C. 2008. Predicting groundwater arsenic contamination in Southeast Asia from surface parameters. *Nat. Geosci.* 1: 536–542.

Environmental Arsenic in a Changing World –
Zhu, Guo, Bhattacharya, Ahmad, Bundschuh & Naidu (Eds)
ISBN 978-1-138-48609-6

Arsenic volume estimates in Holocene clay plug sediments in Bihar, India

S. Kumar, M.E. Donselaar & F. Burgers
Department of Applied Geoscience and Engineering, Delft University of Technology, Delft, The Netherlands

ABSTRACT: Shallow aquifers in the densely-populated areas of the Indo-Gangetic plain are severely polluted with arsenic (As). The occurrence of As and its spatial concentration variability are conditioned by the geomorphological setting of the Holocene floodplain, with the highest concentrations in elevated point-bar surrounded by clay-filled oxbow lakes (clay plugs). The work hypothesis is that the As is locally derived from the clay plugs. In this study, the potential As volume in the clay plugs is calculated, and the migration process of arsenic from clay plug to adjacent point bar is studied. Satellite data in combination with a side-scan sonar survey in present-day oxbow lakes, and As concentrations in sediment from a well were used to calculate the sediment volume of the clay plug, the potential volume of As, and the contact area between the clay plug and adjacent point bar, and to estimate the initial diffusion flux of dissolved As from clay plug to the adjacent point bar.

1 INTRODUCTION

Elevated sandy point bars surrounded by clay-filled oxbow lakes (clay plugs) in Bhojpur and Buxar districts of Bihar (Fig. 1) are hotspots for arsenic (As). Arsenic concentrations in groundwater of the point bars are characterized by a large spatial variation over short distances. Donselaar *et al.* (2016) proposed a generic model in which As is sourced from the clay plugs surrounding the point bars, and subsequently migrates to the groundwater in the permeable point bar sand. The present paper aims to provide insight in the potential As volume in clay plugs of the Holocene Ganges River floodplain, and to present ideas on the migration processes of As from clay plug to adjacent point bar.

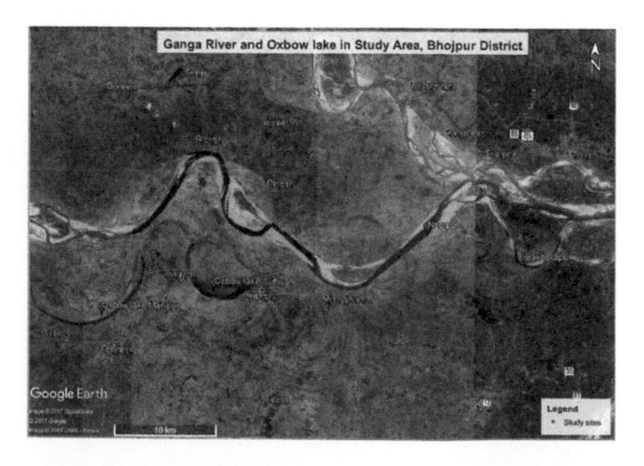

Figure 1. Ganges River and oxbow lakes in the study area.

2 METHODS

Clay plug surface areas were mapped with sentinel satellite data and Google Earth-Pro imagery. Clay plug volumes and contact areas between the clay plug and adjacent point bar were calculated in Matlab by combining the surface areas with bathymetric profiles derived from a side-scan sonar survey in recent oxbow lakes along the Ganges River, Bhojpur District in Bihar (Fig. 2). Arsenic concentrations in sediments from a borehole study in Holocene clay plugs in the Haringhata district (West Bengal, India) of Ghosh *et al.* (2015) were used to calculate As volumes contained in the clay plug deposits.

3 RESULTS AND DISCUSSION

3.1 *Calculation of potential arsenic volumes*

Surface areas of the selected clay plugs ranged from 10^6 to 10^7 m^2. Maximum depth of the oxbow lakes

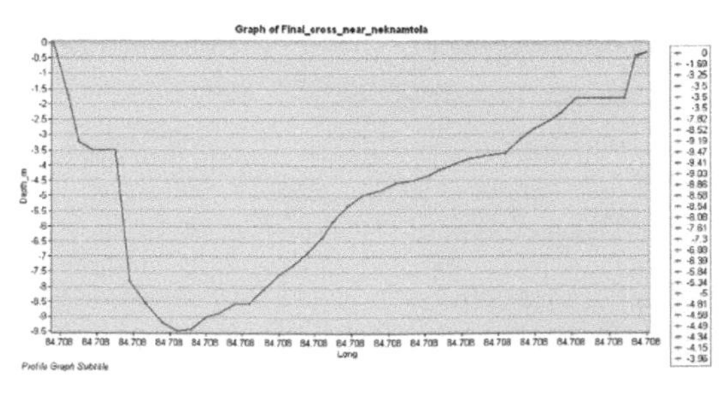

Figure 2. Sonar bathymetric profile of an oxbow lake along the Ganges River in the study area.

in the bathymetric study was 9.47 m. Arsenic content in cores from wells averaged 17.7 $mg\,kg^{-1}$ sediment (Ghosh *et al.*, 2015). The calculated sediment volume of clay plugs ranges from 10^6 to 10^8 m^3. The resultant As volume contained in the clay plug sediment ranges from 10^5 to 10^6 kg. The contact area between the clay plug and the adjacent point bar was in the order of 10^6 to 10^7 m^2.

3.2 *Estimation of diffusion fluxes*

There are no known driving forces in the point bar for the release of As from its solid state and the concentration of soluble As in the pore fluid of the point bar sands is negligible. The discharge depends linearly on the hydraulic conductivity, this soil property is of large influence over time. The discharge also depends on the present As volume. Over time the volume will decrease and thus the discharge will decrease. The discharge decrease is exponentially over time. Equation (1) describes Fick's first law of diffusion process with two assumptions, firstly that the system must be in steady state, and secondly that the medium must be homogeneous.

$$J = -D\nabla C \quad (1)$$

D is the diffusion coefficient, a material specific property, and C is the concentration. With some modifications Fick's first law for transport through a membrane is given by equation (2)

$$J = K\nabla C \quad (2)$$

Where J is diffusion flux [$mol\,m^{-2}s^{-1}$], K the hydraulic conductivity of the medium [$m\ s^{-1}$] and ΔC the difference in concentration [$mol\,m^{-3}$]. The hydraulic conductivity value, which is crucial to the modified version of Fick's first law, was considered to be within the following range: a minimal value of $10^{-8}\,m\,s^{-1}$ and a maximal value of $10^{-7}\,m\,s^{-1}$ (Bear, 1972). Initial diffusion flux was calculated based on Fick's First Law, and ranged between 10^1 to $10^2\,g\,m^{-2}\,y^{-1}$.

3.3 *Discussion*

The presence of high TOC concentrations and lack of water circulation in the deeper part of the oxbow lake (*hypolimnion*) contributes to the development of anoxic conditions that favor the desorption of As from FeOOH (e.g. McArthur *et al.*, 2004). Ghosh *et al.* (2015) reported TOC concentrations of 0.7% in clay plug deposits at shallow depths (6–12 m) in a meandering river geomorphology in Haringhata (Nadia district, West Bengal, India). Advection and diffusion are two processes of As movement from clay plug to point bar. In case of gravitational, compaction-driven expulsion of pore fluid movement, advection occurs while the difference in concentration creating gradient favors diffusion. The hydraulic conductivity is linearly related to the diffusion flux.

4 CONCLUSIONS

Sediment volume of clay plugs in the floodplain along the present-day Ganges River, as calculated from remote sensing data, range from 10^6 to 10^8 m^3. Arsenic concentrations from borehole data in Holocene clay plugs in West Bengal allowed for the estimate of As volume in each of the studied clay plugs. The As content ranges from 10^5 to 10^6 kg per clay plug. Based on Fick's First Law the initial diffusion flux was calculated and ranges from 10 to $10^2\,g\,m^{-2}\,yr^{-1}$. The results of this study are crucial for the dynamic flow modelling of As-contaminated aquifers in the affected areas, and helps to develop guidelines for the location of safe, As-free groundwater extraction.

ACKNOWLEDGEMENTS

Financial support for this study from the NWO-WOTRO research program "Urbanising Deltas of the World" (UDW) is gratefully acknowledged.

REFERENCES

Bear, J. 1972. Dynamics of fluids in porous media. Elsevier, New York, USA.

Donselaar, M.E., Bhatt, A.G. & Ghosh, A.K. 2016. On the relation between fluvio-deltaic flood basin geomorphology and the wide-spread occurrence of arsenic pollution in shallow aquifers. *Sci. Total Environ.* 574: 901–913.

Ghosh, D., Routh, J., Dario, M. & Bhadury, P. 2015. Elemental and biomarker characteristics in a Pleistocene aquifer vulnerable to arsenic contamination in the Bengal Delta Plain, India. *Appl. Geochem.* 61: 87–98.

McArthur, J.M., Banerjee, D.M., Hudson-Edwards, K.A., Mishra, R., Purohit, R., Ravenscroft, P., Cronin, A., Howarth, R.J., Chatterjee, A., Talukder, T., Lowry, D., Houghton, S. & Chadha, D.K. 2004. Natural organic matter in sedimentary basins and its relation to arsenic in anoxic groundwater: the example of West Bengal and its worldwide implications. *Appl. Geochem.* 19: 1255–1293.

Environmental Arsenic in a Changing World –
Zhu, Guo, Bhattacharya, Ahmad, Bundschuh & Naidu (Eds)
ISBN 978-1-138-48609-6

Steady-state groundwater arsenic concentrations in reducing aquifers

B.C. Bostick[1], A.A. Nghiem[1], A. van Geen[1], J. Sun[2], B.J. Mailloux[3], P.H. Viet[4] & P.T.K. Trang[4]
[1]*Barnard College and Lamont-Doherty Earth Observatory, Columbia University, New York, USA*
[2]*University of Western Australia & CSIRO Land and Water, Perth, WA, Australia*
[3]*Department of Environmental Sciences, Barnard College, New York, NY, USA*
[4]*CETASD, Vietnam National University, Hanoi, Vietnam*

ABSTRACT: Arsenic is a common soil and groundwater contaminant in much of the world. Groundwater arsenic is commonly derived from the reductive dissolution of arsenic-bearing iron oxides. Despite considerable efforts, it is still difficult to predict the aqueous concentration of arsenic in the transitional redox environments where arsenic release occurs, or in the reduced sediments through which it is transported. Equilibrium-based partitioning models often no more accurate that simple adsorption isotherms or partition coefficients in these environments. Here, we develop and evaluate a novel kinetics-based approach that incorporates knowledge of the solid-phase, to predict arsenic concentrations. This model defines the steady-state aqueous concentration of arsenic as a function of iron oxide dissolution rate and readsorption of arsenite and arsenate. This model uses measured iron and arsenic redox status to successfully describe As concentrations in redox profiles in heterogeneous Bangladeshi and Vietnamese aquifers. These data imply that the speciation of arsenic, which affects readsorption rates, are the dominant variable controlling As retention, and that iron mineralogy plays an indirect role by affecting the rate of biological reduction.

1 INTRODUCTION

Groundwater arsenic (As) contamination is a global public health problem and also a concern at hundreds of U.S. Superfund sites. A considerable body of research has established that arsenic enters groundwater most commonly through the biological reduction of As-bearing iron oxides. Despite this realization, we are still not able to adequately predict arsenic concentrations at specific sites, or determine how they may be affected as the system evolves. Most predictions are based equilibrium retention by the sediments and predict changes in As concentrations resulting in a change in sediment mineralogy and/or redox status and the change in the quantity of substrate and/or As in the solid-phase. These approaches are broadly consistent with regional trends in that older sediments tend to be oxidized and contain less As due to flushing, but they break down at smaller scales. In this research, we propose that kinetics is the dominant factor that affects As partitioning, and develop a quantitative model to test this hypothesis that is capable of describing As concentrations at specific depths within several depth profiles of dissolved arsenic levels in Vietnam.

1.1 *Modelling*

This simple model is based on the hypothesis that dissolved As is a chemical intermediate between the reductive dissolution that puts it into solution, and adsorption or precipitation processes that remove it from solutions. It assumes that the dissolved As concentrations are at steady-state because retention processes are sufficient that most of the As remains in the solid phase. This model assumes As only resides on Fe(III) minerals, and that the reduction of those minerals releases them proportionally to concentration. The rate of Fe reduction in the sediments can vary due to many factors but under in situ conditions varies strongly with mineralogy (Postma *et al.*, 2010). The rate of readsorption of arsenic depend on the availability (concentration) of iron oxide substrates, and the concentration and oxidation state of As. Arsenate and arsenite are considered independently—the rate of their release is determined by their abundance in the solid-phase. Because adsorption rates for As(V) on Fe(III) oxides are considerably more rapid than As(III) (Kanel *et al.*, 2006), As oxidation state plays a significant, albeit indirect role in regulating As levels.

1.2 *Fe and As speciation*

This model requires knowledge about the solid-phase speciation of both Fe and As. The concentration of solid-phase Fe(II) and Fe(III) oxides is determined using synchrotron-based X-ray absorption spectroscopy (XAS) using established techniques[5]. We convert the fractional speciation determined by XAS to solid phase concentrations of Fe(III) and As(III) or As(V) using X-ray fluorescence (XRF) data collected using a hand-held XRF. We have determined As and Fe speciation every 2 m for several sites where groundwater concentrations also vary with depth. This fine-scale sampling allows us to explain sharp gradients in groundwater arsenic that occur with depth, and to determine the relative roles of Fe(III) and As(V) reduction on regulating As levels.

2 RESULTS AND DISCUSSION

This research is unique in that it allowed us study the redox state in layered aquifers where both redox state and As concentrations varied considerably with depth, and where there were coupled changes in dissolved As levels. In most environments, As(V) and Fe(III) concentrations are high in the surface, and conditions become more reducing with depth. At most of our, Fe(III) concentrations decreased gradually with depth, while As reduction was a sharper transition that occurred below the zone where Fe(III) reduction became apparent. Dissolved As concentrations generally were low at the surface where systems were more oxidized, and there was a sharp gradient between low and high dissolved As concentrations. In more complex layered aquifers systems containing oxidized Pleistocene sediments and reduced Pleistocene sediments, the model was also effective in that dissolved As concentrations oscillated with sediment properties.

The model was able to accurately predict dissolved As levels within a factor of 2 at all depths using measured Fe(III) oxide, As(III) and As(V) solid-phase concentrations, and without any information about groundwater composition. Dissolved As levels mirrored As speciation. Based on these data, we were able to model dissolved As concentrations over a wide range of As and Fe oxidation states given a small number of assumptions about the concentration of Fe and As in the sediments, and the reactivity of the Fe(III) oxides in the system (Fig. 1). Most As concentrations from sediment profiles was effectively described, with the possible exception of some highly reduced sediments.

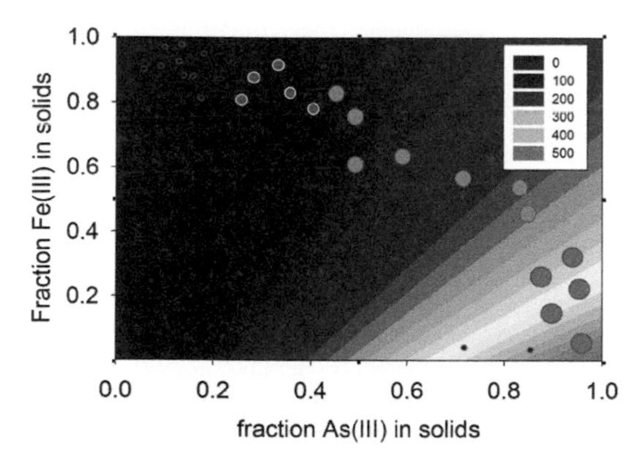

Figure 1. Modeled contour plot of dissolved As concentrations (legend shows color scale) as a function of solid-phase Fe and As oxidation state, with selected data plotted on the figure with symbols sizes and colors proportional to concentration.

3 CONCLUSIONS

The model successfully predicts dissolved As concentrations without refinement of kinetic parameters. This adherence suggests that the assumption that dissolved As can be thought of effectively as a reactive intermediate in the environment.

There are several important applications of this model. First, As concentrations appear to vary with depth because of differences in the rates of retention of As(III) and As(V) rather than differences in their equilibrium adsorption isotherms on aquifer sands. Second, because redox gradients of Fe and As vary from each other spatially, it allows us to identify the relative roles of As and Fe reduction on generating high-As groundwater. Fe gradients are more gradual, and even in most reduced sediments, a considerable fraction of Fe(III) phases remain. Dissolved As concentrations will be highest when reduction rates are fastest, meaning that As concentrations will be elevated most in the presence of reactive Fe oxides such as ferrihydrite. Third, As oxidation state appears to play a much more important role in regulating As levels than does Fe speciation. This model does not consider the rates of As reduction, however, parallel recent research suggests that biologically mediated As(V) reduction is prevalent where As(III) is found. These results suggest that As levels will be highest where there is a combination of high levels of As(III) retained on reactive Fe(III) oxides, and that remediation efforts should concentrate on modulating the rates of retention to be effective.

ACKNOWLEDGEMENTS

U.S. National Institute of Environmental Health Sciences (grant ES010349 and ES009089) and U.S. National Science Foundation (grants 1521356 and 0911557).

REFERENCES

Kanel, S.R., Greneche, J.M. & Choi, H. 2006. Arsenic(V) removal from groundwater using nano scale zero-valent iron as a colloidal reactive barrier material, *Environ. Sci. Technol.* 40: 2045–2050.

Postma, D., Jessen, S., Hue, N.T.M., Mai, T.D., Koch, C.B., Viet, P.H., Nhan P.Q. & Larsena, F. 2010. Mobilization of arsenic and iron from red river floodplain sediments, vietnam. *Geochim. Cosmochim. Acta* 74: 3367–3381.

Environmental Arsenic in a Changing World –
Zhu, Guo, Bhattacharya, Ahmad, Bundschuh & Naidu (Eds)
ISBN 978-1-138-48609-6

Arsenic and polymetallic contaminants in groundwater of the arid regions of South Africa

T.A. Abiye
School of Geosciences, University of the Witwatersrand, Johannesburg, South Africa

ABSTRACT: In South Africa, the majority of rural water supply comes from groundwater hosted within the weathered and fractured crystalline rocks. The main minerals that are responsible for arsenic release into groundwater in South Africa are arsenopyrite (FeAsS), lollingite ($FeAs_2$) and scorodite ($FeAsO_4.2H_2O$), where scorodite is an alteration product of arsenopyrite due to prolonged water-rock interaction process. Owing to the release of arsenic from highly mineralized rocks, its concentration in the groundwater reaches up to 253 $\mu g\ L^{-1}$ (south western part of the country), 6150 $\mu g\ L^{-1}$ (west of Johannesburg), about 500 $\mu g\ L^{-1}$ in the Karoo aquifers. Acid mine drainage is also found to be important source of arsenic and other toxic metals in the environment.

1 INTRODUCTION

In regions that are characterized by climatic aridity and minimal rainfall, groundwater from weathered and fractured crystalline rock aquifers is a primary source of water supply for various community activities. The development of groundwater resources from these aquifers has been regarded as a very important deriver for economic growth of the country owing to its availability at shallow depth ($\approx$40 m) and scarcity of surface water sources often characterized by poor water quality due to arsenic (As) and polymetallic contaminants. In several parts of the world, especially in south-east Asia, high As concentration is prevalent in groundwater from the sedimentary aquifers with enriched organic matter (Ahmed *et al.*, 2004; Bhattacharya *et al.*, 2002a,b, 2004; Mukherjee *et al.*, 2011; Nriagu *et al.*, 2007) related to organic rich sediments. The geology of aquifers that host As in South Africa have undergone large-scale mineralization, where As occurs in sulphide ores such as auriferous, stanniferous and antimonial deposits (Hammerbeck, 1998) that is being released into groundwater through leaching process. Arsenic, U and Se are more mobile under a wider range of naturally occurring groundwater conditions; hence they pose a potentially more widespread problem than other trace constituents (Sami & Druzynski, 2003). Extensive mineralization of rocks is also responsible for the release of toxic metals into the groundwater.

2 MATERIALS AND METHODS

Extensive literature assessment was conducted in order to gather relevant information on the arsenic and toxic metal concentration in the groundwater system of South Africa. In order to get a first-hand information, twenty groundwater samples from granitic aquifers were collected and analyzed for 24 metals, including arsenic. The water samples were diluted in 50 mL polypropylene volumetric flasks with ultra-pure water and acidified with 2% HNO_3 and analyzed by IC-PMS.

3 RESULTS AND DISCUSSION

The geology of the main aquifers in South Africa is represented by weathered and fractured basement rocks, Bushveld igneous complex (basic and felsic intrusive rocks), dykes (pegmatite, syenite and dolerite), the Witwatersrand and Transvaal Supergroup meta-sedimentary rocks (quartzite, sandstone, shale, dolomite), the Karoo and Kalahari sedimentary rocks, meta-volcanic rocks (basalts, andesites and rhyolites) and unconsolidated costal sediments. The main minerals that are responsible for arsenic release into groundwater in South Africa are dominated by arsenopyrite (FeAsS) that has a widespread occurrence in the country, lollingite ($FeAs_2$) and scorodite ($FeAsO_4.2H_2O$). Scordite is a main mineral phase in the withering zone. Arsenopyrite and other sulphide minerals are known to be susceptible to oxidation in near surface environments (Herath *et al.*, 2016), where most boreholes tap water for community supply.

Owing to the release of arsenic from highly mineralized rocks into groundwater, its concentration in the groundwater reaches up to 253 $\mu g\ L^{-1}$ (south western part of the country), 7000 $\mu g\ L^{-1}$, west of Johannesburg (Sami & Druzynski, 2003), about 550 $\mu g\ L^{-1}$ in the Karoo aquifers, within the boreholes having variable depth. In general the arsenic content in the groundwater falls above the WHO drinking water limit (10 $\mu g\ L^{-1}$).

The abandoned gold tailings dams around the city of Johannesburg contain very high metal concentration that release metals into surface and ground water resources. For example, in the tailings dams the content of some metals such as Cr ranges from 170 to 310 mg L^{-1}, Co = 10–240 mg L^{-1}, Cu = 15–254 mg L^{-1} and lead 6–34 mg L^{-1}. High uranium rich mine water with acidic pH decants continuously from abandoned mines into streams (Winde, 2006; Hobbs, 2011; Abiye, 2014; Abiye, *et al.*, 2018). The high concentration of polymetallic contaminants in the groundwater was regulated by oxidation, leaching and precipitation processes besides water-rock interaction. The metal concentration in groundwater is in the order of Fe>Sr>Zn>B>Mn>Ni>Ni>Li>Cr>As>U that falls within the range of 14 μg L^{-1} for U and 3128 μg L^{-1} for Fe in the Namaqualand groundwater (Abiye & Leshomo, 2013). Fe is also the main component in the gold and coal mining regions as a result of acid mine drainage that reaches 890 mg L^{-1} (Abiye *et al.*, 2011). In addition to the leaching of arsenic from sulphide minerals facilitated by oxidation process, oxidation of iron is also believed to contribute for the concentration of arsenic in the groundwater. The lack of aquifer flushing due to low groundwater recharge condition facilitates concentration of metals in the shallow aquifers besides severe evaporation process in the region. This zone is readily accessible by groundwater users for various economic activities.

4 CONCLUSIONS AND RECOMMENDATIONS

South Africa is endowed with a variety of economic minerals both metals and non-metals, which also host minerals responsible for arsenic and other toxic metals release into aquifers. Complex geochemical processes that involve oxidation and precipitation at acidic pH conditions are among few that mobilizes metals from host rocks into groundwater. It was also noted that acid mine drainage in the gold and coal mining areas is the primary media for arsenic and other toxic metal mobilization from sulphides and eventually contaminates water supply systems.

Due to the health related risk associated with arsenic, it is essential to take precautionary measure on boreholes with high content.

ACKNOWLEDGMENT

I would like to thank Prof. Prosun Bhattacharya for his encouragement that helped me to develop this abstract for As2018 conference.

REFERENCES

Abiye, T.A., Mengistu, H. & Demlie, M.B. 2011. Groundwater resource in the crystalline rocks of the Johannesburg area, South Africa. *J. Water Resour. Protection* 3(4): 199–212.

Abiye, T.A. & Leshomo J. 2013. Metal enrichment in the groundwater of the arid environment in South Africa. *Environ. Earth Sci.* 72: 4587–4598.

Abiye, T.A. 2014. Mine water footprint in the Johannesburg area: a review based on existing and measured data. *South Afr. J. Geol.* 117(1): 87–96.

Abiye, T.A, Mkansi, S., Masindi, K. & Leshomo, J. 2018. Effectiveness of wetlands in retaining metals from mine water, South Africa. *J. Water Environ.* (in press).

Ahmed, K.M. Bhattacharya, P., Hasan, M.A., Akhter, S.H., Alam, S.M.M., Bhuyian, M.A.H., Imam, M.B., Khan, A.A. & Sracek, O. 2004. Arsenic contamination in groundwater of alluvial aquifers in Bangladesh: An overview. *Appl. Geochem.* 19(2):181–200.

Bhattacharya, P., Frisbie, S.H., Smith, E., Naidu, R., Jacks, G. & Sarkar, B. 2002a. Arsenic in the Environment: A Global Perspective. In: B.Sarkar (Ed.) *Handbook of Heavy Metals in the Environment*, Marcell Dekker Inc., New York, pp. 147–215.

Bhattacharya, P., Jacks, G., Ahmed, K.M., Khan, A.A. & Routh, J. 2002b. Arsenic in groundwater of the Bengal Delta Plain aquifers in Bangladesh. *Bull. Env. Contam. Toxicol.* 69(4): 538–545.

Bhattacharya, P., Welch, A.H., Ahmed, K.M., Jacks, G. & Naidu, R. 2004. Arsenic in groundwater of sedimentary aquifers. *Appl. Geochem.* 19(2):163–167.

Hammerbeck, E.C.I. 1998. Arsenic. In: T*he Mineral Resources of South Africa: Handbook*, Council for Geoscience, 16, p. 40–45.

Herath, I., Bundschuh, J., Vithanage, M. & Bhattacharya, P. 2016. Geochemical processes for mobilization of arsenic in groundwater. In: P. Bhattacharya, M. Vahter, J. Jarsjö, J. Kumpiene, A. Ahmad, C. Sparrenbom, G. Jacks, M.E. Donselaar, J. Bundschuh, & R. Naidu (eds.) *Arsenic Research and Global Sustainability As 2016*". Interdisciplinary Book Series: "Arsenic in the Environment—Proceedings". Series Editors: J. Bundschuh & P. Bhattacharya, CRC Press/Taylor and Francis (ISBN 978-1-138-02941-5), pp. 23–24.

Hobbs, P.J (Ed.) 2011. Situation assessment of the surface water and groundwater resource environments in the Cradle of Humankind World Heritage Site. Report prepared for the Management Authority. Department of Economic Development. Gauteng Province. South Africa. P424.

Mukherjee, A., Bhattacharya, P. & Fryar, A. E. 2011. Arsenic and other toxic elements in surface and groundwater systems. *Appl. Geochem.* 26(4): 415–420.

Nriagu, J.O., Bhattacharya, P., Mukherjee, A.B., Bundschuh, J., Zevenhoven, R. & Loeppert, R.H. 2007. Arsenic in soil and groundwater: an overview. In: Bhattacharya, P., Mukherjee, A.B., Bundschuh, J., Zevenhoven, R. & Loeppert, R.H. (Eds.) *Arsenic in Soil and Groundwater Environment: Biogeochemical Interactions, Health Effects and Remediation,* Trace Metals and other Contaminants in the Environment, Volume 9, Elsevier B.V. Amsterdam, The Netherlands, pp. 3–60.

Sami, K. & Druzynski, A.L. 2003. Predicted Spatial Distribution of Naturally Occurring Arsenic, Selenium and Uranium in Groundwater in South Africa -Reconnaissance Survey- *WRC Report No. 1236/1/03*.

Winde, F. 2006. Challenges for sustainable water use in dolomitic mining regions of South Africa- a case study of Uranium pollution: part 1 Sources and pathways. *J. Phys. Geogr*. 27 (4): 333–347.

Environmental Arsenic in a Changing World –
Zhu, Guo, Bhattacharya, Ahmad, Bundschuh & Naidu (Eds)
ISBN 978-1-138-48609-6

Effect of recharging water from Meghna River on the arsenic contaminated groundwater

H. Masuda[1], N. Hirai[1] & A.H.M. Selim Reza[2]
[1]*Department of Biology and Geosciences, Osaka City University, Sumiyoshi-ku, Osaka, Japan*
[2]*Department of Geology and Mining, University of Rajshahi, Rajshahi, Bangladesh*

ABSTRACT: Groundwaters and aquifer sediments from Chundpur and its surroundings, along the downstream Meghna River at the meeting with Padma River, Bangladesh, were analyzed to document the mechanism of As contamination. The As was released in association with infiltration of surface water into the shallow groundwater aquifer without strong reduction. In situ chemical weathering and the following desorption at increasing pH would be the primary and secondary processes to release As into the groundwater of the studied area.

1 INTRODUCTION

Delta plain at the south of Bangladesh is one of the most arsenic (As) affected area of the world. In this area, microbial activity is presumed to cause reduction-dissolution of As from the aquifer sediments to groundwater. The reduction-dissolution occurs during the infiltration of surface water into the aquifer. In this study, the geochemistry of groundwaters and aquifer sediments was studied to reveal the role of recharging water from rivers, especially the main channel of Meghna.

2 STUDY AREA AND METHODS

2.1 *Study area and sample collection*

Study area is at the east side of meeting of Meghna (Brahmaputra) and Padma (Ganges) Rivers (Fig. 1). The area is in the modern delta plain. In this area, as similar to the surrounding areas in this delta plain, highly As contaminated groundwater appears in the wells.

Ninty-nine groundwater and five river water samples were collected from the five villages (Fig. 1) in March 2016. The waters were collected in plastic and glass bottles and acidified if it was needed for the analyses. Water temperature, electric conductivity, pH, ORP and DO were measured *in situ*. Aquifer sediments were cored from two sites down to 35m depths using a split barrel sampler in May 2015.

2.2 *Laboratory analyses*

Major and minor element chemistry of water samples were analyzed in the laboratory. The anions were analyzed by ion-chromatography, and major cations and minor elements including As were analyzed by ICP-MS.

Total As and major element chemistry of whole sediment samples were analyzed by ICP-MS after alkali-fusion and X-ray fluorescence photometry respectively. Arsenic and the related elements of the sediments were also analyzed by differentially chemical extraction according to BCR method slightly modified to estimate the phases of mobile As (Rauret *et al.*, 1999).

3 RESULTS AND DISCUSSION

3.1 *Water chemistry*

Major element chemistry was abundantly Ca-HCO_3 type in the three villages close to the Meghna River (MAT, CHU and FAR) and Na-HCO_3 type in the two inland villages (SHA and HAJ). In the all villages, groundwaters were occasionally affected by Na-Cl type water, presumably originated from the household waster waters.

Arsenic concentration is generally high in the groundwaters <70 m depths; the maximum concentration was $0.74\,mg\,L^{-1}$. The groundwaters >100 m depths were almost free from As. Riverwaters contained As $<5\,\mu g\,L^{-1}$. Level of As is high in the HAJ, CHU and FAR villages, of which average As concentration of groundwaters from <70 m depths wells were $0.44\,mg\,L^{-1}$, indicating no relation of As level to the distance from the river. ORPs were the high of the groundwaters from FAR and HAJ compared with those of the other three villages, and the ranges of ORP did not directly related to the As concentration (Fig. 2). Especially, the groundwaters from HAJ gave wide variation of the ORP ($-20\sim-155$ mV) and high range of As ($0.24\sim0.63\,mg\,L^{-1}$ except one; -90 mV

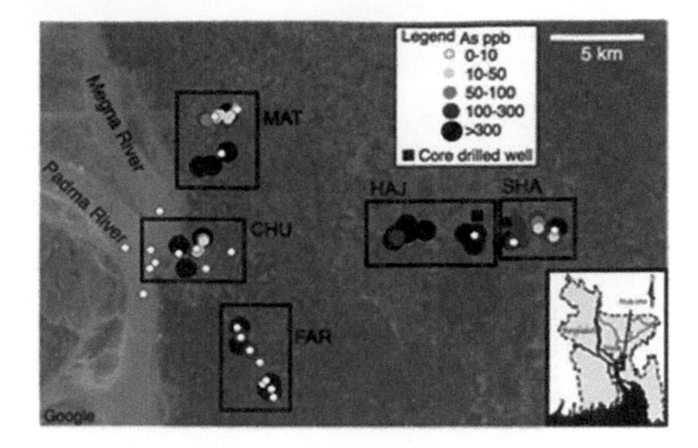

Figure 1. Study area and As concentration of groundwaters of Chandpur and its surroundings.

and 0.006 mg L^{-1} As). The groundwater giving the high ORP (−20 mV) had 0.63 mg L^{-1} As. Iron concentration increased with decreasing ORP (not shown). Thus, in the studied area, reduction-dissolution of Fe-phases does not simply work to increase the As concentrations.

Major ion compositions of the groundwaters containing high As were mostly of Ca-HCO_3 type in the villages MAT, CHU, and FAR, while Ca-HCO_3 type in HAJ and SHA, indicating that the anthropogenic activity does not directly affect releasing As into the groundwaters of this area. Also, the high As groundwaters are rather diluted among the studied groundwater. It is suggested that the As is released from the sediments at the early stage of reaction between the infiltrating groundwater and minerals of the flowing paths.

3.2 *Mineralogy and sequential extracted chemistry*

As concentration had positive relationship to the XRD intensities of clay minerals, chlorite and mica, suggesting that certain amounts of As was adsorbed onto the fine fractions of aquifer sediments.

Two sets of core sediments taken from HAJ down to 23 m depths were analyzed by sequential chemical extraction. The total As concentration of the whole core sediments was within 2 and 8 ppm. Mobile compounds, categorized as weakly adsorbed, reducible and oxidizable fractions were 20 to 60 % of the total As of the sediments. It is notable that the reducible phase, which means the As strongly adsorbed onto and/or fixed in Fe-oxyhydroxides and Mn-oxides, occupies the most abundant host phase of the As, although the mobile phases of As is smaller in the uppermost sediments than the lower ones. Considering increasing As concentration with increasing pH, the level of As of the studied groundwater would be controlled by desorption from the Fe-oxyhydroxides and/or clay fractions

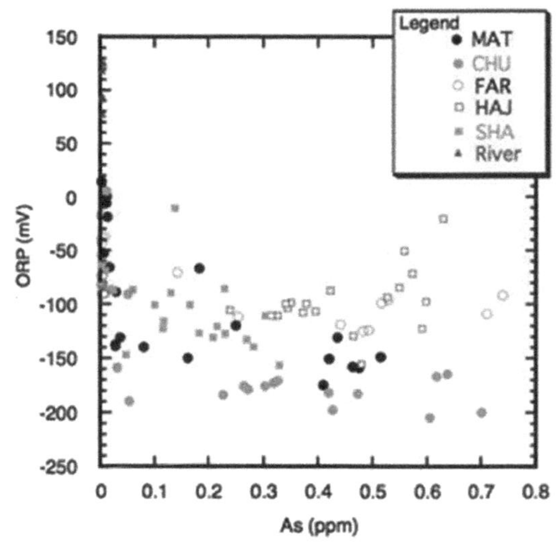

Figure 2. Relationship between As concentration and ORP of groundwaters from Chandpur and its surroundings.

with progressing chemical weathering. Dissolution of the Fe-oxyhydroxides as a host of As may follow the desorption.

4 CONCLUSIONS

Studied groundwaters would contain high As, which was released from the sediments via infiltration of river and/or local meteoric water to promote the chemical weathering and resulting increasing pH to desorb As from the Fe-oxyhydroxides and/or clay minerals as hosts of As in the studied groundwater aquifers. Chemical weathering must be an important mechanism of As releasing into the groundwater in the Neogene aquifers of the studied area.

ACKNOWLEDGEMENTS

We thank to T. Shimonaka and K. Okazaki, Osaka City University, and Drs. A. Marui and M. Ono, AIST, who supported the water analyses.

REFERENCE

Rauret, G., López-Sánchez, J.F., Sahuquillo, A., Rubio R., Davidson, C., Ure, A. & Quevauvillerc, Ph. 1999. Improvement of the BCR three step sequential extraction procedure prior to the certification of new sediment and soil reference materials. *J. Environ. Monit.* 1: 57–61.

Environmental Arsenic in a Changing World –
Zhu, Guo, Bhattacharya, Ahmad, Bundschuh & Naidu (Eds)
ISBN 978-1-138-48609-6

Groundwater arsenic distribution reconnaissance survey in Myanmar

L.A. Richards, G.P. Pincetti Zúñiga & D.A. Polya
School of Earth, Atmospheric and Environmental Sciences and Williamson Research Centre for Molecular Environmental Science, University of Manchester, Manchester, UK

ABSTRACT: Dangerous arsenic concentrations in shallow groundwaters threaten the health and livelihoods of millions of people, particularly in South/Southeast Asia. However, the scope and magnitude of groundwater arsenic hazard is relatively poorly understood in Myanmar as compared to neighboring counties. We undertook a groundwater quality survey across five (hydro)geologically distinct regions of Myanmar, and initial field results indicate elevated arsenic in a number of samples including in areas where previously modeled probability of arsenic concentrations exceeding the provisional WHO guideline was low. Data validation and interpretation of possible geochemical and/or hydrological controls are the subject of ongoing investigation.

1 INTRODUCTION

Millions of people in South/Southeast Asia are chronically exposed to dangerous concentrations of geogenic arsenic in groundwater (e.g. Charlet & Polya, 2006; Smedley & Kinniburgh, 2002 and refs within). Although the scope and magnitude of this problem is reasonably well defined in some areas particularly in countries such as Bangladesh, India, Cambodia, Vietnam and China, widespread international understanding of the distribution of arsenic in Myanmar is much more limited and in some cases is restricted to predictive models based on surface geological parameters (Amini *et al.*, 2008; Winkel *et al.*, 2008), despite elevated levels of arsenic being confirmed in the Ayeyarwady Basin (van Geen *et al.*, 2014). We aim to improve the understanding of arsenic distribution in a cross-country groundwater quality survey incorporating five geologically distinct regions within Myanmar.

2 METHODS/EXPERIMENTAL

2.1 *Field area*

Field sites (n = 85) were located across Myanmar and were broadly based around Maubin (n = 13), Thongwa (n = 11), Loikaw (n = 18), Kalay (n = 23) and Mandalay (n = 20) (Fig. 1). General sampling regions were selected on the basis of (i) encompassing different predicted arsenic hazard categories (Winkel *et al.*, 2008); (ii) encompassing areas of different (hydro)-geological characteristics; (iii) collaboration with local partners. Specific sampling locations within each region were selected with the input of local partners. In brief, the geology is dominated by recent alluvial deposits in the Ayeyarwady basin, Permian limestones in the Shan Plateau, and Cretaceous flysch sedimentary sequences in the Indo-Burman Ranges (Bender, 1983).

2.2 *Sample collection and analysis*

Groundwater samples (n = 85) were collected from existing tube wells and dug wells in December 2017 during which *in-situ* analysis of parameters such as pH, *Eh*, temperature, electrical conductivity, alkalinity, sulfide, ammonium, nitrate, nitrite and fluoride were conducted using methods adapted from previous studies in Cambodia (cf. Richards *et al.*, 2017). Initial field measurements of arsenic were made visually using the ITS Econo-Quick arsenic kit (van Geen *et al.*, 2014).

3 RESULTS AND DISCUSSION

Groundwater arsenic, as measured visually with field kits, ranged from 0–500 $\mu g\,L^{-1}$ (Fig. 2), with the highest concentrations observed in the Ayeyarwady Basin as previously predicted by arsenic hazard models (Winkel *et al.*, 2008) and confirmed by field measurements (van Geen *et al.*, 2014) undertaken within the same region but in different specific areas. Elevated arsenic was also observed in other areas, including the area surrounding Loikaw, an area which was not predicted to be likely to have elevated arsenic by previous predictive models (Winkel *et al.*, 2008). Overall 42% of samples collected (36 of 85) estimated arsenic $\geq 10\,\mu g\,L^{-1}$, including in areas not encompassed by previous arsenic hazard maps, which suggests that it is possible that higher populations and/or additional geographical areas may be at risk from elevated arsenic than previously predicted. Confirming field measurements and linking results with other groundwater geochemical parameters (*e.g.* Fe, HCO_3^-, *Eh*, pH, *etc.*)

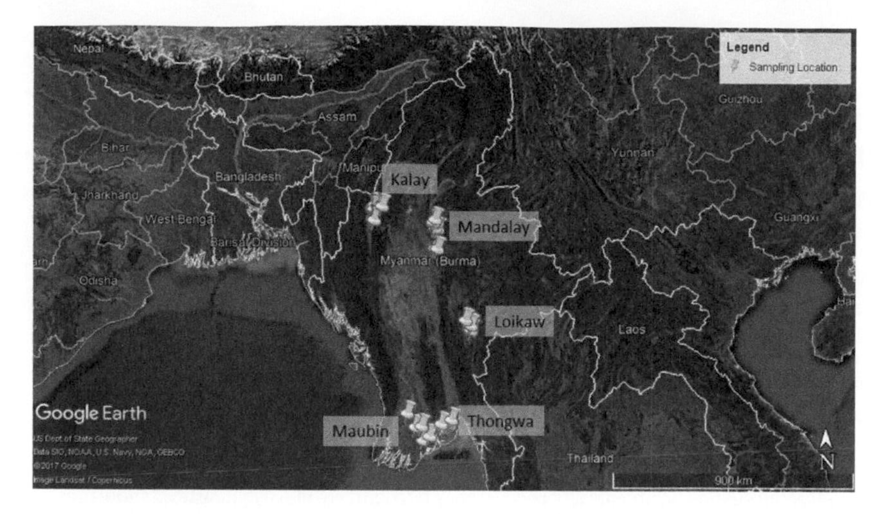

Figure 1. Map showing sampling sites (n = 85) from general field areas based around Maubin (n = 13), Thongwa (n = 11), Loikaw (n = 18), Kalay (n = 23), and Mandalay (n = 20); map adapted from Google Earth with attribution as shown. Samples with arsenic $\geq 50\ \mu g\ L^{-1}$ (n = 9) were located in the Maubin (n = 4), Thongwa (n = 3) and Loikaw (n = 2) sampling areas.

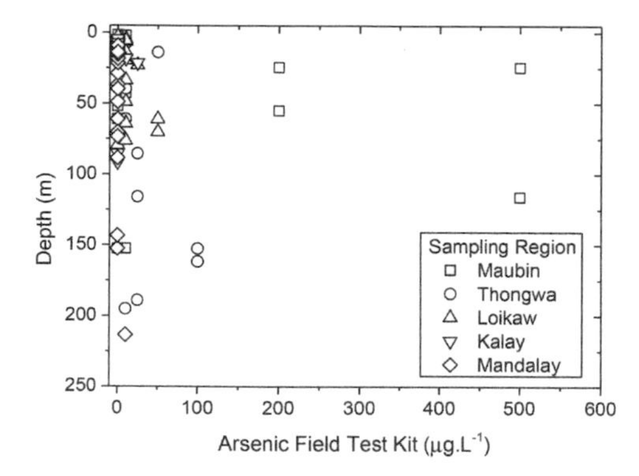

Figure 2. Arsenic (ITS Econo Quick Visual test kit) versus reported depth for each sampling region. Note test kit reports a semi-quantitative range of values assessed visually.

as well as potential (hydro)-geological controls is the subject of ongoing work.

4 CONCLUSIONS

A cross-country groundwater quality survey indicates that elevated levels of arsenic are likely to occur in several distinct, previously under-studied regions of Myanmar. Results will be validated by further, laboratory-based analysis including analysis of major and trace elements and arsenic speciation.

ACKNOWLEDGEMENTS

This research was supported by the Leverhulme Trust (ECF2015-657 to LR), with additional support from an Engineering and Physical Sciences Research Council Improving Diversity Award (to LR, Tun, Gibson and DP). We are very grateful to substantial support especially from Yin Min Tun (eTekkatho Digital Library), project partners and field assistants from Dagon University, Maubin University, Loikaw University, Kalay University and Mandalay University, Helen Downie (University of Manchester) and local authorities and landowners.

REFERENCES

Amini, M., Abbaspour, K.M., Berg, M., Winkel, L., Hug, S.F., Hoehn, E., Yang, H. & Johnson, A. 2008. Statistical modeling of global geogenic arsenic contamination in groundwater. *Environ. Sci. Technol.* 42: 3669–3675.

Bender, F. 1983. *Geology of Burma*: Gebrüder Borntraeger.

Charlet, L. & Polya, D.A. 2006. Arsenic in shallow, reducing groundwaters in southern Asia: An environmental health disaster. *Elements* 2: 91–96.

Richards, L.A., Magnone, D., Sovann, C., Kong, C., Uhlemann, S., Kuras, O., van Dongen, B.E., Ballentine, C.J. & Polya, D.A. 2017. High resolution profile of inorganic aqueous geochemistry and key redox zones in an arsenic bearing aquifer in Cambodia. *Sci. Total Environ.* 590–591: 540–553.

Smedley, P.L. & Kinniburgh, D.G. 2002. A review of the source, behaviour and distribution of arsenic in natural waters. *Appl. Geochem.* 17(5): 517–568.

van Geen, A., Win, K.H., Zaw, T., Naing, W., Mey, J.I., & Mallioux, B. 2014. Confirmation of elevated arsenic levels in groundwater of Myanmar. *Sci. Total Environ.* 478: 21–24.

Winkel, L., Berg, M., Amini, M., Hug, S.J. & Johnson, C.A. 2008. Predicting groundwaters arsenic contamination in Southeast Asia from surface parameters. *Nat. Geosci.* 1: 536–542.

Environmental Arsenic in a Changing World –
Zhu, Guo, Bhattacharya, Ahmad, Bundschuh & Naidu (Eds)
ISBN 978-1-138-48609-6

Potential arsenic contamination in drinking water sources of Tanzania and its link with local geology

J. Ijumulana[1,2], F. Mtalo[1] & P. Bhattacharya[2]
[1] *DAFWAT Research Group, Department of Water Resources Engineering, College of Engineering and Technology, University of Dar es Salaam, Dar es Salaam, Tanzania*
[2] *KTH-International Groundwater Arsenic Research Group, Department of Sustainable Development, Environmental Science and Engineering, KTH Royal Institute of Technology, Stockholm, Sweden*

ABSTRACT: Recent studies on arsenic (As) occurrence particularly in African waters show that several sources of drinking water have elevated concentrations above national and international guidelines. In Tanzania, elevated concentrations of As above the WHO guideline ($10\,\mu g\ L^{-1}$) in Lake Victoria Gold fields is emerging as a threat to public health depending on groundwater and surface water as drinking water sources. In this study, spatial statistics and GIS tools have been used to delineate the relationship between As occurrence and local geological settings. Among the 12 mapped local geological units, the most targeted aquifers for potable water are characterized by granitoids, migmatite, mafic and ultramafic meta-sediments (~50% of water points). The probability of having As levels above the WHO guideline was 0.71 and 0.33 for surface water and groundwater systems respectively.

1 INTRODUCTION

High levels of arsenic (As) have been reported both in surface water and groundwater in several African countries (Ahoulé *et al.*, 2015). The source, distribution and mobilization of As in aqueous environment differs by country and within same country differs by location and are associated with either geogenic or anthropogenic processes. Elevated concentrations of As has been reported in the northern part of Tanzania, particularly around the Lake Victoria Goldfields (LVGF) (Mnali, 2001; Lucca *et al.*, 2017) within Lake Victoria Basin (LVB). In the LVGF, large spatial variability of As occurrence has been identified in terms of concentration as well as speciation which hinders meaningful conclusion on its fate based on the available database (Kassenga & Mato, 2008). The present study aims to investigate: i) the effects of local geological settings on the distribution of As and its concentrations in the drinking water sources; ii) the effects of climate on the variability of As in water sources; and iii) understand the probable links of the adverse health outcomes (viz. cancer cases) due to long term ingestion of inorganic As in drinking water sources in the region. In this abstracts results on spatial variability of arsenic occurrence with respect to local geological setting are presented.

2 MATERIALS AND METHODS

2.1 *Study area*

Lake Victoria Basin, Tanzanian part, is one of the 9 river basins in Tanzania mainland covering area of $119{,}442\,km^2$. The region has little seasonal variation but the eastern sections where Mara region, the study area lies, average only 750–1000 mm of rain. Favourable climatic conditions for agriculture and livestock and the abundance of natural resources have supported the livlihood of the rural population of over 35 million people (Lucca *et al.*, 2017). The geology of the Tanzanian LBV consists of Archean granitoids-greenstone belts hosted in the Tanzanian Craton. More than 80% of rural population depends on groundwater resources for various use.

2.2 *Assessment of drinking water supply points in Lake Victoria Basin with respect to local geology*

More than 80% of rural population in LVB depend on groundwater resources abstracted through boreholes, springs, shallow wells and deep wells. Approximately 50% of abstraction points target aquifers composed of migmatite-granitoid-metasediment complex, metasediments (~22%), sandy, gravelly, silty sediments (~10%), mafic volcanics, meta-basalts, phyllite-greenstone belt with BIF (~9%). The remaining 10% of groundwater abstraction points target aquifers characterized by the 8 remaining local geological units.

2.3 *Water sampling and laboratory analysis*

Water sampling was carried out at the end of dry season during October 2016. A total of 29 water samples were collected, of which 18 samples were taken from groundwater sources and 11 samples from surface

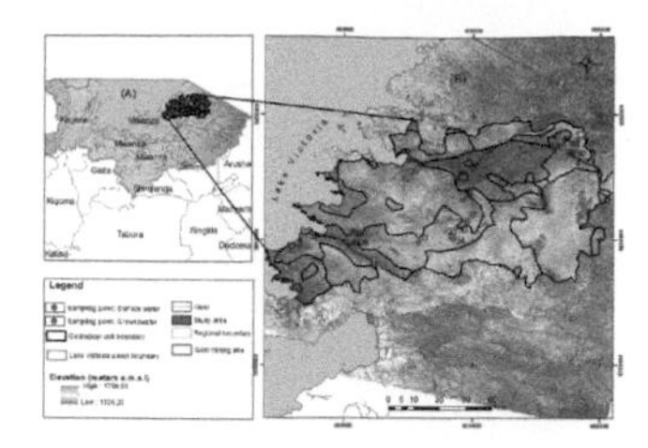

Figure 1. Study area (A) and water sampling locations (B).

water sources were collected (Fig. 1). The physio-chemical parameters such as, pH, temperature (T), electrical conductivity (EC), redox potential (Eh) and elevation (H) were measured in the field. Major anions were analyzed by ion chromatography (IC Dionex DX-120) in the Land and Water Resources Engineering laboratory at KTH Royal Institute of Technology. Major cations were determined by inductively coupled plasma-optical emission spectrometry (ICP-OES) at Linköping University in the Department of Thematic Studies.

2.4 *Creation of spatial database and data analysis*

ArcGIS software was used to create spatial database comprising the location and description of each water sample, physio-chemical parameters and major ions and As. The data analysis part involved calculating and mapping of summary statistics, i.e. minimum, maximum, average, and standard deviation.

3 RESULTS AND DISCUSSION

3.1 *Spatial exploration of water quality parameters with local geological settings*

The collected water samples were from abstraction points targeting aquifers with following sediment types: i) predominantly alluvial and eluvial sediments (aQ) with slightly alkaline pH (7.4) and high Eh (mean +416 mV); ii) migmatite-granitoid-meta-sediment complex (miNA) with neutral pH (7.0) and higher Eh (356.4 mV mean); and iii) volcano-sedimentary complex-Greenstone Belt with banded iron formation (BIF) with approximately neutral pH (6.9). The higher EC values between 715 and 843 μS/cm indicate that aquifer sediments originate from the parent rocks in Tanzanian Craton. Similarly, the higher mean Eh values between 356–416 mV suggest an oxidizing environment in all geologic units.

3.2 *Probability of occurrence of arsenic contamination in groundwater*

The probability of having contaminated aquifers was calculated based on number of samples with arsenic concentrations exceeding WHO guideline value constrained by local geologic units. Figure 2 shows a probability map of potential arsenic contaminated aquifers.

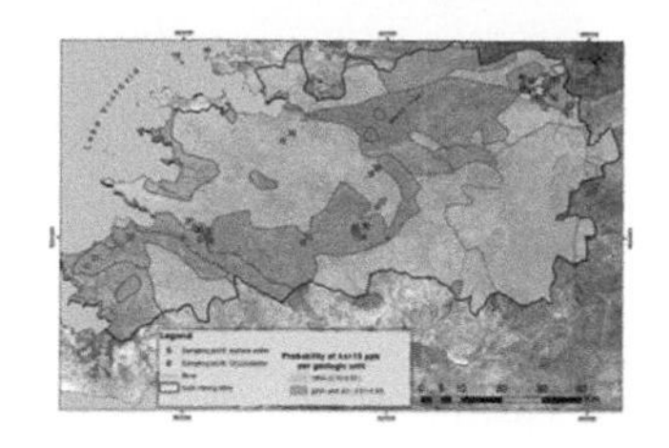

Figure 2. Probability map of arsenic contaminated drinking water sources in Lake Victoria Gold Fields in Mara region.

The most probable aquifers with As levels exceeding 10 $\mu g\ L^{-1}$ are found in the lithologic groups aQ and gsNA rocks/sediments (50–90%). Aquifers in the migmatite-granitoid-meta-sediment complex (miNA) indicate comparatively less likelihood of elevated levels of As in well water. However, this is just a preliminary observation based on the small sample size, and work is currently in progress to link the overall hydrogeochemical characteristics, such as major ions, As and other trace elements with the mapped geological units.

4 CONCLUSIONS AND RECOMMENDATIONS

Arsenic contamination in Lake Victoria Basin is a really problem in drinking water sources. The most targeted aquifers composed of migmatite-granitoid-metasediment complex and metasediments seem to have high levels of arsenic exceeding WHO guideline. The drilling practice during potable water supply should consider the type of geological units and sediments to avoid continual exposure to arsenic toxicity among Lake Victoria Basin communities. The behaviour of excess arsenic needs to be investigated with respect to seasonal variations and depth.

ACKNOWLEDGEMENTS

We acknowledge the Swedish International Development Cooperation Agency (Sida) for supporting the DAFWAT program (Contribution: 51170072).

REFERENCES

Ahoulé, D.G., Lalanne, F., Mendret, J., Brosillon, S. & Maïga, A.H. 2015. Arsenic in African waters: a review. *Wat. Air Soil Poll.* 226(9): 302.

Kassenga, G.R. & Mato, R.R. 2008. Arsenic contamination levels in drinking water sources in mining areas in Lake Victoria Basin, Tanzania, and its removal using stabilized ferralsols. *Int. J. Biol. Chem. Sci.* 2(4): 389–400.

Lucca, E. 2017. Geochemical Investigation of Arsenic in Drinking Water Sources in Proximity of Gold Mining Areas in the Lake Victoria Basin, in Tanzania. MSc Thesis, TRITA SEED-EX 2017:25, KTH Royal Institute of Technology, Sweden, 98p.

Mnali, S. 2001. Assessment of heavy metal pollution in the Lupa gold field, SW Tanzania. *Tanzania J. Sci.* 27(2): 15–22.

Environmental Arsenic in a Changing World –
Zhu, Guo, Bhattacharya, Ahmad, Bundschuh & Naidu (Eds)
ISBN 978-1-138-48609-6

Spatial variability of trace elements with Moran's I Analysis for shallow groundwater quality in the Lower Katari Basin, Bolivian Altiplano

I. Quino[1,2], O. Ramos[1], M. Ormachea[1], J. Quintanilla[3] & P. Bhattacharya[2,3]
[1] *Laboratorio de Hidroquímica, Instituto de Investigaciones Químicas, Universidad Mayor de San Andrés, La Paz, Bolivia*
[2] *KTH-International Groundwater Arsenic Research Group, Department of Sustainable Development, Environmental Science and Engineering, KTH Royal Institute of Technology, Stockholm, Sweden*
[3] *International Center for Applied Climate Science, University of Southern Queensland, Toowoomba, Queensland, Australia*

ABSTRACT: The southeastern part of the Titicaca Lake near the Cohana Bay in the Bolivian Altiplano, has environmental problems caused mainly by urban and industrial wastes upstream of the Katari Basin and by natural geological conditions. This environmental condition has generated an increase in the concentrations of some trace elements in the groundwater. The Moran's I statistic was used with LISA (Local Indicators of Spatial Association) method to know the spatial autocorrelation and the spatial variability of As, Sb, B, Al, Mn and F. Arsenic and antimony are the main pollutants due to natural geological conditions and boron due to the anthropogenic activities. Almost half of all the shallow groundwater samples exceeded the WHO and NB-512 guideline values mainly for antimony, boron and arsenic, whereby the spatial distribution of these trace elements in groundwater raises a significant concern about drinking water quality.

1 INTRODUCTION

Recent studies indicate that the Cohana Bay (Titicaca Lake) has environmental problems caused mainly by urban and industrial wastes upstream of the Lower Katari Basin (Fig. 1) and by natural geological conditions. The study area (91 communities) is located in the southeastern part of the Titicaca Lake, in the Bolivian Altiplano, with an area of 484 km^2 (Fig. 1). Drinking water is extracted from excavated wells (<10 m depth), these wells are also used for irrigation and animal consumption. The objective of this paper is to find the spatial variability using spatial autocorrelation of trace elements (As, Sb, B, Al, Mn and F), considering the hidrogeochemistry, the geographic information systems tools and the water quality of community consumption wells.

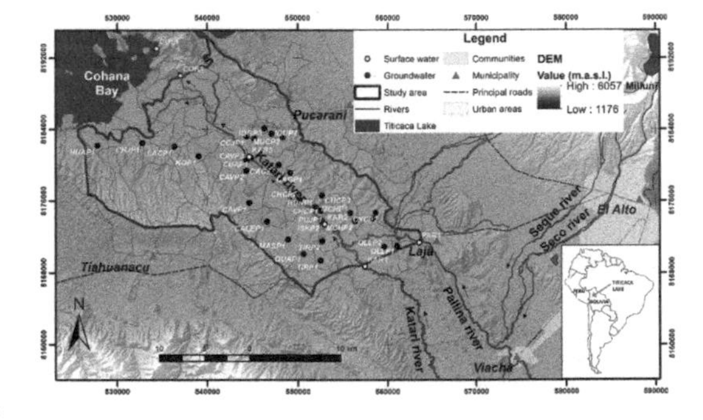

Figure 1. Study area and sampling points.

2 METHODS/EXPERIMENTAL

2.1 *Sampling and laboratory work*

Sampling was carried out during May 2015, 32 groundwater samples (32 communities) and 6 surface water samples were collected. The physicochemical parameters, pH, temperature (T), electrical conductivity (EC), redox potential (Eh) and total dissolved solid (TDS) were measured in the field with a multiparameter HANNA – HI 9828. Major anions were analyzed by Dionex ion chromatograph (ICS 1100) at the Environmental Chemistry Laboratory at Universidad Mayor de San Andrés in La Paz Bolivia. Major cations and TEs were determined by inductively coupled plasma-mass spectrometry at the Mineral Laboratories Canada of the Bureau Veritas Commodities Canada Ltd. in Vancouver Canada.

2.2 *Data analysis*

The Aquachem software (4.0.264 Waterloo Hydrogeologic Inc, 2003) was used to evaluate the analyses results for water samples and the type of water was determined. With the 32 sampling points, an interpolation was made for each TE using the deterministic method Inverse Distance Weighted (IDW). The interpolations were classified, to each of the 91 communities were assigned mean values according to the

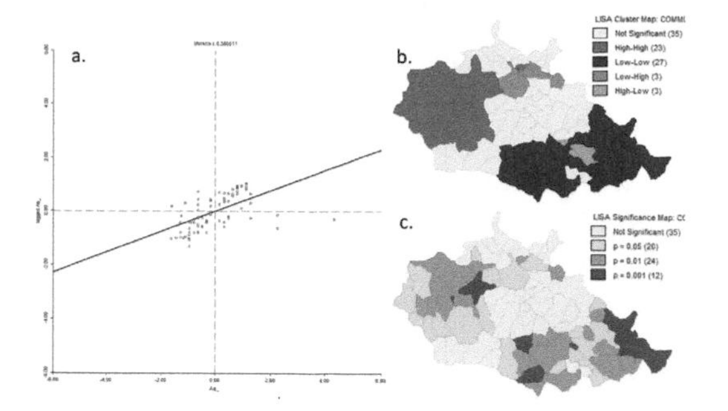

Figure 2. Global Moran's I statistiscal test (a), BiLISA Cluster Map (b) and BiLISA Significance Map (c) for As.

classified interpolation made. The percentage difference (PD) between the measured value and the value predicted by the interpolation was made for the 32 communities where the groundwater samples were taken. All these procedures were done in the ArcGIS 10.2.2 software.

The Moran's I statistic was used with LISA (Local Indicators of Spatial Association) method to know the spatial autocorrelation (SA) of each element. The global spatial dependence analysis (Global Moran's I statistiscal test), the local spatial dependence (BiLISA Cluster Map) and the significant spatial test (BiLISA Significance Map) were made in the GeoDa 1.12.01 software.

3 RESULTS AND DISCUSSION

3.1 *General hydrochemical characteristics*

The pH is slightly alkaline (7.8 mean), the Eh suggesting a moderately oxidizing environment (189.3 mV mean). High EC (279–7984 μS cm^{-1}) due to the lacustrine origin of sediments. 12.5% of the samples are of Ca-HCO_3 water type. 87.5% of the groundwater samples exceed the Bolivian regulation (NB-512) for Sb (5 μg L^{-1}), 56.2% for B (300 μg L^{-1}) and 50% for As considering the WHO guideline values (10 μg L^{-1}).

3.2 *Spatial variability of the trace elements*

The global spatial dependence analysis (Fig. 2a) gave values of Moran's I for As (0.38), Al (0.45), B (0.42), F (0.23), Mn (0.22) and Sb (0.19), these values indicate a positive autocorrelation and statistically significant for all cases, the p-value is 0.001 for the six cases. In the Cluster Map (Fig. 2b), spatial association statistic of Moran's I are presented, there are 23 High-high type communities, which are surrounded by communities with high concentration for As, 17 for Al, 17 for Mn, 16 for B, 16 for F and 13 for Sb, there are also 27 communities of Low-low type, which are communities with low As concentrations that are surrounded by other communities of low concentration of As, for B are 27, 24 for Mn, 17 for F, 16 for Al, and 15 for Sb. In the significance map for As (Fig. 2c), the probabilities of the relationship of contiguity (adjacency) occuring in a random way are shown. 56 significant are highlighted for As, 52 for B, 46 for Mn, 39 for F, 33 for Al and 33 for Sb. With a value that indicates an error probability of 0.001, 0.01 and 0.05 for all cases, in the rest the white color predominates indicating the absence of significance.

The spatial distribution of As shows a positive SA to the northwest (Cohana Bay) of the study area (Fig. 2b). The interaction between surface water and groundwater and the location of the wells around the volcanic formations and their dissolution could be the natural source of As. The B has a positive SA to the southeast of the study area, where the confluence of the Pallina and Katari rivers (alluvial, colluvio – fluvial deposits) is found. The aquifers are superficial and due to the agricultural activity present in the area could develop rapid processes of anthropogenic contamination (Molina *et al.*, 2001). The Sb shows high spatial autocorrelation northeast of the study area where Devonian rocks exist as part of the Bolivian antimony belt of the Eastern Cordillera (Arce-Burgoa & Goldfarb, 2009), this could explain the presence of Sb as the result mainly from weathering of carbonate rocks (Seal *et al.*, 2017) of the Devonian.

4 CONCLUSIONS

The occurrences of Al, B, As, F and Sb are not random and form significant groups in space. More than 50% of the samples exceed the NB-512 and WHO guidelines for Sb, B and As. Sb and As are the main pollutants due to natural geological conditions and boron due to the anthropogenic activities. The spatial distribution of dissolved Sb, B and As concentrations in groundwater raises a significant concern about drinking water quality.

ACKNOWLEDGEMENTS

We are thankful to the Swedish International Development Cooperation Agency (Sida) for the Sweden-Bolivia cooperation program on research capacity development through Contribution: 75000553.

REFERENCES

Arce-Burgoa, O.R. & Goldfarb, R.J. 2009. Metallogeny of Bolivia: *Society of Economic Geologists Newsletter*, 79(1): 8–15.

Molina, S.L., Sanchez, M.F., Pulido, B.A. & Vallejos, I.A. 2001. Consideraciones sobre el boro en las aguas subterraneas del Campo de Dalias (Almeria). *Geogaceta*, 29: 79–82.

Seal, R.R., II, Schulz, K.J. DeYoung, J.H., Jr., *with contributions from* David M. Sutphin, Lawrence J. Drew, James F. Carlin, Jr., & Byron R. Berger, 2017, Antimony, chap. C *of* Schulz, K.J., DeYoung, J.H., Jr., Seal, R.R., II, & Bradley, D.C., eds., Critical mineral resources of the United States—Economic and environmental geology and prospects for future supply: U.S. Geological Survey Professional Paper 1802, p. C1–C17.

Environmental Arsenic in a Changing World –
Zhu, Guo, Bhattacharya, Ahmad, Bundschuh & Naidu (Eds)
ISBN 978-1-138-48609-6

Influence of hydrothermal fluids enriched in As and F on the chemistry of groundwaters of the Duero Basin, Spain

E. Giménez-Forcada[1], S. Timón-Sánchez[1] & M. Vega-Alegre[2]
[1]*Instituto Geológico y Minero de España – IGME, Salamanca, Spain*
[2]*Departamento de Química Analítica, Universidad de Valladolid – UVA, Valladolid, Spain*

ABSTRACT: Chemical and isotopic data of groundwaters from the south edge of the Duero Basin have been interpreted by multivariate statistical analysis including HCA and PCA. The results suggest that waters enriched in arsenic, fluoride and other associated trace elements are alkaline Na-HCO_3 cold-hydrothermal waters, flowing through main faults of the basement.

1 INTRODUCTION

In some areas of the southern area of the Duero Basin (DB), Spain, naturally occurring arsenic and associated trace elements are present in concentrations exceeding the limits established for drinkable water. The study area is located in the Duero Basin where Cenozoic sediments from the basin contrast with the metasedimentary and igneous rocks of the Spanish Central System (SCS) (Fig. 1).

In this range, As and F are present in several rock-forming minerals from the crystalline bedrock. Arsenic is identified in sulfides (arsenopyrite), oxides (magnetite, ilmenite) and ferromagnesian silicates (olivine, pyroxene). Meanwhile, fluorine occurs in primary minerals as biotites, amphiboles, topaz and apatite. Both elements and other trace elements (Mo, V, Cr and U) constitute a suite of incompatible elements, which have difficulty in entering lattice sites of the minerals during the fractional crystallization of magma, and therefore are concentrated in the fluid phase.

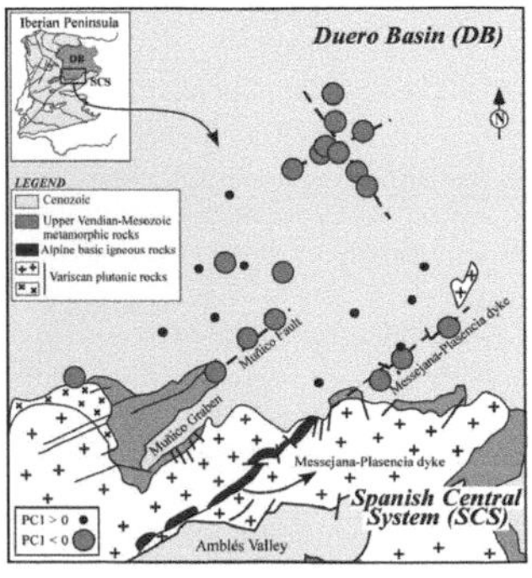

Figure 1. Location of the study area and distribution of PC1 scores.

The aim of this research is to gain knowledge of the geological environment controlling As and F contents in groundwaters of the south edge of the DB, using multivariate statistical tools as Hierarchical Cluster Analyses (HCA) and Principal Components Analyses (PCA) for data interpretation.

2 METHODS

Twenty-one physico-chemical parameters, including temperature, pH, electrical conductivity, redox potential, alkalinity, major anions and cations, natural isotopes and trace elements (As and F, but also Cr, Mo, V and U), were determined in 34 groundwater samples collected from springs, wells and boreholes located in the study area (Ávila province, Spain).

Water temperature, pH, electrical conductivity (EC, 25°C), and oxidation-reduction potential, ORP, were recorded on site. The ORP measurements were corrected for temperature and referenced to the SHE potential. Groundwater samples were filtered (0.45 μm) in situ into polyethylene bottles. Those collected for cation analysis were acidified to 1% v/v with HNO_3 (65%). Major anions and cations and trace elements were determined in the IGME laboratories by standard methods described elsewhere (Giménez-Forcada & Smedley, 2014). Electrical charge imbalances were, in all cases, less than 3%.

3 RESULTS AND DISCUSSION

The HFE-Diagram (Fig. 2) shows that samples with the highest concentrations in arsenic and fluoride correspond mainly to Na-HCO_3 waters.

Correlations of As with F and other hydrochemical variables were uncovered by HCA and PCA. HCA dendrogram shows two main families of variables (Fig. 3). One linked to major chemistry and those associated

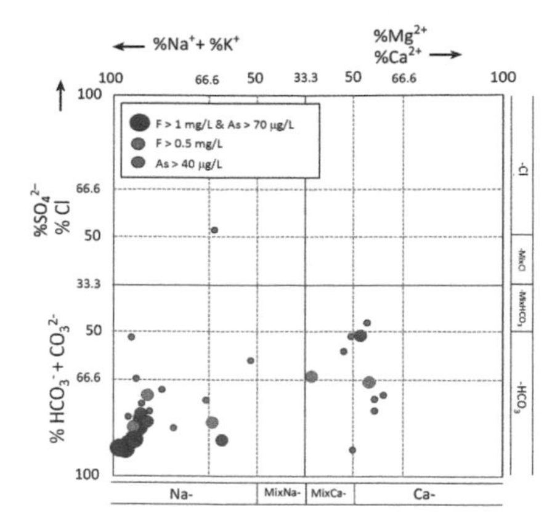

Figure 2. Representation of water samples in the HFE diagram (Giménez-Forcada & Sánchez 2014), modified.

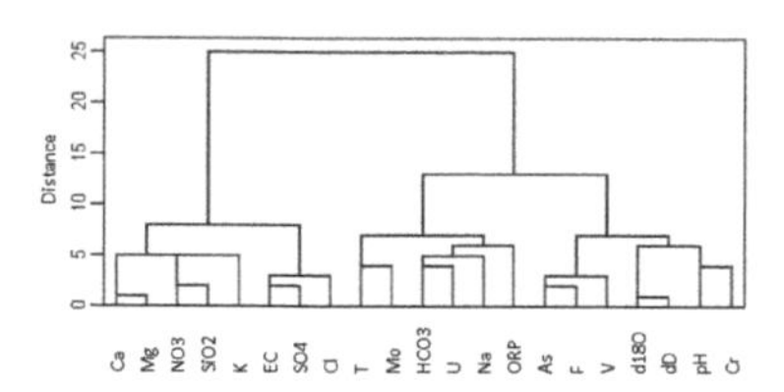

Figure 3. Dendrogram of chemical parameters obtained by HCA using the Ward linkage method.

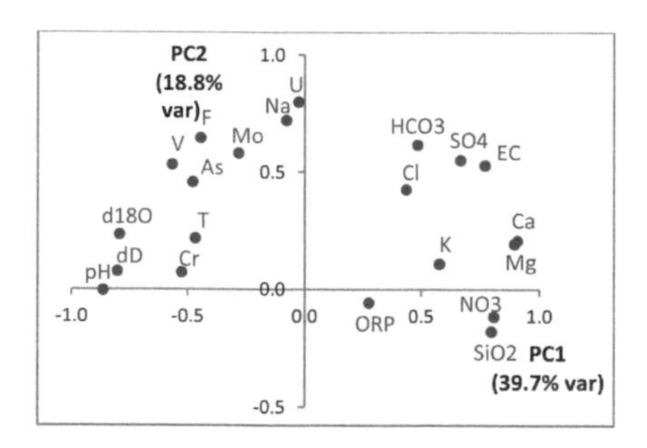

Figure 4. Loadings of the first two principal components obtained by PCA.

with $NaHCO_3$ flows. PCA corroborates this association (Fig. 4). In fact, the first principal component, PC1 (which explains the 39.7% of variance) differences clearly the two main groups referred. These correlations suggest that there are alkaline Na-HCO_3 water flows feeding the basin. These waters, enriched in As, F, V, Cr, U and Mo are characterized by a moderate temperature (cold-hydrothermal waters, 18°C–19°C) and a singular signature of $\delta^{18}O$ and $\delta^{2}H$ (the most negative values of all both parameters).

The correlation between As and other trace elements suggests they have similar geogenic sources and are mobile under similar hydrogeochemical conditions. These sources include igneous-metamorphic bedrocks, mineral occurrences as well as geothermal fluids.

Scores of samples on PC1 have been represented in Figure 4. The highest values are associated to relevant faults and their prolongation in the basin, showing a distribution of As, F and other trace elements controlled by structural features.

Previous works in the area have established the structural control of As distribution in DB, and the correspondence of high contents of F, B and Mo with the highest concentration of As in alkaline Na-HCO_3 groundwaters (Giménez-Forcada & Smedley, 2014).

Fluids flowing through major faults of the DB basement could be a relevant source of As in the study area. Therefore, it seems likely that the occurrence of As and associated trace elements derives at least partially from enriched deep hydrothermal fluids. Arsenic associated with geothermal waters has been reported in several parts of the world (Smedley & Kinniburgh, 2002), and F is recognized as a mobile element under high-temperature conditions and is abundant in hydrothermal solutions (Edmunds & Smedley, 2013).

Without neglecting other sources and processes, the influence of cold-hydrothermal waters enriched in several trace elements and associated with the fissured aquifers from SCS, which in turn form the basement of the DB, could be relevant.

4 CONCLUSIONS

The high concentrations of As and F in groundwaters from the DB south bank could be explained considering inputs of hydrothermal fluids flowing through main faults of the basement. This process does not exclude other possible sources as water-rock interaction processes.

ACKNOWLEDGEMENTS

This work was supported by the Geological Survey of Spain (IGME). HidroGeoTox (Research Project Ref. IGME-2303) and by the Junta de Castilla y León (Research Project Ref. VA291U14/Ref. IGME 2474 – As Cega).

REFERENCES

Edmunds, W.M. & Smedley, P.L. 2013. Fluoride in natural waters. In: O. Selinus, (ed.) *Essentials of Medical Geology, Second Edition*. Springer, 311–336p.

Giménez-Forcada, E. & Sánchez San Román, F.J. 2014. An excel macro to plot the HFE-Diagram to identify sea water intrusion phases. *Groundwater* 53(5): 819–824.

Giménez-Forcada, E. & Smedley, P.L. 2014. Geological factors controlling occurrence and distribution of arsenic in groundwaters from the southern margin of the Duero Basin, Spain. *Environ. Geochem. Hlth*. 36(6):1029–1047.

Smedley, P.L. & Kinniburgh, D.G. 2002. A review of the source, behaviour distribution of arsenic in natural waters. *Appl. Geochem*. 17: 517–568.

1.2 Origin and reactivity of organic matter in high arsenic groundwater systems

Environmental Arsenic in a Changing World –
Zhu, Guo, Bhattacharya, Ahmad, Bundschuh & Naidu (Eds)
ISBN 978-1-138-48609-6

Arsenic methylation and its relationship to abundance and diversity of arsM genes in composting manure

W.W. Zhai, X.J. Tang & J.M. Xu
Institute of Soil and Water Resources and Environmental Science, College of Environmental and Resource Sciences, Zhejiang Provincial Key Laboratory of Agricultural Resources and Environment, Zhejiang University, Hangzhou, P.R. China

ABSTRACT: In this study, two pilot-scale pig manure composting piles were constructed for a systematic investigation of arsenic (As) methylation during manure composting. Microbial community composition, as well as the abundance and diversity of arsM genes were monitored using real-time PCR (qPCR) and amplicon sequencing of both 16S rRNA and arsM genes. Results show an overall accumulation of methylated As occurring during 60 day-composting time. The arsM gene copies increased gradually over time and were correlated positively to the concentrations of methylated As. 16S rRNA gene sequencing and arsM clone library analysis confirmed that the high abundance and diversity of arsM genes shared the same known As-methylating microbes, including *Streptomyces* sp., *Amycolatopsis mediterranei* and *Sphaerobacter thermophiles*, which were likely involved in the methylation process. These results demonstrated that As methylation during manure composting is significant. For the first time, the linkage between As biomethylation and the abundance and diversity of the arsM functional gene in composting manure was established.

1 INTRODUCTION

Arsenic (As)-based feed additives are commonly used in the poultry and livestock industry. Not readily absorbed in animal tissues, almost all the fed As is excreted without attenuation in manure at concentrations up to 300 mg kg^{-1} (Kiranmayi *et al.*, 2015). Methylation of As is normally regarded as one of the main detoxification pathways for As in environment, which is catalyzed by S-adenosylmethionine methyltransferase encoded by *arsM* genes (Qin *et al.*, 2006). Although the mechanism of microbial As methylation is known and arsM genes have been detected in various environments, there remains a limited understanding of how the abundance and diversity of arsM genes correlate with the methylation process during manure composting.

2 METHODS/EXPERIMENTAL

2.1 *Composting experiments and sampling*

Two independent pilot-scale (2.5 m × 1.8 m × 0.75 m in length, width and height) pig manure compost piles were set up in Hangzhou, China, containing on average 1750 ± 60 μg kg^{-1} As (dw, n=3). The first manure compost pile (MC1) was composed of 1,200 kg pig manure and 600 kg sawdust for optimal C/N ratio and water content. The second manure compost pile (MC2) contained 1,200 kg pig manure and 600 kg sawdust mixed with burned rice straw. The moisture contents of the composting sites were maintained at approximately 65% by sprinkling water once every two days. The compost piles were turned over and mixed once every two days in the first month, and once every four days in the second month for aeration. The whole composting process lasted for 60 d, and 2 kg samples were collected on day 1, 5, 15, 25, 35, 45 and 60.

2.2 *Analysis method*

Concentrations of As species were measured by High-Performance Liquid Chromatography Coupled with Inductively Coupled Plasma Mass Spectrometry (HPLC-ICP-MS, NEXION300XX, PerkinElmer, Inc., USA). Total DNA was extracted and copy numbers of arsM gene in the compost samples were estimated by qPCR. 16S rRNA gene was amplified, then sent for sequencing using Illumina Miseq sequencing platform (Miseq, Illumina Inc., USA). Four samples (15-MC1, 15-MC2, 60-MC1 and 60-MC2) were selected for the construction of *arsM* gene clone libraries. The detail were described elsewhere (Zhai *et al.*, 2017).

3 RESULTS AND DISCUSSION

3.1 *The change of As species during composting*

The total concentrations of methylated As species increased more rapidly during the mesophilic and thermophilic phases, while only small increases during the maturing phase. The concentrations of methylated As species represented 37% (MC1) and 35% (MC2) of total As concentration by day 60, clearly indicating As methylation during manure composting. Methylated As species analysis also showed that MMA

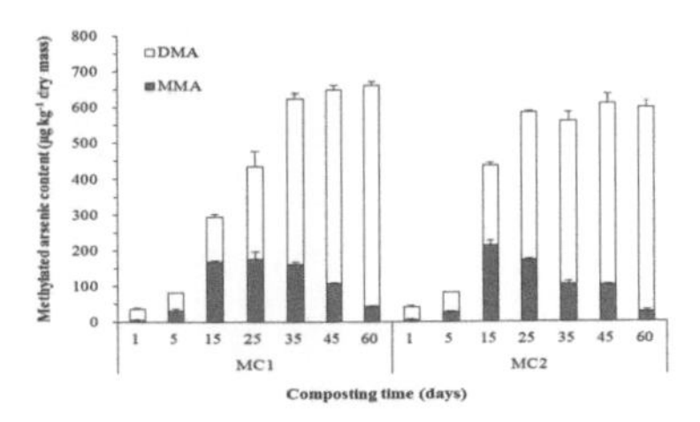

Figure 1. Changes in concentrations of methylated As (MMA and DMA) in the two compost piles.

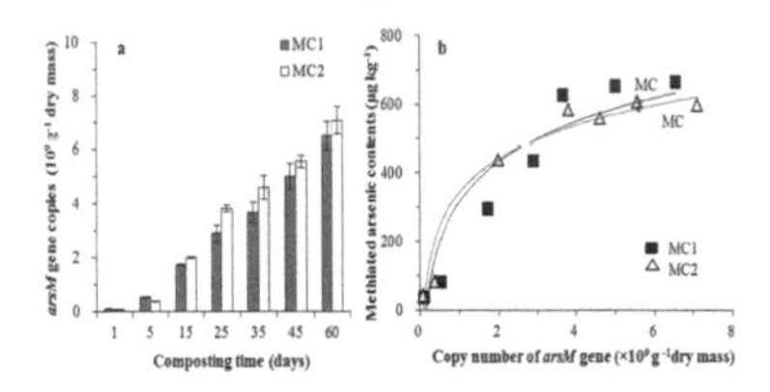

Figure 2. Plot of arsM gene copies (a) and plot of methylated As concentration versus arsM copies in two compost piles (b).

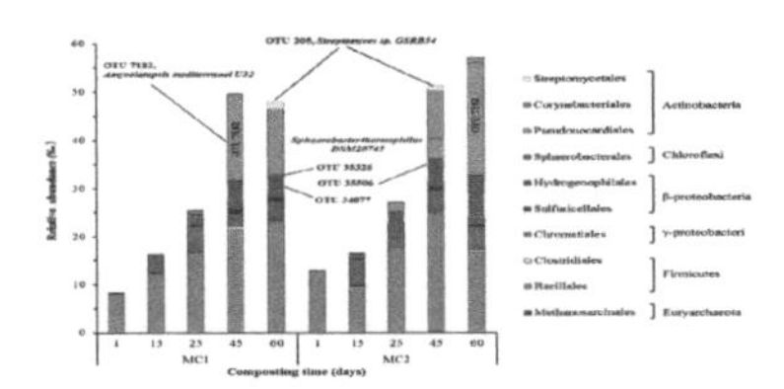

Figure 3. Changes and taxa of selected 16S OTUs related to As methylation.

content in both compost piles peaked during the thermophilic phase (day 5–42) and dropped rapidly during the maturing phase (day 43–60). In contrast, DMA concentrations increased steadily in both compost piles over the composting period (Fig. 1), indicating the transformation of MMA to DMA. And the conversion of DMA to TMA is the rate limiting step in As methylation. Therefore, DMA commonly accumulates in environment samples.

3.2 *Copy numbers of bacterial arsM genes*

The abundance of arsM genes in MC1 and MC2 as a function of composting time increased gradually from $\sim 0.1 \times 10^9$ to $\sim 6.8 \times 10^9$ copies g-1 dry mass in both piles (Fig. 2a). Further, the sum of MMA and DMA concentrations at different time points was found to correlate strongly with the *arsM* gene copy numbers (Fig. 2b). Considering the arsM gene is the key functional gene responsible for microbial As methylation, the positive relationship between the concentrations of methylated As and the copy numbers of arsM genes confirmed the As methylation ability in the manure compost piles.

3.3 *Abundance and biodiversity of arsM genes*

We aligned the 16S rRNA sequences from the compost piles against 16S rRNA sequences from microbes

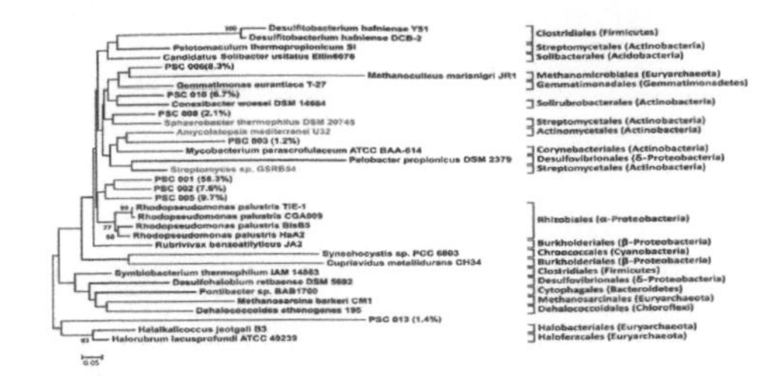

Figure 4. Neighbor-joining analyses of arsM sequences retrieved from composting samples.

containing an arsM gene. Eighty-three OTUs were identified as matching (similarity $\geq$95%) and their relative abundances clearly increased with composting time (Fig. 3). A Neighbor-joining tree of the 8 most abundant *arsM* PSCs ($>$1% relative abundance) was constructed with selected references. Phylogenetic analysis of arsM clone showed the high diversity of arsM gene in the composting samples (Fig. 4). The present study showed putative arsM affiliated with a wide range of phylogenetic taxa that were present in all the composting samples.

4 CONCLUSIONS

In conclusion, two pilot-scale pig manure compost piles were established and an accumulation of methylated As was revealed over the composting process. By qPCR, 16S rRNA sequencing and clone libraries, an increasing in abundance and diversity of arsM genes in composting pig manure were confirmed.

ACKNOWLEDGEMENTS

This work was financially supported by the Provincial Public Technology and Applied Research Projects by Science and Technology Department of Zhejiang Province (2014C33020), the National Key Technology Research and Development Program of the Ministry of Science and Technology of China (2014BAD14B04), Zhejiang Provincial Natural Science Foundation of China (LR13D010001) and Fundamental Research Funds for the Central Universities.

REFERENCES

Kiranmayi, P. M., Asok, A. & Lee, B. 2015. Organoarsenicals in poultry litter: Detection, fate, and toxicity. *Environ. Int.* 75: 68–80.

Qin, J., Rosen, B. P., Zhang, Y., Wang, G.J., Franke, S., & Rensing, C. 2006. Arsenic detoxification and evolution of trimethylarsine gas by a microbial arsenite S-adenosylmethionine methyltransferase. *Proc. Natl. Acad. Sci.* U.S.A. 103(7): 2075–2080.

Zhai, W.W., Wong, M.T., Luo, F., Hashmi, M.Z., Liu, X.M., Edwards, E.A., Tang, X.J. & Xu, J.M. 2017. Arsenic methylation and its relationship to abundance and diversity of arsM genes in composting manure. *Sci. Rep.* 7: 42198.

Environmental Arsenic in a Changing World –
Zhu, Guo, Bhattacharya, Ahmad, Bundschuh & Naidu (Eds)
ISBN 978-1-138-48609-6

Hydrogeological and geochemical comparison of high and low arsenic groundwaters in the Hetao Basin, Inner Mongolia

H.Y. Wang
Institute of Mineralogy and Geochemistry, Karlsruhe Institute of Technology, Karlsruhe, Germany

ABSTRACT: For deeply understanding the arsenic mobilization mechanisms of groundwater in Hetao Basin of Inner Mongolia, sediments and groundwater samples from two multi-level wells (up to 80 m) were analyzed. The results showed that the sediment As content from well with high As concentration (K1) is higher than the well with low As concentration groundwater (K2), with the average As concentrations 14.7 and 12.8 mg kg^{-1} respectively. Interestingly, we found a gray–black peat layer (around 28 m) from the K2, with arsenic concentration 322 mg kg^{-1}, while the TOC content and TS contents are up to 9.6% and 1.8%, respectively. The pyrite also was found in this layer. We conclude that the organic matter triggers the formation of arsenic–sequestering sulphides under strongly reducing conditions, therefore immobilization of As into the sediments. By bonding arsenic in this way, the peat layer occurred in the aquifer plays an active role for arsenic immobilization into the sediments.

1 INTRODUCTION

The Hetao Basin in the Inner Mongolia is one of most serious As-polluted area in China (Guo *et al.*, 2014; Liu *et al.*, 2017). Groundwater with high As concentration (up to 879 $\mu g L^{-1}$) has been widely found in the Hetao Basin, with 76000 people exposed in 35 villages in 2002 (Liu *et al.*, 2017).

For better understanding the mechanisms of As mobilization, we drilled two boleholes up to 80 m in Hangjinhoujin, one countryside highly affected by As pollution. The sediments from different layers were analyzed based on mineralogy and chemistry characters. And the relevance was established between the sediments property and groundwater As concentration.

2 METHODS

2.1 *Study area*

The Hetao Basin is located in between Yellow river to the south and Langshan to the north. Lacustrine deposition and frequent channel changes caused the patchy sediments distribution and allowed organic matter to accumulate in the sediments during Pleistocene and Holocene period. Hangjinhouqi country located in the west of Hetao basin is one of the serious arsenic-affected areas where the groundwater was mainly used for irrigation system.

2.2 *Sediments characterization*

Two boreholes up to 80 m were drilled at the Hangjinhouqi in October of 2015. The XRF analysis was used to determine the mineral phases and major chemical components. The total carbon components (TC) and TS as well as organic matter (TOC) were qualified using carbon-water analyzer. The EC (electric conductivity) value of soluble contents and pH values of sediments were also measured by method of solid-to-liquid ratio of 1:5.

2.3 *Groundwater sampling and analysis*

After sediment sampling, the multi-level wells were installed. The water was pumped from the different depth. Groundwater parameters including temperature (T), pH, redox potential (ORP), electrical conductivity (EC) were measured in the field. The major cations including As and anions were measured by ICP-MS and ICP-OES respectively.

3 RESULTS AND DISCUSSION

3.1 *Sediment geochemistry*

The average sediment grain sizes from K2 borehole are larger than K1, while the medium-coarse sands distribute in 35–40 m and 51–60 m from K2 borehole, they are rarely found in K1. Compared with K1, clay layers are also widely distributed in K2 with yellow-brown layers found in the depth around 16 m, 26 m, 41 m, 51 m, 59 m, 65 m respectively, while only two clay layers appeared in K1 with the depth around 3 m above the aquifer, and 40 m, respectively.

The sediment As contents ranged from 5.8 to 28.9 mg kg^{-1} (Table 1) with higher As contents of clay samples. Therefore, the clay layers intersected in the aquifers can be a major As sink or source. The As contents are well correlated with Fe contents in the sediments, rather than TOC and TS. So the As occurred in the aquifer are mostly from the

Table 1. Chemical compositions of sediment samples collected from two boreholes in the western Hetao basin a: K1, b: K2 (except one sample with As content up to 322 mg kg^{-1}).

a:

Value range	Fe_2O_3 (wt%)	MnO (wt%)	As (mg kg^{-1})	TOC (%)	TIC (%)	TS (mg kg^{-1})
Maximum	7.15	0.120	40.9	0.60	2.15	5125
75% Q	5.76	0.092	21.0	0.37	1.64	512
Median	3.39	0.047	12.1	0.13	0.97	287
25% Q	2.14	0.034	7.71	0.06	0.63	231
Minimum	1.61	0.023	5.82	0.04	0.32	140
Average	3.80	0.060	14.7	0.20	1.14	573
R^2_{Fe}	–	0.93	0.75	0.61	0.80	0.0052
R^2_{As}	–	0.61	–	0.35	0.40	0.0375

b:

Value range	Fe_2O_3 (wt%)	MnO (wt%)	As (mg kg^{-1})	TOC (%)	TIC (%)	TS (mg kg^{-1})
Maximum	6.85	0.109	28.9	0.90	1.92	5438
75% Q	5.17	0.083	16.1	0.47	1.74	685
Median	3.59	0.056	11.9	0.20	1.16	322
25% Q	2.16	0.029	6.59	0.06	0.53	226
Minimum	1.68	0.025	5.03	0.05	0.31	179
Average	3.69	0.058	12.8	0.2	1.10	724
R^2_{Fe}	–	0.97	0.74	0.48	0.7	0.16
R^2_{As}	–	0.66	–	0.50	0.6	0.35

dissolution of Fe/Mn (hydro) oxides or the Fe(Mn) oxides-TOC-As complexes.

The average As content (14.7 mg kg^{-1}) in the K1 borehole is higher than in the K2 (12.7 mg kg^{-1}) (Table 1). It may be an important reason for higher groundwater As concentration from K1 well.

Interestingly, we found a gray-black peat layer in the K2 borehole, The deep sand layer contains trees and plants debris extending from 25.8 to 28.4 m with the As and TOC contents up to 322 mg kg^{-1} and 9.6% respectively. XRD spectra was used to investigate the mineral compositions, revealing that the pyrite minerals appeared in this layer. Under reducing conditions, the S-enriched peat can implicate as a fuel for reductive dissolution of arsenic-bearing iron (hydro) oxides, then accelerate the formation of arsenic-contained pyrite as a result of sufides reacting with the Fe (II).

3.2 *Soluble salts in the sediments*

Generally, the clay and silt samples exhibited higher salinity than the sand. However, the sand layers in the depth of around 80 m in K1, 28 m and 69 m in K2 released high contents of salts due to the high TOC contents (Fig. 1a). The sample EC values near the land surface from K1 are much higher than from K2, indicating that the groundwater from K1 well experienced long time evaporation and caused the salts accumulation in the surface sediments. Contrasting with K1, the sediments from K2 experienced long-term flushing history with low EC values appearing at different aquifer sediments.

Compared with higher sediment As concentration from K1 borehole, the sediments from K2 have higher pH value (Fig. 1a). It may be due to that at alkaline

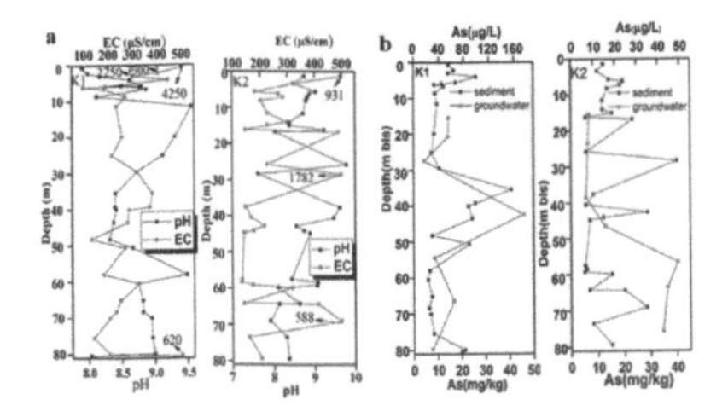

Figure 1. (a) EC value and pH of sediments varies with depth, (b) As concentrations of groundwater and sediments varies with dept.

conditions, dissolved Fe concentrations did not significantly increase upon reduction. Furthermore, the presence of soluble organics under alkaline conditions and the formation of iron oxyhydroxides-organic matter complexes could have retarded Fe reduction and the release of As into solution.

3.3 *Hydrogeochemistry*

The groundwater from K2 well showed higher pH, consisting with higher pH in the sediments. The total dissolved solids (TDS) range from 449 to 3136 mg L^{-1}, and decline with the depth. The groundwater from K1 well experienced long-time water-rock interaction or evaporation process with higher EC value. The groundwater samples from K1 well are mostly Na-Cl-SO_4 or Na-SO_4-Cl type, while samples from K2 well are mostly Na (Ca)-HCO_3-SO_4 type.

The ORP values of both wells are similar under anoxic conditions except the depth of 15–16 m in K2 well. The groundwater As concentrations from different depths of K2 well are lower than 50 μg L^{-1}, and increased with depth (Fig. 1b). Even though the sediment As content in the depth of 28 m from K2 borehole is up to 322 mg L^{-1}, the As concentration in the groundwater according to this depth is lower than 10 μg L^{-1}. Therefore, we conclude that the peat layers accelerate the As immobilization into the sediments.

The groundwater As concentrations in K1 vary with the sediment As contents. The As concentration in the depth of 42.5 m is 174 μg L^{-1} while the As content in the upper clay sediment is up to 25 mg kg^{-1} (Fig. 1b). Under reducing conditions, the DOC released from clay layers above this aquifer may afford the electrons for release of Fe (hydro) oxides.

REFERENCES

Guo, H.M., Wen, D.G., Liu, Z.Y., Jia, Y.F. & Guo, Q. 2014. A review of high arsenic groundwater in mainland and Taiwan, China: distribution, characteristics and geochemical processes. *Appl. Geochem.* 41: 196–217.

Liu, N.J., Deng, Y.M. & Wang, Y. 2017. Arsenic, iron and organic matter in quaternary aquifer sediments from western Hetao Basin, Inner Mongolia. *J. Earth Sci.* 28: 473–483.

Environmental Arsenic in a Changing World –
Zhu, Guo, Bhattacharya, Ahmad, Bundschuh & Naidu (Eds)
ISBN 978-1-138-48609-6

Effects of sediment properties and organic matter on biomobilization of arsenic from aquifer sediments in microcosms

Z. Xie, M. Chen, J. Wang, X. Wei, F. Li, J. Wang & B. Gao
School of Environmental Studies, China University of Geosciences, Wuhan, Hubei, P.R. China

ABSTRACT: In this study, the reducing capacity of the strain *Bacillus cereus* in the two sediments from Datong Basin and Jianghan Plain was investigated by examining the concentrations of As and Fe and the changes of pH and Eh. Results showed that glucose contributed to the mobilization of As and Fe, and As in the sediment from Jianghan Plain was released more easily. It was found that *Bacillus cereus* changed the pH and Eh in the surroundings, and promoted the reductive dissolution of As(V) and Fe(III) in the sediments to enhance As in groundwater.

1 INTRODUCTION

Arsenic (As) is an important element for human carcinogen. More and more people are suffering arsenism for long-term drinking arsenic-contaminated groundwater. Mobilization and enrichment of arsenic in groundwater were affected by several environmental factors.

The objectives of this study were to evaluate the influences of sediment properties and organic matter on bacterial migration of arsenic.

2 MATERIALS AND METHODS

2.1 *Bacterial culture*

The strain *Bacillus cereus* (*B. cereus*) was isolated from the aquifer sediment. The bacterial strain was cultured in modified minimal salt medium (MSM) in an incubator in dark at 25°C for 4 d. Salts supplied per liter of MSM were: 0.14 g KH_2PO_4, 0.50 g KCl, 1.00 g NaCl, 0.13 g $CaCl_2$ and 0.62 g $MgCl_2 \cdot 6H_2O$. The bacterial cells were harvested by centrifugation at 4000 rpm for 20 min at 4°C. The cell pellets were resuspended and cultured in the fresh MSM for the next experiments.

2.2 *Sediment characterization*

Total organic C (TOC), As, Fe, pH and mineral size distribution of sediment particles were determined as the methods described in Simmler *et al.* (2016).

2.3 *Microcosm experiments*

Two sediment samples were collected from high arsenic aquifers at Datong Basin and in Jianghan Plain. The two samples were named as SDB and SJP, respectively. The microcosms were prepared in 500 mL Erlenmeyer flasks with rubber stoppers. In N_2-filled glove box, 10 mL of the bacterium in exponential growth phase were incubated in the Erlenmeyer flask with 0.4 g glucose or 0.5 g fulvic acid, 20 g sieved sediment and 200 mL deionized water. Sediment, Erlenmeyer flask and deionized water were autoclaved prior to use. All the Erlenmeyer flasks were incubated in nitrogen atmosphere, in dark at 25°C. The control groups without organic matter or bacterium were run under the same conditions.

2.4 *Chemical analyses of the aqueous phase*

Before collecting the samples, pH and Eh were measured directly in the water-sediment system in the N_2-filled glove box. Dissolved As(III), As(V), Fe(II) and Fe(III) in the solutions from the microcosms were determined. At regular time intervals, 5 mL of sample were removed from each reaction vessel and centrifuged. As(III) and As(V) were separated from the supernatant by a strong anion exchange column and measured by hydride generation-atomic fluorescence spectrometry. Fe(II) and Fe(III) were determined according to the method of Han *et al.* (2011).

3 RESULTS AND DISCUSSION

3.1 *Sediment characterization*

The content of sand in the two samples was high as shown in Table 1. Nevertheless, sand in SJH had higher percentage. The pH values of SDB and SJP were in slightly alkaline. The contents of TOC in sediments had obvious difference. In addition, the contents of As and Fe in SDB exceeded in SJH. The results indicated that the two sediment samples had different characterization.

3.2 *Speciation and mobilization of As and Fe*

The concentrations of dissolved As(V), As(III), Fe(III) and Fe(II) in aqueous phase in the microcosm were presented in Figure 1. From the figure, we can see that all the concentrations increased over time under the condition of bacterial activities. The concentrations of

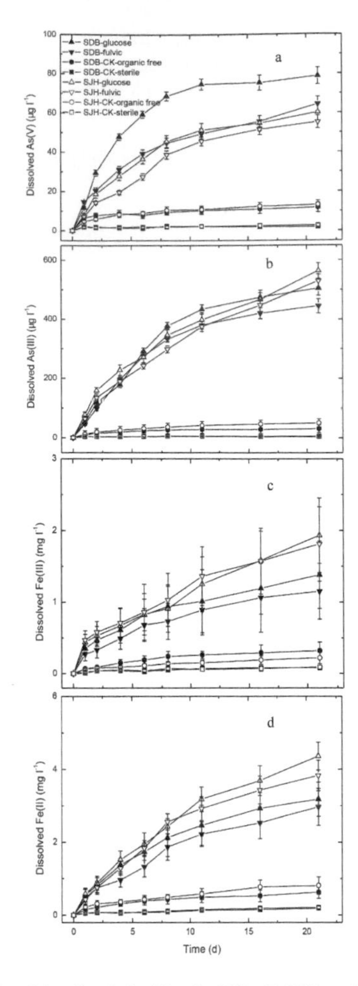

Figure 1. Dissolved As(V), As(III), Fe(III) and Fe(II) in aqueous phase in the microcosm. a) As(V); b) As(III); c) Fe(III); d) Fe(II).

Table 1. Physicochemical characteristics of sediment samples.

	SDB	SJH
pH	7.9	7.6
TOC (%)	0.19	0.56
As (mg kg^{-1})	13.3	9.8
Fe_2O_3 (%)	2.89	2.23
Sand (50–2000 μm) (%)	61.7	76.8
Silt (2–50 μm) (%)	27.7	18.3
Clay (<2 μm) (%)	10.6	4.9

As(III) and Fe(II) exceed those of As(V) and Fe(III) at the same sampling time. Additionally, the concentrations except As(V) in SJH microcosm were higher than in SDB microcosm. Glucose contributes to the mobilization of As and Fe. The results suggested that *B. cereus* promoted the reductive dissolution of As(V) and Fe(III) in the sediments.

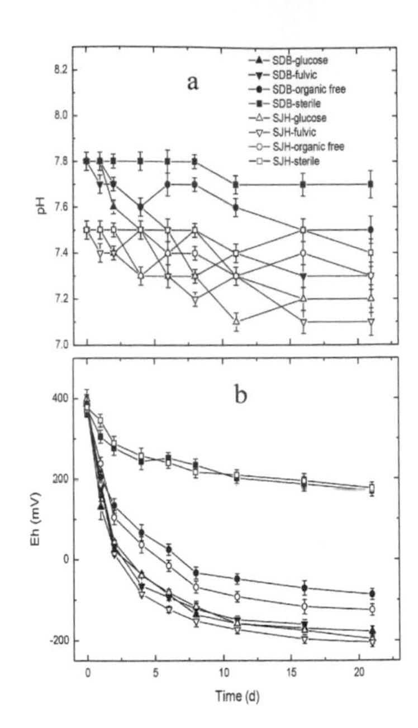

Figure 2. Change of pH and Eh. a) pH; b) Eh.

3.3 *Change of Eh and pH*

It was shown that *B. cereus* affected the pH and Eh of the microcosms microcosms (Fig. 2). The change of pH and Eh contributed to mobilization of As and Fe.

4 CONCLUSIONS

From this study, it was found that the strain *Bacillus cereus* changed the environment factors and promoted the reductive dissolution of As(V) and Fe(III) in the sediments. Then, As and Fe were released into groundwater from sediments.

ACKNOWLEDGEMENTS

This research work was financially supported by National Natural Science Foundation of China (Grant Nos. 41572230 and 41172219), and by the Grant for Innovative Research Groups of the National Natural Science Foundation of China (41521001).

REFERENCES

Han, X., Li, Y.L. & Gu, J.D. 2011. Oxidation of As(III) by MnO_2 in the absence and presence of Fe(II) under acidic conditions. *Geochim. Cosmochim. Acta.* 75: 368–379.

Simmler, M., Suess, E., Christl, I., Kotsev, T. & Kretzschmar, R. 2016. Soil-to-plant transfer of arsenic and phosphorus along a contamination gradient in the mining-impacted Ogosta River floodplain. *Sci. Total Environ.* 572: 742–754.

Environmental Arsenic in a Changing World –
Zhu, Guo, Bhattacharya, Ahmad, Bundschuh & Naidu (Eds)
ISBN 978-1-138-48609-6

Organic acid effect on arsenate bioaccessibility in gastric and alveolar simulated biofluid systems

S.Q. Kong[1,2], R.A. Root[3] & J. Chorover[3]
[1] *School of Environmental Studies and State Key Laboratory of Biogeology and Environmental Geology, China University of Geosciences, Wuhan, P.R. China*
[2] *Laboratory of Basin Hydrology and Wetland Eco-restoration, School of Environmental Studies, China University of Geosciences, Wuhan, P.R. China*
[3] *Department of Soil, Water and Environmental Science, University of Arizona, Tucson, AZ, USA*

ABSTRACT: The risk posed from incidental ingestion of arsenate-contaminated tailings may depend on sorption of arsenate to oxide surfaces in minerals. Popularly organic matter existing in tailings influenced bioaccessibility of arsenate. Arsenate adsorbed minerals were placed in simulated gastric and alveolar (*in vitro*) to ascertain the bioaccessibility of arsenate and changes in arsenate surface speciation caused by the biofluid systems. The effects of organic matter on arsenate bioaccessibility were investigated by adding oxalic and vanllic acid to biofluid systems. These results suggest that the bioaccessibility of arsenate is increased by organic matter and increasing reaction time.

1 INTRODUCTION

Potential exposure to arsenic (As) basically through ingesting or breath from tailings may pose a risk to human health. In physiological systems of the human body, some of the uptake arsenic may become soluble (Meunier *et al.*, 2010). That is the presence or the artificial addition of natural organic matter may influence the mobility of arsenic previously adsorbed onto soil or tailings particle surfaces (Redman *et al.*, 2002; Wang & Mulligan, 2009). The *in vitro* bioassays in this work were carried out in simulated biofluid reacted with several minerals and organic matter. Two synthetic systems included gastric and alveolar simulation biofluid. Ferrihydrite, scorodite and goethite are represented as typical iron oxide minerals in nature. Two representative organic matter, oxalic acid and vanillic acid, represented alkanes and aromatic hydrocarbons. Arsenate was operated in the assays as it is the common arsenic speciation in tailings.

2 METHODS AND EXPERIMENTAL

2.1 *In vitro bioassay*

A physiologically based extraction test (PBET) was used to estimate the bioaccessibility of arsenic from As(V)-containing minerals, as well as from samples mixed with and without natural organic matter.

2.2 *Estimation of arsenic bioaccessibility*

Arsenate bioaccessibility was calculated by dividing the arsenate concentration (μg/g) measured in the *in vitro* gastric solution or the *in vitro* alveolar solution by the total soil arsenate concentration ($\mu g\ g^{-1}$), as described by the following equation (Rodriguez & Basta, 1999; Pouschat & Zagury, 2006):

$$\text{As bioaccessibility (\%)} = (in\ vitro\ \text{As}) / (\text{Total As}) \times 100 \quad (1)$$

3 RESULTS AND DISCUSSION

3.1 *Arsenate bioaccessibility*

The *in vitro* bioassay showed that the bioaccessible As in the As(V)-ferrihydrite, scorodite and As(V)-goethite is intermediate (bioaccessibility: 6.96 to 15.77%). As(V)-ferrihydrite showed the lowest total As release content (G: 26.35 to 56.99 mg L^{-1} and A: 9.23 to 21.89 %), but the highest As bioaccessibility (G: 17.57 to 38.01% and A: 6.16 to 14.60%). In the study, organic carbon concentration (~0.1 wt%) and minerals levels (~1 wt%) suggest that mineralogical characteristics and organic matter content play important roles in the As bioaccessibility of the minerals. The effects of two representative organic matter, oxalic acid and vanillic acid represented alkanes and aromatic hydrocarbons, are similar to different minerals in gastric and alveolar biofluids. It shows that organic acid increases arsenate bioaccessibility and As(V) content in biofluid. With increasing As and Fe concentrations, the relative gastric and alveolar bioaccessibility decreased similarly, despite the different chemical conditions of the two extraction fluids (e.g., pH 1.8 vs 7.4, anaerobic and aerobic).

The sorption behavior of arsenate in soil is highly dependent on organic carbon, while the desorption

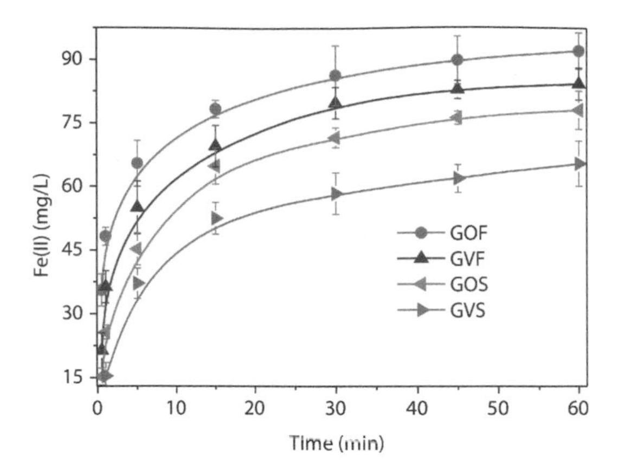

Figure 1. Ferrous concentration in simulated gastric biofluid with time.

is critically influenced by the high concentrations of amorphous and/or crystalline iron oxides. These factors control bioaccessibility after release of As from the crystalline mineral structure. In simulated bio-fluid systems, iron oxides dissolved at specific pH, leading to arsenic release from surface of minerals and formation of ferric hydroxide colloid. A part of released arsenate was re-adsorbed on solid surface through complex with hydroxyl of minerals (Fendorf *et al.*, 1997). Existing of organic matter influenced both the dissolution of minerals and arsenic re-adsorption.

3.2 *Species of Fe and As*

Speciation of As and Fe released were determined for three minerals. There was no detectable As(III) neither in the anaerobic gastric nor in the alveolar fluid. This means that As(V) was not reduced to more toxic As(III) even in anaerobic gastric phase. It is a good phenomenon considering human health. The assay indicated that the Fe that was solubilized during the alveolar phase was totally Fe(III). At the pH of 7.4 of the alveolar phase, Fe(II) would be soluble, but Fe(III) is insoluble. Speciation of Fe in gastric phase shows that detective Fe(II) occurred after organic matter addition (Fig. 1). Fe(II) was the Fe(III) reduction product by the organic substances (Theis & Singer, 1974).

3.3 *Stoichiometry of release Fe and As*

Stoichiometry of release Fe and As uncovered minerals dissolution and arsenic desorption from solid surface. From the results of Fe:As, it shows that the ratios of As(V)-Ferrihydrite and As(V)-goethite in gastric phase were beyond adsorption ratio values, illustrating that As bioaccessibility derived from a dominant process of dissolution minerals. However, in alveolar phase, the ratios implied arsenic release attributed to As desorption from minerals. As for scorodite, a ratio of iron to arsenic nearly 1:1 crystallization mineral, its disintegration led to stoichiometry of release Fe and As very close to 1:1, regardless in gastric and alveolar phases.

Comparison of the same solids in two biofluid phases, shows a stoichiometric increment Fe and As with adding organic matter in gastric phase, and ox-alic acid increased more than vanllic acid. The results means that organic matter inhibited the process of mineral dissolution, while facilitated arsenic desorption. Ionization of organic matter produced hydrogen ions, which inhibited producing hydrogen ions of original gastric phase at the same pH, and then hydroxide ions increased to maintain ion product, leading to increasing ferric hydroxide colloid, and finally improving mineral dissolution. A part of additional organic matter was adsorbed onto the surface of minerals. It was competitive with arsenic on surface activity sites, attributing to arsenic desorption, since organic matter occupied some adsorption sites and inhibited iron release from minerals.

4 CONCLUSIONS

Organic acid increased arsenic bioaccessibility in gastric and alveolar simulated biofluids. Iron species changed to ferrous iron to a certain extent. Arsenate was not transform to arsenite. The ratios of release Fe and As are 1:1 for the three minerals containing arsenate in the biofluids.

ACKNOWLEDGEMENTS

This research was supported by the National Natural Science Foundation of China (no. 41402214).

REFERENCES

Fendorf, S., Eick, M.J., Grossl, P. & Sparks, D.L. 1997. Arsenate and chromate retention mechanisms on goethite .1. Surface structure. *Environ. Sci. Technol.* 31(2): 315–320.

Meunier, L., Walker, S.R., Wragg, J., Parsons, M.B., Koch, I., Jamieson, H.E. & Reimer, K.J. 2010. Effects of soil composition and mineralogy on the bioaccessibility of arsenic from tailings and soil in gold mine districts of Nova Scotia. *Environ. Sci. Technol.* 44(7): 2667–2674.

Pouschat, P. & Zagury, G.J. 2006. In vitro gastrointestinal bioavailability of arsenic in soils collected near CCA-treated utility poles. *Environ. Sci. Technol.* 40(13): 4317–4323.

Redman, A.D., Macalady, D.L. & Ahmann, D. 2002. Natural organic matter affects arsenic speciation and sorption onto hematite. *Environ. Sci. Technol.* 36(13): 2889–2896.

Rodriguez, R.R. & Basta, N.T. 1999. An *in vitro* gastrointestinal method to estimate bioavailable arsenic in contaminated soils and solid media. *Environ. Sci. Technol.* 33(4): 642–649.

Theis, T.L. & Singer, P.C. 1974. Complexation of iron(II) by organic matter and its effect on iron(II) oxygenation. *Environ. Sci. Technol.* 8(6): 569–573.

Wang, S.L. & Mulligan, C.N. 2009. Effect of natural organic matter on arsenic mobilization from mine tailings. *J. Hazard. Mater.* 168(2–3): 721–726.

Environmental Arsenic in a Changing World –
Zhu, Guo, Bhattacharya, Ahmad, Bundschuh & Naidu (Eds)
ISBN 978-1-138-48609-6

Abundance, size distribution and dissolved organic matter binding of arsenic in reducing aquifer

Y.Q. Sun[1,2], H. Lin[3], S.B. Han[4], J. Sun[5], M. Ma[1,2], L.D. Guo[3] & Y. Zheng[2]
[1]*College of Engineering, Peking University, Beijing, China*
[2]*School of Environmental Science and Technology, Southern University of Science and Technology, Shenzhen, China*
[3]*School of Freshwater Sciences, University of Wisconsin-Milwaukee, WI, USA*
[4]*Center for Hydrogeology and Environmental Geology Survey, China Geological Survey, Baoding, China*
[5]*School of Earth Sciences, University of Western Australia, Perth, Western Australia, Australia*

ABSTRACT: The interactions between As and dissolved organic matter (DOM) have significant impacts on the environmental behavior of As. Among various interactions, the binary complexes of As bound to DOM and ternary complexes of As bound to DOM with iron (Fe) bridging, are critical to mobility, speciation, toxicity and bioavailability of As. Flow field flow fractionation (FlFFF) offline with high-resolution inductively coupled plasma mass spectroscopy (HR-ICPMS) were utilized to elucidate the As, Fe and DOM complexation in high arsenic, reducing groundwater from Yinchuan plain. The majority (>70%) of As was truly dissolved with molecular size <0.5 nm; while 13% of As and 37% of Fe were bound to colloidal DOM with molecular size between 16–64 nm. The results imply that As-Fe-DOM complex is distributed as moderately large colloids with molecular size between 16–64 nm. However, artefacts due to changes occurring during sample storage and transportation cannot be ruled out. Thus, field ultrafiltration under N_2 atmosphere will be employed in the future to validate FIFFF results.

1 INTRODUCTION

Exposure to elevated levels of As ($>10\,\mu g\,L^{-1}$) in groundwater has threatened the health of over 100 million people around the world. High As exposure is primarily due to the presence of As sources in aquifers and mobilization of As from sediment to groundwater in microbially mediated reducing environments. Yet the understanding of factors influencing the concentration and mobility of As, especially the role of dissolved organic matter (DOM), is incomplete. DOM is ubiquitous in natural water system and may affect the mobility of As through a number of mechanisms including binding and surface sorption.

McArthur *et al.*, (2001) postulated that microbial degradation of buried deposits peat leads to the reduction of iron (hydro)oxides (FeOOH) and release of Fe coated As in Bangladesh aquifers. Further, Mladenov *et al.*, (2009) suggested that DOM in reducing groundwater can serve to shuttle electrons and enhance the reductive dissolution of iron (oxy)hydroxide minerals, as evidenced by acceleration of the mobilization of dissolved As in Bangladesh aquifers and Gangetic aquifer of late Holocene. Additionally, DOM and As can form complexation which is relevant to As speciation, mobility and toxicity (Wang & Mulligan, 2006). The associations involve bridging metals and deprotonated functional groups of organic matter (i.e. carboxyl, hydroxyl and sulphydryl) (Sharma *et al.*, 2010).

Although many laboratory experiments have found evidence for As-(Fe)-DOM complexes in solution, the kinetic and thermodynamic parameters are difficult to compare due to variable experimental settings in these studies. Further, isolation and determination of this complex in groundwater systems remain scarce. Field (ultra) filtration was conducted for groundwater in the Hetao Basin and the result showed that 20%–40% As is combined with small-size organic colloids and 75% iron is in colloids (Guo *et al.*, 2011). Among various interactions, the binary complexes of As bound to DOM and ternary complexes among As, DOM and Fe, is a mechanism that has emerged to be relevant to mobility of As in reducing aquifers. Here, we explore this possibility as well as isolation of DOM bound of As from Yinchuan Plain where reducing groundwater hosts high levels of As.

2 STUDY AREA AND METHODS

2.1 *Study area*

Our field site is in Yinchuan alluvial plain where As levels reach $210\,\mu g\,L^{-1}$ and correlate with Fe concentrations $>1\,mg\,L^{-1}$ in groundwater in the shallow aquifer (depth 4–40 m). Sediment As and Fe concentrations are also high ($>10\,mg\,kg^{-1}$ and 1% Fe) respectively.

2.2 *Method*

Three groundwater samples were selected for analysis of colloidal size distributions by coupling on-line flow field-flow fractionation (FlFFF) with detectors including UV-absorbance and fluorescence. The two UV-absorbance detectors measured the absorbance at

254 nm and 280 nm respectively, representing the aromatic and protein content. Fluorescence detectors were set for two indexes FLD1 and FLD2 with Ex/Em of 350 nm/450 nm and 275 nm/340 nm, respectively. Selected elements (As, Fe, S, Mn, and Mg) was determined by HR ICP-MS in off-line fraction samples with RSD < 5%.

3 RESULTS AND DISCUSSION

3.1 *Abundance of selected elements in 0.7 μm, 0.4 μm and 0.2 μm filtrates*

For all three groundwater samples, negative ORP values, non-detectable nitrate, and strong sulfide smell during sampling suggest that the groundwater has reached Fe and sulfate reducing stage. Concentrations of selected elements in three samples for the 0.7 μm, 0.4 μm and 0.2 μm filtrates are shown in Table 1. The unusually high Fe concentrations in 0.4 μm filtrates for YC-1 and MLW-7 may have been due to the filter leakage.

3.2 *Abundance of elements from FlFFF*

The asymmetrical FlFFF system (AF2000, Postnova) was equipped with a 0.3 kDa polyether sulfone ultrafiltration membrane and a 0.35 mm spacer with 1 mL of injection volume. A mixed solution with 10 mM NaCl (aq) and 5 mM H_3BO_3 (aq) and pH of 8 (adjusted with NaOH), which was optimized to have the highest recovery and reasonable separation, was used as the carrier solution. Four fractionations with increasing molecular sizes were collected during FlFFF; smallest colloids of 0.5–8 nm, small colloids of 8–16 nm, medium size colloids of 16–64 nm and large colloids of 64–700 nm (Table 1). Concentration in the dissolved fraction (<0.5 nm) was estimated by subtracting the sum of all the colloidal fractions from the total As in <700 nm fraction (Table 1).

Dissolved fractions accounted for the majority of As in YC-1, YC-5 and MLW-7 yet most Fe was found as colloids (Table 1). The medium sized colloidal fraction (16–64 nm) accounted for 12.3% (19.4 $\mu g\,L^{-1}$), 12.3% (29.3 $\mu g\,L^{-1}$) and 13.4% (34.7 $\mu g\,L^{-1}$) of total As concentrations in three samples.

Fractograms from the FlFFF analysis suggested that humic-like DOM (represented by Ex/Em of 350 nm/450 nm) in smaller molecular size, mostly <3 nm but protein-like DOM (characterized by Ex/Em of 275 nm/300 nm) in a wider size spectrum containing both nano-colloidal <4 nm and larger colloidal sizes in the 4–8 nm and >20 nm levels. The different sized DOM components may have different affinity for different forms of As during their interactions, which can be elucidated by simultaneous analysis of both DOM and As in different molecular size fractions. Concerned with sample preservation artefacts, field ultrafiltration under N_2 atmosphere will be employed during future sampling to evaluate reducing groundwater As – organic matter interaction.

Table 1. Concentrations of selected elements of three samples in the 0.7 μm, 0.4 μm and 0.2 μm filtrates (in blue), and 64–700 nm, 16–64 nm, 8–16 nm and 0.5–8 nm fractionations. The concentrations in the dissolved fraction (<0.5 nm) were estimated by subtracting the sum of all the colloidal fractions from the total concentrations in <700 nm fraction.

Sample ID	Size internal (nm)	As ($\mu g\,L^{-1}$)	Fe ($mg\,L^{-1}$)	S ($mg\,L^{-1}$)	Mn ($mg\,L^{-1}$)	Mg ($mg\,L^{-1}$)
YC-1	<700	157.5	1.52	242.9	0.09	97.5
	<400	163.3	2.54	240.2	0.31	97.2
	<200	168.2	1.69	236.6	0.23	95.7
	64–700	10.3	0.71	0.9	0.04	0.57
	16–64	19.4	0.53	2.0	0.03	0.46
	8–16	3.5	0.07	0.5	0.01	0.12
	0.5–8	1.7	0.10	1.1	0.00	0.14
	<0.5	122.6	0.11	238.4	0.01	96.2
YC-5	<700	236.5	0.28	42.6	0.18	30.5
	<400	231.5	0.28	45.2	0.18	31.2
	<200	233.2	0.32	52.5	0.14	35.9
	64–700	12.8	0.08	0.0	0.01	0.08
	16–64	29.1	0.11	0.0	0.01	0.18
	8–16	13.9	0.01	44.6	0.00	0.04
	0.5–8	3.0	0.03	102.0	0.00	0.03
	<0.5	177.7	0.05	−104.0	0.16	30.2
MLW-7	<700	259.5	0.16	0.8	0.17	15.1
	<400	262.3	0.28	0.4	0.16	15.1
	<200	255.0	0.12	0.8	0.04	15.5
	64–700	13.9	0.14	0.8	0.01	0.07
	16–64	34.7	0.80	2.4	0.02	0.36
	8–16	14.9	3.0	0.2	0.12	0.08
	0.5–8	5.5	0.07	0.2	0.01	0.02
	<0.5	190.5	−3.8	−2.9	0.01	14.55

ACKNOWLEDGEMENTS

The study is supported by National Natural Science Foundation (No. 41772265). Special thanks are extended to Tingwen Wu from China Geology Survey.

REFERENCES

Guo, H.M., Zhang, B. & Zhang, Y. 2011. Control of organic and iron colloids on arsenic partition and transport in high arsenic groundwaters in the Hetao basin, Inner Mongolia. *Appl. Geochem.* 26(3): 360–370.

McArthur, J.M., Ravenscroft, P., Safiulla, S. & Thirlwall, M.F. 2001. Arsenic in groundwater: testing pollution mechanisms for sedimentary aquifers in Bangladesh. *Water Resour. Res.* 37(1): 109–117.

Mladenov, N., Zheng, Y., Miller, M.P., Nemergut, D.R., Legg, T., Simone, B., Hageman, C., Rahman, M.M., Ahmed, K.M. & McKnight, D.M. 2009. Dissolved organic matter sources and consequences for iron and arsenic mobilization in Bangladesh aquifers. *Environ. Sci. Technol.* 44(1): 123–128.

Sharma, P., Ofner, J. & Kappler, A. 2010. Formation of binary and ternary colloids and dissolved complexes of organic matter, Fe and As. *Environ. Sci. Technol.* 44(12): 4479–4485.

Wang, S.L. & Mulligan, C.N. 2006. Effect of natural organic matter on arsenic release from soils and sediments into groundwater. *Environ. Geochem. Heal.* 28(3): 197–214.

Environmental Arsenic in a Changing World –
Zhu, Guo, Bhattacharya, Ahmad, Bundschuh & Naidu (Eds)
ISBN 978-1-138-48609-6

Sedimentological controls on the formation of high arsenic aquifers in the central Yangtze River Basin since the Last Glacial Maximum

Y. Deng[1], Y.X. Wang[2] & T. Ma[2]
[1]*Geological Survey, China University of Geosciences, Wuhan, China*
[2]*School of Environmental Studies, China University of Geosciences, Wuhan, China*

ABSTRACT: Understanding the mechanism of arsenic mobilization from sediments to groundwater is important for drinking water supply and water quality management in endemic arsenicosis areas, such as the Jianghan alluvial plain in the middle reaches of the Yangtze river. Sediment samples from three boreholes in Jianghan Plain were collected for bulk geochemistry analysis and OSL dating. Sedimentological processes and palaeoclimatic optima after the Last Glacial Maximum have created favorable conditions for the formation of high-As aquifer systems. Bulk sediment geochemistry results indicated that As was correlated with sulfur in the pre-LGM sediments and by contrast As was correlated with iron in the post-LGM sediments.The intense chemical weathering led to sulfur depletion after the LGM could contribute to the As enrichment in the Holocene and upper Pleistocene aquifer.

1 INTRODUCTION

Geogenic enrichment of arsenic (As) in groundwater has been a topic worldwide concerned over the past decades due to its severe health threat, which impact about one hundred million people. Within river systems draining the Himalaya, tectonic movement, sedimentological processes and palaeoclimatic optima after the Last Glacial Maximum have created favorable conditions for the formation of high-As aquifer systems mainly within the Late Pleistocene-Holocene deposits (McArthur et al., 2011; Wang et al., 2017). The Jianghan Plain located in the middle reach of the Yangtze River Basin in central China has been documented with severe As contamination in the shallow aquifers (Gan et al., 2014; Deng et al., 2018). However, the underlying mechanism responsible for the formation of high arsenic aquifers has not been clearly understood yet. This purpose of this study was to delineate the relationship between the effects of Quaternary sediment deposition responding to sea level change since the Last Glacial Maximum on the arsenic enrichment in the shallow aquifers in the Jianghan Plain.

2 METHODS/EXPERIMENTAL

2.1 *Study area*

Jianghan alluvial plain in Central China is located in the middle Reaches of the Yangtze River, formed by Yangtze River and its largest tributary the Han River (Fig. 1). It is well known as the beautiful and rich land of fish and rice. It has a sub-tropical monsoonal climate with an annual precipitation and evaporation of approximately 1200 mm and 1378 mm, respectively.

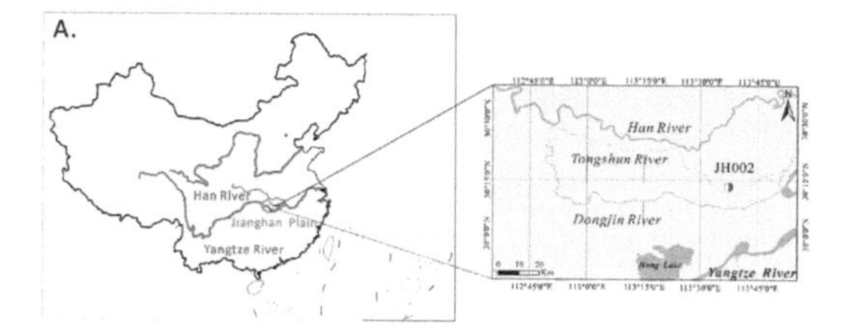

Figure 1. The geographical location of sampling borehole JH002 in the Jianghan Plain.

The Quaternary groundwater systems can be divided into three groups of aquifers. The first group is composed of Holocene alluvial and lacustrine deposits (clay, silt and fine sand) with a thickness of 3–10 m. The second group is composed of late and middle Pleistocene alluvial sediments (medium-coarse sand and gravel, interlaced clay lenses in local area) with a thickness of 30 m; this is the main aquifer for water supply. The third group is composed of early Pleistocene alluvial sediments (fine-medium sand, interlaced clay lenses).

2.2 *Sampling and analysis*

Sediment samples from three boreholes with depths up to 230 m in typical arsenic-affected area were collected for bulk geochemistry analysis and As, Fe and S speciation. Major-element analysis (SiO_2, Fe_2O_3, Al_2O_3, Na_2O, CaO, P_2O_5, and MgO) was done with an XRF spectrometer (RIX2100, Rigaku). Trace elements were determined using ICP-MS after a mixed acid (HNO_3–$HClO_4$–HF) digestion. Total carbon and sulfur, total organic carbon were measured using an Elemental Analyzer (Vario MICRO cube, Elementar).

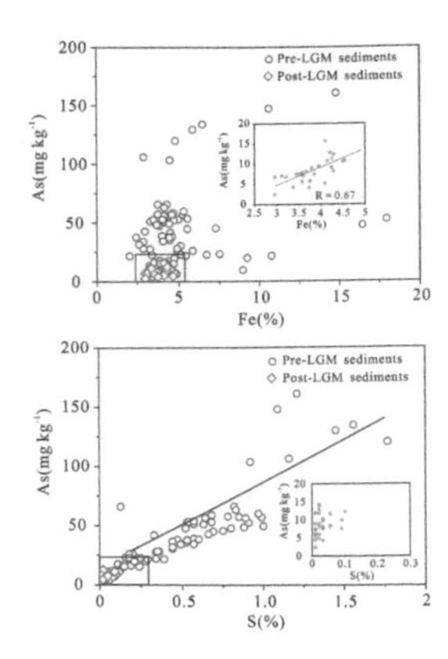

Figure 2. The relationship between arsenic and iron & sulfur in pre-LGM and post-LGM sediments samples of JH002.

We used the optical stimulated luminescence (OSL) dating method to measure the age of 23 sediment samples with the depths from 3.25 to 45.3 m below ground surface (bgs).

3 RESULTS AND DISCUSSION

3.1 *Lithology and bulk geochemistry*

According to the lithology and grain size analysis, the Quaternary aquifers could be divided into the Holocene-upper Pleistocene phreatic aquifer (15–60 m) and middle-lower Pleistocene confined aquifer (>60 m). Bulk sediment geochemistry analysis showed that As, Fe, Mn, P, TOC contents in the sediments of middle-lower Pleistocene aquifer were much higher than those in the Holocene and upper Pleistocene aquifer. Arsenic sequential extraction and SEM-EDS analysis indicated that the shallow sediments possessed an average As content of $9\,\mu g\ g^{-1}$. Arsenic content in the groundwater was up to $2330\,\mu g\ L^{-1}$. The deep aquifer possessed an average As content of $55\,\mu g\ g^{-1}$, and the highest As content in the groundwater was about $100\,\mu g\ L^{-1}$. Arsenic was correlated with sulfur in the pre-LGM sediments ($R = 0.7$, $P < 0.05$) and the pyrite was observed to be the main sink of As in the deep aquifer (>60 m), by contrast arsenic was correlated with iron in the post-LGM sediments ($R = 0.67$, $P < 0.05$) (Fig. 2).

Iron sequential extraction results indicated that Fe mainly existed as iron-oxides in Holocene and upper Pleistocene sediments, while the siderite was the main form of in middle-lower Pleistocene sediments. In addition, abundant Fe in pyrite-form were observed in some sulfur riched samples in middle-lower Pleistocene sediments.

3.2 *Sedimentological controls on the enrichment of arsenic in shallow aquifers since the LGM*

Sea level in the Last Glacial Maximum was 120 m lower than today, which led to depositional break during 40–20 ka. In warming period, the sea level raised immediately, with a quick response of the Yangtze

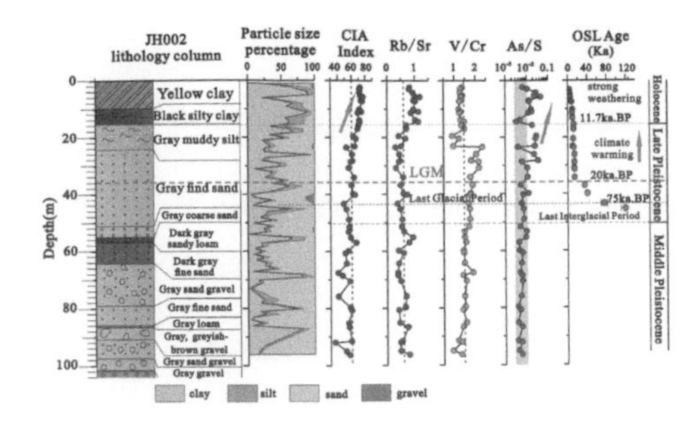

Figure 3. The variation of particle size, CIA index and As/S mole ratio in sediment samples of JH002 borehle.

river. The high CIA, Rb/Sr and V/Cr values indicated an intense chemical weathering and oxidizing sedimentary environment in post-LGM period. The mole ratio of As/S in the post-LGM sediments was >0.01, while in the deep pre-LGM sediments, the As/S kept stable and ranged between 0.0032 and 0.0061 (Fig. 3). Arsenic enriched in the shallow aquifer during post-LGM period may be caused by the intense chemical weathering of arsenopyrite in the sediments after the LGM.

4 CONCLUSIONS

The bulk sediments geochemistry indicated that As was correlated with sulfur in the pre-LGM sediments and by contrast arsenic was correlated with iron in the post-LGM sediments. The intense chemical weathering leading to sulfur depletion after the Last Glacial Maximum could contribute to the As enrichment in the Holocene and upper Pleistocene aquifer.

ACKNOWLEDGEMENTS

The research was financially supported by the National Natural Science Foundation of China (Nos. 41572226 & 41521001), Program of China Geological Survey (Nos. 121201001000150121).

REFERENCES

Gan, Y.Q., Wang, Y.X., Duan Y.H., Deng, Y.M., Guo, X.X. & Ding, X.F. 2014. Hydrogeochemistry and arsenic contami-nation of groundwater in the Jianghan Plain, central China. *J. Geochem. Explor*. 138(3): 81–93.

Deng, Y.M., Zheng, T.L., Wang, Y.X., Liu, L., Jiang, H.C. & Ma, T. 2018. Effect of microbially mediated iron mineral transformation on temporal variation of arsenic in the Pleistocene aquifers of the central Yangtze River basin. *Sci. Total Environ*. 619-620C: 1247–1258.

McArthur, J.M., Nath, B., Banejee, D.M., Purohit, R. & Grassineau, N. 2011. Palaeosol control on groundwater flow and pollutant distribution: the example of arsenic. *Environ. Sci. Technol.* 45: 1376–1383.

Wang, Y.X., Pi, K.F., Fendorf, S., Deng, Y.M. & Xie, X.J. 2017. Sedimentogenesis and hydrobiogeochemistry of high arsenic late pleistocene-holocene aquifer systems. *Earth Sci. Rev.* DOI:10.1016/j.earscirev.2017.10.007

Environmental Arsenic in a Changing World –
Zhu, Guo, Bhattacharya, Ahmad, Bundschuh & Naidu (Eds)
ISBN 978-1-138-48609-6

Roles of dissolved organic matter on seasonal arsenic variation in shallow aquifers of the central Yangtze River Basin by EEM-PARAFAC analysis

X.F. Yuan, Y. Deng & Z.J. Lu
Geological Survey, China University of Geosciences, Wuhan, P.R. China

ABSTRACT: Fluorescence excitation-emission matrix (EEM) analysis was applied to characterize seasonal variation of dissolved organic matter (DOM) in arsenic-affected shallow aquifers of the Jianghan Plain, central China, to better understand the effect of DOM on arsenic transformation during dry season and wet season. It was identified that there are three components of DOM, of which C1 and C3 are fulvic acid-like substances and C2 is humic acid substance, and there is no obvious seasonal variability in C2, and the concentration is slightly higher in the wet season than the dry season. C1 and C3 have the opposite change rules in different seasons, which is mainly reflected in the shallow unconfined water.

1 INTRODUCTION

Naturally arsenic (As) enrichment of groundwater is a subject of great concern, which has been reported from numerous countries worldwide (Keon *et al.*, 2001; Smedley *et al.*, 2002). China is one of the most serious waterborne endemic arsenicosis affected area. Jianghan Plain is a newly discovered arsenic affected area in Southern China. Fluorescence spectroscopy is a simple, sensitive and nondestructive technique that can provide valuable information regarding the molecular structure of dissolved organic matter (DOM) present in natural waters and generate 3-D excitation emission matrices (EEMs) spectra of DOM. PARAFAC can be applied to decompose fluorescence EEMs into different independent groups of fluorescent components, which can then reduce the interference among fluorescent compounds (Engelen *et al.*, 2009; Chen *et al.*, 2010). In this paper, we reveal the temporal changes of DOM and the relationships among DOM and As.

2 METHODS

2.1 *Study area*

Jianghan alluvial plain in Central China is located in the middle Reaches of the Yangtze River, formed by Yangtze River and its largest tributary Han River. A total of 39 monitoring wells with three different depths were installed at 13 sites in Shahu monitoring field site of Jianghan plain. At each site, three monitoring wells with 10 m (A), 25 m (B), and 50 m (C) deep tapped shallow aquifers composed of Quaternary deposits.

2.2 *Sampling and analysis*

Samples were collected in Jan 2016 to Jul 2016 in Shahu monitoring field site of the Jianghan Plain. Some in-situ parameters including T (°C), EC, DO, Eh, pH, S^-, NH_4^+ and Fe^{2+} were monitored and in-situ measured when sampling. Water samples were also collected for further laboratory analysis, including major and trace elements, dissolved organic carbon (DOC) and 3-D fluorescence spectra analysis. Two multivariate statistical techniques including FA and PARAFAC were applied to hydrochemical variables and EEMs of groundwater, respectively.

3 RESULTS AND DISCUSSION

3.1 *Water-level fluctuation and As temporal variation*

The water level between the wet and dry season fluctuates about 1 m. Arsenic has a temporal variation in shallow aquifers, and has a relationship with water level, ORP and DOC (Fig. 1).

3.2 *Fluorescence characteristics of DOM by PARAFAC analysis*

By PARAFAC of samples, it can be identified that there are three components of DOM (Fig. 2), of which C1 and C3 are fulvic acid-like substances and C2 is humic acid substance. The main components of DOM in surface water are C1 and C3, whilst the DOM in groundwater are mostly C1 and C2. C1 and C3 are terrestrial organic matter, the smallest molecule produced in the process of microbial degradation. C2 is the DOM

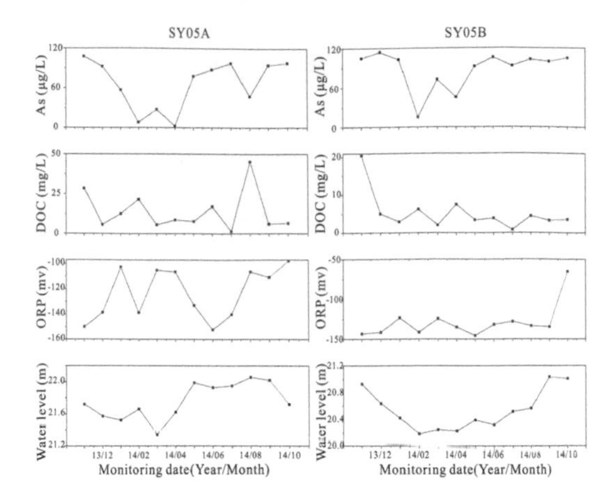

Figure 1. The generalization of As temporal variation with the water level, ORP and DOC in the aquifer of the Jianghan Plain, central Yangtze River Basin.

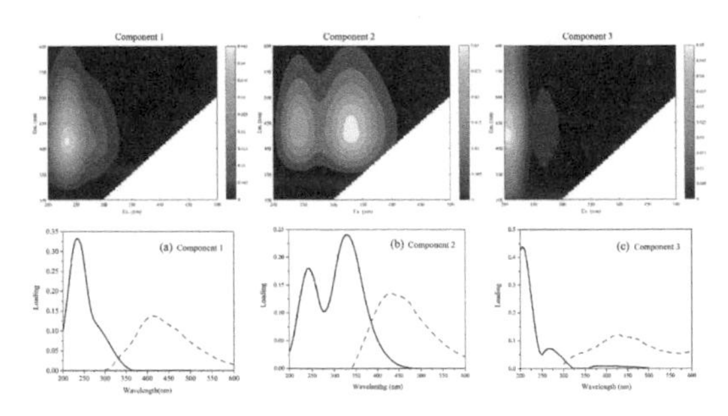

Figure 2. Excitation–emission matrices (EEMs) of groundwater components and spectral characteristics of three-component PARAFAC model validated by split half validation technique.

hybrid component of terrestrial materials, and marine materials and marine organic matter is considered to be the product of microbial utilization. There is no obvious seasonal variability in C2, and the concentration is slightly higher in the wet season than the dry season. C1 and C3 have the opposite change rules in different seasons, which is mainly reflected in the shallow unconfined water. The trend of arsenic concentration changes is the opposite of C1 and C3 (Table 1). The FI index in the sample groundwater is between 1.2 and 1.35, also indicating that the DOM is mainly a terrestrial source, but there is also a mixture of terrestrial source and microbial source, which could come from sediment in the local aquifer.

Table 1. The changes of fluorescence component C1, C2, and C3 in shallow unconfined water and shallow confined water of Fmax value in dry season and wet season.

Monitoring Wells		Dry Season			Wet Season		
		C1	C2	C3	C1	C2	C3
Shallow Unconfined Water	01A	266	126	58	0	140	69
	03A	232	146	70	37	154	96
	05A	132	47	16	121	50	115
	07A	109	33	0	139	54	39
Shallow Confined Water	01B	153	60	42	129	69	119
	03B	151	62	41	130	55	115
	05B	240	119	82	0	113	67
	07B	113	34	0	140	133	88

4 CONCLUSIONS

There is an obvious seasonal variability of As, and its variation has relationship with the water-level, ORP, DOC.There is no obvious seasonal variability in C2, and the concentration is slightly higher in the wet season than the dry season. C1 and C3 have the opposite change rules in different seasons, which is mainly reflected in the 10 m monitoring well.

ACKNOWLEDGEMENTS

The research was financially supported by the National Natural Science Foundation of China (Nos. 41572226 & 41521001), and State Key Laboratory of Biogeology and Environmental Geology (No. 128-GBL21711), China University of Geosciences.

REFERENCES

Chen, M.L., Price, R.M., Yamashita, Y. & Jaffe R. 2010. Comparative study of dissolved organic matter from groundwater and surface water in the Florida coastal Everglades using multidimensional spectrofluorometry combined with multivariate statistics. *Appl. Geochem.* 25:872–880.

Engelen, S., Frosch, S. & Jorgensen, B.M. 2009. A fully robust PARAFAC method for analyzing fluorescence data. *J. Chemom.* 23:124–131.

Keon, N.E., Swartz, C.H., Brabander, D.J., Harvey, C. & Hemond, H.F. 2001. Validation of an arsenic sequential extraction method for evaluating mobility in sediments. *Environ. Sci. Technol.* 35: 2778–2784.

Smedley, P.L. & Kinniburgh, D.G. 2002. A review of the source, behavior and distribution of arsenic in natural waters. *Appl. Geochem.* 17: 517–568.

Environmental Arsenic in a Changing World –
Zhu, Guo, Bhattacharya, Ahmad, Bundschuh & Naidu (Eds)
ISBN 978-1-138-48609-6

Roles of iron and/or arsenic reducing bacteria in controlling the mobilization of arsenic in high arsenic groundwater aquifer

H. Liu, P. Li, Y.H. Wang & Z. Jiang
State Key Laboratory of Biogeology and Environmental Geology, China University of Geosciences, Wuhan, P.R. China

ABSTRACT: To investigate microbial mediated As release and mobilization in the high arsenic groundwater, indigenous iron/As reducing bacteria were isolated from high arsenic groundwater in the Hetao Basin of Inner Mongolia in China. These isolates were identified to be Fe(III) and As(V) reducing (IAR) *Shewanella putrefaciens* IAR-S1, Fe(III) reducing (IR) *S. xiamenensis* IR-S2 and *Klebsiella oxytoca* IR-ZA, and As(V) reducing (AR) *Citrobacter freundii* AR-C1, *Paralostridium bifermentans* AR-P1, *Klebsiella pneumoniae* AR-K1 and *Aeromonas hydrophila* AR-A1. The incubation results showed that strains IAR-S1, IR-S2 and the control *S. oneidensis* MR-1 reduced ferrihydrite quickly and re-immobilized As. Strains AR-C1, AR-P1, AR-K1 and AR-A1 reduces As(V) to As(III), and arsenic (As_{-tot}) was released into the aqueous phase with strain IR-ZA and strain AR-C1.

1 INTRODUCTION

Arsenic (As) as a toxic metalloid element threatens the health of more than 140 million people worldwide. It is well known that microbes play an important role on mediating As-bearing Fe(III) oxyhydroxides reduction and As mobilization. Whether DIR results in As retention or release is still an international debate topic. Therefore, to investigate microbially-mediated DIR affecting on As release and mobilization in high arsenic groundwater environment in Hetao Basin of Inner Mongolia, the primary objectives of this study were to: 1) isolate and characterize the dissimilatory iron and/or arsenic reducing strains from the high arsenic groundwater area of Hetao Basin; 2) investigate the arsenic mobilization from As-bearing ferrihydrite with different strains.

2 METHODS/EXPERIMENTAL

2.1 *Site description and sample collection*

Water samples for bacterial isolation were collected from two tube wells in Hangjinghouqi County of Hetao Basin, where endemic arseniasis is most serious. Groundwater from this region contains high As concentrations up to 1.74 mg L^{-1}.

2.2 *Bacterial isolation and identification Bacterial isolation and identification*

Iron-reducing medium (IRM) or chemically defined medium (CDM) were used to screen iron-reducing and arsenic-reducing bacteria. For isolation, the Hungate roll tube technique was used (Hungate, 1969). All transfers and cultures sampling were performed in an anaerobic chamber with tubes and bottles flushed with O_2-free N_2 gas.

DNA was extracted from over-night cultures using FastDNA® SPIN Kit for Soil according to manufacturer's protocols (Qbiogene, Inc. CA, USA) and 16S rRNA gene was amplified with the primers 27F and 1492R. The 16S rRNA gene sequences were aligned with the closely related sequences in the GenBank database using the BLAST program.

2.3 *Arsenate and ferric reduction assays*

Iron and As(V) reduction activities of isolated strains were assayed with IRM or CDM, respectively, under anaerobic conditions as previously described by Dai *et al.* (2016).

0.8 mL pre-culture of strain and As-bearing ferrihydrite (As/Fe molar ratios was 1%) was transferred into 100 mL bottles with 80 mL basal medium (BM) under strictly anaerobic conditions and incubated at 30°C in dark with 120 rpm. The control experiment was conducted using an autoclaved abiotic sample under the same conditions to confirm only the chemical effect. In order to detect As mobilization with microbial DIR under groundwater conditions of our study area, strains were incubated with artificial groundwater medium (AGM) and natural groundwater instead of BM.

3 RESULTS

3.1 *Isolation and identification*

Three iron reducing strains were obtained and referred to as IAR-S1, IR-S2 and IR-ZA, respectively. BLAST analysis in GenBank identified that strain IAR-S1 belonged to *Shewanella putrefaciens* (99.57%), strain IR-S2 belonged to *Shewanella xiamenensis* (99.36%), and strain IR-ZA belonged to *Klebsiella oxytoca* (99.54%).

Four arsenate reducing strain was obtained and was named as AR-C1, AR-P1, AR-K1 and AR-A1. Phylogenetic analysis places these four strains closest to organisms referring to as *Citrobacter freundii*, *Paralostridium bifermentans*, *Klebsiella pneumoniae* and *Aeromonas hydrophila* with the similarity of 99.93%, 100%, 99.94% and 99.97%, respectively.

3.2 *Iron and arsenic reduction assays*

Three Fe(III)-reducing strains IAR-S1, IR-S2 and IR-ZA performed relatively quick reducing rates in the initial 1.0 day, and then slowed down in the remaining incubation. Strain IAR-S1 has a comparatively strong iron reducing capability and could reduce 20.00 mM Fe(III) into Fe(II) in 1.5 days, while both strain IR-S2 and IR-ZA only could reduce 16 mM and 10 mM in 1.5 days.

In CDM medium, strain AR-C1 could reduce 100% of 1 mM As(V) in 72 h and 75% of 2.5 mM As(V) in 120 h and 23.1% of 5 mM in 48 h. Strain AR-P1 could reduce 100% of 1 mM As(V) in 96 h, 77.6% of 2.5 mM As(V) in 96 h, and 36.9% of 5 mM As(V) in 72 h. Strain AR-K1 could reduce 100% of 1 mM As(V) in 48 h, 55.8% of 2.5 mM As(V) in 84 h, and 33.7% of 5 mM As(V) in 48 h. Strain AR-A1 could reduce 100% of 1 mM As(V) in 48 h, 70.8% of 2.5 mM As(V) in 60 h, and 38.3% of 5 mM As(V) in 48 h. However, strain IR-S2 and IR-ZA did not show any As(V) reducing ability. IAR-S1 produced 1.00 mM As(III) within 1.0 days, which can reduce both Fe(III) and As(V) with glucose as a carbon source.

3.3 *Reduction of As(V)-bearing ferrihydrite*

S. oneidensis MR-1 was chosen as the positive control to detect the As mobilization from As-bearing ferrihydrite by DIR. Under $H_2PO_4^-$ conditions, at the beginning of incubation, the concentration of aqueous total arsenic (As_{-tot}) increased quickly by all strains. However, the decrease of As_{-tot} was observed along with the culture of strain IAR-S1, IR-S2 and MR-1 after the peaks appeared. The fluctuation of As concentration might be due to the aqueous As captured by the secondary minerals gradually. In contrast, As_{-tot} with strain IR-ZA and AR-C1 was gradually increased after 12.0 days incubation. The As release with strain IR-ZA could be attributed to the decrease of the surface area of ferrihydrite with the reductive dissolution. On the other hand, more mobile As(III) produced by the As reducing strain AR-C1 increased the aqueous As.

Strains IAR-S1, IR-S2, IR-ZA AR-C1 and MR-1 incubated in the HCO_3^- conditions, the AGM and the natural groundwater, also performed similar As release and As(III) reduction patterns with those of $H_2PO_4^-$. However, the concentrations of aqueous As_{-tot} released by strains were much lower than those of $H_2PO_4^-$.

3.4 *Effect of As(V)-reduction or As(V) concentration on DIR*

To detect the effect of microbial Fe(III) and As(V) reductions on each other, As(V) reducing strain AR-C1 and Fe(III) reducing strains IR-ZA and IR-S2 (without As(V) reducing capability) were chosen to co-incubate with As-bearing ferrihydrite. Results showed that both Fe(III) and As(V) reductions were improved in the co-incubation. Fe(II) concentrations produced by strains AR-C1 and IR-ZA after 8.0 days was 0.52 mM higher than sole incubation of strain IR-ZA. More 0.33 mM of Fe(II) were detected after 8.0 days co-culture of strains AR-C1 and IR-S2, which is higher than sole incubation of strain IR-S2. Aqueous As_{-tot} concentrations increased with co-incubation of strain IR-ZA and AR-C1 after 18.0 days incubation, which is higher than the separated incubation of these two strains.

The effect of As(V) on DIR was also investigated, through incubation strain IR-S2 with different As/Fe molar ratio of As(V)-bearing ferrihydrite. The quantity of Fe(II) release was found to increase with the As/Fe molar ratio of ferrihydrite. No Fe release or reduction was observed in the blank control.

4 CONCLUSIONS

Our results suggested that the microbial DIR in the groundwater aquifer enhance not only the release but also retention of arsenic. Fe(III) and As(V) reductions mediated by the indigenous bacteria were improved each other. Both microbial As reduction and reductive dissolution of Fe(III) oxyhydroxides were important mechanisms of As enrichment in groundwater aquifers in the Hetao Basin of Inner Mongolia.

ACKNOWLEDGEMENTS

The research work was financially supported by the National Natural Science Foundation of China (No. 41521001 and No.41772260).

REFERENCES

Dai, X.Y., Li, P., Tu, J., Zhang, R., Wei, D.Z., Li, B., Wang, Y.H. & Jiang, Z. 2016. Evidence of arsenic mobilization mediated by an indigenous iron reducing bacterium from high arsenic groundwater aquifer in Hetao Basin of Inner Mongolia, China. *Int. Biodeter. Biodegr.* 128: 22–27.

Hungate, R.E. 1969. Chapter IV A roll tube method for cultivation of strict anaerobes. *Method Microbiol.* 3: 117–132.

Environmental Arsenic in a Changing World –
Zhu, Guo, Bhattacharya, Ahmad, Bundschuh & Naidu (Eds)
ISBN 978-1-138-48609-6

The DOM characteristic in As-affected aquifer of Chaobai River in the North China Plain

Y.F. Jia[1,2] & Y.H. Jiang[1,2]
[1]*State Key Laboratory of Environmental Criteria and Risk Assessment, Chinese Research Academy of Environmental Sciences, Beijing, P.R. China*
[2]*State Environmental Protection Key Laboratory of Simulation and Control of Groundwater Pollution, Chinese Research Academy of Environmental Sciences, Beijing, P.R. China*

ABSTRACT: High As groundwater was found in Chaobai River aquifers of North China Plain (NCP) which poses risk to drinking water resource. Arsenic shows patchy distribution in both horizontal and vertical profiles with concentration of <0.1–304 $\mu g\ L^{-1}$. The enrichment of As is related to reductive dissolution, supported by contradictory distribution of As and NO_3^-, negative correlation between As and SO_4/Cl, and positive and then negative trend between As and Fe. DOM decomposition was evidenced by positive link between As and P. Four florescence components were identified in groundwater DOM including humic acid-like, fulvic acid-like and two kinds of protein-like substances. Arsenic concentration shows negative correlation with humic substances (HS) except for the sample with the highest As. It seems that HS may enhance As mobilization possibly via electron shuttle or complexation effect only when it is in high content. The river water which is characterized with high humic-like substances would infiltrate to supply more HS to groundwater and further promotes As mobilization.

1 INTRODUCTION

High arsenic (As) groundwater is a worldwide environment issue, of which mostly occurred in reducing condition with dissolved organic matter (DOM) prevailing. Reactive DOM is believed to stimulate dissimilatory reduction of iron oxides/hydroxides which is accompanied by As(V) reduction and release of As(III). Recently, humic substance (HS) which is a large proportion of DOM is regarded as electron shuttle to enhance Fe(III) reduction and As release in Bangladesh aquifers (Mladenov *et al.*, 2015). Based on these findings, DOM characterization in high As-affected aquifers seems to be the first step to identify these processes. In this study, Chaobai River Watershed which was found to have high As groundwater was selected to see As mobilization patterns and related DOM characteristic in aquifers.

2 MATERIALS AND METHODS

2.1 *The study area*

The Chaobai River originating in Northern Yanshan Mountain flows through Beijing, Hebei and Tianjin and terminates to the Bohai Sea. It lies in the northern part of North China Plain (NCP) with prevailing semi-arid and semi-humid climates. The annual precipitation and water surface evaporation are around 560 mm and 1100 mm, respectively. Sand and partly gravel prevails in this Holocene porous aquifer. Precipitation and river water infiltration are the main recharge sources for groundwater. The need for drinking water causes a large amount of abstraction further the decline of groundwater table which shows an average rate of 2.8 $m\ yr^{-1}$ in part of area. The study area is located in the upper reach of the river known as the Shunyi district of Beijing. One major drinking water plant is located in this area, indicating an urgent need to deal with the groundwater quality problem and its further evolution.

2.2 *Sampling and analysis methods*

Sixty groundwater samples were collected from wells at depths of 30, 50, 80 and 150 m as well as 6 surface river water samples. Major anions were determined using Ion Chromatography (DX-120, Dionex), major cations, Fe, Mn and As by Inductively Coupled Plasma Atomic Emission Spectroscopy (iCAP 6300, Thermo). Fluorescence spectroscopy was determined using a Hitachi F-7000 fluorescence spectrophotometer. Three-dimensional fluorescence excitation emission matrices (EEMs) were generated by scanning samples over an excitation range of 200–450 nm at 5 nm increments and an emission range of 280–550 nm at 5 nm increments. Hierarchical cluster analysis (HCA) was performed on all samples with Ward's method, which uses the squared Euclidean distance as a similarity measure.

3 RESULTS AND DISCUSSION

3.1 *As mobilization patterns*

Arsenic shows patchy distribution with concentration of <0.1–304 $\mu g L^{-1}$. Eleven of total 60 samples have As >10 $\mu g L^{-1}$, of which 7 above 50 $\mu g L^{-1}$. Arsenic concentration in river water is below the detection limit. High arsenic groundwater was observed in all depth range, indicating a heterogeneous aquifer environment or the different influence of river water infiltration. Nitrate shows high and varied concentration in groundwater (<0.1–90.4 $mg L^{-1}$) as well as the river water (1.43–93.0 $mg L^{-1}$). Contradictory distribution was found between As and NO_3^- with high As groundwater (>10 $\mu g L^{-1}$) have NO_3^- all below 6 $mg L^{-1}$. Although As specie is not determined, it could be concluded that As(III) is more likely to be dominant. Arsenic shows a positive correlation with aqueous Fe when As below 100 $\mu g L^{-1}$, however reverse trend was found between them when As is higher, which is similar with the case in Hetao Basin (Jia *et al.*, 2014). An obvious decline trend of SO_4/Cl ratio was observed when As concentration increases. Total P showed liner and positive correlation with As when As > 10 $\mu g L^{-1}$. These distribution patterns of As and other redox-sensitive components clearly indicate the scenario of reductive dissolution of Fe and related As release (Jia *et al.*, 2017). DOM decomposition plays a critical role in this process and regulating As release.

3.2 *DOM character in high and low As groundwater*

Fluorescence spectroscopy helps to reveal different kind of DOM components. EEMs were delineated into five excitation-emission regions including Region I, II being aromatic protein-like, Region III fulvic acid-like, Region IV soluble microbial by-product-like, and Region V humic acid-like (Chen *et al.*, 2003). The volume of each region represents the cumulative fluorescence response of DOM with similar properties. Generally, As concentration shows positive correlation with total volume of region I, II, and IV, while negative correlation with total volume of region III and V known as HS except for the sample with the highest As. It is dominated by humic acid-like substances, which account for 83.3% of total 5 regions volume. The volume of region V in this sample is the highest value of all samples as well. Therefore, it seems to be ineffective to As mobilization when humic acid-like substances is not in high content. Similar case shows at least 5 $mg L^{-1}$ HS is needed to stimulate microbial ferrihydrite reduction (Jiang *et al.*, 2008).

Four florescence components were identified by EEM spectra coupled with PARAFAC analysis. It shows that components 1 and 2 are originated from humic acid-like, and fulvic acid-like substances, respectively, while components 3 and 4 from protein-like substances. Arsenic concentration shows negative correlation with HS (components 1 and 2) except for the highest As sample. However, no significant pattern were found between As and protein-like substances (components 3 and 4).

4 CONCLUSIONS

This is just a preliminary analysis of DOM in high As aquifers in Chaobai River. The enrichment of As is related to reducing environment attributed by DOM decomposition. However, the specific component of DOM working for this process was not identified. HS may impact As mobilization explained as electron shuttle or complexation effect. However, in this area it may need to be in high content before it works. The concerning thing is that river water which is characterized with high humic-like substances would infiltrate to supply more HS to groundwater, its effect on As mobilization merit further investigation.

ACKNOWLEDGEMENTS

This work was supported by the National Water Pollution Control and Treatment Science and Technology Major Project (No. 2018ZX07109-003, 004).

REFERENCES

Chen, W., Westerhoff, P., Leenheer, J. A. & Booksh, K. 2003. Fluorescence excitation-emission matrix regional integration to quantify spectra for dissolved organic matter. *Environ. Sci. Technol.* 37(24): 5701–5710.

Jia, Y.F., Guo, H.M., Jiang, Y.X., Wu. Y. & Zhou, Y.Z. 2014. Hydrogeochemical zonation and its implication for arsenic mobilization in deep groundwaters near alluvial fans in the Hetao Basin, Inner Mongolia. *J. Hydrol.* 518: 410–420.

Jia, Y.F., Guo, H.M., Xi, B.D., Jiang, Y.H., Zhang, Z., Yuan, R.X., Yi, W.X. & Xue, X.L. 2017. Sources of groundwater salinity and potential impact on arsenic mobility in the western Hetao Basin, Inner Mongolia. *Sci. Total Environ.* 601–602: 691–702.

Jiang, J. & Kappler, A. 2008. Kinetics of microbial and chemical reduction of humic substances: implications for electron shuttling. *Environ. Sci. Technol.* 42(10): 3563–3569.

Mladenov, N., Zheng, Y., Simone, B., Bilinski, T.M., McKnight, D.M., Nemergut, D., Radloff, K.A., Rahman M.M. & Ahmed, K.M. 2015. Dissolved organic matter quality in a shallow aquifer of Bangladesh: implications for arsenic mobility. *Environ. Sci. Technol.* 49(18): 10815–10824.

Environmental Arsenic in a Changing World –
Zhu, Guo, Bhattacharya, Ahmad, Bundschuh & Naidu (Eds)
ISBN 978-1-138-48609-6

Sulfurated fertilizers enhance the microbial dissolution and release of arsenic from soils into groundwater by activating arsenate-respiring prokaryotes

X.C. Zeng, W.X. Shi, W.W. Wu & S.G. Cheng
School of Environmental Studies, China University of Geosciences (Wuhan), Wuhan, Hubei, P.R. China

ABSTRACT: This work aimed to investigate the activity and diversity of the dissimilatory arsenate-respiring prokaryotes (DARPs) in the paddy soils, and the effects of sulfate on the DARPs-catalyzed dissolution and release of arsenic and iron from paddy soils into aqueous phase. We collected arsenic-rich soils from a farmland region of the Xiantao city, Hubei, China. Microcosm assay was used to detect how DARPs catalyze the reduction, dissolution, and release of arsenic and iron from the soils, and how sulfate affects this microbial reaction. HPLC-ICP-MS technique was used to determine the arsenic and iron species. Quantitative PCR was used to measure the arsenate-respiring reductase gene abundances in the microcosms. We found that there are diverse DARPs in the indigenous microorganisms. Microcosm assays indicated that these DARPs efficiently promoted the mobilization, reduction and release of arsenic from soils under anaerobic condition. Remarkably, when sulfate was added into the reactions, the DARPs-mediated arsenic reduction and release were significantly increased. Agricultural activities may significantly promote arsenic-contamination in groundwater through increasing inputs of sulfate into paddy soils. It should be avoid overuses of sulfate fertilizers.

1 INTRODUCTON

Arsenic (As) is a highly toxic metalloid that is widely distributed in the environment. It can exist in organic or inorganic forms (Oremland *et al.*, 2003; Zhu *et al.*, 2014). It is present in more than 200 minerals, usually in combination with sulfur and metals (Kirk & Holm, 2004; Smedley & Kinniburgh, 2002; Ferguson *et al.*, 1972). The most common arsenic-bearing mineral is arsenopyrite (Hao *et al.*, 2014; Savage *et al.*, 2004; Zhu *et al.*, 2008). Arsenic typically occurs in four oxidation states: −3, 0, +3 and +5. The most dominant forms in arsenic-contaminated soils and water are As(III) (arsenite) and As(V) (arsenate). Arsenite is more soluble, mobile and toxic than arsenate. Arsenic compounds have been classified as a carcinogen to humans. Acute high-dose exposure to arsenic may cause severe systemic toxicity and death. Low-dose chronic exposure can result in cancers of various organs and tissues, hyperkeratosis, jaundice, neuropathy, diabetes mellitus, cardiovascular diseases, stroke, lung diseases, hepatotoxicity and other severe diseases (Zhu *et al.*, 2014; Maguffin & Kirk 2015; Singh *et al.*, 2015).

Recently, it was found that high-arsenic groundwater exists in some areas in Jianghan Plain, China. A geochemical survey indicated that approximately 87% of the detected wells contained 10–2330 $\mu g\ L^{-1}$ soluble arsenic. The sediments from the detected sites contained 10.73–136.72 $mg\ L^{-1}$ adsorbed or mineral arsenic (Gan *et al.*, 2014). More recently, we found that the microbial communities from deep sediments, instead of shallow soils, efficiently catalyzed the reduction, dissolution and release of arsenic from insoluble phase into groundwater (Chen *et al.*, 2017). In this study, we aimed to explore how agricultural activities affect the arsenic mobilization and release from shallow soils into groundwater in Jianghan Plain. Because Jianghan Plain is one of the most important food grain production districts in China, sulfate fertilizers are widely used as essential S nutrients for the growth of crops. This significantly increased the sulfate contamination, suggesting that sulfate is one of the major environmental factors in the soils of this region. We found that sulfate has great effects on the indigenous microbial communities-catalyzed arsenic dissolution, reduction and release of arsenic from the paddy soils into groundwater. This finding provided direct evidence that anthropogenic agricultural activity significantly enhanced the arsenic contamination in groundwater.

2 METHODS/EXPERIMENTAL

We collected arsenic-rich soils from a farmland region of the Xiantao city, Hubei, China. Microcosm assay was used to detect how DARPs catalyze the reduction, dissolution, and release of arsenic and iron from the soils, and how sulfate affects this microbial reaction. HPLC-ICP-MS technique was used to determine the arsenic and iron species. Quantitative PCR was used to measure the arsenate-respiring reductase gene abundances in the microcosms.

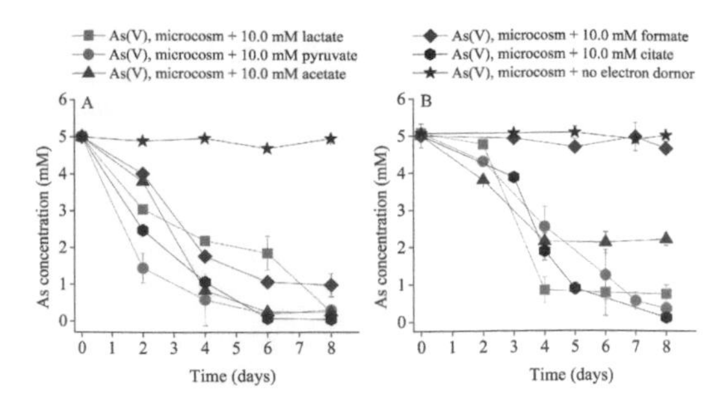

Figure 1. Arsenate-respiring reduction activities of the microbial communities from the sediment samples. Arsenate-respiring curve of the microcosm from the depth of 1.6 m (A) and 4.6 m (B). Active microcosms of the two samples were inoculated into the MM medium amended with 5.0 mM As(V) as the sole electron acceptor and 10.0 mM lactate, formate, pyruvate, acetate or citrate as a electron donor.

3 RESULTS AND DISCUSSION

3.1 *Arsenate-respiring activities of the microorganisms in the soils*

Microcosm assay was used to detect the arsenate-respiring activity of the microbial communities from the soils. The results showed that the arsenate-respiring activities were detectable in all of the two samples that were supplemented with different electron donors, and no significant arsenate-reducing activities were detected in the absence of external electron donors (Fig. 1).

3.2 *Unique diversity of the arsenate-respiring reductase genes from the soils*

To understand the molecular basis of the arsenate-respiring activities of the soils, we explored the molecular diversity of the arsenate-respiring reductase genes present in the microbial communities of the soils by cloning, sequencing and analyzing the *arrA* marker sequence. We identified 27 novel *arrA* genes. Their encoded Arr proteins were referred to as M1 to M18 (from 1.6 m), and N1 to N8 (from 4.6 m). The amino acid sequences of these Arr proteins were each used as queries to search against the GenBank database using the BLAST sever. We found that the Arr proteins from the soils share 71–89% sequence identities with other known microbial Arr proteins, suggesting that a lot of new or new-type DARPs were present in the soils (Fig. 2).

3.3 *Microbial dissolution and release of As(III) and Fe(II) from the shallow soils*

To determine whether the microbial communities in the shallow soils are capable of catalyzing the arsenic release from insoluble phase, we conducted arsenic release assay using the soil samples under anaerobic condition.

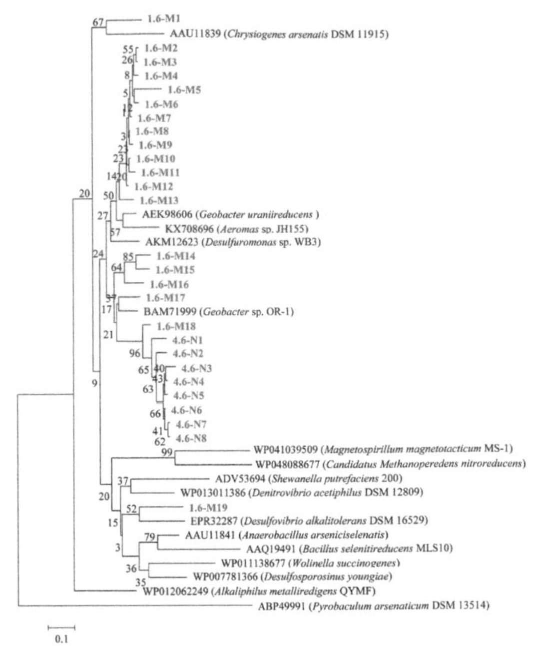

Figure 2. Phylogenetic analysis of arsenate-respiring reductases from the high-arsenic shallow sediments of Jianghan Plain.

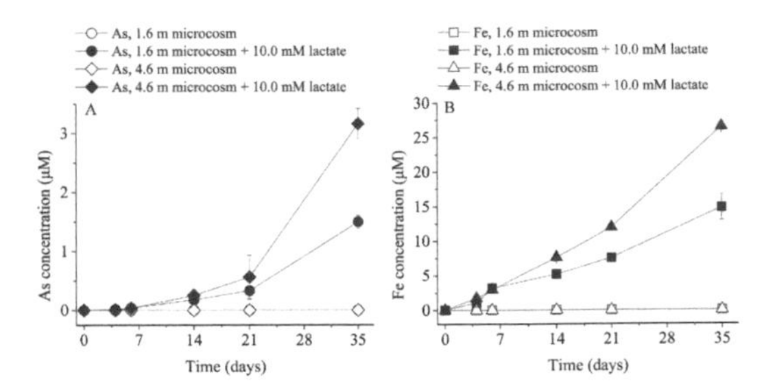

Figure 3. The mobilization, reduction and release of insoluble arsenic and iron catalyzed by microbial communities from the sediment samples using microcosm assay technique. The As(III) (A) and Fe(II) (B) release curves were achieved using the microcosms of the samples from the depth of 1.6 m and 4.6 m. Active microcosms of the two samples were prepared by inoculating each of the samples into the synthetic groundwater supplemented with 10.0 mM lactate.

These results suggest that the microorganisms in the soils of the depths 1.6 and 4.6 m significantly catalyzed the dissolution and release of arsenic and iron from the soils into aqueous phase, and the amount of the released iron was greater than that of the release arsenic. These data also indicated that there were more arsenic and iron released from the soils of 4.6 m than from those of 1.6 m (Fig. 3).

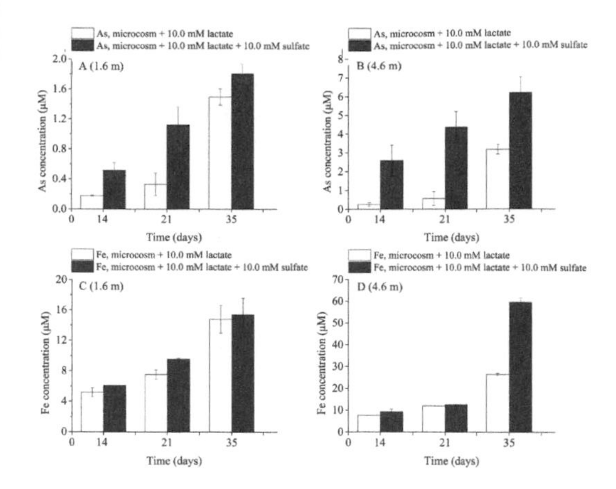

Figure 4. The mobilization, reduction and release of insoluble arsenic and iron catalyzed by microbial communities from the sediment samples in the presence of 10.0 mM sulfate or without sulfate using microcosm assay technique. The As(III) and Fe(II) release curves were achieved using the microcosms of the samples from the depth of 1.6 m (A, C) and 4.6 m (B, D). Active microcosms of the two samples were prepared by inoculating each of the samples into the synthetic groundwater supplemented with 10.0 mM lactate and 10.0 mM sulfate or without sulfate.

3.4 *Sulfate significantly enhanced the microorganisms-mediated release of arsenic and iron*

These results suggested that sulfate significantly enhanced the microorganisms-catalyzed dissolution, reduction and release of arsenic and iron from the soil phase (Fig. 4).

3.5 *Structures of the microbial communities from the soils*

To better understand the microbial basis of the observations, we analyzed the structures of the microbial communities from the soils using illumine high-throughout paired-end sequencing technique. We identified 39 phyla of bacteria.

As shown in Figure 5, the microbial community from the depth of 1.6 m consisted of *Chloroflexi* (38.16% of the total microorganisms), *Proteobacteria* (25.07%), *Acidobacteria* (10.00%), *Gemmatimonadetes* (5.35%), *Latescibacteria* (3.11%), *Aminicenantes* (2.43%), *Nitrospirae* (1.38%), *Actionbacteria* (1.34%), *Firmicutes* (1.00%), *Bacteroidetes* (0.35%), *Spirochaetae* (0.34%), and other microorganisms with less abundance; the microorganisms from the depth of 4.6 m included *Proteobacteria* (50.77%), *Chloroflexi* (15.38%), *Firmicutes* (14.24%), *Nitrospirae* (7.11%), *Bacteroidetes* (6.17%), *Aminicenantes* (2.97%), *Actionbacteria* (2.51%), *Acidobacteria* (1.94%), *Spirochaetae* (1.70%), candidate division Gal 15 (1.53%), *Gemmatimonadetes* (0.43%), *Latescibacteria* (0.30%), and other less abundant microorganisms. This suggests that the microbial community structure from the depth

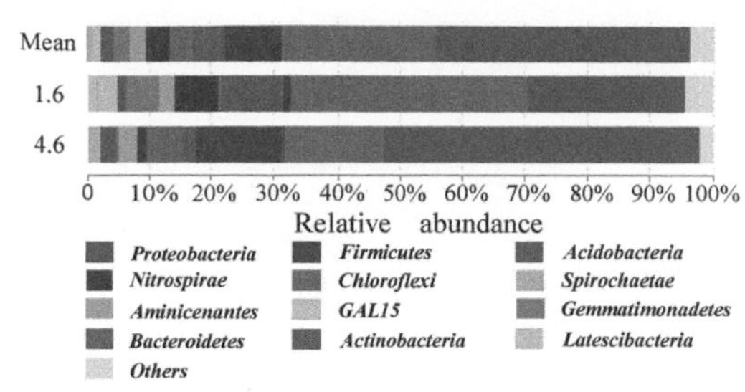

Figure 5. Analysis of the microbial community compositions of the sediment samples at phylum level from the Jianghan Plain.

1.6 m significantly differs from that from the depth of 4.6 m.

4 CONCLUSIONS

This study aimed to explore how arsenic was released from soil phase into groundwater, and how sulfate affects this bioprocess in Jianghan Plain, China. We found that a large diversity of DARPs are present in the shallow soils from the depths of 1.6 and 4.6 m. Microcosm assays indicated that the microbial communities significantly catalyzed the dissolution, reduction and release of arsenic and iron from the soils into aqueous phase. It is interesting to see that addition of 10.0 mM sulfate into the microcosms led to a significant increase of the microorganisms-mediated release of arsenic and iron from the soils. Quantitative PCR analysis for the functional gene abundances suggested that the sulfate-induced increase of microbial release of arsenic and iron was attributed to the significant enhancement of the DARP growth by sulfate. These results suggested that agricultural uses of sulfate fertilizers enhanced arsenic contamination in groundwater. The findings of this study gained new insight into the mechanisms by which the arsenic-contaminated groundwater was formed, and gave direct evidence that agricultural activities may enhance the arsenic contamination in groundwater.

ACKNOWLEDGEMENTS

This work was financially supported by the National Natural Science Foundation of China (grants no. 41472219, 41072181, 41272257 and 41521001).

REFERENCES

Chen, X.M., Zeng, X.C., Wang, J.N., Deng, Y.M., Ma, T., E.G.J., Mu, Y., Yang, Y., Li, H. & Wang, Y.X. 2017. Microbial communities involved in arsenic mobilization and release from the deep sediments into groundwater in Jianghan plain, Central China. *Sci. Total Environ.* 579: 989–999.

Ferguson, J. F. & Gavis, J. 1972. A review of the arsenic cycle in natural waters. *Water Res.* 6: 1259–1274.

Gan, Y.Q., Wang, Y.X., Duan, Y.H., Deng, Y.M., Guo, X.X. & Ding, X.F. 2014. Hydrogeochemistry and arsenic contamination of groundwater in the Jianghan Plain, central China. *J. Geochem. Explor.* 138: 81–93.

Hao, T.W., Xiang, P.Y., Mackey, H.R., Chi, K., Lu, H., Chui, H.K., van Loosdrecht, M.C.M. & Chen, G.H. 2014. A review of biological sulfate conversions in wastewater treatment. *Water Res.* 65: 1–21.

Kirk, M.F., Holm, T.R., Park, J., Jin, Q.S., Sanford, R.A., Fouke, B.W. & Bethke, C.M. 2004. Bacterial sulfate reduction limits natural arsenic contamination in groundwater. *Geology* 32: 953–956.

Maguffin, S.C., Kirk, M.F., Daigle, A.R., Hinkle, S.R. & Jin, Q.S. 2015. Substantial contribution of biomethylation to aquifer arsenic cycling. *Nat. Geosci.* 8: 290–293.

Oremland, R.S. & Stolz, J.F. 2003. The ecology of arsenic. *Science* 300(5621): 939–944.

Savage, K.S., Tingle, T.N., O'Day, P.A., Waychunas, G.A. & Bird, D.K. 2004. Arsenic speciation in pyrite and secondary weathering phases, Mother Lode Gold District, Tuolumne County, California. *Appl. Geochem.* 15(8): 1219–1244.

Smedley, P.L. & Kinniburgh, D.G. 2002. A review of the sources, behavior and distribution of arsenic in natural waters. *Appl. Geochem.* 17: 517–568.

Zhu, W., Young, L.Y., Yee, N., Serfes, M., Rhine, E.D. & Reinfelder, J.R. 2008. Sulfide-driven arsenic mobilization from arsenopyrite and black shale pyrite. *Geochim. Cosmochim. Acta.* 72: 5243–5250.

Zhu, Y.G., Yoshinaga, M., Zhao, F.J. & Rosen, B.P. 2014. Earth Abides Arsenic Biotransformations. *Ann. Rev. Earth Planet. Sci.* 42: 443–467.

1.3 Biogeochemical processes controlling arsenic mobility, redox transformation and climate change impacts

Environmental Arsenic in a Changing World –
Zhu, Guo, Bhattacharya, Ahmad, Bundschuh & Naidu (Eds)
ISBN 978-1-138-48609-6

Structural insight into the catalytic mechanism of arsenate reductase from *Synechocystis* sp. PCC 6803

Y. Yan, J. Ye, X. Zhang, X.M. Xue & Y.G. Zhu
Key Lab of Urban Environment and Health, Institute of Urban Environment, Chinese Academy of Sciences, Xiamen, China

ABSTRACT: Arsenate reductases are the key enzymes in biological arsenic detoxification by catalyzing the intracellular reduction of arsenate to arsenite, and arsenite could be subsequently pumped out of the cells by ArsB or Acr3. The arsenate reductase from cyanobacterium *Synechocystis* sp. Strain PCC 6803 (SynArsC) shows sequence homology with the thioredoxin-dependent arsenate reductase family, while utilizes the glutathione/glutaredoxin system for arsenate reduction. SynArsC is classified as a novel thioredoxin/glutaredoxin hybrid arsenate reductase family. Here we report the crystal structures of SynArsC in the native and phosphate-bound states at 1.37 and 1.55 Å resolutions, respectively. The structures are mostly similar, but also show differences with implications in SynArsC's arsenate reduction mechanism. Our results provide insights into SynArsC's structure-function relationship and its enzymatic mechanism.

1 INTRODUCTION

Arsenic is a toxic metalloid element that causes numerous environment and health problems. Ubiquitous arsenic forced living organisms to evolve different mechanisms for arsenic resistance. A two-step process, arsenate reduction followed by arsenite efflux, is a well-known mechanism of arsenate resistance (Zhu *et al.*, 2017).

Cytoplasmic arsenate reductases (ArsC) are divided into at least three families (Messens & Silver, 2006). The first family is represented by R773ArsC from *Escherichia coli*, which uses glutathione (GSH)/glutaredoxin (Grx) as electron donor and has a single catalytic cysteine. The second family is exemplified by SaArsC from *Staphylococcus aureus*, which uses thioredoxin (Trx) as a reducing system and requires three cysteines for arsenate reduction. The last family is termed as Acr2P, and only present in eukaryotic organisms, such as *Saccharomyces cerevisiae* and *Arabidopsis thaliana*.

Arsenate reductase from *Synechocystis sp.* PCC 6803 (SynArsC) is a novel ArsC belonging to the Trx/Grx hybrid arsenate reductase family. The primary sequence of SynArsC is similar with that of Trx-dependent ArsCs, whereas SynArsC utilized Grx/GSH system for arsenate reduction (López-Maury *et al.*, 2009; Li *et al.*, 2003). Cys8, Cys80 and Cys82 in SynArsC were identified as essential cysteine residues by site-directed mutagenesis, and the Cys80/Cys82 disulfide was detected by equilibrium redox titrations. In order to elucidate the special mechanism of As(V) reduction, we determined the crystal structures of native SynArsC and its complex with phosphate (PO_4^{3-}).

2 METHODS/EXPERIMENTAL

2.1 *Purification and crystallization*

SynArsC with his-tag was overexpressed in *E. coli* Rosetta (DE3), and purified by Ni-NTA agarose column and size-exclusion chromatography. Crystallization trials were set up at 291 K using hanging-drop vapor-diffusion method. The 2 μL hanging drops consisted of 1 μL protein solution and 1 μL reservoir solution were equilibrated against 500 μL reservoir solution (32% (*w*/*v*) PEG 3350, 100 m*M* citric acid, pH 5.5). Needle-like crystals were observed after one week. As the co-crystallization of SynArsC and As(V)/As(III) was unsuccessful, we crystallized SynArsC with PO_4^{3-}, an analog of As(V). The crystals of SynArsC-PO_4^{3-} were obtained in optimized reservoir solution consisting of 28% (*w*/*v*) PEG 3350, 100 m*M* citric acid and pH 4.8 by adding 5 m*M* NaH_2PO_4 into the protein solution. Crystals were mounted and soaked in corresponding reservoir solution supplemented with 20% glycerol and flash-cooled in liquid nitrogen prior to data collection.

2.2 *Data collection, structure determination and refinement*

Diffraction data were collected with a wavelength of 0.9792 Å at 100 K on beamline BL17U of Shanghai Synchrotron Radiation Facility (SSRF). All data were indexed and integrated with *iMosflm*, and scaled with *AIMLESS* from *CCP4* program suite. The structures of native SynArsC and SynArsC-PO_4^{3-} were obtained by the molecular replacement method using *Phaser*. A single structure modified from the 20 conformers of NMR structure of SynArsC (Yu *et al.*, 2011) was used as a search model. The structure models were built using *Coot*, and refined by iterative rounds of *PHENIX* automatically and *Coot* manually.

3 RESULTS AND DISCUSSION

3.1 *Determination of SynArsC structures*

The crystal structure of native SynArsC was determined in space group $P2_12_12_1$, with unit-cell parameters $a = 33.03$, $b = 33.35$, $c = 107.57$ Å. The structure of SynArsC-PO_4^{3-} belonged to space group $C121$, with unit-cell parameters $a = 64.26$, $b = 33.37$, $c = 56.63$ Å, $\beta = 100.2°$. The native SynArsC and SynArsC-PO_4^{3-} structures were refined to resolutions of 1.37 Å and 1.55 Å, respectively. The native SynArsC was modelled and refined with a final R_{work} of 0.166 and an R_{free} of 0.194. The final R_{work} and R_{free} for SynArsC-PO_4^{3-} structure were 0.152 and 0.184, respectively.

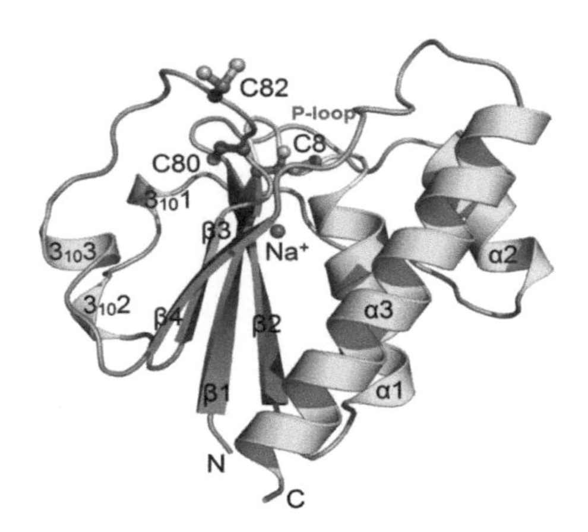

Figure 1. The overall structure of SynArsC shown as cartoon representations. Key residues are shown in stick representation. The P-loop is colored yellow.

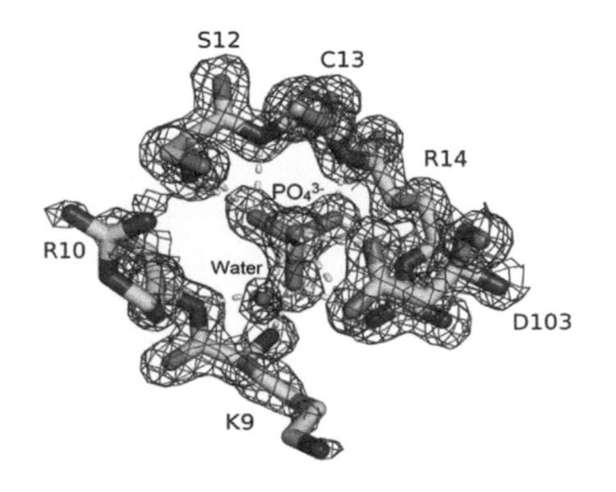

Figure 2. The binding site in the SynArsC-PO_4^{3-} structure. The $2F_{obs} - F_{calc}$ density map was calculated from the refined model contoured at 1.0 σ.

3.2 *Overall structure of SynArsC*

SynArsC showed similar structure with Trx-dependent ArsCs, consisting of a four-stranded and parallel open-twisted β-sheet (β1–4) flanked by three major α-helices (α1–3) and three small 3_{10}-helices (3_{10}1–3) on both sides (Fig. 1).

In the structure of SynArsC, the conserved P-loop connecting the first β-sheet with α-helix is an active site of both phosphatases and arsenate reductases. The first conserved cysteine (cys8) was in the P-loop. However, the other active residues were in unique geometries, with the last two conserved cysteines (cys80 and cys82) locating in the flexible loop and the conserved aspartic acid (Asp103) being in the rigid area.

3.3 *The phosphate-binding site*

PO_4^{3-} is tightly hydrogen-bound to the P-loop in the SynArsC-PO_4^{3-} structure (Fig. 2). The P-loops are almost in the same geometry with Asn11 in the αL conformation with or without binding to PO_4^{3-}. Therefore, the biding of PO_4^{3-} does not remarkably change the conformation of SynArsC. However, it does make the structure more stable. The detailed snapshot also provides insight into the catalytic mechanism of SynArsC.

4 CONCLUSIONS

SynArsC showed similar structure with Trx-coupled ArsCs, but also has its own particularity in structure as a member of new arsenate reductase family.

ACKNOWLEDGEMENTS

This work is supported by the National Natural Science Foundation of China (21507125 and 31270161).

REFERENCES

López-Maury, L., Sánchez-Riego, A.M., Reyes, J.C. & Florencio, F.J. 2009. The glutathione/glutaredoxin system is essential for arsenate reduction in *Synechocystis* sp. strain PCC 6803. *J. Bacteriol.* 191: 3534–3543.

Li, R., Haile, J.D. & Kennelly, P.J. 2003. An arsenate reductase from *Synechocystis* sp. strain PCC 6803 exhibits a novel combination of catalytic characteristics. *J. Bacteriol.* 185: 6780–6789.

Messens, J. & Silver, S. 2006. Arsenate reduction: thiol cascade chemistry with convergent evolution. *J. Mol. Biol.* 362: 1–17.

Yu, C., Xia, B. & Jin, C. 2011. ^1H, ^{13}C and ^{15}N resonance assignments of the arsenate reductase from *Synechocystis* sp. strain PCC 6803. *Biomol. NMR Assignm.* 5: 85–87.

Zhu, Y.G., Xue, X.M., Kappler, A., Rosen, B.P. & Meharg A.A. 2017. Linking genes to microbial biogeochemical cycling: lessons from arsenic. *Environ. Sci. Technol.* 51(13): 7326–7339.

Environmental Arsenic in a Changing World –
Zhu, Guo, Bhattacharya, Ahmad, Bundschuh & Naidu (Eds)
ISBN 978-1-138-48609-6

Impacts of environmental factors on arsenate biotransformation and release in *Microcystis aeruginosa* using Taguchi experimental design

Z.H. Wang[1,2], Z.X. Luo[2] & C.Z. Yan[2]
[1]*College of Chemistry and Environment, Fujian Province Key Laboratory of Modern Analytical Science and Separation Technology, Minnan Normal University, Zhangzhou, China*
[2]*Key Laboratory of Urban Environment and Health, Institute of Urban Environment, Chinese Academy of Sciences, Xiamen, China*

ABSTRACT: We conducted a series of experiments using Taguchi methods to determine optimum conditions for arsenic (As) biotransformation. We found that N is critical for *M. aeruginosa* As(V) biotransformation, particularly with regard to As(III) transformation. Also, As accumulation benefited from low P levels when combined with high N concentrations. Phosphate was second to As(V) as the primary factor to affect As accumulation. Additionally, we found that the small amounts of As that accumulated under low concentrations of As and high P were tightly stored in living algal cells and were easily released after cell death. Our results will be helpful for the understanding, practical applications, and overall control of the key environmental factors, particularly those associated with algal bioremediation for As-polluted water.

1 INTRODUCTION

Algae are widely distributed in aquatic ecosystems and play an important role in arsenic (As) bioaccumulation and biogeochemical cycling (Duncan *et al.*, 2015; Yan *et al.*, 2016). *Microcystis aeruginosa* is generally tolerant to As(V) and exhibits a stronger As bioaccumulation capacity compared to other freshwater algae (Wang *et al.*, 2017). Many abiotic factors affect the metabolic functions of alga contaminated by As, such as As levels, hydrogen ion levels (pH), and key nutrient concentrations of nitrogen (N) and phosphorus (P) in culture media. To date, very limited information is available on how and to what extent environmental factors influence arsenic (As) biotransformation and release in freshwater algae. To further understand environmental factors that impact As(V) uptake, we investigated As biotransformation and release in *M. aeruginosa*, aspects of its growth, intracellular As accumulation in algae cells, and release after algae death. Taguchi methods under their relevant statistical assumptions were applied to determine optimum environmental conditions.

Table 1. Environmental factors of the orthogonal test.

Factor	NO_3^--N (mg L^{-1})	PO_4^{3-}-P (mg L^{-1})	pH	*As*(V) (μM)
Level 1	2	0.02	6	0.1
Level 2	4	0.20	8	1.0
Level 3	10	1.00	10	10.0

Table 2. Experimental L_9 (3^4) orthogonal array.

	Parameters			
Treatment	NO_3^--N (mg L^{-1})	PO_4^{3-}-P (mg L^{-1})	pH	*As*(V) μM
E1	2	0.02	6	0.1
E2	2	0.2	8	1.0
E3	2	1.0	10	10
E4	4	0.02	8	10
E5	4	0.2	10	0.1
E6	4	1.0	6	1.0
E7	10	0.02	10	1.0
E8	10	0.2	6	10
E9	10	1.0	8	0.1

2 METHODS

2.1 *Experimental design*

Three different levels in combination with actual aquatic conditions of each environmental factor were considered (Table 1). Accordingly, we chose an L9 (3^4) orthogonal array, and we obtained experimental conditions (Table 2) by combining Table 1 and the L9 (3^4) orthogonal array. We used experimental data to determine optimal experimental conditions and evaluate experimental results, which we assessed using analysis of variance (ANOVA) and the signal-to-noise (S/N) ratio with biggest characteristics.

2.2 *Batch culture preparation*

The *M. aeruginosa* cultures that we used were incubated in BG-11 without adding additional N and P for 48 h after exponential phases of algal suspension growth were centrifuged and washed twice in sterile Milli-Q water. We separated the above cultures into nine equal parts, which were once again centrifuged and washed in sterile Milli-Q water. They were aseptically transferred to nine sterile 1 L Erlenmeyer flasks

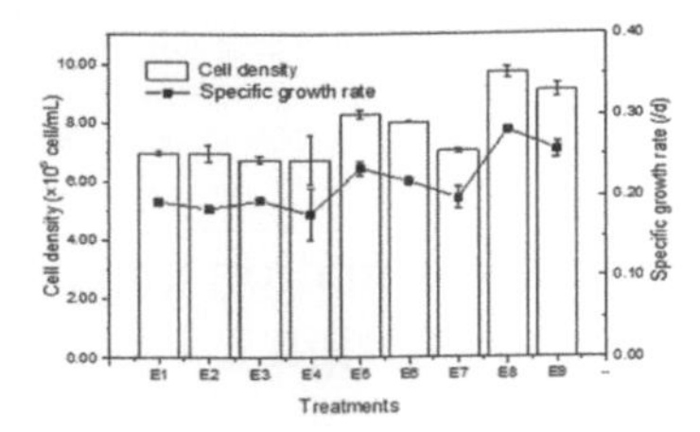

Figure 1. Cell density and the specific growth rate of *M. aeruginosa* under different treatments.

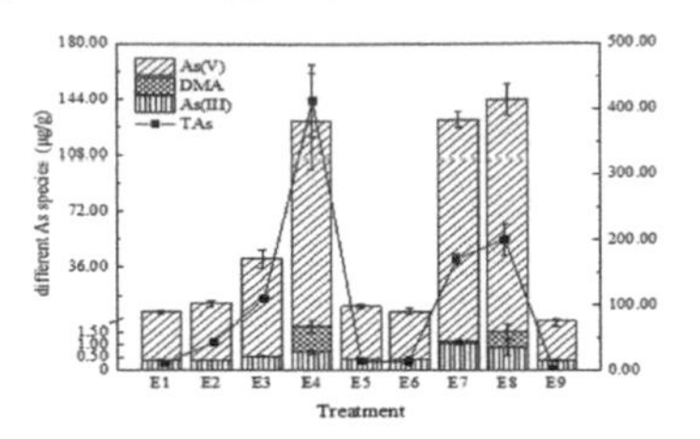

Figure 2. Intracellular total As and different As species concentrations in *M. aeruginosa* under different treatments.

(2 replicates per treatment), containing 250 mL of different autoclaved sterilized modified BG-11 media, according to Table 2. After cultured batch treatments in an illuminated incubator shaker for 96 h, We then harvested approximately 20 mL of the algae via centrifugation at 4500 × g for 10 min. Media was also frozen to determine total As (TAs) and the specific As species.

3 RESULTS AND DISCUSSION

3.1 *Algae growth*

The final cell density and specific growth rate obtained at the conclusion of the 96 h experiment were plotted in Figure 1, which further indicated variation in parameter level combinations. NO_3^--N was found to be the most influential among the four factors on algal growth while PO_4^{3-}-P was the second most influential factor.

3.2 *Intracellular arsenic bioaccumulation*

We ascertained TAs content and As species in algal cells after 96 h in culture media (Fig. 2). High As bioaccumulation and its facilitation by N in media indicated that N could affect algae bioremediation in As polluted water. *M. aeruginosa* could accumulate greater As(V) in cells under high As(V) ambient concentrations, coexisting with low P and high N concentrations. Being key factors affecting intracellular TAs and As(V) accumulation, their similarity in chemical properties between P and As determined that P was second to As in rank and order.

3.3 *Arsenic biotransformation in media*

With the exception of *As*(V), we detected *As* biotransformation and DMA in media after 96 h in algal culture (Fig. 3). Although high *As*(III) and DMA concentrations in media were primarily caused by the high *As*(V) levels, its reduction to *As*(III) in media was

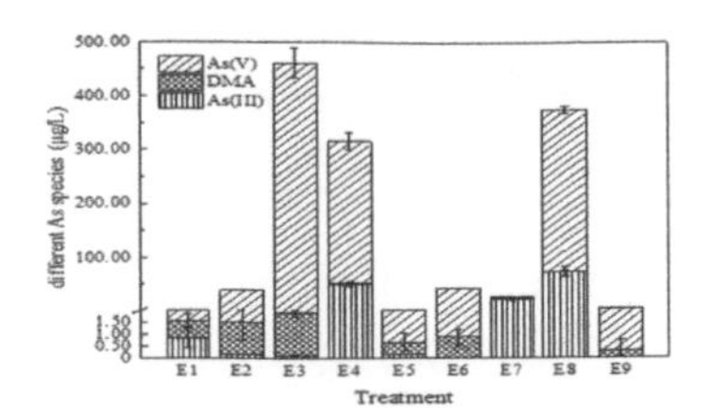

Figure 3. Different As species concentrations in culture media under different treatments.

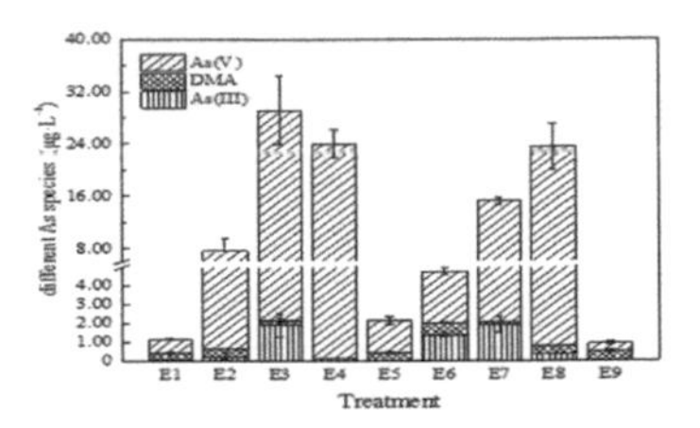

Figure 4. Different As species concentrations in culture media from dead algae cells under different treatments.

inclined to occur in high *N* and low *P* concentrations under slightly acidic environments. As the principal factor, *N* exhibited a significantly positive effect on *As*(III) concentrations in culture media after *As* was introduced.

3.4 *Arsenic release from dead algae*

This study found that As was rapidly released from dead algal cells. Figure 4 shows that approximately 58% to 93% As was released after 8 h resuspension. The high As uptake in algae with increasing initial concentrations of As(V) could result in high As efflux from dead cells. The pH factor yielded principal effects on As(III) and methylated As (OAs) efflux.

ACKNOWLEDGMENTS

This study was jointly supported by the National Nature Science Foundation of China (project nos. 41401552, 41271484 and 21277136) and the Nature Science Foundation of Fujian Province (2016J01691).

REFERENCES

Duncan, E.G., Maher, W.A. & Foster, S.D. 2015. Contribution of arsenic species in unicellular algae to the cycling of arsenic in marine ecosystems. *Environ. Sci. Technol.* 49(1): 33–50.

Wang, Z. H., Luo, Z. X., Yan, C. Z. & Xing, B. S. 2017. Impacts of environmental factors on arsenate biotransformation and release in Microcystis aeruginosa using the Taguchi experimental design approach. *Water Res.* 118: 167–176.

Yan, C.Z., Che, F.F., Zeng, L.Q., Wang, Z.S., Du, M.M., Wei, Q.S., Wang, Z.H., Wang, D.P. & Zhen, Z. 2016. Spatial and seasonal changes of arsenic species in Lake Taihu in relation to eutrophication. *Sci. Total Environ.* 563–564: 496–505.

Environmental Arsenic in a Changing World –
Zhu, Guo, Bhattacharya, Ahmad, Bundschuh & Naidu (Eds)
ISBN 978-1-138-48609-6

Elevated oxidizing compounds influencing the biogeochemistry of arsenic in subsurface environments

W.J. Sun
Department of Civil and Environmental Engineering, Southern Methodist University, Dallas, Texas, USA

ABSTRACT: Arsenic contamination of groundwater and surface water is a worldwide problem, which poses health risks to millions of people in the world. The arsenic in the groundwater is of natural origin, and is released from the weathering of arsenic bearing minerals into the groundwater, owing to the anaerobic conditions of the subsurface. In this study, the oxidation of arsenite (As(III)) to arsenate (As(V)) linked to oxidizing compounds under anoxic conditions was shown to be a widespread microbial activity in anaerobic sludge and sediment samples that were not previously exposed to arsenic contamination. The results indicate that microbial oxidation of As(III) and Fe(II) linked to denitrification resulted in the enhanced immobilization of aqueous arsenic in anaerobic environments by forming Fe(III) (hydr)oxide coated sands with adsorbed As(V). Thus, the elevated oxidizing compounds could play critical roles in influencing the biogeochemistry of arsenic in subsurface environments.

1 INTRODUCTION

Arsenic contamination of groundwater and surface water is a worldwide problem, which poses the health risks to millions of people in the world (Smedley & Kinniburgh, 2002). The arsenic in the groundwater is largely of natural origin, and is released from the weathering of arsenic bearing minerals into the groundwater, owing to the anoxic conditions of the subsurface. As presented in Figure 1, microbiological processes play significant roles in controlling the fate and transport of arsenic in the natural environments. Under reducing conditions, microbial reduction of arsenate (As(V)) to arsenite (As(III)) and ferric (hydr)oxides to soluble Fe(II) are considered as the dominant mechanisms of arsenic mobilization in subsurface environments. On the other hand, if oxidizing conditions can be restored, arsenic can be immobilized by the formation of As(V) and ferric (hydr)oxides. As(V) is more strongly adsorbed than As(III) at circum-neutral conditions by common non-iron metal oxides in sediments such as those of aluminium. Ferric (hydr)oxides have strong affinity for both As(III) and As(V) in circum-neutral environments. Nitrate or (per)chlorate can be considered as alternative oxidants with advantages over elemental oxygen due to their high aqueous solubility and lower chemical reactivity which together enable them to be better dispersed in the saturated subsurface. The main objective of this study was to illustrate the importance of anoxic oxidation of As(III) and Fe(II) linked to oxidizing compounds (nitrate or (per)chlorate) under anoxic conditions in the biogeochemical cycle of arsenic.

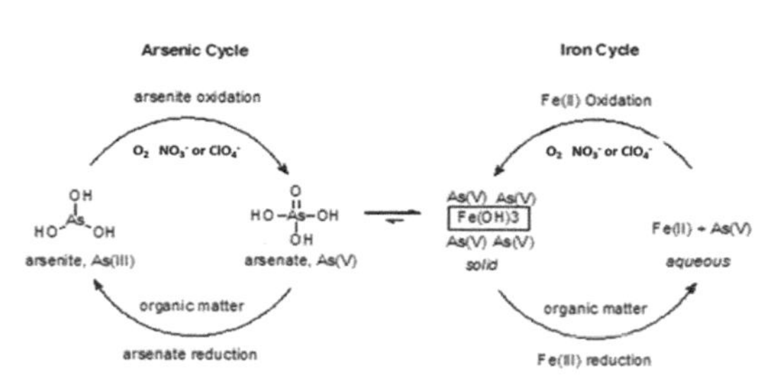

Figure 1. Biogeochemical redox cycles of arsenic and iron.

2 METHODS/EXPERIMENTAL

2.1 *Microorganisms*

Sludge and sediment samples obtained from different sources, including aerobic activated sludge and anaerobically digested sewage sludge, methanogenic granular sludges, and pond sediments, were used as inocula.

2.2 *Batch bioassay*

Batch bioassays were performed under anoxic conditions in serum flasks that were supplied with 120 mL of a basal mineral medium (pH 7.0–7.2) containing bicarbonate as the only carbon source. The medium was also supplemented with As(III) as the electron donor and nitrate (typically 10 mM, unless otherwise specified) as the electron acceptor. Various controls (e.g., abiotic controls, killed sludge controls, controls without electron acceptor) were applied based on the requirements of each experiment. All assays were conducted in triplicate.

Table 1. Summary of microbial As(III) oxidation under denitrifying conditions.

Inoculum		As(V) formation		Time
Name	Sources	With NO_3^-	Without NO_3^-	Days
NGS	Industrial UASB reactor	0.423 ± 0.004	ND	6
EGS	Industrial UASB reactor	ND	ND	ND
ADS	Anaerobic digested sludge	0.425 ± 0.003	ND	13
RAS	Aerobic activated sludge	ND	ND	ND
TDE	Thiosulfate-oxidizing denitrifier	0.416 ± 0.002	ND	10
DPS	Lake sediment	0.415 ± 0.003	ND	6
WCS	Winogradsky column	0.413 ± 0.001	ND	5
PCS	Pinal creek sediment	0.295 ± 0.020	0.298 ± 0.014	>14

ND: Not Detected

2.3 *Column study*

Anoxic As(III) and Fe(II) oxidation under denitrifying conditions was investigated in two glass sand packed bed columns that were fed continuously with synthetic basal medium and inoculated with As(III)-oxidizing denitrifying culture. The treatment column (SF1) was the biologically active column inoculated with chemolithotrophic As(III)-oxidizing denitrifying bacteria and fed with basal medium, As(III) (0.5 mg L^{-1}) and Fe(II) (20 mg L^{-1} Fe) as electron donating substrates, nitrate (155 mg L^{-1} NO_3^-) as the electron acceptor. The control column (SF2) was the same as SF1 but lacked nitrate in the medium.

2.4 *Analytical method*

As(III) and As(V) were analyzed by ion chromatography–inductively coupled plasma–mass spectroscopy (HPLC–ICP–MS). Nitrate and nitrite were analyzed by suppressed conductivity ion chromatography using a Dionex 3000 system. Other analytical determinations (e.g., pH, TSS, VSS) were conducted according to standard methods (APHA, 1999).

3 RESULTS AND DISCUSSION

3.1 *Batch results*

Sediments and sludge samples from environments not known to be contaminated with arsenic were incubated with As(III) either in the presence and absence of NO_3^- and were incubated in the absence of elemental oxygen. As presented in Table 1, six out of the eight inocula tested displayed microbial activity towards the anoxic oxidation of As(III) by nitrate. Five out of the six positive samples showed a dependency on the presence of NO_3^- for the anoxic oxidation of As(III). One of the positive samples, PCS, had this activity both in the presence and absence of added NO_3^-, which may have been due to high levels of oxidized manganese species known to occur in that sediment (Lind & Hem, 1993). The average molar yield of nitrate linked As(III) oxidation was 0.94 ± 0.04 mol As(V) formed mol^{-1} As(III) consumed and the value for the PCS sediment was similar. The results demonstrated that the anoxic oxidation of As(III) to As(V) linked to denitrification is a widespread microbial activity in anaerobic sludge and sediment samples that were not previously exposed to arsenic contamination.

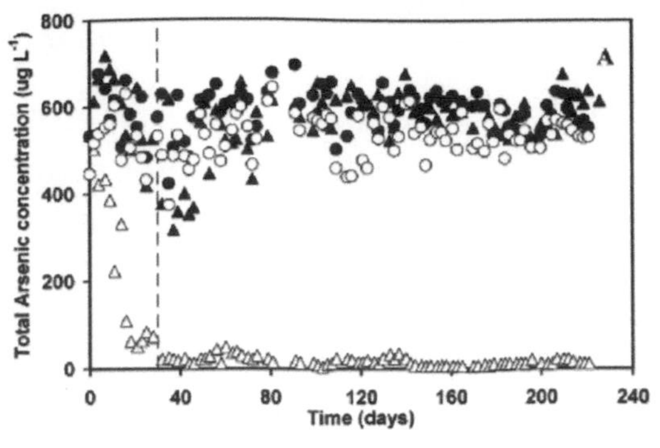

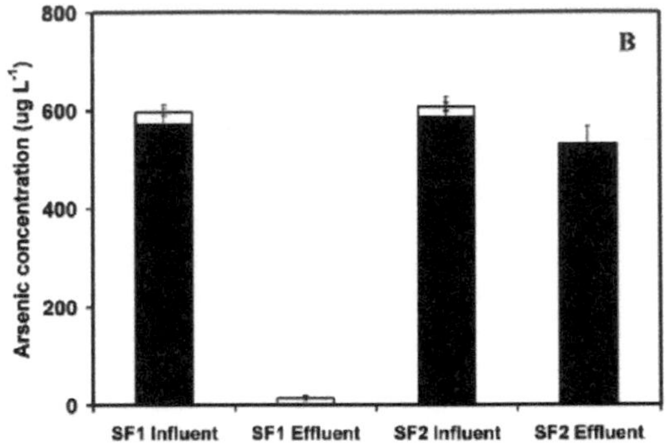

Figure 2. (A) Removal of soluble total As in two sand packed columns fed with 6.67 μM As(III) and 360 μM Fe(II). Column SF1 (fed with 2.5 mM nitrate): (▲) influent, (△) effluent. Column SF2 (without nitrate): (•) influent, (○) effluent. The dashed line indicates the day when the steady state operation was achieved. (B) Arsenic speciation in the influent and effluent of sand packed columns SF1 or SF2: As(III) (solid bars) and As(V) (empty bars).

3.2 *Column results*

The oxidation of As(III) linked to the use of common-occurring nitrate as an electron acceptor may be an important missing link in the biogeochemical cycling of arsenic. Thus, a bioremediation strategy was explored that is based on utilizing nitrate to support the microbial oxidation of Fe(II) and As(III) in the subsurface as a means to immobilize arsenic. Continuous flow columns packed with sand were used to simulate a natural anaerobic groundwater and sediment system with co-occurring As(III) and Fe(II) in the presence (column SF1) or absence (column SF2) of nitrate, respectively.

The time-course of the influent and effluent total soluble arsenic concentrations from the two columns is illustrated in Figure 2A. The result shows that the release of soluble As was greater in SF2 compared to SF1, which is in accordance with the expected adsorption of As on the Fe(III) (hydr)oxides formed from anoxic Fe(II) oxidation. Figure 2B illustrates the average soluble arsenic species in the influent and effluent of columns SF1 and SF2 during the steady state period of operation from day 30 onward. The results show that 99.7% As(III) was eliminated from column SF1 and that it was not recovered as soluble As(V) in the effluent. In contrast, As(III) removal was marginal (9.7%) in the control column lacking nitrate (SF2). The results suggest that no adsorption had occurred, consistent with the low retention of Fe in the column.

4 CONCLUSIONS

Microbial oxidation of As(III) and Fe(II) by denitrifying microorganisms led to the formation of ferric (hydroxides) which adsorbed As(V) formed from As(III)-oxidation. The studies demonstrated that anoxic microbial oxidation of As(III) and Fe(II) linked to denitrification significantly enhance the immobilization of arsenic in the anaerobic subsurface environments and may reduce the risk of arsenic contaminated water to public health and environmental ecological systems.

ACKNOWLEDGEMENTS

The work presented here was funded by a USGS, National Institute for Water Resources 104G Grant (2005AZ114G) and by a grant of the NIEHS-supported Superfund Basic Research Program (NIH ES-04940).

REFERENCES

APHA. 1999. Standard methods for the examination of water and wastewater, twentieth ed. American Public Health Association, Washington, DC.

Lind, C.J. & Hem, J.D. 1993. Manganese minerals and associated fine particulates in the streambed of Pinal Creek, Arizona, USA: a mining-related acid drainage problem. *Appl. Geochem.* 8 (1): 67–80.

Smedley, P.L. & Kinniburgh, D.G. 2002. A review of the source, behaviour and distribution of arsenic in natural waters. *Appl. Geochem.* 17 (5): 517–568.

ISBN 978-1-138-48609-6

Multiple species of arsenic biotransformation occur in *Nostoc* sp. PCC 7120

X.M. Xue & Y.G. Zhu
Key Lab of Urban Environment and Health, Institute of Urban Environment, Chinese Academy of Sciences, Xiamen, China

ABSTRACT: Nostoc cells incubated with arsenite (As(III)) for two weeks were extracted with dichloromethane/methanol (DCM/MeOH) and the extract was partitioned between water and DCM. Arsenic species in aqueous and DCM layers were determined using high performance liquid chromatography – inductively coupled plasma mass spectrometer/electrospray tandem mass spectrometry (HPLC-ICPMS/ESIMSMS). In addition to inorganic arsenic (iAs), the aqueous layer also contained monomethylarsonate (MAs(V)), dimethylarsinate (DMAs(V)), and the two arsenosugars, namely a glycerol arsenosugar (Oxo-Gly) and a phosphate arsenosugar (Oxo-PO_4). Two major arsenosugar phospholipids (AsSugPL982 and AsSugPL984) were detected in DCM fraction. Arsenic in the growth medium was also investigated by HPLC/ICPMS and shown to be present mainly as the inorganic forms As(III) and As(V) accounting for 29%–38% and 29%–57% of the total arsenic, respectively. The total arsenic of methylated arsenic, arsenosugars, and arsenosugar phospholipids in Nostoc cells with increasing As(III) exposure were not markedly different, indicating that the transformation to organoarsenic in Nostoc was not dependent on As(III) concentration in the medium.

1 INTRODUCTION

Arsenic (As) is a ubiquitous and carcinogenic toxic element, and has both acute and chronic toxicity effects on humans. The bioavailability of arsenic and its resultant toxicity are influenced to a great extent by its species.[1] Inorganic arsenic, the major form of arsenic in water and soils, is transformed into organic arsenic species or in reverse in natural biological processes, and microorganisms play a critical role in arsenic biogeochemical cycle (Zhu *et al.*, 2014).

Cyanobacteria are involved in arsenic biogeochemical cycle, and have been reported to have the ability to methylate inorganic arsenic, producing arsenosugars and arsenosugar phospholipids. The previous studies showed that *Nostoc* methylated As(III) to DMAs(V) and TMA(O) (Yin *et al.*, 2011), demethylated MAs(V) and MAs(III) into As(III) (Yan *et al.*, 2015), and produced Oxo-Gly (Miyashita *et al.*, 2012). In this study, HPLC-ICPMS/ESIMS was used to analyze arsenic biotransformation in *Nostoc* in order to understand arsenic biotransformation by *Nostoc* from multiple perspectives.

2 METHODS/EXPERIMENTAL

2.1 *Fractionation of arsenic in Nostoc*

About 30 mg of freeze-dried cells were weighed (to a precision of 0.1 mg) directly into a centrifuge tube (15 mL, polypropylene), 5 mL of a mixture of DCM/MeOH (2 + 1, v/v) was added. The mixture was extracted on a rotary wheel overnight, and centrifuged at 4754 g and 4°C for 15 minutes. 0.5 mL of 1% aqueous NH_4HCO_3 solution was added to the supernatant (~4.5 mL), the solution was separated into an aqueous layer (upper layer, MeOH and H_2O) and DCM-MeOH layer (lower layer).

2.2 *HPLC-ICPMS/ESIMSMS analysis of water-soluble arsenic species and arsenolipids*

Separation was performed under reversed-phase conditions using a Shodex Asahipak C8P-50 4D column (4.6 × 150 mm, 5 μm particle size). The column effluent was split using a passive splitter with 80% being transferred directly to the ESI-MSMS. The remaining 20% of the split flow was transferred to the ICPMS together with a support flow of water containing 1% formic acid and 20 μg L^{-1} Ge, In, Te (0.3 mL min^{-1}) introduced through a T-piece after the splitter.

3 RESULTS AND DISCUSSION

3.1 *Results*

Arsenic species in the aqueous layers were analyzed, and revealed that two arsenosugars were produced by *Nostoc*. The relative proportion of arsenosugars in total aqueous arsenic species in *Nostoc* cells was 0.5%–13% for Oxo-Gly and 0.7%–8% for Oxo-PO_4, and decreased with increasing As(III) exposure.

HPLC/ICPMS analysis of the DCM fraction showed the presence of arsenolipids, and analysis by ESMSMS revealed that the $[M+H]^+$ of two main arsenic-containing compounds were 983 and 985. The chromatographic behavior was the same as that reported previously for arsenosugar phospholipids

($C_{47}H_{88}O_{14}AsP$ (As-PL982) and $C_{47}H_{90}O_{14}AsP$ (As-PL984) in *Synechocystis* sp. PCC 6803 and brown macroalgae.

3.2 *Discussion*

Multiple species of arsenic biotransformation pathways co-occur in *Nostoc*. Microbes have evolved various mechanisms to utilize or detoxify arsenic. Known mechanisms include arsenic redox changes, arsenic methylation and demethylation, As(III) efflux, and the production of complex organoarsenic. Arsenic reduction and efflux are always considered as significant arsenic detoxification in microorganisms. The previous proteomics analysis of *Nostoc* under As(V) stress showed that the expression of two genes, *alr1097* (encoding an As(III) efflux protein; *arsB*) and *alr1105* (encoding an As(V) reductase; *arsC*), was up-regulated, (Pandey, *et al.*, 2012), illustrating that *Nostoc* was capable of performing As(V) uptake, As(V) reduction, and As(III) excretion. Similar results were also found in cyanobacteria *Synechocystis* sp. PCC 6803 and *Microcystis aeruginosa*, some globally significant picocyanobacteria *Prochlorococcus* also were reported to have genomic potential for As(V) reduction and As(III) efflux. (Saunders & Rocap 2015). Yin *et al.* (2011) investigated the ability of arsenic methylation by *Nostoc* and arsenic(III) S-adenosylmethionine methyltransferase (ArsM) from *Nostoc* by chemotrapping volatile TMAs(III). Moreover, our other study showed that *Nostoc* was able to demethylate MAs(III) rapidly to As(III) using ArsI that is a C·As lyase responsible for MAs(III) demethylation, and also could demethylate MAs(V) slowly to As(III), (Yan *et al.*, 2015) suggesting that MAs(V) reduction and MAs(III) demethylation occurred in *Nostoc*. However, MAs(III) oxidation in *Nostoc* cannot be regarded to be catalyzed by an enzyme because ArsH homologous compound oxidizing MAs(III) (Chen, *et al.*, 2015) was not found via blasting against the *Nostoc* proteome with ArsH of *Synechocystis* sp. PCC 6803, (Xue *et al.*, 2014) and we did not ensure that there was As(III) oxidation catalyzed by As(III) oxidase in *Nostoc* because most of As(III) was oxidized by oxygen after long time culture and there was not As(III) oxidase identified in *Nostoc*. In addition, *Nostoc* was found to produce low quantities of arsenosugars and arsenolipids.

An abbreviated biosynthesis pathway of arsenosugar phospholipids in *Nostoc* was hypothesized as described in Figure 1. The methyl groups from S-adenosyl-L-methionine (SAM) are transferred to As(III) (which is absorbed directly by cells or arises from As(V) reduced by ArsC), by ArsM to produce trivalent methylated arsenic. Some of MAs(III) and DMAs(III) bound to ArsM are further methylated into DMAs(III) and TMAs(III). In addition, the adenosyl group from SAM is transferred to DMAs(III) falling off ArsM to generate the key intermediate of arsenosugar synthesis, namely dimethylarsinyladenosine (AsAd). AsAd undergoes glycosidation to produce Oxo-Gly which acts as a precursor to $Oxo\text{-}PO_4$

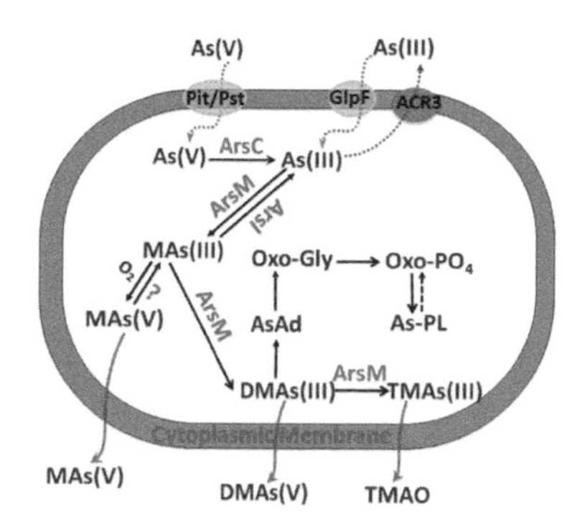

Figure 1. Arsenic metabolism and potential biosynthetic pathways of arsenosugar phospholipids in *Nostoc*.

produced later by cyanobacteria. Finally, fatty cids were added to $Oxo\text{-}PO_4$ to produce arsenosugar phospholipids by unknown enzymes.

In summary, our study revealed that the model organism cyanobacterium *Nostoc* can produce $Oxo\text{-}PO_4$ and arsenosugar phospholipids when exposed to As(III), and the production of complex organoarsenic and arsenic demethylation co-exist in *Nostoc*.

ACKNOWLEDGEMENTS

Our research is supported by the National Natural Science foundation of China (21507125).

REFERENCES

Chen, J., Bhattacharjee H., & Rosen B.P. 2015. ArsH is an organoarsenical oxidase that confers resistance to trivalent forms of the herbicide monosodium methylarsenate and the poultry growth promoter roxarsone. *Mol. Microbiol.* 96(5): 1042–1052.

Miyashita, T., Oda Y., Horiuchi, J., Yin, J.C., Morimoto, T. & Saitoe M. 2012. Mg^{2+} block of Drosophila NMDA receptors is required for long-term memory formation and CREB-dependent gene expression. *Neuron.* 74(5): 887–98.

Pandey, S., Rai, R. & Rai, L.C. 2012. Proteomics combines morphological, physiological and biochemical attributes to unravel the survival strategy of anabaena Sp. PCC7120 under arsenic stress. *J Proteomics.* 75(3): 921–937.

Saunders, J. K. & Rocap, G. 2015. Genomic potential for arsenic efflux and methylation varies among global prochlorococcus populations. *The ISME Journal* 10(1): 197–209.

Xue, X.M., Yan, Y., Xu, H.J., Wang, N., Zhang, X. & Ye, J. 2014. ArsH from synechocystis Sp. PCC 6803 reduces chromate and ferric iron. *FEMS Microbiol Letters* 356 (1): 105–12.

Yan, Y., Ye, J., Xue, X.M. & Zhu, Y.G. 2015. Arsenic demethylation by a C · As cyase in cyanobacterium nostoc Sp. PCC 7120. *Environ. Sci. Technol.* 49: 14350–58.

Yin, X.X., Chen, J., Qin, J., Sun, G.X., Rosen, B.P. & Zhu Y.G. 2011. Biotransformation and volatilization of arsenic by three photosynthetic cyanobacteria. *Plant Physiol.* 156 (3): 1631–38.

Zhu, Y.G., Yoshinaga M., Zhao F.J. & Rosen B.P. 2014. Earth abides arsenic biotransformations. *Annu. Rev. Earth Plan. Sci.* 42(1): 443–67.

Environmental Arsenic in a Changing World –
Zhu, Guo, Bhattacharya, Ahmad, Bundschuh & Naidu (Eds)
ISBN 978-1-138-48609-6

Irrigation activities affecting arsenic mobilization in topsoil in Datong Basin, northern China

Z.Y. Xiao & X.J. Xie
State Key Laboratory of Biogeology and Environmental Geology & School of Environmental Studies, China University of Geosciences, Wuhan, P. R. China

ABSTRACT: The use of high arsenic groundwater for irrigation has been prevailed for decades in the Datong Basin. To reveal the characteristics of As mobilization in topsoil under irrigation activities, a filed plot-experiment has been conducted in the central area of Datong Basin. The irrigation activities promoted Fe(II) oxidizing to Fe(III) (oxyhydr)oxide, and As was adsorbed on Fe(III) (oxyhydr)oxide. As a result, the contents of As and Fe in sediments increased finally. Meanwhile, the irrigation activities led to the increase of soil salinity. Moreover, inverse geochemical modeling suggested that geochemical process was conducted during field experiment. The results illustrated temporal changes in As concentration under irrigation activities.

1 INTRODUCTION

Arsenic (As) contamination of water and soil has been recognized as a grand challenge for humans as well as for fauna and flora (Mandal & Suzuki, 2002). The speciation of arsenic is affected by multiple geochemical processes. Previous studies have demonstrated that the migration and transformation of As in aquifers are controlled by multiple geochemical processes, closely associated with redox environment and Fe (oxyhydr)oxide (Postma *et al.*, 2012). Plenty of agricultural irrigation activities have prevailed in the Datong basin in the past decades. Although high As groundwater in Datong Basin was widely documented in recent years, there are few data of unsaturated As concentrations of soil and soil water. Therefore, the aims of this study are: 1) delineate the composition of soil water and soil sediments under irrigation; 2) understand the arsenic fate in topsoil during irrigation activities.

2 MATERIAL AND METHODS

2.1 *Field experiment*

The field experiment of 2015 and 2016 has been lasting for 3 stages respectively, pre-irrigation before Aug. 10th, irrigation from Aug. 11th to 18th, and post irrigation from Aug. 19th to 24th, 2015, and pre-irrigation before Sep. 5th, irrigation from Sep. 6th to 14th, and post irrigation from Sep 15th to 17th, 2016. The soil water samples have been tapped from 0.5 m and soil sediments samples have been excavated twice in 2015 with the interval of 0.2 m from the depths 0 to 2.0 m on Aug. 10th and Aug. 24th. The irrigated water was from the shallow groundwater aquifer with high As concentration far from the irrigation filed. The chemical and physical parameters including temperature, pH, EC and ORP values were measured on site using an HACH portable meter calibrated before use.

2.2 *Laboratory analysis*

Concentrations of total As were analyzed by hydride generation atomic fluorescence spectrometry (HG-AFS) (AFS-820, Titan). The major cations were determined by ICP-AES (IRIS Intrepid II XSP) and anion concentrations were determined using ion chromatography (IC) (Metrohm 761 Compact). The total As concentrations in the soil samples were dissolved by aqua regia and determined by HG-AFS while total Fe concentrations in the soil samples were digested in the Muffle furnace for 8 h first and dissolved by HCl with 100°C for 6 h, finally measured using o-phenanthroline spectrophotometric analysis method by the spectrophotometry HACH, DR2800.

3 RESULTS AND DISCUSSION

3.1 *Hydrochemistry*

The EC values from the soil samples show the tendency that the conductivity values increase with irrigation on-going (Fig. 1a), while As concentrations show an opposite trend (Fig. 1b).

Moreover, the plots of 2016 vs. 2015 clearly indicate that irrigation activities would result in the salinization and the accumulation of As in topsoil. Intensive extraction of groundwater for agricultural activities and domestic purpose over decades may have made irrigation return flow and salt flushing water become the source of recharge for aquifers at Datong Basin. Meanwhile, the high rate of the evaporation aggravated the phenomena of salinization so that the soil water presented high concentration of salt as well as the groundwater sample of shallow aquifer.

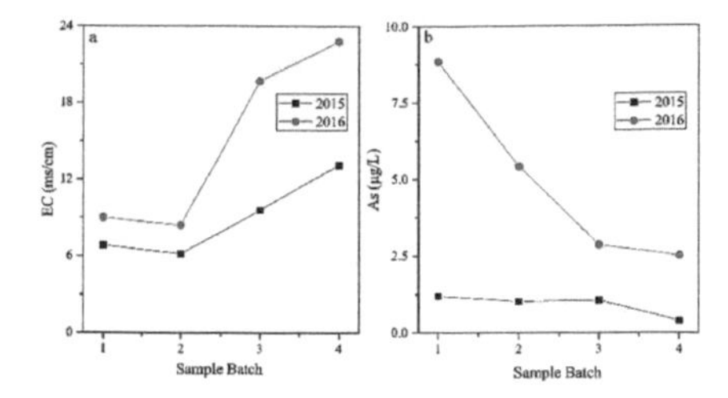

Figure 1. Plots of EC values of 0.5 m from soil water samples of each batch (1 for Inirri-1, 2 for Inirri-2, 3 for Postirri-1 and 4 for Postirri-2).

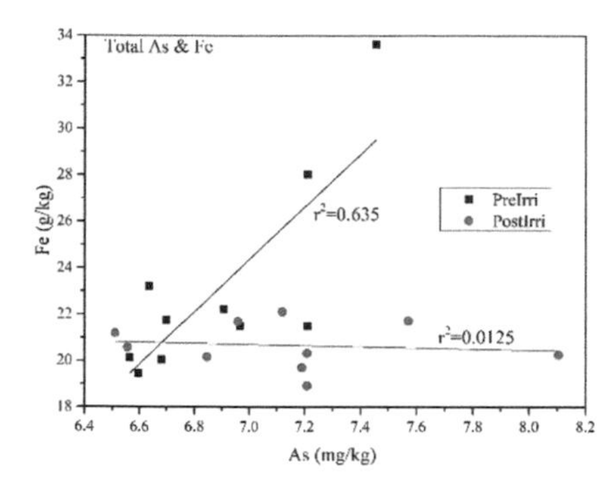

Figure 2. Correlations between total As vs. total Fe from soil sediments analysis.

3.2 *As in sediments*

Geochemical analysis on sediment samples was conducted to exploit potential effects of the forms of As and Fe minerals on As mobilization. Before irrigation activities conducting, the concentrations of total As and Fe show a positive correlation (Figure 2). However, the concentrations of total As and Fe reveal a negative correlation (Figure 2), suggesting that with irrigation on-going As has decoupled with Fe gradually.

3.3 *Inverse geochemical modeling*

Inverse geochemical modeling was made to further analyze the impact of irrigation on As mobilization by code PHREEQC-3 with the wateq4f.database (Charlton & Parkhurst 2011). The soil water chemical compositions of HCO_3^-, SO_4^{2-}, Cl^-, Ca^{2+}, Mg^{2+}, Na^+, Fe, as well as pH were used in the modeling. To ensure the charge balance satisfy the uncertainty limit, the composition Cl^- was chose to adjust the balance. The mineral phases of the modeling were chose from the XRD results and the SI (saturation indices) numbers calculated by PHREEQC-3. Soil water samples of 2015 were chosen for the modeling. The summary of the most optimum choices of the inverse geochemical modeling with phase mole transfers of the minerals and gases is given in Table 1. Modeling results suggest that gypsum and siderite are generally dissolving while calcite, $Fe(OH)_3$(a), FeS(ppt), and CO_2(g) are precipitating. However, H_2S (g) are changed with irrigation activities on-going. The dissolution of H_2S (g) forms

Table 1. The results of inverse geochemical inverse modeling of soil water evolution under irrigation activities.

	Phase mole transfers			
Minerals	PreIrri to InIrri1	InIrri1 to InIrri2	InIrri2 to PostIrri1	PostIrri1 to PostIrri2
Calcite	−0.0098	/	−0.0032	−0.0025
Dolomite	/	/	/	/
Gypsum	0.0098	/	0.0032	0.0025
Siderite	0.0783	0.0251	0.0259	0.0225
$Fe(OH)_3$(a)	−0.0783	/	−0.0259	−0.0200
FeS(ppt)	/	−0.0251	/	−0.0025
CO_2(g)	−0.0685	−0.0251	−0.0227	−0.0200
H_2S(g)	−0.0098	0.0251	−0.0032	/

Positive and negative phase mole transfers indicate dissolution and precipitation, respectively. "/" indicates no phase transfers.

HS- and produces FeS(ppt) in the end. When irrigation, As from soil water showing a decrease trend (Figure 1) indicates that As was adsorbed directed towards Fe sediments since Fe(II) in irrigated water was oxidized to Fe(III) (oxyhydr)oxide (Table 1). Thus, the main geochemical processes during irrigation is oxidization of Fe(II) to Fe(III) (oxyhydr)oxide and adsorption of As in irrigated water with infiltration.

4 CONCLUSIONS

From the hydrochemical properties, sediments results and geochemical modeling, the following characteristics and controlling factors can be inferred.

(1) Irrigation activities would result in the salinization and the accumulation of As in topsoil.
(2) With irrigation on-going, As has decoupled with Fe gradually.
(3) The main geochemical processes indicating from inverse geochemical modeling during irrigation are that Fe(II) was oxidized to Fe(III) (oxyhydr)oxide and As was adsorbed from irrigated water with infiltration.

ACKNOWLEDGEMENTS

The research work was supported by National Natural Science Foundation of China (41372254) and China University of Geosciences (Wuhan).

REFERENCES

Charlton, S. R. & Parkhurst, D. L. 2011. Modules based on the geochemical model PHREEQC for use in scripting and programming languages. *Comput. Geosci.*, 37(10): 1653–1663.

Mandal, B.K. & Suzuki, K.T. 2002. Arsenic round the world: a review. *Talanta* 58(1): 201–235.

Postma, D., Larsen, F., Thai, N. T., Trang, P. T. K., Jakobsen, R., Nhan, P. Q., *along*, T.V., Viet, P.H. & Murray, A.S. 2012. Groundwater arsenic concentrations in vietnam controlled by sediment age. *Nat. Geosci.* 5(9): 656–661.

Environmental Arsenic in a Changing World –
Zhu, Guo, Bhattacharya, Ahmad, Bundschuh & Naidu (Eds)
ISBN 978-1-138-48609-6

Effects of microbial communities on arsenic mobilization and enrichment in groundwater from the Datong Basin, China

L. Yan, X.J. Xie, K.F. Pi, K. Qian, J.X. Li, Z.Y. Chi & Y.X. Wang
School of Environmental Studies & State Key Laboratory of Biogeology and Environmental Geology, China University of Geosciences, Wuhan, China

ABSTRACT: The phospholipid fatty acids (PLFA) as biomarkers of microbial community structures were appointed to elaborate the mobilization of arsenic (As) in groundwater from Datong Basin. The results showed that saturated fatty acids 16:0, 18:0, monounsaturated fatty acids 16:1ω9, 18:1ω9, and cyclopropane fatty acids 17:0, 19:0 had relatively high contents. This may imply the significant impact of dominant *Desulfobacter* and *Clostridium* on the mobilization of As in the shallow groundwater. Sulfate-reducing bacteria and Fe-reducing bacteria can mediate the reductive dissolution of As-bearing Fe oxides/hydroxides to cause the release of As into groundwater.

1 INTRODUCTION

The exposure to high concentrations of arsenic (As) may be fatal to the human health. The genesis of high As groundwater is related to the reductive dissolution of As-bearing iron oxides/hydroxides in the aquifers (Pi *et al.*, 2015). Microbes can mediate different redox reactions, which contributes to the release and enrichment of As in groundwater (Akai *et al.*, 2008). The phospholipid fatty acids (PLFA), forming a major part of the bacterial cell membranes, are regarded to be indicators of living bacterial communities due to their rapid degradation after cell death, which can provide a quantitative measure of the viable biomass and biological diversity of microbial communities (Fang *et al.*, 2006). In the present study, we analyzed lipid biomarkers and the geochemical characteristics of groundwater to illuminate the effects of microbial communities on As mobilization and enrichment in groundwater from the Datong Basin, China.

2 METHODS/EXPERIMENTAL

2.1 *Sampling and geochemical analysis*

The sediment samples were collected from a multilevel groundwater monitoring site in October, 2016 from Shanyin County of the Datong Basin, where As contaminated groundwater occurs. The fresh sediment samples were air-dried, ground, and sieved through a 200 gauge screen. After 0.05 g of the prepared sample was digested using ultra-pure concentrated HNO_3 and HF at 180°C for 24 h, the trace and major elements were tested using inductively coupled plasma mass spectroscopy (ICP-MS) and hydride generation atomic fluorescence spectrometry (HG-AFS), respectively. Total organic carbon (TOC) contents of the sediments were tested using an elemental analyzer (Vario EL cube, Elementar) after the samples were air-dried, ground to 200 mesh and cleaned of inorganic carbon with 0.1 M HCl.

2.2 *PLFA analysis and statistical assay*

A mild alkaline methanolysis method was used to extract total PLFAs, which were then dissolved in 0.5 mL of 1:1 hexane:methyl-*tert* butyl ether and transferred to GC vials for analysis. The correlations of PLFAs contents in the samples were verified via applying principal components analysis using the Past 3.0 software package.

3 RESULTS AND DISCUSSION

3.1 *Sediment geochemistry*

The particle size of the core samples ranged from clay, silt to fine sand. The color of sediments varied from yellow-brown, brown to gray-black, the latter in line with the high TOC levels (0.08 wt.% to 0.41 wt.%, average 0.16 wt.%). The bulk As contents in sediments ranged from 7.20 to 91.6 mg kg^{-1} with an average value of 22.2 mg kg^{-1}, far exceeding the average level of 5–10 mg kg^{-1} in typical modern sediments. The bulk Fe contents varied from 28.4 g/kg to 51.0 g kg^{-1}, with an average of 37.0 g kg^{-1}. Total concentrations of the elements are relatively uniform among all samples, and showed tendency of enrichment in the finer clays. It indicated that trace elements such as As, Cu, Ni were subjected to the isologous geochemical processes in the sediments or had common sources and could be adsorbed by the oxides/hydroxides of Fe/Mn/Al.

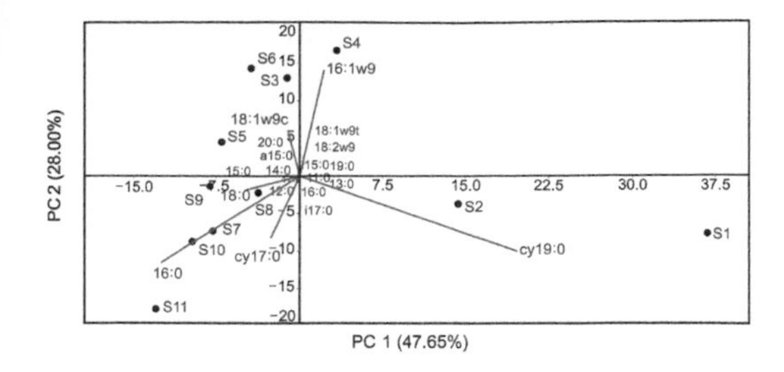

Figure 1. The principal component scattering of PLFAs in sediments from the Datong Basin.

There is a positive correlation between TOC and total As in the bulk sediments ($r^2 = 0.84$), demonstrating As is prone to be enriched in clay sediments with high organic matter content. Dispersed sedimentary organic matter in the aquifer may be easily mobilized to groundwater to promote microbial reactions. Furthermore, the organic matter coated on the surface of Fe minerals can be used as electron donors to fuel in-situ reductive dissolution of Fe(III) oxides/hydroxides, releasing DOM, Fe(II) and As into groundwater.

3.2 *PLFA profiles*

PLFAs with chain length from C_{11} to C_{20} were detected in sediments from Datong Basin. The PLFA profiles were dominated by saturated straight even chain PLFAs (12:0, 14:0, 16:0, 18:0, 20:0), monounsaturated fatty acids (16:1ω9, 18:1ω9), cyclopropane fatty acids (cy17:0, cy19:0), and branched fatty acids (i15:0, a15:0, i16:0, i17:0), whose relative content was 24.31%–90.78%, 11.71%–52.93%, 7.64%–53.87% and 1.69%–4.70%, respectively.

The PLFA profiles of the collected samples were analyzed by principle component analysis (PCA, Fig. 1). PC1 and PC2 accounted for 47.65% and 28.00% of the variances. Factor loading values were the projection of PLFAs on the corresponding ordination axis. The values manifested that 16:0, 18:0, cy19:0, cy17:0 had a strong positive relationship with PC1, which was positive to sulfate-reducing bacteria (SRB: *Desulfobactor*). The 16:1ω9 and 18:1ω9c had good correlation with PC2, which indicated that Fe-reducing bacteria (FeRB: *Clostridium*) were positive to PC2.

3.3 *Specific PLFA analysis*

The contents of arsenic and cyclopropane fatty acids (cy17:0 + cy19:0) showed the negative relationship along the borehole, illustrating high abundant of SRB and FeRB would favor the mobilization of arsenic (Fig. 2). Under strongly reducing conditions, sulfate-reducing bacteria can reduce SO_4^{2-} and produce soluble sulfide. The dissolved sulfide can directly reduce As-bearing Fe(III) oxides/hydroxides. Simultaneously, FeRB may also cause the reductive dissolution of Fe(III) oxides/hydroxides and As release.

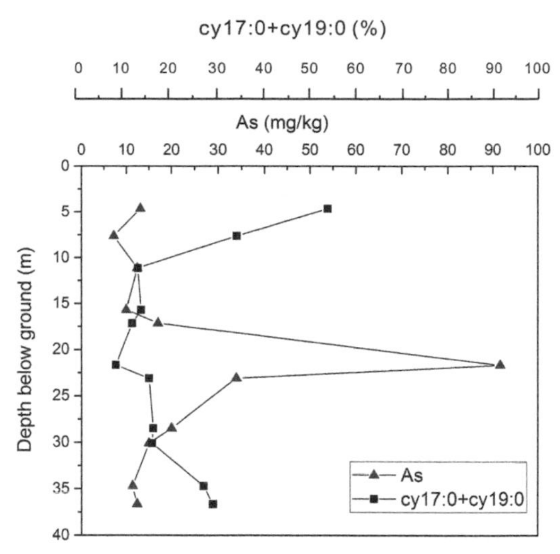

Figure 2. The depth profiles of the contents of arsenic and cyclopropane fatty acids (cy17:0 + cy19:0).

4 CONCLUSIONS

Biomarkers of anaerobia, such as sulfate-reducing bacteria and iron-reducing bacteria, were identified in sediments. The mobilization and enrichment of arsenic may have been controlled and influenced by different microbial communities and processes in shallow groundwater of the Datong Basin. In the high arsenic region, periodic irrigation activities input exogenous organic matter into subsurface, where SRB and FeRB can utilize OM for sulfate and Fe(III) reduction. The production HS^-, could act a reducing agent for iron oxides/hydroxides, expediting the release of the bound arsenic to the groundwater.

ACKNOWLEDGEMENTS

The research work is financially supported by National Natural Science Foundation of China (Nos. 41372254, 41772255).

REFERENCES

Akai, J.J., Kanekiyo, A., Hishida, N., Ogawa, M., Naganuma, T., Fukuhara, H. & Anawer, H.N. 2008. Biogeochemical characterization of bacterial assemblages in relation to release of arsenic from South East Asia (Bangladesh) sediments. *Appl. Geochem.* 23(11):3177–3186.

Fang, J.S., Chan, O., Joeckel, R.M., Huang, Y.S., Wang, Y., Bazylinski, D.A., Moorman, T.B. & Ang Clenment, B.J. 2006. Biomarker analysis of microbial diversity in sediments of a saline groundwater seep of Salt Basin, Nebraska. *Org. Geochem.* 37(8):912–931.

Pi, K., Wang, Y.X., Xie, X.J., Huang, S.B., Yu.Q. & Yu, M. 2015. Geochemical effects of dissolved organic matter biodegradation on arsenic transport in groundwater systems. *J. Geochem. Explor.* 149(149):8–21.

Environmental Arsenic in a Changing World –
Zhu, Guo, Bhattacharya, Ahmad, Bundschuh & Naidu (Eds)
ISBN 978-1-138-48609-6

The effects of the bioanode on the microbial community and element profile in paddy soil

G. Williamson[1] & Z. Chen[2]
[1] *Department of Environmental Science, University of Liverpool, Liverpool, UK*
[2] *Department of Environmental Science, Xi'an Jiaotong-Liverpool University, Suzhou, P.R. China*

ABSTRACT: In paddy soil the reductive dissolution of iron oxide and the availability of organic matter plays an important role in arsenic release under anaerobic conditions. Microbial fuel cells have been shown to reduce organic matter (OM) content and the rate in which this occurs strongly relate to the external resistance applied. In this study we investigated the effects of bioanode operating at different external resistance on the paddy soil microbial community and iron and arsenic concentration. The results show that MFC can be used to reduce soil pore water iron and arsenic concentration and the extent in which this occurs depend on the external resistance applied. The MFC is able to mitigate arsenic release by decreasing organic matter availability. Furthermore, our finding shows that external resistance had a significant influence on the bacterial community composition that develop on the bioanode however only had minimal effect on the community of the bulk soil. These findings suggest that the sMFC can influence the iron and arsenic concentration by reducing OM content and the microbial community that develop in the bioanode vicinity.

1 INTRODUCTION

Paddy fields are some of the most important farmlands because they are used to cultivate rice; a crop that accounts for more than half of the world's caloric and essential micronutrients intake (Fitzgerald *et al.*, 2009). However, today most paddy fields are contaminated with arsenic (As). Moreover, the increase in arsenic bioavailability in paddy fields has been shown to closely correlate with the redox reactivity of iron oxides. Therefore, investigating iron reduction is crucial to understanding arsenic liberation.

Sediment microbial fuel cells (sMFC) are bioelectrochemical system that has the capacity to simultaneously produce electricity and the removal of organic pollutants (Xu *et al.*, 2015). Recent research have reported that the sMFC bioanode can influence the growth of iron reducing bacteria (Bond *et al.*, 2002; Holmes *et al.*, 2004) and can be used as a substitute for Fe oxide during exoelectrogens anaerobic respiration. Therefore, the objectives of this study were to examine effects of the bioanode at difference external resistance on iron oxide reductive dissolution and As mobility. Likewise, the changes in microbial community structure were also eluted at the end of 90 days operational period. This study is expected to provide new insight on the role of the sMFC on the liberation of As in soil pore water.

2 METHODS/EXPERIMENTAL

Briefly, sMFC was constructed using arsenic pollutant from a rice paddy in Hunan South China. Six different treatments (50 ohms, 80 ohms, 200 ohms, 1000 ohms, 2000 ohms and control) were prepared in triplicates. The control was left at open circuit. A data logger was used to record the voltage between the anode and cathode. Each constructed sMFC was equipped with two soil pore water samplers for collecting pore water, with the lower being adjacent to the anode and the upper being 2 cm below the sediment-water interface in the bulk soil. All soil pore water total arsenic and iron concentration were determined by inductively coupled plasma emission spectrometry (ICPMS) (NexlonTM 350x, Pekin Elmer, USA) and atomic absorption spectrometry (AAS) (PinAAcle™ 900, PerkinElmer, USA), respectively, on day 0, 20, 30, 40, 50 and 90. Loss on ignition (LOI) carbon was determined by heating 10 g of soil at 550°C for 4 hrs in a muffle furnace. The microbial community was analyzed after 90 days of operation using 16S rRNA analysis method. The DNA sample will be extracted from the SMFC by using MoBio Laboratories Inc.'s Powersoil DNA isolation Kit according to the manufacturer's instructions. DNA samples was extracted from two points within each sMFC (bulk soil and the anode associated soil).

3 RESULTS AND DISCUSSION

3.1 *Electricity production from sMFC*

The sMFCs were operated at different external resisters of 50, 80, 200, 1000 and 2000 ohms for 90 days. The current of the all sMFC sharply increase during the initial stage regardless of external resistance and reached a maximum current at around 10 days. The highest current produce was observed at an external resistance of 50 ohms (2.4 mA), followed by those of 80 (2.1 mA), 200 (1.6 mA) and 1000 (0.8 mA) ohms. An external load of 2000 ohms produced the lowest current (0.3 mA). However after ca 50 days all the current output irrespective of external load was approximately the same (0.15–0.2 mA). The results

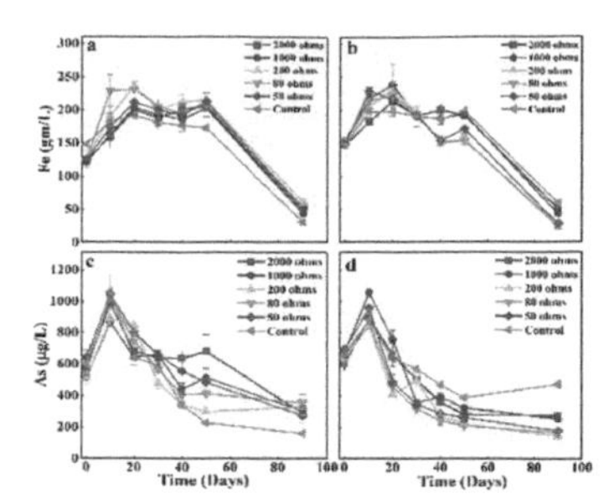

Figure 1. Iron and As variation in soil porewater along with incubation time. a and b represent iron concentration in the bulk soil and anode vicinity, respectively; c and d represent As concentration in the in the bulk soil and anode vicinity, respectively. The error bars represent standard error of measured concentrations of triplicates samples.

obtain here are in accordance with that of Holmes *et al.* (2004) and Hong & Gu (2009), in both studies a maximum current was achieved within 10–20 days which was followed by a gradual decrease current production with time. The different intensity observed in current peak at different external load occurred as a result of higher catalytic activity on the bioanode at lower loads.

3.2 *Organic matter (OM) removal*

The removal efficiencies of soil OM increased with decreasing external resistance. The removal efficiency of OM in the vicinity of the bioanode were 17.7%, 19.7%, 22.6%, 24.6% and 25.1% for the control, 2000 ohms, 1000ohms, 200 ohms, 80 ohms and 50 ohms sMFC, respectively. These finds indicates that external resistance can influence OM removal efficiency (Song *et al.*, 2010).

3.3 *Change in total iron and arsenic*

The iron and As concentration decreased with decreasing external resistance in the bioanode vicinity. Figure 1a–d illustrates the transformation of As and iron in different treatment over the course of the 90 days of incubation. As shown in Figure 1a and b, when sMFC was applied the iron concentration in in all of the treatments to a maximum on day 20 and then gradually declined until the end of the experiment. Arsenic concentration in soil pore water followed a similar trend (Figs. 1c and 1d). At the end of the experiment lower iron and As concentrations were observed in all treatments in the vicinity of the bioanode compared to the control. Furthermore, treatments with lower external resistance (50, 80 and 200 ohms) had significantly lower soil pore water iron and As concentration compared with those at higher external resistance (1000 and 2000 ohms).

However in the bulk soil 2 cm away from the bioanode slightly higher iron and As concentration was observed in the treatments compared to control group. This suggests in the vicinity of the bioanode iron reduction and As mobility in the anoxic paddy soil was limited. This occurred as a result of increase OM removal by the bioanode.

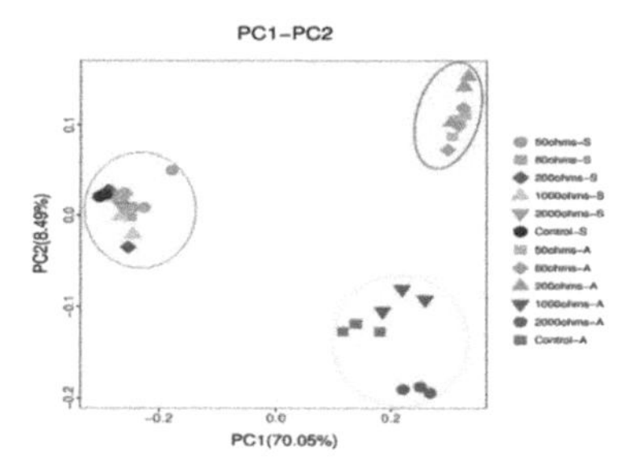

Figure 2. Principal Coordinates Analysis (PCoA) of the sMFC and controls bacterial community composition based on the unweighted UniFrac distance matrix.

3.4 *Microbial community*

The Illumina high through-put sequencing technique was used to investigate the influence of the different external resistance on the microbial community structure. Results shows that the external resistance applied can alter the microbial community that develop in the bioanode associated soil, but had minor effect on the bulk soil community (Fig. 2).

4 CONCLUSIONS

The results obtained in this study demonstrated that operating sMFCs at different external resistance can influence: i) The soil organic matter removal efficiency; ii) The iron and As release in anaerobic paddy soil; and iii) the microbial community that develop on the bioanode.

ACKNOWLEDGEMENTS

This work was supported by the National Science Foundation of China (41571305).

REFERENCES

Bond, D.R., Holmes, D.E., Tender, L.M. & Lovley, D.R. 2002. Electrode reducing microorganisms that harvest energy from marine sediments. *Science* 295(5554): 483–485.

Fitzgerald, M.A., Mccouch, S.R. & Hall, R.D. 2009. Not just a grain of rice: the quest for quality. *Trends Plant Sci.* 14(3): 133–139.

Holmes, D.E., Bond, D.R., O'Neil, R.A., Reimers C.E., Tender, L.R. & Lovley, D.R. 2004. Microbial communities associated with electrodes harvesting electricity from a variety of aquatic sediments. *Microb. Ecol.* 48(2): 178–90.

Hong, Y.G. & Gu, J.D. 2009. Bacterial anaerobic respiration and electron transfer relevant to the biotransformation of pollutants. *Int. Biodeter Biodegr.* 63(8): 973–980.

Song, T. S., Yan, Z.S., Zhao, Z.W. & Jiang, H.L. 2010. Removal of organic matter in freshwater sediment by microbial fuel cells at various external resistances. *J. Chem. Technol. Biot.*

Xu, B.J., Ge, Z. & He, Z. 2015. Sediment microbial fuel cells for wastewater treatment: challenges and opportunities. *Environ. Sci. Water Res. Technol.* 1(3): 279–284.

Environmental Arsenic in a Changing World –
Zhu, Guo, Bhattacharya, Ahmad, Bundschuh & Naidu (Eds)
ISBN 978-1-138-48609-6

Role of carbonate on arsenic mobilization in groundwater

X.B. Gao, P.L. Gong & W.T. Luo
School of Environmental Studies, China University of Geosciences, Wuhan, China

ABSTRACT: As a highly toxic element, arsenic (As) contamination of groundwater causes challenging environmental and health concerns worldwide. Carbonate system has a significant effect on the migration and transformation of trace elements in groundwater. In this study, we do batch adsorption (isothermal, kinetic adsorption) experiments of arsenic with the synthesized calcite. And some characterization methods were employed. The results show that: (1) the adsorption capability was affected by the initial con-centration of arsenic, (2) The adsorption of As(V) on calcite was a comparatively fast reaction within the first 10 h, and when the reaction time reached 72 h, it seemed to have reached a balanced state of sorption and desorption. The research results could help in enriching the scientific understanding of the environmental behavior of synthesized calcite, with great theoretical and practical significance for the further study of the interaction with arsenic in carbonate system and environment.

1 INTRODUCTION

As a highly toxic element, arsenic (As) contamination of groundwater causes challenging environmental and health concerns worldwide (Nordstrom, 2002; Smedley & Kinniburgh, 2002). In recent years, especially in Asia, with the development of industry and agriculture, the arsenic pollution in the groundwater is very serious, which directly threatens the health of the local residents.

As the second major minerals on the earth's surface, carbonates are widely distributed in groundwater system (Monteshernandez *et al.*, 2009). They directly affect the migration and transformation of elements in groundwater, which including arsenic (Yokoyama *et al.*, 2012; Wei *et al.*, 2003).

2 METHODS/EXPERIMENTAL

Calcite were prepared with addition of a mixed solution of $CaCl_2$ (0.5 M L^{-1}), polyacrylic acid and dextran sulfate sodium (to restrain the formation of aragonite), while Na_2CO_3 (0.5 M L^{-1}) solution was added to carry out precipitation reaction.

The batch adsorption (isothermal, kinetic adsorption) experiments were carried out to investigate the interfacial interaction of As(V) on calcite. The concentrations of corresponding arsenic were: 0, 20, 50, 100, 150, 200, 300 μM L^{-1}. We set up three parallel samples) in the isothermal adsorption, the reaction time was 10, 20, 30 min and 1, 3.5, 6, 10.5, 24, 48, 72 h with the same concentration of arsenic (30 μM L^{-1}) in the kinetic adsorption.

Characterization methods, such as Laser Particle Size Analyzer, XRD, FTIR and SEM, were employed to determine the difference of prepared calcite and calcite after their interaction with As(V) in surface morphology, polymorphs and chemical elements (Xu *et al.*, 2013; Leeuw & Parker, 1998).

3 RESULTS AND DISCUSSION

3.1 *Isothermal adsorption*

The isothermal experiment investigated the capacity of calcite on arsenic with different As initial concentrations.

The adsorption capability is affected by the initial concentration of arsenic. When the arsenic content had increased, the amount of arsenic adsorbed on calcite exhibited an increased trend (Fig. 1). The amount of sorbed arsenic increased linearly with solution As concentration.

3.2 *Kinetic adsorption*

The kinetic evaluation (Fig. 2) in this study had shown that the adsorption of As(V) by calcite was a comparatively fast reaction within the first 10 h, followed by a slower and at a relatively constant reaction rate until equilibrium was achieved. After that, the adsorption rate might have decreased. When the reaction time reached 72 h, it seemed to have reached a balanced state of sorption and desorption.

3.3 *Characterization*

The synthesized calcite, had an average particle diameter of approximately 30 μm, ranging from 25 to 45 μm.

Synthesized calcite (Fig. 3a) had good monodispersity without adhesion phenomenon. After adsorption,

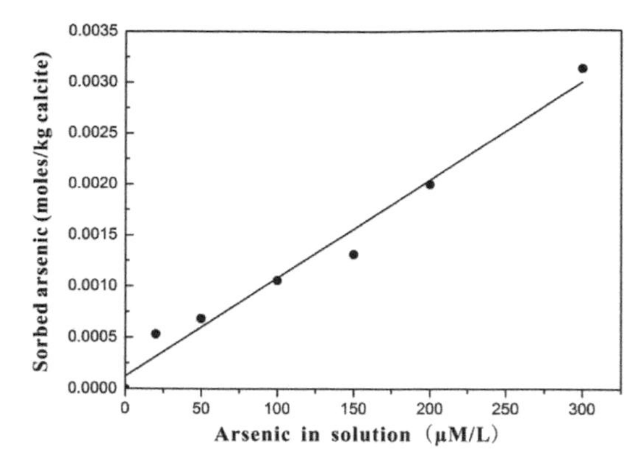

Figure 1. Adsorption isotherms of arsenic on calcite at different initial concentration of arsenic.

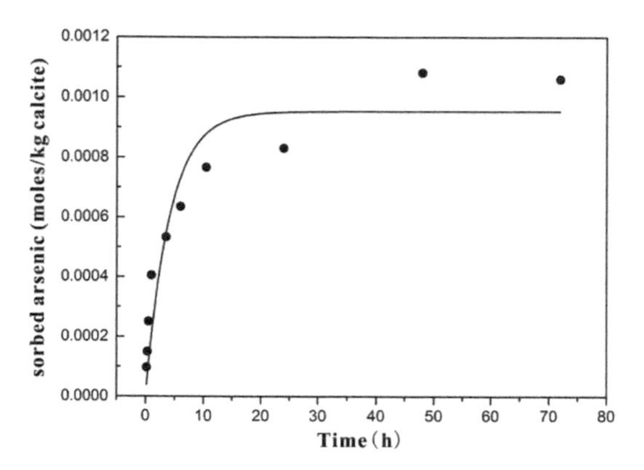

Figure 2. The results of the kinetic uptake experiments as a function of time.

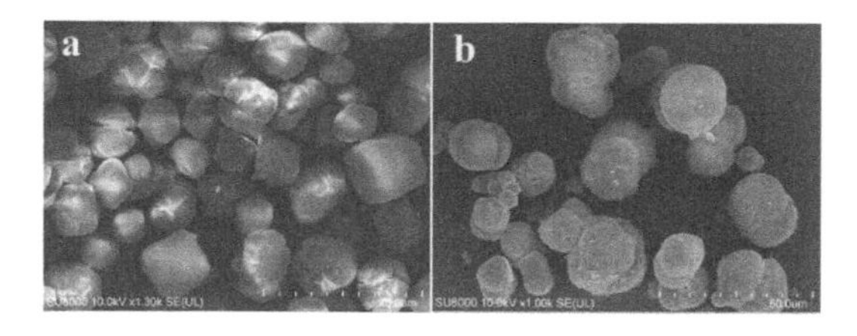

Figure 3. SEM images of (a) Synthesized calcite; (b) calcite after adsorption.

the SEM image (Fig. 3b) showed the increased aggregation and surface pore volume of calcite particles due to the impact of arsenic adsorption. Comparing with Figure 3a and 3b, morphology of calcite got slightly rounded after reaction with arsenic.

4 CONCLUSIONS

The results of this research showed that: (i) The adsorption capability is affected by the initial concentration of arsenic, and the amount of sorbed arsenic increased linearly with solution concentration; (ii) The adsorption of As(V) by calcite was a comparatively fast reaction within the first 10 h, and when the reaction time reached 72 h, it seemed to have reached a balanced state of sorption and desorption.

REFERENCES

de Leeuw, N.H. & Parker, S.C. 1998. Surface structure and morphology of calcium carbonate polymorphs calcite, aragonite, and vaterite: An atomistic approach. *J. Phys. Chem. B* 102(16): 2914–2922.

Montes-Hernandez, G., Concha-Lozano, N., Renard, F. & Quirico, E. 2009. Removal of oxyanions from synthetic wastewater via carbonation process of calcium hydroxide: applied and fundamental aspects. *J. Hazard. Mater.* 166(2–3): 788–795.

Nordstrom, D.K. 2002. Worldwide occurrences of arsenic in ground water. *Science* 296(5576): 2143–2145.

Smedley, P.L. & Kinniburgh, D.G. 2002. A review of the source, behaviour and distribution of arsenic in natural waters. *App. Geochem.* 17(5): 517–568.

Wei, H., Shen, Q., Zhao, Y., Wang, D.J. & Xu, D.F. 2003. Influence of polyvinylpyrrolidone on the precipitation of calcium carbonate and on the transformation of vaterite to calcite. *J. Cryst. Growth* 250(3–4): 516–524.

Xu, J., Yan, C., Zhang, F., Konishi, H., Xu. H. & Teng H.H. 2013. Testing the cation-hydration effect on the crystallization of Ca-Mg-CO_3 systems. *P. Natl. Acad. Sci. USA.* 110(44): 17750–17755.

Yokoyama, Y., Tanaka, K. & Takahashi, Y. 2012. Differences in the immobilization of arsenite and arsenate by calcite. *Geochem. Cosmochim. Acta* 91(91):202–219.

Environmental Arsenic in a Changing World –
Zhu, Guo, Bhattacharya, Ahmad, Bundschuh & Naidu (Eds)
ISBN 978-1-138-48609-6

Effect of symbiotic bacteria on the accumulation and transformation of arsenite by *Chlorella salina*

Y.X. Wang, Q.N. Yu & Y. Ge
Jiangsu Key laboratory of Marine Biology, College of Resources and Environmental Sciences, Nanjing Agricultural University, Nanjing, China

ABSTRACT: Microalgae and bacteria are usually symbiotic in the environment. Algae-bacteria consortia may be potentially applied in wastewater treatment and environmental remediation. Previous research has shown that the algae-bacteria consortia have a strong ability of removing arsenic (As) from solution. However, little is known about the effects of symbiotic bacteria on uptake and metabolism of arsenite (As(III)) by microalgae. In this study, we investigated the effects of symbiotic bacteria on As(III) accumulation and transformation by *Chlorella salina*.

1 INTRODUCTION

Arsenic (As) is a ubiquitously distributed and a non-threshold carcinogen, ranking at the top of the priority list of hazardous substances. Microalgae are widely found in the environment. They are single celled organisms with large cell surface area, strong absorption and adsorption capacity for As, and great potential for the treatment of As contaminated wastewater. Bacteria and microalgae often coexist during the wastewater treatment processes. Although the bacteria may compete with the microalgae for nutrients (Bai *et al.*, 2015), they could also facilitate microalgal harvesting by forming algal-bacterial aggregates. Studies have shown that microalgae metabolism of As(III) includes extracellular oxidation and intracellular methylation, complexation with glutathione (GSH), PCs and stored in the vacuole for detoxification (Jiang *et al.*, 2011). However, little is known about the effects of symbiotic bacteria on As(III) metabolism in microalgae.

2 METHODS/EXPERIMENTAL

Chlorella salina was purchased from Marine Germplasm Bank of China, Chinese Academy of Sciences, Qingdao. *Agrobacterium tumefaciens* and *Erythrobacter citreus* were isolated and identified as the symbiotic bacteria from *C. salina.*

In this study, 0~750 $\mu g\ L^{-1}$ arsenite [As(III)] was set up, and the adsorption, absorption and morphological transformation of As(III) by axenic and non-axenic *C. salina* were measured seven days after treatment.

Determination of total As contents in sterile and non-axenic *C. salina* cells was done through digestion at 120°C. The As species of algal cells were extracted with 1.75% HNO_3. HPLC-HG-AFS was used for the As speciation analyses.

3 RESULTS AND DISCUSSION

Chlorella salina was exposed to a series of concentrations of As(III), however As(V) was the main species in the culture medium and non-axenic algae cells (Figs. 1, 2); but for the axenic *C. salina*, As(III) was the main form in the culture medium (Fig. 3).

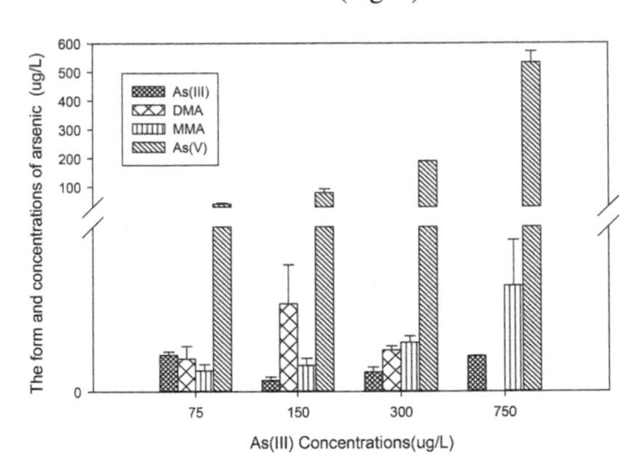

Figure 1. Speciation of As in the culture medium of non-axenic *C. salina*.

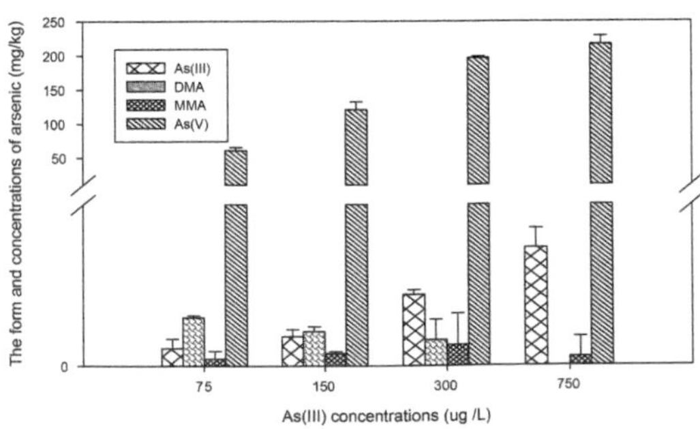

Figure 2. Speciation of As in the non-axenic *C. salina* cells.

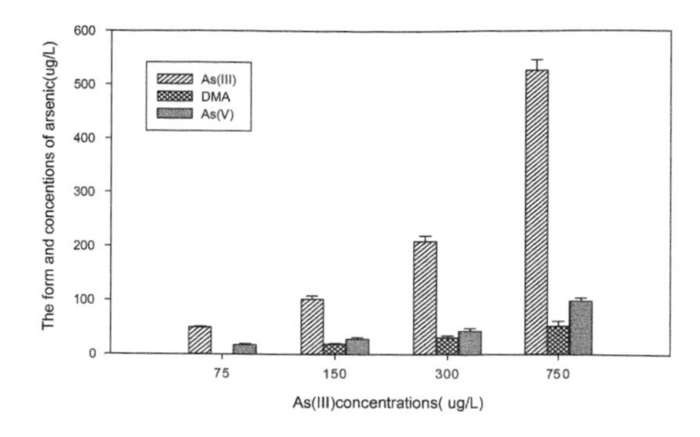

Figure 3. Speciation of As in the culture of axenic *C. salina*.

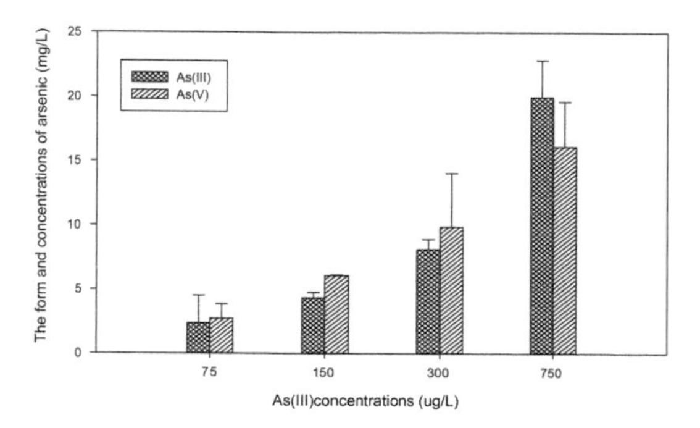

Figure 4. Speciation of As in the axenic *C. salina* cells.

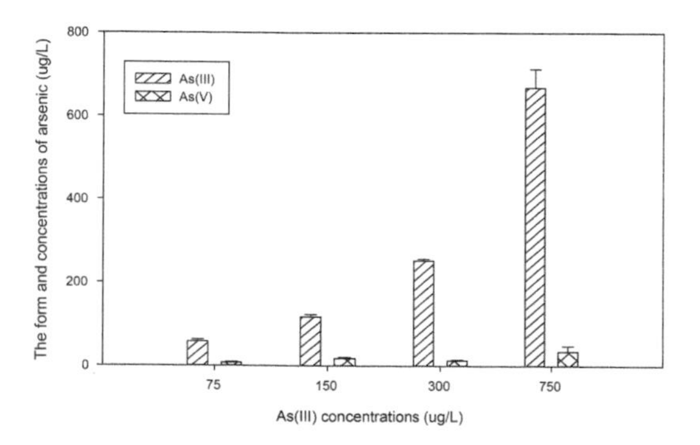

Figure 5. Speciation of As in the culture of *A. tumefaciens*.

As(V) was the main species of intracellular As and a small amount of As(III), MMA and DMA were also detected in these cells. These As species accounted for 2.14%–4.32% of total As in the *C. salina* cells.

There were only As(III) and As(V) forms in the cells of axenic *C. salina*, the ratios of As(V) and As (III) were 44.66%–58.30% and 41.70%–55.34%, respectively (Fig. 4).

As(III) was the main form in the culture medium when the symbiotic bacterium, *A. tumefaciens*, was cultured alone (Fig. 5). The proportion of As(V) was only 7.73%–13.56%, indicating that the bacteria might be an arsenate reduction bacterium.

Results of this study showed that the presence of symbiotic bacteria increased the accumulation and transformation of As by algae cells. After 7 d of treatment with different concentrations of As(III), the removal rates of arsenic in the solution from *C. salina* were 18.92%~55.12%, higher than that of axenic *C. salina* (12.82%~27.43%) and symbiotic bacteria (1.86%~16.19%).

This study showed that bacteria significantly increased the accumulation of arsenic in *C. salina* ($P < 0.05$), and improved the removal of As (III) from water. Symbiotic bacteria increased the amount of EPS on the surface of *C. salina* and increased the binding sites for As, resulting in significantly higher As adsorption by the non-axenic *C. salina*.

In the presence of symbiotic bacteria, As (V) is the major As species in the *C. salina* cells and culture medium, indicating that symbiotic bacteria may increase the expression of the arsenite oxidase gene and lead to the higher oxidation of As (III). Further experiments are being conducted to reveal its mechanisms.

4 CONCLUSIONS

Symbiotic bacteria significantly affected the accumulation of As in the *C. salina* cells. Consortia of the bacteria and *C. salina* could affect the oxidation and methylation of As (III) in the algal cells.

ACKNOWLEDGEMENTS

Financial support from Natural Science Foundation of China (31400450, 31770548) are greatly appreciated.

REFERENCES

Bai, X., Lant, P. & Pratt, S. 2015.The contribution of bacteria to algal growth by carbon cycling. *Biotech. Bioeng.* 112(4): 688–695.

Jiang, Y., Purchase, D., Jones, H. & Garelick H. 2011. Technical Note: Effects of Arsenate (As^{5+}) on growth and production of Glutathione (GSH) and Phytochelatins (PCs) in *Chlorella vulgaris*. *Int. J. Phytorem.* 13(8): 834–844.

Environmental Arsenic in a Changing World –
Zhu, Guo, Bhattacharya, Ahmad, Bundschuh & Naidu (Eds)
ISBN 978-1-138-48609-6

Arsenic transformation mediated by *Pantoea* sp. IMH in spent nZVI waste residue

L. Ye & C.Y. Jing
Research Center for Eco-Environmental Sciences, Chinese Academy of Sciences, Beijing, China

ABSTRACT: Nanoscale zero-valent iron (nZVI) is an effective arsenic (As) scavenger. However, spent nZVI may pose a higher environmental risk than our initial thought in the presence of As-reducing bacteria. Therefore, our motivation was to explore the As redox transformation and release in spent nZVI waste residue in contact with *Pantoea* sp. IMH, an *ars*C gene container. Our incubation results showed that IMH preferentially reduce soluble As(V), not solid-bound As(V), and was innocent in elevating total dissolved As concentrations. μ-XRF and As μ-XANES spectra clearly revealed the heterogeneity and complexity of the inoculated and control samples. Nevertheless, the surface As local coordination was not affected by the presence of IMH as evidenced by similar As-Fe atomic distance (3.32–3.36 Å) and coordination number (1.9) in control and inoculated samples. The Fe XANES results suggested that magnetite in nZVI residue was partly transformed to ferrihydrite, and IMH slowed down the nZVI aging process. IMH distorted Fe local coordination without change its As adsorption capacity as suggested by Mossbauer spectroscopy. Arsenic retention is not inevitably enhanced by in situ formed secondary Fe minerals, but depends on the relative As affinity between the primary and secondary iron minerals.

1 INTRODUCTION

Nanoscale zero-valent iron (nZVI) has been widely used in arsenic (As) remediation because of its large specific surface area and high surface reactivity. nZVI is readily oxidized and the subsequent iron corrosion products are responsible for As adsorption. This As-laden spent nZVI may oxidize and gradually crystallize over time to more ordered forms such as magnetite and other iron minerals. Because of the natural occurrence of iron and its corrosion products, nZVI is designed for direct use and is disposed of in the environment. However, recent studies suggest that the aging of nZVI may facilitate the heavy metal release and the secondary environmental pollution (Calderon & Fullana 2015), especially when spent nZVI is left unattended at treated sites.

Microorganisms play a prominent role in the redox transformation and mobilization of As through the change in As speciation and Fe(III) oxide dissolution (Tian *et al.*, 2015). Bacteria can reduce As(V) either by an anaerobic respiratory pathway mediated by *arr*A genes or by an aerobic detoxification pathway mediated by *ars*C genes. Recent studies adopted the aeration process in nZVI treatment to enhance As removal due to the formation of Fe (hydr)oxides, therefore, the aerobic As-reducing bacteria with *ars*C genes may play a central role in As bio-mobilization. However, whether *ars*C gene containers can directly reduce adsorbed As(V) remains unclear.

The Fe(III) oxide dissolution leads to As mobilization, but subsequently in situ formed secondary iron minerals are reported to re-uptake As. Most previous studies found that As retention is actually enhanced during the formation of secondary Fe minerals. However, limited knowledge is available about the iron transformation in spent nZVI in the presence of *ars*C carriers. In addition, the impact of iron transformation in spent nZVI for As retention is unknown. The desire to decipher the coupled biotransformation of As and nZVI by *ars*C carriers motivates our study.

The purpose of this study was to explore As and Fe redox transformation and release in spent nZVI waste residue in the presence of an aerobic As(V)-reducing bacterium, *Pantoea* sp. IMH. Multiple complementary techniques including incubation experiments, X-ray absorption fine structure (XAFS) and Mössbauer spectroscopy were employed to characterize the reaction process. The insights gained from this study shed new light on the risk assessment of nZVI waste residue in the environment.

2 METHODS/EXPERIMENTAL

2.1 *Incubation experiments*

Spent nZVI waste residue was collected from a rare metal smelting campany. IMH was activated first, and then was inoculated into the culture medium containing 5 g L^{-1} nZVI. The control samples were prepared the same with the incubation samples, except without IMH inoculation. Each incubation and control samples was monitored periodically to determine the number of living cells and the soluble concentration of As(III), As(V), Fe(II), total Fe, anions and other metals.

2.2 *Solid analysis*

The solid samples were freeze-dried and deposited onto a strip of Kapton tape for XAFS analysis. Micro X-ray fluorescence (μ-XRF) and μ-X-ray absorption near edge structure (μ-XANES) spectra were collected at beamline 15U, and the As and Fe K-edge XAFS spectra were collected at beamline 14W at Shanghai Synchrotron Radiation Facility (SSRF), China. XANES spectra were analyzed using LCF fitting and EXAFS data were analyzed using standard FEFF shell fitting approach.

Mössbauer spectra were measured at 293 K with an Austin Science S-600 Mössbauer spectrometry using a γ-ray source of 1.11 GBq^{57}Co/Rh. Measured spectra were fitted to Lorentzian line shapes using standard line shape fitting routines. The software package WinNoroms-for-Igor was used to analyze measured spectra and the rates of all iron speciation were calculated according to peak area.

3 RESULTS AND DISCUSSION

3.1 *Soluble As and Fe speciation*

As(V) was almost completely reduced to As(III) by IMH within 30 h, concomitant with the growth of IMH. Almost no differencee was observed in total dissolved As in control and inoculated samples, suggesting that IMH had a negligible contribution to the As release from nZVI. The results indicate that IMH plays a decisive role in soluble As speciation, but has no impact on the As release. We therefore speculated that IMH should entirely prefer reduce the dissolved As(V) rather than the solid-bound As(V). No significant difference was observed between inoculated and control samples for both Fe(III) ($p = 0.525$) and Fe(II) ($p = 0.848$), implying that IMH facilitated neither Fe(III) precipitation/dissolution nor its reduction. Interestingly, decrease in dissolved As concentration (9% for the control and 12% for the inoculated sample) was not proportional to that of Fe (61% for the control and 62% for the inoculated sample) (Figure 1b), indicating that the fresh precipitated iron oxides did not uptake As by the same ratio as in the initial nZVI residues.

3.2 *Speciation and association of As and Fe*

The speciation of the solid As and Fe was characterized using XANES. Incubation with IMH slightly increased the As(III) percentage from 15% in nZVI residue to 19%, probably due to re-adsorption of dissolved As(III). Compared with the 100% As(III) in the soluble phase due to As(V) bio-reduction, in situ As(V) reduction on solid was inhibited, revealing that IMH prefers to reduce dissolved As(V) rather than adsorbed As(V).

The spatial distribution of As and Fe in solids determined using μ-XRF and the corresponding As μ-XANES spectra at five different spots. The As distribution was linearly correlated with that of Fe in inoculated ($R = 0.97$) and control ($R = 0.95$) samples.

As EXAFS analysis indicated that the As-Fe distance was 3.36 (3.33) Å in the inoculated (control) sample, corresponded to bidentate binuclear corner-sharing As complex on the Fe (hydr)oxides surface.

3.3 *Fe speciation in solid*

To examine the effect of IMH on Fe speciation in solids, XANES analysis was employed. The Fe K-edge XANES analysis indicates that magnetite would be oxidized and transformed to ferrihydrite during the nZVI aging process. IMH slowed down the nZVI aging process as evidenced by the 34% Fe_3O_4 in the inoculated sample compared with zero Fe_3O_4 in the control. To further explore the Fe phase in the presence of IMH, Mössbauer spectroscopy was used to provide insights into the valence state and coordination of Fe atoms. The analysis shows that although IMH did not change the valence state of Fe, it did cause a more distorted polyhedron coordination of Fe atoms.

4 CONCLUSIONS

As(V)-reducing bacteria with *ars*C genes play an important role in the biotransformation and mobility of As in the aerobic environment. *ars*C containers such as strain IMH can readily reduce dissolved As(V) to more toxic As(III) aerobically. The bacterial activity slow down the nZVI aging process and distort the Fe local coordination which may have an impact on the fate and transport of attached pollutants. Furthermore, the As retention is dependent on the relative As affinity between the primary and secondary iron minerals. Therefore, the consequence of secondary Fe mineral formation to As immobilization should be evaluated on a case-by-case basis.

ACKNOWLEDGEMENTS

We acknowledge the financial support of the Strategic Priority Research Program of the Chinese Academy of Sciences (XDB14020201).

REFERENCES

Calderon, B. & Fullana, A. 2015. Heavy metal release due to aging effect during zero valent iron nanoparticles remediation. *Water Res.* 83: 1–9.

Tian, H.X., Shi, Q.T. & Jing, C.Y. 2015. Arsenic biotransformation in solid waste residue: comparison of contributions from bacteria with arsenate and iron reducing pathways. *Environ. Sci. Technol.* 49(4): 2140–2146.

Environmental Arsenic in a Changing World –
Zhu, Guo, Bhattacharya, Ahmad, Bundschuh & Naidu (Eds)
ISBN 978-1-138-48609-6

Land scale biogeography of arsenic biotransformation genes in estuarine wetland

S.Y. Zhang[1], J.Q. Su[2], G.X. Sun[1], Y.F. Yang[3], Y. Zhao[2], J.J. Ding[4], Y.S. Chen[2], Y. Shen[2], G.B. Zhu[5], C. Rensing[2,6] & Y.G. Zhu[1,2]

[1]*State Key Lab of Urban and Regional Ecology, Research Center for Eco-Environmental Sciences, Chinese Academy of Sciences, Beijing, P.R. China*
[2]*Key Laboratory of Urban Environment and Health, Institute of Urban Environment, Chinese Academy of Sciences, Xiamen, P.R. China*
[3]*State Key Joint Laboratory of Environment Simulation and Pollution Control, School of Environment, Tsinghua University, Beijing, P.R. China*
[4]*Key Laboratory of Dryland Agriculture, Ministry of Agriculture, Institute of Environment and Sustainable Development in Agriculture, Beijing, P.R. China*
[5]*Key Laboratory of Drinking Water Science and Technology, Research Center for Eco-Environmental Sciences, Chinese Academy of Sciences, Beijing, P.R. China*
[6]*Fujian Provincial Key Laboratory of Soil Environmental Health and Regulation, College of Resources and the Environment, Fujian Agriculture & Forestry University, Fuzhou, P.R. China*

ABSTRACT: Our study demonstrated that genes involved in As biotransformation were characterized by high abundance and diversity in estuarine sediments at low As levels across Southeastern China. The functional microbial communities showed a significant decrease in similarity along the geographic distance, with higher turnover rates than taxonomic microbial communities based on the similarities of 16S rRNA genes. Further investigation with niche-based models showed that deterministic processes played primary roles in shaping both functional and taxonomic microbial communities. Temperature, pH, total nitrogen concentration, carbon/nitrogen ratio and ferric iron concentration rather than As content in these sediments were significantly linked to functional microbial communities, while sediment temperature and pH were linked to taxonomic microbial communities. We proposed several possible mechanisms to explain these results.

1 INTRODUCTION

It has been documented that microbial genes associated with arsenic (As) biotransformation are widely present in As-rich environments (Zhu *et al.*, 2014). Nonetheless, their presence in natural environment with low As levels remains unclear. To address this issue, we investigated the abundance levels and diversities of Arsenite (As(III)) respiratory oxidation (*aioA*), Arsenate (As(V)) respiratory reduction (*arrA*), As(V) reduction (*arsC*) and As(III) methylation (*arsM*) genes in estuarine sediments at low As levels across Southeastern China to uncover biogeographic patterns at a large spatial scale.

2 EXPERIMENTAL

2.1 *Samples description and DNA extraction*

A total of 70 sediment samples from depths of 0–15 cm were collected in 2013 from 14 estuarine wetlands (five replicates for each sample) distributed in 4 provinces in Southeastern China. DNA was extracted from 0.5 g of the samples using the FastDNA SPIN kit for soil (MP Biomedicals). Quantitative real-time polymerase chain reaction (qPCR) and terminal restriction fragment length polymorphism (T-RFLP) were conducted to investigate the *aioA*, *arrA*, *arsC* and *arsM* genes abundance and diversity, respectively. *arsM* genes were further sequenced on the Illumina Miseq sequencing platform, and yielded a total of 393980 raw read pairs.

2.2 *Data analysis*

Principle coordinates analysis (PCoA) and the Adonis dissimilarity test were conducted to examine the distributions of microbial community compositions involved in As biotransformation. The distance-decay analysis was applied for the *arsM* genes to investigate the declining microbial community similarity with geographic distance. Redundancy analysis (RDA) were chosen to investigate environmental explanatory variables. The variation partitioning analysis resulted in three geographic factors (latitude, longitude and sediment temperature (Tm)) and four soil attributes

(C/N, pH, N, Fe(III)). The rank abundance distribution was examined to test whether niche-based or neutral models best explain the assembly of the microbial community involved in As(III) methylation.

3 RESULTS AND DISCUSSION

3.1 *Prevalence of microbes involved in As biotransformation in estuarine sediments*

Genes involved in As biotransformation were characterized by high abundance and diversity. Among them, *arsC* had the highest abundance among the four As biotransformation genes. The great Oxidation of the Earth resulted in more As(V) in the environment. Additionally, estuaries are adjacent to the marine environment, in which As(V) accumulation by microbes is elevated via the phosphate uptake system due to the low phosphate levels in the marine environment. It is likely that microbes in these environments may acquire more abundant *arsC* gene to detoxify the highly accumulated As in their cytoplasm (Zhang *et al.*, 2013). The abundance of respiratory As(III) oxidizing microbes (*aioA*) and reducing microbes (*arrA*) were lower in sediments but still comparable to those detected in As-contaminated paddy soils (Zhang *et al.*, 2015). The low As levels in these estuarine sediments could explain the lower relative abundance, suggesting that As respiration may not be an important process to either detoxify As or couple it to ATP production due to the lack of available As. The comparable abundant *arsM* genes detected in this study showed great potential for As(III) methylation in the estuarine environment.

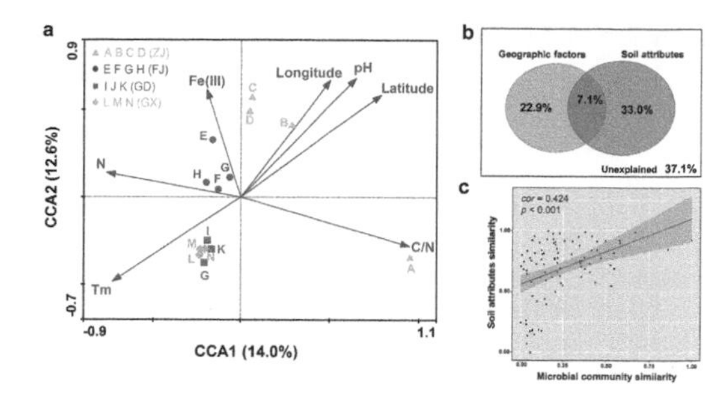

Figure 1. Environment factors contributing to microbial community compositions involved in As(III) methylation.

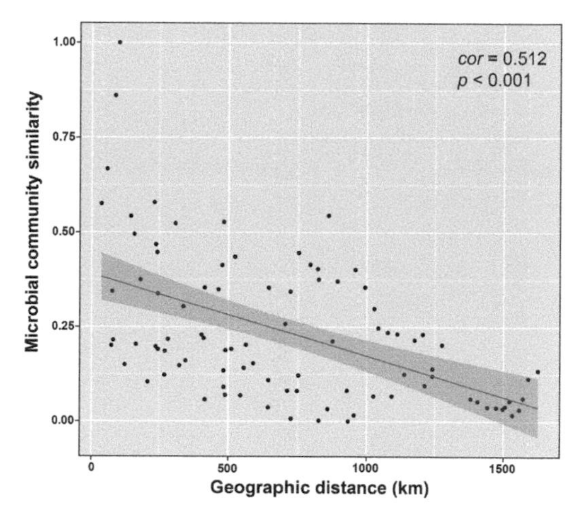

Figure 2. Distance-decay curve for microbial community similarity based on *arsM* gene.

3.2 *Processes and factors driving microbial biogeography associated with As(III) methylation*

Temperature, pH, total nitrogen concentration, carbon/nitrogen ratio and ferric iron concentration rather than As content in these sediments were significantly linked to functional microbial communities (Fig. 1). It is reasonable because As levels in all samples were relatively low and similar. The significant correlations between soil attributes and the microbial community based on *arsM* genes corroborated the concept that the assembly based on niche-based mechanisms is mainly determined by the niche requirements and local habitat conditions. Further investigation with niche-based models also showed that deterministic processes played primary roles in shaping both functional and taxonomic microbial communities.

3.3 *Microbial biogeography: Taxonomic and functional communities*

PCoA of functional microbial communities revealed four clusters for *aioA*, *arrA*, *arsC* and *arsM* genes. The Adonis dissimilarity test corroborated that geographic distribution patterns imposed changes in microbial community compositions involved in As biotransformation. Microbes involved in As(III) methylation displayed a significant decay of microbial community similarity along geographic distance (Fig. 2), with higher turnover rates than taxonomic microbial communities based on the similarities of 16S rRNA genes. The higher fraction of variances (33%) in the microbial community involved in As(III) methylation was explained independently by soil attributes rather than by geographic factors (Fig. 1). It has been shown that metabolic flexibility can be a major predictor for the spatial distribution within microbial communities (Carbonero *et al.*, 2014). We reinforce this viewpoint in our study by showing that the functional composition of microbes is correlated to soil geochemical parameters.

4 CONCLUSIONS

This study is the first to demonstrate the wide presence of As biotransformation genes in an environment with a low As level, which is possibly a legacy of active As metabolism in early life. We further disclosed the biogeographic microbial patterns associated with As biotransformation at a large spatial scale. Since the low As level in these sediments was not strong enough to impose selection on As biotransformation genes, heterogeneity of local soil attributes leading to variable selection for taxonomic microbial communities may become more important in determining functional microbial biogeographic patterns. Compared with microbial taxonomy, environmental selection for functional genes seems to be less complicated, rendering functional microbial communities as an ideal model to study the important role of deterministic processes in shaping microbial biogeographic patterns.

ACKNOWLEDGEMENTS

This study is financially supported by Strategic Priority Research Program of Chinese Academy of Sciences (XDB15020402) and National Natural Science Foundation of China (41430858, 41090282). There is no conflict of interest for this study.

REFERENCES

Carbonero, F., Oakley, B.B. & Purdy, K.J. 2014. Metabolic flexibility as a major predictor of spatial distribution in microbial communities. *PLoS.One* 9(1): e85105.

Zhang, S.Y., Rensing, C. & Zhu, Y.G. 2013. Cyanobacteria-mediated arsenic redox dynamics is regulated by phosphate in aquatic environments. *Environ. Sci. Technol.* 48(2): 994–1000.

Zhang, S. Y., Zhao, F.J., Sun, G.X., Su, J.Q., Yang, X.R. Li, H. & Zhu, Y.G. 2015. Diversity and abundance of arsenic biotransformation genes in paddy soils from Southern China. *Environ. Sci. Technol* 49(7): 4138–4146.

Zhu, Y.G., Yoshinaga, M., Zhao, F.J. & Rosen, B.P. 2014. Earth abides arsenic biotransformations. *Annu. Rev. Earth Planet Sci.* 42: 443–467.

Environmental Arsenic in a Changing World –
Zhu, Guo, Bhattacharya, Ahmad, Bundschuh & Naidu (Eds)
ISBN 978-1-138-48609-6

Arsenic, manganese, and dissolved organic matter biogeochemistry in the Bengal Basin (India) and Southern Pampean plain (Argentina)

M. Vega[1], S. Datta[2], N.N. Sosa[3], H. Kulkarni[2] & M. Berube[2]
[1]*Hydrologic Science and Engineering Program, Colorado School of Mines, Golden, CO, USA*
[2]*Department of Geology, Kansas State University, Manhattan, KS, USA*
[3]*Centro de Investigaciones Geológicas (CONICET-UNLP), Universidad Nacional de La Plata, La Plata, Argentina*

SUMMARY: Arsenic (As) and manganese (Mn) biogeochemistry is intimately linked with dissolved organic matter (DOM) in subsurface aqueous environments. In this study, we explore the role of contrasting DOM quality on the distributions of Mn and As in groundwater. Tubewell and subsurface sediment samples were collected from Murshidabad, West Bengal (WB), India, as well as the Claromecó basin of Argentina. A parallel factor (PARAFAC) model is presented for WB samples that delineate key DOM components in sites characterized by high concentrations of both Mn and As (HMHA), and high Mn concentrations but low As (HMLA). An association between humic-like DOM and elevated As is clear, whereas Mn concentrations are high in the presence of both humic and protein-like DOM. In Claromecó basin groundwaters, an inverse relation between As and Mn is presented and future efforts are planned to incorporate the role of DOM.

1 INTRODUCTION

The arsenic (As) calamity in Southeast Asia has been a significant area of research for decades. More recently, concern over manganese (Mn) in drinking water has inspired work to assess the extent of Mn contamination and associated health impacts. Despite numerous reports of Mn inhibiting the intellectual development of children, increasing infant mortality rates, and causing Parkinson's like neurological disorders, the World Health Organization (WHO) revoked its Mn drinking water limit ($0.4\,mg\,L^{-1}$) because naturally occurring levels were not of health concern (WHO, 2011). We argue that the data needs to be reassessed and the drinking water limit reconsidered.

Numerous papers have explored the relationship between Mn and As in groundwater. Early work demonstrated that Mn oxide reduction can stimulate dissolved As(III) oxidation and immobilization, and later experimentation by Ehlert *et al.* (2016) highlighted that greater Mn-oxide content in contaminated paddy soils lead to increased As retention. Field scale observations noting an inverse relation between dissolved Mn and As in groundwater have been attributed to this phenomenon, as well as variable redox potential (E_h). Yet, few studies to date have assessed the role of DOM in the coupled biogeochemical cycling of Mn and As (Vega *et al.*, 2017).

Dissolved organic matter plays a key role in metal(loid) geochemistry through electron shuttling catalyzing reduction pathways, complexation processes keeping solutes in solution, and labile organic carbon fueling heterotrophic microbial metabolisms. Recent advances in fluorescent and absorbance spectroscopic techniques have permitted qualitative analyses of DOM as it relates to As mobilization, and fluorescence excitation emission matrices (EEMs) coupled with a PARAFAC model revealed associations between DOM and As, but did not include Mn (Kulkarni *et al.*, 2016). The purpose of this study was therefore to investigate the role of DOM in coupled Mn and As biogeochemical cycling by evaluating associations between these metal(loid)s in sites with statistically different DOM composition. This presentation emphasizes recently published work in Murshidabad, India, but also includes ongoing work in the Claromecó river basin of Argentina.

2 METHODS

2.1 *Study area*

Murshidabad, a district in West Bengal, India, is bisected by the north-south flowing River Bhagirathi. West of the river, aquifer sediments are reddish brown and are of Pleistocene age, whereas dark gray Holocene sediments dominate the east. Several studies have reported contrasting As concentrations in the Holocene (high As) and Pleistocene (low As) aquifers, and more recent work has revealed contrasting DOM quality in these distinct systems.

The Claromecó fluvial basin (CFB) is situated in the southern Chaco-pampean plain of Argentina. Late Pleistocene-Holocene aquifer sediments constitute the top of the sedimentary sequence covering the middle and lower basin, and are greenish-brown, mostly volcaniclastic and deposited by fluvio-eolian processes (Sosa *et al.*, 2017).

2.2 *Sampling and analysis*

In Murshidabad, 51 groundwater and 16 subsurface aquifer sediment samples were collected from two sites located on the west side and four sites located on the east side of the river Bhagirathi. Water samples were analyzed for dissolved cations, anions, pH, alkalinity, E_h, DOC, and TDN; DOM was further characterized using EEMs followed by PARAFAC modeling. In the CFB, 29 well water samples were analyzed for dissolved cations, anions, pH, alkalinity, and E_h. Aqua regia digestions were performed on subsurface sediments in both regions to determine bulk concentrations of As, Fe, and Mn; XRF was also employed in the CFB.

3 RESULTS

3.1 *Aquifer metal(loid) geochemistry*

Within Murshidabad, the WHO limits for As (10 μg L^{-1}) and Mn (0.4 mg L^{-1}) were exceeded in 78% and 73% of the surveyed tubewells, respectively. Three sites were characterized by high concentrations of both As (~330 μg L^{-1}) and Mn (~0.8 mg L^{-1}) ["HMHA"], while the remaining three sites had high Mn (~1 mg L^{-1}) but low As (~9 μg L^{-1}) ["HMLA"]. HMLA sites had higher E_h but lower As and Fe concentrations, than HMHA sites. Arsenic concentrations exceeded the WHO limit in all CFB wells (~83 ug L^{-1}), whereas Mn concentrations were consistently low (~0.003 mg L^{-1}). Bulk sediment concentrations for Fe, Mn, and As were statistically similar between HMHA and HMLA sites in Murshidabad, as well as both spatially and with depth in the CFB.

3.2 *DOM-PARAFAC model*

Four DOM components were determined by the PARAFAC model in Murshidabad: (C1) a terrestrial humic-like component; (C2) a humic-like component influenced by agriculture and wastewater; (C3) a protein-like component; and (C4) a microbial humic-like component (Fig. 1). Component 1 was significantly more abundant in HMHA sites, C3 and C4 were significantly more abundant in HMLA sites, and C2 was statistically similar between HMHA and HMLA sites.

4 CONCLUSIONS

The HMHA aquifers have reducing redox potentials that can mobilize Fe, Mn, and As, and the presence of humic-like DOM may be catalyzing these processes. Further, a greater proportion of wells in HMHA sites were saturated with respect to rhodochrosite relative to HMLA sites. This implies a net sink for Mn in HMHA sites, yet Mn concentrations were similar throughout all sites. It is postulated that rhodochrosite precipitation is inhibited by complexation mechanisms with humic-like DOM, keeping Mn in solution. In HMLA aquifers, Mn oxide reduction may be oxidizing As(III) and enhancing the potential re-adsorption of As(V) onto un-reduced or freshly precipitated Fe oxides. The fact that Mn concentrations are high in the presence of both humic-like and protein-like DOM suggests that

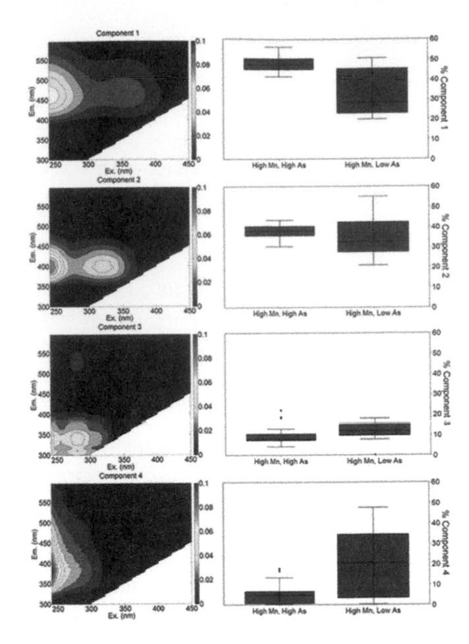

Figure 1. Excitation emission matrices (EEMs) for the four components of the DOM PARAFAC model and their relative distributions in high Mn, high As (HMHA) and high Mn, low As (HMLA) sites.

Mn mobilization may be independent of DOM quality. Future work necessitates experimental validation of the processes in Murshidabad and the application of fluorescence spectroscopic techniques in the CFB to understand the role of DOM quality in a high As, low Mn scenario.

ACKNOWLEDGEMENTS

The authors are grateful to NSF, K-State, CONACYT, and Sigma Xi for making this work possible.

REFERENCES

Ehlert, K., Mikutta, C. & Kretzchmar, R. 2016. Effects of manganese oxide on arsenic reduction and leaching from contaminated floodplain soil. *Environ. Sci. Technol.* 50: 9251–9261.

Kulkarni, H. V., Mladenov, N., Johannesson, K. H. & Datta, S. 2016. Contrasting dissolved organic matter quality in groundwater in Holocene and Pleistocene aquifers and implications for influencing arsenic mobility. *Appl. Geochem.* 77: 194–205.

Sosa, N.N., Zárate, M.A. & Beilinson, E. 2017. Dinámica sedimentaria neógena y cuaternaria continental de la cuenca del arroyo Claromecó, Argentina. *Latin. Am. J. Sediment. Basin. Anal.* 24(2): 1–19.

Vega, M.A., Kulkarni, H.V., Mladenov, N., Johannesson, K., Hettiarachchi, G.M., Bhattacharya, P., Kumar, N., Weeks, J., Galkaduwa, M. & Datta, S. 2017. Biogeochemical controls on the release and accumulation of Mn and As in shallow aquifers, West Bengal, India. *Fron. Environ. Sci.* doi: 10.3389/fenvs.2017.00029.

WHO 2011. Guidelines for Drinking-Water Quality. 4th ed. *World Health Organization, Geneva,* 541p.

Environmental Arsenic in a Changing World –
Zhu, Guo, Bhattacharya, Ahmad, Bundschuh & Naidu (Eds)
ISBN 978-1-138-48609-6

Arsenic mobilization in shallow aquifer of Bengal Delta Plain: role of microbial community and pathogenic bacteria

P. Ghosh & D. Chatterjee
Department of Chemistry, University of Kalyani, Kalyani, West Bengal, India

ABSTRACT: In West Bengal, India, arsenic, iron and microbial contamination has been reported from shallow aquifers (<50 m) of the Bengal Delta Plain (BDP). Six groundwater affected side at Chakdaha Block of Nadia District in West Bengal was selected to study groundwater quality parameter notably arsenic and microbial contamination. The study revealed that arsenic contamination is the highest in site B (up to 0.171 mg L^{-1}), whereas site E has been identified as the lowest (0.089 mg L^{-1}) and rest are in between. Similarly microbial contamination has been the highest in site D and the lowest in site A. Further, high concentrations of arsenic along with microbial contamination have also been found in several sites A, C, D and F. The field scale study indicated simultaneously occurrences of arsenic and microbial contamination, suggesting role of microbial pollution in arsenic mobilization.

1 INTRODUCTION

Arsenic (As) is one of the numerous toxicants that is naturally/anthropogenically introduced into the aquatic ecosystem impairing the water quality and also impairing life of different aquatic species (Biswas *et al.*, 2012). Microorganism has developed several mechanisms that can transform arsenic by extracellular precipitation, intracellular accumulation or oxidation and reduction reaction (Oremland & Stolz, 2003). Interestingly, different arsenic resistant bacteria belonging to enteric group has been isolated from the arsenic prone zone. Some common bacteria Pseudomonas, Geobacter grbicium, Clostridium are isolated from BDP (Islam *et al.*, 2004; Chatterjee *et al.*, 2010, van Geen *et al.*, 2011).

2 METHODS/EXPERIMENTAL

2.1 *Study area*

Ground water samples were collected from six sides (Site- A, B, C, D, E, F) of Bengal Delta Plain (BDP) in Chakdaha Block, West Bengal, India.

2.2 *Sample analysis*

Total arsenic (As) concentration in the samples was measured by hydride generation atomic absorption spectrophotometer (HG.AAS VARIAN-240, detection limit <1 μg L^{-1}). Membrane filtration method was used to determine total coliforms and Thermo tolerant fecal coliforms present in the ground water samples. This microbial population was enumerated on Chromocult Agar and differentiated by standard methods as per US-EPA.

Table 1. Variation in the abundance of arsenic and microbial population in groundwater at the study sites.

Site	[As] %	[MO] %	[As & MO] %
A	81.66	38.75	32.87
B	26.97	34.21	15.13
C	60.37	36.22	27.16
D	23.82	54.04	24.68
E	31.42	32.14	18.21
F	44.55	68.31	34.65

3 RESULTS AND DISCUSSION

3.1 *Results*

Considerable variation in concentration of arsenic in groundwater between the collection sites as well as between different depths were observed. Based on the arsenic content, the total study samples could be set high arsenic content (0.103–0.171 mg L^{-1}).

3.2 *Discussion*

We have observed that in all the six sites under study, high amount arsenic and microbial contamination is present (Fig. 1). Similarly Figure 1c shows that at A, C, D & F sites both arsenic and microbes are present.

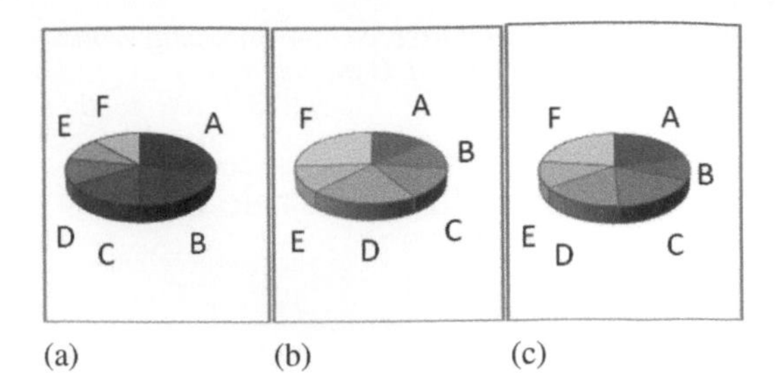

Figure 1. Distribution of a) total arsenic, b) total microbial population, and c) arsenic & microbial population in groundwaters at the study sites A to F.

Another proposed hypothesis is due to the release of arsenic sorbed to aquifer minerals by competitive exchange with microbial phosphate ions that migrates into aquifers due to application of fertilizer to surface soil. In all the situations, the released iron (II) is also oxidized to iron (III) in the presence of oxygen.

4 CONCLUSIONS

Our study indicates that As and microbial activities, specifically fecal coliforms which gains entry into water by fecal contamination. However, too much availability of As into water affects microbial activity and hence shows negative corelation with As-Fe coimmobilization.

ACKNOWLEDGEMENTS

The authors are thankful to the DST-PURSE funding given to the Department of Chemistry, University of Kalyani.

REFERENCES

Biswas, A., Nath, B., Bhattacharya, P., Halder, D.& Kundu, A.K. 2012. Hydrogeochemical contrast between brown and grey sand aquifers in shallow depth of Bengal Basin: Consequences for Sustainable drinking water supply, *Sci.Total Environ.* 431: 402–412.

Chatterjee, D., Halder, D., Majumder, S., Biswas, A. & Nath, B. 2010. Assessment of arsenic exposure from groundwater and rice in Bengal Delta region, West Bengal, India. *Water Res.* 44: 5803–5812.

Islam, F.S., Gault, A.G., Boothman, C., Polya, D.A., Charnock, J.M., Chatterjee, D. & Lloyd, J.R. 2004. Role of metal-reducing bacteria in arsenic release from Bengal delta sediments. *Nature* 430(6995): 68–71.

Oremland, R.S. & Stolz, J.F. 2003. The ecology of arsenic. *Science* 300(5621): 939–944.

van Geen, A., Ahmed, K.M., Akita, Y., Alam, M.J., Culligan, P.J., Emch, M., Escamilla, V., Feighery, J., Ferguson, A.S., Knappett, P., Layton, A.C., Mailloux, B.J., Mcakay, L.D., Mey, J.L., Serre, M.L., Streatfield, P.K., Wu, J. & Yunus, M. 2011. Fecal contamination of shallow tubewells in Bangladesh inversely related to arsenic. *Environ. Sci. Technol.* 45(4): 1199–1205.

Environmental Arsenic in a Changing World –
Zhu, Guo, Bhattacharya, Ahmad, Bundschuh & Naidu (Eds)
ISBN 978-1-138-48609-6

Distribution and hydrogeochemical behavior of arsenic-enriched groundwater in the sedimentary aquifers: Comparison between Datong Basin, China and Kushtia District, Bangladesh

M.-E. Huq, C.L. Su, J.X. Li & R. Liu
School of Environmental Studies, China University of Geosciences (Wuhan), Wuhan, Hubei, P.R. China

ABSTRACT: High arsenic (As) groundwater has become a major global concern due to its widespread occurrence. A comparative hydrogeochemical study was performed on the As-enriched groundwater in Datong Basin, China and Kushtia District, Bangladesh. The elevated As concentrations of Datong Basin ranged from 0.31 to 452 $\mu g L^{-1}$. As enriched groundwater is mainly Na-HCO_3 type water, and characterized by higher pH value, high Na^+, low Ca^{2+}, SO_4^{2-} and NO_3^- along with moderate TDS. The release of As in aquifers resulted from the reduction of As-carrying crystalline iron (Fe) oxide/hydroxides and oxidation of organic matter. The aquifers of Kushtia District, Bangladesh are unconsolidated, alluvial in nature, and developed from Holocene floodplain and Pleistocene deposits. High As (6.04–590.7 $\mu g L^{-1}$) groundwater occurs mainly in shallow aquifers. The Ca-HCO_3 type groundwater is distinguished by circum-neutral pH, medium-high EC, high HCO_3^-, and low content of NO_3^-, SO_4^{2-}, K^+ and Cl^-. The reductive suspension of MnOOH increased the dissolved As loads and redox responsive elements such as SO_4^{2-} and pyrite oxidation acted as the main mechanisms for As release in groundwater. The weak loading of Fe suggests that the release of Fe and As was decoupled in sedimentary aquifers of Kushtia District.

1 INTRODUCTION

High arsenic (As) groundwater is the most crucial natural hazard to public health and the largest environmental hazards in the world (Wang & Wai, 2004). Among the As polluted regions, India, Bangladesh, and China are mostly affected. Currently, millions of people are still consuming As contaminated water above the WHO recommended value (10 $\mu g L^{-1}$) for drinking water both in China and Bangladesh (Bibi *et al.*, 2008; Wang & Deng, 2009). Since last two decades, many studies have been conducted on the causes of As enrichment at Datong Basin, and Bangladesh but the different mechanisms of As release has remain poorly defined. The objectives of this study are (1) to compare the hydrogeochemical characteristics of the elevated As groundwater, and (2) to identify the different mechanisms of As release occurred in the sedimentary aquifers of these two areas.

2 METHODS

2.1 *Study area*

Datong Basin, one of the Cenozoic fault basins of the Shanxi rift system that is situated in the north of Shanxi Province, China and Kushtia District is located in the southwest of Bangladesh were selected as the study area.

2.2 *Sampling and hydrogeochemical analysis*

To investigate the hydrogeochemical properties of groundwater, a total of 132 groundwater samples (83 from Datong Basin and 49 from Kushtia District) were collected from tubewells. All of the water samples were filtered through 0.45 μm membranes and preserved in 50 ml polyethylene bottles for anion, cations and trace elements analysis. Hydride generation-atomic fluorescence spectrometry (HG-AFS) (Titan AFS 830) inductively coupled plasma atomic emission spectrometry (ICP-AES) and Ion Chromatography (IC) (Metrohm-761 Compact IC) were used to measure total dissolved As, major ions and concentration of trace elements respectively.

3 RESULTS AND DISCUSSION

Different mechanisms are responsible for As enrichment in the groundwater of each study area. Factor analysis (FA) was performed on the hydrochemical data to explore the causes of elevated As occurrence in the groundwater of these two study areas. In Datong Basin, China, $F1$ explains 20% of total variance including strong positive loadings of Ca^{2+}, Mg^{2+}, Na^+, Cl^-, SO_4^{2-}, Fe, Cu, Se, Li, B, Sr and V with the negative loading of EC, while others show weak loading (Table 1). The positive loading of Cl^- can be derived due to evaporation and may increase along the flow path from recharge to discharge area. $F2$ accounts for 19.79% of total variance, with strong positive loading of pH, B, and Ba and negative loading of Ca^{2+}. In Datong Basin, high As mostly occurs with the pH value exceeding 7.95. $F3$ covers 16.28% of the total variance. Ba is an indicator for a marine groundwater environment, so this factor may be associated with a marine groundwater environment due

Table 1. Factor loadings of different chemical parameters of groundwater from the two study sites.

	Datong Basin, China			Kushtia, Bangladesh		
Components	*F*1	*F*2	*F*3	*F*1	*F*2	*F*3
As	−.311	.318	−.141	−.055	.146	.108
Ca^{2+}	**.602**	**−.659**	.245	.161	.075	**.884**
K^{+}	.441	−.225	.234	**.680**	.085	−.049
Mg^{2+}	**.760**	−.468	−.097	**.507**	**.548**	.253
Na^{+}	**.798**	−.077	**−.521**	**.817**	−.138	.217
Cl^{-}	**.798**	−.077	**−.521**	**.888**	.197	.105
NO_3^-	.407	−.149	−.049	−.004	.048	−.104
SO_4^{2-}	**.806**	−.365	−.271	**.848**	.048	−.002
pH	−.308	**.504**	−.484	−.017	**−.557**	−.049
EC	**−.590**	.103	.243	.347	**.675**	.215
T	−.148	.112	.024	092	.181	.217
Depth	−.480	.147	.032	−.167	−.090	−.188
HCO_3^-	.477	.409	**−.593**	.193	.213	**.806**
TDS	.370	−.254	−.382	.487	.207	**.683**
Cr	.026	−.025	.250	.008	−.185	.050
Mn	.481	.264	.343	−.169	−.129	**.783**
Fe	**.751**	−.276	.407	−.197	.205	−.191
Ni	.370	.117	.398	−.008	−.010	.039
Cu	**.822**	.308	.133	−.203	−.097	.012
Pb	.467	.365	−.245	.042	**−.785**	−.122
Se	**.710**	.425	.241	−.321	.100	.304
Li	**.876**	.182	.159	−.010	**−.778**	.063
B	**.684**	**.619**	−.148	.362	**.580**	.075
Ba	.282	**.612**	**.570**	**.552**	.018	.269
Sr	**.851**	−.092	.258	.415	.307	**.575**
V	**.641**	.416	.121	−.015	**−.736**	−.104
Variability (%)	20.00	19.79	16.28	16.06	13.8	12.9
Cumulative (%)	20.00	39.79	56.07	16.06	29.9	42.8

to the strong positive loading of Ba (0.57). In Kushtia District, Bangladesh in *F*1 strong loading of Mg^{2+} suggests that it may come from the weathering of silicate or carbonate suspension. The excessive alkalinity may serve as an extra source of Mg^{2+} and the balance of Cl^- and SO_4^{2-}. In *F*2, the strong negative loading of pH suggests that the pH-dependent reactions are not the predominant features of As enrichment in groundwater. *F*3 shows a positive loading of Ca^{2+} that is favorable to the development of the alkali-type soil. Regular and long-term irrigation with Ca^{2+} enriched water in this area makes the soil plastic as well as humid in the wet season and form clod soil crust in the dry season. HCO_3^- also shows a positive strong loading indicating three main weathering factors (evaporation, silicate, and carbonate) which increase the As concentration (Table 1).

The negative loading of HCO_3^- in Datong Basin, indicates that the alkaline environment is associated with the As mobilization (Table 1 and Figure 1a). The comparison of HCO_3^- and dissolved Fe suggests that Fe reduction plays a vital role in As release (Figure 1c). HCO_3^- and SO_4^{2-} are considered to induce As desorption from the Fe mineral shells, and their interactions in the present study reveal that the reactions of redox are the major controlling mechanisms to release As in aquifer system.

The weak loading of Fe (Table 1) of Kushtia District, Bangladesh may be ascribed to the re-precipitation of suspended Fe, suggesting that the release of Fe and As is decoupled or influenced by other geochemical

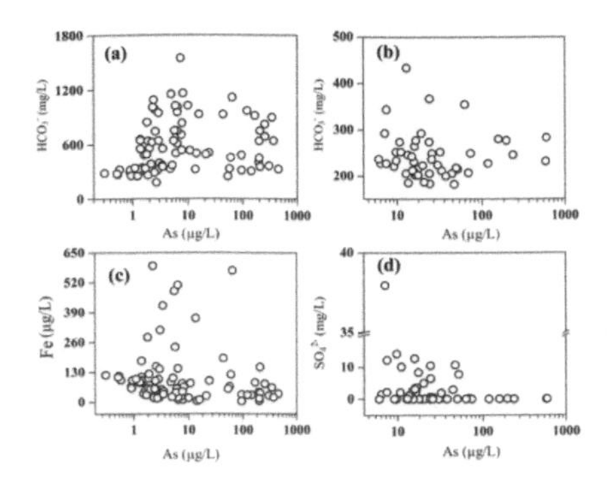

Figure 1. Correlation plots of As with (a) HCO_3^- for Datong Basin (b) HCO_3^- for Kushtia (c) Fe for Datong Basin and (d) SO_4^{2-} for Kushtia.

mechanisms (Radloff *et al.*, 2007). The strong positive loading of Mn (Table 1) indicates that the As is predominantly released with the bacterial reductive suspension of Mn oxy-hydroxides. The influence of HCO_3^- might dominate the As release of the aquifers of Kushtia District in reduction dissolution (Figure 1b). The significant loading of SO_4^{2-} demonstrates that the pyrite oxidation is the key mechanism to As discharge in groundwater (Figure 1d).

4 CONCLUSIONS

The association of As and Fe indicates that the iron-bearing minerals might be the key carrier of elevated As in aquifers. High alkaline situation, As release from sediments with high pH environment, and As enrichment under reducing condition are the responsible factors that control and mobilize As in groundwater from sedimentary aquifers of Datong Basin China.In Kushtia district high As mainly occurred in the shallow aquifer indicating that the shallow parts are recharged/infiltrated by the surface water supply. Low concentrations of NO_3^- and SO_4^{2-} reveal an absolute reducing situation. The high concentration of HCO_3^- maybe related to microbial mediated Fe(III) reduction of organic matter, which creates the reducing condition and releases As in the aquifers. The strong positive loading of Mn suggests that the suspension of MnOOH might be responsible for dissolved As loads.

REFERENCES

Bibi, M.H., Ahmed, F. & Ishiga, H. 2008. Geochemical study of arsenic concentrations in groundwater of the Meghna River Delta, Bangladesh. *J. Geochem. Explor*. 97: 43–58.

Radloff, K.A., Cheng, Z.Q., Rahman, M.W., Ahmed, K.M., Mailloux, B.J., Juhl, A.R., Schlosser, P. & Van Geen, A. 2007. Mobilization of arsenic during one-year incubations of grey aquifer sands from Araihazar, Bangladesh. *Environ. Sci. Technol.* 41: 3639–3645.

Wang, J.S. & Wai, C.M. 2004. Arsenic in drinking water-a global environmental problem. *J. Chem. Educ.* 81: 207.

Wang, Y.X. & Deng, Y.M. 2009. Environmental geochemistry of high-arsenic aquifer systems. Heavy Metal in the Environment.

Environmental Arsenic in a Changing World –
Zhu, Guo, Bhattacharya, Ahmad, Bundschuh & Naidu (Eds)
ISBN 978-1-138-48609-6

Unravelling the role of microorganisms in arsenic mobilization using metagenomic techniques

J.R. Lloyd[1], E.T. Gnanaprakasam[1], N. Bassil[1], B.E. van Dongen[1], L.A. Richards[1], D.A. Polya[1], B.J. Mailloux[2], B.C. Bostick[2] & A. van Geen[2]
[1]*School of Earth and Environmental Sciences, The University of Manchester, UK*
[2]*Barnard College and Lamont-Doherty Earth Observatory, Columbia University, New York, NY, USA*

ABSTRACT: The contamination of groundwaters, abstracted for drinking and irrigation, by sediment-derived arsenic, threatens the health of tens of millions worldwide. Microbial processes are accepted as playing a key role in arsenic mobilisation from sediments into groundwaters, but the precise biogeochemical mechanisms remain a subject of debate. A combination of field investigations, coupled to laboratory experimentation with sediment samples using "microcosm" approaches, has provided a significant body of evidence supporting a key role for anaerobic metal-reducing bacteria in the reductive mobilization of arsenic in aquifers in West Bengal, Cambodia, Vietnam and Bangladesh. The application of high-throughput next generation sequencing, combined with metagenomic reconstructions and other "omics" techniques from the life sciences is shedding new insight into the processes at play, and identifying new mircoorganisms and coupled biogeochemical processes that control the solubility of arsenic in Asian aquifers.

1 MICROBIALLY DRIVEN MOBILIZATION OF ARSENIC IN AQUIFERS: MICROCOSM STUDIES

Field studies have shown that contamination of groundwater with naturally occurring arsenic in the subsurface poses a global public health crisis in many countries including Mexico, China, Hungary, Argentina, Chile, Cambodia, India (West Bengal), and Bangladesh. Although the mechanism of arsenic release from these sediments has been a topic of intense debate, there is broad consensus that microbially-mediated reduction of assemblages comprising arsenic (most likely as arsenate) sorbed to ferric oxyhydroxides is the dominant mechanism for the mobilization of arsenic into these groundwaters (Akai *et al.*, 2004; Islam *et al.*, 2004). For example, an early "microcosm"-based study from our group in Manchester provided direct evidence for the role of indigenous metal-reducing bacteria in the formation of toxic, mobile As(III) in sediments from the Ganges Delta (Islam *et al.*, 2004). This laboratory study showed that the addition of acetate to anaerobic sediments, as a proxy for organic matter and a potential electron donor for metal reduction, resulted in stimulation of microbial reduction of Fe(III) followed by As(V) reduction and the subsequent release of As(III), presumably by As(V)-respiring bacteria that were previously respiring Fe(III). Microbial communities responsible for metal reduction and As(III) mobilization in the stimulated anaerobic sediment were analyzed using molecular (PCR) and cultivation-dependent techniques. Both approaches confirmed an increase in numbers of metal-reducing bacteria, principally *Geobacter* species. Later work confirmed that *Geobacter* species are able to couple to oxidation of organic matter to the reduction of both Fe(III) oxides and As(V) that may be associated with the iron minerals within aquifer sediments.

Microbial communities in the subsurface are complex, and although they may become less diverse in laboratory microcosm incubations that are designed to mimic the biogeochemical conditions that support maximal levels of arsenic mobilization, identifying the causative organisms is challenging. One approach to identify the metabolically active causative organisms is to use stable isotope probing (SIP), which can link the active fraction of a microbial community to a particular biogeochemical process. This technique has been used to identify As(V)-respiring bacteria in Cambodian aquifer sediments implicated in the reductive mobilization of arsenic (Rowland *et al.*, 2007). 13C-labeled acetate was added to microcosm incubations, and promoted the reduction of the As(V) present in the sediments. PCR analysis of the "heavy" labeled DNA that was synthesized by the active fraction of the microbial community within the sediments (and separated from unlabeled "light" background nucleic acids by ultracentrifugation) led to the detection of known arsenate-respiring bacteria *Desulfotomaculum* sp. and *Desulfosporosinus* sp. from their characteristic 16S rRNA gene signatures. When 10 mM As(V) was added, to enrich for As(V)-respiring bacteria, an organism closely related to the arsenate-reducing bacterium *Sulfurospirillum* strain NP4 was identified. This organism was also closely related to clones identified previously in West Bengal sediments associated with high arsenic concentrations. Functional gene analysis of sediments amended with 13C-labeled acetate and As(V) that targeted the As(V) respiratory reductase gene (*arrA*) using highly specific PCR primers, identified gene sequences most closely related to those found in *S. barnesii* and *G. uraniireducens*. Subsequent high-throughput pyrosequencing of 13C amended, heavy-labelled DNA from similar sediment incubations, have further emphasized the potential importance of organisms affiliated with known *Geobacter* species. Here, organisms most closely related to *G. uraniireducens* dominated and carried copies of the *arrA* As(V) respiratory reductase gene, implicated in mediating As-release in laboratory incubations.

2 IDENTIFYING AS-MOBILISING BACTERIA VIA CORRELATION ANALYSIS OF IN SITU GEOCHEMICAL PROFILES & HIGH-THROUGHPUT MICROBIAL COMMUNITY SEQUENCING

Recent improvements in DNA sequencing and analysis, alongside the availability of carefully extracted and stored field samples, has made it possible to explore the in situ microbial ecology of sediments with high groundwater arsenic concentrations.

With colleagues in Barnard College and Columbia University (New York, USA) and the University of Dhaka (Bangladesh) we recently used a comprehensive suite of state-of-the art molecular techniques to better constrain the relationship between indigenous microbial communities and the iron and arsenic mineral phases present in sediments at two well-characterized arsenic-impacted aquifers in Bangladesh (Gnanaprakasam *et al.*, 2017). First, we examined the hydrogeologies of the sediments and water samples from two contrasting sites (sites F and B), followed by analyses of the composition of bacterial communities (and key functional genes that could impact the As speciation) using next generation sequencing. We then explored correlations between the microbial community compositions and the data obtained from the hydrogeological analyses of the sediment cores and relevant water samples. At both sites, arsenate [As(V)] was the major species of As present in sediments at depths with low aqueous As concentrations, while most sediment As was arsenite [As(III)] at depths with elevated aqueous As concentrations. This was consistent with a role for the microbial As(V) reduction in mobilizing arsenic. 16S rRNA gene analysis indicated that the arsenic-rich sediments were colonized by diverse bacterial communities implicated in both dissimilatory Fe(III) and As(V) reduction, while detailed correlation analyses suggested the involvement of phylogenetic groups not normally associated with As mobilization. Again, *Geobacter* species known to reduce both Fe(III) and As(V) and implicated in arsenic mobilization in previous studies were detected throughout and could play a critical role in arsenic release. In addition, Spearman rank correlations identified new phylogenetic groups that could be linked directly or indirectly with As(V) reduction and mobilization, and identification of their potential role in such processes warrants further investigation. However, given the complex microbial communities identified and the relatively low abundance of known metal-reducing bacteria, the organisms causing these problems are likely to constitute a relatively minor component of the microbial communities. The mobilization of arsenic in carbon-stimulated microcosms and pure culture lab experiments described previously are typically rapid (on the order of weeks). It is likely that under in situ conditions, these processes take longer to deliver the arsenic into the aqueous phase, consistent with higher arsenic waters that are several decades old. This would also be consistent with lower loadings of bioavailable organic material expected at depth and the relatively low abundance of metal-reducing bacteria in the microbial communities detected. The precise nature of the organic matter fueling metal reduction at depth clearly requires further investigation, as does the role of other potential electron donors, such as ammonium and methane, in such processes. This is a focus of our current work, alongside detailed metagenomic investigations of the microbial communities present in high arsenic aquifers in Cambodia, Bangladesh and Vietnam. These studies are already yielding molecular clues that we hope will lead to a better understanding of the microbial process that lead to arsenic mobilization in aquifers, and perhaps more importantly, competing microbial processes that may mitigate As contamination.

ACKNOWLEDGEMENTS

Funding from the Royal Society, Natural Environment Research Council (grant NE/P01304X/1) and NIEHS Superfund Research Program (grant P42 ES010349) are gratefully acknowledged.

REFERENCES

Akai, J., Lzumi, K., Fukuhara, H., Masuda, H., Nakano, S., Yoshimura, T., Ohfuji, H., Anawar, H.M. & Akai, K. 2004. Mineralogical and geomicrobiological investigations on groundwater arsenic enrichment in Bangladesh. *Appl. Geochem.* 19(2):215–230.

Gnanaprakasam, E.T., Lloyd, J.R., Boothman, C., Ahmed, K. M., Choudhury, I., Bostick, B.C., Geen A.V. & Mailloux B.J. 2017. Microbial community structure and arsenic biogeochemistry in two arsenic-impacted aquifers in Bangladesh. *Mbio* 8: e01326-17.

Islam, L.N., Nabi, A.M., Rahman, M.M., Khan, M.A. & Kazi, A.I. 2004. Association of clinical complications with nutritional status and the prevalence of Leukopenia among arsenic patients in Bangladesh. *Int. J. Environ. Res. Public. Health.* 1(2): 74–82.

Rowland, H., Pederick, R., Polya, D.A., Pancost, R.D., van Dongen, B.E., Gault, A.G., Vaughan, D.J., Bryant, C., Anderson, B. & Lloyd, J.R., 2007. The control of organic matter on microbially mediated iron reduction and arsenic release in shallow alluvial aquifers, Cambodia. *Geobiol.* DOI: 10.1111/j.1472-4669.2007.00136.x.

Environmental Arsenic in a Changing World –
Zhu, Guo, Bhattacharya, Ahmad, Bundschuh & Naidu (Eds)
ISBN 978-1-138-48609-6

Understanding arsenic evolution in a shallow, reducing aquifer in the lower Mekong basin, Cambodia using geochemical tracers

L.A. Richards[1], D. Magnone[1,2], J. Sültenfuß[3], A.J. Boyce[4], C. Bryant[5], C. Sovann[6], B.E. van Dongen[1], D.J. Lapworth[7], D.C. Gooddy[7], L. Chambers[8,9], C.J. Ballentine[10] & D.A. Polya[1]

[1]*School of Earth, Atmospheric and Environmental Sciences and Williamson Research Centre for Molecular Environmental Science, University of Manchester, Manchester, UK*
[2]*School of Geography, University of Lincoln, Lincoln, UK*
[3]*Institute of Environmental Physics, University of Bremen, Bremen, Germany*
[4]*Scottish Universities Environmental Research Centre, East Kilbride, UK*
[5]*NERC Radiocarbon Facility, East Kilbride, UK*
[6]*Department of Environmental Science, Royal University of Phnom Penh, Cambodia*
[7]*British Geological Survey, Wallingford, UK*
[8]*Lancaster Environment Centre, Lancaster University, Lancaster, UK*
[9]*Department of Civil and Environmental Engineering, University of Strathclyde, Glasgow, UK*
[10]*Department of Earth Sciences, University of Oxford, Oxford, UK*

ABSTRACT: Millions of people globally are exposed to dangerous arsenic concentrations in groundwater-derived drinking water sources. Arsenic mobilization in circum-Himalayan groundwater is widely attributed to the reductive dissolution of iron minerals containing arsenic, although a number of questions remain the fundamental mechanisms of this process. In a high resolution study site in Kandal Province, Cambodia, the evolution of groundwater geochemistry was evaluated along dominant groundwater flowpaths in a heavily arsenic-contaminated aquifer using a suite of geochemical tracers (*e.g*. As, Fe, SO_4, $\delta^{18}O$, $\delta^{2}H$, $^{3}H/^{3}He$, ^{14}C) to understand the processes which may contribute to arsenic mobilization in shallow, reducing aquifers typical to Southeast Asia.

1 INTRODUCTION

Millions of people in South/Southeast Asia are chronically exposed to groundwater containing dangerous concentrations of geogenic arsenic, widely attributed to reductive dissolution of iron minerals requiring metal reducing bacteria (Islam *et al.*, 2004) and bioavailable organic matter. Fundamental questions regarding the (bio)geochemical controls on this process remain contentious and important in understanding future secular changes in arsenic hazard (Harvey *et al.*, 2002; Polya & Charlet, 2009). We aim to improve the understanding of the controls on arsenic mobilization in a well-characterized shallow, reducing, arsenic-affected aquifer in Cambodia.

2 METHODS/EXPERIMENTAL

2.1 *Field area*

The field sites located in an area heavily affected by arsenic in Kien Svay (Kandal Province, Cambodia) (Berg *et al.*, 2007; Lawson *et al.*, 2013; Polizzotto *et al.*, 2008; Polya *et al.*, 2003; Sovann & Polya, 2014). are oriented broadly parallel with major inferred groundwater flowpaths along clay or sand-dominated transects (Richards *et al.*, 2017a), initially characterized using electrical resistivity tomography (Uhlemann *et al.*, 2017).

2.2 *Sample collection and analysis*

Groundwater sample collection and inorganic geochemical analysis was previously described (Richards *et al.*, 2017a). In brief, major and trace elements were analyzed using inductively coupled plasma atomic emission spectrometry (ICP-AES), inductively coupled plasma mass spectrometry (ICP-MS) and ion chromatography (Richards et al., 2017a); aqueous organic matter was characterized using fluorescence spectroscopy (Richards *et al.*, under review and references within); stable isotope (δD and $\delta^{18}O$) analyses (Richards *et al.*, 2018) using mass spectrometry techniques (Donnelly *et al.*, 2001); analysis of ^{3}H, He and Ne isotopes (Richards *et al.*, 2017b) was conducted (Sültenfuß *et al.*, 2009) and apparent ^{3}H-^{3}He ages were derived as previously described (Sültenfuß *et al.*, 2011); groundwater ^{14}C-total organic carbon (TOC) and ^{14}C-total inorganic carbon (TIC) was prepared and analysed with mass spectrometry (Xu *et al.*, 2004).

3 RESULTS AND DISCUSSION

Groundwater arsenic is highly heterogeneous and generally consistent with mobilization by reductive dissolution of iron (hydr)oxides (Richards *et al.*, 2017a). Stable isotope signatures indicate that high arsenic groundwater can have differing recharge contributions (Richards *et al.*, 2018). Apparent ^{3}H-^{3}He ages indicate most groundwaters are modern and a dominant vertical hydrological control within the aquifer (Richards et al., 2017b). The application of other tracers will also be discussed. The combined information derived from various tracers provides the conceptual framework for improved understanding of the controls on arsenic mobilization within the aquifer.

4 CONCLUSIONS

Various geochemical including isotopic tracers such as As, Fe, SO_4, $\delta^{18}O$, $\delta^{2}H$, $^{3}H/^{3}He$, ^{14}C and organic matter fluorescence properties are used to understand arsenic evolution within a shallow, reducing aquifer in Cambodia. The suite of tracers allows increased understanding of the dominant controls on groundwater arsenic concentrations in this aquifer and the results presented may be more widely applicable to other similar arsenic-affected aquifers.

ACKNOWLEDGEMENTS

This research was supported by a NERC Standard Research Grant (NE/J023833/1 to DP, BvD and CB), a NERC PhD studentship (NE/L501591/1 to DM) and a Leverhulme Trust Early Career Fellowship (ECF2015-657 to LR). Stable isotope analysis was supported by the NERC Isotope Geosciences Facility (IP-1505-1114) and radiocarbon analysis by the NERC Radiocarbon Facility NRCF010001 (1814.0414, 1834.0714 and 1906.0415). We are very grateful to field assistants (Royal University of Phnom Penh & Royal University of Agriculture, Cambodia), Resources Development International – Cambodia, local landowners and local drilling team.

REFERENCES

Berg, M., Stengel, C., Trang, P.T.K., Viet, P.H., Sampson, M. L., Leng, M., Samreth, S., & Fredericks, D. 2007. Magnitude of arsenic pollution in the Mekong and Red River Deltas – Cambodia and Vietnam. *Sci. Total Environ.* 372(2–3): 413–425.

Donnelly, T., Waldron, S., Tait, A., Dougans, J., & Bearhop, S. 2001. Hydrogen isotope analysis of natural abundance and deuterium-enriched waters by reduction over chromium on-line to a dynamic dual inlet isotope-ratio mass spectrometer. *Rapid Commun. Mass Sp.* 15: 1297–1303.

Harvey, C.F., Swartz, C.H., Badruzzaman, A.B.M., Keon-Blute, N., Yu, W., Ashraf Ali, M., Jay, J., Beckie, R., Niedan, V., Brabander, D., Oates, P.M., Ashfaque, K.N., Islam, S., Hemond, H.F., & Ahmed, M.F. 2002. Arsenic mobility and groundwater extraction in Bangladesh. *Science* 298: 1602–1606.

Islam, F.S., Gault, A.G., Boothman, C., Polya, D.A., Charnock, J.M., Chatterjee, D., & Lloyd, J.R. 2004. Role of metal-reducing bacteria in arsenic release from Bengal delta sediments. *Naure,* 430: 68–71.

Lawson, M., Polya, D.A., Boyce, A.J., Bryant, C., Mondal, D., Shantz, A., & Ballentine, C.J. 2013. Pond-derived organic carbon driving changes in arsenic hazard found in Asian groundwaters. *Environ. Sci. Technol.* 47: 7085–7094.

Polizzotto, M.L., Kocar, B.D., Benner, S.G., Sampson, M., & Fendorf, S. 2008. Near-surface wetland sediments as a source of arsenic release to ground water in Asia. *Nature* 454: 505–508.

Polya, D.A., & Charlet, L. 2009. Rising arsenic risk? *Nature Geosci* 2(6): 383–384.

Polya, D.A., Gault, A.G., Bourne, N.J., Lythgoe, P.R., & Cooke, D.A. 2003. Coupled HPLC-ICP-MS analysis indicates highly hazardous concentrations of dissolved arsenic species in Cambodian groundwaters. *RSC Special Publication* 288: 127–140.

Richards, L.A., Lapworth, D.J., Magnone, D., Gooddy, D.C., Chambers, L., Williams, P.J., van Dongen, B.E., & Polya, D.A. (under review). Dissolved organic matter tracers in an arsenic bearing aquifer in Cambodia: a fluorescence spectroscopy study. *submitted to Geosci. Front.*

Richards, L.A., Magnone, D., Boyce, A.J., Casanueva-Marenco, M.J., van Dongen, B.E., Ballentine, C.J., & Polya, D.A. 2018. Delineating sources of groundwater recharge in an arsenic-affected Holocene aquifer in Cambodia using stable isotope-based mixing models. *J. Hydrol.* 557: 321–334.

Richards, L.A., Magnone, D., Sovann, C., Kong, C., Uhlemann, S., Kuras, O., van Dongen, B.E., Ballentine, C.J., & Polya, D.A. 2017a. High resolution profile of inorganic aqueous geochemistry and key redox zones in an arsenic bearing aquifer in Cambodia. *Sci. Total Environ.* 590–591: 540–553.

Richards, L.A., Sültenfuß, J., Ballentine, C.J., Magnone, D., van Dongen, B.E., Sovann, C., & Polya, D.A. 2017b. Tritium tracers of rapid surface water ingression into arsenic-bearing aquifers in the Lower Mekong Basin, Cambodia. *Procedia Earth Planet Sci.* 17C: 849–852.

Sovann, C., & Polya, D.A. 2014. Improved groundwater geogenic arsenic hazard map for Cambodia. *Environ. Chem.* 11(5): 595–607.

Sültenfuß, J., Purtschert, R., & Führböter, J.F. 2011. Age structure and recharge conditions of a coastal aquifer (northern Germany) investigated with ^{39}Ar, ^{14}C, ^{3}H, He isotopes and Ne. *Hydrogeol. J.* 19: 221–236.

Sültenfuß, J., Roether, W., & Rhein, M. 2009. The Bremen mass spectrometric facility for the measurement of helium isotopes, neon, and tritium in water. *Isot. Environ. Healt. S.* 45(2): 83–95.

Uhlemann, S., Kuras, O., Richards, L.A., Naden, E., & Polya, D.A. 2017. Electrical Resistivity Tomography determines the spatial distribution of clay layer thickness and aquifer vulnerability, Kandal Province, Cambodia. *J. Asian Earth Sci.* 147: 402–414.

Xu, S., Anderson, R., Bryant, C., Cook, G.T., Dougans, A., Freeman, S., Naysmith, P., Schnabel, C., & Scott, E.M. 2004. Capabilities of the New SUERC 5MV AMS Facility for ^{14}C Dating. *Radiocarbon* 46(1): 59–64.

Environmental Arsenic in a Changing World –
Zhu, Guo, Bhattacharya, Ahmad, Bundschuh & Naidu (Eds)
ISBN 978-1-138-48609-6

Significance of arsenic resistant prokaryotes in climate change perspective

A.B.M.R. Islam[1,2], S.A. Ahmad[1], M. Alauddin[3] & K. Tazaki[4]
[1]*Department of Occupational and Environmental Health, Bangladesh University of Health Sciences, Dhaka, Bangladesh*
[2]*Coordination of Environment and Life Line (CELL), Next International Co. Ltd, Sangenjaya, Tokyo, Japan*
[3]*Department of Chemistry, Wagner College, Staten Island, New York, USA*
[4]*Department of Earth Sciences, Faculty of Science, Kanazawa University, Kanazawa, Ishikawa, Japan*

ABSTRACT: The purpose of this study was to observe symbiotic prokaryotes under the optical and scanning electron microscope for their morphology and multiple roles including arsenic immobilization in the natural environment. Microbial mats proliferating near to the arsenic contaminated (141 to 997 μg L^{-1}) tube-wells were collected for this study. The results of microscopic observation showed that the symbiotic prokaryotes in microbial mats were mainly consisting of *Coccus, Bacillus, and Filamentous* microbes associated with photosynthetic *Cyanobacteria;* enabling to accumulate high concentrations of As (550 mg kg^{-1}). ED-XRF analysis also confirmed the presence of As in microbial mats associated with Na, Mg, Al, Si, P, S, K, Ca, Ti, Mn, Fe, Zn and Sr. Henceforth, it can be concluded that the symbiotic prokaryotes can play a significant role in enabling to uptake arsenic from tube-well water, and release O_2 in the ambient air simultaneously. Our findings are very encouraging for further research to promote a sustainable mitigation effort for provision of As safe drinking water and a CO_2 reduced safer environment.

1 INTRODUCTION

Arsenic (As) toxicity is a global health problem affecting millions of people through drinking water. Besides global warming is an important factor for climate change due to increasing CO_2 in the atmosphere. Therefore As contamination of ground water and climate change are receiving attentions from global community as important issues of contemporary research for the sake of public health and environmental disaster. Arsenic, the metalloid poses serious health threats for human when it exceeds the WHO guideline value 10 μg L^{-1} in drinking water. However, some symbiotic prokaryotes are enabling to obtain energy even in an extreme or toxic environment and are intimately involved with the biogeochemical cycling of metals or metalloids for their own niche (Oremland & Stolz, 2003). It is well known that certain prokaryotes such as Cyanobacteria are famous for their oxygenic photosynthesis (Badqer & Price, 2003). Besides this microbial mats consisting of microbes are capable of accumulate As and produce biominerals in geo-aqua-eco system (Tazaki et al., 2003; Islam et al., 2014). The aim of this study was to observe symbiotic prokaryotes under the optical and scanning electron microscope for their multiple roles; such as controls on arsenic immobilization and photo synthetic activities.

2 METHODS AND MATERIALS

2.1 *Materials*

Various colored microbial mats proliferating near to the arsenic contaminated tube-wells were collected from Hazigonj Upazila, a highly As contaminated areas in Bangladesh (BGS/DPHE, 2001). Tube-well waters were circum pH with a high concentration of As (141 to 997 μg L^{-1}), much higher than that of WHO standard. A part of sample was fixed with 1% gluteraldehyde and carried to the Kanazawa University laboratory for further analysis and observations.

2.2 *Observations and analyses*

Optical microscopic observations were carried out by an epifluorescence microscope (Nikon EFD3, equipped with Digital camera: COOLPIX995). Episcopic and DAPI stained samples were observed in this study. Scanning electron microscopic (JEOL JSM-5200LV) technique was applied for micro morphology of symbiotic prokaryotes. Besides this, to determine the elemental concentrations; dried microbial mats were analyzed by ED-XRF (JEOL JSX 3201). Especially, Neutron activation analytical technique was applied for determining the arsenic concentration in microbial mats.

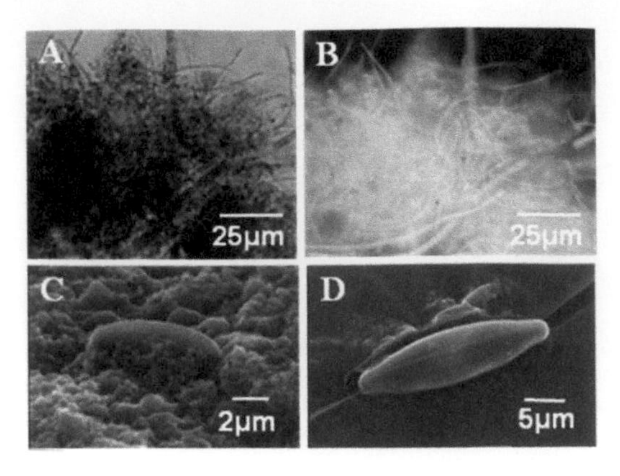

Figure 1. Micrographs show the presence of Coccus, Bacillus, Filamentous and Cyanobacteria in As rich microbial mats.

3 RESULTS AND DISCUSSION

3.1 *Microscopic observations*

The results of microscopic observation are shown in Figure 1. Optical micrograph indicated the colony of symbiotic prokaryotes in microbial mats, mainly consisting of autotrophs and heterotrophs, such as *Coccus, Bacillus, and Filamentous* microbes associated with *Cyanobacteria* (A). DAPI stained epifluorescence micrograph showed the fluorescent blue and red part indicating DNA in prokaryotic cells, photosynthetic pigments in chromatophores of photo-autophytes including *Cyanobacteria* (B). SEM micrograph confirmed the presence of As resisting *Coccus* and *Bacillus* types of prokaryotes associated with diatom in a fixed sample (C and D). In this study, we found vigorous proliferations of As resistant *Cyanobacteria* in microbial mats, those can reduce CO_2 in the environment as they are only photosynthetic prokaryotes able to produce Oxygen (Badqer & Price, 2003). Furthermore it is reported that *Cyanobacteria* are responsible for roughly 10 % of global photosynthetic activities (Mangan *et al.*, (2016).

3.2 *As accumulation in microbial mats*

The profile of microbial metal accumulation is presented in Table 1. The semi quantitative analytical data showed the presence of As associated with Na, Mg, Al, Si, P, S, K, Ca, Ti, Mn, Fe, Zn and Sr in all microbial mats. Intimate involvement of microbe metal interactions might play the key role in immobilizing the corresponding metals or metalloids from the As contaminated tube-well water sources (Tazaki *et al.*, 2003). However, Neutron activation analysis (NAA) data showed a high concentration of As (550 mg kg^{-1}) in microbial mats, which is several thousand times higher than that of corresponding tube well water (Islam *et al.*, 2014).

Table 1. Elemental concentrations in different microbial mats.

El	D-Gy	LG	DG	LB 1	LB 2	LB 3	Gy
Na	nd	nd	0.56	nd	nd	nd	nd
Mg	1.65	4.72	1.76	1.41	1.29	2.15	1.19
Al	5.55	3.18	8.38	3.00	3.3	6.24	4.3
Si	40.2	32.86	51.20	19.34	28.22	46.08	28.34
P.	2.63	3.00	2.84	4.47	4.35	3.32	5.63
S	1.47	2.40	1.08	1.78	3.62	1.40	2.64
K	7.56	3.64	7.99	4.99	9.17	7.63	6.86
Ca	13.83	10.98	7.42	17.28	17.03	7.77	12.45
Ti	2.71	2.88	2.47	1.46	2.95	3.33	3.37
Mn	0.93	4.00	0.32	0.86	1.30	1.11	2.66
Fe	22.89	30.05	15.95	44.43	28.22	20.35	31.23
Zn	0.33	1.89	nd	0.36	0.18	0.16	0.68
As	0.16	0.26	0.02	0.46	0.21	0.18	0.50
Sr	0.1	0.17	nd	0.15	0.12	0.08	0.15
	100	100.03	99.99	99.99	99.96	99.89	100 wt%

El; Elements, D-Gy; Dim Grey, LG; Light Green, DG; Deep Green, LB 1; Light Brown 1, LB 2; Light Brown 2, LB 3; Light Brown 3, Gy; Grey, nd; not detected.

4 CONCLUSIONS

Symbiotic prokaryotes can play a significant role in enabling to uptake arsenic from well water, and release of O_2 in the atmosphere. Further research are needed to develop a sustainable mitigation effort for provision of As free safe drinking water and a CO_2 reduced safer environment.

ACKNOWLEDGEMENTS

We thank Ministry of Education, Science and Culture, Japan for supporting this study. We are thankful to Dr. Wahid Uddin Ahmed and GETE Research Center, Bangladesh for additional support.

REFERENCES

BGS/DPHE 2001. Arsenic contamination of groundwater in Bangladesh, Volume1: ReportWC/00/19, Keyworth: British Geological Survey.

Badqer, M.R. & Price, G.D. 2003. CO_2 concentrating mechanisms in Cyanobacteria: Molecular components, their diversity and evolution, *J. Exp. Biol.* 54(383): 609–622.

Islam, A.B.M.R., Jalil, M.A. & Tazaki, K. 2014. Roles of microbial community on arsenic removal in drinking – water. *Chem. Express* 3(3): 85–96.

Mangan, N.M., Flamholz, A., Hood, R.D., Milo, R. & Savage, D.F. 2016. pH determines the energetic efficiency of the cyanobacterial CO_2 concentrating Mechanism, *Proc. Natl. Acad. Sci. USA.* 113(36): E5354–E5362.

Oremland, R.S. & Stolz, J.F. 2003. The ecology of arsenic, *Science* 300: 939–944.

Tazaki, K., Islam, A.B.M.R., Nagai, K. & Kurihara, T. 2003. $FeAs_2$ biomineralization on encrusted bacteria in hot springs: an ecological role of symbiotic bacteria, *Can. J. Earth Sci.* 40: 1725–1738.

Environmental Arsenic in a Changing World –
Zhu, Guo, Bhattacharya, Ahmad, Bundschuh & Naidu (Eds)
ISBN 978-1-138-48609-6

Microbial study related with the arsenic hydrogeochemistry of the Xichú River in Guanajuato, Mexico

U.E. Rodríguez Castrejón, A.H. Serafín Muñoz, C. Cano Canchola & A. Álvarez Vargas
Departamento de Biología, División de Ciencias Naturales y Exactas, Universidad de Guanajuato, Campus Guanajuato, México

ABSTRACT: Microorganisms play an important role in many surface and near-surface geochemical environments. Partial physicochemical characterization of Xichu River waters showed high concentrations of As (98 $\mu g L^{-1}$) in the zone of the "Aurora" mine probably due to arsenic release into the environment through metal sulfide oxidation. Several microorganisms strains resistant or tolerant to 20mM of As(III) or As(V) were isolated from the river water, sediments and/or biofilms. The microorganisms likely interact with arsenic through mechanisms such as sorption, mobilization, precipitation and redox reactions, understanding of which may facilitate the development of new bio-treatments.

1 INTRODUCTION

Arsenic (As) is toxic to humans and other organisms. Its mobility in the environment occurs through natural processes, for instance, weathering reactions, biological activity and volcanic emissions, as well as through anthropogenic processes (Hu & Gao, 2008). Arsenic can affect exploitation of water, agricultural development and sustainable use of soil, directly affecting socioeconomic growth in affected areas (Litter *et al.*, 2009).

Environmental geochemistry includes a consideration of geological, mineralogical and chemical processes that introduce and control concentrations, reactions and mobility of potentially toxic elements (PTE) in the environment (Hudson *et al.*, 2011). Microorganisms interact with the arsenic through mechanisms such as adsorption, redox reactions and precipitation, and may play a major role in determining the rate of releasing of arsenic in geochemical environments.

2 METHODS/EXPERIMENTAL

2.1 *Water samples processing*

Water samples were collected from the Xichú river, which contains high concentrations of arsenic as well as tons of tailings from the "Aurora" mine. The sample site is located inside the "Sierra Gorda of Guanajuato", a Mex. Biosphere and federal natural reserve in the state of Guanajuato, Mexico.

Samples were taken at 7 sites either in April, September or both months: Before Jales (BJ) Jales (J), After Jales (AJ), Entry Jales (EJ), Waterhole (WO), Mixture (M) and Laja River (LR).

In-situ analyses were carried out for: alkalinity, electric conductivity, dissolved oxygen and temperature with a potentiometer (Thermo Scientific). Furthermore, sulfates, sulfides, ferrous ion, nitrate, and nitrites were determined in the laboratory with HACH equipment. Metals (Pb, Cd, Zn, Mn, Cu) and arsenic were determined by atomic absorption spectrometry.

2.2 *Microbiological analysis*

The isolation of microorganisms from the each sample was carried out on nutritive agar (NA) as a growth medium amended with sodium arsenite (As(III)) and sodium arsenate (As(V)) in increasing concentrations to 20 mM to obtain microorganisms (tolerant or resistant) to both arsenic salts under aerobic or anaerobic conditions. Microorganisms tolerant or resistant to (As(III)) or (As(V)) with capacities to grow under aerobic or anaerobic conditions were morphologically characterized for its molecular identification by using comparisons of 16S ribosomal gene sequences. Other characterization that will be done is the detection of genes involved with its arsenic interaction.

3 RESULTS AND DISCUSSION

3.1 *Water samples*

Water analyses are summarized in Table 1. The highest arsenic concentration found was 98 $\mu g L^{-1}$, the lowest 0 $\mu g L^{-1}$, with the higher values near the "Aurora" mine as reported by Salas (214), and the lower values being found at sampling sites more distant from the "Aurora" mine and also during the summer. These trends reflect a point source discharge diluted with distance (Drahota *et al.*, 2013).

Table 1. Water Analysis Results. ND = not determined.

	April Samples					September Samples					
Parameters	*AJ*	*J*	*EJ*	*DJ*	*OA*	*AJ*	*J*	*EJ*	*DJ*	*MZ*	*RL*
SO_4^{2-} mg L^{-1}	110	133	145	160	123	12.3	17.6	12	15.3	24	18
pH	9.37	9.55	9.50	10.19	7.50	8.43	8.38	8.34	8.66	8.60	8.77
S^{-2} mg L^{-1}	0.00	0.00	0.00	0.00	0.00	0.09	0.15	0.14	0.10	0.14	0.07
EC $\mu S\,cm^{-1}$	630	957	653	661	647	156	157	157	161	189	231
DO mg L^{-1}	5.7	4.68	4.18	4.45	4.2	2.08	1.67	0.8	1.6	1.2	0.96
T(°C)	20.2	20.5	22.6	17.821.9	27.1	17.8	17.7	17.7	18.1	21.1	22.9
AsT $\mu g\,L^{-1}$	35	40	50	98	8.9	4	5	5	9	3	0
Fe^{2+}mg L^{-1}	0.04	0.036	0.042	0.08	0.022	0.34	0.39	0.73	0.34	0.56	0.24
Alk mg $CaCO_3$ L^{-1}	ND	ND	ND	ND	ND	6	7	6	7	9	12
NO^{-3} mg L^{-1}	9.6	1.25	6.8	0.9	4.0	2.6	2.9	2.1	4.4	5.8	5.3

3.2 *Microbiology analysis*

We isolated 50 arsenic tolerant strains with different phenotypes: Different Gram, cell morphology, colonial morphology (colors blue, yellow, orange, red and withe), arsenite or arsenate tolerance, specific ability to growth on aerobic or anaerobic conditions. We note that Shakya *et al.* (2012) has previously reported microorganisms able to tolerate similar concentration of As, viz. 13 mM of arsenate and 10 mM of arsenite.

Strains isolated from water were typically more tolerant of arsenite than those from sediment of arsenate, whilst strains isolated from biofilms had similar reistance to arsenite and arsenate. The fact that the microorganism growed in both species: As (III) and As (V) could be influenced for the physicochemical characteristics in the study zone, where it is even seen as a reducing environment and would favor the present of As (III) in water, the arsenate could be in the sediments. Both species can appears in neutral pH in oxic and anoxic environments (Nordstrom, 2009).

The importance of having isolated microorganism aerobics and anaerobic facultative that grow in presence of arsenic (III) and (V) perhaps indicate that the bacterial are using some arsenic detoxification mechanism through the arsenite oxidation (Oremland & Stolz, 2003), these hypothesis is due to that the strains coming from water samples had a better growth in presence of As(III). The presence of As(III) can be maintained in oxygenic water by biological reduction of As(V) (Santini & Ward, 2012). The arsenate reduction in neutrals environments of pH are catalyzed majority by microorganism including production of energy and arsenic detoxification (Huang, 2009).

4 CONCLUSIONS

High concentrations of arsenic in the Xichu River during both the spring and summer represent a significant hazard to citizens that live near mining waste or drink the river water.

The discovery of various microorganisms with tolerance to all of As(III), As(V) or both species suggests that both these arsenic species may be present in the natural environment – a hypothesis that could be checked by appropriate speciation analysis. The discovery of four types of microorganism, aerobic and anaerobic strict and facultative, suggests different mechanisms of interaction between microorganisms and environmental arsenic.

ACKNOWLEDGEMENTS

University of Guanajuato CONACYT (Consejo Nacional de Ciencia y Tecnología).

REFERENCES

Drahota, P., Nováková, B., Matoušsek, T., Mihaljevič, M., Rohovec J. & Filippi, M. 2013. Diel variation of arsenic, molybdenum and antimony in a stream draining a natural As geochemical anomaly. *Appl. Geochem.* 31:84–93.

Hu, Z.C. & Gao, S. 2008. Upper crustal abundances of trace elements: A revision and update. *Chem. Geol.* 253: 205–221.

Huang, A.H., Teplitski, M., Rathinasabapathi, B. & Ma, L. 2009. Characterization of arsenic-resistant bacterial comunities in the rhizosphere of an arsenic hyperaccumulator *Peteris vittata L. Can. J. Microbiol.* 56(3): 263–46.

Hudson, K.A., Jamieson, H.E. & Lottermoser, B.G. 2011. Mine wastes: past, present and future. *Elements* 7(6): 375–380.

Litter, M.I., Armienta, M.A., & Farías, S.S. 2009. Iberoarsen. Metodologías analíticas para la determinación y especiación de arsénico en aguas y suelos. Argentina: CYTED.

Nordstrom, D. K. 2002. Public health-Worldwide occurrences of arsenic in ground water. *Science* 296: 2143–2145.

Oremland, R.S. & Stolz, J.F. 2003. The ecology of arsenic. *Science* 300: 939–44.

Salas, E.F. 2014. Geoquímica y Mineralogía de Jales en Mina Aurora, Xichú, Guanajuato, México, D. F. Thesis.

Santini, J.M. & Ward, S.A. (Eds.) 2012. The Metabolism of Arsenite. In: Arsenic in the Environment 5. CRC Press, Boca Raton, Florida, 189 p.

Shakya, S., Pradhan, B, Smith, I., Shrestha, J. & Tuladhar, S. 2012. Isoltion and characterization of aerobic culturable arsenic-resistant bacteria from surface water and groundwater of Rautahat District, Nepal. *J. Environ. Manag.* 95: S20–S25.

1.4 Arsenic and other trace elements in groundwater of China

Environmental Arsenic in a Changing World –
Zhu, Guo, Bhattacharya, Ahmad, Bundschuh & Naidu (Eds)
ISBN 978-1-138-48609-6

Nano-TiO_2 both increases and decreases arsenic toxicity: Evidence from different aquatic animal experiments

Z.X. Luo[1,2], Z.H. Wang[2], F. Yang[1] & C.Z. Yan[1]
[1]*Key Laboratory of Urban Environment and Health, Institute of Urban Environment, Chinese Academy of Sciences, Xiamen, P.R. China*
[2]*College of Chemistry and Environment, Fujian Province Key Laboratory of Modern Analytical Science and Separation Technology, Minnan Normal University, Zhangzhou, P.R. China*

ABSTRACT: Titanium dioxide nanoparticles (nano-TiO_2) are usually applied to treat arsenic-polluted drinking water. Additionally, wide application of nano-TiO_2 can enter into the environment inevitably. Furthermore, titanium dioxide nanoparticles (nano-TiO_2) can adsorb ambient pollutants to modify their bioavailability in aquatic organisms, as well as their toxicity of its ambient pollutants. This study investigated arsenic accumulation, subcellular distribution and toxicity on two aquatic animals (*Daphnia magna* in freshwater and *Artemia salina* in salt water). Nano-TiO_2 acts as a positive carrier, significantly facilitating D. magna's ability to uptake As(V) as well as *Artemia salina*. In *Artemia salina*, As percentage in biologically active metal (BAM) fractions significantly decreased after the addition of 10 $mg\,L^{-1}$ of nano-TiO_2, while the fractional percentage of As increased in cellular debris. These lower As proportions in the sensitive fractions of cells indicate As(V) toxicity inhibition. Higher As(V) EC_{50} values in nauplii (by a magnitude from 1.97 to 2.76) compared to the control (nano-TiO_2 free) as we increased nano-TiO_2 concentrations from 1 to 1000 $mg\,L^{-1}$, indicates that nano-TiO_2 could alleviate As(V) toxicity in *A. salina* nauplii by enhancing efflux and decreasing the proportion of As in the sensitive fractions of cells. In contrast, in *D. magna*, even though As accumulation increased with increasing nano-TiO_2 concentrations in *D. magna*, As(V) toxicity associated with nano-TiO_2 exhibited a dual effect. Compared to the control, the increased As was mainly distributed in BDM (biologically detoxified metal), but Ti was mainly distributed in MSF (metal-sensitive fractions) with increasing nano-TiO_2 levels. Differences in subcellular distribution demonstrated that adsorbed As(V) on nano-TiO_2 could dissociate itself and be transported separately, which results in increased toxicity at higher nano-TiO_2 concentrations. Decreased As(V) toxicity associated with lower nano-TiO_2 concentrations results from unaffected As levels in MSFs (when compared to the control), where several As components continued to be adsorbed on nano-TiO_2. Accordingly, arsenic toxicity on aquatic animals by nano-TiO_2 depends on animal species and the behaviors of the co-contaminants inside, especially their assimilation in the digestive tract. More attention should be paid to the influence of nano-TiO_2on As(V) assimilation in the digestive tract.

1 INTRODUCTION

Arsenic (*As*) is a metalloid included in the priority list of hazardous substances. This is a result of its high toxicity and wide distribution. Potential health hazards and environmental impacts of manufactured nanoparticles (MNPs) have become a significant concern with the rapid development of nanotechnology. It has been well established that nano-TiO_2 has the potential to facilitate the entry of co-contaminants adsorbed by nanoparticles into aquatic organisms and to subsequently promote potential toxic effects (Li *et al.*, 2016; Yan *et al.*, 2017). However, little available information describes the impact of nanoparticles on biokinetics of various co-contaminants in aquatic organisms, particularly for the spatial and subcellular distribution of co-contaminants associated with nanoparticles.

Titanium dioxide nanoparticles (nano-TiO_2) are usually applied to treat polluted drinking water by arsenic (Yan *et al.*, 2017). Additionally, wide application of nano-TiO_2 can enter into the environment inevitably. Most previous relevant studies observed that *As*(V) is strongly adsorbed to TiO_2 nanoparticles and could subsequently affect chemical accumulation. Arsenic toxicity on aquatic animals by nano-TiO_2 depends on animal species and the behaviors of the co-contaminants inside, especially their assimilation in the digestive tract. This study investigated arsenic

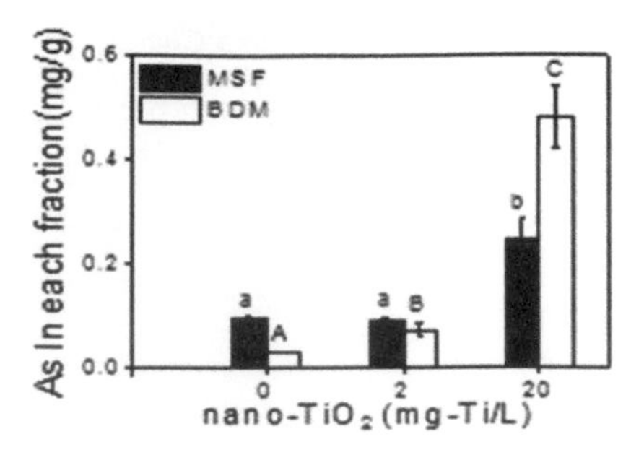

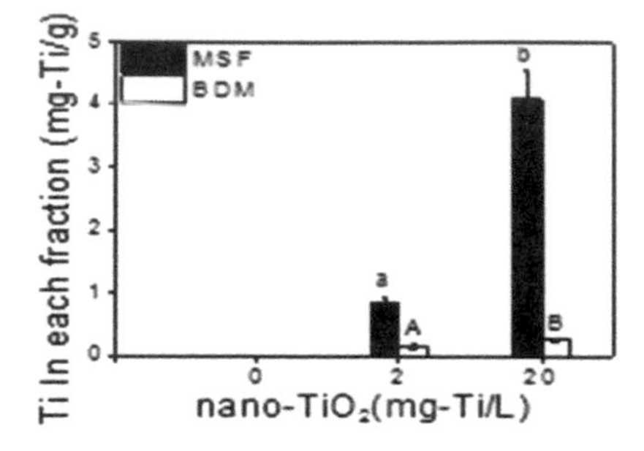

Figure 1. Subcellular distribution in *D. magna*.

accumulation, subcellular distribution and toxicity on two aquatic animals (*Daphnia magna* in freshwater and *Artemia salina* in salt water).

2 METHODS

2.1 *Subcellular distribution*

We determined subcellular partitioning of total *As* and *Ti* in *two aquatic animals* body tissues using the methods of differential centrifugation (Wang *et al.*, 2017). We obtained a total of five different fractions, including cellular debris (containing cell membranes), organelles (containing nuclear, mitochondrial, microsomes, and lysosomes), heat-denatured protein (HDP, containing enzymes), heat-stable protein (HSP, or metallothionein-like proteins), and metal-rich granules (MRG). Results are expressed in mg (dry weight) per individual.

3 RESULTS AND DISCUSSION

3.1 *Subcellular distribution of arsenic and titanium*

The dissociation of arsenic from nano-TiO_2 of $20\,mg\,L^{-1}$ was observed due to the differences of

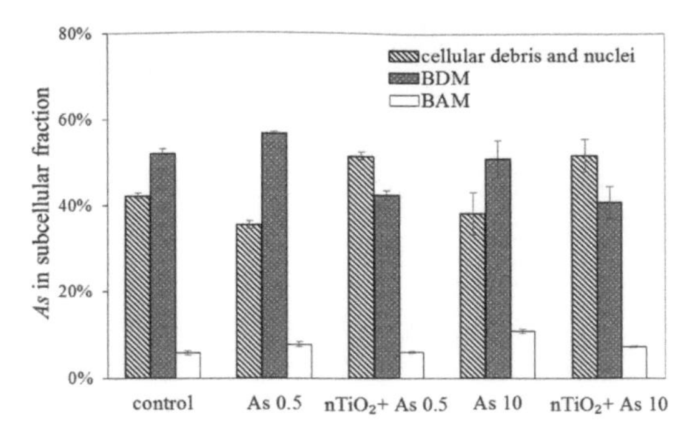

Figure 2. Subcellular distribution in *A. salina*.

subcellular distribution of arsenic and titanium in *D. magna*, resulting in the higher toxicity of arsenate in higher levels of nano-TiO_2 (Fig. 1).

However, decreased As(V) toxicity associated with lower nano-TiO_2 concentrations results from unaffected As levels in MSFs (when compared to the control), where several As components continued to be adsorbed by nano-TiO_2 (Fig. 1).

Arsenic percentage in biologically active metal (BAM) fractions significantly decreased, decreasing *As*(V) toxicity on *A. salina* (Fig. 2).

ACKNOWLEDGMENTS

This study was jointly supported by the National Nature Science Foundation of China (project nos. 41401552, 41271484 and 21277136) and the Nature Science Foundation of Fujian Province (2016J01691).

REFERENCES

Li, M., Luo, Z., Yan, Y., Wang, Z., Chi, Q., Yan, C. & Xing, B.S. 2016. Arsenate accumulation, distribution, and toxicity associated with titanium dioxide nanoparticles in *Daphnia magna*. *Environ. Sci. Technol.* 50(17): 9636–9643.

Wang, Z. H., Luo, Z. X., Yan, C. Z. & Xing, B. S. 2017. Impacts of environmental factors on arsenate biotransformation and release in *Microcystis aeruginosa* using the Taguchi experimental design approach. *Water Res.* 118: 167–176.

Yan, C. Z., Yang, F., Wang, Z. S., Wang, Q. Q., Seitz, F. & Luo, Z. X. 2017. Changes in arsenate bioaccumulation, subcellular distribution, depuration, and toxicity in *Artemia salina nauplii* in the presence of titanium dioxide nanoparticles. *Environ. Sci. Nano.* 4(6): 1365–1376.

Environmental Arsenic in a Changing World –
Zhu, Guo, Bhattacharya, Ahmad, Bundschuh & Naidu (Eds)
ISBN 978-1-138-48609-6

Interfacial interaction of arsenic(V) with Mg-containing calcite and calcite

P.L. Gong, X.B. Gao, Q.Q. Yi, W.T. Luo, X. Zhang & F. Liu
School of Environmental Studies, China University of Geosciences, Wuhan, P.R. China

ABSTRACT: Carbonate system has a significant effect on the migration and transformation of trace elements in groundwater. But impure calcite (such as Mg-containing calcite) is the main occurrence pattern of carbonate in nature groundwater system. In this study, batch adsorption experiments were carried out to investigate the interfacial interaction of arsenic(V) on Mg-calcite and calcite. Characterization methods were employed to compare the changes after reaction. The adsorption reactions showed that the adsorption capacity of Mg-containing calcite was higher than pure calcite.

1 INTRODUCTION

Carbonate system has a significant effect on migration and transformation of trace elements in groundwater (Yokoyama *et al.*, 2012; Stollenwerk, 2003). There are several relevant researches on surface adsorption and co-precipitation of the pure calcite (Monteshernandez *et al.*, 2009). However, it is not well understood about interaction between non-pure calcite and arsenic (As). In natural groundwater system, there is no pure calcite, impure calcite (such as Mg-containing calcite) is the main occurrence pattern of carbonate. In this study, on behalf of impure calcite, interfacial interaction of As(V) with Mg-containing calcite was simulated.

2 METHODS/EXPERIMENTAL

Mg-calcite was prepared with calcium to magnesium ratios, which is 9:1 (Long *et al.*, 2011, 2012; Xu *et al.*, 2013). Then batch adsorption (isothermal, kinetic adsorption) experiments were carried out to investigate the interfacial interaction of As(V) on Mg-calcite and calcite. Characterization methods, such as Laser Particle Size Analyzer, XRD, FTIR and SEM, were employed to determine the Mg-calcite, calcite, and after their interaction with As(V) in surface morphology, polymorphs and chemical elements.

3 RESULTS AND DISCUSSION

On this basis, the characteristics and mechanism of As(V) adsorption on Mg-Calcite were investigated.

3.1 *Characteristic of calcite*

Mg-calcite prepared by the precipitation method was nearly spherical and had a particle size of about 25–40 μm. With the increase of the magnesium content in the initial solution, the grain size increased slightly and the crystallinity decreased (Fig. 1).

Figure 1. FE-SEM Images of (a) Ca_9Mg_1, (b) Calcite (Ca_nMg_m refers to the Mg-calcite, prepared with the initial Ca:Mg molar ratio of n:m.).

3.2 *Adsorption experiments*

Isothermal batch adsorption experiments (24 h) of Mg-calcite and calcite with different As(V) concentrations (0~300 μM), followed Freundlich adsorption isotherm model. Kinetic adsorption of Mg-calcite and calcite with 30 μM As(V) with different reaction times (10 min~120 h), which was a comparatively fast reaction within the first 10 h, followed by a slower and at a relatively constant reaction rate until equilibrium was achieved, and the adsorption behavior was in accordance with the second-order kinetic adsorption equation.

3.3 *Adsorption capacity*

Adsorption reactions showed that the adsorption capacity of Mg-calcite is of the order, calcite < Ca_9Mg_1; the maximum arsenic adsorption potential of Ca_9Mg_1 was 0.00344 mol kg^{-1} calcite, which was higher than that of calcite (Fig. 2). This phenomenon is mainly due to the impact of integrated Mg in crystal lattice of calcite, which could result in a

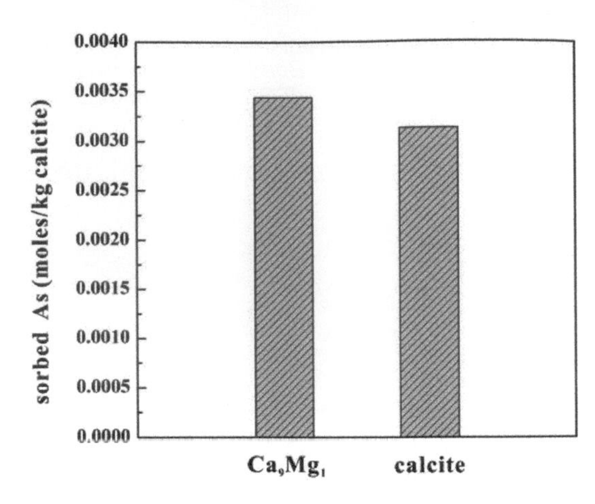

Figure 2. Adsorption isotherms at initial arsenic concentration of 300 μM.

mineral surface with higher positive charge and greater propensity for the adsorption of negatively charged species of arsenic.

4 CONCLUSIONS

The results of this research showed that the interaction of As(V) with Mg-calcite can be directly impacted by Mg content. The presence of Mg in the solution had led to the morphology and phase transformation from calcite to other calcium carbonate forms (such as aragonite and vaterite) in the process of precipitation.

The adsorption capacity of Mg-containing calcite was higher than pure calcite. So, in the arsenic polluted groundwater, such as that in the Northern China, the adsorption capacity of the aquifer medium for As(V) is enhanced when the content of Mg-calcite increases in the aquifer sediments.

REFERENCES

Long, X., Ma, Y. & Qi, L. 2011. In vitro synthesis of high Mg calcite under ambient conditions and its implication for biomineralization process. *Cryst. Growth Des.* 11(11): 2866–2873.

Long, X., Nasse, M.J. & Ma, Y. 2012. From synthetic to biogenic Mg-containing calcites: a comparative study using FTIR microspectroscopy. *Phys. Chem. Chem. Phys.* 14(7): 2255.

Monteshernandez, G., Conchalozano, N. & Renard, F. 2009. Removal of oxyanions from synthetic wastewater via carbonation process of calcium hydroxide: applied and fundamental aspects. *J. Hazard Mater.* 166(2–3): 788.

Stollenwerk, K.G. 2003. Geochemical processes controlling transport of arsenic in groundwater: A review of adsorption. *Arsenic in Ground Water. Springer US.* 67–100.

Xu, J., Yan, C. & Zhang, F. 2013. Testing the cation-hydration effect on the crystallization of Ca-Mg-CO_3 systems. *Proc. Natl. Acad. Sci. USA.* 110(44): 17750–177505.

Yokoyama, Y., Tanaka, K. & Takahashi, Y. 2012. Differences in the immobilization of arsenite and arsenate by calcite. *Geochim. Cosmochim. Acta* 91(91):202–219.

Environmental Arsenic in a Changing World –
Zhu, Guo, Bhattacharya, Ahmad, Bundschuh & Naidu (Eds)
ISBN 978-1-138-48609-6

Temporal dynamics of microbial community structure and its effect on arsenic mobilization and transformation in Quaternary aquifers of the central Yangtze River Basin

T.L. Zheng[1], Y. Deng[1], H.C. Jiang[2] & Y.X. Wang[2]
[1]*Geological Survey, China University of Geosciences, Wuhan, P.R. China*
[2]*State Key Laboratory of Biogeology and Environmental Geology, China University of Geosciences, Wuhan, P.R. China*

ABSTRACT: To elaborate the seasonal variation of groundwater arsenic (As) and its underlying regulating biogeochemical processes, long term biogeochemical monitoring and sediment incubation experiments were established in the aquifers from the Jianghan Plain, central Yangtze River basin. The results indicated that the groundwater As concentration exhibited significant seasonal variations (exceeding $800\,\mu g\ L^{-1}$) between the pre-monsoon (January, February and March) and monsoon (June, July and August) season, which was regulated by the redox status fluctuation in the groundwater. The correlation between the Fe^{2+} and As concentrations in the groundwater from different seasons suggested the As mobilization in the aquifers was controlled by the iron mineral transformation. Furthermore, the incubation experiments with the sediments collected from different depths in the same aquifers indicated that the variation of microbial community was correlated with the released As concentration ($R = 0.7$, $P < 0.05$) and the iron-reducing bacteria, including *Pseudomonas, Clostridium* and *Geobacter*, were the main drivers for the As mobilization from the sediments at 26 m and 36 m depth. In addition, the bacterial sulfate reducing process could affect the As temporal variation in groundwater through forming the Fe-sulfide minerals to scavenge the As in aqueous phase. Those results could provide new insights into the mechanism of As mobilization and seasonal As variation in groundwater systems.

1 INTRODUCTION

Natural elevation of arsenic (As) in groundwater poses serious health risk in various parts of the world. A prominent problem in the As-affected groundwater, which remains not well understood, is that the As concentrations vary temporally in groundwater. Recently, the shallow Quaternary aquifers in the Jianghan Plain, central Yangtze River Basin, has been reported with natural enrichment and seasonal variation of As in the groundwater, however little is known about the underlying regulating processes. Thus long term biogeochemical monitoring and anaerobic incubation experiments from aquifer sediments of different depths were conducted to reveal the underlying biogeochemical processes responsible for the As mobilization and temporal variation.

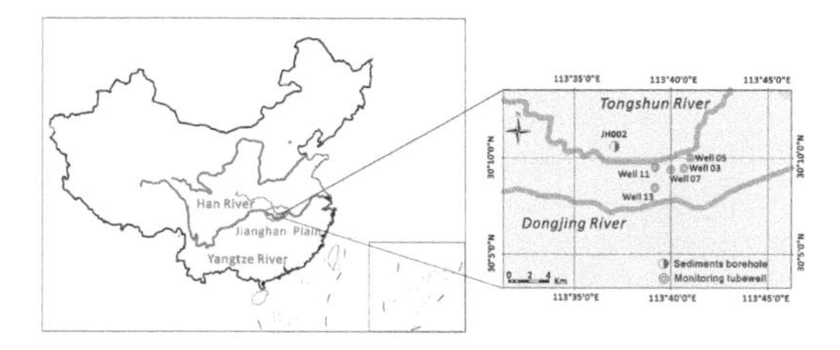

Figure 1. The geographical location of sampling borehole JH002 and monitoring wells 03, 05, 07, 11 and 13.

2 METHODS AND MATERIALS

2.1 *Study area*

Jianghan Plain is a semi-closed Quaternary basin with the subtropical monsoon climate, and it is formed by the alluvial sediments of the Yangtze River and its largest tributary Han River (Fig. 1). The Quaternary aquifer systems in the Jianghan Plain could be divided into two groups: 1) the Holocene unconfined aquifer, which is composed of Holocene alluvial and lacustrine deposits (clay, silt and fine sand) in the depth of 3–15 m below the ground surface (bgs); 2) the late and middle Pleistocene confined aquifer, which is composed of alluvial sediments (medium coarse sand and gravel with clay lens) in the depth of 15–80 m bgs.

2.2 *Sampling and analysis*

The pH, Eh, SES and DO of groundwater were analyzed on site, and the groundwater samples of As, As(III)/As(V), cations, anions and total dissolved carbon were analyzed in laboratory. In addition, the sediment geochemistry characteristics were comprehensively analyzed by the X-ray Fluorescence Spectroscopy (XRF-1800, Shimadzu, Japan), X-Ray Diffractometer (D8-Focus, Germany) and the procedure of sequential extraction (Keon *et al.*, 2001).

The microbial community dynamics and variation of microbial functional genes were analyzed by the Illumina high-throughput sequencing and quantitative PCR techniques. The details of sample collection and analysis were described in a recent study (Deng *et al.*, 2018).

3 RESULTS AND DISCUSSION

3.1 *Seasonal variation of microbial community compositions responding to the changing groundwater level and As concentration*

The results indicated that the microbial community compositions exhibited significant seasonal variation between the monsoon season and pre-monsoon season, in which the *α-Proteobacteria* decreased from pre-monsoon season to monsoon season and the *γ-Proteobacteria* increased from pre-monsoon season to monsoon season (Fig. 2). The correralation between the microbial compostion dynamics and groundwater level fluctuation indicated that the observed seasonal variation of microbial community may be significantly affected by the varying ground water level. This observation may be ascribed to the fact that the groundwater level fluctuation changed the redox status in the aquifers, and supsequently altered the anaerobic and aerobic metabolism in the microbial community.

Furthermore, the corresponding seasonal variations of Fe(II) and As concentrations were observed in the groundwater, suggesting that the As concentration variation in the aquifer is associated with the Fe mineral transformations. Meanwhile, it is notable that the positively correlative variations between *Geobacter* sp. and As/Fe(II) were observed in the groundwater, while negative correlation between the *Sideroxydans* sp. and As/Fe(II) was also observed. Those results suggested that the microbial functional groups could facilitate the As transformation in aquifers through affecting the redox cycle of iron.

3.2 *Microbial functional groups accounting for the As transformation and temporal variation*

In order to further understand the underlying regulating biogeochemical processes, aquifer sediment incubation experiments were conducted with sediments from different depths (Deng *et al.*, 2018). The results indicated that the As mobilization was significantly associated with the reductive dissolution of iron oxide minerals and the variation of microbial community was correlated with As mobilization ($R = 0.7$, $P < 0.05$) in JH26 and JH36. The OTUs associated with FeRB, including *Pseudomonas, Pedobacter, Paenibacillus, Cellulomonas, Clostridium, Thiobacillus* and *Geobacter*, were predominant in the microbial community, indicating those FeRB were the main drivers for As mobilization from the aquifer sediments.

In addition, the detection of *dsrB* gene and the elimination of As in the late cultural period indicated that the bacteria sulfate reduction could scavenge the dissolved As in the solution. This observation was in

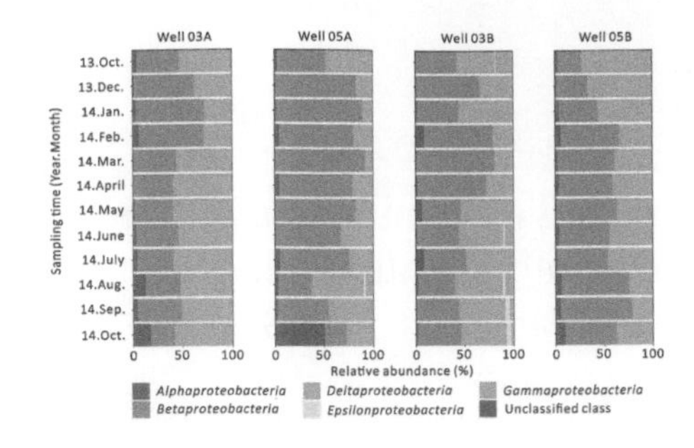

Figure 2. The seasonal variation of microbial community composition in the shallow groundwater from central Yangtze River Basin.

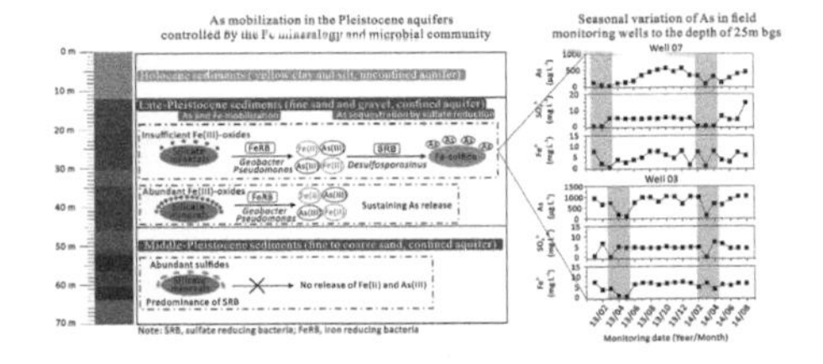

Figure 3. The generalization of As mobilization and temporal variation controlled by the Fe mineralogy and microbial community in the aquifer sediments from the Jianghan Plain, central Yangtze River Basin.

consistent with our field monitoring results in the pre-monsoon season, indicating that the microbial sulfate reduction process may also contribute to the temporal decrease of As concentration (Fig. 3).

4 CONCLUSIONS

The seasonal variation of As in aquifers could be affected by the variation of microbial community structures, and the As mobilization was mainly controlled by the distribution of microbial community and Fe mineralogy in the aquifer sediments.

ACKNOWLEDGEMENTS

The research was financially supported by the National Natural Science Foundation of China (Nos. 41572226 & 41521001), Program of China Geological Survey (Nos. 121201001000150121), and State Key Laboratory of Biogeology and Environmental Geology, China University of Geosciences.

REFERENCES

Deng, Y., Zheng, T., Wang, Y., Liu, L., Jiang, H. & Ma, T. 2018. Effect of microbially mediated iron mineral transformation on temporal variation of arsenic in the Pleistocene aquifers of the central Yangtze River basin. *Sci. Total Environ.* 619:1247–1258.

Keon, N.E., Swartz, C.H., Brabander, D.J., Harvey, C. & Hemond, H.F. 2001. Validation of an arsenic sequential extraction method for evaluating mobility in sediments. *Environ. Sci. Technol.* 35(13): 2778–2784.

Environmental Arsenic in a Changing World –
Zhu, Guo, Bhattacharya, Ahmad, Bundschuh & Naidu (Eds)
ISBN 978-1-138-48609-6

Variation of Extracellular Polymeric Substances (EPS) of *Chlamydomonas reinhardtii* under arsenic stress

C.H. Li[1], J.Y. Zhang[1], Q.N. Yu[1], B.B. Dong[1], C.H. Zhang[2] & Y. Ge[2]
[1]*College of Resources and Environmental Sciences, Jiangsu Provincial Key Laboratory of Marine Biology, Nanjing Agricultural University, Nanjing, P.R. China*
[2]*Demonstration Laboratory of Elements and Life Science Research, Laboratory Centre of Life Science, Nanjing Agricultural University, Nanjing, P.R. China*

ABSTRACT: Extracellular Polymeric Substances (EPS) are an important component of microorganisms. In this paper, we first compared the effectiveness of EPS extraction methods for amodel green algae (*Chlamydomonas reinhardtii*), and then investigated variations of EPS under arsenic (As) stress. Components of the algal EPS with or without arsenite and arsenate treatments were analyzed. The results showed that the different EPS extraction methods ranked as follows: NaOH > heating > cation exchange resin > EDTA > centrifugation. However, NaOH extraction caused more cell disruption than the heating approach. The optimal extraction was heating at 50°C, under which the amount of EPS was $40.06 \pm 0.297\ mg\ g^{-1}$. Under $20\ \mu g\ L^{-1}$ As(V) and $200\ \mu g\ L^{-1}$ As(III), the amounts of EPS were $41.35\ mg\ g^{-1}$ and $38.63\ mg\ g^{-1}$ respectively, compared to control ($46.4\ mg\ g^{-1}$), indicating that the EPS production of *C. reinhardtii* was not affected by As.

1 INTRODUCTION

Extracellular polymeric substances (EPS) are the products with a high molecular weight, and they usually secreted by microorganisms (such as algae). The EPS typically are composed of polysaccharides, proteins and nucleic acids. Additionally, as a crucial substance adhesion to cell surface, EPS has a variety of biological roles on algal cells due to its abundant functional groups (e.g., −COOH, −NH etc.). EPS forms an important extracellular protective barrier on the algae cell surface, which can prevent toxicants. Moreover, those toxicants (e.g., Pb^{2+}, Cd^{2+}) also affect the production and composition of EPS. Research have shown that some microalgae has a strong ability to adsorb As and have the potential in removing the As pollution from water. However, the relation between microalgal EPS and As was not clear. Therefore, the objectives of this study were: (1) to establish a suitable method for the extraction and analysis of extracellular polymers substances (EPS) from *C. reinhardtii*, and (2) to quantify andanalyze the algal EPS under various As treatments.

2 METHODS AND MATERIALS

2.1 *Cultivation of algae and EPS extraction*

The green alga *C. reinhardtii* in this study was purchased from the Institute of Hydrobiology, Chinese Academy of Sciences. EPS was extracted by the following methods: centrifugation, heating, NaOH, EDTA, CER treatments, respectively.

2.2 *Effect of arsenic on EPS compositions and content*

Upon exposure to different As(V) and As(III) concentrations, EPS from algae cultures was extracted using heating method. Proteins in EPS extract were determined by the modified Bradford method. The polysaccharides content in EPS was measured using the phenol-sulfuric method. Nucleic acids were determined by the diphenylamine colorimetric method.

3 RESULTS AND DISCUSSION

3.1 *Comparison of EPS extraction methods*

Table 1 showed that the amounts of EPS ranked as follows: NaOH >heating >cation exchange resin >EDTA >centrifugation. When NaOH ($0.05\ mol\ L^{-1}$) was used, the amount of EPS reached $70.42\ mg\ g^{-1}$. However, this method caused higher cell disruption (DNA $0.94\ mg\ g^{-1}$) than others (Table 1). When the heating temperature was 50°C, the amount of EPS was $40.06 \pm 0.30\ mg\ g^{-1}$. Thus, heating method was chosen for EPS extraction from *C. reinhardtii.*

3.2 *Variations of EPS contents and compositions under arsenic stress*

The amount of EPS by the microalgae under the stress of different concentrations of As(V) in the 4-d cultivation are shown in Figure 1. When *C. reinhardtii* was inoculated in TAP media with 20, 100, 200, $500\ mg\ L^{-1}$ of As(V), the EPS production at the 4-d

Table 1. Contents of EPS components measured by different extraction methods.

Extraction method	Yield of EPS (mg·g^{-1} DW)			
	polysaccharide	protein	DNA	DNA (%)
Centrifugation	0.53 ± 0.04d	ND	ND	ND
Heating	25.48 ± 0.21b	14.02 ± 0.4b	0.56 ± 0.03b	1.41 ± 0.071b
NaOH	28.71 ± 1.24a	40.77 ± 0.18a	0.94 ± 0.12a	1.36 ± 0.202b
EDTA	13.74 ± 1.38c	1.26 ± 0.09d	0.38 ± 0.04c	2.35 ± 0.353a
CER	27.88 ± 5.16a	7.21 ± 0.14c	ND	ND

Note: Data in table are means ± standard deviations. Different letters indicate significant difference between different extraction methods ($P < 0.05$). ND means not detected or below the detection limit. Data are means ± SD (n = 3).

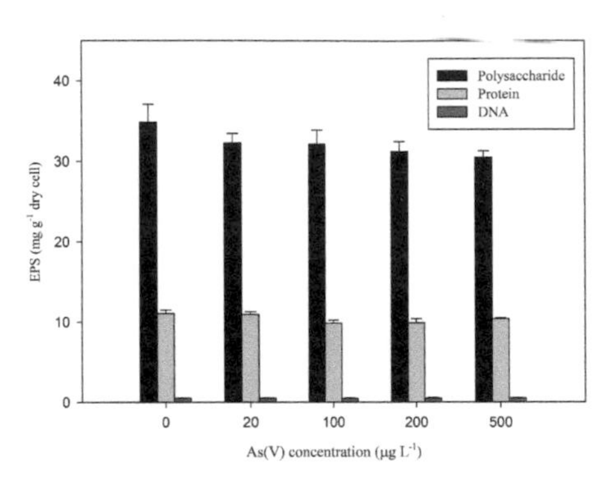

Figure 1. Variations in EPS from *C. reinhardtii* after exposure to different concentrations of As(V). Data are means ± SD (n = 3).

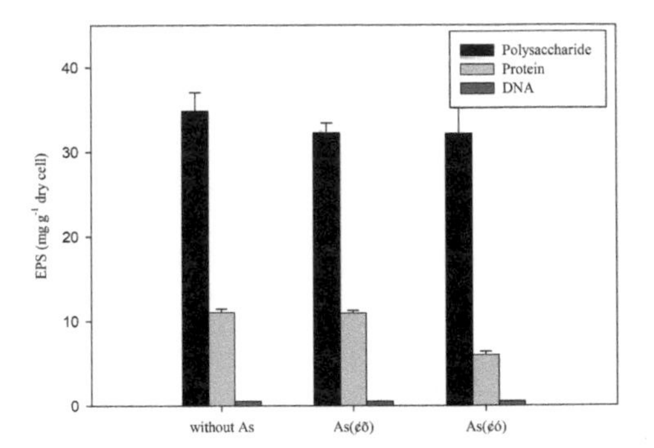

Figure 2. Variations of EPS from *C. reinhardtii* after exposure to arsenate (20 μg L^{-1}) and arsenite (200 μg L^{-1}). Data are means ± SD (n = 3).

decreased to ~43.7, 42.37, 41.54, 41.35 (mg g^{-1} dry cell), respectively, compared to the cultures in the TAP media without arsenate (the control).

The total amount of EPS of *C. reinhardtii* decreased from 46.4 mg g^{-1} (control) to 38.63 mg g^{-1} with 200 μg L^{-1} As (III) (Fig. 2). In this case, the protein of EPSsignificantly reduced by 45.72% (Fig. 2), which was much more than that with 20 μg L^{-1} As(V).

The interaction of EPS with As has been suggested to decrease As toxicity to microorganisms. For example, *Rhizobium* strain VBCK1062 isolated from *Vigna radiata* plants showed an increase in EPS production when exposed to 5 mM As(V), however, there was a decline with concentration of As(V) higher than 5 mM (Deepika *et al.* 2016). On the contrary, the present study showed that EPS was decreased upon various As treatments, suggesting that the relationship between EPS and As is complicatedand needs further investigations.

4 CONCLUSIONS

1) Heating method (50°C, 3 h) was considered as the suitable EPS extraction method with high extraction efficiency (40.06 mg g^{-1}), simple process and little cell disruption.
2) The amount of EPS production slightly decreased 5.82–11.53% when exposed to various arsenate and arsenite treatments, suggesting that EPS and As interactions are complex and more studies are needed.

ACKNOWLEDGEMENTS

Financial support from Natural Science Foundation of China (31770548) is greatly appreciated.

REFERENCES

Dai, Y.F., Xiao, Y., Zhang, E.H., Liu, L.D. & Qiu, L. 2016. Effective methods for extracting extracellular polymeric substances from Shewanella oneidensis MR-1. *Water Sci. Technol.* 74(12): 2987–2996.

Deepika, K.V., Raghuram, M., Kariali, E. & Bramhachari, PV. 2016. Biological responses of symbiotic *Rhizobium radiobacter* strain VBCK1062 to the arsenic contaminated rhizosphere soils of mung bean. *Ecotox. Environ. Safe.* 134: 1–10.

Taylor, C., Matzke, M., Kroll, A., Read, D.S. & Svendsen, C. 2016. Toxic interactions of different silver forms with freshwater green algae and cyanobacteria and their effects on mechanistic endpoints and the production of extracellular polymeric substances. *Environ. Sci-Nano* 3(2): 396–408.

Wang, Y., Wang, S., Xu, P., Liu, C. & Liu, M. 2015. Review of arsenic speciation, toxicity and metabolism in microalgae. *Rev. Environ. Sci. Bio.* 14(3): 427–451.

Environmental Arsenic in a Changing World –
Zhu, Guo, Bhattacharya, Ahmad, Bundschuh & Naidu (Eds)
ISBN 978-1-138-48609-6

The connection of manganese and arsenic in unconfined groundwater and shallow confined groundwater of Jianghan plain, China

X. Yu & Y. Deng
Geological Survey, China University of Geosciences, Wuhan, P.R. China

ABSTRACT: The mean content of arsenic (As) in the unconfined groundwater of Jianghan plain is 91.5 $\mu g\,L^{-1}$, and the average content of manganese is 1.19 $mg\,L^{-1}$. In the shallow confined groundwater (15–60 m), the content of arsenic is 40.34 $\mu g\,L^{-1}$, the average content of manganese is 0.59 mg L^{-1}. With the concentration of manganese rises, the concentration of As drops. In the monitoring field of Jianghan plain, we find that the concentration of manganese has seasonal change, and this change has a tendency to inhabit the reduction of arsenic. This can be explained that the reduction of manganese consumed the reducer of system.

1 INTRODUCTION

Naturally arsenic (As) enrichment of groundwater is a subject of great concern, which has been reported from numerous countries worldwide (Bai *et al.*, 2016). In addition to Bangladesh, China is one of the countries with the worst endemic arsenic poisoning. The contents of arsenic and manganese in the Jianghan plain are very high. In particular, manganese oxides have been shown to oxidize aqueous As(III) and Fe(II), enhance sorption of the oxidize As(V), and enhance sequestration of arsenic in the sadiment (Bai *et al.*, 2014).

2 METHODS/EXPERIMENTAL

2.1 *Study area*

Jianghan plain is a plain formed by the alluvial deposits of the Yangtze river and the Han River. At the high arsenic groundwater monitoring site in Shahu town, 10 m, 25 m and 50 m of monitoring wells are monitored through hydrogeological drilling. The 10-meter monitoring (A) well is located in the shallow pore aquifer, 25 m (B) well in the shallow confined aquifer.

2.2 *Sampling and analysis*

Temperature, pH, EC, and Eh were measured on site using a portable pH, EC, and Eh meter in-situ (HACH HQ40D, USA). NH_4-N, Fe^{2+} and sulfide were measured on site using a portable spectrophotometer (HACH2800). Concentrations of HCO_3^- were measured within 24 h using acid-base titration methods. Total concentration of dissolved ions (Ca, Mg, Na, K, Fe, and Mn) was determined using an inductively coupled plasma atomic emission spectrometer (ICP-AES) (IRIS Intrepid II XSP, USA). Anions such as Cl^-, NO_3^-, and SO_4^{2-} were determined using an ion chromatograph.

3 RESULTS AND DISCUSSION

3.1 *Spatial distribution of manganese and its influence on the arsenic content of groundwater*

The mean content of arsenic in the unconfined groundwater of Jianghan plain is 91.5 $\mu g\,L^{-1}$, and the average content of manganese is 1.19 $mg\,L^{-1}$ (Fig. 1). In the shallow confined groundwater (15-60 m), the content of arsenic is 40.34 $\mu g\,L^{-1}$, the average content of manganese is 0.59 $mg\,L^{-1}$ (Fig. 2). The concentration of manganese had two peaks at 5 m and 15 m. Arsenic reaches its peak at about 20 m. With the concentration of manganese rises, the concentration of arsenic drops. There is a negative correlation between the changes of arsenic and manganese. Seasonal variation of manganese and the relation to arsenic (Deng *et al.*, 2014).

Both the concentration of manganese and arsenic in unconfined groundwater under the shallow un-fined

Table 1. Statistics of groundwater chemical data in Jianghang plain.

Index	Depth Unit	0–15 m			15–60 m		
		Max	Min	Mean	Max	Min	Mean
pH		5.58	8.15	7.07	6.17	8.3	6.93
Eh	mV	254.3	171.5	−145.1	360.5	152.1	2.827
Ca	$mg\,L^{-1}$	3.54	92.4	113.38	0.27	96.2	85.39
Mg	$mg\,L^{-1}$	1.041	60.3	26.37	0.871	49.5	20.81
Na	$mg\,L^{-1}$	0.598	48.8	26.65	0.355	46.1	16.36
HCO_3^-	$mg\,L^{-1}$	50.1	918	493	170.6	983	548.3
SO_4^{2-}	$mg\,L^{-1}$	0	99.6	45.33	0	86.7	4.63
Fe	$mg\,L^{-1}$	0	9.66	2.45	0	9.97	5.7
Mn	$mg\,L^{-1}$	0	7.43	1.19	0	5.32	0.59
As	$\mu g\,L^{-1}$	0	91.5	15.12	0	96.2	40.34
S^{2-}	$\mu g\,L^{-1}$	0	490	11.93	0	224	6.68

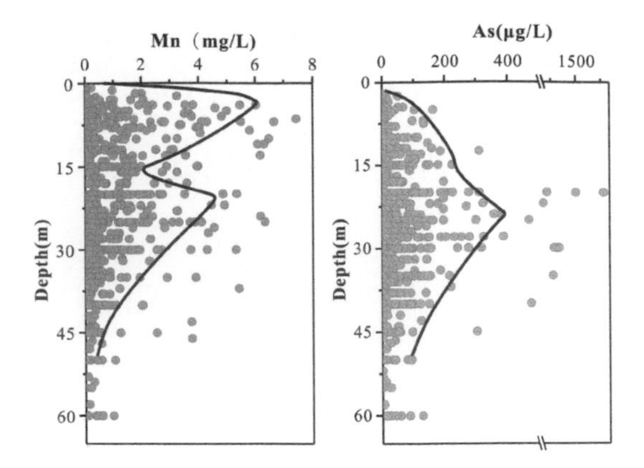

Figure 1. The concentration changes of arsenic and manganese in the unconfined groundwater and shallow confined groundwater of Jianghang plain.

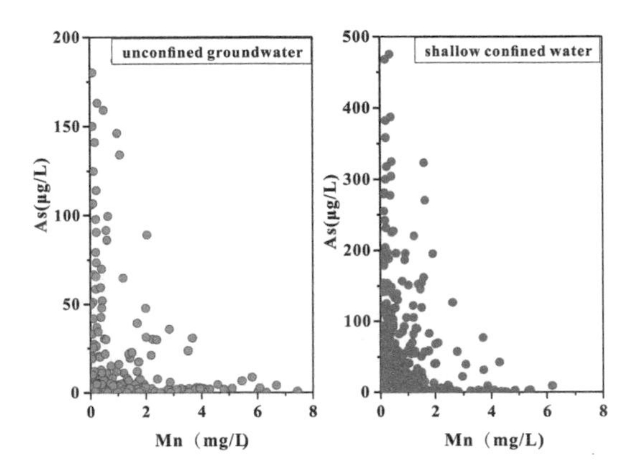

Figure 2. The relationship between arsenic and manganese in groundwater of Jianghang plain.

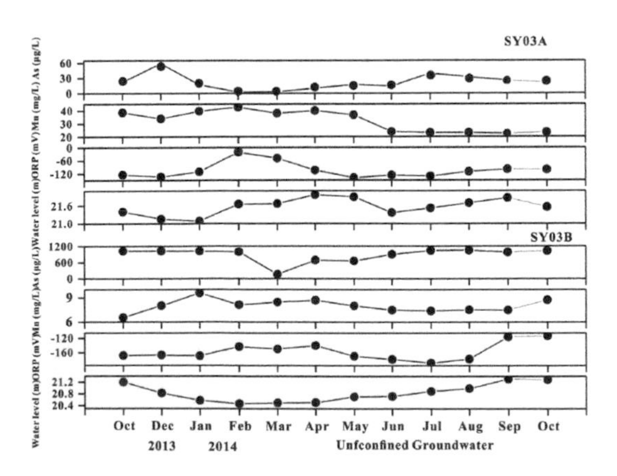

Figure 3. Temporal variation of chemical indices in confined groundwater from October 2013 to October 2014.

groundwater had obvious seasonal changes. The concentration of arsenic increased significantly during the period from June to September, and the concentration of arsenic decreased significantly after the dry period

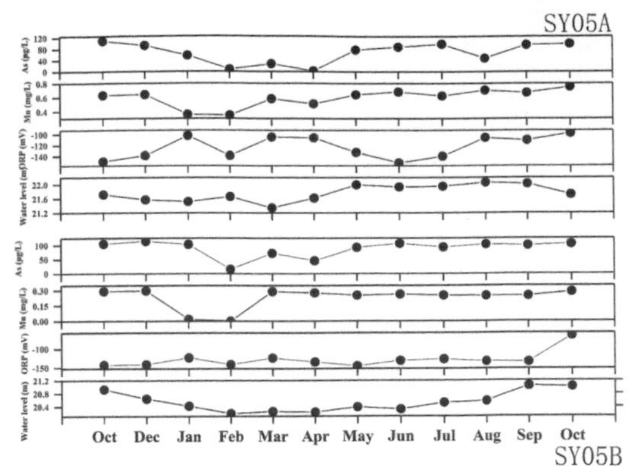

Figure 4. Temporal variation of chemical indexs in shallow cofined groundwater from October 2013 to October 2014.

(Fig. 3). The content of manganese rises sharply from the end of the harvest period and falls sharply around the dry period and starts to recover sharply in March (Fig. 4). When the concentration of arsenic rose, the concentration of manganese rise slowly. These data indicate the concentration of manganese can control the release of arsenic.

4 CONCLUSIONS

In the unconfined groundwater and shallow confined groundwater of Jianghang plain, the concentration of manganese has a certain influence on the changes of arsenic. There is a negative correlation between the changes of arsenic and manganese. When the concentration of manganese in the groundwater is increased, the concentration of arsenic decreases. The seasonal variation in the concentration of manganese and arsenic are different, they have the opposite trend in the selling period. It may be that the reduction of manganese consumed the reducer of system and inhibited the release of arsenic.

ACKNOWLEDGEMENTS

The research was financially supported jointly by National Natural Science Foundation of China (41102153, 41120124003).

REFERENCES

Bai, Y.H., Yang, T.T., Liang, J.S. & Qu, J.H. 2016. The role of biogenic Fe-Mn oxides formed in situ for arsenic oxidation and adsorption in aquatic ecosystems. *Water Res.* 98: 119–127.

Deng, Y.M., Li, H.J., Wang, Y.X., Duan, Y.H. & Gan, Y.Q. 2014. Temporal variability of groundwater chemistry and relationship with water-table fluctuation in the Jianghan plain, central China. *Proc. Earth Planet. Sci.* 10:100–103.

Environmental Arsenic in a Changing World –
Zhu, Guo, Bhattacharya, Ahmad, Bundschuh & Naidu (Eds)
ISBN 978-1-138-48609-6

Functions and unique diversity of genes and microorganisms involved in arsenic methylation in the arsenic-rich shallow and deep sediments of Jianghan Plain

X.B. Zhu, X.C. Zeng, Y. Yang, W.X. Shi & X.M. Chen
State Key Laboratory of Biogeology and Environmental Geology and Department of Biological Science and Technology, School of Environmental Studies, China University of Geosciences, Wuhan, P.R. China

ABSTRACT: Almost nothing is known about the activities and diversities of microbial communities involved in As methylation in arsenic-rich shallow and deep sediments; the correlations between As biomethylation and environmental parameters also remain to be elucidated. To address these issues, we collected 9 arsenic-rich sediment samples from the depths of 1, 30, 65, 95, 114, 135, 175, 200 and 223 m in Jianghan Plain. Microcosm assay indicated that the microbial communities in all of the sediment samples significantly catalyzed arsenic methylation. We identified 90 unique *arsM* genes from the eight samples, all of which code for new or new-type ArsMs, suggesting that As-methylating microorganisms are widely distributed in the samples from shallow to deep sediments. To determine whether biomethylation of As occurs in the sediments under natural geochemical conditions, we conducted microcosm assays without exogenous As and carbons. After 80.0 days of incubation, approximately 4.5–15.5 $\mu g\,L^{-1}$ DMAsV were detected in all of the microcosms with the exception of that from 30.0 m, and 2.0–9.0 $\mu g\,L^{-1}$ MMAsV were detected in the microcosms of 65, 135, 200, and 230 m; moreover, approximately 36.4–97.2 $\mu g\,L^{-1}$ soluble As(V) were detected from the nine sediment samples. This suggests that approximately 5.3%, 0%, 8.1%, 28.9%, 18.0%, 8.7%, 13.8%, 10.2% and 14.9% of total dissolved As were methylated by the microbial communities in the sediment samples from 1, 30, 65, 95, 114, 135, 175, 200 and 223 m without exogenous As and carbons, respectively. The concentrations of biogenic DMAsV show significant positive correlations with the depths of sediments, and negative correlations with the environmental NH_4^+ and NaCl, but show no significant correlations with other environmental parameters, including NO_3^-, SO_4^{2+}, TOC, TON, Fe, Sb, Cu, K, Ca, Mg, Mn and Al. This work for the first time revealed the activities and diversities of As-methylating microbes in the arsenic-rich shallow and deep sediments, and helps to better understand the microorganisms-mediated biogeochemical cycles of arsenic in arsenic-rich shallow and deep sediments.

1 INTRODUCTON

Arsenic (As) is widely distributed in the Earth's crust at an average concentration of 2.0 $mg\,kg^{-1}$. It occurs in more than 200 minerals, usually in combination with sulfur and metals (Nordstrom, 2002). Arsenic-contaminated groundwater exists in more than 70 countries worldwide, including Bangladesh, India, Pakistan, Burma, Nepal, Vietnam, Cambodia, China and United States (Kirk *et al.*, 2004; Gu *et al.*, 2012; Shao *et al.*, 2016). Natural processes, biochemical reactions and anthropogenic activities are responsible for the dissolution and release of arsenic from minerals into groundwater (Fendorf *et al.*, 2010). Many bacteria, archaea, fungi, and animals are able to methylate As (Zhu *et al.*, 2014). Arsenic methylation was catalyzed by As(III)-S-adenosylmethionine methyltransferase (ArsM) (Ajees *et al.*, 2012). The first arsM gene was cloned from the soil bacterium Rhodopseudomonas palustris (Qin *et al.*, 2006). It codes for a protein with 283 amino acid residues. ArsM is present in both prokaryotic and eukaryotic microorganisms (Zhu *et al.*, 2017). Purified ArsM can convert As(III) into DMAsV, TMAO, and volatile TMAsIII (Chen et al., 2014; Jia *et al.*, 2013; Slyemi *et al.*, 2012; Zhu et al., 2014). The arsM gene can be used as a molecular marker for investigations of the diversities of As(III)-methylating microbes (Qin *et al.*, 2006, 2009). Recently, some single bacterial strains with significant As(III)-methylating activities were isolated from paddy soils, wastewater ponds, and microbial mats (Huang et al., 2016; Kuramata *et al.*, 2015; Wang et al., 2016; Zhang *et al.*, 2015b). Functional analyses suggest that ArsMs play a role in the detoxification of As(III) and could be exploited in bioremediation of arsenic-contaminated groundwater (Huang *et al.*, 2015; Saunders *et al.*, 2016).

This study aimed to explore the microbial communities-catalyzed methylation of arsenic in the arsenic-rich sediments in Jianghan Plain. To the best of our knowledge, this is the first effort to characterize the functions and diversities of microbial communities involved in arsenic methylation, and the correlations between arsenic biomethylation and environmental parameters in the severely contaminated shallow and deep sediments. This work gained a new understanding of the microorganisms-mediated biogeochemical cycles of arsenic in arsenic-rich sediments.

2 METHODS/EXPERIMENTAL

2.1 *Sampling*

The sampling site (113°36′35.028″E, 30°8′34.944″N) was near a paddy field, located in the Jiahe village that is affiliated to the Shahu town of Xiantao city, Hubei province, China (Fig. 1a). Sediment samples were collected from the depths of 1, 30, 65, 95, 114, 135, 175, 200, and 223 m. The external layers of the sediment cores were carefully removed to avoid contaminations and oxidation by oxygen. The samples were placed into sterilized tubes, and kept on ice immediately. All of the samples were transported into the laboratory in 12.0 hours.

2.2 *Geochemical analyses*

To determine the total arsenic content, one gram of dried sample powders was mixed with 20.0 mL of aqua regia. After 2.0 hours of incubation at 100°C, the mixture was centrifuged, and the supernatant was collected for determination of the arsenic concentration using the atomic fluorescence spectrometry (AFS) (AFS-9600, Haiguang, Beijing, China). Soluble arsenic species, including As(III), As(V), MMAsV and DMAsV, were determined by high-performance liquid chromatography linked to atomic fluorescence spectrometry (HPLC-ICP-MS) (LC-20A, Shimadzu, Japan; ELAN, DRC-e, PerkinElmer, USA).

2.3 *Determination of As-methylating activities of the sediment samples*

To detect the As-methylating activity of the microbial community from sediment samples,active microcosms were prepared in triplicate by inoculating 8.0 g of samples into 20.0 mL of simulated groundwater (Wang et al., 2017) amended with 0.23 mM As(III) and 20.0 mM lactate in a 50-mL flask. All of the mixtures were incubated at 30°C without shaking. After 80 days, approximately 1.0 mL of cultures was removed from each flask for determination of the concentrations of $MMAs^V$ and $DMAs^V$ using HPLC-ICP-MS.

2.4 *Amplification, cloning and analysis of arsM genes from genomic DNAs of the samples*

To detect the diversity of As-methylating microbes from the samples, three pairs of primers were used to amplify arsM genes from the genomic DNA of each sample. The primers were listed in Table 1. Metagenomic DNA was extracted from the samples using the MiniBEST bacterial genomic DNA extraction kit (TaKaRa, Japan) (Plenge *et al.*, 2016). PCR products were gel purified using the E.Z.N.A gel extraction kit (Omega, USA). Purified DNA was ligated into a T vector. An arsM gene library was generated by introducing the recombinant plasmids into Escherichia coli DH5α competent cells. All of the clones from the library were sequenced and analyzed as described previously (Zhong *et al.*, 2017).

2.5 *Methylated As release assay in the absence of exogenous As and carbons*

To determine whether microbial communities from the sediment samples catalyze arsenic methylation under natural conditions, we prepared microcosms without addition of exogenous As and carbon sources. Eight grams of the sediment samples were mixed with 20.0 mL of simulated groundwater in a 50-mL flask. All of the microcosms were incubated at 30°C for 80.0 days without shaking. After 80.0 days of incubation, approximately 1.0 mL of slurries was removed from the flaks for measurement of the concentrations of As(III), As(V), $MMAs^V$ and $DMAs^V$.

2.6 *Analysis of the correlations between methylated As and geochemical parameters*

The Spearman's Rank Correlation Coefficient was used to determine the correlations between the concentrations of methylated As species produced by microbial communities, and the geochemical parameters in the environment. The SPSS software was used for statistical analyses. Correlations were considered to be statistically significant at 95% confidence level ($P < 0.05$).

3 RESULTS AND DISCUSSION

3.1 *Characterization of the sampling site*

We collected nine sediment samples from the depths of 1, 30, 65, 95, 114, 135, 175, 200, and 223 m, respectively. Figure 1b shows the appearances of the sediment samples. Geochemical analyses indicated that the sediments 180 contain high contents of total As (ranging from 6.74 to 27.9 mg/kg) and soluble As (ranging from 1.9 to 100.7 μg/L). The arsenic contents showed no correlations with the depths of the sediments. The concentrations of total arsenic also showed no correlations with those of soluble arsenic; this suggests that multiple factors controlled the arsenic dissolution and release from the sediments. The sediment samples contain 0.27–8.37 g/kg total organic carbons (TOC) and 0.18–1.26 g/kg total organic nitrogen (TON); these substances could provide essential carbon and nitrogen sources for the growth of microorganisms in the sediments. The sediments also contain relatively high contents of sulfate (ranging from 14.53 to 863.87 mg/kg), and relatively low contents of ammonium (ranging

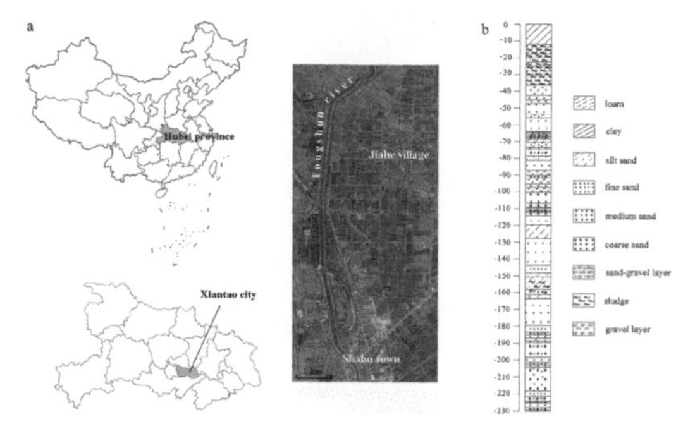

Figure 1. Illustrations of the sampling site and sample appearances. (a) Geographical map of the sampling location in the Jiahe village of the Shahu town in Jianghan Plain. (b) The appearances of the sediment samples collected from the depths of 1 to 230 m.

from 3.12 to 52.77 mg/kg) and nitrate (ranging from 0.23 to 3.03 mg/kg).

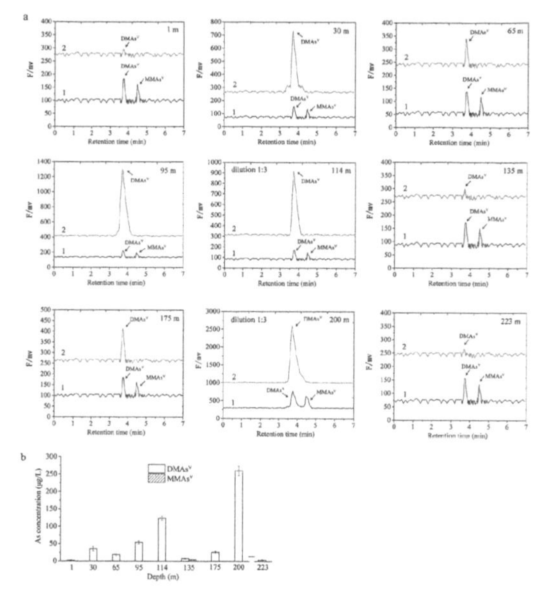

Figure 2. Arsenc methylation activities of the microbial communities of the nine samples collected from the arsenic-rich shallow and deep sediments in Jianghan Plain. (a) HPLC-ICP-MS chromatograms of the methylated As species produced by the microbial communities of the samples from 1 to 223 m after 80.0 days of incubation. Curve 1, standards of DMAsV and MMAsV; curve 2, DMAsV and MMAsV concentrations in the microcosms amended with 0.23 mM As(III) as the substrates of ArsM enzymes, and 20.0 mM lactate as the carbon sources after 80.0 days of incubation. (b) Concentrations of the methylated As in the microcosms of the samples from 1 to 223 m after 80.0 days of incubation.

3.2 *As-methylating activities of the microbial communities from the sediments*

Microcosm assay was used to determine the Asmethylating activities of the nine sediment samples. The microcosms were amended with 0.23 mM As(III) as the substrates of ArsM enzymes, and 20.0 mM lactate as the carbon sources. The results showed that no detectable amounts of methylated arsenic species were observed in the autoclaved sediment slurries. In contrast, when the sediment microcosms were not autoclaved, approximately 2.06, 35.53, 18.83, 53.99, 121.86, 7.54, 7.00, 260.38, and 3.02 μg/L $DMAs^V$ were detected in the microcosms from 1,30, 65, 95, 114, 135, 175, 200, and 223 m, respectively (Figs. 2a and 2b). No significant amounts of $MMAs^V$ were detectable in all of the sediment samples with the exception of that from the depth of 135 m, in which 3.30 μg/L of biogenic $MMAs^V$ was produced. This suggests that the microbial communities in the nine sediment samples were able to significantly catalyze As methylation, and the dominant products were $DMAs^V$.

3.3 *Unique diversities of As-methylating microbes in the sediment samples*

To understand the microbial basis of the As-methylating activities of the nine sediment samples, we explored the diversities of the As-methylating microbes present in the microbial communities by cloning, sequencing and analyzing the *arsM* genes from the metagenomic DNAs of the samples. The results showed that we identified 90 different ArsM proteins from the microbial communities of the eight samples, including 20, 22, 8, 10, 4, 5, 13, and 7 different ArsMs from the samples of 1, 30, 65, 95, 114, 135, 200, and 223 m, respectively (Fig. 3a). A phylogenetic tree was constructed based on the multiple alignments of the ArsM proteins from this study and their closely related ArsMs from other known microorganisms (Fig. 3b). An ArsM sequence from archaea was chosen as the outgroup.

3.4 *Methylated As release from the sediments*

To detect whether the microbial communities-catalyzed As methylation occurs under natural conditions, we performed methylated As release assay using the microcosms in the absence of any exogenous organic carbons and As(III) compounds. We found that approximately 5.0, 30.0, 55.5, 126.1, 5.7, 25.2, 260.3, 9.0 μg/L $DMAs^V$ were produced in the slurries of the samples from the depths 30, 65, 95, 114, 135, 175, 200, and 223 m, respectively, and no significant amount of $DMAs^V$ were detected from the microcosm of 30 m (Fig. 4a). We also examined the concentrations of soluble As in the sediment samples after 80.0 days of incubation. The results showed that approximately 85.05, 18.69, 86.64, 36.36, 39, 97.2, 50.79, 151.53 and 84.09 μg/L of soluble As(V) were detected in the slurries of the samples from 1, 30, 65, 95, 114, 135, 175, 200 and 223 m, respectively (Fig. 4b). This suggests that approximately 5.29%, 0%, 8.08%, 28.87%, 17.95%, 8.74%, 13.78%, 10.23% and 14.87% of total

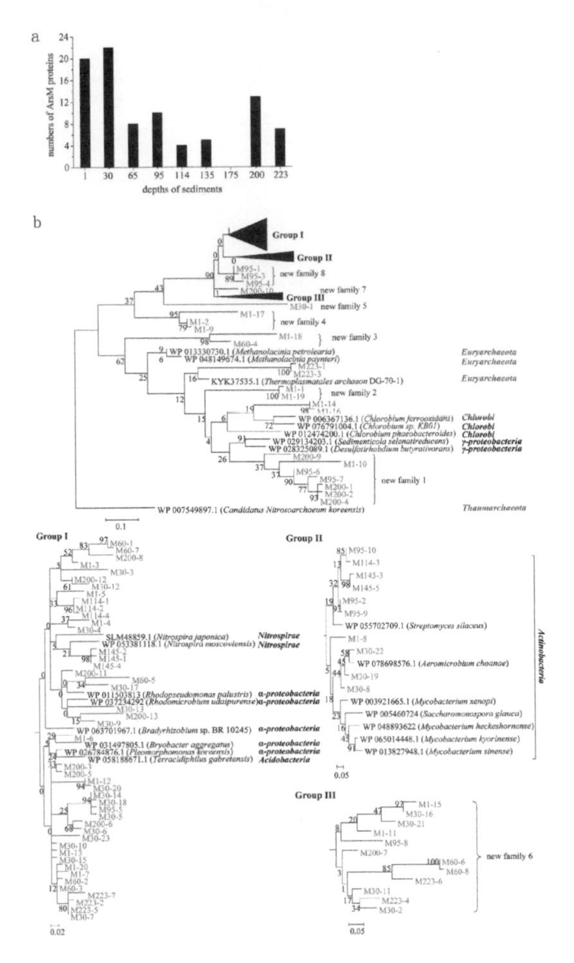

Figure 3. Characterization of the ArsM proteins identified from microbial communities of the sediment samples from Jianghan Plain. (a) The numbers of different ArsM proteins identified from the microbial community of each sample from the depths of 1 to 223 m. (b) Maximum likelihood tree constructed from the multiple sequence alignments of the obtained ArsM proteins from the sediment samples and their closely related homologues from other known microorganisms. All of the ArsM sequences in this study are shown in red; numbers on the branches are bootstrap values based on 1000 replicates; only the bootstrap values larger than 50% are shown; the scale bar represents the average number of substitutions per site.

soluble As were methylated by the microbial communities in the microcosms of the samples from 1, 30, 65, 95, 114, 135, 175, 200 and 223 m, respectively.

3.5 *Correlations of the methylated As species and some geochemical parameters*

The Spearman's Rank Correlation Coefficient was used to discover the correlations between methylated As species and geochemical parameters. We found that the concentrations of biogenic DMAsV from the sediments of different depths show significant positive correlations with the depths of the sediments

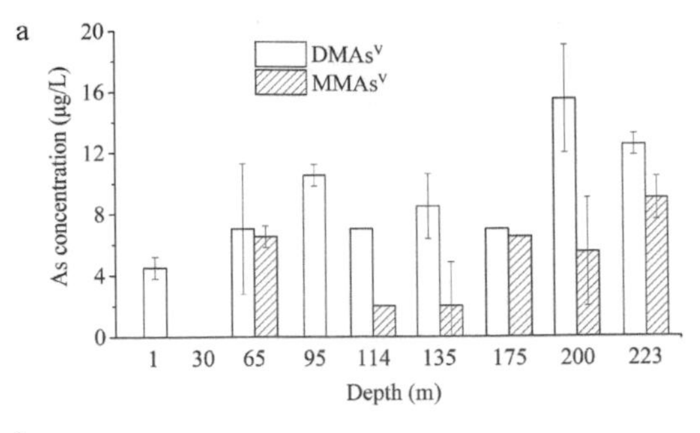

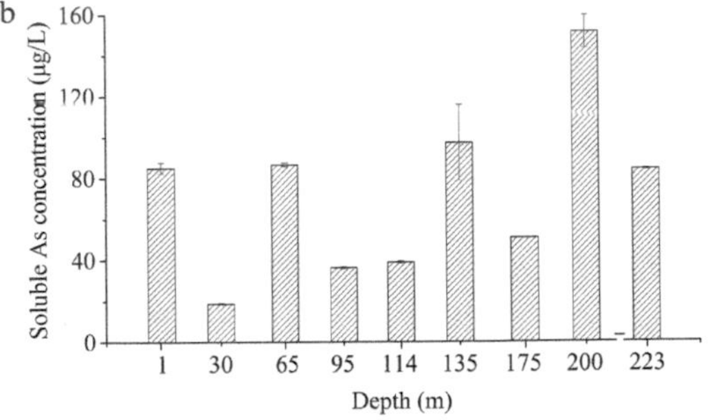

Figure 4. The microbial communities-catalyzed methylation and dissolution of arsenic in the microcosms prepared from the sediment samples of the depths 1 to 223 m in the absence of exogenous As and carbons. (a) Concentrations of biogenic methylated As after 80.0 days of incubation. (b) Concentrations of biogenic soluble As after 80.0 days of incubation.

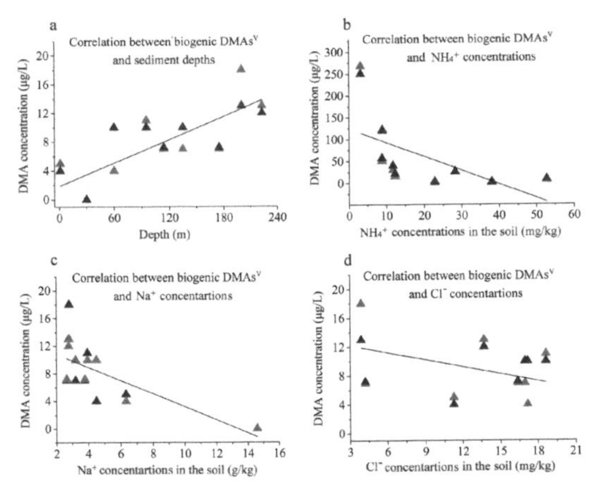

Figure 5. Correlations between arsenic biomethylation and some environmental parameters. (a) Correlation between biogenic $DMAs^V$ and sediment depths ($r = 0.78565$, $p = 0.01209$); (b) Correlation between biogenic $DMAs^V$ and concentrations of environmental NH_4^+ ($r = -0.60782$, $p = 0.0825$); (c) Correlation between biogenic $DMAs^V$ and concentrations of environmental Na^+ ($r = -0.78529$, $p = 0.01215$); (d) Correlation between biogenic $DMAs^V$ and concentrations of environmental Cl^- ($r = -0.67712$, $p = 0.04512$).

($r = 0.78565; p = 0.01209$) (Fig. 5a), and negative correlations with the contents of NH_4^+ ($r = -0.60782$; $p = 0.0825$), Na^+ ($r = -0.78529$; $p = 0.01215$), and Cl^- ($r = -0.67712$; $p = 0.04512$) in the environment (Fig. 5b, 5c and 5d).

4 CONCLUSIONS

This study aimed to explore how arsenic was released from soil phase into groundwater, and how sulfate affects this bioprocess in Jianghan Plain, China. We found that a large diversity of DARPs are present in the shallow soils from the depths of 1.6 and 4.6 m. Microcosm assays indicated that the microbial communities significantly catalyzed the dissolution, reduction and release of arsenic and iron from the soils into aqueous phase. It is interesting to see that addition of 10.0 mM sulfate into the microcosms led to a significant increase of the microorganisms-mediated release of arsenic and iron from the soils. Quantitative PCR analysis for the functional gene abundances suggested that the sulfate-induced increase of microbial release of arsenic and iron was attributed to the significant enhancement of the DARP growth by sulfate. These results suggested that agricultural uses of sulfate fertilizers enhanced arsenic contamination in groundwater. The findings of this study gained new insight into the mechanisms by which the arsenic-contaminated groundwater was formed, and gave direct evidence that agricultural activities may enhance the arsenic contamination in groundwater.

ACKNOWLEDGEMENTS

This work was financially supported by the National Natural Science Foundation of China (grants no. 41472219, 41072181, 41272257 and 41521001).

REFERENCES

Chen, X., Zeng, X.C., Wang, J., Deng, Y., Ma, T.E.G. & Wang, Y. 2017. Microbial communities involved in arsenic mobilization and release from the deep sediments into groundwater in Jianghan plain, Central China. *Sci. Total Environ.* 579: 989–999.

Ferguson, J.F. & Gavis, J. 1972. A review of the arsenic cycle in natural waters. *Water. Res.* 6: 1259–1274.

Gan, Y., Wang, Y., Duan, Y., Deng, Y., Guo, X. & Ding, X. 2014. Hydrogeochemistry and arsenic contamination of groundwater in the Jianghan Plain, central China. *J. Geochem. Explor.* 138: 81–93.

Hao, T.W., Xiang, P.Y., Mackey, H.R., Chi, K., Lu, H., Chui, H.K. & Chen, G.H. 2014. A review of biological sulfate conversions in wastewater treatment. *Water Res.* 65: 1–21.

Kirk, M.F., Holm, T.R., Park, J., Jin, Q., Sanford, R.A., Fouke, B.W. & Bethke, C.M. 2004. Bacterial sulfate reduction limits natural arsenic contamination in groundwater. *Geology* 32: 953–956.

Maguffin, S.C., Kirk, M.F., Daigle, A.R., Hinkle, S.R. & Jin, Q. 2015. Substantial contribution of biomethylation to aquifer arsenic cycling. *Nat. Geosci.* 8: 290–293.

Savage, K.S., Tingle, T.N., O'Day, P.A., Waychunas, G.A. & Bird, D.K. 2000. Arsenic speciation in pyrite and secondary weathering phases, Mother Lode gold district, Tuolumne County, California. *Appl. Geochem.* 15(8): 1219–1244.

Smedley, P.L. & Kinniburgh, D.G. 2002. A review of the sources, behavior and distribution of arsenic in natural waters. *Appl. Geochem.* 17: 517–568.

Zhu, W., Young, L.Y., Yee, N., Serfes, M., Rhine, E.D. & Reinfelder, J.R. 2008. Sulfide-driven arsenic mobilization from arsenopyrite and black shale pyrite. *Geochim. Cosmochim. Acta* 72: 5243–5250.

Zhu, Y.G., Yoshinaga, M., Zhao, F.J. & Rosen, B.P. 2014. Earth abides arsenic biotransformations. *Ann. Rev. Earth Planet. Sci.* 42: 443–467.

Environmental Arsenic in a Changing World –
Zhu, Guo, Bhattacharya, Ahmad, Bundschuh & Naidu (Eds)
ISBN 978-1-138-48609-6

Characteristics and mechanisms of arsenic behavior during the microbial oxidation-reduction of iron

W. Xiu[1], H.M. Guo[1,2], X.N. Yu[2], W.J. Yuan[2] & T.T. Ke[2]
[1] *State Key Laboratory of Biogeology and Environmental Geology, China University of Geosciences, Beijing, P.R. China*
[2] *School of Water Resources and Environment, China University of Geosciences (Beijing), Beijing, P.R. China*

ABSTRACT: The (bio)geochemical cycling of arsenic (As) and iron (Fe) is generally coupled in high As groundwater. However, characteristics and mechanisms of As sequestration-release behavior during formation-reduction of biogenic Fe(III) minerals (bio-FeM) are poorly understood. The *Pseudogulbenkiania* sp. strain 2002 induced the formation of binary bio-FeM, facilitating As immobilization compared to single Fe(III) minerals. Net As release was found during reduction of binary bio-FeM by *Shewanella oneidensis* strain MR-1, which was affected by secondary iron mineral formation and competitive adsorption. We suggested that bio-FeM deserve more attention due to their outstanding potential roles in controlling As behavior in aquatic systems.

1 INTRODUCTION

High arsenic (As) groundwater (As concentration > $10\,\mu g\,L^{-1}$) is one of the major worldwide concerns. Arsenic (bio)geochemical cycling is often coupled to the cycling of iron (Fe) (Zhu *et al.*, 2017). For instance, natural ferric (hydro)oxides, mainly consisting of various biogenic Fe(III) minerals (bio-FeM), possess high sorption reactivity, which have the potential to co-precipitate or adsorb As (Feris, 2005; Xiu *et al.*, 2015; Xiu *et al.*, 2016; Sowers *et al.*, 2017). Whereas, dissimilarly reductive dissolution of Fe(III) (hydro)oxides is of utmost importance in controlling As enrichment in groundwater under reducing condition (Islam *et al.*, 2004; Postma *et al.*, 2010; Schaefer *et al.*, 2017). However, characteristics and mechanisms of As sequestration-release behavior during the formation-reduction of biogenic Fe(III) minerals are poorly understood. The Objectives of this study are to (i) characterize As sequestration by biogenic Fe(III) minerals as well as related mechanisms of As sequestration; (ii) explore the character and mechanisms of As release during bio-reduction of biogenic Fe(III) minerals.

2 METHODS/EXPERIMENTAL

2.1 *Sources of microbes*

The anaerobic Fe-oxidizer, *Pseudogulbenkiania* sp. strain 2002, was retrieved from lab stock (20% glycerol at −80°C, Geomicrobiology Lab at the China University of Geosciences, Beijing) and routinely cultivated at 30°C using freshwater basal medium with nitrate and acetate as electron acceptor and donor, respectively.

The facultative dissimilarly iron reducer, *Shewanella oneidensis* MR-1 was routinely cultured aerobically in Luria-Bertani (LB) broth (pH = 7.0) at 30°C. Once cell growth reached the mid to late log phase, centrifuged in anaerobic chamber (filled with N_2/H_2, 92.5%/7.5% (v/v), Coy Laboratory Products, USA), washed three times with deoxygenated bicarbonate buffer ($2\,g\,L^{-1}$ $NaHCO_3$ and $0.08\,g\,L^{-1}$ KCl, pH = 7.0) to remove residual LB, and resuspended in buffer to a final concentration of $\sim 2 \times 10^8$ cells mL^{-1}. In selected treatments, AQDS (Sigma) was added separately.

2.2 *Experiential setups*

To obtain the As-containing bio-FeM, anaerobic cultures with $1\,mg\,L^{-1}$ As(V) and approximately initial 3.0 mM Fe(II) (initial molar ratio of Fe(II)/As(V) around 225) were incubated with later log-phase cells of strain 2002 (5%,v/v). The bio-FeM was sampled by repeated centrifugation (1000 rpm for 10 min), following by rinsing with deionized water, dried in an anaerobic glove box until analyzed within one week. Reduction of bio-FeM was conducted in bicarbonate buffer containing bio-FeM ($\sim 100\,mg\,mL^{-1}$) and MR-1 cells ($\sim 2 \times 10^8$ cells mL^{-1}). Tubes were purged with N_2/CO_2 gas mix (80:20) and sealed with thick butyl rubber stoppers. The control consisted of one tube that received 1 mL of sterile bicarbonate buffer in place of MR-1 cells.

2.3 *Analytical methods*

Suspension were anoxically taken at different time intervals and divided into two equivalents. One was filtered with 0.22 μm membrane filter, and analyzed for dissolved As species, dissolved Fe species. The other

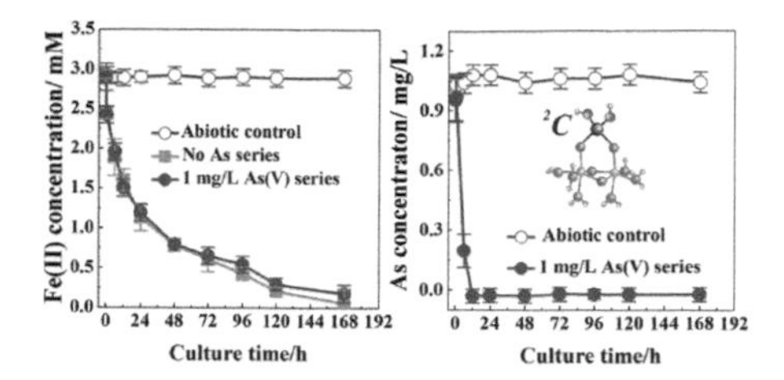

Figure 1. Changes of dissolved As and Fe(II) during strain 2002-induced Fe(II) oxidation: (left) changes of dissolved Fe(II); (right) changes of dissolved As.

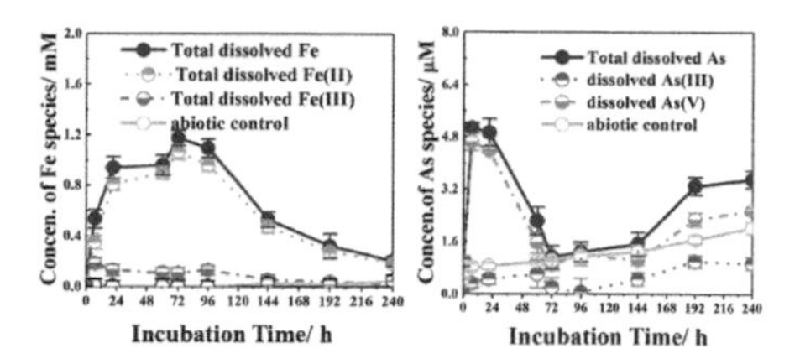

Figure 2. Changes of dissolved As and Fe(II) during MR-1-induced Fe(III) reduction: (left) changes of dissolved Fe species; (right) changes of dissolved As species.

was used to monitor the progress of Fe(III) reduction, following the 1,10-phenanthroline assay. The bio-FeM were sampled and analyzed by scanning electron microscopy (SEM), X-ray photoelectron spectroscopy (XPS), frontier transform IR spectra (FTIR), X-ray diffraction (XRD), and As and Fe K-edge XANES and EXAFS.

3 RESULTS AND DISCUSSION

3.1 *As immobilization during the formation of biogenic Fe(III) minerals*

Results showed that a rapid decrease in Fe(II) concentration was observed in early stage (before 24 h) and thereafter, Fe(II) concentrations dropped relatively placid after 24 h, forming the mixture of two-line ferrihydrite and goethite. Simultaneously, As(V) was efficiently removed from solutions through adsorption onto and/or co-precipitation with biogenic ferrihydrite-goethite biminerals (Fig. 1), forming binulear bidentate corner sharing As-Fe complexes (^{2}C). The $As_{immobilized}/Fe_{precipitated}$ at the end of incubation (168 h) achieved 0.005, which was higher than that in poorly crystalline or even amorphous biogenic Fe(III) oxides under similar condition. No detectable redox transformations of As(V) were observed.

3.2 *Reduction of biogenic Fe(III) minerals*

Result showed that Fe(III) reduction occurred when log phase of MR-1 was reached. A rapid increase in Fe(II) concentration was observed in early stage (<72 h), and thereafter dropped, possibly due to the formation of secondary iron minerals. Simultaneously, As release occurred before 24 h (mainly As(V) resulting from iron reduction), and then released As was re-immobilized into secondary iron minerals, and finally As was mobilized from secondary iron minerals possibly via competitive adsorption) (Fig. 2).

4 CONCLUSIONS

The *Pseudogulbenkiania* sp. strain 2002 induced the formation of binary bio-FeM, facilitating As immobilization compared to single Fe(III) minerals. Net As release was found during reduction of binary bio-FeM by *Shewanella oneidensis* strain MR-1, affected by secondary iron mineral and competitive adsorption.

ACKNOWLEDGEMENTS

The study is financially supported by the National Natural Science Foundation of China (Nos. 41702272, 41672225, 41222020 and 41172224) and the Fundamental Research Funds for the Central Universities (No. 53200759026, 2652013028).

REFERENCES

Ferris, F.G. 2005. Biogeochemical properties of bacteriogenic iron oxides. *Geomicrobiol. J.* 22(3-4): 79–85.

Guo, H., Ren, Y., Liu, Q., Zhao, K. & Li, Y., 2013. Enhancement of arsenic adsorption during mineral transformation from siderite to goethite: mechanism and application. *Environ. Sci. Technol.* 47(2): 1009–1016.

Hohmann, C., Morin, G., Ona-Nguema, G., Guigner, J.M., Brown Jr, G.E. & Kappler, A. 2011. Molecular-level modes of As binding to Fe (III)(oxyhydr) oxides precipitated by the anaerobic nitrate-reducing Fe (II)-oxidizing Acidovorax sp. strain BoFeN1. *Geochim. Cosmochim. Acta* 75(17): 4699–4712.

Islam, F.S., Gault, A.G., Boothman, C., Polya, D.A., Charnock, J.M., Chatterjee, D. & Lloyd, J.R. 2004. Role of metal-reducing bacteria in arsenic release from Bengal delta sediments. *Nature* 430(6995): 68.

Postma, D., Jessen, S., Hue, N.T.M., Duc, M.T., Koch, C.B., Viet, P.H., Nhan, P.Q. & Larsen, F. 2010. Mobilization of arsenic and iron from Red River floodplain sediments, Vietnam. *Geochim. Cosmochim. Acta* 74(12): 3367–3381.

Schaefer, M.V., Guo, X., Gan, Y., Benner, S.G., Griffin, A.M., Gorski, C.A., Wang, Y. & Fendorf, S. 2017. Redox controls on arsenic enrichment and release from aquifer sediments in central Yangtze River Basin. *Geochim. Cosmochim. Acta* 204: 104–119.

Sowers, T.D., Harrington, J.M., Polizzotto, M.L. & Duckworth, O.W. 2017. Sorption of arsenic to biogenic iron (oxyhydr) oxides produced in circumneutral environments. *Geochim. Cosmochim. Acta* 198: 194–207.

Xiu, W., Guo, H., Liu, Q., Liu, Z. & Zhang, B. 2015. Arsenic removal and transformation by *Pseudomonas* sp. strain GE-1-induced ferrihydrite: co-precipitation versus adsorption. *Water Air Soil Poll.* 226(6): 167.

Xiu, W., Guo, H., Shen, J., Liu, S., Ding, S., Hou, W., Ma, J. & Dong, H. 2016. Stimulation of Fe (II) oxidation, biogenic lepidocrocite formation, and arsenic immobilization by *Pseudogulbenkiania* sp. strain 2002. *Environ. Sci. Technol.* 50(12): 6449–6458.

Zhu, Y.G., Xue, X.M., Kappler, A., Rosen, B.P. & Meharg, A.A. 2017. Linking genes to microbial biogeochemical cycling: lessons from arsenic. *Environl. Sci. Technol.* 51(13): 7326–7339.

Environmental Arsenic in a Changing World –
Zhu, Guo, Bhattacharya, Ahmad, Bundschuh & Naidu (Eds)
ISBN 978-1-138-48609-6

Microbial community in high arsenic groundwater aquifers from Hetao Plain of Inner Mongolia, China

Y.H. Wang, P. Li, Z. Jiang, H. Liu, D.Z. Wei & H.L. Wang
State Key Laboratory of Biogeology and Environmental Geology, China University of Geosciences, Wuhan, P. R. China

ABSTRACT: The microbial community composition in high arsenic groundwater aquifers was investigated by 454 pyrosequencing and illumina MiSeq sequencing approaches. Both groundwater and sediment samples were divided into low and high arsenic groups according to principal component analysis and hierarchical clustering. Microbial communities were significantly different among samples with different geochemistry. Predominant populations were *Acinetobacter*, *Pseudomonas*, *Psychrobacter* and *Alishewanella* in high reducing area, *Alishewanella*, *Psychrobacter*, *Methylotenera* and *Crenothrix* in agricultural area, and *Thiobacillus*, *Pseudomonas* and *Hydrogenophaga* in high arsenic sediments. Arsenic, TOC, NH_4^+, Fe(II) and SO_4^{2-} were important factors shaping indigenous microbial communities.

1 INTRODUCTION

Arsenic (As) contamination in groundwater aquifers is a serious environmental issue in many countries. Results of previous studies showed that As release and mobilization were usually controlled by a series of microbially mediated reactions and geochemical processes. However, microbial communities in groundwater and sediments with different geochemistry have yet to be fully understood.

2 METHODS/EXPERIMENTAL

2.1 *Sample collection and geochemical analysis*

Hangjinhouqi County is one of the most serious high arsenic groundwater threatened area located in the west of Hetao Plain. Groundwater samples from high reducing area and along agricultural drainage channels and sediments from three boreholes in Hangjinhouqi County were collected. Microbial samples were collected by filtering of 15L fresh waters through 0.2 μm filters. Geochemical parameters were measured in situ or sampled according to our previous studies (Wang *et al.*, 2016).

2.2 *454 pyrosequencing and illumina MiSeq sequencing*

Pyrosequencing of the V4 region of the 16S rRNA gene was carried out from the 515F-end of the amplicons with Roche (454) genome sequencer FLX+ system (454 Life Sciences, USA) at SeqWright Inc (Houston, USA). Illumina sequencing were conducted by an Illumina MiSeq 2000 instrument at the Yale Center for Genome Analysis. All raw data analysis was performed using the QIIME software package, version 1.5.0.

2.3 *Statistical analysis*

All statistical analysis in this study was performed, based on genus-level OTUs at the 97% similarity level under the package of Vegan in R.

3 RESULTS AND DISCUSSION

3.1 *Geochemistry*

Samples with different geochemistry presented distinct biogeochemical characteristics and redox properties. Groundwater and sediment samples could be divided into well-defined high and low As groups by Hierarchical clustering and PCoA analyses of geochemical parameters. Arsenic concentrations of groundwater samples along agricultural drainage channels showed positive correlations with NH_4^+ and TOC, while showed no obvious correlations with Fe(II) and SO_4^{2-}, differencing from those samples from strongly reducing areas which were characterized with low concentrations of SO_4^{2-}, negative ORP and high ratios of Fe(II) to Fe(III). In contrast, high As sediments in this study were under weak reducing conditions where As (III)/As (V) increased with depth from 0.02 to 0.52 and SO_4^{2-}/Total S ranged between 0.001 and 0.75.

3.2 *Microbial richness and diversity*

The microbial community diversities of groundwater from highly reducing area and sediments were analyzed by 454 pyrosequencing. A total of 233,704 reads were obtained for 39 samples after removal of low-quality and chimeric sequences. A variety of taxa were observed at the 97% OTU level, with 12–267 observed and 23–355 predicted OTUs (based on Chao1) and coverage values ranging from 31.3% to 81.6%. Richness and diversity indexes of these samples showed no significant correlations with the concentrations of As or any other geochemical variables. Illumina sequencing was used to examine the microbial communities of groundwater from agricultural drainage channels. Sequencing results revealed that a total of 329–2823 OTUs were observed at the 97% OTU level. Microbial richness and diversity of high arsenic groundwater samples along the drainage channels were lower than those of low arsenic groundwater samples, but higher than those of high arsenic groundwaters from strongly reducing areas, implying that irrigation and drainage activities might affect local geochemical environment and thus enhance the diversity of microorganisms.

3.3 *Microbial compositions in high As groundwater aquifers*

Microbial community compositions at phylum level showed no distinct difference between high and low As groundwater samples as well as sediment samples. For groundwater samples from highly reducing areas and sediment samples, *Acinetobacter*, *Psychrobacter* and *Alishewanella* were more abundant in groundwater than sediments. The average abundances of *Thiobacillus*, *Pseudomonas*, *Hydrogenophaga*, *Enterobacteriaceae*, *Sulfuricurvum* and *Arthrobacter* were higher in sediments than groundwater samples. It was worth noting there were two populations *Acinetobacter* and *Thiobacillus* which highly contributed to the dissimilarities of these two kinds of samples (average abundance 30.45% and 12.31%, respectively). *Acinetobacter* was distinctly dominated in groundwater with the markedly high relative abundance (62.41%), and *Thiobacillus* was significantly abundant (average abundance 24.62%) in sediments, while being absent in most of the groundwater (average abundance 0.01%). The archaeal abundance was mostly lower than 1%. Methanogens were distinctly predominant in most of the high arsenic groundwater, while similar results haven't been found in high arsenic sediments.

Illumina results of groundwater samples from agricultural drainage channels indicated that the bacterial community was composed of 62 phyla. Of these, only 9 phyla dominated each community, among which *Proteobacteria* was the most dominant group (32.02–86.50%), followed by *Firmicutes* (0.16–18.48%), *Actinobacteria* (0.34–12.08%) and *Nitrospirae* (0–22.64%). Archaeal populations accounted only for 0.01–8.69% of the microbial populations. All detected archaeal populations belonged to the phyla *Euryarchaeota* and *Crenarchaeota* which represented by the classes of *Parvarchaea*, *Methanomicrobia* and *Methanobacteria*. These methanogenic populations might provide even stronger reducing conditions and thereby accelerate As release in groundwater aquifers in the Hetao Plain. However, the irrigation and drainage activities in agricultural area could bring more oxygen and the oxidizing conditions might weaken the role of methanogenic populations. At the genus level, *Alishewanella*, *Psychrobacter*, *Pseudomonadaceae*, *Methylotenera*, *Crenothrix* and *Comamonadaceae* were dominant communities in high As groundwater samples from agricultural drainage channels with high average abundances. These dominant populations indicated that As release in groundwater along agricultural drainage channels was correlated with microbial carbon, nitrogen and iron reactions. Different from samples from the highly reducing area, many oxidizing microbial populations such as *Methylotenera* and *Crenothrix* were found in the present study, indicating the possible influence of geochemical difference on microbial composition.

3.4 *Microbial community structure in relation to geochemistry*

Co-inertia analysis was carried out in order to reveal the relative importance of geochemical vectors that affect microbial community structures. Results revealed that geochemical variables had a significant influence on their microbial community composition. Arsenic is one of the key environmental factors that contribute to the difference in the geochemistry and microbial community structure in both groundwater and sediments. Besides, geochemical parameters including As, TOC, SO_4^{2-}, SO_4^{2-}/TS and Fe^{2+} were important factors causing the difference of the microbial community in groundwater samples from highly reducing area and sediment samples. Geochemical variables such as As, TOC, NH_4^+, Fe and ORP might also affect the microbial composition from agricultural drainage channels. These important geochemical factors shaping the microbial community structure were different from those groundwater samples of highly reducing area and sediment samples in the Hetao Plain (Li *et al.*, 2015).

ACKNOWLEDGEMENTS

This study was funded by National Natural Science Foundation of China (No. 41372348, 41521001).

REFERENCES

Li, P., Wang, Y., Dai, X., Zhang, R., Jiang, Z., Jiang, D., Wang, S., Jiang, H., Wang, Y. & Dong, H. 2015. Microbial community in high arsenic shallow groundwater aquifers in Hetao Basin of Inner Mongolia, China. *PloS one* 10(5): e0125844.

Wang, Y., Li, P., Jiang, Z., Sinkkonen, A., Wang, S., Tu, J., Wei, D., Dong, H. & Wang, Y. 2016. Microbial community of high arsenic groundwater in agricultural irrigation area of Hetao Plain, Inner Mongolia. *Front. Microbiol.* 7: 1917.

1.5 Spatial and temporal evolution of arsenic in mine waste and tailings

Environmental Arsenic in a Changing World –
Zhu, Guo, Bhattacharya, Ahmad, Bundschuh & Naidu (Eds)
ISBN 978-1-138-48609-6

Application of stable isotopes on bioaccumulation and trophic transfer of arsenic in aquatic organisms around a closed realgar mine

F. Yang[1,2,3] & C.Y. Wei[1,2]
[1] *Institute of Geographic Sciences and Natural Resources Research, Chinese Academy of Sciences, Beijing, P.R. China*
[2] *Key Laboratory of Land Surface Pattern and Simulation, Chinese Academy of Sciences, Beijing, P.R. China*
[3] *University of Chinese Academy of Sciences, Beijing, P.R. China*

ABSTRACT: Arsenic (As) bioaccmulation and trophic transfer were studied in various species of aquatic organisms from the streams and a reservior (Zaoshi Lake) around Shimen realgar mine, which had been closed since 2001. The As concentrations in top soils, sediments and water were decreasing with the distance from the central mining sites, indicating the As mining activities had still produced eminent impact on As contamination around the mine. Throughout the sampling sites, the As in biota was ranked in an order of tadpole > crab > frog > loach > other fishes. The $\delta^{13}C$ and $\delta^{15}N$ values in the same species of organisms varied greatly with sites, suggesting the impact of diet variation on the trophic levels of the same species in the food chain. The trophic levels for various species of organisms were determined as from 1.25 to 3.76 based on $\delta^{15}N$ values. In streams, As levels were much higher in lower trophic organisms, but were quite stable at higher trophic levels, indicating no bio-magnification of As in the food chain, while in the reservoir organisms had much lower arsenic concentrations but obvious biomagnification trend through the food webs. Moreover, the As speciation, including iAs(III), iAs(V), MMA, DMA, AsB and other species were analyzed to explore the biotransformation of As and the sources and bio-transfer of organic As in the food chain. It was found that the organic As species increased with trophic levels in streams, while in reservoir they showed opposite patterns.

1 INTRODCTION

Arsenic (As) is a pervasive environmental toxicant andu carcinogen that can pose significant health risks to humans and animals (Mandal & Suzuki, 2002; Smedley & Kinniburgh, 2002). As toxicity depends on not only its total concentration but also its chemical speciation (Ng, 2005). As loading from mines has long been a public concern, with serious health consequences for residents living in As-contaminated areas. To explore the As accumulation and transformation throughout the food webs, surface water, sediment, soil, aquatic organisms were collected from Zaoshi Reservior used as a reference site and five study sites along the streams where metal contaminants had been discharged from the abandoned mine. The findings of this research could provide important information for understanding the mechanisms of As bioaccumulation in As-contaminated environments and for assessing the potential environmental impacts of elevated As concentrations on mining activities at the realgar mine area.

2 METHODS

2.1 *Total arsenic analysis*

The total As content in samples was determined by ICP-MS (DRC-e, PerkinElmer, Waltham, USA.) following acid digestion. Initially, about 200–300 mg freeze-dried powders were directly weighed into a 25 mL graduated vessels. After which 5 mL of concentrated HNO_3 was slowly added to the samples, and then left overnight in order to digest gently covered with a perforated glass stopper. In the following morning, the vessel was carried out on a hot plate at 120°C–140°C for complete digestion until the solution became transparent and about 1 mL solution remained. After cooling, about 2 mL H_2O_2 was added and heated for an hour, and then transferred to a 20-mL tube. The tube was brought to volume with ultra-pure water for analysis. To determine the As concentration in soil samples, a portion (0.3 ± 0.02 g) was weighed and digested with 10 mL HNO_3-$HClO_4$ (v/v 9:1) at 180°C on an electric hot plate.

2.2 *Determination of As species*

The dried samples (100–300 mg) was weighed into a 50 mL polyethyene tube and extracted using a methanol-water mixture (15 mL, 1 + 1 v/v). The tubes were then placed on a shaker at 150 rpm under room temperature overnight. Then the extracted solution was sonicated in an ultrasonic bath for 30 min before centrifugation at 3000 rpm for 15 min. The supernatant was transferred into a 25 mL tubes, while the residue was re-extracted with 5 mL of 1:1 methanol: water mixture following the former procedure. The mixture

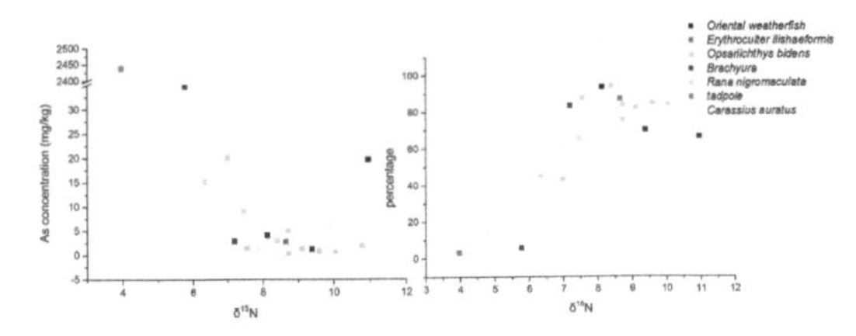

Figure 1. Total arsenic (left) and arsenic speciation (right) of aquatic organisms from the contaminated streams.

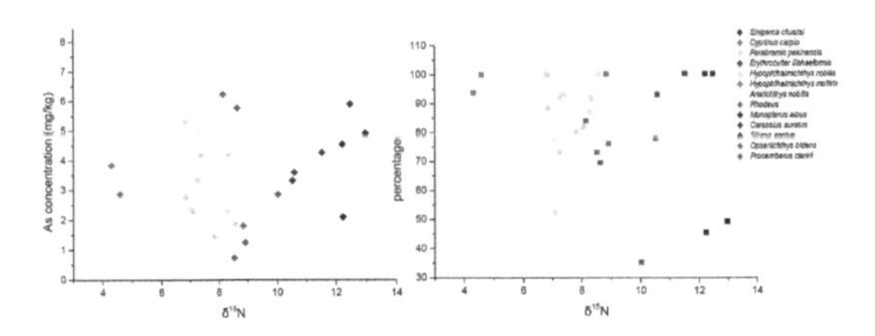

Figure 2. Total arsenic (left) and arsenic speciation (right) of aquatic organisms from the reservoir.

was maintained shaking for 1 h, sonicated for 30 min and centrifuged for 20 min. The two combined extracts were mixed and evaporated using a pressured N_2 gas-blowing concentrator until less than 5 mL solution remained. Before analysis, the samples were filtered with a 0.45 μm disposable syringe filters into HPLC vials and stored at 4°C.

3 RESULTS AND DISCUSSION

Analysis of carbon (δ^{13}C) and nitrogen (δ^{15}N) stables isotopes provides the numerical characterization of trophic level, and as such, we combined it with As to evaluate the biomagnification/biodilution within a freshwater ecosystem. The total As in surface soil, sediments and water were in the range of 35–5000 mg kg^{-1}, 43–4543 mg kg^{-1} and 5–3293 μg L^{-1}, respectively (Figs. 1 and 2). The As concentrations were decreasing with the distance from the central mining sites, indicating that the As mining activities had produced eminent impact on As contamination in Shimen.

The main As compounds identified by HPLC-ICP-MS were inorganic As (arsenite, As(III), and arsenate, As(V)), dimethylarsinic acid (DMA) and monomethylarsonic acid (MMA), while arsenobetaine (AsB) and arsenocholine (AsC) were mostly present as minor species. In our study, the proportions of DMA and MMA and the As(III) accounts for higher proportions in the muscle tissue compared to those in the water.

The δ^{13}C and δ^{15}N values in the same species of organisms varied greatly with sites, which suggest the impact of diet difference on the trophic levels of the same species in the food chain. The trophic levels for various species of organisms were determined as from 1.25 to 3.76 based on δ^{15}N values. In streams, As levels were much higher in lower trophic organisms, but were quite stable at higher trophic levels, indicating no bio-magnification of As in the food chain, however, the reservoir showed slight arsenic concentrations and obvious biomagnification through the food webs. It is noteworthy that MMA and DMA were the dominant forms of methylated As in the higher trophic level in contaminated streams. The same trends results were reported in the team group previously, which indicating that the lower trophic levels of organisms have a greater ability to accumulate arsenic and higher trophic levels of organisms have a greater ability to methylate arsenic. But the opposite pattern was found in the organisms in reservoir. The reason and mechanisms need further researches.

4 CONCLUSIONS

Arsenic contamination from mines is obvious in freshwaters. There are a number of reports documented biodilution of As with increasing trophic levels in freshwater food webs which can be related to many factors. This study gives a full picture of As characterization in both aquatic and terrestrial ecosystem in the As-contaminated area in China. However, a contrast pattern can be found in both total arsenic and its species. No conclusive evidence from the field study has been presented yet, thus further studies are needed.

ACKNOWLEDGEMENTS

This study was financially supported by the National Natural Science Foundation of China (Grant No. 41571470).

REFERENCES

Mandal, B.K. & Suzuki, K.T. 2002. Arsenic round the world: a review. *Talanta.* 58(1): 201–235.

Ng, J.C. 2005. Environmental contamination of arsenic and its toxicological impact on humans. *Environ. Chem.* 2(2): 146–160.

Smedley, P.L. & Kinniburgh, D.G. 2002. A review of the source, behaviour and distribution of arsenic in natural waters. *Appl. Geochem.* 17(5): 517–568.

Environmental Arsenic in a Changing World –
Zhu, Guo, Bhattacharya, Ahmad, Bundschuh & Naidu (Eds)
ISBN 978-1-138-48609-6

Arsenic characteristics in the terrestrial environment in the vicinity of the Shimen realgar mine, China

C.Y. Wei
Key Laboratory of Land Surface Pattern and Simulation, Institute of Geographic Sciences and Natural Resources Research, Chinese Academy of Sciences, Beijing, P.R. China

ABSTRACT: In this study, multiple types of samples, including soils, plants, litter and soil invertebrates, were collected from a former arsenic (As) mine in China. The total As concentrations in the soils, earthworms, litter and the aboveground portions of grass from the contaminated area followed the decreasing order of 83–2224 mg kg^{-1}, 31–430 mg kg^{-1}, 1–62 mg kg^{-1} and 2–23 mg kg^{-1}, respectively. XANES analysis revealed that the predominant form of As in the soils was arsenate (As(V)). In the grass and litter of the native plant community, inorganic As species were the main species, while minor amounts of DMA, MMA were also detected by HPLC-ICP-MS. The major As species extracted from earthworms were inorganic and AsB was the only organic species present in the earthworm samples, although at low proportions. The internal bioconversion of other As species is hypothesized to contribute greatly to the formation and accumulation of AsB in earthworms, although the direct external absorption of organic As from soils might be another source. This study sheds light on the potential sources of complex organoarsenicals, such as AsB, in terrestrial organisms.

1 INTRODUCTION

Arsenic (As) is a toxic metalloid element that is widely distributed in the environment (Ng, 2005). The As toxicity is known to depend on the chemical species (Ng, 2005). In general, inorganic As species are more toxic than organic forms. Therefore, it is essential to monitor both the total As concentration and As speciation in environmental and biological samples. Generally, As can enter plants through their root systems and can be dispersed by leaf litter, posing a threat to soil invertebrates living in As-contaminated soils (Wang *et al.*, 2016). To the best of our knowledge, the literature has primarily focused on the relationships between soils and plants or earthworms in terrestrial ecosystems, but such a limited scope is insufficient (Kramar *et al.*, 2017). To further understand the As cycling in the terrestrial environment, comprehensive studies on As bioaccumulation and biotransformation via a soil-plant-leaf litter-terrestrial fauna approach are highly desired.

2 METHODS AND MATERIALS

2.1 *Sampling*

The Shimen realgar mine (29°38′11″–29°38′43″N, 111°2′06″–111°2′23″E) is located in Shimen County, Hunan Province, China. This mine contains the largest reserves of realgar in Asia. In this study, eleven sites in the contaminated area (around the smelter and tailings dam), while six sites in the uncontaminated area (>4 km from the upper part of the smelter) were established at different distances from the center of the mine.

2.2 *Total arsenic analysis*

Approximately 0.05 g of each soil sample was digested with 10 ml of 5:1 HNO_3:$HClO_4$, and approximately 0.2 g of each biological sample was digested with 10 ml of 9:1 HNO_3:$HClO_4$ in acid-cleaned vessels. The test tubes were left overnight and then heated at 140°C on an electric hot plate. The samples were analyzed using an AFS-8230 spectrometer (Beijing Titian Co., China).

2.3 *Arsenic speciation analysis*

Arsenic speciation was analyzed in aqueous methanol extracts of the dried samples and of the reference materials as described in the literature (Ciardullo *et al.*, 2010) with some modifications. The K-edge XANES spectra of As were obtained at the 1W1B beam line at the Beijing Synchrotron Radiation Facility (BSRF), Beijing, China.

3 RESULTS AND DISCUSSION

3.1 *Arsenic characterization in soils*

In contaminated areas, the soil pH varied markedly from 4.51 to 7.98, while the content of TOM ranged from 18.05 to 49.33 g kg^{-1}. In the uncontaminated areas, little variation in soil pH (6.21–6.74) was found, and the TOM content varied from 16 to 26 g kg^{-1}. The As concentrations in soils collected from uncontaminated areas were in the range of 8–23 mg kg^{-1}, which were much lower than those in contaminated areas (61–2224 mg kg^{-1}).

3.2 *Arsenic concentration in the native plants*

In general, the As concentrations in this study displayed a pattern of leaves > stems, similar to the results of other studies (Tsipoura *et al.*, 2017). The As uptake by plants was found to differ among species; the BF in terrestrial plants, excluding two *Pteris* species, varied from 0.0003 to 0.20, which was in the same range as that reported in the literature. The fern species *Pteris vittata* and *Pteris cretica* collected from contaminated sites showed the greatest As concentrations among the sampled plants, with up to 584 mg kg^{-1} and 391 mg kg^{-1} As in their aboveground parts, respectively. In this study, the As concentrations in corn much higher than those from the uncontaminated area. Similarly, the indigenous *Miscanthus sinensis* and *Conyza Canadensis*. Therefore, the past As mining and smelting activities appear to still be causing As accumulation in crops and indigenous plants.

3.3 *Arsenic concentration in the soil-grass-litter-invertebrate system*

In general, the As level increased in earthworms with increasing As concentration in their surroundings in this study, which is in agreement with previous studies. BFs tended to decrease with increasing As concentrations in soils, suggesting that As is mainly autoregulated and can be sequestered by earthworms. Earthworms accumulated more As than did the grass and litter in this study. Significant positive correlations were found between the As concentrations in the soils and earthworms, the earthworms and leaf litter, and the soils and leaf litter, suggesting close relationships among the As inputs in the soil-litter-earthworm system.

3.4 *Arsenic transformation in the terrestrial environment*

In the present study, As(V) was the most dominant form in the soils, as determined by XANES analysis, while As(III) was the predominant As species present in many terrestrial plant species in this study. Therefore, the As(V) taken up by the plants was converted to As(III). Notably, various organic As in plant suggested organoarsenic species taken up from soils. The main As species extracted from earthworms was inorganic As. AsB was the only organoarsenic compound detected in earthworms. In this study, no or trace amounts of AsB were detected, while other organic As species, such as AsC, DMA, and MMA, and an unknown As species appeared at relatively high percentages in plants or leaf litter. Therefore, the AsB in earthworms does not largely come from ambient environmental sources, but a result of the biotransformation by earthworms themselves (Langdon *et al.*, 2003; Button *et al.*, 2009).

4 CONCLUSIONS

In this study, the total As concentration and As speciation were studied in different environmental media from both contaminated and uncontaminated area. The aboveground parts of the local plant community in this mining area contained low As concentrations. With increasing As concentration in the environment, the concentration of As in earthworms generally increased, while the BFs decreased in the present study. The most abundant As compounds found in plants, leaf litter and grass were inorganic As species, while a variety of organoarsenic species were also detected. The high proportion of inorganic As in earthworms suggested dietary sources, while the appearance of AsB as the only organoarsenic species was largely the result of biotransformation from other As species.

ACKNOWLEDGEMENTS

This study was financially supported by the National Natural Science Foundation of China (Grant No. 41571470).

REFERENCES

Button, M., Jenkin, G.R.T., Harrington, C.F. & Watts, M.J. 2009. Arsenic biotransformation in earthworms from contaminated soils. *J. Environ. Monitor.* 11: 1484–1191.

Ciardullo, S., Aureli, F., Raggi, A. & Cubadda, F. 2010. Arsenic speciation in freshwater fish: Focus on extraction and mass balance. *Talanta* 81: 213–221.

Kramar, U., Norra, S., Berner, Z., Kiczka, M. & Chandrasekharam, D. 2017. On the distribution and speciation of arsenic in the soil-plant-system of a rice field in West-Bengal, India: A μ-synchrotron techniques based case study. *Appl. Geochem.* 77: 4–14.

Langdon, C.J., Piearce, T.G., Meharg, A.A. & Semple, K.T. 2003. Interactions between earthworms and arsenic in the soil environment: A review. *Environ. Pollut.* 124: 361–373.

Ng, J.C. 2005. Environmental contamination of arsenic and its toxicological impact on humans. *Environ. Chem.* 2: 146–160.

Tsipoura, N., Burger, J., Niles, L., Dey, A., Gochfeld, M., Peck, M. & Mizrahi, D. 2017. Metal levels in shorebird feathers and blood during migration through Delaware Bay. *Arch. Environ. Con. Toxicol.* 72: 562–574.

Wang, Z.F., Cui, Z.J., Liu, L., Ma, Q.C. & Xu, X.M. 2016. Toxicological and biochemical responses of the earthworm *Eisenia fetida* exposed to contaminated soil: Effects of arsenic species. *Chemosphere* 154: 161–170.

Environmental Arsenic in a Changing World –
Zhu, Guo, Bhattacharya, Ahmad, Bundschuh & Naidu (Eds)
ISBN 978-1-138-48609-6

Hydrochemical characteristics and the genesis of high arsenic groundwater in the ecotone between polymetallic sulfide mining area and irrigated agricultural area

Y.H. Dong[1,2], J.L. Li[1] & T. Ma[2]
[1]*School of Environmental Sciences, China University of Geosciences, Wuhan, P.R. China*
[2]*School of Water Resources and Environmental Engineering, East China University of Technology, Nanchang, P.R. China*

ABSTRACT: The goal of this study was to expand existing knowledge to genesis of high arsenic groundwater and to reveal the influence of anthropogenic practices on high arsenic aquifer system, finally to provide the comprehensive theory for developing appropriate water treatment technologies. High arsenic groundwater is distributed at depths of 10–40 m in the north of Hanjinghouqi county along the Yin Mountains where is the ecotone of the polymetallic sulfide mining area and the irrigation area. Dissolved As concentrations in groundwater are in the range of 0.5 μg L^{-1} to 764.8 μg L^{-1} with an average of 161.6 μg L^{-1}. 66.88% of (103) groundwater samples exceed the standard for allowable arsenic concentration 10 μg L^{-1} in drinking water. High arsenic in the ecotone is geological origin and influenced by anthropogenic activities including irrigation and mining.

1 INTRODUCTION

An ecotone is a transition zone between two biomes (Fridley *et al.*, 2009) or adjacent patches (Ward *et al.*, 1999). It is where two communities meet and integrate (Pearl *et al.*, 2011). It may be narrow or wide, and it may be local (the zone between a field and forest) or regional (the transition between forest and grassland ecosystems). An ecotone may appear on the ground as a gradual blending of the two communities across a broad area, or it may manifest itself as a sharp boundary line.

The ecotone between polymetallic sulfide mining area and irrigated agricultural area is a mosaic transition zone between two areas affected by human activities. It is used in this study to consider the exchange, mixing and interaction of matter and exchange of energy in surface water, groundwater and sediments between polymetallic sulfide mining area and irrigation area. In this ecotone, migration and interaction of substances among surface water, groundwater, sediments and rocks are susceptible to interference from artificial factors, while formation mechanism of groundwater becomes extremely complicated.

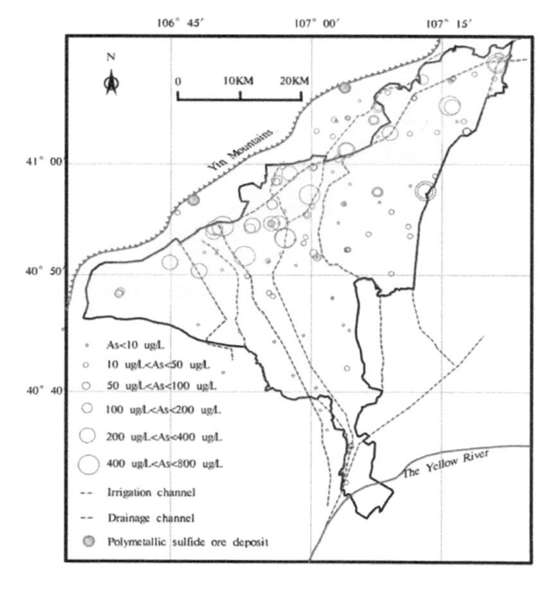

Figure 1. The ecotone between polymetallic sulfide mining area and irrigated agricultural area (in light).

2 METHODS

A summary of 103 groundwater samples were collected from the study area Hangjinhouqi County and analyzed for chemical compositions (Fig. 1). The pH, EC and ORP of solution obtained were monitored by portable parallel analyzer (HQ40d, HACH). The solutions were filtered through membrane filters with 0.45-μm pore diameter for chemical analysis. Then aqueous samples were added with 1–2 drops of hydrochloric acid for total As content analysis by Atomic Fluorescence Spectrometry (AFS-820, Titan Instruments). Concentrations of the total Fe, sulfate (calculated as SO_4^{2-}), and dissolvable sulfide (calculated as S^{2-}) in filtered solutions were analyzed by portable spectrophotometer (DR2800, HACH).

3 RESULTS AND DISCUSSION

3.1 *Groundwater flow field*

Groundwater from the irrigated agricultural area and mining area converges in the ecotone, and flows along the main channel from southwest to east. Water level of a well in the ecotone (the front of the alluvial fans) varied from 0.2 to 8.2 m below the land surface, and from 0.2 to 3.3 m below the land surface in the flat plain region. The fluctuation pattern of water level was different between the ecotone and the irrigated agricultural area. That is mainly because the ecotone is far from the Yellow River, where the groundwater is extracted for irrigation, the water level rose when the irrigation ceased and stabilized during winter season without irrigation (Guo *et al.*, 2013).

3.2 *Groundwater hydrochemistry*

In the ecotone, it is obvious that groundwater has a high electrical conductivity, indicating the more dissolved solid than the mining area and irrigation area. Groundwater also has a high average concentration of total dissolved arsenic. The hydrochemical types are complicated in this area (Table 1).

3.3 *Genesis of high arsenic groundwater in the ecotone*

The enrichment processes mainly contain two aspects, groundwater dynamic and groundwater hydrochemical processes. The hydrodynamic processes in the ecotone contains mixing process among the groundwater from the polymetallic sulfide mining area, groundwater from the irrigated agricultural area, and/or water from the main drainage channel in horizontal direction; and the groundwater evaporation the vertical direction. Hydrochemical processes mainly refer to cation exchange, weathering of silicate minerals, reductive dissolution of Fe/Mn-oxides/hydroxides, and sulfate reduction. Moreover, groundwater nitrate pollution is an essential factor influencing the hydrochemical processes of arsenic enrichment.

4 CONCLUSIONS

The groundwater hydrochemistry and distribution of groundwater arsenic in the ecotone were discussed to better understand the groundwater dynamic field and hydrochemical field. The genesis of high arsenic groundwater was revealed.

Table 1. The chemical compositions of groundwater in the ecotone.

Index	Unit	Max	Min	Average
T	°C	21.0	10.1	13.2
pH		9.2	6.5	7.9
EC	$\mu S\ cm^{-1}$	106300	763	3595
ORP	mV	253.0	−169.8	−30.2
DO	$mg\ L^{-1}$	7.0	1.2	3.5
HCO_3^-	$mg\ L^{-1}$	1276.2	141.0	615.2
CO_3^{2-}	$mg\ L^{-1}$	77.3	0.0	7.8
Total dissolved As	$\mu g\ L^{-1}$	764.8	0.5	161.6
As(III)	$\mu g\ L^{-1}$	541.1	5.3	134.5
As(V)	$\mu g\ L^{-1}$	231.0	4.4	39.8
As(p)	$\mu g\ L^{-1}$	197.3	2.6	37.2
Ca^{2+}	$mg\ L^{-1}$	457.2	2.4	90.6
K^+	$mg\ L^{-1}$	61.9	1.2	16.8
Mg^{2+}	$mg\ L^{-1}$	506.9	3.8	116.4
Na^+	$mg\ L^{-1}$	2462.9	33.2	467.6
Cl^-	$mg\ L^{-1}$	4495.3	46.3	638.1
SO_4^{2-}	$mg\ L^{-1}$	1475.7	BLD	363.8
NO_3^-	$mg\ L^{-1}$	190.8	BLD	13.7
F^-	$mg\ L^{-1}$	9.45	BLD	1.49
Fe	$mg\ L^{-1}$	6.90	BLD	0.85
Mn	$mg\ L^{-1}$	0.68	BLD	0.14
TP	$mg\ L^{-1}$	0.56	BLD	0.05
DOC	$mg\ L^{-1}$	77.1	0.4	13.4
NH_4^+	$mg\ L^{-1}$	13.4	BLD	3.0
NO_2^-	$mg\ L^{-1}$	0.93	BLD	0.04
Dissolved sulfide	$\mu g\ L^{-1}$	1206.4	BLD	46.1
PO_4^{3-}	$mg\ L^{-1}$	4.36	0.02	0.28
Fe^{2+}	$mg\ L^{-1}$	2.42	0.02	0.56

Note: BLD: below limit of detection.

ACKNOWLEDGEMENTS

This work was financially supported by the National Natural Science Foundation of China (No. 41372252) and IAEA Coordinated Research Project (No. R21122/R0).

REFERENCES

Fridley, J.D., Senft, A.R. & Peet, R.K. 2009. Vegetation structure of field margins and adjacent forests in agricultural landscapes of the North Carolina piedmont. *Castanea* 74(4): 327–339.

Guo, H.M., Zhang, Y., Jia, Y.F., Zhao, K., Li, Y. & Tang X.H. 2013. Dynamic behaviors of water levels and arsenic concentration in shallow groundwater from the Hetao Basin, Inner Mongolia, *J. Geochem. Explor.* 135: 130–140.

Pearl, S.E., Berg, L.R. & Martin, D.W. 2011. *Biology*. Belmont, California.

Ward, J.V., Tockner, K. & Schiemer, F. 2010. Biodiversity of floodplain river ecosystems: Ecotones and connectivity. *River Res. Appl.* 15: 125–139.

Environmental Arsenic in a Changing World –
Zhu, Guo, Bhattacharya, Ahmad, Bundschuh & Naidu (Eds)
ISBN 978-1-138-48609-6

Potential threat of arsenic contamination of water sources from gold mining activities in Lake Victoria areas, Tanzania

R.R.A.M. Mato & G.R. Kassenga
Department of Environmental Science and Management, Ardhi University, Dar es Salaam, Tanzania

ABSTRACT: Occurrence of arsenic in drinking water sources is among the contemporary challenging water quality problems that need immediate attention worldwide. Both natural and human induced arsenic sources are potentially contaminating the water resources and thus exposing millions of people to serious public health risks including cancer. In some areas, mining activities, especially gold mining, have been reported to induce elevated levels of arsenic in adjacent water sources. In most developing countries like Tanzania most lack scientific information on the extent of arsenic levels in water sources. A rapid assessment conducted to determine occurrence of arsenic contamination in water sources in the Lake Victoria mining areas in Tanzania has showed elevated levels of arsenic in some areas. Arsenic levels as high as $70\,\mu g\,L^{-1}$ were observed and more than 40% of the water samples taken from the areas had arsenic levels equal to or exceeding the Tanzania Drinking Water Quality Standards threshold value of $10\,\mu g\,L^{-1}$. Areas with high arsenic levels included Maswa, Shinyanga and Musoma districts. The arsenic mobilization into water sources is suspected to be due to oxidation of arsenopyrite after exposure of the ore to the air partly due to mining activities. This linkage makes mining activities potential threat to quality of drinking water sources in the Lake Victoria mining areas, a reason which triggers search for safe water sources. This paper discusses the potential health risks that exists in the Lake Victoria mining areas due to exposure to arsenic contamination and appropriate measures that can be undertaken to protect the health of general public.

1 INTRODUCTION

Arsenic is a known water contaminant, which can have devastating effects on health (WHO, 2000). Both acute and chronic health effects can be manifested depending on the arsenic dosage and time duration of exposure. Some countries like Bangladesh experiences serious arsenic contamination (Halsey, 2000). The WHO (2000) has set the Drinking Water Guideline for arsenic at $10\,\mu g\,L^{-1}$. However, drinking water arsenic limits vary from country to country based on different regulatory limits set by the governments.

Arsenic naturally occurs primarily as arsenopyrite (FeAsS), which is the most abundant arsenic mineral, dominantly in mineral veins (Halsey, 2000). Its occurrence in water sources thus has a close linkage with mining or mineral extractions. According to the BGS (2001), arsenic mobilizes in the local environment as a result of arsenopyrite oxidation, induced (or exacerbated) by the mining activity. High arsenic concentrations have been reported in soils and river waters close to the gold mining activity in Ghana (BGS, 2001; Bhattacharya *et al.*, 2012; Kassenga & Mato, 2008; Smedley, 1996). Arsenic contamination from mining activities has also been identified in numerous areas of the USA where concentrations of up to $48{,}000\,\mu g\,L^{-1}$ in groundwater have been reported (Welch *et al.*, 2000). In Bangladesh concentration of arsenic in groundwater has been reported to exceed $2{,}000\,\mu g\,L^{-1}$ (Halsey, 2000).

In Tanzania, arsenic contamination in water sources has scarcely been reported. Kassenga and Mato (2008) reported elevated levels of arsenic in Lake Victoria region. Earlier, the Department of Geology, University of Dar es Salaam, Tanzania (1994) reported presence of arsenic concentrations ranging between 0.5 and $379\,\mu g\,L^{-1}$ in some water and sediment samples collected from the Lake Victoria mining areas. The suspected source of arsenic pollution in the water and sediment samples was identified to be oxidation of arsenopyrite in mine tailings dumped into the rivers during gold panning.

Despite intensive mining activities in the Lake Victoria areas, there is no comprehensive study to determine the ingress of arsenic in water sources. The Lake Victoria Basin is the leading gold producing area in Tanzania. There are a number of goldfields in the area (collectively known as Lake Victoria Goldfields) and they are located in the Archean Nyanzian greenstone belt east and south of Lake Victoria in northwest Tanzania (Department of Geology, 1994). Mining is done at both small scale (artisanal) and large scale. The large scale gold mining companies include North Mara Gold Mine; Geita Gold Mine, Kahama Gold mine, and Buzwagi Gold mine just to mention a few. However, there are numerous artisanal gold mining activities spatially spread throughout the Lake Victoria

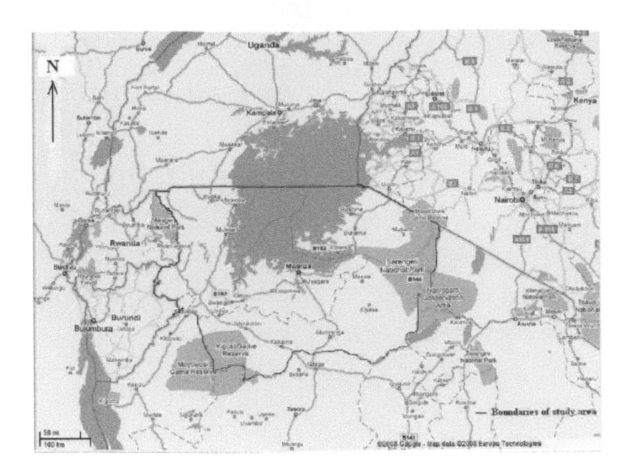

Figure 1. Location map of Tanzania showing the study area.

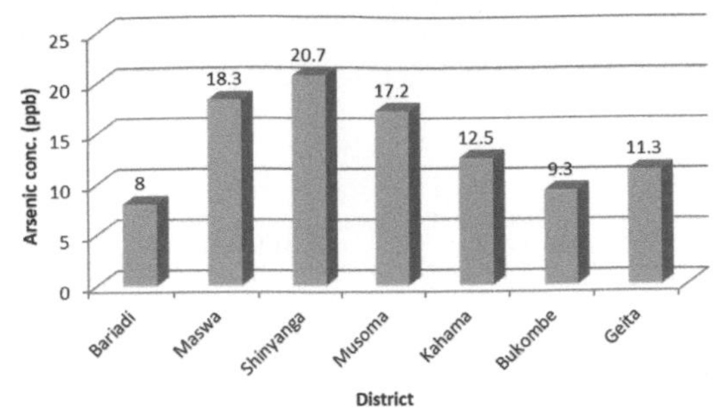

Figure 2. Measured average concentration levels of arsenic in selected water sources in the districts of the Lake Victoria mining areas, Tanzania.

region. Both open cast and underground mining can expose the gold ores containing arsenic mineral to oxidation that lead to contamination of the groundwater sources. Probably management of tailing dumps contribute more to the arsenic pollution of the surface water sources. Such pollution exposes more than 12 million people in the Lake Victoria areas to a range of health risks.

Tanzania has adopted the WHO tolerant limit of $10\,\mu g\,L^{-1}$ (TZS 789: 2003). However, arsenic in Tanzania has not been an important or common variable for measurement in routine water quality analysis. Thus, its distribution in Tanzania is not yet clearly known. A rapid assessment was conducted in 2008 with the objective of determining the levels of arsenic contamination in drinking water sources in mining areas of the Lake Victoria Basin in Tanzania and information collected is part of this presentation.

2 METHODS

2.1 *Study area*

The study, which was conducted in 2008 in the Lake Victoria Basin in Tanzania, which covered some parts of the regions of Shinyanga, Mwanza, Geitia, Simiyu, and Mara. The study involved conducting field measurements of arsenic in selected water sources in the districts of Bariadi and Meatu (Simiyu Region); Bukombe, Maswa and Kahama (Shinyanga Region); Musoma and Serengeti (Mara Region); Geita (Geita Region) and Mwanza Urban, Sengerema, and Kwimba (Mwanza Region), as shown in Figure 1.

The total population in the Lake Victoria zone exceeds 12 million people, which is approximately 25% of the entire population of Tanzania (URT, 2012).

2.2 *Analytical methods*

The study involved carrying out in-situ measurement of arsenic levels in selected water sources in the districts. Only important water sources used for community water supply were selected. A total of 96 drinking water sources were selected and analysed for arsenic. The field studies involved onsite measurements of concentrations of arsenic and other associated parameters like iron and phosphorus. Field analysis of arsenic was done using Arsenic Test Kit (Hach Company, Loveland, CO), which has a detection limit of $5\,\mu g\,L^{-1}$. The equipment uses visual comparison test and is ideal for use almost anywhere that trace amounts of total inorganic arsenic must be quantified. This kit uses safe, easy-to-handle reagents packaged in unit doses, with a test strip to determine the final result.

In this method, hydrogen sulfide is first oxidized to sulfate to prevent interference, and the oxidizing environment is then neutralized. Sulfonic acid and powdered zinc react to create strong reducing conditions in which inorganic arsenic is reduced to arsine gas (AsH_3). The arsine gas then reacts with mercuric bromide in the test strip to form mixed arsenic/mercury halogenides that discolor the test strip. The color ranges from yellow through tan to brown, depending on the concentration.

3 RESULTS AND DISCUSSION

3.1 *Arsenic contamination levels*

As earlier reported by Kassenga and Mato (2008), arsenic was detected in 58% of the water sources measured. This means that they had arsenic concentration exceeding $5\,\mu g\,L^{-1}$, which was the detection limit of the equipment used. The actual data indicated a range of 5 to $70\,\mu g\,L^{-1}$ with Maswa district showing the highest and Bariadi the lowest concentrations (see Figure 2). Moreover, more than 40% of the analysed water sources had arsenic concentrations exceeding the Tanzania Drinking Water Quality Standards threshold value, which is $10\,\mu g\,L^{-1}$.

Despite of showing high values of arsenic, Maswa district had no large scale mining activity, though artisanal mining has been going on in the areas for a long time. The districts of Kahama and Geita, which had large gold mines (Kaham and Geita gold mines), indicated a fairly low arsenic values of 5–30 $\mu g L^{-1}$. The lowest arsenic contamination levels (5–10 $\mu g L^{-1}$) were obtained in Bukombe and Bariadi districts, which did not have large scale mining activities. Nevertheless, arsenic contamination in the study area was observed to be less serious compared to the situation in other parts of the world reported elsewhere like Ghana, Bangladesh and USA (Welch *et al.*, 2000; Smith *et al.*, 2000; Smedley, 1996; Mukherjee *et al.*, 2006).

Kassenga & Mato (2008) argued that the fact that the observed arsenic concentrations in the study area in Tanzania are inconsequential compared to those reported in other gold mining areas should not be the reason for complacency. Efforts should be done to protect the rapidly increasing population in the Lake Victoria areas especially that arsenic is carcinogenic even at very low concentrations and the effect is cumulative (even at the WHO guideline of 10 $\mu g L^{-1}$, the health risk is still 0.2%).

Probably the most serious threat of arsenic contamination in water sources is increased mining activities in the Lake Victoria basin due to improved mining technologies and involvement of many more people in the sector. The current government drive towards industrialization will expand more the small and large scale mining activities that can be preceded by intensive mineral exploration activities. Unfortunately there exist inadequate environmental management practices of the tailing dumps and exploration holes. The problem is pronounced for small and medium scale mining companies which do not have sufficient resources to install appropriate mitigation measures to manage tailing dumps. Since there are no comprehensive studies, which have been undertaken to reveal the status of arsenic contamination of water sources in relation to on-going mining activities, it is difficult to tell the extent of the problem. However, referring to experiences in other countries, there is a serious potential public health risks due to arsenic contamination. The problem can be magnified by the fact that in the Lake Basin more than 67% of the households live more than a kilometre from a safe drinking water source (URT, 2002); and, that the Lake Victoria is generally dry with high evaporation rates (Sutcliffe & Petersen, 2007) thus there is a possibility of concentrating the contaminant in water sources.

The conclusion by Kassenga & Mato (2008) is that if left unattended the arsenic contamination problem in the Lake Victoria mining areas in Tanzania could sooner than later reach a crisis level. Since comprehensive studies to map arsenic contamination levels could prove to be costly and time demanding, a search for safe water sources can be a sustainable solution. Similar approaches have been adopted for other water sources contaminants by identifying sources with low or safe arsenic levels.

4 CONCLUSION

Arsenic contamination in Lake Victoria mining areas in Tanzania is a real public health threat. The problem is exacerbated by increased rates of mining activities in the areas and ever increasing population exposed to arsenic contamination. Searching for alternative safe water sources could save more lives especially in the poor rural areas.

REFERENCES

Bhattacharya, P., Sracek, O., Eldvall, B., Asklund, R., Barmen, G., Jacks, G., Koku, J., Gustafsson, J.E., Singh, N. & Brokking, B.B. 2012. Hydrogeochemical Hydrogeochemical study on the contamination of water resources in a part of Tarkwa mining area, Western Ghana. *J. Afr. Ear. Sci.* 66–67: 72–84.

Department of Geology. 1994. Study on mercury and other heavy metal pollution in gold mining areas around Lake Victoria. (Unpublished report). University of Dar es Salaam, Dar es Salaam.

Halsey, P.M. 2000. Arsenic Contamination Study of Drinking Water in Nepal. PhD Thesis. Massachusetts Institute of Technology, Boston. p. 73.

Kassenga, G.R. & Mato, R.R.A.M. 2008. Arsenic contamination levels in drinking water sources in mining areas in Lake Victoria Basin, Tanzania, and its removal using stabilized ferralsols. *Int. J. Biol. Chem. Sci.* 2(4): 389–400.

Mukherjee, A., Sengupta, M.K., Hossain, M.A., Ahamed, S., Das, B., Nayak, B, Lodh, D., Rahman, M.M. & Chakraborti, D. 2006. Arsenic contamination in groundwater: a global perspective with emphasis on the Asian Scenario. *J. Heal. Popul. Nutr.* 24(2): 142–163.

Smedley, P.L. 1996. Arsenic in rural groundwater in Ghana. *J. Afr. Earth. Sci.* 22: 459–470.

Smith, A.H., Lingas, E.O. & Rahman, M. 2000. Contamination of drinking water by arsenic in Bangladesh: a public health emergency. *Bulletin of the World Health Organization* 78(9): 1093–1103.

Sutcliffe, J.V. & Petersen, G. 2007. Lake Victoria: derivation of a corrected natural water level series. *Hydrol. Sci. J.* 52(6): 1316–1321.

URT (United Republic of Tanzania). 2002. Population and Housing Census. National Bureau of Statistics: Dar es Salaam.

URT (United Republic of Tanzania). 2012. Population and Housing Census. National Bureau of Statistics: Dar es Salaam.

Welch, A.H., Westjohn, D.B., Helsel, D.R. & Wanty, R.B. 2000. Arsenic in ground water of the United States: Occurrence and geochemistry. *Ground Water* 38: 589–604.

World Health Organisation (WHO). 2001. *Arsenic and Arsenic Compounds*. Second edition. Environmental Health Criteria 224, World Health Organization: Geneva.

1.6 Arsenic mobility and fate in contaminated soils and sediments

Environmental Arsenic in a Changing World –
Zhu, Guo, Bhattacharya, Ahmad, Bundschuh & Naidu (Eds)
ISBN 978-1-138-48609-6

Effect of humic acid on microbial arsenic reduction in anoxic paddy soil

X.M. Li, J.T. Qiao & F.B. Li
Guangdong Key Laboratory of Integrated Agro-environmental Pollution Control and Management,
Guangdong Institute of Eco-Environmental Science & Technology, Guangzhou, P. R. China

ABSTRACT: In this study, humic acid (HA) was extracted from peat soil and added to an arsenic-contaminated paddy soil for microcosm experiments to investigate its effect on microbial As(V) reduction in the anoxic paddy soil. The concentration of dissolved As(III) with amendment of HA increased gradually over time and, at the end of the incubation, reached 4.2 mg L^{-1}, which is three times that of the control without amendment. Total RNAs were extracted to profile the potentially active microbial communities and suggested that HA increased the abundance of *Variovorax*, *Azoarcus*, and *Anaeromyxobacter* sp. relative to the control. Transcriptions of the respiratory As(V)-reducing gene *arrA* (10^6 copies mL^{-1}) were significantly higher than those of the cytoplasmic As(V)-reducing gene *arsC* (10^4 copies mL^{-1}). The addition of HA increased the transcripts of *arsC* only during Day 2–6, but increased those of total bacteria and *arrA* during the whole time-course incubation. Our results indicated that the HA extracted from soil enhanced As(V) reduction primarily via up-regulated expression of respiratory As(V)-reducing gene.

1 INTRODUCTION

Arsenic (As) in paddy fields is of great concern due to its high bioavailability to rice and subsequent bioaccumulation in human via dietary intake of rice. Flooding conditions in paddy fields can enhance As(V) reduction, driven by cytoplasmic As(V)-reducing and respiratory As(V)-reducing microorganisms. Humic substances make up the largest single class of dissolved organic matter in the natural environment and can significantly influence microorganism activity and arsenic fate in soils. However, little is known about its effect on microbial As(V) reduction in anoxic paddy soil. In this study, humic acid (HA) was extracted and added to arsenic-contaminated paddy soil to investigate its effect on arsenic transformation, the associated microbial community and transcription levels of As(V)-reducing genes.

2 EXPERIMENTAL METHODS

2.1 *Microcosm experiment and Arsenic speciation*

The arsenic-contaminated soil was collected from a paddy field in Shantou City (Guangdong Province, PRC). HA was extracted from a peat soil (View Sino International Ltd) using the standard procedure described by International Humic Substances Society (Swift, 1996), and then freeze-dried before use.

Anaerobic microcosm experiments were conducted in serum vials with 70 mL $Na_2CO_3/NaHCO_3$ solution (30 mM, pH 7.3), 7 g (wet weight) paddy soil, 1 mL l^{-1} trace element solution, 1 mL l^{-1} vitamin solution with and without 1% HA at 30°C in an anaerobic chamber in the dark. At intervals, dissolved As(III)/As(V) was determined from the supernatant of each culture using atomic florescence spectroscopy (SA-20, Jitian Inc., Beijing, China), and the pellets were used for RNA extraction.

2.2 *RNA extraction and microbial community characterization*

Total RNA was extracted using a MoBio PowerSoil™ Total RNA Isolation Kit (Mio Bio Laboratories, USA). The RNA was reverse transcribed using a PrimeScript RT reagent kit with gDNA Eraser (Takara, Shiga, Japan). The cDNA samples were amplified with primers specific to V4 region of 16S rRNA gene using primers 515F and 806R for further high-throughput sequencing. Data from 16S rRNA amplicon libraries were processed using the Quantitative Insights Into Microbial Ecology (QIIME 1.8.0) toolkit.

2.3 *Transcription and statistical analyses*

The cDNA was amplified with primers specific to the *arsC* gene using primers amlt-42-F/amlt-376-R and smrc-42-F/smrc-376R. PCR amplification of the *arrA* gene was achieved with primers *arrA*-CV1F/*arrA*-CV1R. Quantification of transcripts of 16S rRNA, *arrA* and *arsC* genes were performed on an iQ™ 5 Multicolor Real-Time PCR Detection System (Bio-Rad Laboratories, USA) using SYBR Green I detection method. Statistical significance of differences between HA amended and control microcosms was determined via one-way analysis of variance (ANOVA) followed by Duncan's multiple range tests in SPSS 22.0, and with $P < 0.05$ considered to be statistically significant.

3 RESULTS AND DISCUSSION

3.1 *Arsenic transformation*

At Day 0, concentrations of dissolved As(III) were very low (<0.001 mg l^{-1}) in both HA-amended and control microcosms. In the control without amendment, dissolved As(III) began to increase after Day 6 and reached 1.7 ± 0.1 mg L^{-1} at the end of the incubation (Day 30). With the HA amendment, the concentrations of dissolved As(III) increased after Day 0 and at the end were 4.2 ± 0.2 mg L^{-1}, which is three times that of the control. Concentrations of dissolved As(V) showed a similar trend in both HA-amended and control treatments: firstly increasing from 0.2 ± 0.1 mg L^{-1} on Day 0 to 0.4 ± 0.1 mg L^{-1} on Day 6, and then decreasing over time during the rest of the incubation. These results indicated that As(V) from the solid phase could be released into soil solution and reduced to As(III) in the soil without any amendment, while the addition of HA could facilitate the reduction of As(V) (mainly in solid phase) resulting in the enhancement of As(III) release into the soil solution.

3.2 *Composition of active bacterial communities*

In both HA-amended and control treatments, *Variovorax* followed by *Azoarcus* were the most abundant species during the incubation. The abundance of *Variovorax* sp. in the control increased from 1–8% during Day 0–2 to 20–22% during the rest of incubation, suggesting that *Variovorax* sp. could be stimulated in anoxic paddy soil without any amendment. In contrast, during the whole incubation time, the abundance of *Variovorax* maintained a high level in the HA-amemded microcosm (17–28%). The abundance of *Azoarcus* sp. in the HA-amended treatment was 6–10% during Day 0–2, which was higher than those in the control (0–4%). However, its abundance in both the HA-amended and control treatments decreased in a similar manner as time elapsed during the rest of incubation. Compared with control (<1%), HA had a specific stimulation on the abundance of *Anaeromyxobacter* and its abundance increased gradually over time from $1 \pm 0.03\%$ on Day 0 to $4 \pm 0.1\%$ on Day 30.

V. paradoxus has been reported to be resistant to many metals/metalloids including arsenic (Belimov *et al.*, 2005), and the *ars* operon has been found in the whole-genome of *V. paradoxus* S110 (Han *et al.*, 2011). Some *V. paradoxus* can grow with HA as the sole carbon source (Filip & Bielek, 2002). *Azoarcus* sp. has not shown any ability of As(V) reduction so far, but its genome analysis reveals the existence of genes related to the *ars* operon (Faoro *et al.*, 2017). *Anaeromyxobacter* sp. has been frequently detected in arsenic-contaminated areas and *Anaeromyxobacter* sp. PSR-1 can use As(V) as electron acceptor for repiration (Kudo *et al.*, 2013). Our results suggested that both *Variovorax* and *Azoarcus* sp. were the dominant indigenous microorganisms in the anoxic paddy soil, and HA addition only stimulated the *Variovorax* and *Azoarcus* sp. at the beginning of incubation, but stimulated the *Anaeromyxobacter* sp. at the later stages of incubation.

3.3 *Transcription of functional genes*

Transcriptions of total bacteria increased rapidly from 0.1×10^8 copies mL^{-1} on Day 0 to 10×10^8 and 5×10^8 copies mL^{-1} on Day 2 in the HA-amended and control treatments, respectively, and then decreased over time to 7×10^8 and 2×10^8 copies mL^{-1} on Day 30. The transcriptions of the *arrA* gene in the control were constant at 0.2×10^6-0.7×10^6 copies mL^{-1} during Day 2–30, while those in the HA-amended treatment increased gradually from 0.4×10^6 copies mL^{-1} on Day 2 to 2×10^6 copies mL^{-1} on Day 30. The differences of transcriptions of total bacteria and *arrA* genes between these two treatments were significant ($P < 0.05$). However, transcriptions of *arsC* gene (3.3×10^4 and 4×10^4 copies mL^{-1}) in the HA-amended treatment were significantly higher than those (2×10^4 and 1.5×10^4 copies mL^{-1}) in the control only on Day 2 and Day 6, respectively.

4 CONCLUSIONS

The HA extracted from peat soil can significantly facilitate the reduction of As(V) and release of As(III) into paddy soil solution, probably due to the increase in abundance of *Variovorax*, *Azoarcus*, and *Anaeromyxobacter* sp. as well as the respiratory As(V)-reducing gene *arrA*.

ACKNOWLEDGEMENTS

This work was supported by the National Science Foundation of China (41330857 and 41571130052) and the Guangdong Natural Science Funds for Distinguished Young Scholars (2017A030306010).

REFERENCES

Belimov, A.A., Hontzeas, N. & Safronova, V.I. 2005. Cadmium-tolerant plant growth-promoting bacteria associated with the roots of Indian mustard (Brassica juncea L. Czern.). *Soil Biol. Biochem.* 37(2): 241–250.

Faoro, H., Rene, M.R. & Battistoni, F. 2017. The oil-contaminated soil diazotroph *Azoarcus olearius* DQS-4(T) is genetically and phenotypically similar to the model grass endophyte *Azoarcus* sp. BH72. *Environ. Microbiol. Rep.* 9(3): 223–238.

Filip, Z. & Bielek, P. 2002. Susceptibility of humic acids from soils with various contents of metals to microbial utilization and transformation. *Biol. Fert. Soils.* 36(6): 426–433.

Han, J.I., Choi, H.K. & Lee, S.W. 2011. Complete genome sequence of the metabolically versatile plant growth-promoting endophyte *Variovorax paradoxus* S110. *J. Bacteriol.* 193(5): 1183–1190.

Kudo, K., Yamaguchi, N. & Makino, T. 2013. Release of arsenic from soil by a novel dissimilatory arsenate-reducing bacterium, *Anaeromyxobacter* sp. strain PSR-1. *Appl. Environ. Microbiol.* 79(15): 4635–4642.

Swift, R.S. 1996. Organic matter characterization. In D.L. Sparks (eds). *Methods of Soil Analysis. Part 3. Chemical methods.* Soil Science Society of America Madison: WI. pp. 1018–1020.

Environmental Arsenic in a Changing World –
Zhu, Guo, Bhattacharya, Ahmad, Bundschuh & Naidu (Eds)
ISBN 978-1-138-48609-6

Arsenic relative bioavailability in contaminated soils: comparison of animal models, dosing schemes, and biological endpoints

H.B. Li[1], J. Li[1], L.Q. Ma[1] & A.L. Juhasz[2]
[1]*State Key Laboratory of Pollution Control and Resource Reuse, School of the Environment, Nanjing University, Nanjing, P.R. China*
[2]*Future Industries Institute, University of South Australia, Mawson Lakes, South Australia, Australia*

ABSTRACT: Different animals and biomarkers have been used to measure arsenic (As) relative bioavailability (RBA) in contaminated soils. However, there is a lack of comparison of As-RBA based on different animals (i.e., swine and mouse) and biomarkers [area under the blood As concentration curve (AUC) after a single gavaged dose vs. steady state As urinary excretion (SSUE) and As accumulation in liver and kidneys after multiple doses via diet]. In this study, As-RBA in 12 As-contaminated soils with known As-RBA via swine blood AUC model were measured using mouse blood AUC, SSUE, liver and kidneys analyses. Arsenic-RBA for the four mouse assays ranged 2.8–61%, 3.6–64%, 3.9–74%, and 3.4–61%, respectively. Compared to swine blood AUC assay (7.0–81%), though well correlated ($R^2 = 0.83$), the mouse blood AUC assay generally yielded lower values. Similarly, strong correlations were observed between mouse blood AUC and mouse SSUE (R2 = 0.86) and between urine, liver, and kidneys ($R^2 = 0.75$–0.89), suggesting As-RBA values were congruent among different animals and endpoints. Selection of animals and biomarkers would not significantly influence the role of *in vivo* assays to validate *in vitro* assays. Based on its simplicity, mouse liver or kidney assay following repeated doses of soil-amended diet is recommended for future As-RBA studies.

1 INTRODUCTION

Reliable assessment of human health risks from the ingestion of As-contaminated soil depends not only on total As concentration, but also its bioavailability. As such, various animal bioassays have been developed to quantify As relative bioavailability (RBA, relative to the absorption of sodium arsenate) in contaminated soils. As a measure of As-RBA, different biomarkers (As concentration in blood, kidneys, liver, or urine) have been used to determine As absorption following a single gavaged dose or multiple repeated doses of As-contaminated soil via diet. In addition to different animal models, varying feeding schemes have been employed to measure As-RBA in contaminated soil. While different animal models and feeding schemes have been used to assess As-RBA, comparative studies detailing the influences of these parameters on As-RBA measurement are lacking.

2 METHODS/EXPERIMENTAL

2.1 *Arsenic-contaminated soils*

Twelve contaminated soils (42 to 1114 mg As kg^{-1}) that had been measured for As-RBA using swine incorporating a single gavaged dose and area under blood As time curve (AUC) analysis were used in this study (Juhasz *et al.*, 2009).

2.2 *Determination of As relative bioavailability using mouse bioassays*

Arsenic-RBA in the soils was determined using mouse bioassays using different dosing regimens and biological endpoints. Initially, As-RBA was assessed in fasted animals following administration of a single gavaged dose of sodium arsenate and soil suspension. Sodium arsenate was used as the reference dose. Blood samples of control and treated mice were collected at 4, 8, 16, 24, and 48 h following gavage for As analysis to develop the blood As time curve. Arsenic-RBA was calculated by dividing the AUC following soil administration by the AUC following sodium arsenate administration following dose normalization.

In addition to AUC analysis, As-RBA in contaminated soils was also assessed using steady state As urinary excretion and As accumulation in liver and kidneys following multiple repeated As doses via diet. Prior to *in vivo* assays, soil and sodium arsenate were incorporated into the mouse basal diet. Each soil- or arsenate-amended diet (3 g each mouse per day) was provided to mice (n = 3) daily at 9 am for 10 d. Following 8 d of exposure, mice were transferred to metabolic cages and urine collected each morning for the last two days of exposure and analyzed for As. At the end of the 10-d exposure period, the mice were sacrificed, and the liver and kidneys samples were collected for As analysis. When urine was used as the endpoint, As-RBA in soil was calculated as the ratio of the urinary As

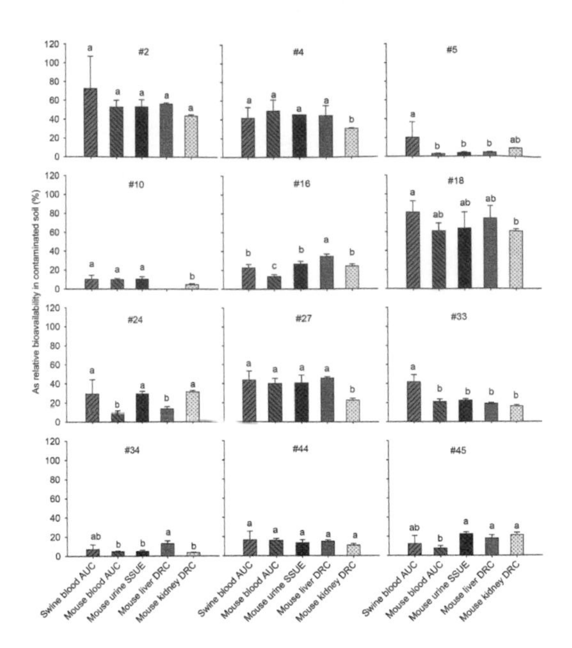

Figure 1. Arsenic relative bioavailability in 12 contaminated soils measured using swine blood AUC and mouse blood AUC models following a single gavage dose to fasted animals, mouse steady state urinary excretion (SSUE), and mouse liver and kidney dose response curve (DRC) models following multiple doses to soil-amended diets for 10 d.

excretion fraction for soil exposure to that for sodium arsenate exposure. When liver or kidneys were used as the endpoint, Arsenic-RBA was calculated as the ratio of dose normalized As accumulation in liver/kidneys following 10-d exposure to soil-amended diet to that following 10-d arsenate-amended diet exposure.

3 RESULTS AND DISCUSSION

3.1 *Comparison of As-RBA by swine and mouse blood AUC bioassays*

Arsenic RBA in the 12-contaminated soils varied from 2.8 to 61% using the mouse AUC assay (Fig. 1). Compared to the swine AUC model, the mouse AUC assay generally yielded lower As-RBA for 11 out of the 12 soils. However, the differences between the two animal models were insignificant for most samples. A strong linear relationship ($R^2 = 0.83$) was observed for As-RBA based on swine and mouse models. The difference in As-RBA obtained by swine and mouse models was mainly caused by their different physiological structures, which affect As absorption.

3.2 *Comparison of As-RBA by different endpoints of mouse bioassays*

When urine was used as the biomarker of As exposure in mice, As-RBA in contaminated soils ranged from 3.6 to 64%. When mouse liver and kidneys were used as exposure end points, As-RBA in contaminated soils ranged from 3.9 to 74% and 3.4 to 61%, respectively, with generally lower values using kidneys. There was a significant correlation between As-RBA using kidneys and liver with $R^2 = 0.75$. In addition, As-RBA determined using mouse liver and kidneys was correlated with that determined using urine as the endpoint ($R^2 = 0.88$–0.89).

3.3 *Comparison of As-RBA by different dosing approaches of mouse bioassays*

For AUC (single gavage dose) and SSUE (multiple doses via diet consumption) models, a strong linear relationship was observed ($R^2 = 0.86$). Similarly, strong linear correlations ($R^2 = 0.88$ and 0.69) were found between As-RBA determined using the mouse AUC and mouse liver or kidney assays.

4 CONCLUSIONS

The results suggested that selection of different animals probably would not significantly influence the role of *in vivo* assays to validate *in vitro* assays. Though swine is the preferred animal model for As-RBA determination, mice can be used as a surrogate animal model due to its significant correlation with swine model in addition to its low cost and ease of handling.

In addition, different dosing approaches and endpoints did not significantly affect the *in vivo-in vitro* correlations. Compared to the single gavaged dose, which needs to establish the blood As time curve using large animal sizes, the repeated soil dose via diet offered the advantage of small numbers of animals required to determine As-RBA. For multiple dosing schemes, As in liver or kidneys is a simpler biological endpoint compared to urine. In short, based on this study, the mouse liver or kidney assay following repeated doses of soil-amended diet to mice is recommended for future As-RBA studies.

ACKNOWLEDGEMENTS

This work was supported by National Natural Science Foundation of China (21507057; 41673101).

REFERENCE

Juhasz, A.L., Weber, J., Smith, E., Naidu, R., Rees, M., Rofe, A., Kuchel, T. & Sansom, L. 2009. Assessment of four commonly employed in vitro arsenic bioaccessibility assays for predicting in vivo arsenic relative bioavailability in contaminated soils. *Environ. Sci. Technol.* 43(24): 9887–9894.

Environmental Arsenic in a Changing World –
Zhu, Guo, Bhattacharya, Ahmad, Bundschuh & Naidu (Eds)
ISBN 978-1-138-48609-6

Facilitated release of arsenic from polluted sediment in Plateau Lakeshore Wetland by phosphorus input

R. Zhao[1,2], L. Hou[1,2] & Y.G. Liu[1,2]
[1]*Research Institute of Rural Sewage Treatment, Southwest Forestry University, Kunming, P.R. China*
[2]*College of Ecology and Soil & Water Conservation, Southwest Forestry University, Kunming, P.R. China*

ABSTRACT: The effect of exogenous phosphorus on the release of arsenic in the sediment of plateau lakeshore wetland deserves further study. The sediment samples in Yangzonghai Lakeshore Wetland which is located in Yunnan Province was selected as the research object, and the batch experiment method was carried out to monitor the release kinetics of arsenic. It was found that phosphorus input would facilitate the release of arsenic in the short period (less than 3 d), but the release would tend to be minimum after long period (30 d). After the input of phosphorus, the bioavailability of arsenic in the sediment would increase, whereas the sorption capacity for phosphorus decreased.

1 INTRODUCTION

Arsenic (As) contamination is a worldwide environmental problem, which attracts concerns from many scientists as its high carcinogenic activities (Bhattacharya *et al.*, 2007; Islam *et al.*, 2000). Plateau Lake under the arsenic pollution is worth more research due to the high risk to the people in the catchment and also the difficulties of control. Lakeshore wetland is the natural screen for the retention of contaminants, especially the two-side effect of the sediment, which could both be the sink and the source of the contaminants. When the exogenous phosphorus entered into the arsenic polluted wetland, what would happen for the lake environment deserves much more researches.

2 MATERIALS & METHODS

2.1 *Sediment sample*

Yangzonghai Lakeshore Wetland which is located in Yunnan Province was selected as the research object, and the surface layers (0–10 cm) of the sediment samples were collected. Several basic properties of the sediment was determined, such as pH, sediment texture, contents of organic matter, total phosphorus and arsenic.

2.2 *Release of arsenic by phosphorus input*

The experiment was carried out by batch method. The sediment sample was freeze-dried and sieved with 1 mm mesh. Specific amount of sediments was put into a series of vials, and different concentrations of phosphorus (KH_2PO_4) solutions were added into the vials as 0, 1, 5, 10, 20, 40, 60, and 100 mg L^{-1}. The release kinetics of arsenic were monitored at the following time points as 1, 2, 4, 10, 24, 48, 72, 168, 360, and 720 h. The concentrations of arsenic both in the water and sediment was measured, and the sequential extraction method was used for the determination of speciation of arsenic and inorganic phosphorus in the sediments. Duplicates were conducted for all the experiment and sterilization was avoided as the real simulation of wetland environment was designed.

Table 1. Selected properties of arsenic polluted sediment.

Properties	Values
pH	8.13
Sand (%)	37.2
Silt (%)	55.4
Clay (%)	7.41
OM (g kg^{-1})	25.7 ± 0.811
TP (mg kg^{-1})	473 ± 48.4
TAs (mg kg^{-1})	19.8 ± 3.66

3 RESULTS AND DISCUSSION

3.1 *Basic properties of the sediment samples*

Some of the selected properties of the arsenic polluted sediments were listed in Table 1. More than half of the total arsenic existing in the residue state (10.2 ± 0.108 mg kg^{-1}).

3.2 *Facilitated release of arsenic*

Release of arsenic was observed after the input of phosphorus, and it changed significantly when compared the short and long periods (Fig. 1). For the short period, the release of arsenic increased with the

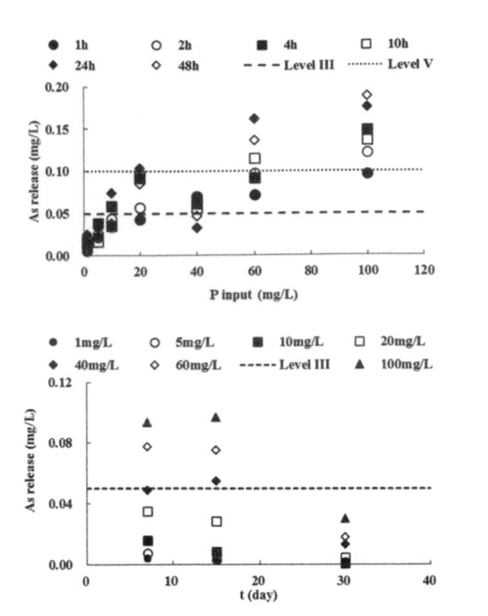

Figure 1. Facilitated release of arsenic from sediment by phosphorus input, the above is for short period (less than 3 days) and the below if for relatively long period (30 days).

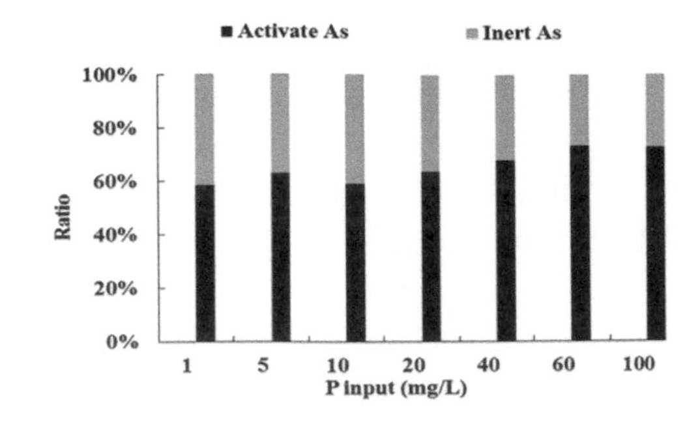

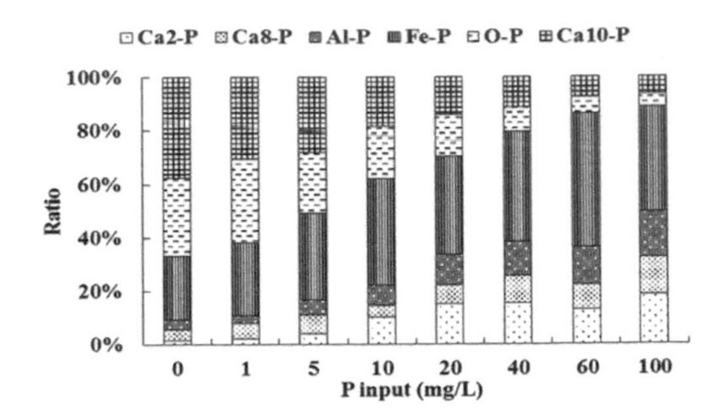

Figure 2. Speciation of arsenic and inorganic phosphorus in the sediment after phosphorus input for 30 d.

increasing concentrations of input amount of phosphorus, and when the concentration of phosphorus increased to 10 mg L^{-1}, the concentration of arsenic in the water exceeded the level III limit for environmental quality standard for surface water after 4 h. When the concentration went to 60 mg L^{-1}, the concentration of arsenic exceeded the level V limit after 10 h.

Interestingly, for the relatively long period, the concentration of arsenic in the water all showed the trends of decreasing. After 30 d, the concentrations of arsenic all went below the level III limit, regardless of the phosphorus input amount. The results might attributed to the counter balance of arsenic between water and sediment phases, and the microbes might play a significant role which deserved further exploration.

3.3 *Speciation of arsenic and phosphorus in the sediment*

The speciation of the arsenic and inorganic phosphorus was monitored after the input of phosphorus. The data at 30 days were showed in Figure 2. After the input of phosphorus, the activities of the arsenic in the sediment showed a shift trend from inert to activate, and the ratio changed from 51.51% to less than 40%, indicating that phosphorus input increase the relative activities of arsenic in the sediment. And the ratio of activate part of arsenic increased with the increasing input of phosphorus.

On the other hand, the ratio of inorganic phosphorus didn't changed much compared with the original sediment, although the total amount of phosphorus of the sediment increased (data not shown). Another obvious trend was shown in Figure 2 that the ratio of activated phosphorus of the sediment increased with the increasing input of exogenous phosphorus, which indicates that the worse eutrophication not only increased the bioavailability of the arsenic in the sediment but also decreased its phosphorus sorption capability (Rubinos *et al.*, 2011).

4 CONCLUSIONS

Phosphorus input would facilitated the release of arsenic in the short period (less than 3 d), but the release would tend to be minimum after long period (30 d). After the input of phosphorus, the bioavailability of arsenic in the sediment would increase, whereas the sorption capacity for phosphorus decreased.

ACKNOWLEDGEMENTS

This project was supported by the National Natural Science Foundation of China (grants 31560147 and 51469030).

REFERENCES

Bhattacharya, P., Welch, A. H. & Stollenwerk, K. G. 2007. Arsenic in the environment: biology and chemistry. *Sci. Total Environ.* 379(2): 109–120.

Islam, F.S., Gault, A.G., Boothman, C., Polya, D.A., Charnock, J.M., Chatterjee, D. & Lloyd, J.R. 2000. Role of metal-reducing bacteria in arsenic release from Bengal delta sediments. *Nature.* 430(6995): 68–71.

Rubinos, D.A., Iglesias, L. & Diazfierros, F. 2011. Interacting effect of pH, phosphate and time on the release of arsenic from polluted river sediments. *Aquat. Geochem.* 17(3): 281–306.

Environmental Arsenic in a Changing World –
Zhu, Guo, Bhattacharya, Ahmad, Bundschuh & Naidu (Eds)
ISBN 978-1-138-48609-6

Quantification of arsenic adsorption and oxidation on manganese oxides

B. Rathi[1], O. Cirpka[1], J. Sun[2], J. Jamieson[2], A. Siade[2], H. Prommer[2] & M. Zhu[3]
[1]*Centre for Applied Geoscience, University Tübingen, Tübingen, Germany*
[2]*University of Western Australia and CSIRO Land and Water, Perth, Australia*
[3]*Department of Ecosystem Science and Management, University of Wyoming, Laramie, WY, USA*

ABSTRACT: Oxidation of arsenite (As(III)) by manganese (Mn) oxide minerals commonly found as coatings on aquifer sediments can be an important process controlling arsenic mobility in many aquifers. To date the mechanistic details of As(III) oxidation process have only been described qualitatively in a number of studies, however, these mechanisms vary markedly with respect to the intermediate reactions involved and the products formed. We carried out a detailed geochemical analysis of key literature datasets by translating the known reaction mechanisms into conceptual models of sequentially increasing complexities. These conceptual models were then tested for their feasibility in a numerical modelling framework. The results of this modelling exposed significant limitations in the current conceptual models. Although none of the model simulations produced accurate fit to the data, we were able to ascertain that As(III) oxidation by Mn oxides is most likely a two-step process in which the rate of first oxidation step was dependent only on a small fraction of the Mn oxide initially present in the system. The second rate of oxidation step was slower than first and was responsible for producing Mn(II) ions. The conceptual models can be further improved with data on solid phase characterisation of Mn mineralogy and speciation of adsorbed species during the experiments. This modelling framework provides a good foundation for assessing the influence of other geochemical factors on As(III) oxidation in future research.

1 INTRODUCTION

Arsenite, As(III), mobility and toxicity can be dramatically reduced due to oxidation by Mn oxides that are ubiquitous in soils and sediments (Post, 1999). Despite a large set of studies investigating As(III) oxidation by both laboratory synthesized and natural Mn oxides, the details of the proposed reaction mechanisms vary markedly with respect to the intermediate steps involved, the contribution of surface reactions, and the surface properties of Mn oxides (e.g., Oscarson *et al.*, 1983; Scott & Morgan, 1995; Nesbitt *et al.*, 1998). Furthermore, the extent of As(III) oxidation can be affected by a wide range of geochemical factors, including the solution pH, competing solutes (Fe(II), phosphate, etc.) and specific Mn mineralogy. So far, only a few attempts have been made to quantify As(III) oxidation by Mn oxides (e.g., Amirbahman *et al.*, 2006; Rathi *et al.*, 2017) and they do not follow a rigorous process-based modelling approach to interrogate the intermediate surface reactions and the prevailing geochemical conditions that influence the extent and rate of As(III) oxidation. This study aims to review the known As(III) oxidation mechanisms and to develop and evaluate suitable process-based modelling approaches against the experimental data available in the literature.

2 METHODS

2.1 *Dataset and modelling approach*

Datasets from both batch (Manning *et al.*, 2002) and flow (Lafferty *et al.*, 2010) experiments were used to test the conceptual models. Batch experiments were conducted at pH 6.50 with acid-Birnessite. Flow experiments were conducted at pH 7.20 with delta-Birnessite. Both types of Birnessites differ in their crystal size and hence the surface area. Both experiments provided data on aq. As(III) and arsenate, As(V), concentrations in additional to aqueous Mn(II) concentrations from flow experiments. The geochemical modelling code PHREEQC (Parkhurst & Appelo, 1999) and reactive transport modelling code PHT3D (Prommer *et al.*, 2003) were used to analyze experiment data. The adsorption reactions of As(III), As(V) and Mn(II) were modelled using surface complexation approach and redox transformations between As(III) and As(V) were assumed to be kinetically controlled.

2.2 *Conceptual models*

The current understanding of the oxidation process was evaluated by translating three specific sets of

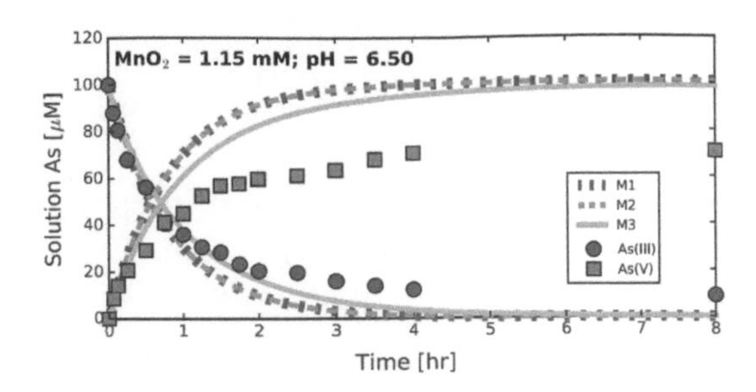

Figure 1. Simulation results of models M1 to M3 for data from laboratory batch experiments (Manning *et al.*, 2002).

processes into separate model variants and by investigating the replicability of the model output to the available data. The simplest model variant, in which oxidation was considered to occur as a one-step process, followed the reaction

$$MnO_2 + H_3AsO_3 + H^+ \rightarrow Mn^{2+} + H_2AsO_4^- + H_2O \quad (1)$$

and was defined as the base model (M1). Subsequently, one or more degrees of complexity were sequentially added to consider kinetically controlled oxidation in two steps by mineral MnO_2 (s) and an intermediate product $MnOOH^*$ (s), either without (model M2) or with adsorption of aqueous species (model M3).

3 RESULTS AND DISCUSSION

3.1 *Results of models M1 and M2*

Model M1 results failed to capture the observed temporal concentration changes of all aqueous species in both batch and flow experiments (Figs. 1 and 2). The simulations of both experiments matched observed As(III) and As(V) concentrations initially followed by being underestimated for As (III) and overestimated for As(V) in the remaining duration. Aq. Mn(II) was overestimated while following an As(V):Mn(II) ratio of 1:1 throughout M1 simulation. These results suggest that the one-step oxidation model lacked key processes, which would slow As(III) adsorption over time and cause retention of a fraction of As(V) and Mn(II) on Mn oxides.

3.2 *Numerical implementation of model M2*

Model M2 incorporates two-step As(III) oxidation reactions with MnO_2 (s) and $MnOOH^*$ (s) (Nesbitt *et al.*, 1998):

$$2Mn^{IV}O_2(s) + H_3AsO_3 + H_2O \rightarrow 2Mn^{III}OOH^*(s) + H_3AsO_4 + 2H^+ \quad (2)$$

and

$$2Mn^{III}OOH^*(s) + H_3AsO_3 + 4H^+ \rightarrow 2Mn^{2+} + H_3AsO_4 + 3H_2O \quad (3)$$

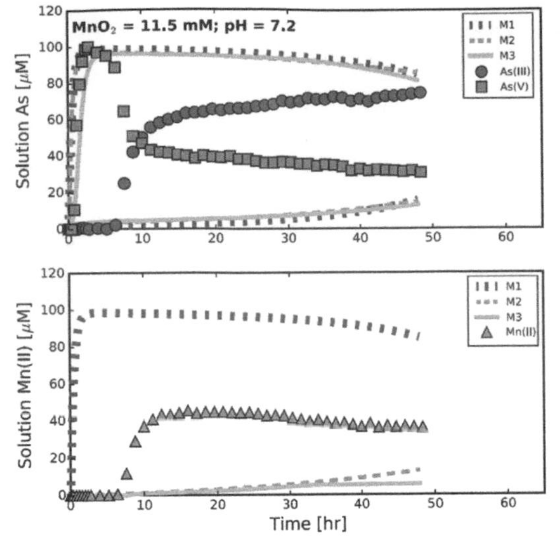

Figure 2. Simulation results of models M1 to M3 for data from laboratory flow experiment (Lafferty *et al.*, 2010).

M2 simulations produced identical results to M1 except for Mn(II) concentrations in the flow experiment where the delayed appearance of Mn(II) was captured better than M1 (Figs. 1 and 2). M2 results show that oxidation was predominantly caused by MnO_2(s) suggesting that only a fraction of the total MnO_2 (s) was required for As(III) oxidation at the onset of every experiment.

3.3 *Numerical implementation of model M3*

Model M3 was also setup by modifying model M2 to include adsorption reactions of As(III) and As(V), and the oxidation rates dependent only on the concentrations of adsorbed As(III). MnO_2 (s) and $MnOOH^*$(s) were both assumed to provide surface sites $>Mn^{IV}$ and $>Mn^{III}$, respectively:

$$2(>Mn^{IV} - OH) + H_3As^{III}O_3 + H_2O \rightarrow 2(>Mn^{III} - OH) + HAs^VO_4^{2-} + 2H^+ \quad (4)$$

and

and $2(>Mn^{III} - OH) + H_3As^{III}O_3 \rightarrow 2Mn^{2+} + HAs^VO_4^{2-} + H_2O + 4H^+$ (5).

The results of the M3 yielded a slightly improved calibration of the aq. As(III) and As(V) concentrations observed in the batch experiment, but failed to reproduce the results of the flow experiment (Figs. 1 and 2). Oxidation was predominantly caused by MnO_2 (s) and surface passivation was minor.

4 CONCLUSIONS

The numerical implementation of all conceptual models (M1 to M3) showed some major limitations in the reactions mechanisms proposed in the literature. None of the models could simulate the observed arsenic or Mn(II) data and point towards the possibility that only a limited fraction of the Mn oxides effectively participates in the reactions. A detailed information on the solid phase characterisation of Mn mineralogy and speciation of adsorbed species could provide more data to improve our understanding of As(III) oxidation mechanism.

The numerical model that simulates As(III) oxidation by Mn oxides is relevant for a wide range of scenarios in natural groundwater systems. High abundance of Mn oxide coatings on natural sediments inadvertently impose significant influence on arsenic mobility in groundwater. This study has explored interaction between arsenic and pure Mn oxides and provides a good foundation for assessing the influence of other solution species or chemical factors in future research. Development of a modelling framework based on adsorption and surface oxidation mechanisms supports a more refined understanding of the complex interaction between arsenic and Mn oxides.

ACKNOWLEDGEMENTS

We acknowledge the financial support provided by UWA and CSIRO through postgraduate scholarship and travel funding.

REFERENCES

Amirbahman, A., Kent, D.B., Curtis, G.P. & Davis, J.A. 2006. Kinetics of sorption and abiotic oxidation of arsenic(III) by aquifer materials. *Geochim. Cosmochim. Acta* 70: 533–547.

Lafferty, B.J., Ginder-Vogel, M. & Sparks, D.L. 2010. Arsenite oxidation by a poorly crystalline manganese-oxide 1. stirred-flow experiments. *Environ. Sci. Technol.* 44: 8460–8466.

Manning, B.A., Fendorf, S.E., Bostick, B. & Suarez, D.L. 2002. Arsenic(III) oxidation and Arsenic(V) adsorption reactions on synthetic birnessite. *Environ. Sci. Technol.* 36: 976–981.

Nesbitt, H.W., Canning, G.W. & Bancroft, G.M. 1998. XPS study of reductive dissolution of 7Å-birnessite by H3AsO3, with constraints on reaction mechanism. *Geochim. Cosmo-chim. Acta* 62: 2097–2110.

Oscarson, D.W., Huang, P.M., Liaw, W.K. & Hammer, U.T. 1983. Kinetics of oxidation of arsenite by various manganese dioxides. *Soil Sci. Soc. Am. J.* 47: 644–648.

Parkhurst, D.L. & Appelo, C.A.J. 1999. User's guide to PHREEQC (version 2) a computer program for speciation, batch-reaction, one-dimensional transport, and inverse geo-chemical calculations. Open-File Reports Vol. XIV, U.S. Geological Survey: Earth Science Information Center.

Post, J.E. 1999. Manganese oxide minerals: Crystal structures and economic and environmental significance. *Proc. Natl. Acad. Sci. U.S.A.* 96: 3447–3454.

Prommer, H., Barry, D.A. & Zheng, C. 2003. MOD-FLOW/MT3DMS-based reactive multicomponent transport modeling. *Ground Water*. 41: 247–257.

Rathi, B., Neidhardt, H., Berg, M., Siade, A. & Prommer, H. 2017. Processes governing arsenic retardation on Pleistocene sediments: Adsorption experiments and model-based analysis. *Water Resour. Res.* 53: 4344–4360.

Scott, M.J. & Morgan, J.J. 1995. Reactions at Oxide Surfaces. 1. Oxidation of As(III) by Synthetic Birnessite. *Environ. Sci. Technol.* 29: 1898–1905.

Environmental Arsenic in a Changing World –
Zhu, Guo, Bhattacharya, Ahmad, Bundschuh & Naidu (Eds)
ISBN 978-1-138-48609-6

A novel MAs(III)-selective ArsR transcriptional repressor

J. Chen, V.S. Nadar & B.P. Rosen
Department of Cellular Biology and Pharmacology, Herbert Wertheim College of Medicine, Florida International University, Miami, Florida, USA

ABSTRACT: All heavy metals can lead to toxicity and oxidative stress when taken up in excessive amounts, imposing a serious threat to the environment and human health. Expression of genes for resistance to heavy metals and metalloids is frequently transcriptionally regulated by the toxic ions themselves. Arsenic is a ubiquitous, naturally occurring toxic metalloid widely distributed in soil and groundwater. Microbes biotransform both pentavalent (As(V)) and trivalet (As(III)) arsenic into more the toxic methylated metabolites methylarsenite acid (MAs(III)) and dimethylarsenite (DMAs(III)). Environmental arsenic is sensed by members of the ArsR/SmtB family of metalloregulatory proteins. The arsR gene is autoregulated and is typically part of an operon that contains other ars genes involved in arsenic detoxification. To date, every identified ArsR is regulated by inorganic As(III). Here we described a novel ArsR from Shewanella putrefaciens that is specific for MAs(III) and does not respond to As(III). SpArsR has three conversed cysteine: Cys83, Cys101 and Cys102. Substitutions of Cys83 with serine has no effect for ArsR activity. However, mutation of C101S and C102S mutants lost MAs(III) binding affinity, which indicates that these two cysteines are required for MAs(III) binding. SpArsR can be converted into an As(III)-responsive repressor by introduction of an additional cysteine that allows for 3-coordinate As(III) binding. Our results indicate that SpArsR evolved selectivity for MAs(III) over As(III) in order to control expression of genes for MAs(III) detoxification.

1 INTRODUCTION

Nearly every organism has genetic mechanisms for arsenic resistance. Their genes are usually organized in *ars* operons, which are found either on the chromosome or on plasmids (Zhu *et al.*, 2014). Typically *ars* operons are regulated by an ArsR As(III)-responsive transcriptional repressor. Recently a parallel biocycle for organoarsenicals has been identified (Chen *et al.*, 2014). In response to this environmental pressure, other bacteria have evolved mechanisms to detoxify MAs(III). These organoarsenical detoxification genes are in *ars* operons regulated by a homodimeric As(III)-responsive ArsR repressor. Three different ArsRs have been identified that have three types of As(III) binding sites (Ordóñez *et al.*, 2008). The ArsR repressor encoded by *Escherichia coli* plasmid R773 (termed a Type 1 site). Two other ArsR orthologs with different As(III) binding sites were identified in the *ars* operon of *Acidithiobacillus ferrooxidans* and *Corynebacterium glutamicum*, respectively. Each ArsR has a high-affinity As(III) binding site composed of three cysteine residues at spatially distinct locations in their three-dimensional structures (Qin *et al.*, 2007). In this study we identified a novel MAs(III)-responsive ArsR that regulates expression of the *arsP* and *arsH* MAs(III) resistance genes in *Shewanella putrefaciens* 200. SpArsR is induced most effectively by MAs(III), with little response to As(III). SpArsR appears to have two conversed Cys101 and Cys102, which were involved in MAs(III) induction. We postulate that SpArsR evolved selectivity for MAs(III) in order to respond to environmental MAs(III) produced by other soil microbes.

2 METHODS/EXPERIMENTAL

2.1 *Biosensor construction*

To analysis the ArsR regulation by arsenicals, a biosensor with *gfp* reporter was constructed. For expression of *SpArsR* from *S. putrefaciens*, plasmid pBAD-*SpArsR* was constructed. The *gfp* reporter is under control of the *SparsP* promoter and generated plasmid pACYC184-*ParsP-gfp*. Transcriptional activity of the biosensor was estimated from arsenical-responsive expression of *gfp*. Expression of *gfp* was assayed from the fluorescence of cells.

2.2 *Mutagenesis of the SpArsR gene*

SparsR mutations were generated by site-directed mutagenesis using a Quick Change mutagenesis kit. The codons for residues Cys83, Cys101 and Cys102 were changed to serine codons, generating three different single cysteine SpArsR mutants.

3 RESULTS AND DISCUSSION

3.1 *SpArsR has a MAs(III) binding site*

In ArsRs, cysteine thiolates usually form three-coordinate complexes with As(III). In AfArsR, Cys95, Cys96 and Cys102 form the three-coordinate As(III) binding site. In the homology model of SpArsR,

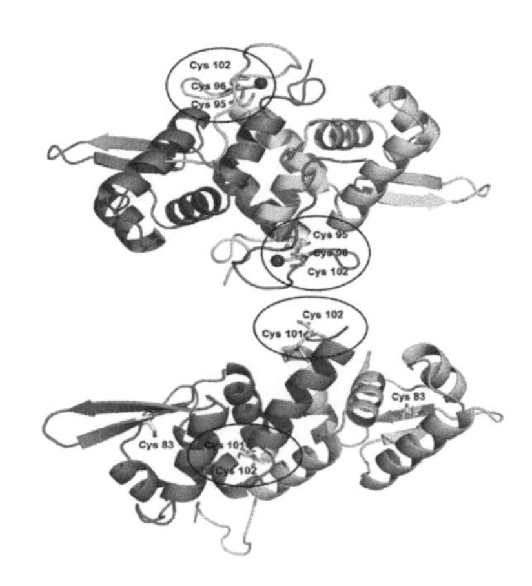

Figure 1. Homology modeling of As(III)-bound AfArsR and MAs(III)-bound SpArsR.

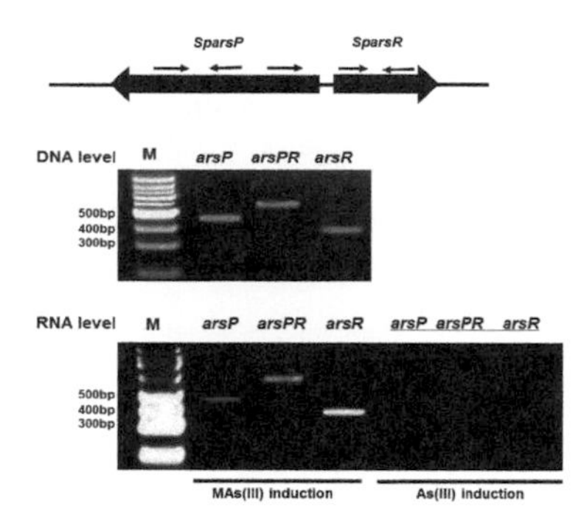

Figure 2. *SparsR* and *SparsP* are linked transcriptionally by MAs(III) induction.

Cys101 and Cys102 are superimposible with AfArsR Cys95 and Cys96. Cys83, which is found in putative orthologs but not in AfArsR, is located distant from Cys101 and Cys102 and would not be expected to contribute to inducer binding (Fig. 1).

3.2 *SpArsR is a MAs(III)-selective repressor*

Qualitative reverse transcription-PCR (RT-PCR) was used to assess *SparsR* and *SparsP* expression following exposure to As(III) or MAs(III). There was no detectable expression of either *SparsR* or *SparsP* expression following exposure to As(III) (Fig. 2). In contrast, both were expressed at high levels following exposure to MAs(III). Results showed that *SparsR* is induced by MAs(III) and not As(III) and regulates a gene for MAs(III) detoxification.

3.3 *The MAs(III) binding site in SpArsR*

Elevated *gfp* expression indicates that SpArsR is highly selective for MAs(III) binding and derepresses *gfp* expression. Cys83 is not required for MAs(III) response. Both Cys101 and Cys102 are involved in MAs(III) binding. Cells with the wild type SparsR biosensor responded only to MAs(III), and cellular *gfp* derepression was proportional to the concentration of

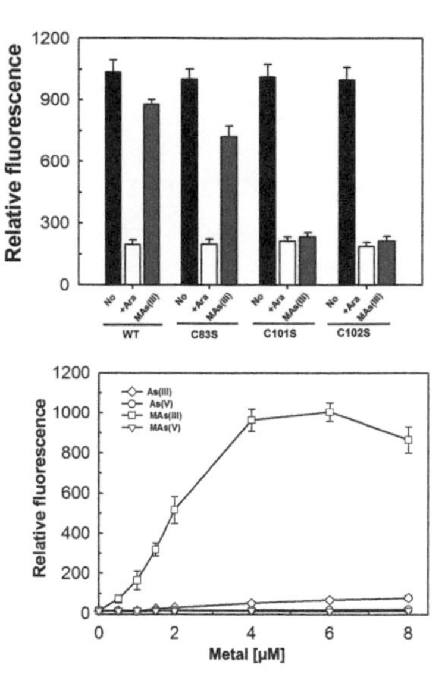

Figure 3. Binding of MAs(III) to SpArsR involves specific cysteine residues.

MAs(III). Cells with the wild type SpArsR biosensor were unresponsive to MAs(V) or As(V) and exhibited only a very low response to As(III) (Fig. 3).

4 CONCLUSION

The MAs(III) binding site of SpArsR appears to have evolved from an ancestor common to the AfArsR As(III) binding site. The SpArsR MAs(III) binding site appears to be the result of a loss-of-function mutation where the third As(III) ligand is no longer present. In reality, though, it is a gain-of-function mutation to provide specificity for MAs(III) that allows for transcriptional regulation of MAs(III) detoxification mechanisms.

ACKNOWLEDGEMENTS

This work was supported by National Institute of General Medical Sciences and NIH grant R01 GM55425.

REFERENCES

Chen, J., Sun, S., Li, C.Z., Zhu, Y.G. & Rosen, B.P. 2014. Biosensor for organoarsenical herbicides and growth promoters. *Environ. Sci. Technol.* 48: 1141–1147.

Ordóñez, E., Thiyagarajan, S., Cook, J.D., Stemmler, T.L., Gil, J.A., Mateos, L.M. & Rosen, B.P. 2008. Evolution of metal(loid) binding sites in transcriptional regulators. *J. Biol. Chem.* 283: 25706–25714.

Qin, J., Fu, H.L., Ye, J., Bencze, K.Z., Stemmler, T.L., Rawlings, D.E. & Rosen, B.P. 2007. Convergent evolution of a new arsenic binding site in the ArsR/SmtB family of metalloregulators. *J. Biol. Chem.* 282: 34346–34355.

Zhu, Y.G., Yoshinaga, M., Zhao F.J. & Rosen, B.P. 2014. Earth abides arsenic biotransformations. *Annu. Rev. Earth Planet. Sci.* 42: 443–467.

Environmental Arsenic in a Changing World –
Zhu, Guo, Bhattacharya, Ahmad, Bundschuh & Naidu (Eds)
ISBN 978-1-138-48609-6

A preliminary investigation on adsorption behavior of As(III) and As(V) in Jianghan Plain

S. Wang[1], R. Ma[1,2], Y.F. Liu[1] & J.Y. Wang[1]
[1]*School of Environmental Studies, China University of Geosciences, Wuhan, P.R. China*
[2]*Laboratory of Basin Hydrology and Wetland Eco-restoration, China University of Geosciences, Wuhan, P.R. China*

ABSTRACT: High arsenic groundwater has posed a severe health threat to the local people in Jianghan Plain. Several studies have been conducted to illuminate the factors and mechanism controlling seasonal variations in As concentrations. However, the quantitatively description of As adsorption has not been studied and this impedes the explanation of As reactive transport and fate in the study area. In this study, a series of adsorption experiments on natural sediments were conducted to investigate the characteristics of surface complexation reactions. The results showed that As(III) and As(V) have different adsorption behaviors on natural sediments. The HCO_3^- competitive adsorption has negligible effect on As(V) adsorption in artificial groundwater (AGW), while it has significant influences on As(III) adsorption. With the increase of pH, adsorbed As(III) concentration increased and As(V) concentration decreased. The results from this study strongly suggested that different adsorption behavior of As(III) and As(V) should be considered as a important process for controlling the seasonal variation in arsenic concentration in Jianghan Plain.

1 INTRODUCTION

Jianghan plain, which is located in the middle reaches of Yangtze River, was found to be widely distributed with high-arsenic groundwater. This posed a severe health threat to local people. Several studies have been conducted to investigate the distribution of high arsenic (As) groundwater and explore the factors and mechanism that control the seasonal variations in As concentrations. However, the As reactive transport has not been qualitatively simulated until know, which impedes the further understanding of As reactive transport processes. Adsorption and desorption play an important role in controlling the arsenic reactive transport. The behavior of adsorption/desorption and its quantitatively modeling is of great importance for the coupled flow- reactive transport modeling of arsenic and thus understanding its fate in Jianghan Plain.

Hence, the aims of this research is to characterize the surface complexation behavior of As(III) and As(V) on natural sediments in Jianghan Plain by laboratory batch experiments, and to develop the surface complexation model.

2 METHODS/EXPERIMENTAL

2.1 *Study site and sediments collection*

Jianghan Plain is an alluvial plain created by Yangtze and Han rivers. It is comprised of the central and southern regions of Hubei Province. Arsenic concentration measured in the sediments and groundwater is much higher than the global average. Meanwhile, adsorption and desorption are main factors for controlling arsenic concentrations under various geochemical conditions (Duan *et al.*, 2015; Rathi *et al.*, 2017). The study site, Shahu site, is located within the Jianghan Plain as shown in Figure 1.

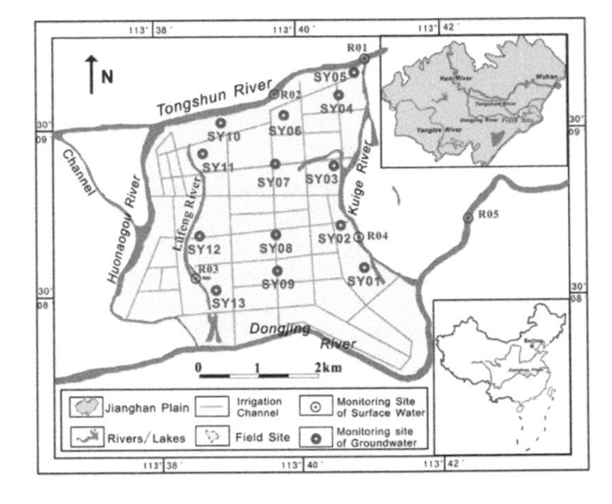

Figure 1. Location of the monitoring site and the sediments is collected at SY07 (cited from Duan *et al.*, 2015).

2.2 *Adsorption experiments*

A series of batch experiments were conducted to investigate the adsorption/desorption behavior of As(III) and As(V) on sediments in Jianghan Plain with varying

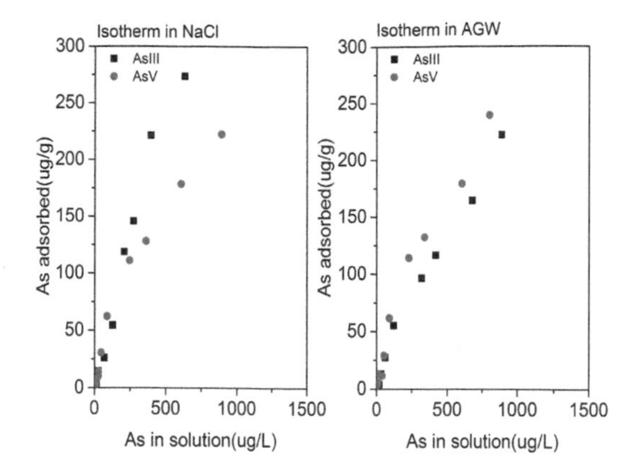

Figure 2. Adsorption isotherm for As(III) and As(V) on sediments in NaCl and AGW solution.

initial arsenic concentration, solution pH and HCO_3 concentration. Sediments were mixed with solutions which were representative of in-situ groundwater geochemistry, and then were placed on a rotating mixer for an equilibration time of 168 hours. All As(III) adsorption experiments were conducted in an anaerobic environment.

2.3 *Geochemical modeling*

A generalized surface complexation model was proposed to model the arsenic adsorption and desorption. PHREEQC (version 3) software was employed to simulate the equilibrium surface complexation reactions and other related reactions (Davies *et al.*, 1998). All ions, As(III), As(V) and some related minor elements such as Fe and Mn were taken into consideration during modeling (Rathi *et al.*, 2017).

3 RESULTS AND DISCUSSION

3.1 *Arsenic adsorption*

The As(III) adsorption was greater than As(V) in NaCl solution as shown in Figure 2. In comparison, more As(V) was absorbed than As(III) in AGW solution. The As(V) adsorption in AGW was similar to that in NaCl, while more As(III) was adsorbed in NaCl solution. This may be caused by competitive adsorption. The results from HCO_3^- competitive adsorption experiments further evidenced this hypothesis. Figure 3 shows that the adsorbed As(III) increased with the increase in HCO_3^- concentration, while the adsorbed As(V) almost remained unconstant with the increase in HCO_3^- concentrations. The degree of the influences of HCO_3^- competitive adsorption on As(V) and As(III) was different.

The reactions for major ions were directly taken from PHREEQC standard database and the water complexation reactions and their equilibrium constants were taken from Wateq database. A non-electrostatic generalized composite surface complexation model (GC-SCM) is proposed as shown in Table 1. The model was preliminarily developed and ran to obtain the equilibrium constants.

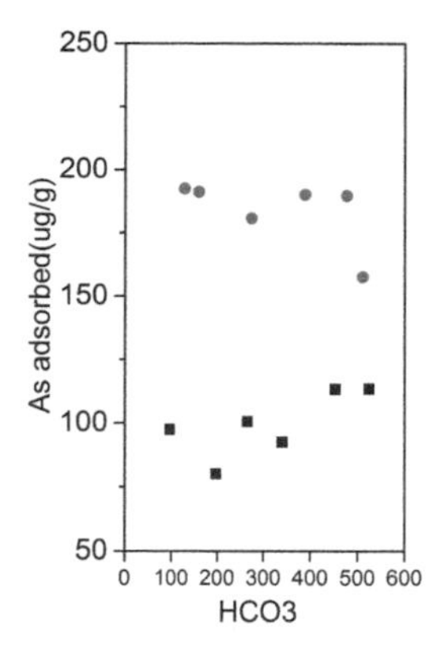

Figure 3. Effect of bicarbonate on adsorption of As(III) and As(V) on different sediments surface.

Table 1. Parameters for SCM of experimental data.

Solution species	Reactions	Equilibrium constant
H^+ and	$Jh\_cOH + H^+ = Jh\_cOH_2^+$	pK_c^1
OH^-	$Jh\_cOH + = Jh\_cO^- + H^+$	pK_c^2
As(III)	$Jh\_cOH + AsO_3^{-3} + 3H^+ = Jh\_cH_2AsO_3 + H_2O$	$\log K_c^1$
	$Jh\_cOH + AsO_3^{-3} + 2H^+ = Jh\_cHAsO_3^- + H_2O$	$\log K_c^2$
	$Jh\_cOH + AsO_4^{-3} + H^+ = Jh\_cAsO_3^{2-} + H_2O$	$\log K_c^3$
As(V)	$Jh\_sOH + AsO_4^{-3} + 3H^+ = Jh\_sH_2AsO_4 + H_2O$	$\log K_c^4$
	$Jh\_sOH + AsO_4^{-3} + 2H^+ = Jh\_sHAsO_4^- + H_2O$	$\log K_c^5$
	$Jh\_sOH + AsO_4^{-3} + H^+ = Jh\_sAsO_4^{2-} + H_2O$	$\log K_c^6$
HCO_3^-	$Jh\_sOH + CO_3^{2-} + 2H^+ = Jh\_sCO_3^- + H_2O$	$\log K_c^7$
	$Jh\_cOH + CO_3^{2-} + 2H^+ = Jh\_cHCO_3 + H_2O$	$\log K_c^8$

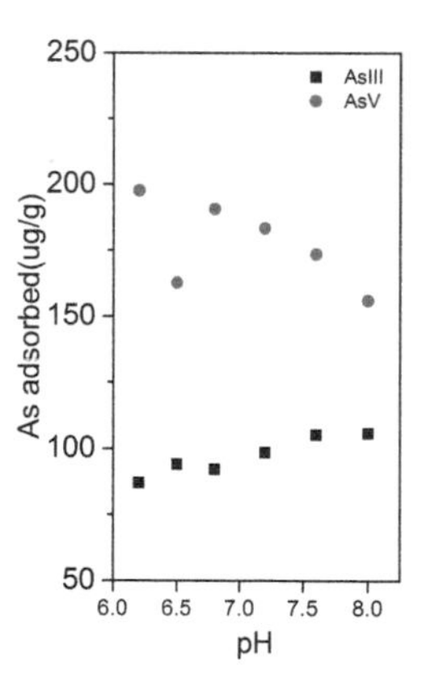

Figure 4. Effect of pH on adsorption of As(III) and As(V) on different sediments surface.

3.2 *Effects of pH*

pH has a relatively significant influence on both As(III) and As(V) adsorption as indicated in Figure 4. While the more As(III) was absorbed as the pH increased, the adsorbed As(V) become less with the increase in pH.

4 CONCLUSIONS

The different adsorption behaviors of As(III) and As(V) was shown by this study. The experimental results showed that effect of the HCO_3^- competitive adsorption on As(V) adsorption was minor. However, it has a significant impact of As(III) adsorption. While the increase in pH enhances the As(III) adsorption, it has negative influences on As(V) adsorption. The seasonal variation in arsenic concentration was observed in the study area and the different valence state of As dominated at different seasons. The results from this study strongly implied that different adsorption behavior of As(III) and As(V) should be considered as an important process for controlling the seasonal variation in arsenic concentration in Jianghan Plain.

ACKNOWLEDGEMENTS

This research is financially supported by China Geological Survey (No. 12120114069301).

REFERENCES

Davis, J.A., Coston, J.A., Kent, D.B. & Fuller, C.C. 1998. Application of the surface complexation concept to complex mineral assemblages. *Environ. Sci. Technol.* 32 (19): 2820–2828.

Duan, Y., Gan, Y., Wang, Y., Deng, Y., Guo, X. & Dong, C. 2015. Temporal variation of groundwater level and arsenic concentration at Jianghan Plain, central China. *J. Geochem. Explor.* 149: 106–119.

Rathi, B., Neidhardt, H., Berg, M., Siade, A. & Prommer, H. 2017. Processes governing arsenic retardation on Pleistocene sediments: Adsorption experiments and model-based analysis. *Water Resour. Res.* 53: 4344–4360.

Environmental Arsenic in a Changing World –
Zhu, Guo, Bhattacharya, Ahmad, Bundschuh & Naidu (Eds)
ISBN 978-1-138-48609-6

Surface complexation modeling of arsenic mobilization from goethite: Interpretation of in-situ experiments in a sedimentary basin of Inner Mongolia, China

L. Stolze[1], D. Zhang[2], H.M. Guo[2] & M. Rolle[1]
[1]*Department of Environmental Engineering, Technical University of Denmark, Lyngby, Denmark*
[2]*School of Water Resources and Environment, China University of Geosciences, Beijing, P.R. China*

ABSTRACT: We present and compare conceptual and numerical modeling approaches for a quantitative interpretation of in-situ experiments that consisted in monitoring the temporal change of adsorbed-As concentration by incubating As-loaded goethite coated sand in the groundwater. We employed the diffuse double layer (DDL) and charge distribution multisite complexation (CD-MUSIC) models to simulate the measured As-sorbed concentration and to investigate the effects of multicomponent adsorption processes on As mobility.

1 INTRODUCTION

Sorption competition and mineral-oxides reductive transformation have been recognized as predominant mechanisms controlling the release and mobility of arsenic in Asia (e.g., Neidhart *et al.*, 2014). Despite the considerable effort toward understanding the geochemical mechanisms controlling the occurrence of arsenic, numerical modeling of these processes remain rare and mostly limited to well-controlled laboratory experiments (Rawson *et al.*, 2016).

Surface complexation models (SCMs) have been widely applied as a powerful alternative to traditional empirical sorption modeling approaches (Dzombak & Morel, 1990; Hiemstra & Van Riemsdijk, 1996). However, these models are typically developed and/or applied to simple aqueous systems rather than complex environmental groundwater conditions.

Field experiments were recently performed in an As contaminated basin in Inner Mongolia, China (Guo *et al.*, 2011). These experiments provided some new insights on the As mechanisms of release from Fe-(oxyhydr-)oxides under in-stu conditions (Zhang *et al.*, 2017).

We applied two surface complexation modeling approaches available in the geochemical code PHREEQC (i.e. DDL and CD-MUSIC) to simulate the desorption of As from goethite in order to provide quantitative interpretation of the in-situ experiments as well as to assess the performances of the 2 SCMs approaches under environmental multicomponent aqueous conditions.

2 METHODS/EXPERIMENTAL

A reactive transport model was developed in a Matlab® environment using the iPhreeqc module (Charlton & Parkhurst, 2011). The transport of groundwater solutes and all geochemical reactions occurring at the surface-aqueous interface were simulated through stepwise batch-type reactions occurring within single reactor cells. Such reactors represent the small containers encapsulating the As-loaded goethite-coated sand in the groundwater wells. The modeling implementation was based on data and results from the in-situ experiments performed in 7 wells (Zhang *et al.*, 2017).

We assumed that multicomponent sorption competition was the only geochemical process leading to the release of As from goethite. This was hypothesized by (Zhang *et al.*, 2017) according to the observed steady Fe-content and absence of other Fe-oxide phases rather than goethite in the in-situ experiments. Sorption reactions on goethite were modeled using the diffuse double layer approach (DDL) by Dzombak & Morel (1990) and the charge distribution multisite complexation (CD MUSIC) model (Hiemstra & Van Riemsdijk, 1996). The two distinct SCM approaches were applied in order to allow a critical comparison of the results and to assess the performances of the models when used for complex aqueous systems. For each SCM, a set of protonation/deprotonation and surface complexation reactions with their relative affinity constants (logK) was selected based on a thorough literature review.

The parameters (i.e. affinity constants, surface area, site densities, time step) were calibrated through a joint inversion procedure by minimizing the root mean squared error (RMSE) between predicted values and measured sorbed As onto goethite. Parameters calibration was performed (i) for each well individually and (ii) for the 7 wells simultaneously by parallelizing the simulation in order to find a unique set of parameters describing the experimental observations.

3 RESULTS AND DISCUSSION

A single set of parameters for both SCMs could provide satisfying fits of the measured adsorption As for all wells. However, simulation results showed significant differences applying the DDL and CD-MUSIC models respectively.

Figure 1 shows the result for a selected well and presents the temporal decrease of As from the surface of goethite predicted by the 2 SCMs where affinity constants were calibrated (CD-MUSIC 1 and DDL 1) or set equal to the best estimates reported in the literature. Without calibrating the affinity constants, the DDL model provides results closer to the measured data compared to the CD-MUSIC model. However, when calibration to the in-situ observations

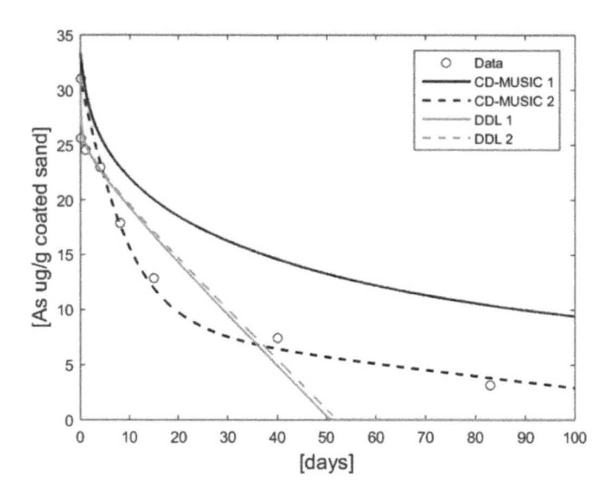

Figure 1. Temporal change in adsorbed As concentrations in a groundwater well. Dashed and full lines are results predicted by the SCMs. 1: parameters fixed to best estimates values re-ported in the literature. 2: parameters calibrated within range of values reported in the literature.

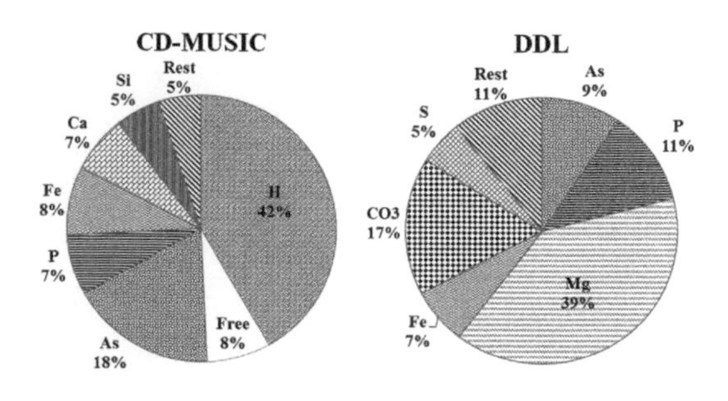

Figure 2. Surface composition at t = 37 days predicted by the CD-MUSIC 1 and DDL 1 model.

was implemented, the CD-MUSIC allowed matching the measurements, whereas the performance of the DDL were not significantly improved (CD_MUSIC 2 and DDL 2).

The surface compositions predicted by the two SCMs were compared at times when the simulated adsorbed As was found equal between the DDL and the CD-MUSIC. Figure 2 shows that the two SCMs present significant difference in modeling the competition of the aqueous species for surface sites in the considered multicomponent aqueous system. The surface composition is dominated by protons in the CD-MUSIC model whereas magnesium is the dominant adsorbed specie in the DDL. The CD-MUSIC model does not predict any sorption of carbonate or sulfur unlike the DDL, but suggests sorption of silica and calcium.

4 CONCLUSIONS

Reactive transport modeling implementing surface complexation with the DDL and the CD-MUSIC models allowed us obtaining a reasonable agreement with the measured As-adsorbed concentrations in multicomponent aqueous conditions. However, the higher flexibility of the CD-MUSIC model provides a superior capability in fitting the experimental datasets with a unique set of parameters. Furthermore, the role of the aqueous species in the sorption competition significantly differs between the predictions of the DDL and the CD-MUSIC models.

REFERENCES

Charlton, S. & Parkhurst, D. 2011. Modules based on the geochemical model PHREEQC for use in scripting and programming languages. *Computat. Geosci.* 37: 1653–1663.

Dzombak, D. & Morel, F.M.M. 1990. *Surface Complexation Modeling- Hydrous Ferric Oxide*. New York: Wiley.

Guo, H., Zhang, B., Li, Y., Berner, Z., Tang, X., Norra, S. & Stüben, D. 2011. Hydrogeological and biogeochemical constrains of arsenic mobilization in shallow aquifers from the Hetao basin, Inner Mongolia. *Environ. Pollut.* 159: 876–883.

Hiemstra, T. & Van Riemsdijk, W. 1996. A surface structural approach to ion adsorption: the charge distribution (CD) model. *J. Colloid Interf. Sci.* 179: 488–508.

Neidhart, H., Berner, Z.A., Freikowski, D., Biswas, A., Majumder, S., Winter, J., Gallert, C., Chatterjee, D. & Norra, S. 2014. Organic carbon induced mobilization of iron and manganese in a West Bengal aquifer and the muted response of groundwater arsenic concentrations. *Chem. Geol.* 367: 51–62.

Rawson, J., Prommer, H., Siade, A., Carr, J., Berg, M., Davis, J.A. & Fendorf, S. 2016. Numerical modeling of arsenic mobility during reductive iron-mineral transformations. *Environ. Sci. Technol.* 50(5): 2459–2467.

Zhang, D., Guo, H., Xiu, W., Ni, P., Zheng, H. & Wei, C. 2017. In-Situ Mobilization and Transformation of Iron Oxides-Adsorbed Arsenate in Natural Groundwater. *J. Hazard. Mater.* 321: 228–237.

Environmental Arsenic in a Changing World –
Zhu, Guo, Bhattacharya, Ahmad, Bundschuh & Naidu (Eds)
ISBN 978-1-138-48609-6

Effects of carbonate and Fe(II) on As(III) adsorption and oxidation on hydrous manganese oxide

Z.X. Zhao[1], S.F. Wang[2] & Y.F. Jia[2]
[1]*College of Chemistry and Chemical Engineering, Xinjiang Normal University, Urumqi, China*
[2]*Institute of Applied Ecology, Chinese Academy of Sciences, Shenyang, China*

ABSTRACT: Manganese oxide plays an important role in redox transformation of arsenic in the aquatic environment. High levels of carbonate and Fe(II) were found as major coexistent ions with arsenic, but their influences on oxidation of arsenite (As(III)) to arsenate (As(V)) by manganese oxide were not fully understood. This work researched oxidation of As(III) on hydrous manganese oxide (HMO) and investigated the effects of carbonate and Fe(II).

1 INTRODUCTION

Manganese oxide is an important oxidant for the transformation of arsenic speciation in the aquatic environment (Ying *et al.*, 2012). It was also proposed as an oxidant for the treatment of arsenic contaminated groundwater (Xu *et al.*, 2011). As(III) oxidation by manganese oxide may be influenced by coexistent ions, such as Ca(II) and Zn(II) (Lafferty *et al.*, 2011). In anaerobic groundwater condition, the concentration of Fe(II) was higher than that of heavy metals (Appelo *et al.*, 2002). However, little is known about its influence on As(III) oxidation by manganese oxide. Fe(II) was reactive and readily oxidized to ferric oxide by MnO_2 (Han *et al.*, 2011), hence may compete with As(III) for the oxidant.

2 METHODS/EXPERIMENTAL

2.1 *As(III) adsorption and oxidation*

As(III) adsorption and oxidation experiments were performed by contacting As(III) solution with HMO suspensions in batch reactors at 20°C, with the background electrolyte of 0.02 mol L^{-1} NaCl. 0.1 mL of 1 g L^{-1} As(III) stock solution was added to 50 mL HMO suspension with the pH pre-adjusted to target value and the initial concentration of 2 mg L^{-1} As(III) was obtained. The pH of the mixture was maintained constant using 0.05 mol L^{-1} HCl and/or NaOH solutions.

2.2 *Determination of arsenic and iron concentrations*

The concentrations of As(III) and As(T) were measured using hydride-generation atomic fluorescence spectrophotometer (AFS-2202E, Haiguang Corp., Beijing) with a detection limit of 0.01 μg L^{-1} and duplication analysis agreed within 5%. The concentration of Mn(II) was determined using a flame atomic absorption spectrophotometer (AA240, Varian) with a detection limit of 0.05 mg L^{-1} and duplication analysis agreed within 7%.

3 RESULTS AND DISCUSSION

3.1 *Kinetic studies of As(III) adsorption and oxidation on HMO*

The effect of carbonate and Fe(II) on As(III) adsorption and oxidation on hydrous manganese oxide was observed in kinetic studies under Mn/As = 8.5 and pH = 7.0 conditions. Adsorption and oxidation of As(III) in HMO system with carbonate and Fe(II) was compared with pure HMO system (see Fig. 1). The concentrations of As(III) and As(V) in aqueous and solid phase in pure HMO system without coexistent carbonate and Fe(II) were showed in Figure 1. The concentration of As(III) in aqueous phase dropped from 2.0 mg L^{-1} to 1.7 mg L^{-1} within 1 h and decreased slowly to a stable level (~1.5 mg L^{-1}) after 48 h reaction. The concentration of As(V) in aqueous phase increased to 0.25 mg L^{-1} after 48 h reaction. In solid phase, few of As(III) could be detected during 48 h reaction time. The concentration of aqueous As(V) increased to 0.2 mg L^{-1} within 1 h and reached to 0.25 mg L^{-1} after 48 h reaction time. The As(V) in solid increased to 0.3 mg L^{-1} after 48 h reaction time. The results indicated that 30% of added As(III) was oxidized to As(V). About 55% of oxidized As(V) was adsorbed and few of As(III) was adsorbed on HMO. This implied that at the neutral pH HMO has strong

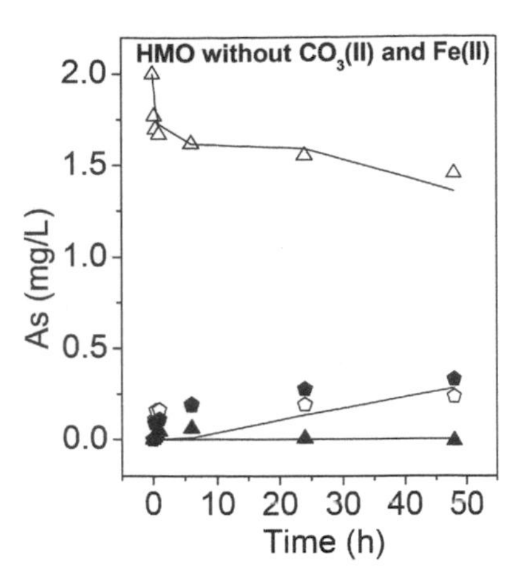

Figure 1. Distribution of As(III) and As(V) (triangle and pentagon respectively) in aqueous and solid phase (open and solid respectively) by contacting As(III) with HMO as a function of time at Mn/As = 8.5 and pH 7.0.

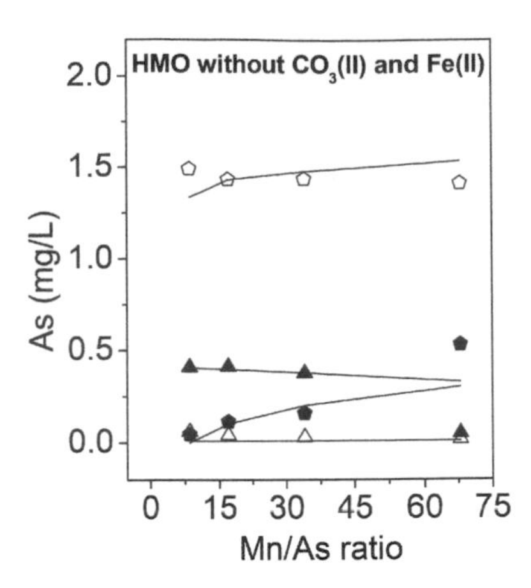

Figure 2. Distribution of As(III) and As(V) (triangle and pentagon respectively) in aqueous and solid phase (open and solid respectively) by contacting As(III) with HMO as a function of Mn/As ratio at pH 3.0. Solid lines represent model results.

oxidation capacity for As(III) and high adsorption capacity for As(V).

Carbonate and Fe(II) showed distinct effect on kinetics of As(III) adsorption and oxidation on HMO. The addition of carbonate only slightly increased As(V) adsorption at the beginning of the reaction. The addition of Fe(II) increased As(III) oxidation significantly and 60% of As(III) was oxidized to As(V). More As(III) and As(V) were adsorbed on HMO in the presence of Fe(II). The maximum value of As(III) concentration in solid phase at the beginning of the reaction indicated that adsorption and oxidation of As(III) on HMO were competitive process, which controlled the concentrations of As(III) and As(V) in aqueous and solid phase. The addition of carbonate and Fe(II) at the same time mainly exhibited the effect of Fe(II) on adsorption and oxidation of As(III) on HMO.

3.2 *Equilibrium studies under acidic pH condition*

The effect of carbonate and Fe(II) on As(III) adsorption and oxidation on hydrous manganese oxide was observed in equilibrium studies under pH = 3.0 condition. Adsorption and oxidation of As(III) in HMO system with carbonate and Fe(II) was compared with pure HMO system (see Fig. 2). The concentrations of As(III) and As(V) in aqueous and solid phase in pure HMO system without coexistent carbonate and Fe(II) are shown in Figure 2. Under pH = 3.0 condition, both of the As(III) in aqueous and solid phase could be ignored. The concentration of As(V) in aqueous and solid phase increased to 1.4 and 0.6 mg L^{-1} at Mn/As ratio is 68. This indicated that the influence of pH on As (III) oxidation was great. Under acidic condition As (III) oxidation increased remarkably even at low Mn/As ratio and presented great independence with Mn/As ratio. The reaction was largely driven by the proton concentration (Villalobos *et al.*, 2014). The produced As(V) existed primarily in aqueous phase and presented more in solid phase at higher Mn/As ratios.

3.3 *Equilibrium studies under neutral pH condition*

The effect of carbonate and Fe(II) on As(III) adsorption and oxidation on hydrous manganese oxide was observed in equilibrium studies under pH = 7.0 condition. Adsorption and oxidation of As(III) in HMO system with carbonate and Fe(II) was compared with pure HMO system (see Fig. 3). The concentrations of As(III) and As(V) in aqueous and solid phase in pure HMO system without coexistent carbonate and Fe(II) are presented in Figure 3. Under pH = 7.0 condition, As(III) concentration in aqueous phase decreased to 0.8 mg L^{-1} at higher Mn/As ratio. The concentration of As(V) in aqueous and solid phase increased to 0.8 and 0.5 mg L^{-1} at Mn/As ratio is 68. Under neutral condition As (III) oxidation increased with Mn/As ratio and presented great dependence with Mn/As ratio. Half of produced As(V) existed in aqueous phase at all four Mn/As ratios.

In HMO system with carbonate, fewer of As(V) adsorbed in solid phase than that in pure HMO system at higher Mn/As ratio. This was consistent with the results under acidic pH condition. In HMO system with Fe(II), more As (III) was oxidized to As(V) than that in pure HMO system. Moreover, more As(III) and As(V) was adsorbed in solid phase than that in pure HMO system. This suggested that addition of Fe(II) could increased As(III) oxidation and As(III) and As(V) adsorption significantly. There were two reason for these results. One reason was that Fe(II) was

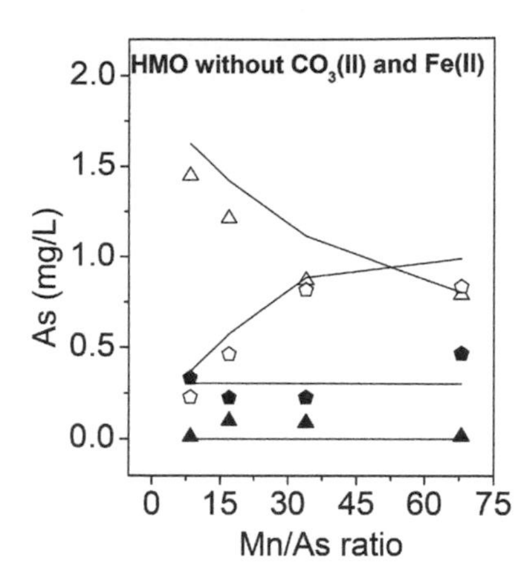

Figure 3. Distribution of As(III) and As(V) (triangle and pentagon respectively) in aqueous and solid phase (open and solid respectively) by contacting As(III) with HMO as a function of Mn/As ratio at pH 7.0.

oxidized to Fe(III) oxide at neutral pH and supplied new sites for As(V) adsorption (Ying *et al.*, 2012).

4 CONCLUSIONS

(1) Addition of Fe(II) decreased oxidation of As(III) on HMO at acidic pH, while increased As(III) oxidation at neutral and basic pH. This indicated that different mechanisms may be involved in the presence of Fe(II). Formation of amorphous ferric arsenate or ferric oxide at different pH is responsible for this.
(2) Carbonate ions decreased As(V) adsorption at neutral pH under higher Mn/As ratio. Precipitation of $MnCO_3$ was the major reason because its existence on HMO decreased As(V) adsorption on HMO. Formation of $MnCO_3$ also immobilized manganese and decreased the concentration of Mn(II) in aqueous phase. This has important environmental implications in protecting the groundwater from secondary manganese contamination during remediation of arsenic contaminated groundwater using HMO.

ACKNOWLEDGEMENTS

This work was financially supported by the National Natural Science Foundation of China (NO. 41203071). We sincerely thank Prof. Shiqiang Wei and his students at NSRL and Prof. Yidong Zhao and his students at BSRL for their assistances with XANES measurements.

REFERENCES

Appelo, C.A.J., Van der Weiden, M.J.J., Tournassat, C. & Charlet, L. 2002. Surface complexation of ferrous iron and carbonate on ferrihydrite and the mobilization of arsenic. *Environ. Sci. Technol.* 36(14): 3096–3103.

Han, X., Li, Y.L. & Gu, J.D. 2011. Oxidation of As(III) by MnO2 in the absence and presence of Fe(II) under acidic conditions. *Geochim. Cosmochim. Acta* 75(2): 368–379.

Lafferty, B., Ginder-Vogel, M.A. & Sparks, D.L. 2011. Arsenite oxidation by a poorly-crystalline manganese oxide. 3. arsenic and manganese desorption. *Environ. Sci. Technol.* 45(21): 9218–9223.

Villalobos, M., Escobar-Quiroza, I. N. & Salazar-Camachoa, C. 2014. The influence of particle size and structure on the sorption and oxidation behavior of birnessite: I. Adsorption of As(V) and oxidation of As(III). *Geochimi. Cosmochimi. Acta* 125: 564–581.

Xu, W., Wang, H., Liu, R., Zhao, X. & Qu, J. 2011. Arsenic release from arsenic-bearing Fe-Mn binary oxide: Effects of Eh condition. *Chemosphere* 83(7): 1020–1027.

Ying, S.C., Kocar, B.D. & Fendorf, S. 2012. Oxidation and competitive retention of arsenic between iron- and manganese oxides. *Geochimi. Cosmochim. Acta* 96(11): 294–303.

Environmental Arsenic in a Changing World –
Zhu, Guo, Bhattacharya, Ahmad, Bundschuh & Naidu (Eds)
ISBN 978-1-138-48609-6

Effects of the seeding phenomenon in the scorodite precipitation process

J.S. Villalobos, J.C. Bravo & F.P. Mella
EcoMetales Limited, Providencia, Región Metropolitana, Santiago de Chile, Chile

ABSTRACT: In the copper mining industry, the impurity of the highest concentration is arsenic, with ore grades of up to 0.3% in mine minerals. During the copper production process via sulphide, arsenic is concentrated simultaneously with the copper closing the treatment cycle in copper smelting. Because of its toxic, mutagenic and carcinogenic characteristics, it should precipitate and be disposed of as a stable residue, in order to prevent it from infiltrating into the soil, and thus into the ground water. Ecometales has designed a process at industrial scale to stabilize arsenic from flue dust into the form of scorodite ($FeAsO_4*2H_2O$), a highly stable compound. This process comprises the stages of primary leaching, oxidation and precipitation of arsenic in the form of scorodite. The present study aims at evaluating an optimization of the scorodite precipitation stage by using seeds in the primary precipitation stage. In this sense, laboratory tests were carried out simulating the industrial circuit. Parameters such as the precipitation kinetics and arsenic grade in the precipitated residue were verified. By using the seeds, a substantial improvement was obtained in: the precipitation kinetics (from 24 to 16 h) and in the arsenic grade in the obtained residue from 9 to 13%. The percentage of scorodite in the residue grew from 30 to 41%.

1 INTRODUCTION

In the copper mining industry, the impurity of the highest concentration is arsenic, with ore grades of up to 0.3% in mine minerals (Filippou *et al.*, 2007). During the copper production process via sulphide, arsenic (As) is concentrated simultaneously with the copper closing the treatment cycle in copper smelting. Because of its toxic, mutagenic and carcinogenic characteristics, it should precipitate and be disposed of as a stable residue, in order to prevent it from infiltrating into the soil, and thus into the groundwater.

Ecometales has designed a process, which is unique at industrial scale, to stabilize arsenic from flue dust into the form of scorodite ($FeAsO_4 \cdot 2H_2O$), a highly stable compound (Demopoulos, 1995). The residues treated correspond to flue dust, refinery effluents and effluents from the acid plant. The process comprises the stages of primary leaching, oxidation of Pregnant Leaching Solution (PLS) and precipitation of arsenic in the form of scorodite.

Currently the plant processes around 155 tons d^{-1} of flue dust and 500 m^3/d^{-1} of refinery effluent, generating 130 tons d^{-1} of solid waste of which only 30% represents scorodite and 70% gypsum, residues which are disposed of in a deposit of hazardous industrial waste (Kruusn & Errrr, 1988).

The low scorodite concentration in the precipitated residue is a problem since it reduces the storage capacity of the deposit. Therefore, an increase in scorodite grade in the precipitate is required.

The present study aims at optimizing the primary scorodite precipitation stage by using seeds. In this sense, precipitation tests were carried out in a laboratory simulating the industrial circuit. Finally, parameters such as the precipitation kinetics and arsenic grade in the precipitated residue were verified.

2 METOHODS/EXPERIMENTAL

Precipitation kinetics tests were performed in 130 L reactors in batch mode, simulating an industrial process of two stages: a primary and a secondary.

The materials used for the tests such as scorodite (to be used as seeds), oxidized PLS, solution of ferric sulphate, limestone slurry, neutral water and acidified water, were sampled from the industrial plant and previously characterized.

2.1 *Primary precipitation tests*

Oxidized PLS was poured into an agitated 130 L reactor and the temperature was established at 80°C. The ferric solution was added adjusting the Fe^{+3}/As^{+5} ratio to 1. Three tests were carried out: one without adding seeds and with an initial solid concentration of 1% p/p (PP1), one with addition of seeds (initial solid concentration of 5% p/p) of a scorodite sample obtained from the previous test (PP2) and one with addition of seeds (initial solid concentration of 5% p/p) of a scorodite sample obtained from the industrial plant.

The oxidized PLS slurry was left to react for a period of 24 h and every 4 h samples were taken from the pulp. They were filtered and sent to be analyzed for As_T, As^{+5}, Fe_T, Fe^{+3}, H^+ both in solid and liquid phases.

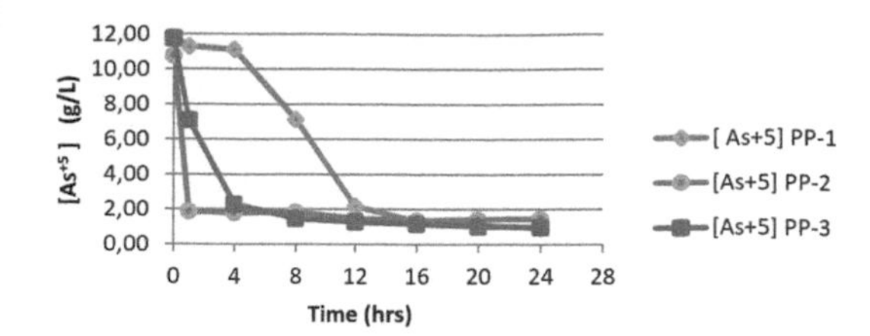

Figure 1. Concentration of As^{5+} in the final PLS versus time for tests PP-1, PP-2 and PP-3.

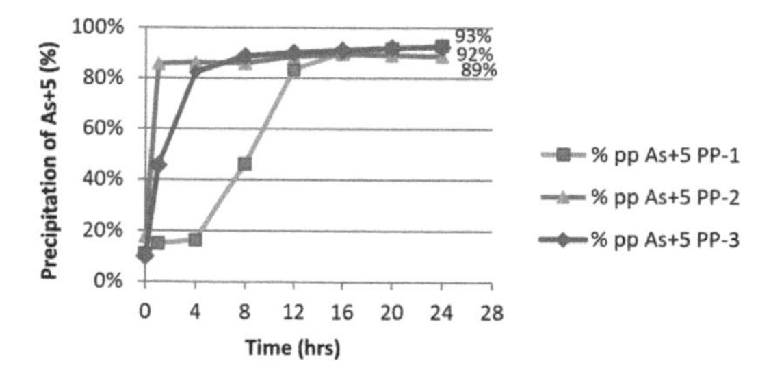

Figure 2. Percentage of precipitation accumulated As^{5+} versus time for tests PP-1, PP-2 and PP-3.

2.2 *Secondary precipitation tests*

The test was continued in the same reactor with the same parameters of the primary leaching test. However, the acid concentration was adjusted to 20 g L^{-1}, by adding limestone slurry. The initial solid concentration varied from 5 to 3 and 8% (PS-1, PS-2 and PS-3, respectively). Then, the oxidized PLS slurry was left to react from a period of 16 h and every 4 h samples were taken from the pulp. They were filtered and sent to be analyzed for As_T, As^{+5}, Fe_T, Fe^{+3}, H^+ both in solid and liquid phases.

3 RESULTS

3.1 *Primary precipitation tests*

Figure 1 shows that for PP-1 a value of 0.931 g L^{-1} was achieved, PP-2 a value of 1.459 g L^{-1} and for PP-3 0.98 g L^{-1}. Taking into consideration this variable, the best results were obtained in tests PP-1 and PP-3.

This may be due to the fact that in SC-3, industrial plant scorodite was used, which could contain small amounts of unreacted limestone, which would enhance the precipitation of As^{5+}, which did not happen in the case of PP-2, where the seed generated from PP-1 (without addition of limestone) was used.

Figure 2 shows that the kinetics for the tests with seeding (PP-2 and PP-3) is much faster than for the test without seeding (PP-1), mainly due to the phenomenon of heterogeneous nucleation (contribution of surface), which facilitates the precipitation.

When reaching 24 h residence time, precipitations of 93% are reached for the test without seeding (PP-1), 89% for the test with seeds of the laboratory test (PP-2) and 92% for the test with seeds of the industrial plant (PP-3). It can be observed that the flat line of the curves (maximum precipitation) is reached after 16 h.

Table 1. Summary of results obtained in tests PS-1, PS-2 and PS-3.

	Precipitation As^{+5} (%)		As^{+5} in PLS (g L^{-1})		As_T in pptate (%)	Scorodite in pptate (%)
N°	t = 16 h	t = 32 h	t = 16 h	t = 32 h	t = 32 h	t = 32 h
PS-1	93.0	99.0	1.74	0.11	13.17	41
PS-2	96.0	99.5	0.53	0.05	13.16	41
PS-3	96.0	99.5	0.57	0.20	13.4	41

3.2 *Secondary precipitation tests*

Table 1 shows that the PLS in the PS-1 test has a final concentration of As^{+5} of 0.11 g L^{-1} which means that 99% of the As^{+5} managed to be precipitated. This condition generates a precipitate with an As_T concentration of 13.17, equivalent to a scorodite concentration in the precipitate of 41%. This presents an improvement to the plant condition (without seed-ing) of only 30%.

For all tests an As^{5+} precipitation of close to 99% is achieved, which thus confirms that the initial solid concentration does not affect the As^{5+} precipitation percentage.

In the secondary stage there is no large degree of progress in the precipitation reaction due to the fact that everything precipitates that had not precipitated in the first stage, which represents a small portion. However, the last stage is necessary for compliance with the process parameters ([As_T] $\leq$ 1 g L^{-1} in final PLS).

4 CONCLUSIONS

(1) When incorporating the seeding, the As^{+5} precipitation kinetics in the first stage improved from 24 to 16 h, thus achieving an average As^{5+} precipitation of 90%.
(2) At the end of the second precipitation stage, an As^{5+} precipitation of 99% was received, with a scorodite concentration in the precipitate of 41%.
(3) For both stages, the initial solid concentration does not affect the arsenic degree or the As^{5+} precipitation percentage As^{5+}.

REFERENCES

Demopoulos, G.P. 1995. Precipitation of crystalline scorodite ($FeAsO_4{\cdot}2H_2O$) from chloride solutions. *Hydrometallurgy* 38(3): 245–261.

Filippou, D., St-Germain, P. & Grammatikopoulos, T. 2007. Recovery of metal values from copper-arsenic minerals and other related resources, Min. Proc. *Extr. Metallurgy Rev.* 28(4): 247–298.

Kruusn, E. & Errrr, Y.A. 1988. Solubility and stability of scorodite, $FeAsO_4{\cdot}2H_2O$: New data and further discussion, *American Mineralogist*, 73, 850–854.

Environmental Arsenic in a Changing World –
Zhu, Guo, Bhattacharya, Ahmad, Bundschuh & Naidu (Eds)
ISBN 978-1-138-48609-6

Effect of microbial sulfate reduction on arsenic mobilization in aquifer sediments from the Jianghan Plain, Central Yangtze River Basin

J. Gao[1], Y. Deng[1], T.I. Zheng[1] & H.C. Jiang[2]
[1]*Geological Survey, China University of Geosciences, Wuhan, China*
[2]*State Key Laboratory of Biogeology and Environmental Geology, China University of Geosciences, Wuhan, China*

ABSTRACT: The natural enrichment of arsenic (As) in the shallow groundwater of the Jianghan Plain, central Yangtze River Basin, has been reported recently. However, the mechanism of As mobilization in the aquifer is not clearly understood yet. To evaluate the effect of microbial sulfate-reducing activities on As mobilization in the aquifer, the sediment samples were collected from 20.75 m at the typical As-rich depth of shallow aquifer. The results indicated that 3 $mg\,kg^{-1}$ As released from the sediment with acetate amended or without amendment. It was notable that, in the late cultural period, the Fe(II) was scavenged in the aqueous phase along with the decrease of sulfate, suggesting that the Fe-sulfide mineral may be generated in the incubation systems. Correspondingly, high abundance of sulfate-reducing bacteria *Desulfomicrobium* were observed, indicating the occurrence of bacteria sulfate reduction. Those results suggested that the As was mobilized under the sulfate reducing condition, and furthermore, the released As was not immobilized by the re-generation of Fe-sulfide minerals. These results can provide new insights for the mobilization mechanism of As in aquifer systems.

1 INTRODUCTION

Natural enrichment of arsenic (As) in groundwater has caused serious healthy and environmental crisis in many parts of the world. The microbially mediated reductive dissolution of iron oxide minerals is widely accepted as the main mechanism of As mobilization in subsurface aquifers (Fendorf *et al.*, 2010). However, recently, the sulfur redox cycle has been recognized with important influences on the As mobilization, in which indigenous microbes play crucial roles in facilitating As transformation in high As groundwater systems (Huang, 2014).

On one hand, microbial sulfate-reducing process is recognized as a mechanism for sequestering As via forming As-sulfide minerals directly or forming iron-sulfide minerals through co-precipitation and/or adsorption (Burton *et al.*, 2014); on the other hand, the sulfide produced by bacteria can reduce As-bearing iron oxides to release As and form sulfarsenate and sulfarsenite further to enhance As solubility (Sun *et al.*, 2016).

2 METHODS/EXPERIMENTAL

2.1 *Experimental materials and methods*

The experiment were conducted in an anaerobic glovebox under N_2 atmosphere. Sediments were homogenized before use and the microcosm slurries were generated by mixing 5 g of sediment with 50 mL artificial groundwater (10 $mmol\,L^{-1}$ PIPES, 7.9 $mmol\,L^{-1}$ NaCl, 2.7 $mmol\,L^{-1}$ KCl, 0.3 $mmol\,L^{-1}$ $MgSO_4$, 0.4 $mmol\,L^{-1}$ $CaCl_2 \cdot 2H_2O$, pH = 7.1).

The microcosm batches were set up into three groups in duplicate: 1) the first group was supplemented with $mol\,L^{-1}$ acetate to stimulate the microbial activities; 2) the second group was not amended; 3) the third group was sterilized twice by autoclave at 121°C, 250 Kpa for 20 min. The aqueous subsamples were filtered through 0.22 μm syringe filters and then collected for the analysis of Fe(II), As(III), As(V), As(total), anions and cations. The remaining solids in the microcosms through centrifuging were collected and stored at −75°C for molecular biological analysis.

2.2 *Sampling and analysis*

The pH and Eh of solutions were measure by a portable gel-filled probe (HQ40D, HACH, Japan). Fe(II) was measured using a Spectronic 601 spectrophotometer (UV-1750, Shimadzu, Japan) with the ferrozine method at 562 nm wavelength. The total As was determined using an atomic fluorescence spectrophotometer (HG-AFS-930, Haiguang, China). The As speciation was measured by the liquid chromatography-hydride generation-atomic fluorescence spectrometry (LC-HG-AFS, Haiguang, China). The major cations were measured using an inductively coupled plasma optical emission spectrometer (ICP-OES) (iCAP 6000 series, Thermo Fisher Scientific, USA). The anions including F^-, Cl^-, NO_3^-, and SO_4^{2-} were determined using an ion chromatograph (ICS-2100, Thermo Fisher Scientific, USA).

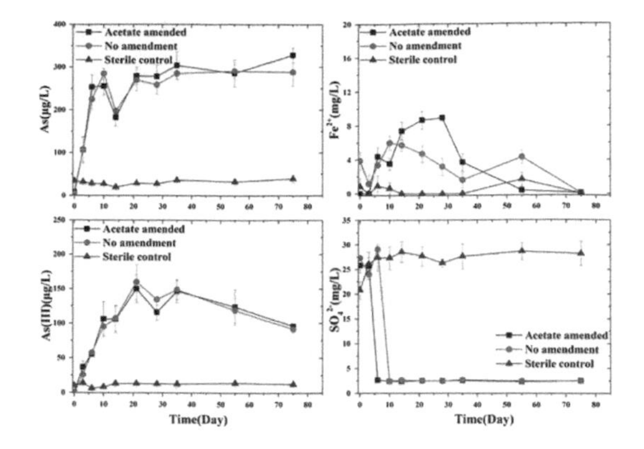

Figure 1. Dissolved arsenic, arsenite, iron and sulfate concentrations during the incubation.

The microbial community dynamics were analyzed by the illumine high-throughput sequencing.

3 RESULTS AND DISCUSSION

3.1 *The mobilization and retention of As and Fe(II) during incubation*

The results indicated 3 mg kg^{-1} As is released from the sediment. From the first to tenth day, the As concentration increased correspondingly with the increase of Fe(II), during which sulfate was reduced completely. The As concentration decreased from the fourteenth day and then remained stable from the twenty-first day (Fig. 1).

In the late culture period, the Fe(II) was scavenged in the aqueous phase along with the decrease of sulfate. However, the As concentration remained stable. These results suggested that the Fe-sulfide mineral may be generated in the incubation systems. As was mobilized under the sulfate reducing condition, and furthermore, the released As was not immobilized by the re-generation of Fe-sulfide minerals.

3.2 *The role of functional microbial groups involved in arsenic mobilization*

According to the results of the illumina high-throughput sequencing, the sulfate-reducing bacteria, Desulfomicrobium, and the iron-reducing bacteria, Geobacter, were the predominant functional bacteria in the microbial community, which can affect the As variation (Fig. 2). In the late culture period, the methane producing bacteria became the main functional bacteria, which have competitive impact with the sulfate-reducing bacteria.

From the first to tenth day, the As concentration increased. Correspondingly, high abundance of sulfate-reducing bacteria Desulfomicrobium were observed, indicating the occurrence of bacterial sulfate reduction. These results can provide direct evidence and insights for the cause of As-rich groundwater in Jianghan plain.

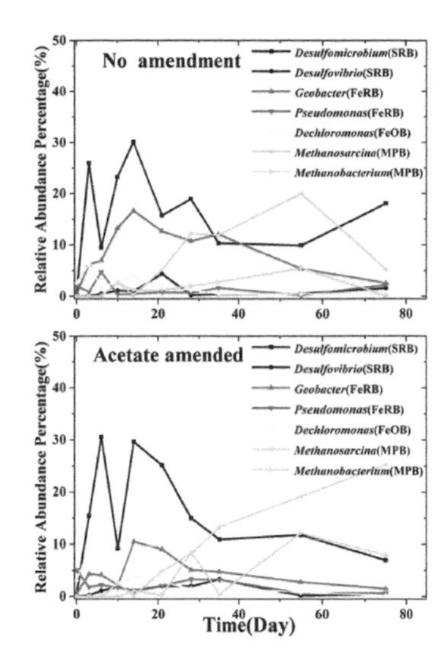

Figure 2. The variation of functional microbial groups involved in arsenic mobilization.

4 CONCLUSIONS

In this study, 3 mg kg^{-1} As was released from the sediment with acetate amended or without amendment. The microbial sulfate-reducing activities had important influences on the As mobilization, in which indigenous microbes, the sulfate-reducing bacteria, *Desulfomicrobium*, played crucial roles and the Fe-sulfide mineral may be generated in the incubation systems. The As was mobilized under the sulfate reducing condition.

ACKNOWLEDGEMENTS

This research was financially supported by the National Natural Science Foundation of China (Nos. 41572226 & 41521001), Program of China Geological Survey (Nos. 12120114069301 & 12120100100015 0121), and State Key Laboratory of Biogeology and Environmental Geology (No.128-GBL21711), China University of Geosciences.

REFERENCES

Burton, E.D., Johnston, S.G. & Kocar, B.D. 2014. Arsenic mobility during flooding of contaminated soil: the effect of microbial sulfate reduction. *Environ Sci Technol.* 48(23): 13660–13667.

Fendorf, S., Michael, H.A. & Geen, A.V. 2010. Spatial and Temporal Variations of Groundwater Arsenic in South and Southeast Asia. *Science* 328(5982): 1123–1127.

Huang, J.H. 2014. Impact of microorganisms on arsenic Biogeochemistry: A Review. *Water, Air Soil Poll.* 225(2): 1848.

Sun, J., Quicksall, A.N., Chillrud, S.N., Mailloux, B.J. & Bostick, B.C. 2016. Arsenic mobilization from sediments in microcosms under sulfate reduction. *Chemosphere* 153: 254–261.

Environmental Arsenic in a Changing World –
Zhu, Guo, Bhattacharya, Ahmad, Bundschuh & Naidu (Eds)
ISBN 978-1-138-48609-6

Assessing arsenic ecotoxicity in tropical soils for regulatory purposes: Which endpoints are more appropriate?

L.R.G. Guilherme[1], G.C. Martins[1], C. Oliveira[1], E.S. Penido[2], T. Natal-da-Luz[3] & J.P. Sousa[3]
[1]*Department of Soil Science, Federal University of Lavras, Minas Gerais, Brazil*
[2]*Department of Chemistry, Federal University of Lavras, Minas Gerais, Brazil*
[3]*Centre for Functional Ecology, Department of Life Sciences, University of Coimbra, Coimbra, Portugal*

ABSTRACT: Ecotoxicological tests with plants are often used to help regulators setting soil screening values of many contaminants, arsenic (As) included. Yet, the measurement of a comprehensive set of endpoints in representative plant species is very time-consuming and hard to achieve in a single test. This study evaluated the best-suited endpoints to assess As toxicity and contamination effects in crop plants grown in oxidic soils, with a focus on Tropical agroecosystems. Our aim is to recommend an approach for setting soil screening values that uses arsenic ecotoxicological studies chosen on the basis of the sensitivity and reliability of endpoints in plant growth tests. From the measured endpoints, the most sensitive ones were: first germination count > relative leaf area > total dry mass > germination speed index, while the most reliable endpoints were: first germination count > total dry mass = germination speed index > plant height. The species *Phaseolus vulgaris* and *Zea mays* were the most and least sensitive to arsenic toxicity, respectively. The use of extractable As concentrations for toxicity characterization allows estimating more realistic soil screening values for arsenic in Tropical soils.

1 INTRODUCTION

Arsenic (As) is a metalloid widely distributed throughout Earth, and it is known to be toxic to most living organisms. It has been ranked number one in the ATSDR's Substance Priority List of hazardous substances for the last 20 years, posing the most significant potential threat to human health and the environment. Because of that, arsenic accumulation and effects in plants should be taken into account when evaluating its toxicity.

Different plant species have distinctive capacities to absorb, accumulate, and tolerate As, which also depends on soil properties. Furthermore, arsenic absorption by plants is quite variable, since each species has different levels of metabolic tolerance and detoxification mechanisms to arsenic compounds (Yoon *et al.*, 2015).

For this reason, laboratory ecotoxicological tests using different standard species as test organisms have been used to characterize the risk associated to specific substances in different soils. However, it is not always possible to measure many plants endpoints in a single test. Therefore, it is important to define which endpoints are most relevant to evaluate As toxicity.

Developing ecotoxicological studies for As contributes for the development of a better database to be used in decision-making actions related to environmental issues. Therefore, the objective of this study was to evaluate the best-suited endpoints to assess As toxicity and contamination effects in crop plants, based on the sensitivity and reliability of endpoints in plant growth tests.

For this purpose, eleven endpoints were evaluated in six plant species, which were exposed to gradients of increasing As concentrations in two natural soils and one artificial tropical soil.

2 METHODS/EXPERIMENTAL

2.1 *Soil sampling and test plants*

A Red-yellow Latosol – Oxisol (21°17′08″E, 44°47′43″N) and a Haplic Cambisol – Inceptsol (21°13′46″E, 44°59′10″N), and an tropical artificial soil – TAS, which was produced by mixing dry kaolinite clay (20%), fine sand (70%), and coconut fiber (10%), were used in the experiments. Soil samples were air-dried, sieved to 2 mm and characterized. Maximum water holding capacity was determined as defined by ISO 11274. Soil As was determined using the Mehlich-1 extractant. The maximum adsorption capacity of each soil was determined as defined by Campos *et al.* (2007). Arsenic levels were measured by graphite furnace absorption spectroscopy (Perkin Elmer – AAnalyst™). Attributes of the soils are shown in Table 1.

Six commercial crop species were chosen: monocotyledonous: maize (*Zea mays*), rice (*Oryza sativa*), sorghum (*Sorghum bicolor*); and eudicotyledonous: beans (*Phaseolus vulgaris*), sunflower (*Helianthus annuus*), and radish (*Raphanus sativus*).

2.2 *Greenhouse experimental design*

The experiments were conducted under greenhouse with natural light and ambient temperature (25 ± 3°C), controlled by an electronic ventilation system. Tests with plants were based on procedures described in OECD Test n° 208. Three days prior to experiments, soils were fertilized (N = 100 mg; P = 100 mg; K = 50 mg; Ca = 37 mg; Mg = 15 mg; S = 25 mg; B = 0.25 mg; Cu = 0.75 mg; Zn = 2.5 mg; Mn = 5 mg; and Mo = 0.1 mg). The following concentration gradients: 0, 8, 14.5, 26, 46.5, 84, 150 and 270 mg As kg^{-1} were used to cover soil quality guidelines values adopted in Brazil. The doses were directly added to soils, keeping 50% of the water holding capacity of each soil. The tests began with seeds sowing and lasted until 21 days after ≥50% of the seeds from control units had germinated. The endpoints considered in each test for analysis of arsenic toxicity were: germination speed index (GSI), final germination count (FnC), total germination (TG), plant survival (PS), completely expanded leaves (CEL), soil plant analysis development (SPAD), relative leaf area (RLA), stem diameter (SD), plant height (PH) and; and total dry mass (TDM). At the end of the experimental period, plants were cut at the base and dried at 60°C until constant weight, for the subsequent determination of dry weight and As concentration.

The (total) arsenic concentration for 50% of effect (EC_{50}) was estimated for each endpoint. To evaluate the adequacy of the endpoints, the parameters were characterized according to their sensitivity and reliability. Endpoints were ranked in two sequences for each soil and test species: i) descending order of sensitivity (the lower for higher EC_{50} value); and, ii) descending order of reliability (the lower for higher amplitude of 95% confidence intervals of the EC_{50} estimated). The endpoints with the lower mean ranks were considered the most sensitive or reliable. Therefore, these endpoints were selected for evaluating the contamination effects of As in crop plants. For these selected endpoints, EC_{50} values were estimated with total and phytoavailable arsenic concentrations measured in soil samples.

2.3 *Chemical analysis of arsenic in soil and plants*

Dried samples were digested using HNO_3 (65%) in a microwave oven (Mars 5, CEM Corporation, Matthews, NC, USA) for total As contents. Phytoavailable contents were measured by adding 0.5 g of sample in 5 mL of a Mehlich-1 solution (0.05 mol L^{-1} HCl + 0.0125 mol L^{-1} H_2SO_4), followed by shaking (5 min) and filtration after 16 hours (0.45-μm filter). Arsenic concentrations were determined by atomic absorption spectrophotometry with electrothermal atomization using graphite furnace (Perkin Elmer – AAnalyst™800, Norwalk, CT, 171 USA).

Certified reference materials containing matrices compatible with the samples were used: plant material 180 BCR-482 No. 638 and for soil samples SRM 2711a and SRM 2710a. The recovery rate were 76.5%, 98.2% and 101% for BCR-482, SRM 2711a, and SRM 2710a, respectively.

Table 1. Physical and chemical properties of soils.

	ATS	Oxisol	Inceptsol
pH (H_2O)	5.2	4.4	4.6
OM/%	7.84	0.24	1.87
CEC/$cmol_c\ dm^{-3}$	2.27	0.27	2.01
P-rem/$mg\ L^{-1}$	36.1	6.84	4.31
As/$mg\ kg^{-1}$	<0.01	0.026	<0.01
Qe/$mg\ As\ kg^{-1}$	nd	714.3	1,667
WHC/%	74.0	41.0	59.0
Clay/%	19	26	33
Silt/%	8	8	66
Sand/%	73	66	19
Fe_2O_3/%	2.86	3.84	25.1

nd = not detected; Qe = maximum adsorption capacity; WHC = water holding capacity.

Table 2. Germination and growth endpoints sorted in descending order of sensitivity or reliability and respective average ranks (Rk) for all test species, considering all soils.

Sensitivity		Reliability	
Endpoint	Rk	Endpoint	Rk
FrC	2.3	FrC	3.9
RLA	4.3	TDM	4.9
TDM	4.7	GSI	4.9
GSI	5.1	PH	5.0
PH	5.5	FnC	5.3
FnC	5.6	TG	5.4
TG	6.7	RLA	6.2
PS	7.3	CEL	7.0
SD	7.4	PS	7.2
CEL	7.5	SD	7.5
SPAD	9.0	SPAD	8.3

3 RESULTS AND DISCUSSION

Considering all soils, the most sensitive endpoints were FrC, RLA, and TDM, while GSI, PH, and FnC had a median sensitivity. SPAD was always the least sensitive endpoint (Table 2).

Germination is the first attribute that may be influenced by contaminants (Li *et al.*, 2007). The early germination (FrC) is often the most sensitive endpoint, because some of the defense mechanisms of the plant are still not developed (Liu *et al.*, 2005). Endpoints related to germination (FrC, FnC, and TG) had contrasting sensitivity and reliability, as apparently, there are concentrations that only retard germination. After germination, subsequent processes become affected by soil contamination, which reflected in the development of plant and growth endpoints (Srivastava *et al.*, 2013). Another effect of As in plants is related to chlorophyll degradation, but in the present study, this phenomenon did not occur intensively.

Table 3. Total and phytoavailable arsenic content (respective recoveries over the nominal content) in the soils.

Total (mg kg^{-1})			
Dose	TAS	Oxisol	Inceptisol
0	3.1	2.43	1.68
8	7.18 (90%)	5.57 (70%)	7.87 (98%)
14.5	14.3 (99%)	12.0 (83%)	7.47 (51%)
26	25.3 (97%)	21.8 (84%)	22.7 (87%)
46.5	40.6 (87%)	40.1 (86%)	32.3 (64%)
84	80.4 (96%)	69.2 (82%)	87.4 (104%)
150	148 (99%)	134 (89%)	113 (76%)
270	260 (97%)	247 (92%)	257 (95%)
Phytoavailable (mg kg^{-1})			
Dose	TAS	Oxisol	Inceptisol
0	0.33	0.37	0.21
8	1.17 (15%)	0.64 (8%)	0.10 (1%)
14.5	2.82 (19%)	1.72 (12%)	0.21 (1%)
26	4.62 (18%)	3.34 (13%)	0.48 (2%)
46.5	18.8 (40%)	7.47 (16%)	1.17 (2%)
84	29.3 (35%)	14.8 (15%)	1.42 (2%)
150	56.9 (38%)	31.6 (21%)	3.69 (2%)
270	145 (54%)	78.4 (29%)	11.6 (4%)

Our data showed a tendency of smaller confidence intervals in EC_{50} values estimated by FrC, followed by TDM = GSI, PH, and FnC endpoints. The EC_{50} values estimated by SPAD, SD, and PS had generally the lowest reliability (Table 2).

Total arsenic concentrations had good recoveries (Table 3), which confirms that that the test species were exposed to a broad concentration gradient. Regarding phytoavailable As concentrations, the highest recovery percentages were observed for the TAS and the lowest for the Inceptisol. The pattern of availability observed in each test soil was related to the specific attributes of each one (Table 1) and corroborates with the relative toxicities found among the three tested soils (Table 4).

The estimated EC_{50} using total As concentrations show the highest differences among contrasting soils, while the lowest differences were observed for EC_{50} estimated using phytoavailable As concentrations. These results reinforce the use of available levels for regulatory purposes, because they reduce differences among toxicity data on contrasting soils.

From the species studied, *P. vulgaris* was the most sensitive and *Z. mays* was the least sensitive to arsenic. Both species also accumulated less arsenic in the tissues (Table 5). The species *H. annuus* and *R. sativus* had the highest arsenic accumulation, which is a reflex of adaptive strategies to the contaminant.

4 CONCLUSIONS

The most sensitive endpoints were: first germination count > relative leaf area > total dry mass > germination speed index, while the most reliable endpoints were: first germination count > total dry mass = germination speed index > plant height. Amongst the evaluated plant species, *P. vulgaris* and *Z. mays* were the most and least sensitive to arsenic toxicity, respectively. The use of extractable As concentrations for toxicity characterization allows estimating more realistic toxic values among different soils.

Table 4. EC50 values (with corresponding 95% confidence intervals) for the effects on total dry mass of the test species cultivated in contrasting soils. EC50 estimations were based on total and phytoavailable arsenic concentrations and are expressed in mg kg^{-1}.

Total			
	ATS	Inceptisol	Δ
Rice	24.0 (19.6–28.5)	189 (156–223)	165
Maize	49.9 (24.5–75.3)	>257	207
Sorghum	38.9 (28.6–49.3)	183 (153–213)	144
Common-bean	<7.18	46.7 (38.8–54.7)	39.5
Sunflower	27.1 (21.2–33.0)	> 257	229
Radish	41.9 (36.7–47.2)	>257	215
Phytoavailable			
	ATS	Inceptisol	Δ
Rice	4.45 (3.63–5.28)	7.80 (6.10–9.51)	3.30
Maize	22.8 (15.8–29.7)	>11.6	−11.2
Sorghum	8.55 (3.30–13.7)	6.94 (5.24–8.64)	−1.61
Common-bean	<1.17	1.34 (1.24–1.45)	0.17
Sunflower	5.69 (4.48–6.89)	>11.6	5.91
Radish	19.2 (17.1–21.3)	>11.6	−7.60

Note: Δ = EC_{50} (Inceptisol) −EC_{50} (ATS).

Table 5. Arsenic content in the shoot of the species cultivated at the dose 26 mg As kg^{-1}. The values are expressed in mg kg^{-1}.

	ATS	Oxisol	Inceptisol
Rice	24.9 ± 0.13	3.09 ± 0.58	0.89 ± 0.09
Maize	2.54 ± 2.72	0.65 ± 0.14	0.87 ± 0.07
Sorghum	1.95 ± 0.18	1.22 ± 0.19	0.81 ± 0.16
Beans	n.d.	1.72 ± 0.54	0.66 ± 0.31
Sunflower	13.9 ± 2.45	5.69 ± 0.68	0.79 ± 0.15
Radish	14.1 ± 1.71	10.3 ± 1.17	1.81 ± 1.04
As-Total	25.3 ± 1.23	21.8 ± 1.13	22.7 ± 0.41
As-Phytoavailable	4.62 ± 1.83	3.34 ± 0.35	0.48 ± 0.10

ACKNOWLEDGEMENTS

The authors thank the funding agencies Minas Gerais State Research Foundation (FAPEMIG), National Council for Scientific and Technological Development (CNPq Grant# 140081/2014-3), and Coordination for the Improvement of Higher Education Personnel

(CAPES Grant# BEX3204/15-4) for the financial support.

REFERENCES

Campos, M.L., Guilherme, L.R.G., Lopes, R.S., Antunes, A.S., Marques, J.J.G. de S. e M. & Curi, N. 2007. Teor e capacidade máxima de adsorção de arsênio em Latossolos brasileiros. *Rev. Bras. Ciência do Solo* 31: 1311–1318.

Li, C., Feng, S., Shao, Y., Jiang, L., Lu, X. & Hou, X. 2007. Effects of arsenic on seed germination and physiological activities of wheat seedlings. *J. Environ. Sci.* 19: 725–732.

Liu, X., Zhang, S., Shan, X. & Zhu, Y.G. 2005. Toxicity of arsenate and arsenite on germination, seedling growth and amylolytic activity of wheat. *Chemosphere* 61: 293–301.

Srivastava, S., Akkarakaran, J.J., Suprasanna, P. & D'Souza, S.F. 2013. Response of adenine and pyridine metabolism during germination and early seedling growth under arsenic stress in Brassica juncea. *Acta Physiol. Plant* 35: 1081–1091.

Yoon, Y., Lee, W.M. & An, Y.J. 2015. Phytotoxicity of arsenic compounds on crop plant seedlings. *Environ. Sci. Pollut. Res.* 22: 11047–11056.

Environmental Arsenic in a Changing World –
Zhu, Guo, Bhattacharya, Ahmad, Bundschuh & Naidu (Eds)
ISBN 978-1-138-48609-6

Arsenic and trace metal mobility in alum shale areas in Sweden

G. Jacks[1] & P. Bhattacharya[2]
[1]*Division of Water and Environmental Engineering, Department of Sustainable Development, Environmental Science and Engineering, KTH Royal Institute of Technology, Stockholm, Sweden*
[2]*KTH International Groundwater Arsenic Research Group, Department of Sustainable Development, Environmental Science and Engineering, KTH Royal Institute of Technology, Stockholm, Sweden*

ABSTRACT: Alum shales found in Sweden are enriched in As and trace metals like Mo, Zn, Cd, Ni and U with the redox conditions being crucial for the mobilization of those elements. The shales have been mined, e.g. for uranium. Uranium is mobilized under oxidizing conditions, while arsenic is mobilized under reducing conditions. This plays a role under natural conditions in the alum shale but more so in the tailings after mining. An investigation of a natural area indicates small environmental risks while the utilization of the shales has left tailings that tend to leak into the environment.

1 INTRODUCTION

Black organic rich shales formed in deep sea environments under anoxic conditions are enriched in arsenic and a number of trace metals. Alum shales of Cambrian age are occurring in several sites in Sweden. They are present in southern Sweden and are outcropping in the eastern rim of the Caledonian mountains in northern Sweden. Similar shales are present in the British Isles and in North America in the Appalachian mountains. They have been used in generally minor projects for several purposes, extraction of alum for leather tanning, as fuel for production of lime mortar, for oil production (Sundin, 2008) and more recently to extract uranium. The mobilization of arsenic and the variety of trace metals are of concern under natural conditions, but many reports indicate that the tailings left after mining are of more serious concern. Several sites have been the subject to remediation with varying results. The experiences from this will be reviewed here as well as conditions in areas where no utilization has so far taken place. The main field work in this presentation has been done in an area in the Caledonian mountains where about 11% of the today globally known uranium occurrence is present.

This site is considered to be representative for areas where no mining has taken place while the literature survey is viewed upon as examples where the shale has been utilized.

2 METHODS/EXPERIMENTAL

Soils, plants, groundwater and surface water have been sampled. Water solutes were speciated by in-situ dialysis to reveal the forms in which trace elements were mobile. Water was analyzed by ICP-MS to get a large range of trace elements. The speciation was also done by the thermodynamic program Visual MINTEQ (Gustafsson, 2016).

A large number of publications are available from alum shale areas in southern Sweden which are illustrative for conditions where utilization of the shales has taken place.

3 RESULTS AND DISCUSSION

3.1 *The Viken area in the Caledonian mountains*

The soil in the area was not podzolised and had a neutral pH due to the Ordovician limestone surrounding the shale and mixed into the surface soil during the last Ice Age. The arsenic concentration in the soil profile down to 0.4 m was similar to the concentration in the parent material, the shale, indicating the arsenic was adsorbed onto ferric oxy-hydroxides at the site. The immobility of the arsenic in this site is supported by the low content in surface water and groundwater. The reason for this is likely to be the pronounced topography of the area that did not support any wetland formation where reducing conditions would exist. The lack of wetland means that there were no sinks for uranium. The concentrations of trace metals in soil indicate a substantial loss of U, Mo, Zn and Ni (Table 1) also verified by the elevated concentrations of notably U and Mo in groundwater and surface water.

Uranium was noted taken up by plant while molybdenum was high especially in red clover, common fodder specie in the area. This could cause secondary copper deficiency which has been observed in alum

Table 1. Metals and metalloids in shale, soil and crust.

Material	Mo	Cu	Zn	As	Ni	U	V
Shale	313	127	408	70	475	186	2133
Soil	139	58	201	65	103	27	480
Soil*	1.2	60	70	2	85	2.5	60

* Global average as per Alloway (2015).

shale areas in southern Sweden (Frank 2004). Uranium in water was high, $>100\,\mu g\ L^{-1}$, however in the form of calcium-carbonate-uranyl complexes, which are considered not to be toxic (Prat *et al.*, 2009).

3.2 *Review of literature on alum shale sites in Southern Sweden*

The review of the literature regarding the environmental conditions in alum shale sites in southern Sweden indicates more of problems. At Ranstad in SW Sweden about 1.5 million tons of shale was mined and extracted for uranium. A recent review (SWAPO, 2005) shows that there is considerable pollution in receiving water courses of uranium. Arsenic in drainage water from the water filled open pit is low and arsenic is attached to ferric hydroxides and retained in the surface sediments (Widerlund & Ingri, 1995). In surface water a sizeable transport of arsenic is possible only in water with an elevated content of DOC and ferric precipitates forming particulates or colloids (Jacks *et al.*, 2013).

In another site on the island of Öland shale has been processed by burning (Lavergren *et al.*, 2009). It has been observed that the in situ shale does not form acid drainage in response to weathering which is seen in the burnt shale. Considerable spread of metals are seen in connection to the burnt shale.

Arsenic and uranium in alum shale areas have been investigated in groundwater in dug and drilled wells (Jemander, 2008). Arsenic and uranium was in 30 wells below the respective health limits at $10\,\mu g\ L^{-1}$ and $30\,\mu g\ L^{-1}$. Only one well had $20\,\mu g\ L^{-1}$ of U. Low concentrations of iron indicated that the well water was under oxidizing regime providing good adsorption of arsenic. Even though the alum shales have a high content of organic matter, in the order of 25%, the organics are quite refractory and the formation of reducing conditions less common. The Caledonian shales have been subject to thermal alteration up to about 300°C (Snäll, 1988).

4 CONCLUSIONS

The investigation of the alum shale area in the Caledonian mountains shows that there is currently no environmental risks, possibly a risk for molybdenosis, secondary copper deficiency in animals. However, the use of concentrate supplemented with copper is likely to eliminate that risk. The uranium in groundwater and surface water is in a non-toxic form. The review of the report from sites, where the alum shale has been mined and utilized for different purposes, shows considerably more of problems. There are plans for large scale mining of the alum shale amounting to 700 million tons at the here investigated site. A considerable problem would be the tailings from the bacterial leaching of the shale, still containing 15–25% of the metals and the arsenic (Bhatti, 2015). If deposited so that the tailings are oxygenated, uranium and molybdenum will escape while a lowland deposition under reducing conditions would favor the escape of arsenic. Arsenic in surface water is in the boreal climate transported in particulate or colloidal form attached to ferric hydroxides and humic matter (Jacks *et al.*, 2013). These forms of arsenic does not seem to be bioavailable.

ACKNOWLEDGEMENTS

The investigation of the Caledonian alum shale site was funded by the foundation ÅForsk, Stockholm in Sweden, which is thankfully acknowledged.

REFERENCES

Alloway, B. J. 2015. Heavy metals in soils. *Mineral. Mag.* 55(8): 1318–1324.

Bhatti, T. M. 2015. Bioleaching of organic carbon rich polymetallic black shale. *Hydrometallurgy* 157: 246–255.

Frank, A. 2004. A review of the "mysterious" disease in Swedish moose, related to molybdenosis and disturbances in copper metabolism. *Biol. Trace Elem. Res.* 102(1–3): 143–159.

Gustafsson, J.P. 2016. Visual MINTEQ ver. 3.1. URL: https://vminteq.lwr.kth.se.

Jacks, G., Slejkovec, Z., Mörth, M. & Bhattacharya, P. 2013. Redox cycling of arsenic along the water pathways in a sulphidic metasediment area, N. Sweden. *Appl. Geochem.* 35: 35-43.

Jemander, L. 2008. Metals in an alum shale area in the county of Östergötland. M Sc thesis. Umeå Uinversity, Dept. of Natural Geopgraphy 32 pp.

Lavergren, U., Åström, M. E., Falk, H. & Bergbäck, B. 2009. Metal dispersion in groundwater in an area with natural and processed black shale. *Appl. Geochem.* 24: 359–369.

Prat, O., Vercouter, T., Ansoborio, E., Fichet, P., Kurttio, P. & Salonen, L. 2009. Uranium speciation in drinking water from drilled wells in southern Finland and its potential link to health effects. *Environ. Sci. Technol.* 43: 3941–3945.

Snäll, S. 1988. Mineralogy and maturity of alum shales of south-central Jämtland, Sweden. *Swedish Geological Survey, Ser* C 818. 46 pp.

Sundin, B. 2008. The Swedish shale-oil industry during the first half of the 19th century. Available at: www.diva-portal.org/smash/get/diva2:274308.pdf.

SVAPO. 2005. Ranstad: Environmental and health risk assessment of the Tranebärssjön and tailings area. Report (In Swedish). 51 pp.

Widerlund, A. & Ingri, J. 1995. Early diagenesis of arsenic in sediments of the Kalix estuary, northern Sweden. *Chem. Geol.* 125: 185–196.

1.7 Arsenic in dust and road deposits

Environmental Arsenic in a Changing World –
Zhu, Guo, Bhattacharya, Ahmad, Bundschuh & Naidu (Eds)
ISBN 978-1-138-48609-6

Trends in antimony pollution near exposed traffic nodes: comparison with arsenic

B. Dousova[1], M. Lhotka[1], V. Machovic[1], F. Buzek[2], B. Cejkova[2] & I. Jackova[2]
[1]*University of Chemistry and Technology Prague, Prague, Czech Republic*
[2]*Czech Geological Survey, Prague, Czech Republic*

ABSTRACT: The significant part of antimony (Sb) contamination can be associated with extremely loaded traffic areas, where braking vehicles produce the abrasion dust containing up to 5% wt. of Sb_2S_3. Heavy loaded cross-roads from three different regions of the Czech Republic (Central Europe) were monitored for Sb content in road dusts, topsoils and reference soils during the two-year season (2015–2017). The concentration of antimony exceeded up to 60 times the background values and varied from 5 to 70 $\mu g\,g^{-1}$ in topsoils, and from 70 to 160 $\mu g\,g^{-1}$ in road dusts with the preferable binding to fine particle fraction (<0.1 mm).

1 INTRODUCTION

Antimony (Sb) is a global environmental contaminant and one of the elements of increasing environmental significance. While arsenic (As) represents a historical toxic substance monitored in the long term, antimony is a typical emerging contaminant that has been classified as a serious problem for future generations (CEC, 1998). The significant part of Sb contamination can be associated with extremely loaded traffic areas, where braking vehicles produce the abrasion dust containing up to 5% wt. of Sb_2S_3. Antimony released during braking process is mostly bound in fine atmospheric particulate matter, i.e. in PM_{10} (ca 32 μg Sb/braking/car) and $PM_{2.5}$ (ca 22 μg Sb/braking/car) (Iijima *et al.*, 2008), a minor part remains in non-exhaust traffic emmissions (road dust). Roadside topsoils accumulate large amounts of vehicle emissions, so the concentrations of antimony in them can reach more than 60 times of background value. As with arsenic, the soil pH/Eh importantly determine antimony species in soils. Arsenic has been found as pentavalent arsenate, AsO_4^{3-} in the oxidic surface zone (0–3 cm) and as trivalent arsenite, AsO_3^{3-} under slightly reducing conditions (9–12 cm) (Wilson *et al.*, 2010), while antimony as pentavalent hydrated antimonate, $Sb(OH)_6^-$ represents the major oxidation state over a wide Eh range (Mitsunobu *et al.*, 2008).

The binding of both As and Sb oxyanions in soils has been mostly controlled by the complexation with iron (Fe^{III}), whereas Sb exhibits significantly higher affinity to natural organic matter (NOM) compare to As (Mitsunobu *et al.*, 2008).

The aim of this study was to monitore heavy loaded cross-roads from three different regions of the Czech Republic (Central, South and East Bohemia) for antimony content in road dusts and topsoils during the two-year season (2015–2017). The same samples were tested for As content to the evaluation of current contamination trends in exposed urban areas.

2 EXPERIMENTAL

2.1 *Sampling*

Road dusts and urban topsoils (0–5 cm) were collected periodically (4 sampling data during 2015–2017) at busy cross-roads in three big cities in the Czech Republic (Prague (P) in middle Bohemia, České Budějovice (CB) in southern Bohemia and industrial Ostrava (O) in eastern Bohemia) and kept in PVC packages. Topsoils were taken as three sub-samples from the area of 20 × 20 cm, mixed together, quartered, dried and sieved to <0.315–0.1 mm and <0.1 mm fractions, respectively. Road dusts were collected close to roads exposed to heavy traffic circulation, mixed together, dried and sieved to <0.315–0.1 mm and <0.1 mm fractions, respectively.

2.2 *Sample processing*

The fractions of solid samples were analysed for major components by XRF (Rigaku NEX QC), for the specific surface area as S_{BET} (Micromeritics ASAP 2020) and for TOC using elementary analysis (Fisons 1108). Finally, all samples were mineralized by microwave digestion (Berghof MWS-2 speedwave) with HCl and HNO_3 (1:3).

The concentration of Sb in aqueous solutions was determined by HG-AFS using PSA 10.055 Millennium Excalibur. The samples were pre-treated with a solution of HCl (36% v/w) and KI (50%) with ascorbic acid (10%). The declared detection limit was 0.05 ppm

Table 1. Average Sb concentration ($\mu g\ g^{-1}$) in topsoil and road dust fractions.

	Urban soils		Road dust	
Locality	<0.1*)	0.1–0.315	<0.1	0.1–0.315
P	13.6	5.5	45.3	29.9
CB	53.7	23.1	164.6	89.1
O	5.6	2.5	12.8	6.5

*) particle size in mm

and the standard deviation was experimentally determined as 2.5%.

3 RESULTS AND DISCUSSION

3.1 *Antimony content in topsoils and road dust*

The data in Table 1 illustrate a major binding of Sb in fine particles (more than double the concentrations compare to the coarser fraction), and a significantly higher content of Sb in road dust compare to soils. While the concentration ranged from units to tens of $\mu g\ g^{-1}$, in road dust it reached up to tens to hundreds of $\mu g\ g^{-1}$ in topsoils. Evident local variability in Sb contamination (CB >> P > O) can be explained by the different traffic intensity at the sampling sites.

3.2 *Antimony correlation with Fe and OM*

In all aspects, antimony demonstrated better correlation with organic matter than with Fe. Very good Sb/C_{org} correlation ($R^2 = 0.893$) was found in topsoils, less apparent correlation factor ($R^2 = 0.702$) could be assigned to Sb/C_{org} in road dust. Unlike arsenic, that exhibited high positive correlation with Fe in different natural solids (Dousova *et al.*, 2012), Sb/Fe correlation in topsoils was worse ($R^2 = 0.594$) and even none Sb/Fe correlation was found in road dust.

3.3 *Important trends in Sb and As pollution in urban areas*

The contamination impact of brake abrasion (Fig. 1) illustrated decreasing trend in Sb content from source brake abrasion over the road dust and topsoils to the reference soils, which were collected in places without transport, as close as possible to the sampling sites. Nevertheless, the minimum Sb concentration measured in the reference soils exceeded the declared background value (0.1–1.9 $\mu g\ g^{-1}$) markedly.

Thanks to brake pads abrasion and growing usage of PET bottles the world antimony production sharply increased over the last 40 years (USGS, 2017). Conversely, in most of European countries arsenic emissions declined by 70–85% following the considerable decrease of atmospheric emissions in Central Europe from 1985–2000. Over the 20-year time span As concentrations in vertical atmospheric deposition decreased 8 times, those in interception deposition decreased 16 times (Doušová *et al.*, 2007).

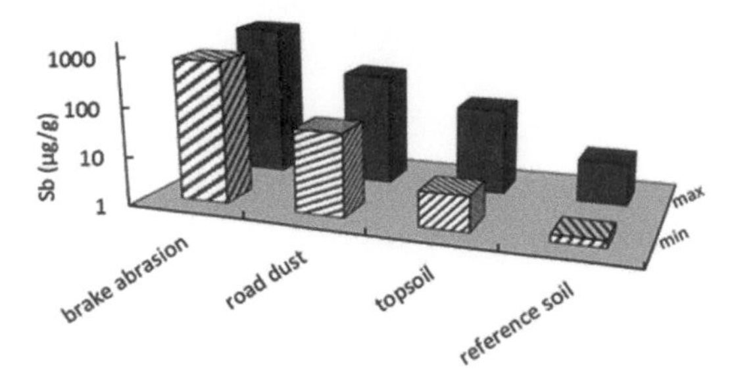

Figure 1. Contamination effect of source brake abrasion.

4 CONCLUSIONS

Brake abrasion dust represents one of the major sources of antimony pollution. In exposed localities antimony concentration exceeded up to 60 times the background value and decreased in the order: brake abrasion ($\approx 10^3\ \mu g\ g^{-1}$) > road dust ($\approx 10^2\ \mu g\ g^{-1}$) > topspoils ($\approx 10^1\ \mu g\ g^{-1}$) $\gg$ reference soils ($<1\ \mu g\ g^{-1}$) being mostly accumulated in fine particles. Contrary to arsenic, antimony correlated better with OM than with Fe in topsoils, its straight correlation with Fe/OM in road dust was not observed. While arsenic contamination has recently decreased thanks to a massive decline of arsenic emissions, antimony contamination indicates a dangerous progress due to growing automotive traffic.

ACKNOWLEDGEMENTS

This work was part of projects 16-13778S (Grant Agency of Czech Republic).

REFERENCES

CEC, 1998. Evaluating the cost and supply of alternatives to MTB in Californias's reformulated gasoline. Draft Report. California Energy Commission.

Doušová, B., Erbanová, L., & Novák, M. 2007. Arsenic in artmospheric deposition at the Czech-Polish border: Two sampling campaigns 20 years apart. *Sci. Total Environ.* 387: 185–193.

Dousova, B., Buzek, F., Rothwell, J., Krejcova, S. & Lhotka, M. 2012. Adsorption behavior of arsenic relating to different natural solids: soils, stream sediments and peats. *Sci. Total Environ.* 433: 456–461.

Iijima, A., Sato, K., Yano, K., Kato, M., Kozawa, K. & Furuta, N. 2008. Emission factor for antimony in brake abrasion dusts as one of the major atmospheric antimony sources. *Environ. Sci. Technol.* 42: 2937–2942.

Mitsunobu, S., Takahashi, Y. & Sakai, Y. 2008. Abiotic reduction of antimony(v) by green rust ($Fe_4(II)Fe_2(III)(OH)_{12}$ $SO_4 . 3H_2O$). *Chemosphere.* 70: 942–947.

USGS. 2017. Mineral commodity summaries 2017: United States Geological Survey 202p. (https://doi.org/10.3133/70180197).

Wilson, S.C., Lockwood, P.V., Ashley, P.M. & Tighe, M. 2010. The chemistry and behaviour of antimony in the soil environment with comparisons to arsenic: a critical review. *Environ. Pollut.*, 158: 1169–1181.

Environmental Arsenic in a Changing World –
Zhu, Guo, Bhattacharya, Ahmad, Bundschuh & Naidu (Eds)
ISBN 978-1-138-48609-6

Arsenic contaminated dust and the mud accident in Fundão (Brazil)

C.S. Fernandes, A.C. Santos, G.R. Santos & M.C. Teixeira
Environmental Engineering Graduating Program (ProAmb), Pharmacy Department (DEFAR), School of Pharmacy, Federal University of Ouro Preto, MG, Brazil

ABSTRACT: On November 5th, 2015, Fundão tailings dam in Mariana (Minas Gerais) collapsed. This was considered the biggest environmental disaster in Brazil and one of the largest in the world. Mud wastes (more than 50 million m^3) were dumped downstream of the Doce River basin. Hundreds of thousands of people living along the rivers have been affected. The aim of this study was to characterize dust samples collected after the mud deposition. Although iron mining wastes were classified as "not-hazardous" materials, our chemical analyses results pointed out the occurrence of potentially hazardous elements as As, Al, Mn and Pb at very high levels in the collected particulate material.

1 INTRODUCTION

Minas Gerais state faced its worst environmental disaster in November 2015: the rupture of a dike of Fundão tailings dam (owned by the mining company Samarco) at Mariana. This accident was the most devastating environmental disaster in the Brazilian history. So far, the magnitude of the impacts is unknown, as well as how much time it will take to repair all the damages generated by this tragedy. Iron mining wastes (sludge) were dumped downstream of Doce River basin (DRB), affecting all the basin ecosystem and also some coastal beaches located more than 660 km far from the accident site. Six months after the disaster, with the beginning of the dry season, people still living within the affected areas, started complaining because of the dust generated during wastes removal process. This study aimed to collect and characterize dust samples since: (i) the accident took place in an arsenic contaminated area, (ii) arsenic mobilization is totally likely; (iii) it is well known that arsenic and other contaminants may be carried not only by water but also by soil and dust and; (iv) the negative effects of particulate material (PM) to human health are not negligible and, (v) PM inhalation is a function of their size, the smaller the particle, the easier its inhalation and the higher would be its potential toxicity (Segura *et al.*, 2016).

2 METHODS/EXPERIMENTAL

2.1 *Reagents, materials, software and equipment*

Analytical grade reagents; disposable plastic flasks, buckets and bags; glass flasks; sieves. Software: Excell, ArcGIS 10.1 and QGIS 2.18. Equipment: Mechanical sieve agitator, analytical scale, hot plate, Inductively Coupled Plasma-Optical Emission Spectrometer (ICP-OES, Varian 725-ES), Total Reflection X-Ray Fluorescence–TXRF spectrometer (S2 Picofox, Brucker); Zetasizer (Nano, Malvern); portable global positioning syatem-GPS (Garmin).

2.2 *Sampling and sample characterization*

Sampling points were distributed along the course of the Gualaxo do Norte River (the first affected by mud). Plastic buckets (2 L, 1.80 m high) were distributed on the river banks (±70 km), between Ouro Preto (OP), Mariana (M) and Barra Longa (BL) municipalities, at the very beginning of the dry season (June, 2016). A dried mud sample (GST) was collected as a reference material. After a 4-month interval, 23 dust samples were retrieved, dried (room temperature) and sieved ($\leq$0.088 mm = PM). GST and dust samples with size distribution between 0.037–0.2 mm were used for quantitative chemical analysis (ICP_OES) after hot plate acid digestion. Dust and GST particles smaller than 0.037 mm were used for the determination of the Zeta potential and polydispersity index and chemical composition was accessed by TXRF.

3 RESULTS AND DISCUSSION

3.1 *Dust samples – Physical properties*

Point 1 (S 20°16.847′, W 43°28.185′) was located at OP, upstream of the dam. All other points were easily accessible and had been purposely distributed at high traffic areas (used for mud removal). Points

Table 1. Granulometric distribution – particulate matter (PM).

	Granulometric distribution (PM) [%]			
	x<0.044 [mm]	0.044<x<0.053 [mm]	0.053<x<0.088 [mm]	x>0.088 [mm]
1	25,70	24,59	24,84	24,86
2	24,55	23,75	24,18	27,52
3	25,19	24,06	24,19	26,56
4	22,95	21,55	22,40	33,09
5	23,60	23,41	23,30	29,69
6	24,27	23,58	24,11	28,04
7	25,12	23,92	24,82	26,13
8	26,61	21,61	22,90	28,88
9	28,02	23,87	24,19	23,92
10	17,37	23,97	25,89	32,77
11	20,41	22,45	23,11	34,03
12	26,25	23,83	23,94	25,98
13	26,20	24,35	24,70	24,75
14	25,67	24,11	24,50	25,72
15	24,47	24,49	24,33	26,71
16	33,56	–	33,10	33,34
17	25,59	24,69	24,65	25,07
19	27,86	24,15	21,12	26,87
20	10,30	21,07	25,20	43,43
21	25,79	24,73	24,84	24,64
22	25,82	24,51	24,71	24,96
23	25,95	24,38	24,49	25,18
GST	9.44	28.80	53.01	8.75

GST, dried mud deposited and collected as a reference material.

2–10 were located in Mariana while the other 16 points were distributed at BL. Points 10 and 23 were close to residences affected by mud, on the banks of the Gualaxo do Norte River (S 20°18.142′, W 43°14.987′ and S 20°16.949′, W 43°02.558′). Granulometric distribution of collected samples is shown in Table 1. Approximately 10 g of dust was deposited in each collector (2 L, 0.021 m^2), representing an average deposition of ±476 g m^2. Of particular note are points 7 (S 20°17.578′ W 43°15.160′) and 23, were 38.241 and 63.4 g of dust, were collected. Fine GST particles showed a diameter of 0.8555 μm and a polidispersive index of 0.624 and a Zeta potential of −20.7 mV. Diameter and the negative charge of GST particles should be considered for the prediction of its physiological and toxicological behavior. As approximately 50% of the GST material was considered as fine particles and some particles were smaller than 1 μm, those particles would be easily inhalable and absorbable.

3.2 *Dust samples – Chemical composition*

Elemental analysis of GST and dust particles pointed to the presence of some potentially toxic trace elements such as As, Al, Cu, Mn, Pb and Zn, besides Fe (the main element). Our results showed that chemical composition vary depending on particle size. In the case of arsenic, GST average arsenic content was 23.34 mg As kg^{-1}. Dust samples averaged arsenic content was 49.47 mg As kg^{-1}. The highest As content was found in sample 3 (S 20°16.257′, W 43°17.802′), 194.52 mg As kg^{-1}, mainly found in the particles sized between 0.053 and 0.088 mm (462 mg kg^{-1}). Sample 21 (S 20°16.949′, W 43°02.558′) presented the second higher As content, 139.35 mg As kg^{-1} and in that case, As was concentrated in the very fine particles (481 mg kg^{-1}). After the disaster, some reports were issued with the objective of ascertaining the situation of the affected area, these studies were carried out by both governmental and non-governmental organizations. Results of the analyzes of the water and sediment samples collected in the Doce River indicated contamination rates by As, Mn, Pb, Al and Fe, above the values allowed by Brazilian legislation (BRASIL Resolução 2005). The same elements were found in both mud samples collected immediately after the event (Segura *et al.*, 2016) and dust samples (present study). Such data, however, differ from those historically observed prior to the accident. Thus, changes in both air and water quality may be related to flooding by tailings sludge, as well as possible environmental changes and disturbances to the health of the inhabitants of affected areas (IGAM, 2015).

4 CONCLUSIONS

Even if in their origin, iron mining wastes were free of potentially dangerous elements like arsenic, it seems that the energy from the mud flooding was so intense that caused the re-suspension and re-solubilization of some very old sediments deposited at the river banks for centuries. Chemical elements from sediments were carried with the mud and are now mixed with the waste. Toxic elements mobilized during the artisanal gold mining are at the surface again and some of those particles are small enough to be classified as PM and therefore, being inhaled by human.

ACKNOWLEDGEMENTS

Authors are grateful to LGQA/DEGEO/UFOP and LCMEM/DEQUI/UFOP for ICP-OES and TXRF chemical analyses, respectively.

REFERENCES

BRASIL Resolução. 2005. CONAMA N° 357. pp. 58–63. www.mma.gov.br/port/conama/res/res05/res35705.pdf.

IGAM. 2015. Technical report-Acompanhamento da Qualidade das Águas do Rio Doce Após o Rompimento da Barragem da Samarco no distrito de Bento Rodrigues. Belo Horizonte: Instituto Mineiro de Gestão das Águas. 49p. (www.cbhdoce.org.br/wp-content/uploads/2016/06/Relatorio_Qualidade_15dez2015.pdf).

Segura, F.R., Nunes, E.A., Paniz, F.P., Paulelli, A.C.C., Rodrigues, G.B., Braga, G.Ú.L., Pedreira Filho, W.D.R., Barbosa Jr., F., Cerchiaro, G., Silva, F.F. & Batista, B.L. 2016. Potential risks of the residue from Samarcos's mine dam burst (Bento Rodrigues, Brazil). *Environ. Pollut.* 218: 813–825.

1.8 Advances and challenges in arsenic analysis in solid and aqueous matrix

Environmental Arsenic in a Changing World –
Zhu, Guo, Bhattacharya, Ahmad, Bundschuh & Naidu (Eds)
ISBN 978-1-138-48609-6

Effects of arsenite oxidation on metabolic pathways and the roles of the regulator AioR in *Agrobacterium tumefaciens* GW4

G.J. Wang, K.X. Shi, Q. Wang & X. Fan
State Key Laboratory of Agricultural Microbiology, College of Life Science and Technology, Huazhong Agricultural University, Wuhan, P.R. China

ABSTRACT: To determine the related metabolic pathways mediated by As(III) oxidation and whether AioR regulates other cellular responses to As(III), proteomic and genetic analyses were performed on the heterotrophic arsenite-oxidizing *Agrobacterium tumefaciens* GW4. The results showed that AioR was the main regulator for As(III) oxidation, and As(III) oxidation altered several cellular processes, especially phosphate, cell wall and carbon metabolism. In addition, the synchronous regulation of As resistance/oxidation proteins and phosphate transport proteins indicated that the metabolism of As was highly associated with that of phosphate. AioR suppressed the expression of the pst2/pho2 system and up-regulated the pst1/pho1 system to transport As and phosphate economically. AioR is the key driver of strain GW4 to adapt well in the As(III)-enriched sediment environment.

1 INTRODUCTION

Agrobacterium tumefaciens GW4 is a highly As(III) resistant [minimal inhibitory concentration (MIC) = 25 mM] and As(III)-oxidizing bacterium isolated from As-enriched groundwater sediments (Fan *et al.*, 2008). Unlike most of the heterotrophic As(III)-oxidizing bacteria using As(III) oxidation as an As detoxification process, the As(III) oxidation of strain GW4 enhanced the bacterial growth, and the strain showed positive chemotaxis toward As(III) (Shi *et al.*, 2017; Wang *et al.*, 2015). However, the mutant strain GW4-Δ*aioR* failed to demonstrate increased growth, and its As(III) oxidation and As(III) chemotaxis phenotypes were both disrupted (Shi *et al.*, 2017). It appeared that the effect of As(III) oxidation in strain GW4 was different from that in all the well-recognized heterotrophic and chemoautotrophic As(III)-oxidizing strains. We speculated that As(III) may be involved with several different metabolism pathways, and AioR may regulate other cellular functions besides As(III) oxidation.

Thus, in this study, we developed *A. tumefaciens* GW4 as a model to understand the alteration of global metabolism pathways with As(III) oxidation, and the regulatory roles of AioR. The methods includes isobaric tags for relative and absolute quantitation (iTRAQ) proteomics, gene knock out, and complementation and gene transcription analyses.

2 MATERIALS AND METHODS

2.1 *A. tumefaciens culture*

GW4 was grown at 28°C in $MMNH_4$ medium containing 55 mM mannitol as the primary carbon source and modified to contain 0.1 mM phosphate (Liu *et al.*, 2012). Total protein was extracted from the control, the As(III) treated strain GW4 and GW4-Δ*aioR* cells. iTRAQ proteomics was analyzed using AB SCIEX nanoLC-MS/MS (Triple TOF 5600 plus).

2.2 *Gene knock out and complementation of aioR*

Gene knock out and complementation were performed as described in Wang *et al.* (2015) and Shi *et al.* (2017) and generated the mutant strain GW4-Δ*aioR* and the complemented strain GW4-Δ*aioR*-C.

2.3 *Quantitative RT-PCR analysis*

Overnight cultures ($OD_{600} = 0.7$–0.8) were inoculated into 100 mL $MMNH_4$ medium. RNA extraction and

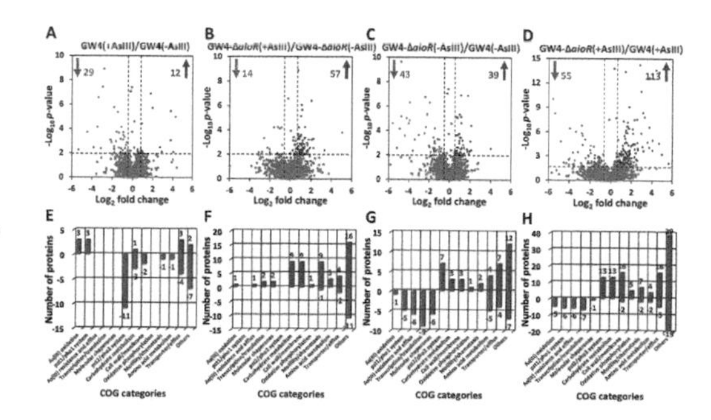

Figure 1. iTRAQ-based proteomic analysis of global perturbation of *A. tumefaciens* GW4 cellular metabolism in response to As(III) exposure. Gray vertical lines denote the 1.6 and −1.5 fold change designated cut-off. Horizontal gray lines denote a statistical significant *p*-value ≤0.01. (E-H) Classification of proteins in COG categories. Red bars represent the up-regulated proteins and green bars represent the down-regulated proteins.

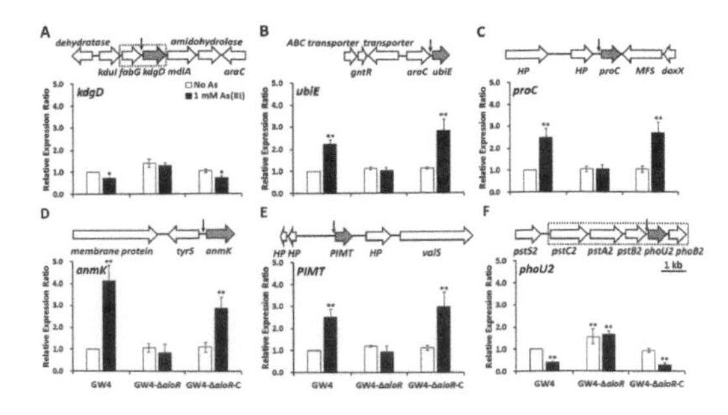

Figure 2. Influence of As(III) and AioR on the transcription of the *kdgD* (A), *ubiE* (B), *proC* (C), *anmK* (D), *PIMT* (E) and *phoU2* (F) genes. The locations of the arrows are the AioR putative binding sites. The gene clusters in the dashed boxes are co-transcribed (data not shown).

Quantitative RT-PCR were performed as described before (Liu *et al.*, 2012; Wang *et al.*, 2015).

3 RESULTS AND DISCUSSION

3.1 *Proteiomics*

iTRAQ was compared in four treatments, GW4 (+AsIII) / GW4 (-AsIII), GW4-Δ*aioR* (+AsIII) / GW4-Δ*aioR* (-AsIII), GW4-Δ*aioR* (-AsIII) / GW4 (-AsIII) and GW4-Δ*aioR* (+AsIII) / GW4 (+AsIII). A total of 41, 71, 82 and 168 differentially expressed proteins were identified, respectively (Fig. 1).

3.2 *Quantitative RT-PCR analysis*

Gene transcription levels of the wild type GW4, GW4-Δ*aioR* and GW4-Δ*aioR*-C showed that AioR affected the transcription of *kdgD, ubiE, proC, anmK, PIMT* and *phoU2* genes (Fig. 2). EMSA analysis also showed that AioR interacted with the promoter region of these genes (data not shown). These genes are related to several metabolic pathways.

3.3 *The regulatory cellular networks controlled by AioR*

Based on our previous studies (Wang *et al.*, 2015; Shi *et al.*, 2017), and the above data, we suggested that the regulatory cellular networks controlled by AioR are As resistance (ars operon), phosphate metabolism (*pst/pho* system), TCA cycle, cell wall/membrane, amino acid metabolism and motility/chemotaxis. AioR regulates several cellular processes. AioR regulates the expression of AioBA to oxidize As(III) (1.1), and regulates the expression of AioE to transfer the electron from AioBA to AioE with the generation of NADH (1.2). AioR could also regulate the expression of UbiE to maintain the electron transport within oxidative phosphorylation complex (1.2). AioR suppress the processes of glycolysis and TCA cycle by regulating KdgD (1.3). AioR also regulates the expression of Ars, AnmK, ProC and PIMT to participate in

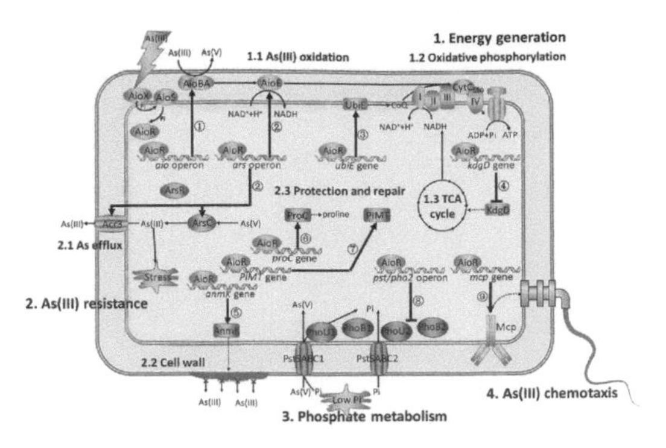

Figure 3. A general view showing the regulatory cellular networks controlled by AioR.

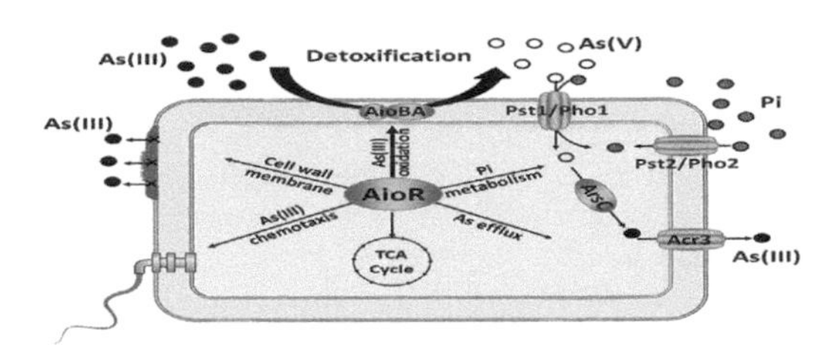

Figure 4. A general view showing of the roles of arsenic regulator AioR.

As efflux (2.1), recycling of peptidoglycan (2.2), synthesis of proline (2.3) and PIMT repair system (2.4) to enhance As(III) resistance. Besides, AioR regulates PhoU2-PhoB2 in phosphate metabolism and regulates Mcp in As(III) chemotaxis (Fig. 3).

In addition, the synchronous regulation of As resistance/oxidation proteins and phosphate transport proteins indicated that the metabolism of As was highly associated with that of phosphate. AioR suppressed the expression of the *pst2/pho2* system and up-regulated the *pst1/pho1* system to transport As and phosphate economically (Fig. 4).

4 CONCLUSIONS

A. tumefaciens GW4 represents a new type of heterotrophic As(III)-oxidizing bacteria that can use As(III) to enhance growth, and so which is different from most of the known heterotrophic As(III)-oxidizing bacteria using As(III) oxidation as a detoxification process. Our results show that cellular responses to As(III) are delicate, and AioR mainly drives As(III) oxidation. Additionally, AioR could regulate multiple genes involved in phosphate metabolism, As(III) resistance, and energy acquisition. AioR suppressed the expression of the *pst2/pho2* system and up-regulated the *pst1/pho1* system to transport As and phosphate economically. The regulatory cellular networks controlled by AioR provide the high As(III) resistance and better adaptation for strain GW4 to live in the As-rich sediment environment.

ACKNOWLEDGEMENTS

The present study was supported by the National Natural Science Foundation of China (31670108). During the revision of this abstract, a part of the results are accepted by Environmental Pollution (in press).

REFERENCES

Fan, H., Su, C., Wang, Y., Yao, J., Zhao, K., Wang, Y. & Wang, G. 2008. Sedimentary arsenite-oxidizing and arsenate-reducing bacteria associated with high arsenic groundwater from Shanyin, Northwestern China. *J. Appl. Microbiol.* 105: 529–539.

Liu, G., Liu, M., Kim, E.H., Maaty, W.S., Bothner, B., Lei, B., Rensing, C., Wang, G. & McDermott, T.R. 2012. A periplasmic arsenite-binding protein involved in regulating arsenite oxidation. *Environ. Microbiol.* 14: 1624–1634.

Shi, K., Fan, X., Qiao, Z., Han, Y., McDermott, T.R., Wang, Q. & Wang, G. 2017. Arsenite oxidation regulator AioR regulates bacterial chemotaxis towards arsenite in Agrobacterium tumefaciens GW4. *Sci. Rep.* 7: 43252.

Wang, Q., Qin, D., Zhang, S., Wang, L., Li, J., Rensing, C., McDermott, T.R. & Wang, G. 2015. Fate of arsenate following arsenite oxidation in Agrobacterium tumefaciens GW4. *Environ. Microbiol.* 17: 1926–1940.

Environmental Arsenic in a Changing World –
Zhu, Guo, Bhattacharya, Ahmad, Bundschuh & Naidu (Eds)
ISBN 978-1-138-48609-6

Sulfur-arsenic interactions and formation of thioarsenic complexes in the environment

I. Herath[1], J. Bundschuh[1] & P. Bhattacharya[2]
[1]*Faculty of Health, Engineering and Sciences, University of Southern Queensland, Toowoomba, Queensland, Australia*
[2]*KTH-International Groundwater Arsenic Research Group, Department of Sustainable Development, Environmental Science and Engineering, KTH Royal Institute of Technology, Stockholm, Sweden, and International Center for Applied Climate Science, The University of Southern Queensland, Toowoomba, Queensland, Australia*

ABSTRACT: Thiolated arsenic compounds are the sulfur analogous substructures of oxo-arsenicals as the arsinoyl (As=O) is substituted by an arsinothioyl (As=S) group. Relatively brief history of thioarsenic research, mostly in the current decade has endeavored to understand their consequences in the natural environment. However, thioarsenic related aspects have by far not attached much research concern on global scale compared to other arsenic species. This paper provides a critical overview on formation mechanisms of thioarsenicals and their chemistry aiming to direct future research towards thioasenic mitigation strategies. Sulfur-arsenite/arsenate interactions and dissolution of arsenic sulfide minerals are the main mechanisms that involve in the formation of different thioarsenic species in the natural environment. The formation and chemical analysis of thioarsenicals in soil and sediments are highly unknown. Future research needs to be more inclined towards in determining the molecular structure of unknown thioarsenic complexes in various environmental suites.

1 INTRODUCTION

Arsenic (As) and sulfur (S) commonly co-exist in the environment due to their strong interrelationship in the biogeochemical cycles. Arsenic species where the oxygen-bonded As is substituted by sulfur, thereby forming As-SH and/or As=S substructures are known as thioarsenic compounds (Guo *et al.*, 2017). Occurrence of thioarsenic species has been identified as a vital parameter to better understand the geologic and geochemical mechanisms of As enrichment in sulfide-rich waters. Therefore, a holistic understanding of the formation, speciation and chemical analysis of thioarsenic species is crucial for investigating the fate of As in different environmental, geological, and biological systems. However, the definite chemical nature of many thioarsenic compounds still remains a subject of debate and hence, it is urgently necessary to develop proper analytical protocols for resolving this dispute in order to understand mechanisms of As speciation, geochemistry, and mobility in the natural environment. The present paper provides an overview of the formation mechanisms of thiolated As complexes and their chemistry in order to direct future research towards thioarsenic mitigation strategies.

2 DISCUSSION

2.1 *Formation of thioarsenic species*

Thioarsenic species are formed as a result of geochemical interactions between As and sulfur-bearing compounds (sulfides; S^{2-}/HS^{-}) in favorably reducing or anoxic environments (Härtig & Planer-Friedrich 2012). Laboratory as well as field scale studies have demonstrated the formation pathways of thioarsenic complexes in sulfidic water containing both inorganic arsenate and arsenite. Oxidation of arsenite in sulfidic environments and dissolution of arsenic sulfide minerals have been recognized as the most significant geochemical triggers for the formation of thioarsenic species in the environment.

2.2 *Formation of thioarsenic species by sulfur-arsenite interactions*

Thioarsenic complexes in the form of thioarsenate are formed through a series of redox reactions between arsenite and sulfur while donating non-bonding electron pair of arsenite to sulfur atom. The formation of thioarsenate takes place through the deprotonation of arsenite in the presence of aqueous sulphide (H_2S), where arsenite is partially oxidized to monothioarsenate (reaction 1).

$$5H_3AsO_3 + 3H_2S \rightarrow 2As + 3H_2AsO_3S^- + 6H_2O + 3H^+ \quad (1)$$

In natural environments, such as some ground water and geothermal fluids, this reaction occurs due to presence of electron acceptors such as sulfate and sulfur which are produced from H_2S by bacterial activities. The monothioarsenate produced from reaction (1) is further converted to dithioarsenate in the presence of excess sulfide (reaction 2).

$$H_2AsO_3S^- + H_2S \rightarrow H_2AsO_2S_2^- + H_2O \quad (2)$$

Microbial activities of sulfate-reducing bacteria, such as *Desulfotomaculum* can strongly influence the formation of thioarsenate from the oxidation of arsenite in sulfidic geothermal water (Wu *et al.*, 2017). In sulfidic aquifers under anoxic conditions, *Desulfotomaculum* bacterium can oxidize arsenite to arsenate which can react with S^{2-}/HS^-, thereby producing thioarsenates ($AsO_{4-x}S_x^{2-}$ with x = 1−4) along with mono-, di-, and tri-thioarsenate as dominant species (reaction 3).

$$[HAs(V)O_4]^{2-} + [HS]^- \rightarrow [HAs(V)S^{2-}xO_{(4-x)}]^{2-} + [OH]^- \quad (3)$$

2.3 *Formation of thioarsenicals by dissolution of arsenic sulfide minerals*

Arsenic bearing sulfide minerals, specifically orpiment (As_2S_3), enargite (Cu_3AsS_4) and arsenopyrite (As-Fe-S) are important sources of the formation of various thioarsenic species in the environment. Dissolution of these mineral phases may trigger the release and formation of thioarsenic compounds which play a crucial role in As cycle in aqueous sulfidic systems.

Orpiment dissolution

The dissolution of amorphous As_2S_3 in sulfide deficient and sulfide rich solutions has been investigated at pH 4.4 ± 0.4 and a range of temperature from 25 to 90°C. In sulfide-deficient solutions ($H_2S < 10^{-3}$ M), H_3AsO_3 is the main product of the dissolution of As_2S_3 and the dissolution reaction can be expressed as:

$$12As_2S_3 + 3H_2O \Leftrightarrow H_3AsO_3 + 32H_2S \quad (4)$$

In sulfide rich solutions ($H_2S > 10^{-3.8}$ M), the solubility of As_2S_3 is controlled by the concentration of H_2S and the dissolution As_2S_3 may result $H_2As_3S_6^-$ as the dominant species of As (reaction 5). This trimeric thioarsenite is the solubility controlling species of amorphous or crystalline orpiment in sulfur rich solutions.

$$3As_2S_3 + 3H_2S \Leftrightarrow 2H_2As_3S_6^- + 2H^+ \quad (5)$$

Arsenopyrite dissolution

The oxidative dissolution of arsenopyrite in alkaline media produces dominantly monothioarsenates at (ΣAs/ΣS) of 0.8–1.3 (reaction 6) (Suess & Planer-Friedrich 2012).

$$FeAsS + 6OH^- \rightarrow Fe(OH)_3 + H_3AsO_3S + 6e^- (6) \quad (6)$$

The formation of monothioarsenate from arsenopyrite dissolution is governed by a physisorption mechanism of hydroxide under both oxic and anoxic conditions. First, arsenopyrite reacts with dissolved oxygen and subsequently hydroxyl group is adsorbed via a physisorption mechanism (reaction 7–9).

$$FeAsS.O_2 + H_2O + 2e^- \rightarrow FeAsS + H_2O_2 + OH^- \quad (7)$$

$$FeAsS.OH^- \rightarrow Fe(OH)AsS + e^- \quad (8)$$

$$Fe(OH)AsS^- \leftrightarrow FeAs(OH)S \quad (9)$$

As a result of these electron transfer reactions, a complex of As-OH-S is formed and thereby resulting in a surface complex of monothioarsenate (reaction 10).

$$[Fe(OH)AsS] \rightarrow FeAs(OH)S\ ] + OH^- \leftrightarrow Fe(OH)_3 + As(OH)_3S + 5e^- \quad (10)$$

Apart from the occurrence of thioarsenicals in environmental and geological systems, various thioarsenic complexes have also been detected in biological systems. Micro-biota living in human and animal gut can enhance the production of inorganic as well as methylated thioarsenate compounds due to complexation reactions between thiol groups and ingested arsenates in the gut. The replacement of OH- groups of methylated arsenic species by SH- (thiolation) can occur in H_2S producing gut microbiota within the distal gastrointestinal tract. For instance, highly toxic dimethylmonothioarsenate (LC50 – 10.7 in cultured A431 human epidermoid carcinoma cells) is formed by the thiolation of dimethylarsenate during the metabolism of dimethylarsenate in human urine.

3 CONCLUSIONS

This review focuses on the existing knowledge in relation to the formation mechanisms of thioarsenic species in the natural systems. Sulfur-arsenite/arsenate interactions and dissolution of arsenic sulfide minerals are the main mechanisms that involve in the formation of different thioarsenic species in the natural environment. The formation of thioarsenic complexes may lead to increase the solubility, mobility and bioavailability of As in the environment. More future researches are essential on thioarsenic related aspects in clarifying formation mechanisms, and developing upgraded analytical instrumentation to establish a sustainable arsenic mitigation on a global scale.

REFERENCES

Guo, Q., Planer-Friedrich, B., Liu, M., Li, J., Zhou, C. & Wang, Y. 2017. Arsenic and thioarsenic species in the hot springs of the Rehai magmatic geothermal system, Tengchong volcanic region, China. *Chem. Geol.* 453: 12–20.

Härtig, C. & Planer-Friedrich, B. 2012. Thioarsenate Transformation by Filamentous Microbial Mats Thriving in an Alkaline, Sulfidic Hot Spring. *Environ. Sci. Technol.* 46(8): 4348–4356.

Suess, E. & Planer-Friedrich, B. 2012. Thioarsenate formation upon dissolution of orpiment and arsenopyrite. *Chemosphere* 89(11): 1390–1398.

Wu, G., Huang, L., Jiang, H., Peng, Y., Guo, W., Chen, Z., She, W., Guo, Q. & Dong, H. 2017. Thioarsenate Formation Coupled with Anaerobic Arsenite Oxidation by a Sulfate-Reducing Bacterium Isolated from a Hot Spring. *Front. Microbiol.* 8(1336).

Environmental Arsenic in a Changing World –
Zhu, Guo, Bhattacharya, Ahmad, Bundschuh & Naidu (Eds)
ISBN 978-1-138-48609-6

Application of spectral gamma and magnetic susceptibility in an As-bearing loessic aquifer, Argentina

L. Sierra[1], S. Dietrich[1], P.A. Weinzettel[1], P.A. Bea[1], L. Cacciabue[1], M.L. Gómez Samus[2] & N.N. Sosa[3]
[1]*Instituto de Hidrología de Llanuras "Dr. Eduardo J. Usunoff" (IHLLA), Azul, Buenos Aires, Argentina*
[2]*Lab. de Entrenamiento Multidisciplinario para la Investigación Tecnológica (LEMIT), La Plata, Argentina*
[3]*Centro de Investigaciones Geológicas (CIG), La Plata, Argentina*

ABSTRACT: Application of Natural Gamma Spectroscopy (NGS) borehole logging and magnetic susceptibility (χ) on aquifer samples are compared with the identified mineralogy and trace elemental analyses. Acceptable correlation between ICP analyses (K and Th) and NGS signals has been found. However, some observed discrepancies on them could be related to the different exploratory volumes of each technique. Magnetic Susceptibility measurements present better correlations between the superparamagnetic signal (χdf) and As contents, probably linked to the pedogenetic origin of such particles.

1 INTRODUCTION

Geophysical measurements are easily implemented and non-destructive techniques that allow to obtain key features of aquifers. In fact, Natural Gamma Spectroscopy (NGS) borehole logging is a mineralogical log that improves the results of classical gamma ray log (GR) (Svendsen *et al.*, 2001). This tool estimates the ^{40}K, ^{232}Th and ^{238}U contents based on their radioactive activity (Killeen *et al.*, 2015). Scanlon *et al.* (2009) used GR logging to identify volcanic ash layers and related with As content in water. NGS can be used in cased wells thus expanding the exploratory possibilities. On the other hand, environmental magnetism is useful for studying sedimentary and post-depositional processes of magnetic minerals (Walden *et al.*, 1999). Magnetic susceptibility (χ) is related to the presence of magnetic minerals. Frequency dependent susceptibility (χ_{df}) signal may be increased by the relative abundance of superparamagnetic particles (SP) like ultrafine magnetite/maghemite. This pedogenic minerals are very common in soils and paleosols of pampean loessic sediments (Gómez Samus *et al.*, 2017). The objective of this work was to assess the NGS and χ logging signals, and relate them with mineralogy and As-bearing phases in the sediments of the aquifer.

2 MATERIALS AND METHODS

The Chaco-Pampean Plain in Argentina is one of the largest flatlands regions (1×10^6 km^2) with high As groundwater concentrations (Nicolli *et al.*, 2012). The study site is located in Tres Arroyos city, Buenos Aires province (−38.367°, −60.246°). The hosted aquifer consists locally of 120 m thick reworked aeolian mantle (loess) of Neogene to Quaternary age. It is composed of loessic sandy silts that are intermixed paleosols and calcrete horizons. As content in 1 m-averaged samples is 3.5–6.3 mg kg^{-1}. Groundwater is alkaline and dominates the Na-HCO_3 type. Arsenic content ranges between 10 and 234 μg L^{-1} (Sierra *et al.*, 2016). The volcanic glass shards are suggested as the main potential As source in this region. Secondary minerals such as clays and Fe-Mn oxy-hydroxides may be considered as additional sources of As (Sosa *et al.*, 2015).

2.1 *Sediment sampling and analysis*

Thirty samples for X-ray diffraction (XRD) and trace elemental analyses (ICP-MS, aqua-regia acid digestion) on each 1 m-mixed sediments were used.

2.2 *Geophysical measurement*

Magnetic sampling was performed on cores and detritus samples. Bulk samples were air-dried, milled and placed in 10 cc plastic boxes. The χ was measured in low (χ_{lf}) and high frequency (χ_{hf}) by a MS Bartington device with MS2B sensor and the frequency dependent susceptibility ($\chi_{df} = \chi_{lf} - \chi_{hf}$), was calculated. NGS logging was performed with a Robertson Geologging equipment using static measurements at 50 cm interval with 1 min counting time.

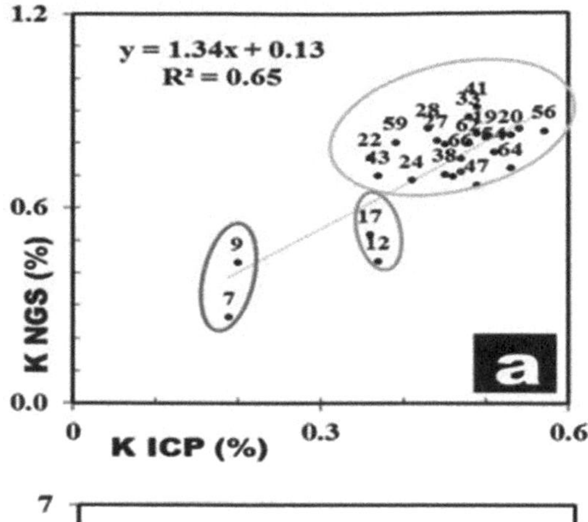

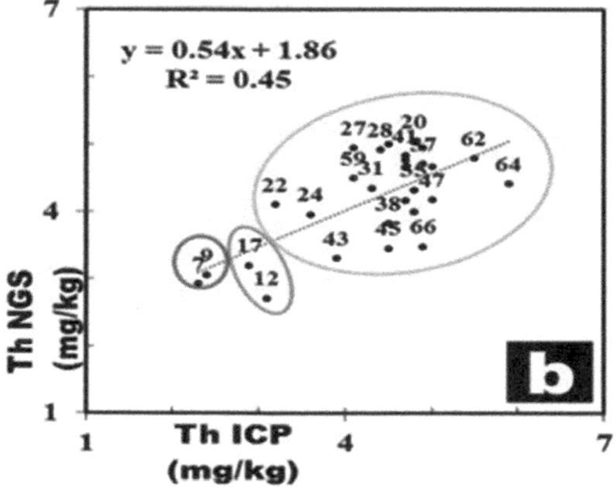

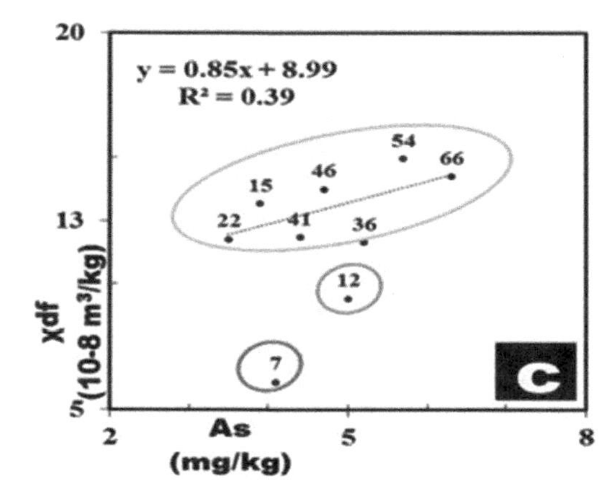

Figure 1. a) and b) Scatter plots NGS vs ICP content of a) K and Th NGS vs ICP content. b) χ_{df} vs As content. Three sections are marked: Red (low χ_{df} calcareous sands), Blue (variable χ_{df} calcretes) and Orange (high χ_{df} clayey silts).

3 RESULTS AND DISCUSSION

3.1 *Natural Gamma Spectroscopy*

Correlation coefficients between NGS log and ICP trace elements analyses (Fig. 1) showed acceptable values for K ($r^2 = 0.65$) and Th ($r^2 = 0.45$). No correlation was observed for U due to a) low concentrations in sediments. Good correlation was also observed between Fe content and K_{NGS} ($r^2 = 0.60$) and Th_{NGS} log (r^2: 0.41). This can be explained by the relative abundance of K-rich micas like illite and I/S interlayer in associations with Fe-oxyhydroxides. The NGS log can be divided in three sections (Figure 1). The abundance of calcite and quartz minerals lower GR, K_{NGS} and Th_{NGS} signals of the calcareous sands and calcretes between 5 and 17 meters below surface (mbs). High As calcrete at 12 mbs may be to surface or precipitation process that retain As on calcite.

The lower section is in the range 20–70 mbs. This older sediments (upper Miocene?) are enriched in K and Th. sandy silt loessic sediments palaeosols with intercalations of clayey silt and low As-Fe calcretes.

3.2 *Magnetic susceptibility*

Magnetic Susceptibility (χ_{lf} and χ_{hf}) correlation with As is low maybe due to multiple mineral sources. However, better correlations were achieved using χ_{df} with Fe ($r^2 = 0.72$) and As ($r^2 = 0.39$). This can be explained by the relative importance of SP magnetic particles, which are typical of paleosols of this region (Gómez Samus et al., 2017). The correlation of χ_{df} with radioelements shows also acceptable values for a) ICP analyses: K_{ICP} ($r^2 = 0.73$) and Th_{ICP} ($r^2 = 0.75$) due to the presence of SP magnetic particles with K-clays. b) NGS logging: K_{NGS} ($r^2 = 0.64$) and Th_{NGS} ($r^2 = 0.35$) lower correlations may be related to sample volumes.

4 CONCLUSIONS

Comparison of NGS geophysical log and ICP trace element analysis in sediments showed acceptable correlations that may be useful to characterizes As-bearing mineral in the aquifers. χ measurements presents better correlations between χ_{df} and As content because of the presence of superparamagnetic particles (SP) because of the pedogenic origin. Disparities are attributable to the different sample volumes.

ACKNOWLEDGEMENTS

We thank PICT1805/2014 (ANPCyT) and PID 0075/2011 (COHIFE) for supporting this research. We also acknowledge IHLLA Lab and Institute of Physics Arroyo Seco (IFAS) for analyses. Special thanks to the reviewers.

REFERENCES

Gómez Samus, M., Rico, Y. & Bidegain, J. 2017. Magnetostratigraphy and magnetic parameters in Quaternary sequences of Balcarce, Argentina. a contribution to understand the magnetic behaviour in cenozoic sediments of South America. *J. Geo. Res.* 13: 66–82.

Killeen, P. 2015. Tools and Techniques: Radiometric Methods. In: *Schubert, G. Treatise on geophysics*. Elsevier. (11): 447–524

Nicolli, H.B., Bundschuh, J., Blanco, M., Tujchneider, O., Panarello, H.O., Dapeña, C. & Rusansky, J.E. 2012. Arsenic and associated trace-elements in groundwater from the Chaco-Pampeano plain, Argentina: Results from 100 years of research. *Sci. Total Environ.* 429: 36–56.

Scanlon, B., Nicot, J., Reedy, R., Kurtzman, D., Mukherjee, A. & Nordstrom, D. 2009. Elevated naturally occurring arsenic in a semiarid oxidizing system, Southern High Plains aquifer, Texas, USA. *Appl. Geochem.* 24(11): 2061–2071.

Sierra, L., Cacciabue, L., Dietrich, S., Weinzettel, P. & Bea, S. 2016. Arsenic in groundwater and sediments in a loessic aquifer, Argentina. In *Arsenic Research and Global Sustainability: Proc. of the Sixth Int. Congress on Arsenic in the Environment (As2016),* Stockholm, Sweden (p. 92).

Sosa, N., Datta, S. & Zarate, M. 2015. Arsenic concentrations in soils and sediments of the southern Pampean Plain, within Claromecó River Basin (Argentina). In *AGU Fall Meeting.*

Svendsen, J. & Hartley, N. 2001. Comparison between outcrop-spectral gamma ray logging and whole rock geochemistry: implications for quantitative reservoir characterisation in continental sequences. *Mar. Pet. Geol.* 18(6): 657–670.

Walden J., Oldfield, F. & Smith, J.P. 1999. Environmental magnetism: a practical guide. *Technical Guide, No 6. Quaternary Research Association, London*, 214 pp.

Environmental Arsenic in a Changing World –
Zhu, Guo, Bhattacharya, Ahmad, Bundschuh & Naidu (Eds)
ISBN 978-1-138-48609-6

Speciation of arsenic in sediment and groundwater

W.T. Luo, X.B. Gao, Y. Li, X. Zhang & P.L. Gong
School of Environmental Studies, China University of Geosciences, Wuhan, P.R. China

ABSTRACT: Yuncheng basin is an area contaminated by arsenic (As). The species of As present in the study area need to be better characterized. In this study, species of As in sediment was investigated using sequential extraction method. Ion Chromatography – Hydride Generation – Atomic fluorescence Spectrometry (IC-HG-AFS) was used to determine the species of As in aquifer sediments and surrounding shallow groundwater. The instrumental system of ion chromatography(IC) couples with HG-AFS proved to be an effective method for analysis of the total As and the As speciation in sediments.

1 INTRODUCTION

Arsenic (As) is a toxic trace element presents in soils and groundwaters throughout the world (Javed *et al.*, 2014; Korte, 1991; Huang *et al.*, 2008). As one of the areas contaminated by As, Yuncheng basin shows a slightly higher As in surface soils than the background values of soils in China (Wei *et al.*, 1991). It belongs to the typical basin-derived high As groundwater with As contents in groundwater up to 50 $\mu g\ L^{-1}$. The toxicity of As varies with different species (e.g. As(III), As(V), monomethylarsonic acid (MMA) and dimethylarsenic acid (MMA)). The inorganic As species are believed to be more toxic than organic species. The forms of As present in the study area need to be better characterized. This study analyzed the species of As in groundwater and bore-drilling sediments in high-As groundwater area in Qiji County, Yuncheng basin.

2 METHODS/EXPERIMENTAL

Sediment samples were collected from one borehole in the Yuncheng Basin, Shanxi Province in 2015. Thirty one groundwater samples were collected from domestic water wells with depth between 40 and 65 m. Sequential extraction was performed to speciate As in sediment samples to get the six As fractions (Water-soluble fraction, Exchangeable fraction, Bound to carbonates or specifically adsorbed fraction, Bound to Iron and Manganese Oxide fraction, Bound to Organic Matter, Residue fraction). Different species of As were determined by IC-HG-AFS (ICS2100, Thermo Scientific, USA; RGF-6300, Bohui, Beijing, China) method (Fig. 1).

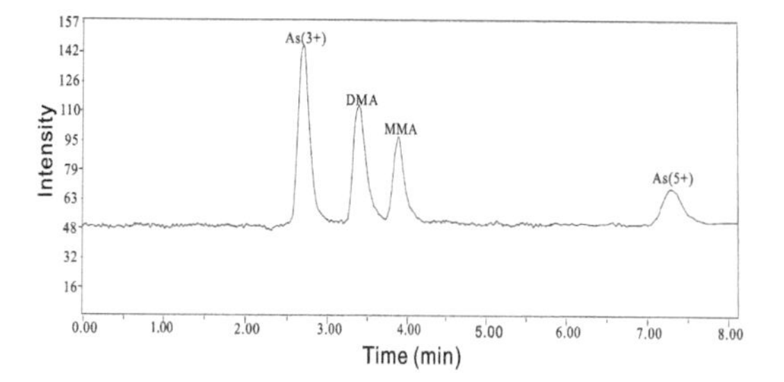

Figure 1. Chromatogram of IC-HG-AFS analyses for a standard solution of four arsenic species at 10 $\mu g\ L^{-1}$ concentration for each species. Intensity counts per second. The retention times of As(III), DMA, MMA, and As(V) is 2.54, 3.26,3.73, and 7.07 min, respectively.

3 RESULTS AND DISCUSSION

3.1 *Speciation of arsenic in sediment*

The chromatogram of the species analysis (Fig. 2) showed that As(V) and As(III) were detected in all As fractures, while organic As (DMA and MMA) was not found in fractions 1, 3 and 4. Exchangeable fraction is the loosely bound state. Only a little organic As (DMA and MMA) was found in fraction 2. The weight percent of As(III) and As(V) was 74.66% and 7.75% respectively, indicating that the migration of As(III) is stronger than As(V) in loosely bound state. The contents of these two fractions in the sediments were limited, accounting for a less impact on groundwater As pollution. Inorganic As remained dominant in fractions 3 and 4. Moreover, obviously increasing content of As(V) was observed in fraction 3, while the proportion of As(III) reduced relatively.

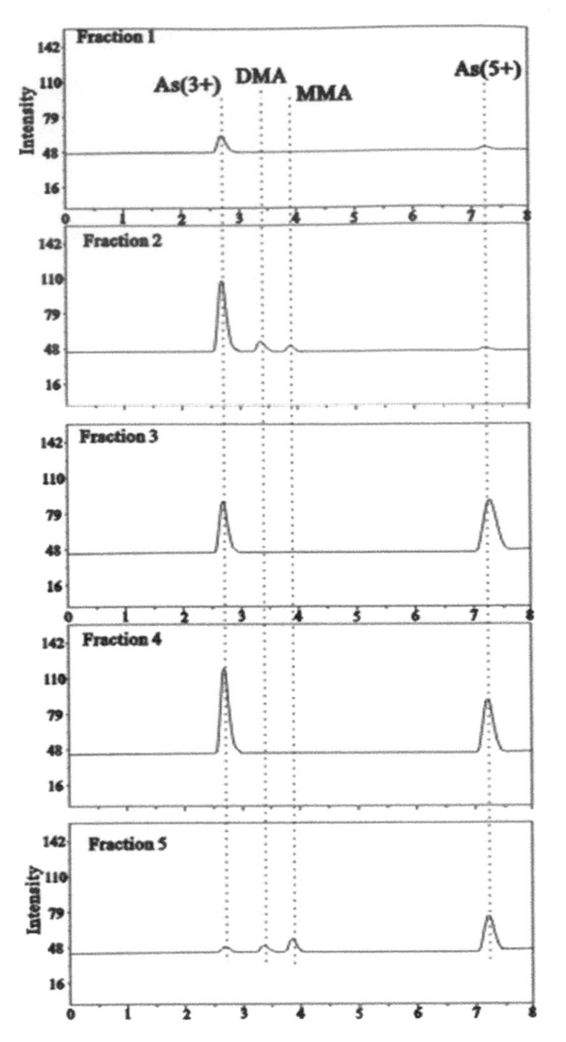

Figure 2. The chromatogram of the arsenic species in sediment samples.

The As(V) content reached its maximum in fraction 4. The weight percentage of As(V) in fraction 4 was 75.94%. The results showed that bounding to carbonates or specifically adsorbed fraction and bounding to iron and manganese oxide fraction are mainly related to As(V). In other words, the dominant species of As in the strongly bound state was As(V). Strongly bound As influenced the groundwater mainly through reductive dissolution and adsorption/desorption. Organic As (DMA and MMA) with a relatively low content was detected in fraction 5. Inorganic As remained dominant in this fraction.

3.2 *Speciation of arsenic in groundwater*

Total As contents in groundwater samples were above the World Health Organization's standard of $10\,\mu g\,L^{-1}$, expect for the QJ05 ($8.59\,\mu g\,L^{-1}$). The content of As in groundwater ranged from $8.59\,\mu g\,L^{-1}$ to $19.40\,\mu g\,L^{-1}$. Arsenic species analysis revealed that organic As is very low in groundwater and inorganic As is the dominate part. As to the inorganic As, As(V) was abundant, whereas As(III) was relatively low in all samples.

4 CONCLUSIONS

The instrumental system of ion chromatography (IC) couples with HG-AFS proves to be an effective method to analyze the As speciation in sediments and groundwater.

REFERENCES

Huang, C.L., Zheng, P. & Chen, Y.L. 2008. Change of the arsenic content in soils of the Linfen – Yuncheng basin, Shanxi, China. *Geol. Bull. China* 27(2):246–251.

Javed, M.B., Kachanoski, G. & Siddique, T. 2014. Arsenic fractionation and mineralogical characterization of sediments in the Cold Lake area of Alberta, Canada. *Sci. Total Environ.* 500–501: 181–190.

Korte, N. 1991. Naturally-occurring Arsenic in groundwaters of the mid western United State. *Environ. Geol. Water Sci.* 18(2): 137–141.

Wei, F.S., Chen, J.S. & Wu, Y.Y. 1991. Research of the soil environmental background value in China. *Environ. Sci.* 12(4): 12–19.

Environmental Arsenic in a Changing World –
Zhu, Guo, Bhattacharya, Ahmad, Bundschuh & Naidu (Eds)
ISBN 978-1-138-48609-6

Total arsenic and inorganic arsenic speciation and their correlation with fluoride, iron and manganese levels in groundwater intended for human consumption in Uruguay

I. Machado[1], V. Bühl[1] & N. Mañay[2]
[1]*Analytical Chemistry, Faculty of Chemistry, DEC, Universidad de la República (UdelaR), Montevideo, Uruguay*
[2]*Toxicology, Faculty of Chemistry, DEC, Universidad de la República (UdelaR), Montevideo, Uruguay*

ABSTRACT: The aim of this study is to develop analytical methodologies for the assessment of total arsenic (As) and inorganic As species, fluoride, iron and manganese in groundwater samples used for human consumption in Uruguay, and to evaluate the possible correlations between them. For this, 50 samples of groundwater from north Uruguay were analyzed. The concentration ranges found were: tAs (0.94–58.9) $\mu g\,L^{-1}$, F^- (0.168–1.528) $mg\,L^{-1}$, Fe (2.1–95.5) $\mu g\,L^{-1}$, and Mn (0.12–138.6) $\mu g\,L^{-1}$. More than half of the samples presented tAs concentration above WHO As limits for drinking water, with the corresponding risks for public health. Pearson correlations between tAs and F^-, Fe and Mn were studied, resulting in strong positive correlations for As/F^- and As/Fe. Preliminary results showed higher As(III) levels in the samples, which is in agreement with the reducing conditions of the aquifers.

1 INTRODUCTION

Arsenic (As) levels in natural waters have been cited in different environments, although the highest concentrations were found in groundwaters, mainly linked to natural geochemical processes (Smedley & Kinninburgh, 2002). In reducing aquifers, like those found in Uruguay, As(III) generally constitutes a high proportion of total As (tAs). Its mobilization is usually originated by desorption of mineral oxides and the reductive dissolution of Fe and Mn oxides. Since inorganic As (iAs) species present higher toxicity, especially As (III), it is important to carry out speciation analysis (Smedley *et al.*, 2001).

Geogenic As in groundwater surveillance is recently conducted in Uruguay, and different aquifers of the country showed As levels above those recommended by WHO guidelines for drinking water (10 $\mu g\,L^{-1}$) (WHO, 2017). The aim of this study is the development and optimization of analytical methodologies for the determination and assessment of tAs and iAs species, fluoride, iron and manganese in groundwater samples used for human consumption in Uruguay, as well as to evaluate the possible correlations between tAs and iAs with the mentioned inorganic parameters, of which, no systematic control studies are carried out.

2 EXPERIMENTAL

2.1 *Analytical determinations*

For analytical determinations of tAs, Fe and Mn by ET-AAS, an atomic absorption spectrometer (iCE 3500, Thermo Scientific) equipped with a graphite furnace atomizer and Zeeman-based correction was employed. Argon (Ar 99.99%, Linde) was used as purge and protective gas. The analytical lines used were: 193.7 nm (As), 302.1 nm (Fe) and 279.5 nm (Mn) and the signal used for quantification was integrated absorbance (peak-area). For all the determinations, pyrolytically coated graphite tubes (Thermo Scientific) were used (Table 1). For As determination palladium matrix modifier (Merck) was used (5 μg) (Table 2).

For iAs speciation analysis, a continuous flow hydride generation system made *in-house* was employed. Ar (99.99%, Linde) was used as carrier gas, with a flow rate of 75 $mL\,min^{-1}$, controlled by means of a rotameter (Cole Parmer). Two T-pieces (0.8 mm inner bore) were

Table 1. ET-AAS temperature programs for tAs, Fe and Mn.

Parameter	Temperature (°C)	Ramp rate (°C s^{-1})	Hold time (s)
Drying	100	10	$50^a/30^{b,c}$
Pyrolysis	$1200^a/1100^b/900^c$	150	20
Atomization	$2200^a/2100^b/1800^c$	0	3
Cleaning	2600	0	3

a-As, b-Fe, c-Mn. Ar flow: 0.2 $mL\,min^{-1}$, except during atomization step. Injection volume: 30 μL.

Table 2. Main figures of merit obtained by ET-AAS.

Parameter	As	Fe	Mn
Linearity ($\mu g\,L^{-1}$)	up to 20.0	up to 100	up to 3.0
LOD (3σ, n = 10) ($\mu g\,L^{-1}$)	0.27	0.54	0.032
LOQ (10σ, n = 10) ($\mu g\,L^{-1}$)	0.90	1.8	0.11
Precision (RSD, n = 10)		≤10%	
Trueness (compared to Certified value, n = 6)*	98–102%	99–101%	99–103%

*CRM NIST: SRM 1643f Trace elements in water.

used to merge sample flow with reductant flow and, downstream, to merge the reaction mixture flow with Ar flow. The outlet from a second T-piece was connected to a 3-mL internal volume gas–liquid separator with a forced outlet. Sample and reductant solutions were delivered by a peristaltic pump (RP-1 Dynamax). Sample and reductant flow rates were 4.0 mL min^{-1} and 1.5 mL min^{-1} respectively. The system was coupled to a flame atomic absorption spectrometer (AAnalyst 200, Perkin Elmer). An EDL lamp (Perkin Elmer) operated at 193.7 nm was used. The quartz T-tube cell with a path-length of 165 mm and a diameter of 12 mm, was heated to 980°C by an acetylene-air flame. The previous separation of the species was achieved by a liquid chromatograph (LC-20AT Prominence Shimadzu) working in isocratic mode, equipped with a Dionex IonPac AS22 (4 × 250 mm) column, using 60 mM $NH_4H_2PO_4$ solution adjusted to pH 5.8 as mobile phase at 1.5 mL min^{-1}.

Fluoride (F^-) concentrations were measured by a combined fluoride electrode connected to a pH/ion meter (Orion VersaStar Thermo Scientific). Total ionic strength adjustment buffer TISAB II (Orion Thermo Scientific) was used. For each measurement, 50 mL of sample were mixed with 10 mL of TISAB II solution.

2.2 *Reagents*

All chemicals used were of analytical reagent grade and all the solutions were prepared with ultrapure water of 18.2 MΩcm resistivity (ASTM Type I).

All glassware was soaked overnight in 10% (v/v) HNO_3 before use. For the preparation of calibration curves, commercial standard solutions (1000 mg L^{-1}) of As(V), Fe and Mn (Merck) were used, as well as sodium arsenite ($NaAsO_2$ 99.0%, Sigma-Aldrich) and sodium fluoride (NaF 99.5%, Merck).

2.3 *Samples*

A total of 50 groundwater samples were collected from different points of the north part of Uruguay at depths between 40 and 50 m. The pH of the samples was adjusted with HCl to 2.0 in order to prevent interchanges between As species. Samples were stored in polypropylene bottles at 4–5°C.

3 RESULTS

3.1 *tAs, iAs, F^-, Fe and Mn levels in the samples*

The tAs concentration values found in the groundwater samples, ranged from 0.94 to 58.9 μg L^{-1}. More than half of the samples (27 samples) were above those recommended by WHO (10 μg L^{-1}) for drinking water. The high concentrated sample (58.9 μg L^{-1}) was taken from an area located at the northwest part of the country called "Salto Grande" which corresponds to a zone of thermal waters that emanates at 45°C from a depth of 1295 m. This relatively high natural concentration may be associated only to mineral components of the geological framework, since there is no mining activity in this area.

The iAs determinations are still on process, but preliminary results showed higher As(III) levels in the samples, which is in agreement with the reducing conditions of the aquifers.

Total concentration values of Fe and Mn in the samples, ranged from 2.1 to 95.5 μg L^{-1} and from 0.12 to 138.6 μg L^{-1} respectively. F^- values ranged from 0.168 to 1.528 mg L^{-1}. The sample from "Salto Grande" also presented the highest F^- value, exceeding the limit of 1.5 mg L^{-1} recommended by WHO. This could be related to natural geochemical processes such as the ion exchange with the increase of the precipitation of fluorite (CaF_2).

3.2 *Correlation of tAs with F^-, Fe and Mn levels*

Pearson correlation analysis of tAs with F^-, Fe and Mn concentrations in the samples were performed, resulting in highly significant positive correlation coefficients ($p < 0.05$) for As/F^- ($r = 0.876$) and As/Fe ($r = 0.652$). However, a negative correlation coefficient was found for As/Mn ($r = -0.206$).

4 CONCLUSIONS

Groundwater analysis showed that in some sampled points As and F^- concentrations were above those of WHO guidelines, being the highest values found in samples from the northwest part of the country. Also, strong positive Pearson correlations were found for these two elements.

The lack of knowledge of As in groundwater issues and its use as drinking water for human and animal consumption in Uruguay is a major area of interest in Medical Geology research in Uruguay, as there is also a lack of epidemiological studies on As exposure health risks. Arsenic levels in groundwater and its relationship with other elements, should be deeply studied to prevent long-term health effects.

The results of this ongoing study will contribute not only to the understanding of the As situation of groundwater in Uruguay, but also to expand the existing information on the distribution of As species in the regions of the world.

ACKNOWLEDGEMENTS

The authors thank PEDECIBA-Química.

REFERENCES

Smedley, P.L., Kinniburgh, D.G., Huq, I., Zhen-dong, L. & Nicolli, H.B. 2001. In W.R. Chappell, C.O. Abernathy & R.L. Calderon (eds), *Arsenic Exposure and Health Effects IV*: 467–450. New York: Elsevier Science Ltd.

Smedley, P.L. & Kinniburgh, D.G. 2002. A review of the source, behavior and distribution of arsenic in natural waters. *Appl. Geochem.* 17: 517–568.

WHO. 2017. Guidelines for drinking-water quality: Fourth edition incorporating the first addendum. Geneva.

Environmental Arsenic in a Changing World –
Zhu, Guo, Bhattacharya, Ahmad, Bundschuh & Naidu (Eds)
ISBN 978-1-138-48609-6

Optimization of the high pressure leaching of Complex Copper Concentrates of Codelco using a process simulator

N.P. Werth
EcoMetales Limited, Providencia, Santiago, Chile

ABSTRACT: In recent years, arsenic (As) content in copper concentrates has increased along with environmental restrictions for their transport and processing. Under this scenery is being developed the Complex Concentrate Leaching Project (PLCC), which allows the treatment of copper concentrates with high As content (>0.5%) through a high pressure (2873 kPa) and temperature (220°C) leaching process within an autoclave. The addition of high purity oxygen and cooling water to control the temperature in the different autoclave compartments is possible to dissolve over 99% of copper and As compounds. Dissolved As is then precipitated as an arsenical residue. The PLS is sent to a solvent extraction plant to produce copper cathodes, where it is previously mixed with heap/dump PLS solutions. On the other hand, the arsenical residue is stored on a deposit or sent to a silver recovery process. The gases from the autoclave, slurry depressurization and slurry cooling are sent to a gas cleaning system, where water and entrainments are recovered before emitting the gas to the atmosphere. A process simulator was developed in order to assist the review of the feasibility study and analyze different process configurations. The purpose of this work is to describe the process simulator developed for the PLCC project and present some simulation results for different process configuration analyzed during the feasibility of the PLCC project, which permitted to improve the process and reduce the CAPEX and OPEX.

1 INTRODUCTION

Between years 2000 and 2015 world copper production from concentrates has increased approximately from eleven to fifteen millions of metric tonnes per year. In 2014, Chile produced 26.5% of the total copper from concentrates and in the future it is expected that copper production in Chile will rise, mainly associated to production of concentrates (Fig. 1).

In recent years arsenic (As) content in copper concentrates has increased along with environmental restrictions for their handling, transport and processing. China has set a benchmark maximum of 0.5% for the As content in copper concentrates and in the future. It is expected that this value is going to be more restrictive.

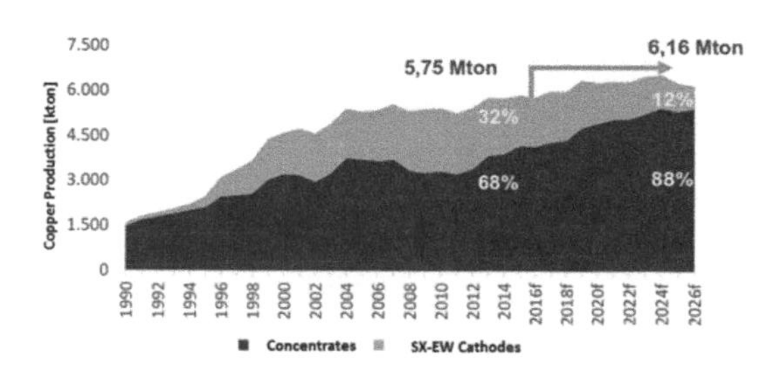

Figure 1. Copper production in Chile.

Under this scenery, the Complex Concentrate Leaching Project (PLCC) has developed leaching process using autoclave technology which allows the treatment of copper concentrates with high As content (>0.5%) of Codelco through a high pressure (2873 kPa) and temperature (220°C) (Marsden & Brewer, 2003; Parra *et al.*, 2017).

This work describes the process simulator that was developed in order to assist the review of the feasibility study and analyze different process configurations which allowed reducing the CAPEX and OPEX of the project.

2 METHODS/EXPERIMENTAL

During the years 2014 and 2016, it was developed the prefeasibility and feasibility studies of the PLCC project, respectively. In both stages, batch and pilot campaigns were done in order to define the operational and design parameters for the high pressure leaching and solid/liquid separation stages (Marsden, 2007; McDonald & Muir, 2007).

Together with these tests, a process simulator was developed using Matlab/Simulink® software including the mass and heat balance of all the stages of the process. This permitted to review the calculations done by the engineering company (Hatch) and to analyze some changes in the flowsheet in order to improve the OPEX and CAPEX of the project.

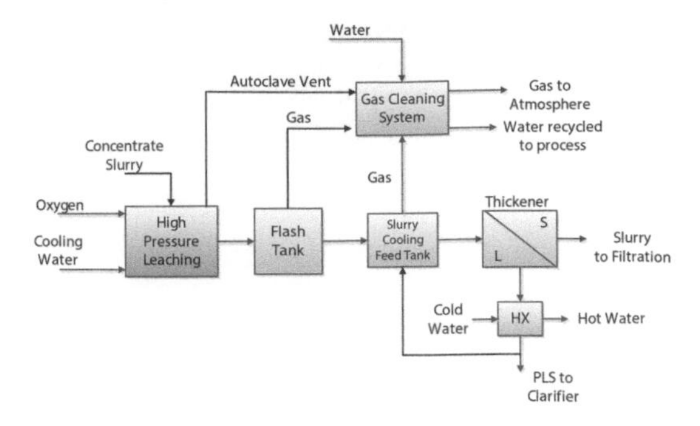

Figure 2. Prefeasibility simplified process flow diagram.

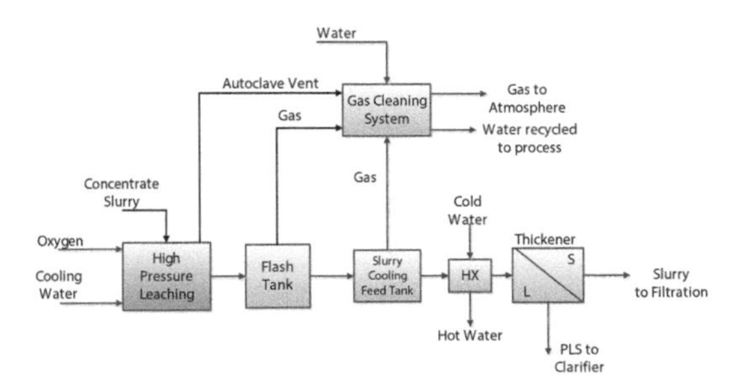

Figure 3. Feasibility simplified process flow diagram.

The chemical reactions and their extent inside the autoclave were defined and adjusted to represent the results obtained for the different processes and feed conditions. All the parameters for the rest of the processes were adjusted according to the specifications done in the engineering design.

3 RESULTS

During the prefeasibility, it was seen that in order to minimize the As redissolution after the high pressure leaching in the autoclave, the slurry has to be cooled to 40°C. To do this, it was defined to recirculate part of the cooled slurry after the thickening stage to the seal tank (Figure 2). This permitted to cool the flash tank slurry from 90°C to 71°C. A heat exchange stage was considered after the thickener, which allowed cooling the slurry from 69°C to 40°C.

During the feasibility, another option was analyzed, which considered cooling the slurry before the thickener (Figure 3). A new simulation was done considering this new flowsheet, and it was seen that doing this modification the flowrate to the thickener was reduced 42%, which allowed reducing the size of the slurry cooling feed tank and thickener. This modification permitted to reduce the CAPEX in 3 MUS$. Also this rapid cooling allowed improving the As precipitation as ferric arsenate from 85% to 90%.

Another point analyzed during the feasibility was the water make-up in the process. During the prefeasibility, it was defined to use cooling towers to cool the water used in: i) gas cleaning system, ii) slurry cooling system, and iii) oxygen plant. Most of the make-up water of the process was due to the loss of water (evaporation) in the cooling towers, so during the feasibility the use of dry coolers was analyzed, instead of cooling towers. A new simulation was done considering the use of only dry coolers and it was seen that make-up water was reduced 90%. This modification allowed to reduce the OPEX in 2.3 MUS$ y^{-1}.

4 CONCLUSION

The development of a simulator for the high pressure leaching of copper concentrates with high As contents of Codelco allowed to review the engineering of the project and to analyze different process configurations, which allowed reducing the CAPEX, OPEX and improving the As precipitation.

REFERENCES

Marsden, J. & Brewer, R. 2003. Hydrometallurgical processing of copper concentrates by Phelps Dodge at the Bagdad mine in Arizona. *Proceedings of the ALTA copper conference. Melbourne, Australia 2003. ALTA Metallurgical Services.*

Marsden, J. 2007. Medium-temperature pressure leaching of copper concentrates Part I: Chemistry and initial process development. *Miner. Metall. Proc.* 24(4): 193–204.

McDonald, R.G. & Muir, D.M. 2007. Pressure Oxidation leaching of chalcopyrite. Part I: Comparison of high and low temperature reaction kinetics and products. *Hydrometallurgy* 86: 191–205.

Parra, N., Acuña, M. & Fuentes, O. 2017. PLCC Project: Autoclave Technology Applied to Complex Copper Concentrates. *Hydroprocess ICMSE 2017, 9th International Seminar on Process Hydrometallurgy.* Santiago, Chile.

Environmental Arsenic in a Changing World –
Zhu, Guo, Bhattacharya, Ahmad, Bundschuh & Naidu (Eds)
ISBN 978-1-138-48609-6

Comparative genomic analysis reveals organization, function and evolution of *ars* genes in *Pantoea* spp.

L.Y. Wang & C.Y. Jing
Research Center for Eco-Environmental Sciences, Chinese Academy of Sciences, Beijing, P.R. China

ABSTRACT: Numerous genes are involved in various strategies to resist toxic arsenic (As). However, the As resistance strategy in genus *Pantoea* is poorly understood. In this study, a comparative genome analysis of 23 *Pantoea* genomes was conducted. Two vertical genetic *arsC*-like genes without any contribution to As resistance were found to exist in the 23 *Pantoea* strains. Besides the two *arsC*-like genes, As resistance gene clusters *arsRBC* or *arsRBCH* were found in 15 *Pantoea* genomes. These *ars* clusters were found to be acquired by horizontal gene transfer (HGT) from sources related to *Franconibacter helveticus*, *Serratia marcescens*, and *Citrobacter freundii*. During the history of evolution, the *ars* clusters were acquired more than once in some species, and were lost in some strains, producing strains without As resistance capability. This study revealed the organization, distribution and the complex evolutionary history of As resistance genes in *Pantoea* spp. The insights gained in this study improved our understanding on the As resistance strategy of *Pantoea* spp. and its roles in the biogeochemical cycling of As.

1 INTRODUCTION

A series of arsenite (As(III)) oxidizing bacteria have been successfully isolated and employed to transform As(III) to arsenate (As(V)) for As remediation, because As(V) is much more strongly adsorbed than As(III). Considering the ubiquitous existence of As(III) oxidizing bacteria in groundwater, we hypothesize that the indigenous As(III) oxidizing bacteria in groundwater should affect the speciation of adsorbed As in filters.

In traditional molecular biology research, As resistance traits are revealed primarily based on the cultivation of a specific strain, and it is impossible to study the As strategy of all strains in a genus. Nevertheless, the genomic sequence of a strain contains nearly all of the genetic information. Therefore, fundamental knowledge such as the phylogenetic, the genetic traits of As resistance and its evolutionary history can be obtained through comparative genomic analysis (Colston *et al.*, 2014).

Pantoea is a genus of Gram-negative, facultative anaerobic bacteria. Currently, the genus contains twenty-six species. In 2013, the strain *Pantoea* sp. IMH was an isolate that reported firstly as the strain having the As resistance capability within *Pantoea* species (Wu *et al.*, 2013). Further, we sequenced the genome of *Pantoea* sp. IMH and found two *ars* clusters (*arsR1B1C1H1* and *arsR2B2C2H2*) co-contributing to its As resistance (Wang *et al.*, 2016). However, the evolutionary history and genetic traits of As resistance in genus *Pantoea* are not fully understood.

Herein, we present the first study of the genetic traits of As resistance in *Pantoea* spp., as well as their evolutionary history. Two vertically transmitted *arsC*-like genes without any contribution to As resistance were found to exist in the 23 *Pantoea* strains. Besides these two *arsC*-like genes, As resistance gene clusters *arsRBC* or *arsRBCH* were found in 15 *Pantoea* genomes. These *ars* clusters were acquired by horizontal gene transfer (HGT) from sources related to *Franconibacter helveticus*, *Serratia marcescens,* and *Citrobacter freundii.* The insights gained in this study improve our understanding on the complex evolutionary history of As resistance genes and their roles in the biogeochemical cycling of As.

2 METHODS/EXPERIMENTAL

2.1 *Comparative genomics*

All of the orthologous pairs between *Pantoea* test genomes were identified by Pan Genome Analysis Pipeline (PGAP). The common dataset of shared genes among test strains was defined as their core genome. The total set of genes with test genomes was defined as the pan genome. The set of genes in each strain not shared with other strains was defined as the unique genes.

2.2 *Strains, plasmids and culture conditions*

E. coli and *Pantoea* strains were grown in LB medium (per liter contains: 10 g tryptone, 5 g yeast and 10 g NaCl) or LB plates (LB medium with w/v 1.5% agar) at 30°C. When appropriate, antibiotics were added at the following concentration: 100 $\mu g\, mL^{-1}$ ampicillin. Resistance to As species was tested by plating serial dilutions of cultures of each strain onto agar plates containing filtered sodium arsenate (Na_3AsO_4).

2.3 *Construction of the recombinant plasmids and E. coli strains*

A 3.86 kb *Bam*HI-*Xba*I DNA fragment containing the complete *ars1* cluster of *P. stewartii* S301, a 3.43 kb *Bam*HI-*Xba*I DNA fragment containing the complete *ars2* gene cluster of *P. agglomerans* Tx10, a 860 bp *Bam*HI-*Xba*I DNA fragment containing the complete *arsC1*-like gene of *P. stewartii* DC283 and a 942 bp *Bam*HI-*Xba*I DNA fragment containing the complete *arsC2*-like gene of *P. stewartii* DC283 were PCR amplified with primers. The above PCR products were ligated to the *Bam*HI-*Xba*I site of plasmid pUC18, yielding plasmids pUC18-ars1, pUC18-ars2, pUC18-arsC1-like and pUC18-arsC2-like. Then the plasmids were transferred to *E. coli* AW3110, yielding the recombinant *E. coli* AW3110-ars1, *E. coli* AW3110-ars2, *E. coli* AW3110-arsC1-like and *E. coli* AW3110-arsC2-like strains, respectively.

3 RESULTS AND DISCUSSION

3.1 *Distribution and organization of As-related genes in Pantoea genomes*

The *ars* genes in a genome are prone to group together as *ars* clusters (*arsRBC* and *arsRBCH*). Although comparison of the COG assignments of 23 genomes revealed that the DNA sequences between homologous genes within these *ars* clusters are conserved, some variations exist in DNA sequences, which can be divided into two sub-groups (*ars1* and *ars2*). Unlike the two *ars* clusters in *Pantoea* sp. IMH, only one *ars* cluster, either *ars1* or *ars2*, was observed in other strains. The *ars* gene clusters generally exhibited more than 80% identity within each sub-group and about 54% identity between two sub-groups. Actually, the *ars* clusters were not detected in 8 strains including Sc1, BL1, 9140, DC283, MP7, C91, ND04 and FF5. Moreover, two *arsC*-like genes with only 25% homology (*arsC1*-like and *arsC2*-like) were found in the 23 genomes. Based on the different *ars* genes distributions, the 23 strains were categorized into 4 sub-groups.

3.2 *Evolution and the origin of ars clusters*

The distribution and organization of *ars* genes in *Pantoea* raise a question as to their evolution. We detected the G + C content of *ars* clusters and their corresponding genomes. The results indicated that these *ars* clusters may be acquired in *Pantoea* strains by HGT. To further elucidate the evolution of the *ars* gene clusters, we compared the chromosomal regions flanking the *ars* gene clusters among the 23 *Pantoea* strains and found that the genes in the upstream and downstream regions were conserved among strains of the same species.

To gain insights into the origin of *ars* genes clusters in *Pantoea*, a NJ phylogenetic tree was constructed based on the ArsRBC protein sequences. The results imply that the *ars1* cluster may be acquired via HGT from *Franconibacter helveticus*, and *ars2* from *Serratia marcescens and Citrobacter freundii* in early evolutionary history.

3.3 *Functional analysis of ars gene and arsC-like genes*

The growth of the yielded recombinant *E. coli* strains *E. coli*-ars1 and *E. coli*-ars2, was tested in 5 mM concentration As(V). Both *E. coli*-ars1 and *E. coli*-ars2 survived in 5 mM As(V), and *E. coli*-ars1 grew better than *E. coli*-ars2. This result suggests that both the *ars1* and *ars2* clusters enabled *E. coli* AW3110 to resist As, and *ars1* seemed to have a more effective As resistance capability than *ars2*. Neither the *arsC1*-like nor *arsC2*-like gene enables *E. coli* AW3110 to resist As. In line with the alignment result, the function analysis demonstrates that these two *arsC*-like genes do not contribute to As resistance.

4 CONCLUSIONS

Arsenic resistance gene clusters *arsRBC* or *arsRBCH* were found in most of *Pantoea* genomes. These *ars* clusters were found to be acquired by horizontal gene transfer (HGT) from sources related to *Franconibacter helveticus*, *Serratia marcescens*, and *Citrobacter freundii*. During the history of evolution, the *ars* clusters were acquired more than once in some species, and were lost in some strains, producing strains without As resistance capability. Our results improved the understanding of the As resistance strategy in *Pantoea* spp.

ACKNOWLEDGEMENTS

We acknowledge the financial support of the National Basic Research Program of China (2015CB932003), the Strategic Priority Research Program of the Chinese Academy of Sciences (XDB14020302), and the National Natural Science Foundation of China (41373123, 41425016, 41503094, and 21321004). We thank Yongguan Zhu for the strain *E. coli* AW3110.

REFERENCES

Colston, S. M., Fullmer, M. S. & Beka, L. 2014. Bioinformatic genome comparisons for taxonomic and phylogenetic assignments using aeromonas as a test case. *MBio*. 5(6): 2150–7511.

Wang, L., Zhuang, X., Zhuang, G. & Jing, C. 2016. Arsenic resistance strategy in *Pantoea* sp. IMH: organization, function and evolution of *ars* genes. *Sci. Rep.* 6, 39195.

Wu, Q., Du, J., Zhuang, G. & Jing, C. 2013. *Bacillus* sp. SXB and *Pantoea* sp. IMH, aerobic As(V)-reducing bacteria isolated from arsenic-contaminated soil. *J. App. Microbiol.* 114: 713–721.

Environmental Arsenic in a Changing World –
Zhu, Guo, Bhattacharya, Ahmad, Bundschuh & Naidu (Eds)
ISBN 978-1-138-48609-6

Thioarsenic compounds exist in the drinking groundwater

J.H. Liang[1] & Z. Chen[2]
[1]*Key Laboratory of Karst Ecosystem and Treatment of Rocky Desertification, Institute of Karst Geology, Chinese Academy of Geological Sciences, Guilin, P.R. China*
[2]*Department of Environmental Science, Xi'an Jiaotong-Liverpool University, Suzhou, P.R. China*

ABSTRACT: Arsenic (As) polluted groundwater in Northern China was used as drinking water source and caused severe health problems. Characterization of As speciation is crucial to understand the health risk of As and its biogeochemical behaviors in groundwater. In this study, groundwater samples were collected from 26 wells in Northern China. Arsenic species in the groundwater were measured by high-performance liquid chromatography (HPLC) and inductively coupled plasma mass spectrometry (ICP-MS). Thioarsenate, one of the As species that is seldom reported in drinking groundwater, was detected in two-third of the sampling wells, even in the samples with total As concentration lower than $10\,\mu g{\cdot}L^{-1}$. Furthermore, the occurrence of thioarsenate in groundwater samples was dependent on the pH of groundwater, and thioarsenate was transformed to arsenite below a pH value of 8.2. The study demonstrated thioarsenate was prevalent in the alkaline drinking groundwater and would be a new As exposure pathway for the people and livestock living in the As-rich area.

1 INTRODUCTION

Arsenic (As) thiolation occurs under alkaline reducing conditions with high sulfur concentration (Chen *et al.*, 2005). The formation, stability and toxicity of thioarsenate have been investigated in the laboratory (Rader *et al.*, 2004; Stauder *et al.*, 2005; Stucker *et al.*, 2014). Thioarsenate was found in the alkaline geothermal waters with high sulfur content (Planer-Friedrich *et al.*, 2007) and groundwater from the contaminated site with industrial As (Wallschlager & Stadey, 2007). Besides, thioarsenates were detected and contributed up to 33% of the As speciation in stream of an As-enriched peatland (Thomas Arrigo *et al.*, 2016). Although there are some reports of thioarsenate in the natural groundwater (Stauder *et al.*, 2005), it remains unclear that whether the thioarsenate exits in drinking groundwater.

Although the majority of As species in the groundwater in Northern China were identified as As(III) and As(V), the occurrence and formation of other As species are not well understood. In this study, the objectives were to clarify the As species profile with updated preservation methods and reveal the key environmental factors affecting the speciation of As in groundwater.

2 METHODS

2.1 *Sampling areas*

The sampling was conducted in the southern Datong basin of Shanxi province (Shanyin country) of Northern China where groundwater is the most heavily contaminated by As in China. The studied area is the Cenozoic basin located at the Shanxi Rift System. The pH values of groundwater we investigated were documented under partial alkaline condition.

2.2 *Laboratory analysis*

Arsenic species in fresh groundwater were analyzed using high-performance liquid chromatography (HPLC, 1100, Agilent, USA) and inductively coupled plasma mass spectrometry (ICP-MS, Agilent 7500, USA). A PRP-X100 anion-exchange column (250×4.1 mm, $10\,\mu m$) (Hamilton Co., Reno, NV, USA) was used for separation of As(III), As(V) and thioarsenate at room temperature (25°C) with the mobile phase of 10 mM $(NH_4)_2HPO_4$ and 10 mM NH_4NO_3 (pH 6.2).

The samples were also analyzed by HPLC-ICP-MS with size exclusive column (Shodex OHpak SB-802.5 HQ, $30\,cm \times 8.0\,mm \times 6.0\,\mu m$, Showa Denko America Inc., New York, NY, USA) to test the potential complexation of As with other large molecules. (Liu *et al.*, 2011). The total As in filtered samples was detected by ICP-MS.

3 RESULTS AND DISCUSSION

3.1 *Determination of As species*

The total As concentrations of 90% groundwater samples exceeded $10\,\mu g{\cdot}L^{-1}$, which is WHO guideline value for As for drinking water. The highest concentration of As was $703.7\,\mu g{\cdot}L^{-1}$. No methylated As

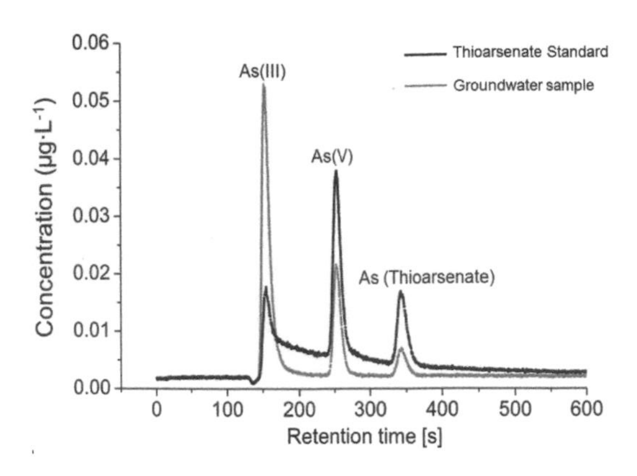

Figure 1. Arsenic speciation in field sample (gray line) and thioarsenate standard (black line) from HPLC-ICP-MS chromatograms. Elution order of arsenic speciation: As(III), As(V), As (Thioarsenate).

and NOM (Natural Organic Matter)-As complexation was detected in the groundwater samples. The As methylation was well observed in high organic matter conditions, like paddy field, but was not found in groundwater. As(III) and As(V) were dominant in all the samples.

One unidentified As compound, with the exception of As(III) and As(V), was observed in most samples and the retention time of its peak was 346 s, same to thioarsenate standard (Fig. 1). Thioarsenate is unstable under acidic condition and has been widely reported in high sulfur condition (Jia *et al.*, 2015). Although structure of the putative thioarsenate was not determined in this study because of its low concentration in field samples, the alkaline environment and high sulfur level in the investigated groundwater of the study sites provides a suitable geochemical condition for the thioarsenate formation. Other studies showed the concentrations of SO_4^{2-} in the investigated area varied from $0.02–3121\,mg{\cdot}L^{-1}$ with the geological heterogeneity.

3.2 *Occurrence of thioarsenate*

The standard thioarsenate was artificially synthesized by mixing arsenite and sulfide, and measured by $(NH_4)_2HPO_4$ mobile phase at pHs 6.2, 7.2, 8.2 and 9.2. Along with the pH gradient, the retention time of thioarsenate was gradually reduced from 435 s at pH 9.2 to 342 s at pH 6.2. The pH variation in mobile phase resulted in the change of the percentage of As(III), As(V) and thioarsenate. In the investigated pH range, the percentage of As(V) was slightly varied. Thioarsenate was converted to As(III) at low pH. When the pH of mobile phase decreased from 8.2 to 7.2 and 6.2, the percentage of As(III) increased from 18% to 23% and 40%, and in the meantime, the thioarsenate contents dropped from 37% to 31% and 18% respectively. The results indicated that monothioarsenate was transformed into arsenite at the pH lower than 8.2 in the groundwater.

4 CONCLUSIONS

The results suggested that thioarsenate would be a new As exposure pathway for human health, especially for the people living in the As-contaminated area of Northern China. The conversion between thioarsenate and arsenite was found at a pH threshold of 8.2. The sensitivity of thioarsenate to pH provides a way to control its formation or degradation. This study improved our understanding of As species and interconversion of As in the natural drinking groundwater. Further investigation is required to elucidate the formation mechanism of thioarsenate in alkali groundwater and reveal the consequence when people or animals are exposed to thioarsenate through mouth.

ACKNOWLEDGEMENTS

Financial support was obtained from Cooperation @epfl (SDC-EPFL fund, project: Relevance of arsenic complexed with organic matter in Chinese groundwater) and the Natural Science Foundation of China (No. 41371459 and No. 41571305).

REFERENCES

Chen, Z., Zhu, Y.G., Liu, W.J. & Meharg, A.A. 2005. Direct evidence showing the effect of root surface iron plaque on arsenite and arsenate uptake into rice (Oryza sativa) roots. *New Phytol.* 165: 91–97.

Jia, Y., Bao, P. & Zhu, Y.G. 2015. Arsenic bioavailability to rice plant in paddy soil: influence of microbial sulfate reduction. *J. Soils Sediments* 15: 1960–1967.

Planer-Friedrich, B., London, J., McCleskey, R.B., Nordstrom, D.K. & Wallschlager, D. 2007. Thioarsenates in geothermal waters of yellow stone national park: Determination, preservation, and geochemical importance. *Environ. Sci. Technol.* 41: 5245–5251.

Stucker, V.K., Silverman, D.R., Williams, K.H., Sharp, J.O. & Ranville, J.F. 2014. Thioarsenic species associated with increased arsenic release during biostimulated subsurface sulfate reduction. *Environ. Sci. Technol.* 48: 13367–13375.

Thomas Arrigo, L.K., Mikutta, C., Lohmayer, R., Planer-Friedrich, B. & Kretzschmar, R. 2016. Sulfidization of organic freshwater flocs from a minerotrophic peatland: speciation changes of iron, sulfur, and arsenic. *Environ. Sci. Technol.* 50: 3607–3616.

Wallschlager, D. & Stadey, C.J. 2007. Determination of (oxy)thioarsenates in sulfidic waters. *Anal. Chem.* 79:3873–3880.

Environmental Arsenic in a Changing World –
Zhu, Guo, Bhattacharya, Ahmad, Bundschuh & Naidu (Eds)
ISBN 978-1-138-48609-6

Arsenic speciation of groundwater and agricultural soils in central Gangetic basin, India

M. Kumar[1,2] & AL. Ramanathan[2]
[1]*Department of Environmental Science, School of Earth, Environment & Space Studies, Central University of Haryana, Jant Pali, Mahendergarh, India*
[2]*School of Environmental Sciences, Jawaharlal Nehru University, New Delhi, India*

ABSTRACT: The current study was performed to estimate the amount of inorganic forms [arsenite, As(III) and arsenate, As(V) of arsenic (As) present in groundwater (n = 18) and agricultural soils from eleven locations in the central Gangetic basin, India. Water samples were speciated using a disposable cartridge, while a microwave assisted method was used to obtain As species in agricultural soil samples. The estimation of As species concentration was performed using ion chromatography (IC) coupled with inductively coupled plasma mass spectrometer (ICP-MS) in solution matrix. Approximately 73% of the groundwater samples (n = 18) show As(III) as the dominant species, while 27% reveals As(V) was the dominant species. Groundwater (80%) samples exceeded the World Health Organization (WHO) guideline value ($10\,\mu g\,L^{-1}$) of As. The concentration of As(III) in agricultural soil samples varies from not detectable to $40\,\mu g\,kg^{-1}$ and As(V) was observed as the major species (ranging from 1050 to $6835\,\mu g\,kg^{-1}$) while the total As concentration varied from 3528 to $14{,}690\,\mu g\,kg^{-1}$. Arsenate (V) species dominate in oxygen-rich environments and well-drained soils, whereas in the reducing conditions, such as regularly flooded soils, As(III) is the stable oxidation state.

1 INTRODUCTION

Arsenic (As) is a common element which forms many compounds in the environment and biological system as well. Nowadays, natural occurrence of As has been reported in almost all the region of South East Asia and other countries around the world like; Mexico, Argentina, Poland and Canada (Chen *et al.*, 2006). The determination of the total As concentration alone is insufficient for many environmental exposure scenarios, but the determination of the species is important in order to accurate assessments of environmental impact and human health risk (Rahman *et al.*, 2009; Vassileva *et al.*, 2001), the toxicity and bio-availability of As compounds depend on the chemical form of the As (Gong *et al.*, 2002) and cycling of As in different environmental conditions like lake (Zheng *et al.*, 2003; Kumar *et al.*, 2018). Current study reported As level and speciation in water, agriculture fields soil using ion chromatography (IC) coupled with inductively coupled plasma mass spectrometer (ICP-MS) in solution matrix in Gangetic basin, India.

2 METHODS

Water samples (18) were collected from two blocks Mohiuddin Nagar and Mohanpur of the district Samastipur, Bihar during June 2015 to know As-speciation ratio in reducing environment of central Gangetic basin. Agricultural soil samples were collected from 11 different locations of the As-affected area. Generally, shallow tubewells water used for irrigation in this area, which is contaminated with geogenic As contamination. A microwave digester (CEM, MARS 6) having 42 digester vessels were used to digest all soil and sediment samples. Microwave-assisted extraction technique (1 M orthophosphoric acid) was used for the extraction of the As species from agricultural soils (Kumar *et al.*, 2016a).

3 RESULTS AND DISCUSSION

Arsenic speciation for groundwater are shown in Table 1. Approximately 73% (14 out of 18) of the samples shows As(III) as dominant species while only 27% (5 out of 18) shows As(V) as dominant species. Arsenite (III) for all samples was 63.8% and As(V) was 36.2%. It is expected for groundwaters where reducing environment occurs in the aquifers.

Many studies also have been found arsenite as the primary arsenic species present in central Gangetic plain (Kumar *et al.*, 2016b), West Bengal and Bangladesh (Kim *et al.*, 2003). Groundwaters mainly have inorganic As (Elci *et al.*, 2008; Hughes *et al.*, 2011), hence the sum of the two species (AsIII and AsV) will be equal to the total As (Table 1).

To know the level of the As in the agricultural soil which is irrigated by water with As concentration

Table 1. Groundwater sample As species in central Gangetic basin.

S. No.	As(III)	As(V)	As(t)	As(III)/As(V)
1	57.8	42.19	29.44	1.37
2	95.8	4.16	81.42	23.02
3	44.9	55.09	35.14	0.82
4	84.6	15.38	5.98	5.50
5	91.8	8.23	12.64	11.15
6	76.7	23.29	1.47	3.29
7	88.1	11.94	2.62	7.38
8	8.1	91.90	5.59	0.09
9	67.6	32.39	14.82	2.09
10	70.9	29.11	19.20	2.43
11	57.7	42.29	1.35	1.36
12	87.5	12.46	2.79	7.03
13	42.9	57.09	16.66	0.75
14	58.9	41.06	6.95	1.44
15	94.6	5.43	1.34	17.41
16	14.78	85.23	2.65	0.17
17	11.6	88.40	23.50	0.13
18	79.6	20.38	104.7	3.91

As(III) and As(V) represented in (%) and As(total) presented in $\mu g\,L^{-1}$.

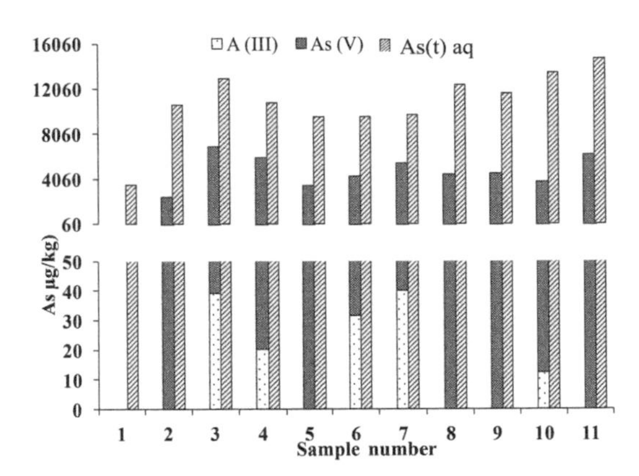

Figure 1. As speciation in agricultural soil samples of central Gangetic basin.

levels of $>10\,\mu g\,L^{-1}$. Tubewell water was not collected from the corresponding agricultural soil samples. The total concentration of As observed in agricultural soils ranged from 3527–14690 $\mu g\,kg^{-1}$ with the percent extractable concentration of 22.8 to 55.9 $\mu g\,kg^{-1}$ (Fig. 1). The concentration of As(III) and As(V) varied from bdl– 40.07 $\mu g\,kg^{-1}$ and 1050–6835 $\mu g\,kg^{-1}$ respectively. As(V) was detected in almost all the agricultural soil samples, while As(III) was detected only in 6 samples having a lower range of (0.1–40 $mg\,kg^{-1}$) only. Arsenate [As(V)] species dominated in oxygen-rich environments and well-drained soils, whereas in the reducing conditions, such as regularly flooded soils, arsenite [As(III)] is the stable oxidation state, but elemental As and arsine can also be present in strongly reducing environments (Vicky-Singh *et al.,* 2010).

4 CONCLUSIONS

The study revealed that As(III) dominated in groundwater, while As(V) in the agricultural filed soil. The dominance of As(III) indicates the reducing conditions of the aquifers in the central Gangetic basin. Groundwater (80%) samples exceeded the World Health Organization (WHO) guideline value ($10\,\mu g\,L^{-1}$) of As. The concentration of As(III) in agricultural soil samples was comparatively very low, while As(V) was observed as the dominating species. It was also observed that As(V) species dominated in oxygen-rich environments and well-drained soils, whereas As(III) dominated in the regularly flooded locations.

REFERENCES

Chen, Z., Akter, K.F., Rahman, M.M. & Naidu, R. 2006. Speciation of arsenic by ion chromatography inductively coupled plasma mass spectrometry using ammonium eluents. *J. Sep. Sci.* 29: 2671–2676.

Elci, L., Divrikli, U. & Soylak, M. 2008. Inorganic arsenic speciation in various water samples with GFAAS using coprecipitation. *Int. J. Environ. Anal. Chem.* 88: 711–723.

Gong, Z., Lu, X., Ma, M., Watt, C. & Le, X.C. 2002. Arsenic speciation analysis. *Talanta* 58:77–96.

Hughes, M.F., Beck, B.D., Chen, Y., Lewis, A.S. & Thomas, D.J. 2011. Arsenic exposure and toxicology: a historical perspective. *Toxicol. Sci.* 123:305–332.

Kim, M.J., Nriagu, J. & Haack, S. 2003. Arsenic behavior in newly drilled wells. *Chemosphere* 52: 623–633.

Kumar, M., Ramanathan, AL., Rahman, M.M. & Naidu, R. 2016a. Concentrations of inorganic arsenic in groundwater, agricultural soils and subsurface sediments from the middle Gangetic plain of Bihar, India. *Sci. Total Environ.* 573, 1103–1114.

Kumar, M., Rahman, M.M., Ramanathan, A.L. & Naidu, R. 2016b. Arsenic and other elements in drinking water and dietary components from the middle Gangetic plain of Bihar, India: health risk index. *Sci. Total Environ.* 539: 125–134.

Kumar, M. & Ramanathan, A.L. 2018. Vertical Geochemical Variations and Speciation Studies of As, Fe, Mn, Zn, and Cu in the Sediments of the Central Gangetic Basin: Sequential Extraction and Statistical Approach. *Int. J. Environ. Res. Publ. Health* 15(2): 183.

Rahman, M.M., Chen, Z. & Naidu, R. 2009. Extraction of arsenic species in soils using microwave-assisted extraction detected by ion chromatography coupled to inductively coupled plasma mass spectrometry, *Environ. Geochem. Hlth.* 31: 93–102.

Vassileva, E., Becker, A. & Broekaert, J. 2001. Determination of arsenic and selenium species in groundwater and soil extracts by ion chromatography coupled to inductively coupled plasma mass spectrometry. *Anal. Chim. Acta* 441: 135–146.

Vicky-Singh, Brar, M.S., Preeti-Sharma & Malhi, S.S. 2010. Arsenic in Water, Soil, and Rice Plants in the Indo-Gangetic Plains of Northwestern India. *Commun. Soil Sci. Plant Anal.* 41: 1350–1360.

Zheng, J., Hintelmann, H., Dimock, B. & Dzurko, M.S. 2003. Speciation of arsenic in water, sediment, and plants of the Moira watershed, Canada, using HPLC coupled to high resolution ICP–MS. *Anal. Bioanal. Chem.* 377: 14–24.

Environmental Arsenic in a Changing World –
Zhu, Guo, Bhattacharya, Ahmad, Bundschuh & Naidu (Eds)
ISBN 978-1-138-48609-6

Single-particle identification of trace arsenic constituents in environmental samples

V.S.T. Ciminelli[1,2], I.D. Delbem[2,3], E. Freitas[2,3], M. Morais[4] & M. Gasparon[2,5]
[1] *Universidade Federal de Minas Gerais-UFMG, Department of Metallurgical and Materials Engineering, Belo Horizonte, Brazil*
[2] *National Institute of Science and Technology on Mineral Resources, Water and Biodiversity, INCT-Acqua, Brazil*
[3] *Universidade Federal de Minas Gerais-UFMG, Center of Microscopy, Belo Horizonte, Brazil*
[4] *Kinross Brasil Mineração, Brazil*
[5] *The University of Queensland, School of Earth and Environmental Sciences, St Lucia, Australia*

ABSTRACT: An analytical protocol was developed to identify arsenic (As) in soil samples and in the $\leq 10\,\mu m$ fraction of surface dust samples (fine surface dust-FSD). Single-particle identification of trace As constituents was undertaken by combining scanning electron microscopy with automated image analyses and high-resolution, transmission electron microscopy. Two forms of As association with iron and aluminum nanoscale phases were identified. In the predominant one, As was identified in oriented aggregates formed by crystalline nanoparticles of Fe-(hydr)oxides. In the FSD samples, As was additionally detected in an assembly of hematite and goethite nanocrystals forming larger particles of few hundreds of nanometers, often entangled with phyllosilicates. These mixed phases carried various elements, such as P, Ba, Pb, among others. Even rare As-bearing phases (e.g., 1 to 9 particles out of approx. 30,000 particles analyzed), such as arsenopyrite and ferric arsenate, and possibly scorodite were identified in some samples. The developed analytical protocol brings a novel and practical contribution to As speciation in environmental samples.

1 INTRODUCTION

Precise, single particle characterization of arsenic (As)-bearing phases in environmental samples is not a simple task despite the advances in analytical techniques. Bulk X-ray absorption spectroscopy has been applied to identify the molecular environment of As in various matrices for more than two decades (Foster et al., 1998). Theoretical modeling combined with spectroscopic techniques has further advanced the understanding of the mechanisms of As fixation in the environment (Ladeira *et al.,* 2001). Micro-X-ray fluorescence combined with microfocused-X-ray absorption spectroscopy has enabled in situ characterization of As with spatial resolution usually down to the micrometer level (Ono *et al.,* 2015). Nevertheless, none of the aforementioned methods provides the spatial resolution necessary to investigate highly heterogeneous nanoscale phases in environmental samples, down to a few nanometres, or allow statistically sound quantification of As-bearing phases.

To overcome these limitations, our group has combined scanning electron microscopy with automated image analysis and high-resolution transmission electron microscopy (HRTEM). An analytical protocol has been developed and applied to soil and fine ($\leq 10\,\mu m$ fraction) surface dust samples collected in a gold mining region. The results will demonstrate that the developed analytical protocol brings a novel and practical contribution to As speciation in environmental samples.

2 METHODS/EXPERIMENTAL

2.1 *Sampling, sample preparation and analysis*

The collection of surface soil (0–20 cm) samples was undertaken following local and international protocols (USEPA, 1991) in four geological units and four classes of soils, comprising areas of gold mineralization and areas representing the region's background. The samples were collected in areas with no indication of anthropogenic activities. The bulk samples were oven-dried at 40°C for 12 hours then disaggregated, split into sub-samples and sieved at 2 mm, and then finely-ground ($<44\,\mu m$) for chemical analyses and particle characterization by transmission electron microscopy TEM.

Surface dust sampling was conducted in the residential area (where citizens may be regularly exposed to resuspended dust) near the gold mine operation in two campaigns (dry and wet seasons). These samples possibly represent a combination of fugitive dust from construction and excavation sites (including the mine site), natural geological background and baseline associated with local and regional industrial and agricultural activities. The collection and storage followed the procedures described by USEPA (1991). Aiming to assess the respiratory exposure, five surface dust samples were sieved to obtain the $\leq 10\,\mu m$ fraction (Fine surface dust – FSD). Two samples collected from the crushing area within the mine site were also sieved at $10\,\mu m$. The sieving apparatus consisted of adapting an Ultrasonic Sieving (HK Technologies, USA) to a mechanical vibrator (Cleveland Vibrator Company – Model: VJ-1212, USA).

The acid extractable As content in the soil samples was determined following digestion with aqua regia using a microwave-assisted (Ethos, Milestone) digestion procedure (USEPA, 2007). Arsenic was analyzed by inductively coupled plasma optical emission spectrometry (ICP-OES) (Perkin Elmer Optima 7300DV) or ICP-MS (Agilent 7500cs, CA, USA). Two standard reference materials (NIST SRM 2710a and CANMET/ CCRMP-Till-3) were analyzed together with each batch of 10 soil samples. Duplicates and analytical blanks were analyzed as well. Arsenic recoveries ranged from 84 to 101%. All blank extractions returned values below the method detection limits.

2.2 *Electron Microscopy analyses*

The characterization of As-bearing phases and quantitative mineralogy were based on single particle analyses using a FEI Quanta 650 Field Emission Gun Scanning Electron Microscope (FEG-SEM) equipped with two Bruker Quantax X-Flash 5010 energy dispersion X-ray (EDX) detectors and FEI's Mineral Liberation Analyzer-MLA for data acquisition and process. The grain-based X-ray mapping (GXMAP) measurement mode was applied to the analyses of polished sections and loose particles. In this mode, a series of backscattered electron (BSE) images is collected. Identification of mineral grains by MLA is based on BSE image segmentation and collection of EDX-spectra of the particles/grains. Collected EDX-spectra are then classified using a pre-defined list of mineral spectra collected by the user. The method has a resolution of grain size down to $0.1–0.2\,\mu m$ (Gu *et al.*, 2003). For the TEM analyses each powder sample was dispersed in Milli-Q water in Eppendorf tubes and sonicated in ultrasound bath. A drop of each suspension was placed on carbon coated Cu-TEM grids (300 mesh) and left to dry in a desiccator. The analysis was performed using High Resolution TEM (HRTEM), Scanning TEM (STEM), EDX spectroscopy and Electron Energy-Loss Spectroscopy (EELS) using a FEI TEM Tecnai G2-20 (200 kV).

3 RESULTS AND DISCUSSION

3.1 *As-bearing phases*

Table 1 shows the main mineral phases in the soil samples according to the analyses carried out by the MLA. The main phases (>2 wt.%) are the silicates: quartz (SiO_2), mica/clay minerals and microcline ($KAlSi_3O_8$) and others. Goethite (FeO(OH)) and hematite (α-Fe_2O_3) are the main Fe-(hydr)oxides. The MLA tool allows for quantitative single particle analysis of a large number of grains. The total number of particles analyzed per sample ranged from 33,864 to 79,330, thus providing good statistics. The differences in the number of particles reflect differences in particle size distribution for a fixed measuring time. Large variations in the content of Fe-(hydr)oxides from 2% to 44% are observed. In general, the As concentration increases with the increase of the concentration of Fe-(hydr)oxides, with a significant As-enrichment in the coarse fraction.

Table 1. Main mineral phases (%) and arsenic minerals (number of particles) identified by MLA in three soil samples.

Samples	S0 <2 mm	S1 <2 mm	S4 >2 mm	 <2 mm
Quartz	21.6	49.1	26.6	28.2
Other silicates	72.1	39.8	27.6	55.1
FeOx-As free	1.7	1.4	22.6	6.7
FeOx-As	0.6	4.1	21.3	4.4
Ilmenite	n.d.	2.7	n.d.	4.2
Others (<2%wt)	4.0	2.9	1.9	1.4
Number of particles				
Arsenopyrite	n.d.	4	n.d.	n.d.
Scorodite	n.d.	1	n.d.	n.d
FeOx-As	738	2,901	11,702	1,968
Total particles	79,330	41,831	33,864	59,477
As(mg kg^{-1})	411	1,560	7,556	1,355

n.d. = not detected.

Arsenic is found mainly associated with Fe-(hydr)oxides, with rare arsenopyrite and ferric arsenate, likely scorodite. Arsenopyrite and scorodite are the main As phases in the local sulfide and oxidized ore bodies, respectively. The relatively low number arsenopyrite particles and other sulfides (not shown) is consistent with the low bulk sulfur concentration (range of <100 to 288 mg kg^{-1}). The results indicate a small contribution of sulfides and arsenates from the mineralized lithologies to the bulk soil chemistry. The quantification of the mineral phases determined by the MLA was shown to be consistent with the chemical analyses (not shown) of the major elements (iron and silicon) within 15% variation.

A BSE-SEM image of polished sections prepared from the soil samples show typical mineral associations found in the soil samples: quartz inclusion and the intergrowth of phyllosilicate lamellae (muscovite) with Fe-(hydr)oxides (goethite or hematite) (Fig. 1).

The MLA was also employed to analyze the mineral phases in the FSD samples. The method allowed a good reading of particles with diameter smaller than $10\,\mu m$. The main constituents were again the silicate

Table 2. Main mineral phases (%) and As minerals (no. particles) identified by MLA in four FSD samples.

Samples	FC <2 mm	F2 <2 mm	F17 >2 mm	F19 <2 mm
Mica/Clay	85.0	66.0	58.0	56.6
Quartz	9.7	18.2	7.9	12.0
Fe-Ox	1.6	3.8	4.9	9.0
Organic matter	0.4	2.4	4.9	4.0
Other silicates	1.8	4.5	14.2	10.0
Carbonates	1.4	2.7	7.3	6.0
Others	0.1	2.4	2.8	2.4
Number of particles				
Arsenopyrite	8	1	n.d.	1
Scorodite	9	1	3	2
FeOx-As	109	240	150	211
As mixed phases	47	17	60	32
Tot. no. particles	35,624	36,856	35,144	38,517
As (mg kg^{-1})	279	445	212	265

n.d. = not detected.

minerals, such as quartz, muscovite and other clay minerals; Fe oxy-hydroxides; carbonates and organic matter – identified by the typical morphology and the level of carbon – associated with low concentrations of elements such as Si, Al and Mg.

In the FSD samples As is found in five phases: (i) arsenopyrite, (ii) scorodite (iii) iron oxy-hydroxide, (iv) mixed phases and (v) muscovite/clay. Iron oxy-hydroxides were identified as the major As-bearing phase in all the samples followed by the As-bearing mixed phases. According to the EDX analyses, these mixed phases contain Al (2.44 to 47.31%), Si (1 to 4.02%), P (0.47 to 34.29%), Ca (0.47 to 7.34%), Fe (0.18 to 23.60%), As (0.06 to 7.23%), Ba (0.11 to 6.01%), Pb (0.14 to 25.64%), O (50.67 to 78.80%), whereas Na, Ti, Cl, S, K, V, Cr, Cu, Sr, Cd appear as minors. No specific known mineral phase could be assigned to these phases. Arsenopyrite and scorodite are rare. In the case of FSD samples, the preparation of polished sections for quantitative MLA analyses is not possible due to the fine particle size of the material and though the numbers could not be fully validated by chemical analyses, there is a clear and consistent trend when comparing both methods. The technique allows for the identification of As-bearing phases from a significant population of particulate material (up to 30,000 particles), in a heterogenous, complex mineral assembly, which would not have been possible by the common techniques applied to particulate analyses. The refinement of the method for quantitative analyses of FSD is under development.

The main As carriers – aggregates and mixed phases – were further investigated by TEM analysis. Based on d_{hkl} spaces, the aggregates (present in both soil and FSD samples) were identified as goethite and hematite. Freitas *et al.* (2015) investigated As- enriched Fe-Al-oxisols after their use as liners in disposal facilities of sulfide tailings. The

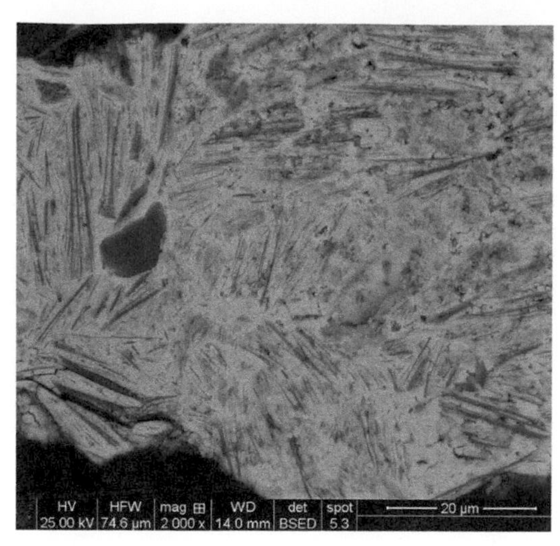

Figure 1. BSE-SEM images of polished sections showing the inclusion of quartz and intergrowth of phyllosilicate lamellae (muscovite) with Fe-(hydr)oxides (goethite) in soil samples.

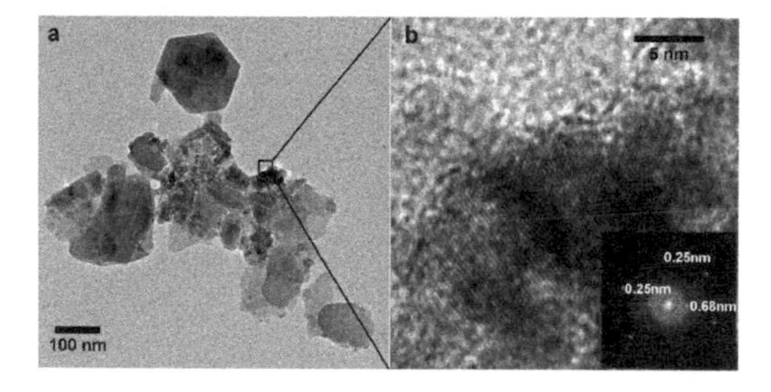

Figure 2. Phyllosilicates entangled with mixed phases in FSD samples (left), and a HRTEM image of the oriented aggregate of nanoparticles (right).

results demonstrated that As was present in oriented aggregates formed by crystalline nanoparticles of Fe-(hydr)oxides. The same pattern was found in the soil samples collected in the present study. In addition to oriented aggregates described before, TEM analyses showed that the mixed phases are essentially an assembly of hematite and goethite nanocrystals that form larger particles of a few hundreds of nanometers. These mineral phases were identified by HRTEM image analysis and selected area diffraction- SAD. Aluminum was also detected.

The nanoparticles of Fe-(hydr)oxides are often entangled with phyllosilicates (Figure 2). The TEM-EDS analysis has not shown the presence of other elements apart from O, Fe, Al, and As in probed particles. The occurrence of other elements such as P and Si observed in the SEM-EDS data might be due to the larger interaction volume (a few micrometers size), around the target particle, where the X-rays are emitted from thicker samples.

4 CONCLUSIONS

An analytical protocol, combining scanning electron microscopy with automated image analysis and high-resolution transmission electron microscopy, was developed and applied to soil and fine ($\leq 10\,\mu m$ fraction) surface dust samples. This protocol allows for a statistically sound quantification of As-bearing phases with the spatial resolution necessary to investigate highly heterogeneous nanoscale phases in environmental samples, down to a few nanometres. Arsenic was mainly identified in oriented aggregates formed by crystalline nanoparticles of Fe-(hydr)oxides. In the FSD samples, As was additionally detected in an assembly of hematite and goethite nanocrystals forming larger particles of a few hundreds of nanometers, and often entangled with phyllosilicates. Rare arsenopyrite and scorodite were identified in the samples.

ACKNOWLEDGEMENTS

The authors are grateful to the Brazilian government agencies – CNPq, FAPEMIG and CAPES, and to Kinross Brasil Mineração (KBM) for financial support and the PVE fellowship from the Science Without Borders program to M. Gasparon. The authors also acknowledge the Center of Microscopy/UFMG. Prof. Jack Ng and Dr. Lixia Qi at UQ are gratefully acknowledged for hosting M. Morais in the UQ Laboratory. The authors also thank Dr Claudia L. Caldeira, Patricia Lopes and Filipe A.T. Alves at UFMG for the ICP-OES analyses, and Dr. Marcus Manoel Fernandes for the soil samples.

REFERENCES

Foster, A.L., Brown, Jr. G.E., Tingle, T.N. & Parks, G.A. 1998. Quantitative As speciation in mine tailings using X-ray absorption spectroscopy. *Am. Mineral.* 83: 553–68.

Freitas, E.T.F., Montoro, L.A., Gasparon, M. & Ciminelli, V.S.T. 2015. Natural attenuation of arsenic in the environment by immobilization in nanostructured hematite. *Chemosphere* 138: 340–347.

Gu, Y. 2003. Automated scanning electron microscope based mineral liberation analysis. *J. Miner. Mater. Character. Eng.* 2: 33–41.

Ladeira, A.C.Q., Ciminelli, V.S.T., Alves, M.C.M. & Duarte, H.A. 2001. Mechanism of anion retention from EXAFS and Density Functional Calculations: Arsenic (V) adsorbed on Gibbsite. *Geochim. Cosmochim. Acta* 65(8): 1211–1217.

Mason, B.J. 1992. Preparation of soil sampling protocols: sampling techniques and strategies (No. PB-92-220532/XAB). Nevada Univ., Las Vegas, NV (United States). Environmental Research Center.

Ono, F.B., Tappero, R., Sparks, D. & Guilherme, L.R.G. 2015. Investigation of arsenic species in tailings and windblown dust from a gold mining area. *Environ. Sci. Poll. Res.* 23(1): 638–647.

USEPA–Environmental Protection Agency. 1991. Compendum of ERT waste sampling procedures. Emergency Response Division. EPA/540/P-91/008. Oswer Directive 9360.4-07.

USEPA–United States Environmental Protection Agency 2007. Microwave assisted acid digestion of sediments, sludges, soils and oils – Method 3051a. Revision 1.

Section 2: Arsenic in a changing agricultural ecosystem and food chain effects

2.1 Processes and pathways of arsenic in agricultural ecosystems

Environmental Arsenic in a Changing World –
Zhu, Guo, Bhattacharya, Ahmad, Bundschuh & Naidu (Eds)
ISBN 978-1-138-48609-6

Derivation of soil thresholds for arsenic applying species sensitivity distribution

C.F. Ding & X.X. Wang
Key Laboratory of Soil Environment and Pollution Remediation, Institute of Soil Science, Chinese Academy of Sciences, Nanjing, P.R. China

ABSTRACT: It is essential to establish an accurate soil threshold for the implementation of soil management practices. This study takes root vegetable as an example to derive soil thresholds for arsenic (As) based on the food quality standard using species sensitivity distribution (SSD). A soil type-specific bioconcentration factor (BCF, ratio of As concentration in plant to that in soil) generated from soil with a proper As concentration gradient was calculated and applied in the derivation of soil thresholds instead of a generic BCF value to minimize the uncertainty. The derived soil thresholds were dependent on the combination of soil pH and iron oxide (Fe_{OX}) content.

1 INTRODUCTION

Generally, soil thresholds are calculated from the maximum levels for potentially toxic elements in food products for the various crops and the soil-plant transfer model (de Vries & McLaughlin, 2013). The food quality standards (FQSs) for contaminants are set according to different food categories, such as cereals, leafy vegetables, root vegetables, fruits, beans, nuts, etc. (Shao *et al.*, 2014). However, the current soil quality standards (SQSs) of potentially toxic elements used for agriculture in many countries around the world do not consider the diverse crop species and cultivars and the effects of soil properties, and therefore may be either over- or under-conservative, resulting in unnecessary economical or ecological costs (Recatalá *et al.*, 2010; Zhao *et al.*, 2015). Therefore, it is urgent to revise and improve the SQSs.

Species sensitivity distributions (SSDs) are commonly used to derive threshold values for contaminants. After estimation of the SSD parameters using statistical extrapolation methods, a hazard concentration from the fifth percentile of the distribution (HC5) is calculated. The HC5 is the concentration at which less than 5% of the species within an ecosystem is expected to be affected and is often used for deriving environmental quality standards (Korsman *et al.*, 2016). Very few studies have applied the SSD methodology to derive soil thresholds for As due to the lack of data generated from different crop species and different soil types.

Therefore, with a focus on widely consumed root vegetables, this study aims to derive soil thresholds for As based on the food quality standard using SSD method.

2 PROCEDURE AND METHODS

2.1 *Framework for deriving soil thresholds for As*

In brief, the procedures of deriving soil thresholds for As included, first, the investigation of the sensitivity distributions of different vegetable cultivars for accumulating As. The second step is the construction of the prediction model for As transfer from soil to vegetable. The third step is the verification of cross-species extrapolation of the prediction model and the normalization of As bioaccumulation data. Finally, the SSD curves were constructed and HC5 values were calculated, and then the prediction model for HC5 was developed as a function of soil properties.

2.2 *Investigation of sensitivity distributions and development of prediction model*

A total of twenty-one soils covering a wide variation in soil properties were collected throughout China. Firstly, two typical soils, an acidic Ferralsols (pH 4.84) and a neutral Cambisols (pH 6.93), were used to test the sensitivity variations of different cultivars for accumulating soil As. Three species of root vegetables, radish (*Raphanus sativus* L.), carrot (*Daucus carota* L.), and potato (*Solanum tuberosum* L.) were used in the experiment; four representative cultivars of each species were selected. Then, to minimize the effect of soil properties on the bioaccumulation data, the soil-plant transfer model was developed and used as the normalization relationship from carrot cultivar New Kuroda grown in the twenty-one soils. Uncontaminated soils were spiked with soluble As salt ($Na_3AsO_4 \cdot 12H_2O$).

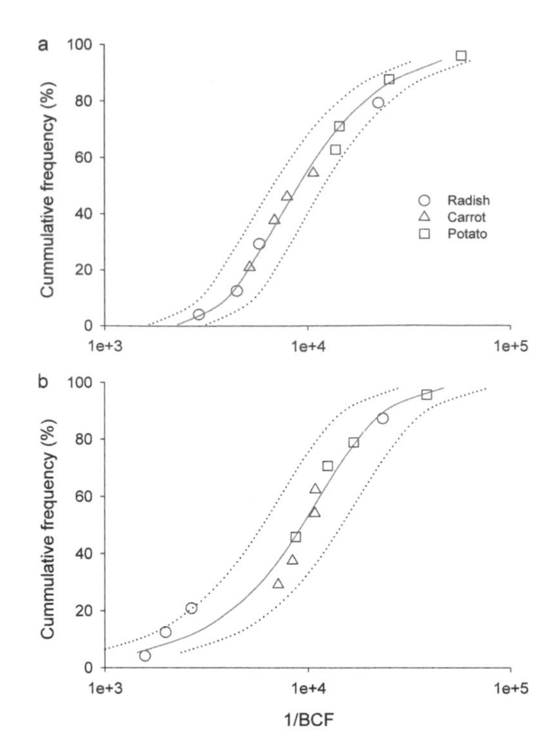

Figure 1. Species sensitivity distributions of the twelve vegetable cultivars grown in Ferralsols (a) and Cambisols (b) fitted by Burr Type III distribution.

Three treatments were applied, including the control, low-As (As1, 15–30 mg kg^{-1}), and high-As (As2, 30–60 mg kg^{-1}). After aging for three months, vegetable plants were cultured under regular farming management style. Three replicates were tested per treatment.

3 RESULTS AND DISCUSSION

The SSD curves (Fig. 1) were constructed with Burr Type III distribution based on the accumulation data of different vegetable cultivars grown in Ferrosols and Cambosols. The geometric means of BCFs under three As treatments were calculated and taken to represent the sensitivity for the same vegetable cultivar. The SSD curves showed that the twelve vegetable cultivars exhibited sensitivity variations for As accumulation, with radishes and potatoes in both soils being the most and least sensitive, respectively.

The normalization relationship was shown as log [BCF] = 0.13pH–1.39log [Fe_{OX}] – 1.98 with $R^2 = 0.67$. The model was subsequently used for normalizing all individual BCF values of the twelve cultivars cultivated in Ferralsols and Cambisols to soil conditions with different combinations of soil pH (4.5–9.0) and Fe_{OX} (15–65 g kg^{-1}).

The critical soil concentration for each cultivar under different soil conditions after normalization was back calculated from the corresponding BCF value and the food quality standard of As (0.5 mg kg^{-1}, fresh weight). Then the hazardous concentrations (HC5) were calculated from the Burr Type III fitted SSD models. The calculation formula for HC5 was provided in Figure 2. The results suggested that the current SQS were only valid for soils with limited combinations of soil pH and Fe_{OX} content.

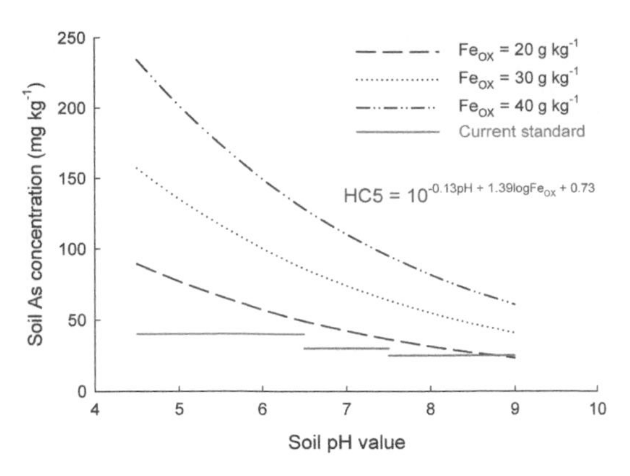

Figure 2. Comparison of the derived soil As thresholds with current Chinese soil quality standard.

4 CONCLUSIONS

This study adopted the SSD methodology to derive soil thresholds for As in view of the food quality standard while taking into account the influences of soil properties. The approach proposed here is widely applicable to other crops as well as other trace elements that have the potential to cause food safety issues.

ACKNOWLEDGEMENTS

This work was supported by the National Key Research and Development Program of China (2016YFD0800400).

REFERENCES

de Vries, W. & McLaughlin, M.J. 2013. Modeling the cadmium balance in Australian agricultural systems in view of potential impacts on food and water quality. *Sci. Total Environ.* 461–462: 240–257.

Korsman, J.C., Schipper, A.M. & Hendriks, A.J. 2016. Dietary toxicity thresholds and ecological risks for birds and mammals based on species sensitivity distributions. *Environ. Sci. Technol.* 50(19): 10644–10652.

Recatalá, L., Sánchez, J., Arbelo, C. & Sacristán, D. 2010. Testing the validity of a Cd soil quality standard in representative Mediterranean agricultural soils under an accumulator crop. *Sci. Total Environ.* 409(1): 9–18.

Shao, Y., Wang, J., Chen, X. & Wu, Y.N. 2014. The consolidation of food contaminants standards in China. *Food Control* 43: 213–216.

Zhao, F.J., Ma, Y.B., Zhu, Y.G., Tang, Z. & McGrath, S.P. 2015. Soil contamination in China: current status and mitigation strategies. *Environ. Sci. Technol.* 49(2): 750–759.

Environmental Arsenic in a Changing World –
Zhu, Guo, Bhattacharya, Ahmad, Bundschuh & Naidu (Eds)
ISBN 978-1-138-48609-6

Spatial variation of arsenic in irrigation well water from three flood plains (Ravi, Chenab and Jhelum) of Punjab, Pakistan

A. Javed[1,2], Z.U. Baig[2], A. Farooqi[2] & A. van Geen[3]
[1]*Department of Earth and Environmental Sciences, Bahria University, Islamabad, Pakistan*
[2]*Environmental Geochemistry Laboratory, Department of Environmental Sciences, Faculty of Biological Sciences, Quaid-i-Azam University, Islamabad, Pakistan*
[3]*Lamont Doherty Earth Observatory, Columbia University, Palisades, New York, NY, USA*

ABSTRACT: Research work was conducted in six rice growing districts of Punjab, covering three flood plains (Ravi, Chenab and Jhelum) to investigate the geographic variations of arsenic (As) in irrigation well water. In field, irrigation well water was tested using ITS Field Kit which was later analyzed by ICP-MS at Columbia University, New York, USA. Results indicate that Ravi flood plain is different from Chenab and Jhelum Flood plain; with Ravi floodplain had a higher As concentration in irrigation wells than those from the Chenab and Jhelum Flood plain and high As wells was concentrated near the river bank. The results of this study form a strong basis for the researchers in designing detailed studies on the accumulation of As in paddy soil and possible uptake by rice.

1 INTRODUCTION

Arsenic (As) contamination of groundwater is a well-recognized problem in Southeast Asia. Studies have shown that wells distribution of As is highly variable geographically and wells located near the rivers have high As levels (Berg *et al.*, 2007). According to results of first blanket testing in Pakistan (by Abida Farooqi, USAID/HEC funded project) and in Indian side by Chander Kumar Singh (PEER funded project), As concentrations in drinking water are higher in the areas along the Ravi Flood plain whereas concentrations were low along Satluj, Chenab and Jhelum Flood plain. Therefore, in this study we hypothesized that the concentration of As in groundwater used for irrigation may also vary from one flood plain to another within a same region. Therefore, the objective of this study is the screening of irrigation wells for As that would form a basis for the researchers in designing detailed studies on the accumulation of As in paddy soil and possible uptake by rice.

2 METHODS/EXPERIMENTAL

2.1 *Study area and sampling*

In this study, assessment of As in irrigation wells was carried out in six rice growing districts (Narowal, Mandi Bahauddin, Gujrat, Gujranwala, Sialkot and Lahore) (Fig. 1). Mandi Bahauddin and Gujrat District lies between Jhelum and Chenab Flood plain whereas Narowal and Lahore are on Ravi Flood plain and Sailkot district lies between Ravi and Chenab flood plain.

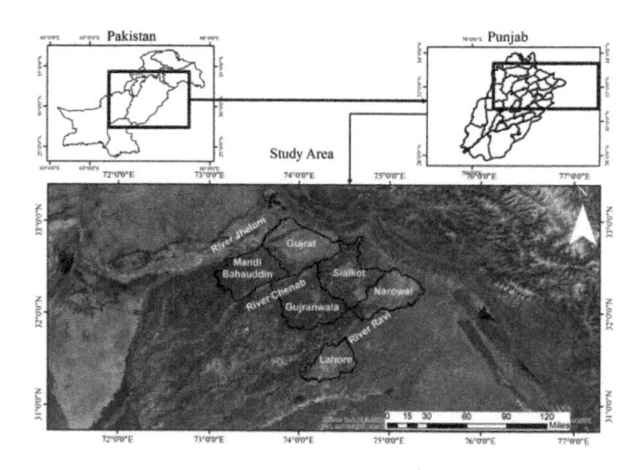

Figure 1. Map of the study area showing six rice growing districts of Punjab.

2.2 *Analysis*

Irrigation water samples were analyzed for As using portable ITS Arsenic Econo-Quick Kit, which relies on the generation of arsine gas and visual detection on a strip impregnated with mercuric bromide and the standard reaction time was maintained at 10 min. All the data were uploaded in survey CTO. For quality control, the standards of known quantity were prepared and tested every day before the start of work. The same well water samples were also collected for

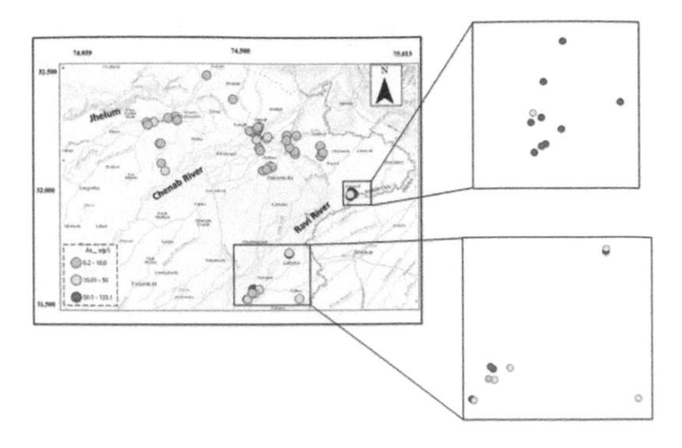

Figure 2. Spatial distribution of arsenic in irrigation wells by ICP-MS. Red dot shows the arsenic concentration >50 $\mu g L^{-1}$.

measurement of As at Lamont Doherty Earth Observatory (LDEO) Laboratory of Columbia University, New York, USA. Water was collected in 20 mL clean plastic vial and was transported to Laboratory. Water was analyzed for As by inductively coupled plasma mass spectrometry (ICP-MS) following 1:10 dilution. Before analysis, water samples were acidified with 1% HNO_3. The detection limit of the method for dissolved As is 0.1 $\mu g L^{-1}$. Three standards were used (NIST 1640 A) with As value 8.75 $\mu g L^{-1}$, NIST (1643 F) with As value 56.85 $\mu g L^{-1}$ and LDEO standard with As value 430 $\mu g L^{-1}$) along with method and vial blanks. We also validated this field kit method by comparing kit As measurements with ICP-MS measurements on 60 water samples and found good co-relation ($R^2 = 0.786$) across 1–102 $\mu g L^{-1}$ range of concentrations.

3 RESULTS AND DISCUSSION

The geographic variations of As in all three flood plains is shown in Figure 2. Results indicate that Ravi flood plain is different from Chenab and Jhelum flood plain, The Ravi floodplain had a higher As concentration (ranged from 7.27 to 135 $\mu g L^{-1}$) in irrigation wells than those from the Chenab (ranged from 0.2335 to 27.75 $\mu g L^{-1}$) and Jhelum Flood plain (ranged from 0.2335 to 21.86 $\mu g L^{-1}$). In Jhelum and Chenab flood plains none of the wells were found to have As concentration greater than 50 $\mu g L^{-1}$ whereas in Ravi flood plain, 65% of the well exceeds 50 $\mu g L^{-1}$ and 10% exceeds 100 $\mu g L^{-1}$. Similar observations were also reported for groundwater As concentrations across the different geomorphological units of the Bangladesh with low concentrations of As in the higher altitudePleistocene terraces, and at high concentrations in Holocene floodplains (Ravenscroft *et al.*, 2005). The explanation for this is that Pleistocene sediments are more highly weathered and leached of As. A study on the source of As in the Holocene/Pleistocene sediments from the Terai plain of Nepal (that stratigraphically resemble Pakistani sediments) proposed a number of complex processes which can explain the differences in As concentration between Holocene and Pleistocene sediments (Guillot *et al.*, 2015).

In addition, spatial analysis showed that wells with higher concentrations of As were concentrated near the river bank. As the distance from River increased, the concentration of As in groundwater decreased. Stahl *et al.* (2016) suggests that aquifers throughout South and Southeast Asia may be vulnerable to As contamination where riverine recharge flows through recently deposited sediments. Prior to the onset of substantial groundwater pumping, As released in the pore-water of reactive riverbed sediments would have discharged into rivers and would not have influenced groundwater concentrations further inland. With large-scale groundwater pumping throughout South and Southeast Asia, it is likely that many aquifers are now experiencing net inflow from nearby rivers (Berg *et al.*, 2007). This entire phenomenon could possibly explain the geographic variation in As concentration of groundwater in the study area.

4 CONCLUSIONS

Results indicate that Ravi flood plain is different from Chenab and Jhelum, with Ravi floodplain had a higher As concentration in irrigation wells than those from the Chenab and Jhelum Flood plain. The results of this study form a strong basis for the researchers in designing detailed studies on the accumulation of As in paddy soil and possible uptake by rice.

ACKNOWLEDGEMENTS

The author acknowledges the Columbia University, New York, USA, Higher Education Commission (HEC), Pakistan and US-Pakistan Centers for Advanced Studies in Water, Mehran University of Engineering and Technology (MUET), Jamshoro for financial and technical support.

REFERENCES

Berg, M., Stengel, C., Pham, T.K., Pham, H.V., Sampson, M. L. & Leng, M. 2007. Magnitude of arsenic pollution in the Mekong and Red River Deltas-Cambodia and Vietnam. *Sci. Total Environ.* 372(2–3): 413–425.

Guillot, S., Garçon, M., Weinman, B., Gajurel, A., Tisserand, D., France-Lanord, C., van Geen, A., Chakraborty, S., Huyghe, P., Upreti, B.N. & Charlet, L. 2015. Origin of arsenic in Late Pleistocene to Holocene sediments in the Nawalparasi district (Terai, Nepal). *Environ. Earth Sci.* 74(3): 2571–2593.

Ravenscroft, P., Burgess, W.G., Ahmed, K.M., Burren, M. & Perrin, J. 2005. Arsenic in groundwater of the Bengal Basin, Bangladesh: distribution, field relations, and hydrogeological setting. *Hydrogeol. J.* 13(5–6): 727–751.

Stahl, M.O., Harvey, C.F., van Geen, A., Sun, J., Thi Kim Trang, P., Mai Lan, V. & Bostick, B.C. 2016. River bank geomorphology controls groundwater arsenic concentrations in aquifers adjacent to the Red River, Hanoi Vietnam. *Water Resour. Res.* 52(8): 6321–6334.

Environmental Arsenic in a Changing World –
Zhu, Guo, Bhattacharya, Ahmad, Bundschuh & Naidu (Eds)
ISBN 978-1-138-48609-6

Occurrence of arsenic in agricultural soils from the Chaco-Pampean plain (Argentina)

C.V. Alvarez Gonçalvez[1,2,3], F.E. Arellano[1,2,3], A. Fernández Cirelli[1,2,3] & A.L. Pérez Carrera[1,2,3]
[1] *Instituto de Investigaciones en Producción Animal UBA-CONICET (INPA), CONICET – Universidad de Buenos Aires, Buenos Aires, Argentina*
[2] *Universidad de Buenos Aires. Centro de Estudios Transdisciplinarios del Agua (CETA), Universidad de Buenos Aires, Buenos Aires, Argentina*
[3] *Universidad de Buenos Aires. Cátedra de Química Orgánica de Biomoléculas, Universidad de Buenos Aires, Buenos Aires, Argentina*

ABSTRACT: Arsenic pollution is naturally present in groundwater and soils of the main agriculture production area of Argentine. Arsenic in soils, can be taken up and bioaccumulated in plant, and may be bio-transferred to agri-food and humans. All samples collected shown values between 18.6 and 33.1 mg kg^{-1} of total As. Range percentage of organic matter in analyzed samples were from 6.7 to 24.9%. No significant correlation ($p > 0.05$) between total As level and organic matter was found. The As concentration in the analyzed samples were above the limits considered in Argentine. However, all samples were into the range previously reported by other authors.

1 INTRODUCTION

Arsenic (As) pollution affects several countries around the world. One of the most affected countries is Argentina, where As is naturally present in groundwater and soils of the main agriculture production areas (Bundschuh *et al.*, 2012). It is known that presence of As in soils has an adverse impact on ecosystems (Bhattacharya *et al.*, 2007; Huang, 2016). Arsenic in soils, can be taken up and bioaccumulated in plants. Also, livestock intake As from environmental matrices and food. Thereby, from animal and plant tissues, As may be bio-transferred to agri-food and humans, which implicates a potential risk to health.

Furthermore, it is known that organic matter present in soils affects the mobility of As species in the environment, being able enhance the release of As from soils facilitating As leaching into the groundwater (Wang & Mulligan, 2006), and increasing their bioavailability (Huang *et al.*, 2006).

The aim of this study is to assess total arsenic level in agricultural soils destined to fodder implantation, and further analyze the relationship between arsenic levels in soil and the organic matter present in the soil.

2 MATERIAL AND METHODS

2.1 *Study area and sampling*

This study was carried out in the Pampean Plain (Argentina). The study area is one of the most important livestock production regions. The samples zones are located in the west of Buenos Aires province and in the southeast of Cordoba province.

Thirty fodders fields from medium size farms were taken account for this study. For a preliminary analysis 14 samples were random select. All samples were collected from in plastic bags. In all cases soils samples were taken up from areas destined to the implantation of fodder and correspond to surface soil (0 to 20 cm underground). In each farm, samples were collected along transect in the field at regular intervals. Soil portions were mixed into the plastic bag, and were kept in the dark and refrigerated to 4°C until analysis.

2.2 *Chemical analysis*

Organic matter percentage was determinate by gravimetric procedures. Arsenic determination was according ISO 11466 (1995), soils were digested using a mixture of nitric acid and hydrochloric acid (1:3) at 120°C during 2 h with reflux. Arsenic was determined by emission spectroscopy in an inductively coupled plasma optical emission spectrometer (ICP-OES, Optima 2000, Perkin Elmer).

3 RESULTS AND DISCUSSION

3.1 *Arsenic level in soils*

Usually, arsenic in soils is 1 to 40 mg kg^{-1} (Kabata & Pendias, 1992). Canadian Council of Ministers of the Environment (2001), considered in the Soil Quality Guideline for Environmental Health (SQG_E)

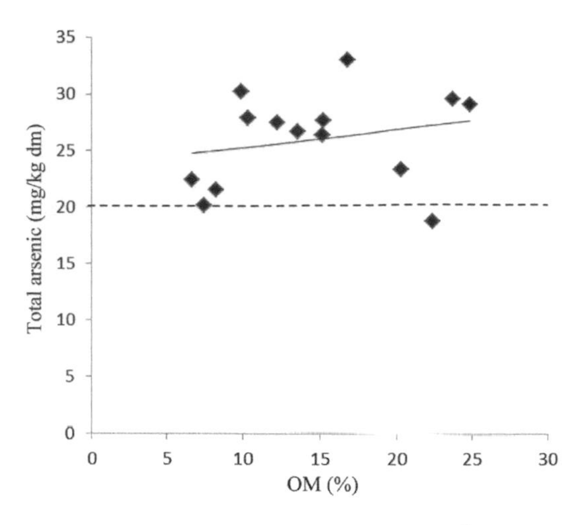

Figure 1. Relationship between total As ($mg\,kg^{-1}$ dry mass) and organic matter percentage (OM%) in soil samples. Continued line represents correlation between them. Dotted line represents maximum recommended level ($20\,mg\,kg^{-1}$) to total arsenic in soils destined to agricultural activities in Argentina.

for agricultural soil a maximum level of $17\,mg\,kg^{-1}$ for inorganic arsenic. Recommended values for total arsenic content in agricultural soils in Argentina is $20\,mg\,kg^{-1}$ (Ley 24.051, Decreto 831/93). All samples analyzed shown values between 18.6 and 33.1 $mg\,kg^{-1}$ of total As, with a mean $\pm$ standard deviation (SD) of 26.0 ± 4.2. Taking on account the recommended values for total As in Argentina, it was observed that 93% of samples exceeded this guideline.

3.2 *Total As level and organic matter*

Relationship between total As level and organic matter amount was assessed. Range percentages of organic matter in analyzed samples were from 6.7 to 24.9% with a mean $\pm$SD of 14.8 ± 6.1. No significant correlation ($p > 0.05$) between total As level and organic matter was found (Fig. 1).

4 CONCLUSIONS

The total As concentration in the analyzed samples were above the limits considered in Argentine. However, all samples were into the range previously reported by other authors. Therefore, it is necessary to determine the As species and its bioavailability present in those soils to evaluate the potential risk from the extensive use of this matrix.

On the other hand, organic matter from the first layer did not show evidence of correlation with As. Thus, to evaluate the organic matter effect in As mobility, it is suggested to evaluate the interaction of As with lower layers of soil.

ACKNOWLEDGEMENTS

The authors are grateful to the University of Buenos Aires and CONICET for financial support.

REFERENCES

Bhattacharya, P., Welch, A.H., Stollenwerk, K.G., McLaughlin, M.J., Bundschuh, J. & Panaullah, G. 2007. Arsenic in the environment: biology and chemistry. *Sci. Tot. Environ.* 379: 109–120.

Bundschuh, J., Litter, M.I., Parvez, F., Román-Ross, G., Nicolli, H.B., Jean, J.S., Liu, C.W., López, D., Armienta, M.A., Guilherme, L.R.G., Cuevas, A.G., Cornejo, L., Cumbal, L. & Toujaguez, R. 2012. One century of arsenic exposure in Latin America: a review of history and occurrence from 14 countries. *Sci. Tot. Environ.* 429: 2–35.

Canadian Council of Ministers of the Environment. 2001. Canadian Soil Quality Guidelines for the Protection of Environmental and Human Health: Arsenic (inorganic) (1997). Updated In: Canadian Environmental quality guidelines, 1999. Canadian Council of Ministers of the Environment, Winnipeg.

Decreto 831/93. Decreto Reglamentario de la ley 24.051. Argentina.

Huang, J.H. 2016. Arsenic trophodynamics along the food chain/web of different ecosystems: a review. *Chem. Ecol.* 32(9): 803–823.

Huang, R.Q., Gao, S.F., Wang, W.L., Staunton, S. & Wang G. 2006. Soil arsenic availability and transfer of soil arsenic to crops in suburban areas in Fujian Province, southeast China. *Sci. Tot. Environ.* 15(2–3): 531–541.

ISO 11466. 1995. Soil Quality – Extraction of trace element soluble in aqua regia.

Kabata, P.A. & Pendias, H. 1992. *Trace elements in soils and plants*. Second edition, CRC Press, Boca Raton, Ann Arbor, London, pp. 203–209.

Ley 24.051 Ley Nacional de Residuos Peligrosos. Argentina.

Wang, S. & Mulligan, C.N. 2006. Effect of natural organic matter on arsenic release from soils and sediments into groundwater. *Environ. Geochem. Health* 28(3): 197–214.

Environmental Arsenic in a Changing World –
Zhu, Guo, Bhattacharya, Ahmad, Bundschuh & Naidu (Eds)
ISBN 978-1-138-48609-6

Arsenic and antimony concentrations in Chinese typical farmland soils

A. Parvez[1], C.L. Ma[1], Q.Y. Zhong[1], C.Y. Lin[1], Y.B. Ma[2] & M.C. He[1]
[1]*State Key Laboratory of Water Environment Simulation, School of Environment, Beijing Normal University, Beijing, P.R. China*
[2]*Institute of Agricultural Resources and Regional Planning, Chinese Academy of Agricultural Sciences, Beijing, P.R. China*

ABSTRACT: The pollution extent by heavy metals in soils is dependent on their total contents, chemical speciation, sources, and some soil physicochemical factors affected their geochemical behaviors. The difficulty in characterization of soil, and current lack of ecological risk assessment methods for heavy metals in farmland soils in China are to be addressed. This study aimed are to investigate the distribution of Arsenic (As) and Antimony (Sb) in different soils collected from 21 provinces of China, and to assess the total content of As and Sb in the soil samples. The present study evaluates the understanding of differences in soil bio-toxicity effects against different heavy metals. The total content of As and Sb in different soils was extracted by double acid digestion (HNO_3 and HCl). The concentration was measured using HG-AFS technique. It provides the basic information about the presence of these toxic metals in different types of soils around China.

1 INTRODUCTION

Arsenic is a commonly encountered contaminant in the environments because of its release during industrial, mining, and agricultural activities (Smedley & Kinniburgh, 2002). The toxicity and mobility of arsenic in soil environments is a function of its oxidation state and its sorption to soil minerals and organic matter (Charlet *et al.*, 2011). Antimony is a toxic element, and excess intake results in many diseases in humans, such as cancers, cardiovascular disease, liver disease and respiratory disease (WHO, 2003). China has the most abundant Sb resources of any country in the World (He *et al.*, 2012). Soil contamination by As and Sb is suspected to cause health problems in some regions in China. For example, residents of antimony mining area in the south Guizhou province in China are subjected to long-term Sb exposure, and liver cirrhosis is the primary cause of death (Cen *et al.*, 2007).

2 EXPERIMENTAL METHODS

2.1 *Soil sampling and extraction method*

Soil samples from 21 provinces of China were air-dried, sifted to remove stone and debris, then ground to pass through a 20-mesh sieve (0.99 mm in particle size). Approximately 0.1 g of this fine powder from each sample was accurately weighed into a glass tube (25 mL) containing 3 mL concentrated HNO_3 and 1 mL HCl and left overnight. Samples will be digested for 2 h in a water bath at 90°C, after which the supernatants were obtained via centrifugation and filtration. As and Sb was measured using a hydride generation atomic fluorescence spectrometer (HG-AFS) (Wei *et al.*, 2011). Blanks, duplicated samples and soil standard reference materials (Centre for Reference Materials, China) were used to verify the accuracy of the analysis.

2.2 *Analytical technique*

Numerous analytical techniques and experimental approaches have been proposed in the last two decades to identify and measure As and Sb aimed at obtaining reliable results and correct evaluations in the field of environmental chemistry. Atomic fluorescence spectroscopy coupled to HG has received increasing attention because its suitability for As and Sb determination at trace levels due to its high sensitivity, wide dynamic range (4–6 orders of magnitude), simplicity and very low instrumental cost (Chen *et al.*, 2003; Miravet *et al.*, 2005; Zhang *et al.*, 2005).

The QA/QC included the calibration, blanks and using the reference materials. Initial calibration was performed before all the samples. The calibration curve yielded correlation coefficient of 0.998. QC samples were prepared and analyzed with each set of soil samples for As and Sb detection. The samples were gone through the same sample preparation method in the same manner to ensure the evaluation of extraction procedure. The recovery percentage for reference materials ranged from 88% to 97%.

3 RESULTS AND DISCUSSION

3.1 *Arsenic*

The concentrations of As are in the range of 1.688 and 10.28 mg kg^{-1} (mean 5.575 mg kg^{-1}). The soil of Xinjiang soil is presenting a highest concentration of As (10.28 mg kg^{-1}) among the different farmland soils of China. The minimum amount of As is found in the sample of Jilin soil (1.688 mg kg^{-1}). The binding of As via ternary complexes and mobile colloids has been considered the most important but not the sole driver of As mobility in soils. The total content of As gives the information about the presence of As concentration as a whole. However, the environmental availability and toxicological effects of As depend on its speciation, the knowledge of the oxidation state of As in soils is essential for the risk assessment.

3.2 *Antimony*

Soil is the main plant-developing medium; it is also a major contaminant sink in various ways: fertilizers and amendments inputs, aerial deposition, and water percolation. Therefore, the concentration of Sb is varying in different provinces of China (0.091–0.521 mg kg^{-1}). The highest concentration of Sb was found in the soil of Anhui province (0.521 mg kg^{-1}). The soil of Chongqing city showed the minimum concentration of Sb as shown in Table 1.

Table 1. Concentration of arsenic and antimony in different soils (mg kg^{-1}).

Province	Sb	As
Hainan	0.100	8.560
Guangdong	0.274	6.770
Hunan	0.266	3.779
Yunnan	0.331	4.735
Jiangsu	0.207	3.746
Zhejiang	0.128	5.290
Anhui	0.521	2.267
Jiangxi	0.260	6.052
Sichuan	0.211	4.393
Chongqing	0.091	6.119
Shanxi	0.224	4.427
Shandong	0.140	6.431
Shanxi	0.372	9.809
Hubei	0.309	7.140
Henan	0.170	5.898
Hebbi	0.173	7.420
Beijing	0.351	3.734
Ningxia	0.252	3.295
Xinjiang	0.150	10.28
Jilin	0.179	1.688
Heilongjiang	0.213	5.245

4 CONCLUSIONS

The present study gathers the current state of knowledge on As and Sb concentrations in farmland soils of China. The concentration may vary according to the difference in physico-chemical properties of soil. The study helped to develop the soil heavy metal risk assessment methods that will contribute to China's farmland soil protection and food safety. It has an important social value, which can provide technical guidance for farmland soil remediation.

ACKNOWLEDGEMENTS

This work was supported by the project of "Research on Migration/Transformation and Safety Threshold of Heavy Metals in Farmland Systems", National Key Research and Development Program of China (2016YFD0800405) and the National Natural Science Foundation of China (21477008).

REFERENCES

Cen, R.G., Li, B., Wei, S.Y., Mo, X.J. & Zhang, L. 2007. Investigation on correlation between chronic antimony poisoning and liver fibrosis. *Labeled Immunoass.Clin. Med.* 14(2): 106–107.

Charlet, L., Morin, G., Rose, J., Wang, Y., Auffan, M., Burnol, A. & Fernandez-Martinez, A. 2011. Reactivity at (nano) particle-water interfaces, redox processes, and arsenic transport in the environment, C.R. *Geosci.* 343(2–3): 123–139.

Chen, B., Krachler, M. & Shotyk, W. 2003. Determination of antimony in plant and peat samples by hydride generation-atomic fluorescence spectrometry (HG-AFS). *J. Anal. Atomic Spectrom.* 18(10): 1256–1262.

He, M., Wang, X., Wu, F. & Fu, Z. 2012. Antimony pollution in China. *Sci. Total Environ.* 421–422: 41–50.

Miravet, R., Bonilla, E., Lopez-Sanchez, J.F., & Rubio, R. 2005. Antimony speciation in terrestrial plants. Comparative studies on extraction methods. *J. Environ. Monitor.* 7(12): 1207–1213.

Smedley, P.L. & Kinniburgh, D.G. 2002. A review of the source, behaviour, and distribution of arsenic in natural waters. *Appl. Geochem.* 17(5): 517–568.

Wei, C., Deng, Q., Wu, F., Fu, Z. & Xu, L. 2011. Arsenic, antimony, and bismuth uptake and accumulation by plants in an old antimony mine, China. *Biol. Trace Elem. Res.* 144(1–3): 1150–1158.

WHO 2003. Antimony in Drinking-Water: Background Document for Preparation of WHO Guidelines for Drinking-Water Quality. WHO/SDE/WSH/03.04/74. *World Health Organization*, Geneva.

Zhang, W.B., Gan, W.E. & Lin, X.Q. 2005. Electrochemical hydride generation atomic fluorescence spectrometry for the simultaneous determination of arsenic and antimony in Chinese medicine samples. *Anal. Chim. Acta* 539(1): 335–340.

Environmental Arsenic in a Changing World –
Zhu, Guo, Bhattacharya, Ahmad, Bundschuh & Naidu (Eds)
ISBN 978-1-138-48609-6

Characterization of an agricultural site historically polluted by the destruction of arsenic-containing chemical weapons

F. Battaglia-Brunet[1], J. Lions[1], C. Joulian[1], N. Devau[1], M. Charron[1], D. Hube[1], J. Hellal[1], M. Le Guédard[2], I. Jordan[3], P. Bhattacharya[4], K. Loukola-Ruskeeniemi[5], T. Tarvainen[5] & J. Kaija[5]
[1]*BRGM, Orléans, France*
[2]*LEB Aquitaine Transfert, Villenave d'Ornon, France*
[3]*GEOS Ingenieurgesellschaft mbH, Halsbrücke, Germany*
[4]*KTH-Groundwater Arsenic Research Group, Department of Sustainable Development, Environmental Science and Engineering, KTH Royal Institute of Technology, Stockholm, Sweden*
[5]*Geological Survey of Finland, Espoo, Finland*

ABSTRACT: Arsenic in agricultural soils may represent a risk for crop quality and surrounding water resources. In the frame of AgriAs project, "Evaluation and management of Arsenic contamination in agricultural soil and water", a former chemical weapons destruction site converted into agricultural land was characterized. The objective of this study was to identify possible links between arsenic concentration and speciation and bio-indicators informing about the bioavailability of the toxic element. Plants lipidic bio-indicator Omega-3 Index showed that toxicity of the soil was not directly correlated with arsenic concentration. Conversely, arsenic level in the soil samples seemed to influence the density of microbes transforming As species.

1 INTRODUCTION

Arsenic (As) and its compounds are toxic and ubiquitous in the environment. It has been estimated that over 220 million people are exposed to As from drinking water or food (Naujokas *et al.*, 2013). Arsenic in agricultural soils and water, and its subsequent entering into the food chain cause potential risk to human health. Indeed, As can accumulate in crop plants such as corn or barley. High As concentrations may also reduce crop yields. Production, destruction and storage of As-rich chemical weapons have increased As concentrations in the European environment. The present study, performed in the framework of the WaterJPI AgriAs project, is focused on the elucidation of the links between As concentration and speciation in soils and a range of bio-indicators that could be helpful to understand the behavior and bio-availability of arsenic in agricultural soils.

2 METHODS/EXPERIMENTAL

2.1 *Site description and sampling*

The sampling site is a former chemical ammunition breaking-down facility of the interwar period converted into agricultural land near Verdun, France. The field is located 25 km north-east from Verdun. The field was used as a pasture in 2002 and was cultivated from 2002 to 2015. Since 2015, it is a fallow ground because cultivation was forbidden when the pollution was detected. The sampling was performed in May 2017. Many plant species could be observed. The diversity was lower in the most polluted places but plants were observed even in those places. Soils were sampled in a reference zone, far from the polluted area, and along a three-zones transect representing a gradient of As pollution. Soils were sampled in the 0–20 cm layer. Each sample was taken as a composite of 5 points from 3 m × 3 m squares.

2.2 *Chemical analyzes*

Arsenic concentration in the soil surfaces was determined on site with X-ray fluorescence (NITON) apparatus. Speciation of arsenic was performed by HPLC-ICP-MS after H_3PO_4 extraction. Diphenylarsinc acid was determined by HPLC-DAD and apolar organoarsenicals by GC/MS.

2.3 *Bio-indicators*

As(III)-oxidizing and As(V)-reducing microbial concentrations were evaluated by Most Probable Number methods (5 tubes), using specific media in microplates. As(III) was visualized in the wells by adding pyrrolidine dithio-carbamate (Thouin *et al.*, 2016). Plants were collected in the four zones described previously. On the reference, low and medium zones, three and five species of plants were collected to measure As bioavailability and the lipidic bioindicator Omega-3 Index. Arsenic bioavailability was analyzed from collected plants by ICP-MS after mineralization by acid

Table 1. Concentration of arsenic species in the soil samples. ND: not detected

Zones	As(V) mg kg^{-1}	As(III) mg kg^{-1}	Diphelylarsinic acid mg kg^{-1}	Triphenylarsnine mg kg^{-1}
Reference	15	0.3	ND	ND
Transect Low	20	0.2	ND	ND
Transect Medium	192	1.4	ND	ND
Transect High	756	14.4	2.19	0.95

attack. For the measurement of the lipidic bioindicator, the leaf fatty acid composition was analyzed by GC-FID on the plants collected on the site but also on lettuces grown under controlled conditions on the four collected soils.

3 RESULTS AND DISCUSSION

3.1 *Arsenic in soil*

Arsenic concentration varied from 15 mg kg^{-1} in the reference not polluted area of the site, to 775 mg kg^{-1} in the highly polluted spot (Table 1).

The main arsenic form was arsenate (As(V)), however As(III) proportion was higher in the polluted spot. Two organic molecules derived from the weapons were detected only in the highly polluted zone: diphenylarsinic acid and thriphenylarsine.

3.2 *Bio-indicators*

The collected plants were *Brassica napus*, *Plantago major*, *Carduus* sp. in the reference zone; *Brassica napus, Plantago major, Carduus sp.*, and *Chamaemelum nobilis* in the low-pollution zone; *Brassica napus, Plantago major, Carduus sp.* and *Chamaemelum nobilis* in the medium-polluted zone; *Carduus sp., and Chamaemelum nobilis* in the high-pollution zone. Unidentified species were also collected in all zones except the highly-polluted one. Arsenic was more phytoavailable on the medium polluted soil than on the reference and the low polluted soils for all the three species collected. On the highly polluted soil, only one species could be analyzed. For this species, the As content was higher in the highly polluted zone. However, regardless of the zones, the As content in plants was very low compared to plants found on other As-polluted sites. In both site plants and lettuces, the lipidic bioindicator was lower on the medium polluted soil. It seems that the medium polluted soil was the most toxic for plants. This result might be interpreted in the light of further soil characterizations. As(III)-oxidizing microbes were more numerous in the highly polluted soils than

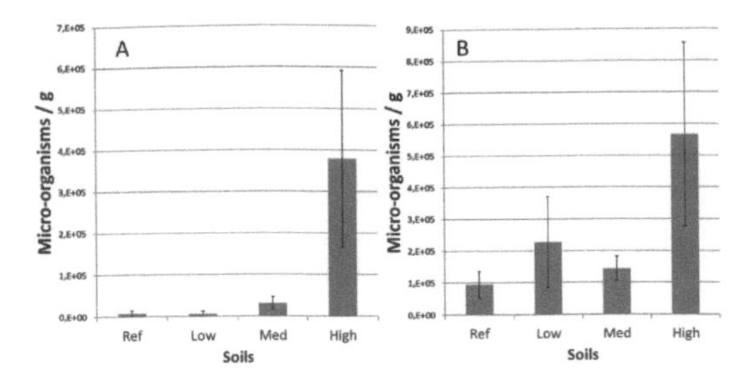

Figure 1. Evaluation by MPN methods of As(III)-oxidizing (A) and As(V)-reducing (B) microorganisms in the soil samples.

in the other samples. The same tendency was observed for As(V)-reducing microbes, but less significantly (Fig. 1).

4 CONCLUSIONS

Whereas the studied site was polluted one century ago, and was used for agriculture for years, impact of the pollution level is still observable on the diversity of wild plants. Arsenic level seems to exert influence on the soil composition in As-transforming microbes. These results suggest that arsenic was sufficiently bio-available to exert a selection pressure on the microbial communities. Next experimental steps will include the study of the influence of agricultural practices on the fate of As, through a biogeochemical approach including modelling, and connected with risk assessment.

ACKNOWLEDGEMENTS

AgriAs is co-funded by EU and the Academy of Finland, L'Agence Nationale de la Recherche, Bundesministerium für Ernährung und Landwirtschaft and Forskningsrådet FORMAS under the ERA-NET Cofund WaterWorks2015 Call. This ERA-NET is an integral part of the 2016 Joint Activities developed by the Water Challenges for a Changing World Joint Programme Initiative (Water JPI).

REFERENCES

Naujokas, M.F., Anderson, B., Ahsan, H., Vasken Aposhian, H., Graziano, J.H., Thompson, C. & Suk, W.A. 2013. The broad scope of health effects from chronic arsenic exposure: update on a worldwide public health problem. *Environ. Health Persp.* 121, 295–302.

Thouin, H., Le Forestier, L., Gautret, P., Hube, D., Laperche, V., Dupraz, S., & Battaglia-Brunet, F. 2016. Characterization and mobility of arsenic and heavy metals in soils polluted by the destruction of arsenic-containing shells from the Great War. *Sci. Total Environ.* 550: 658–669.

Environmental Arsenic in a Changing World –
Zhu, Guo, Bhattacharya, Ahmad, Bundschuh & Naidu (Eds)
ISBN 978-1-138-48609-6

Absorption and distribution of phosphorus from Typha under arsenic

W. Ren, G.Y. Yang, W. Yan, L. Hou & Y.G. Liu
Research Institute of Rural Sewage Treatment and College of Ecology and Soil & Water Conservation, Southwest Forestry University, Kunming, P.R. China

ABSTRACT: In order to reveal absorption and distribution of phosphorus under from under arsenic. By the method of arsenic addition, the concentration gradient of arsenic in the sediment was set 0, 50, 100, 150, 200, 400 and 600 mg kg^{-1}. To investigate the absorption and distribution of different parts of Typha at different arsenic concentrations. The results show: (1) With the increase of arsenic concentration in the sediments, Typha on phosphorus absorption was first increased and then decreased. The absorption of arsenic also increased first, then decreased and tended to be stable, and the absorption of phosphorus and arsenic reached the peak under 150 mg kg^{-1} stress. Add different concentrations of arsenic affected the absorption of phosphorus and arsenic in Typha, may be related to the biomass of plants. (2) The uptake of phosphorus in the aerial part of Cattail was significantly higher than that of the root, while the absorption and accumulation of arsenic were opposite.

1 INTRODUCTION

Arsenic and arsenic compounds are common environmental pollutants, endangering the ecological environment, and seriously threatening human health (Mandal & Suzuki, 2002). Phosphorus is a necessary nutrient element for plant growth. It is also an important limiting element for eutrophication in lakes. In view of the high risk of eutrophication in the plateau lakes. However, the lake area is vast and the hydrological conditions are complex. It is difficult to realize the large-scale restore by physical and chemical methods. Phytoremediation is the ecological treatment method which is concerned at present. Wetland plants can remove pollutants in water by absorption, adsorption and enrichment (Tu & Ma, 2003). The less studies on the distribution and absorption of phosphorus and arsenic. Pot culture and gradient arsenic stress were used in this study. The absorption and distribution of phosphorus were studied under arsenic stress. The relationship between arsenic and phosphorus in wetland plants was analyzed in order to provide theoretical basis and technical support for phytoremediation of plateau lakes with eutrophication and heavy metal arsenic pollution.

2 MATERIALS & METHODS

The cattail as the research object, through the pot culture and adding different concentrations of arsenic by way of indoor simulation experiment. Set the arsenic pollution in the sediment concentration: 0, 50, 100, 150, 200, 400, 600 mg kg^{-1}, were roughly the same biomass and growth status of seedlings planted in good soil for pollution stress, and unified equal amount of garden fertilizer, according to evaporation of soil moisture, not to keep the water regularly. The soil is submerged. Plants were harvested after 45 d plants, and determination of plant height, root length, upper (lower) dry weight and the content of arsenic, explore typical wetland plants Typha in plants under different concentrations of arsenic uptake and distribution of various parts of the.

Table 1. Basic properties of the tested soil.

Property	Value
pH	3.97
TP (g kg^{-1})	0.57 ± 0.004
TAs (mg kg^{-1})	8.39 ± 0.64
TCa (g kg^{-1})	2.86 ± 0.58
OM (g kg^{-1})	28.84 ± 0.81

3 RESULTS AND DISCUSSION

3.1 *Basic properties of the tested soil*

Some of tested soil sediments were listed in Table 1.

3.2 *Effects of arsenic pollution on the growth of Typha habitat*

Different arsenic pollution habitat Typha growth showed increased before decreasing rule. The high content of arsenic in cattail plant 0∼150 mg kg^{-1} soil increased volatility, and reached the peak at 150 mg kg^{-1} in the high pollution, arsenic pollution decreased significantly (Fig. 1). Arsenic pollution in the environment, underground cattail than shoot the degree of toxicity was significantly increased, root length reached maximum in the arsenic content of 50 mg kg^{-1}, higher than 50 mg kg^{-1} showed a downward trend, and decreased significantly.

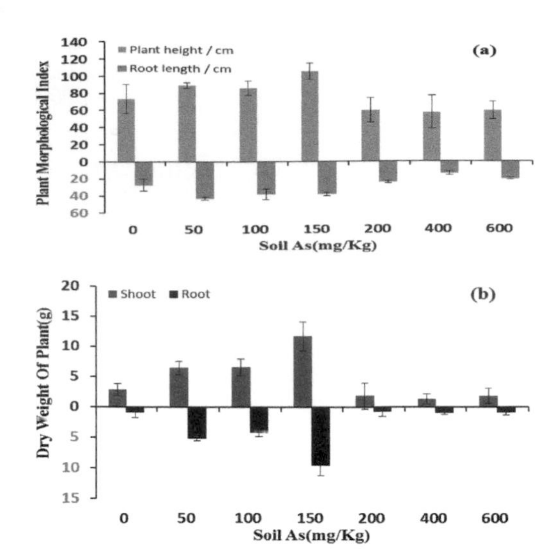

Figure 1. Effects of arsenic pollution on the growth of Typha habitat: a) Plant morphological index, b) Dry weight of plant.

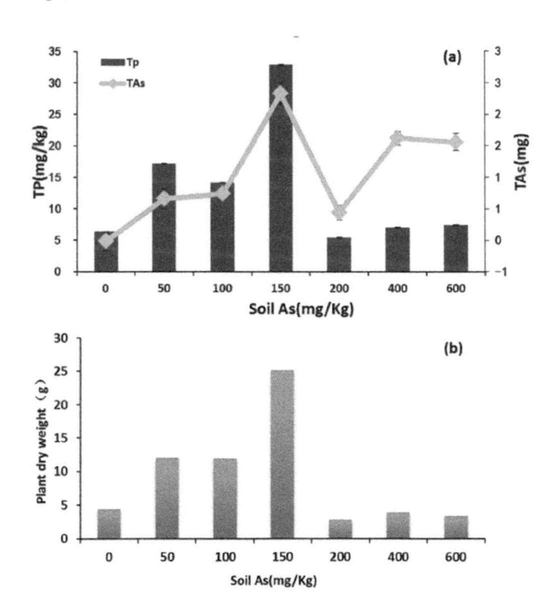

Figure 2. Effects of arsenic on the absorption of phosphorus and arsenic in cedar and its effect on the biomass of cattail: a) Absorption of phosphorus and arsenic from cattails, b) Dry weight of persimmon plants.

3.3 *Absorption of phosphorus and arsenic and its effect on the biomass of Typha angustifolia in arsenic contaminated habitats*

With the increase of arsenic concentration of exogenous pollutants, the absorption of phosphorus and arsenic increased first and then decreased, and the absorption of phosphorus arsenic reached the peak value in the soil with arsenic content of 150 mg kg^{-1} (Fig. 2) – synergistic effect. This is consistent with the law of dry weight change of the plant, indicating

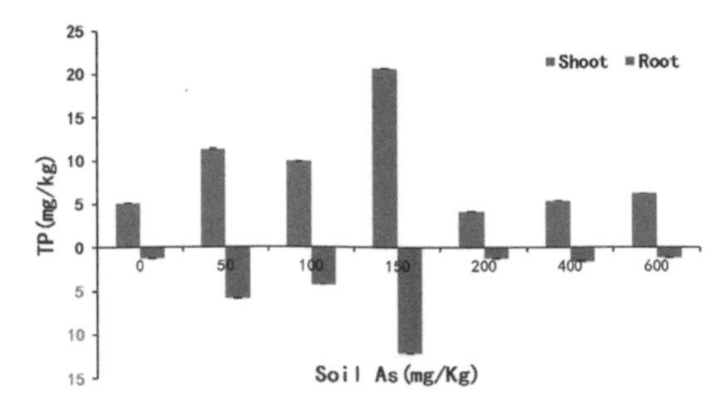

Figure 3. Distribution of phosphorus from Typha under arsenic.

that the addition of different concentrations of arsenic affects the absorption of phosphorus arsenic, which may be related to the biomass of the plant.

3.4 *Distribution of phosphorus from Typha under arsenic*

With the increase of arsenic concentration of exogenous pollutants, the uptake of phosphorus was higher than that of roots (Fig. 3). The transport coefficient of cedar is higher, indicating that cattail can absorb phosphorus quickly, which is less affected by the change of arsenic concentration. In the arsenic polluted habitat, the enrichment ability of phosphorus in the upper part and the underground part of cattail was weak.

4 CONCLUSIONS

1) With the increase of arsenic concentration in the sediments, on phosphorus and arsenic absorption was first increased and then decreased. Add different concentrations of arsenic affected the absorption of phosphorus and arsenic in Typha, may be related to the biomass of plants.
2) Typha on the part of the absorption of phosphorus accumulation was significantly higher than that of the root.
3) The phosphorus transport coefficient of different parts of Typha angustifolia was higher, while the arsenic was lower.

ACKNOWLEDGEMENTS

This project was supported by the National Natural Science Foundation of China (grants 31560147 and 51469030).

REFERENCES

Mandal, B.K & Suzuki, K.T. 2002. Arsenic round the world: a review. *Talanta* 58(1): 201–235.

Tu, S. & Ma, L.Q. 2003. Interactive effects of pH, arsenic and phosphorus on uptake of As and P and growth of the arsenic hyper accumulator *Pteris vittata* L. under hydroponic conditions. *Environ. Exper. Bot.* 50(3): 243–251.

Environmental Arsenic in a Changing World –
Zhu, Guo, Bhattacharya, Ahmad, Bundschuh & Naidu (Eds)
ISBN 978-1-138-48609-6

Adsorption of arsenic by birnessite-loaded biochar in water and soil

H.Y. Wang, P. Chen & G.-X. Sun
State Key Laboratory of Urban and Regional Ecology, Research Center for Eco-Environmental Sciences, Chinese Academy of Sciences, Beijing, P.R. China

ABSTRACT: A novel biochar was successfully prepared through modification of birnessite to improve its capability to adsorb As in water and soil. The saturated adsorption capabilities of birnessite-loaded biochar (BRB) for As(III) and As(V) were as large as 3621 and 2381 mg kg^{-1} (calculated by Langmuir isotherm model), much higher than for the corresponding non-loaded biochar (no sorption of As). In comparison with control, after 6 weeks' incubation in contaminated soil, BRB showed potential for reducing As concentration under flooded condition in pore water, respectively. These results suggested that BRB could be used as an effective sorbent for simultaneous immobilization of heavy metals, especially As in environmental and agricultural systems.

1 INTRODUCTION

Adsorption is one of the most commonly used methods for removal of arsenic (As) from the aqueous phase or for immobilization of them in soil because of their non-degradable properties and tendency for bio-accumulation in the food chain. Biochar has few or no sorption abilities for anionic forms such as As, or even enhanced reduction and release of As from soil because of the increase in organic matter and pH in the soil after biochar addition (Beesley *et al.*, 2011).

Several studies have reported on the modification of biochars to enhance their sorption abilities for As from aqueous solutions. Birnessite is one of most abundant manganese (Mn) oxides in terrestrial environments, and many studies have noted its high sorption capabilities for As. Biochar can provide an effective surface for birnessite coating and might be a promising media for increasing As removal. In this work, we speculated that biochar could be loaded with birnessite and that this environmentally friendly material could be a suitable adsorbent for simultaneously removing As from aqueous solutions or immobilizing both of them in soils. The sorption mechanism of As by birnessite was unclear.

The objectives of this work were to: (1) prepare and characterize the birnessite-loaded biochar; (2) explore the capabilities and mechanisms of birnessite-loaded biochar for sorption of As (As(V) or As(III)).

2 METHODS/EXPERIMENTAL

2.1 *Preparation of birnessite-loaded rice husk biochar*

Rice husk was pyrolyzed in a muffle furnace under an N2 environment by holding the peak temperature at 600°C for 4 h. Unloaded rice husk biochar (RB) was used as the control.

Birnessite was loaded onto RB using the $KMnO_4$ precipitation method described by McKenzie (1971) and Feng (2007) and designated as BRB. Biochar (10 g) was added to the $KMnO_4$ solution (0.4 M, 100 mL) and stirred for 2 h, then boiled for 20 min, followed by drop addition of 6.6 mL concentrated HCl as reducer at the rate of 0.7 mL min^{-1}, then boiled for another 10 min. The resulting biochar-birnessite composites were rinsed thoroughly with deionized water, then oven dried at 60°C for about 12 h.

2.2 *Sorption kinetics for Cd, As(III) and As(V)*

Sorption kinetics of As(III) (5 mg L^{-1}), As(V) (6 mg L^{-1}), and Cd (26 mg L^{-1}) by BRB were investigated. Briefly, BRB (0.05 g) was added into 10 mL of sorbate solutions (As(III), As(V)) in 15 mL centrifuge tubes at room temperature (25°C), with 0.01 mol L^{-1} Centrifuge tubes containing the BRB and solutions were placed onto a rotary shaker and shaken at 150 rpm. At each sampling time (1, 2, 4, 8, 12, 24, 48 h), three replicates were taken out and immediately filtered through 0.45 μm nylon membrane filters. The concentrations of As(III) and As(V) in supernatants were determined using ICP-OES or ICP-MS. The initial pH values of the solutions were adjusted to 6.0 ± 0.2 with 0.01 mol L^{-1} HCl or NaOH.

3 RESULTS AND DISCUSSION

3.1 *Characterization of BRB*

Compared with RB, RB loaded with birnessite resulted in more than 2× higher pore volume than unmodified RB, while the BET surface area decreased by more than two thirds, indicating that birnessite changed the pore size distribution of the RB. The SEM image of BRB was different from that of RB, with randomly

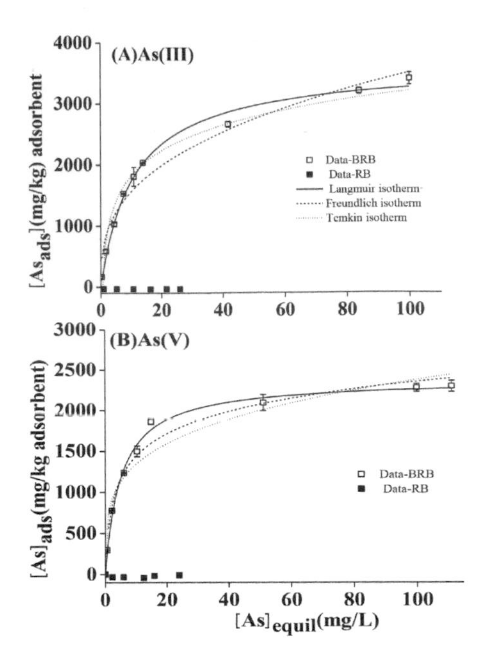

Figure 1. As(III) (A) and As(V) (B) sorption isotherm and fitted models for BRB.

stacked birnessite on the surface of the BRB; the poorly crystallized birnessite was further confirmed by XRD analysis. The XPS wide-scan spectra of BRB showed a substantial increase in Mn content compared with the wide-scan spectra of RB.

3.2 *As(III), As(V) and Cd sorption kinetics and isotherms*

As(III) and As(V) sorption onto BRB were biphasic, with a rapid initial phase over the first few hours followed by a much slower sorption phase, indicating that more than one mechanism affected the sorption processes. Pseudo-first-order, pseudo-second-order, and Elovich models were tested for their abilities to simulate the sorption kinetics data. The results showed that the pseudo-second-order kinetic model was better than the pseudo-first-order model for describing all data ($R^2 > 0.92$), indicating that binuclear adsorption was dominant. The Elovich model, an empirical equation, was the best to describe kinetic data with coefficients of correlation (R^2) above 0.96. Based on the adsorption isotherm, BRB exhibited superior abilities to adsorb As(III) and As(V) and the maximum sorption capacities obtained from the Langmuir model were 3621 mg kg^{-1} for As(III) and 2381 mg kg^{-1} for As(V), while RB had no ability to adsorb As (Fig. 1). The sorption capacity of BRB was much higher for As(III) than for As(V), probably because the oxidation of As(III) by synthetic birnessite was accompanied by Mn(II) and As(V) release, and dissolution of Mn(II) created more fresh reaction sites for As(V) on birnessite surfaces.

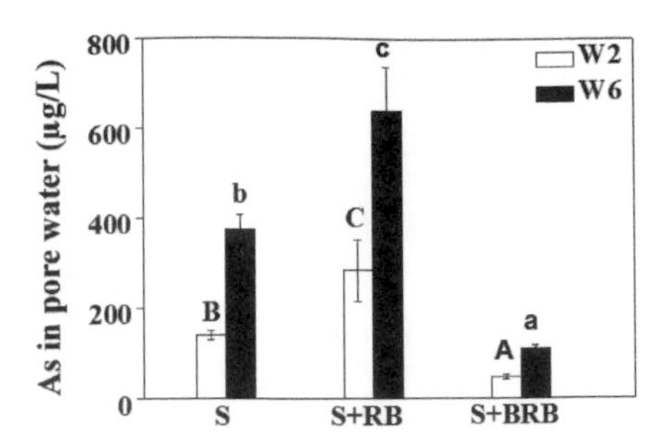

Figure 2. Concentrations of As in pore water of the treatment of soil (S), soil + rice husk char (S + RB), soil + birnessite-loaded rice husk biochar (S + BRB) under and flooded condition in the second week (W2) and sixth week (W6).

3.3 *Application of BRB in soil under flooded conditions*

The As concentration in pore water of the control increased by 1.7 times from week 2 to week 6. At W6, the As concentration in the BRB treatment (111 µg L^{-1}) was 5.76-fold lower than that in the RB treatment and 3.39-fold lower than in the control treatment (Fig. 2). This experiment suggests that the application of BRB could effectively reduce the As bioavailability in soil under flooded conditions.

4 CONCLUSIONS

A novel biochar was prepared by modification with birnessite. Birnessite-loaded biochar enhanced its ability to remove As from aquatic solution and soil and exhibited a hybrid adsorption property to simultaneously remove As and Cd. These results suggested that birnessite-loaded biochar has potential for remediation of heavy metal contaminated water and soil, especially in the case of multi-metal contamination.

ACKNOWLEDGEMENTS

This project was financially supported by the Natural Science Foundation of China (No. 41371459), the State Key Program of Natural Science Foundation of China (No. 41330853, 41430858).

REFERENCES

Beesley, L., Jiménez, E.M., Gomez-Eyles, J.L., Harris, E. Robinson, B. & Sizmur, T. 2011. A review of biochars' potential role in the remediation, revegetation and restoration of contaminated soils. *Environ. Pollut.* 159(12): 3269–3282.

Feng, X.H., Zhai, L.M., Tan, W.F., Liu, F. & He, J.Z. 2007. Adsorption and redox reactions of heavy metals on synthesized Mn oxide minerals. *Environ. Pollut.* 147(2): 366–373.

McKenzie, R.M. 1971. The synthesis of birnessite, cryptomelane, and some other oxides and hydroxides of manganese. *Mineral. Mag.* 38(296): 493–502.

Environmental Arsenic in a Changing World –
Zhu, Guo, Bhattacharya, Ahmad, Bundschuh & Naidu (Eds)
ISBN 978-1-138-48609-6

The application of organics promotes arsenic methylation in paddy soils

G.L. Duan[1], Y.P. Yang[1], X.Y. Yi[1] & Y.G. Zhu[1,2]
[1]*Research Center for Eco-Environmental Sciences, Chinese Academy of Sciences, Beijing, P.R. China*
[2]*Institute of Urban Environment, Chinese Academy of Sciences, Xiamen, P.R. China*

ABSTRACT: Arsenic (As) accumulation in rice grains poses critical health risk for populations whose staple food is rice. Organics could affect As transformation and volatilization in soil, and subsequently affect As accumulation and speciation in rice grains. Straw application significantly promoted As release from soil and As methylation, also enhanced As volatilization from soil. However, straw application significantly increased DMA accumulation in polished rice grains. Straw biochar application significantly decreased inorganic As concentration in polished grains, but did not significantly affect As methylation and DMA accumulation. Fulvic acid addition significantly increased As bio-availability in the soil, and high level of humic acid addition significantly enhanced As volatilization.

1 INTRODUCTION

Arsenic (As) is a toxic metalloid, and inorganic As is classified as a class 1 and non-threshold carcinogen (Smith *et al.*, 2002). Arsenic ubiquitously exists in the environment, and paddy fields are widely contaminated with As (Zhao *et al.*, 2015). Arsenic in paddy soil can be readily taken up by rice roots, and accumulated to rice grains, the staple food for more than half of the world's population and two thirds of the Chinese population. It has been reported that rice consumption constitutes a major source of dietary intake of inorganic As for populations whose staple food is rice (Li *et al.*, 2011). Therefore, it is of great significance to reduce As accumulation in rice grains (Zhu *et al.*, 2014).

Different As speciation has different toxicity, generally, inorganic As (such as arsenite and arsenate) is more toxic than organic As (such as MMA and DMA). In rice grains, arsenite and DMA are the predominant arsenic species. It has been demonstrated that rice plants lack the ability to methylate arsenic, and DMA in rice grain is derived from the soil through microorganism-mediated methylation (Lomax *et al.*, 2012). Therefore, promoting the process of arsenic methylation in paddy, and subsequently decrease the proportion of arsenite in rice gains are of great importance for safe rice production at As contaminated fields. In addition, the last product of As methylation is TMA, which is volatile gas. Thus, enhancing As volatilization could be a potential strategy for As bioremediation of paddy soil.

2 METHODS/EXPERIMENTAL

2.1 *Rice cultivation*

Rice straw is the main by-product of rice production. It has been reported that straw incorporation could enhance As methylation. Therefore, pot experiments were carried out to investigate the effects of straw incorporation on As volatilization. For this experiment, 4 kg As contaminated soil, which was sampled from Qiyang city (Hunan province, China), The As concentration of the soil was 84.7 mg kg^{-1}. The treatments were as follows: (1) Control (No straw biochar nor straw); (2) 1% straw biochar; (3) 2% straw biochar; (4) 4% straw biochar; (5) 1% straw; (6) 2% straw; (7) 4% straw. Fertilization and watering were managed according to the local practices. After grain mature, rice grains were harvested for As concentration and speciation analysis by HPLC-ICP-MS.

2.2 *Soil incubation with organics*

To understand how organics affect As transformation in soil, humic acid and fulvic acid were incubated with As contaminated soil, and the dynamic As transformation and volatilization was monitored. In this experiment, 200 g soil (the same as pot experiment) was mixed with humic acid and fulvic acid at 5 levels: 0, 2, 5, 10, 15 mg kg^{-1} soil. Then 240 mL deionized water was added to the flask and keeping a water layer about 2–3 cm above the soil surface during the experiment. Every week, the volatilized gas was sampled.

As in the volatilized gas was trapped and the amount was analyzed.

2.3 *Volatile As collection*

Arsenic volatilizing from each pot was collected with silica gel rinsed with 2% $AgNO_3$ for one week at 4 different growing stages. After sampling, the trapping tubes were collected, and then the silica gel in the trapping tubes was sampled and digested. Total As concentration was measured by ICP-MS.

3 RESULTS AND DISCUSSION

3.1 *Straw application increased DMA accumulation in polished rice grains*

Straw application significantly increased total As concentration in polished grains and husk, while decreased total As in straw, and no significant effects on As accumulation in root. In polished grains, total As concentration could be increased by 4-fold by 4% straw application. Straw biochar application had no significant effect on total As concentration in polished grains.

In polished grains, As(V), As(III) and dimethylarsenate (DMA) were detected. Straw application significantly increased DMA concentration in polished grains. At the treatment of 4% straw application, DMA concentration in polished grains increased by about 8-fold when compared to control. Straw biochar application significantly decreased inorganic As concentration in polished grains but did not significantly affect DMA concentration.

3.2 *Straw application enhanced As volatilization from paddy soil*

During the whole rice-growing season, small amounts of volatile As were detected in the control and straw biochar treatments, and straw biochar did not affect As volatilization, when compared with the control. However, straw application significantly enhanced the As volatilization from paddy soil. At seedling stage, As volatilization apparently increased with increasing dose of straw added to soil. Compared with control, the treatment of 4% straw application increased As volatilization by 166, 64 and 12 times at seedling, tillering, and heading stages, respectively. Arsenic volatilization reached the maximum rate at seedling stage, then showed a decreasing trend with the rice growth stage. At maturing stage, the fluxes of volatile As were less than one-tenth of that at seedling stage.

3.3 *Addition of humic acid and fulvic acid promoted As methylation in paddy soil*

Without organics addition, As concentration in soil solution was increased with the incubation time, and reached maximum concentration at day 14, then declined. With addition of humic acid, As concentration in soil solution showed similar dynamic pattern to control. However, with addition of fulvic acid, As concentration in soil solution increased within 14 days incubation, and then kept at high levels until the experiment was finished after 28 days incubation.

Humic acid and fulvic acid addition increased As bioavailability in soil. Total As concentration in soil solution was increased by organics addition, especially by fulvic acid addition. When soil was incubated with $10\,mg\,kg^{-1}$ fulvic acid for 28 days, arsenite concentration in soil solution increased by about 5-fold, when compared with control. Arsenite was the dominated As species in soil solution.

Humic acid and fulvic acid addition enhanced As methylation in the soil. Arsenic volatilization was significantly enhanced by higher humic acid ($>5\,mg\,kg^{-1}$) addition, and lower fulvic acid ($<10\,mg\,kg^{-1}$) addition. With the same addition level, the fluxes of volatile As collected from humic acid treatments were higher than that from fulvic acid treatments. MMA and DMA concentration in soil solution were significantly increased by fulvic acid addition. These results suggested that, compare to humic acid, fulvic acid was more efficient for activating As in soil, but the activated As in soil solution was not readily for As volatilization.

4 CONCLUSIONS

Straw application promoted As release from soil and As methylation, enhanced As volatilization, but increased DMA accumulation in polished rice grains. Fulvic acid addition significantly increased As bioavailability in the soil, but the activated As in soil solution was not readily for As volatilization.

ACKNOWLEDGEMENTS

This research was funded by the National Natural Science Foundation of China (grant no. 41371458 and 21677157), and the special fund for agro-scientific research in the public interest (grant no. 201403015).

REFERENCES

Li, G., Sun, G.X., Williams, P.N., Nunes, L. & Zhu, Y.G. 2011. Inorganic arsenic in Chinese food and its cancer risk. *Environ. Int.* 37(7): 1219–1225.

Lomax, C., Liu, W.J., Wu, L.Y., Xue, K., Xiong, J., Zhou, J., McGrath, S.P., Meharg, A., Miller, A.J. & Zhao, F.J. 2012. Methylated arsenic species in plants originate from soil microorganisms. *New Phytol.* 193(3): 665–672.

Smith, A.H., Lopipero, P.A., Bates, M.N. & Steinmaus, C.M. 2002. Public health – arsenic epidemiology and drinking water standards. *Science* 296: 2145–2146.

Zhao, F.J., Ma, Y., Zhu, Y.G., Tang, Z. & McGrath, S.P. 2015. Soil contamination in China: current status and mitigation strategies. *Environ. Sci. Technol.* 49(2): 750–759.

Zhu, Y.G., Yoshinaga, M., Zhao, F.J. & Rosen, B.P. 2014. Earth abides arsenic biotransformations. *Annu. Rev. Earth Planet. Sci.* 42: 443–467.

Environmental Arsenic in a Changing World –
Zhu, Guo, Bhattacharya, Ahmad, Bundschuh & Naidu (Eds)
ISBN 978-1-138-48609-6

Traceability of arsenic in agricultural water in Irrigation District 005, Mexico

M.C. Valles-Aragón[1] & M.L. Ballinas-Casarrubias[2]
[1]*Faculty of Agrotechnological Science, Autonomous University of Chihuahua, Chihuahua, Chih., Mexico*
[2]*Faculty of Chemical Sciences, Autonomous University of Chihuahua, Chihuahua, Chih., Mexico*

ABSTRACT: The Irrigation District 005 is an agricultural territory in Chihuahua, Mexico where has been detected As in groundwater. So, the objective was to determine As in agricultural wells, and the metalloid traceability to soil and crops. 119 irrigation wells were sampled, 140 samples of soil, 32 onion and 37 chili plants were collected. The As concentration was determined in all matrices by HG-AAS. 11% of the wells exceeded the maximum permissible level (MPL) in Mexico of As in water for agricultural use (100 μg L^{-1}), the highest concentration was 576 μg L^{-1}. 13% of the soil samples exceeded the MPL for agricultural use (22 mg kg^{-1}), the maximum concentration was 57.7 mg kg^{-1}. The highest As presence in crops was in root, 7.4 mg kg^{-1} in chili and 28.8 mg kg^{-1} in onion. However, 0.16 mg kg^{-1} in chili crop and 0.38 mg kg^{-1} in onion crop were also identified.

1 INTRODUCTION

In Mexico, many rural localities, have water sources contaminated with arsenic (As), among them localities from Chihuahua State (Rivera-Huerta *et al.*, 2008). Several studies support this fact, Espino-Valdes *et al.* (2009) determined at the south-center region of Chihuahua State 72% of the samples exceeded the maximum permissible level (MPL) of As in water for human consumption of 25 μg L^{-1} established by the Mexican regulation (SSA 1994). Olmos-Márquez (2011) determined the As concentration in the feeding of 116 plants of reverse osmosis in municipalities of Chihuahua, 73% exceeded the MPL of water for human consumption. Vega-Gleason (2013) estimated the As interval in water between 0.5 and 500 μg L^{-1} in the south region of Chihuahua State. González-Horta *et al.* (2015) reported high As levels in water for human consumption, as well as 49% of the sampled people with high levels of As in urine. The Irrigation District 005 (ID005) located in the south-central region of Chihuahua State in Mexico, is a territory that has a representative agricultural economy (INEGI 2013). Due this, it is considered that crops irrigated with contaminated water with As may be a contribution to the daily intake of As (Santra *et al.*, 2013). So, the objective was to determine the As concentration in agricultural wells, and the traceability of this metalloid to soil and plants for human consumption.

2 METHODS/EXPERIMENTAL

2.1 *Water, soil and plant sampling*

The sampling was conducted in the ID005, two water sampling were performed, the first on summer (Sampling 1) and the second on autumn (Sampling 2). In the first sampling, water from 65 wells was collected, in the second from 54 wells. For soil sampling, were considered the zones that exceeded the MPL according to Mexican regulations for As in water for agricultural use (CONAGUA 2016). Zones were systematic sampled by quadrants, each quadrant had 24 sampling points at 250 m of distance between them, in an area of 100 ha. Giving a total of 140 soil samples. The plants collection was in the harvest time from May to July. 32 samples of onion, and 37 of chili plant were collected in the same quadrants.

2.2 *Sample preparation and As determination*

The water samples before the analysis were filtered with Whatman No.2 and Millipore filters. Soil and plant samples were dried at 45°C, and digested in a Mars 6 microwave. EPA-3052 digestion method was used for soil and EPA-3051 for plants. As determination in all matrices was done by a Perkin Elmer AAnalyst700 Atomic Absorption Spectrometry equipment, coupled to FIAS 100 Hydride Generator. The detection limit was 3.12 μg L^{-1}, samples were analyzed in triplicates, using controls and standards.

3 RESULTS AND DISCUSSION

3.1 *As in water*

During sampling 1, 9% of the wells exceeded the MPL of As for agricultural water of 100 μg L^{-1} (CONAGUA 2016) and 38% the MPL for human consumption (25 μg L^{-1}) (SSA 1994), the maximum concentration was 335 μg L^{-1}. In Sampling 2, 13% of the wells exceeded the MPL for agricultural irrigation and 33% the one for human consumption. The maximum concentration of As was 576 μg L^{-1} (Table 1).

Table 1. Arsenic in water of the Irrigation District 005.

Samples	MPL agricultural use $>100\,\mu g\,L^{-1}$	MPL human use $>25\,\mu g\,L^{-1}$	Not detectable $<3\,\mu g\,L^{-1}$	Interval $\mu g\,L^{-1}$
Global n = 119	11% n = 13	36% n = 43	43% n = 51	<3–576
Sampling 1 n = 65	9% n = 6	38% n = 25	45% n = 29	<3–335
Sampling 2 n = 54	13% n = 7	33% n = 18	41% n = 22	<3–576

Table 2. Arsenic in soil in the Irrigation District 005.

Samples	MPL agricultural $>22\,mg\,kg^{-1}$	Interval $mg\,kg^{-1}$
Global n = 140	13% n = 18	4.3–57.7
MM33 n = 24	8% n = 2	4.4–57.7
MM35 n = 24	0% n = 0	4.3–15.6
MM38 n = 24	0% n = 0	5.8–13.5
MR14 n = 24	17% n = 4	7.9–35.6
MJ01 n = 20	10% n = 2	8.3–23.5
MJ02 n = 24	42% n = 10	8.7–47.5

Table 3. Arsenic in chili and onion plants.

Crop	Root	Stem	Leaf	Chili/onion
Jalapeño	0.3213	0.2236	0.5325	0.1583
Serrano	7.4327	0.0000	0.6992	0.0546
Chilaca	1.2110	0.2143	1.1862	0.0000
Onion	28.8492	0.8069	–	0.3814

3.2 *As in soil*

The soil results were done in 6 sampling sites classified as: MJ01, MJ02, MM33, MM35, MM38 and MR14. The soil with the highest As concentrations was MJ02 with 42% of the samples that exceeded the MPL of $22\,mg\,kg^{-1}$ (SEMARNAT/SSA 2004), for As in agricultural land (Table 2). However, it was observed that in all soil samples, there was As presence, which evidences the accumulation of it by water irrigation. However, it is emphasized that the global average of As concentration in soil is $6.83\,mg\,kg^{-1}$ (Kabata-Pendias 2011).

3.3 *As in plant*

The analysis of chili of different species jalapeño, serrano and chilaca showed a higher As concentration in root with 0.32, 7.4 and $1.2\,mg\,kg^{-1}$ of As, respectively. In leaf with an average concentration of 0.53, 0.7 and $1.2\,mg\,kg^{-1}$, respectively. Although, the edible part of chili presented the lowest As levels, with an average of 0.16 and $0.05\,mg\,kg^{-1}$ in jalapeño and serrano, without presence in chilaca (Table 3).

The analysis of onion showed the existence of a higher As concentration in root with an average of $28.8\,mg\,kg^{-1}$. However, it was observed that it was accumulated in the onion bulb in a greater quantity than in the chili plants. An average As concentration of $0.38\,mg\,kg^{-1}$ was determined (Table 3).

4 CONCLUSIONS

As is an element present in the ID005, it is at levels that can cause an agricultural risk for production, it can even mean a public health risk, especially if it is used as water for human consumption for agricultural communities. The As presence in soil is relevant although in most cases the MPL are not exceeded, evidences the traceability of the metalloid from water to soil. Moreover, being As present in water and soil, it can be absorbed by the plant; although it is mostly concentrated in the root of the plants, chili and onion can store arsenic in the edible part too.

ACKNOWLEDGEMENTS

To the National Council of Science and Technology of Mexico (CONACYT) for the financial support.

REFERENCES

CONAGUA. 2016. Ley federal de derechos disposiciones aplicables en materia de aguas nacionales. Mexico.

Espino-Valdes, M.S., Barrera-Prieto, Y. & Herrera-Peraza, E. 2009. Presencia de arsénico en la seccion norte del acuífero Meoqui-Delicias del Estado de Chihuahua. *Tecnociencia Chihuahua* 3(1): 8–18.

González-Horta, C., Ballinas-Casarrubias, L., Sánchez-Ramírez, B., Ishida, M., Barrera-Hernández, A., Gutiérrez-Torres, D., Zacarias, O., Saunders, R., Drobná, Z., Mendez, M., García-Vargas, G., Loomis, D., Stýblo, M. & Del Razo, L.M.A. 2015. Concurrent exposure to arsenic and fluoride from drinking water in Chihuahua, Mexico. *Int. J. Environ. Res. Public Health* 12(5): 4587–4601.

INEGI. 2013. Instituto Nacional de Estadistica y Geografía. http://www.inegi.org.mx/ (accessed 29-Diciembre-2015).

Kabata-Pendias, A. 2011. *Trace Elements in Soils and Plants*. Fourth edition. Boca Raton, New York: CRC Press.

Olmos-Márquez, M.A. 2011. Remoción de As en el agua por fitorremediación con *Eleocharis macrostachya* en humedales construidos de flujo subsuperficial. Chihuahua, CIMAV, Mexico.

Rivera-Huerta, M., Martín-Domínguez, A., Piña-Soberanis, M., Pérez-Castrejón, P., García-Espinosa, J. 2008. La electrocoagulación: una alternativa para el tratamiento de agua contaminada con As. IMTA, Mexico.

Santra, S.C., Samal, A.C., Bhattacharya, P., Banerjee, S., Biswas, A. & Majumdar, J. 2013. Arsenic in food chain and community health risk: a study in Gangetic West Bengal. *Procedia Environ. Sci.* 18(1): 2–13.

SEMARNAT/SSA. 2004. NOM-147-SEMARNAT/SSA1. Secretaría de Medio Ambiente y Recursos Naturales/ Secretaría de Salud, Mexico.

SSA. 1994. NOM-127-SSA1. Secretaría de Salubridad y Asistencia, Mexico.

Vega-Gleason, S. 2013. Riesgo sanitario ambiental por la presencia de arsénico y floruros en los acuíferos de México. CONAGUA, Mexico.

Environmental Arsenic in a Changing World –
Zhu, Guo, Bhattacharya, Ahmad, Bundschuh & Naidu (Eds)
ISBN 978-1-138-48609-6

Arsenic data availability in agricultural soils and waters in Europe

T. Hatakka & T. Tarvainen
Geological Survey of Finland, Espoo, Finland

ABSTRACT: The AgriAs project – Evaluation and management of arsenic contamination in agricultural soil and water has compiled a summary of European-wide databases and publications on As concentrations in agricultural soils and related water bodies, As concentrations in crops together with a list of gaps in data and knowledge. The literature review and the AgriAs questionnaire revealed that there is a large amount of nationwide or large-scale data on arsenic concentrations in soil, surface water and/or groundwater from many countries. However, none of the countries has published regional maps of arsenic concentrations in crops. According to the literature review and the questionnaire, no national-scale data on As concentrations in groundwater, stream water, sediment or topsoil are available from Belgium, Denmark, Estonia, Greece, Latvia, Moldova and Romania. Moreover, there are no data from European-wide mapping programmes for two countries: Moldova and Turkey. There is no up-to-date map of arsenic concentrations in European groundwater related to agricultural sites. There are quite extensive European-wide datasets on As concentrations in agricultural soil, but more detailed regional mapping at the national level is needed, especially in those areas where anomalously high As concentrations in topsoil have been discovered. According to the AgriAs studies, European-wide data, as well as the nationwide data, on As concentrations in crops are almost completely lacking. The AgriAs project provides the European Union with reliable data on the existing risks of arsenic exposure through agriculture, a complete summary of existing tools available for As remediation as well as an array of tools for ecotoxicity and bioavailability assessment.

1 INTRODUCTION

Arsenic is a toxic and carcinogenic substance. According to WHO, the greatest threat to public health from arsenic originates from contaminated groundwater. Food is another notable pathway for As exposure in humans. There are numerous arsenic-related publications and reports, but many of them focus on arsenic problems in South-East Asia or are limited to groundwater or contaminated soil. The AgriAs (Water-JPI 2017) project has focused on European data on agricultural soils and the quality of related surface water and groundwater.

2 METHODS

The AgriAs (*http://projects.gtk.fi/AgriAs/index.html*) Task Assessment of data availability on As concentrations in water, soil and crops in Europe has summarized European-wide databases and publications on As concentrations in soil and water. This was followed by a literature review and a questionnaire on national-level data sources of As concentrations in agricultural soil and water. A web-based AgriAs questionnaire on national and large-scale regional data sources on arsenic in soil, surface water, groundwater and crops in Europe was sent in May 2017 to 116 organizations from 23 countries. The selected organizations represented environmental authorities and research organizations working on water quality, (agricultural) soil and environmental issues. The general findings concerning As concentrations in crops were summarized from the literature. Following the assessment of data availability, a list of major data gaps was reported.

3 RESULTS AND DISCUSSION

The AgriAs project has concentrated on European publications, reports and data related to arsenic. Reimann *et al.* (2009) have provided a comprehensive summary of the European-wide availability of data on As concentrations in soil and water. More up-to-date information on national and regional data sources were identified from the AgriAs questionnaire. The questionnaire revealed that regional-scale data are available on arsenic concentrations in soil and surface water. These data can provide a detailed insight into the European-wide anomalies found in the FOREGS (Salminen *et al.*, 2005), Baltic Soil Survey (Reimann *et al.*, 2003), GEMAS (Reimann *et al.*, 2015; Tarvainen *et al.*, 2015) and LUCAS surveys (Tóth *et al.*, 2015a, 2015b). The European-wide arsenic data availability is presented in Table 1. In addition, regional and national data are available on arsenic concentrations in groundwater, which have been lacking from the European-wide geochemical mapping programmes. Unfortunately, no publicly available

Table 1. The European-wide arsenic data availability in the Water JPI member countries in Europe. SW = Stream water, SD = Sediment, TS = Topsoil. The letters in bold indicate the maximum concentrations of As are relatively high (SW: As Max >7 $\mu g L^{-1}$, SD and TS: As Max >25 $mg kg^{-1}$, LUCAS TS: As Max >50 $mg kg^{-1}$). White = no data available. FOREGS = Geochemical Atlas of Europe (Salminen *et al.*, 2005), BSS = Baltic Soil Survey (Reimann *et al.*, 2003), GEMAS = Geochemistry of European agricultural and pasture soils (Reimann *et al.*, 2015), LUCAS = Lucas-topsoil survey (Toth *et al.*, 2015a, 2015b).

Country	FOREGS Agricultural Soil Catchments	BSS	GEMAS Agricultural Soil	LUCAS Agricultural Soil
Austria	SW, SD, TS		**TS**	**TS**
Belgium			TS	TS
Cyprus			TS	TS
Denmark			**TS**	TS
Estonia	SW, SD, TS	TS	TS	TS
Finland		TS	**TS**	TS
France	**SW**, **SD**, **TS**		**TS**	**TS**
Germany	SW, **SD**, TS	TS	**TS**	**TS**
Greece			**TS**	**TS**
Hungary	SW, SD, TS		**TS**	**TS**
Ireland			**TS**	**TS**
Italy	**SW**, SD, **TS**		**TS**	**TS**
Latvia	SW, SD, TS	TS	TS	TS
Moldova				
The Netherland			TS	**TS**
Norway		TS	TS	**TS**
Poland	SW, SD, TS	TS	TS	
Portugal	SW, **SD**, **TS**		**TS**	**TS**
Romania				TS
Spain	**SW**, SD, **TS**		**TS**	TS
Sweden		TS	TS	**TS**
Turkey				
United Kingdom	SW, **SD**, TS		**TS**	**TS**

national or regional datasets were reported on arsenic concentrations in crops.

According to the literature review and the questionnaire, no national-scale data on As concentrations in groundwater, stream water, sediment or topsoil are available from Belgium, Denmark, Estonia, Greece, Latvia, Moldova or Romania. There are no data from European-wide mapping programmes for two countries: Moldova and Turkey.

4 CONCLUSIONS

There are quite extensive European-wide datasets on As concentrations in agricultural soil, but more detailed regional mapping at the national level is needed, especially in those areas where anomalously high As concentrations in topsoil have been discovered. According the AgriAs questionnaire and literature study, European-wide data as well as nationwide data on As concentrations in crops are entirely lacking. There is no up-to-date map of arsenic concentrations in European groundwater related to agricultural sites. European-wide or large-scale regional databases very seldom combine arsenic concentrations in agricultural topsoil with concentrations in adjacent surface water or groundwater.

The AgriAs project provides the European Union with reliable data on the existing risks of arsenic exposure through agriculture, a complete summary of existing tools available for As remediation as well as an array of tools for ecotoxicity and bioavailability assessment.

ACKNOWLEDGEMENTS

AgriAs is co-funded by EU and the Academy of Finland, L'Agence Nationale de la Recherché (France), Bundesministerium für Ernährung und Landwirtschaft (Germany) and Forskningsrådet FORMAS (Sweden) under the ERA-NET Cofund WaterWorks2015 Call.

REFERENCES

Reimann, C., Siewers, U., Tarvainen, T., Bityukova, L., Eriksson, J., Gilucis, A., Gregorauskiene, V., Lukashev, V., Matinian, N.N. & Pasieczna, A. 2003. Agricultural Soils in Northern Europe: A Geochemical Atlas. Geologisches Jahrbuch, Sonderhefte, Reihe D, Heft SD 5. Schweizerbart'sche Verlagsbuchhandlung, Stuttgart.

Reimann, C., Matschullat, J., Birke, M. & Salminen, R. 2009. Arsenic distribution in the environment: the effects of scale. *Appl. Geochem.* 24(7), 1147–1167.

Reimann, C., Birke, M., Demetriades, A., Filzmoser, P. & O'Connor, P. (eds) 2015. Chemistry of Europe's Agricultural Soils. Part A: Methodology and Interpretation of the GEMAS Data Set. Geologisches Jahrbuch Reihe B Heft 102. Schweizerbart'sche Verlagsbuchhandlung, Stuttgart.

Salminen, R., Batista, M.J. & Bidovec, M. 2005. Geochemical Atlas of Europe. Part 1 – Background Information, Methodology and Maps. Geological Survey of Finland, Espoo, Finland.

Tarvainen, T, Birke, M. & Reimann, C. 2015. Arsenic anomalies in European agricultural and grazing land soil. pp. 81–88. In: C. Reimann, M. Birke, A. Demetriades, P. Filzmoser & P. O'Connor (eds) Chemistry of Europe's Agricultural Soils. Part B: General Background Information and Further Analysis of the GEMAS Data Set. Geologisches Jahrbuch Reihe B Heft 103. Schweizerbart'sche Verlagsbuchhandlung, Stuttgart.

Tóth, G. Hermann, T., Szatmári, G. & Pásztor, L. 2015a. Maps of heavy metals in the soils of the European Union and proposed priority areas for detailed assessment. *Sci. Total Environ.* 565: 1054–1062.

Tóth, G., Hermann, T., Da Silva, M.R. & Montanarella, L. 2015b. Heavy metals in agricultural soils of the European Union with implications for food safety. *Environ. Int.* 88: 299–309.

Water-JPI. 2017. Water JPI, Challenges for a changing world. http://www.waterjpi.eu/index.php?option=com_content&view=article&id=79&Itemid=686

Environmental Arsenic in a Changing World –
Zhu, Guo, Bhattacharya, Ahmad, Bundschuh & Naidu (Eds)
ISBN 978-1-138-48609-6

Arsenic in cattle: Evaluation of possible exposure biomarkers

C.V. Alvarez-Gonçalvez, F.E. Arellano, A. Fernández-Cirelli & A.L. Pérez Carrera
Instituto de Investigaciones en Producción Animal UBA-CONICET (INPA), CONICET – Universidad de Buenos Aires, Buenos Aires, Argentina
Centro de Estudios Transdisciplinarios del Agua (CETA), Universidad de Buenos Aires, Buenos Aires, Argentina
Cátedra de Química Orgánica de Biomoléculas, Universidad de Buenos Aires, Buenos Aires, Argentina

ABSTRACT: Arsenic pollution is naturally present in groundwater and soils of the main agriculture production areas in Argentina. Arsenic can be taken up and bioaccumulated by forages, and may be bio-transferred to animals, food and humans. Blood and hair are generally used in humans as biomarkers but there is a lack of information in livestock. This study shows that As levels in cattle hair from As affected area were significantly higher than in the control area. In analyzed blood samples no significant differences were found between control and As affected areas. Blood appear not to be an effective biomarker in cattle such as it has been determinate for human but hair may indicate chronic exposure in cattle.

1 INTRODUCTION

Arsenic (As) pollution affects several countries around the world. One of the most affected countries is Argentina, where As is naturally present in groundwater and soils of the main agriculture production areas (Bundschuh *et al.*, 2012). Arsenic can be taken up and bioaccumulated in plants and livestock, which means a potential risk to human health through agri-food.

Presence of toxic substances in blood is a known exposure biomarker (Lowry *et al.*, 1989). On the other hand, in the case of As, its levels in cattle hair might be used as another biomarker exposure. Even though, it is commonly used for humans, there is a severe lack of knowledge for other animal species.

The aim of this study is to assess total arsenic levels in soils and drinking water from cattle farms; and to analyze the relationship with blood and hair As content in cattle.

2 MATERIAL AND METHODS

This study was carried out in the Pampean Plain (Argentina). The study area is one of the most important livestock production regions. The samples zones were located in Buenos Aires province (Control area) and in the southeast of Cordoba province (As affected area).

Soil and drinking water samples were taken up from livestock production systems and dairy farms. Soil samples correspond to surface soil (0 to 20 cm deep). Samples were collected along transect in the field at regular intervals. Arsenic determination was according to ISO 11466 (1995). Soils were digested using a mixture of nitric and hydrochloric acids (1:3) at 120°C during 2 h with reflux. Water samples correspond to groundwater used for cattle. They were collected in polypropylene tubes, acidified at 20% with HNO_3 and conserved refrigerated at 4°C. Blood and hair samples were collected from adult and female Holando cows. The samples were digested with HNO_3 at hot plate. Arsenic in all samples were determined by inductively coupled plasma optical emission spectrometry (ICP-OES, Optima 2000, Perkin Elmer).

3 RESULTS AND DISCUSSION

Usually, arsenic in soils ranged between 1 to 95 mg kg^{-1} (Kabata & Pendias, 2001), and in groundwater (main water source for animal drinking in the Pampean Plain) arsenic is documented in the literature up to values of 5000 $\mu g\,L^{-1}$ (Ravenscroft *et al.*, 2009). Maximum total As content for animal drinking water and soil destined to agricultural activities in Argentina are 0.5 mg L^{-1} and 20 mg kg^{-1} respectively (Law 24.051), but the international guidelines recommend a maximum level of As in animal drinking water 200 $\mu g\,L^{-1}$ (FAO, 1985) and in agricultural soil 17 mg kg^{-1} (CCME, 2001). The As levels for all samples analyzed in this study are shown in Table 1.

It is known that drinking water is the main As source for cattle (Perez Carrera *et al.*, 2012, 2016), the levels obtained through forage and fodder or accidental soil take up is relatively low. In the analyzed soil samples, As levels were below the detection limit (LD = 3.5 mg kg^{-1}).

Regarding to As level in hair, in all samples from control area, were below 1000 ng g^{-1}. In affected areas

Table 1. Arsenic levels (min–max.) for water, cattle's hair and blood in affected and control areas.

Area	Water $\mu g L^{-1}$	Hair As $ng g^{-1}$	Blood As $\mu g L^{-1}$
Affected area	*51–268*	<LD–1520	<LD–*75*
Control area	*<LD–50*	<LD–520	<LD

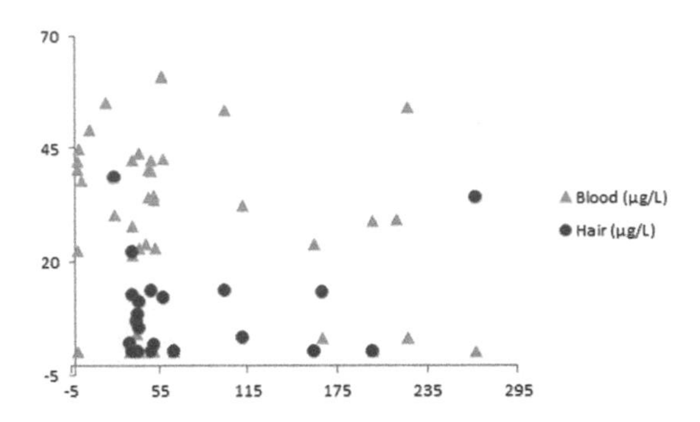

Figure 1. Relationship between total arsenic ($\mu g L^{-1}$) in blood and hair with water As concentration. X axis show As levels in water while Y axis shows As levels in blood and hair.

As hair levels were between <LD–1520. In cattle there are no enough data to determine a confident baseline value. These results are according to previous studies (Pérez Carrera & Fernandez Cirelli, 2012).

Arsenic levels in cattle blood were between <LD–75 $\mu g L^{-1}$. These values are below than those previously reported by Rana *et al.* (2008, 2010) in West Bengal, India, that inform mean levels in blood of 284 $\mu g L^{-1}$. Relationship between total As level in drinking water and As levels in blood and hair should be of relevance. However, no significant correlation ($p > 0.05$) was observed. Results are shown in Figure 1.

4 CONCLUSIONS

Total As concentration in 50% of water samples exceeded the recommended value for cattle (200 $\mu g L^{-1}$) in the affected area. Blood appear not to be an effective biomarker for chronic exposure in cattle. Regards hair values in affected area were significantly higher than values determined in the control area and may be used as exposure biomarkers.

ACKNOWLEDGEMENTS

The authors are grateful to the University of Buenos Aires and CONICET for financial support.

REFERENCES

Bundschuh, J., Litter, M.I., Parvez, F., Román-Ross, G., Nicolli, H.B., Jean, J.-S., Liu, C.-W., López, D., Armienta, M.A., Guilherme, L.R.G., Cuevas, A.G., Cornejo, L., Cumbal, L. & Toujaguez, R. 2012. One century of arsenic exposure in Latin America: a review of history and occurrence from 14 countries. *Sci. Total Environ.* 429: 2–35.

CCME-Canadian Council of Ministers of the Environment. 2001. Canadian soil quality guidelines for the protection of the environmental and human health: Arsenic (inorganic) (1997). Updated in: Canadian environmental quality guidelines, 1999, Canadian Council of Ministers of the Environment, Winnipeg.

FAO-Food and Agriculture Organization 1985. Water quality for agriculture. Rome: Food and Agriculture Organization of the United Nations.

ISO 11466. 1995. Soil Quality – Extraction of trace element soluble in aqua regia.

Kabata P.A. & Pendias, H. 2010. *Trace Elements in Soils and Plants*. Third edition, CRC Press, Boca Raton, Ann Arbor, London.

Lowry, L.K. 1995. Role of biomarkers of exposure in the assessment of health risks. Toxicology letters 77(1–3): 31–38.

Perez-Carrera, A.L. & Fernandez-Cirelli, A. 2012. Arsenic exposure in cows from high contaminated area in the Chaco Pampean plain, Argentina. Proceedings of the 4th International Congress on Arsenic in the Environment, 22–27 July 2012, Cairns, Australia.

Pérez-Carrera, A., Alvarez-Gonçalvez, C.V. & Fernández-Cirelli, A. 2016. Transference factors as a tool for the estimation of arsenic milk concentration. *Environ. Sci. Pollut. Res.* 23(16): 16329–16335.

Rana, T., Sarkar, S., Mandal, T.K., Bhattyacharyya, K. & Roy, A. 2008. Arsenic residue in blood, urine and faeces samples from cattle in the Nadia district of West Bengal in India. *Internet J. Vet. Med.* 4(1).

Rana, T., Bera, A.K., Das, S., Bhattacharya, D., Bandyopadhyay, S., Pan, D. & Das, S.K. 2010. Effect of chronic intake of arsenic-contaminated water on blood oxidative stress indices in cattle in an arsenic-affected zone. *Ecotoxicol. Environ. Saf.* 73(6), 1327–1332.

Ravenscroft P., Brammer H. & Richards K. *Arsenic Pollution: A Global Synthesis*. West Sussex, UK: John Wiley & Sons; 2009.

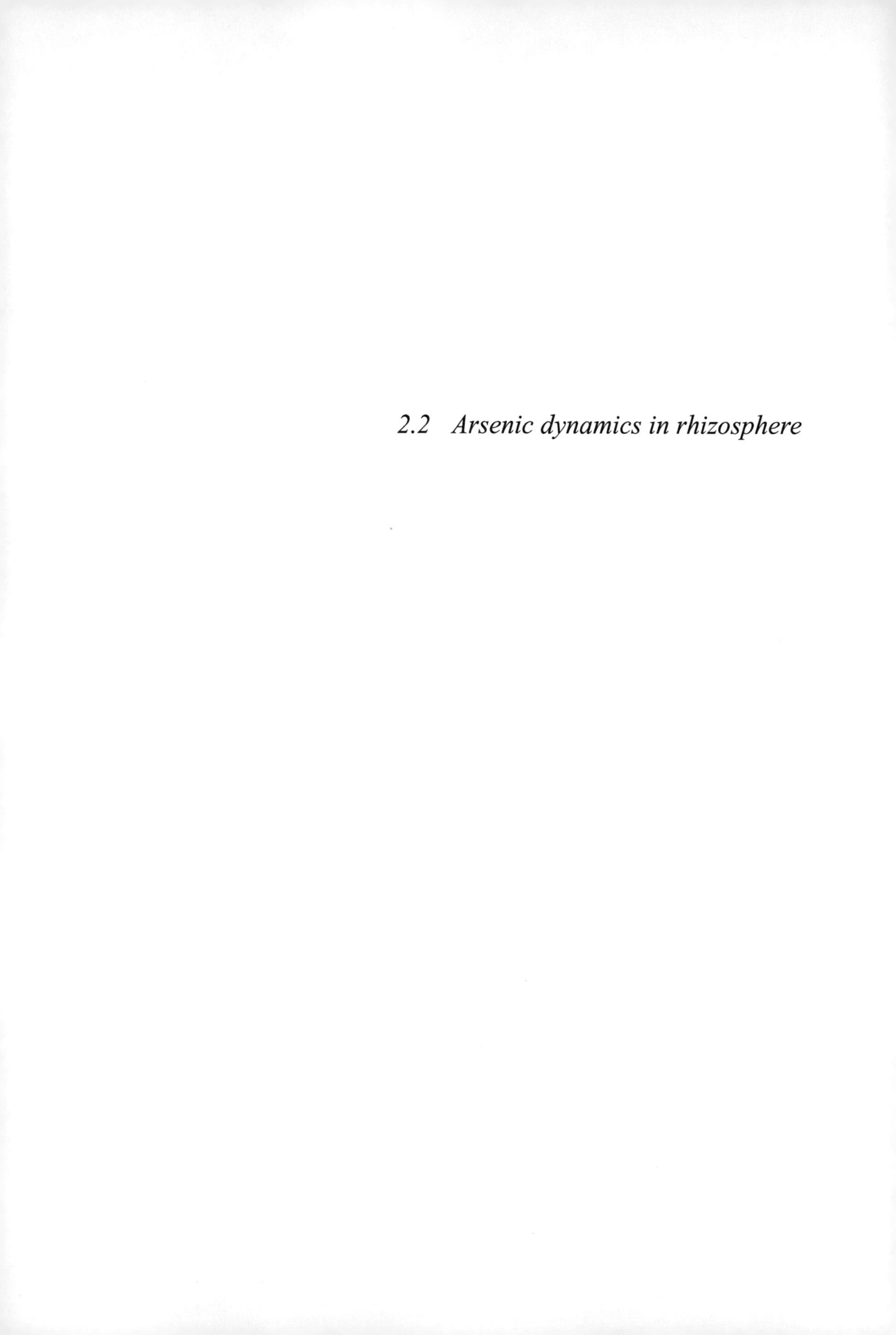

2.2 Arsenic dynamics in rhizosphere

Environmental Arsenic in a Changing World –
Zhu, Guo, Bhattacharya, Ahmad, Bundschuh & Naidu (Eds)
ISBN 978-1-138-48609-6

Bioelectrochemical arsenite oxidation in rice rhizosphere in plant-microbial fuel cells

X.Q. Wang, Y.H. Lv, C.P. Liu, F.B. Li & Y.H. Du
Guangdong Key Laboratory of Integrated Agro-environmental Pollution Control and Management, Guangdong Institute of Eco-Environmental Science & Technology, Guangzhou, P.R. China

ABSTRACT: The potential for arsenite removal using plant-microbial fuel cells (PMFC) was evaluated. In such a system, O_2 is excreted by living rice plant roots and Fe(II) play important roles in initiating Fenton reactions. Fenton reagents, including H_2O_2, Fe(II), were generated in-situ without an electricity supply. The HO· produced in the Fenton reaction were capable of oxidizing arsenite apparently and efficiently. The results might provide a new insight for As(III) oxidation in rice rhizosphere.

1 INTRODUCTION

Arsenic (As) is commonly presented as trivalent (arsenite) and pentavalent (arsenate) species in aquatic and soil systems, and the former is much more toxic and mobile than the latter. In flooded paddy soils, Fe(III) (hydr)oxides can be reduced to Fe(II), and arsenate adsorbed on the minerals can then be released and reduced to arsenite (Mitsunobu *et al.*, 2006). At the same time, O_2 released from the surfaces of rice root can directly oxidize Fe(II) in rhizosphere to Fe(III) (hydr)oxides (Colmer *et al.*, 2003). Recently, Wang *et al.* (2014) used a 'bio-electro-Fenton' reaction system successfully to oxidize and remove arsenite at neutral pH. In wet plant rhizosphere, microorganism activities can generate electricity from the biodegradation of plant root exudates, rhizodeposites and other organic matters. Therefore, we hypothesize that, arsenite can be oxidized to arsenate simultaneously by reactive oxygen species such as HO· via Fenton-like reactions (Wang *et al.*, 2014) and finally incorporated onto Fe(III) (hydr)oxides. As a result of As immobilization, the uptake of As by rice plants may be decreased, e.g., As accumulation in rice may be alleviated.

Consequently, a PMFC was constructed to demonstrate the feasibility of our hypothesis. Arsenite removal efficiencies and kinetics in the PMFC were investigated. The arsenite removal mechanisms in the PMFC were also explored. It should be valuable for understanding the biogeochemical processes of paddies under flooding conditions.

2 METHODS/EXPERIMENTAL

PMFC configuration was constructed according to a previously published protocol (Wang *et al.*, 2014). Differently, the PMFC in present study enlarged both the anode and cathode chambers to 600 mL. The PMFC experiments were operated in recycling batch mode. During the experiments, a pure culture of *Shewanella putrefaciens* SP200 was used as the biocatalyst in the sterilized anode chamber with 20 mM lactate as the electron donor. Three rice seedlings were planted in the cathode electrode. Rice roots were anticipated to expand along the chamber and be in close contact with the carbon felt/γ-FeOOH composite cathode. The growth medium contained a 100 mM phosphate buffer-based nutrient solution (pH 7.0). 1129 $\mu g\,L^{-1}$ arsenite concentration was used to study its removal efficiency by the PMFC. During the startup, an external resistance of 1000 Ω was used to connect the anode and cathode, and the MFC was operated at a controlled temperature of 30°C. The cell voltages were recorded by a 16-channel voltage collection instrument (AD8223, China).

3 RESULTS AND DISCUSSION

3.1 *Arsenite removal in the PMFC*

Although arsenite concentrations in the control reactors under open-circuit conditions decreased lower than that in the PMFC, high arsenite removal efficiencies were found at the end of all the treatments (Fig. 1). In the cathode chamber of the PMFC, more than 90% arsenite was depleted within 20 h. A small amount of arsenate was detected at 1.5 h, which increased firstly and reached a peak value at 24 h and then decreased over the subsequent time period. No arsenate was found in the two control experiments. The high arsenite removal efficiency in the two control reactors might be attributed to the adsorption by γ-FeOOH pasted on the carbon felt. Arsenite removal kinetics in the three reactors were also investigated and were found to follow first-order kinetics with high coefficients of determination ($R^2 > 0.92$, $p < 0.0001$).

The arsenite removal rate in the PMFC was much higher than those in the control experiments (Table 1). These results indicate that the bioelectrochemical process in the PMFC plays an important role in arsenite

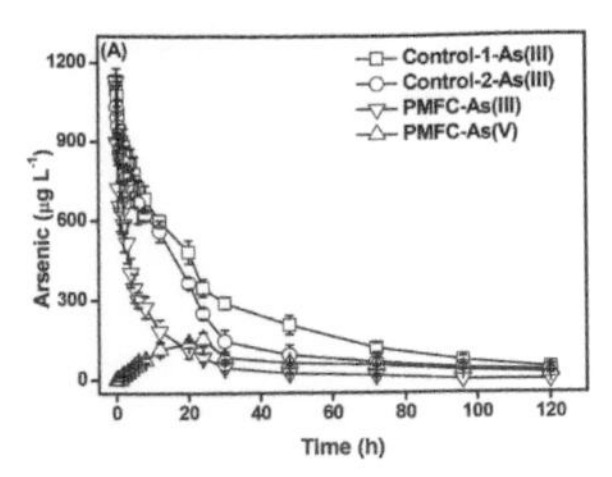

Figure 1. Variation of arsenic concentrations in the PMFC under closed and opened circuit conditions.

Table 1. Performance of PMFC for arsenite removal under various conditions.

	k_{obs} (h^{-1})	Fe(II) (mg L^{-1})	H_2O_2 (mg L^{-1})	OCE (%)
Control-1	0.0271	0.57 ± 0.14		
Control-2	0.0313	0.61 ± 0.02		
PVDF	0.1107	0.88 ± 0.13	1.58 ± 0.26	68.4

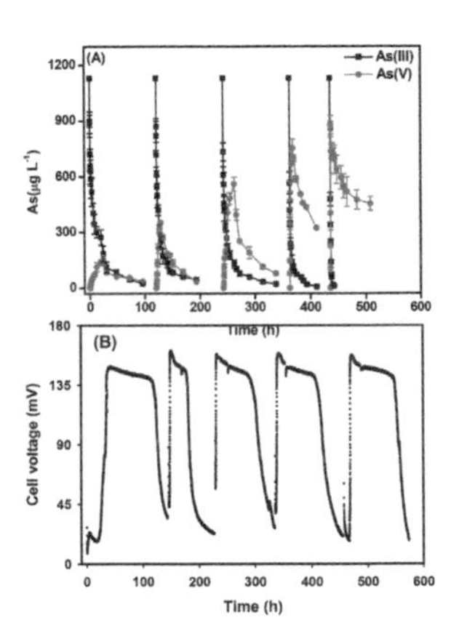

Figure 2. Performances of long-term operation of the PMFC.

oxidation, and the rice plants in the PMFC can stimulate this process. The value of oxidation current efficiency was determined to be 68.41% in the first 60 h, which was comparable with that of a previous study for Cr(VI) reduction using a PMFC (Habibul *et al.*, 2016).

The repeatable adsorption and oxidation experiments were run for five cycles. Figure 2A shows As species in the aqueous solution of the cathode chamber and the cell voltage of the PMFC over the five cycles. The cycles promoted the depletion rate of arsenite and generation rate of arsenate even if 2 g of γ-FeOOH was not maintained in the PMFC during the entire experiment. The apparent rate constants from the first to the fifth run were calculated to be 0.11 ± 0.02 ($R^2 = 0.95$), 0.22 ± 0.02 ($R^2 = 0.90$), 0.50 ± 0.03 ($R^2 = 0.91$), 0.85 ± 0.13 ($R^2 = 0.91$) and 0.71 ± 0.12 ($R^2 = 0.93$) h^{-1}. The dissolved arsenate increased more and more rapidly from the first cycle to the fifth cycle, implying that the oxidation process plays a more important role than the adsorption process in arsenite depletion. The cell voltage reached a plateau and then started to decrease due to the consumption of lactate in the anode chamber (Fig. 2B) in all the five cycles. These results implied that the root oxygen loss of rice plants prevail for arsenite oxidation in the rhizosphere.

3.2 *Mechanism of arsenite removal in the PMFC and environmental implications*

The increase of arsenate in the cathode chamber indicates that arsenite can be oxidized directly in the cathode coupled with oxidation of organics (lactate) in the anode of the PMFC. The reaction efficiency is highly dependent upon the ROL of living rice plants in the cathode chamber. In addition, the much higher removal rates of arsenite in the PMFC compared with those of the two controls also indicate that the bio-electro-chemical process for arsenite removal is more rapid. As Wang *et al.* (2014) suggested, this arsenite removal process is important in natural system such as the rhizosphere of rice plants in flooding paddies. In such a system, O_2 could be reduced to H_2O_2 via accepting electricity generated by microorganism activities, combined with Fe(II) produced by Fe(III) (hydr)oxides reductive dissolution, the Fenton reaction was initiated and the resulted HO· was active for arsenite oxidation.

4 CONCLUSIONS

O_2 excreted by rice roots can accept electrons derived from microorganism activities and participate in Fenton or Fenton like reactions, which products benefit arsenite oxidation and combination on Fe(III) (hydr)oxides. In summary, the results indicate that bioelectrochemical process plays important roles in arsenite removal in paddy soil-rice systems.

ACKNOWLEDGEMENTS

This work was financially supported by the National Natural Science Foundation of China (41201504) and Guangdong academy of sciences special funds for the talent with high degree and special technical ability (2017GDASCX-0829).

REFERENCES

Colmer, T.D. 2003. Long-distance transport of gases in plants: a perspective on internal aeration and radial oxygen loss from roots. *Plant Cell. Environ.* 26(1): 17–36.

Habibul, N., Yi, H., Wang, Y.K., Wei, C., Yu, H.Q. & Sheng, G.P. 2016. Bioelectrochemical chromium(VI) removal in plant-microbial fuel cells. *Environ. Sci. Technol.* 50(7): 3882–3889.

Mitsunobu, S., Harada, T. & Takahashi, Y. 2006. Comparison of antimony behavior with that of arsenic under various soil redox conditions. *Environ. Sci. Technol.* 40 (23): 7270–7276.

Wang, X., Liu, C., Yuan, Y. & Li, F. 2014. Arsenite oxidation and removal driven by a bio-electro-Fenton process under neutral pH conditions. *J. Hazard. Mater.* 275(2): 200–209.

Environmental Arsenic in a Changing World –
Zhu, Guo, Bhattacharya, Ahmad, Bundschuh & Naidu (Eds)
ISBN 978-1-138-48609-6

Thioarsenate formation in paddy soils

J. Wang[1], M. Romani[2], M. Martin[3] & B. Planer-Friedrich[1]
[1]*Environmental Geochemistry, Bayreuth University, Bayreuth, Germany*
[2]*Ente Nazionale Risi, Milan, Italy*
[3]*Dipartimento di Scienze Agrarie, Forestali e Alimentari, University of Torino, Turin, Italy*

ABSTRACT: Despite the significance of sulfur in controlling arsenic cycling, thioarsenate formation has not been studied in paddy soils, yet. Here, we combined lab incubation experiments with field porewater sampling from two Italian paddy soils to assess the geochemical significance of thioarsenate. Incubation data showed formation of both inorganic and methylated thioarsenates under flooded conditions, even in the absence of extra sulfate addition. Percentages of total arsenic were 12.5% and 18.3%, respectively. All species were also discovered after draining the systems for 4 days and re-flooding. Field sampling at rice flowering stage further confirmed the existence of inorganic (average 22.3%) and methylated thioarsenates (average 2.2%) under natural conditions. In summary, since our results conclusively prove the occurrence and quantitative importance of thioarsenate in paddy soils, their environmental fate urgently needs to be addressed.

1 INTRODUCTION

Elevated arsenic (As) concentrations in rice grains pose a serious health risk for over half the global population (Meharg *et al.*, 2009). Over the past years, knowledge about arsenite, arsenate, and methylated arsenates has increased tremendously. However, the formation of soluble arsenic-sulfur complex, so-called thio-arsenates ($As^{V}S^{0}S_{n}^{-II}O_{4-n}^{3-}$, with n = 0–3), has never been studied in rice agroecosystems, despite the long-recognized significance of reduced sulfur ligands in controlling aqueous metal speciation (Boulegue *et al.*, 1982). The neglectance of considering thioarsenates is partially explained because acidification, which is both used for stabilization as well as for chromatographic separation, leads to precipitation of AsS-minerals (Smieja & Wilkin, 2003) and transformation of thioarsenates (Planer-Friedrich & Wallschläger, 2009). Using flash-freezing for sample stabilization and chromatographic separation in an alkaline eluent as analytical method (Planer-Friedrich *et al.*, 2007), we have very recently gathered preliminary evidence from Italian and French rice fields revealing that inorganic and methylated thioarsenate do occur under natural conditions. In the present study, two Italian paddy soils were selected to elucidate the formation of thioarsenate via lab incubations. Porewater samples were also taken in the open fields at rice flowering stage to further assess the geochemical significance of thioarsenate in rice agroecosystems.

2 METHODS/EXPERIMENTAL

The paddy soils were sampled from two farms "Cascina Veronica" and "Cascina Formazzo" in Mortara (Italy). For each microcosm, 400 g fresh soil (2 mm sieved) were premixed with 2.5 g rice straw, then 250 mL tap water was added. Microcosms (500 mL polyethylene bottles) were equipped with rhizonsamplers and connected to a Teflon shut-off valve. No additional sulfate was added during the first 20 days of flooding. Thereafter, soils were drained and oxidation was allowed to occur for 4 days. Oxidized soils were then reflooded for 14 days with tap water containing 1 mM $(NH_4)_2SO_4$, thus to mimic the sulfate fertilizer application practice in Italy. Additionally, three porewater samples from each farm were taken at rice flowering stage. Porewater E_H and pH were measured following standard protocol. Iron was measured photometrically by the ferrozine assay and free sulfide by the methylene blue method. Aqueous arsenic and sulfur speciation was analyzed by anion exchange chromatography (Dionex ICS-3000 SP) ICP-MS without suppressor, following the method of Wallschläger and London (2007).

3 RESULTS AND DISCUSSION

3.1 *Thioarsenate formation under flooding-drainage-reflooding*

Besides arsenite, arsenate, monomethylarsenate (MMA), and dimethylarsenate (DMA), nine other As species were detected in both soils after flooding. Three of those species were confirmed as inorganic thioarsenates (thioAs), specifically mono- (MTA), di- (DTA) and trithioarsenate (TTA). Another three were identified as methylated thioarsenates (MethylthioAs), namely monomethylmonothioarsenate (MMMTA), monomethyldithioarsenate

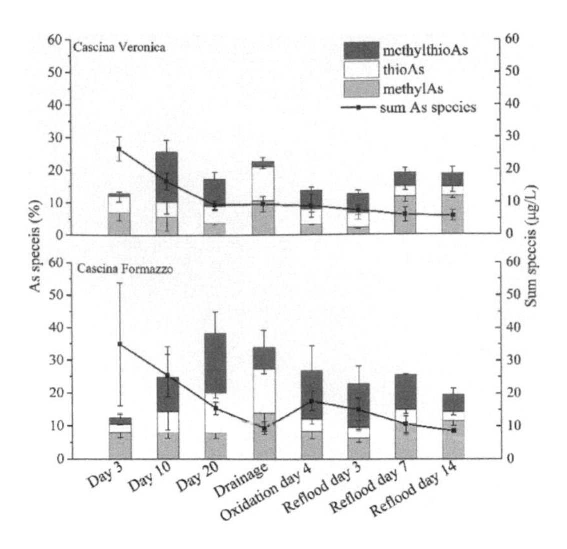

Figure 1. Porewater As speciation under flooding-drainage-reflooding cycling.

(MMDTA), and dimethyldithioarsenate (DMDTA). Additionally, we detected 3 unknown peaks with general abundance <1%.

Relative abundance of thioAs peaked after 20 days flooding in both Veronica and Formazzo soils (Fig. 1) with 5.3 ± 1.3% and 12.5 ± 1.2%, respectively. Relative abundance of methylthioAs peaked at day 10 for Veronica and day 20 for Formazzo (Fig. 1), with 15.2 ± 3.7% and 18.3 ± 6.5%, respectively.

Interestingly, higher percentages of thioAs were found in the drainage when compared that of porewater at day 20 in both soils (Fig. 1). This observation could indicate less absorption of thioAs to the soil solid phase, probably due to less affinity of thioAs to Fe minerals as suggested by previous studies (Couture *et al.*, 2013). All species were also discovered after reflooding, though in different shares. Since free sulfide was below detection limit under flooded conditions, we speculate it is elemental sulfur formation which determines As thiolation (Planer-Friedrich *et al.*, 2015).

3.2 *Arsenic speciation in field samples*

MethylthioAs species were of minor quantitative importance, except in sample Formazzo-1, with an average percentage of 2.2%. The main As species in Veronica porewater were arsenate (34.4 ± 5.9%), arsenite (18.4 ± 6.0%), MTA (17.1 ± 1.1%), DTA (13.0 ± 3.3%), DMA (12.8 ± 6.8%) and MMA (1.8 ± 1.2%). In Formazzo, the main porewater As species detected were arsenite (59.8 ± 7.9%), arsenate (19.1 ± 5.4%), DTA (10.9 ± 7.7%), DMA (3.5 ± 1.8%), MTA (1.4 ± 0.8%), and MMA (1.8 ± 2.3%). TTA (1.3 ± 0.9). The percentage of the thioAs (average 22.3%) was found to outweight that of methylate arsenate (average 9.7%) in all the porewater samples (Fig. 2).

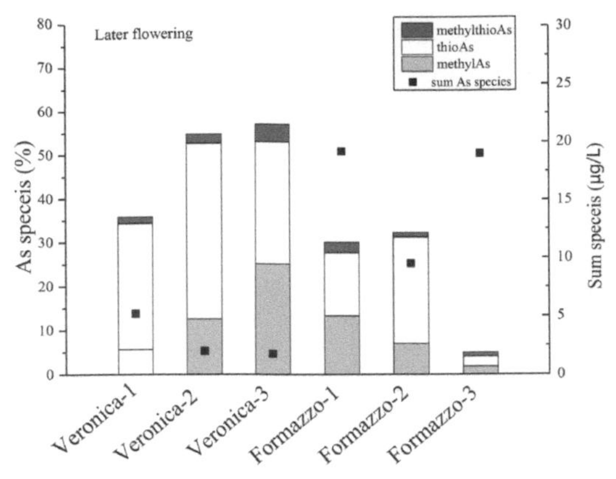

Figure 2. Porewater As speciation at rice flowering stage at fields.

4 CONCLUSIONS

Our lab experimental and field sampling data conclusively prove the occurrence and quantitative importance of inorganic and methylated thioarsenates in rice agroecosystems. Given the ubiquitous natural occurrence of arsenic and sulfur in reduced paddy soils, As thiolation needs to be considered in future As cycling studies.

REFERENCES

Boulegue, J., Lord, C.J. & Church, T.M. 1982. Sulfur speciation and associated trace metals (Fe, Cu) in the pore waters of Great Marsh, Delaware. *Geochim. Cosmochim. Acta* 46(30): 453–464.

Couture, R.-M., Rose, J., Kumar, N., Mitchell, K., Wallschläger, D. & Van Cappellen, P. 2013. Sorption of arsenite, arsenate, and thioarsenates to iron oxides and iron sulfides: a kinetic and spectroscopic investigation. *Environ. Sci. Technol.* 47(11): 5652–5659.

Meharg, A.A., Williams, P.N., Adomako, E., Lawgali, Y.Y., Deacon, C., Villada, A., Cambell, R.C., Sun, G., Zhu, Y.-G. & Feldmann, J. 2009. Geographical variation in total and inorganic arsenic content of polished (white) rice. *Environ. Sci. Technol.* 43(5): 1612–1617.

Planer-Friedrich, B. & Wallschläger, D. 2009. A critical investigation of hydride generation-based arsenic speciation in sulfidic waters. *Environ. Sci. Technol.* 43(13): 5007–5013.

Planer-Friedrich, B., London, J., McCleskey, R.B., Nordstrom, D.K. & Wallschläger, D. 2007. Thioarsenates in geothermal waters of Yellowstone national park: determination, preservation, and geochemical importance. *Environ. Sci. Technol.* 41(15): 5245–5251.

Planer-Friedrich, B., Hartig, C., Lohmayer, R., Suess, E., McCann, S. & Oremland, R. 2015. Anaerobic chemolithotrophic growth of the haloalkaliphilic bacterium strain MLMS-1 by disproportionation of monothioarsenate. *Environ. Sci. Technol.* 49(11): 6554–6563.

Smieja, J.A. & Wilkin, R.T. 2003. Preservation of sulfidic waters containing dissolved As(III). *J. Environ. Monit* 5(6): 913–916.

Wallschläger, D. & London, J. 2007. Determination of methylated arsenic-sulfur compounds in groundwater. *Environ. Sci. Technol.* 42(1): 228–234.

Environmental Arsenic in a Changing World –
Zhu, Guo, Bhattacharya, Ahmad, Bundschuh & Naidu (Eds)
ISBN 978-1-138-48609-6

The role of radial oxygen loss on the flux of arsenic and other elements in rice rhizosphere

D. Yin[1], J. Luo[1], W. Fang[1] & P.N. Williams[2]
[1]*State Key Laboratory of Pollution Control and Resource Reuse, School of the Environment, Nanjing University, Jiangsu, P.R. China*
[2]*Institute for Global Food Security, Queen's University Belfast, Belfast, UK*

ABSTRACT: In this study, we combined diffusive gradients in thin films (DGT) with planar optode (PO), a two-dimensional (2D) in situ chemical techniques for investigating the geochemical behaviors of arsenic in rice rhizosphere soil and bulk soil at high spatial resolution (sub-mm). We had observed three distinctive regions in rice rhizosphere: soil-water interfaces (SWI, O+), rhizosphere aerobic soils (O+) and bulk anaerobic soils (O−). The mobility of arsenic and other elements is greater in rhizosphere than bulk zone, flux maxima for As, Fe, P, Pb had also been observed around root tips. Our results indicate rice rhizosphere is a special unit to gather oxygen and affect metals mobility, both flux maxima for metals and radial oxygen loss from root tips are common existed in rice rhizosphere. We have provided new evidence for the importance of rhizosphere oxidation and coupled diffusion in modulating arsenic mobilization and dispersion, showing microniches are important geochemical phenomena exploited by rice plants to acquire metals or nutrients.

1 INTRODUCTION

Rhizosphere processes affect the fate of arsenic and other elements in rice rhizosphere. It is important and complicated for managing transformation of arsenic and other elements to plants. There exist complexity of root exudates and highly microbial communities in rhizosphere. (Lee *et al.*, 2013) Williams *et al.* (2014) had developed a new sandwich sensor technology, consisting of a pH/O_2-sensitive planar optode, overlain by an ultrathin DGT layer for measuring As and other elements. We need a better understanding of micro/macroniches in trace elements uptake in rice. We still want to know whether those findings are common in other environment, such as higher level of arsenic or with other technologies.

2 EXPERIMENTAL

2.1 *Method summary*

An ultrathin DGT layer is exposed to the soil and backed by a planar optode as described in Williams *et al.* (2014). After rice was cultivated in the rhizotrons, the sandwich sensors were deployed in the rice rhizosphere. The DGT binding gel could record the locally induced As and other elements flux and the O_2 planar optode can resolve the O_2 concentration dynamics for the same location.

2.2 *Chemical mapping*

Three ultrathin DGT resin gel used in the sensors: suspended particulate reagent-iminodiacetate (SPR-IDA) (Warnken *et al.*, 2004), precipitated ferrihydrite (PF) (Luo *et al.*, 2010) and precipitated zirconia (PZ) (Guan *et al.*, 2015). After deployed 24 h in the rhizotrons, the gels were retrieved and rinsed with MQ water, dried using a gel drier (Bio-Rad model 543) and then carefully fixed onto glass plates using double-sided adhesive tape prior laser ablation (LA) combined with ICP-MS analysis. ICP-MS was used to record ^{13}C, ^{75}As, ^{82}Se, ^{121}Sb, ^{31}P, ^{184}W, ^{98}Mo, ^{57}Fe, and ^{115}In signals. Sampling soil and pore water when harvesting the plant using a compartmented design: soil-water interfaces, bulk anaerobic soils and rhizosphere aerobic soils. We used color ratiometric planar optode sensors to achieve oxygen images in the rhizotrons (Larsen *et al.*, 2011).

3 RESULTS AND DISCUSSION

3.1 *As and other elements flux in rice rhizosphere*

Various location of Fe, As, P, Pb, Mn in different distance to rice roots. DGT-measured flux is determined by the concentration adjacent to the device surface, localized maxima in the DGT-flux reflect localized maxima in the elements concentration, indicating a

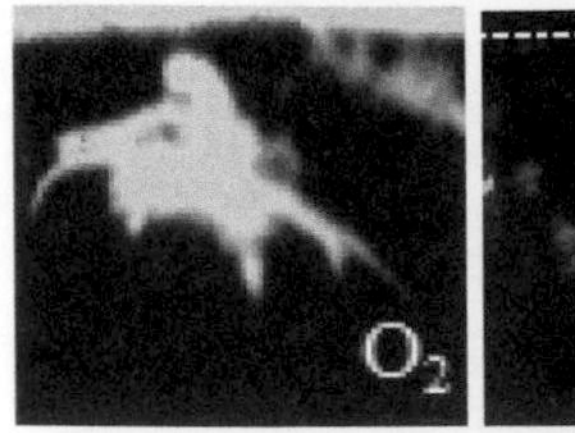

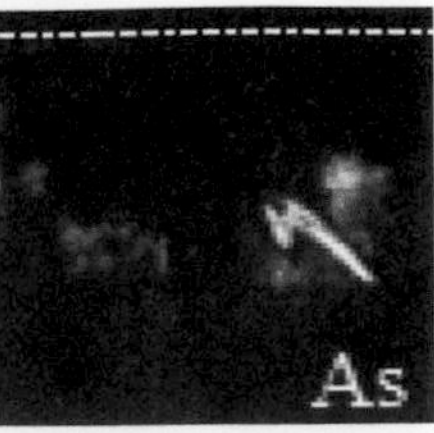

Figure 1. Visualization of O2, Fe, As, P, Pb, Mn around a set of rice roots. The metal fluxes (f_{DGT}, pg cm^{-2} s^{-1}) and oxygen concentration (percent air saturation) increased sequentially with the color scale shown from blue to white.

localized process of mobilization (Williams *et al.*, 2014). High-resolution imaging of DGT-measured elements flux around rice roots was conducted to obtain insight into the spatial variability of metals (Fig. 1). The oxygen images from rice rhizosphere and the corresponding elements fluxes from different SPR-IDA DGT gel. As, P, Pb and Fe(II) mobilization is low in where high oxygen around tap roots, which may due to Fe plaque promote these metals deposition onto root surfaces, resulting in decrease of metals availability. Wu *et al.* (2011) and Mei *et al.* (2009) had noted that there was a significant negatively correlation between ROL and As tolerance and accumulation in rice, which may be due to the effects of ROL on Fe plaque formation.

As and V, Mn, Fe, Co, Cu concentration in SWI are higher than the other two areas, may indicate that SWI could be a sink of these metals. This could be because the redox potential changes over soil depth under submerged condition, as a consequence of elements transport and accumulate in SWI. As accumulation in the topsoil relates to freshly precipitated Fe/Mn oxides. This oxidized layer acted as a sink for As, Fe, Mn, V, Co, Cu and might explain the depletion occurred in soil pore water monitored as revealed by the samplers located at the soil-water interface.

3.2 *Consistency of As with P and Fe*

As mobilization was consistent with P and Fe, there are flux maximum exist in SWI and root apexes. As and P belong to the same chemical group and have comparable dissociation constants for their acids and solubility products for their salts, resulting in physicochemical similarity in soil. (Adriano, 2001) It is not surprising for the co-occurrence of As and Fe flux maxima. Arsenic is readily retained on Fe oxides, and their release often coincides. Meanwhile, with the redox threshold for the Fe(III)/Fe(II) transition lying close to the arsenate/arsenite boundary, the transformation to arsenite would favor a more rapid and extensive desorption. (Takahashi *et al.*, 2004).

4 CONCLUSIONS

There is similar transport of As, Zn and Mn across the root plaque, but the location of the sub-surface resupply is well defined. Clearly the root apexes act as a sink for As, Fe(II), V, Mn, Co, Cu, metal remobilized in the tips of root and at the surface of water-soil. Flux maxima for As and some other elements are quite common, but not all roots will show this phenomenon.

ACKNOWLEDGEMENTS

This work was funded by the National Natural Science Foundation of China (NSFC) (No. 41771271 and 21477053), the NSFC and Newton Fund/Royal Society (No. 21511130063 and R1504GFS).

REFERENCES

Adriano, D.C. 2001. Arsenic. In: *Trace Elements in Terrestrial Environments*. Springer, New York, 219–261.

Guan, D.X., Williams, P.N., Luo, J., Zheng, J.L., Xu, H.C., Cai, C. & Ma, L.Q. 2015. Novel precipitated zirconia-based DGT technique for high-resolution imaging of oxyanions in waters and sediments. *Environ. Sci. Technol.* 49(6): 3653–3661.

Larsen, M., Borisov, S.M., Grunwald, B., Klimant, I. & Glud, R.N. 2011. A simple and inexpensive high resolution color ratiometric planar optode imaging approach: application to oxygen and pH sensing. *Limnol. Oceanogr. Meth.* 9(9): 348–360.

Lee, Y.J., Mynampati, K., Drautz, D., Arumugam, K., Williams, R. & Schuster, S. 2013. Understanding aquatic rhizosphere processes through metabolomics and metagenomics approach. *EGU General Assembly*, 15.

Luo, J., Zhang, H., Santner, J. & Davison, W. 2010. Performance characteristics of diffusive gradients in thin films equipped with a binding gel layer containing precipitated ferrihydrite for measuring arsenic (V), selenium (VI), vanadium (V), and antimony (V). *Anal. Chem.* 82(21): 8903–8909.

Mei, X.Q., Ye, Z.H. & Wong, M.H. 2009. The relationship of root porosity and radial oxygen loss on arsenic tolerance and uptake in rice grains and straw. *Environ. Pollut.* 157(8): 2550–2557.

Takahashi, Y., Minamikawa, R., Hattori, K.H., Kurishima, K., Kihou, N. & Yuita, K. 2004. Arsenic behavior in paddy fields during the cycle of flooded and non-flooded periods. *Environ. Sci. Technol.* 38(4): 1038–1044.

Warnken, K.W., Zhang, H., & Davison, W. 2004. Performance characteristics of suspended particulate reagent-iminodiacetate as a binding agent for diffusive gradients in thin films. *Anal. Chim. Acta* 508(1): 41–51.

Williams, P.N., Santner, J., Larsen, M., Lehto, N.J., Oburger, E., Wenzel, W. & Zhang, H. 2014. Localized flux maxima of arsenic, lead, and iron around root apices in flooded lowland rice. *Environ. Sci. Technol.* 48(15): 8498–8506.

Wu, C., Ye, Z., Shu, W., Zhu, Y. & Wong, M. 2011. Arsenic accumulation and speciation in rice are affected by root aeration and variation of genotypes. *J. Exp. Bot.* 62(8): 2889–2898.

2.3 Microbial ecology of arsenic biotransformation in soils

Environmental Arsenic in a Changing World –
Zhu, Guo, Bhattacharya, Ahmad, Bundschuh & Naidu (Eds)
ISBN 978-1-138-48609-6

Microbial transformation of arsenic in Bengal floodplain

H. Afroz, A.A. Meharg & C. Meharg
Institute for Global Food Security, Queens University Belfast, Belfast, UK

ABSTRACT: Methylation is an important biotransformation process that limits the toxicity of arsenic (As) in soil. Here we investigated how geomorphology and paddy management influences As speciation, and the abundance and diversity of the arsM gene responsible for bacterial methylation of arsenic. Soil samples collected from paddy and non-paddy fields of Holocene and Pleistocene regions of Bangladesh were incubated under anaerobic conditions to identify how these treatments affected As speciation in soil solution as well as to investigate the changes in relative arsM copy no and diversity in soil. The Holocene soil had higher concentration of soil solution arsenic species (inorganic arsenic, dimethylarsinic acid (DMA), trimethylarsenic oxide (TMAO), with qPCR showing higher copy numbers of both 16S and arsM in Holocene soil compared to Pleistocene soil. Lower soil Eh may explain the higher arsM copy number in Holocene soil, with arsenic methylation known to be increased under anaerobic conditions. The higher pH in Holocene soil may also explain the increase in 16S copy number, with bacteria known to be less abundant in acidic soils. Further to that amplicon sequencing showed an increased species richness (chao1) and diversity (Simpson) for both 16S and arsM in Holocene compare to Pleistocene soil. PiCrust analysis of the 16S amplicon results showed the presence of arsenic metabolism related genes some of which were increased in Holocene compared to Pleistocene soil. The results showed that presence of As soil chemistry strongly correlates with arsenic transformation and copy number of arsenic metabolizing genes and bacterial as well as arsM gene diversity.

1 INTRODUCTION

Paddy soils in the Holocene regions of Bangladesh is elevated in inorganic arsenic (iAs), while Pleistocene soils are lower in iAS, most likely due to its and paddy management practices, such as flooding of irrigation of Holocene soils with As contaminated waters (Lu, *et al.*, 2009). Several studies found that, Holocene floodplain contained higher total As compared to Pleistocene terrace soil (Chowdhury *et al.*, 2017; Lu, *et al.*, 2009; William *et al.*, 2011). This is because, Holocene soils are less weathered and the parent material are still undergoing subsequent weathering process while the Pleistocene soils are highly weathered (William *et al.*, 2011). As methylation is an important biotransformation process that catalyzes toxic iAs to less toxic methylated As species such as MMA (monomethylarsonic acid), DMA and TMAO. The rate of methylation depends on As species and the abundance and diversity of arsM gene in soil (Jia *et al.*, 2014; Zhao *et al.*, 2013). In this study, we investigated the effect of geomorphology and paddy management system on As speciation and diversity of arsM gene in Bangladeshi soil.

2 MATERIAL AND METHODS

2.1 *Soil sample collection incubation*

A total of 20 soil samples were collected from paddy and adjacent non-paddy fields across Holocene (Araihazar Upazila) and Pleistocene regions (Madhupur Upazila) of Bangladesh. Field moist soils were incubated for 2 weeks under anaerobic condition to stimulate the activity of microbial population in soil. After two weeks, soil pore water and soils were collected and both soil and pore water immediately frozen and stored at $-20°$ until subsequent analysis. pH and Eh was recorded in replicated microcosms by inserting an Eh/pH meter inti the soil at harvest. Arsenic species in 0.4-micron Millipore membrane filtered soil pore waters (iAs, DMA and TMAO) were determined using a Dionex IC chromatographic system interfaced with ICP-MS (ICS-5000 DC, Thermo Scientific).

2.2 *DNA extraction, qPCR and Amplicon sequencing*

Soil DNA was extracted using the Powerlyzer® PowersSoil® DNA isolation kit (MOBIO Laboratories, Inc.) following the manufacturer's instructions. Soil DNA was amplified for bacterial 16S and arsM gene using specific primer pairs 16S_1369F/16S_1492R (Jia *et al.*, 2014) and arsMF1/arsMR2 (Jia *et al.*, 2013), respectively and relative copy no was measured by qPCR on the Eppendorf mastercycler (Realplex^4, Hamburg, Germany) using SYBR green (PrimerDesign, USA). Following purification and quantification soil DNA samples were sent to Centre for Genomic Research institute, University of Liverpool for amplicon sequencing of the 16S rDNA and arsM gene on the Illumina Miseq using 250 paired end sequencing.

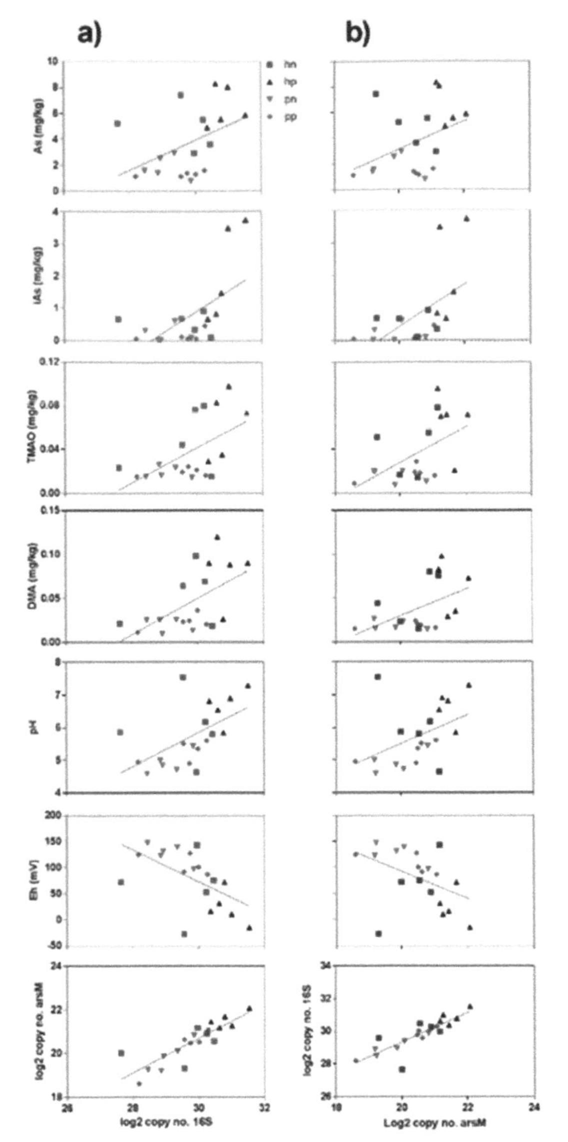

Figure 1. Linear Relationships for bacterial 16S (a) and arsM (b) gene copy no. versus soil As, soil solution As species, pH and Eh. The line on each graph is the regression line for this parameter. "hn" represents Holocene non-paddy "hp" represents Holocene paddy, "pn" Pleistocene non-paddy, "pp" Pleistocene paddy.

Sequences were analyzed using QIIME version 1.8 (Caporaso *et al.*, 2010) followed by statistical analysis in R.

2.3 *Characterization of soils*

The soil samples were oven dried (70°C for 48 h) to constant weight and analyzed for t-As using ICP-MS.

2.4 *Statistical analysis*

Statistical analysis was performed in Minitab version 16 (Minitab, PA, USA) and Graphpad prism7.

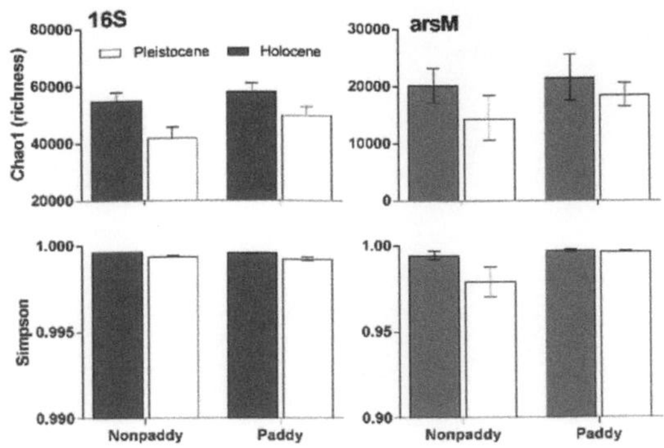

Figure 2. Alpha diversity (Richness and Simpson evenness) in bacterial 16S and arsM gene. Significant differences ($p < 0.05$) was observed for geomorphology (Holocene/Pleistocene).

3 RESULTS AND DISCUSSION

3.1 *Soil solution As species, pH, Eh and qPCR*

The concentration of soil As, soil solution As species and relative copy no of bacterial 16S and arsM gene were higher in Holocene soils compared to Pleistocene soils. The linear relationship between bacterial 16S and arsM gene copy no with chemical parameters (soil As, soil solution As species, pH and Eh) is shown in Figure 1.

The strong linear relationships of 16S and arsM gene copy with soil pH and Eh suggests that, these are the key factors that regulating the DNA copy no and As speciation in soil. High redox potential and acidic pH in Pleistocene soils leads to lower arsM gene copy no in that soils. Because arsM gene is redox sensitive and more active under anaerobic condition (Jia *et al.*, 2013, 2014). In contrast, no significant Paddy management effect was observed in this study. The Pleistocene soils either from paddy and non-paddy soils were less variable in different fields while the Holocene non-paddy soils were more variable.

3.2 *Diversity of Bacterial 16S and arsM gene*

Alpha diversity measures showed higher species richness (chao1) and overall diversity (Simpson) for both 16S and arsM gene in Holocene compared to Pleistocene soil. While this effect was significant for both 16S and arsM, it was more pronounced in arsM gene (Fig. 2).

4 CONCLUSION

Overall, the concentration of soil As, pore water As species and the abundance of arsM and other As metabolism related genes was shown to vary in two different geomorphic units of Bangladesh, with chemical and biological parameters correlating. Future investigations will focus on metagenomics analysis to further explore these results.

ACKNOWLEDGEMENTS

The research work was funded by Commonwealth Scholarship Commission, UK.

REFERENCES

Caporaso, J.G., Kuczynski, J., Stombaugh, J., Bittinger, K., Bushman, F.D., Costello, E.K., Fierer, N., Peña, A.G., Goodrich, J.K., Gordon, J.I., Huttley, Kelley, S.T., Knights, D., Koenig, J.E., Ley, R.E., Lozupone, C.A., McDonald, D., Muegge, B.D., Pirrung, M., Reeder, J., Sevinsky, J.R., Turnbaugh, P.J., Walters, W.A., Widmann, J., Yatsunenko, T., Zaneveld, J. & Knight, R. 2010. QIIME allows analysis of high-throughput community sequencing data. *Nat. Methods* 7(5): 335–336.

Chowdhury, M.T.A., Deacon, C.M., Jones, G.D., Huq, S.M.I., Williams, P.N., Hoque, A.F.M.M., Winkel, L.H.E., Price, A.H., Norton, G.J. & Meharg, A.A. 2017. Arsenic in Bangladeshi soils related to physiographic region, paddy management, and mirco- and macro-elemental status. *Sci. Total Environ.* 590: 406–415.

Jia, Y., Huang, H., Zhong, M., Wang, F., Zhang, L. & Zhu, Y. 2013. Microbial arsenic methylation in soil and rice rhizosphere. *Environ. Sci. Technol.* 47(7): 3141–3148.

Jia, Y., Huang, H., Chen, Z. & Zhu, Y. 2014. Arsenic uptake by rice is influenced by microbe-mediated arsenic redox changes in the rhizosphere. *Environ. Sci. Technol.* 48(2): 1001–1007.

Lu, Y., Adomako, E.E., Solaiman, A.R.M., Islam, M.R., Deacon, C., Williams, P.N., Rahman, G.K.M.M. & Meharg, A.A. 2009. Baseline soil variation is a major factor in arsenic accumulation in Bengal delta paddy rice. *Environ. Sci. Technol.* 43(6): 1724–1729.

Williams, P. N., Zhang, H., Davison, W., Meharg, A. A., Hossain, M., Norton, G. J., Brammer, H. & Islam, R.M. 2011. Organic matter-solid phase interactions are critical for predicting arsenic release and plant uptake in Bangladesh paddy soils. *Environ. Sci. Technol.* 45(14): 6080–6087.

Zhao, F., Harris, E., Yan, J., Ma, J., Wu, L., Liu, W., McGrath, S.P., Zhou, J.Z. & Zhu, Y.G. 2013. Arsenic methylation in soils and its relationship with microbial arsM abundance and diversity, and as speciation in rice. *Environ. Sci. Technol.* 47(13): 7147–7154.

Environmental Arsenic in a Changing World –
Zhu, Guo, Bhattacharya, Ahmad, Bundschuh & Naidu (Eds)
ISBN 978-1-138-48609-6

Concurrent methylation and demethylation of arsenic in fungal cells

S.M. Su, X.B. Zeng, L.Y. Bai, Y.N. Wang & C.X. Wu
Institute of Environment and Sustainable Development in Agriculture, Chinese Academy of Agricultural Sciences/Key Laboratory of Agro-Environment, Ministry of Agriculture, Beijing, P.R. China

ABSTRACT: Microbial methylation and demethylation are central to arsenic's (As) biogeochemical cycling. Here, the transformations of monomethylarsonic acid (MMA(V)) (50 mg L^{-1}) for 15 days in cells of As-methylating fungi, *Fusarium oxysporum* CZ-8F1, *Penicillium janthinellum* SM-12F4, and *Trichoderma asperellum* SM-12F1, were evaluated, and trace concentrations of As(III) and As(V) were observed in fungal cell extracts. Trace amounts of DMA(V) were also detected in MMA(V) and *P. janthinellum* SM-12F4 incubations. In situ X-ray absorption near edge structure (XANES) indicated that after exposure to MMA(V) (500 mg L^{-1}) for 15 days, 28.6–48.6% of accumulated As in fungal cells was DMA(V), followed by 18.4–30.3% from As(V), 0–28.1% from As(III), and 4.8–28.9% from MMA(V). The concurrent methylation and demethylation of As occurs in fungal cells. The findings of this study will develop our understandings of microorganisms that drive As speciation transformation.

1 INTRODUCTION

Microbial methylation and volatilization of arsenic (As) drives As biogeochemical cycling ecosystems. Notably, organic matter, moisture, and pH all regulate As methylation and volatilization. Demethylation of methylarsenicals has been also widely studied in soils. However, there are very limited studies associated with As demethylation in microbial cells. Furthermore, some As-methylating microorganisms are capable of demethylation of methyarsenicals. *NsarsI* cloned from As-methylating *Nostoc* sp. PCC 7120 encodes a C·As lyase responsible for MMA(III) demethylation (Yan *et al.*, 2015). However, whether methylation and demethylation can be triggered concurrently in the same microbial cell remains unknown. Three fungal strains, *F. oxysporum* CZ-8F1, *P. janthinellum* SM-12F4, and *T. asperellum* SM-12F1, were shown to be capable of As(III) or As(V) methylation, with MMA(V) and DMA(V) generated as products (Su *et al.*, 2012). In this study, after exposure to MMA(V), As speciation transformation in *F. oxysporum* CZ-8F1, *P. janthinellum* SM-12F4, and *T. asperellum* SM-12F1 cells was evaluated using in *situ* X-ray absorption spectroscopy (XAS) and ion exchange chromatographic separation.

2 METHODS/EXPERIMENTAL

2.1 *As speciation in fungal culture system using ion exchange chromatography separation*

Briefly, after incubation with the fungal spore suspension (0.2 mL, 10^4 cfu mL^{-1}) in potato-glucose-peptone (PGP) medium (spiked with MMA(V) of 50 mg L^{-1}) for 1, 3, 5, 10, and 15 days, the culture medium was removed via centrifugation (12,000 *g*, 15 min). A mixture of buffer solution (0.1 M KH_2PO_4/K_2HPO_4, pH 7.0, 10 mL) and ultrapure water (10 mL) was used to wash the compact fungal cells. Arsenic speciation in fungal cells was extracted using tetramethylammonium hydroxide (TMAH, 25%, 0.3 mL) and ultrapure water (4.7 mL) by grinding, centrifugation (12,000 *g*, 15 min), and filtration (diameter, 0.20-μm). Arsenic speciation analysis was conducted using high performance liquid chromatography-hydride generation-atomic fluorescence spectrometry (HPLC-HG-AFS, SA-10, Titan Instruments, Beijing, China). PGP medium spiked with MMA(V) and without fungi was used to investigate the natural variance of MMA(V). Three replicates were run for each sampling time. No As was detected in the spore suspension used as the fungal inoculum.

2.2 *As speciation in fungal cells using in-situ XAS*

Fungal cells were harvested after cultivation for 15 days in PGP medium spiked with 500 mg L^{-1} MMA(V) (CH_4AsNaO_3, Strem Chemicals, Inc. Newburyport, USA). Sample preparation and analysis using X-ray absorption near edge structure (XANES) was performed as previously reported (Su *et al.*, 2012). For As distribution mapping using X-ray fluorescence (XRF), each point was scanned for 2 s using a silicon/lithium (Si/Li) detector with a spot size of 3.08 × 2.78 μm^2 and a step size of 5 μm. For As speciation determination using XANES, each site was scanned for 6 s using a spot size of 3.08 × 2.78 μm^2 from 11,850 to 11,920 eV with a 0.5 eV step size. During XANES analysis, As standard compounds were prepared with high-purity (>94.5%) chemicals.

3 RESULTS AND DISCUSSION

3.1 *As speciation in fungal cells using ion exchange chromatographic separation*

MMA(V), a small amount of inorganic As, and DMA(V) were observed in fungal cells after being exposed to MMA(V). MMA(V) contents in the fungal cells initially increased and then decreased with cultivation time. Furthermore, a small amount of As(III), As(V), or DMA(V) were also observed in fungal cells. For *T. asperellum* SM-12F1, the converse varying trends for As(III) and As(V) were observed with time. When cultivation time was prolonged to 15 days, the contents of As(III) and As(V) were 0.6 and 3.8 $\mu g\, g^{-1}$, respectively. For *P. janthinellum* SM-12F4, DMA(V) of 0.4 and 28.7 $\mu g\, g^{-1}$ at a cultivation time of 2 and 3 days was observed, respectively. As(III) and As(V) contents reached 1.5 and 0.9 $\mu g\, g^{-1}$ when the cultivation time extended to 15 days, respectively. For *F. oxysporum* CZ-8F1, As(III) and As(V) of 0.3 and 1.2 $\mu g\, g^{-1}$ were observed at a cultivation time of 15 days, respectively. This indicates that MMA(V) demethylation into As(III) and As(V) and methylation into DMA occurred in fungal cells after exposure to MMA(V) (data see Su *et al.*, 2017).

3.2 *As speciation in fungal cells using in-situ XAS*

Arsenic accumulation in fungal cells was easily observed using XRF, as shown in Figure 1, after exposure to 500 mg L^{-1} of MMA(V) for 15 days. The XANES spectra for each fungal strain and the standards are shown in Figure 1D. Evaluation of the XANES spectra beyond the absorption edge shows differences in the 11,862–11,875 eV regions of the three fungal strains. DMA(V) and inorganic As were simultaneously observed in fungal cells after MMA(V) exposure. For *T. asperellum* SM-12F1, 48.6% of the accumulated As was DMA(V), followed by 28.1% As(III), 18.4% As(V), and 4.8% MMA(V). For *P. janthinellum* SM-12F4, 41.1% of the accumulated As was DMA(V), followed by 30.3% As(V), and 28.9% MMA. No As(III) was observed. However, for *F. oxysporum* CZ-8F1, 29.0%% of the accumulated As was As(V), followed by 28.6% DMA(V), 26.3% MMA(V), and 16.2% As(III). MMA(V) remethylation and demethylation could be simultaneously triggered in fungal cells (data see Su *et al.*, 2017).

The concurrent methylation and demethylation of As may exist in fungal hyphae. The concurrent methylation and demethylation of As is supported by our observations using the ion exchange chromatographic separation and *in situ* XANES. Comparatively, XANES analysis over-estimated the methylation of MMA(V). The differences between both methods might be attributed to: (I) As exposure condition, as higher As exposure might help to emerge the methylated arsenide in biological cells; (II) measurement scales, XANES analysis is used to determine As species in a particular area of a sample. While the typical method produced the average value of As in samples; (III) sample preparation, since samples for XANES analysis are preserved in their native state. However, the mixed cellular sample is acquired by cellular lysis and centrifugation in advance of HPLC-HG-AFS analysis. Notably, these differences do not interfere with our observation that the concurrent methylation and demethylation of As exists in fungal cells. The typical method coupled with *in situ* XANES is recommended to detect As species in samples.

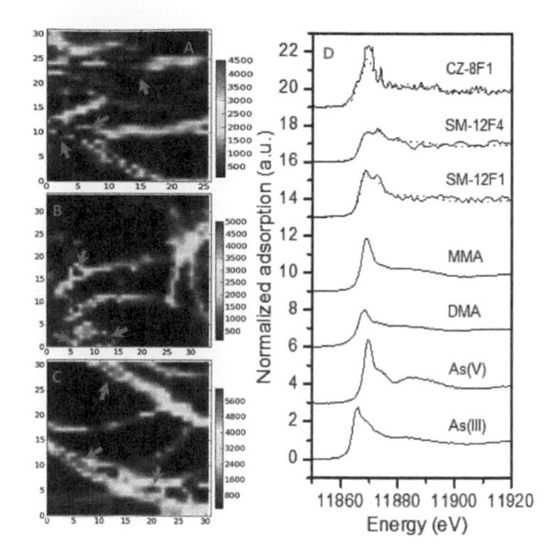

Figure 1. X-ray fluorescence (XRF) images and X-ray absorption near edge structure (XANES) spectra of three fungal strains after exposure to 500 mg L^{-1} of MMA(V) for 15 days. A, B, and C: XRF images for SM-12F1, SM-12F4, and CZ-8F1; D, XANES spectra of four references and three fungal strains.

ACKNOWLEDGEMENTS

The authors thank for financial support from the National Foundation of Natural Science of China, Grant No. 41671328, and the Yong Elite Scientist Sponsorship Program by the China Association for Science and Technology.

REFERENCES

Su, S.M., Zeng, X.B., Li, L.F., Duan, R., Bai, L.Y., Li, A.G., Wang, J. & Jiang, S. 2012. Arsenate reduction and methylation in the cells of *Trichoderma asperellum* SM-12F1, *Penicillium janthinellum* SM-12F4, and *Fusarium oxysporum* CZ-8F1 investigated with X-ray absorption near edge structure. *J. Hazard. Mater.* 243: 364–367.

Su, S.M., Zeng, X.B., Bai, L.Y., Wang, Y.N., Zhang, L.L., Li, M.S. & Wu, C.X. 2017. Concurrent methylation and demethylation of arsenic by fungi and their differential expression in the protoplasm proteome. *Environ. Pollut.* 225: 620–657.

Yan, Y., Ye, J., Xue, X.M. & Zhu, Y.G. 2015. Arsenic demethylation by a C·As lyase in cyanobacterium *Nostoc* sp. PCC 7120. *Environ. Sci. Technol.* 49: 14350–14358.

Environmental Arsenic in a Changing World –
Zhu, Guo, Bhattacharya, Ahmad, Bundschuh & Naidu (Eds)
ISBN 978-1-138-48609-6

Response of soil microbial communities to elevated antimony and arsenic contamination indicates the relationship between the innate microbiota and contaminant fractions

W. Sun[1], B.Q. Li[1], Z.X. Xu[1], E.Z. Xiao[2] & T.F. Xiao[2]

[1]*Guangdong Key Laboratory of Agricultural Environment Pollution Integrated Control, Guangdong Institute of Eco-Environmental Science & Technology, Guangzhou, P.R. China*

[2]*Key Laboratory of Water Quality and Conservation in the Pearl River Delta, Ministry of Education, School of Environmental Science and Engineering, Guangzhou University, Guangzhou, P.R. China*

ABSTRACT: We selected two sites in Southwest China with different levels of Sb and As contamination to study interactions among various Sb and As fractions and the soil microbiota, with a focus on the microbial response to metalloid contamination. Comprehensive geochemical analyses and 16S rRNA gene amplicon sequencing demonstrated distinct soil taxonomic inventories depending on Sb and As contamination levels. In addition, metagenomics revealed the potential metabolic pathways of the soil microbial ecosystems.

1 INTRODUCTION

The extensive mining and smelting activities at the Banpo Sb mine, Southwest China, have made this area one of the most severely Sb-contaminated areas in the world. Elevated concentrations of Sb and contamination of both Sb and As have been observed in surface waters, aquatic sediments, and tailing dumps with effects on their bacterial communities (Sun *et al.*, 2016a, 2016b). Therefore, these sites provided excellent natural laboratories to study the microbial response to different concentrations of Sb and As (Xiao *et al.*, 2016a, 2016b). However, the innate terrestrial soil bacterial microbiota of these sites has not yet been examined. In this study, we obtained soil samples from two contaminated sites with various levels of contamination. The aims were to: i) examine the impact of Sb and As contamination on the soil microbial community composition and diversity, ii) elucidate the correlation between soil microbial assemblages with environmental factors, especially those directly linked to Sb and As contamination, and iii) explore the potential metabolic pathways of the soil microbial ecosystems.

2 METHODS/EXPERIMENTAL

The study sites are located in eastern Dushan County, Guizhou Province, southwest China, with different levels of Sb and As contamination. Comprehensive geochemical analyses and 16S rRNA gene amplicon sequencing demonstrated distinct soil taxonomic inventories depending on Sb and As contamination levels. We used Random Forest (RF) and Stochastic Gradient Boosting (SGB) analyses to estimate the contributions of environmental factors and quantify the strengths of connections. Shotgun metagenomic libraries from five soil samples were constructed on an Illumina Hiseq 4000 platform to explore the potential metabolic pathways of the soil microbial ecosystems.

3 RESULTS AND DISCUSSION

3.1 *Correlation between environmental factors and microbial communities*

SGB and RF models were used to interpret the relative importance of environmental variables to the community diversity as expressed by Shannon index. As(V)-C, nitrate, C/N, total N, and Sb(V)-C were

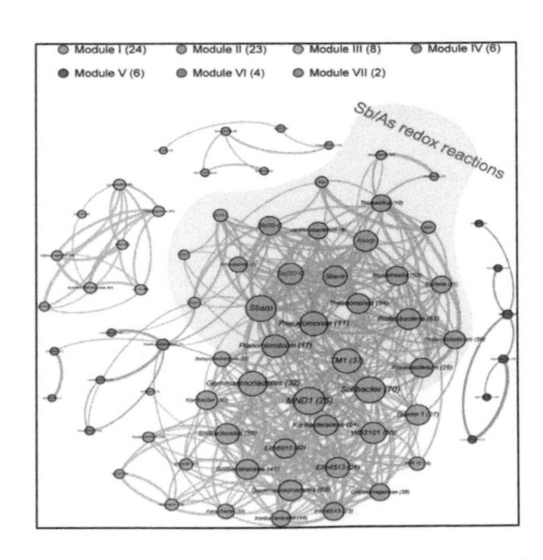

Figure 1. Co-occurrence network showing the correlation between bacterial taxa (OTUs) and contaminant fractions as nodes colored by modularity class.

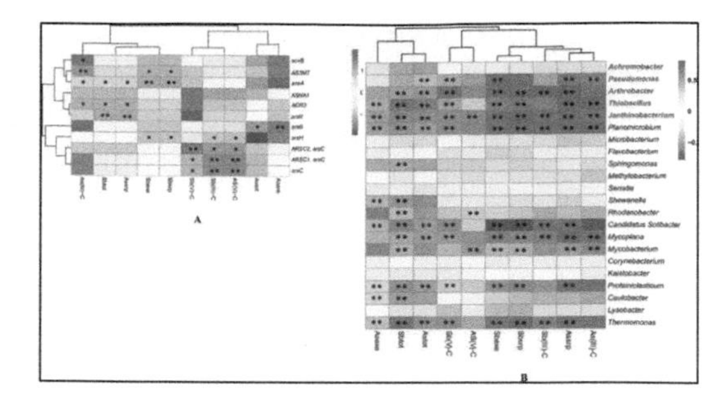

Figure 2. Heatmap of Spearman's rank correlations coefficients between the contaminant fractions and (A) the annotated As-related genes as indicated in the bracket; (B) abundant genera (red genera were identified as "core" genera).

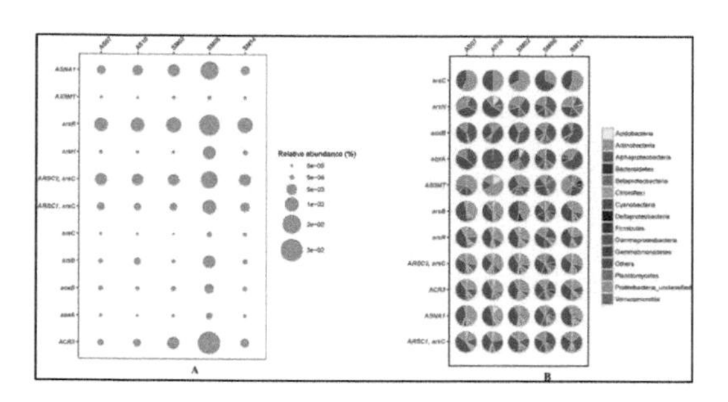

Figure 3. Percentage of reads annotated to Arsenic-related genes for five shotgun metagenomics sequencing libraries.

among the top five factors influencing the Shannon index by SGB. The co-occurrence network further demonstrated correlations between individual phylotypes and contaminant fractions (Fig. 1). Among these identified genera, many contain members relating to Sb and As redox reactions such as *Arthrobacter*, *Janthinobacterium*, *Thiobacillus*, and *Pseudomonas*.

3.2 *Microbial response to contamination at the genus level*

All of the five selected core genera have members with reported capabilities in As redox transformations and Sb resistance. In addition, four core genera were classified in module II of the co-occurrence network, and three core genera were significantly enriched at the more contaminated site. The correlation of these "core" genera with multiple contaminant fractions suggests their selection by Sb/As contamination (Fig. 2).

3.3 *Metabolic functions revealed by metagenomics*

Genes related to *aoxA*, *aoxB* and As(V) reductase (*ARSC1* and *ARSC2*) were detected in all libraries, suggesting possible biogeochemical As redox reactions mediated by the microbial communities in these soils (Fig. 3). Because As and Sb share structural similarities, microorganisms may use similar metabolic pathways to transform both As and Sb. In addition to the presence of As-related genes across all samples, we observed significant positive correlations among As-related genes and various contaminant fractions – not only of As fractions, but those of Sb as well.

4 CONCLUSIONS

Stochastic gradient boosting indicated that citric acid extractable Sb(V) and As(V) contributed 5% and 15%, respectively, to influencing the community diversity. Random forest predicted that low concentrations of Sb(V) and As(V) could enhance the community diversity but generally, the Sb and As contamination impairs microbial diversity. Co-occurrence network analysis indicated a strong correlation between the indigenous microbial communities and various Sb and As fractions. A number of taxa were identified as core genera due to their elevated abundances and positive correlation with contaminant fractions (total Sb and As concentrations, bioavailable Sb and As extractable fractions, and Sb and As redox species). Shotgun metagenomics indicated that Sb and As biogeochemical redox reactions may exist in contaminated soils. All these observations suggest the potential for bioremediation of Sb- and As-contaminated soils.

ACKNOWLEDGEMENTS

This research was funded by the Public Welfare Foundation of the Ministry of Water Resources of China (201501011), the High-level Leading Talent Introduction Program of GDAS (2016GDASRC-0103), the Natural Science Foundation of China (41103080; 41473124; 41420104007), the Opening Fund of the State Key Laboratory of Environmental Geochemistry (SKLEG2016907), and SPICC Program (2016GDASPT-0105).

REFERENCES

Sun, W., Xiao, E., Dong, Y., Tang, S, Krumins, V., Ning, Z., Sun, M., Zhao, Y., Wu, S. & Xiao, T. 2016a. Profiling microbial community in a watershed heavily contaminated by an active antimony (Sb) mine in Southwest China. *Sci. Total Environ.* 550: 297–308.

Sun, W., Xiao, E., Kalin, M., Krumins, V., Dong, Y., Ning, Z., Liu, T., Sun, M., Zhao, Y., Wu, S., Mao, J. & Xiao, T. 2016b. Remediation of antimony-rich mine waters: assessment of antimony removal and shifts in the microbial community of an onsite field-scale bioreactor. *Environ. Pollut.* 215: 213–222.

Xiao, E., Krumins, V., Dong, Y., Xiao, T., Ning, Z., Xiao, Q. & Sun, W. 2016a. Microbial diversity and community structure in an antimony-rich tailings dump. *Appl. Microbiol. Biotechnol.* 100(7): 1–13.

Xiao, E., Krumins, V., Tang, S., Xiao, T., Ning, Z., Lan, X. & Sun, W. 2016b. Correlating microbial community profiles with geochemical conditions in a watershed heavily contaminated by an antimony tailing pond. *Environ. Pollut.* 215: 141–153.

Environmental Arsenic in a Changing World –
Zhu, Guo, Bhattacharya, Ahmad, Bundschuh & Naidu (Eds)
ISBN 978-1-138-48609-6

Transformation of roxarsone by *Enterobacter sp.* CZ-1 isolated from an arsenic-contaminated paddy soil

K. Huang, F. Gao & F.-J. Zhao
College of Resources and Environmental Sciences,
Nanjing Agricultural University, Nanjing, P.R. China

ABSTRACT: Roxarsone [Rox(V)] is widely used as a feed additive in poultry industries. This arsenic-containing compound may be degraded by microbes to release toxic inorganic arsenic in the environment. To date, most of the studies of microbial mediated Rox(V) degradation have focused on anaerobic microorganisms. Here, we isolated a pure cultured aerobic Rox(V)-degrading bacterial strain, CZ-1, from an arsenic-contaminated paddy soil in China. On the basis of 16S rRNA gene sequence, strain CZ-1 was classified as a member of genus *Enterobacter*. This strain could efficiently degrade Rox(V), converting all of 10 μM Rox(V) within 24 h in liquid culture. Four metabolites including n-acetyl-4-hydroxy-m-arsanilic acid (N-AHPAA[V]), 3-amino-4-hydroxyphenylarsonic acid (3-AHPAA[V]), arsenite (As[III]) and an unknown arsenic compound were detected and identified by HPLC-ICP-MS and HPLC-LC-MS. N-AHPAA(V) was the main product, likely to be formed from acetylation of 3-AHPAA(V). Based on these results, a novel degradation pathway of Rox(V) by *Enterobacter*. sp CZ-1 is proposed.

1 INTRODUCTION

The extensive use of the organoarsenical 3-nitro-4-hydroxybenzene arsonic acid (roxarsone; Rox[V]) as a feed additive in the poultry industry can lead to increasing arsenic contamination of soil and water environments (Nachman *et al.*, 2005; O'Conner *et al.*, 2005). Previous studies using chicken feces and sewage sludge show that microbes play significant roles in degrading Rox(V) in the environment (Cortinas *et al.*, 2006; Stolz *et al.*, 2007). Stolz *et al.* (2007) reported that *Clostridium* species were mainly responsible for Rox(V) degradation under anaerobic conditions in their enrichment cultures obtained from chicken litter. They also isolated an anaerobic Rox(V)-degrading bacterium *Clostridium* OhILAs (Stolz *et al.*, 2007). Two other anaerobic bacteria, *Shewanella oneidensis* MR-1 and *Shewanella putrefaciens* CN32, have also been found to possess the Rox(V) degradation ability (Chen *et al.*, 2016; Han *et al.*, 2017). Roxarsone biodegradation activity of an aerobic microbial consortium has also been demonstrated (Guzman-Fierro *et al.*, 2015), but to date no strains of aerobic microbes capable of Ros(V) degradation have been isolated. The degradation pathway under aerobic conditions remains unknown. In this study, we isolated and characterized an aerobic Rox(V)-degrading strain of *Enterobacter*, named CZ-1, and propose a novel biodegradation pathway.

2 MATERIAL AND METHODS

2.1 *Isolation and characterization of a Rox(V)-degrading bacterium*

One strain with a Rox(V)-degrading ability was isolated from an As-contaminated paddy soil and named CZ-1. Its ability for Rox(V) degradation was determined quantitatively. The cell morphology was observed by SEM (S-3000N, Hitachi). The 16S rRNA gene was amplified by PCR and sequenced in Genscript Co. Ltd. The neighbor-joining phylogenetic tree was constructed by MEGA 5.0.

2.2 *Chemical analysis*

The metabolites produced in Rox(V) degradation were analyzed by HPLC (series 200; PerkinElmer)-ICP-MS (NexION 300D; PerkinElmer) using an anion-exchange column (Hamilton PRP-X100; 250 mm by 4.6 mm) eluted with the mobile phase [A: 50 mM $(NH_4)_2HPO_4$; B: water] in gradient elution and LC-MS consisting of an Agilent 1200 series HPLC coupled to an Agilent 6410B triple quadrupole mass spectrometer using an Agilent Eclipse Plus C18 column (2.1 × 150 mm, 3.5 μm) with the mobile phase [95% solution A (0.1% formic acid) and 5% solution B (acetonitrile)] at a flow rate of 0.2 mL min^{-1}.

3 RESULTS AND DISCUSSION

3.1 *Strain isolation and characterization*

One strain, designated strain CZ-1, capable of degrading Rox(V) was isolated from an arsenic-contaminated paddy soil in Chenzhou city, Hunan province, China. After 24 h of inoculation, strain CZ-1 was able to degrade all of 10 μM Rox(V) in $ST10^{-1}$ medium. Cells were Gram-negative, aerobic, and motile. A phylogenetic tree based on the 16S rRNA gene sequence of strain CZ-1 was constructed. Strain CZ-1 was closely related to *Enterobacter* sp. X72, with a sequence similarity score of 99%, suggesting that strain CZ-1 is an *Enterobacter* species.

3.2 *Identification of the metabolites during Rox(V) degradation*

The products of the degradation of Rox(V) by strain CZ-1 was first analyzed by HPLC-ICP-MS. The HPLC-ICP-MS analysis showed that nearly all of Rox(V) (10 μM) in the culture medium was converted to four As-containing metabolites designated compounds 1, 2, 3 and 4. Compound 4 was found to be the main product. Compounds 1, 2 and 4 were preliminarily identified as As(III), 3-AHPAA(V) and N-AHPAA(V), respectively, while compound 3 was still unknown that did not correspond to several common phenylarsine compounds such as 4-hydroxyphenylarsonic acid (HPAA[V]), 4-aminophenylarsonic acid (4-APAA[V]), 2-aminophenylarsonic acid (2-APAA[V]) and carbarsone. When treated with 30% H_2O_2, compound 1 was transformed to As(V), providing further evidence that compound 1 was As(III). In addition, the LC-MS analysis demonstrated the presence of 3-AHPAA(V) and N-AHPAA(V), confirming the results from the HPLC-ICP-MS analysis.

On the basis of above results, we propose a novel transformation pathway of Rox(V) mediated by strain CZ-1, which is different from the putative degradation pathways in *Alkaliphilus oremlandii* strain OhILAs (Thomas *et al.*, 2014) and *Shewanella putrefaciens* strain CN32 (Han *et al.*, 2017). In the new pathway, the nitro group in Rox(V) is first reduced to amino group to form the compound 3-AHPAA(V), and the latter is then acetylated into N-AHPAA(V). N-AHPAA (V) was found to be stable.

4 CONCLUSIONS

To our best knowledge, strain CZ-1 is the first aerobic bacterium with a Rox(V)-degrading ability isolated from an arsenic-contaminated paddy soil. A novel transformation pathway of Rox(V) by this strain is proposed.

ACKNOWLEDGEMENTS

This study was supported by the Natural Science Foundation of China (grant 41703069 and 41330853), the Natural Science Foundation of Jiangsu Province (grant BK 20170723).

REFERENCES

Chen, G.W., Ke, Z.C., Liang, T.F., Liu, L. & Wang, G. 2016. *Shewanella oneidensis* MR-1-induced Fe (III) reduction facilitates roxarsone transformation. *PLoS One* 11(4): e0154017.

Cortinas, I., Field, J.A., Kopplin, M., Garbarino, J.R., Gandolfi, A.J. & Sierra-Alvarez, R. 2006. Anaerobic biotransformation of roxarsone and related N-substituted phenylarsonic acids. *Environ. Sci. Technol.* 40(9): 2951–2957.

Guzman-Fierro, V.G., Moraga, R., Leon, C.G., Campos, V.L., Smith, C. & Mondaca, M.A. 2015. Isolation and characterization of an aerobic bacterial consortium able to degrade roxarsone. *Int. J. Environ. Sci. Technol.* 12(4): 1353–1362.

Han, J.C., Zhang, F., Cheng, L., Mu, Y., Liu, D. F., Li, W. W. & Yu H.Q. 2017. Rapid release of arsenite from roxarsone bioreduction by exoelectrogenic bacteria. *Environ. Sci. Technol. Lett.* 4(8): 350–355.

Nachman, K.E., Graham, J.P., Price, L.B. & Silbergeld, E.K. 2005. Arsenic: a roadblock to potential animal waste management solutions. *Environ. Health Perspect.* 113(9): 1123–1124.

O'Conner, R., O'Conner, M., Irgolic, K., Sabrsula, J., Gurleyuk, H., Brunette, R., Howard, C., Garcia, J., Brien, J., Brien, J. & Brien, J. 2005. Transformation, air transport, and human impact of arsenic in poultry litter. *Environ. Forensics* 6(1): 83–89.

Stolz, J. F., Perera, E., Kilonzo, B., Kail, B., Crable, B., Fisher, E., Ranganathan, M., Wormer, L. & Basu, P. 2007. Biotransformation of 3-nitro-4-hydroxybenzene arsonic acid (Roxarsone) and release of inorganic arsenic by *Clostridium species*. *Environ. Sci. Technol.* 41(3): 818–823.

Thomas, J.A., Chovanec, P., Stolz, J.F. & Basu, P. 2014. Mapping the protein profile involved in the biotransformation of organoarsenicals using an arsenic metabolizing bacterium. *Metallomics* 6(10): 1958–1969.

Environmental Arsenic in a Changing World –
Zhu, Guo, Bhattacharya, Ahmad, Bundschuh & Naidu (Eds)
ISBN 978-1-138-48609-6

Functional microbial communities in high arsenic groundwater

P. Li, Z. Jiang & Y.X. Wang
China University of Geosciences, Wuhan, P.R. China

ABSTRACT: Microbial functional potential in high arsenic (As) groundwater remains largely unknown. In this study, the microbial community functional composition of high arsenic groundwater samples from Hetao Basin, China, was investigated using integrated methods including DGGE, gene clone library, qPCR, and GeoChip. The results showed that As-related genes (*arsC* and *arrA*), sulfate-related genes (*dsrA* and *dsrB*), and methanogen genes (*mcrA*) were correlated with As, SO_4^{2-}, or CH_4 concentrations in groundwater, respectively. Arsenic, total organic content, SO_4^{2-}, NH_4^+, oxidation-reduction potential (ORP), and pH were important factors shaping the functional microbial community structure. Alkaline and reducing conditions associated with microbially-mediated geochemical processes could be linked to As enrichment in groundwater.

1 INTRODUCTION

Arsenic (As) in groundwater is a serious environmental issue due to its widespread distribution and high toxicity, which threatens the health of millions of people worldwide. Previous studies showed that As mobilization and transformation could be ascribed to the complex interactions between microbes and geochemical processes and that microbes play key roles in driving the biogeochemical cycle in high As groundwater aquifers (Li *et al.*, 2014, 2017; Wang *et al.*, 2015). The integrated role of various functional microbial populations on As mobilization in groundwater remained unknown. To investigate the functional potential microbial community structures in high As groundwater aquifers, we used a series of technologies including DGGE, gene clone library, qPCR, and GeoChip, which targets functional gene families involved in As, C, and S geochemical cycling.

2 METHODS

2.1 *Sample collection and geochemistry measurements*

Groundwater samples were collected from tubewells with a depth range of 20–30 m from Hetao Basin, Inner Mongolia. Microbial samples were collected by on-line filtering of 5–10 L water through 0.22-μm filters. Water pH, electrolytic conductivity (EC), oxidation-reduction potential (ORP), total dissolved solids (TDS), dissolved oxygen (DO), ammonium, sulfide, ferrous iron (Fe(II)), and total iron were measured in the field using portable hand-held meters. The measurements of cation, anion, dissolved organic carton (DOC), total carbon (TC), total nitrogen (TN), CH_4, and As species were performed in the laboratory.

2.2 *DNA extraction and functional gene detection*

DNA for all groundwater samples was extracted using FastDNA spin kits for soil according to the manufacturer's manual. DNA concentrations were quantified by PicoGreen, using a FLUO star Optima instrument. The *dsrB* gene structure were investigated using denaturing gradient gel electrophoresis (DGGE). The functional communities of *mcrA* genes and arsenic related genes including *aioA*, *arsC,* and *arrA* were detected with gene clone library method. N, C, S and As related genes were qualified by qPCR and geochip. For geochip detection, ~500 ng of DNA was labeled with Cy. The labeled DNA was re-suspended in hybridization solution, and then hybridized in an Agilent hybridization oven at 67°C for 24 h. After hybridization, the arrays were scanned with a NimbleGen MS200 Microarray Scanner. The images were extracted by the Agilent Feature Extraction program. Poor quality spots with a signal-to-noise ratio of less than 2.0 were removed before statistical analysis. The positive signals were normalized within each sample and across all samples.

2.3 *Statistical analysis*

Data were further analyzed by the Vegan package in R 3.1.1 (http://www.r-project.org/). Hierarchical clustering was performed with CLUSTER 3.0 were visualized in TREEVIEW. Canonical correspondence analysis (CCA) were performed to link microbial communities to environmental variables.

3 RESULTS AND DISCUSSION

3.1 *Geochemistry*

Groundwater samples could be divided into well-defined high and low As groups based on geochemistry. Samples with high As concentrations generally had low concentrations of sulfate, negative ORP, and relatively high concentrations of As(III), Fe(II), CH_4, NH_4^+, and TOC contents. These geochemical characteristics indicated that alkaline and reducing conditions could be linked to natural As enrichment in groundwater.

3.2 *Functional microbial communities*

Analysis of *aioA* gene clone library found that As-oxidizing bacteria in high As groundwater of Hetao Plain belonged to *Rhodoferax ferrireducens*, *Leptothrix* sp., *Acidovorax* sp., *Pseudomonas*, *Acinetobacter*, *Xanthobacter autotrophicus*, *Bradyrhizobium* sp., *Bosea* sp., *Aminobacter* sp. and *Nitrobacter hamburgensis*. As-reducing populations in high As groundwater mainly belonged to *Geobacter* sp., *Desulfosporosinus* sp., *Chrysiogenes* and *Desulfurispirillum*. Most of these populations were found capable of dissimilatory arsenate reducing or sulfur oxidizing in previously studies.

Clone library result of *mcrA* gene indicated methanogens in high As groundwater were mainly composed of *Methanomicrobia*, *Methanobacteria* and the uncultured group, among which 83% clones belonged to *Methanomicrobia* including *Methanomicrobiales* and *Methanosarcinales*.

Sulfate reducing bacteria in high As groundwater were mainly dominated by *Desulfotomaculum*, *Desulfobulbus*, *Desulfosarcina* and *Desulfobacca*. Most of the functional populations were found capable of both arsenate reducing and sulfur reducing.

3.3 *The abundance of functional genes*

The number of detected As related genes *arsC* ($r = 0.462$, $p < 0.01$) and *arrA* ($r = 0.486$, $p < 0.01$) have positive correlations with As concentration. More *arsC*/*arsB* genes were detected than the other genes, such as *arrA* (around 5–8 folds), indicating that the most common detoxification mechanism was reduction of As(V) to As(III) rather than respiratory arsenate reduction in the high arsenic groundwater aquifers. Besides, As detoxification genes *arsC* and *arsB*, and the respiratory arsenate (As(V)) reductase gene *arrA* had higher relative abundances in the high As group than low As group.

There were positive correlations between As and sulfur reduction genes (*dsrB* and As: $r = 0.327$, $p < 0.05$). Positive correlations were also observed between methanogenesis genes and As and methane concentrations (*mcrA*: $r = 0.549$, $p < 0.05$; $r = 0.314$, $p < 0.05$, respectively). These results indicated that sulfur reduction and methanogenesis might be an important metabolic process and accelerate As release and accumulation in high arsenic aquifers.

3.4 *Functional microbial community structure in relation to geochemistry*

Significant differences were observed in functional microbial structure between low and high As groups ($r = 0.4776$, $p = 0.007$). Six environmental variables, As_{Tot}, SO_4^{2-}, NH_4^+, pH, ORP, and TOC were the most significant environmental variables shaping the microbial community structure ($p = 0.021$). Group 1 (SO_4^{2-}, ORP) explained 22.63% ($p = 0.010$), and group 2 (As_{Tot}, TOC and NH_4^+) explained 24.98% ($p = 0.010$). The pH (group 3) independently explained 17.64% ($p = 0.005$) of the observed variation. About 58.80% of the community functional variation remained unexplained by the above selected variables.

4 CONCLUSIONS

Our results reveal that environmental variables account for the majority of variation in microbial functional potential and provides evidence for possible links between As, S, and C related microbial functional gene abundances and As geochemistry. Alkaline and reducing conditions associated with microbially-mediated geochemical processes could be linked to As enrichment in groundwater in this study area.

ACKNOWLEDGEMENTS

This work was financially supported by National Natural Science Foundation of China (grant numbers 41372348, 41521001).

REFERENCES

Li, P., Li, B., Webster, G., Wang, Y., Jiang, D., Dai, X., Jiang, Z., Dong, H. & Wang, Y. 2014. Abundance and diversity of sulfate-reducing bacteria in high arsenic shallow aquifers. *Geomicrobiol. J.* 31(9): 802–812.

Li, P., Jiang, Z., Wang, Y., Deng, Y., Van Nostrand, J.D., Yuan, T., Liu, H., Wei, D. & Zhou, J. 2017. Analysis of the functional gene structure and metabolic potential of microbial community in high arsenic groundwater. *Water Res.* 123: 268–276.

Wang, Y., Li, P., Dai, X., Zhang, R., Jiang, Z., Jiang, D. & Wang, Y. 2015. Abundance and diversity of methanogens: potential role in high arsenic groundwater in Hetao plain of inner Mongolia, China. *Sci. Total Environ.* 515: 153–161.

2.4 Molecular mechanisms of plant arsenic uptake

Environmental Arsenic in a Changing World –
Zhu, Guo, Bhattacharya, Ahmad, Bundschuh & Naidu (Eds)
ISBN 978-1-138-48609-6

Heterologous expression of PvACR3;1 decreased arsenic accumulation in plant shoots

Y.S. Chen, Y. Cao, C.Y. Hua, M.R. Jia, J.W. Fu, Y.H. Han, X. Liu & L.Q. Ma
State Key Lab of Pollution Control and Resource Reuse, School of the Environment, Nanjing University, Jiangsu, P.R. China

ABSTRACT: Arsenic (As) is a toxic carcinogen so it is crucial to decrease As accumulation in crops to reduce its risk to human health. Arsenite (As(III)) antiporter ACR3 protein is critical for As metabolism in organisms, but it is lost in flowering plants. Here, a novel ACR3 gene from As hyperaccumulator *Pteris vittata*, PvACR3;1, was cloned and expressed in *Saccharomyces cerevisiae* (yeast), *Arabidopsis thaliana* (model plant), and *Nicotiana tabacum* (tobacco). Yeast experiments showed that PvACR3;1 functioned as an As(III)-antiporter to mediate As(III) efflux to an external medium. At 5 μM As(III), PvACR3;1 transgenic Arabidopsis accumulated 14–29% higher As in the roots and 55–61% lower As in the shoots compared to WT control, showing lower As translocation. Besides, transgenic tobacco under 5 μM As(III) or As(V) (arsenate) also showed similar results, indicating that expressing PvACR3;1 gene increased As retention in plant roots. Moreover, observation of PvACR3;1-green fluorescent protein fusions in transgenic Arabidopsis showed that PvACR3;1 protein localized to the vacuolar membrane, indicating that PvACR3;1 mediated As(III) sequestration into vacuoles, consistent with increased root As. Thus, our study provides a potential strategy to limit As accumulation in plant shoots, shedding light on engineering low-As crops to improve food safety.

1 INTRODUCTION

In yeast, As(III) transporter ACR3 (Arsenic Compounds Resistance 3) is localized to the plasma membrane to export As(III) out of the cell (Wysocki *et al.*, 1997). Its homologues exist in plants including moss, lycophytes, ferns, and gymnosperms, but not angiosperms (Indriolo *et al.*, 2010). In As-hyperaccumulator *Pteris vittata*, two ACR3 homologues, PvACR3 and PvACR3;1, were reported, with PvACR3 being localized to the vacuolar membrane and likely efflux of As(III) into the vacuole for sequestration in *P. vittata* (Indriolo *et al.*, 2010). However, in transgenic Arabidopsis, PvACR3 localizes to the plasma membrane and its heterologous expression increases As(III) efflux and As(III) translocation (Chen *et al.*, 2013; Wang *et al.*, 2017). Although PvACR3;1 was reported by Indriolo *et al.* (2010), it was not investigated in their study so its function is unclear. Here, in this study, we successfully cloned the PvACR3;1 full length coding sequence (CDS) from *P. vittata* and tested its function following expression in transgenic yeast and in transgenic plants.

2 METHODS/EXPERIMENTAL

2.1 *Yeast growth assays*

PvACR3;1 CDS was cloned into the GAL1 promoter cassette of yeast vector pAG413GAL-ccdB by recombination. Then it was transformed into Δ*acr*3 mutant with BY4741 background. Yeast transformants that expressing PvACR3;1 were then selected for growth assays.

2.2 *Transgenic plant generation, plant growth and As determination*

PvACR3;1 CDS was cloned into the CaMV 35S promoter cassette of plant expression vector pSN1301 by recombination. Agrobacterium strain C58 was transformed with the binary vector by electroporation. The Agrobacterium culture was used to transform *Arabidopsis thaliana* and tobacco, respectively. For analysis of As accumulation in transgenic plants, plants were cultivated under different As(III) or As(V) treatments. Total As concentrations were determined by inductively coupled plasma mass spectrometry (ICP-MS).

3 RESULTS AND DISCUSSION

3.1 *PvACR3;1 encodes a functional As(III) antiporter in yeast*

Expression of PvACR3;1 enhanced the tolerance of Δ*acr*3 to both As(III) and As(V) and effectively suppressed As-sensitive phenotype. Arsenic determination showed that expressing PvACR3;1 significantly reduced As accumulation in the yeast after growing under As(III) or As(V). Considering As(V) can be reduced to As(III) in yeast cells, we concluded

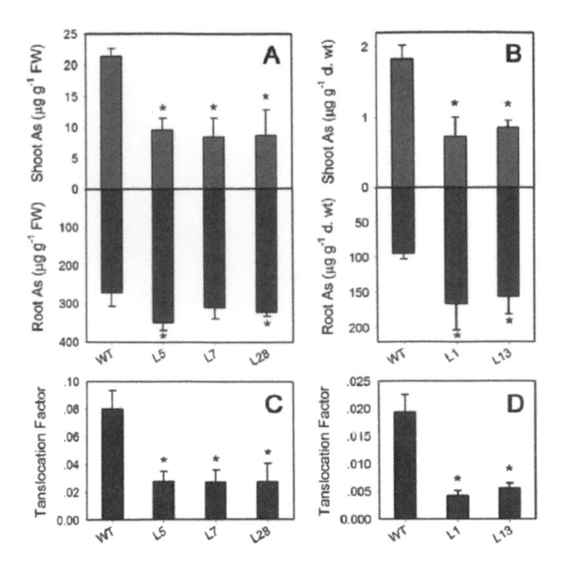

Figure 1. Arsenic (As) accumulation and translocation in PvACR3;1 transgenic plants under 5 μM arsenite (As(III)) treatments.

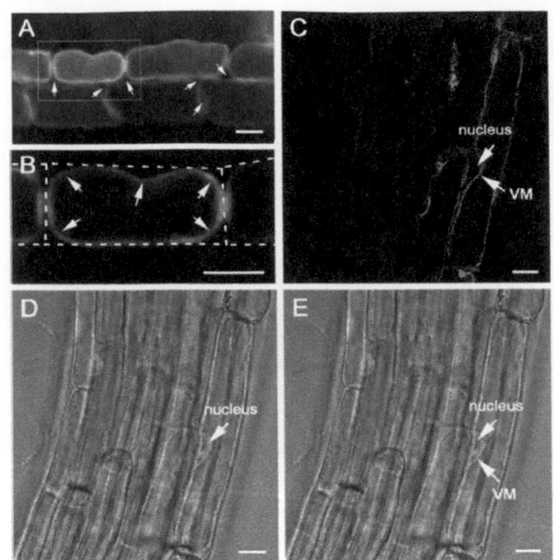

Figure 2. Subcellular localization of PvACR3;1 in plants through the observation of GFP fluorescence of PvACR3;1-GFP fusions.

that the obtained PvACR3;1 was a functional As(III) antiporter, which played an important role in As(III) efflux to the external medium across the yeast plasma membrane.

3.2 *Expressing PvACR3;1 decreased shoot As accumulation in plants under As(III) exposure*

PvACR3;1 gene was transformed into model plant Arabidopsis and tobacco and three transgenic Arabidopsis lines (L5, L7, and L28) and two transgenic tobacco lines (L1 and L13) were obtained. At 5 μM As(III), both transgenic Arabidopsis and transgenic tobacco lines accumulated higher As in the roots and lower As in the shoots, compared to the WT controls (Fig. 1A,B). As a result, As translocation dropped by ~66% in transgenic Arabidopsis L5, L7, and L28 and by 71–78% in transgenic tobacco L1 and L13, compared to WT controls (Fig. 1C,D). These results showed that heterologous expression of PvACR3;1 decreased As(III) translocation in plants, probably by retaining As(III) in plant roots.

3.3 *Subcellular localization of PvACR3;1 in plants*

The subcellular location of PvACR3;1 in plants is critical for its physiological function. Thus, vector expressing (C-terminal) PvACR3;1-GFP fusions was transformed into transgenic Arabidopsis to visualize GFP fluorescence. GFP signal of PvACR3;1-GFP showed a clear localization of GFP to the indented region of the rectangular cell (Fig. 2A,B). In addition, the PvACR3;1-GFP also showed a clear localization of GFP delineating the nucleus at the root cell, unambiguously demonstrating a vacuolar location (Fig. 2C–E), indicating that PvACR3;1 may mediate As(III) sequestration in transgenic plant roots, thus increasing As accumulation in the roots and subsequently decreasing As(III) xylem loading and As translocation to plant shoots.

4 CONCLUSIONS

PvACR3;1 localized to vacuolar membrane when heterologous expressed in plants and mediated As(III) sequestration into vacuoles in the roots, therefore reducing As accumulation in plant shoots. Thus, our work provides a potential strategy to decrease As accumulation in plant shoots for food safety.

ACKNOWLEDGEMENTS

This work was supported by Jiangsu Provincial Natural Science Foundation (Grant No. BK20160649), and the National Natural Science Foundation of China (Grant No. 21707068, 21637002).

REFERENCES

Chen, Y., Xu, W., Shen, H., Yan, H., He, Z. & Ma, M. 2013. Engineering arsenic tolerance and hyperaccumulation in plants for phytoremediation by a *PvACR3* transgenic approach. *Environ. Sci. Technol.* 47(16): 9355–9362.

Indriolo, E., Na, G., Ellis, D., Salt, D.E. & Banks, J.A. 2010. A vacuolar arsenite transporter necessary for arsenic tolerance in the arsenic hyperaccumulating fern *Pteris vittata* is missing in flowering plants. *Plant Cell* 22(6): 2045–2057.

Wang, C., Na, G., Bermejo, E.S., Chen, Y., Banks, J.A., Salt, D.E. & Zhao, F.J. 2018. Dissecting the components controlling root-to-shoot arsenic translocation in *Arabidopsis thaliana*. *New Phytol.* 217(1):206–218.

Wysocki, R., Bobrowicz, P. & Ulaszewski, S. 1997. The *Saccharomyces cerevisiae* ACR3 gene encodes a putative membrane protein involved in arsenite transport. *J. Biol. Chem.* 272(48): 30061–30066.

Environmental Arsenic in a Changing World –
Zhu, Guo, Bhattacharya, Ahmad, Bundschuh & Naidu (Eds)
ISBN 978-1-138-48609-6

Exploration of biochemical properties of soil and groundwater in arsenic affected blocks of Murshidabad district and isolation of potential arsenic resistant bacteria

S. Ahmed, A. Basu, D. Mandal, I. Saha & M. Biswas
Sripat Singh College, Jiaganj, Murshidabad, India

ABSTRACT: Arsenic is one of the major contaminants of soil and groundwater, responsible for a number of health hazards. Various blocks of Murshidabad district show arsenic concentration above the maximum permissible limit in soil and water samples. In the present study, we have focused on some highly arsenic contaminated regions of Murshidabad district. We have analyzed various biochemical parameters of soil and water samples of this district. The soils of these regions show high alkalinity and the groundwater samples also exhibits a basic pH. The total dissolved solids of these water samples varied from 200 mg L^{-1} to 300 mg L^{-1}. Also, the bacterial load in the water samples was extremely high. Since, groundwater is a source of drinking water in the blocks of Murshidabad district, consumption of such high amount of total dissolved solids would take a toll on the detoxifying and excretory system of the body i.e., on the hepatic and renal systems. The uncharacterized microorganisms in the soil could be fecal coliforms or other pathogenic bacteria, and their consumption would have serious health consequences. After serial dilutions (10^6 to 10^9 folds) of the groundwater and the soil samples, we could still isolate some bacteria thriving in these samples. Since, the soil and groundwater of these regions are highly arsenic contaminated, bacterial colonies isolated after serial dilutions could be potentially arsenic resistant bacteria.

1 INTRODUCTION

Arsenic toxicity is one of the serious problems both from national and global perspective. Arsenic is a toxic metal, which contaminates soil and groundwater (Hendryx, 2009; Huq *et al.*, 2006; Pais & Benton, 1997). In India arsenic contamination of soil and groundwater is a severe problem in lower Gangetic plain and Ganga-Brahmaputra deltaic region (Rahman *et al.*, 2005). U.S. Environmental Protection Agency prescribed the maximum permissible limit of arsenic in drinking water to be 10 $\mu g\,L^{-1}$ (EPA, 2006). However, World Health Organization notified that in absence of any alternate source of drinking water, the maximum permissible limit of arsenic would be 50 $\mu g\,L^{-1}$. In countries like Canada and Australia, this permissible limit is 5 $\mu g\,L^{-1}$ and 7 $\mu g\,L^{-1}$, respectively. Arsenic is released in the soil by natural biogeochemical cycles, from there it leaches into the groundwater by natural processes. Arsenic enters the body through water and food (Kapaj *et al.*, 2006). There are evidences that staple food crops like rice have accumulated arsenic. This leads to biomagnification of arsenic within the food chain. West Bengal is one of the severely arsenic affected states of India. Within West Bengal nine districts are affected by arsenic toxicity. Murshidabad is one of the severely arsenic affected districts of West Bengal. The river Ganga separates it from Bangladesh. Eastern bank of Bhagirathi river constitutes more arsenic contaminated blocks of Murshidabad district (64.7% above 10 $\mu g\,L^{-1}$ and 32.5% above 50 $\mu g\,L^{-1}$) compared to the western bank of the river (30.1% above 10 $\mu g\,L^{-1}$ and 11.7% above 50 $\mu g\,L^{-1}$). In this study, our main aim is to analyze the conditions of soil and groundwater in highly arsenic contaminated regions of Murshidabad district (Fig. 1). For this purpose, we have qualitatively and quantitatively checked various parameters of soil & groundwater samples from these affected regions. Also, we have attempted the isolation of potential arsenic resistant bacteria from these regions.

2 METHODS/EXPERIMENTAL

2.1 *Measurement of pH of soil and groundwater samples*

Soil samples were diluted to obtain a concentration of 1 mg soil/1 mL distilled water. Soil samples were mixed well with water by constant stirring so, that the soluble solutes i.e. ions and electrolytes come in the polar phase. The particulate matter in the soil (soil sediments) was allowed to settle down. The pH of the solution was measured at 25°C. Similarly, the pH of the water samples was measured at the same temperature.

2.2 *Estimation of bacterial load in the soil and groundwater samples and isolation of potential arsenic resistant bacteria*

Water samples, and soil samples dissolved in water was plated in a LB-agar plate to estimate the bacterial load in the soil and groundwater samples.

Figure 1. Soil samples collected from a) Mahisasthali village.

Further, serial dilutions of water and soil samples were made and these diluted stocks were plated in LB-agar plates for isolation of single colonies of potential arsenic resistant bacteria.

3 RESULTS AND DISCUSSION

3.1 *Estimation of pH and total dissolved solids in the soil and groundwater samples from severely arsenic affected regions of Murshidabad district*

Water samples were collected from Bhagobangola I, Hariharpara, Chunakhali and Asrampara, which are the regions amongst the severely arsenic affected blocks of Murshidabad district. Soil samples were also collected from Bhagobangola I, Hariharpara and Chunakhali. pH of the soil and water samples was measured at 25°C. Both soil and water samples from the four arsenic contaminated blocks were found to be alkaline in nature. Soil sample of Hariharpara showed the maximum pH (8.37) amongst all the other soil samples. Among the water samples, water sample from Bhagobangola I showed the maximum pH (7.60). All the soil samples showed pH above 8. Total dissolved solids (TDS) in the water samples were found to be very high. TDS in water samples from Hariharpara was 200 mg L^{-1}. The other three sources showed TDS concentration of 300 mg L^{-1}.

3.2 *Isolation of potential arsenic resistant bacteria from the soil and groundwater samples*

Next, with an aim to isolate arsenic resistant bacteria from these samples, soil and water samples from aforementioned arsenic affected blocks were plated with and without dilutions on LB agar plate. Huge number of bacterial colonies was observed when undiluted soil and water samples were plated. As PHED had marked these blocks as severely arsenic contaminated, we wanted to identify arsenic resistant microorganisms from soil and water samples of these blocks. Therefore, we isolated bacteria from these soil and water samples by serially diluting the samples and spreading them on LB agar plate. Single bacterial colonies could be observed in the plates. Considering the huge arsenic toxicity of the aforementioned regions, these colonies could be categorized as colonies of potential arsenic resistant bacteria.

4 CONCLUSIONS

We have concentrated our study on four blocks of Murshidabad district. These were Bhagobangola I, Hariharpara, Chunakhali and Asrampara. All of these blocks had been marked by PHED (Public Health and Engineering Department) as severely arsenic affected blocks. We collected soil and water samples from these areas and measured the pH and total dissolved solids (TDS). Both soil and water samples of these areas were found to be very alkaline. Soil & water samples from the above-mentioned arsenic affected blocks were plated with and without dilutions on LB-agar plate. With serial dilutions single bacterial colonies could be obtained on the LB-agar plate. Considering the huge arsenic toxicity of these regions, the soil and water microbiota obtained from these regions could be potentially arsenic resistant. Huge number of bacterial colonies was observed when undiluted soil and water samples were plated. Therefore, people of these blocks were drinking water contaminated with millions of bacteria. As characterization of these bacteria was not done therefore, it could only be assumed that these water samples might be contaminated with toxic bacteria including fecal coliforms and other pathogenic bacteria.

ACKNOWLEDGEMENTS

The authors are thankful to Sripat Singh College for providing laboratory facilities & Department of Biotechnology, Government of West Bengal for the financial assistance.

REFERENCES

EPA. 2006. Drinking Water Requirements for States and Public Water System, EPA.

Hendryx, M. 2009. Mortality from heart, respiratory and kidney disease in coal mining areas of Appalachia. *Int. Arch. Occup. Environ. Health* 82(2): 243–249.

Huq, S.M.I., Joardar, J.C., Parvin, S., Correll, R. & Naidu, R. 2006. Arsenic contamination in food chain: transfer of arsenic into food materials through groundwater irrigation. *J. Health Popul. Nutr.* 24(3): 305–316.

Kapaj, S, Peterson, H, Liber, K & Bhattacharya, P. 2006. Health effects from chronic arsenic poisoning – a review. *J. Environ. Sci. Health A* 41(10): 2399–2428.

Pais, I.J & Benton, J.R.J. 1997. *The Hand Book of Trace Elements*. Publishing by: St. Luice Press, Boca Raton, FL.

Rahman, M.M., Sengupta, M.K., Ahamed, S., Lodh, D., Das, B., Hossain, M.A., Nayak, B., Mukherjee, A., Chakraborti, D., Mukherjee, S.C., Pati, A., Saha, K.S., Palit, S.K., Kaies, I., Barua, A.K. & Asad, K.A. 2005. Murshidabad-one of the nine groundwater arsenic-affected districts of West Bengal, India. part I: magnitude of contamination and population at risk. *Clin. Toxicol.* 43(7): 823–834.

Environmental Arsenic in a Changing World –
Zhu, Guo, Bhattacharya, Ahmad, Bundschuh & Naidu (Eds)
ISBN 978-1-138-48609-6

An effective rhizoinoculation restraints arsenic translocation in peanut and maize plants exposed to a realistic groundwater metalloid dose

J.M. Peralta[1], C.N. Travaglia[1], R.A. Gil[2], A. Furlan[1], S. Castro[1] & E.C. Bianucci[1]
[1]*Departamento de Ciencias Naturales, Facultad de Ciencias Exactas, Físico-Químicas y Naturales, Universidad Nacional de Río Cuarto, Córdoba, Argentina*
[2]*Laboratorio de Espectrometría de Masas – Instituto de Química de San Luis (CCT-San Luis), Área de Química Analítica, Universidad Nacional de San Luis, San Luis, Argentina*

ABSTRACT: Groundwater with high arsenic (As) concentration constitutes a serious problem for crops, since roots can accumulate the metalloid acting as the first stage of As distribution in the trophic chain. The aim of this research was to elucidate the impact of a realistic As(V) dose in peanut and maize plants and to determine the contribution of plant growth promoting bacteria (PGPB) to metalloid translocation in both crops. The results obtained revealed that rhizoinoculation of plants exposed to metalloid, contributed not only to improve growth but also to reduce As transport to shoots. Hence, inoculation of peanut and maize with the correct PGPB partner prevents metalloid translocation in plants avoiding possible fruit contamination.

1 INTRODUCTION

Arsenic (As) is a harmful metalloid that impacts on crops acting as the first stage of As distribution in the trophic chain. In Córdoba province (Argentina), 90% of the region is affected by groundwater As concentrations that exceeds the maximum allowed level in drinking water (FAO, 2015), being arsenate (As(V)) the prevalent arsenic form (Blarasin *et al.*, 2014). Peanut and maize plants constitute important crops in Córdoba, as they represent approximately 90% of the Argentinean production. These crops establish interactions with plant growth promoting bacteria (PGPB) improving yield (Glick, 1995). Given the As amounts determined in groundwater, the metalloid absorption by crop plants could represent an agricultural and human health problem. Therefore, our experiments were conducted to elucidate the impact of a realistic As(V) dose in peanut and maize plants growth and to determine the contribution of PGPB to metalloid translocation in both plants.

2 METHOD/EXPERIMENTAL

2.1 *Bacterial strains*

Bradyrhizobium sp. SEMIA6144 and *Bradyrhizobium* sp. C-145 strains were obtained from MIRCEN (Brazil) and INTA (Argentina), respectively. *Azospirillum brasilense* AZ39 and *A. brasilense* CD (ATTCC 29710) were provided by IMIZA-INTA (Argentina) and EMBRAPA (Brasil), respectively. Bradyrhizobial strains were cultivated in liquid YEM medium (Vincent, 1970) and *Azospirillum* strains in liquid Nfb medium (Dobereiner, 1988).

2.2 *Plant material and experimental design*

Peanut and maize seeds were supplied by "El Carmen S.A" and DEKALB respectively. Seeds were surface sterilized (Vincent, 1970) and pre-germinated seeds were transferred to a Leonard Jar system containing sterile substrate sand:perlite (2:1) with Hoagland's nutrient solution (Hoagland and Arnon, 1950) devoid of As(V) (control) or containing 3 μM As(V), supplied as $Na_2HAsO_4.7H_2O$ (the metalloid concentration found in groundwater of some areas of Córdoba). Plants were divided in two groups (non-inoculated and inoculated) and grown in a controlled environment for 30 days. At harvest, growth and nodulation variables and nitrogen content (by Kjeldahl method) were determined.

2.3 *Arsenic accumulation and translocation in plant tissues*

Metalloid concentration was determined in peanut and maize shoots, roots and nodules (legume) by using an inductively coupled plasma mass spectrometry (ICP-MS) (Sobrino-Plata *et al.*, 2009). The translocation factor (TF) was calculated as Singh and Agrawal (2007).

2.4 *Statistical analysis*

Experiments were conducted in a completely randomized design and repeated three times. The data were analyzed using ANOVA and Duncan's test at P > 0.05. Prior to the test of significance, normality and homogeneity of variance were verified using the modified Shapiro-Wilk and Levene tests, respectively.

3 RESULTS AND DISCUSSION

3.1 *Arsenic impact on peanut and maize growth*

Peanut plants exposed to As(V) showed a significant reduction in shoot dry weight only in non-inoculated plants, without differences in the root dry weight (Table 1). Nodule number and dry weight of peanut plants were reduced by As. Nevertheless, the comparison between bradyrhizobial strains revealed that *Bradyrhizobium* sp. C-145 had a better behavior than *Bradyrhizobium* sp. SEMIA6144 since these variables were significantly higher in both control and treated condition. In the control treatment, nitrogen content was similar in plants inoculated with either bradyrhizobial strain, and significantly greater than in non-inoculated plants. However, As(V) addition caused a reduction of nitrogen content only in plants inoculated with *Bradyrhizobium* sp. SEMIA6144. In maize plants, all growth variables as well as nitrogen content were decreased by metalloid addition irrespective of the inoculation condition tested. Remarkably, *Azospirillum* strains allowed greater bi-ological nitrogen fixation (BNF) in plants, as determined by nitrogen content, compared to non-inoculated plants even when the crops were exposed to the metalloid (Table 1). In support of our results a significant reduction of soybean, peanut, lupin and maize growth were observed when exposed to As(V), being the generation of reactive oxygen species the main reason of cellular toxicity (Bianucci *et al.*, 2017, 2018; Finnegan & Chen, 2012; Lu *et al.*, 2017; Vázquez-Reina *et al.*, 2005). In addition, reduction of nodulation variables was also observed in lupin and soybean plants exposed to As (Bianucci *et al.*, 2018; Carpena *et al.*, 2006). It is known that the interaction established between peanut or maize with PGPB, as used in this research, is important since it fulfils a plant's N demand via the BNF process (Glick, 1995). However, reports evaluating the impact of As in the symbiotic interaction between peanut and rhizobia is scarce, even more in the maize-PGPB interaction. According to our results, it is possible to suggest that the inoculation of peanut with *bradyrhizobia* or maize with *Azospirillum* contribute to enhance growth variables in crops exposed to a realistic As groundwater dose. In this sense, *Bradyrhizobium* sp. C-145 and *A. brasilense* Az39 strains had the better behavior among tested strains.

3.2 *Arsenic accumulation and translocation in peanut and maize plants*

Arsenic accumulation by peanut plants was mainly detected in roots, followed by nodules and finally in shoots, regardless of the inoculation condition (Table 2). Metalloid content in shoots of plants

Table 1. Impact of arsenate on peanut and maize growth.

		Shoot dry weight (g)		Root dry weight (g)		Nodule number plant^{-1}		Nodule dry weight (g plant^{-1})		Nitrogen content (mg g^{-1} dry weight)	
		Control	As(V)	Control	As(V)	Control	As(V)	Control	As(V)	Control	As(V)
Peanut	non-inoculated	0.93±0.05^{A1}	0.75±0.02^{A2}	0.21±0.004^{A1}	0.20±0.041^{B1}	nd	nd	nd	nd	17.09±2.75^{A1}	16.19±1.26^{A1}
	Bradyrhizobium sp. SEMIA 6144	0.92±0.14^{A1}	0.75±0.03^{A1}	0.23±0.018^{A1}	0.24±0.009^{A1}	27.70±1.10^{A1}	21.20±1.00^{A2}	0.04±0.004^{A1}	0.03±0.001^{A2}	26.24±2.49^{B1}	17.01±0.40^{A2}
	Bradyrhizobium sp. C-145	1.05±0.14^{A1}	0.85±0.04^{A1}	0.21±0.012^{A1}	0.23±0.006^{A1}	54.00±4.70^{B1}	25.20±2.00^{B2}	0.06±0.003^{B1}	0.03±0.001^{A2}	26.03±0.48^{B1}	25.22±1.13^{B1}
Maize	non-inoculated	0.22±0.01^{B1}	0.17±0.01^{B2}	0.43±0.02^{B1}	0.28±0.02^{A2}	nd	nd	nd	nd	17.93±0.55^{C1}	13.61±0.85^{B2}
	Azospirillum brasilense Az39	0.31±0.01^{A1}	0.24±0.02^{A2}	0.46±0.05^{AB1}	0.27±0.03^{A2}	nd	nd	nd	nd	25.27±1.09^{A1}	16.48±0.21^{A2}
	Azospirillum brasilense CD	0.28±0.02^{A1}	0.22±0.01^{A2}	0.55±0.03^{A1}	0.24±0.01^{A2}	nd	nd	nd	nd	21.28±0.65^{B1}	15.63±0.44^{A2}

Data represent the mean ± SE(n = 10) Different letters in each column indicate significant differences among inoculation conditions for each treatment and different numbers on each row indicate significant differences among treatment for each inoculation condition according to the Duncan's test (P < 0.05). Nd: not determined.

Table 2. Arsenic accumulation and translocation in peanut and maize plants exposed to As(V).

		Arsenic accumulation ($\mu g\ g^{-1}$ dry weight)			
		Shoot	Root	Nodule	TF
Peanut	non-inoculated	18.64±0.62^{b}	64.45±1.22^{c}	nd	0.29±0.007^{a}
	Bradyrhizobium sp. SEMIA 6144	15.04±0.39^{b}	178.85±4.00^{a}	130.70±10.85^{a}	0.08±0.002^{b}
	Bradyrhizobium sp. C-145	34.41±2.25^{a}	121.47±4.34^{b}	81.70±1.24^{b}	0.28±0.017^{a}
Maize	non-inoculated	9.14±0.39^{a}	7.21±0.16^{c}	nd	1.26±0.046^{a}
	Azospirillum brasilense Az39	7.52±0.34^{b}	14.19±0.55^{b}	nd	0.53±0.015^{b}
	Azospirillum brasilense CD	10.16±0.23^{a}	33.27±0.83^{a}	nd	0.30±0.003^{c}

Data represent the mean ± SE (n = 10) Different letters in each column indicate significant differences among inoculation conditions for each tissue according to the Duncan's test (P < 0,05). Nd: not determined.

inoculated with *Bradyrhizobium* sp. C-145 was higher than that found in other inoculation conditions. On the other hand, roots and nodules of plants inoculated with *Bradyrhizobium* sp. SEMIA6144 presented the highest As content. Interestingly, inoculated plants revealed higher As content compared to non-inoculated plants in below-ground organs. Regarding As translocation from roots to shoots, inoculation of peanut plants with *Bradyrhizobium* sp. SEMIA6144 revealed a significant lower TF among inoculation conditions proved (Table 2).

In inoculated maize plants, metalloid accumulation was mainly detected in roots compared with shoots. On the contrary, non-inoculated plants accumulated higher As content on shoot than roots. The comparison between inoculated strains revealed that shoot and root As content of plants inoculated with *A. brasilense* CD was higher than that found with *A. brasilense* Az39. As observed in peanut plants, inoculated maize plants presented higher As content in roots compared to non-inoculated plants. Regarding As translocation from roots to shoots, inoculation of maize plants decreased metalloid TF in a significant way compared to non-inoculated plants. Metalloid distribution pattern in the studied crops are in agreement with that found in non-hyperaccumulating legumes (Bianucci *et al.*, 2017, 2018; Mandal *et al.*, 2008) and in poaceae (Derlicková *et al.*, 2013). Taking into account the results presented, *Bradyrhizobium* sp. SEMIA6144 could constitute a promising inoculant for peanut plant in order to decrease As translocation to shoots avoiding fruit contamination. Although, maize inoculation with both tested PGPB avoid metalloid translocation to aerial part of the plant, *A. brasilense* Az39 also promoted growth. Thus, it is possible to suggest that the selection of the best PGPB-plant interaction that not only improve growth but also reduce As accumulation in shoots and therefore in the harvest product, could be a promising biotechnological tool to be used in contaminated field.

4 CONCLUSIONS

Arsenic negatively impacts on peanut and maize growth and the inoculation with PGPB represents an effective and promising strategy to improve plant development by restraining metalloid translocation to edible parts. In this sense, the rhizoinoculation with *Bradyrhizobium* sp. SEMIA6144 in peanut plants and *Azospirillum* Az39 in maize, results a biotechnological approach to prevent metalloid translocation avoiding As distribution in the trophic chain.

Thus to prepare a strategy for safe tubewell installation leading to the development of a method combining hydrogeological suitability and social mapping which allowed to optimize the locations for safe well installations.

ACKNOWLEDGEMENTS

This research was supported by Secretaría de Ciencia y Técnica de la Universidad Nacional de Río Cuarto (SECYT-UNRC), Consejo Nacional de Investigaciones Científicas y Técnicas (CONICET) and Fondo para la Investigación Científica y Tecnológica (FONCYT) PICT 2014-0956.

REFERENCES

Bianucci, E., Furlan, A., Tordable, M.C., Hernández, L.E., Carpena-Ruiz, R.O. & Castro, S. 2017. Antioxidant responses of peanut roots exposed to realistic groundwater doses of arsenate: identification of glutathione S-transferase as a suitable biomarker for metalloid toxicity. *Chemosphere* 181: 551–561.

Bianucci, E. Godoy, A., Furlan, A., Peralta, J.M., Hernández, L., Carpena-Ruiz, R.O & Castro S. 2018. Arsenic toxicity in soybean alleviated by a symbiotic species of *Bradyrhizobium. Symbiosis* 74(3): 167–176.

Blarasin, M., Cabrera, A. & Matteoda, E. 2014. Aguas subterráneas de la provincia de Córdoba. UniRío. Universidad Nacional de Río Cuarto. Argentina.

Carpena, R., Esteban, E., Lucena, J.J., Peñalosa, S., Vázquez, P., Zornoza, P. & Gárate, A. 2006. Simbiosis y fitorrecuperación de suelos. *Fijación de Nitrógeno: Fundamentos y Aplicaciones.* Sociedad Española de Fijación del Nitrógeno, pp. 255–268.

Döbereiner, J. 1988. Isolation and identification of root associated diazotrophs. *Plant Sci.* 110: 207–212.

Drličková, G., Vaculík, M., Matejkovic, P. & Lux, A. 2013. Bioavailability and toxicity of arsenic in maize (*Zea mays*

L.) grown in contaminated soils. *Bull. Environ. Contam. Toxicol.* 91: 235–223.

FAO. 2015. http://faostat.fao.org.

Finnegan, P. & Chen, W. 2012. Arsenic toxicity: the effects on plant metabolism. *Front. Physiol.* 3: 182.

Glick, B.R. 1995. The enhancement of plant growth by free-living bacteria. *Can. J. Microbiol.* 41(2): 109–117.

Hoagland, D. & Arnon, D. 1950. The water-culture method for growing plants without soil. California Agricultural Experiment Station.

Lu, H.-D., Xue, J.-Q. & Guo, D.-W. 2017. Efficacy of planting date adjustment as a cultivation strategy to cope with drought stress and increase rainfed maize yield and water-use efficiency. *Agric. Water Manag.* 179: 227–235.

Mandal, S.M., Pati, B.R., Das, A.K. & Ghosh, A.K. 2008. Characterization of a symbiotically effective *Rhizobium* resistant to arsenic: isolated from the root nodules of *Vigna mungo* (L.) Hepper grown in an arsenic-contaminated field. *J. Gen. Appl. Microbiol.* 54(2): 93–99.

Singh, R. & Agrawal, M. 2007. Effects of sewage sludge amendment on heavy metal accumulation and consequent responses of *Beta vulgaris* plants. *Chemosphere* 67(11): 2229–2240.

Sobrino-Plata, J., Ortega-Villasante, C., Flores-Cáceres, M.L., Escobar, C., Del Campo, F.F. & Hernández, L.E. 2009. Differential alterations of antioxidant defenses as bioindicators of mercury and cadmium toxicity in alfalfa. *Chemosphere* 77(7): 946–954.

Vásquez-Reina, S., Esteban, E. & Goldsbrough, P. 2005. Arsenate-induced phytochelatins in white lupin: influence of phosphate status. *Physiol. Plant.* 124(1): 41–49.

Vincent, J. 1970. A manual for the practical study of the root-nodule bacteria. In: *A Manual for the Practical Study of the Root-Nodule Bacteria.* Blackwell Scientific Publications Ltd, pp. 73–97.

2.5 Speciation and toxicity of arsenic in food chain

Environmental Arsenic in a Changing World –
Zhu, Guo, Bhattacharya, Ahmad, Bundschuh & Naidu (Eds)
ISBN 978-1-138-48609-6

Arsenic speciation in soil-water system and their uptake by rice (*Oryza sativa*)

P. Kumarathilaka[1], J. Bundschuh[1], S. Seneweera[2] & A.A. Meharg[3]
[1]*School of Civil Engineering and Surveying, Faculty of Health, Engineering and Sciences, University of Southern Queensland, Toowoomba, QLD, Australia*
[2]*Center for Crop Health, Faculty of Health, Engineering and Sciences, University of Southern Queensland, Toowoomba, QLD, Australia*
[3]*Institute for Global Food Security, Queen's University Belfast, Belfast, UK*

ABSTRACT: Rice is the main staple food for billions people worldwide. Geogenic and anthropogenic sources lead to concentration of As, in particular, above the plow pan of paddy soils. Flooded conditions increase the As concentrations in paddy soil-water system above those of non-flooded conditions. Formation of rice root plaque, availability of metal (hydr)oxides (i.e., Fe and Mn), dissolved organic carbon (DOC), competing ions (i.e., P and Si), and microorganisms play an important role in As speciation and distribution in rice rhizosphere and subsequent uptake by rice plant. Arsenic species are acquired by rice roots through the pathways for nutrients and metabolized via a variety of mechanisms. A number of metabolic pathways including As(III) efflux, and As-thiol complexation and sequestration may decrease As burden in rice tissues.

1 INTRODUCTION

Arsenic (As) is a carcinogen and related to human health concern more than any other toxic element or compound. Arsenic mainly associates with sulfur (S) rich minerals such as arsenopyrites (FeAsS), realgar (As_4S_4), and orpiment (As_2S_3). Over the past few decades, millions of people all over the world have been suffering from chronic diseases and deaths, primarily in response to the As-contaminated drinking water. Groundwater As contamination in Bangladesh is recognized as the most drastic mass poisoning in history. Therefore, recommended and regulatory limits established by various authorities have been revised. World Health Organization (WHO) guideline value for total As in drinking water was reduced in 1993 from 50 to $10\,\mu g\,L^{-1}$.

Another important concern regarding source of As exposure is rice (*Oryza sativa*) (Adeyemi *et al.*, 2017; Chen *et al.*, 2016). Rice is cultivated in over 100 countries in the world with distinctive cropping seasons. Rice cultivation is managed in a special way to ensure wet cultivation of rice and the management practices consist: (a) plowing to homogenize of waterlogged top soils (b) submerging (flooding) the top soil during the cropping period (c) soil is drained and dried before the harvest. Natural and anthropogenic activities lead to concentrate and distribution of As in paddy environment. More importantly, rice grains can contain approximately 10 times as much as baseline total As as other cereals (Halder *et al.*, 2014; Ma *et al.*, 2016; Tenni *et al.*, 2017).

2 DISCUSSION

2.1 *Arsenic bioavailability in rhizosphere*

Paddy management practices (flooding and draining) mainly alter the redox chemistry of paddy soils. Iron and Mn co-exist in rice root plaques under flooded conditions. However, Fe is the main element followed by Mn as secondary in the plaques since Fe(III)-hydro(oxides) precipitate at lower redox potentials than Mn oxides. Radial oxygen loss (ROL) via the root aerenchyma oxidizes the As(III) in rhizosphere and resulted As(V) is sequestrated in Fe plaque. Therefore, As(V)/As(III) ratio is high in Fe plaque whereas the As in the soil solution is mainly present as As(III) which is highly mobile.

Dissolved organic carbon (DOC) derived from decomposed plant and animal products plays an important role in As dynamics. Dissolved organic carbon can promote the reduction of Fe(III) (hydr)oxides to release As. The presence of the competing anions such as phosphorus (P) and silicon (Si) influence the As mobility in paddy soil. Phosphate (PO_4^{3-}) is an analog to As(V) and can influence on As(V) distribution in soil and subsequent uptake. Similarly, Si competes with As(III) for retention sites on soil mineral surfaces.

Aerobic and anaerobic microorganisms in rhizosphere greatly mediate As speciation through assimilatory and dissimilatory pathways. Phylogenetically diverse *aioA* genes mediate As(III) oxidation to the As(V) whereas *arsC* and *arrA* genes reduce As(V) to As(III) in the detoxification and respiratory pathways, respectively. Microbial transformation of inorganic As species by *arsM* genes produces methylated arsenicals such as dimethylarsinic acid (DMA(V)), monomethylarsonoic acid (MMA(V)). Therefore, methylated As species present in rice plants are the product of rhizosphere microorganisms.

2.2 *Arsenic species uptake and translocation*

Arsenic species such as As(III), As(V), MMA(V), and DMA(V) in rice rhizosphere are taken up and translocated up into the stalks and grains. Key transporters including nodulin 26-like intrinsic proteins (NIPs) and PO_4^{3-} transporter genes (OsPHTs) mainly involve the uptake of inorganic and methylated As species. In rice roots, Lsi1 can mediate influx of silicic acid [$Si(OH)_4$] and as well as As(III). Following the uptake, As(III) efflux is a key mechanism and release portion of As(III) into the external environment minimizing cellular As burden. In addition, the formation of As(III)-thiol complexes and sequestration in vacuoles decrease the As(III) translocation.

Phosphate transporter, OsPHT1;8 (OsPT8), is found to have a high affinity for PO_4^{3-} and As(V) uptake in rice. Following the uptake, As(V) readily reduces into the As(III) and the reduction process is stimulated by As(V) reductase enzymes. It has been suggested that Lsi1 mediates the uptake of undissociated MMA(V) and DMA(V) in rice plants. In rice roots, MMA(V) reduces into MMA(III) which is then bind with thiols. So far, there have been no reports of volatile As species released by rice plant under typical growing conditions unless rice plants are treated with trimethylarsines (TMAs(V)O).

Natural Resistance-Associated Macrophage Protein (NRAMP) transporter, OsNRAMP1, and Lsi2 mediate As(III) efflux in the direction of xylem. Putative peptide transporter in rice, OsPTR7, involves in the long-distance (root to grain) translocation of methylated As species in rice. Interestingly, the phloem transportation acts as the primary route of transport to rice grains for both inorganic and methylated As species.

Despite the rice genotypes, location, and season, total As concentrations in rice tissues decrease in the order of root $\gg$ shoot $>$ straw $>$ husk $>$ grain. The highest translocation efficiency (caryopsis-to-root) in rice is gained by DMA(V) followed by MMA(V) and inorganic As species. The lack of DMA(V)-PC complexation enhances the DMA(V) to be readily moved between root and shoots. Table 1 summarizes the concentration of different As species in rice grains in various localities. The pattern of inorganic and methylated As loading into the grain is time-dependent.

Table 1. Arsenic speciation in market-base rice in different localities (Adeyemi *et al.*, 2017; Chen *et al.*, 2016; Halder *et al.*, 2014; Ma *et al.*, 2016; Tenni *et al.*, 2017). Food and Agricultural Organization (FAO) recommended level for inorganic As in rice is 200 $\mu g\, kg^{-1}$.

	Concentration ($\mu g\, kg^{-1}$)			
Country	As(III)	As(V)	DMA(V)	MMA(V)
Bangladesh	129	66	14	2
China	107.1	4.7	9.6	0.3
Taiwan	61.6	4.3	12.1	2.7
Italy	91	8.2	55	–
Thailand	81.4	3.7	29	<2
South Korea	80	5	30	2
Australia	178*		68	–
USA	98*		157	–
India	50*		10	–
Nigeria	47*		11.5	0.3

*Sum of As(III) and As(V).

3 CONCLUSIONS

Physico-chemical properties together with biological processes contribute to the speciation and bioavailability of As in the rice rhizosphere. Number of transporters and enzymes is associated with As uptake, translocation and transform in rice tissues. Various metabolic pathways may decrease the total As concentration in rice grains. Application of agronomical, physico-chemical, and breeding approaches would further be needed to produce rice for human consumption, particularly in As prone areas.

REFERENCES

Adeyemi, J.A., Adedire, C.O., Martins-Junior, A.d.C., Paulelli, A.C., Awopetu, A.F., Segura, F.R., de Oliveira-Souza, V.C., Batista, B.L. & Barbosa Jr, F. 2017. Arsenic speciation in rice consumed in south-western Nigeria, and estimation of dietary intake of arsenic species through rice consumption. *Environ. Toxicol. Chem.* 99(5–6): 1–8.

Chen, H.L., Lee, C.C., Huang, W.J., Huang, H.T., Wu, Y.C., Hsu, Y.C. & Kao, Y.T. 2016. Arsenic speciation in rice and risk assessment of inorganic arsenic in Taiwan population. *Environ. Sci. Pollut. Res.* 23(5): 4481–4488.

Halder, D., Biswas, A., Šlejkovec, Z., Chatterjee, D., Nriagu, J., Jacks, G. & Bhattacharya, P. 2014. Arsenic species in raw and cooked rice: implications for human health in rural Bengal. *Sci. Total Environ.* 497: 200–208.

Ma, L., Wang, L., Jia, Y. & Yang, Z. 2016. Arsenic speciation in locally grown rice grains from Hunan province, China: spatial distribution and potential health risk. *Sci. Total Environ.* 557: 438–444.

Tenni, D., Martin, M., Barberis, E., Beone, G.M., Miniotti, E., Sodano, M., Zanzo, E., Fontanella, M.C. & Romani, M. 2017. Total As and As speciation in Italian rice as related to producing areas and paddy soils properties. *J. Agr. Food Chem.* 65(17): 3443–3452.

Environmental Arsenic in a Changing World –
Zhu, Guo, Bhattacharya, Ahmad, Bundschuh & Naidu (Eds)
ISBN 978-1-138-48609-6

Application of nanofilms for arsenic speciation using surface-enhanced Raman spectroscopy (SERS)

V. Liamtsau[1,2] & Y. Cai[1,2]
[1]*Department of Chemistry and Biochemistry, Florida International University, Miami, FL, USA*
[2]*Southeast Environmental Research Center, Florida International University, Miami, FL, USA*

ABSTRACT: The novel analytical method based on combination of "coffee ring" effect and Surface-enhanced Raman spectroscopy (SERS) has been developed for the speciation of thiolated arsenicals. The gold nanofilm (AuNF) has been fabricated by the coating of glass surface with 3-aminopropyltrimetoxysiloxane (APTMS) and citrate derived negatively charged gold nanoparticles. After the deposition of thioarsenicals buffer solution onto AuNF, following solvent evaporation, ring stamp was formed. Due to the difference in charges of molecules, DMMTA(V) and DMDTA(V) travelled different distances. Overall, the "coffee ring" effect has a decent potential for the speciation of arsenic compounds.

1 INTRODUCTION

The "coffee ring" effect has been previously employed primarily for the analytes preconcentration, however with proper glass surface modification it could have a potential for the separation of arsenic species. Dimethylmonothioarsinic acid (DMMTA(V)) and dimethyldithioarsinic acid (DMDTA(V)) are thioarsenicals that has been widely detected in animal urine after exposure to iAs(III) and DMA(V) (Fig. 1). Pentavalent thioarsenicals can bind to the proteins in contrast to iAs(III) and DMA(V), thus the toxicities of thioarsenicals are similar with that of iAs(III) and are much higher than those of DMA(V). So, high toxicity and protein binding capacity demonstrate that thioarsenicals might be the key arsenic metabolites related to the As overall toxicity. As a result, it is vital to develop a method that can precisely determine the exact amount of thioarsenicals in biological systems.

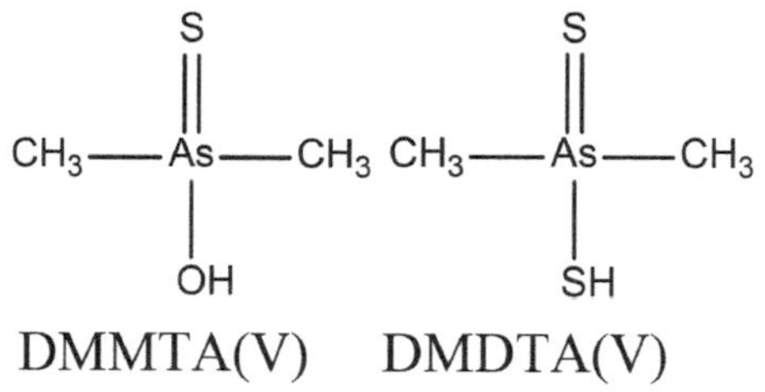

S S
CH_3—As—CH_3 CH_3—As—CH_3
OH SH
DMMTA(V) DMDTA(V)

Figure 1. The structures of Dimethylmonothioarsinic (DMMTA(V)) acid and dimethyldithioarsinic acid (DMDTA(V)).

2 METHODS/EXPERIMENTAL

DMMTA(V) was synthesized according Cullen *et al.* (2016), the molar ratio of Na_2S/H_2SO_4/DMA was 1.6:1.6:1.

DMDTA(V) was synthesized by two different methods. The first method reported by Suzuki *et al.* (2004) provided the acidic form of DMDTA(V) in contrast to the second approach that has been described by Fricke *et al.* (2005) and allowed to synthesize the salt form of DMDTA(V). The molar ratio for these two methods of Na_2S/H_2SO_4/DMA(V) was 7.5:7.5:1.

The preparation of a gold nanofilm includes two stages, synthesis of citrate-coated gold and gold nanoparticles (AuNPs) and coating nanoparticles onto silicon wafers. AuNPs were synthesized by Turkevich method. The fabrication of AuNF was performed by silanization of glass substrates by 3-aminopropyltrimetoxysiloxane (APTMS), following the deposition of negatively charged gold nanoparticles onto the positively charged amino groups silanized surface.

DMMTA(V) and DMDTA(V) solutions (10^{-4} M) solutions of thioarsenicals were prepared by dissolving thioarsenicals in buffer, following the deposition of 2 μL of DMMTA(V) and DMDTA(V) buffer solution onto the AuNF. Once the droplet was completely dried and a ring-shaped stain was formed on the AuNF, SERS signals were obtained from the center to the edge of the dried droplet.

3 RESULTS AND DISCUSSION

Raman spectra of solid thioarsenicals are shown in Figures 2 and Figures 3.

Since charged and neutral species would have various interactions with negatively charged SERS

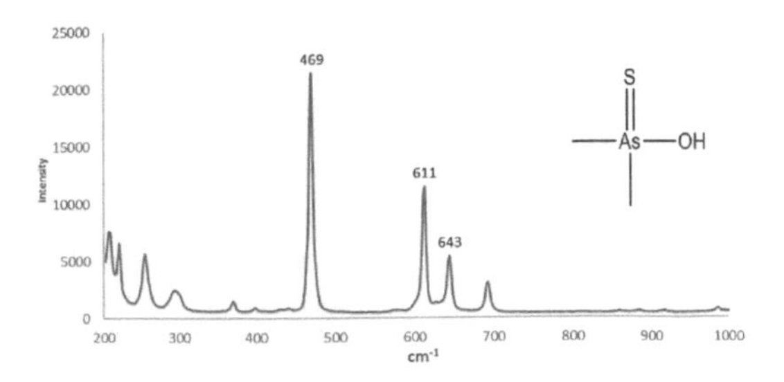

Figure 2. Raman spectroscopy of DMMTA(V).

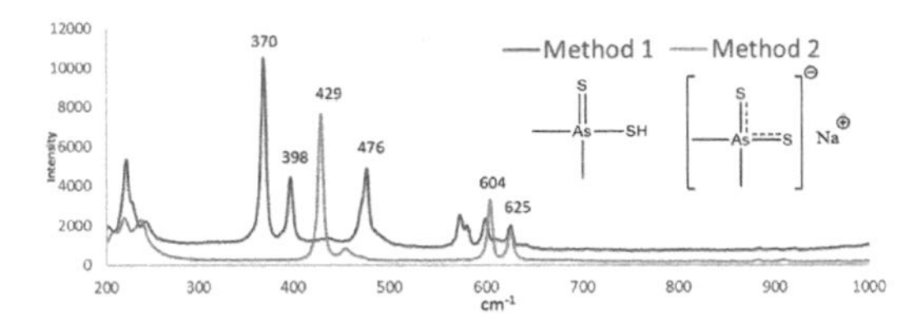

Figure 3. Raman spectroscopy of DMDTA(V).

Table 1. Raman vibrational frequencies of DMMTA(V) and DMDTA(V).

Arsenical	Assignment	Experimental Raman frequency (cm^{-1})
DMMTA(V)	As = S	469
	As – C	641
	As – O	643
DMDTA(V)	As – S	370,398
	As ÷ S	429
	s C – As – C	573,604
	a C – As – C	625

surface, their travelling distances might be different (Table 1). AuNF was applied for the separation of DMMTA(V) and DMDTA(V).

From the graph it is clear that As–S vibrations shift from 429 cm^{-1} to 417 cm^{-1} for DMDTA(V) and from 469 cm^{-1} to 465 cm^{-1} for DMMTA(V) respectively (Fig. 4), thus, thioarsenicals were adsorbed onto the AuNF and charge transfer between the molecules and AuNF surface occurred. However, it is not clear whether a capillary force that drown thioarsenicals to the edges of evaporating droplet or attraction/repulsion interaction between analytes and the AuNF surface had the major impact onto the separation of thioarsenicals.

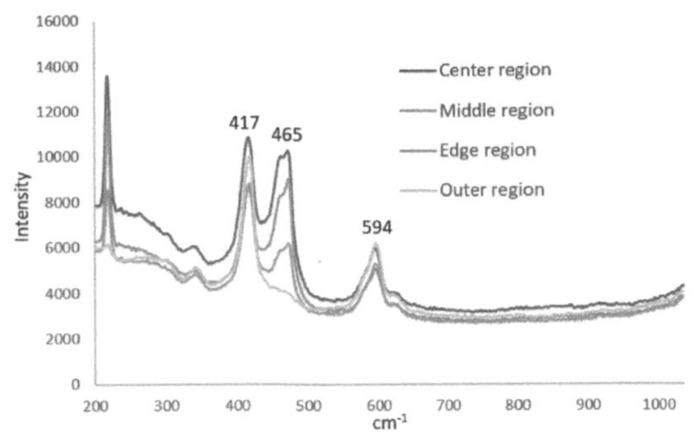

Figure 4. SERS of 10^{-4} M DMMTA(V) and DMDTA(V) buffer solution separated by "coffee ring" effect.

4 CONCLUSIONS

DMMTA(V) and DMDTA(V) were separated using "coffee ring" effect and identified by SERS. Nanofilms are versatile tools, facile to fabricate and easy to operate under necessary conditions. Overall, nanofilms had a decent potential for the separation of arsenic species.

ACKNOWLEDGEMENTS

I would like to thank Dr. Guangliang Liu and Dr. Changjun Fan from the Department of Chemistry and Biochemistry at FIU for their faith in my project.

REFERENCES

Cullen, W.R., Liu, Q., Liu, X., McKnight-Whitford, A., Peng, H., Popowich, A., Yan, X., Zhang, Q., Fricke, M., Sun., H. & Le, C. 2016. Methylated and thiolated arsenic species for environmental and health research – a review on synthesis and characterization. *J. Environ. Sci.* 49: 7–27.

Fricke, M.W., Zeller, M., Sun, H., Lai, V W.-M., Cullen, W.R., Shoemaker, J.A., Witkowski, M.R. & Creed, J.T. 2005. Chromatographic separation and identification of products from the reaction of dimethylarsinic acid with hydrogen sulfide. *Chem. Res. Toxicol.* 18(12): 1821–1829.

Suzuki, K.T., Mandal, B.K., Katagiri, A., Sakuma, Y., Kawakami, A., Ogra, Y., Yamaguchi, K., Sei, Y., Yamanaka, K., Anzai, K., Ohmichi, M., Takayama, H. & Aimi, N. 2004. Dimethylthioarsenicals as arsenic metabolites and their chemical preparations. *Chem. Res. Toxicol.* 17(7): 914–921.

Environmental Arsenic in a Changing World –
Zhu, Guo, Bhattacharya, Ahmad, Bundschuh & Naidu (Eds)
ISBN 978-1-138-48609-6

Interannual variability of dissolved and rice grain concentrations of arsenic and cadmium in paddy fields subjected to different water managements

T. Honma[1], K. Nakamura[2], T. Makino[2] & H. Katou[2]
[1]*Niigata Agricultural Research Institute, Nagaoka, Japan*
[2]*Institute for Agro-Environmental Sciences, NARO, Tsukuba, Japan*

ABSTRACT: Interannual variability of dissolved arsenic (As) and cadmium (Cd) concentrations as well as those in rice grain was investigated in paddy fields subjected to different water managements. While substantial variability was observed in the As and Cd concentrations, unique relationships existed between dissolved and rice grain concentrations across different years and water managements. The results suggest that interannual variability of dissolved As and Cd concentrations was responsible for the variability in the concentrations in rice grain and should be taken into account in risk assessment.

1 INTRODUCTION

Arsenic and Cd accumulation in rice grain is of serious concern to human health. Because of anaerobic conditions prevailing in rice growing fields, rice can accumulate a higher concentration of inorganic arsenic (iAs), a human carcinogen, than other cereal crops. The uptake of As and Cd by rice is sensitive to redox conditions in soil and affected by water management during growth period. Reducing conditions produced by flooding lead to increase in As solubility through reductive dissolution of As-sorbing iron (III) (hydr)oxides and decrease in Cd solubility through formation of hardly soluble cadmium sulfide, while oxidizing conditions produced by drainage promote immobilization of As and solubilization of Cd. Honma *et al.* (2016) conducted field experiments to compare the effects of intermittent irrigation with different irrigation intervals. They found that concentrations of As and Cd in rice grain were linearly related to dissolved concentrations during postheading 3 weeks and suggested an optimal Eh of −73 mV for simultaneously suppressing rice grain As and Cd concentrations. The objectives of the present study were to investigate (i) to what extents dissolved and rice grain As and Cd concentrations are interannually variable under the same water managements, and (ii) whether the relationships between dissolved As and Cd and rice grain iAs and Cd, as well as the optimal Eh, are influenced by the interannual variability.

2 METHODS/EXPERIMENTAL

2.1 *Field experiments*

Field experiments were conducted in 2013 through 2015 in a paddy field on an alluvial plain in central Japan. The soil, classified as a Typic Hydraquent (Soil Survey Staff, 2014), had a total carbon content of 16.2 g kg^{-1}, 1 M HCl-extractable As of 2.49 mg kg^{-1}, and 0.1 M HCl-extractable Cd of 0.84 mg kg^{-1}. The textural composition was 52% sand (0.2–2 mm), 30% silt (0.02–0.2 mm), and 18% clay (<2 mm), with soil pH of 5.8 measured at a soil:water ratio of 1:2.5.

Seedlings of rice (*Oryza sativa* L. cv. Koshihikari) were transplanted in mid-May. The rice plants were grown under flooded conditions for approximately 5 weeks, followed by 14 days of midseason drainage. Thereafter, different water managements were practiced during preheading 3 weeks and postheading 3 weeks. They included (i) flooded (FLD), (ii) intermittent irrigations F3D3, F3D5, and F3D7, where numerals before the letters ‘F’ and ‘D’ designate the length (in days) of flooding and drainage in irrigation cycle, respectively, and (iii) rainfed (RFD). The water management F3D7 was not conducted in 2013 whereas F3D5 and F3D7 were not conducted in 2015. The field was drained for harvest in early September.

2.2 *Plant and soil analysis*

Rice grains air-dried to 15% of moisture and remaining on the 1.85 mm-sieve were used for analysis. Inorganic As(III) and As(V) concentrations in the grains were determined by inductively coupled mass spectroscopy (ICP-MS) according to the methods of Nishimura *et al.* (2010) and Baba *et al.* (2014) after digestion with dilute HNO_3. Total Cd concentration was determined by flow injection (FI)-ICP-MS after digestion with HNO_3 followed by H_2O_2.

Soil redox potential (Eh) was measured in duplicate with platinum electrodes installed at a depth of 15 cm in the field. Soil solution samples were collected using solution samplers installed at the same depth, at intervals of 1–2 weeks from mid-June to early September.

Immediately after collection, the samples were acidified with HNO_3 and analyzed for total As and Cd by FI-ICP-MS.

3 RESULTS AND DISCUSSION

3.1 *Dissolved As and Cd*

In Figure 1, dissolved As and Cd concentrations averaged over postheading 3 weeks are plotted against soil Eh averaged over the same period for each water management in the three years. General trends of sharp decrease of dissolved As with soil Eh, particularly at Eh above −100 mV, and gradual increase of dissolved Cd with Eh were evident across different years and water managements. However, both dissolved As and Cd concentrations showed considerable interannual variability, particularly in the plots where the concentrations were high (i.e., dissolved As in FLD and F3D3; dissolved Cd in F3D3 and RFD). The optimal soil Eh for simultaneously suppressing dissolved As and Cd concentrations was identified using the "trade-off value" proposed by Honma *et al.* (2016) and found at approximately −50 mV. This value was in reasonable agreement with the previously reported value of −73 mV in the same field by Honma *et al.* (2016). Among the water managements compared in this study, intermittent irrigation F3D5 was most effective in simultaneously suppressing dissolved As and Cd.

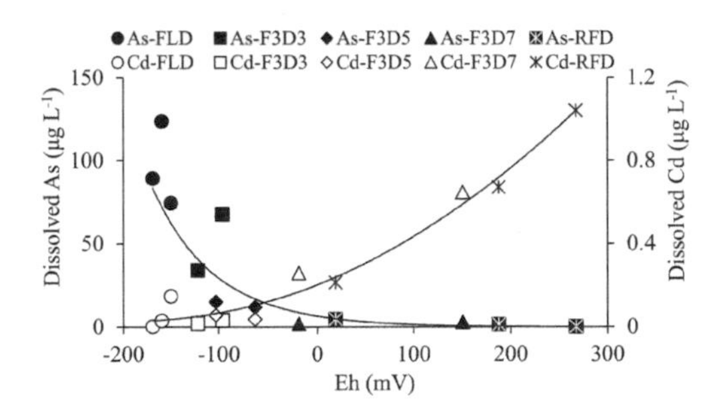

Figure 1. Relationships between dissolved As and Cd concentrations and soil Eh. Each symbol represents the average value for postheading 3 weeks in different water managements and years.

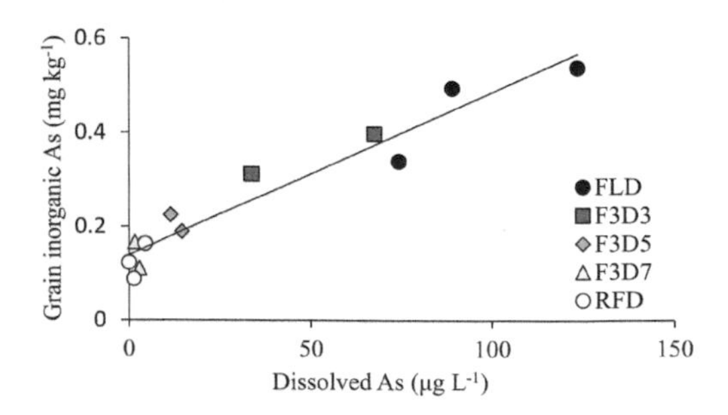

Figure 2. Relationship between inorganic As concentration in rice grain and dissolved As concentration across different water managements and years.

3.2 *Inorganic As and Cd in rice grain and their relations to dissolved As and Cd*

Inorganic As and Cd concentrations in rice grain also showed considerable interannual variability. Nonetheless, across different water managements and years, there was a linear relationship between the average dissolved As concentration in the postheading 3 weeks and iAs concentration in rice grain (Fig. 2). A linear relationship was also observed between the average dissolved Cd concentration in the same period and rice grain Cd concentration. These results suggest that interannual variability of dissolved As and Cd concentrations was responsible for the variability in the iAs and Cd concentrations in rice grain. A larger interannual variability was found as the iAs and Cd concentrations increased, and this should be taken into account in risk assessment.

4 CONCLUSIONS

Both dissolved As and Cd concentrations in paddy soil and iAs and Cd concentrations in rice grain showed a substantial interannual variability under the same water managements. However, there were unique linear relationships for As and Cd, across different water managements and years, between average dissolved concentrations during postheading 3 weeks, and concentrations in rice grain. This suggests that interannual variability of dissolved concentrations in response to variable weather and redox conditions was responsible for the variability in the concentrations in rice grain. A larger interannual variability is expected under water managements conductive to higher iAs and Cd concentrations, and this should be taken into account in risk assessment.

ACKNOWLEDGEMENTS

This work was supported by a grant from the Ministry of Agriculture, Forestry, and Fisheries of the Japanese Government (Research project for improving food safety and animal health As-240).

REFERENCES

Baba, K., Arao, T., Yamaguchi, N., Watanabe, E., Eun. H. & Ishizaka, M. 2014. Chromatographic separation of arsenic species with pentafluorophenyl column and application to rice. *J. Chromatogr. A 1354*: 109–116.

Honma, T., Ohba, H., Kaneko-Kadokura, A., Makino, T., Nakamura, K. & Katou, H. 2016. Optimal soil Eh, pH, and water management for simultaneously minimizing arsenic and cadmium concentrations in rice grains. *Environ. Sci. Technol.* 50(8): 4178–4185.

Nishimura, T., Hamano-Nagaoka, M., Sakakibara, N., Abe, T., Maekawa, Y. & Yonetani, T. 2010. Determination method for total arsenic and partial-digestion method with nitric acid for inorganic arsenic speciation in several varieties of rice. *J. Food Hyg. Soc. Jpn.* 51(4): 178–181.

Soil Survey Staff. 2014. Keys to Soil Taxonomy, Twelfth Edition. U.S. Department of Agriculture, Natural Resources Conservation Service.

Environmental Arsenic in a Changing World –
Zhu, Guo, Bhattacharya, Ahmad, Bundschuh & Naidu (Eds)
ISBN 978-1-138-48609-6

The effects of different arsenic species in relation to straighthead disease in rice

H.P. Martin[1], W. Maher[2], M. Ellwood[1], E. Duncan[3], P. Snell[4] & F. Krikowa[2]
[1]*Research School of Earth Sciences, Australian National University, Canberra, ACT, Australia*
[2]*Institute for Applied Ecology, University of Canberra, Canberra, ACT, Australia*
[3]*Future Industries Institute, University of South Australia, Adelaide, SA, Australia*
[4]*Department of Primary Industries, Yanco Agricultural Institute, Orange, NSW, Australia*

ABSTRACT: The effects of inorganic arsenic on plants, the mechanisms involved in the uptake and transport of arsenic and how inorganic arsenic enters food chains are well documented. Regulatory limits have been established to control the inorganic arsenic concentrations in certain foods including rice. There is, however, a knowledge gap with respect to dimethylarsenic concentrations. In this study rice was grown hydroponically and exposed to varying DMA concentrations. High levels of DMA were detrimental to rice plants whereby plants showed symptoms consistent with Straighthead disease, a disease that results in dramatic yield losses.

1 INTRODUCTION

Straighthead is a physiological disorder in rice that results in sterile florets and poorly developed panicles, leading to the head of the rice plant remaining upright when it reaches maturity (Dunn & Dunn, 2012). Straighthead disease has the potential to generate in substantial yield declines. In Australia straighthead is expected to cost the rice industry in excess of \$1-million per year. Although this cost could be much larger due to straighthead commonly being confused for cold-weather sterility. The exact cause of straighthead is unknown, however, studies have revealed a strong correlation between symptoms of the disease and elevated arsenic concentrations especially DMA within the plant and soil. The role arsenic plays with respect to straighthead disease is unknown, however, the presence of both inorganic (Rahman *et al.*, 2008), and DMA (Yan *et al.*, 2008) have been shown to increase the frequency and severity of straighthead symptoms in rice.

When investigating the relationship between arsenic speciation and total arsenic accumulation within rice plants, studies have found that the inorganic arsenic follows a hyperbolic pattern and levels out around 0.2 $mg\,kg^{-1}$. DMA, however, does not follow this pattern and as a result has the potential to accumulate at much higher concentrations within the plant (Zhao *et al.*, 2013). The uptake pathways and metabolism of DMA are largely unknown.

2 METHODS/EXPERIMENTAL DESIGN

2.1 *Seed germination and plant culture*

Rice seeds were surface sterilized utilizing a technique adapted by Kim (2005). Seeds were washed with

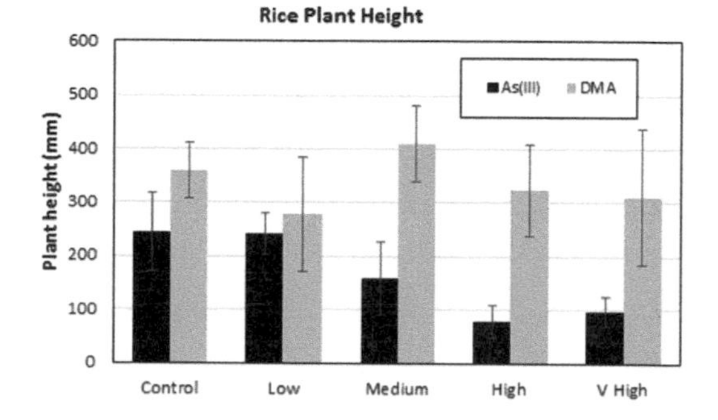

Figure 1. Effects of As(III) and DMA on plant height.

10% H_2O_2 for 10 minutes followed by 70% ethanol for 5 minutes followed by deionized water (DI). The seeds are than placed in DI water and transferred to an incubator at 30°C for 48 hr.

Seedlings were transplanted into purpose built hydroponic systems. The rice plants were grown using nutrient solution, exposing the rice to a half strength solution for the first 14 days, followed by full strength solution thereafter. The nutrient solution was aerated throughout the entire experiment, with the nutrient solution replaced weekly. The pH was kept between 5.5 and 5.8, adjusted daily using 1 $mol\,L^{-1}$ HCL and 1 $mol\,L^{-1}$ NaOH.

Rice plants were exposed to either DMA or As(III) at four different concentrations; Low 0.8 $\mu\,mol\,L^{-1}$, Medium 1.8 $\mu\,mol\,L^{-1}$, High 3.5 $\mu\,mol\,L^{-1}$ and Very High 6.7 $\mu\,mol\,L^{-1}$. Each treatment was conducted in triplicate. Samples were grown for a total of 38 days.

Rice root and shoot samples where oven dried at 70°C until constant weight. Plants were digested using a low-volume microwave procedure developed. Approximately 0.2 g of dried plant samples into Teflon

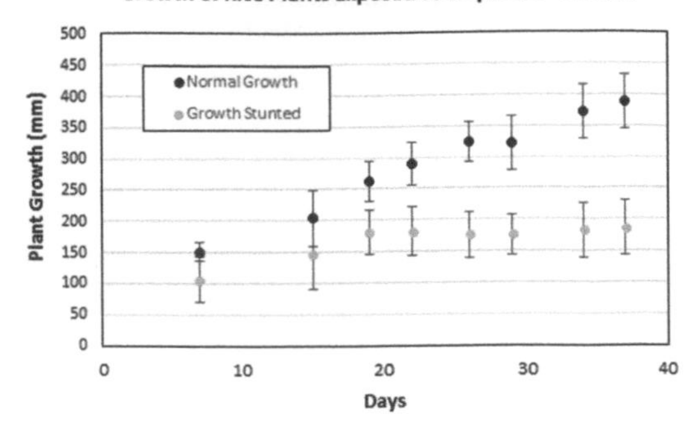

Figure 2. The grow rates of the plants of the of rice plants exposed to 6.7 μmol L^{-1}, showing the different growth rates of plants within the same treatments.

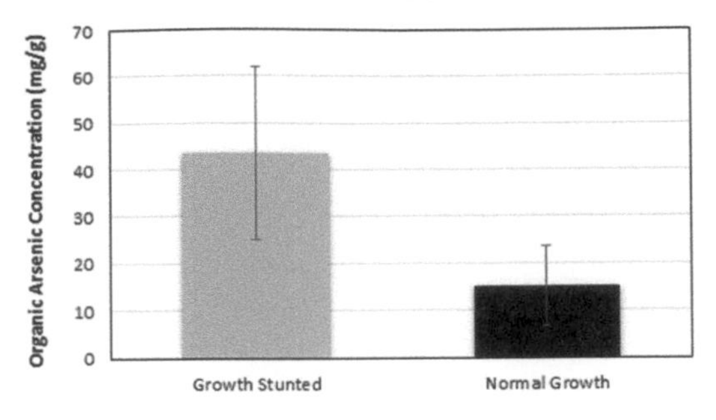

Figure 3. The concentration of arsenic in the rice shoots exposed to 6.7 μmol L^{-1} of DMA.

closed digestion vessels with 2 ml concentrated ultra-pure nitric acid and 1 mL hydrogen peroxide. Samples were then microwaved using a preprogrammed digestion procedure. After digestion, samples are cooled at room temperature and the digests diluted to 10 mL with DI water for analysis by Inductively Coupled Plasms Mass Spectrometry (ICP-MS).

3 RESULTS AND DISCUSSION

3.1 *Plant growth*

Rice plants exposed to As(III) were observed to reduce in plant height with increasing As(III) exposure ($p < 0.05$). No significant differences in growth between the DMA treatments was observed, although there was significant variability in the data (Fig. 1).

3.2 *DMA effected rice plants*

The large variation in heights for DMA exposed plants was investigated further by looking at individual plant growth rates within treatments. For individuals exposed to 6.7 μ mol L^{-1} DMA, these was a clear difference with two distinct populations; plants that exhibited normal growth and plants with stunted growth (Fig. 2). Plants were determined to have stunted growth if their growth rate was less than 50% of the control. Growth rate divergence occurred fourteen days after transplanting. Similar trends were observed across all the DMA treatments.

When the total arsenic concentrations between the two groups (Normal Growth and Stunted Growth) were compared, distinct differences were also observed (Fig. 3). The plants with stunted growth had more arsenic per gram of plant tissue with arsenic concentrations within the shoots nearly double of that of the plants with normal growth (Normal growth $15 \pm 8\,\mu g\,g^{-1}$, Stunted growth $43 \pm 8\,\mu g\,g^{-1}$).

This indicates that DMA does not affect all plants, only a few in each treatment. Close inspection revealed that the reduced growth plants were not dead, even though growth had stopped. For most treatments, this occurred between days 7 and 20 after transplanting. Rice plants exposed to DMA showed similar symptoms to plants affected by pathogens, with only a select few plants being affected.

4 CONCLUSIONS

This study highlighted that DMA has a larger effect on the rice plants than previously thought. While DMA is relatively non-toxic to humans, it still affects the plant differently than inorganic arsenic. There is, however, still considerable uncertainty regarding the uptake and metabolism of DMA. This study has highlighted that rice plants metabolize DMA differently to inorganic arsenic. With growing global concern of arsenic contamination in the environment, DMA could pose a real concern for crop health and productivity.

ACKNOWLEDGEMENTS

Ecochem UC for lab use and data analysis.

Australian Government Research Training Program ANU OCG Travel Scholarship.

REFERENCES

Dunn, B.W. & Dunn, T.S. 2012. Influence of soil type on severity of straighthead in rice. *Commun. Soil Sci. Plant Anal.* 43(12): 1705–1719.

Kim, D.W., Rakwal, R., Agrawal, G.K., Jung, Y.H., Shibato, J., Jwa, N.-S., Iwahashi, Y., Iwahashi, H., Kim, D.H., Shim, I-S. & Usui, K. 2005. A hydroponic rice seedling culture model system for investigating proteome of salt stress in rice leaf. *Electrophorseis* 26(23): 4521–4539.

Rahman, M.A., Hasegawa, H. & Rahman, M.M. 2008. Straighthead disease of rice (*Oryza sativa* L.) induced by arsenic toxicity. *Environ. Exper. Bot.* 62(1): 54–59.

Yan, W.G., Agrama, H.A. & Slaton, N.A. 2008. Soil and plant minerals associated with rice straighthead disorder induced by arsenic. *Agron. J.* 100(6): 1655–1661.

Zhao, F.J., Zhu, Y.G. & Meharg, A.A. 2013. Methylated arsenic species in rice: geographical variation, origin, and uptake mechanisms. *Environ. Sci. Technol.* 47(9): 3957–3966.

Environmental Arsenic in a Changing World –
Zhu, Guo, Bhattacharya, Ahmad, Bundschuh & Naidu (Eds)
ISBN 978-1-138-48609-6

Effects of foliar application of silicon on uptake and translocation of arsenite and DMA in rice (*Oryza sativa*)

W.-J. Liu, Y. Sun & Q.-L. Zhao
College of Resources and Environmental Science, Hebei Agricultural University, Baoding, P.R. China

ABSTRACT: A hydroponic experiment was conducted to explore the effects of foliar application of silicon on the uptake and translocation of As(III) and DMA in rice. The result showed that the ratio of foliar application of Si concentration and the As concentration in the solution at 100:1 decreased significantly arsenic concentrations in the phloem sap and the rice shoots. However, when DMA supplied, Si/As ratio at 200:1 increased significantly arsenic and silicon concentrations in rice phloem ($P < 0.05$). In addition, there was a positive correlation between arsenic concentrations and silicon concentration in rice phloem ($R = 0.856$, $P < 0.01$, $n = 12$). Regardless of whether As(III) or DMA was supplied to rice, foliar application of low concentration of silicon (Si/As at 100:1) could decrease arsenic concentration and translocation in rice seedlings.

1 INTRODUCTION

Arsenite (As(III)) is the predominant species of As in soil solution from flooded paddy fields. As(III) as $H_3AsO_3^0$ is an analogue of silicic acid ($H_4SiO_4^0$) ($pH < 8$) and can be taken up by rice roots through the silicic acid transport system (Ma *et al.*, 2008). The similar situation had happened to Di-methylated arsenic (DMA) into rice roots (Li *et al.*, 2009). However, application Si in the submerged soil would increase As(III) and DMA mobility (Lee *et al.*, 2014; Liu *et al.*, 2014). Therefore, we want to know that the impact of foliar application Si with the different concentrations on the uptake and translocation of As(III) and DMA in rice.

2 METHODS/EXPERIMENTAL

2.1 *Plant growth*

Rice (*Oryza sativa* L. cv. Italica Carolina, the Japonica type, an early flowering cultivar) seeds were sterilized in 30% H_2O_2 solution for 15 min followed by thorough washing with de-ionized water. The seeds were germinated in moist quartz sand. After 3 weeks, rice seedlings were transplanted to 500 mL PVC pots (7.5 cm diameter and height 14 cm, one plant per pot) containing half-strength Kimura B solution. The composition of the nutrient solution was 0.091 mM KNO_3, 0.183 mM $Ca(NO_3)_2$, 0.274 mM $MgSO_4$, 0.091 mM KH_2PO_4, 0.183 mM $(NH_4)_2SO_4$, 0.5 μM $MnCl_2$, 3 μM H_3BO_3, 0.1 μM $(NH_4)_6Mo_7O_{24}$, 0.4 μM $ZnSO_4$, 0.2 μM $CuSO_4$, 20 μM NaFe(III)-EDTA (pH adjusted to 5.5 with KOH). Nutrient solutions were renewed once every 3 d. The growth conditions were 14 h photoperiod with a light intensity of 280 $\mu mol\, m^{-2}s^{-1}$ and day/night temperatures were maintained at 28/25°C.

2.2 *Experiment treatments*

Rice plants at heading stage were exposed to the half-strength Kimura B nutrient solution containing 10 μM As(III) or 10 μM DMA. Different levels of 0 (CK), 1000 (D_{Si}), 2000 (G_{Si}) μM Si as silicic acid were sprayed on the rice shoots at 10:10 am every day and the exposure of As and Si would last for 72 h. Thus, the ratios of Si to As(Si/As) would be 100:1 and 200:1. Silicic acid was prepared by passing potassium silicate through a cation-exchange resin (Amberlite IR-120B, H^+ form).

After 72 h of As and Si exposure, phloem sap, rice shoot and rice root were collected for total arsenic detection.

2.3 *Total As analysis*

Shoots (0.2 g FW) and roots (approx. 0.1 g FW) samples were digested in 5 mL HNO_3 (Liu *et al.*, 2004b). Reagent blank and certified reference material (GBW07603 from the National Research Center for Standard Materials in China) were included to verify the accuracy and precision of the digestion procedure and subsequent analysis. Total As concentrations in the samples (phloem sap, rice shoot and root) were determined atomic by fluorescence spectrometry (AFS9600, Beijing Haiguang Instrument Co., Beijing, China).

2.4 *Statistical analysis*

The significance of treatment effects was determined by analysis of variance (ANOVA) using statistical package SPSS 17.0 for windows. Tukey's multiple comparison tests were carried out according to the LSD (least significance difference) values.

3 RESULTS AND DISCUSSION

3.1 *Arsenic in phloem sap influenced by foliar application of Si*

For As(III) treatment, the foliar application of Si concentration at 1000 μM (Si/As ratio at 100:1) decreased significantly As levels in phloem sap ($P < 0.05$) (Fig. 1). However, when DMA applied in nutrient solution, the foliar application of Si increased As concentrations in phloem sap significantly. There was a positive correlation between As concentrations and silicon levels in rice phloem only for rice exposed to medium with DMA ($R = 0.856$, $P < 0.01$, $n = 12$).

3.2 *Arsenic concentrations in rice root and shoot influenced by Si foliar application*

When As(III) supplied in nutrient solution, Si foliar application did not influence As accumulation in root while 1 mM Si addition decreased As levels in root significantly (Table 1). For DMA treatment, both 1 mM and 2 mM silicic acid foliar application reduced arsenic accumulation in rice roots and shoots. It indicated that foliar application Si could mediate As transported into rice grain under DMA in the growth medium and foliar Si application would be better to control As accumulation in rice plants for DMA treatment than that of As(III) supply.

3.3 *Arsenic translocation affected by Si foliar application*

As translocation factor ranged from ~0.03–0.05 for As(III) treatment while from ~0.17–0.19 for DMA treatment (Fig. 2), which meant DMA had higher efficiency to transport from root to shoot than that of

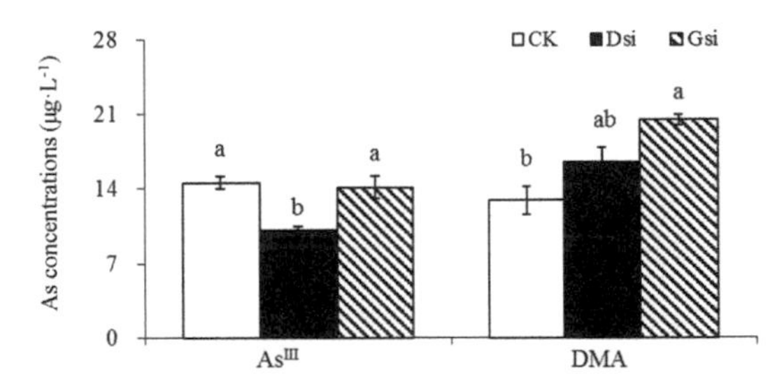

Figure 1. As concentration in rice phloem sap under the different Si concentration.

Table 1. As concentration in rice roots and shoots under the different treatments.

	CK	Dsi	Gsi
	Root-As (mg/kg)		
As(III)	216.3 ± 2.03b	304.2 ± 8.53a	202.0 ± 9.08b
DMA	24.4 ± 0.41a	19.5 ± 1.25b	16.0 ± 1.21c
	Shoot-As (mg/kg)		
As(III)	10.5 ± 0.07a	9.01 ± 0.19b	10.6 ± 0.07a
DMA	4.96 ± 0.34a	3.23 ± 0.39b	2.99 ± 0.25b

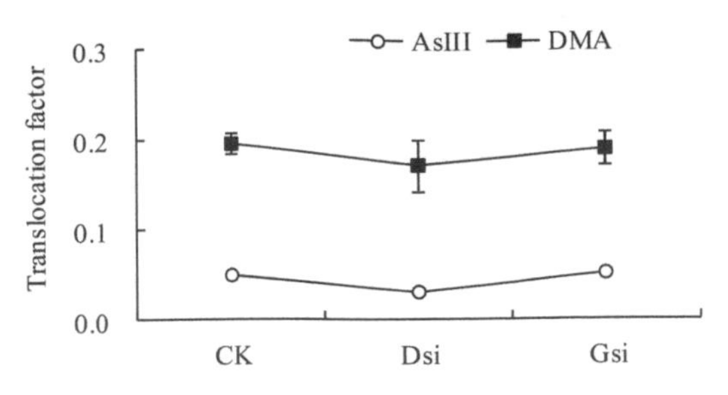

Figure 2. The translocation factor of arsenic in rice root to shoot under different Si treatments.

As(III). When 1 mM Si was sprayed on the rice shoots, the ratios of arsenic concentrations from roots to shoots were lowest regardless of whether As(III) or DMA in the growth medium.

4 CONCLUSIONS

The ratio of foliar application of Si concentration to As concentration in the solution at 100:1 decreased significantly As levels in phloem sap and rice shoots for arsenite treatment. When DMA supplied, Si/As ratio at 200:1 increased significantly arsenic and silicon concentrations in rice phloem ($P < 0.05$). In addition, there was a positive correlation between As concentrations and silicon levels in rice phloem ($R = 0.856$, $P < 0.01$, $n = 12$). Whether As(III) or DMA was supplied to rice, foliar application of low concentration of silicon (Si/As at 100:1) decrease arsenic concentration and translocation in rice.

ACKNOWLEDGEMENTS

This research was supported by the Natural Science Foundation of China (41471398). We thank Professor Fangjie Zhao of Agricultural University of Nanjing and Rothamsted Research for providing rice seeds. We also sincerely thank Dr. Ben Simon for improving this manuscript in language.

REFERENCES

Lee, C.H., Huang, H.H., Syu, C.H., Lin, T.H. & Lee, D.Y. 2014. Increase of As release and phytotoxicity to rice seedlings in As-contaminated paddy soils by Si fertilizer application. *J. Hazard. Mater.* 276: 253–261.

Li, R.Y., Ago, Y., Liu, W.J., Mitani, N., Feldmann, J., McGrath, S.P., Ma, J.F. & Zhao, F.J. 2009. The rice aquaporin Lsi1 mediates uptake of methylated arsenic species. *Plant Physiol.* 150(4): 2071–2080.

Liu, W.J., McGrath, S.P. & Zhao, F.J., 2014. Silicon has opposite effects on the accumulation of inorganic and methylated arsenic species in rice. *Plant Soil* 376(1–2): 423–431.

Ma, J.F., Yamaji, N., Mitani, N., Xu, X.Y., Su, Y.H., McGrath, S.P. & Zhao, F.J. 2008. Transporters of arsenite in rice and their role in arsenic accumulation in rice grain. *Proc. Natl. Acad. Sci. U S A* 105(29): 9931–9935.

Environmental Arsenic in a Changing World –
Zhu, Guo, Bhattacharya, Ahmad, Bundschuh & Naidu (Eds)
ISBN 978-1-138-48609-6

Persistence and plant uptake of methylarsenic in continuously- and intermittently-flooded rice paddies

S.C. Maguffin[1,2], M.C. Reid[1,2], A. McClung[1,2] & J. Rohila[1,2]
[1]*Civil & Environmental Engineering, Cornell University, Ithaca, NY, USA*
[2]*Dale Bumpers National Rice Research Center, Stuttgart, AR, USA*

ABSTRACT: Rice grown in the South-Central USA tends to accumulate a larger fraction of methylarsenic species than rice grown in other regions, and the reason for these patterns are not clear. The goal of this study was to determine how soil management conditions influence the persistence and plant uptake of methylarsenic species in rice paddy soils, using field-scale experimental plots in Arkansas, USA. We expect methylated arsenic to persist in pore water and be more abundant in biomass in continuously flooded grown rice.

1 INTRODUCTION

Rice accumulates arsenic (As) primarily as inorganic arsenic and the methylarsenic species dimethylarsinic acid (DMAs(V)) (Ma *et al.*, 2008). Rice grown in the South-Central United States tends to accumulate a larger fraction of DMAs(V) than rice grown in other regions. The reasons for these patterns remain unclear, and these uncertainties underscore a broader lack of knowledge around how the persistence and plant availability of methylarsenic species in rice paddy soils are governed by soil environmental conditions (Zhao *et al.*, 2013).

The goal of this study was to determine the persistence and fate of As originating from the methylarsenic-based pesticide monosodium methanearsonate (MSMA) in field-scale experimental rice plots cultivated under different soil management conditions. The specific objectives of this study were: (1) Monitor MSMA in pore waters and associated with soil minerals to determine the effect of irrigation regime on plant-available methylarsenic species; and (2) Track As abundance and speciation in rice plant tissues at different stages of maturity to examine links between belowground As dynamics and As accumulation in plant tissues, with a special focus on rice grains.

2 METHODS AND EXPERIMENTAL DESIGN

2.1 *Experimental design*

A field monitoring study was conducted during the 2017 growing season at the United States Department of Agriculture Dale Bumpers National Rice Research Center in Stuttgart, Arkansas, USA, as part of a multi-decade study on the impacts of MSMA amendment on As uptake into rice. A 2 × 2 factorial experimental design (Fig. 1) was used to test the effects of irrigation regime (continuously-flooded (CF) vs. alternate wetting and drying (AWD)) and MSMA amendment (1.1 mg-As kg^{-1} – applied annually for more than 20 years) on As speciation in pore water as well as As associated with soil solid phases.

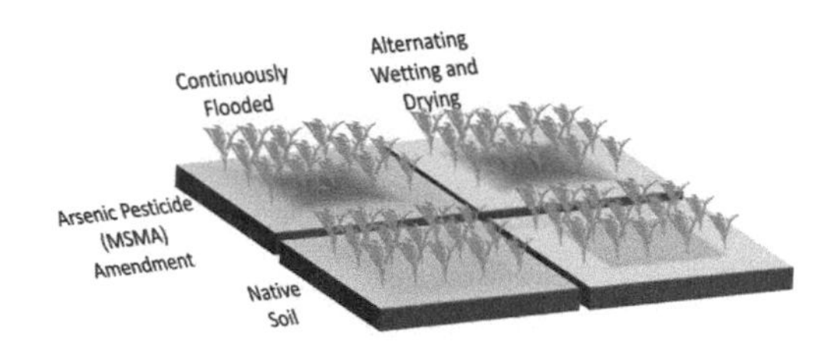

Figure 1. Conceptual illustration of the 2 × 2 factorial experimental design at USDA-ARS Dale Bumpers Research Service. The red represents the application of the MSMA pesticide onto the rice paddy.

2.2 *Soil solution sampling*

Three pore water sampling locations were established in each experimental field. Each location had a Rhizon pore water sampler at 10 cm and 25 cm depth. Pore water samplers were equipped with 0.15 μM filters and were anaerobically sampled bi-weekly starting 3 weeks after plots were planted using a vacuumed, autoclaved, acid-washed bottle. In the MSMA-amended AWD experiment, we sampled pore water at a higher frequency during a mid-season wetting event to more closely examine the effects of soil wetting and drying on aqueous chemistry and arsenic speciation.

Pore water arsenic speciation was determined using HG-CT-GC-AAS. Pore water samples were also analyzed for inorganic anions using ion chromatography, total elemental analysis using ICP-OES, and dissolved organic carbon (DOC) using a total organic carbon analyzer. Soil temperature moisture were logged using in situ probes.

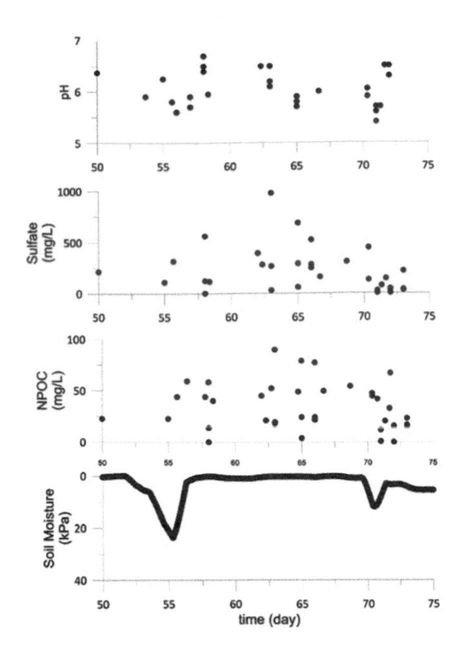

Figure 2. Mid-growing season high resolution aqueous sampling data from the AWD-MSMA-amended field's shallow wells. Alongside regular sampling, high-frequency sampling was conducted to more clearly resolve arsenic redox and speciation chemistry during specific alternating wetting and drying events.

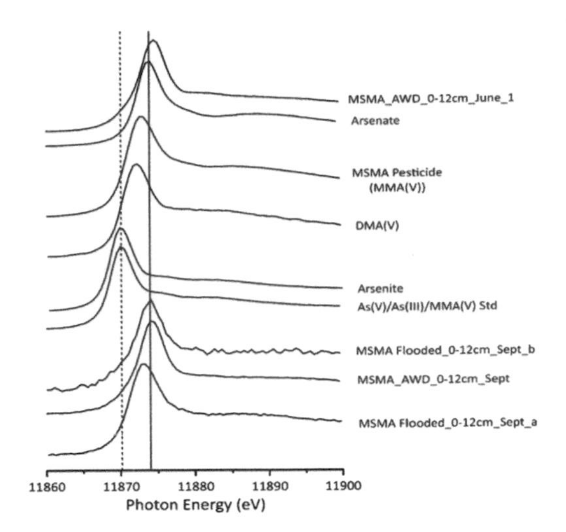

Figure 3. XANES spectra of reference standards and soil samples from flooded and AWD fields. The solid vertical line indicates the arsenate white line energy. The dotted vertical line indicates the arsenite white line energy (not all data shown).

2.3 *Solid phase analysis*

Soil was collected using a soil probe and anaerobically preserved in heat sealed BoPET bags flushed with research grade argon gas. Total elemental analysis was performed via XRF and by acid digestion followed by ICP-OES. As speciation was determined using X-ray absorption near-edge structure (XANES) at the Cornell High Energy Synchrotron Source (CHESS).

3 EXPECTED RESULTS AND DISCUSSION

3.1 *Pore water*

Initial results indicate a difference in pore water chemistry between irrigation methods. Soil solution chemistry was more greatly affected by AWD controls at 10 cm depth than at 25 cm. In the MSMA-AWD experiment, high resolution sampling highlighted the effect soil moisture has on pH, sulfate, and DOC (Fig. 2). An increase in soil moisture correlated strongly with an increase in sulfate and DOC before sulfate reducing conditions were reached and both decreased (Fig. 2). Pore water As analyses have not been completed at the time of abstract submission.

3.2 *Solid phase*

XANES analysis was employed to speciate the arsenic within soil at the beginning and end of the growing season (Fig. 3). Importantly, arsenic in soil from the MSMA-flooded experiment collected at the end of the growing season was different than arsenate's white line energy; this indicates that MSMA was not completely demethylated to arsenite and/or arsenite was not completely oxidized to arsenate. This was not the case for results from MSMA-AWD experiments and thus suggests a persistence of MSMA in flooded paddy soils is possible.

4 CONCLUSIONS

Data from arsenic analyses, combined with pore water and solid phase chemistry, will provide a detailed characterization of methylation chemistry in rice paddy soils as a function of irrigation controls and background arsenic concentrations. Expected results will highlight differences in arsenic speciation among experimental designs.

ACKNOWLEDGEMENTS

This work has been funded through the National Science Foundation Postdoctoral Fellowship Program and Cornell University.

REFERENCES

Ma, J.F., Yamaji, N., Mitani, N., Xu, X.-Y., Su, Y.-H., McGrath, S. P. & Zhao, F.-J. 2008. Transporters of arsenite in rice and their role in arsenic accumulation in rice grain. *Proc. Natl. Acad. Sci. U.S. A.* 105(29): 9931–9935.

Zhao, F.J., Zhu, Y.G., & Meharg, A.A. 2013. Methylated arsenic species in rice: geographical variation, origin, and uptake mechanisms. *Environ. Sci. Technol.* 47(9): 3957–3966.

Environmental Arsenic in a Changing World –
Zhu, Guo, Bhattacharya, Ahmad, Bundschuh & Naidu (Eds)
ISBN 978-1-138-48609-6

Geographical variation of arsenic in rice from Bangladesh: Cancer risk

S. Islam[1,2,3], M.M. Rahman[1,2] & R. Naidu[1,2]
[1] *Global Centre for Environmental Remediation (GCER), Faculty of Science, The University of Newcastle, Callaghan, NSW, Australia*
[2] *Cooperative Research Centre for Contamination Assessment and Remediation of the Environment (CRC CARE), The University of Newcastle, Callaghan, NSW, Australia*
[3] *Department of Soil Science, Bangladesh Agricultural University, Mymensingh, Bangladesh*

ABSTRACT: This study analyzed 965 rice samples collected by household survey from Bangladesh to determine the distribution of arsenic (As), geographical variation, daily intake and the potential cancer risk form consuming rice. The results showed that As content in rice grain ranged between 3–680 μg/kg^{-1} with the highest fraction being 98.6% of inorganic As. The daily intake of inorganic As from rice ranged between 0.38 to 1.92 μg/kg^{-1} body weight (BW). The incremental lifetime cancer risk (ILCR) for individuals due to the consumption of rice varied between 0.57×10^{-3} and 2.88×10^{-3} in different locations higher than the US EPA threshold.

1 INTRODUCTION

Arsenic is recognized as a toxic element and has been classified as a Group I human carcinogen. Arsenic exposure to humans mainly occurs via drinking As-contaminated water and consumption of food crops grown in irrigated agricultural lands where As-contaminated groundwater is used for irrigation. In addition to supplying drinking water, As-contaminated groundwater is used extensively for the irrigation of crops in Bangladesh and West Bengal, India, especially for the cultivation of paddy rice during the dry season. As a result, food crops accumulate elevated levels of As so that it poses a potential threat to human health. Rice is the main dietary staple for many populations worldwide due to its nutritional value. Arsenic exposure from rice intake can be considerable and intake of inorganic As via this route is a significant risk factor for cancers, especially for the population who depend on rice diet substantively (Mondal & Polya, 2008). In most developing countries, paddy rice is grown under flooded conditions and for this reason, the accumulation of As in rice grain is significant. Rice is more effective in As accumulation than other cereal crops such as wheat and barley (Williams *et al.*, 2007). The objective of this study is to assess the distribution of As in rice from different locations of Bangladesh to understand the geographical variation and the potential cancer risk from consuming rice.

2 METHODS

For this study, we collected and analyzed 965 rice samples form 73 Upazilas (sub-districts) from 20 districts (Barisal, Chandpur, Comilla, Dhaka, Dinajpur, Faridpur, Gaibandha, Gopalganj, Lakshmipur, Lalmonirhat, Mymensingh, Naogaon, Narayanganj, Nawabganj, Netrokona, Noakhali, Rajshahi, Sherpur and Sylhet) of Bangladesh. Districts were selected based on the As concentrations (severely, moderately and less affected) in groundwater. The sampling sites are shown in Figure 1. The detail study results were reported elsewhere (Islam *et al.*, 2017a, 2017b). Both dry (Boro) and wet (Aman) season's rice were included in this study. Most of the rice were local varieties, some high yielding variety and hybrid varieties were also included. Digestion of the rice samples for total As analysis was carried out employing the method used by Rahman *et al.* (2009). Inductively coupled plasma mass spectrometer (ICP-MS) (Agilent Technologies, Tokyo, Japan), was used to determine the amount of total As. For As speciation a liquid chromatography system (model 1100, Agilent Technologies, Tokyo, Japan) equipped with a guard and Hamilton PRP-X100 separation column, coupled with ICP-MS was used. Appropriate CRM was also used to verify the results of total As and As species in rice.

3 RESULTS AND DISCUSSION

Mean and median values of total As concentrations in rice grains are 126 and 107 μg kg^{-1}, respectively, and the As level range between 3.4 and 680 μg kg^{-1}, dry weight. These total As results are comparable to those previously found from Boro rice (range, 40–910 μg kg^{-1}), and Aman rice (range, <40–920 μg kg^{-1}) samples in 25 districts throughout Bangladesh (Williams *et al.*, 2006). Rice collected

Table 1. Arsenic concentration and varietal differences in Bangladesh.

Rice genotypes	Concentration of As ($\mu g\ kg^{-1}$)
BRRI dhan33	137 ± 11
Binadhan-7	166 ± 07
BRRI dhan30	179 ± 12
BR22	211 ± 08
BR26	217 ± 05
BR24	219 ± 02
BRRI dhan47	104 ± 09
BR14	152 ± 05
BRRI dhan50	164 ± 04
BRRI dhan29	166 ± 01
BRRI dhan58	174 ± 09
BRRI dhan28	174 ± 08
Super hybrid	210 ± 11
Sonar Bangla hybrid	285 ± 16

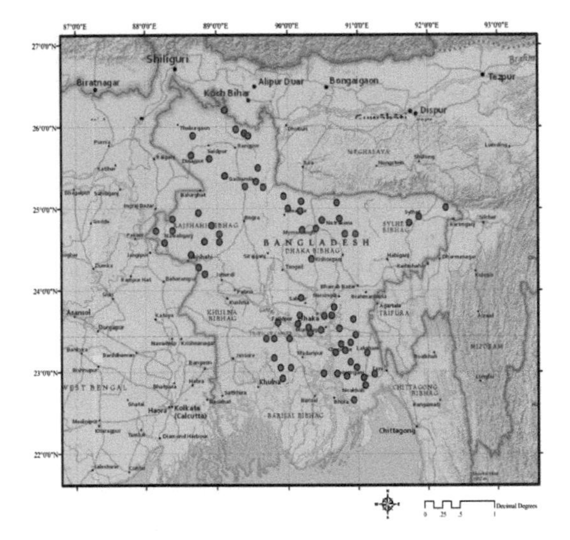

Figure 1. Sampling location in Bangladesh (red dots shows Upazilas) (Islam *et al.*, 2017a).

from the markets showed that As concentrations ranged from 180–310 $\mu g\ kg^{-1}$ in Aman rice and 210–270 $\mu g\ kg^{-1}$ in Boro rice (Williams *et al.*, 2006). Arsenic species results showed that inorganic As is predominant over organic As which ranges from 59 to 98% of the total As. Our results indicate that the aromatic rice varieties had less inorganic As compared to non-aromatic rice. The varietal differences on total As in rice grain in this study are summarized in Table 1.

About 47% of variation was accounted for by differences among the 73 upazilas, while 71% was due to differences between districts (Islam *et al.*, 2017a). Much of the variation between samples could therefore be attributed to differences between districts (Fig. 2). The highest average daily intake of inorganic As from rice consumption in Narayanganj district was

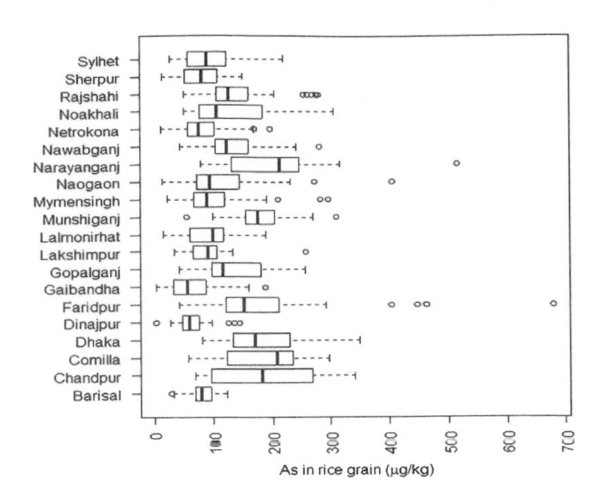

Figure 2. Statistical box plot of the concentrations of As in rice grain in 20 districts of Bangladesh (Islam *et al.*, 2017a).

1.92 $\mu g\ kg^{-1}$ BW. The average incremental lifetime cancer risk (ILCR) value of inorganic As for consumers in Narayanganj district was 2.88×10^{-3}, 288 times higher than the acceptable limit (Islam *et al.*, 2017a).

4 CONCLUSIONS

The study reveals that rice poses a significant risk to the local inhabitants. Further work is necessary to reduce the As levels from rice as well as to minimize the human exposure risk and potential cancer risk.

REFERENCES

Islam, S., Rahman, M.M., Islam, M.R. & Naidu, R.. 2017a. Geographical variation and age-related dietary exposure to arsenic in rice from Bangladesh. *Sci. Total Environ.* 601: 122–131.

Islam, S., Rahman, M.M., Islam, M.R. & Naidu, R. 2017b. Inorganic arsenic in rice and cancer risk. Clean Up conference, Melbourne, Australia, 12–15 Sep.

Mondal, D. & Polya, D.A. 2008. Rice is a major exposure route for arsenic in Chakdaha block, Nadia district, West Bengal, India: a probabilistic risk assessment. *App. Geochem.* 23(11): 2987–2998.

Rahman, M., Owens, G. & Naidu, R.. 2009. Arsenic levels in rice grain and assessment of daily dietary intake of arsenic from rice in arsenic-contaminated regions of Bangladesh – implications to groundwater irrigation. *Environ. Geochem. Health* 31(1): 179–187.

Williams, P.N., Islam, M.R., Adomako, E.E., Raab, A,. Hossain, S.A., Zhu, Y.G., Feldmann, J. & Meharg, A.A. 2006. Increase in rice grain arsenic for regions of Bangladesh irrigating paddies with elevated arsenic in groundwaters. *Environ. Sci. Technol.* 40(16): 4903–4908.

Williams, P.N., Raab, A., Feldmann, J. & Meharg, A.A., 2007. Market basket survey shows elevated levels of As in South Central U.S. Processed rice compared to California: consequences for human dietary exposure. *Environ. Sci. Technol.* 41(7): 2178–2183.

Environmental Arsenic in a Changing World –
Zhu, Guo, Bhattacharya, Ahmad, Bundschuh & Naidu (Eds)
ISBN 978-1-138-48609-6

Translocation of arsenic in food chain: A case study from villages in Gangetic basin, India

D. Ghosh[1] & J. Routh[2]
[1]*Centre for Earth Sciences, Indian Institute of Science, Bangalore, India*
[2]*Department of Thematic Studies – Environmental Change, Linköping University, Linköping, Sweden*

ABSTRACT: Arsenic (As) contamination of groundwater is a worldwide concern and West Bengal in India is affected by this problem. Different food crops may accumulate As differently, and studies often show conflicting results. A previous study from Karimpur block, Nadia district, West Bengal indicated high concentration of As in two tube wells. Groundwater from these wells is used for irrigating nearby agricultural fields. The aim of this study was therefore to analyze As and other metals concentration in different vegetables and rice growing in these fields. In addition, animal fodder plants and cow milk was also analyzed to investigate the possibility of transfer of arsenic through food chain. The samples were analyzed using a microwave digestion procedure followed by inductively coupled plasma mass spectrometry. Animal fodder roots showed high concentration of As compared to other samples. The results also showed possible translocation of Cd along with As from root to shoot, in animal fodder. The milk samples showed high concentrations of As and Pb.

1 INTRODUCTION

In many parts of the world contamination of groundwater from various inorganic and organic pollutants is a crucial problem. Over 150 million people worldwide suffer from arsenic (As) poisoning (Stroud *et al.*, 2011). Over the past two decades extensive research efforts have been launched to investigate the role of As on humans (Bhattacharya *et al.*, 1997). Research on the health effects has shown that changes in skin pigmentation and skin keratosis are two common consequences of exposure to high As levels in humans.

Groundwater in India is major source of drinking water. Irrigation using As contaminated aquifers could lead to translocation and biomagnification of As through various agricultural activities particularly through cereals that are consumed. There is very little information on As levels in vegetables, fruits and other food products from areas which are known As 'hot-spots' (Bhattacharya *et al.*, 1997; Ghosh *et al.*, 2015, 2017). Previous studies conclude there are still many uncertainties about As uptake by crops. This is due to the variations that exist between the different types of vegetation, soil types and land use practices (Kabata-Pendias & Pendias, 1989). Moreover, toxicity and availability of As in plants depend on many factors such as: hydrology, adsorption capacity, translocation in plants etc.

The focus in various studies is on arsenic contaminated water and its effects on human, however, it is also important to investigate the link between irrigation with contaminated water and accumulation of As in food crops It is also important to trace how As is transferred through the food chain involving; water-soil – food crops – humans, and impacts human health.

2 METHODS

2.1 *Sampling*

Sampling was done for wet (May–November) and dry (January–May) harvesting of crops in 2014, dry 2015 and wet and dry periods in 2016. The following villages of Nadia district of West Bengal, India: Haringhata, Gayeshpur, and Birohi (Chakdah Block), Rehematpur and Jamaipara (Karimpur Block), Balarampur, Sashipur, Donapur, and Kamgachia (Ranaghat Block). There were earlier reports of As contamination in the groundwater from these aquifers. Young and mature vegetable samples were collected from the field along with the soil sample and irrigation water samples from the respective fields. In addition, animal fodder (*Sorghum)* plants and milk samples were collected from these villages.

2.2 *Extraction and quantification*

The samples were lyophilized and powdered. About 0.5 g of each sample was used to extract the trace metals using microwave assisted technique with H_2O_2 and HNO_3. Blanks and replicates were performed to ensure quality control of the analytical procedures. The SMR 1547 – Peach leaf from National Institute of Standards and Technology, BCR Reference Material Nr. 62 OLEA EUROPAEA Commission of the European Communities and Certified Reference Material: NIMT/UOE/FM/001 Inorganic Elements in Peat were used to correct the methodological errors if any.

Trace metal analyses in the extracts was done on a Perkin Elmer ICP-Mass Spectrometer NexION 300D.

3 RESULTS AND DISCUSSION

The vegetable samples collected during the dry harvest period in all three years had higher accumulation of As, and other heavy metals like Pb, Cd, and Zn, in comparison to the monsoon crops. The possible reason could be the water source used for irrigation. Crops grown during the dry harvest period are irrigated using groundwater, whereas crops during the wet period are irrigated using rainwater (Shrivastava *et al.*, 2015). However, statistical analyses did not show any co-relation between As and trace metal accumulation within the vegetables. The highest As accumulation was observed in rice (both polished and unpolished varieties) and animal fodder plants. The specific reasons are unknown. However, a possible explanation could be that phosphate transporters found in cell membrane wall of the family *Poaceae*, easily translocates As, which has stoichiometric similarity to it.

The sampling stations were divided into two groups, high and low As rich aquifers. Milk and vegetable samples showed a clear grouping with respect to their sampling site. However, accumulation of As also occurred in milk samples collected during both seasons from all stations.

The ability of translocation characteristics of various metals within a plant can be calculated using translocation factor (TF). Results >1 would indicate translocation of the selected metal (Zabin & Howladar, 2015).

$$TF = (As_{shoot}) / (As_{root}) \quad (1)$$

Transportation characteristics of various trace metals were calculated for animal fodder samples. The results indicated generally low TF values. Other than As (TF = 1.93), only one trace metal showed indication of possible translocation from root to shoot i.e. Cd with a TF value of 1.40. For the rest of the trace metals, the indication of translocation from root to shoot was generally less and none of the selected trace metals exceeded the TF value of 1. Chromium and Co had the lowest TF value (0.01) for both animal fodder samples.

In general, Cu has low mobility and it mostly accumulates in roots (Kabata-Pendias & Pendias, 1989). Consistent with this, the results of TF values for Cu (0.25 and 0.33) indicated there is no translocation of Cu from root to shoot. Both translocation factor and bioconcentration factors are needed to evaluate the specific potential by a plants for phytoremediation (Usman *et al.*, 2013).

4 CONCLUSIONS

Overall, the study shows the translocation of As and other trace metals into higher animals through the food chain. The contaminated groundwater used for irrigation leads to accumulation of As in animal fodder, which is consumed by cattle and accumulates in milk. Human is the ultimate consumer of all these products including the different food crops and milk. This most likely leads to the different As-related health problems in the area. Drinking water is one of the major way of As entering into human body, and many measures have been taken to reduce the contamination levels in drinking water. More research is required to understand the As accumulation pathways and how can As accumulation be reduced by using GMO crops or by pre-treating irrigation water by either chemical treatment or phyto-remediation.

ACKNOWLEDGEMENTS

We thank the Swedish Research Link- Asia Program for funding the research. Indira Pasic, Molly Aylesbury, and Mårten Dario helped us with the laboratory analyses.

REFERENCES

Stroud, J.L., Norton, G.J., Islam, M.R., Dasgupta, T., White, R.P., Price, A.H., Meharg, A.A., McGrath, S.P. & Zhao, F.J. 2011. The dynamics of arsenic in four paddy fields in the Bengal delta. *Environ. Pollut.* 159: 947–953.

Bhattacharya, P., Chatterjee, D. & Jacks, G. 1997. Occurrence of arsenic contaminated groundwater in alluvial aquifers from Delta Plains, Eastern India: options for safe drinking water supply. *Int. J. Water Resour. Manag.* 13, 79–82.

Ghosh, D., Routh, J., Därio, M. & Bhadury, P. 2015. Elemental and biomarker characteristics in a Pleistocene aquifer vulnerable to arsenic contamination in the Bengal Delta Plain, India. *Appl. Geochem.* 61:87–98.

Ghosh, D., Routh, J. & Bhadury, P. 2017. Sub-surface biogeochemical characteristics and its effect on arsenic cycling in the Holocene grey sand aquifers of the Lower Bengal Basin. *Front. Environ. Sci.* 5:82.

Shrivastava, A., Ghosh, D., Dash, A. & Bose, S. 2015. Arsenic Contamination in Soil and Sediment in India: Sources, Effects, and Remediation. *Curr. Pollution Rep.*: 35–46.

Kabata-Pendias, A. & Pendias, H. 1989. *Trace elements in soils and plants*; CRC Press: Boca Raton, FL.

Usman, A.R.A., Alkredaa, R.S. & Al-Wabel, M.I. 2013. Heavy metal contamination in sediments and mangroves from the coast of Red Sea: Avicennia marina as potential metal bioaccumulator. *Ecotoxicol. Environ. Saf.* 97: 263–270.

Zabin, S.A. & Howladar, S.M. 2015. Accumulation of Cu, Ni and Pb in selected native plants growing naturally in sediments of water reservoir dams, Albaha Region, KSA.Nature and Science 13(3): 11–17.

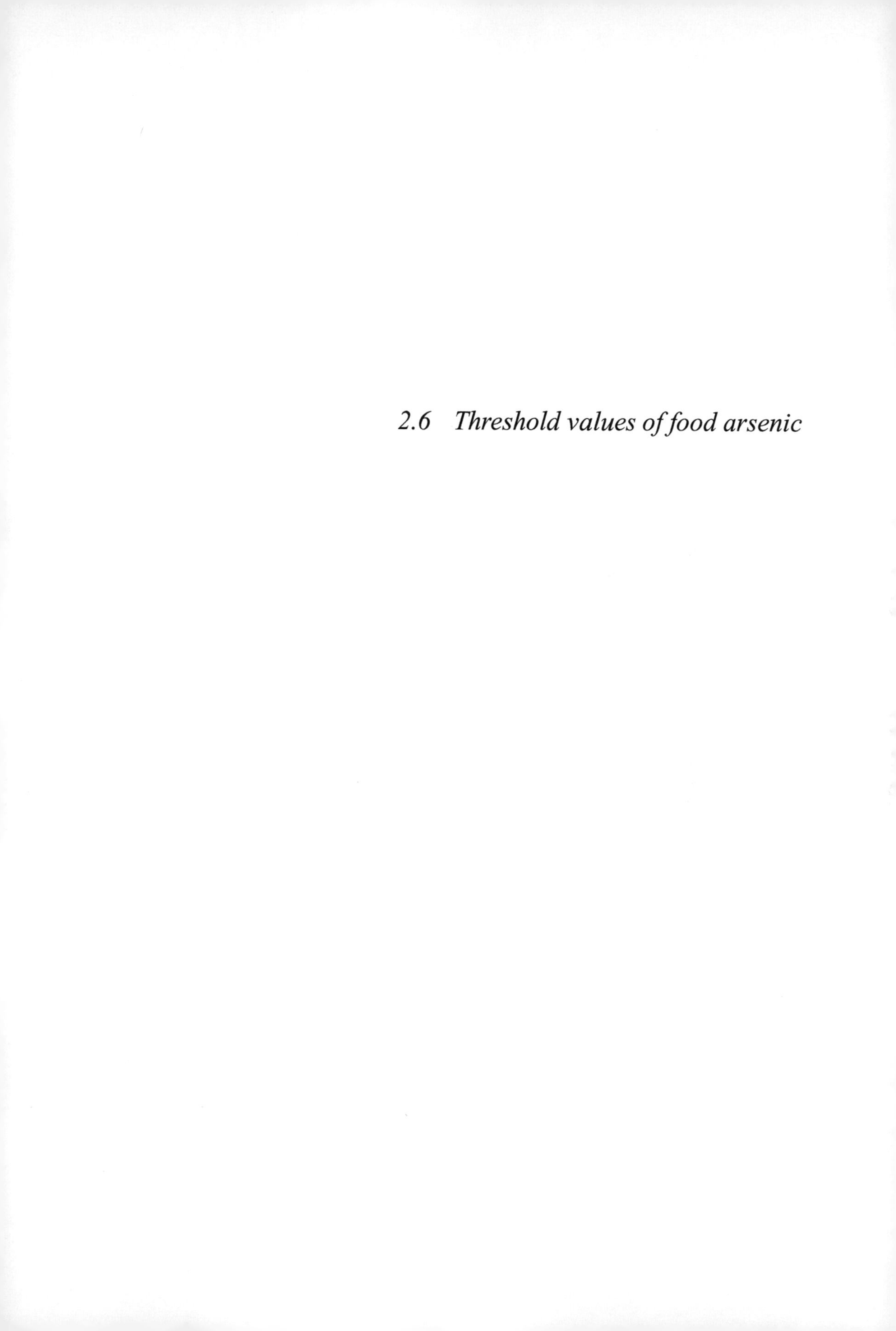

2.6 Threshold values of food arsenic

Environmental Arsenic in a Changing World –
Zhu, Guo, Bhattacharya, Ahmad, Bundschuh & Naidu (Eds)
ISBN 978-1-138-48609-6

Bioaccessibility and arsenic speciation in carrots, beets and quinoa from a contaminated area of Chile

I. Pizarro[1], M. Gómez-Gómez[2], D. Román[2] & M.A. Palacios[2]
[1]*Facultad de Ciencias Básicas, Universidad de Antofagasta, Antofagasta, Chile*
[2]*Department of Analytical Chemistry, Faculty of Chemistry, Universidad Complutense de Madrid, Madrid, Spain*

ABSTRACT: The study is focused on locally-grown, such as carrots, beets and quinoa from the As-polluted Chiu Chiu area in Northern Chile. For the vegetables, were investigated: i) Their total As, Cu, Pb, Cr, Cd and Mn content ii) Arsenic speciation by LC-ICPMS analysis; iii) Arsenic bioaccessibility iv) Arsenic species present in the extracts and v) Arsenic dietary exposure. In vitro gastrointestinal digestion of As from carrots and beets, whereas efficiency was about 40% for quinoa. For carrots, only As(III) and As(V) species were found, being their concentration levels similar. In the case of quinoa, around 85% of the element was present as As(V).

1 INTRODUCTION

The Northern Chilean economy has been mainly supported by the exploitation of copper mining resources and mineral smelting plants. Arsenic can be present in minerals, associated with copper and other heavy metals. In addition, the Loa river water is also naturally contaminated by As. The concentration of As in the water of the Loa hydrographic basin changes dramatically over the course of the river. Assessment of the contribution to human diet of As and other contaminant elements in food requires studying not only their total element content but also their absorption rates in the gastrointestinal tract (Khouzam *et al.*, 2011; Pizarro *et al.*, 2003). A way to evaluate the absorbable fraction in trace element bioaccessibility studies is the "in vitro" simulation of the digestion process.

2 METHODS/EXPERIMENTAL

2.1 *Materials and methods*

An ICPMS, (HP-4500, Agilent Technologies, Analytical System, Tokyo, Japan) operating under normal multi-element tuning conditions was used for total arsenic and other metals analysis. For arsenic speciation, an ICPMS has been used as detector system after LC species separation. Chilean vegetable samples were collected from 3 different places of the Chiu Chiu Chilean area. Carrots and beets were peeled with a plastic knife and the edible part was treated in a titanium blender until a homogeneous mash was achieved. A fraction of about 2 kg of flesh were lyophilized. For sample mineralization, about 0.5 g of the vegetables was digested in PTFE vessels with 5 mL of concentrated HNO_3 and 2 mL of H_2O_2.

2.2 *Extraction procedures and in vitro digestion*

For arsenic speciation by LC-ICPMS, two extraction procedures have been employed: Methanol: water (1:1) and enzymatic protease-lipase extraction. Arsenic species were identified in the different extracts. Samples were digested following a laboratory-simulated three-step digestion process: Salivary digestion, about 1.0 g of lyophilized sample was mixed with 5 mL of salivary juice. Gastric digestion, 10 mL of the prepared gastric juice was added to the solid fraction after salivary digestion and intestinal digestion, 10 mL of the prepared intestinal juice was added to the solid fraction obtained in the gastric digestion.

3 RESULTS AND DISCUSSION

3.1 *As and Cr, Cu, Mn, Pb and Cd analysis in carrots, beets and quinoa*

When vegetables are grown in contaminated soils, the bioavailable fraction of the elements present in the soil may migrate to the vegetable through intake mechanisms where microorganisms and the vegetal physiology are frequently involved.

Table 1 shows the concentrations of As, Cd, Cu, Pb, Mn and Cr in the lyophilized flesh of carrots, beets and in dried quinoa growing in the Chiu Chiu area; the concentrations of the same elements in local vegetables from Spain; and the metal concentration ratios between Chilean and Spanish samples. The concentrations of all tested elements in the Spanish samples fall within those levels reported for uncontaminated vegetables. The arsenic concentrations found in lyophilized carrots, beets and quinoa were 0.52,

Table 1. Total As, Cd, Pb, Cu, Mn and Cr content in Chilean and Spanish lyophilized vegetables. Results expressed in mg kg^{-1}.

Vegetables		As	Cd	Pb	Cu	Mn
Carrots	Flesh-C	0,52 ±0.04	0,05 ±0,01	0,12 ± 0,02	7,75 ± 0,91	4,75 ± 0,51
	Flesh-S	0,02 ± 0,01	0.07 ±0,0 1	0.09 ± 0,01	8,82 ± 0,59	2,28 ± 0,18
	Factor	**24**	**0.7**	**1.3**	**0.9**	**2**
	Peel-C	1,62 ±0,02	0,08 ± 0.01	0,21 ± 0.02	16,3 ± 1,5	9,95 ± 0,20
	Peel -S	0,15 ± 0,01	0,12 ± 0,02	0,23 ± 0,03	17,3 ± 0,9	13,7 ± 0,9
	Factor	**11**	**0.7**	**0.9**	0.9	**0.7**
Beets	Flesh-C	0,62 ± 0.05	0,09 ±0,00	0,36 ±0,01	6,71 ± 0,95	8,38 ±0.16
	Flesh-S	0,02±0,00	0.05±0,00	<LD	3,82±2,1	3,84±0,16
	Factor	**26**	**1.8**	(0,006) --	**1,7**	**2,2**
	Peel-C	3,20±0,06	0,11 ± 0,01	0,31 ± 0,05	28,6 ± 0,5	21,8 ± 0,2
	Peel-S	0,22±0,01	0,03±0,01	<LD	3,8±1,2	11,1± 0,3
	Factor	14	3,6	(0,008) --	**7**	**2**
Quinoa –C		0,20 ± 0,02	0,38 ± 0,07	0,04 ± 0,02	7,89 ± 0,73	13,5 ± 2,4
Quinoa -S		0.01± 0,01	<LD	0,09 ± 0,01	1,14 ± 0,16	4,11 ± 0,11
Factor		**20**	--	**0,4**	7	**3**

C: Chilean; S: Spanish. Factor: Metal concentration in Chilean/Spanish samples.

0.62 and 0.20 mg kg^{-1}, respectively. The most studied vegetable, the As concentration permitted by the Codex Alimentarious is 0.3 mg kg^{-1} for total As and 0.2 mg kg^{-1} of inorganic As. The WHO recommendation for rice is a limit of 1.0 mg As kg^{-1}. Chile has established a maximum of 0.5 mg As kg^{-1} for cereals. Therefore, the concentration found in the analyzed vegetables did not greatly exceed those levels allowed by the Chilean law.

Previous studies performed a decade ago in carrots growing in the same area showed total As levels around 50 mg kg^{-1}, a hundred times higher than those obtained nowadays soil cleaning and water decontamination treatments carried out nowadays, could be responsible for the decreasing trend found. The methodology employed for the analysis of most of the elements has been validated with CRMs of vegetables.

3.2 *Arsenic speciation in carrots, beets and quinoa*

In carrots, only inorganic As(III) and As(V) species were found, at similar concentration levels. For quinoa, 82% of the As was present as As(V), being about 15% as As(III). No other species were found in the water: methanol (1:1) fraction for both vegetables. Terrestrial plants in general are known to present very low efficiency mechanisms of biomethylation. Probably the contamination of these soils makes it difficult for soil microorganisms to develop biomethylation mechanisms that might produce methylated species available for absorption through the vegetables roots. Beets showed a different behavior, with inorganic As(V) (about 20%) and several unknown As species (representing about 70% of the total As), with retention time of 4.5 min, being present (data not shown). Sucrose is the sugar extracted from beets; therefore As-sucrose by-products might be the unidentified species. The results obtained in the present study, show a lack of methylated species probably due to the lower arsenic concentration than that previously found. However, the proportion of As(III) and As(V) species are in a similar range.

Table 2. Bioaccessibility of As in the digestion process of Chilean samples by ICP-MS.

Vegetables		As	Cd	Pb	Cu	Mn
Carrots	Flesh-C	0,52 ±0.04	0,05 ±0,01	0,12 ± 0,02	7,75 ± 0,91	4,75 ± 0,51
	Flesh-S	0,02 ± 0,01	0.07 ±0,0 1	0.09 ± 0,01	8,82 ± 0,59	2,28 ± 0,18
	Factor	**24**	**0.7**	**1.3**	**0.9**	**2**
	Peel-C	1,62 ±0,02	0,08 ± 0.01	0,21 ± 0.02	16,3 ± 1,5	9,95 ± 0,20
	Peel -S	0,15 ± 0,01	0,12 ± 0,02	0,23 ± 0,03	17,3 ± 0,9	13,7 ± 0,9
	Factor	**11**	**0.7**	**0.9**	0.9	**0.7**
Beets	Flesh-C	0,62 ± 0.05	0,09 ±0,00	0,36 ±0,01	6,71 ± 0,95	8,38 ±0.16
	Flesh-S	0,02±0,00	0.05±0,00	<LD	3,82±2,1	3,84±0,16
	Factor	**26**	**1.8**	(0,006) --	**1,7**	**2,2**
	Peel-C	3,20±0,06	0,11 ± 0,01	0,31 ± 0,05	28,6 ± 0,5	21,8 ± 0,2
	Peel-S	0,22±0,01	0,03±0,01	<LD	3,8±1,2	11,1± 0,3
	Factor	14	3,6	(0,008) --	**7**	**2**
Quinoa –C		0,20 ± 0,02	0,38 ± 0,07	0,04 ± 0,02	7,89 ± 0,73	13,5 ± 2,4
Quinoa -S		0.01± 0,01	<LD	0,09 ± 0,01	1,14 ± 0,16	4,11 ± 0,11
Factor		**20**	--	**0,4**	7	**3**

Results expressed in lyophilized vegetables (n = 3).

3.3 *Bioaccessibility of As in carrots, beets and quinoa*

Compared to in vivo studies, the in vitro sequential gastrointestinal fluid extraction is a relatively simple, yet effective approach to estimate the bioaccessibility of dietary arsenic. Table 2 shows: the total As concentrations in the different extracts of the in vitro gastrointestinal digestion process.

No significant transformation of the original As species found in the methanolic or enzymatic solutions seems to take place during "in vitro" gastrointestinal digestion of the vegetables.

4 CONCLUSIONS

Inorganic As(III) and As(V) were the only species present in carrots and quinoa. Regarding the toxicological risk, quinoa is the vegetable to be able to accumulate the highest amount of inorganic As, but its bioaccessibility is only about 40%. On the contrary, for carrots and beets the As bioaccessibility is about 100%. However, since the main As species present in beets are probably arsenosugars, presumably less toxic.

ACKNOWLEDGEMENTS

The authors thank Universidad Complutense de Madrid, Spain (007/ 11 VIII-2011 (S2013/ABI-3028), the European FEDER program and the project MECE-SUP U.A., Chile, for financial support.

REFERENCES

Khouzam, R.B., Pohl, P. & Lobinski, R. 2011. Bioaccessibility of essential elements from white cheese, bread, fruit and vegetables. *Talanta* 30(86):425–428.

Pizarro, I., Gómez, M., Cámara, C. & Palacios, M.A. 2003. Arsenic speciation in environmental and biological samples. Extraction and stability studies. *Anal. Chim. Acta* 495(1–2): 85–98.

Environmental Arsenic in a Changing World –
Zhu, Guo, Bhattacharya, Ahmad, Bundschuh & Naidu (Eds)
ISBN 978-1-138-48609-6

Contribution of (factors) cooking water, raw rice, and traditional cooking method on cooked rice arsenic level

D. Chatterjee & U. Mandal
Department of Chemistry, University of Kalyani, Kalyani, Nadia, West Bengal, India

ABSTRACT: Rice consumption has now been established as a well-recognized exposure source of arsenic for many subpopulations. The aim is to investigate the various factors that might exert control on the final arsenic content in cooked rice following the indigenous cooking practice pursued by the rural villagers of West Bengal. It was found that the use of arsenic-rich groundwater for cooking elevates the arsenic concentration in cooked rice (up to 162% than the raw rice), making the rural population of West Bengal particularly vulnerable to arsenic poisoning through rice consumption. Results of our study show that for the cooking method employed, rice variety, background arsenic concentration in raw rice and cooking water arsenic concentration are important predisposing factors that direct the accumulation of arsenic in cooked form.

1 INTRODUCTION

In the last few decades, a lot of articles have been published which show that millions of people worldwide are exposed to high levels of arsenic (As) from drinking contaminated water (Nriagu *et al.*, 2007). The major point of concern is that rice is the dominant food source for the world population, particularly in developing Asian countries, where an average of 72.8% of the daily calorie intake per capita are provided from rice intake. Since rice has the potential to accumulate more As than other food crops, rice intake has provided a facile route for As exposure (Chatterjee *et al.*, 2010; Meharg *et al.*, 2009). The risk posed from rice depends upon both the amount of rice consumed and the concentration of inorganic As in rice grains, moderated by gut bioavailability. In the present study, rice samples were cooked in individual households in rural Bengal affected by endemic arsenocosis, following the traditional cooking method adopted by the villagers. The use of As-free cooking water in decreasing As content in cooked rice was explored so as to evaluate the reduction in As exposure for the population.

2 MATERIALS AND METHODS

2.1 *Study area*

The study area is located in Chaku-danga village of Chakdaha block in Nadia district, West Bengal. The agricultural system is highly dependent upon groundwater to meet the water requirement for irrigation of crop.

2.2 *Rice cooking and sample collection*

For this study, rice was cooked in individual households following the procedure adopted by rural villagers. Generally, rice was washed 3–4 times with water and the washed rice was soaked in excess water (10–12 times the weight of raw rice) for 15–20 minutes and finally cooked. After cooking, the excess starch water (gruel) was discarded by tilting the pan against the lid. The entire cooking procedure as described above was done using the supplied Milli-Q water.

2.3 *Sample analysis*

Cooking water, raw rice, starch water, and cooked rice were analyzed for As by hydride generation atomic absorption spectroscopy (HG-AAS) (Varian AA240).

2.4 *Statistical analysis*

SPSS statistical software, version 17.0 by IBM was used for data analysis. Independent variables such as rice variety, As concentration in raw rice and cooking water were tested for multiple linear regression analysis with As content in cooked rice. Statistical significance was indicated by values of $p < 0.05$.

3 RESULTS AND DISCUSSION

Our initial baseline survey revealed that Ratna, Satabdi, Pratik, Parijat, Ranjit, Swarna are the most common rice varieties that are used by the villagers in the study area. The median As concentration in raw rice (uncooked) for the six rice varieties investigated in this study was 228 μg kg^{-1} ranging between 105–510 μg kg^{-1}. Categorizing the collected raw rice As concentration with respect to rice varieties, it was found that Swarna had the highest grain As concentration (median: 380 μg As kg^{-1}; range: 270–510 μg As kg^{-1}) followed by Ranjit (median: 221 μg As kg^{-1}; range: 206–360 μg As kg^{-1}), Parijat (median: 205 μg As kg^{-1}; range: 140–370 μg As kg^{-1}), Pratik (median: 194 μg As kg^{-1}; range: 155–358 μg As kg^{-1}), Ratna (median: 179 μg As kg^{-1}; range: 130–235 μg As kg^{-1}), while Satabdi (median: 163 μg As kg^{-1}; range: 105–356 μg As kg^{-1}) had the lowest As concentration. The variability and distribution of total As in the classified rice grains have been shown by Box and Whisker plot.

Studies have shown that there can be severe variation in the final As concentration in cooked rice from the initial As concentration in raw rice. This study indicates among the six rice varieties studied, Swarna variety (short bold) has the maximum As concentration compared to the other rice varieties. However, the discrepancies in As concentration with grain size is somewhat perturbed when plotted for the cooked form indicating that the As concentration in cooked rice may vary greatly from the raw rice, depending upon cooking water and pattern and rice varieties. Results of our study indicate that As in cooking water can be a very important determinant for As in cooked rice. Elevated concentration of As in cooking water can result in the influx of As in cooked rice, thereby causing an increase in As content in the cooked form. Moreover, rice variety also influences the amount of As influx in the cooked rice samples, thereby being a principal factor controlling the cooked rice As concentration. Cooking rice with water with low As concentration can therefore be seen as a "decontamination" process as it tends to leach-out As from rice grain into the gruel and thereby ultimately decreasing the As exposure from rice intake. The results confirm the findings of the past laboratory based studies which have shown reduction in As content in cooked rice following the use of low As water.

4 CONCLUSIONS

This study further suggests that, provided As-safe cooking water are used, the traditional cooking practice can actually be beneficial for the rural people of Bengal as the cooking procedure and the large volume of low As water enables the contaminated rice with sufficient contact time to leach As from the rice grains into the starch water, thereby food grain highly elevated with As may be made suitable for human consumption. Therefore, although strategies on decreasing the As content in raw rice grains should be promoted, As in cooked rice is the actual point of convergence for the dietary exposure for the population and attempts to limit the As in cooked rice must be taken into consideration to successfully combat the menace of As.

ACKNOWLEDGEMENTS

We would also like to thank UGC–SAP program and DST-FIST program to the Department of Chemistry, University of Kalyani.

REFERENCES

Chatterjee, D., Halder, D., Majumder, S., Biswas, A., Nath, B., Bhattacharya, P., Bhowmick, S., Mukherjee-Goswami, A., Saha, D., Maity, P.B., Chatterjee, D., Mukherjee, A. & Bundschuh, J. 2010. Assessment of arsenic exposure from groundwater and rice in Bengal Delta Region, West Bengal, India. *Water Res.* 44(19): 5803–5812.

Meharg, A.A., Williams, P.N., Adomako, E., Lawgali, Y.Y., Deacon, C., Villada, A., Cambell, R., Sun, G.X., Zhu, Y.G., Feldmann, J., Raab, A., Zhao, F.J., Islam, M., Hossain, S. & Yanai, J. 2009. Geographical variation in total and inorganic arsenic content of polished (white) rice. *Environ. Sci. Technol.* 43(5): 1612–1617.

Nriagu, J., Bhattacharya, P., Mukherjee, A., Bundschuh, J., Zevenhoven, R. & Loeppert, R. 2007. Arsenic in soil and groundwater: an overview. In: Bhattacharya, P., Mukherjee, A.B., Bundschuh, J., Zevenhoven, R., Loeppert, R.H. (eds), *Arsenic in Soil and Groundwater Environment*. Elsevier, Amsterdam, pp. 3–60.

Environmental Arsenic in a Changing World –
Zhu, Guo, Bhattacharya, Ahmad, Bundschuh & Naidu (Eds)
ISBN 978-1-138-48609-6

Study of arsenic in drinking water and food in three different districts of Bihar, India

P.K. Sharma[1], S. Suman[1], R. Kumar[1], A.K. Ghosh[1], D. Mondal[2], N. Bose[3] & S.K. Singh[4]
[1]*Mahavir Cancer Institute and Research Centre, Patna, Bihar, India*
[2]*School of Environment & Life Sciences, University of Salford, UK*
[3]*Department of Geography, A.N. College, Patna, India*
[4]*Department of Environment and Water Management, A.N. College, Patna, India*

ABSTRACT: Arsenic contamination in groundwater is a serious and widespread global public health problem and India is one of the highly arsenic exposed countries with maximum impact in the Ganga Meghna Brahmaputra plains. A total of 18 districts out of 38 districts of the state of Bihar are highly affected with groundwater arsenic poisoning. In the present study, three arsenic affected districts were taken selected for survey. The study deals with the survey of the affected village in the district, with groundwater arsenic quantification with special attention to determine the concentration of arsenic in food material. The maximum arsenic contamination in water sample was 107.1 $\mu g\,L^{-1}$ in Saran while 1257 $\mu g\,kg^{-1}$ in rice in Bhagalpur, 567.5 $\mu g\,kg^{-1}$.

1 INTRODUCTION

Arsenic in the recent times has caused health hazards in a huge population inhabiting Ganga-Meghna-Brahmaputra basin. The entire Ganga-Meghna Brahmaputra (GMB) plains has high arsenic concentrations ($>10\,\mu g\,L^{-1}$) in groundwater. Populations residing in these plains are exposed to arsenic which has led to many health-related problems like keratosis, melanosis and even cancer (Ahamed *et al.*, 2006).

There is no effectively remedial action of chronic arsenicosis. An assessment of As and other elements in the food components was conducted to understand the actual disease burden caused by ingestion of food in people residing in the middle Gangetic plain (Kumar *et al.*, 2016).

In Bihar, presently 18 districts out of a total of 38 are affected by arsenic and it is assumed that a population of about 5 million are consuming arsenic contaminated water with more than $50\,\mu g\,L^{-1}$ (Singh *et al.*, 2014). Bhojpur, Buxer and Bhagalpur was the worst affected district among others. The highest value of arsenic ($1861\,\mu g\,L^{-1}$) was detected in the water of the tube-well at the Bhojpur area. All three confirmed sources of arsenic; water, soil and food with elevated arsenic level is posing a serious threat to the residing communities and hence their health is at risk (Singh *et al.*, 2011).

2 METHODS

2.1 *Study area*

Samples were collected from three different arsenic affected district of Bihar., namely Bhagalpur (located on southern bank of Ganges river), Saran (situated near the junction of Ghaghara and Ganga river) and Patna (located on southern bank of Ganges river). These districts were selected as a part of our ongoing NUTRI-SAM project supported by DST-UKIERI.

2.2 *Sample collection and arsenic estimation*

Water sample and three types of food samples like raw rice, wheat and potato were collected in a sterile polyethylene bag. Arsenic was estimated by atomic absorption spectroscopy (Perkin Elmer PinAAcle 900T) after processing and digestion.

2.3 *Statistical analysis*

To determine the significant mean, correlation between the various data, standard deviation SPSS software (IBM, version 22) was used.

3 RESULTS AND DISCUSSION

3.1 *Arsenic profile in water samples*

The arsenic concentration (Table 1) in drinking water was elevated in some of the tested samples. The maximum arsenic concentration in groundwater sample reported was $107.10\,\mu g\,L^{-1}$ in Saran District with the mean value $44.13\,\mu g\,L^{-1}$ and standard deviation 38.07 and the lowest arsenic concentration in groundwater reported was $50.60\,\mu g\,L^{-1}$ in Patna district with the mean value is $26.87\,\mu g\,L^{-1}$ and the standard deviation is 19.13.

Table 1. Arsenic in water samples.

District	Maximum ($\mu g\ L^{-1}$)	Mean ($\mu g\ L^{-1}$)	Standard deviation
Bhagalpur (N = 14)	99.5	30.5	43.2
Saran (N = 14)	107.1	44.1	38.1
Patna (N = 14)	50.6	26.9	19.1

Table 2. Arsenic in raw rice samples.

District	Maximum ($\mu g\ kg^{-1}$)	Mean ($\mu g\ kg^{-1}$)	Standard deviation
Bhagalpur (N = 13)	1257.5	108.3	345.8
Patna (N = 14)	66.3	48.7	10.5

Table 3. Arsenic in wheat.

District	Maximum ($\mu g\ kg^{-1}$)	Mean ($\mu g\ kg^{-1}$)	Standard deviation
Bhagalpur (N = 11)	567.5	340.6	161.4
Saran (N = 12)	175.0	57.9	57.8
Patna (N = 10)	215.5	135.3	46.3

Table 4. Arsenic in Potato samples.

District	Maximum ($\mu g\ kg^{-1}$)	Mean ($\mu g\ kg^{-1}$)	Standard deviation
Bhagalpur (N = 9)	826.0	558.6	127.4
Saran (N = 14)	195.7	123.0	38.0
Patna (N = 10)	170.1	143.6	28.6

3.2 *Arsenic profile in food samples*

For the total arsenic in analyzed samples of raw rice (Table 2), the highest arsenic concentration was in 1257.5 $\mu g\ kg^{-1}$ in Bhagalpur district with mean value 108.34 $\mu g\ kg^{-1}$ and the standard deviation is 345.78. In Patna district arsenic in raw rice was in very low concentration, maximum value was found was 66.30 $\mu g\ kg^{-1}$ with mean value 48.68 $\mu g\ kg^{-1}$ and the standard deviation is 10.49.

In total analyzed wheat samples (Table 3) the highest arsenic concentration was 567.5 $\mu g\ kg^{-1}$ in Bhagalpur district with mean value 340.6 $\mu g\ kg^{-1}$ and standard deviation 161.4 In Patna district arsenic in wheat was in moderate concentration, maximum value was 215.5 $\mu g\ kg^{-1}$ with mean value 135.3 $\mu g\ kg^{-1}$ and standard deviation 46.32 The lowest arsenic concentration in wheat is in Saran district i.e. 175 $\mu g\ kg^{-1}$ with the mean value 57.9 and standard deviation 57.8.

In total analyzed potato samples (Table 4) the highest arsenic concentration was in 826 $\mu g\ kg^{-1}$ in Bhagalpur district with the mean value 558.7 $\mu g/kg$ and standard deviation 127.40 In Saran district arsenic in wheat is in moderate concentration, maximum value was 195.7 $\mu g\ kg^{-1}$ with mean value 123.0 $\mu g\ kg^{-1}$ and the standard deviation is 37.98. The lowest arsenic concentration in wheat is in Patna district i.e. 170.1 $\mu g\ kg^{-1}$ with the mean value 143.6 and standard deviation 28.6.

Arsenic in food items (Rice, Wheat and Potato) was not correlated with the concentration in drinking water. This was due to the fact that the food procured is not grown locally. The water used for irrigation is different from the water used for the drinking.

4 CONCLUSIONS

The mean arsenic concentration in water samples of three entire selected region was not so much. Most of the samples were within the range ($<50\ \mu g\ L^{-1}$). The arsenic concentration in food samples (rice, wheat and potato) of Bhagalpur was found to be high in comparison to Patna and Saran district. The arsenic content in our wheat samples were within the range of 490–1150 $\mu g\ kg^{-1}$. No correlation was observed in food and drinking water samples of the study area.

ACKNOWLEDGEMENTS

This study is funded by DST-UKIERI Thematic partnership project. We thank all participants.

REFERENCES

Ahamed, S., Sengupta, M.K., Mukherjee, A., Hossain, M.A., Das, B., Nayak, B., Pal, A., Mukherjee, S.C., Pati, S. & Dutta, R.N. 2006. Arsenic groundwater contamination and its health effects in the state of Uttar Pradesh (UP) in upper and middle Ganga plain, India: a severe danger. *Sci. Total Environ.* 370(2–3): 310–322.

Kumar, M., Rahman, M.M., Ramanathan, A.L. & Naidu, R. 2016. Arsenic and other elements in drinking water and dietary components from the middle Gangetic plain of Bihar, India: health risk index. *Sci. Total Environ.* 539: 125–134.

Singh, S.K. & Ghosh, A.K. 2011. Entry of arsenic into food material-a case study. *World Appl. Sci. J.* 13(2): 385–390.

Singh, S., Ghosh, A., Kumar, A., Kislay, K., Kumar, C., Tiwari, R., Parwez, R., Kumar, N. & Imam, M. 2014. Groundwater arsenic contamination and associated health risks in Bihar, India. *Int. J. Environ. Res.* 8(1): 49–60.

Environmental Arsenic in a Changing World –
Zhu, Guo, Bhattacharya, Ahmad, Bundschuh & Naidu (Eds)
ISBN 978-1-138-48609-6

Effects of arsenic on growth, saponins and flavoids yield of *Panax notoginseng* (Burk. F.H. Chen) and its control measurements

N. Li, S.C. Cheng, X.Y. Mei, Q. Li & Y.Q. Zu
College of Resources and Environment, Yunnan Agricultural University, Kunming, P.R. China

ABSTRACT: Arsenic stress inhibited the growth of *Panax notogensing*. The citric acid and phosphorus fertilizer can promote the growth of *notogensing*, increase the biomass and morphological index, improve the efficacy of saponins and flavonoid contents. They also make it release of soil arsenic, increasing arsenic content in *Panax notogensing* to exceed the standard, phosphorus is not conducive to the release of arsenic in soil, reducing the absorption of arsenic in plants.

1 INTRODUCTION

Arsenic (As) is a naturally toxic element and widely distributed in the environment, and it is cancerogenic to humans. Environmental As pollution in soil has become more and more severe problem (Dudka & Adriano, 1997; Zhou *et al.*, 2016). High soil arsenic background value, which was mean 65.6 mg kg^{-1} in the range of 6.9 to 242.0 mg kg^{-1}, was found in Wenshan province. the main production area of *Panax notoginseng* (Burk. F.H. Chen), which has been cultivated for importantly traditional medicine in China, *Panax notoginseng* is an herbal species in the genus of Panax, family of Araliaceae (Dahui *et al.*, 2014; Guo *et al.*, 2010). The mining activities and unreasonable fertilization could cause arsenic content exceeding the standard and affect the growth and quality of *P. notogensing*.

2 METHODS/EXPERIMENTAL

Field experiments were carried out in Dalongtan village (22°39′N, 103°06′E, elevation 2251 m), Xundian County, Kunming prefecture, Yunnan province, China. Each plot was 3 m^2 in size. Soil collected from the 0–15 cm soil layer was mixed with the As stock solutions with As(V) 0, 20, 80, 140, 200, 260 mg kg^{-1} (prepared using H_3AsO_4). The As treated soil was incubated for two weeks. Soil before As treatment had a pH 5.32, organic matter content of 16.77 g kg^{-1}, total N content of 0.60 g kg^{-1}, total P content of 1.20 g kg^{-1} and As content at 12.00 mg kg^{-1}. Soil type was upland red soil with clay loam texture. Plot and pot experiments were conducted to study effects of citric acid (0, 4, 6, 8 mmol kg^{-1}) and calcium magnesium phosphate fertilizer (0, 60, 120, 180 mg kg^{-1}) on growth, photosynthesis and medicinal components flavonoids and spanins of *P. notogensing* under arsenic stress.

3 RESULTS AND DISCUSSION

The results showed that: (1) The biomass and plant growth under As 140 mg kg^{-1} treatment significantly decreased compared with control (0 mg kg^{-1} As treatment). For biennial *P. notoginseng*, arsenic stress had no significant effect on the contents of saponins. For triennial *P. notoginseng*, there was a negative correlation between content of saponins and content of arsenic ($P < 0.05$). There were no significant changes in PAL, SS and CHS enzyme activities of the *P. notoginseng* enzymes in the biosynthesis of flavonoids in triennial *P. notoginseng*. (2) Arsenic concentrations in main root under 140 mg kg^{-1} and 8 mmol kg^{-1} citric acid treatment reached the highest and increased by 30% coping with non citric acid treatment. Citric acid promoted the synthesis of photosynthetic pigments, increased the content of photosynthetic pigments, decreased the content of malondialdehyde (MDA). Citric acid effectively improved the flavonoids accumulation (Fig. 1). Under As 140 mg kg^{-1} treatment, flavonoid accumulation increased by 29% with 8 mmol kg^{-1} citric acid application. (3) The content of arsenic in 6 mmol kg^{-1} and 8 mmol kg^{-1} citric acid leaching solution was significantly higher 21% and 33% than that of non citric acid leaching solution, respectively. Compared with non citric acid, contents of F2 (specifically adsorbed arsenic), F3 (weak crystalline iron and aluminum oxide hydrate combination fraction) and F4 (crystalline hydrated iron aluminum oxide with arsenic) of 8 mmol kg^{-1} citric acid increased by 11%, 33% and

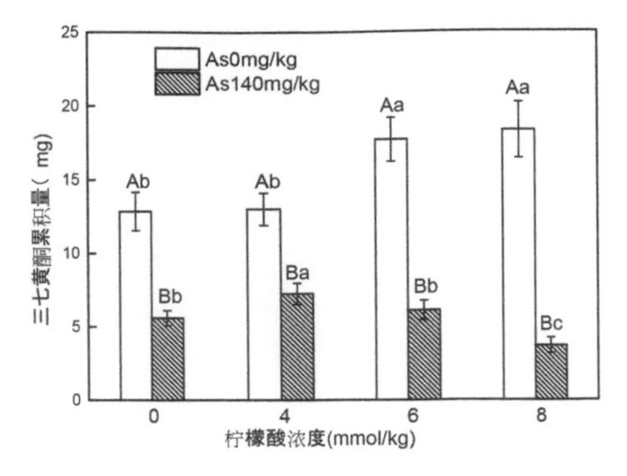

Figure 1. Effect of citric acid on the flavonoids accumulation under As 140 mg kg^{-1} treatment.

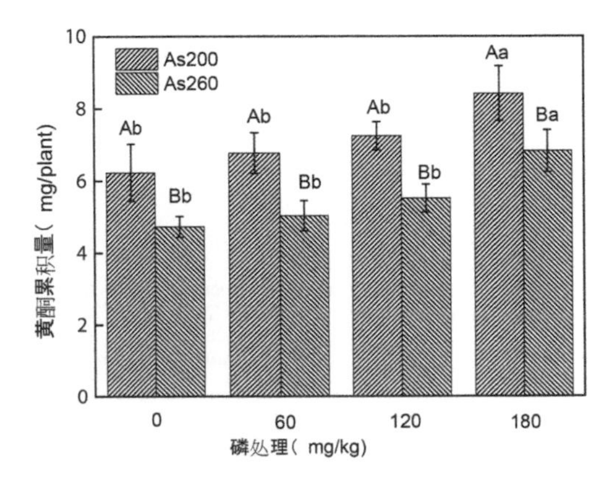

Figure 2. Effect of different P levels on flavonoids accumulation of *P. notoginseng* under As stress.

32%, respectively. (4) Phosphate fertilizer could significantly improve the *P. notogensing* biomass under As concentration of 200 mg kg^{-1} and 260 mg kg^{-1} and phosphate fertilizer 180 mg kg^{-1} treatment, biomass significantly increased by 56% and 17%. Plant heights were significantly increased by 5% and 26%, respectively; the leaf area increased by 33% and 41%. Flavonoids accumulation increased under different P treatment levels (Fig. 2).

4 CONCLUSIONS

Arsenic stress inhibited the growth of *Panax notogensing*, the citric acid and phosphorus fertilizer can promote the growth of *notogensing*, increased the biomass and morphological index, improve the efficacy of saponins and flavonoid content, but also easy to make soil arsenic release, lead arsenic content exceed the standard, phosphorus is not conducive to the release of arsenic in soil thus, reduce the absorption of arsenic in plants.

ACKNOWLEDGEMENTS

The study was supported by the National Natural Science Foundation of China (Grant Nos. 41261096 and 31560163).

REFERENCES

Dahui, L., Na, X., Li, W., Xiuming, C., Lanping, G., Zhihui, Z., Jiajin, W. & Ye, Y. 2014. Effects of different cleaning treatments on heavy metal removal of Panax notoginseng (Burk) FH Chen. *Food Addit. Contam. Part A* 31(12): 2004–2013.

Dudka, S. & Adriano, D.C., 1997. Environmental impacts of metal ore mining and processing: a review. *J. Environ. Qual.* 26(3): 590–602.

Guo, H.B., Cui, X.M., An, N., & Cai, G.P., 2010. Sanqi Ginseng (*Panax notoginseng* (Burkill) F.H. Chen) in China: distribution, cultivation and variations. *Genet. Resour. Crop Ev.* 57(3): 453–460.

Zhou, Q., Guo, J.J., He, C.T., Shen, C., Huang, Y.Y., Chen, J.X., Guo, J.H., Yuan, J.G. & Yang, Z.Y. 2016. Comparative transcriptome analysis between low- and high – cadmium – accumulating genotypes of Pakchoi (*Brassica chinensis* L.) in response to cadmium stress. *Environ. Sci. Technol.* 50(12): 6485–6494.

Section 3: Health impacts of environmental arsenic

3.1 Exposure and epidemiology of arsenic impacts on human health

Environmental Arsenic in a Changing World –
Zhu, Guo, Bhattacharya, Ahmad, Bundschuh & Naidu (Eds)
ISBN 978-1-138-48609-6

Arsenic, DNA damage, and cancers of bladder and kidney – a long-term follow-up of residents in arseniasis endemic area of north-eastern Taiwan

S.-L. Wang[1], S.-F. Tsai[1], L.-I. Hsu[2], C.-J. Chen[2], K.-H. Hsu[3] & H.-Y. Chiou[4]
[1]*National Institute of Environmental Heath Sciences, National Health Research Institutes, Miaoli, Taiwan*
[2]*Genomics Research Centre, Academia Sinica, Taipei, Taiwan*
[3]*Department of Health Care Management, Chang-Gung University, Taoyuan, Taiwan*
[4]*Department of Public Health, Taipei Medical University, Taipei, Taiwan*

ABSTRACT: The aim is to assess dose-response association between arsenic exposure, DNA damage biomarkers, and the incidence of bladder and kidney cancers. A total of 8102 men and women from 3901 households have been enrolled in 1991–1994, and followed in 2011–2014. The data collected included well water consumption, habits of cigarette smoking, alcohol consumption, exercise and diet, through standardized personal interview. The individual urinary arsenic species were quantified using high-performance liquid chromatography–inductively coupled plasma/mass spectrometry (HPLC-ICP/MS). For assessment of oxidative and methylated DNA lesions and depletion, urinary 8-oxo-7,8-dihydro-2′-deoxyguanosine (8-oxodG) and N7-methylguanine (N7-MeG) were measured respectively, using liquid chromatography/tandem mass spectrometry (LC-MS/MS). The National Cancer Registry Data using the pathology finding defined bladder and kidney cancers. Urinary levels of the two DNA adduct increased significantly with increasing urinary arsenic level (iAs+MMA+DMA) in both men ($\beta = 0.82$, $\beta = 0.34$ for 8-oxodG and N7-MeG, respectively, $p < 0.0001$ for both) and women ($\beta = 1.03$, 0.38, $p < 0.0001$) adjusted for potential confounders. Incidence rate of bladder and kidney cancer using person-years tended to be the highest for higher urinary inorganic and methylated arsenic with higher DNA adduct level than the medians. It is suggested that subjects with high arsenic exposure experienced further cancer risk with high level of DNA damage biomarker.

1 INTRODUCTION

Arsenic is ubiquitous and becoming one of the largest environmental health concerns in the area where groundwater is needed as source of drinking water, or arsenic is prevalent in air pollutants. High inorganic arsenic exposure has been found to be related to various cancers and cardiovascular diseases (Lee & Yu, 2016; Rorbach-Dolata *et al.*, 2015; Yager *et al.*, 2013). The aim of this study is to assess dose-response association between arsenic exposure, DNA damage biomarkers, and the incidence of bladder and kidney cancers.

2 METHODS

2.1 *Subjects*

A total of 8102 men and women from 3901 households have been enrolled in 1991–1994 and followed in 2011–2014. The data collected included well water consumption, habits of cigarette smoking, alcohol consumption, exercise and diet, and personal and family history of major diseases through standardized personal interview.

2.2 *Measurements*

The individual urinary arsenic species were quantified using high-performance liquid chromatography–inductively coupled plasma/mass spectrometry (HPLC-ICP/MS).

For assessment of oxidative and methylated DNA lesions and depletion, urinary 8-oxo-7,8-dihydro-2′-deoxyguanosine (8-oxodG) and N7-methylguanine (N7-MeG) were measured respectively, using liquid chromatography/tandem mass spectrometry (LC-MS/MS). The National Cancer Registry Data using the pathology finding defined bladder and kidney cancers.

3 RESULTS AND DISCUSSION

Urinary levels of the two DNA adduct increased significantly with increasing urinary arsenic level (iAs+MMA+DMA) in both men ($\beta = 0.82$, $\beta = 0.34$ for 8-oxodG and N7-MeG, respectively, $p < 0.0001$ for both) and women ($\beta = 1.03$, 0.38, $p < 0.0001$) adjusted for potential confounders of age and cigarette smoking. Incidence rate of bladder and kidney cancer using person-years tended to be the highest for higher urinary inorganic and methylated arsenic with higher DNA adduct level than the medians.

4 CONCLUSIONS

It is suggested that subjects with high arsenic exposure experienced further cancer risk with high level of DNA damage biomarker.

ACKNOWLEDGEMENTS

Grant from Ministry of Science and Technology (MOST) Taiwan.

REFERENCES

Lee, C.H. & Yu, H.S. 2016. Role of mitochondria, ROS, and DNA damage in arsenic induced carcinogenesis. *Front Bioscan (Schol Ed)*8: 312–320.

Rorbach-Dolata, A., Marchewka, Z. & Piwowar, A. 2015. The biochemical carcinogenesis of selected heavy metals in bladder cancer. *Postepy Biochem.* 61(2): 176–182.

Yager, J.W., Gentry, P.R., Thomas, R.S., Pluta, L., Efremenko, A., Black, M., Arnold, L.L., McKim, J.M., Wilga, P. & Gill, G. 2013. Evaluation of gene expression changes in human primary uroepithelial cells following 24-hr exposures to inorganic arsenic and its methylated metabolites. *Environ. Mol. Mutagen.* 54(2): 82–98.

Environmental Arsenic in a Changing World –
Zhu, Guo, Bhattacharya, Ahmad, Bundschuh & Naidu (Eds)
ISBN 978-1-138-48609-6

Arsinothricin: a novel arsenic-containing antibiotic

M. Yoshinaga[1], V.S. Nadar[1], J. Chen[1], D.S. Dheeman[1], B.P. Rosen[1],
M. Kuramata[2] & S. Ishikawa[2]
[1]*Herbert Wertheim College of Medicine, Florida International University, Miami, FL, USA*
[2]*Institute for Agro-Environmental Sciences, NARO, Tsukuba, Ibaraki, Japan*

ABSTRACT: The emergence and spread of bacterial resistance highlights the urgent need for new antibiotics. Organoarsenicals have been used as antimicrobials since Paul Ehrlich's salvarsan. Recently a soil bacterium was shown to produce the organoarsenical arsinothricin (AST), a mimetic of the herbicidal antibiotic phosphinothricin (PT). We demonstrate that AST is an effective broad-spectrum antibiotic, showing that bacteria have acquired the ability to utilize environmental arsenic to produce a potent antimicrobial. With every new antibiotic, resistance inevitably arises. The *arsN* gene, widely distributed in *ars* operons, confers PT resistance. The functional linkage of *arsN* to arsenic detoxification was unclear. Here we show that *arsN* selectively confers resistance to AST. Crystal structures of the ArsN *N*-acetyltransferase shed light on the mechanism of selectivity. These results may lead to development of a new class of antimicrobials and ArsN inhibitors.

1 INTRODUCTION

New antibiotics are urgently needed because the emergence of resistance has rendered nearly every clinically used antibiotic ineffectual. Recently a new arsenic-containing natural product, arsinothricin (AST), was shown to be produced by the rhizosphere microbe *Burkholderia* sp. GSRB05 (Kuramata *et al.*, 2016). AST is an arsenic mimetic of the *Streptomyces* phosphonate antibiotic phosphinothricin (PT). The use of arsenicals as antimicrobial and anticancer agents is well-established (Jolliffe, 1993). Here we show that AST is a potent broad-spectrum antibiotic and propose that AST has the potential to be the progenitor of a new class of organoarsenical antibiotics and provide a countermeasure to the return of the pre-antibiotic era.

2 METHODS

2.1 *In vivo resistance assays*

Cells were grown in LB medium to log phase, following which the cells were 100-time diluted in M9 medium containing arsenite (As(III)), methylarsenite (MAs(III)), AST or PT. Resistance was determined from the A_{600} after 24 h incubation.

2.2 *Gene cloning, protein purification and crystallization and enzyme assays*

Pseudomonas putida KT2440 *arsN* (*PparsN*) and the *pat* gene encoding PT N-acetyltransferase from *Streptomyces viridochromeogenes* (*Svpat*) were cloned into pBAD expression vector and transformed into *E. coli* TOP10. His-tag PpArsN and SvPAT were purified from the induced cells using a Ni-NTA column chromatography. Crystals of apo- and AST-bound PpArsN were grown at room temperature. X-ray data were collected on beamline 22ID at the Advanced Photon Source, Argonne National laboratory. The structures were solved by molecular replacement method using PDB ID: 1YVO as a template. The enzymatic activities and kinetic parameters of the purified PpArsN and SvPAT for PT and AST were determined (Thompson *et al.*, 1987).

3 RESULTS AND DISCUSSION

3.1 *AST is a potent broad-spectrum antibiotic*

AST is effective on a variety of Gram-negative and Gram-positive bacterial species at a thirtieth of the concentration of PT, suggesting that AST is a much more effective broad-spectrum antibiotic than PT (Fig. 1). Two bacterial species were resistant to AST. *Burkholdeira* sp. GSRB05 is the producer of AST, so it is not unreasonable that the species has a mechanism of resistance to AST and PT. As discussed below, the *arsN* gene confers resistance in *P. putida* KT2440. In *E. coli* AW3110 Δ*arsRBC*, AST was found to be considerably more inhibitory than As(III), and its inhibitory activity was similar to that of highly toxic MAs(III). These results suggest that AST-synthesizing bacteria use environmental inorganic arsenic to produce and secrete AST to kill off competitors.

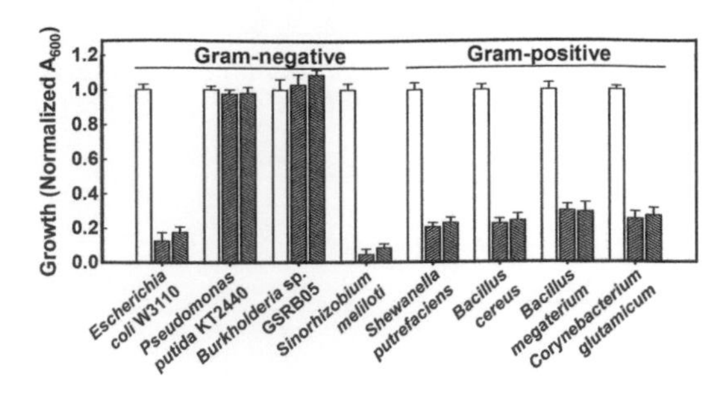

Figure 1. AST is a broad-spectrum antibiotic. Cells were cultured in the absence (left) or presence of 25 μM AST (middle) or 800 μM PT (right). Data are the mean ± SE (n − 3).

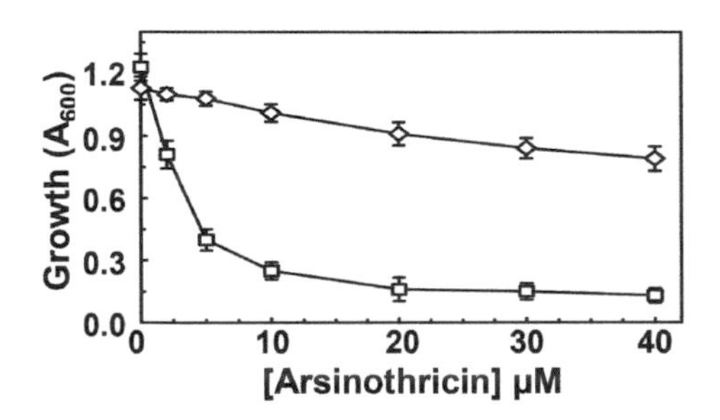

Figure 2. PpArsN confers resistance to AST. Strains: *E. coli* AW3110 bearing control plasmid (□) or plasmid carrying the *PparsN* gene (◇). Data are the mean ± SE (n = 3).

3.2 *ArsN confers AST resistance*

With every antibiotic, resistance inevitably arises. PT resistance is conferred by PT *N*-acetyltransferase (PAT). A gene termed *arsN* is found in many arsenic resistance (*ars*) operons that has been annotated as encoding a PAT. ArsN from *P. putida* KT2440 (PpArsN) has been demonstrated to inactivate PT by *N*-acetylation (4). Why an enzyme for PT resistance should be in *ars* operons was a mystery. The identification of AST as a natural product suggested that the biological function of ArsN could be as an AST resistance. We examined the ability of ArsN to confer AST resistance (Fig. 2).

AW3110 is sensitive to AST, whereas AW3110 expressing *PparsN* became resistant to AST, supporting our hypothesis that ArsN confers AST resistance. Heterologous expression of *PparsN* also conferred PT resistance. These results are consistent with the results of *P. putida* (Páez-Espino *et al.*, 2015) (Fig. 1). In addition, we demonstrated that purified PpArsN is more selective for AST over PT. We compared this with the substrate selectivity of purified PAT from the PT-synthesizing *S. viridochromeogenes* (SvPAT), whose natural substrate is PT. SvPAT showed comparable activity with PT and AST. Thus ArsN, which is selective for AST, is different from PAT, which shows no selectivity.

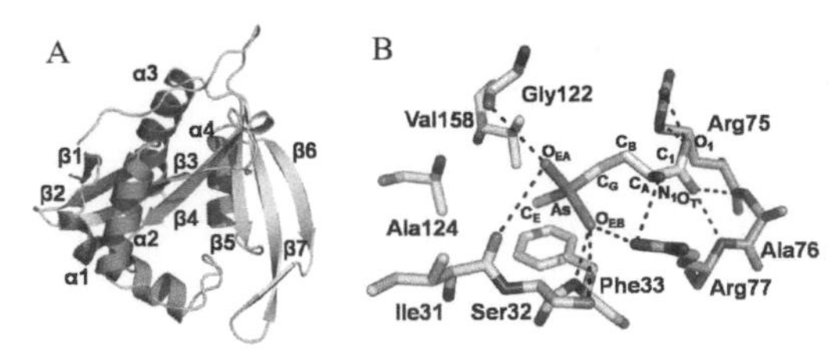

Figure 3. Structure of PpArsN. A: Overall fold of apo-PpArsN. α helices are shown in dark green. B: Interaction of AST with PpArsN. The AST binding site is formed by residues from both the chains A (cyan) and B (green) of the dimer.

3.3 *Crystal structure of ArsN*

To elucidate the mechanism of PpArsN resistance and its selectivity for AST, we solved structures of apo- and AST-bound PpArsN. PpArsN belongs to the GCN5 *N*-acetyltransferase (GNAT) superfamily and like other GNAT superfamily members, the overall conformation of PpArsN is an α/β three-layer sandwich fold (Fig. 3A). PpArsN forms an asymmetric homodimer, as does PAT. Although PpArsN shares low protein sequence identity (~35%) with PAT, the backbone structures of these proteins are well conserved. PpArsN has two AST-binding sites, which are asymmetrically formed by amino acid residues from both Chains A and B. Both active sites are composed of nine residues: six residues from one chain and three residues from the other (Fig. 3B). Comparison of the structures of AST-bound PpArsN and PT- and CoA-bound PAT from *S. hygroscopicus* (ShPAT) (PDB ID: 5T7E) (Christ *et al.*, 2017) demonstrates similarities of the active sites of these two related *N*-ace-tyltransferases.

The residues involved in AST binding in PpArsN are conserved well in ShPAT, with several conservative replacements. Surprisingly, sets of amino acid residues used by ShPAT to interact with each chemical moiety in PT are different from those used by PpArsN to interact with the corresponding chemical moiety in AST. As a result, while the arsenic atom of AST overlaps with the phosphorus atom of PT, the orientation of the amino group of PT is inclined to 95° towards the acetyl-CoA binding site with respect to the AST binding site. The minor difference in composition of amino acid residues forming the substrate binding site between ArsN and PAT may lead in part to the variation of their substrate selectivity.

4 CONCLUSIONS

Our results demonstrate that the novel organoarsenical natural product AST is a broad-spectrum antibiotic much more potent than its phosphorus mimetic PT. We also show that ArsN selectively confers resistance to AST over PT, elucidating the functional linkage of ArsN to arsenic detoxification. Our ArsN crystal structures will be useful to design new drugs that can evade resistance or enhance AST effectiveness as antibiotics.

ACKNOWLEDGEMENTS

This work was supported by NIH grant R01 GM055425 to BPR, Herbert Wertheim College of Medicine Pilot Fund (#800008403) to MY, and a JSPS KAKENHI grant (No. 23380044) to SI.

REFERENCES

Christ, B., Hochstrasser, R., Guyer, L., Francisco, R., Aubry, S., Hörtensteiner, S. & Weng, J.-K. 2017. Non-specific activities of the major herbicide-resistance gene BAR. *Nat. Plants* 3(12): 937.

Jolliffe, D.M. 1993. A history of the use of arsenicals in man. *J. Roy. Soc. Med.* 86(5): 287–289.

Kuramata, M., Sakakibara, F., Kataoka, R., Yamazaki, K., Baba, K., Ishizaka, M., Hiradate, S., Kamo, T. & Ishikawa, S. 2016. Arsinothricin, a novel organoarsenic species produced by a rice rhizosphere bacterium. *Environ. Chem.* 13(4): 723–731.

Páez-Espino, A.D., Chavarría, M. & Lorenzo, V. 2015. The two paralogue *phoN* (phosphinothricin acetyl transferase) genes of *Pseudomonas putida* encode functionally different proteins. *Environ. Microbiol.* 17(9): 3330–3340.

Thompson, C.J., Movva, N.R., Tizard, R., Crameri, R., Davies, J.E., Lauwereys, M. & Botterman, J. 1987. Characterization of the herbicide-resistance gene *bar* from *Streptomyces hygroscopicus*. *EMBO J.* 6(9): 2519–2523.

Environmental Arsenic in a Changing World –
Zhu, Guo, Bhattacharya, Ahmad, Bundschuh & Naidu (Eds)
ISBN 978-1-138-48609-6

Arsenic induces Th1/Th2 imbalance in immune and non-immune organs

Y.Y. Guo[1], L. Zhao[1], S. Yang[1], G.F. Sun[1], B. Li[1], X.X. Duan[2] & J.L. Li[3]
[1]*Environment and Non-Communicable Disease Research Center, Key Laboratory of Arsenic-related Biological Effects and Prevention and Treatment in Liaoning Province, School of Public Health, China Medical University, Shenyang, P.R. China*
[2]*Department of Toxicology, School of Public Health, Shenyang Medical College, Shenyang, Liaoning, P.R. China*
[3]*Department of Occupational and Environmental Health, Key Laboratory of Occupational Health and Safety for Coal Industry in Hebei Province, School of Public Health, North China University of Science and Technology, Tangshan, Hebei, P.R. China*

ABSTRACT: CD4+ T helper (Th) cells, especially well balanced Th1/Th2 responses, are essential regulators of the immune responses and inflammatory diseases. Arsenic is a known immunosuppressive metalloid, but its detailed effects on immune and non-immune organs are poorly understood. In the present study, mice were treated with 2.5, 5 and 10 mg kg^{-1} $NaAsO_2$ intra-gastrically for 24 h. We found that arsenic exposure led to the persistent aberration of the inherent capability of Th cells to differentiate into Th1 and Th2 cells. These results might provide novel therapeutic strategies on arsenic-induced immune related diseases.

1 INTRODUCTION

Arsenic is the number one contaminant of concern for human health worldwide according to the Agency for Toxic Substances and Disease Registry (ATSDR, 2007). Micromolar concentrations of arsenic could not only reduce the differentiation of human peripheral blood mononuclear cells (PBMCs) into functional macrophages (Lemarie *et al.*, 2006a), but also reverse the major phenotypic and genetic features of human mature macrophages (Lemarie *et al.*, 2006b). What's more, the balance of the earliest determined Th cell subsets, Th1 and Th2, play important roles in immune function. In the present study, we set up arsenic exposure models by treating mice with an oral administration of 2.5, 5 and 10 mg kg^{-1} $NaAsO_2$, and determined the Th1/Th2 differentiation in spleen, thymus and lung. The present study was undertaken to comprehensively assess the immunomodulatory effect of arsenic exposure in vivo.

2 METHODS/EXPERIMENTAL

2.1 *Animals and experimental procedures*

Forty female C57BL/6 mice (weighing 18–22 g, 6–7 weeks old) were obtained from the Center for Experimental Animals at China Medical University (Shenyang, China). Mice were exposed to environmentally relevant concentrations of $NaAsO_2$ (2.5, 5 and 10 mg kg^{-1}) intragastrically for 24 h, respectively. At autopsy, the entire spleen, thymus and lung were promptly removed, weighed, and stored at −80°C for future use.

2.2 *Determination of tissue arsenic levels*

Spleen, thymus and lung were homogenized on ice with 500 μL deionized water and 0.05 g tissues. Determination of arsenic species, including iAs, monomethylarsonic acid (MMA) and dimethylarsinic acid (DMA), were determined by HPLC-HG-AFS (SA-10 Atomic Fluorescence Species Analyzer, Titan Co, Beijing). Total arsenic (T-As) levels were then calculated by summing up the levels of iAs, MMA and DMA. All samples were analyzed in triplicate, and the results were expressed as mean ± SD (n = 4).

2.3 *Total RNA isolation and qPCR analysis*

Total RNA of spleen, thymus and lung from experimental mice was isolated using a Trizol Reagent (Invitrogen). Real-time PCR was conducted using a two-step method with an ABI 7500 Real-Time PCR System (ABI, USA).

2.4 *Statistical analysis*

Data were presented as mean ± standard deviation (SD). Statistical significance was determined by ANOVA. P values of less than 0.05 were considered as statistically significant.

Table 1. Total arsenic (T-As, $\mu g\ As\,g^{-1}$ tissue) levels and weights in spleen, thymus and lung of control and different experimental mice.

$NaAsO_2$ ($mg\,kg^{-1}$)	T-As in spleen	T-As in thymus	T-As in lung	Spleen weight	Thymus weight	Lung weight
control	<LD	<LD	<LD	0.07 ± 0.01	0.07 ± 0.01	0.13 ± 0.01
2.5	0.95 ± 0.08	1.21 ± 0.46	0.35 ± 0.02	0.06 ± 0.01*	0.05 ± 0.01*	0.14 ± 0.01
5	1.51 ± 0.06	1.82 ± 0.72	0.91 ± 0.17	0.05 ± 0.01*	0.04 ± 0.01*	0.13 ± 0.01
10	2.73 ± 0.31	3.01 ± 0.56	2.2 ± 0.29	0.05 ± 0.01*	0.04 ± 0.01*	0.13 ± 0.01

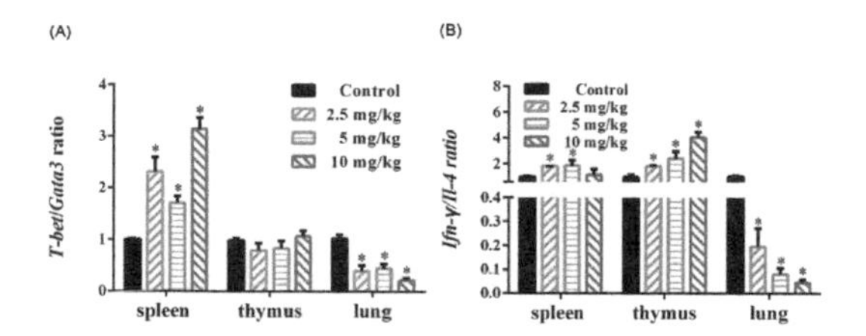

Figure 1. Effects of arsenic exposure on Th1/Th2 differentiation in spleen, thymus and lung of control and different experimental mice. T-bet/Gata3 ratio and Ifn-γ/Il-4 ratio were determined and calculated, respectively. * denoted $p < 0.05$ compared with control.

3 RESULTS AND DISCUSSION

3.1 *T-As levels and weights of spleen, thymus and lung by arsenic exposure*

In our study, no significant difference of body weight, as well as general status were observed among all the study mice, suggesting no obvious growth inhibition of the experimental mice (data not shown). We detected T-As levels at 12 h and found that T-As levels were significantly dose-dependent in different arsenic-treatment group (Table 1). We only found that the splenic and thymic weights in all arsenic-treated groups were consistently lower than in control group, suggesting arsenic induced immunotoxicity in mice.

3.2 *Arsenic exposure regulates Th1/Th2 differentiation in spleen, thymus and lung*

Recent years, the Th1/Th2 immune balance is thought to exert critical roles in stabilize microenvironment in vivo (Moriyama and Nakamura, 2005). Our results here observed an increase in T-bet/Gata3 ratio of spleen, and a decrease in lung. Ifn-γ/Il-4 ratio were also moderately upregulated both in spleen and thymus, as well as significantly decreased in lung (Fig. 1). Our results indicate that arsenic could induce an imbalance of Th1/Th2 differentiation and Th1-polarized immune responses in spleen and thymus, while Th2-polarized immune responses in lung.

4 CONCLUSIONS

In conclusion, our present study demonstrated that arsenic-induced immune changes are associated with modulation of Th1/Th2 differentiation in spleen, thymus and lung of C57BL/6 mice.

ACKNOWLEDGEMENTS

This study was supported by grants from National Natural Science Foundation China (NSFC, No. 81673114) and Key Laboratory Basic Research Funds from Liaoning Education Department (LS201607).

REFERENCES

ATSDR, 2007. Toxicological Profile for Arsenic (Update). Agency for Toxic Substances and Disease Registry. Atlanta, GA.

Lemarie, A., Morzadec, C., Merino, D., Bourdonnay, E., Fardel, O. & Vernhet, L. 2006a. Human macrophages constitute targets for immunotoxic inorganic arsenic. *J. Immunol.* 177 (5): 3019–3027.

Lemarie, A., Morzadec, C., Merino, D., Micheau, O., Fardel, O. & Vernhet, L. 2006b. Arsenic trioxide induces apoptosis of human monocytes during macrophagic differentiation through nuclear factor-KappaB-related survival pathway down-regulation. *J. Pharmacol. Exp. Ther.* 316(1): 304–314.

Moriyama, M. & Nakamura, S. 2016. Th1/Th2 immune balance and other T helper subsets in IgG4-related disease. *Curr Top. Microbiol. Immunol.* 401: 75–83.

Environmental Arsenic in a Changing World –
Zhu, Guo, Bhattacharya, Ahmad, Bundschuh & Naidu (Eds)
ISBN 978-1-138-48609-6

Thiolation in arsenic metabolism: a chemical perspective

C. Fan[1], G. Liu[1] & Y. Cai[1,2]
[1]*Department of Chemistry and Biochemistry, Florida International University, Miami, FL, USA*
[2]*Southeast Environmental Research Center, Florida International University, Miami, FL, USA*

ABSTRACT: In recent years, methylated thioarsenicals have been widely detected in various biological and environmental matrices, suggesting their broad involvement and biological implications in arsenic metabolism. However, very little is known about the formation mechanisms of methylated thioarsenicals and the relation between arsenic methylation and thiolation processes. It is timely and necessary to summarize and synthesize the reported information on thiolated arsenicals for an improved understanding of arsenic thiolation. To this end, we examined the proposed formation pathways of methylated oxoarsenicals and thioarsenicals from a chemical perspective, and proposed a new arsenic metabolic scheme in which arsenic thiolation is integrated with methylation processes (instead of being separated from methylation as currently reported), followed lastly by discussion on the biological implications of the new scheme of arsenic metabolism. This informative review on arsenic thiolation from the chemical perspective will be helpful to better understand the arsenic metabolism at the molecular level and the toxicological effects of arsenic species.

1 INTRODUCTION

Arsenic is a toxic metalloid widely distributed in the environment and poses health risk to human beings. On the other hand, arsenic has been recognized in successful treatment of cancers (e.g., multiple myeloma). The double-edged effects of arsenic are highly dependent on the arsenic species and their metabolic processes in living organisms. Arsenic metabolism is a complex process involving a series of biochemical reactions including oxidation, reduction, methylation, and thiolation. Methylation process was considered a major arsenic metabolic pathway, and the end products would be methylated oxyarsenicals (Cullen, 2014; Dheeman *et al.*, 2014). Methylated thioarsenicals, referring to pentavalent species in this review unless stated otherwise and including monomethylmonothioarsinic acid (MMMTA(V)), dimethylmonothioarsinic acid (DMMTA(V)) and dimethyldithioarsinic acid (DMDTA(V)), were recently found as common arsenic metabolites and may be of important biological implications, as they were demonstrated to be more toxic than their oxy-counterparts, possibly because of their capability of binding proteins. The arsenic metabolic pathways proposed previously focused predominantly on methylated oxoarsenicals, but very limited studies reported sporadic, inconsistent information on thiolation process independent of arsenic methylation (Sun *et al.*, 2016; Wang *et al.*, 2015). It is urgently needed to review the reported information on the formation of methylated thioarsenicals and more importantly, to use a holistic approach to systematically consider both methylation and thiolation processes during arsenic metabolism.

2 FORMATION PATHWAYS OF METHYLATED OXOARSENICALS AND THIOARSENICALS

Previously proposed arsenic methylation pathways including classic oxidative methylation, reductive methylation, and the processes involving protein-bound arsenicals were reviewed from a chemical perspective. We found that the oxidative methylation pathway is more chemically plausible, regardless of the arsenic binding to proteins or not. Pentavalent arsenicals seem to be obligated intermediates or exist in the transition state during the methylation reactions. These proposed arsenic methylation pathways were ended up with forming methylated oxoarsenicals, contributing little to explain the increasingly common detection of thioarsenicals. Arsenic thiolation process was regarded as a separated pathway from methylation process, and methylated thioarsenicals were presumably produced through 1) thiolation of methylated arsenicals (free or protein-bound) and 2) thiolation of iAs following successive methylation. These chemical reactions were studied, and direct thiolation of pentavalent arsenic was found to be more likely to occur.

3 A NEWLY PROPOSED FORMATION PATHWAY OF METHYLATED THIOARSENICALS

A new formation pathway of methylated thioarsenicals accompanying methylation process was proposed (Fig. 1). The new arsenic metabolic pathway takes into account thiolation and methylation simultaneously while considering the fact that most arsenic

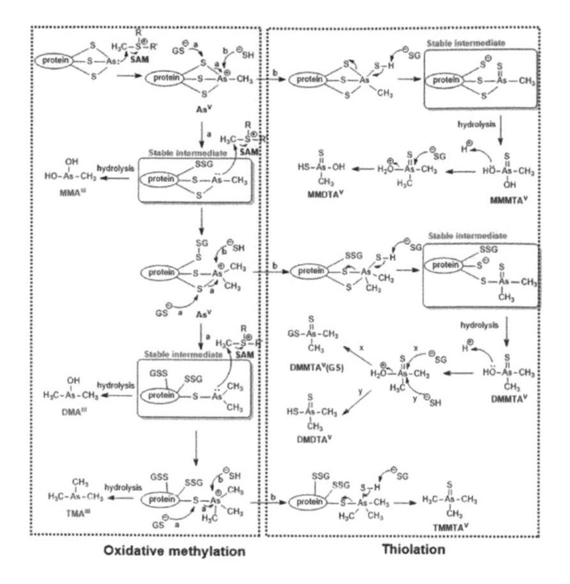

Figure 1. Proposed arsenic metabolic scheme integrating methylation with thiolation.

are bound to proteins in the biological matrix before being excreted. Trivalent inorganic arsenicals undergo an oxidative methylation in the presence of S-adenosylmethionine (SAM) while they are bound to proteins or enzymes. After accepting a methyl group from SAM, the arsenicals are oxidized to pentavalent state while still being bound to proteins. We proposed that the resulting intermediates could undergo reduction (process a) or thiolation (process b) to form stable protein-bound products. The reduction process yields trivalent protein-bound arsenicals in the presence of reductive reagents, such as glutathione (GSH) (process a). The thiolation process could be a competitive reaction to form stable pentavalent protein-bound arsenicals to keep their oxidative state (process b). Reduction and thiolation processes produce protein-bound methylated trivalent arsenicals and pentavalent thioarsenicals, respectively. These two types of protein-bound arsenicals are stable and crucial intermediates in arsenic metabolism.

Although lack of techniques to directly detect the protein-bound arsenic species, the evidence from previous studies related to the formation of the proposed intermediates was reviewed in this article. For example, $DMMTA^V$ was demonstrated to form a stable As-S bond with GSH and the product, DMMTA(V)(GS), as a stable arsenic metabolite, was found in *Brassica oleracea* extracts after exposure to dimethylarsinic acid (DMA(V)) as well as in human cancer cell line treated with darinaparsin (Raab *et al.*, 2007; Yehiayan *et al.*, 2014).

4 IMPLICATIONS OF THE PROPOSED MECHANISM

According to the new pathway, methylated thioarsenicals can be produced through reaction of sulfide with not only free arsenic metabolites but also protein-bound arsenic intermediates. Thiolation reaction occurs competitively with methylation reaction, and methylated thioarsenicals should be important metabolites rather than occasional by-products. After uptake, most arsenic species are bound to proteins in biological systems. The protein-bound arsenicals used to be considered as only in the trivalent forms. Pentavalent thioarsenicals also can bound to proteins to form stable complexes. Therefore, free thioarsenicals, just like free trivalent arsenicals, should be usually found in low concentrations and could be easily ignored. Thioarsenicals can convert to oxyarsenicals during sampling, storage, preparation, and analysis, which could cause underestimation of the amount of thioarsenicals. Methylation of inorganic arsenic was previously regarded as a detoxification pathway due to the formation of less toxic monomethylarsonic acid (MMA(V)) and DMA(V), until their trivalent forms, monomethylarsonous acid (MMA(III)) and dimethylarsinous acid (DMA(III)), were detected. Similarly, pentavalent arsenic was considered to be less toxic because of low binding affinity to proteins, however, MMMTA(V) and DMMTA(V) were demonstrated to be highly toxic. The high toxicity, on the other hand, could make thioarsenicals potential anticancer drug candidates. The newly proposed arsenic metabolic pathway could provide valuable information for study of arsenic-based anticancer medicines.

REFERENCES

Cullen, W.R. 2014. Chemical mechanism of arsenic biomethylation. *Chem. Res. Toxicol.* 27(4): 457–461.

Dheeman, D.S., Packianathan, C., Pillai, J.K. & Rosen, B.P. 2014. Pathway of human AS3MT arsenic methylation. *Chem. Res. Toxicol.* 27(11): 1979–1989.

Raab, A., Wright, S.H., Jaspars, M., Meharg, A.A. & Feldmann, J. 2007. Pentavalent arsenic can bind to biomolecules. *Angew. Chem.* 119: 2648–2651.

Sun, Y., Liu, G. & Cai, Y. 2016. Thiolated arsenicals in arsenic metabolism: occurrence, formation, and biological implications. *J. Environ. Sci.* 49: 59–73.

Wang, Q.Q., Thomas, D.J. & Naranmandura, H. 2015. Importance of being thiomethylated: formation, fate, and effects of methylated thioarsenicals. *Chem. Res. Toxicol.* 28(3): 281–289.

Yehiayan, L., Stice, S., Liu, G., Matulis, S., Boise, L.H. &Cai, Y. 2014. Dimethylarsinothioyl glutathione as a metabolite in human multiple myeloma cell lines upon exposure to darinaparsin. *Chem. Res. Toxicol.* 27(5): 754–764.

Environmental Arsenic in a Changing World –
Zhu, Guo, Bhattacharya, Ahmad, Bundschuh & Naidu (Eds)
ISBN 978-1-138-48609-6

AS3MT polymorphisms, arsenic metabolites and pregnancy

A. Stajnko[1,2], Z. Šlejkovec[1], D. Mazej[1], M. Horvat[1,2] & I. Falnoga[1]
[1]*Department of Environmental Sciences, Jožef Stefan Institute, Ljubljana, Slovenia*
[2]*Jožef Stefan International Postgraduate School, Ljubljana, Slovenia*

ABSTRACT: Arsenic methylating enzyme (AS3MT) genetic variations and their role in variability of As metabolism were evaluated in population of pregnant/non-pregnant women exposed to low levels of As. We have tested the associations between genotypes/haplotype of seven *AS3MT* SNPs and As metabolites in urine (methylation efficiency). Significant associations of individual SNPs and protective haplotype with arsenic metabolites have been confirmed among non-pregnant women. On the contrary, these associations have been almost absent among pregnant women – most possibly a counsequence of physiological changes in pregnancy, and/or impact of external DMA due to higher seafood intake. Both factors can influence As metabolism.

1 INTRODUCTION

Arsenic (III oxidation state) methyltrasferase (AS3MT) is an enzyme generating methylated (MMA) and dimethylated (DMA) arsenicals and therefore it has important role in human arsenic metabolism. Its genetic variations can have important impact on individual and population variability of As metabolism. Various studies, mainly conducted in populations exposed to high levels of As in drinking water, point on beneficial *AS3MT* genetic variants (protective haplotype) resulting in more efficient As methylation (lower AsIII% and MMA% and higher DMA%)[1,5]. However in recent study, Xu et al. (2016) emphasised that the extent of the influence of *AS3MT* variants on As metabolism may also depends on levels of exposure to As, or better – on levels of inorganic arsenic (iAs). Therefore, in present study the influence of 7 *AS3MT* variants and pregnancy status on As metabolite patterns (urine) was evaluated in population of women exposed to low levels of As.

2 METHODS/EXPERIMENTAL

2.1 *Study population*

Study was conducted on Croatian-Slovenian population divided in a group of pregnant ($n = 136$, average age of 30 ± 5; third trimester of pregnancy) and a group with non-pregnant women ($n = 209$; average age of 39 ± 4). A small subset of non-pregnant women ($n = 33$) was consisted of women previously participating in pregnant group; they were re-sampled a few years after pregnancy. All participants were recruited between 2006 and 2017 within prospective birth cohort PHIME and follow up survey CROME-LIFE+.

2.2 *Analytical methods*

Total levels of metalloids in blood (As), plasma (selenium) and in urine (As) were measured using ICP-MS method, while arsenic metabolites (MMA, DMA and AsIII) in urine were determined by HPLC-HG-AFS. The percentage of metabolites were determined based on their sum (MMA+DMA+AsIII, U-SAs). Methylation efficiency was evaluated by first methylation step (PMI, MMA/AsIII), second methylation step (SMI, DMA/MMA) and total methylation rate (TMI, ((DMA+MMA)/AsIII). DNA was isolated from peripheral blood and genotyped for 7 *AS3MT* (rs7085104, rs3740400, rs3740393, rs3740390, rs11191439, rs10748835, rs1046778) and one AQP9 (rs2414539) by pre-designed TaqMan SNP Genotyping assays and qPCR method.

2.3 *Statistical methods*

The differences between groups and associations of genotypes/haplotypes with arsenic metabolites were tested by non-parametric Kruskal-Wallis test. Associations were further adjusted for age, weight, height, BMI, gestational age, parity, seafood intake, U-As, P-Se, smoking and AQP9 SNP in multivariable linear regression models. Statistical analysis was performed by STATA12/SE software and construction of haplotypes with Haploview 4.2.

3 RESULTS AND DISCUSSION

3.1 *As exposure and urine metabolites*

Arsenic exposure in studied population was low; geometric mean (GM) of total As was below 2.6 ng/g for blood and below 23 ng/ml for urine. Analysis of As urine metabolites and methylation efficiency

Table 1. As metabolites in urine and methylation efficiency.

	Pregnant women	Non-pregnant women	
N	136	33*	176
AsIII%	3.83 (0.13–71.4)	4.58 (0.63–24.0)	5.78 (0.31–29.6)
MMA%	5.13 (0.26–98.0)	9.94 (0.62–28.6)	10.9 (0.74–40.0)
DMA%	82.7 (1.34–99.6)	80.1 (57.1–98.1)	77.1 (40.0–98.9)
PMI	1.34 (0.01–146)	2.17 (0.33–6)	2.08 (0.16–19.6)
SMI	16.1 (0.01–384)	8.05 (2.0–158)	8.58 (1.0–98.7)
TMI	24.1 (0.04–770)	20.4 (3.17–156)	15.9 (2.38–325)

B-blood, U-urine, U-SAs (MMA+DMA+AsIII); data presented as GM (range), * women from pregnant group re-sampled during follow up study.

(Table 1) showed significantly higher DMA% (and SMI) accompanied with lower MMA% (and PMI) for pregnant group in comparison with non-pregnant women. Results are consistent with previous studies and indicate more efficient and changed As methylation in pregnancy, what could be partially explained by hormonal changes (estrogen up-regulates other methylation pathways) and dietary supplementation (folic acid increases urine DMA%)[3], TMI was higher in pregnant women, but the difference was lower when compared to re-sampled non-pregnant women. Explanation could be possible higher (DMA containing)-seafood intake in re-sampled subgroup[5].

3.2 *AS3MT genotypes/haplotypes and As metabolites*

After adjustment, the significant associations of individual *AS3MT* SNPs with As metabolites were found for 4 SNPs (rs7085104, rs3740400, rs11191439 and rs10748835) among pregnant women and for 6 out of 7 SNPs among non-pregnant women (exception was rs11191439; re-sampled women were excluded due to missing data). Interestingly, when associations were estimated among participants with blood-As level above median ($\geq$2.7 ng/g and $\geq$0.53 ng/ml for pregnant and non-pregnant women, respectively) the associations became stronger for non-pregnant group. On other hand, three associations were lost for pregnant group; only rs11191439 stayed associated with As metabolites. Obtained results are generally consistent with literature data where protective alleles (6 minor alleles, and one common – rs11191439) were found associated with more efficient methylation (lower AsIII% and MMA% and higher DMA%, SMI and TMI).

Metabolites were also marginally affected by AQP9 SNP.

Due to high LD (linkage disequilibrium) for SNPs, we constructed *AS3MT* haplotypes. Calculated frequency of protective haplotype was 9.5 % what is similar to the frequencies reported for European countries with comparable low iAs exposure (10–14%; Slovakia, Hungary and Romania)[2]; percentage was much lower from frequencies reported for populations adapted to historic exposure to high iAs levels in drinking water (50–70%; Argentina, Peru, Chile)[1,4].

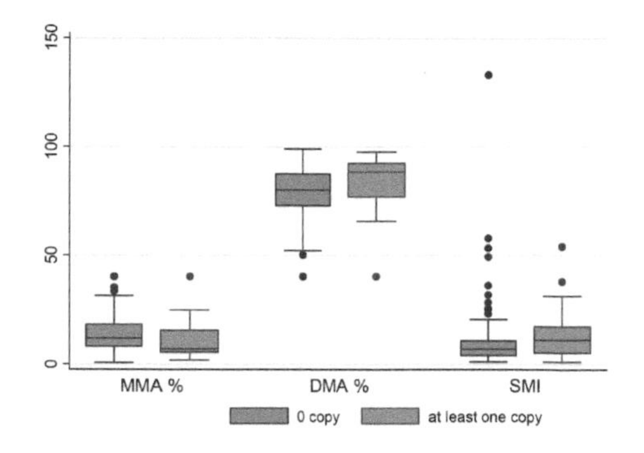

Figure 1. MMA%, DMA% and SMI in non-pregnant women stratified by the presence of protective haplotype.

In present study a protective haplotype was significantly associated with lower MMA% and with higher DMA% and SMI in group of non-pregnant women (Figure 1), while among pregnant group no significant associations were observed.

4 CONCLUSIONS

Our results indicate that pregnancy, due to physiological changes, and DMA containing seafood may had significant influence on As metabolites and therefore, blur the associations with genetic variants at such low level of exposure. On the other hand, among non-pregnant women – with more stable physiology, significant associations with As metabolites were found for separate SNPs as well as for protective haplotype.

ACKNOWLEDGEMENTS

Study was supported by Slovenian Research Agency (P1 043), and PHIME, CROME-LIFE+, HEALS projects.

REFERENCES

Apata M. et al., 2017, Human adaptation to arsenic in Andean populations of the Atacama Desert, *Am J Phys Antropol*, 163:192–199.

Engström K. et al., 2015, Genetic Variation in Arsenic (+3 Oxidation State) Methyltransferase (AS3MT), ArsenicMetabolismand Risk of Basal Cell Carcinoma in a European Population, *Environ Mol Mutagenesis* 56:60–69.

Li et al., 2008, Nutritional status has marginal influence on the metabolism of inorganic arsenic in pregnant Bangladeshi Women. *Envirn Health Perspect,* 116(3):315–321.

Schlebush M.C. et al., 2013, Possible positive selection for an arsenic-protective haplotype in humans, *Environ Health Perspect*, 121(1):53–58.

Taylor et al., 2017, Human exposure to organic arsenic species rom seafood, *Science of the total Environ*, 580: 266–282.

Xu et al., 2016, Associations between variants in Arsenic (+3 Oxidation State) Methyltransferase (AS3MT) and urinary metabolites of Inorganic Arsenic: Role of Exosure level. *Tox Sci*: 153(1):112–123.

Environmental Arsenic in a Changing World –
Zhu, Guo, Bhattacharya, Ahmad, Bundschuh & Naidu (Eds)
ISBN 978-1-138-48609-6

Prevalence of cancer incidences in Bihar, India due to poisoning in groundwater

A. Kumar, R. Kumar, M. Ali & A.K. Ghosh
Mahavir Cancer Institute and Research Centre, Patna, Bihar, India

ABSTRACT: Arsenic (As) poisoning in groundwater in Bihar (India) has caused various health related issues in the village population. In State of Bihar, India, it is estimated that more than 5 million people are drinking water with As concentrations greater than 10 $\mu g L^{-1}$ (WHO permissible limit), and presently, the groundwater As contamination has been confirmed in 18 districts of the state of Bihar as per our study. The As contamination in the groundwater was maximum observed in Buxar district in the Simri village with maximum level of 1929 $\mu g L^{-1}$. Apart from the symptoms of arsenicosis, the other symptoms observed were general body weakness, loss of appetite, anemia, cough or bronchitis, constipation, change in the mood or behavior, mental stress, hormonal imbalance etc. Our study has confirmed very high As contamination in the groundwater as well as in their biological samples of As exposed population. The present study deals with the evaluation of As contamination in groundwater and its correlation with incidences of cancer cases in As hit areas of Bihar. This will enable to make a strategic policy to control the cancer incidences in these areas with As contaminated aquifers.

1 INTRODUCTION

In the Gangetic flood plain region of Bihar, the arsenic (As) poisoning in groundwater has led to various health related problems in the population. The population exhibited typical symptoms of arsenicosis like hyperkeratosis in sole and palm, melanosis, pigmentations in skin, general body weakness, loss of appetite, anemia, cough or bronchitis, constipation, change in the mood or behavior, mental stress, hormonal imbalance etc. The incidences of cancer cases have increased many folds in these As hotspot areas in recent times. Approximately, 80,000 new cancer cases are reported yearly from this state and our institute – Mahavir Cancer Sansthan & Research Centre, Patna, Bihar alone registered approximately 26,000 new cancer patients in 2016–17. The cancer cases reported are squamous cell carcinoma of skin, liver and gall bladder cancer, renal cell carcinoma, breast cancer, melanoma, leukemia, lymphoma etc. The present study evaluates the correlation between As contaminations in groundwater and cancer incidences.

2 METHODS/EXPERIMENTAL

2.1 *Ethical approval*

The research work was approved by the IEC (Institutional Ethics Committee) of the institute as the work was on human subjects. The survey work was carried out from February of 2016 till date when the cancer patients from the village reported in our cancer research center.

2.2 *Location*

The study was done in the As hit villages of Buxar, Saran, Patna and Samastipur villages of state Bihar, India.

2.3 *Sample collection & survey*

Cancer patients were selected from the As hit village for the study to know their blood As levels as well as hair As levels.

Apart from this, random 400 cancer patient's blood samples were collected from the pathology department of the institute and were assayed to know their blood As levels.

The water samples were collected from the cancer patient's households. After collection, all the samples were digested using concentrated HNO_3 on hot plate under fume hood and total As concentration (inorganic form) was estimated as per the protocol of (NIOSH, 1994) through graphite furnace atomic absorption spectrophotometer (Pinnacle 900T, Perkin Elmer, Singapore). Simultaneously, health assessment of the population was also done through a health survey questionnaire proforma. For determining the exact location of the hand pump, hand held Global Positioning System (GPS) receivers (Garmin etrex10, of USA) with an estimated accuracy of ≈10 m were utilized.

3 RESULTS AND DISCUSSION

3.1 *Arsenic assessment in cancer patient's blood samples*

The study shows novel findings ever explored in this area. The maximum As concentration in blood samples of cancer patients observed was 461.2 $\mu g\,L^{-1}$ (Fig. 1).

3.2 *Health assessment*

The rural population exhibited the typical symptoms of arsenicosis like hyperkeratosis in palm and sole, melanosis in palm and sole, blackening of tongue, skin irritation, anemia, gastritis, constipation, loss of appetite, bronchitis & cough, etc. The cancer incidences were very severely found along with high As concentration in their water samples. The cancer cases recorded were squamous cell carcinoma of skin, renal cell carcinoma, ovarian cancer, breast cancer, skin melanoma etc. in the village population (Fig. 2).

The maximum As concentration in the hair samples were observed in 6.3 $mg\,kg^{-1}$ (Fig. 2B) while in hand pump water was 826.2 $\mu g\,L^{-1}$ As determination in urine, hair and nail are considered as most reliable indicator of exposure. Hair samples are used as a biomarker for As exposure because inorganic As and dimethylarsinic acid are stored in hair root and thus reflect past exposure (Yoshida *et al.*, 2004). Elevated As levels in hair indicates the past exposure of 6–12 months (NRC, 2001).

Various molecular pathways have been deciphered which indicates that As is one of the toxic agent which is causing cancer in the population under study. Arsenic also has the affinity to directly bind sulfhydryl (SH) moieties and conversion of SH to S-S by free radicals (ROS) leading to cause severe toxicity at the cellular level. Although, interaction between As and thiol groups may occur in 200 known human proteins (Abernathy *et al.*, 1999).

The other health related assessments showed hormonal imbalance in the population. The rural population exhibited elevated levels of serum estrogen while decreased levels of serum testosterone levels denotes that the As contamination in groundwater and its consumption by the rural population has caused severe health problems to them (Kumar *et al.*, 2015, 2016). The entire study reveals that high As concentration in groundwater has somehow led to cause cancer incidences in the As hit area. Map of the occurrence of As contamination in groundwater and the corresponding incidences of cancer in Bihar is presented in Figure 3.

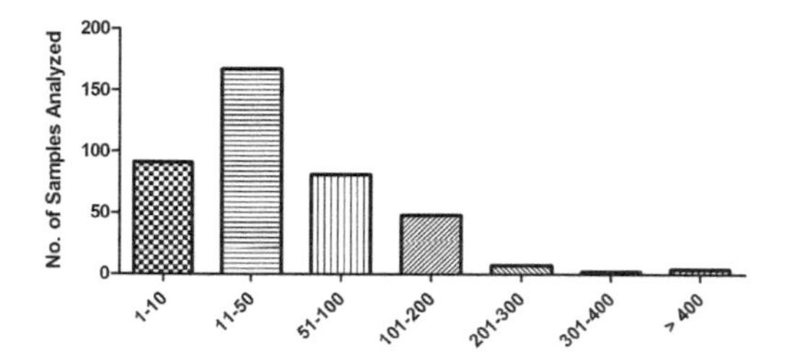

Figure 1. Arsenic concentration in blood samples of cancer patients analyses through GF-AAS (ANOVA-Dunnett's Test, $P < 0.05$).

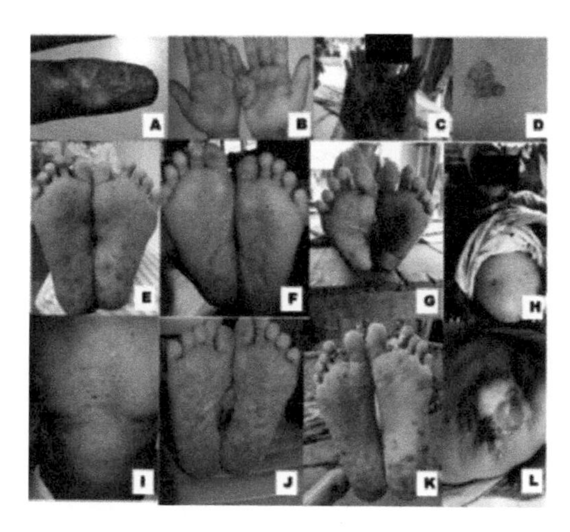

Figure 2. Showing cancer cases in the population along with the arsenicosis symptoms (A, B, E & F – Squamous cell carcinoma, C, G & L – Breast cancer, D – Skin melanoma. I & J – Renal cell carcinoma, H & K – Ovarian cancer).

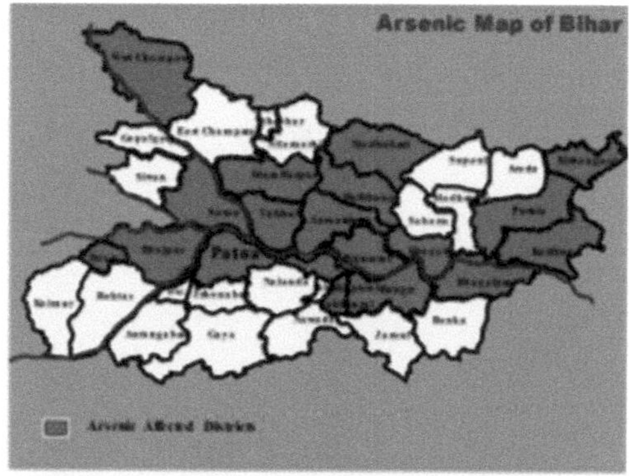

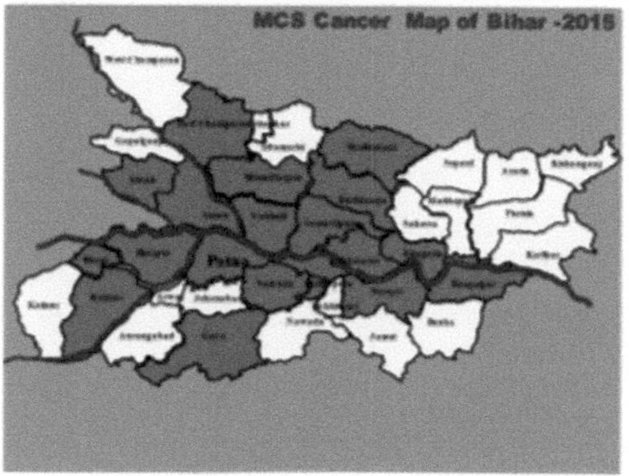

Figure 3. Comparison between arsenic map of Bihar, India and cancer map of Bihar.

4 CONCLUSIONS

The study thus concludes that high As contamination in the groundwater has led to cause incidences of cancer in these hotspot areas. Many As induced health related problems were also observed in the population like keratosis, melanosis, leuco-melanosis, hormonal imbalance. The As map of the state with the cancer map are almost similar correlates the present study.

ACKNOWLEDGEMENTS

The authors acknowledge support extended by Mahavir Cancer Sansthan and Research Centre, Patna for providing laboratory and other infrastructural facilities for this study.

REFERENCES

Abernathy, C.O., Liu, Y.P., Longfellow, D., Aposhian, H.V., Beck, B., Fowler, B., Goyer, R., Menzer, R., Rossman, T., Thompson, C. & Waalkes, M. 1999. Arsenic: health effects, mechanisms of actions, and research issues. *Environ. Health Perspect.* 107(7): 593–597.

Kumar, A., Ali, M., Rahman, S.M., Iqubal, A.M., Anand, G., Niraj, P., Shankar, P. & Kumar, R. 2015. Ground water arsenic poisoning in "Tilak Rai Ka Hatta" village of Buxar district, Bihar, India causing severe health hazards and hormonal imbalance. *J. Environ. Anal. Toxicol.* 5(4): 290. 1.

Kumar, A., Rahman, M.S., Iqubal, M.A., Ali, M., Niraj, P.K., Anand, G. & Kumar, P. 2016. Ground water arsenic contamination: a local survey in India. *Int. J. Prev. Med.* 7: 100.

NRC. 2001. Arsenic in drinking water: 2001 update. The National Academies Press, Washington, DC.

Yoshida, T., Yamauchi, H. & Fan, S.G. 2004. Chronic health effects in people exposed to arsenic via the drinking water: dose-response relationships in review. *Toxicol. Appl. Pharmacol.* 198(3): 243–252.

Environmental Arsenic in a Changing World –
Zhu, Guo, Bhattacharya, Ahmad, Bundschuh & Naidu (Eds)
ISBN 978-1-138-48609-6

Arsenic exposure from drinking water and the occurrence of micro- and macrovascular complications of type 2 diabetes

D.D. Jovanović[1], K. Paunović[2], D.D. Manojlović[3] & Z. Rasic-Milutinović[4]
[1]*Institute of Public Health of Serbia "Dr Milan Jovanović Batut", Belgrade, Serbia*
[2]*Institute of Hygiene and Medical Ecology, Faculty of Medicine, University of Belgrade, Belgrade, Serbia*
[3]*Institute of Chemistry, Technology and Metallurgy, Center of Chemistry, Belgrade, Serbia*
[4]*Departments of Endocrinology, University Hospital Zemun, Belgrade, Serbia*

ABSTRACT: The highest arsenic concentrations in Serbia are measured in drinking water in the Vojvodina Region. The research was designed as a cross-sectional cohort study comprising exposed and unexposed patients with the type 2 diabetes to arsenic in drinking water in Zrenjanin municipality, Serbia. Median life time arsenic exposure from drinking water in exposed group was calculated at 181.00 μg day^{-1} (ranged 1.15–1369.51 μg day^{-1}. Multivariate logistic regression model showed significantly higher odds ratio for the occurrence of myocardial infarction and stroke at arsenic concentrations above 10 μg L^{-1}, while life time arsenic exposure significantly contributed to the occurrence of heart failure. These results support the hypothesis that exposure to arsenic in drinking water may play a role in the occurrence of micro- and macrovascular complications of type 2 diabetes.

1 INTRODUCTION

Arsenic is not uniquely present in waters in Serbia. Vojvodina region on the north belongs to the southern part of the Pannonian Basin, which contains high concentrations of naturally occurring arsenic (Dangic, 2007; Varsanyi and Kovacs, 2006). The highest arsenic concentrations in Serbia are measured in drinking water in the Middle Banat region within Vojvodina (Jovanović *et al.*, 2011). Prevalence of type 2 diabetes in Serbia is approximately 600 thousand persons, or 8.2% of the total population, similarly to European prevalence (Sicree *et al.*, 2009). The aim of this study was to explore the association between exposure to arsenic in drinking water and the occurrence of macro- and microvascular complications of type 2 diabetes in Zrenjanin municipality, Serbia.

2 METHOD

The research was designed as a cross-sectional cohort study comprising 187 patients with type 2 diabetes divided in two groups (i.e. exposed and unexposed) according to the arsenic concentration in drinking water. The cut-off point for drinking water arsenic concentrations was the national standard of 10 μgAs L^{-1}. The research was conducted in Primary Care Center "Bosko Vrebalov" in town Zrenjanin. Each patient was undergone following measurements: body weight and height, waist and hip circumferences, blood pressure and hair arsenic concentrations. The data on age, gender, educational level, family and personal history of diabetes type 2, drinking water source and drinking water habits, smoking status, dietary habits, physical activity, as well as the level of stress were obtained through the questionnaire. In addition, data on existence and history of macro- and microvascular complications, lipid and glucose status were obtained from clinical records for each patient.

The following macrovascular complications of type 2 diabetes were included in the study: angina pectoris, myocardial infarction, chronic heart failure and stroke. Diabetic foot, retinopathy, nephropathy and poly-neuropathy were observed as microvascular complications.

Life time arsenic exposure was calculated based on daily water consumption, and arsenic concentration in drinking water, from water supplies at the current and previous places of residence.

3 RESULTS AND DISCUSSION

Mean arsenic concentrations in drinking water from public water supply systems in 171 waters consumed by exposed patients was 125.19 ± 85.18 μg L^{-1} (median value 95.00 μg L^{-1}, range 11.00–349.00 μg L^{-1}), while mean arsenic concentrations consumed by unexposed patients was 1.62 ± 1.82 μg L^{-1} (median value 1.00 μg L^{-1}, range 0.13–10.00 μg L^{-1}).

Median life time arsenic exposure from drinking water in exposed group was calculated at 181.00 μg day^{-1} (ranged 1.15–1369.51 μg day^{-1}). Multivariate logistic regression model was performed including age, gender, BMI, duration of diabetes, exposure to arsenic above 10 μg L^{-1} in drinking water and life time arsenic exposure as independent variables. This analysis showed significantly higher odds ratio for the occurrence of myocardial infarction and stroke at arsenic concentrations above 10 μg L^{-1}, while life time arsenic exposure significantly contributed to the occurrence of heart failure (Table 1). The same model was performed for microvascular

Table 1. Odds ratio (95% confidence interval) for the occurrence of macrovascular complications in relation to arsenic exposure.

Macrovascular complications	Exposure to arsenic in drinking water	Life time arsenic exposure
Angina pectoris	1.004 (0.517–1.950)	
Myocardial infarction	3.817 (1.322–11.020)	
Heart failure		5.328 (1.831–15.506)
Stroke	3.992 (1.073–14.849)	

Adjusted for age, gender, body mass index and duration of diabetes.

Table 2. Odds ratio (95% confidence interval) for the occurrence of microvascular complications in relation to arsenic exposure.

Microvascular complications	Exposure to arsenic in drinking water	Life time arsenic exposure
Polyneuropathy		3.259 (1.364–7.788)
Retinopathy		2.353 (1.038–5.338)
Nephropathy	1.456 (0.475–4.460)	
Diabetic foot		5.853 (1.203–28.471)

Adjusted for age, gender, body mass index and duration of diabetes.

complications, showing significantly higher odds ratio for the occurrence of poly-neuropathy, retinopathy and diabetic foot after consumption of 181.00 $\mu g\ day^{-1}$ of drinking water arsenic, calculated as life time arsenic exposure (Table 2).

A study conducted in Hungary, Romania and Slovakia reported life time arsenic dose rate of 20.8–65.6 $\mu g\ day^{-1}$ for the exposed study population in Hungary (Hough *et al.*, 2010). This is much below calculated life time arsenic exposure in our study, therefore the data on long term arsenic exposure cannot be compared between these two studies, due to different method of their calculations. The higher prevalence of microvascular and macrovascular diseases (20.0% and 25.3%, respectively) was observed in arseniasis-endemic areas for diabetics than in the nonendemic areas in Taiwan (Wang *et al.*, 2003). One of the studies which investigated the dose-dependent effect of arsenic ingestion on the prevalence of microvascular diseases, especially the kidney and nervous system, was conducted in Taiwan in 2005. It showed a dose-dependant increase in the prevalence of microvascular diseases at arsenic exposure via drinking water over 300 $\mu g\ L^{-1}$ and an increasing trend in subjects who in the same time had diabetes (Chiou *et al.*, 2005). This study also suggested that there was a higher threshold of arsenic in water for the occurrence of neuropathy ($\geq$600 $\mu g\ L^{-1}$) in healthy subjects compared to a lower threshold ($\geq$300 $\mu g\ L^{-1}$) in patients with diabetes. The results of our study indicated that significantly lover concentrations of arsenic in drinking water (around 100 $\mu g\ L^{-1}$) may increase the risk of diabetic neuropathy, diabetic foot and retinopathy. Further investigations are needed to explore association between arsenic exposure from drinking water and the occurrence of micro- and macrovascular complications of type 2 diabetes.

4 CONCLUSIONS

This cross-sectional study showed that arsenic exposure from drinking water may increase risk for the occurrence of micro- and macrovascular complications of type 2 diabetes. Long term arsenic exposure from drinking water may increase risk of the occurrence of acute myocardial infarction, heart failure, stroke, diabetic foot, retinopathy and poly-neuropathy, considered as microvascular and macrovascular complications of type 2 diabetes. These results support the hypothesis that exposure to arsenic in drinking water may play a role in the occurrence of micro- and macrovascular complications of type 2 diabetes.

ACKNOWLEDGEMENTS

All authors wish to thank their collaborators in the Institute of Public Health of Zrenjanin, whose supports are greatly acknowledged.

REFERENCES

Chiou, J.-M., Wang, S.-L., Chen, C.-J., Deng, C.-R., Lin, W. & Tai, T.-Y. 2005. Arsenic ingestion and increased microvascular disease risk: observations from the south-western arseniasis-endemic area in Taiwan. *Int. J. Epidemiol.* 34(4): 936–943.

Dangic, A. 2007. Arsenic in surface- and groundwater in cen-tral parts of the Balkan Peninsula (SE Europe). In: Bhattacharya, P., Mukherjee, A.B.B., Bundschuh, J., Zevenhoven, R., Loeppert, R.H. (eds). *Arsenic in Soil and Groundwater Environment – Biogeochemical Interactions, Health Effects and Remediation.* Elsevier, Amsterdam.

Hough, R.L., Fletcher, T., Leonardi, G.S., Goessler, W., Gnagnarella, P., Clemens, F., Gurzau, E., Koppova, K., Rudnai, P. & Kumar, R. 2010. Lifetime exposure to arsenic in residential drinking water in Central Europe. *Int. Arch. Occup. Environ. Health* 83(5): 471–481.

James, K.A., Marshall, J.A., Hokanson, J.E., Meliker, J.R., Zerbe, G.O. & Byers, T.E. 2013. A case-cohort study examining lifetime exposure to inorganic arsenic in drinking water and diabetes mellitus. *Environ. Res.* 123: 33–38.

Jovanović, D., Jakovljević, B., Rašić-Milutinović, Z., Paunović, K., Peković, G. & Knezević, T. 2011. Arsenic occurrence in drinking water supply systems in ten municipalities in Vojvodina region, Serbia. *Environ. Res.* 111(2): 315–318.

Sicree, R., Shaw, J.E. & Zimmet, P.Z. 2009. The global burden of diabetes. In: Gan, D. (ed). *Diabetes Atlas.* 4th ed. International Diabetes Federation, Brussels.

Varsanyi I. & Kovacs, L.O. 2006. Arsenic, iron and organic matter in sediments and groundwater in the Pannonian Basin, Hungary. *Appl. Geochem* 21(6): 949–963.

Wang, S.-L., Chiou, J.-M., Chen, C.-J., Tseng, C.-H., Chou, W.-L., Wang, C.-C., Wu, T.-N. & Chang, L.W. 2003. Prevalence of non-insulin-dependent diabetes mellitus and related vascular diseases in southwestern arseniasis-endemic and nonendemic areas in Taiwan. *Environ. Health Persp.* 111(2): 155–159.

Environmental Arsenic in a Changing World –
Zhu, Guo, Bhattacharya, Ahmad, Bundschuh & Naidu (Eds)
ISBN 978-1-138-48609-6

Chronic arsenic exposure, endothelial dysfunction and risk of cardiovascular diseases

K. Hossain[1], S. Himeno[2], M.S. Islam[1,3], A. Rahman[1] & M.M. Hasibuzzaman[1]
[1]*Department of Biochemistry and Molecular Biology, University of Rajshahi, Rajshahi, Bangladesh*
[2]*Laboratory of Molecular Nutrition and Toxicology, Faculty of Pharmaceutical Sciences, Tokushima Bunri University, Tokushima, Japan*
[3]*Department of Applied Nutrition and Food Technology, Islamic University, Kushtia, Bangladesh*

ABSTRACT: Cardiovascular diseases (CVDs) are the major causes of arsenic-related morbidity and mortality. Hypertension is a major risk of CVDs. Atherosclerosis is the main biochemical event of all forms of CVDs including hypertension. Atherosclerosis is a complex biochemical process in which several molecules are involved. However, underlying mechanism of arsenic exposure-related atherosclerosis leading CVDs especially hypertension has not yet been clearly understood. In our research, we have focused on the pathophysiology of arsenic-induced atherosclerosis leading to CVDs recruiting human subjects from arsenic-endemic and non-endemic rural areas in Bangladesh. We found that arsenic exposure metrics were positively associated with the elevated levels of diastolic (DBP) and systolic blood pressure (SBP) of the study subjects. We have also analyzed several circulating molecules related to vascular endothelial dysfunction and atherosclerosis including vasoconstrictors, and inflammatory, oxidative and adhesion molecules. We have demonstrated the relationship of arsenic exposure with these molecules. Our results suggest that exposure to arsenic is associated endothelial dysfunction probably through its pro-oxidative and pro-inflammatory properties. Our results also suggest that chronic exposure to arsenic impairs the vascular homeostasis leading to the development of hypertension and other forms CVDs. Thus, endothelial dysfunction and related biochemical events are the hallmarks of arsenic-induced cardiovascular diseases.

1 INTRODUCTION

Exposure to arsenic through drinking water is a major threat to the public health in many countries especially Bangladesh. Consumption of arsenic-contaminated water has turned into an environmental tragedy causing thousands of deaths in Bangladesh. Approximately 80–100 million people are at risk of arsenic toxicity in the country because of the consumption of higher concentrations of arsenic through drinking water than the permissive limit ($>10\,\mu g\,L^{-1}$) set by World Health Organization (WHO). Chronic exposure to arsenic has been associated with CVDs (Huda *et al.*, 2014; Karim *et al.*, 2013; Rahman *et al.*, 1999). Hypertension is a risk of CVDs. However, mechanism of arsenic exposure-related CVDs especially hypertension has not yet been clearly understood. We therefore, explored the underlying mechanism of arsenic-induced CVDs especially hypertension recruiting human subjects from arsenic-endemic and non-endemic rural areas in Bangladesh.

2 METHODS

Study areas and study subjects were selected from the several villages in Bangladesh as we described previously (Ali *et al.*, 2010; Islam *et al.*, 2011, 2015; Karim *et al.*, 2010, 2013; Rahman *et al.*, 2015). Arsenic in the subjects' drinking water, hair and nails were determined by inductively coupled plasma mass spectroscopy (ICP-MS). Circulating molecules were analyzed by bioanalyzer or microplate reader through commercially available kits. The standard protocol recommended by WHO was used for the measurement of blood pressure. Hypertension was defined as an SBP of $\geq$140 mmHg and a DBP of $\geq$90 mmHg on three repeated measurements. Statistical analysis for this study was done with Statistical Package for the Social Sciences (SPSS version 17.0).

3 RESULTS AND DISCUSSION

We found that subjects' drinking water, hair and nail arsenic concentrations were positively and significantly associated with DBP and SBP (Table 1) which indicated that chronic exposure to arsenic was associated with the elevation of blood pressure leading to hypertension. Endothelial function plays an inevitable role in vascular homeostasis. Endothelial dysfunction or damage is the major physiopathological mechanism implicated in the development of CVDs including hypertension. The presence of endothelial dysfunction is considered as clinical syndrome that is associated with and predicts an increased risk of CVDs (McLenachan *et al.*, 1990). Thrombomodulin (TM) is widely distributed on endothelial cell surface of the blood vessels as a high affinity receptor for thrombin. Soluble TM (sTM), a proteolytically cleavage form of

Table 1. Correlation of arsenic exposure metrics with subjects DBP and SBP levels.

Variables	Water As	Hair As	Nail As
DBP	$r_s = 0.332$*	$r_s = 0.320$*	$r_s = 0.293$*
SBP	$r_s = 0.289$*	$r_s = 0.279$*	$r_s = 0.247$*

As = Arsenic; *$p < 0.001$. No. of study subjects for DBP and SBP were 321.

thromobomodulin (TM) in blood represents the degree of endothelial damage. In our study (Hassibuzzaman *et al.*, 2017), we found that subjects' drinking water, hair and nails arsenic concentrations were positively associated with serum sTM levels. Previously, we also reported that arsenic exposure metrics were positively associated with the levels of plasma big endothelin-1 (Big ET-1) (Hossain *et al.*, 2012). Big ET-1 is a precursor form of well-known vasoconstrictor endothelin-1 (ET-1) and a marker of endothelial dysfunction. Positive association of arsenic exposure metrics with sTM and BigET-1 levels clearly indicate that chronic exposure to arsenic causes endothelial dysfunction. We also found that the hypertensive patients had significantly higher levels of sTM and Big ET-1 than the normotensive counterpart. Another important circulating molecule related to blood pressure or hypertension is angiotension II. Angiotension II, an important component of rennin-angiotensin system has strong vasoconstrictive activity. It can also increase the aldosterone level that acts on kidney tubule to reabsorb sodium and water. Reabsorption of sodium and water ultimately increases the blood volume causing elevation of blood pressure. Angiotensin II is produced from angiotensin I by angiotensin converting enzyme (ACE). Intriguingly, in our preliminary experiment, we found that arsenic exposure metrics were positively associated with circulating ACE levels. Positive association between arsenic exposure and ACE indicates that chronic exposure to arsenic causes dysfunction of rennin-angiotensin system. Our preliminary results also indicate that like sTM and Big ET-1, the study subjects who were hypertensive had a significantly higher level of ACE than the normotensive subjects. Our previous studies (Huda *et al.*, 2014; Karim *et al.*, 2013) demonstrated that chronic arsenic exposure metrics were positively associated with circulating inflammatory molecules (C-reactive protein and uric acid), adhesion molecules (intercellular adhesion molecule-1and vascular cell adhesion molecule-1), and inversely associated with anti-inflammatory and anti-oxidative molecule (high density lipoprotein cholesterol-C). Increased levels of pro-inflammatory, pro-oxidative and adhesion molecules, and decreased levels of anti-inflammatory molecules have been reported to be implicated in the development of atherosclerosis leading CVDs through endothelial dysfunction. Arsenic exposure-related endothelial dysfunction and elevated levels of vasoconstrictors and proatherogenic molecules may be the important features of arsenic-induced CVDs and hypertension.

4 CONCLUSIONS

Our results demonstrated that chronic exposure to arsenic was positively associated DBP and SBP. Associations of arsenic exposure with several circulating proatherogenic molecules indicate that chronic exposure to arsenic causes vascular endothelial dysfunction. Our results also suggest that arsenic exposure shift the vascular microenvironment toward vasoconstriction by increasing circulating vasoconstrictors. Taken together, our results reveal that endothelial dysfunctions and constriction of vascular system are the hallmarks of arsenic-induced CVDs especially hypertension.

REFERENCES

Ali, N., Hoque, M.A., Haque, A., Salam, K.A., Karim, M.R., Rahman, A., Islam, K., Saud, Z.A., Khalek, M.A. & Akhand, A.A. 2010. Association between arsenic exposure and plasma cholinesterase activity: a population based study in Bangladesh. *Environ. Health* 9(1): 36.

Hasibuzzaman, M., Hossain, S., Islam, M.S., Rahman, A., Anjum, A., Hossain, F., Mohanto, N.C., Karim, M.R., Hoque, M.M. & Saud, Z.A. 2017. Association between arsenic exposure and soluble thrombomodulin: a cross sectional study in Bangladesh. *PloS One* 12(4): e0175154.

Hossain, E., Islam, K., Yeasmin, F., Karim, M.R., Rahman, M., Agarwal, S., Hossain, S., Aziz, A., Al Mamun, A. & Sheikh, A. 2012. Elevated levels of plasma Big endothelin-1 and its relation to hypertension and skin lesions in individuals exposed to arsenic. *Toxicol. Appl. Pharm.* 259(2): 187–194.

Huda, N., Hossain, S., Rahman, M., Karim, M.R., Islam, K., Al Mamun, A., Hossain, M.I., Mohanto, N.C., Alam, S. & Aktar, S. 2014. Elevated levels of plasma uric acid and its relation to hypertension in arsenic-endemic human individuals in Bangladesh. *Toxicol. Appl. Pharm.* 281(1): 11–18.

Islam, K., Haque, A., Karim, R., Fajol, A., Hossain, E., Salam, K.A., Ali, N., Saud, Z.A., Rahman, M. & Rahman, M. 2011. Dose-response relationship between arsenic exposure and the serum enzymes for liver function tests in the individuals exposed to arsenic: a cross sectional study in Bangladesh. *Environ. Health* 10(1): 64.

Islam, M.S., Mohanto, N.C., Karim, M.R., Aktar, S., Hoque, M.M., Rahman, A., Jahan, M., Khatun, R., Aziz, A. & Salam, K.A. 2015. Elevated concentrations of serum matrix metalloproteinase-2 and-9 and their associations with circulating markers of cardiovascular diseases in chronic arsenic-exposed individuals. *Environ. Health* 14(1): 92.

Karim, M.R., Salam, K.A., Hossain, E., Islam, K., Ali, N., Haque, A., Saud, Z.A., Yeasmin, T., Hossain, M. & Miyataka, H. 2010. Interaction between chronic arsenic exposure via drinking water and plasma lactate dehydrogenase activity. *Sci. Total Environ.* 409(2): 278–283.

Karim, M.R., Rahman, M., Islam, K., Mamun, A.A., Hossain, S., Hossain, E., Aziz, A., Yeasmin, F., Agarwal, S. & Hossain, M.I. 2013. Increases in oxidized low-density lipo-protein and other inflammatory and adhesion molecules with a concomitant decrease in high-density lipoprotein in the individuals exposed to arsenic in Bangladesh. *Toxicol. Sci.* 135(1): 17–25.

McLenachan, J.M., Vita, J., Fish, D., Treasure, C.B., Cox, D.A., Ganz, P. & Selwyn, A.P. 1990. Early evidence of endothelial vasodilator dysfunction at coronary branch points. *Circulation* 82(4): 1169–1173.

Rahman, M., Al Mamun, A., Karim, M.R., Islam, K., Al Amin, H., Hossain, S., Hossain, M.I., Saud, Z.A., Noman, A.S.M. & Miyataka, H. 2015. Associations of total arsenic in drinking water, hair and nails with serum vascular endothelial growth factor in arsenic-endemic individuals in Bangladesh. *Chemosphere* 120: 336–342.

Rahman, M., Tondel, M., Ahmad, S.A., Chowdhury, I.A., Faruquee, M.H. & Axelson, O. 1999. Hypertension and arsenic exposure in Bangladesh. *Hypertension* 33(1): 74–78.

Environmental Arsenic in a Changing World –
Zhu, Guo, Bhattacharya, Ahmad, Bundschuh & Naidu (Eds)
ISBN 978-1-138-48609-6

Arsenic exposure, lung function, vitamin D and immune modulation in the Health Effects of Arsenic Longitudinal Study (HEALS) cohort

F. Parvez[1], M. Eunus[2], F.T. Lauer[3], T. Islam[2], C. Olopade[4], X. Liu[1], V. Slavkovich[1], H. Ahsan[4], J.H. Graziano[1] & S.W. Burchiel[3]
[1]*Mailman School of Public Health, Columbia University, New York City, NY, USA*
[2]*Columbia University Arsenic Research Project, Dhaka, Bangladesh*
[3]*College of Pharmacy, University of New Mexico, Albuquerque, NM, USA*
[4]*Departments of Health Studies, Medicine and Human Genetics and Cancer Research Center, The University of Chicago, Chicago, IL, USA*

ABSTRACT: Epidemiologic research, including our own, has demonstrated associations of arsenic (As) exposure with impaired lung function. While an overwhelming number of recent reports demonstrates a protective role of vitamin D (vit. D) on the respiratory system, its role on lung function in an As exposed population has not been evaluated. In an ongoing study in the Health Effects of Arsenic Longitudinal Study (HEALS) in Bangladesh, we have observed a positive association between serum vit. D and lung function among people exposed to As. Our data show a strong adverse association between water As and lung function among individuals with low vit. D ($<=15\,ng\,mL^{-1}$). The findings suggest that people with adequate levels of vitamin D are less likely to experience impairment of lung function associated with As. Our research has important public health implications in As exposed populations around the globe.

1 INTRODUCTION

Chronic exposure to water arsenic (WAs) has been linked to non-malignant respiratory outcomes including respiratory symptoms, airway epithelial damage, impaired lung function, respiratory infections, prevalence and mortality from COPD. Impaired lung function has also been observed in populations exposed to As from Chile, India and Pakistan. In a longitudinal analysis, we have observed a significant decline of FEV1 (-80.6 mL, $p < 0.01$) and FVC (-97.3 mL, $p = 0.02$) in the study participants ($N = 950$) exposed to WAs $>97\,\mu g\,L^{-1}$ as compared to those exposed to $<19\,\mu g\,L^{-1}$. Notably emerging evidence suggests a protective effect of Vitamin D (vit. D) on lung function, however, a role of circulating vit. D on lung function in As exposed populations has not been evaluated.

We are conducting an epidemiological investigation to examine the effects of long term As exposure, at low-to-moderate levels, on lung function and immune function as well as a possible protective role of vit. D on As-related lung function in the Health Effects of Arsenic Longitudinal Study (HEALS) participants in Araihazar, Bangladesh.

2 MATERIALS AND METHODS

The HEALS cohort was established in Araihazar in 2000, Bangladesh, to prospectively examine health effects of As exposure from drinking water among ~35,000 adults and their children. Since recruitment, we have been following HEALS participants for their health status using a number of procedures including active quarterly in person visits by research assistants, bi-annual visits by physicians, and regular surveillance by village health workers. During the follow-up visits all cohort members were evaluated by trained physicians using standard protocol to ascertain lung disorders.

For this analysis, we utilized preliminary findings from an ongoing investigation. The aims of the project are to evaluate effects of long terms exposure to As on lung function, immune function and serum vit. D in a group of adults ($N = 630$) in the HEALS cohort.

2.1 *Assessment of exposure and study outcomes*

Assessment of exposure: Water, urine and urinary As were measured by high-resolution inductively-coupled plasma mass spectrometry (HR-ICP-MS).

Assessment of study outcomes: We conducted pre- and post-bronchodilator administered *Lung function tests (PFT)* using a portable spirometer (Koko, nSpire Health, Longmont, CA) and were scored based on acceptability criteria developed according to ATS guidelines. We have included only post-bronchodilator PFT data for this analysis.

Vitamin D (25-hydroxyvitamin D2 and 25-hydroxyvitamin D3) were measured simultaneously in serum using Ultra-Performance Liquid Chromatography combined with tandem mass spectrometry (LC-MS/MS) following liquid-liquid extraction.

T-cell function was determined by activating T-cell with anti-CD3/anti-CD28 in peripheral blood mononuclear cell (PBMC) samples.

3 RESULTS AND DISCUSSION

We have conducted an interim analysis in 231 individuals to assess the relationship between As exposure and lung function, vit. D and immune function. At the Arsenic Conference in July we will present findings utilizing a larger dataset.

In the current analysis, we observe an inverse association between As exposure and lung function, particularly among smokers and males. Our analysis shows that WAs exposure was inversely associated with Forced Expiratory Volume in 1 sec (FEV_1) ($r = -0.20$, $p < 0.05$) and Forced Vital Capacity (FVC) ($r = -0.23, p = 0.02$) among smokers. The relationships among non-smokers were as follows: FEV_1 ($r = -0.15, p = 0.06$) and FVC ($r = -0.16, p = 0.04$). Among males we observed an association between WAs and lung function [FEV1 ($r = -0.16, p = 0.09$) and FVC ($r = -0.19, p = 0.04$)]. Similar relationships were also observed with urinary As (UAs).

We have found a positive association between vit. D and FEV_1 ($r = 0.22$, $p < 0.01$) and FVC ($r - 0.23$, $p < 0.01$). In contrast, WAs ($r = -0.15, p = 0.09$) and UAs ($r = -0.38$, $p = 0.006$) were found to significantly inhibit serum vit. D. Importantly, our data show a strong negative association between WAs and FEV_1 ($r = -0.46, p = 0.09$) and FVC ($r = -0.54, p = 0.04$) and FVC among individuals with low serum vit. D ($<= 15$ ng/ml) as compared to those with high vit. D level (> 15 ng/ml) for FEV_1 ($r = -0.13, p = 0.18$) and FVC ($r = -0.05, p = 0.57$).

Additionally, we have observed that a number of cytokines (IL-4, IL-8, IL-10, IL-17α, IL-1β and TNFα) were significantly inhibited by WAs. IL-17α was one of the most affected cytokines by WAs ($r = -0.40$, $p < 0.07$) and UAs ($r = -0.43$, $p < 0.05$). This association was found to be stronger among smokers. A positive correlation between lung function and some pro-inflammatory cytokines (IL-2, IL-4, IL-8, and TNFα) was also observed.

4 CONCLUSIONS

To our knowledge, our study is the first to demonstrate a protective effect of vit. D on lung function in a population chronically exposed water As. The study has several strengths including multiple measures of As exposure including water and urine samples, and measurement of post-bronchodilator lung function tests to assist in establishing the presence of reversible obstructive lung disease. Our analysis also reveals an As-associated immune modulation which may explain part of As related non-malignant respiratory outcomes. Findings from our study may have important public health implication in As exposed populations around the globe.

ACKNOWLEDGEMENTS

This work was supported by US NIEHS grants P42 ES10349, P30 ES 09089, R01 ES019968-02S1 and R01 ES023888. Thanks to our field staff and the study participants.

REFERENCES

Ahsan, H., Chen, Y., Parvez, F., Argos, M., Hussain, A.I., Momotaj, H., Levy, D., van Geen, A., Howe, G., Graziano, J. 2006. Health Effects of Arsenic Longitudinal Study (HEALS): description of a multidisciplinary epidemiologic investigation. *J. Expo. Sci. Environ. Epidemiol.* 16(2): 191–205.

Burchiel, S.W., Lauer, F.T., Beswick, E.J., Gandolfi, A.J., Parvez, F., Liu, K.J. & Hudson, L.G. 2014. Differential susceptibility of human peripheral blood T cells to suppression by environmental levels of sodium arsenite and monomethylarsonous acid. *PLoS ONE* 9(10): e109192.

Dauphine, D.C., Ferreccio, C., Guntur, S., Yuan, Y., Hammond, S.K., Balmes, J., Smith, A.H. & Steinmaus, C. 2011. Lung function in adults following in utero and childhood exposure to arsenic in drinking water: Preliminary findings. *Int. Arch. Occup. Environ. Health* 84: 591–600.

Farzan, S.F., Korrick, S., Li, Z., Enelow, R., Gandolfi, A.J., Madan, J., Nadeau, K. & Karagas, M.R. 2013. In utero arsenic exposure and infant infection in a United States cohort: a prospective study. *Environ. Res.* 126: 24–30.

Kunisaki, K.M., Niewoehner, D.E., Singh, R.J. & Connett, J.E. 2011. Vitamin D status and longitudinal lung function decline in the lung health study. *Eur. Respir. J.* 37: 238–243.

Lange NE, Sparrow D, Vokonas P, Litonjua AA. 2012. Vitamin d deficiency, smoking, and lung function in the normative aging study. *Am. J. Respir. Crit. Care Med.* 186: 616 621.

Lehouck, A., Mathieu, C., Carremans, C., Baeke, F., Verhaegen, J., Van Eldere, J., Decallonne, B., Bouillon, R., Decramer, M. & Janssens, W. 2012. High doses of vitamin D to reduce exacerbations in chronic obstructive pulmonary disease: A randomized trial. *Ann. Intern. Med.* 156: 105–114.

Mazumder, D.N., Steinmaus, C., Bhattacharya, P., von Ehrenstein, O.S., Ghosh, N., Gotway, M., Sil, A., Balmes, J.R., Haque, R., Hira-Smith, M.M. & Smith, A.H. 2005. Bronchiectasis in persons with skin lesions resulting from arsenic in drinking water. *Epidemiology* 16(6): 760–765.

Nafees, A.A., Kazi, A., Fatmi, Z., Irfan, M., Ali, A. & Kayama, F. 2011. Lung function decrement with arsenic exposure to drinking groundwater along river indus: A comparative cross-sectional study. *Environ. Geochem. Health* 33: 203–216.

Parvez, F., Chen, Y., Brandt-Rauf, P.W., Slavkovich, V., Islam, T., Ahmed, A., Argos, M., Hassan, R., Yunus, M., Haque, S.E., Balac, O., Graziano, J.H. & Ahsan, H. 2010. A prospective study of respiratory symptoms associated

with chronic arsenic exposure in bangladesh: Findings from the health effects of arsenic longitudinal study (HEALS). *Thorax* 65: 528–533.
Parvez, F., Chen, Y., Yunus, M., Olopade, C., Segers, S., Slavkovich, V., Argos, M., Hasan, R., Ahmed, A., Islam, T., Akter, M.M., Graziano, J.H., Ahsan, H. 2013. Arsenic Exposure and Impaired Lung Function: Findings from a Large Population-based Prospective Cohort Study. *Am. J. Resp. Crit. Care* 188(7): 813–819.
Smith, A.H., Marshall, G., Yuan, Y., Ferreccio, C., Liaw, J., von Ehrenstein, O., Steinmaus, C., Bates, M.N. & Selvin, S. 2006. Increased mortality from lung cancer and bronchiectasis in young adults after exposure to arsenic in utero and in early childhood. *Environ. Health Perspect.* 114: 1293–1296.
von Ehrenstein, O.S., Mazumder, D.N., Yuan, Y., Samanta, S., Balmes, J., Sil, A., Ghosh, N., Hira-Smith, M., Haque, R., Purushothamam, R., Lahiri, S., Das, S. & Smith, A.H. 2005. Decrements in lung function related to arsenic in drinking water in West Bengal, India. *Am. J. Epidemiol.* 162:
533–541.
Wu, A.C., Tantisira, K., Li, L., Fuhlbrigge, A.L., Weiss, S.T. & Litonjua, A. 2012. Effect of vitamin D and inhaled corticosteroid treatment on lung function in children. *Am. J. Respir. Crit. Care Med.* 186: 508–513.

Environmental Arsenic in a Changing World – Zhu, Guo, Bhattacharya, Ahmad, Bundschuh & Naidu (Eds) ISBN 978-1-138-48609-6

Arsenic in drinking water and childhood mortality: A 13-year follow-up findings

M. Rahman[1], N. Sohel[2] & M. Yunus[3]
[1]*Research and Evaluation Division, BRAC, Mohakhali, Dhaka, Bangladesh*
[2]*Department of Epidemiology and Biostatistics, McMaster University, Hamilton, Canada*
[3]*icddrb, Dhaka, Bangladesh*

ABSTRACT: Around 51 million children globally are exposed to elevated levels of arsenic in drinking water However; the extent to which exposure is related to deaths from cancer at a young age is unknown. We assembled a cohort of 58,406 children aged 5–18 years from Health and Demographic Surveillance System of icddrb in Bangladesh and followed during 2003–2015. The follow-up period was 543415 person-years. We observed a significant association between childhood cancer mortality and arsenic exposure in the highest exposure tertile (HR = 2.70, 95% CI = 1·25–5.83). Arsenic exposure was associated with substantial increased risk of deaths at young age from cancers where child (5–11 years) had a higher risk of death compared to adolescent (12–18 years).

1 INTRODUCTION

Our previous study (Rahman *et al.*, 2007, 2009) followed up childhood residents of Matlab for 7 years and found a significant dose-response relationship between cumulative arsenic exposure and all cancers and cardiovascular combined deaths. Therefore, this study continued with same population of 58,406 Matlab residents who had been followed up for an average of 13 years in an effort to elucidate the dose-response relationship between ingested arsenic exposure and mortality risks. The large number of study participants, longer period of follow-up with more incident cases, and wide range of arsenic exposure levels provided us with a unique opportunity to further investigate the association. However, further aim included analyzing the modifying effects of surviving children and attained education that have been exposed to arsenic through the consumption of tubewell water.

2 METHODS/EXPERIMENTAL

We assembled a population-based cohort study on childhood mortality with excess drinking arsenic water registered in Matlab Health Demographic Surveillance System (HDSS) from January 01, 2003 to December 31, 2015. All childhood deaths (age 5–18 years at baseline) were ascertained. Cardiovascular causes of death were defined as ischemic heart diseases, I20–I25 (International Classification of Disease, 10th Revision (ICD-10)), and cerebrovascular diseases (stroke), I60–I69 (ICD-10) and cancer C1-C99. Causes of deaths were identified from routine VA conducted by specially trained field staff of HDSS who were unaware of the arsenic exposure of the household members.

Follow-up time in person-years was calculated as the number of days between the baseline interview and date of death, out migration, or report of being alive on December 31, 2015 whichever came first. Participants with an accident-related cause of death such as road traffic accident, drowning or other accidental deaths or being alive were censored. The association in baseline and average exposures were calculated for each individual. The association between maternal age at birth, sex, education, asset score, and individual level arsenic exposures were analyzed in order to assess possible associations ($P \leq 0.10$), using chi square or Spearman's correlation coefficients was considered appropriate for the data. The mortality risks of arsenic exposures were estimated by Cox proportional hazards models, adjusting for potential confounders. Factors crudely associated with mortality at 5% level of significance, were included in the model.

3 RESULTS AND DISCUSSION

Of the 58,406 children, 313 died during follow-up. Baseline arsenic exposure was associated with increased all cancer cause of childhood mortality (Table 1). Children were exposed in current arsenic well water ($>138.71\,\mu g\,L^{-1}$) had 2.32 times higher

Table 1. Association between baseline and average exposures and childhood mortalities, hazard ratios (HRs) and 95% confidence interval (95% CI).

Exposure variables	aHRs 95% CI, baseline exposure	aHR and 95% CI, average exposure
$\leq$1.10	1.00	1.00
1.11–138.71	1.89 (0.83–4.28)	0.97 (0.40–2.40)
>138.71	2.70 (1.25–5.83)	2.32 (1.09–4.93)

chance of deaths than those who were exposed to $\leq 1.10\,\mu g\,L^{-1}$ level of arsenic contamination after adjusting age, sex, education attainment and SES.

We noted a higher mortality rate among both cancer and non-cancer children deaths (8.92% and 7.03%, respectively) that had no educational attainment. An earlier study reported that ensuring arsenic free drinking water would increase educational attainment among the Bangladeshi males (Murray & Sharmin, 2015). And also education is a known factor that correlates with the populations' perception about arsenic poisoning. We also demonstrated that the highest mortality rate (9.09 per 100,000 person-years) in 2015 was prevalent among the cancer suffered children in those who exposed $>138.71\,\mu g\,L^{-1}$.

Multiple studies conducted in the arsenic contaminated countries around the world found positive link with various types of cancers including urinary bladder, lung, kidney, hematolymphatic malignancy, skin cancers followed by mortality who are drinking arsenic contaminated water (Axelson *et al.*, 1978; Hopenhayn-Rich *et al.*, 1998; Karagas, 2010; Lewis *et al.*, 1999; Smith *et al.*, 1998; Wu *et al.*, 1989). We found only five childhoods cancer studies relevant to our study, however, these all were ecological and made no inference of all-cause mortality risks (Infante-Rivard *et al.*, 2001; Liaw *et al.*, 2008; Moore *et al.*, 2002; Rubin *et al.*, 2007). All of these studies reported a non-significant relative risk to various cancers including lymphoblastic leukemia from Canada cohort, all types of cancers combined from Chile and Nevada in United States (Infante-Rivard *et al.*, 2001; Liaw *et al.*, 2008; Moore *et al.*, 2002).

4 CONCLUSIONS

Despite some limitation, this study could have been done because of uniqueness of Matlab population having individual level As exposure data and the independent prospective demographic surveillance system covering 0.5 million population for about five decades. Further studies are needed with a longer follow up period to investigate if the risk is further enhanced and causal link becomes clearer though it might be hard enough to find similar population-elsewhere with exposure data for investigating the risk of death due to As exposure at young age.

ACKNOWLEDGEMENTS

The Authors would like to thank Professor Joseph H. Graziano, Mailman School of Public Health, Columbia University, New York for reviewing the revised draft to critically review the child development part. The AsMat study was conducted at icddrb with the support of Swedish International Development Agency (Sida) World Health Organization (WHO) and United States of Agency for International Development (USAID).

REFERENCES

Axelson, O., Dahlgren, E., Jansson, C.D. & Rehnlund, S.O. 1978. Arsenic exposure and mortality: a case-referent study from a Swedish copper smelter. *Br. J. Ind. Med.* 35: 8–15.

Hopenhayn-Rich, C., Biggs, M.L. & Smith, A.H. 1998. Lung and kidney cancer mortality associated with arsenic in drinking water in Cordoba, Argentina. *Int. J. Epidemiol.* 27: 561–569.

Infante-Rivard, C., Olson, E., Jacques, L. & Ayotte, P. 2001. Drinking water contaminants and childhood leukemia. *Epidemiology* 12: 13–19.

Karagas, M.R. 2010. Arsenic-related mortality in Bangladesh. *Lancet* 376: 213–214.

Lewis, D.R., Southwick, J.W., Ouellet-Hellstrom, R., Rench, J. & Calderon, R.L. 1999. Drinking water arsenic in Utah: a cohort mortality study. *Environ. Health Perspect.* 107: 359–365.

Liaw, J., Marshall, G., Yuan, Y., Ferreccio, C., Steinmaus, C. & Smith, A.H. 2008. Increased childhood liver cancer mortality and arsenic in drinking water in northern Chile. *Cancer Epidemiol. Prev. Biom.* 17: 1982–1987.

Moore, L.E., Lu, M. & Smith, A.H. 2002. Childhood cancer incidence and arsenic exposure in drinking water in Nevada. *Arch. Environ. Health* 57(3): 201-206.

Murray, M.P. & Sharmin, R. 2015. Groundwater arsenic and education attainment in Bangladesh. *J. Health Popul. Nutr.* 33: 20.

Rahman, A., Vahter, M., Ekstrom, E.C., Rahman, M., Golam Mustafa, A.H., Wahed, M.A., Yunus, M. & Persson, L.Å. 2007. Association of arsenic exposure during pregnancy with fetal loss and infant death: a cohort study in Bangladesh. *Am. J. Epidemiol.* 165(12): 1389–1396.

Rahman, A., Vahter, M., Smith, A.H., Nermell, B., Yunus, M., El Arifeen, S., Persson, L.Å. & Ekström, E.C. 2009. Arsenic exposure during pregnancy and size at birth: a prospective cohort study in Bangladesh. *Am. J. Epidemiol.* 169(3): 304–312.

Rubin, C.S., Holmes, A.K., Belson, M.G, Jones, R.L., Flanders, W.D., Kieszak, S.M., Osterloh, J., Luber, G.E., Blount, B.C., Barr, D.B., Steinberg, K.K., Satten, G.A., McGeehin, M.A. & Todd, R.L. 2007. Investigating childhood leukemia in Churchill County, Nevada. *Environ. Health Perspect.* 115(1): 151-157.

Smith, A.H., Goycolea, M., Haque, R. & Biggs, M.L. 1998. Marked increase in bladder and lung cancer mortality in a region of Northern Chile due to arsenic in drinking water. *Am. J. Epidemiol.* 147: 660–669.

Wu, M.-M., Kuo, T.-L., Hwang, Y.-H. & Chen, C.-J. 1989. Dose-response relation between arsenic concentration in well water and mortality from cancers and vascular diseases. *Am. J. Epidemiol.* 130: 1123–1132.

3.2 Genetic predisposition of chronic arsenic poisoning

Environmental Arsenic in a Changing World –
Zhu, Guo, Bhattacharya, Ahmad, Bundschuh & Naidu (Eds)
ISBN 978-1-138-48609-6

OsRCS3 functions as a cytosolic *O*-acetylserine(thiol)lyase and regulates arsenic accumulation in rice

C. Wang, Z. Tang & F.-J. Zhao
College of Resources and Environmental Science, Nanjing Agricultural University, Nanjing, P.R. China

ABSTRACT: Reduction of the levels of arsenic in rice shoot is important challenge for agriculture. OsRCS3 is identified as an *O*-acetylserine(thiol)lyase in present study. *OsRCS3* is able to complement an *Escherichia coli* cys-auxotroph mutant.OsRCS3 is localized to the cytoplasm and the expression of *OsRCS3* in roots was up-regulated by arsenic and cadmium treatment. In the presence of As(V), *Osrcs3* mutants showed significantly lower levels of Cys, GSH, PC_2 in roots compared with wild-type. Knockout of OsRCS3 reduced As accumulation in the roots but increased As accumulation in shoots. We conclude that OsRCS3 is the cytosolic O-acetylserine(thiol)lyase that plays an important role in non-protein thiol synthesis and in restricting As accumulation in rice shoots.

1 INTRODUCTION

Complexation of arsenite by thiols is an important mechanism of arsenic (As) detoxification in non-hyperaccumulating plants. Plants could decrease the translocation of As from roots to shoots by increasing the synthesis of a range of thiol-containing compounds in roots (Liu *et al.*, 2010). Cysteine biosynthesis is a central metabolic pathway to incorporate inorganic sulfur into organic molecules. In plants, cysteine is synthesized from *O*-acetyl-L-serine and sulfide by *O*-acetylserine(thiol)lyase (OAS-TL). OAS-TL activity is present in the cytosol, plastids and mitochondria (Hell *et al.*, 2002). In the present study, we investigated the As accumulation of rice (*Oryza sativa*) T-DNA insertion mutants for *OsRCS3*.

2 METHODS/EXPERIMENTAL

2.1 *Functional complementation of OsRCS3 in Escherichia coli*

The full-length coding sequence of *OsRCS3 (Os03g0747800)* was cloned into the prokaryotic expression vector pET-29a. The cysteine auxotroph *E. coli* strain NK3 transformed with the vector or pET-29a were cultured at 37°C for 2-d.

2.2 *RNA extraction and transcriptional analysis by qRT-PCR*

Four-week-old rice seedlings (cv. *Nipponbare* (NPB)) grown in hydroponics were exposed to 10 μM of As(III/V) or cadmium for 24 h. Total RNA were extracted from roots. Rice *Actin* and *HistoneH3* were used as reference genes.

2.3 *Subcellular localization of OsRCS3*

The OsRCS3 full coding cDNA was cloned into pSAT6-EGFP-C1 vector between *HindIII* and *BamHI*. Then *35S:eGFP:OsRCS3* fragment was subcloned into the vector pRCS2-ocs-nptII with *PI-PspI*. The protoplast was isolated from the tobacco leaves transformed with the two-expression vector and observed using a confocal laser scanning microscope (Zeiss).

2.4 *Analysis of total As content and non-protein thiol levels*

Four-week-old *Osrcs3* mutants (Tos17) and wild-type plants were exposed to 2 μM As(V) for 3 days. Phosphate was withheld in the As(V) experiments to facilitate As(V) uptake. Total As and non-protein thiol content were analyzed, as described previously (Wang *et al.*, 2018).

3 RESULTS AND DISCUSSION

3.1 *OsRCS3 functions as O-acetylserine(thiol)lyase*

To test if *OsRCS3* is able to synthesize cysteine, we expressed the gene in the cysteine synthase-deficient *E. coli* strain NK3. The mutant strain transformed empty vector was unable to grow on M9 medium without cysteine. While heterologous expression of *OsRCS3* restored the growth of the *E. coli* strain NK3 on the medium.

3.2 *OsRCS3 was localized in the cytoplasm*

To determine the subcellular localization of the OsRCS3, we isolated protoplasts from the transgenic tobacco leaves expressing *OsRCS3* under the control of cauliflower mosaic virus 35S promoter. Analysis

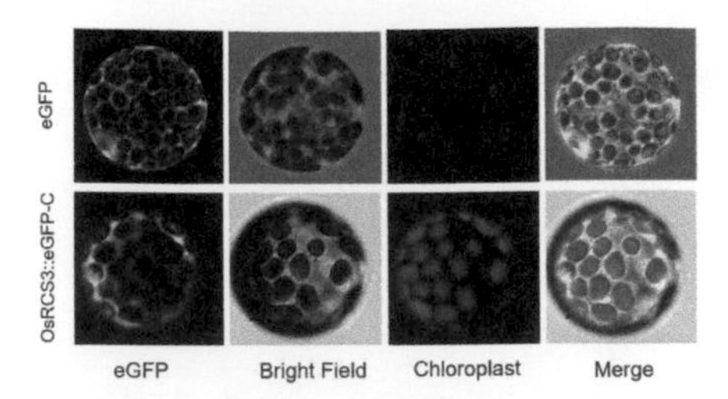

Figure 1. Subcellular localization of OsRCS3 in cytoplasm.

of OsRCS3-GFP-expressing protoplast by confocal microscopy showed that the green fluorescence was confined to cytoplasm. Thus, we confirmed that OsRCS3 is a cytosolic *O*-acetylserine(thiol)lyase (Fig. 1).

3.3 *Arsenic and cadmium-inducible expression of OsRCS3*

The transcription level of the OAS-TL genes was examined by qRT-PCR from 4-week-old roots treated with or without arsenic (III/V) or cadmium for 24 h. A high level of expression was observed for *OsRCS3* and *OsRCS1 (Os12g0625000)*. In contrast, the expression of other *OAS-TL* genes was low. Expression of *OsRCS3* increased after As(III/V) and cadmium treatment.

3.4 *Cytosolic OsRCS3 predominantly contributes to the cellular OAS-TL activity in root*

To investigate the contribution of OsRCS3 to the cellular OAS-TL activity, the OAS-TL activities in the roots and shoots of the *Osrcs3* mutants were determined using crude protein extracts of the mutants and wild-type. We obtained three independent knockout lines of *OsRCS3* mutants from the Rice *Tos17* insertion mutant database. Analysis using RT-PCR showed that the expressions of *OsRCS3* were abolished in these mutants.

In the root, significant reductions in the OAS-TL activity were observed in three *Osrcs3* mutants, NG6001, NG5746, and ND9034. The OAS-TL activity in shoot displayed lower levels than that in root and showed no difference between *Osrcs3* and wild-type. These results indicated that cytosolic OsRCS3 showed predominant OAS-TL activity in the root (Fig. 2).

3.5 *The total As and thiol levels in Osrcs3 mutants*

To determine the contribution of OsRCS3 to cysteine synthesis and subsequent GSH and PCs synthesis in vivo, thiol levels and total As of plants grown in hydroponics treated with 2 μM As(V) for 3-d were measured in the shoots and roots of *Osrcs3* mutants and compared with those of the wild type. The *Osrcs3* mutants showed significant reduction both in the total As content and Cys, GSH, PC_2 levels in the roots (Fig. 3). In the shoots, total As content and PC_2 level were significantly increased in the mutants.

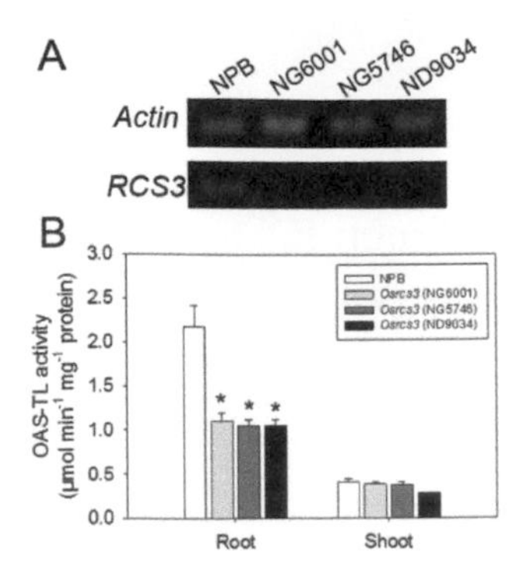

Figure 2. Expression of *OsRCS3* (A) and OAS-TL activity (B) in *Osrcs3* mutants and wild-type in hydroponics without arsenic.

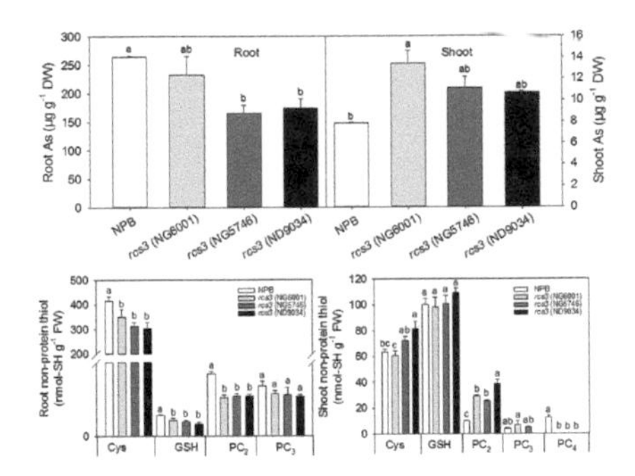

Figure 3. Total As content (A) and thiol levels (B) in roots of *Osrcs3* mutants and wild-type in hydroponics with 2 μM As(V) for 3 days.

4 CONCLUSIONS

The results indicate that OsRCS3 serves as a cytosolic O-acetylserine(thiol)lyase that plays an important role in non-protein thiol synthesis and in restricting As accumulation in rice shoots.

ACKNOWLEDGEMENTS

We thank Professor Jian Feng Ma for the provision of *Osrcs3* mutants. This work was supported by the Natural Science Foundation of China (31520103914).

REFERENCES

Hell, R., Jost, R., Berkowitz, O. & Wirtz, M. 2002. Molecular and biochemical analysis of the enzymes of cysteine biosynthesis in the plant *Arabidopsis thaliana. Amino Acids* 22(3): 245–257.

Liu, W., Wood, B.A., Raab, A, McGrath, S. P., Zhao, F. & Feldmann, J. 2010. Complexation of arsenite with phytochelatins reduces arsenite efflux and translocation from roots to shoots in Arabidopsis. *Plant Physiol.* 152(4): 2211–2221.

Wang, C., Na, G., Bermejo, E.S., Chen, Y., Banks, J.A., Salt, D.E. & Zhao, F.J. 2018. Dissecting the components controlling root-to-shoot arsenic translocation in *Arabidopsis thaliana. New Phytol.* 217(1): 206–218.

Environmental Arsenic in a Changing World –
Zhu, Guo, Bhattacharya, Ahmad, Bundschuh & Naidu (Eds)
ISBN 978-1-138-48609-6

Epigenomic alterations in the individuals exposed to arsenic through drinking water in West Bengal, India

A.K. Giri, D. Chatterjee & N. Banerjee
Molecular Genetics Division, CSIR-Indian Institute of Chemical Biology, Kolkata, India

ABSTRACT: We have evaluated epigenetic changes in the arsenic exposed population at West Bengal, India. We observed that the arsenic exposed individuals had changes in the promoter methylation status of tumor suppressive and DNA repair genes. The downstream effectors mainly proteins were also altered in case of arsenic exposure. Global miRNA profiling showed several altered miRNAs in arsenic exposed samples. Of the above altered miRNAs, a significant number was associated with arsenic induced senescence and diseases like skin lesions, skin cancer, respiratory distress and peripheral neuropathy.

1 INTRODUCTION

Arsenic is not mutagenic but carcinogenic and human. There is a strong believe that arsenic follows the epigenetic pathways to induce cancer in humans. DNA methylation is an epigenetic modification of DNA that is tightly regulated in mammalian development and is responsible for maintaining the normal functioning of the adult organism (Schaefer *et al.*, 2007). Micro RNAs (miRNA) are small 19–25 nucleotides long noncoding RNA molecules that functions in controlling gene expression post-transcriptionally by destabilizing the transcribed mRNA or translational repression (Filipowicz *et al.*, 2008). Evidence showed that epigenetic modifications including DNA methylation and altered micro RNA expression patterns contribute to carcinogenesis (Watanabe *et al.*, 2008). So here we have investigated the epigenetic alterations in arsenic exposed population in West Bengal, India.

2 METHODS

2.1 *Selection of study subjects and sample collection*

Study subjects were recruited from the highly arsenic-affected Murshidabad district of West Bengal. The control subjects were chosen from the arsenic unaffected district of East Midnapur. All the study subjects were age and sex matched with similar socio-economic status. The selection criteria were followed as described in Ghosh *et al.* (2007). Samples were collected only from those subjects who provided informed consent to participate in the study. This study was conducted in accord with the Helsinki II Declaration and approved by the Institutional Ethics Committee of CSIR-Indian Institute of Chemical Biology. The collected samples included drinking water and urine for exposure assessment by Atomic Absorption Spectrometer and blood samples were used to isolate DNA, RNA and protein for subsequent studies.

2.2 *Identification of skin lesions and peripheral neuropathy*

An expert dermatologist identified the characteristics of arsenic-induced skin lesions and helped in the recruitment of exposed study participants. Arsenic-induced neuronal problems were recorded by an expert neurologist for the various types of clinical manifestations including the power and deep tendon reflexes, calf tenderness, pressure and pain as detailed previously (Paul *et al.*, 2013). The studied participants, clinically confirmed to have peripheral neuropathy by the neurologist, were brought for confirmatory electrophysiological studies such as nerve conduction velocity test (NCV) and electromyography (EMG) test.

2.3 *DNA promoter methylation, quantitative real time PCR and Western Blotting*

Briefly bisulfite modification of DNA was performed and then subjected to methylation specific PCR for the promoter regions of the three above mentioned genes separately. RNA was converted to complementary DNA (cDNA) using RevertAid H Minus First Strand cDNA Synthesis Kit and then subjected to gene or microRNA expression analysis. Protein was isolated from the peripheral blood mononuclear cells and cell culture pellet using RIPA lysis reagent and then subjected to western blot for the respected proteins.

3 RESULTS AND DISCUSSION

3.1 *Promoter methylation status is altered and regulates the gene expression status*

The tumor suppressor genes namely DAPK and p16 showed significant hypermethylation in the individuals with skin lesion (WSL) compared with the individuals without skin lesion (WOSL). It also shows that the individuals with DAPK and p16 hypermethylation had higher risk of developing skin lesions

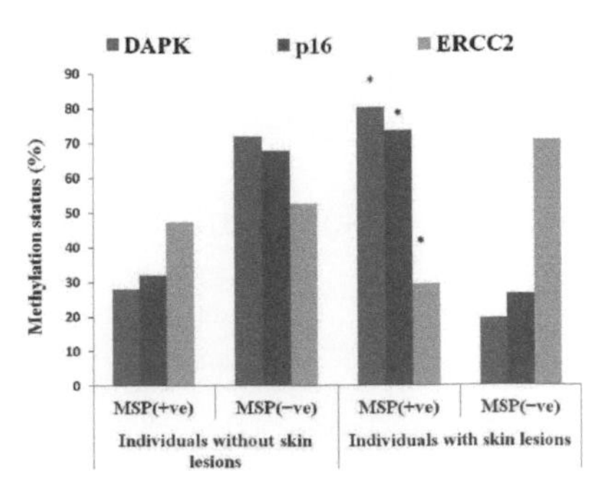

Figure 1. Promoter methylation index (%) of DAPK, p16 and ERCC2 in human study population; *p < 0.5.

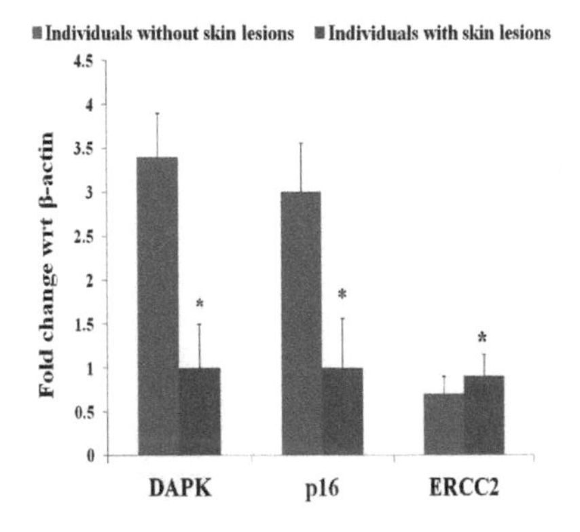

Figure 2. Alteration of protein levels in the differentially altered promoter. *p < 0.5.

as is predicted from the OR values (Banerjee *et al.*, 2013). In humans, arsenic exposure led to hypomethylated promoters of ERCC2, and this was significantly highest in the exposed WSL compared to the referents (Fig. 1). Tumor suppressor proteins, DAPK and p16 was found to downregulated significantly whereas the DNA repair related protein i.e. ERCC2 was up-regulated in the exposed population compared to the reference group (Fig. 2).

3.2 *miRNA levels are associated with arsenic induced diseases*

We performed global miRNA profiling and found that miRNA's involved in arsenic induced senescence and diseases like skin lesions and peripheral neuropathy are altered. Several senescence related miRNAs like miR-21, miR-34a, miR-29a, 126, 141 and 424 was up-regulated in arsenic exposed individuals. This suggests that epigenetic mediators like miRNAs are involved in cellular processes perturbed by arsenic. Results showed that miR-21 was up-regulated in arsenic exposed individuals and contributes to arsenic induced skin lesions, skin cancers and respiratory distress. Similarly, miR-29a was strongly associated with the occurrence of peripheral neuropathy by regulating its downstream effector PMP22 (Banerjee *et al.*, 2017; Chatterjee *et al.*, 2018). *In vitro* studies on different cell lines treated with varying doses of sodium arsenite validated the above observed results from arsenic exposed individuals.

4 CONCLUSIONS

Our study reveals that epigenetic mechanisms contribute to arsenic altered cellular processes like DNA repair ability, tumor suppressive mechanisms and senescence as well as diseases like skin lesions, skin cancer, respiratory distress and peripheral neuropathy. Further molecular insights into the epigenetic mechanisms might provide novel therapeutic strategies for better management of arsenic induced diseases.

ACKNOWLEDGEMENTS

The author acknowledges Council of Scientific and Industrial Research (CSIR), New Delhi, India for funding the research works and field study.

REFERENCES

Banerjee, N., Paul, S., Sau, T.J., Das, J.K., Bandyopadhyay, A., Banerjee, S. & Giri, A.K. 2013. Epigenetic modifications of *DAPK* and *p16* genes contribute to arsenic-induced skin lesions and nondermatological health effects. *Toxicol. Sci.* 135(2): 300–308.

Banerjee, N., Bandyopadhyay, A.K., Dutta, S., Das, J.K., Chowdhury, T.R., Bandyopadhyay, A. & Giri, A.K. 2017. Increased microRNA 21 expression contributes to arsenic induced skin lesions, skin cancers and respiratory distress in chronically exposed individuals. *Toxicology* 378: 10–16.

Chatterjee, D., Bandyopadhyay, A., Sarma, N., Basu, S., Roychowdhury, T., Roy, S.S. & Giri, A.K. 2018. Role of microRNAs in senescence and its contribution to peripheral neuropathy in the arsenic exposed population of West Bengal, India. *Environ. Pollut.* 233: 596–603.

Filipowicz, W., Bhattacharyya, S.N. & Sonenberg, N. 2008. Mechanisms of posttranscriptional regulation by microRNAs: are the answers in sight? *Nat. Rev. Genet.* 9(2): 102.

Ghosh, P., Banerjee, M., De Chaudhuri, S., Chowdhury, R., Das, J.K., Mukherjee, A., Sarkar, A.K., Mondal, L., Baidya, K. & Sau, T.J. 2007. Comparison of health effects between individuals with and without skin lesions in the population exposed to arsenic through drinking water in West Bengal, India. *J. Exp. Sci. Environ. Epidemiol.* 17(3): 215–223.

Paul, S., Das, N., Bhattacharjee, P., Banerjee, M., Das, J.K., Sarma, N., Sarkar, A., Bandyopadhyay, A.K., Sau, T.J. & Basu, S. 2013. Arsenic-induced toxicity and carcinogenicity: a two-wave cross-sectional study in arsenicosis individuals in West Bengal, India. *J. Exp. Sci. Environ. Epidemiol.* 23(2): 156–162.

Schaefer, C.-B., Ooi, S.-K., Bestor, T.-H. & Bourc'his, D. 2007. Epigenetic decisions in mammalian germ cells. *Science.* 316(5823): 398–399.

Watanabe, M., Ogawa, Y., Itoh, K., Koiwa, T., Kadin, M.E., Watanabe, T., Okayasu, I., Higashihara, M. & Horie, R. 2008. Hypomethylation of CD30 CpG islands with aberrant JunB expression drives CD30 induction in Hodgkin lymphoma and anaplastic large cell lymphoma. *Lab. Invest.* 88(1): 48–57.

Environmental Arsenic in a Changing World –
Zhu, Guo, Bhattacharya, Ahmad, Bundschuh & Naidu (Eds)
ISBN 978-1-138-48609-6

Identification of arsenic susceptibility by using the micronucleus assay and Single Nucleotide Polymorphisms (SNP)

A.K. Bandyopadhyay[1], D. Chatterjee[2] & A.K. Giri[2]
[1]*Health Point Multispecialty Hospital, Kolkata, India*
[2]*Molecular Genetics Division, CSIR-Indian Institute of Chemical Biology, Kolkata, India*

ABSTRACT: About 30 million people in the State of West Bengal, India, are exposed to very high amount of arsenic in their drinking water. Although a large number of individuals are exposed to arsenic through drinking water, but only 10–12% individuals showed arsenic induced specific skin lesions. For this reason, it is believed that genetic variations among the arsenic exposed individuals might be responsible for this arsenic susceptibility and carcinogenicity. We have tried to identify the arsenic susceptible individuals by assessing the genetic damage as measured by micronucleus assay in human lymphocytes and Single Nucleotide Polymorphisms (SNP) in both arsenic induced skin symptomatic and asymptomatic individuals exposed to similar arsenic contaminated water.

1 INTRODUCTION

About 150 million people are exposed to arsenic through different sources throughout the World. In West Bengal the groundwater of 9 districts out of 16 districts have very high concentrations of arsenic which people are drinking for more than 2 decades. Skin lesions are the hallmark sign of arsenic toxicity. Hence, we attempted a combinatorial approach of assessing single nucleotide polymorphisms (SNPs) in the p53 (tumor suppressive) and ERCC2 (DNA repair) gene with genetic damage as measured by micronuclei formation in lymphocytes, to predict the risk of skin lesions and skin cancer upon arsenic exposure.

2 METHODS

2.1 *Selection of study subjects*

Recruitment of the arsenic exposed individuals were done from three highly arsenic-affected districts of West Bengal i.e. North 24 Parganas, Nadia and Murshidabad. The control subjects were chosen from East Midnapur district of West Bengal with little or no history of arsenic exposure. All the study subjects were age and sex matched with similar socio-economic status. The selection criteria were followed as described in Beau *et al.* (2004). Samples were collected from those subjects who provided informed consent to participate in the study. This study was conducted in accord with the Helsinki II Declaration and approved by the Institutional Ethics Committee of CSIR-Indian Institute of Chemical Biology. Drinking water and urine were collected for exposure assessment by Atomic Absorption Spectrometer and blood samples were used to isolate DNA and Micronucleus assay.

2.2 *Genotyping and Micronucleus assay from lymphocytes*

DNA extraction from blood was carried out and PCR was done using primers that flanked the p53 codon 72 Arg/Pro polymorphic locus and ERCC2 codon 751 Lysine/glutamine and then subjected to restriction digestion to carry out genotyping of the study samples. Lymphocyte culture was done according to Basu *et al.* (2004). Micronucleus assay was done according to established protocol and Giemsa stained, scored for MN under the microscope (Basu *et al.*, 2004).

3 RESULTS AND DISCUSSION

3.1 *Genetic variations and susceptibility in Arsenic exposed population*

PCR, sequence-BLAST and RFLP analysis were used to identify the key SNPs in DNA-repair and tumor suppressor pathways have been associated with increased incidence of arsenic-induced skin lesions. Table 1 tabulates the SNPs that may be responsible towards arsenic induced skin lesions and premalignant form of skin lesions in the population exposed to arsenic through drinking water (Banerjee *et al.*, 2007; De Chaudhuri *et al.*, 2006).

Table 1. SNPs in different genes studied in this population.

Gene of interest	Polymorphism	Genotype	Risk/protection OR [95% C.I.]
p53	Arg72Pro (G > C)	GG	1.99 [1.28–3.08]
ERCC2	Lys751Gln (A > C)	AA	4.8 [2.78–8.3]

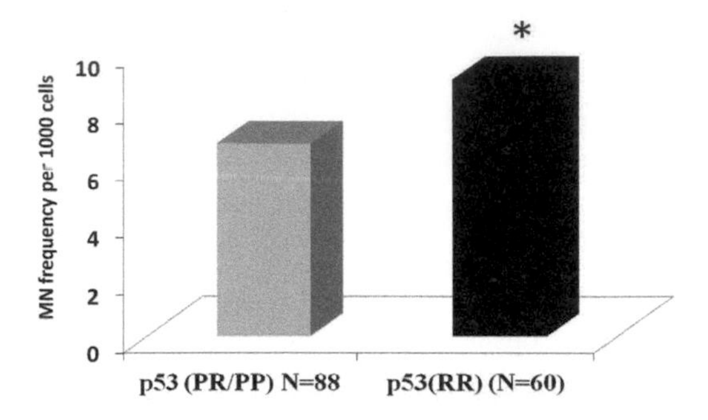

Figure 1. Comparison of the genetic damage (MN) in the risk genotype (RR) and non-risk genotype (PR/PP) in both the study groups; $p < 0.01$.

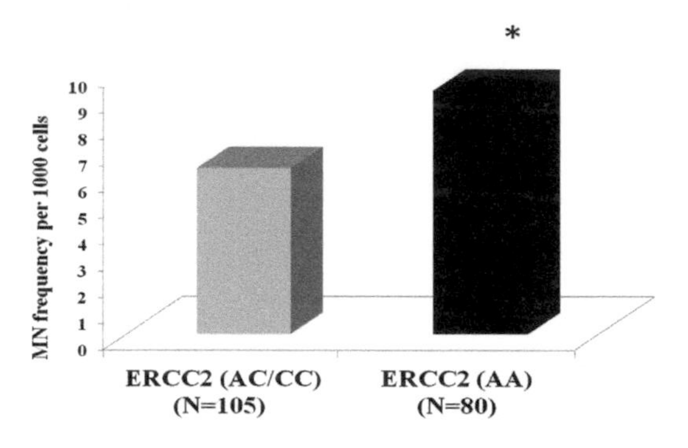

Figure 2. Comparison of the genetic damage (MN) in the risk genotype (AA) and non-risk genotype (AC/CC) in both the study groups; $p < 0.01$.

3.2 *Association of genetic damage with genetic variants*

Frequency of MN/1000 cells in the individuals exhibiting hyperkeratosis and those devoid of any arsenic-specific skin lesion in the risk genotype has been depicted in Figure 1 and Figure 2.

4 CONCLUSIONS

In this study, we speculate that, may be, it is the contribution of the p53 Arg/Arg and ERCC2 Lys/Lys genotype individuals that is responsible for increased MN frequency in the arsenic induced skin lesion group. Hence a combined evaluation of the SNPs and genetic damage as measured by MN assay might be a good predictor of the occurrence of skin lesions.

ACKNOWLEDGEMENTS

The author is extremely grateful to Dr. AKG for providing facilities to carry out the work and CSIR-SRFship to D.C.

REFERENCES

Banerjee, M., Sarkar, J., Das, J.K., Mukherjee, A., Sarkar, A.K., Mondal, L. & Giri, A.K. 2007. Polymorphism in the ERCC2 codon 751 is associated with arsenic-induced premalignant hyperkeratosis and significant chromosome aberrations. *Carcinogenesis* 28(3): 672–676.

Basu, A., Ghosh, P., Das, J.K., Banerjee, A., Ray, K. & Giri, A.K. 2004. Micronuclei as biomarkers of carcinogen exposure in populations exposed to arsenic through drinking water in West Bengal, India: a comparative study in three cell types. *Cancer Epidemiol. Prev. Biomar.* 13(5): 820–827.

De Chaudhuri, S., Mahata, J., Das, J.K., Mukherjee, A., Ghosh, P., Sau, T.J., Mondal, L., Basu, S., Giri, A.K. & Roychoudhury, S. 2006. Association of specific *p53* polymorphisms with keratosis in individuals exposed to arsenic through drinking water in West Bengal, India. *Mutat. Res.* 601(1): 102–112.

Environmental Arsenic in a Changing World –
Zhu, Guo, Bhattacharya, Ahmad, Bundschuh & Naidu (Eds)
ISBN 978-1-138-48609-6

Alternative splicing of arsenic (III oxidation state) methyltransferase

D. Sumi & S. Himeno
Faculty of Pharmaceutical Sciences, Tokushima Bunri University, Tokushima, Japan

ABSTRACT: We identified two splicing variants of the human arsenic (III oxidation state) methyltransferase (AS3MT) gene in HepG2 cells. One splicing variant was an exon-3 skipping (Δ3) form which produced a premature stop codon, and the other was an exon-4 and -5 skipping (Δ4,5) form which produced a 31.1 kDa AS3MT protein, lack of methyltransferase activity. In addition, we found that exposure of HepG2 cells to hydrogen peroxide (H2O2) resulted in increased levels of a novel spliced form skipping exon-3 to exon-10 (Δ3–10). These data suggest that abnormal alternative splicing of AS3MT mRNA may affect arsenic methylation ability and the splicing of AS3MT pre-mRNA was disconcerted by oxidative stress.

1 INTRODUCTION

In mammals, arsenic (III oxidation state) methyltransferase (AS3MT) catalyzes the formation of monomethylarsonate (MMA(V)) and dimethylarsinic acid (DMA(V)) from arsenite (As(III)) and monomethylarsonous acid (MMA(III)), respectively, as a substrate. When exposed to As(III) contained in the diet or drinking water, AS3MT knockout mice showed high systemic toxicity in the bladder epithelium, mild acute inflammation in the liver, and hydronephrosis in the kidneys, and the incidences of these tissue damages in AS3MT knockout mice were higher than those in WT mice. Although these reports indicate that AS3MT activity plays an important role in the modulation of arsenic toxicity, little is known about the mechanisms underlying the regulation of AS3MT activity and of AS3MT gene expression (Sumi *et al.*, 2011).

Alternative splicing generates more than two mRNAs by alteration in the location and combination at the splicing sites, resulting in variant isoforms of the protein translated from a single gene. Environmental chemicals that are known to cause oxidative stress, such as paraquat and arsenic, were shown to impair control over mRNA splicing, resulting in the deregulation of the survival of motor neurons (SMN) and the induction of DNA damage in gene 45α (GADD45α). It has been reported that hydrogen peroxide (H_2O_2) stimulates alternative splicing of hypoxanthine guanine phosphoribosyl transferase (HPRT) and soluble guanylyl cyclase (sGC). Thus, it is apparent that oxidative stress causes splicing abnormalities on specific mRNAs. However, it remains unknown whether the control of splicing of AS3MT mRNA is vulnerable to oxidative stress.

We identified two splicing variants of the human AS3MT gene in human HepG2 cells. One of the two splicing variants of human AS3MT does not have methyltransferase activity. In addition, we found a novel splicing variant of AS3MT mRNA triggered by H_2O_2 (Sumi *et al.*, 2016).

2 METHODS/EXPERIMENTAL

2.1 *Cell culture*

HepG2 cells were obtained from ATCC and were cultured at 37°C in a humidified atmosphere of 5% CO_2 using Dulbecco's modified Eagle's medium containing 10% fetal calf serum, penicillin (100 U mL^{-1}), and streptomycin (100 $\mu g\,mL^{-1}$).

2.2 *Determination of arsenic methylation activity*

The reacted samples and cell lysates suspended in 150 mM Tris-HNO_3 were then sonicated and incubated at 70°C for 30 min, and H_2O_2 was added to a final concentration of 10% at room temperature for 3 h for the oxidation of arsenic metabolites. These samples were centrifuged at 15,000 × g for 10 min at 4°C, and the supernatant was applied to an Amicon YM-3 centrifugal filter at 15,000 × g for 20 min at 4°C. The eluate (5 μL) was separated by a nanospace HPLC system on a Capcell Pak C18 MGII (1.0 mm i.d. ×150 mm long) using 5 mM tetrabutylammonium hydroxide, 3 mM malonic acid, and 4% methanol, as a mobile phase with a flow rate of 200 μL/min. The eluates from the HPLC column were directly introduced into the ICP-MS spray chamber (ICP-MS 7700x).

3 RESULTS AND DISCUSSION

3.1 *Splicing variants of human AS3MT mRNA and activity*

When we carried out RT-PCR with total RNA extracted from HepG2 cells and the primers containing the initiation and termination codons of human AS3MT,

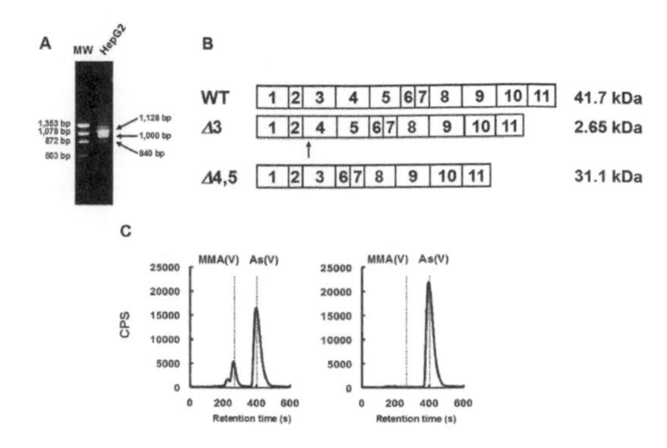

Figure 1. Splicing variants of human AS3MT mRNA and activity. A: RT-PCR was performed in HepG2 cells. B: The diagram of the full-length, Δ3, and Δ4,5 splicing variants of AS3MT mRNA. The arrow indicates the site of premature stop codon in Δ3 AS3MT mRNA. C: Arsenic speciation in samples reacted with recombinant WT or Δ4,5 AS3MT protein, SAM, GSH and As(III).

we found that three PCR products were amplified (Fig. 1A), suggestive of alternative splicing. Therefore, we determined DNA sequences of the three PCR products following the ligation of each PCR product to the sequencing vector. Results obtained from DNA sequence revealed that 1,128, 1,000, and 840 bp cDNAs represent a full-length (WT) AS3MT mRNA, a splicing variant in which exon-3 is deleted (Δ3 AS3MT), and a splicing variant in which both exon-4 and -5 are deleted (Δ4,5 AS3MT), respectively (Fig. 1B). The WT AS3MT protein was calculated to be 41.7 kDa (375 amino acids), while the Δ4,5 AS3MT protein was 31.1 kDa (279 amino acids). On the other hand, the Δ3 AS3MT protein was calculated to be 2.65 kDa (23 amino acids) due to the appearance of a premature stop codon derived by frame shift (Fig. 1B, arrow).

In order to investigate whether the product of Δ4,5 AS3MT mRNA possesses arsenic methyltransferase activity, purified protein was prepared. Arsenic speciation with HPLC-ICP-MS showed a clear peak of MMA(V) after the incubation of As(III) with the recombinant WT AS3MT protein (Fig. 1C, left). However, the peaks of methylated arsenic species were not detected after the incubation with the recombinant Δ4,5 AS3MT protein (Fig. 1C, right). These results indicate that the Δ4,5 AS3MT protein lacks arsenic methyltransferase activity.

3.2 *A novel spliced form of human AS3MT mRNA triggered by H_2O_2 exposure*

In order to determine whether oxidative stress induces alternative splicing on AS3MT mRNA, total RNA was extracted from HepG2 cells exposed to H_2O_2. We carried out RT-PCR with total RNA using primers, located at exon 2 and exon 11 on AS3MT mRNA, respectively. As shown in Figure 2, the mRNA levels of the spliced form of Δ3 were not increased by exposure to H_2O_2, but rather decreased. However, we detected an increase

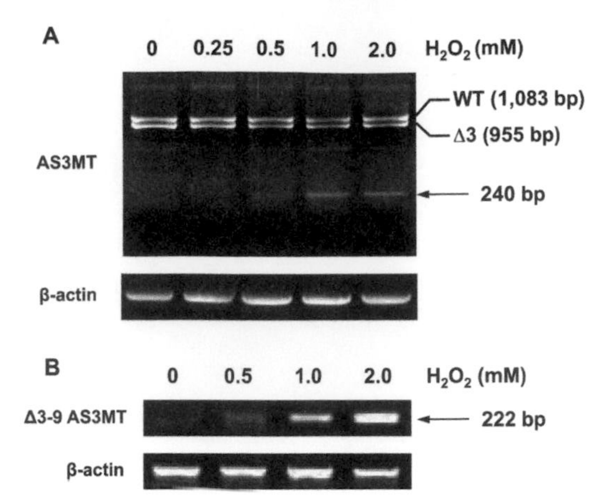

Figure 2. A novel spliced form of human AS3MT mRNA triggered by H_2O_2 exposure. HepG2 cells were exposed to indicated concentrations of H_2O_2 for 6 h. RT-PCR was performed.

of a novel band at 240 bp in an H_2O_2-concentration-dependent manner. DNA sequencing revealed that this band is derived from the AS3MT mRNA skipping the exons from 3 to 9 (Δ3–9). The appearance of the Δ3–9 splicing variant was confirmed by using the specific primers located at the junction of exons 2 and 10 (Fig. 2B).

4 CONCLUSIONS

We found that alternative splicing of AS3MT mRNA in basal condition and splicing abnormalities in AS3MT mRNA were caused by H_2O_2, suggesting that the splicing of AS3MT mRNA is vulnerable to oxidative stress. It has been reported that the expression of AS3MT plays an important role in arsenic metabolism and toxicity manifestation among people living in arsenic-contaminated areas. Further studies are required to evaluate alternative splicing of the AS3MT gene among different populations and the roles of such splicing in individual differences in the capacity for arsenic methylation.

ACKNOWLEDGEMENTS

This work was supported by JSPS KAKENHI Grant Number JP26670068.

REFERENCES

Sumi, D., Fukushima K., Miyataka, H., & Himeno, S. 2011. Alternative splicing variants of human arsenic (+3 oxidation state) methyltransferase. *Biochem. Biophys. Res. Commun.* 415(1): 48–53.

Sumi, D., Takeda, C., Yasuoka, D., & Himeno S. 2016. Hydrogen peroxide triggers a novel alternative splicing of arsenic (+3 oxidation state) methyltransferase gene. *Biochem. Biophys. Res. Commun.* 480(1): 18–22.

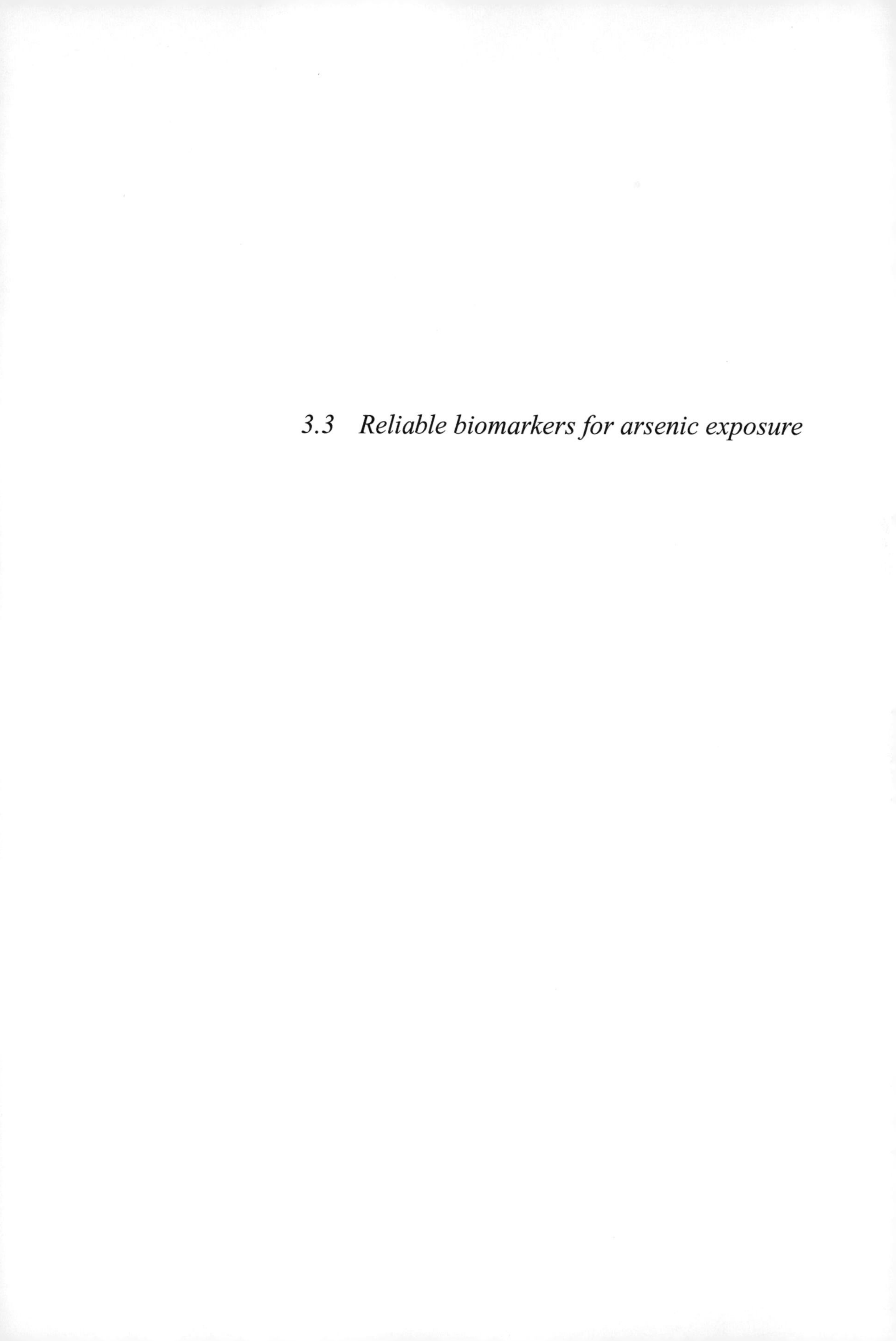

3.3 Reliable biomarkers for arsenic exposure

Environmental Arsenic in a Changing World –
Zhu, Guo, Bhattacharya, Ahmad, Bundschuh & Naidu (Eds)
ISBN 978-1-138-48609-6

Drinking water arsenic exposure, thyroid hormone biomarkers and neurobehavioral outcomes in adolescents

K.M. Khan[1], F. Parvez[2], L. Kamendulis[1], B. Hocevar[1], T. Zoeller[3] & J.H. Graziano[2]
[1]*Department of Environmental and Occupational Health, School of Public Health, Indiana University, Bloomington, Indiana, USA*
Department of Environmental and Occupational Health, Indiana University Bloomington, Bloomington, IN, USA
[2]*Department of Environmental Health, Columbia University New York, New York, NY, USA*
[3]*Department of Biology, University of Massachusetts Amherst, Amherst, MA, USA*

ABSTRACT: Exposure to arsenic (As) from drinking water has been linked to neurocognitive impairment, however, the physiological mechanisms of As-induced neurotoxicity are not known. Emerging evidence from animal studies suggest disruption of the endocrine system in early life as a plausible mechanism. We examined neurobehavioral performance and serum thyroid hormone (TH) among adolescents living in Araihazar, Bangladesh who have been consistently exposed to water arsenic (WAs) above and below 10 $\mu g\,L^{-1}$ since *in utero*. Mothers of these children are participants of the Health Effects of Arsenic Longitudinal Study (HEALS). We have observed inverse relationships between As exposure and TH biomarkers as well as between TH and multiple neurobehavioral outcomes in these adolescents. Findings from this study suggest that As-induced neurotoxicity may be mediated via disruption of thyroid hormone.

1 INTRODUCTION

Chronic arsenic (As) exposure has been recognized as a major environmental health problem that affects multiple organs and biological systems including the developing brain in children (Das *et al.*, 2012: Gong *et al.*, 2015; Wasserman *et al.*, 2007). Thyroid hormones (TH), on the other hand, are key determinants of the development and maturation of central nervous system (CNS) from the pre-natal period to the middle of childhood (Brent, 2010; Gilbert *et al.*, 2012; Parvez *et al.*, 2011; Rosado *et al.*, 2007; Rovet, 2014; Roy *et al.*, 2011). Therefore, one potential mechanism linking As exposure to neurocognitive function is the disruption of thyroid hormone (TH) action. Arsenic appears to interfere with thyroid function by lowering blood levels of thyroid hormone (Kahn *et al.*, 2014; Wasserman *et al.*, 2014; Zoeller *et al.*, 2014). The goal of the present study was to examine the relationships between water As (WAs), four serum TH biomarkers such as free thyroxine (fT4), thyroid stimulating hormone (TSH), total triiodothyronine (tT3) and antibodies to thyroperoxidase (TPOAb) and neurodevelopmental outcomes measured by the Behavioral Assessment and Research System BARS computer test battery in adolescents in Bangladesh who have known WAs exposure history from birth to adolescence.

2 MATERIALS AND METHODS

We recruited 32 healthy adolescents (16 boys and 16 girls) aged between 15 and 17 years whose mothers have been participating in the Health Effects of Arsenic Longitudinal Study (HEALS), since 2000 in Araihazar, Bangladesh. Every mother brings water for all family members and therefore, the sources of drinking water for mother and her child remain the same. Currently, we have WAs and Urinary As (UAs) data available for the mothers covering at least five critical stages of childhood: birth, age 3, 6, 9, and 12, which helped us estimate lifetime WAs exposure. Half of these children were consistently exposed to WAs above the EPA/WHO MCL of 10 $\mu g\,L^{-1}$ from prenatal to adolescence; the remaining half had consistent low lifetime WAs exposure ($<10\,\mu g\,L^{-1}$). We analyzed their serum samples for thyroid biomarkers at the IUB Toxicology Laboratory, urine samples for UAs at Trace Metal Core Laboratory of Columbia University and also evaluated their neurobehavioral performance using the Behavioral Assessment and Research System (BARS) test battery. The computer-based tests in BARS measure attention, response speed, coordination and memory by recording time and accuracy of the responses of the participants even if they do not have prior experience in computer use.

3 RESULTS

3.1 *Arsenic exposure and thyroid biomarkers*

There was no significant difference between high and low As exposed children in terms of age, BMI and blood pressure. We observed that concurrent WAs and UAs were negatively associated with fT_4. At the same time, both WAs and UAs were positively and significantly correlated with TPOAb, which is consistent with earlier reports in adults (Table 1).

Table 1. Spearman correlation coefficients for the relationships between lifetime (In utero and early life) As exposure and serum thyroid biomarkers.

Exposure Variables	Total T_3	Free T_4	TSH	TPOAb
Urinary As (μg/g Cr)	0.05	−0.10	0.01	0.43**
Water As (ppb)	0.02	−0.16	−0.06	0.36*

**p-value = 0.015; *p-value = 0.04

Table 2. Spearman correlation coefficients for the associations between serum thyroid biomarkers and NB outcomes.

NB Outcomes	Total T3	Free T4	TSH	TPOAb
Digit span F	−0.14	**0.31**	**−0.10**	**−0.19**
Digit span R	−0.03	**0.21**	**−0.03**	**−0.25**
Symbol digit latency (ms)	**−0.05**	**−0.13**	**0.18**	**0.20**
Reaction time latency (ms)	**−0.06**	0.09	**0.41***	**0.18**
Continuous performance latency (ms)	**−0.28**	0.38	**0.22**	**0.15**
Continuous performance d-prime	−0.34	−0.09	0.12	0.22
Match-to-sample (correct count)	**0.23**	**0.16**	**−0.36***	0.04
Match-to-sample (correct latency ms)	−0.23	−0.11	0.03	−0.04

Associations demonstrating inverse relationship of serum thyroid biomarkers with NB outcomes are in bold text; *p < 0.05

3.2 *Arsenic exposure and NB outcomes*

With a shorter form of the BARS test battery containing five tests we demonstrated that children exposed to high WAs had poorer neurobehavioral (NB) performance in six out of eight outcome variables (data not shown). This observation showing a negative impact of As on multiple measures NB performance is consistent with the findings of previous epidemiological studies.

3.3 *Thyroid biomarkers and NB outcomes*

We examined the correlations between concurrent thyroid function biomarkers and NB outcomes. Significant negative association between TSH and match-to-sample correct count ($p = 0.04$) and significant positive association TSH and reaction time latency ($p = 0.02$) were observed. Majority of the associations of NB outcomes with TH biomarkers are consistent with the hypothesis that increased TSH and TPOAb and decreased tT3 and fT4 predict decline of cognitive functions in adolescents (Table 2).

4 CONCLUSIONS

Overall, our findings demonstrate that thyroid system disruption by WAs can be revealed by decreased free thyroxine (fT4) or total triiodothyronine (tT3) and increased serum thyroid stimulating hormone (TSH) and antibodies to thyroperoxidase (TPOAb) in adolescents, which in turn may lead to neurocognitive deficits in children.

ACKNOWLEDGEMENTS

This study was funded by the Faculty Research Grant Program (FRGP) of IU School of Public Health, US NIEHS Grants P42 ES 10349 and facilitated by the HEALS staff in Araihazar, Bangladesh.

REFERENCES

Brent, G.A. 2010. Environmental exposures and autoimmune thyroid disease. *Thyroid* 20(7): 755–761.

Das, N., Paul, S., Chatterjee, D., Banerjee, N., Majumder, N.S., Sarma, N., Sau, T.J., Basu, S., Banerjee, S., Majumder, P., Bandyopadhyay, A.K., States, J.C. & Giri, A.K..2012. Arsenic exposure through drinking water increases the risk of liver and cardiovascular diseases in the population of West Bengal, India. *BMC Public Health* 12(1): 639.

Gilbert, M.E., Rovet, J., Chen, Z. & Koibuchi, N. 2012. Developmental thyroid hormone disruption: prevalence, environmental contaminants and neurodevelopmental consequences. *Neurotoxicology* 33(4): 842–852.

Gong, G., Basom, J., Mattevada, S. & Onger, F. 2015. Association of hypothyroidism with low-level arsenic exposure in rural West Texas. *Environ. Res.* 138: 154–160.

Kahn, L.G., Liu, X.H., Rajovic, B., Popovac, D., Oberfield, S., Graziano, J.H. & Factor-Litvak P. 2014. Blood lead concentration and thyroid function during pregnancy: results from the Yugoslavia prospective study of environmental lead exposure. *Environ. Health Persp.* 122(10): 1134–1140.

Parvez, F., Wasserman, G.A., Factor-Litvak, P., Liu, X., Slavkovich, V., Siddique, A.B., Sultana, R., Sultana, R., Islam, T., Levy, D., Mey, J.L., Geen, A.V., Khan, K., Kline, J., Ahsan, H. & Graziano, J.H. 2011. Arsenic exposure and motor function among children in Bangladesh. *Environ. Health Pers.* 119(11): 1665–1670.

Rosado, J.L., Ronquillo, D., Kordas, K., Rojas, O., Alatorre, J., Lopez, P., Garcia-Vargas, G., Caamaño, M.dC., Cebrián, M.E. & Stoltzfus, R.J. 2007. Arsenic exposure and cognitive performance in Mexican schoolchildren. *Environ. Health Pers.* 115(9): 1371.

Rovet, J.F. 2014. The role of thyroid hormones for brain development and cognitive function. *Endocrin. Dev.*26: 26–43.

Roy, A., Kordas, K., Lopez, P., Rosado, J.L., Cebrian, M.E., Vargas, G.G., Ronquillo, D. & Stoltzfus, R.J. 2011. Association between arsenic exposure and behavior among first-graders from Torreon, Mexico. *Environ. Res.* 111(5): 670–676.

Wasserman, G.A., Liu, X., Parvez, F., Ahsan, H., Factor-Litvak, P., Kline, J., van Geen, A., Slavkovich, V., LoIacono, N.J. & Levy, D. 2007. Water arsenic exposure and intellectual function in 6-year-old children in Araihazar, Bangladesh. *Environ. Health Pers.* 115(2): 285–289.

Wasserman, G.A., Liu, X., LoIacono, N.J., Kline, J., Factor-Litvak, P., van Geen, A., Mey, J.L., Levy, D., Abramson, R. & Schwartz, A. 2014. A cross-sectional study of well water arsenic and child IQ in Maine schoolchildren. *Environ. Health* 13(1): 23.

Zoeller, R.T. & Rovet, J. 2014. Timing of thyroid hormone action in the developing brain: clinical observations and experimental findings. *J Neuroendocrinol* 16(10): 809–818.

Environmental Arsenic in a Changing World –
Zhu, Guo, Bhattacharya, Ahmad, Bundschuh & Naidu (Eds)
ISBN 978-1-138-48609-6

Arsenic exposure in the Canadian general population: levels of arsenic species measured in urine, and associated demographic, lifestyle or dietary factors

A. St-Amand, S. Karthikeyan, M. Guay, R. Charron, A. Vezina & K. Werry
Health Canada

ABSTRACT: In Canada, the primary source of exposure to arsenic is food, followed by drinking water, soil, and air. Measuring inorganic arsenic in urine provides a measure of internal dose integrating all sources and routes of exposure. Urine samples were collected through the Canadian Health Measures Survey (CHMS) in 2009 to 2011 from 2538 respondents and analyzed for various inorganic related arsenic species (arsenite, arsenate, monomethylarsonic acid, dimethylarsinic acid, arsenocholine and arsenobetaine). The geometric mean of urinary dimethylarsinic acid (DMA) in the Canadian population was 3.5 (95% CI: 3.0–4.0) μg As L^{-1}. Concentrations were significantly higher in children than in adults. Canadians who eat rice once or more per day have higher urinary concentrations. No association was found with sources of drinking water.

1 INTRODUCTION

Urinary data for inorganic arsenic provides a measure of internal dose integrating all sources and routes of exposure. This study describes nationally-representative concentrations of inorganic related arsenic species in urine in the Canadian population as measured in the Canadian Health Measure Survey (CHMS) 2009–2011. Potential factors influencing exposure data such as age, sex, smoking status as well as dietary factors such as drinking water source, fish, rice and juice consumption are examined.

2 METHODS/EXPERIMENTAL

2.1 *Data collection and laboratory analyses*

The CHMS is designed as a cross-sectional survey conducted over 2-year cycle. Urine samples from 2538 Canadians from 3 to 79 years of age were collected in 18 sites across Canada from August 2009 to November 2011. Urine samples were diluted in ammonium carbonate and analyzed for arsenite (III oxidation state), arsenate (V oxidation state), monomethylarsonic acid (MMA), dimethylarsinic acid (DMA), and arsenocholine and arsenobetaine combined using ultra performance liquid chromatography (UPLC) on a Waters Acquity UPLC coupled to ICP-MS on a Varian 820-MS (INSPQ, 2009).

2.2 *Statistical analysis*

The data were analyzed with SAS 9.2 and SUDAAN 10.0.1 software. All analyses were weighed using the CHMS cycle 2 survey weights (environmental urine subsample) in order to be representative of the Canadian population. Variance estimates were produced using bootstrap weights, taking into account the 13 degrees of freedom. Concentrations that were below LOD were assigned a value of LOD/2. For each biomarker, descriptive statistics (geometric means and selected percentiles with their associated 95% confidence intervals) were calculated on the volumetric concentrations.

3 RESULTS AND DISCUSSION

3.1 *Descriptive statistics*

Descriptive statistics of arsenic species and metabolites for the Canadian population are presented in Table 1. Inorganic arsenic species and metabolites were detected in 24.4% of samples as arsenite, 0.5% as arsenate, 27% as MMA and 96.2% as DMA. The geometric mean of DMA in the Canadian population was 3.5 (95%CI: 3.0–4.0) μg As L^{-1}.

3.2 *Simple regression model*

Models were developed for the sum of urinary MMA, DMA and arsenite. The results of simple regression models show that concentrations of the sum were significantly associated with rice consumption (higher concentrations in individuals with more than twice per week consumption), fish and shellfish consumption (higher concentrations in fish and shellfish eaters), detection of arsenobetaine and arsenocholine (higher

Table 1. Geometric means and 95th percentiles of urinary speciated arsenic concentrations in $\mu g As L^{-1}$ (95% confidence interval) for the Canadian population aged 3–79 years, Canadian Health Measures Survey cycle 2, 2009–2011.

Species	Sample size	% detects[a]	Geometric mean[b]	95th percentile
Arsenite	2537	245	—	2.7[c] (1.3–4.0)
Arsenate	2538	<1	—	<LOD
MMA	2538	27	—	1.6 (1.2–2.0)
DMA	2538	96	3.5 (3.0–4.0)	16[c] (7.3–26)
Arsenocholine, arsenobetaine	2538	51	—	48[c] (30–67)

DMA: dimethylarsinic acid; MMA: monomethylarsinic acid.
[a]Limits of detection were: $0.7 \mu g As L^{-1}$ for arsenite, arsenate, MMA and DMA, and 1.5 for arsenocholine/arsenobetaine, respectively.
[b]If detection was $\leq 60\%$, the geometric mean was not calculated.
[c]Data are used with caution as coefficient of variation is between 16.6% to 33.3%.

concentrations in individuals with detected levels), and ethnic background.

3.3 *Multiple regression model*

Multiple regression models for the sum of MMA and DMA show that LSGM concentrations were significantly higher in 3–5 and 6–11 years old than in 20–79 years old adults (3–5 vs 20–79 p-value = 0.01, 6–11 vs 20–79 p-value < 0.0006). The R^2 was 0.50, meaning that 50% of the variability in the sum of MMA and DMA was explained by the model.

4 CONCLUSIONS

This nationally representative dataset reflects background exposure for the Canadian general population which is consistent with levels measured in the US population (Caldwell *et al.*, 2009). Models indicate that sources of exposure to inorganic arsenic in the diet are quite varied but that rice and seafood are among the larger contributors. However, there are others such as vegetables, fruits and fruit juices, other grain products that can contribute to dietary inorganic arsenic exposure. Approximately a third of dietary inorganic arsenic exposure is methylated arsenic. A high proportion DMA in urine could be from direct ingestion and not inorganic arsenic metabolism.

ACKNOWLEDGEMENTS

This study was funded under the Government of Canada's Chemicals Management Plan. We would like to thank Statistics Canada for planning and Centre de Toxicologie du Québec for conducting the laboratory analyses.

REFERENCES

Caldwell, K.L., Jones, R.L., Verdon, C.P., Jarrett, J.M., Caudill, S.P. & Osterloh, J.D. 2009. Levels of urinary total and speciated arsenic in the US population: National Health and Nutrition Examination Survey 2003–2004. *J. Expo. Sci. Env. Epid.* 19(1): 59–68.

INSPQ 2009. Analytical method for the determination of arsenic species in urine by ultra-performance liquid chromatography coupled to argon plasma induced mass spectrometry (HPLC-ICP-MS) (M-585), condensed version for CHMS. Laboratoire de Toxicologie, Québec, QC.

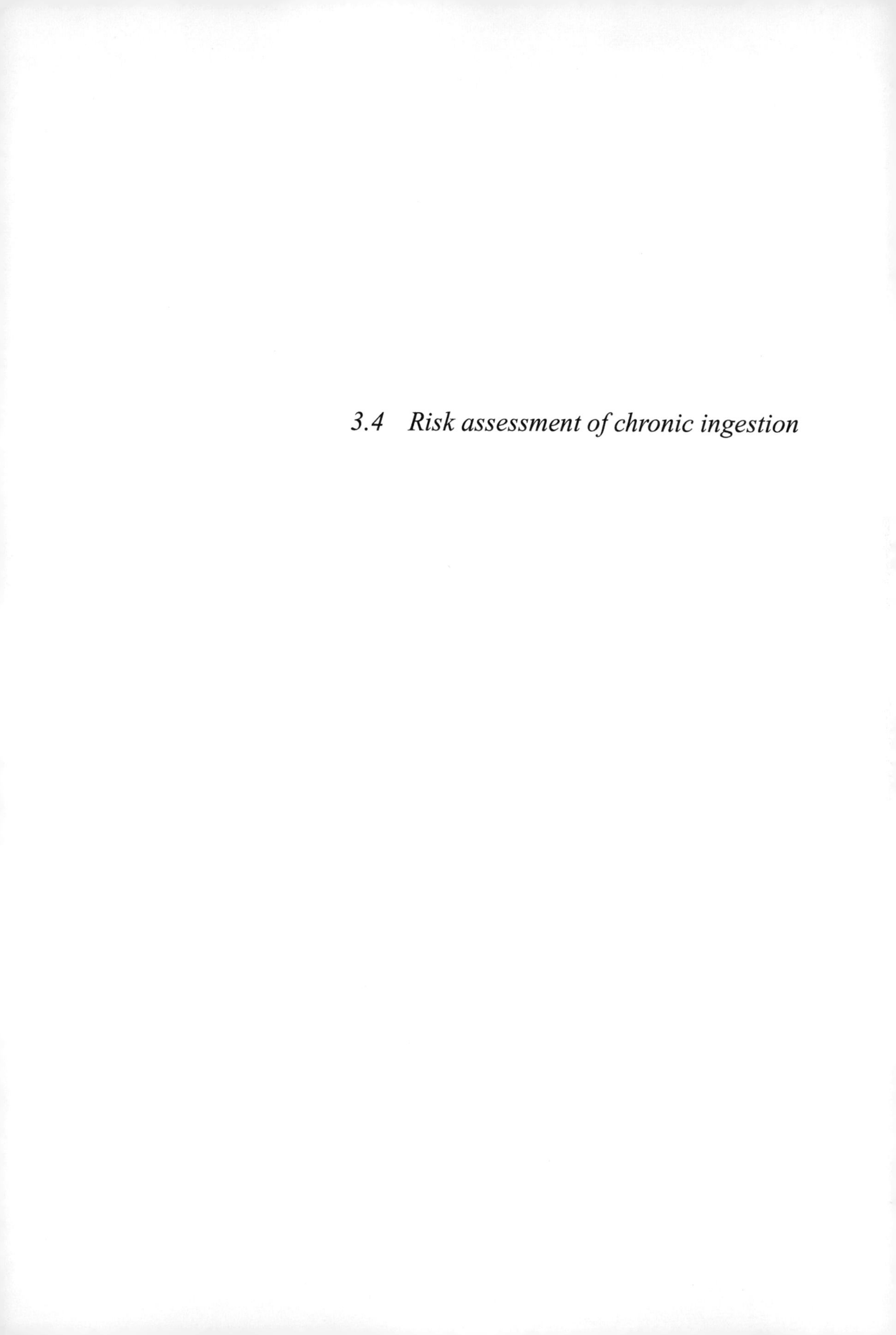

3.4 Risk assessment of chronic ingestion

Environmental Arsenic in a Changing World –
Zhu, Guo, Bhattacharya, Ahmad, Bundschuh & Naidu (Eds)
ISBN 978-1-138-48609-6

Beyond the wells: role of diet on arsenic induced toxicity in exposed populations of Bihar, India

D. Mondal[1], S. Suman[2], P. Sharma[2], S. Kumari[2], R. Kumar[2], A.K. Ghosh[2], N. Bose[3], S.K. Singh[4], H. Matthews[5] & P.A. Cook[5]
[1]*School of Environment & Life Sciences, University of Salford, Salford, UK*
[2]*Mahavir Cancer Institute and Research Center, Patna, India*
[3]*Department of Geography, A.N. College, Patna, India*
[4]*Department of Environment and Water Management, A.N. College, Patna, India*
[5]*School of Health Sciences, University of Salford, Salford, UK*

ABSTRACT: Drinking water is the major route of arsenic exposure but dietary factors can modulate the arsenic induced toxicity. We aim to determine the dietary correlates of arsenic exposure and factors that can act as host defense. As a part of an on-going research project here, we present with the preliminary findings from survey of a study population with known arsenic exposure from drinking water in Bihar, India.

1 INTRODUCTION

Arsenic levels in drinking water as high as $1900\,\mu g\,L^{-1}$ has been recently detected in one of the villages from arsenic affected districts (18 out of 38) of Bihar where more than 10 million people are facing contamination from drinking water (Thakur & Gupta, 2006). Diet is known to modulate the risk of arsenic induced toxicity, for example diet rich in gourds and root vegetables along with increased diversity of food is found to reduce the risk of arsenical skin lesions (Pierce *et al.*, 2010). On the other hand, malnutrition, which is highly prevalent in Bihar (NFHS, 2016), is known to increase susceptibility of arsenical skin lesions (Mazumder *et al.*, 2011). The aim of this study is to determine the dietary correlates of arsenic induced toxicity in Bihar.

2 METHODS

2.1 *Study area*

Samples (N = 42) were collected from residents of arsenic contaminated areas (reported concentration $>10\,\mu g\,L^{-1}$) from Bhagalpur (Rannuchak and Nandgola), Banka (Rajpura and Kakna), Saran (Sabalpur) and Patna (Haldi Chapara) districts of Bihar, India as a part of ongoing Nutri-SAM project.

2.2 *Data collection*

After taking informed consent, data on socio-economic, nutritional and health status were collected using questionnaires. Food consumption pattern and nutrient intake was determined using food frequency questionnaire and 24-hour recall methods, respectively. Nutrient content was calculated using Indian Food Composition Table – 2017 (Longvah *et al.*, 2017). Blood, water, hair, rice (raw and cooked), wheat and potato samples were collected and arsenic content was estimated by Atomic Absorption Spectrophotometer (PerkinElmer PinAAcle 900 T) using APHA protocol.

2.3 *Statistical analysis*

Descriptive data as mean ± standard deviation along with comparison tests, association and correlations are presented using SPSS (IBM, version 22) and Stata (version 11.2, StataCorp, USA).

3 RESULTS AND DISCUSSION

3.1 *Demographic and nutritional status*

There were significant demographic differences between male and female participants (Table 1) along with difference in nutrient intake. Overall, 23% participants were underweight, 14% were overweight and 9% were found to be obese. Most of the participants were married, having extended family (more than six members) but had a common kitchen. None consume alcohol (prohibited by law in Bihar), but 38% were found to take some form of tobacco such as *gutkha* or chewing leaves.

Table 1. Demography and nutrient intake (mean ± SD).

	Overall	Male (n = 20, 48%)	Female (n = 22, 52%)	p value
Age (years)	52 ± 16	58 ± 17	46 ± 15	**0.018**
BMI ($kg\,m^{-2}$)	22.9 ± 6.6	20.5 ± 3.4	25.2 ± 7.0	**0.010**
W/H Ratio	0.8 ± 0.07	0.89 ± 0.02	0.85 ± 0.01	0.169
Blood Pressure (normal %)	76	75	77	0.863
Energy ($kcal\,d^{-1}$)	1667 ± 655	1928 ± 628	1430 ± 127	**0.012**
Protein ($gm\,d^{-1}$)	51 ± 18.5	58 ± 4	43 ± 4	**0.006**
CHO ($gm\,d^{-1}$)	281 ± 114	332 ± 24	234 ± 21	**0.004**
Fat ($gm\,d^{-1}$)	48.5 ± 28.0	52.8 ± 7.1	44.7 ± 5.1	0.357
Dietary fibre	41.1 ± 22.1	49.1 ± 5.0	33.8 ± 4.2	**0.023**

Table 2. Summary of arsenic concentrations.

Sample (N, unit)	Mean	S.D	Maximum
Water (42, $\mu g L^{-1}$)	33.86	34.99	107.0
Blood (26, $\mu g L^{-1}$)	12.32	24.77	104.78
Hair (23, $\mu g kg^{-1}$)	807.62	689.76	2480.40
Raw rice (41, $\mu g kg^{-1}$)	50.98	194.65	1257.50
Cooked rice (18, $\mu g kg^{-1}$)	139.61	386.16	1654.50
Wheat (33, $\mu g kg^{-1}$)	175.57	158.01	567.50
Potato (33, $\mu g kg^{-1}$)	248.05	205.56	826.00

Table 3. Correlates of skin manifestation.

	Yes (n = 10, 24%)	No (n = 32, 76%)	p value
Gender (number)	M = 5; F = 5	M = 15; F = 17	0.863
Age (years)	48 ± 17	53 ± 16	0.415
BMI ($kg m^{-2}$)	22 ± 3	23 ± 6	0.781
W/H Ratio	0.85 ± 0.04	0.88 ± 0.08	0.575
Energy ($kcal d^{-1}$)	1820 ± 942	1619 ± 548	0.403
Protein ($gm d^{-1}$)	52 ± 23	50 ± 17	0.712
CHO ($gm d^{-1}$)	323 ± 170	268 ± 91	0.191
Fat ($gm d^{-1}$)	53 ± 24	36 ± 29	0.504
Dietary fibre	43 ± 32	40 ± 18	0.646
Water ($\mu g kg^{-1}$)*	39 ± 40	32 ± 33	0.562
Raw rice ($\mu g kg^{-1}$)*	156 ± 413	21.3 ± 25.4	0.065
Wheat ($\mu g kg^{-1}$)*	355 ± 161	97 ± 66	**0.000**
Potato ($\mu g kg^{-1}$)*	616 ± 111	166 ± 106	**0.000**
Blood ($\mu g kg^{-1}$)*	73 ± 44	7 ± 15	**0.000**
Hair ($\mu g kg^{-1}$)*	1301 ± 686	525 ± 528	**0.007**

*Measured arsenic concentrations.

67% of the respondents were found to skip breakfasts every day and only 14% had four or more meals a day. The main sources of nutrients were cereals such as wheat and rice with 87% participants having pulses every day. Milk intake in the study population was 67% and 9.5% consumed curd every day. Most of participants consume green leafy, roots and tubers and other vegetables everyday whereas only 17% had fruits regularly. Use of refined sugar was common (81%) while *jaggery* and carbonated beverages were consumed occasionally.

3.2 *Arsenic, diet and health*

The overall arsenic concentration (Table 2) in drinking water was less than the Indian permissible limit of $\mu g L^{-1}$ but higher than the WHO limit of $10 \mu g L^{-1}$ and Saran district had the highest concentration (n = 14; mean ± SD: $44 \pm 38 \mu g L^{-1}$; max: $107 \mu g L^{-1}$). Arsenic in food items (rice, wheat and potato) was not correlated with concentrations in drinking water. This could be because either the food is sourced from markets or the water used for irrigation is different to water used for drinking.

There was no significant difference in demography and nutrient intake between participants with skin manifestations (pigmentation or lesions, a hall mark for arsenic toxicity, 24%) compared to participants without skin symptoms (Table 3).

Participants with skin manifestations had significantly high concentrations of arsenic in blood and were consuming higher arsenic from food items significantly so for wheat and potato (Table 3).

Table 4. Association between arsenic in food and in biomarkers.

	Blood		Hair	
Arsenic in	r	p	r	P
Water	−0.20	0.32	0.69**	<0.001
Cooked rice	0.93**	<0.001	0.49	0.17
Wheat	0.59**	0.007	0.16	0.52
Potato	0.93**	<0.001	0.53*	0.03
Energy intake/day	0.43*	0.02	−0.14	0.54
Protein intake/day	0.38	0.05	−0.14	0.55
CHO intake/day	0.49*	0.01	−0.04	0.85
Dietary fiber intake/day	0.40*	0.04	−0.12	0.59

Though there was no relationship between drinking water and blood arsenic, statistically significant positive correlation was found between arsenic in cooked rice, wheat, potato, energy intake, protein intake, carbohydrates intake, dietary fiber intake and arsenic in blood. A strong relationship between water and hair arsenic was observed and statistically significant positive correlation was observed between arsenic in water and potato with arsenic in hair samples (Table 4).

4 CONCLUSIONS

Arsenic in food items (wheat and potato) and nutrient intake is found to be related with blood arsenic levels and the blood arsenic is found to be significantly higher is participants with skin manifestations, compared to those without any skin symptoms. Both potato and wheat are found to be significant route of arsenic exposure in this population. This is a part of an ongoing study and further results will be presented at the conference.

ACKNOWLEDGEMENTS

This study is funded by the DST-UKIERI Thematic partnership project. We thank all the participants.

REFERENCES

Longvah, T., Ananthan, R., Bhaskarachary, K. & Venkaiah, K. 2017. Indian food composition table. *National Institute of Nutrition (ICMR), Hyderabad, India.*

Mazumder, D.G. & Dasgupta, U.B. 2011. Chronic arsenic toxicity: studies in West Bengal, India. *Kaohsiung J. Med. Sci.* 27(9): 360–370.

NFHS. 2016. State fact sheet-Bihar. Ministry of health and family welfare, *Government of India*. 1–4.

Pierce, B.L., Argos, M., Chen, Y., Melkonian, S., Parvez, F., Islam, T., Ahmed, A., Hasan, R., Rathouz, P.J. & Ahsan, H. 2010. Arsenic exposure, dietary patterns, and skin lesion risk in Bangladesh: a prospective study. *Am. J. Epidemiol.* 173(3): 345–354.

Thakur, B.K. & Gupta, V. 2016. Arsenic concentration in drinking water of Bihar. *J Water Sanit. Hyg. Dev.*, 06(2): 331–341.

Environmental Arsenic in a Changing World –
Zhu, Guo, Bhattacharya, Ahmad, Bundschuh & Naidu (Eds)
ISBN 978-1-138-48609-6

Effects of folate on arsenic methylation pattern and methionine cycle in sub chronic arsenic-exposed mice

D. Wang, Y. Li, B. Li, L. Lin & G.F. Sun
Research Center of Environment and Non-Communicable Disease, School of Public Health, China Medical University, Shenyang, P.R. China

ABSTRACT: Arsenic compounds are serious harm to human health. Skin damages, cardiovascular disease, diabetes and cancer may cause by chronic arsenic exposure. Arsenic metabolism pattern is an important index of arsenic poisoning, our previous studies have shown that exogenous factors play a more important role in arsenic poisoning than environmental factors. Based on this, the focus of this research is folic acid which are important cofactors in methionine cycle. In the present study, we establish high arsenic exposure animal model and then applying interference factors. Our results show that As levels were significantly dose-dependent in different experimental groups, as well as moderate increase of SAH, decrease of SAM and SAM/SAH were found in the blood by sub chronic arsenic administration. We also observed the involvement of folate in promoting arsenic methylation in vivo. Furthermore, SAM level increased after the treatment of folate.

1 INTRODUCTION

Inorganic arsenic (iAs) is a known toxicant and carcinogen. Worldwide arsenic exposure has become a threat to human health. The severity of arsenic toxicity after exposure is related to the speed of arsenic metabolism and clearance. The methylation capacity plays an important role in the metabolism of arsenic. Multiple studies demonstrate that protective effects of folate against arsenic-induced oxidative injuries in vivo. However, the effect of folate on arsenic metabolism and methionine cycle remains unclear. In view of the continued exposure to arsenic and associated health risk, we are trying to provide the experimental evidences to counteract and relieve the chronic injuries of arsenic by means of folate.

2 METHODS/EXPERIMENTAL

2.1 *Animals and treatment*

Totally seventy mice were randomized into seven groups of ten mice each and were treated as the following details:

Group I control mice.
Group II treated with $NaAsO_2$ in drinking water alone ($10\,mg\,L^{-1}$, free access to drinking water, 6 weeks).
Group III treated with $NaAsO_2$ in drinking water alone ($50\,mg\,L^{-1}$, free access to drinking water, 6 weeks).
Group IV treated with $NaAsO_2$ in drinking water alone ($100\,mg\,L^{-1}$, free access to drinking water, 6 weeks).
Group V treated with arsenic (as in group II) plus folate ($10\,mg\,kg^{-1}$, gavage, twice one week for 6 weeks).
Group VI treated with arsenic (as in group III) plus folate ($10\,mg\,kg^{-1}$, gavage, twice one week for 6 weeks).
Group VII treated with arsenic (as in group IV) plus folate ($10\,mg\,kg^{-1}$, gavage, twicc one week for 6 weeks).

The day before the end of the experiment, the mice were housed into the metabolism cages (two mice per cage) to collected 24 h cumulative urine sufficiently for analysis of the urine arsenic and its metabolites. Then at the end day of the treatment, all mice were weighed and killed by ether anesthesia. Blood was collected through eyeball extirpating into non-heparinized and heparinized vials for serum separation and the subsequent biochemical assays, respectively.

2.2 *Determination of SAM and SAH*

Based on protein precipitation, SAH and SAM in the RBCs samples were determined by HPLC with ultraviolet detection and the detector wavelength was 254 nm. HPLC analytical column was Thermo Syncronis C 18 WP ($150\,mm \times 4.6\,mm$, $5\,\mu m$) at 30°C. The separation was carried out with the mobile phase of $5\,mmol\,L^{-1}$ sodium heptanesulfonate in $50\,mmol\,L^{-1}$ sodium phosphate buffer ($pH = 4.3$)-methanol (80:20, v/v) at $1.0\,mL\,min^{-1}$. The assay utilized a quantification method of external standard.

2.3 *Determination of arsenic metabolites*

Arsenic species (iAs, MMA, DMA and TMA) determination in urine were performed by an atomic

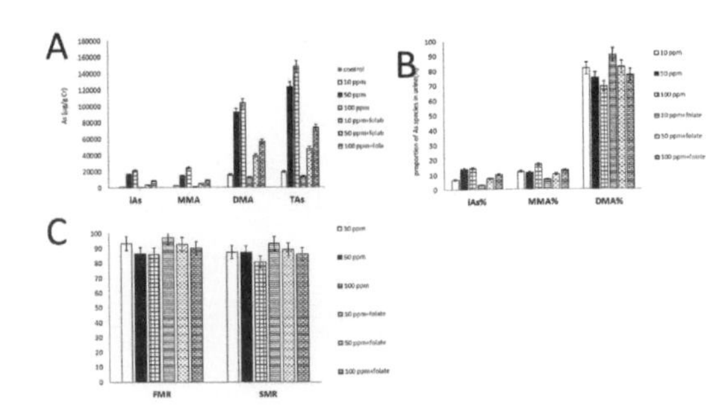

Figure 1. Arsenic concentrations and profiles in the urine of mice of arsenic exposure. (A) Concentrations of iAs, MMA and DMA in urine. The TAs was calculated by summing up the concentrations of iAs, MMA and DMA. (B) The proportions of urine arsenic metabolites. (C) The arsenic methylation indices. FMR, first methylation ratio; SMR, secondary methylation ratio. n = 6–8. Values are mean ± SEM.

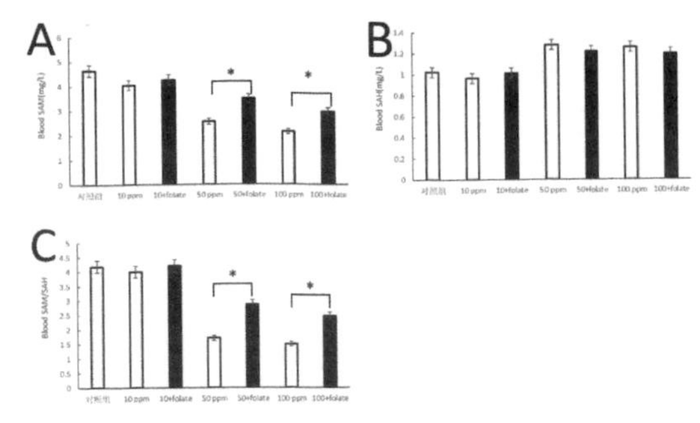

Figure 2. SAM and SAH concentrations, SAM/SAH in the blood of mice of arsenic exposure. (A) Concentrations of SAM in blood. (B) Concentrations of SAH in blood. (C) The ratio of SAM and SAH. n = 6–8. Values are mean ± SEM.

absorption spectrophotometer (AA-6800, Shimadzu Co. Kyoto, Japan) with an arsenic speciation pretreatment system (ASA-2sp, Shimadzu Co. Kyoto, Japan) as described previously (Sun *et al.*, 2007).

3 RESULTS AND DISCUSSION

In this study, we determined the levels of T-As, as well as various arsenic metabolites in urine, to explore the potential effects of folate on arsenic methylation and urinary excretion. As shown in Figure 1A, urinary arsenic levels decreased significantly following folate treatment. But a clear enhancement of both urinary DMA% and two methylation indices (PMR and SMR), as well as the reduction of urinary iAs% in Figure 1B and C. Our results here suggested that facilitation of arsenic methylation may be involved in the protective effects of folate.

The concentrations of SAM and SAH, as the substrate and product of essential cellular methyltransferase reactions, are important metabolic indicators of the cellular methylation capacity. Therefore, we compared SAM and SAH concentrations among Arsenic-treated groups, folate-treated groups and control.

Figure 2 show a significant decrease in SAM. This resulted in a significant decrease in the SAM/SAH ratio following arsenic treatment compared with that of the control and significant increase occurred after folate treated. SAH has a higher affinity for the methyltransferase active site compared with that of the SAM precursor. Pathological SAH accumulation can decrease the SAM/SAH ratio and inhibit most cellular methyltransferase activity, which may induce DNA hypomethylation. Thus, our results suggest that arsenic decreased methylation capacity of mice, however, these can be relieved by folate. These maybe the reason why promoting arsenic methylation occurred.

4 CONCLUSIONS

In summary, As levels were significantly dose-dependent in different experimental groups, as well as moderate increase of SAH, decrease of SAM and SAM/SAH were found in the blood by sub chronic arsenic administration. We also observed the involvement of folate in promoting arsenic methylation and SAM level in vivo. Our study confirmed the antagonistic roles of folate to counteract inorganic arsenic-induced hypomethylation in vivo, and suggested that enhanced methionine cycle might be valuable for the protective effects of folate against arsenic intoxication. This provides a potential useful chemopreventive dietary component for human populations were threaten with arseniasis.

ACKNOWLEDGEMENTS

This study was supported by grants from National Natural Science Foundation of China (NSFC) (Nos. 81502764).

REFERENCE

Sun, G., Xu, Y., Li, X., Jin, Y., Li, B. & Sun, X. 2007. Urinary arsenic metabolites in children and adults exposed to arsenic in drinking water in Inner Mongolia, China. Environ Health Perspect 2007;115(4):648–652.

Environmental Arsenic in a Changing World –
Zhu, Guo, Bhattacharya, Ahmad, Bundschuh & Naidu (Eds)
ISBN 978-1-138-48609-6

Health risk assessment of arsenic dispersion from mining in Mount Isa

J. Zheng[1], B.N. Noller[1], T. Huynh[1], R. Taga[1,2], J.C. Ng[2,3], V. Diacomanolis[2] & H.H. Harris[3]
[1] *Centre for Mined Land Rehabilitation, Sustainable Minerals Institute, The University of Queensland, Brisbane, QLD, Australia*
[2] *Queensland Alliance for Environmental Health Sciences (QAEHS), The University of Queensland, Brisbane, QLD, Australia*
[3] *Cooperative Research Centre for Contamination Assessment and Remediation of the Environment (CRC CARE), Newcastle, NSW, Australia*
[4] *Department of Chemistry, Adelaide University, Adelaide, SA, Australia*

ABSTRACT: The study identified that arsenic (As) exposure to the population of Mount Isa is of limited significance but surface tailings and fall out (city) may be higher risk because their % bioaccessibilities are higher than those of other sources. Total As concentrations ($\mu g\,m^{-2}$) in surface wipes indicate a need to clean surfaces where children regularly contact. Airborne particulate concentrations were not significant for the sampling period.

1 INTRODUCTION

The Lead Pathway Air Study (2007–2012) has also examined sources of arsenic (As) and provided an understanding of human exposure at Mount Isa from mining and processing activities (Noller *et al.*, 2017). If not properly managed, dispersion of metals and metalloids from mine activities may cause adverse effects on the environment and human health. The National Environment Protection Measures (NEPMs) guidelines for soil contamination in Australia (NEPC, 2013) identify a need to undertake further Tier II risk assessment of the site if the Health Investigation Levels (HIL) are exceeded. In the absence of site specific data, the NEPMs may not provide accurate close out criteria for mined land. The aim of this study is to use a risk assessment process to assess any potential health effects from As sources in Mount Isa city including from mining and mineral processing, natural and other sources of exposure to the local population.

2 METHODS/EXPERIMENTAL

Samples of ores and processed materials were collected from Mount Isa Mines, and soil and dust from the nearby city (Noller *et al.*, 2017). Samples initially sieved to <2 mm for measurement of total concentrations (aqua regia digest) were also sieved to <250 μm for bioaccessibility (gastro-intestinal tract simulation) measurement using the physiologically-based extraction test (Ng *et al.*, 2015; Ruby *et al.*, 1996). Air particulates and surface wipes were also collected (Noller *et al.*, 2017). Arsenic concentrations were determined in digest solutions by ICP-MS. Particle size analysis was undertaken by Malvern Mastersizer (2000). Hazard assessment, exposure assessment and risk characterization of arsenic followed the Australian enHealth (2012) health risk assessment framework.

3 RESULTS AND DISCUSSION

Table 1 gives results for total As and bioaccessible-adjusted concentrations ($mg\,kg^{-1}$) in soil, minerals and dusts from Mount Isa and environs. Whilst some total As concentrations exceed NEPM HILs, only surface tailings and fall out (city) exceed HIL A for bioaccessibility adjusted As concentration (Table 1). Total As concentrations ($\mu g\,m^{-2}$) in surface wipes at Mount Isa city (Table 2) show that floor and carpet surfaces, based on an indicative guideline, have much lower concentrations than window sill and trough and external surfaces which are not cleaned in houses and regularly as floors and carpets. These data indicate a need to clean surfaces that children regularly contact. Table 3 gives airborne concentrations of As ($\mu g\,m^{-3}$) in outdoor and indoor air at the Mount Isa residential area and show that the As guideline for total suspended particulates was not exceeded for the period of sample collections. Bioaccessibility (% BAc, in-vitro) is generally a more conservative approach to estimate bioavailability of contaminated wastes such as from mining. Figure 1 shows the particle size distribution of samples from the mine site and residential area as cumulative (%). All samples have 50% of volume in the range 10–110 μm except for surface tailings with 50% of volume being 1 μm but bimodal. This data

Table 1. Total As concentrations and bioaccessible As in soil and dust samples from Mount Isa.

Sample	Total As ($mg\,kg^{-1}$) Mean ± sd (n*)	Bioaccessibility (BAc) Mean ± sd (n*) Adjusted ($mg\,kg^{-1}$)	BAc %
Background geology			
Outcrops	31 ± 27 (13)	3 ± 2.1 (5)	9 ± 6 (5)
Mineralization	497 ± 197 (4)	7 (1)	1 (1)
Mining and mineral processing			
Mine site dust	1320 ± 4040 (72)	25 ± 23 (16)	8 ± 6 (16)
Haul road dust	291 ± 3 (9)	32 ± 27 (9)	11 ± 4 (9)
Tailings surface	361 ± 31 (7)	112 ± 96 (7)	31 ± 28 (7)
Smelter dust	375 ± 525 (16)	30 ± 42 (22)	8 ± 5 (22)
City residential area			
Garden soil (<250 μm)	12 ± 9 (76)	2 ± 2 (76)	15 ± 8 (76)
Garden soil (<10 μm)	39 ± 28 (74)	11 ± 11 (74)	28 ± 15 (74)
Footpath	9 ± 6 (14)	1.9 ± 0.9 (14)	27 ± 14 (14)
Fallout dust	291 ± 248 (15)	130 ± 90 (15)	47 ± 23 (15)
Roof gutter dust	1139 ± 1511 (18)	52 ± 54 (18)	11 ± 10 (18)
Carpet dust	46 ± 26 (6)	9 ± 7 (6)	22 ± 10 (6)
Australian soil contamination guidelines (NEPC 2013)			
HIL A (residential/ garden)	100	–	–
HIL C (public open space)	300	–	–

*n = number of samples.

Table 2. Total As concentrations ($\mu g\,m^{-2}$) in surface wipes.

Sample	Total As ($\mu g\,m^{-2}$) mean ± sd (n*)
Floor wipe	17 ± 48 (176)
Carpet wipe	13 ± 4 (68)
Window sill	102 ± 235 (130)
Window trough	2163 ± 5140 (47)
Roof wipe	1247 ± 2834 (134)
Veranda wipe	146 ± 248 (84)
Guideline (Brookhaven National Laboratory 2014)	
Surface wipe for As	1500 $\mu g\,m^{-2}$*

*n = number of samples; Criteria type – Housekeeping – all.

supports that tailings surface material and city fall out may be higher risk because their % BAc's (Table 1) are 31 and 47, respectively.

4 CONCLUSIONS

The study identified that As exposure to the population of Mount Isa is low but surface tailings and fall out (city) may be higher risk because their % BAc's are higher than those of other sources. Total As concentrations ($\mu g\,m^{-2}$) in surface wipes indicate a need to clean surfaces where children regularly contact.

Table 3. Airborne As concentrations ($\mu g\,m^{-3}$) in Mount Isa.

Sample	Total As ($\mu g\,m^{-3}$) Mean ± sd (n*)
City outdoor	
(2009)	0.005 ± 0.008 (12)
(2010)	0.005 ± 0.008 (12)
(2011)	0.005 ± 0.008 (12)
City Indoor	
(2010)	0.004 ± 0.004 (12)
Guideline (NEPC 2002)	
As in total suspended particles ('TSP')*	0.006 $\mu g\,m^{-3}$ (1 year averaging period)

*n = number of samples.

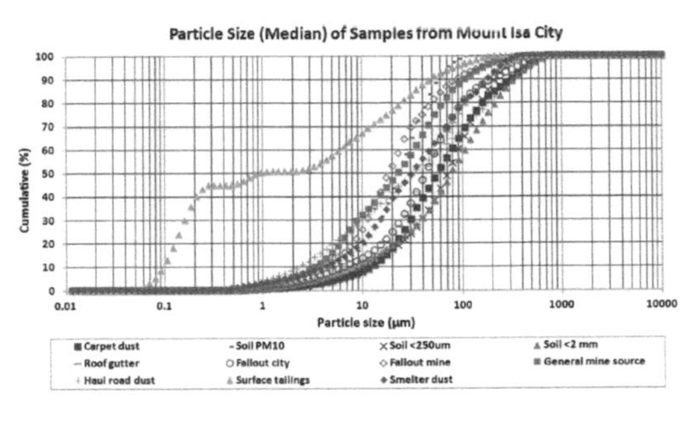

Figure 1. Particle size distribution of samples from the mine site and residential area as cumulative (%).

Airborne particulate concentrations were not significant for the sampling period.

ACKNOWLEDGEMENTS

Glencore- Mount Isa Mines funded the study.

REFERENCES

enHealth 2012. Environmental health risk assessment. Guidelines for assessing human health risks from environmental hazards. The Environmental Health Committee.

NEPC 2002. Ambient air quality standards. National Environment Protection Council. Canberra.

NEPC 2013. National Environmental Protection (Assessment of Site Contamination) Measures, National Environment Protection Council, Adelaide.

Ng, J.C., Juhasz, A., Smith, E. & Naidu, R. 2015. Assessing bioavailability and bioaccessibility of metals and metalloids. *Environ. Sci. Pollut. R.* 22(12): 8802–8825.

Noller, B., Zheng, J., Huynh, T., Ng, J., Diacomanolis, V., Taga, R. & Harris, H. 2017. Lead Pathways Study – Air. Health Risk Assessment of Contaminants to Mount Isa City. 7 February 2017 Mount Isa Mines Limited, Mount Isa. pp 1–414 plus appendices. http://www.mountisamines.com.au/EN/sustainability/Pages/LEADPATHWAYSSTUDYPORTAL.aspx.

Ruby, M.V., Davis, A., Schoof, R., Eberle, S. & Sellstone, C.M. 1996. Estimation of lead and arsenic bioavailability using a physiologically based extraction test. *Environ. Sci. Technol.* 30(2): 422–430.

Environmental Arsenic in a Changing World –
Zhu, Guo, Bhattacharya, Ahmad, Bundschuh & Naidu (Eds)
ISBN 978-1-138-48609-6

Assessing drinking water quality at high dependent point of sources and potential health risk of massive population: A view from Tala Upazila of Satkhira District in Bangladesh

R. Saha, N.C. Dey & M. Rahman
BRAC Research and Evaluation Division, BRAC Centre, Dhaka, Bangladesh

ABSTRACT: Drinking water is the important indicator to assess environmental health. Aim of the present research is to investigate water quality at highly dependent drinking water sources and population exposed to potential contamination. Tala *upazila* of Satkhira district in Bangladesh was selected as study area. Drinking water temperature, EC, Fe, As, total coliform, *E. coli,* fecal coliform were tested following standard procedure in 649 highly dependent drinking water sources including DTW, STW and PSF. Besides, corresponding 260 dependent households' coliform bacteria were tested and semi-structured questionnaire survey was conducted. According to WHO, 99%, 83% and 74% water sources exceeded the drinking water standard for EC, Fe and As, respectively. WQI suggested that majority (77%) of highly dependent drinking water sources were unsuitable for drinking and 40% population (0.12 million) in the study area were exposed to potential health. Most frequently people were suffered from fever, diarrhea and high blood pressure and usually they were spending $ 3–13 per month for their health-related expenditure. Regular water testing facilities, keeping water sources at a safe distance from contamination, practicing hygiene behavior, etc. could be taken for improving situation.

1 INTRODUCTION

Safe drinking water is enormously essential requirement for humans as well as all living organisms (Das and Acharya, 2003) and potential indicator of environmental health. Groundwater is one of the most important sources of drinking water to many people around the world (Chatterjee *et al.*, 2010) and also used for various purposes (Khan *et al.*, 2003; Sargaonkar & Deshpande, 2003). Bangladesh is a densely populated country with 139 million population (BBS, 2014). About 20% of total area (about 29000 km^2) is considered as coastal zone, especially characterized by influence of tidal waters, salinity intrusion and cyclones/storm surges, etc. (MoWR 2005). In coastal area, 53% aquifer area has been affected by saline water intrusion from Bay of Bengal (MoFDM, 2005). In Bangladesh, shallow aquifer in *Meghna* river basin and coastal plain are extremely As enriched and, in this area, more than 80% TW were As contaminated (Ahmed *et al.*, 2004). High Fe concentration, turbidity and to some extent bacterial contamination are also associated with the water sources/extraction devices in the south-western coastal areas. According to WHO, about 80% of human diseases are caused by contaminated water. Thus, dependent drinking water sources need to be free from high level elemental concentration and pathogenic microorganism (Javed *et al.*, 2014) and it is essential to obtain precise and appropriate information on water quality for essential life-saving and other beneficial uses (Donadkar *et al.*, 2016; Gadhave *et al.*, 2008; Ilangeswaran *et al.*, 2009; Shyamala *et al.*, 2008; Talawat & Chandel, 2008). Overall aim of the study was to examine water quality at high dependent water sources and estimating populations exposed to potential contamination. Study findings will provide sufficient information of drinking water quality of highly dependent point of sources which will help to reduce risk of potential acute and chronic diseases in massive population.

2 METHODS/EXPERIMENTAL

2.1 *Subsection*

Tala *upazila* of Satkhira district, in the coastal area of Bangladesh, has been selected as the study area because incidences of frequent natural calamities and scarcity of safe drinking water (Fig. 1). A total of 649 high dependent drinking water sources were identified including deep tube-well (DTW), shallow tube-well (STW) and pond sand filter (PSF). *In-situ* measurements were done for temperature and electrical conductivity (EC) by Hach Pocket Pro High Range Conductivity Tester; total iron (Fe) by Hach Iron Test Kit-Model IR-B Color Disc; arsenic (As) by Hach Dual Range Arsenic Test Kit-Test Strip EZ and coliform bacteria (total coliform, *E. coli* and fecal coliform) by membrane filtration technique based Hach Microbiology Test Kit. Weighed arithmetic water quality index (WQI) was used to calculate suitability of

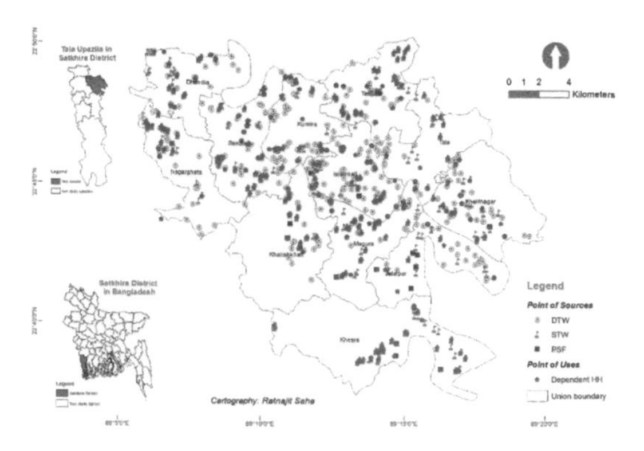

Figure 1. Location of water sampling PoS and corresponding PoU in the study area.

Table 1. Basic statistics of water quality parameters.

WQ Para	Unit	Range	Mean ± SD	Ex. WHO Stand (%)
High dependent drinking water PoS				
Temp	°C	25.8–36.6	30.53 ± 1.38	100
EC	$\mu S\,cm^{-1}$	165–8715	2494.14 ± 2192.95	100
Fe	$mg\,L^{-1}$	0–18	3.07 ± 2.90	83
As	$\mu g\,L^{-1}$	0–500	61.69 ± 67.66	74
TC	CFU/100 mL	0–208	16.81 ± 32.67	38
E. coli	CFU/100 mL	0–160	3.90 ± 15.79	24
FC	CFU/100 mL	0–212	25.25 ± 44.97	45
Corresponding dependent HHs' PoU				
TC	CFU/100 mL	0–220	27.47 ± 42.72	54
E. coli	CFU/100 mL	0–166	5.17 ± 15.03	35
FC	CFU/100 mL	0–228	32.05 ± 46.21	55

drinking water. As, Fe, EC and temperature were tested in 649 high dependent water sources and coliform bacteria were tested in only 377 water sources and 260 corresponding dependent households. At the corresponding households (HH) a semi-structured questionnaire survey was conducted.

3 RESULTS AND DISCUSSION

3.1 *Water quality status at high dependent PoS*

All the PoS exceeded WHO drinking water standard for temperature and EC (Table 1). Range of Fe levels was 0–18 $mg\,L^{-1}$ with low variability and 83 water sources crossed drinking water standard. Minimum and maximum As concentrations were 0 and 500 $\mu g\,L^{-1}$ (mean 61.69) and the variation of tested results was quite high (SD ± 67.66). According to WHO standard, three-fourth (74%) of highly dependent water sources exceeded standard limit. The content of coliform bacteria and the proportion of bacterial contamination was found to be much higher in the corresponding dependent HH than dependent water sources.

EC is a good measure of dissolved solids in water. It is dependent on concentrations of ions and nutrient

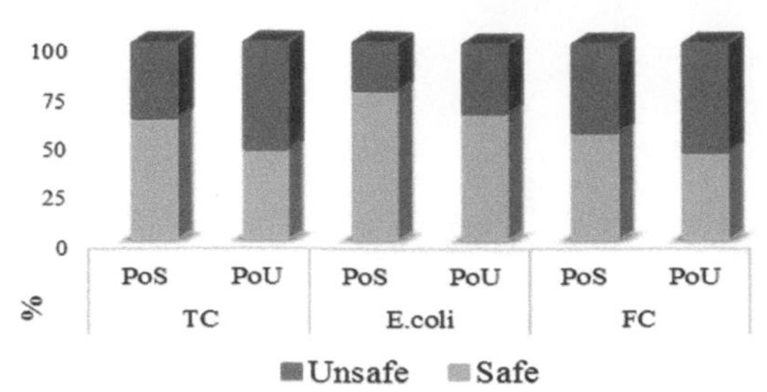

Figure 2. Status of coliform bacteria in PoS and corresponding PoU.

status (Gupta *et al.*, 2009). To survive, groundwater is uplifted for drinking water and high level of salinity threatening environmental health (Ahmed, 2006; Minar *et al.*, 2013; National Water Policy, 1999). Excess amount is responsible for the objectionable taste of drinking water (Gupta *et al.*, 2009). High Fe concentration is understood to be non-toxic and health impacts are not widely documented (Merrill *et al.*, 2011). Besides, intake of As through drinking water and the food chain could be a potential health hazard for the local population (Rahman *et al.*, 2008). Because As is transported by blood to different organs in the body, mainly in the form of monomethylarsonic acid, bladder cancer risk is increased 2.7 times if As concentration exceeded 10 $\mu g\,L^{-1}$ in drinking water (Akter *et al.*, 2016) and it has a variety of adverse health effects in the form of acute and chronic exposures such as dermal changes, respiratory, pulmonary, cardiovascular, renal, neurological and other carcinogenic effects (Mandal & Suzuki, 2002).

3.2 *Contamination pathway from PoS to PoU*

In all tested Point of Uses (PoU), range of TC was 0–220 CFU/100 mL (mean 27.47) and proportion of contaminated PoU (50%) was much higher than corresponding PoS (45%) (Fig. 2). Besides, minimum and maximum values of *E. coli* were 0 and 166 CFU/100 mL (mean 5.17). *E. coli* contaminated PoU (55%) was much higher than corresponding PoS (40%). In all tested PoU, FC range were 0–228 CFU/100 mL (mean 32.05). Similarly, large proportion of PoU (60%) were FC contaminated which was about double of the corresponding contaminated PoS (32%).

Coliform bacteria have many adverse effects on public health such as urinary tract infections, cystitis and kidney infections and these bacteria are also responsible for environmental degradation, for instance, *E. coil* is responsible for damaging vegetables (Javed *et al.*, 2014; CCME, 2009).

3.3 *Water Quality Index (WQI) in PoS*

Range of calculated value of WQI was from 0.8 to 7352 (mean 200). In all tested water sources, 89 (14%)-excellent, 90 (14%)-good, 41 (6%)-poor, 27 (4%)-very poor and 402 (62%)-unsuitable for drinking. More

Table 2. Population exposed to potential water borne diseases in the study area.

Dependency classification/ Exp population	No. of PoS poor-unsuitable	No of dependent HH	Population exposed (5 per HH)	Population exposed (million)
Fairly high	146	1941	9705	0.01
Moderately high	160	5013	25065	0.03
High dependent	122	8507	42535	0.04
Most high dep	42	7975	39875	0.04
Total	470	23436	117180	0.12

than half of DTW (52%) were good to excellent and almost all STW (98%) and PSF (94%) were unsuitable as drinking water sources.

3.4 *Population exposed to potential diseases and approaching for solving*

According to WQI, majority of usable PoS were unsuitable for drinking. HH dependency basis calculation suggested that about 0.12 million people (40% population of the study area) were directly consuming contaminated water (Table 2). As a consequence, massive number of population were at potential health risk of acute and chronic diseases. In the last 15 days of survey date, dependent HH members most frequently suffered from fever, diarrhea, dysentery, colds and high blood pressure. Dependent HH regularly spent significant amount of money ($3–13 per month/HH) for health-related expenditure, even though more than half of them were poor. To get safe drinking water, majority of dependent HH (62%) expressed willingness to pay. Among them, 58% people wanted to pay $ 1 per month and only 20% HH showed willingness to pay $ 2.6 per month.

Once groundwater is contaminated, its quality cannot be restored, so it is highly necessary to have a concern to protect and manage groundwater quality (Singh *et al.*, 2012). It is imperative to investigate water quality in the dependent water sources. Regular monitoring is needed for securing public health and protecting the groundwater system from further deterioration (Rahman *et al.*, 2000; Ramakrishnaiah *et al.*, 2009). To improve the health risk situation and ensure safe drinking water to rural communities, immediate steps should be taken.

4 CONCLUSIONS

In the studied area, a majority of highly dependent drinking water sources have been contaminated by arsenic, iron, salinity and coliform bacteria. Therefore, inadequate safe drinking water options have been observed and an increasing number of people are demanding water from single water sources. Study findings revealed that one third of the population in the studied area are exposed to potential health risks because a majority of drinking water sources are unsuitable for drinking. To improve the safe water crisis situation, certain steps might be effective such as keeping a safe separation of source of contamination from water sources, water quality testing facilities at regular intervals, awareness raising, practicing hygiene behavior, providing adequate health service facilities, water treatment plants and sufficient safe drinking water options for local communities, etc.

ACKNOWLEDGEMENTS

The authors sincerely acknowledging BRAC Water, Sanitation and Hygiene (WASH) Programme for providing financial support and BRAC Research and Evaluation Division for providing scope to conduct this study. Special thanks go to the research assistants who participated in *in-situ* experiment.

REFERENCES

Ahmed, A.U. 2006. Bangladesh: Climate Change Impacts and Vulnerability (A Synthesis). *Dhaka: Climate Change Cell, Department of Environment, Comprehensive Disaster Management Programme, Government of the People's Republic of Bangladesh.*

Ahmed, K.M., Bhattacharya, P., Hasan, M.A., Akhter, S.H., Alam, S.M., Bhuyian, M.H., Imam, M.B., Khan, A.A. & Sracek, O. 2004. Arsenic enrichment in groundwater of the alluvial aquifers in Bangladesh: an overview. *Appl. Geochem.* 19(2): 181–200.

Akter, T., Jhohura, F.T., Akter, F., Chowdhury, T.R., Mistry, S.K., Dey, D., Barua, M.K., Islam, M.A. & Rahman, M. 2016. Water Quality Index for measuring drinking water quality in rural Bangladesh: a cross-sectional study. *J. Health Popul. Nutr.* 35(1): 4.

BBS. 2014. Bangladesh Population and Housing Census 2011, National Report, Volume 2, Union Statistics. Dhaka: Bangladesh Bureau of Statistics, Statistics and Information Division, Ministry of Planning, *Government of the People's Republic of Bangladesh*, pp: 349.

Chatterjee, R., Tarafder, G. & Paul, S. 2010. Groundwater quality assessment of Dhanbad district, Jharkhand, India. *B. Eng. Geol. Environ.* 69(1): 137–141.

Das, J. & Acharya, B.C. 2003. Hydrology and assessment of lotic water quality in Cuttack City, India. *Water Air Soil Poll.* 150(1–4):163–175.

Donadkar, D.K., Rahangdale, P.K. & Gour, K. 2016. Assessment of ground water quality in and around Gadchiroli district. *Int. J. Eng. Res. Technol.* 5(04):626–631.

Gadhave, A., Thorat, D. & Uphade, B. 2008. Water quality parameters of ground water near industrial areas, Shrirampur (M.S). *Rasayan J. Chem.* 1(4):853–855.

Gupta, P., Vishwakarma, M. & Rawtani, P.M. 2009. Assessment of water quality parameters of Kerwa Dam for drinking suitability. *Int. J. Theor. Appl. Sci.* 1(2): 53–55.

Ilangeswaran, D., Kumar, R. & Kannan D. 2009. Assessment of quality of groundwater in Kandarvakottai and Karambakudi areas of Pudukkottai district, Tamilnadu, India. *E-J Chem.* 6(3): 898–904.

Javed, F., Ahmed, M.N., Shah, H.U., Iqbal, M.S., Wahid, A. & Ahmad, S.S. 2014. Effects of seasonal variations on physicochemical properties and concentrations of faecal coliform in River Kabul. *World Appl. Sci. J.* 29(1): 142–149.

Khan, F., Husain, T. & Lumb, A. 2003. Water quality evaluation and trend analysis in selected watersheds of the Atlantic region of Canada. *Environ. Monit. Assess.* 88(1–3): 221–248.

Mandal, B.K. & Suzuki, K.T. 2002. Arsenic round the world: a review. Talanta 58(1): 201–235.

Merrill, R.D., Shamim, A.A., Ali, H., Jahan, N., Labrique, A.B., Schulze, K., Christian, P. & West Jr, K.P. 2011. Iron status of women is associated with the iron concentration of potable groundwater in rural Bangladesh-3. *J. Nutr.* 141(5): 944–949.

Minar, M., Hossain, M.B. & Shamsuddin, M. 2013. Climate change and coastal zone of Bangladesh: vulnerability, resilience and adaptability. *Middle-East J. Sci. Res.* 13(1): 114–120.

MoFDM 2005. Building resilient future. Dhaka: Ministry of Food and Disaster Management, Government of Bangladesh MoWR 2005. Coastal zone policy, 2005. Dhaka: Ministry of Water Resources, Government of Bangladesh National Water Policy 1999. Dhaka: Ministry of Water Resources, Government of the People's Republic of Bangladesh.

Rahman, M.A., Hasegawa, H., Rahman, M., Miah, M. & Tasmin, A. 2008. Straighthead disease of rice (Oryza sativa L.) induced by arsenic toxicity. *Environ. Exp. Bot.* 62(1): 54–59.

Rahman, M.M., Hassan, M.Q., Islam, M.S. & Shamsad, S. 2000. Environmental impact assessment on water quality deterioration caused by the decreased Ganges outflow and saline water intrusion in south-western Bangladesh. *Environ. Geol.*, 40(1–2): 31–40.

Ramakrishnaiah, C.R., Sadashivaiah, C. & Ranganna, G. 2009. Assessment of water quality index for the groundwater in Tumkur Taluk, Karnataka State, India. *E-J Chem.* 6(2): 523–530.

Shyamala R., Shanthi M. & Lalitha P. 2008. Physicochemical analysis of borewell water samples of Telungupalayam area in Coimbatore district, Tamilnadu, India. *E-J Chem.* 5(4): 924–929.

Singh, M., Jha, D. & Jadoun, J. 2012. Assessment of physicochemical status of groundwater samples of Dholpur district, Rajasthan. India. *Int. J. Chem.* 4(4): 96.

Talawat R.K. & Chandel C.P.S. 2008. Quality of groundwater of Jaipur city, Rajasthan (India) and its suitability for domestic and irrigation purpose. *Appl. Ecol. Environ. Res.* 6(2): 79–88.

Environmental Arsenic in a Changing World –
Zhu, Guo, Bhattacharya, Ahmad, Bundschuh & Naidu (Eds)
ISBN 978-1-138-48609-6

Health impact of chronic arsenic exposure in the population of Gyaspur-Majhi village, Patna, Bihar India

S. Abhinav, S. Navin, P. Shankar, M.S. Rahman, R. Kumar, P.K. Niraj, A. Kumar & A.K. Ghosh
Mahavir Cancer Institute and Research Centre, Patna, Bihar, India

ABSTRACT: Arsenic contamination in the groundwater is a serious and widespread global public health problem and India is one of the highly arsenic exposed country with maximum impact in the Ganga-Meghna-Brahmaputra (GMB) plains. A total of 18 districts out of 38 districts of the state of Bihar are highly affected with groundwater arsenic poisoning. Gyaspur-Mahaji village (a flood plain region of river Ganga) of Patna district was under taken for the study on the basis of large number of skin and breast cancer cases along with typical symptoms of arsenicosis diagnosed in Mahavir Cancer Institute and Research Centre, Patna. The study deals with the survey of the entire village with groundwater and blood arsenic estimation with special attention to know the health status of the arsenic exposed population. The study revealed high arsenic concentration in the groundwater in the village. Arsenic in the blood samples indicates biomarker for chronic arsenic exposure. The overall health status of the village people was very poor but the most unfortunate part was the appearance of skin cancer and breast cancer in many of the arsenic exposed population of this village. Although the number of such persons was few, it was still alarming as more persons may be impacted if they continue to drink arsenic contaminated water in future.

1 INTRODUCTION

Arsenic in the recent times has caused serious health hazards in a huge population inhabiting Ganga-Meghna- Brahmaputra plains. The entire Ganga-Meghna Bramhaputra (GMB) plains has high arsenic concentrations ($<10\,\mu g\,L^{-1}$) in the groundwater. Population residing in these plains are exposed to arsenic which has led to many health-related problems like keratosis, melanosis and different types of cancer (Yunus *et al.*, 2016). In Bihar, presently 18 districts out of 38 districts are affected by arsenic poisoning and it is assumed that a population of about 5 million are consuming arsenic contaminated groundwater with arsenic concentration more than $50\,\mu g\,L^{-1}$ (Singh *et al.*, 2014). In the present study, Gyaspur-Mahaji village (a flood plain region of river Ganga) of Patna district was under taken for the study. The study deals with the entire survey of the village along with groundwater and blood arsenic estimation. The health status of the chronic arsenic exposed population was the major objective of the study.

2 MATERIALS AND METHODS

2.1 *Location and ethical approval*

The present study was undertaken at Gyaspur-Mahaji village (N25°30′02.3″ E085°27′14.2″) of Patna, Bihar, India. The population of the Gyaspur- Mahazi village is approximately 3,380 with male population 1,970 and female population 1410. There are approximately 509 households (Census, 2011) in the village. The research work was approved by the IEC (Institutional Ethics Committee) of the institute as the work was on human subjects. The survey was carried out in February, 2016.

2.2 *Collection and analysis of total arsenic*

Altogether, 58 groundwater samples from hand tube-wells (HTWs) were randomly collected in duplicates from each household situated at every 50 m of distance in the village. The blood samples of the 58 subjects were also collected from the same households. After the collection, all the samples were estimated as per the standardized protocol through Graphite Furnace Atomic Absorption Spectrophotometer (Pinnacle 900T, Perkin Elmer, Singapore) at Mahavir Cancer Institute and Research Centre, Patna, Bihar, India. Data were analyzed with statistical software (Graphpad Prism 5) and values were expressed as Mean ± SEM.

2.3 *Health assessment*

The suspected people of the four villages included in this study were exclusively interrogated and examined for arsenic related diseases and other health consequences to know their present health status. For this a questionnaire method was prepared, utilized and health related data were extensively collected (Singh *et al.*, 2014).

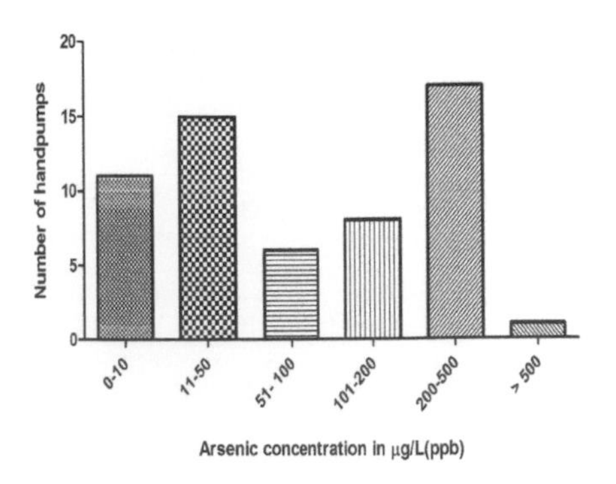

Figure 1. Arsenic concentration in groundwater samples of HTWs of village Gyaspur-Majhi.

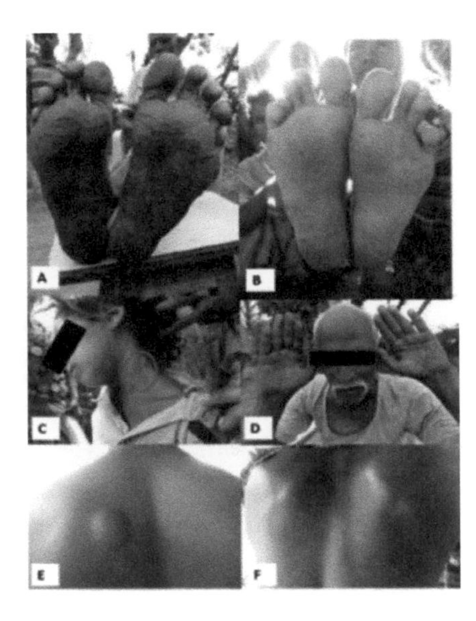

Figure 2. A. Hyperkeratosis of sole in 60 yrs old male (As in drinking water 270.5 $\mu g\,L^{-1}$); B. Hyperkeratosis of sole in 14 yrs old girl (As in drinking water 368.5 $\mu g\,L^{-1}$); C. Cervical Nodes in 10 yr old girl (As in drinking water 368.5 $\mu g\,L^{-1}$); D. Hyperkeratosis of palm in 60 yrs old male (As in drinking water 270.5 $\mu g\,L^{-1}$); E. Lump at back of 35 yrs old male (As in drinking water 361 $\mu g\,L^{-1}$); F. Severe melanosis at back of 47 yrs old male (As in drinking water 104.4 $\mu g\,L^{-1}$).

3 RESULTS AND DISCUSSION

3.1 *Groundwater and blood arsenic assessment*

The analysis of 58 water samples from the entire village represented high prevalence of arsenic contamination in the groundwater. The maximum arsenic concentration in groundwater sample was 826.4 $\mu g\,L^{-1}$. The groundwater samples showed more than 10 $\mu g\,L^{-1}$ arsenic concentration in about 89% of the samples analyzed. Only 11% samples had normal levels (below 10 $\mu g\,L^{-1}$), 17% had levels between 11–50 $\mu g\,L^{-1}$ while rest 83% samples were highly arsenic contaminated water samples (Fig. 1).

The total analyzed 58 blood samples showed arsenic concentration in their blood with 41% below 1 $\mu g\,L^{-1}$ whereas 59% subjects had the levels more than the permissible limit. The maximum arsenic concentration in blood sample of the village subject reported was 64.98 $\mu g\,L^{-1}$ which was extremely very high.

3.2 *Health assessments*

Persons in this village exhibited typical symptoms of arsenicosis like hyperkeratosis in sole and palm, hyperpigmentation in palm (Fig. 2). Many of them exhibited hyperpigmentation (spotted pigmentation) on their whole body, melanosis, cervical nodes on their neck region and tumor at the back (Fig. 2). The most unfortunate part of the study were the cases of cancer, where the first subject (male) reported in our Cancer center with typical symptoms of arsenicosis as hyperkeratosis in sole and palm with rain drop pigmentation over his entire body. The second subject (female) reported with medullary carcinoma of breast. She also exhibited with typical symptoms of arsenicosis as hyperkeratosis on sole and palm.

4 CONCLUSIONS

From the study, it can be concluded that the magnitude of arsenic groundwater contamination is severe in the Gyaspur-Majhi village. Prolonged use of arsenic contaminated drinking water leads to various health complications including cancer. Presence of arsenic in the blood samples of the subjects of all the age groups are very alarming. It reflects that there is no any or limited alternate safe source of drinking, cooking and irrigation water supply in these areas. Furthermore, the future health risk can be controlled through supply of safe drinking water.

ACKNOWLEDGEMENTS

The authors acknowledge support extended by Mahavir Cancer Institute and Research Centre, Patna, Bihar, India for providing laboratory and other infrastructural facilities for this study.

REFERENCES

Census. 2011. Interim Report of Population Census of India. (http://www.censusindia.gov.in/).

Singh, S.K., Ghosh, A.K., Kumar, A., Kislay, K., Kumar, C., Tiwari, R.R., Parwez, R., Kumar, N. & Imam, M.D. 2014. Groundwater arsenic contamination and associated health risks in Bihar. India. *Int. J. Environ. Res.* 8(1): 49–60.

Yunus, F.M., Khan, S., Chowdhury, P., Milton, A.H., Hussain, S. & Rahman, M. 2016. A review of groundwater arsenic contamination in Bangladesh: the millennium development goal era and beyond. *Int. J. Environ. Res. Public Health* 13(2): 215.

Environmental Arsenic in a Changing World –
Zhu, Guo, Bhattacharya, Ahmad, Bundschuh & Naidu (Eds)
ISBN 978-1-138-48609-6

Health risks related to seafood consumption and arsenic speciation in fish and shellfish from North Sea (Southern Bight) and Açu Port area (Brazil)

P. Baisch[1], Y. Gao[2], N.V. Larebeke[2], W. Baeyens[2], M. Leermakers[2], N. Mirlean[1] & F.M. Júnior[3]
[1]*Laboratório de Oceanografia Geológica, Instituto de Oceanografia, Universidade Federal do Rio Grande (FURG), Rio Grande, RS, Brazil*
[2]*Analytical, Environmental and Geochemical Department (AMGC), Vrije Universiteit Brussel, Brussels, Belgium*
[3]*Laboratório de Ensaios Farmacológicos e Toxicológicos, Instituto de Ciências Biológicas, Universidade Federal do Rio Grande (FURG), Rio Grande, RS, Brazil*

ABSTRACT: In both the North Sea and the Açu Port (Brazil) coastal areas, high As concentrations were observed in water, soil and sediments. Therefore, the impact of this contamination on fish and shellfish species bought from local fishermen was studied. Total As was assessed by ICP-MS while toxic As was assessed by ICPMS-HPLC and HG-AFS. Several fish species had average Total As concentrations above 1 mg g^{-1}, but the highest concentrations were found in less spotted dogfish, lemon sole and whelks from the North Sea. Toxic As fractions were high in scallops but rarely exceeded 2% in other species. Considering consumption of 150 g, only 3 samples exceeded the total daily intake. Using mean toxic As concentrations for each species, lifetime cancer risk values at the actual global seafood consumption rate of 54 g/day are above 10^{-4} for whelks, scallops, dogfish, ray and lemon sole.

1 INTRODUCTION

The North Sea and the Açu Port area (Brazil) ecosystems have been highly impacted by As with sources of different nature. In the North Sea, the average dissolved As concentration is around 0.75 mg L^{-1}, but river input can substantially modify the As baseline.

In the Açu Port area, high As levels in small lakes and marine sediments originate from calcareous biota and iron oxyhydroxides rich in arsenic. Arsenic levels in lake sediments vary between 20 and 270 mg kg^{-1}, while marine sediments contain around 50 mg kg^{-1} (Mirlean & Baisch, 2016).

This study aimed at examining if high As levels in both marine ecosystems led to high As levels in fish, at comparing total daily intake of As from fish consumption and at estimating the individual life-time cancer risk due to seafood consumption.

2 METHODS

In the North Sea, ray, dogfish, lemon sole, pouting, pollack, brill, cod, whiting and ling (fish species), besides whelks and scallops (shellfish species) were studied. Samples included 79 fish and 37 shellfish. In the Açu Port, located in Brazil's Atlantic coast, big tooth corvina, catfish and shorthead drum (N = 7) were studied. In the North Sea, fish and shellfish were collected in 2009–2010 whereas, in the Açu Port, fish were caught in 2013. Methods used for total As determination in fish comprised microwave assisted mineralization and ICP-MS detection. Arsenic species of fish and shellfish were selectively determined by liquid chromatography-inductively coupled plasma – mass spectrometry (ICPMS-HPLC). Separation of As species was performed by an anion exchange column with ammonium phosphate solution as mobile phase. The following arsenic compounds were determined: As(III), As(V), MMA and DMA, which is called toxic As fraction "As-Tox". Arsenic speciation in samples from the North Sea was also determined by hydride generation-atomic fluorescence spectrometry (HG-AFS). All analytical methods comply with quality assurance/quality control procedures.

3 RESULTS AND DISCUSSION

3.1 *Arsenic in fish and shellfish*

Species from the North Sea, i.e., pouting, whiting, cod, brill, ling and pollack, seemed to show similar total and toxic arsenic levels in their tissues. Many fish species under study had average concentrations above 1 μg g^{-1}, levels which need to be further investigated. Three North Sea species (less spotted dogfish, lemon sole and whelks) were found to be clearly more contaminated than the others. Average Total As levels in thornback ray, pelagic fish group and scallops (North Sea species) and tilefish, namorado sandperch and pink conger (Rio de Janeiro coastal fish species) ranged from 1 to 13 μg g^{-1}. It indicates that, for several fish species, comparable levels were observed in both marine areas. However, additional studies of As speciation in fish are critical for the assessment of consumption risk because arsenobetaine, a non-toxic form, is the major As species in marine fish. In the

Açu Port area (RJ-Brazil), average total arsenic concentrations in drum, catfish and corvina fish species ranged from 0.73 to 8.9 $\mu g\, g^{-1}$. Average toxic arsenic levels are high in less spotted dogfish, whelk, scallop and lemon sole (from 0.15 to 0.46 $\mu g\, g^{-1}$) while in bigtooth corvina, catfish, pelagic fish group and short-head drum, levels ranged between 0.023 and 0.067 $\mu g\, g^{-1}$. In fish, the highest arsenic (total and toxic) levels were found in dogfish and lemon sole, both from the North Sea, but dogfish was also the longest (632 mm) fish species under study. Correlations between toxic and total As levels in fish muscle were generally good with a correlation coefficient r equal to 0.78 for all fish species and 0.75 for whelks. Correlation was not good ($r = 0.33$) for scallops, only. The highest toxic arsenic fractions (toxic arsenic conc./total arsenic conc.) were found in great scallops (8.9%) and bigtooth corvina (3.2%), whereas the lowest ones were found in less spotted dogfish (0.6%) and lemon sole (0.7%). Data showed that toxic arsenic fractions in fish and shellfish were small by comparison with those found in surrounding water and sediment.

3.2 *Cancer risk assessment*

Cancer risk assessments were expressed as Individual lifetime cancer risk (TR). This assessment is defined as the daily average intake per kg of body weight, multiplied by an element-specific factor, the so-called slope factor (SF). TR can be calculated by:

$$TR = \frac{CR * Cf * EF * ED * SF}{BW * AT}$$

where CR is consumption rate of seafood, CF is the 95th upper confidence limit of the toxic as concentration mean for all samples, EF is the exposure frequency, ED is exposure time, BW is body weight and AT is averaging time. In the case of cancer effects, consumption of 13.5 g day^{-1} seafood with an average toxic As concentration shows, only for whelks and lemon sole individual lifetime cancer risks that are higher than the acceptable limit. Those seafood species were also the ones that exceeded the permissible dose calculation (Table 1). However, fish consumption in the world increased from 13.5 (USEPA,1997) to 54 g day^{-1} (FAO, 2016) in 20 years. Considering seafood consumption of 54 g day^{-1}, whelks, scallops, dogfish, ray and lemon sole exceeded the acceptable limit in case average toxic As values are used; Besides, the pelagic fish group joins those species when the 95th percentile of the toxic arsenic levels is used. In addition, when estimating cancer risks, one should take into account that arsenic is only one of the multiple carcinogens to which humans are exposed to. The fact that seafood essentially contains arsenobetaine, a non-toxic arsenic form, is a reassuring element. But the presence of arsenic in seafood, which is similar to inorganic As in terms of metabolite formation and tissue accumulation, should be considered. In addition, after seafood storage and cooking, toxic arsenic metabolites can be formed.

Table 1. Lifetime cancer risks of consumption of seafood species for 2 consumption rates (13.5 g day 1 TR2 and 54 g day 1 TR1) and 2 concentrations (mean and 95th percentile). Consumption As-Tox concentration 95th upper confidence limit (1); 95th percentile (2).

Species	n	As Tox Mean	TR1 mean	TR2 mean	As Tox 95th perc (2)	TR1 95th perc (2)
				mg / kg (ww)		
Whelk	10	0.46	5.32E-04	1.33E-04	1.10	1.27E-03
Scallop	27	0.23	2.65E-04	6.63E-05	0.52	6.03E-04
Dogfish	10	0.29	3.38E-04	8.44E-05	0.63	7.23E-04
Ray	9	0.15	1.69E-04	4.22E-05	0.23	2.65E-04
Pelagic Fish Group	30	0.07	7.71E-05	1.93E-05	0.11	1.25E-04
Lemon Sole	30	0.35	4.05E-04	1.01E-04	1.00	1.16E-03
Drum	3	0.031	3.59E-05	8.97E-06	0.05	5.44E-05
Catfish	2	0.029	3.36E-05	8.39E-06	0.03	3.82E-05
Corvina	2	0.023	2.66E-05	6.65E-06	0.04	4.51E-05
Consumption As (1)	TR1 54 g / day		TR2 13.5 g / day			
Exposure Frequency				365 days / year		
Exposure Duration				70 years		
Slope Factor (SF)				1.5 mg / kg / day		
Body weight				70 kg		
Average time				70 years		
Consumption		TR1 54 g / day		TR2 13.5 g / day		
As-Tox concentration		95th upper confidence limit				
Exposure Frequency		365 days / year				
Exposure Duration		70 years				
Slope Factor SF		1.5 mg / kg / day				
body weight		70 kg				
averaging time		70 years				

4 CONCLUSIONS

Several fish species had average total As concentrations above 1 $\mu g\, g^{-1}$ wet weight, but the highest levels, i. e., 50, 49 and 50 $\mu g\, g^{-1}$, were found in less spotted dogfish, lemon sole and whelks from the North Sea, respectively. High Total As levels correspond to high toxic As levels, except for scallops having increased toxic As concentrations. toxic As fractions were high in scallops (10%) but rarely exceeded 2% in the other species. Toxic arsenic fractions in liver samples were 2–4-fold higher than the ones in muscle. Considering consumption of 150 g seafood, only 3 samples exceeded the provisional total daily intake of 2 mg As kg^{-1}. However, cancer risks are non-negligible. Using mean toxic As concentrations for each of the different fish and shellfish species under study, lifetime cancer risk values at the actual global seafood consumption rate of 54 g day^{-1} were above 10^{-4} for whelks, scallops, dogfish, ray and lemon sole.

ACKNOWLEDGEMENTS

The authors thank the CB&I Company for the support in the field work in the Açu Port area.

REFERENCES

FAO, 2016. The State of World Fisheries and Aquaculture, p. 204.

Mirlean, N. & Baisch, P. 2016. Arsenic in Brazilian tropical coastal zone. Proceedings 6° Intern. Congress on Arsenic in the Environment. CRC Press, Boca Raton, FL, pp. 169–170.

USEPA. 1997. Exposure Factors Handbook. Update to Exposure Factors Handbook (Chapter 10): *Intake of fish and shellfish*. EPA/600/8e89/043.

Environmental Arsenic in a Changing World –
Zhu, Guo, Bhattacharya, Ahmad, Bundschuh & Naidu (Eds)
ISBN 978-1-138-48609-6

A dietary intervention in Bangladesh to counteract arsenic toxicity

J.E.G. Smits[1], R.M. Krohn[1], A. Vandenberg[2] & R. Raqib[3]
[1]*Faculty of Veterinary Medicine, University of Calgary, Calgary, Alberta, Canada*
[2]*Department of Plant Sciences, University of Saskatchewan, Saskatoon, Canada*
[3]*Nutritional Biochemistry Laboratory, Dhaka, Bangladesh*

ABSTRACT: This 6-month clinical trial tests whether high-selenium lentils, as a whole food solution, can improve the health of arsenic-exposed Bangladeshi villagers. The study entails 400 participants in two treatment groups. All participating households have tubewell water containing $\geq 100\,\mu g\,L^{-1}$, but over 50% are $>250\,\mu g\,L^{-1}$. In this double-blind study, one group is daily consuming high-selenium lentils from the Canadian prairies, the other, low-selenium lentils grown in another ecozone. At the onset, mid-term, and end of the trial, samples (blood, urine, stool, hair) are collected, and health examinations include testing lung inflammation, body weight and blood pressure. The major outcome will be arsenic excretion in urine and feces, and arsenic deposition in hair. Secondary outcomes also include antioxidant status, and blood lipid profile.

1 INTRODUCTION

Up to 100 million people worldwide are chronically exposed to dangerously high concentrations of arsenic (As) in their drinking water and food supply. Bangladesh is facing contamination of groundwater by As. Since tubewell water contamination by As was discovered in the 1990s, As in drinking water has been reduced by 40%, yet approximately 45 million Bangladeshis remain at risk from As concentrations greater than the WHO guideline value of $10\,\mu g\,L^{-1}$ (ppb) in the well water (WHO Fact sheet N°372, December 2012) and greater than the Bangladeshi limit of $50\,\mu g\,L^{-1}$. Micronutrient-deficient (e.g. selenium) soils lead to less nutritious local foods, which becomes a partner in crime, exacerbating the toxic burden incurred by As exposure.

Selenium (Se) is an important trace element and an antagonist to As toxicity. Se pills are currently used to treat arsenicosis. But pills are not often well received by people and are costly for low-income families. Lentils are a common food in Bangladesh. These lentils are rich in Se, 425–672 $\mu g\,kg^{-1}$, which is mostly in the highly bioavailable form, L-selenomethionine (Thavarajah *et al.*, 2007).

The purpose of this work is to determine whether high-Se lentils in the daily diet can decrease the body burden of As in exposed Bangladeshis, when compared to the treatment group on low-Se lentils.

2 METHODS/EXPERIMENTAL

This dietary trial is a parallel, randomized, controlled trial (Fig. 1). Four hundred Bangladeshis (approximately 80 families) from Shahrasti area, who have As

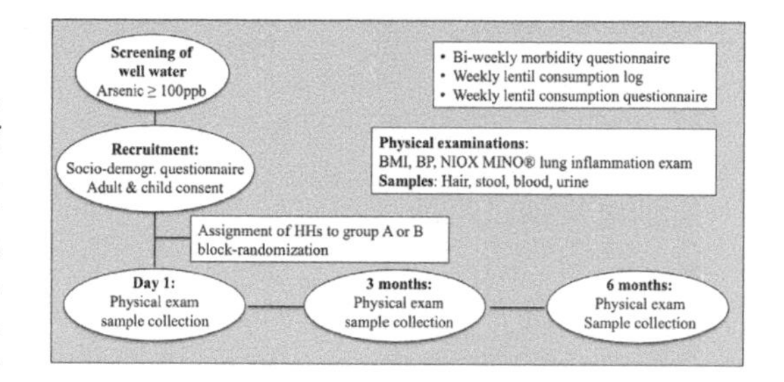

Figure 1. Flow chart of the trial procedures.

levels in the household tubewells $\geq 100\,\mu g\,L^{-1}$, were randomly assigned to one of two treatment groups. Eligibility criteria included apparently healthy people aged 14 to 75 years old. In this double-blind study, one group will have high-selenium lentils, the other will consume low-Se lentils.

Each participant is required to eat 65 g of lentils every day for 6 months. At the beginning, after 3 months, and end of the trial, blood, urine and stool, plus hair (day 1 and at 6 months only) samples are collected. The health examination includes assessment of acute lung inflammation, body weight, height, and blood pressure. The major outcome is As excretion in urine and feces, as well as As deposition in hair. Secondary outcomes include antioxidant status, lipid profile, lung inflammation status and blood pressure.

Arsenic levels in hair, stool and urine samples are measured by Hydride Generation Atomic Absorption Spectrometry (HGAAS). Oxidative stress and antioxidant status is determined by measuring 8-OHdG, a major product of ROS-induced oxidative stress, in urine and plasma using ELISA kits. Selenium-dependent glutathione peroxidase uses to

Table 1. Nutritional analysis of low- and high-Se lentils.

		High Se lentils	Low Se lentils
Macronutrients			
Protein	% by weight	26.22	27.73
Starch	% by weight	38.00	37.00
TDF	% by weight	8.48	6.66
Fat	% by weight	0.78	0.77
Ash	% by weight	2.77	3.16
Phytochemicals			
Phytic acid	$g\,kg^{-1}$	0.61	0.72
Minerals			
Calcium	$mg\,kg^{-1}$	327.88	377.51
Copper	$mg\,kg^{-1}$	9.34	11.44
Iron	$mg\,kg^{-1}$	75.75	65.3
Zinc	$mg\,kg^{-1}$	42.15	51.9
Selenium	$mg\,kg^{-1}$	0.854	0.029
Arsenic	$mg\,kg^{-1}$	<0.001	<0.001

scavenge reactive oxygen species (ROS), resulting in the production of oxidized GSSG (33). The antioxidant status based on reduced glutathione (GSH) and the oxidized form, GSSG, is determined by the ratio of GSH:GSSG using an ELISA. Triglyceride, total cholesterol, high-density-lipoprotein (HDL), and low-density-lipoprotein (LDL)-cholesterol plus acute phase proteins are measured using a Biochemistry Analyzer Cobas c 311.

3 RESULTS AND DISCUSSION

The human dietary study in Bangladesh is under way and early results from the spring of 2016 will be presented. We tested tubewells of 102 households in the Shahrasti region for As. Over 90% had As $>100\,\mu g\,L^{-1}$. The high- and low-Se lentils were grown in the summer of 2014. The macro- and micronutrient content has been determined (Table 1).

We have 4 experimental studies from laboratory animals that showed clear benefits from the high selenium diets.

In rats exposed to As in drinking water, immune function, blood biochemistry, and oxidative stress biomarkers were examined. Liver damage, increased oxidative stress, decreased blood levels of protective antioxidants, and suppressed antibody mediated immunity were the most sensitive health changes from As exposure (Nain & Smits, 2010). Next, we exposed young, growing rats for 4 months to environmentally realistic levels of As in drinking water. Rats on high Se diets had higher blood stores of the major antioxidant, glutathione, As-suppressed antibody response was reversed, and body burdens of As were reduced, evident through lower arsenic in the kidneys, and higher fecal and urinary arsenic excretion in animals on high Se diets (Sah & Smits, 2012).

In an atherosclerotic mouse model, mice were fed high- or low-Se lentil diets for 13 wk while consuming low levels of As in their drinking water ($200\,\mu g\,L^{-1}$). Arsenic-triggered plaques were formed in the aorta. Plaque development was significantly reduced or completely abolished in aortas of mice on the high Se lentil diet. Our study also indicated that Se deficiency plus As exposure resulted in higher low-density lipoproteins (LDLs), the so-called 'bad' cholesterol (Krohn et al., 2016).

4 CONCLUSIONS

The responses from our animal studies, plus those in which Se pills have been advocated for improving health status in As exposed people, are the driving force behind our human dietary study.

Physiological responses and health benefits in humans on the high Se lentil diets will be evident through blood pressure changes, favorable changes in blood triglyceride composition, decreased markers of oxidative stress, plus increased excretion of arsenic, compared with the group on the low Se lentils.

ACKNOWLEDGEMENTS

We thank Gene Arganosa and Barry Goetz for the lentil nutrient analysis.

REFERENCES

Krohn, R., Lemaire, M., Negro Silva, L., Mann, K. & Smits. J.E. 2016. High-selenium lentil diet protects against arsenic-induced atherosclerosis in mice. *J. Nutr. Biochem.* 27: 9–15.

Nain, S. & Smits, J.E.G. 2010. Pathological, immunological and biochemical biomarkers of sub-chronic arsenic toxicity in experimental rats. *Environ. Toxicol.* 27: 244–254.

Sah, S. & Smits, J.E. 2012. Dietary selenium fortification: a potential solution to chronic arsenic toxicity. *Toxicol. Environ. Chem.* 94(7): 1–13.

Thavarajah, D., Vandenberg, A., George, G. & Pickering, I. 2007. Selenium species in selenium-rich lentils from Saskatchewan. *J. Agric. Food Chem.* 55(18): 7337–7341.

WHO 2012. Fact sheet N° 372, December 2012, World Health Organoization, Geneva.

Environmental Arsenic in a Changing World –
Zhu, Guo, Bhattacharya, Ahmad, Bundschuh & Naidu (Eds)
ISBN 978-1-138-48609-6

Arsenic contamination in drinking water from groundwater sources and health risk assessment in the Republic of Dagestan, Russia

T.O. Abdulmutalimova[1], B.A. Revich[2] & O.M. Ramazanov[3]
[1]*Institute of Geology, Dagestan Center of Science, Russian Academy of Sciences, Makhachkala, Dagestan, Russia*
[2]*Institute for Forecasting of the Russian Academy of Science, Moscow, Russia*
[3]*Institute of Geothermal Problems, Dagestan Center of Science, Russian Academy of Sciences, Makhachkala, Dagestan, Russia*

ABSTRACT: In this study the data of arsenic content in groundwater of the Republic of Dagestan are presented. Due to naturally occurring arsenic in the sediments the groundwaters, used by population for drinking, content high level of arsenic compound. The contamination level varied from 1 to 500 $\mu g\,L^{-1}$ with an average arsenic concentration of 190 $\mu g\,L^{-1}$. 90% of water samples are above the WHO standard of 10 $\mu g\,L^{-1}$. The high arsenic concentrations found in the artesian water samples (15.8% above 10 $\mu g\,L^{-1}$ and 84.2% above 50 $\mu g\,L^{-1}$) indicate that more than 500 thousand of people consuming untreated groundwater might be at a considerable risk of chronic arsenic poisoning. The evaluation of cancer risks from oral exposure to As were found to be above the acceptable U.S.EPA and WHO cancer health risk range of 1×10^{-6} to 1×10^{-4}.

1 INTRODUCTION

Groundwater is a significant source for portable water for millions of people worldwide (Bhattacharya *et al.*, 2006). Naturally occurring groundwater As contamination has been reported in several countries of the world, with concentration levels exceeding the WHO drinking water guideline value of 10 $\mu g\,L^{-1}$ (WHO, 2001; 2012).

Previous studies had revealed that groundwater, used by population as drinking water, in some areas of the Dagestan Republic also contained high levels of As (Kurbanova *et al.*, 2013) exposing a significant population to health risk. However, the extent of risk and the extent of As contamination in groundwater sources is unknown. A human health risk assessment exposed to arsenic in drinking water was carried out in this study, in line with US EPA risk assessment guidelines. The aim of this study is to show spatial distribution of As in this region and human health risk assessment of chronic oral exposure.

2 METHODS

2.1 *Water collection, arsenic analysis and exposure*

703 water samples were collected from the arsenic-contaminated groundwater in the northern districts of Dagestan Republic. During sampling, the plastic containers were rinsed several times with the sample before the final sample was taken. The water samples were acidified by adding 1 mL of 10% analytical grade HNO_3 and brought to the laboratory. Water samples were analyzed for total As by flow-injection hydride generation-atomic absorbtion spectrophotometry (FI-HG-AAS).

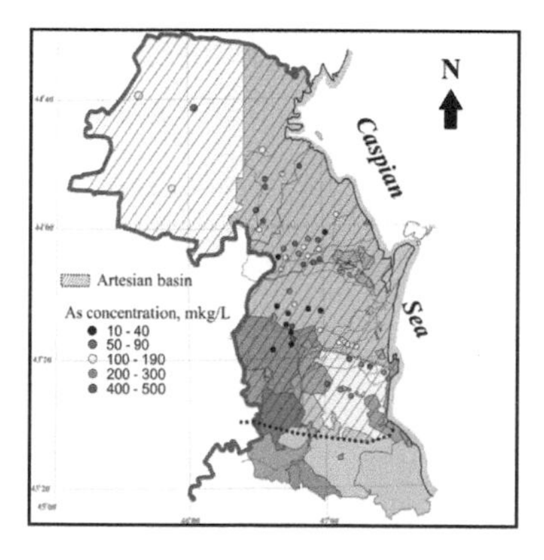

Figure 1. Map of artesian basin with As concentration in different villages of the Dagestan Republic, M 1:500000.

2.2 *Human health risk assessment: Calculation of carcinogenic health risk*

More than 200 million people in 105 countries around the world are exposed to arsenic at concentrations much higher than the guideline value, and therefore the public health priority should be to reduce exposure for these people. In this study cancer health risk for population was evaluated using human health risk assessment method (WHO, 2010). The exposure route evaluated is oral. According to IARC (1987; 2004), arsenic has been classified as a class 1 human carcinogen, based on sufficient evidence from human epidemiological data that exposure to arsenic causes several forms of cancers. Carcinogenic risks for oral exposure to As were calculated by using following equation:

Cancer health risk (CR_{oral}) = (ADD) $\times$ CSF (Eqn. 1)

Table 1. Association between exposed population and As level in drinking water.

Area	The level of arsenic in drinking water, $\mu g\ L^{-1}$	Population
1	10–40	167134
2	50–90	16985
3	100–190	108147
4	200–300	9023
5	400–500	8444

where ADD is the average daily dose of As in water via oral exposure route in the study area by population; CSF is the cancer slope factor for oral exposure to As which is $1.5\ mg\ kg^{-1}\ d^{-1}$.

3 RESULTS AND DISCUSSION

3.1 *Spatial distribution of arsenic in drinking water*

The concentration of As in water samples collected in this study ranged from 10 to $500\ \mu g\ L^{-1}$ with a mean value $190\ \mu g\ L^{-1}$ (Fig. 1). The observed differences in As concentrations are due to spatial variation in the regional hydrogeochemical parameters which control the mobilization in groundwater (Bhattacharya *et al.*, 2006).

Most of the water samples (97%) contained As at levels higher than WHO MCL value of $10\ \mu g\ L^{-1}$. 4.8% water samples contained As at the levels 400–$500\ \mu g\ L^{-1}$. 53.9% of population use drinking water with the level of As 10–$40\ \mu g\ L^{-1}$. Groundwater with the high concentration of As (400–$500\ \mu g\ L^{-1}$) used by 3% of the northern districts population. The level of As in the water is detected 20 times or more in 12 villages with the total population of about 16 thousand of people (Table 1).

3.2 *Cancer risk results*

Cancer health risk is defined as the incremental probability that cancer may develop during life due to chronic As exposure. According to U.S.EPA exposure factors and the International Agency Research on Cancer (IARC), As is the metal among the list of metals that are likely to be carcinogenic via oral ingestion with contaminated water[3].

The results of cancer risk for resident adults from the study area exposed to As via oral ingestion of water are presented in Table 2.

In the study area, ADD for oral exposure ranged from $0.0003\ mg\ kg^{-1}\ d^{-1}$ to $0.014\ mg\ kg^{-1}\ d^{-1}$. It has been shown that the lifetime individual cancer risks were at the minimum As concentration ($10\ \mu g\ L^{-1}$) – 4.3×10^{-4}; at maximum As concentration ($500\ \mu g\ L^{-1}$) – 2.1×10^{-3}, respectively, with a mean of $190\ \mu g\ L^{-1}$ – 8.1×10^{-3}. The evaluation of cancer risks from exposure to As in this study were found to be above the acceptable U.S.EPA and WHO cancer health risk (1×10^{-6} to 1×10^{-4}) and not permissible for population. The annual cancer health risks for population were from 1 to 95 additional cases of possible occurrence of cancer.

Table 2. Cancer risk for exposed population in the Dagestan Republic.

Level of arsenic in drinking water ($\mu g\ L^{-1}$)	Individual cancer health risk
10–40	4.29×10^{-4}–1.71×10^{-3}
50–90	2.14×10^{-3}–3.86×10^{-3}
100–190	4.29×10^{-3}–8.14×10^{-3}
200–300	8.57×10^{-3}–1.29×10^{-2}
400–500	1.71×10^{-2}–2.14×10^{-2}

4 CONCLUSIONS

The results of the water samples analysis revealed high concentrations of As in groundwater, used by population for drinking. 97% of water samples showed As concentration higher than WHO and safe levels. Health risk assessment models indicated that oral exposure to As by consumption of groundwater presented potential carcinogenic risks and the population of the north districts of the Dagestan republic in the study area are at risk of developing cancerous diseases due to As exposure via drinking water. Further research should be directed to case-control epidemiological studies. in the areas of Dagestan with elevated As in groundwater.

REFERENCES

Bhattacharya, P., Claesson, M., Bundschuh, J., Sracek, O., Fagerberg, J., Jacks, G., Martin, R.A., Storniolo, A.R. & Thir, J.M. 2006. Distribution and mobility of arsenic in the Rio Dulce alluvial aquifers in Santiago del Estero Province, Argentina. *Sci. Total Environ.* 358(1–3): 97–120.

IARC 1987. Arsenic and arsenic compounds. In: Overall Evaluations of Carcinogenicity. *IARC Monographs on the Evaluation of Carcinogenic Risk of Chemicals to Humans*, Suppl. 7, Lyon, France.

IARC 2004. Some drinking-water disinfectants and contaminants, including arsenic. *Monographs on the Evaluation of Carcinogenic Risks to Humans*. Volume 84 WHO; Lyon, France.

Kurbanova, L.M., Samedov, S.G., Gazaliev, I.M. & Abdulmutalimova, T.O. 2013. Arsenic in the groundwaters of the north Dagestan artesian basin. *Geochem. Int.* 51(3): 237–239.

WHO 2001. Environmental health criteria 224, arsenic and arsenic compounds. Inter-organization programme for the sound management of chemicals. World Health Organization, Geneva.

WHO 2010. Human health risk assessment toolkit: chemical hazards. World Health Organization, Geneva.

WHO 2012. Arsenic Fact Sheet no 372. World Health Organization, Geneva.

3.5 Multi-metal synergies in chronic exposure cases

Environmental Arsenic in a Changing World –
Zhu, Guo, Bhattacharya, Ahmad, Bundschuh & Naidu (Eds)
ISBN 978-1-138-48609-6

Interaction of polyaromatic hydrocarbons and metals on bioaccessibility and toxicity of arsenic

J.C. Ng, Q. Xia, S. Muthusamy, V. Lal & C. Peng
Queensland Alliance for Environmental Health Sciences (QAEHS), The University of Queensland, Brisbane, QLD, Australia

ABSTRACT: We investigated the interaction effects of cadmium, lead and four PAHs on the toxicity of arsenic and each other. A simulated in vitro human digestive system and human liver cell-based assays were employed to determine their respective bioaccessibility, uptake and toxicity of individual toxicants and mixtures of up to all seven chemicals. Under most environmental conditions, additive effect can be assumed when assessing bioaccessibility of mixed arsenic, cadmium and lead with or without polyaromatic hydrocarbons whilst hepatic uptake was less than additive for cadmium mixed with arsenic/lead/pyrene/ benzo[a]pyrene. Toxicity in HepG2 cells including cytotoxicity, oxidative stress, genotoxicity and AhR (aryl hydrocarbon receptor) activation were determined. The interaction amongst mixtures vary depending on toxic endpoints, concentration and number of individual chemicals in the mixtures. Therefore, the traditional concentration addition or independent action model for health risk assessment may over or under estimate the risk of mixtures. This study has gained understanding of mixture interaction and helps to refine current risk assessment of mixed contaminants.

1 INTRODUCTION

Arsenic (As) is often found to co-exist with cadmium (Cd), lead (Pb) and organic toxicants such as polyaromatic hydrocarbons (PAHs) in contaminated sites (e.g. former gasworks sites). Information required for the risk assessment of arsenic and these other contaminants in this type of mixtures is lacking. A better understanding of potential interaction effects of mixtures with regard to their bioaccessibility and toxicity would inform the risk assessment process. A simulated in vitro human digestive system (Unified BARGE (Bioaccessibility Research Group of Europe) method (which is known as UBM)) and HepG2 cell-based assays were employed to measure the bioaccessibility, uptake and toxicity of individual toxicants and mixtures of up to all seven chemicals. Selected toxicants included As, Cd, Pb, benzo[a]pyrene (B[a]P), naphthalene (Nap), phenanthrene (Phe) and pyrene (Pyr).

2 METHODS/EXPERIMENTAL

2.1 *Bioaccessibility and bioavailability*

United BARGE Method (UBM) (Denys *et al.*, 2012) was used to measure bioaccessibility (BAC) of individual chemicals and in combinations of mixtures. Uptake by human liver cells (HepG2) is treated as a surrogate for bioavailability estimate.

2.2 *Toxicity interaction effects*

Toxicity interaction effects of individual chemicals and in mixtures were studied using HepG2 cells. Biological end points included cytotoxicity (MTS assay), genotoxicity (micronucleus assay), oxidative stress (Nrf2 antioxidant pathway as determined in the ARE reporter-HepG2 cells) and AhR activation as measured by CAFLUX assay.

Individual dose-response curves obtained were used to selected dosages for the mixture experiments. A factorial design for combination of dosages was used for the interaction studies of mixed toxicants (Muthusamy *et al.*, 2016a, 2016b).

Potential interaction effects between As, Cd, Pb and PAHs are illustrated in Figure 1. This study is aimed to investigate the relationship between bioaccessibility, cell uptake and toxicity effects of some biological endpoints.

3 RESULTS AND DISCUSSION

3.1 *Bioaccessibility and uptake in HepG2 cells*

Bioaccessibility of As, Cd and Pb was not affected by PYR or B[a]P in spiked and aged soils. However, after UBM-extracted As, Cd and Pb progressed to target organ cells (HepG2), uptake of Cd was inhibited by PYR or B[a]P probably due to the more damaged cell membrane whilst uptake of As and Pb was not affected. Arsenic and Pb interfered the sorption of Cd in soils, leading to an increase in Cd intestinal bioaccessibility. Pure solution results were in agreement with UBM extracted mixtures. Taken together, bioaccessibility of As, Cd and Pb was not likely to be affected by PAHs due to distinct behavior of inorganic and organic contaminants in soils and solutions whilst accumulation

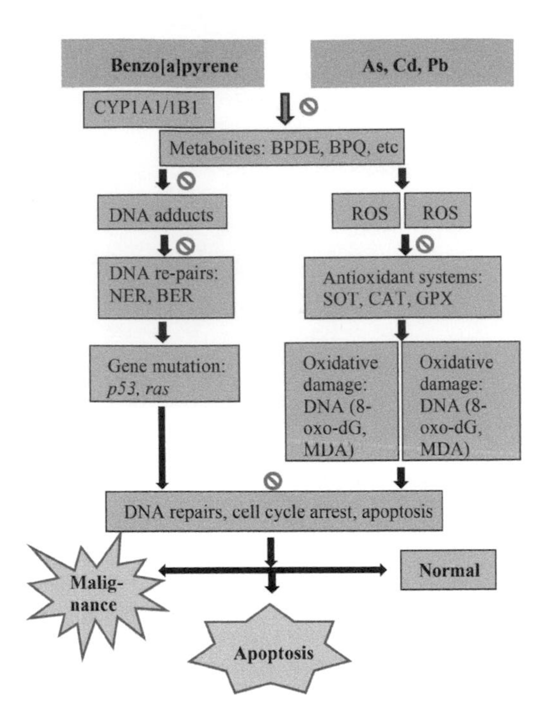

Figure 1. Potential interaction between As, Cd, Pb and PAHs using benzo[a]pyrene as an example, where ⊘ indicates inhibition.

of metal/metalloid at toxic concentrations (e.g. Cd in this study) in hepatocytes might be inhibited by PAHs.

3.2 *HepG2 cell toxicity*

The cytotoxicity and interaction effects of binary to quaternary mixtures of B[a]P, As, Cd and Pb were found to be better predicted by using the combination index (CI)-isobologram method. The predicted interaction between these mixtures was synergism, additivity and antagonism at low, medium and higher concentration combinations of mixtures, respectively.

The mixture effect of the four selected PAHs and As, Cd and Pb on the Nrf2 antioxidant pathway (an indicator of oxidative stress response) was determined in the ARE reporter-HepG2 cells. The mixture study was carried out for binary, ternary, quaternary and seven-chemical combinations and the mixture effects were predicted using the CA model.

Binary or ternary mixtures of As, Cd and Pb increased MN formation and AhR induced by B[a]P and the mixtures of PAHs decreased MN formation and AhR induction by B[a]P. The mixtures at higher order, quaternary and seven chemical combinations of As, Cd, Pb and PAHs did not increase the genotoxicity of B[a]P.

Toxicity studies with the bioaccessible fractions (UBM extracts) showed that the concentrations of toxicants were insufficient to induce cytotoxicity. The Nrf2 activation assay is more sensitive and hence useful to determine these chemicals response from UBM extracts with relatively low concentrations. The results from soil experiments showed that UBM method is useful to determine the bioaccessibility of metals but did not give good results for B[a]P. The combination of bioaccessibility and bioassays could be useful for risk assessment of chemicals.

4 CONCLUSIONS

Under most environmental conditions, additive effect can be assumed when assessing bioaccessibility of mixed arsenic, cadmium and lead with or without polyaromatic hydrocarbons whilst hepatic uptake was less than additive for cadmium mixed with arsenic/lead/pyrene/benzo[a]pyrene. Bioaccessibility was also influenced by soil properties. Taken together, interaction should be interpreted individually for different physiological processes.

For toxicity evaluation, it was found that the CI method (Chou & Talalay, 1984) could be better for predicting the mixture toxicity compared to that of concentration addition (CA) and independent action (IA) models. Whereas CA model may be appropriate for oxidative stress response assessment.

The interaction between these mixtures vary depends on toxic endpoints, concentration and number of individual chemicals in the mixtures. Therefore, the traditional CA or IA model for health risk assessment may over or under estimate the risk of mixtures. Further mechanistic studies are required to understand the observed effects in various toxic endpoints.

ACKNOWLEDGEMENTS

The research was funded by CRC CARE (project no. 3.1.01-11/12), the Chinese Science Council and UQ IPRS scholarships. QAEHS incorporating the former National Research Centre for Environmental Toxicology (Entox) is a partnership between Queensland Health and the University of Queensland.

REFERENCES

Chou, T.C. & Talalay, P. 1984. Quantitative analysis of dose-effect relationships: the combined effects of multiple drugs or enzyme inhibitors. *Adv. Enzyme. Regul.* 22:27–55.

Denys, S., Caboche, J., Tack, K., Rychen, G., Wragg, J., Cave, M., Jondreville, C. & Feidt, C. 2012. In vivo validation of the unified BARGE method to assess the bioaccessibility of arsenic, antimony, cadmium, and lead in Soils. *Environ. Sci. Technol.* 46(11): 6252–6260.

Muthusamy, S., Peng, C. & Ng, J. 2016a. The binary, ternary and quaternary mixture toxicity of benzo[a]pyrene, arsenic, cadmium and lead in HepG2 cells. *Toxicol. Res.* 5(2): 703–713.

Muthusamy, S., Peng, C. & Ng, J. 2016b. Effects of multi-component mixtures of polyaromatic hydrocarbons and heavy metal/loid(s) effects on Nrf2-Antioxidant Response Element (ARE) pathway in ARE reporter–HepG2 cells. *Toxicol. Res.* 5(4): 1160–1171.

Environmental Arsenic in a Changing World –
Zhu, Guo, Bhattacharya, Ahmad, Bundschuh & Naidu (Eds)
ISBN 978-1-138-48609-6

Contamination of arsenic and heavy metals in coal exploitation area

R. Sharma[1], A. Yadav[1], S. Ramteke[1], S. Chakradhari[1], K.S. Patel[1],
L. Lata[2], H. Milosh[2], P. Li[3], J. Allen[3] & W. Corns[3]
[1] *School of Studies in Chemistry/Environmental Science, Pt. Ravishankar Shukla University, Raipur, India*
[2] *Department of Soil Science/Geology, Maria Curie-Skłodowska University, Lublin, Poland*
[3] *PS Analytical Ltd, Arthur House, Orpington, Kent, UK*

ABSTRACT: The coal is a dirty fuel, containing As and other heavy metals (HMs) at the trace levels. Several millions tons of coals are exploited in the Korba basin, CG, India to generate electricity. In this work, distribution of As and other HMs i.e. Cr, Mn, Fe, Ni, Cu, Zn, Cd, Pb and Hg in the surface soil and sediment are described. Three metals i.e. Mn, Fe and Ni occurred at higher concentrations, ranging ($n = 30$) from 2.3–6.4, 0.08–0.22 and 0.04–0.16% with mean value ($p = 0.05$) of 4.3 ± 0.4, 0.14 ± 0.02 and 0.08 ± 0.01%, respectively. The concentration of elements i.e. As, Cr, Cu, Zn, Cd, Pb and Hg in the soil was ranged ($n = 30$) from 49–164, 30–78, 44–131, 87–220, 0.11–0.56, 72–194 and 0.11–0.39 mg kg^{-1} with mean value ($p = 0.05$) of ($p = 0.05$) 106 ± 11, 51 ± 5, 86 ± 8, 156 ± 10, 0.33 ± 0.04, 130 ± 10 and 0.22 ± 0.03 mg kg^{-1}, respectively. The concentration of the As in the sediments ($n = 26$) was ranged from 36–154 mg kg^{-1} with mean value of 93 ± 12 mg kg^{-1}. The toxic inorganic As(III) and As(V) species are found to exist in the soil and sediment. The concentration variations, pollution indices and sources of the contaminants in the geomedia are discussed.

1 INTRODUCTION

Pollution of urban geomedia is of a great public health interest due to receiving of large amounts of pollutants from multiple sources including industrial wastes, vehicle emissions, coal and biomass burnings, etc. (Li *et al.*, 2012; Obaidy *et al.*, 2013; Plyaskina & Ladonin, 2009). India is the third-largest producer of coal in the world. Coal is a naturally occurring combustible material contains elemental carbon, sulfur, hydrocarbons, trace metals, etc. (Chou, 2012). Several environmental issues i.e. acid mine drainage, deposition of toxic compounds, environmental pollution, halting of acid rain, health hazards, storage of solid waste, etc. were observed due to coal burnings (Agrawal *et al.*, 2010; Bhuiyan *et al.*, 2010; Guttikunda *et al.*, 2014; Pandey *et al.*, 2011; Sengupta *et al.*, 2010; Sheoran *et al.*, 2011; Singh *et al.*, 2010).

Several millions tons of coal is mined out and burnt in the Korba basin for generation of electricity by pouring the effluents into the environment. The whole environment is covered by the fly ash and black carbon (BC). The rain and groundwater are acidic with high content of metals and fluoride Patel *et al.*, 2001; 2016). The HMs contamination of the Korba basin has not been carried out so far. Hence, in this work, the contamination of surface soil and sediment of the Korba basin with elements i.e. BC, As, Fe, Cr, Mn, Ni, Cu, Zn, Cd, Pb and Hg is described. The concentration variations, enrichments and sources of the metals in the soil and sediment are discussed.

2 METHODS AND EXPERIMENTAL

2.1 *Study area and sampling*

Korba coalfield (22.35°N and 82.68°E) is located in the Chhattisgarh state, India in the basin of the Hasdeo river, extending over ≈530 km^2. The population of Korba area is ≈ 1.0 million. Several coal mines are in operation with annual production of ≈3 BT coal annually since year 1960. A huge amount of coal (≈20 MT yr^{-1}) is consumed by various units of thermal power plants to produce 6000 MW electricity with emission of ≈6 MT ash into the environment. The Asia biggest aluminum plant (3.2×10^5 TPY Aluminum smelter) is also in the operation in the Korba area.

The soil and sediment samples were collected from 30 and 26 locations of Korba area lie over ≈500 km^2 areas (Fig. 1). One kilogram of sample from each site (0–10 cm) was collected in a clean polyethylene container during January, 2011–2017 as prescribed in the literature (Tan, 2005). For depth profile studies, the soil samples were collected at depth of 0–10, 11–20 and 21–30 cm.

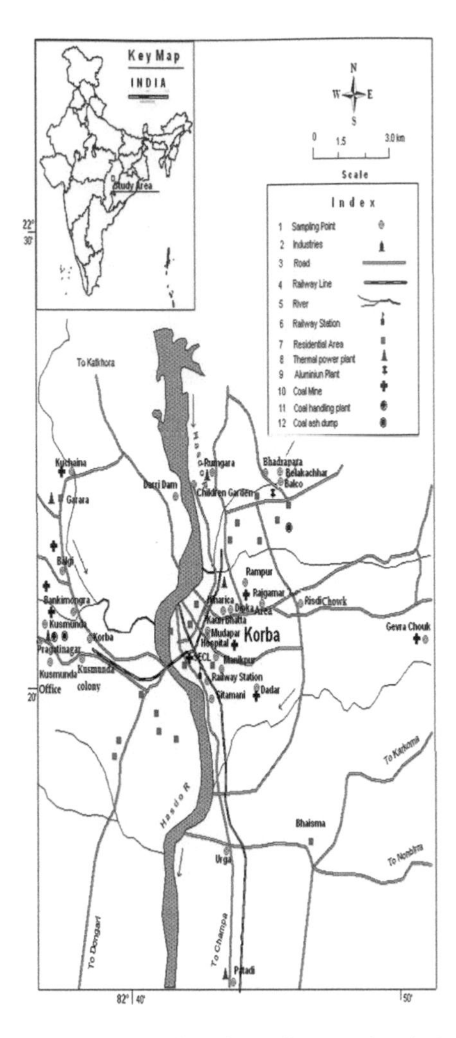

Figure 1. Representation of sampling locations in Korba basin.

2.2 *Analysis*

The soil samples were dried, milled and particles of ≤1 mm were sieved out. A 10 g sample with 20 mL deionized water in a 100-mL conical flask was agitated in an ultrasonic bath for 6 h. The pH values of the extract were measured by the Hanna pH meter type-HI991300.

The weighed amount of soil sample (0.25 g) was digested with acids (3 mL HCl and 1 mL HNO_3) in closed system with P/T MARS CEM (Varian Company) microwave oven. The acid extract was used for monitoring of the elements.

The CHNSO–IRMS Analyzer, SV Instruments Analytica Pvt. Ltd. was used for analysis of the black or elemental carbon (BC or EC). The total carbon (TC) in the soil sample was oxidized at 1020°C with O_2 into CO_2 with constant helium flow.

The analytical techniques i.e. Varian ICP-OES-700-ES, GF-AAS SpectrAA 220 and CV-AAS SpectrAA 55B were used for monitoring of metals i.e. Cr, Mn, Fe, Ni, Cu, Zn and Pb; As and Cd; and Hg in the soil extract, respectively. The standard soil sample (NCS DC 73382 CRM) was used for the quality control.

A 0.2 g of sediment sample was taken into a 10-mL Teflon centrifuge tube by subsequent addition of 5 mL of 0.5 M H_3PO_4 and kept overnight. A 2 mL of filtered solution was diluted to 10 mL with sodium phosphate to use as HPLC mobile phase. The As-species (i.e. As(III), As(V), MMA and DMA) were quantified by using the technique i.e. HPLC-AFS

The pollution indices i.e. enrichment factor (EF), contamination factor (CF) and pollution load index (PLI) are used to determine element contents in the soil samples with respect to the base line concentration. These relate the concentration of an element, X, to a crustal element (e.g. Al) in the soil sample, and this ratio is then normalized to the ratio of those elements in the earth's crust. The following equations were used for the calculation of the pollution indices (Sinex & Helz, 1981; Tomlinson *et al.*, 1980):

$$EF = \{[X_s]/[Al_s]\}/\{[X_e]/[Al_e]\}$$
$$CF = \{[X_s]/[X_e]\}$$
$$PLI = (CF_1 \times CF_2 \times CF_3 \times CF_4 \ldots\ldots\ldots CF_n)^{l/n}$$

where, symbols: X_s, X_e, Al_s and Al_e denote concentration of metal and Al in the soil and earth crust, respectively.

The IBM SPSS Statistics 23 software was used for the preparation of the dendrogram.

3 RESULTS AND DISCUSSION

3.1 *pH of extract*

The soil and sediment were colored, ranging from brown to blackish. The pH values of the soil and sediment extracts (n = 30) ranged from 5.4–7.5 and from 5.4–8.1 with mean value (p = 0.05) of 6.6 ± 0.2 and 6.6 ± 0.3, respectively. The extracts were observed to be slightly acidic, may be due to high chloride and sulfate contents. The lowest pH value was seen in the Korba city, may be due to the high anthropogenic activities.

3.2 *Distribution of black carbon*

The BC concentration in the soil and sediment was ranged from 3.4–7.9 and 3.6–14.0% with mean value (p = 0.05) of 5.6 ± 0.4 and 9.2 ± 1.0%, respectively. The higher loading of the BC in the sediment samples was observed, may be due to its transport by the runoff water. The BC content in the geomedia of the studied area was found to be higher than reported in other regions, probably due to huge coal burning (Han *et al.*, 2009; He & Zhang, 2009; Muri *et al.*, 2002).

3.3 *Distribution of elements*

The metals i.e. Mn, Fe and Ni occurred in the soil of Korba basin at high levels, ranging (n = 30) from 2.3–6.4, 0.08–0.22 and 0.04–0.16 % with mean value (p = 0.05) of 4.3 ± 0.4, 0.14 ± 0.02 and 0.08 ± 0.01%, respectively. The concentration

Table 1. Distribution of major metals in soil, mg kg^{-1}.

S. No.	Location	As	Cd	Pb	Hg
1	Niharica	146	0.47	163	0.21
2	Kuan Bhatta	80	0.26	119	0.18
3	Railway Station	78	0.11	95	0.11
4	Sitamani	140	0.43	126	0.23
5	Rajgamar	164	0.56	189	0.39
6	Rumgara	60	0.21	126	0.24
7	Darri dam	96	0.27	116	0.19
8	Chaildren Garden	61	0.35	116	0.20
9	Dipka-I	61	0.27	117	0.14
10	Dipika-II	49	0.21	137	0.23
11	Gevera Chowk	102	0.31	114	0.22
12	Kusmunda-I	96	0.17	127	0.21
13	Kusmunda-II	106	0.29	126	0.25
14	Kusmunada-III	130	0.41	154	0.25
15	Korba	145	0.54	190	0.37
16	Banki mongra	120	0.21	110	0.26
17	Balgi	111	0.38	88	0.25
18	Kuchaina	97	0.25	72	0.13
19	Balco	134	0.44	168	0.22
20	bhadrapara	86	0.33	106	0.11
21	Risdi Chowk	83	0.34	102	0.13
22	Manikpur	79	0.30	96	0.15
23	Dadar	94	0.39	138	0.24
24	SECL	98	0.41	140	0.21
25	Mudapar	105	0.38	119	0.24
26	Rampur, PWD	101	0.32	125	0.22
27	SECL Hospital	149	0.49	160	0.31
28	Belakachar	131	0.29	149	0.30
29	Urga	134	0.21	129	0.26
30	Patadi	163	0.48	194	0.38

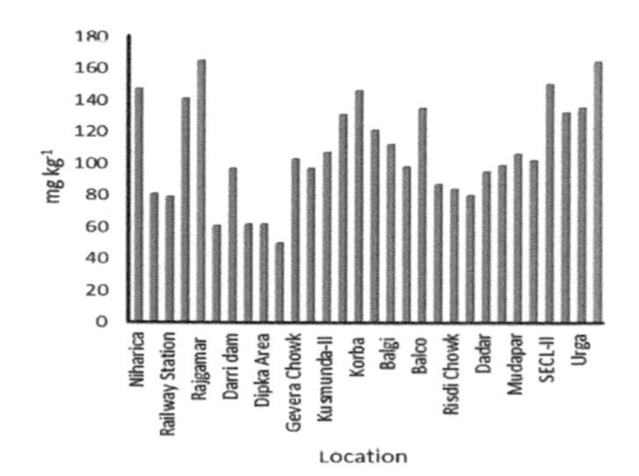

Figure 2. Distribution of As in surface soil.

of elements i.e. As, Cr, Cu, Zn, Cd, Pb and Hg ranged (n = 30) from 49–164, 30–78, 44–131, 87–220, 0.11–0.56, 72–194 and 0.11–0.39 mg kg^{-1} with mean value (p = 0.05) of 106 ± 11, 51 ± 5, 86 ± 8, 156 ± 10, 0.33 ± 0.04, 130 ± 10 and 0.22 ± 0.03 mg kg^{-1}, respectively (Table 1).

The concentration of elements i.e. As, Cr, Cu, Zn, Cd, Pb and Hg in the sediments ranged from 36–154, 29–7, 18–92, 42–294, 0.14–1.19, 26–127 and 0.12–0.82 mg kg^{-1} with mean value of 93 ± 12, 49 ± 5, 49 ± 8, 142 ± 28, 0.59 ± 0.11, 72 ± 13 and 0.44 ± 0.08 mg kg^{-1}, respectively (Table 2). The lower concentration of the elements (except Cd and Hg) in the sediment than the soil was observed.

Table 2. Distribution of heavy metals in sediments.

S. No.	Location	As	Cd	Pb	Hg
1	Shakti Nagar	63	0.14	33	0.16
2	Gevra, Dipka	96	0.24	37	0.17
3	PN, Dipka	98	0.27	49	0.23
4	Banki, Dipka	36	0.52	47	0.25
5	Delwadih	100	0.63	81	0.31
6	Shingali	82	0.36	78	0.28
7	Kusmunda	92	0.50	92	0.36
8	Rajgamar-3	65	0.45	74	0.26
9	Mudapar	114	0.65	35	0.21
10	PN, Darri	138	0.82	106	0.61
11	Darri west	127	0.95	127	0.77
12	Jamnipali	154	1.19	82	0.63
13	Gopalpur	149	1.18	120	0.82
14	HTPP, Darri	111	0.84	112	0.74
15	Manuikpur-1	127	0.77	107	0.63
16	Manikpur-2	51	0.39	31	0.30
17	Dader-1	79	0.66	54	0.54
18	Dader-2	99	0.72	87	0.65
19	Kudarikhar	43	0.42	37	0.33
20	Naktikhar	93	0.64	75	0.53
21	Danras-1	98	0.62	53	0.46
22	Danras-2	41	0.20	26	0.12
23	SN-Balco	47	0.22	48	0.32
24	Pathadi	79	0.96	125	0.71
25	Dhendheni	69	0.62	108	0.61
26	Sukhri	30	0.35	35	0.31

They occurred in the following increasing order: Hg < Cd < Cr < Cu < As < Pb < Zn < Ni < Mn < Fe in the soil. The highest concentration of As was seen near the point sources i.e. thermal power plant, coal mine, urban area, etc. as shown in Figure 2. The concentration of As in the geomedia of studied area was several folds (>10) higher than permissible limit of 5 mg kg^{-1}. The concentration of the As and heavy metals in the soil and sediment of the Korba basin was higher than values reported in other region of the country and world, probably due to huge coal mining and burning (Dahal *et al.*, 2008; Ilwon *et al.*, 2003; Kwon *et al.*, 2017; Muller, 1969; Sheela *et al.*, 2012; Rudnick & Gao, 2003; Shrivastava *et al.*, 2014; Whitmore *et al.*, 2008).

3.4 *Vertical distribution of elements*

The concentration of elements (i.e. As, Cr, Ni, Cu, Zn, Cd and Hg) was increased as the depth profile was increased up to 30 cm unlikely to Mn, Fe and Pb, may be due to their less binding with the organic compounds (Fig. 3).

3.5 *Temporal variation of elements*

The increased temporal variation of As over periods: 2011–2015 was observed due to continuous mining and burning of the coals (Fig. 4). The rate of temporal increase in the As concentration was ≈6% in the geomedia due to continuous coal burning. At least 15% higher concentration of the As in the soil with respect

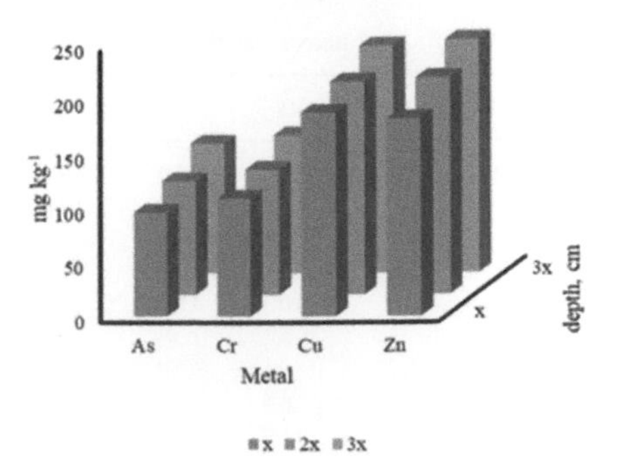

Figure 3. Verticle distribution of As and other metals in soil, x = 0–10, 2x = 10–20, 3x = 20–30 cm.

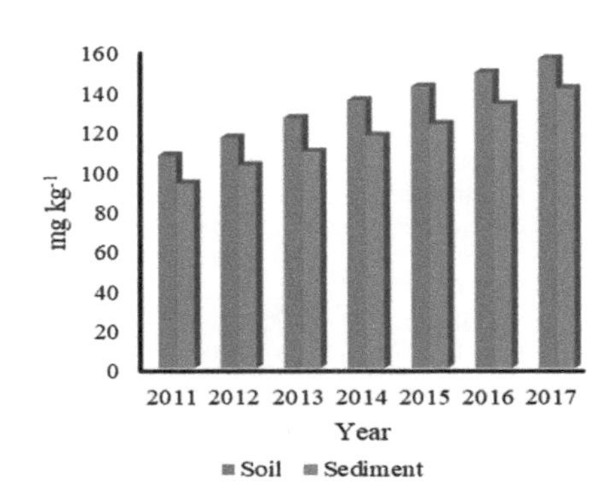

Figure 4. Temporal variation of As in soil and sediment.

Table 3. Speciation of arsenic in sediments, mg kg^{-1}.

S. No.	As(III)	DMA	MMA	As(V)	AsT	EE, %
S1	1.08	–	–	14.0	15.08	27.1
S2	–	–	–	9.08	9.08	110.7
SE1	0.1	–	–	2.10	2.20	73.3
SE2	0.15	–	–	6.0	6.15	130.8
SD3	–	–	–	2.40	2.4	75

EE = Extraction efficiency (100%).

the sediment was observed due to its washout by the aqueous media.

3.6 *Speciation of arsenic*

The As(III), As(V), MMA and DMA were quantified in the soil and sediment extracts by using technique i.e. HPLC-AFS (Table 3). The organic species i.e. MMA and DMA were not detected in the soil and sediment extracts. The whole As in soil and sediment samples of studied area was found in the toxic inorganic forms.

3.7 *Sources*

The correlation coefficient matrix for BC and nine metals in the soil is presented in Table 4. A fair correlation (r = 0.58–0.82) of the metals with BC was observed, may be due to deposition by multiple sources i.e. coal burning, fly ash, alumina roasting, etc. Similarly, fair correlation coefficient (0.39–0.80) among the metals was marked, showing deposition by the multiple sources.

Table 4. Correlation coefficient matrix of elements.

	BC	Fe	Ni	Cr	Cu	Zn	Pb	Hg	Cd	As
BC	1.00									
Fe	0.58	1.00								
Ni	0.59	0.41	1.00							
Cr	0.70	0.55	0.39	1.00						
Cu	0.74	0.60	0.56	0.73	1.00					
Zn	0.74	0.64	0.45	0.56	0.75	1.00				
Pb	0.81	0.77	0.59	0.59	0.72	0.72	1.00			
Hg	0.76	0.55	0.52	0.52	0.65	0.67	0.80	1.00		
Cd	0.82	0.68	0.57	0.57	0.58	0.67	0.69	0.59	1.00	
As	0.78	0.42	0.49	0.49	0.61	0.64	0.67	0.72	0.68	1.00

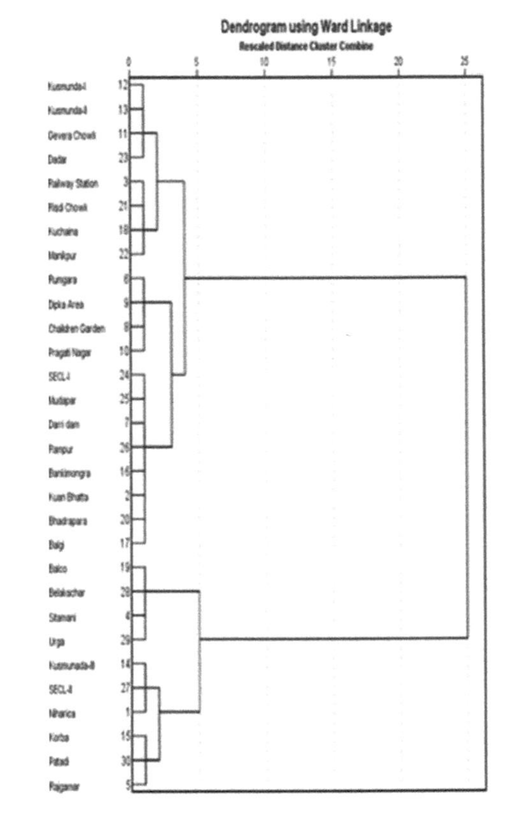

Figure 5. Dendrogram for group linkage of chemical characteristics of locations.

A dendrogram to know linkage between similar groups is plotted, using sum of total concentration of three metals i.e. As, Zn and Pb as discriminating factor. The dendrogram was categorized into two groups: group-I and -II with inclusion of 20 and10 locations as shown in Figure 5. The higher concentration of the As+Zn+Pb (>400 mg kg^{-1}) was seen in group-II, may be due to loading of point sources (i.e. coal mines, aluminum plant and thermal power plant) effluents.

3.8 *Enrichment*

The background concentration of Al, Fe, Mn, Cr, Zn, Ni, Cu, Pb, As, Cd and Hg reported was 8.2%, 3.9%, 775 mg kg^{-1}, 92 mg kg^{-1}, 67 mg kg^{-1}, 47 mg kg^{-1}, 28 mg kg^{-1}, 17 mg kg^{-1}, 4.8 mg kg^{-1}, 0.09 mg kg^{-1} and 0.05 mg kg^{-1}, respectively. The elements i.e. Ni and As; P, Cd, Pb and Hg; and S, Cl, Mn, Cu and Zn were highly ($20 \geq EF < 40$), significantly ($5 \geq EF < 20$) and moderately ($2 \geq EF < 5$) were enriched in the soil, respectively. Similarly, the soil was highly ($CF \geq 6$), considerably ($3 \geq CF < 6$) and moderately ($1 \geq CF < 3$) contaminated with As, Ni and Pb; S, Cl, P, Cu, Cd and Hg and Mn, Fe and Zn, respectively. The PLI value for heavy metals i.e. As, Ni, Pb, Cd and Hg was found to be 8.6 indicating their high contamination in the soil.

4 CONCLUSIONS

The enormous coal burning tends to increase the concentration of elements i.e. As, Zn, Cd, Pb and Hg in the geomedia of the Korba basin. The highest pollution indices for elements i.e. As, Ni and Pb were recorded. The vertical transport of elements i.e. As, Cr, Cu, Ni, Zn, Cd and Hg via the soil media was enhanced with increasing soil depth profile, may be due to less binding with the organic compounds. The whole As exists in the toxic inorganic forms i.e. As(III) and As(V) in the geomedia.

ACKNOWLEDGEMENTS

We are thankful to the UGC, New Delhi for awarding BSR grant to KSP.

REFERENCES

Agrawal, P., Mittal, A., Prakash, R., Kumar, M., Singh, T.B. & Tripathi, S.K. 2010. Assessment of contamination of soil due to heavy metals around coal fired thermal power plants at Singrauli region of India. *Bull. Environ. Contam. Toxicol.* 85(2): 219–223.

Bhuiyan, M.A.H., Parvez, L., Islam, M.A., Dampare, S.B. & Suzuki, S. 2010. Heavy metal pollution of coal mine-affected agricultural soils in the northern part of Bangladesh. *J. Hazard. Mater*. 173(1–3): 384–392.

Chou, C.L. 2012. Sulfur in coals: a review of geochemistry and origins. *Int. J. Coal Geol.* 100(10): 1–13.

Dahal, B.M., Fuerhacker, M., Mentler, A., Karki, K.B., Shrestha, R.R. & Blum, W.E.H. 2008. Arsenic contamination of soils and agricultural plants through irrigation water in Nepal. *Environ. Pollut.* 155(1): 57–163.

Guttikunda, S.K. & Jawahar, P. 2014. Atmospheric emissions and pollution from the coal-fired thermal power plants in India. *Atmos. Environ*. 92: 449–460.

Han, Y.M., Cao, J.J., Chow, J.C., Watson, J.G., An, Z.S. & Liu, S.X. 2009. Elemental carbon in urban soils and road dusts in Xi'an, China and its implication for air pollution. *Atmos. Environ.* 43(15): 2464–2470.

He, Y. & Zhang, G.L. 2009. Historical record of black carbon in urban soils and its environmental implications. *Environ. Pollut.* 157(10): 2684–2688.

Ilwon, K., Joo, S.A., Young, S.P. & Kyoung-Woong, K. 2003. Arsenic contamination of soils and sediments from tailings in the vicinity of Myungbong Au mine, Korea. *Chem. Spec. Bioavailab.* 15(3): 67–74.

Kwon, J.C., Nejad, Z.D. & Jung, M.C. 2017. Arsenic and heavy metals in paddy soil and polished rice contaminated by mining activities in Korea. *Catena* 148: 92–100.

Li, X., Liu, L., Wang, Y., Luo, G., Chen, X., Yang, X., Gao, B. & He, X. 2012. Integrated assessment of heavy metal contamination in sediments from a coastal industrial basin, NE China. *PLOS One* 7(6): e39690.

Muller, G. 1969. Index of geo-accumulation in sediments of the Rhine River. *Geo. J.* 2(3): 108–118.

Muri, G., Cermelj, B., Faganeli, J. & Brancelj, A. 2002. Black carbon in Slovenian alpine lacustrine sediments. *Chemosphere* 46(8): 1225–1234.

Obaidy, A.H.M.J.Al. & Mashhadi, A.A.M.Al. 2013. Heavy metal contaminations in urban soil within Baghdad city, Iraq. *J. Environ. Protec.* 4(1): 72–82.

Pandey, V.C., Singh, J.S., Singh, R.P., Singh, N. & Yunus, M. 2011. Arsenic hazards in coal ?y ash and its fate in Indian scenario. *Resour. Conserv. Recyl.* 55(9): 819–835.

Patel, K.S., Shukla, A., Tripathi, A.N. & Hoffmann, P. 2001. Heavy metal concentrations of precipitation in east Madhya Pradesh of India. *Water Air Soil Pollut.* 130(1–4): 463–468.

Patel, K.S., Yadav, A., Rajhans, K., Ramteke, S., Sharma, R., Wysocka, I. & Jaron, I. 2016. Exposure of fluoride in coal basin. *Int. J. Clean Coal Energy* 5(1): 1–12.

Plyaskina, O.V. & Ladonin, D.V. 2009. Heavy metal pollution of urban soils. *Eura. Soil Sci.* 42(7): 816–823.

Rudnick, R.L. & Gao, S. 2003. Composition of the continental crust. In 'The Crust' (Rudnick, R. L., ed) Treatise on Geochemistry, (Holland, H.D. and Turekian, K.K., eds), Elsevier- Pergamon, Oxford, pp. 1–64.

Sengupta, S., Chatterjee, T., Ghosh, P.B. & Saha, T. 2010. Heavy metal accumulation in agricultural soils around a coal fired thermal power plant (Farakka) in India. *J. Environ. Sci. Eng.* 52(4): 299–306.

Sheela, A.M., Letha, J., Joseph, S. & Thomas, J. 2012. Assessment of heavy metal contamination in coastal lake sediments associated with urbanization: Southern Kerala, India. *Lakes Reserv. Res. Manag.* 17(2): 97–112.

Sheoran, V., Sheoran, A.S. & Tholia, N.K. 2011. Acid mine drainage: an overview of Indian mining industry. *Int. J. Earth Sci. Eng.* 4(6): 1075–1086.

Shrivastava, A., Barla, A., Yadav, H. & Bose, S. 2014. Arsenic contamination in shallow groundwater and agricultural soil of Chakdaha block, West Bengal, India. *Front. Environ. Sci.* 2: 1–9.

Sinex, S.A. & Helz, G.R. 1981. Regional geochemistry of trace elements in Chesapeake Bay sediments. *Environ. Geol.* 3(6): 315–323.

Singh, R., Singh, D.P., Kumar, N., Bhargava, S.K. & Barman, S.C. 2010. Accumulation and translocation of heavy metals in soil and plants from fly ash contaminated area. *J. Environ. Bio.* 31(4): 421–430.

Tan, K.H. 2005. *Soil Sampling, Preparation and Analysis*. 2nd ed., CRC Press, Boca Raton, FL.

Tomlinson, D.L., Wilson, J.G., Harris, C.R. & Jeffrey, D.W. 1980. Problem in assessment of heavy-metal levels in estuaries and the formation of a pollution index. *Helgoländer Meeres.* 33(1–4): 566–575.

Whitmore, T.J., Riedinger-Whitmore, M.A., Smoak, J.M., Kolasa, K.V., Goddard, E.A. & Bindler, R. 2008. Arsenic contamination of lake sediments in Florida: evidence of herbicide mobility from watershed soils. *J. Paleolimnol.* 40(3): 869–884.

Environmental Arsenic in a Changing World – Zhu, Guo, Bhattacharya, Ahmad, Bundschuh & Naidu (Eds)
ISBN 978-1-138-48609-6

Contamination of water, soil and plant with arsenic and heavy metals

A. Yadav[1], K.S. Patel[1], L. Lata[2], H. Milosh[2], P. Li[3], J. Allen[3] & W. Corns[3]
[1]*School of Studies in Chemistry/Environmental Science, Pt. Ravishankar Shukla University, Raipur, India*
[2]*Department of Soil Science/Geology, Maria Curie-Skłodowska University, Lublin, Poland*
[3]*PS Analytical Ltd, Arthur House, Orpington, Kent, UK*

ABSTRACT: The environment of Ambagarh Chowki, Rajnandgaon, central India is contaminated with As at excessive levels. The domestic animals are suffering with the diseases due to intake of the contaminated water and food. In this work, the contamination of the water, soil and plant leaf with As and other heavy metals (HMs) i.e. Cr, Mn, Cu, Zn and Pb in the village Koudikasa are described. The concentration of As in the 45 dried leaves were ranged from 0.3–27 mg kg^{-1} with mean value ($p = 0.05$) of 5.6 ± 1.4 mg kg^{-1}. The toxic inorganic arsenic species are found to be dominated. The bioindicators for the arsenic and other HMs contaminations are highlighted.

1 INTRODUCTION

Arsenic, a carcinogenic environmental and occupational pollutant is a menace to all of us due to human poison (Hong *et al.*, 2014). The contamination of the environments with As and other heavy metals (HMs) in the several regions of the World were reported (Bhattacharya *et al.*, 2010; Chaurasia *et al.*, 2012; Pandey *et al.*, 2002, 2007; Sawidis *et al.*, 2011). However, the As contamination occurred at elevated levels in the environment of Ambagarh tehsil, Rajnandgaon district, Chhattisgarh over 3000 km^{-2} (Pandey *et al.*, 2007; Sawidis *et al.*, 2011). Arsenic in this area is causing serious health hazards in humans by entering through water and food (Pandey *et al.*, 2007). Hence, in this work, the contamination of water, soil, plant leaf and animal stool with As and other HMs i.e. Cr, Mn, Cu, Zn and Pb in the contaminated village is described.

2 METHODS/EXPERIMENTAL

2.1 *Study area and sampling*

The study area is the Koudikasa village, Ambagarh Chowki, Rajnandgaon, CG, India (20.78209°N 80.74117°E) where there are 10,000 humans and domestic animals i.e. cow, buffalo, goat and sheep. The trees and animal stool were reported as bioindicators to know environmental contaminations of the metals (Gupta, 2013; Sawidis *et al.*, 2011). The water samples were collected from 5 locations i.e. tank and canal, hand pump and agricultural field (0–10 cm) of the Kourikasa village in February, 2017 by using established methodologies. The leaves of forty five plants were collected manually. They were washed with deionized water and dried in the sunlight. The solid samples were died, crushed and sieved out particles of <0.1 mm.

2.2 *Analysis*

The physical characteristics (i.e. pH, conductivity and TDS) of the water was measured at the spot. A 0.25 g of the solid sample was digested with acids (3 mL HCl and 1 mL HNO_3) in closed system with P/TMARS CEM (Varian Company) microwave oven. The acid extract was used for monitoring the metals with the GF-AAS and ICP-AES. The speciation of arsenic was quantified by the AFS technique.

3 RESULTS AND DISCUSSION

3.1 *Physicochemical characteristics*

The surface water (i.e. tank, reservoir and canal) and groundwater (GW)were used for drinking purposes by the animals i.e. cow, buffalo, goat, sheep, etc. The concentrations of As, Cr, Mn, Cu, Zn and Pb in the GW ($n = 5$) was ranged from 0.030–0.048, 0.019–0.061, 0.089–0.152, 0.027–0.040, 0.07–0.11 and 0.005–0.015 $\mu g\ L^{-1}$ with men value ($p = 0.05$) of 0.038 ± 0.06, 0.120 ± 0.021, 0.032 ± 0.005, 0.096 ± 0.015 and 0.009 ± 0.003 $\mu g\ L^{-1}$, respectively. The higher concentration of As and other HMs in the GW was observed, may be due to their mineralization from the bed rocks. Several folds higher concentration of As in the water was observed than prescribed value of 10 $\mu g\ L^{-1}$. The higher As concentration in the water of the studied area than other regions of the country was found (Chaurasia *et al.*, 2012).

The concentration of As, Cr, Mn, Cu, Zn and Pb ($n = 5$) in the soils was ranged from 58–302, 45–73, 778–986, 64–105, 76–114 and 17–41 mg kg^{-1} with mean value ($p = 0.05$) of 207 ± 97, 60 ± 10, 893 ± 72, 84 ± 15, 94 ± 13 and 30 ± 8 mg kg^{-1}, respectively. The concentration of As in the soil was several folds higher than the background levels of 5.0 mg kg^{-1}.

The remarkably high As concentration in the soil of the studied area was marked.

The concentrations of the elements i.e. As, Cr, Mn, Cu, Zn and Pb in 45 dried leaves were ranged from 0.3–27, 3.3–25, 18–159, 9.1–62, 29–238 and 0.8–8.0 mg kg^{-1} with mean value ($p = 0.05$) of 5.6 ± 1.4, 11 ± 2, 89 ± 10, 27 ± 4, 86 ± 12 and 3.7 ± 0.6 mg kg^{-1}, respectively. All leaves were contaminated with As at 56 ± 14 times higher than the permissible limit of 0.1 mg kg^{-1} (FAO/WHO, 2011). Similarly, all leaves were loaded with the Zn, Cr and Pb at least 4 ± 1, 6 ± 2 and 17 ± 4 times higher than permissible limit of 20, 2.3 and 0.3 mg kg^{-1}, respectively. The higher concentration of As in the plant leaves of the studied area than values reported in other regions of the country and the world was observed (Bhattacharya *et al.*, 2010; Pandey *et al.*, 2002, 2007; Sawidis *et al.*, 2011).

3.2 *Speciation*

In general, inorganic forms of As are more toxic in the environment than organic forms and, among inorganic forms, arsenite is more toxic than arsenate. In the soil samples, all As was in the inorganic forms: As(III) and As(V). In the stool samples, lower concentrations of the non-toxic, DMA (3.5%) and MMA (1%) were detected.

3.3 *Correlation and sources*

The correlation coefficient (r^2) of the elements i.e. As, Cr, Mn, Cu, Zn and Pb among their mean concentrations in the SW, GW, soil and leaf samples was computed. They had good correlation ($r^2 = >0.91$) among themselves except Zn which showed partial correlations, indicating origin from the similar sources i.e. arsenopyrite leaching, etc.

Some species i.e. *D. bulbifera, A. concinna, T. foenumgraecum, L. purpureus, M. oleifera, T. grandis, H. sabdariffa, A. esculentus and A. marmelos*, are seemed to be as Zn hyper accumulator species as the ratio of $[Zn_{leaf}]/[Zn_{soil}]$ was found > 1.0. Similarly, the highest accumulation of other elements i.e. As, Cr, Mn, Cu and Pb was marked with leaf of plants i.e. *Vigna unguiculata, Diospyros melanoxylon, Mangifera indica, Hibiscus sabdariffa and Zingiber officinale*, respectively which could be assessed as bioindicators for the respective elements.

3.4 *Toxicity*

The concentrations of As, Cr, Mn, Cu, Zn and Pb were ranged from 41–60, 33–50, 212–301, 840–1220, 429–630 and 39–84 mg kg^{-1} with mean value of 51 ± 7, 39 ± 6, 254 ± 34, 1003 ± 120, 529 ± 71 and 63 ± 14 mg kg^{-1}, respectively. The highest concentration of As was marked in the goat stool, may be due to high intake of the green leaves. The elements i.e. As, Cr, Mn, Cu, Zn and Pb were enriched at least 9, 3, 3, 37, 6 and 17-folds higher than the respective leaf value. The higher loading of three metals i.e. Cu, Zn and Pb in the stool samples of the domestic animals than the soil sample was observed, may be due to preconcentration from the contaminated leaves. The skin cancers and fibrosis in the cow and sheep are seen in the studied area.

4 CONCLUSIONS

The whole environment (i.e. water, soil and plant) of Koudikasa village are contaminated with arsenic at the tremendous high level. Arsenic existed mostly in the inorganic As(V) form. The plant leaves i.e. *Vigna unguiculata, Dalbergia melanoxylon, Mangifera indica, Hibiscus sabdariffa, Moringa oleifera* and *Zingiber officinale* were as bio-indicator for assessing pollution of metals i.e. As, Cr, Mn, Cu, Zn and Pb, respectively. The stool samples of the domestic animals were contaminated with As at extremely high levels (>50 mg kg^{-1}). They are enriched from 3–37-folds higher in the stool sample with respect to the leaf mean value. The skin cancers are observed in the animals of the studied area.

ACKNOWLEDGEMENTS

The UGC, New Delhi is greatly acknowledged for awarding the BSR grant to KSP.

REFERENCES

Bhattacharya, P., Samal, A.C., Majumdar, J. & Santra, S.C. 2010. Arsenic contamination in rice, wheat, pulses, and vegetables: a study in an arsenic affected area of west Bengal, India. *Water Air Soil Pollut.* 213(1–4): 3–13.

Chaurasia, N., Mishra, A. & Pandey, S.K. 2012. Finger print of arsenic contaminated water in India – a review. *J. Forensic Res.* 3: 172.

FAO/WHO 2011. Food Standards programme Codex committee on contaminants in foods Food. *CF/5 INF/1*, pp. 1–89.

Gupta, V. 2013. Mammalian feces as bio-indicator of heavy metal contamination in Bikaner zoological garden, Rajasthan, India. *Res. J. Animal, Veterinary and Fishery Sci.* 1(5): 10–15.

Hong, Y.S., Song, K.H. & Chung, J.Y. 2014. Health effects of chronic arsenic exposure. *J. Prev. Med. Public Health.* 47(5): 245–252.

Pandey P.K., Yadav S., Nair S. & Bhui, A. 2002. Arsenic contamination of the environment: a new perspective from central-east India. *Environ. Int.* 28(4): 235–245.

Pandey P.K., Yadav, S. & Pandey, M. 2007. Human arsenic poisoning issues in central-east Indian locations: biomarkers and biochemical monitoring. *Int. J. Environ. Res. Publ. Health* 4(1): 15–22.

Patel K.S., Shrivas K., Brandt R., Jakubowski, N., Corns, W. & Hoffmann, P. 2005. Arsenic contamination in water, soil, sediment and rice of central India. *Environ. Geochem. Health* 27(2): 131–145.

Sawidis, T., Breuste, J., Mitrovic, M., Pavlovic, P. & Tsigaridas, K. 2011. Trees as bioindicator of heavy metal pollution in three European cities. *Environ. Pollut.* 159(12): 3560–3570.

3.6 Assessment of global burden of arsenic in drinking water and health care systems for exposed population

Environmental Arsenic in a Changing World –
Zhu, Guo, Bhattacharya, Ahmad, Bundschuh & Naidu (Eds)
ISBN 978-1-138-48609-6

Pharmacodynamic study of the selenium-mediated arsenic excretion in arsenicosis patients in Bangladesh

M. Alauddin[1], R. Cekovic[1], S. Alauddin[1], L. Bolevic[1], S. Saha[2], J.E. Spallholz[3], P.F. LaPorte[4], S. Ahmed[5], H. Ahsan[6], J. Gailer[7], O. Ponomarenko[8], I.J. Pickering[8], S.P. Singh[8] & G.N. George[8]
[1]*Department of Chemistry, Wagner College, Staten Island, New York, USA*
[2]*BIRDEM Hospital, Dhaka, Bangladesh*
[3]*Division of Nutritional Sciences, Texas Tech University, Lubbock, Texas, USA*
[4]*Division of Hematology-Oncology, University of California Los Angeles, Los Angeles, California, USA*
[5]*Institute of Child and Mother Health, Sk Hospital, Dhaka, Bangladesh*
[6]*Department of Health Sciences, University of Chicago, Chicago, USA*
[7]*Department of Chemistry, University of Calgary, Calgary, Canada*
[8]*Department of Geological Sciences, University of Saskatchewan, Canada*

ABSTRACT: Since 1938 several investigators have confirmed antagonistic interaction whereby the administration of selenite mitigates arsenite toxicity in mammals. More recently a number of animal model and human studies have revealed that the bimolecular basis of this antagonism involves the glutathione mediated formation of major metabolite seleno-bis(S-glutathionyl)arsinium [(GS)2AsSe]- ion which is rapidly excreted through bile. To investigate whether the oral supplementation of sodium selenite will enhance the fecal excretion of arsenites in human, a limited pharmacodynamics study was carried out involving ten arsenicosis patients. However, data from five patients are reported here. Patients were screened for pre-existing hepatic diseases and they were excluded from the study. Subjects were selected based on the arsenic level in their household drinking water, hair, nail samples and diffuse melanosis symptoms. Patients received ^{77}Se-labeled sodium selenite as oral supplement, which they ingested together with their arsenic-containing drinking water. Total selenium levels increased in both urine and feces following the co-administration of sodium selenite (800 μg). The patients receiving a placebo showed no change in their selenium excretion. The increase in selenium levels in feces and arsenic to some extent after dosing agrees with our hypothesis that selenium supplementation promotes co-excretion of arsenic and selenium through formation of [(GS)2AsSe]- in the bile.

1 INTRODUCTION

Since the discovery of unsafe levels of arsenic in groundwater in Bangladesh in 1993, a number of reports have indicated that approximately 70–80 million people are being exposed to this metalloid to concentrations in drinking water that are above the World Health Organization (WHO) permissible level of 10 $\mu g\,L^{-1}$ from shallow aquifers contaminated by geogenic arsenic (Mukherjee & Bhattacharya, 2001; Kinniburgh & Smedley 2001; Ahmed *et al.*, 2006). Chronic exposure of humans to inorganic arsenic leads to arsenicosis with pathological symptoms including skin melanosis and keratosis, endocrine diseases and cancers of the lungs, the liver, the kidneys and the bladder (Kapaj *et al.*, 2006).

1.1 *Antagonism of Arsenic and selenium*

During 1930–1950, several animal model studies indicated that arsenite can detoxify selenium in form of selenite and/or seleniferous wheat in rats, dogs, cattles (Moxon *et al.*, 1938). The therapeutic benefit of selenium for countering the toxicity of inorganic arsenic stems from a number of studies that have been carried out over the past 80 years. The oxy-anions of arsenic and selenium are arsenite and selenite – are toxic to animals if given separately. However, if both are administered together they mutually de-toxify each other. This antagonistic interaction has been linked to the in vivo formation of the seleno-bis(S-glutathionyl)arsinium [(GS)2AsSe]-. Synchrotron-based X-ray absorption (XAS) spectroscopy has recently revealed the formation of the seleno-bis(S-glutathionyl)arsinium anion in hepatocytes of rabbits and its subsequent excretion in the bile (Gailer *et al.*, 2002, 2004). In addition, studies indicate that multidrug resistance protein 2 (MRP2) transports [(GS)2AsSe]- mediates its excretion from the liver to the bile (Carew & Leslie, 2010).

During 2006 to 2009 we have conducted a 48-week, 821 patient, randomized, double-blinded placebo controlled Phase III clinical trial in Bangladesh to investigate the effectiveness of daily oral dose of 200 μg selenite in combating arsenic toxicity. Our study suggested that the employed selenium dose was inadequate to reduce the prevalent condition of melanosis and other arsenicosis symptoms since their arsenic intake through drinking water and food were quite high (Alauddin *et al.*, 2012). However, our recent studies which involved the administration of selenite and arsenite to rats and hamsters conclusively revealed the presence of [(GS)2AsSe]- in the bile (George *et al.*, 2016; Ponomarenko *et al.*, 2017).

1.2 *Objectives of the current study*

A number of our previous studies using rabbits, rats, hamsters and human have demonstrated that the co-administration of selenite will promote the hepatobiliary co-excretion of arsenic and selenium in the form of the [(GS)2AsSe]- ion. We hypothesize that the dietary supplementation of selenium may promote the formation of the arsenic-selenium-glutathione conjugate which is excreted in the bile and therefore increases the excretion of arsenic via feces. The present study is a follow up is a follow up of the previous pharmacodynamics study involving five patients. Our objectives were to monitor the excretion of total arsenic and selenium via feces and urine in five patients that were provided with daily meals and drinking water which they would normally consume in their household. Our aim was to assess the effectiveness of a selenium supplement in the excretion of arsenic from the body of arsenicosis patients.

2 EXPERIMENTAL

2.1 *Study design*

This is a limited study which involved five arsenicosis patients from Laksham, Shahrasty in Bangladesh. The subjects were patients which were recruited following defined selection criteria including a) arsenic exposure based on arsenic in their drinking water source, and b) normal kidney and liver function. The study protocol was explained to all patients and signed consents was obtained from all participants. All participants stayed in a private clinic in Laksham for 10 days where they were given fixed food regimen prepared with the drinking water from their homes. On the sixth day they received either a 800 μg Se-77 labeled sodium selenite dose or a placebo, which was drinking water that did not contain selenite. The stable isotope Se-77 labelled sodium selenite was used as supplement to differentiate native selenium and selenium that was administered to patients. All urine and fecal matter were collected from all participants every day to account for all excretions. Feces were weighed, volumes of urine were measured and all excretions were recorded during the patients 10 day stay in the clinic. Data for 5th thru 8th day are presented in this paper. The urine samples and fecal matters were pooled together separately for each day for each participant. All food samples for breakfast, lunch, dinner were weighed for each participant every day and recorded. A portion of breakfast, lunch, dinner, snacks were homogenized in distilled deionized water and aliquots were preserved in vials for arsenic and selenium analysis. All water, fruit juice, tea intake for each patient were recorded daily. All pooled feces samples were weighed and an weighed portion was homogenized in distilled deionized water and aliquots were preserved at −4°C in a freezer before further analysis in the laboratory. The aliquots of food, water, patient samples such as blood, urine and feces were transported to the USA and stored immediately in a −80°C freezer prior to analysis.

2.2 *Reagents and standards used for the chemical analysis of all biological tissues for total arsenic and selenium*

Homogenized feces samples were oven dried at 95°C for an hour and dry weight was obtained. The dried fecal matter were digested in high purity nitric acid and diluted with deionized water to a fixed volume. Arsenic (As), selenium (Se) in urine and fecal matter were determined by the DRC-ICP-MS technique while the arsenic, selenium in food and water samples were determined by the graphite furnace atomic absorption spectroscopy (GF-AAS) (Perkin Elmer Model AAnalyst 800) with Zeeman background correction and electrodeless discharge lamps (EDL). Matrix modifier for Se analysis were prepared from a solution containing 1% (w/v) $Ni(NO_3)_2.6H_2O$, 2% (w/v) $Mg(NO_3)_2.6H_2O$ and 0.1% (v/v) Triton x-100 while the matrix modifier for As analysis consisted of 1% (w/v) $Ni(NO_3)_2.6H_2O$ and 0.1% Triton X-100. For all matrix modifiers, high purity (99.999%) $Ni(NO_3)_2.6H_2O$, 2% (w/v) $Mg(NO_3)_2.6H_2O$ (Sigma-Aldrich Co., USA) 18 $M\Omega cm^{-1}$ water were used. Stock As, Se standard solutions (Perkin Elmer Co., USA) were used to calibrate AAS. The National Institute of Standards and Technology (NIST) standard reference materials (SRM 1640), served as control reference material for As, Se analysis in water and food samples. The method detection limit for As, Se analysis in water and urine was 1.0 $\mu g\,L^{-1}$, with an RSD of 2.5%.

3 RESULTS

3.1 *Fecal and urinary Se, As data*

Arsenic and selenium in fecal matter from all (5) patients collected before Se dosing and after Se dosing are shown in Table 1. Patients 1, 2 and 3 received selenium supplements and patients 4 and 5 received placebo (drinking water without selenium dose) on day 6. Here we report data for day 5 thru day 8 for brief observation for the fate of selenium administered to patients.

Table 1. Daily As, Se intake and pooled fecal and urinary excretion (μg) by patients .

	Day	In As	Ex As	In Se78	Ex Se78	In Se77	Ex Se77
Pat1	5	2026	1825	39.4	53.9	0	74.6
	6	1948	1688	24.2	49.8	800	54.6
	7	2161	1221	30.7	50.1	0	9985
	8	1803	1224	19.7	46.6	0	4132
Pat2	5	1400	374	39.6	21	0	23.3
	6	1146	602	23.9	30.5	800	31.9
	7	1488	497	38.7	31.3	0	4352
	8	1401	877	19.4	69.4	0	7328
Pat3	5	1846	694	39.4	27.7	0	31.9
	6	1906	1237	26.5	38.9	800	44.6
	7	1771	391	31.9	17.2	0	4318
	8	1835	732	19.7	16.6	0	1995
Pat4	5	1804	2435	59.9	52.8	0	56.6
	6	1774	1853	35.9	52.4	0	57.6
	7	1767	1199	64.2	32.8	0	63.1
	8	1801	1958	23.8	77.9	0	88.5
Pat5	5	1344	646	38.9	25.6	0	26.6
	6	1210	852	23.4	29.4	0	33.7
	7	1416	632	37.2	25.8	0	28.7
	8	1634	867	19.4	33.9	0	46.2

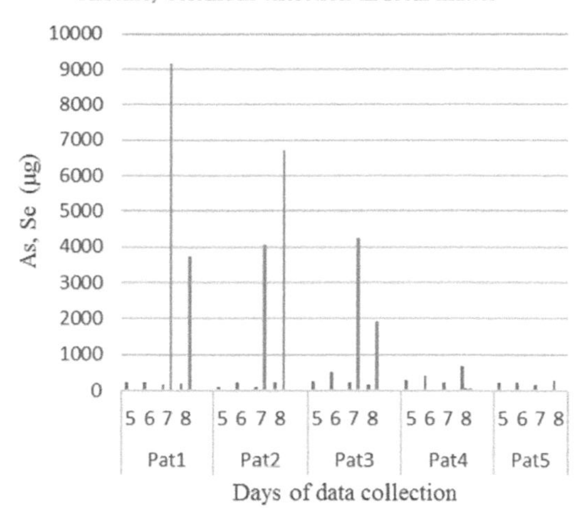

Figure 2. Fecal excretion of total As and Se by patients.

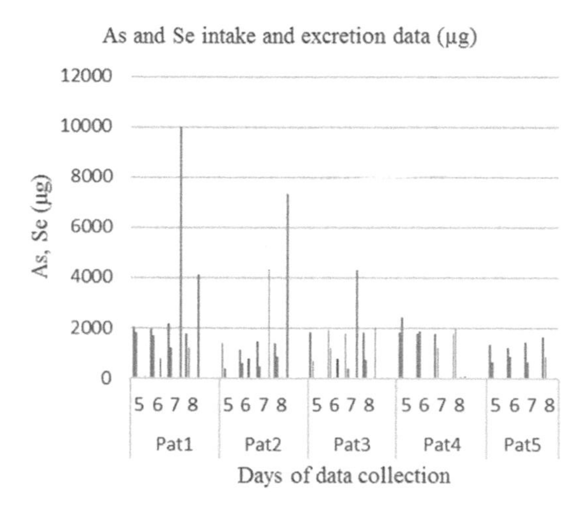

Figure 1. As, Se intake and excretion (μg) for patients 1–5.

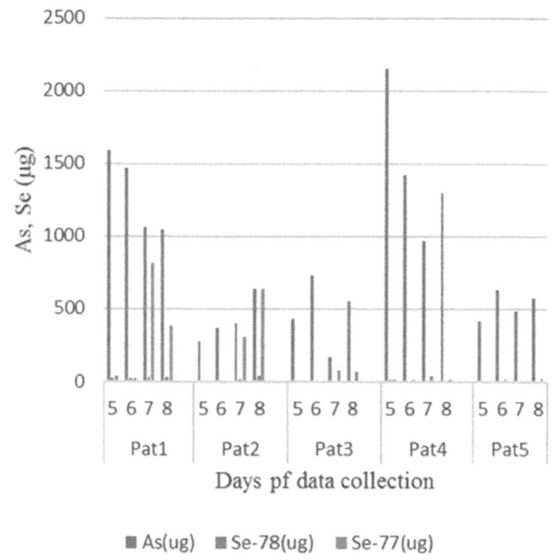

Figure 3. Urinary excretion of total As and Se by patients.

3.2 *Intake and excretion of As, Se*

In Table 1 total As, Se intake by a patient are calculated from the As, Se levels in their daily food intake and their drinking water consumed by each patient. The total excretion of As, Se have been determined from the As, Se levels in urine and fecal matter and the pooled amount of urine and fecal matter for each patient for each day. Patients 1–3 received Se supplementation on day 6. Patients 4 and 5 did not receive selenium supplement.

The As and Se in fecal excretion are high, especially in fecal matter for the supplement group (Table 1) as expected. Data for only 2 pre-dose days and 2 post-dose days have been reported here for a brief overview of the pharmacodynamics study. Detailed outcome involving all ten patients and analysis of patient samples from all 10 days will be reported later. The present data indicate that (a) selenium increases in fecal matter in supplement group on the 7th and 8th days at the appropriate interval 6–12 hours after Se ingestion and (b) in every case fecal arsenic excretion increases after Se ingestion. The total amount of Se-77 excretion is much higher in all three patients in the supplement group than the intake amount. The urinary As, Se excretion increased for patient 2, 3 in the supplement group. High urinary As excretion in patient 4 might be indicative of other physiologic condition for this patient.

4 CONCLUSIONS

The urine, fecal matter samples from the remaining 5 patients in the same batch are being analyzed and the data will be reported later in a separate paper. The current data obtained from a limited number of patients are indicative of potential pathway for removal of body burden arsenic by Se supplementation. Currently a pharmacodynamics study involving a larger number of patients is under way by our group.

CLINICAL TRIAL REGISTRY

Clinical Trial.gov Identifier: NCT02377635.

ACKNOWLEDGEMENTS

The study has been supported by grant from Grand Challenges Canada to the University of Saskatchewan. The authors thank R. Gerads from the Brooks Applied Laboratory, Bothell, Washington, USA for Se measurements in samples through ICP-MS.

REFERENCES

Ahmed, M.F., Ahuja, S., Alauddin, M., Hug, S.J., Lloyd, J.R., Pfaff, A., Pichler, T., Saltikov, C., Stute, M. & van Geen, A. 2006. Epidemiology: ensuring safe drinking water in Bangladesh. *Science* 314: 1687–1688.

Alauddin, M., Wheaton, T., Valencia, M., Stekolchik, E., Spallholz, J.E., La Porte, P.F., Ahmed, S., Chakaraborty, B., Bhattacharjee, M., Zakaria, A.B.M., Sultana, S., George, G.N., Pickering, I.J. & Gailer, J. 2012. Clinical trial involving selenium supplementation to counter arsenic toxicity among rural population in Bangladesh In: J.C. Ng, B.N. Noller, R. Naidu, J. Bundschuh & P. Bhattacharya (Eds.) *Understanding the Geological and Medical Interface of Arsenic As 2012:* Proceedings of the 4th International Congress on the Arsenic in the Environment. Taylor and Francis Group, CRC Press, London, UK. pp. 143–147.

Carew, M.W. & Leslie, E.M. 2010. Selenium-dependent and independent transport of arsenic by human multidrug resistance protein 2 (MRP2/ABCC2): implications for the mutual detoxification of arsenic and selenium. *Carcinogenesis* 31: 1450–1455.

Gailer, J., George, G.N., Pickering, I.J., Prince, R.C., Younis, H.S. & Winzerling, J.J. 2002. Biliary excretion of $[(GS)_2AsSe]^-$ after intravenous injection of of rabbits with arsenite and selenite. *Chem. Res. Toxicol.* 15: 1466–1471.

Gailer, J., Ruprecht, I., Reitmeir, P., Benker, B. & Schramel, P. 2004. Mobilization of exogenous and endogenous selenium to bile after the intravenous administration of environmentally relevant doses of arsenite to rabbits. *Appl. Organomet. Chem.* 18: 670–675.

George, G.N., Gailer, J., Ponomarenko, O., La Porte, P.F., Strait, K., Alauddin, M., Ahsan, H., Ahmed, S., Spallholz, J.E. & Pickering, I.J. 2016. Observation of the selenobis-(S-glutathionyl) arsinium anion in rat bile. *Inorg. Biochem* 158: 24–29.

Kapaj, S., Peterson, H., Liber, K. & Bhattacharya, P. 2006. Human health effects from chronic arsenic poisoning-a review. *J. Environ. Sci. Health Part A* 41(10): 2399–2428.

Kinniburgh, D.G. & Smedley, P.I. (Eds) 2001. Arsenic contamination of groundwater in Bangladesh, Final Report (BGS Technical Report WC/00/19, British Geological Survey, Keyworth, UK.

Mukherjee, A.B. & Bhattacharya, P. 2001. Arsenic in groundwater in the Bengal delta Plain, slow poisoning in Bangladesh. *Environ. Res.* 9(3): 189–220.

Moxon, A.L. 1938. The effect of arsenic on the toxicity of seleneferous grains. *Science* 88, 81.

Ponomarenko, O., La Porte, P.F., Singh, S.P., Langan, G., Fleming, D.E.B., Spallholz, J.E., Alauddin, M., Ahsan, H., Ahmed, S., Gailer, J., George, G.N. & Pickering, I.J. 2017. Selenium-mediated arsenic excretion in mammals: a synchrotron-based study of whole-body distribution and tissue-specific chemistry. *Metallomics* 9: 1585–1595.

Environmental Arsenic in a Changing World –
Zhu, Guo, Bhattacharya, Ahmad, Bundschuh & Naidu (Eds)
ISBN 978-1-138-48609-6

Exposure levels to various arsenic species and their associated factors in Korean adults

J.D. Park[1], I.G. Kang[1], S.G. Lee[1], B.S. Choi[1], H. Kim[2] & H.J. Kwon[3]
[1]*College of Medicine, Chung-Ang University, Seoul, South Korea*
[2]*Chungbuk National University, Cheongju, South Korea*
[3]*Dankook University, Cheongju, South Korea*

ABSTRACT: Arsenic (As) has been known a human carcinogen, which is widely distributed in the environment. This study was performed to evaluate the exposure level to arsenic and to determine contributing factors to human exposure of arsenic in Korean adults. Diet intake and urine sample were obtained from 2,044 study subjects older than 19 years of age. Human exposure to the arsenic was assessed by using the concentrations of various arsenic species such as inorganic As (As(V), As(III)), MMA, DMA, arsenobetaine in urine. The concentration of arsenic species in urine was analyzed by HPLC-ICP-MS method. In the arsenic speciation analysis, the geometric mean concentrations were 41.36 $\mu g\ L^{-1}$ of arsenobetaine, 24.72 $\mu g\ L^{-1}$ of DMA, 1.42 $\mu g\ L^{-1}$ of MMA, 2.96 $\mu g\ L^{-1}$ of inorganic As. The levels of various arsenic species in urine were influenced generally by age, smoking, inhabitant area and seafood intake. The consumption of seafood was positively correlated with some of arsenic species such as inorganic As, DMA and arsenobetaine. These findings suggest that the seafood might be a one of major source to the arsenic exposure in Korean adults, and overconsumption of seafood could increase human exposure to inorganic arsenic as well as organic.

1 INTRODUCTION

Arsenic (As) has been known one of major environmental pollutants which is ubiquitously distributed in the environment. Arsenic occurred from anthropogenic source such as mining, industry, agricultural uses as well as from naturally in soil, rock and volcanic eruption (ATSDR, 2007). There are several forms of arsenic. The toxicities of arsenic are quite different from their chemical form and valence, inorganic is much toxic than organic and trivalent is more toxic compared to pentavalent. Inorganic arsenic, arsenate (As(V)) and arsenite (As(III)), are metabolized in the body to the much less toxic metabolites, monomethylarsonic acid (MMA) and dimethylarsinic acid (DMA), through reduction and methylation process, however, which are more toxic than organoarsenics such as arsenobetaine (AsB), arsenocholine (AC) and arsenosugars (AS). Organic arsenics are rapidly excreted in urine without metabolism in the body, which are essentially non-toxic. Human exposure to arsenic can be estimated from based on arsenic levels in blood, hair, nails, and urine (Choi *et al.*, 2010; Meliker *et al.*, 2006). Arsenic concentration in urine is being used as the most valuable biomarker which reflects the exposure to arsenic within the last several days (Navas-Acien *et al.*, 2011; Saoudi *et al.*, 2012) . However, several forms of arsenic such as organoarsenic as well as inorganic arsenic and their metabolites, are excreted in urine. So, it is necessary to determine various arsenic species respectively for the risk assessment in human arsenic exposure. Several studies reported that seafood consumption increased DMA excretion in urine and suggested that kind of organoarsenic in seafood could be metabolized to DMA in the body. Accordingly, it is questionable yet whether overconsumption of seafood could be hazardous to human health or not. Seafood is one of favorable food in Korean. However, there are very limited data for arsenic species analyses in the urine of Korean.

In this study, we performed speciation analysis of arsenic in urine and analyzed the relations between seafood consumption and various arsenic species in urine, which will be useful to do the risk assessment for arsenic in human exposure.

2 MATERIAL AND METHODS

Total of 2,044 study subjects, 888 males and 1,156 females were recruited, who were 19 years old or older. They had not been exposed to As occupationally. We sampled study subjects by probability sampling methods stratified by sex and age from the 102 sampling sites in Korea. Written informed consent was obtained from all study subjects participated in this study. We had conducted 1:1 interviews to investigate demographic characteristics and diet information during the last 24 h. Diet study was performed by the 24-h recall method. Also, we asked whether seafood intake within 72 hours before this study. Then, urine samples were collected and stored at −70°C until As analysis. The speciation analysis of As, such as As(V), As(III), MMA, DMA, arsenobetaine in urine was performed by HPLC-ICP-MS (HPLC pump PerkinElmer Series 200, PerkinElmer NEXION 300S). The levels of various arsenic in urine were presented as arithmetic mean, geometric mean, median and value at 95 percentiles. The concentrations of arsenic in urine were distributed log-normally rather than normal distribution, which were log-transformed for the statistical analyses. And statistical analyses were performed with SAS version 9.2 (SAS institute Inc., Cary, NC, USA). The level of statistical significance was set at $p < 0.05$.

3 RESULTS AND DISCUSSION

The concentrations of geometric mean of various arsenic species in urine were 2.96 μg L^{-1} of InAs (As(III) + As(V)), 1.42 μg L^{-1} of MMA, 24.72 μg L^{-1} of DMA, 41.36 μg L^{-1} of AsB, and also were 30.90 μg L^{-1} of TmetAs (As(III) + As(V) + MMA + DMA) and 84.67 μg L^{-1} of TsumAs (As(III) + As(V) + MMA + DMA + AB) in Korean adults (Table 1).

Urinary As level in Korean adult is similar or less than Taiwan, Japan, China, while is higher compare to United State, France, Germany and United Kingdom (Fig. 1).

In general, contributing factors to the concentrations of arsenic species in urine were determined as age, coastal area inhabitant, and seafood consumption (Table 2).

Table 1. Mean concentrations of various As species in urine (μg L^{-1}).

		InAs[a]	MMA[b]	DMA[c]	AsB[d]	TmetAs[e]	TsumAs[f]
Total	**AM ±**	6.02 ±	2.22 ±	34.11 ±	98.15 ±	42.35 ±	140.49 ±
(n =	**SD**	13.75	2.12	39.35	208.37	50.99	221.62
2044)	**GM**	2.96	1.42	24.72	41.36	30.90	84.67
	(GSD)	(3.33)	(3.47)	(2.25)	(4.20)	(2.18)	(2.63)
	Median	2.95	1.73	25.29	43.82	31.88	82.85
	P95	19.1	5.62	83.11	337.96	100.85	407.07
	PTsum	4.20%	1.60%	24.30%	70.00%	30.00%	100.00%

*a: As(III) + As(V), b: monomethylarsonic acid, c: dimethylarsinic acid, d: arsenobetaine, e: As(III) + As(V) + MMA + DMA, f: As(III) + As(V) + MMA + DMA + AsB, PTsum: percentage in relation to Tsum, AM: arithmetic mean, SD: standard deviation, GM: geometric mean, GSD: geometric standard deviation, P95: value at 95 percentile. * TmetAs: As(III) + As(V) + MMA + DMA, TsumAs: As(III) + As(V) + MMA + DMA + AsB.

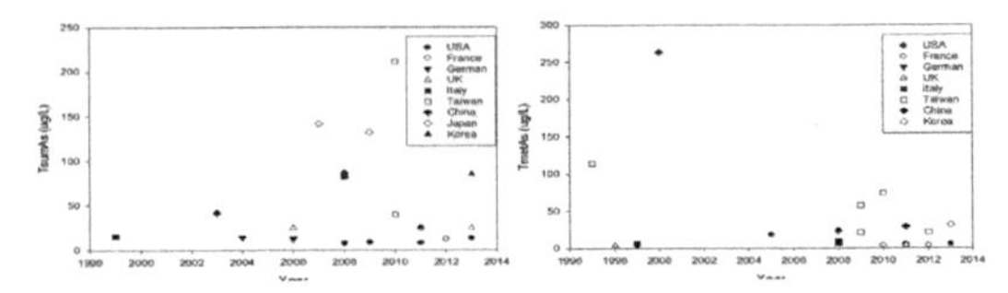

Figure 1. International comparisons of arsenic concentration in urine (μg/L) (TmetAs, TsumAs).

No relations were observed between total food consumption and arsenic levels in urine. However, the consumption of fish/shellfish was positively correlated with InAs, DMA, arsenobetaine, TmetAs and TsumAs. The consumption of seaweeds was also correlated positively with InAs, DMA, TmetAs and TsumAs, but not arsenobetaine.

4 CONCLUSIONS

Urinary As level in Korean adult is similar or less than Taiwan, Japan, China, while is higher compare to United State, France, Germany and United Kingdom. Our data suggests that seafood was primary cause of arsenic exposure among Korean adults, and overconsumption of seafood could increase human exposure level to the inorganic As as well as organic As.

ACKNOWLEDGEMENTS

This study was supported by a Grant (14162MFDS654) from the Ministry of Food and Drug Safety in 2014.

REFERENCES

Choi, B.S., Choi, S.J., Kim, D.W., Huang, M., Kim, N.Y., Park, K.S., Kim, C.Y., Lee, H.M., Yum, Y.N., Han, E.S., Kang, T.S., Yu, I.J. & Park, J.D. 2010. Effects of repeated seafood consumption on urinary excretion of arsenic species by volunteers. *Arch. Environ. Contam. Toxicol.* 58(1): 222–229.

Meliker, J.R., Franzblau, A., Slotnick, M.J. & Nriagu, J.O. 2006. Major contributors to inorganic arsenic intake in southeastern Michigan. *Int. J. Hyg. Environ. Health* 209(5): 399–411.

Navas-Acien, A., Francesconi, K.A., Silbergeld, E.K. & Guallar, E. 2011. Seafood intake and urine concentrations of total arsenic, dimethylarsinate and arsenobetaine in the US population. *Environ. Res.* 111(1): 110–118.

Saoudi, A., Zeghnoun, A., Bidondo, M.L., Garnier, R., Cirimele, V., Persoons, R. & Fréry, N. 2012. Urinary arsenic levels in the French adult population: the French National Nutrition and Health Study, 2006–2007. *Sci. Total Environ.* 433: 206–215.

Table 2. Mean concentrations of various As species in urine by demographic characteristics (μg/L).

		N	InAs[a]	MMA[b]	DMA[c]	AsB[d]	TmetAs[e]	TsumAs[f]
Gender	**Male**	888	2.97 (3.19)	1.50 (3.59)	24.24 (2.21)	45.55 (3.87)	30.37 (2.16)	87.68 (2.59)
	Female	1156	2.95 (3.45)	1.36 (3.38)	25.10 (2.28)	38.40 (4.45)	31.31 (2.21)	82.4 (2.66)
Age (yrs)	**19–29**	355	2.76 (3.25)	1.45 (3.10)	20.61 (2.39)	29.10 (4.84)	26.56 (2.26)	66.42 (2.74)
	30–39	360	3.10 (3.07)	1.31 (3.78)	23.72 (2.34)	37.83 (4.03)	30.09 (2.17)	78.67 (2.57)
	40–49	450	3.20 (3.13)	1.55 (3.25)	26.82 (2.17)	47.08 (3.99)	33.19 (2.14)	92.89 (2.56)
	50–59	475	2.78 (3.33)	1.48 (3.28)	26.22 (2.08)	46.65 (4.01)	32.29 (2.07)	92.17 (2.61)
	60–	404	2.97 (3.88)	1.30 (4.02)	25.64 (2.27)	45.83 (4.07)	31.68 (2.29)	91.35 (2.62)
Inhabitant area	**Inland**	1541	2.77 (3.31)	1.46 (3.34)	23.88 (2.21)	35.24 (4.13)	29.86 (2.13)	76.13 (2.50)
	Costal	503	3.63(3.33)	1.30 (3.88)	27.49 (2.36)	67.54 (3.97)	34.30(2.33)	117.30 (2.86)
Seafood intake†	**No**	734	2.61 (3.40)	1.34 (3.83)	20.97 (2.28)	29.18(4.46)	26.58 (2.20)	65.62 (2.60)
	Yes	1295	3.17 (3.26)	1.46 (3.29)	27.15 (2.20)	50.71 (3.90)	33.65 (2.15)	98.12 (2.57)

†: seafood intake during the last 3 days before this study. Data are presented as geometric mean and geometric standard deviation, a: As(III) + As(V), b: monomethylarsonic acid, c: dimethylarsinic acid, d: arsenobetaine, e: As(III) + As(V) + MMA + DMA, f: As(III) + As(V) + MMA + DMA + AsB.

Environmental Arsenic in a Changing World –
Zhu, Guo, Bhattacharya, Ahmad, Bundschuh & Naidu (Eds)
ISBN 978-1-138-48609-6

A medical geology perspective of arsenic as a poison and medicinal agents

J.A. Centeno
Division of Biology, Chemistry and Materials Science, US Food and Drug Administration, Washington DC, USA

ABSTRACT: The public health concerns for environmental exposure to arsenic has been recognized for decades. However, recent human activities have resulted in eve greater arsenic exposures and the potential increase for chronic arsenic poisoning on a worldwide basis.

1 INTRODUCTION

The public health concerns for environmental exposure to arsenic has been recognized for decades. However, recent human activities have resulted in eve greater arsenic exposures and the potential increase for chronic arsenic poisoning on a worldwide basis. The natural sources of arsenic exposure vary from burning of arsenic-rich coal (China) and mining activities (Malaysia, Japan) to the ingestion of arsenic-contaminated drinking water (Taiwan, The Philippines, Mexico, Chile). The groundwater arsenic contamination in Bangladesh and the West Bengal Delta of India has received the greatest international attention because of the large number of people potentially exposed and the high prevalence of arsenic-induced disease (Selinus *et al.*, 2013).

2 MEDICAL GEOLOGY – THE SCIENCE THAT DEALS WITH THE IMPACTS OF NATURAL GEOLOGIC MATERIALS

Medical geology – the science that deals with the impacts of natural geologic materials and processes on animal and human health – is aimed at increasing the interactions between geoscientists, environmental and biomedical communities, and by stimulating increased research collaboration among these disciplines (Centeno, 2008; Centeno *et al.*, 2016; Smith *et al.*, 2010; Zikovsky & Chah, 1990) Medical geologists are a group of scientists that are primarily interested in outbreaks of disease in which the characteristics of the local geological constituents contribute to the occurrence of various disease states. For the most part, diseases of interest have often included the effects of deficiency or toxicity of a variety of metallic and non-metallic elements on various systemic organs.

3 MEDICAL GEOLOGY – THE SCIENCE THAT COLLABORATES ON A WIDE RANGE OF ENVIRONMENTAL HEALTH PROBLEMS SEEKING CAUSES AND SOLUTIONS LS

Medical geologists are scientists (geoscientists, biomedical/public health scientists, chemists, toxicologists, epidemiologists, hydrologists, geographers, etc.) who generally collaborate on a wide range of environmental health problems seeking causes and solutions. Among these problems are the health impacts of geogenic (natural) dusts, naturally occurring elements in surface water, groundwater and soil, geologic processes such as volcanoes, erosions, earthquakes, tsunamis, etc., occupational exposure to natural materials and natural radiation, and long-term effects of exposure to oncogenic elements, the most prominent example being arsenic.

Medical geologists try to determine the sources, transport and fate of potentially harmful trace elements such as arsenic, fluorine, selenium, copper and other metals. They try to determine the pathways of exposure and produce maps that illustrate local, regional and/or global geologic and geochemical factors and their relationship to existing or potential health problems. A good example of collaborative research on medical geology is the arsenic issue in Bangladesh and West Bengal, India. In this region, medical geologists are working together to determine the source of the high arsenic levels in well water that put at risk the health of thousands of people in this region.

This presentation will focus on providing a medical geology perspective on arsenic as an environmental and medicinal agent. The discussion explores problems inferring risk and disease causation from natural exposures to arsenic, particularly for chronic outcomes, and will argue for the importance of the ecological perspective in assessing pathogenesis. Additionally, the potential beneficial aspects of arsenic as a medicinal agent will be explored.

REFERENCES

Centeno, J.A. 2008. Foreword: 10th anniversary review: natural disasters and their long-term impacts on the health of communities. *J. Environ. Monit.* 10(2): 166–166.

Centeno, J.A., Finkelman, R.B. & Selinus, O. 2016. Medical geology: impacts of the natural environment on public health. *Geosciences* 6(1): 8.

Selinus, O., Alloway, B.J., Centeno, J.A., Finkelman, R.B., Fuge, R., Lindh, U. & Smedley, P. 2013. *Essentials of medical geology*. Springer, New York.

Smith, A.H., Lingas, E.O. & Rahman, M. 2010. Contamination of drinking-water by arsenic in Bangladesh: a public health emergency. *Bull. World Health Organ.* 78(9): 1093–1103.

Zikovsky, L. & Chah, B. 1990. The lognormal distribution of radon concentration in groundwater. *Groundwater* 28: 673–676.

Section 4: Technologies for arsenic immobilization and clean water blueprints

4.1 Adsorption and co-precipitation for arsenic removal

Environmental Arsenic in a Changing World –
Zhu, Guo, Bhattacharya, Ahmad, Bundschuh & Naidu (Eds)
ISBN 978-1-138-48609-6

Interaction of arsenic with co-precipitated Fe(II,III) (hydr)oxides

C.M. van Genuchten[1,2], T. Behrends[1], P. Kraal[1], S.L.S. Stipp[2] & K. Dideriksen[2]
[1]*Department of Earth Sciences-Geochemistry, Faculty of Geosciences, Utrecht University, Utrecht, The Netherlands*
[2]*Nano-Science Center, Department of Chemistry, University of Copenhagen, Copenhagen, Denmark*

ABSTRACT: Iron electrocoagulation (EC) is a low-cost arsenic (As) treatment method that is able to produce Fe(II,III) phases that bind As effectively, such as green rust (GR) and magnetite (Mag). We compared the As removal performance of coprecipitated GR, Mag and lepidocrocite (Lp) produced by Fe(0) EC. The As(III) removal efficiency for these phases was >60%, increasing in order of GR < Mag ≤ Lp, and all three Fe phases removed >95% of the initial As(V) (0.5 to 11 $mg\,L^{-1}$). Spectroscopic data of EC samples revealed that As was able to substitute for tetrahedral Fe sites in Mag, associating with multiple Fe atoms, while As formed binuclear corner sharing geometries on GR and Lp surfaces. The formation of multinuclear As complexes during Mag production by Fe(0) EC can explain the low fraction of As remobilized from the solids during extractions with 2.5 mM PO_4^{3-} or 0.01 M NaOH. Taken together, our results show that Mag is the optimum phase for As removal in Fe(0) EC field treatment.

1 INTRODUCTION

Chronic exposure to arsenic (As) in groundwater used for drinking threatens the lives of millions of people worldwide. Iron electrocoagulation (Fe(0) EC) is a low cost As removal method based on the production of Fe(II) by Fe(0) electrolysis, followed by (partial) Fe(II) oxidation to form Fe precipitates that bind As. The mixed valent Fe(II,III) (hydr)oxide phases, green rust (GR) and magnetite (Mag), can be produced easily and at low cost by Fe(0) EC. The formation of these minerals can improve the performance of Fe(0) EC systems by improving particle separation and by binding As in highly stable multi-nuclear complexes (Wang et al., 2014). However, the As redox and sorption behavior during Fe(II,III) (hydr)oxide formation by Fe(0) EC must be understood before promoting these minerals in future EC systems.

In this work, we combined wet chemical measurements with X-ray absorption spectroscopy to determine the As removal efficiency and As speciation during GR, Mag and lepidocrocite (Lp) formation. These results can be used to optimize the design and operation of Fe(0) EC systems for As removal.

2 METHODS

2.1 *Arsenic removal by Fe(0) EC*

For the As removal experiments, we generated GR, Mag and Lp by EC following our previous approaches (Wang et al., 2014). Briefly, GR and Mag were produced by adding 3 mM Fe(II) by Fe(0) EC at 300 μM Fe(II) min^{-1} to solutions open to the atmosphere that contained 10 mM dissolved inorganic carbon (to form GR) or NaCl (to form Mag) at pH 7.5 and 9. Lepidocrocite was generated in a NaCl electrolyte by adding 3 mM Fe(II) by EC at 30 μM Fe(II) min^{-1}. In each experiment, the electrolyte solution initially contained either As(III) or As(V) at concentrations ranging from 0.5 to 11 $mg\,L^{-1}$. Immediately following the electrolysis stage, an aliquot of the suspension was passed through 0.2 μm filters and acidified with HNO_3. Arsenic in the filtrate was measured by GF-AAS (Perkin Elmer AAnalyst 800).

2.2 *X-ray absorption spectroscopy*

Samples for As K-edge X-ray absorption spectroscopy (XAS) were prepared by filtering particle suspensions in an anaerobic chamber and loading the solids into custom sample holders. Spectra were collected at 77°K in fluorescence mode at beam line BM-26A (DUBBLE) at ESRF (Grenoble, FR) and beam line 4-1 at SSRL (Menlo Park, USA). The As oxidation state in EC samples was quantified by linear combination fits using As reference minerals. The As uptake mode was determined with shell-by-shell fits of the Fourier transformed EXAFS spectra.

2.3 *Extraction experiments*

Extraction experiments were performed by filtering As(V)-bearing suspensions of GR, Mag and Lp and adding the solids to various solutions: 0.1 M NaCl, 0.01 M HCl (pH 2), 2.5 mM PO_4^{3-} and 0.01 M NaOH (pH 12). Aliquots of each extraction were filtered after 1 h and 1 d. The mobilization of As from the EC solids was measured by GF-AAS.

3 RESULTS AND DISCUSSION

3.1 *Arsenic removal*

The formation of Mag and Lp (3 mM Fe) resulted in >90% removal of initial As(III) ranging from 0.5 to 11 mg L^{-1} at pH 7.5 and 9 (Fig. 1). By contrast, GR was less effective at removing As(III), with ≈82% removal of 0.5 mg L^{-1} As(III) at pH 7.5 and 9. While As(III) removal varied with Fe(II,III) mineral and pH, As(V) removal was >95% for GR, Mag and Lp regardless of pH.

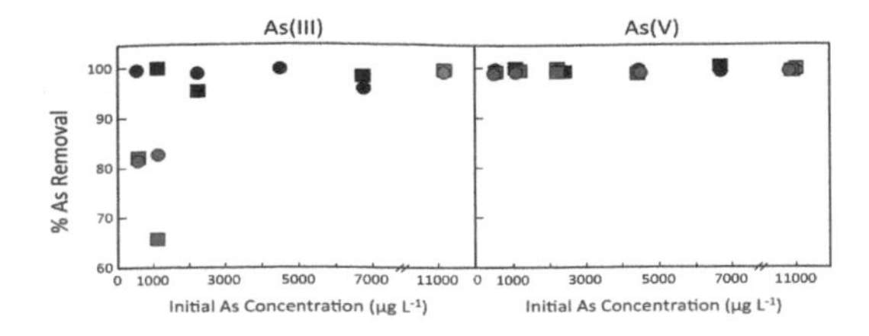

Figure 1. Removal of As(III) and As(V) during the formation of GR (green symbols), Mag (black symbols) and Lp (orange symbols) by Fe(0) EC. The squares and circles represent experiments at pH 7.5 and 9.

3.2 *As K-edge XANES and EXAFS spectroscopy*

Comparing the As K-edge XANES spectra across EC samples (Fig. 2) shows that the extent of As(III) oxidation increases during GR and Mag coprecipitation when pH increases from 7.5 to 9. In addition, decreasing the rate of Fe(II) production, which produced Fe(III) phases instead of Fe(II,III) (hydr)oxides, increased As(III) oxidation significantly. For example, at pH 7.5 and an As/Fe ratio of 5 mol%, producing GR (300 μM Fe(II) min^{-1}) oxidized only 17% As(III), whereas generating Lp (30 μM Fe(II) min^{-1}) oxidized 60% As(III). No evidence for As(V) reduction during coprecipitation with GR and Mag (data not shown), nor during adsorption to presynthesized GR and Mag (Fig. 2A) was observed. Although we found little difference between the extent of As(III) oxidation when generating GR and Mag, the EC Mag samples showed distinct post edge features (arrows in Fig. 2B) that were absent in the XANES spectra for all other EC samples and adsorption references, which indicates a unique As coordination environment.

Consistent with the XANES spectra, the EXAFS spectra and corresponding Fourier transforms (data not shown) of the coprecipitated EC Mag samples were unique among the data set. The second shell peak of the Fourier transformed EC Mag samples had considerably larger amplitude relative to those of the adsorption standards and GR and Lp EC samples. The second shell peak amplitude and position of the EC Mag samples matched that of a synthetic Mag reference with incorporated As(V), which suggests at least partial incorporation of As into EC-generated Mag. Shell-by-shell fits of the EC Mag samples confirmed the existence of multinuclear As coordination geometries, with fit-derived As-Fe coordination numbers ($N_{As-Fe1} = 5.3$, $N_{As-Fe2} = 3.1$) and interatomic distances ($R_{As-Fe1} = 3.44$ Å, $R_{As-Fe2} = 3.65$ Å) consistent with partial As substitution for tetrahedral Fe in Mag.

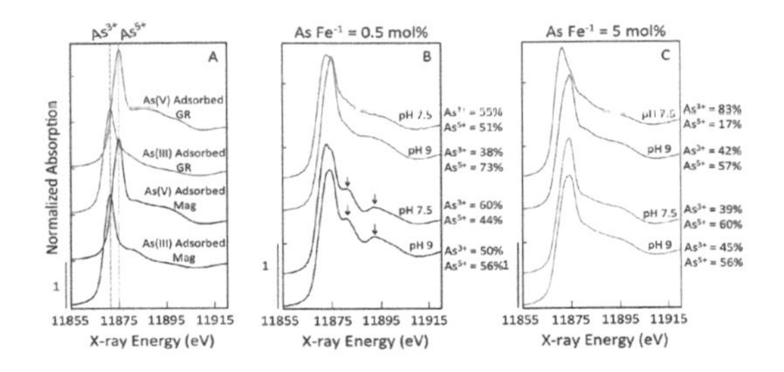

Figure 2. As K-edge XANES spectra of (A) adsorption references and (B-C) EC samples containing As(III) concentrations of 1.1 and 11 mg L^{-1} (As/Fe = 0.5 to 5 mol%). GR samples are shown in green, Mag in black and Lp in orange.

3.3 *Arsenic remobilization*

No As(V) was mobilized from any EC generated sample in the 0.1 M NaCl and pH 2 extractions, but different As(V) fractions were extracted from the PO_4^{3-} and pH 12 solutions, depending on the Fe phase (data not shown). We found that 2.5 mM PO_4^{3-} mobilized 22 to 25% of As(V) bound to GR and Lp, while only 12% was extracted from Mag. Furthermore, only 20% of As(V) initially bound to Mag was extracted at pH 12 compared to nearly 50% for Lp. The lower fraction of As(V) mobilized from Mag relative to GR and Lp in the PO_4^{3-} and pH 12 extractions can be explained by the partial incorporation of As(V) into EC generated Mag, which was detected by EXAFS spectroscopy.

4 CONCLUSIONS

Combining the As(III) and As(V) removal experiments with the EXAFS results and chemical extractions suggests that As treatment performance in the field can be optimized by generating Mag rather than GR or Lp. The benefits of producing Mag are further supported by the ease of separating Mag particle suspensions using low strength magnetic fields.

ACKNOWLEDGEMENTS

We acknowledge Dipanjan Banerjee (ESRF) and Ryan Davis (SSRL) for assistance during XAS data collection. CMvG acknowledges funding support by a NWO Veni grant (Project No. 14400).

REFERENCE

Wang, Y., Morin, G., Ona-Nguema, G. & Brown, G.E. 2014. Arsenic(III) and arsenic(V) speciation during transformation of lepidocrocite to magnetite. *Environ. Sci. Technol.* 48(24): 14282–14290.

Environmental Arsenic in a Changing World –
Zhu, Guo, Bhattacharya, Ahmad, Bundschuh & Naidu (Eds)
ISBN 978-1-138-48609-6

Adsorptive removal of arsenic by calcined Mg-Fe-(CO_3) LDH: An artificial neural network model

M.K. Yadav[1], A.K. Gupta[2], P.S. Ghosal[2], A. Mukherjee[3] & I.S. Chauhan[4]
[1]*School of Environmental Science and Engineering, Indian Institute of Technology, Kharagpur, India*
[2]*Environmental Engineering Division, Department of Civil Engineering, Indian Institute of Technology, Kharagpur, India*
[3]*Department of Geology and Geophysics, Indian Institute of Technology, Kharagpur, India*
[4]*Department of Civil Engineering, Indian Institute of Technology (BHU) Varanasi, Varanasi, India*

ABSTRACT: The multivariate modeling of adsorptive removal of arsenic from aqueous solution by a calcined Mg-Fe-(CO_3) layer double hydroxide, synthesized by a co-precipitation method at a low supersaturation, was conducted by an artificial neural network (ANN). The major influencing parameters of the adsorption process, i.e., adsorbent dose (0.25–4 g L^{-1}), reaction time (2–240 min), pH of the solution (3–12) and agitation rate (80–220 rpm) were varied through a 'one variable at a time' (OVAT) experiment to assess their individual effect on the arsenic removal efficiency. The OVAT experimental data were used for multivariate modeling through a feed forward ANN network with back propagation algorithm. The optimized network showed a correlation coefficient for the training, validation, testing and overall process above 0.99 and the mean square of error as 0.996. The analysis of variance conducted on the predicted values from the model and the actual experimental value exhibited a high F value of and low p value less than 0.001, which showed the applicability of ANN model in delineating the adsorption process of arsenic removal.

1 INTRODUCTION

Arsenic is classified as a Class I human carcinogen by the International Agency for Research on Cancer (IARC). The geogenic pollutant is one of the major concerns in water research by affecting more than 200 million people worldwide. Adsorption is considered to be a preferred choice for arsenic removal and layered double hydroxide is one of the efficient adsorbent in this field. The adsorption process is affected by several parameters, such as adsorbent dose, reaction time, pH, agitation rate. Generally, the univariate modeling through OVAT experiments are performed for assessment of the influence of the adsorption process parameters suggesting their individual effect. However, the multivariate modeling of the process parameters explores the interactive effect of different parameter and predicts the overall optima of the system. In the present study, the calcined Mg-Fe-(CO_3) LDH was prepared and employed for the removal of arsenic. The role of the major influencing factors for the arsenic adsorption, i.e., adsorbent dose, reaction time, pH of the solution and agitation rate was assessed by conducting OVAT experiment. The data of the OVAT experiment was used as an input to ANN model. The multivariate modeling through ANN was introduced as a prediction model for the overall adsorption system.

2 EXPERIMENTAL

2.1 *Synthesis of the adsorbent*

A co-precipitation method at a low supersaturation followed by calcination was used for the preparation of the calcined Mg-Fe-(CO_3) LDH. The effect of the preparation process parameter, i.e., molar ratio of the tri and bi-valent metal ion, formation pH and calcination temperature was optimized by multivariate optimization by a 3^3 factorial design followed by response surface method and reported in our previous study (Yadav *et al.*, 2017). The material prepared at optimized condition, i.e., molar ratio of 2, pH 12 and calcination temperature as 300°C, was used as the adsorbent in the present work.

2.2 *Adsorption experiments and modeling*

The OVAT experiments were conducted for the sorption study in a BOD incubator shaker. The major influencing factor, i.e., adsorbent dose, reaction time, pH of the solution and agitation rate was varied from 0.25–4 g L^{-1}, 2–240 min, 3–12, 80–220 rpm, respectively. The univariate and multivariate modeling was done on the OVAT experimental data.

3 RESULTS AND DISCUSSION

3.1 *Univariate study*

The individual effect of the adsorption parameter was presented in Figure 1. The removal efficiency showed an increasing trend with adsorbent dose and the saturation was attained at 4 g L^{-1} dose showing insignificant change in removal efficiency afterwards. The kinetic plot showed a very first uptake and the equilibrium was reached around 240 min. The pH of the solution showed an antagonistic effect on removal efficiency. The drastic decrease of removal efficiency was

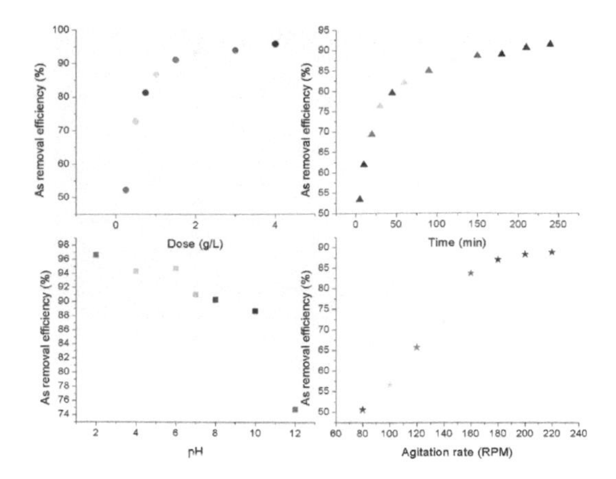

Figure 1. Variation of the removal efficiency with adsorbent dose, time, pH and agitation rate.

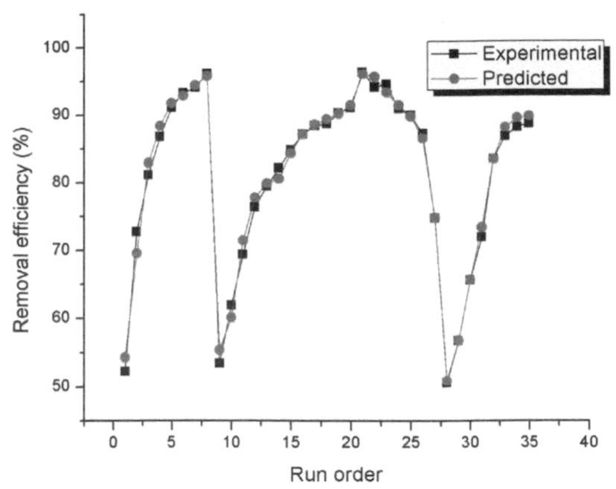

Figure 2. Experimental and predicted values of the removal efficiency.

observed above pH 10. The agitation rate enhanced the removal efficiency up to 200 rpm; the increase in removal efficiency was insignificant afterwards.

3.2 *Multivariate modeling through ANN*

The multivariate optimization was performed by ANN with a feed forward network. The ANN architecture was comprised of one input layer, one hidden layer and one output layer (Ghosal *et al.*, 2016). The OVAT experimental data was used as an input and the removal efficiency was obtained as the output for the combined system. The ANN analysis was performed in MATLAB, 2010a (The MathWorks, INC.) environment. The training algorithm used in this study was Levenberg-Marquardt optimization and the network was analyzed through the back-propagation method. The hyperbolic tangent sigmoid function was chosen as the non-linear neural transfer function in the hidden layer and as follows:

$$O = f(U) = \frac{2}{\left(1+e^{(-2u)}\right)-1} \quad (1)$$

The number of neurons in the hidden layer was optimized by rigorous iteration based on the performance indicator as root mean square of error (RMSE) and the correlation coefficient (R). The variation of RMSE with the number of neurons was conducted and the optimized network was framed with eight neurons in the hidden layer with a RMSE of 1.181. The value of R for the training, testing, validation and overall model for the optimized network was found as 0.999, 0.99, 0.996 and 0.994, respectively.

The network was simulated for the entire set of the experimental data and the predicted values were calculated. The actual experimental values and the predicted values from the model were presented in Figure 2. The ANOVA for the predicted and the actual values showed the adequacy of the model with a high F value of 4391 and a low p value less than 0.001 (Table 1). The ANN model was found to be a significant in terms of goodness of fit with a high R^2 value of 0.993 and adjusted R^2 of 0.992.

Table 1. ANOVA for predicted values from ANN model and the experimental values for removal efficiency.

Model	*SS*	*MS*	*F*	*p-value*
Regression	5954	5954	4391	<0.001
Residual	44.7	1.4		
Total	5999			

4 CONCLUSIONS

A calcined Mg-Fe-(CO_3) LDH was prepared by a co-precipitation method followed by calcination and employed for the adsorptive removal of arsenic. The effect of the adsorption process parameters was analyzed by conducting OVAT experiments. The kinetics of adsorption exhibited a fast uptake at initial stage and equilibrium was reached at 240 min. The pH and agitation rate showed negative and positive influence on removal efficiency. The arsenic removal efficiency was more than 95% for an adsorbent dose of 4 g L^{-1}. The multivariate modeling through ANN showed a significant prediction capacity with R^2 and adjusted R^2 more than 0.99. The importance of the present study of the multivariate modeling lies in its usefulness in the related field.

REFERENCES

Ghosal, P.S. & Gupta, A.K. 2016. Enhanced efficiency of ANN using non-linear regression for modeling adsorptive removal of fluoride by calcined Ca-Al-(NO_3)-LDH, *J. Mol. Liq*. 222: 564–570.

Yadav, M.K., Gupta, A.K., Ghosal, P.S. & Mukherjee, A. 2017. pH mediated facile preparation of hydrotalcite based adsorbent for enhanced arsenite and arsenate removal: insights on physicochemical properties and adsorption mechanism, *J. Mol. Liq*. 240: 240–252.

Environmental Arsenic in a Changing World –
Zhu, Guo, Bhattacharya, Ahmad, Bundschuh & Naidu (Eds)
ISBN 978-1-138-48609-6

Influence of Fe(II), Fe(III) and Al on arsenic speciation in treatment of contaminated water by Fe and Al (hydr)oxides co-precipitation

I.C. Filardi Vasques[1], R. Welmer Veloso[2] & J.W. Vargas de Mello[3]
[1]*Universidade Federal de Lavras, Lavras, MG, Brazil*
[2]*Instituto Federal de Rondônia, Lavras, MG, Brazil*
[3]*Universidade Federal de Viçosa, Lavras, MG, Brazil*

ABSTRACT: Arsenic is considered a highly toxic metalloid, but the toxicity depends on its speciation. Inorganic species are considered more toxic than organic ones and the As(III) more toxic than As(V). This study evaluated the speciation of soluble arsenic after treatment of contaminated water by co-precipitation of Fe and Al (hydr)oxides. The results showed that the valence of Fe and presence of Al play an important role on As speciation in soluble phases resulting from water treatment.

1 INTRODUCTION

Arsenic is considered a very toxic metalloid that is usually found associated to gold deposits. Acid mine drainage (AMD) is among the major environmental problems in base metal exploitations. Therefore, contamination of water, soil and sediments with As associated to AMD is a major concern in mining industry. This study aimed to evaluate arsenite and arsenate removal from highly contaminated water and redox changes due to co-precipitation of Fe(II), Fe(III) and Al (hydr)oxides.

2 EXPERIMENTAL

2.1 *Fe and Al (hydr)oxides synthesis*

Synthesis of Fe and Al (hydr)oxides were performed following Schwertmann and Cornell (2000) guidelines with some adaptations in order to approach AMD. Iron and Al (hydr)oxides were precipitated from sulfates salts of Fe(II), Fe(III) and Al in a solution containing high concentrations of As ($500\,mg\,L^{-1}$). Three Fe:Al molar ratios were considered (100:0, 80:20 and 60:40), in order to obtain pure Fe (hydr)oxides, Al substituted Fe (hydr)oxides and Al (hydr)oxides. The pH was buffered to 8 with KOH in the beginning of the experiments and whenever necessary.

2.2 *Arsenic speciation*

Arsenic concentrations in supernatants were monitored by hydride generation atomic fluorescence spectrometry (AFS-HG) during the experimental period (84 days). At the end of the experimental period, speciation of soluble arsenic in supernatants was performed by HPLC-AFS-HG.

3 RESULTS AND DISCUSSION

3.1 *Removal efficiency*

Efficiencies of As removal from water reached high percentages just after precipitation (>93%), for both As species, with a slightly lower efficiency for arsenite (AsIII) treated with Fe(II) in the presence of Al (Table 1).

3.2 *Arsenic speciation*

Goethite, lepidocrocite and gibbsite were identified by XRD in precipitates from Fe(II). No significant differences were identified in Fe and Al (hydr)oxides after synthesis with arsenite or arsenate (Fig. 1).

In these treatments, soluble As(III) was oxidized to As(V), mainly in the presence of Al, suggesting that Al favors As oxidation (Table 2).

Table 1. Efficiencies for As (III and V) removal from water by co-precipitation of Fe and Al (hydr)oxides at different Fe:Al molar ratios 2 h after precipitation.

Treatments	Fe:Al	Efficiency mean after 2 h
Fe(II): Al: As(V)	100:0	99.98 a
Fe(III): Al: As(V)	100:0	99.98 a
Fe(III): Al: As(III)	100:0	99.98 a
Fe(III): Al: As(V)	80:20	99.98 a
Fe(III): Al: As(V)	60:40	99.98 a
Fe(III): Al: As(III)	80:20	99.96 a
Fe(II): Al: As(III)	100:0	99.93 a
Fe(III): Al: As(III)	60:40	99.90 a
Fe(II): Al: As(V)	80:20	99.89 a
Fe(II): Al: As(V)	60:40	99.89 a
Fe(II): Al: As(III)	80:20	99.33 a
Fe(II): Al: As(III)	60:40	93.49 b

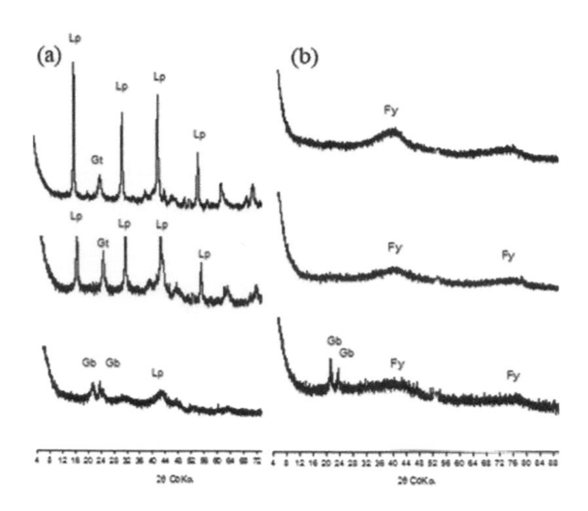

Figure 1. Diffractograms of precipitates from Fe(II) treatments (a) and Fe(III) treatments (b), with the three Fe:Al molar ratios 100:0, 80:20 and 60:40. (Fy: ferrihydrite, Gb: gibbsite, Gt: goethite, Lp: lepidocrocite).

Table 2. Soluble arsenite and arsenate concentrations in supernatants at the end of incubation period for treatments with 500 mg L^{-1} of As(III) or As(V).

Arsenic valence	Fe(II):Al ratio	Total [As] ($\mu g\ L^{-1}$)	[As (III)] ($\mu g\ L^{-1}$)	[As (V)] ($\mu g\ L^{-1}$)
As(III)	100:00	1352	818	534
	80:20	232	25	207
	60:40	441	20	421
As(V)	100:00	198	43	155
	80:20	83	22	61
	60:40	140	0	140
	Fe(III):Al	Total [As] ($\mu g\ L^{-1}$)	[As (III)]	[As (V)]
As(III)	100:00	161	137	25
	80:20	330	330	0
	60:40	663	660	3
As(V)	100:00	77	70	7
	80:20	135	106	30
	60:40	112	110	2

This behavior is similar to oxidation of Fe(II) in the presence of Al, as it favors precipitation of goethite in detriment to magnetite (Mello *et al.*, 2018). On the other hand, ferrihydrite and gibbsite were identified in precipitates from Fe(III) treatments (Fig. 1), but As(III) was not oxidized to As(V). Differently from Fe(II), Al did not favor As oxidation in Fe(III) system. This difference can be related to Fenton Reactions that take place in Fe(II) systems, leading to Fe oxidation, but not in Fe(III) systems (Ona-Nguema *et al.*, 2010). On the other hand, As(V) was reduced to As(III) in a higher extent in Fe(III) treatments, but apparently, there are not reported reasons for that.

4 CONCLUSIONS

Considering the higher toxicity of arsenite compared to arsenate, the main contribution of this study was to demonstrate the As(III) oxidation in treatments with Fe(II), in the presence of Al. On the contrary, As(V) was reduced to As(III) in Fe(III) treatments.

ACKNOWLEDGEMENTS

The authors would like to thank CAPES, CNPq, FAPEMIG and INCT Acqua.

REFERENCES

Mello, J.W.V., Gasparon, M. & Silva, J. 2018. Effectiveness of arsenic co-precipitation with Fe-Al (hydr)oxides for the treatment of contaminated water. *Braz. J. Soil Sci.* 42: e0170261.

Ona-Nguema, G., Morin, G., Wang, Y., Foster, A. L., Juillot, F. & Calas, G. 2010. XANES evidence for rapid arsenic (III) oxidation at magnetite and ferrihydrite surfaces by dissolved O_2 via Fe^{2+}-mediated reactions. *Environ. Sci. Technol.* 44(14): 5416–5422.

Schwertmann, U. & Cornell, R.M. 2000. *Iron Oxides in the Laboratory. John Wiley & Sons.*

Environmental Arsenic in a Changing World –
Zhu, Guo, Bhattacharya, Ahmad, Bundschuh & Naidu (Eds)
ISBN 978-1-138-48609-6

Sorption of toxic oxyanions to modified kaolines

M. Lhotka & B. Dousova
University of Chemistry and Technology Prague, Prague, Czech Republic

ABSTRACT: Natural clay materials are considered effective and perspective sorbents for economical end ecological reasons. In general, aluminosilicates are not selective sorbents for anionic contaminants thanks to a low pH (ZPC). Their simple surface pre-treatment with Fe (II) can significantly change the surface charge and thus strongly improve the affinity to anionic contaminants. This work deals with a sorption of toxic arsenic, antimony and selenium oxyanions on calcinated, Fe-modified, kaolin. The natural kaolin calcinated at 550°C was treated by rehydroxylation in autoclave with 0.6 M solution of $FeCl_2$. The concentration of initial As/Sb/Se solutions was $2{\cdot}10^{-4}$ M. The main result of this work is the comparison of sorption efficiency and sorption capacity of modified sorbent from the point of view of the rehydration time and the type of oxyanionic particle.

1 INTRODUCTION

Arsenic, antimony and selenium are among the toxic elements occurring in small quantities in the water and in the earth's crust. The degree of toxicity of these elements depends on the oxidation degree and form of occurrence. The As, Sb, Se and their compounds leads to both acute and chronic poisoning leading to death. The use of kaolinite and rehydroxylated kaolinite as sorbents of these contaminants belongs to very efficient methods. Under the hydrothermal treatment, kaolinite is transformed into semicrystalline and very reactive metakaolinite, and the rehydroxylation of metakaolinite to kaolinite is possible. Rocha *et al.* (1990) studied the rehydration of metakaolinite to kaolinite which was heated at 155–250°C for 1–14 days and concluded that the amorphous metakaolinite can be transformed to crystalline form. The specific surface area was highly increased during the rehydroxylation. Lhotka *et al.* (2012) studied the rehydroxylation of metakaolinite to kaolinite which was heated at 150–250°C for 1–14 days. He found that the rehydroxylation of metakaolinite to kaolinite was strongly dependent on temperature and time of the hydrothermal process. The optimum transformation from the point of view of the surface properties was observed after longer-term autoclaving (4–7 days) at 175°C, when the specific surface S_{BET} of raw kaolinite increased more than three times. Oxyanions showed a higher adsorption affinity for Fe-modified sorbents. During the interaction of raw clay and Fe salt solution, ion-exchangeable Fe^{3+} particles in amorphous and/or poorly crystalline form have been fixed on the clay surface forming active adsorption sites. Substantial variability in growing Fe phases (hydrated Fe_2O_3, non-specific Fe^{3+} species, ferrihydrite) resulted from using different types of aluminosilicate carrier and the treatment conditions (Dousova *et al.*, 2009). The aim of this work was to prepare kaoline based sorbents using rehydroxylated method and to compare the sorption efficiency and sorption capacity of modified sorbents from the point of view of rehydroxylation time and the type of oxyanionic particle.

2 METHODS/EXPERIMENTAL

2.1 *Preparation of modified kaolinites*

The crystalline kaolinite from West Bohemia was used for the preparation of modified sorbents. This kaolinite was calcined at 650°C for 3 h and converted to metakaolinite. For the preparation of rehydroxyled kaolinite, 8 g of metakaolinite was mixed with 30 g of water. The suspension was stirred for 2 min at room temperature and then was inserted into autoclave. The autoclave was heated at 175°C for 4, 7, 10 and 14 days. In next experiments, 30 g of 0.6 M $FeCl_2$ was used instead of water. After removal from autoclave, the suspension was filtered off, washed with distilled water and dried at 100°C for 24 h.

2.2 *Analytical methods*

The model solutions of oxoanions As(III), As(V), Se(IV), Se(VI), Sb(V) with a concentration of $2{\cdot}10^{-4}$ M were used. The suspension of model solution and sorbent (6 g L^{-1}) was shaken in sealed polyethylene bottle at room temperature for 24 h. The product was filtered off; the filtrate was analyzed for residual As concentration and pH value. Equilibrium adsorption isotherms of nitrogen were measured at 77 K using static volumetric adsorption systems (TriFlex analyzer, Micromeritics). The adsorption isotherms were fitted in the BET specific surface area and the pore size distribution by the DFT and BJH method. The concentration of As, Sb and Se in aqueous solutions

Table 1. Specific surface area of raw kaolinite, metakaolinite and rehydroxylated sorbents (temperature of rehydroxylation – 175°C).

sample	S_{BET} ($m^2\ g^{-1}$)
Kaolinite (K)	15.84
Metakaolinite (MK)	17.81
S4 (4 days)	98.57
S7 (7 days)	103.10
S4 (4 day-Fe)	70.21
S7 (7 days-Fe)	73.81
S10 (10 days-Fe)	71.05
S14 (14 days-Fe)	69.71

Table 2. As(V)/Sb(V)/Se(IV) adsorption on Fe-modified rehydroxylated kaolinite under the various rehydroxylation time.

	As(III)		As(V)		Sb(V)		Se(IV)	
sample	q*	ε**	q	ε	q	ε	q	ε
K4	0.07	57	0.14	99	0.18	96	0.07	100
K7	0.08	59	0.14	99	0.24	97	0.09	99
K10	0.09	56	0.12	100	0.10	98	0.07	99
K14	0.08	61	0.08	99	0.23	98	0.11	99

*maximum adsorption capacity (mmol g^{-1})
**maximum sorption efficiency (%).

was determined by HG-AFS using PSA 10.055 Millennium Excalibur.

3 RESULTS AND DISCUSSION

3.1 *Textural properties of sorbents*

The samples of sorbents were prepared under different reaction conditions. The surface area of newly prepared sorbents was much larger than that of raw kaolinite and metakaolinite (from 15.8 to ~103.1 $m^2\ g^{-1}$). At the rehydration temperature of 175°C the maximum surface area was achieved. Using the maximum S_{BET} at 175°C (Table 1), this temperature was used for the Fe treatment of samples. The comparison of S_{BET} of sorbents prepared at 175°C with and without Fe ions is shown in Table 1 Fe treated kaolinites reached the maximum S_{BET} after four days reaction time. The decrease of their S_{BET} value compare to Fe-free rehydrated samples corresponded to the binding of Fe^{3+} particles on the surface, forming active adsorption sites (Lhotka *et al.*, 2012).

3.2 *Sorption of arsenic, antimony and selenium*

The experimental series for the characterization of As/Sb/Se adsorption on Fe-modified kaolinite (175°C, 4–14 days) were performed with 8 samples of varying sorbent dosage for each reaction time. This arrangement enabled to evaluate sorption parameters according to the Langmuir model. The comparison of

Table 3. Theoretical sorption capacities Q of Fe-modified kaolinite.

	Theoretical sorption capacity Q (mmol g^{-1})			
Oxyanion	4 days	7 days	10 days	14 days
As(III)	–	–	–	–
As(V)	0.09	0.10	0.08	0.08
Se(IV)	0.06	0.08	0.06	0.07
Se(VI)	0.04	0.03		
Sb(V)	0.11	0.16	0.10	0.14

adsorption efficiency ε and maximum sorption capacities q for investigated oxyanions are summarized in Table 2. Sorption parameters evaluated according to the Langmuir model are shown in Table 3.

4 CONCLUSIONS

The aim of this work was the sorption of toxic oxyanions of arsenic, antimony and selenium to $FeCl_2$ treated rehydroxylated kaolines. Kaolines were divided into 4, 7, 10 and 14 days according to rehydroxylation time. Comparison of sorptive capacities has shown that the sorbent rehydroxylated for 7 days achieved the highest sorption capacities (0.03–0.16 mmol g^{-1}), and thus the best sorption properties. This was also confirmed by the comparison of the Langmuir isotherms for As(V) sorbed on kaolin with different rehydroxylation times. By comparing the sorption capacities for separate oxyanions Sb(V) indicated the best sorption capacity (and the highest affinity to sorbent). The affinity of the oxyanions to the sorbents decreases in the range of Sb(V) > As(V) > Se(IV) > Se(VI) > As(III). The As(III) adsorption did not followed either Langmuir or Freundlich model. In term of adsorption efficiency, the sorbent showed almost 100% removal of all oxyanions except As(III), which reached values of about 60%.

ACKNOWLEDGEMENTS

This work was part of projects 16-13778S (Grant Agency of Czech Republic).

REFERENCES

Doušová, B., Fuitová, L., Grygar, T., Machovič, V., Koloušek, D., Herzogová, L. & Lhotka, M. 2009. Modified aluminosilicates as low-cost sorbents of As(III) from anoxic groundwater. *J. Hazard. Mater.* 165(1–3): 134–140.

Lhotka, M., Doušova, B. & Machovič, V. 2012. Preparation of modified sorbents from rehydrated clay minerals. *Clay Miner.* 47(2): 251–258.

Rocha, J., Adams, J.M. & Klinowski J. 1990. The rehydration of metakaolinite to kaolinite: evidence from solid-state NMR and cognate techniques. *J. Solid State Chem.* 89(2): 260–274.

Environmental Arsenic in a Changing World – Zhu, Guo, Bhattacharya, Ahmad, Bundschuh & Naidu (Eds)
ISBN 978-1-138-48609-6

Efficient removal of arsenic species by green rust sulfate (GR_{SO4})

J.P.H. Perez[1,2], H.M. Freeman[1], J.A. Schuessler[1] & L.G. Benning[1,2]
[1]*Helmholtz Centre Potsdam – GFZ German Research Centre for Geosciences, Potsdam, Germany*
[2]*Institute of Geological Sciences, Department of Earth Sciences, Free University of Berlin, Berlin, Germany*

ABSTRACT: The interfacial reactivity between green rust sulfate (GR_{SO4}) and arsenic (As) species were investigated using batch adsorption experiments. GR_{SO4} is capable of adsorbing up to 155 and 88 mg g^{-1} of As(III) and As(V), respectively. The value of pH greatly affects adsorption, with As(III) removal 2.5 times higher at alkaline pH, while As(V) removal higher at circum-neutral conditions. Competing groundwater species such as magnesium (Mg^{2+}) or phosphate (PO_4^{3-}) reduce removal efficiency. Our results show that GR_{SO4} is a stable, efficient and environmentally-relevant mineral substrate for As sequestration in anoxic groundwater systems.

1 INTRODUCTION

Due to its wide-spread distribution, toxicity and mobility, arsenic (As) contamination of groundwater resources remains a significant problem worldwide. Although conventional remediation techniques (e.g., oxidation, coagulation-flocculation, ion exchange) are widely employed for clean up, they are often too costly and environmentally unsuitable (Nicomel *et al.*, 2016).

In contrast, adsorption-based strategies are promising alternatives for groundwater treatment due to their facile implementation, cost-effectiveness and high removal efficiency (Leus *et al.*, 2017). However, interfacial reactions between (redox-active) mineral surfaces and contaminants such as As need to be quantitatively assessed in order to maximize their potential for subsurface remediation. Moreover, there is an imminent challenge regarding the development and testing of adequate (redox-active) mineral phases that have high metal-specific uptake capacities, strong binding affinities and excellent stabilities.

Herein, we report an in-depth investigation on the interfacial interactions between freshly precipitated green rust sulfate (GR_{SO4}) and aqueous As species. We tested the influence of pH, adsorbent loading, ionic strength and presence of potentially interfering ions and evaluated the performance of GR_{SO4} for the removal of As(III) and As(V).

2 EXPERIMENTAL SECTION

2.1 *Synthesis and characterization of GR_{SO4}*

GR_{SO4} was synthesized in an anaerobic chamber (95% N_2, 5% H_2) by co-precipitating stoichiometric amounts of $(NH_4)_2Fe(SO_4)_2{\cdot}6H_2O$ and $Fe_2(SO_4)_3$ with NaOH (Géhin *et al.*, 2002). The obtained GR_{SO4} was characterized by X-ray powder diffraction (XRPD), inductively coupled plasma optical emission spectroscopy (ICP-OES), transmission electron microscopy (TEM) and electron energy loss spectroscopy (EELS).

2.2 *Adsorption experiments*

Batch adsorption experiments were done at room temperature, with GR_{SO4} suspensions reacted at pH 7 to 9 with 10 mg L^{-1} aqueous As(III) and As(V) through shaking at 250 rpm for 24 h. The resulting samples were filtered, supernatants acidified and the separated solid and aqueous phased stored at 4°C until analysis.

3 RESULTS AND DISCUSSION

3.1 *Influence of adsorption parameters*

As(V) was efficiently removed (>99%) at all tested pH values. However, the As(III) removal increased significantly from 50 at pH 7 to 95% at pH 9, likely due to the possible formation of multi-nuclear As(III) complexes on the GR_{SO4} surface (Ona-Nguema *et al.*, 2009; Wang *et al.*, 2010).

Removal efficiency decreased as ionic strength increased from 0.005 to 0.5 M, with As(V) removal dropping by ~10% at high ionic strength (from >99 to 90%). On the other hand, inhibition of As(III) removal was more apparent, with removal efficiency decreasing from 59 to 38%. This reduction in As removal may be caused by the presence of potentially interfering species in the bulk solution. Hence, competitive adsorption experiments were conducted using binary solutions of As with calcium (Ca^{2+}), magnesium (Mg^{2+}) and phosphorous (PO_4^{3-}).

The competitive adsorption results suggest that Mg^{2+} and PO_4^{3-} can inhibit the adsorption of As on the GR_{SO4} surface, while Ca^{2+} is not affecting As adsorption. In the presence of Mg^{2+}, As(III) and As(V) removal decreased by ~7 and ~22%, respectively. Mg^{2+} can block the access of As to the surface active sites of GR_{SO4} due to the formation of outer-sphere hydrated complexes (Lightstone *et al.*, 2001). On the other hand, co-existing PO_4^{3-} species resulted in the reduction of the removal efficiency by ~7 and ~25% for As(III) and As(V), respectively. This is a consequence of the competition between PO_4^{3-} and As species for surface binding sites due to their similar structure (Bocher *et al.*, 2004).

3.2 *Adsorption kinetics and isotherms*

Adsorption kinetic data showed that a pseudo-2nd order kinetic model gives the best fit ($R^2 \geq 0.9990$) and that the uptake rate of As(III) and As(V) were 4 and 11 times faster at alkaline conditions compared to circum-neutral pH.

The equilibrium adsorption data were fitted to the Langmuir and Freundlich isotherms. Based on the model fits, As adsorption onto GR_{SO4} can be best described by a Langmuir model, a result that supports previous X-ray absorption spectroscopic studies (Jönsson and Sherman, 2008; Wang *et al.*, 2010) that suggested that As species form monodentate mononuclear and bidentate binuclear inner-sphere complexes at GR particle edges.

Using the Langmuir isotherm model, the maximum adsorption capacity (q_{max}) for As onto GR_{SO4} at pH 7 and 8–9 was evaluated. Our results show that at alkaline pH, GR_{SO4} can adsorb up to 155 mg As(III) per g of solid, a value 2.5 times higher than at neutral pH. On the contrary, a higher adsorption capacity was observed for As(V) at neutral pH (88 mg g^{-1}) compared to alkaline pH (67 mg g^{-1}).

3.3 *Environmental implications on the fate and mobility of As*

We show that GR_{SO4} has an excellent ability to adsorb As species and has the highest sorption capacity of any iron (oxyhydr)oxides found in surface and near-surface environments, including ferrihydrite, goethite, hematite, maghemite and magnetite. In terms of surface coverages, GR_{SO4} outperforms all of these phases (Fig. 1), and more importantly, it is also one of the very few redox active mineral phases that exhibits unusually high As(III) coverage.

Finally, we show that GR_{SO4} remains stable post-As adsorption at typical contaminated groundwater conditions (~10 mg L^{-1}). Long term monitoring (three months) of GR_{SO4} samples showed no structural and morphological changes, as confirmed by XRD analysis and TEM imaging. This was supported by analyses of aqueous As in the reacting supernatant, which showed no significant re-release of adsorbed As into the aqueous phase.

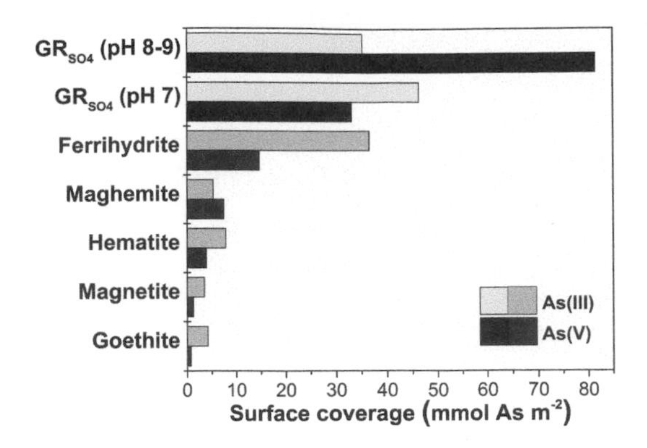

Figure 1. Comparison of As surface coverages of GR_{SO4} (this study) with other iron (oxyhydr)oxides.

4 CONCLUSIONS

We quantified the interfacial reactivity between GR_{SO4} and As species. Using batch adsorption experiments, we examined the influence of various critical environmental parameters on As removal. We successfully demonstrated that GR_{SO4} is a highly effective and stable mineral adsorbent compared to other iron (oxyhydr)oxide phases at the concentration of As-contaminated groundwater (~10 mg L^{-1}). GR_{SO4} demonstrated an exceptional As sorption reactivity which makes it a potentially novel and environmentally-relevant mineral substrate for the sequestration of As in reduced groundwater systems. Overall, our results provide important insights as to how uptake and release of As in contaminated groundwaters can be influenced by the presence of redox-active GR phases.

ACKNOWLEDGEMENTS

This project has received funding from the European Union's Horizon 2020 research and innovation programme under the Marie Sklodowska-Curie grant agreement No. 675219 and the German Helmholtz Recruiting Initiative funding to LGB and HMF.

REFERENCES

Bocher, F., Géhin, A., Ruby, C., Ghanbaja, J., Abdelmoula, M. & Génin, J.-M.R. 2004. Coprecipitation of Fe(II–III) hydroxycarbonate green rust stabilised by phosphate adsorption. *Solid State Sci.* 6(1): 117–124.

Géhin, A., Ruby, C., Abdelmoula, M., Benali, O., Ghanbaja, J., Refait, P. & Génin, J.-M.R. 2002. Synthesis of Fe(II–III) hydroxysulphate green rust by coprecipitation. *Solid State Sci.* 4(1): 61–66.

Jönsson, J. & Sherman, D.M. 2008. Sorption of As(III) and As(V) to siderite, green rust (fougerite) and magnetite: implications for arsenic release in anoxic groundwaters. *Chem. Geol.* 255(1–2): 173–181.

Leus, K., Perez, J.P.H., Folens, K., Meledina, M., Van Tendeloo, G., Du Laing, G. & Van Der Voort, P. 2017. UiO-66-$(SH)_2$ as stable, selective and regenerable adsorbent for the removal of mercury from water under environmentally-relevant conditions. *Faraday Discuss.* 201: 145–161.

Lightstone, F.C., Schwegler, E., Hood, R.Q., Gygi, F. & Galli, G. (2001). A first principles molecular dynamics simulation of the hydrated magnesium ion. *Chem. Phys. Lett.* 343(5–6): 549–555.

Nicomel, N.R., Leus, K., Folens, K., Van Der Voort, P. & Du Laing, G. 2016. Technologies for arsenic removal from water: current status and future perspectives. *Int. J. Environ. Res. Publ. Health* 13(1): 62.

Ona-Nguema, G., Morin, G., Wang, Y., Menguy, N., Juillot, F., Olivi, L., Aquilanti, G., Abdelmoula, M., Ruby, C., Bargar, J.R., Guyot, F., Calas, G. & Brown, Jr, G.E. 2009. Arsenite sequestration at the surface of nano-$Fe(OH)_2$, ferrous-carbonate hydroxide, and green-rust after bioreduction of arsenic-sorbed lepidocrocite by *Shewanella putrefaciens*. *Geochim. Cosmochim. Acta* 73(5): 1359–1381.

Wang, Y., Morin, G., Ona-Nguema, G., Juillot, F., Guyot, F., Calas, G. & Brown, G.E. 2010. Evidence for different surface speciation of arsenite and arsenate on green rust: an EXAFS and XANES Study. *Environ. Sci. Technol.* 44(1): 109–115.

Environmental Arsenic in a Changing World –
Zhu, Guo, Bhattacharya, Ahmad, Bundschuh & Naidu (Eds)
ISBN 978-1-138-48609-6

Arsenite removal from water by an iron-bearing Layered Double Hydroxide (LDH)

Q.H. Guo & Y.W. Cao
State Key Laboratory of Biogeology and Environmental Geology & School of Environmental Studies, China University of Geosciences, Hubei, P.R. China

ABSTRACT: A Mg/Fe-based Layered Double Hydroxide (LDH) intercalated with chloride (Nano iowaite), was synthesized to evaluate its performance for arsenite removal from water. It has a much higher arsenite uptake capacity than other LDHs that are commonly used for water dearsenication. The surface adsorption of the solution arsenite onto the iowaite samples and the anion exchange of the arsenite in solution with chloride, which is originally in the iowaite interlayers, are the primary mechanisms for the uptake of arsenite by iowaite.

1 INTRODUCTION

Arsenic was ranked among the top 20 highest priority harmful substances by the Agency for Toxic Substances and Disease Registry, but nearly 50 millions of people worldwide are suffering from the hazards of drinking high-arsenic waters. Therefore, the treatment of both arsenic-containing wastewater and high-arsenic drinking water has received substantial attention in countries where arsenic problems occur. In recent years, the use of layered double metal hydroxides (LDHs) for purifying arsenic-containing waters has been reported (Goh *et al.*, 2009; Guo & Tian, 2013). LDHs, which are also called anion clays, were first discovered as a natural mineral (hydrotalcite) by Hohsteter in 1842 and have a general chemical formula of $[M^{2+}_{1-x}M^{3+}_{x}(OH)_2](A^{n-})_{x/n}\cdot mH_2O$, where M^{2+} is a divalent cation (typically Mg^{2+}, Co^{2+}, Ni^{2+}, Zn^{2+}, or Mn^{2+}), M^{3+} is a trivalent cation (typically Al^{3+}, Fe^{3+}, Cr^{3+}, or Ga^{3+}), and A^{n-} is an anion (typically OH^-, Cl^-, NO_3^-, CO_3^{2-}, or SO_4^{2-}) (Rives and Ulibarri, 1999). These hydroxides have large surface areas, huge numbers of exchangeable anions within their interlayer regions, and relatively weak interlayer bonding (Zaneva & Stanimirova, 2004).

In this study, an iron-bearing LDH, Nano-iowaite, was tested for its arsenite uptake capacity in view that among major arsenic species, arsenite is much more toxic than the others. The experimental results indicate that Nano-iowaite performed better than many other commonly used sorbents with respect to solution dearsenication under similar experimental conditions (e.g. the mass ratio of solution arsenic to sorbent). Iron-bearing LDH is promising to be used in arsenic removal for either natural arsenic-rich waters or contaminated high-arsenic waters.

2 METHODS

2.1 *Synthesis of iowaite*

The nanocrystalline Mg-Fe(III) LDH (denoted by Nano-iowaite) was synthesized via fast coprecipitation followed by hydrothermal treatment. A mixed salt solution comprising 20 mmol of $MgCl_2\cdot 6H_2O$ and 10 mmol of $FeCl_3\cdot 6H_2O$ was promptly added to 50 mL of 1.4 M NaOH with nitrogen purging and vigorous stirring. The iowaite slurry was centrifuged at 4000 r min^{-1} for 10 min, washed three times with Nanopure water, and finally dispersed in Nanopure water. The resulting aqueous suspension was then hydrothermally treated in a stainless-steel autoclave at 80°C for 4 h. After air-cooling, a stable homogeneous iowaite suspension (the product) was further freeze-dried at −20°C to avoid aggregation. The synthesized iowaite samples were kept in a desiccator to prevent the possible sorption of atmospheric CO_2 and moisture.

2.2 *Batch tests*

The arsenite removal experiments were performed using the batch sorption technique by reacting 100 mL arsenite solutions of various initial concentrations (from 0.15 to 750 mg L^{-1}) with 0.2 g of synthetic iowaite. During the experiments, the sample bottles were sealed and placed in a constant-temperature water bath shaker at 25°C for a predetermined period. After the sorption experiments, the solution samples were decanted from the bottles, centrifuged and filtered through a 0.02-μm microporous membrane.

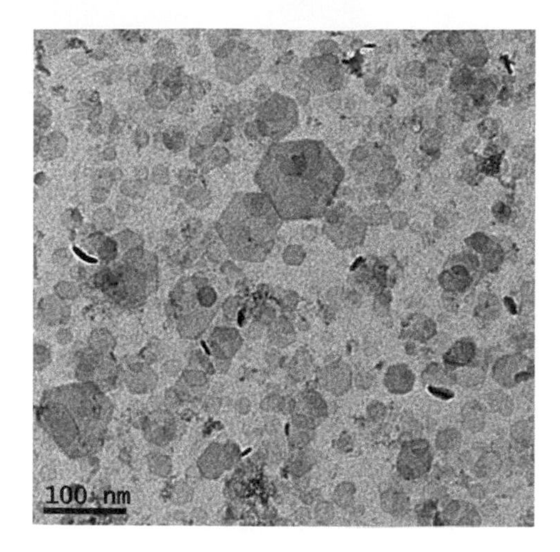

Figure 1. TEM image of synthetic Nano-iowaite.

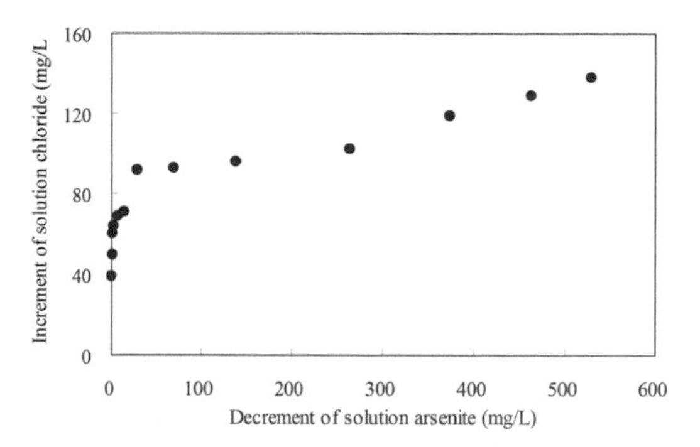

Figure 2. Variations in the increment of chloride vs. the decrement of arsenite in the solutions reacted with Nano-iowaite.

3 RESULTS AND DISCUSSION

3.1 *Characteristics of synthetic iowaite*

Elemental analysis reveals that the Mg/Fe molar ratio of synthetic Nano-iowaite is 2.57, almost the same as its designed ratio (i.e., the molar ratio [2.50] of their starting salts). Nano-iowaite has a narrow particle size distribution (166–675 nm with an equivalent hydrodynamic diameter of 386 nm), as demonstrated by the PCS analysis, and its BET specific surface area is 124.6 $m^2\ g^{-1}$. TEM analysis shows that the stable suspensions of Nano-iowaite are monodispersed nanoparticles with a well-shaped hexagonal form (Fig. 1).

3.2 *Arsenite removal by iowaite and the possible mechanisms involved*

The sorption isotherms of arsenite sorbed on Nano-iowaite can be well fitted by the Freundlich and Langmuir models. By extrapolating the Langmuir equation, the sorption maxima of Nano-iowaite for arsenite was calculated to be 263.2 $mg\ g^{-1}$. This high capacity of arsenite uptake can be attributed to the narrow particle size distribution and large surface area of Nano-iowaite.

According to the analyses of arsenite and chloride in solution after the sorption reaction, the increment of the solution chloride concentration is positively related to the decrement of the solution arsenite concentration (Fig. 2). Thus, the anion exchange between the arsenite in solution and the chloride that was originally intercalated into the interlayers of the iowaite samples was the primary mechanism for the uptake of arsenite by iowaite. The sorbed arsenite basically binds to the iowaite surface or within its interlayers by forming an outer-sphere complex; that is, the driving force for arsenite sorption onto/into iowaite is likely the Coulombic attraction between arsenite and the positively charged layers of iowaite. However, it cannot be ruled out that the inner-sphere complexation of arsenite with certain functional groups on the layers of iowaite may also occur and form a more stable and stronger arsenite bond.

4 CONCLUSIONS

The experimental work conducted in this study indicates that iowaite, an iron-bearing LDH, can be used to effectively remove arsenite from aqueous solution. The nanocrystallization of iowaite via fast coprecipitation followed by hydrothermal treatment helps to enhance its dearsenication capacity. Nano-iowaite performed better than many other commonly used sorbents with respect to solution dearsenication under similar experimental conditions.

ACKNOWLEDGEMENTS

This study was financially supported by the National Natural Science Foundation of China (grant numbers 41572335 and 41772370) and the research program of State Key Laboratory of Biogeology and Environmental Geology of China (No. GBL11505).

REFERENCES

Goh, K.-H., Lim, T.-T. & Dong, Z. 2009. Enhanced arsenic removal by hydrothermally treated nanocrystalline Mg/Al layered double hydroxide with nitrate intercalation. *Environ. Sci. Technol.* 43(7): 2537–2543.

Guo, Q. & Tian, J. 2013. Removal of fluoride and arsenate from aqueous solution by hydrocalumite via precipitation and anion exchange. *Chem. Eng. J.* 231: 121–131.

Rives, V. & Ulibarri, M.A.A. 1999. Layered double hydroxides (LDH) intercalated with metal coordination compounds and oxometalates. *Coord. Chem. Rev.* 181(1): 61–120.

Zaneva, S. & Stanimirova, T. 2004. Crystal chemistry, classification position and nomenclature of Layered Double Hydroxides. Bulgarian Geological Society, *Annual Scientific Conference "Geology 2004"*, pp. 110–112.

Environmental Arsenic in a Changing World –
Zhu, Guo, Bhattacharya, Ahmad, Bundschuh & Naidu (Eds)
ISBN 978-1-138-48609-6

Effect of Ca^{2+} and PO_4^{3-} on As(III) and As(V) adsorption at goethite-water interface: Experiments and modeling

Y.X. Deng, L.P. Weng & Y.T. Li
Agro-Environmental Protection Institute, Ministry of Agriculture, Tianjin, P.R. China

ABSTRACT: Adsorption process is a key reaction that controls the bioavailability and mobility of arsenic in the environment. So far, studies regarding As adsorption to minerals in multiple component systems are still limited. In this study, we investigated the effect of Ca^{2+} as the major cation and PO_4^{3-} as the major anion on As(III) or As(V) adsorption at goethite-water interface over a wide pH range. Data of the adsorption experiments were compared to those calculated with the CD-MUSIC model. Results show that Ca^{2+} promoted As(V) adsorption at high pH, but slightly influenced (increased or decreased) As(III) adsorption. Phosphate competed strongly with the adsorption of both As(III) and As(V), especially at relatively low pH for As(III) and high pH for As(V). The combined effect of Ca^{2+} and PO_4^{3-} on As adsorption results in an increased As(III) adsorption with the increase of pH (3–10), whereas for As(V) a maximum adsorption can be seen around pH 6. The CD-MUSIC model can predict As adsorption in the complex systems by considering electrostatic interactions and site competition. Competition with PO_4^{3-} leads to a decrease of the bidentate As(III) and As(V) surface species. Our results can provide understanding needed in the risk assessment and remediation of As contaminated soils and water.

1 INTRODUCTION

The predominant As species in water and soil are As(III) and As(V). Arsenic adsorption to minerals controls to a large extent its bioavailability and mobility. The adsorption process is significantly influenced by pH and co-occurrence of other ions such as Ca^{2+} and PO_4^{3-} (Kanematsu *et al.*, 2008, Stachowicz *et al.*, 2010). However, there are so far very few studies that investigate simultaneously the effect of Ca^{2+} and PO_4^{3-} on As(III) or As(V) adsorption under different pH. Similarly, the ability of the CD-MUSIC model, which is an advanced surface complexation model (Hiemstra & Van Riemsdijk, 2006), to predict As adsorption in multiple component systems has not been sufficiently tested. The objectives of this study are: (1) to reveal the effect of Ca^{2+} and PO_4^{3-} on As(III) or As(V) adsorption to goethite under various pH; (2) to validate the reliability of CD-MUSIC model prediction for the above-mentioned systems; and (3) to discuss Ca^{2+} and PO_4^{3-} effect on the surface speciation of adsorbed As on goethite based on the CD-MUSIC model.

2 METHODS

Goethite material was prepared according to Hiemstra *et al.* (1999). The specific surface area and PZC of goethite are 89 $m^2\ g^{-1}$ and 9.2 respectively.

Adsorption experiments were carried out in gas-tight polyethylene centrifuge tubes, to which appropriate amounts of stock solutions were added under N_2 atmosphere. For all treatments, the finial concentration of goethite was 1.1 $g\ L^{-1}$, of NaCl was 0.1 M. For both As(III) and As(V), treatments of As, As + Ca, As + P and As + Ca + P were prepared and the pH was adjusted with acid and base to a range between 3.5 and 11.5. The prepared suspensions were equilibrated on a horizontal shaker at 25°C for 3 days. The suspensions final pH was measured, then centrifuged and filtered through 0.45 μm membranes. The filtrates were analyzed on ICP-MS for As concentration and on ICP-OES for Ca and P concentrations. The CD-MUSIC model parameters were adopted from the literatures (Stachowicz *et al.*, 2007, 2008), except that the logK values for As(III), As(V), Ca^{2+} and PO_4^{3-} were optimized in the current study using data of single adsorbate experiments. The model calculations were carried out using the computer program ECOSAT.

3 RESULTS AND DISCUSSION

In the As(III)-Ca-P system, the competitive effect of PO_4^{3-} on As(III) adsorption is very strong especially at pH < 10, whereas the Ca^{2+} effect is only visible at pH > 6 (Fig. 1a). Interestingly, there is a decrease rather than an increase of As(III) adsorption at pH about 6–10, when Ca^{2+} is added to the As(III)-P system. This phenomenon can be attributed to the increase of PO_4^{3-} adsorption in the presence of Ca^{2+} as a result of synergistic electrostatic interactions, which consequently leads to a decrease of As(III) adsorption due to site competition between As(III) and PO_4^{3-}.

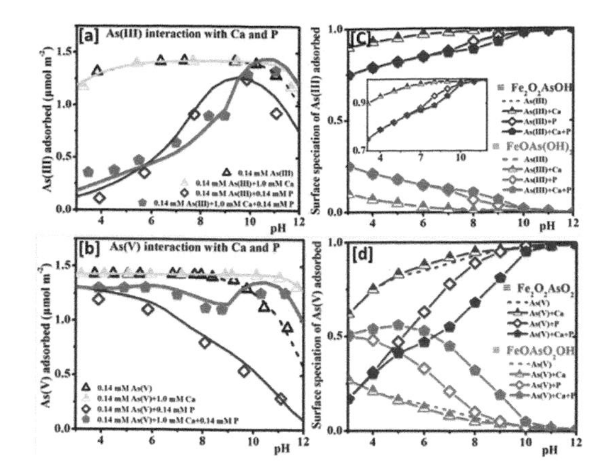

Figure 1. Influence of Ca^{2+} and PO_4^{3-} on the adsorption (a, b) and surface speciation (c, d) of As(III) or As(V) to goethite. (a, b) Symbols are experimental data, and lines are CD-MUSIC model predictions; (c, d) all lines are calculated with the CD-MUSIC model.

In the As(V)-Ca-P system, the magnitudes of the competitive effect of PO_4^{3-} and the synergistic effect of Ca^{2+} on As(V) adsorption are comparable at pH 5–9, whereas the Ca^{2+} effect is more significant at higher pH and PO_4^{3-} effect is more significant at lower pH (Fig. 1b). When PO_4^{3-} is present, the influence of Ca^{2+} appears at a lower pH compared with the As(V)-Ca system, which is caused by the more negative charge and potential at goethite surface in the presence of PO_4^{3-}.

The CD-MUSIC model well predicted adsorption features of both As(III) and As(V) in different systems (Figs. 1c, d). However, in order to achieve good agreements between calculated and measured soluble Ca^{2+} and PO_4^{3-} concentrations, two types of calcium phosphate precipitates need to be included, i.e. $Ca_2HPO_4(OH)_2$ (hydroxyspodiosite) and $Ca_5(PO_4)_3OH$ (hydroxyapatite) with fitted solubility product ($\log K_{so}$) of −32.9 and −47.0 respectively. The Ca^{2+} and PO_4^{3-} effect on As(III) and As(V) adsorption is accounted in the model via electrostatic interactions and site competition. The weaker Ca^{2+} effect on As(III) than on As(V) adsorption can be explained by the weaker electrostatic interactions between Ca^{2+} and As(III).

In the CD-MUSIC modeling, adsorbed As(III) is represented using two inner-sphere complexes, i.e., a bidentate $\equiv Fe_2O_2AsOH$ and a monodentate $\equiv FeOAs(OH)_2$ surface species; and adsorbed As(V) is represented using three inner-sphere complexes, i.e. a bidentate $\equiv Fe_2O_2AsO_2$, a protonated bidentate $\equiv Fe_2O_2AsOOH$, and a protonated monodentate $\equiv FeOAsO_2OH$ surface species (Stachowicz *et al.*, 2008). Our model calculations show that the bidentate surface complex of adsorbed As(III) and As(V) is greatly suppressed in the presence of PO_4^{3-} (Fig. 1c, 1d), whereas the effect of Ca^{2+} on the surface speciation of As(III) or As(V) in the As-Ca systems is quite slight (in Fig. 1c and 1d, the lines of As-Ca treatments coincide with those of As treatments). In the As(III)/As(V)-P systems, the presence of Ca^{2+} decreased the bidentate complex of As(III) and As(V), especially for As(V) due to site competition as a result of enhanced As(V) and PO_4^{3-} adsorption.

4 CONCLUSIONS

This study demonstrated that the presence of PO_4^{3-} and Ca^{2+} influences not only the magnitude of adsorption of As(III) and As(V), but also their adsorption pH dependency. In the natural environment (pH 3.5–8.5), when As is in the form of As(III), the adsorption of As will increase with the increase of pH; When As is present in the form of As(V), a maximum adsorption around pH 6 can be expected when Ca^{2+} is present. The increase of PO_4^{3-} concentration will decrease both As(III) and As(V) adsorption, whereas an increase of Ca^{2+} concentration will increase As(V) adsorption, but it may increase or decrease As(III) adsorption depending on conditions. Increasing soil pH by liming is an effective approach to decrease the mobility and bioavailability of heavy metals such as Cd. Our analysis indicates that the mobility and bioavailability of As can possibly be decreased by liming in dry land soils when the pH remains below 6, whereas the opposite effect may take place when the pH is above 6.

The CD-MUSIC model is a reliable tool to understand and predict the adsorption behavior of As on iron (hydr)oxide minerals.

ACKNOWLEDGEMENTS

This work was supported by the National Key Research and Development Program of China (2016YFD0800304), and the National Natural Science Foundation of China (No. U1401234).

REFERENCES

Hiemstra, T. & Van Riemsdijk, W.H. 1999. Surface structural ion adsorption modeling of competitive binding of oxyanions by metal (hydr)oxides. *J. Colloid. Interf. Sci.* 210(1): 182–193.

Hiemstra, T. & Van Riemsdijk, W.H., 2006. On the relationship between charge distribution, surface hydration, and the structure of the interface of metal hydroxides. *J. Colloid. Interf. Sci.* 301(1): 1–18.

Kanematsu, M., Young, T.M., Fukushi, K., Green, P.G. & Darby, J.L. 2010. Extended triple layer modeling of arsenate and phosphate adsorption on a goethite-based granular porous adsorbent. *Environ. Sci. Technol.* 44(9): 3388–3394.

Stachowicz, M., Hiemstra, T. & van Riemsdijk, W.H. 2007. Arsenic-bicarbonate interaction on goethite particles. *Environ. Sci. Technol.* 41(16): 5620–5625.

Stachowicz, M., Hiemstra, T. & van Riemsdijk, W.H. 2008. Multi-competitive interaction of As(III) and As(V) oxyanions with Ca^{2+}, Mg^{2+}, PO_4^{3-}, and CO_3^{2-} ions on goethite. *J. Colloid. Interf. Sci.* 320(2): 400–414.

Environmental Arsenic in a Changing World –
Zhu, Guo, Bhattacharya, Ahmad, Bundschuh & Naidu (Eds)
ISBN 978-1-138-48609-6

An insight into As(III) adsorption behavior on β-cyclodextrin functionalized hydrous ferric oxide: Synthesis & characterization

I. Saha[1], K. Gupta[2], S. Ahmed[1], D. Chatterjee[3] & U.C. Ghosh[2]
[1]*Department of Chemistry, Sripat Singh College, Murshidabad, India*
[2]*Department of Chemistry, Presidency University, Kolkata, India*
[3]*Department of Chemistry, University of Kalyani, Kalyani, West Bengal, India*

ABSTRACT: The particular study explores the adsorption potential of As(III) on β-cyclodextrin functionalized hydrous ferric oxide (HCC). This is characterized by XRD, FESEM, AFM, XPS, BET, surface site concentration and FTIR. The modification of hydrous ferric oxide (HFO) surface by β-cyclodextrin provides copious –OH groups which in turn enhances As(III) adsorption on HCC compared to HFO. The adsorption remains almost constant in pH range 3–8 which decreases at higher pH (>8) and followed monolayer and pseudo first order kinetics. It is spontaneous at 303 K with increasing entropy and decreasing enthalpy. Thus HCC is found to be a more efficient adsorbent media compared to HFO.

1 INTRODUCTION

The metalloid arsenic (As) which has been inflicting havoc on mankind for the past few decades and has assumed global proportions transcending geographical boundaries and communities, has targeted water which is the support system of a civilization as the medium for its transportation into the environment and biological systems which explains the countless deaths it has left behind in its wake (Biswas *et al.*, 1998; Smedley & Kinniburgh, 2002). The cases of As poisoning through drinking water can be traced long back in the world history (Ritchie, 1961; Vaughan, 2006; Wang *et al.*, 2007). In 1898, the first cases of skin cancer were observed among population consuming As contaminated water in Poland. Since then it has been conclusively established that unabated ingestion of As via drinking water causes a variety of carcinogenic effects of skin, liver, kidney and other organs which ultimately leads to death. All these incidents have led to World Health Organization (WHO) and other national agencies to fix the permissible limit to 10 $\mu g L^{-1}$ of As in drinking water. The principle source of As contamination is groundwater and the majority of rural and semi urban population in vast parts of the globe depend on this form of water for their drinking water and other household needs. The problem associated with As contamination is severe in Bangladesh and major parts of West Bengal, India. In West Bengal, more than 6 million people from 12 districts covering 111 blocks are worst affected and consuming drinking water with As concentration >50 $mg L^{-1}$ (Mazumder *et al.*, 2010). The present investigations provide promising results demonstrating that HFO modified by β-cyclodextrin would be a better scavenger of As(III) from contaminated groundwater as against conventional pure HFO.

2 METHODS/EXPERIMENTAL

A ferric chloride solution was prepared in 0.1 M HCl and mixed with an aqueous solution of β-cyclodextrin. The mixture was warmed at 90°C accompanied by gradual addition of dilute NH_3 solution with constant stirring till the pH reached around 7. The dark brown precipitate formed was aged for 48 h, filtered, and washed repeatedly with deionized water. The filtered mass obtained was dried at 100°C into an air oven. The fully dried mass was ground in a mortar and sieved to obtain a particle size between 52–100 mesh (250–150 microns) and subsequently used for adsorption studies.

3 RESULTS AND DISCUSSION

Figure 1A shows the XPS spectra of HCC. The peak at 711.9 eV (spectrum 1B) is due to the 2p electron of Fe and peak at 532.0 eV (spectrum 1C) value indicates the O 1s electron. The peak at 286.1 eV showed the presence of C-OH/C-O-C bonds on the HCC surface which clearly reveals that HFO was modified with β-cyclodextrin.

As(III) adsorption was observed to slightly decrease up to pH_i 8.0 for HCC. Above this pH_i value the adsorption decreased markedly up to pH 11.0. At pH_i 7.0, 65.8% As(III) uptake was observed for pure HFO whereas for HCC 81.5% of the total As(III) were removed from solution. When the initial solution pH was raised to 9.0 the As(III) adsorption dropped down to 66.3% for HCC. The pH_{ZPC} value of HCC lied within the range 7.6–7.9 indicating that below this pH range the surface charge of HCC was positive and above it surface becomes negative. At low pH the most dominant form of As(III) species present in the solution is $As(OH)_3$ ($pK_1 = 9.2$). Due to the presence of large number of surface –OH groups the neutral $As(OH)_3$ species form strong H-bond with HCC surface at low pH region.

4 CONCLUSIONS

Hydrous ferric oxide (HFO) has proven to be a potent adsorbent for dissolved As(III) in contaminated water. However, surface modification of HFO with β-cyclodextrin by a simple and economical route shows large enhancement of As(III) scavenging power. The systematic As(III) adsorption by this material has showed that the optimum pH and equilibrium contact time are $\sim 7.0 \pm 0.1$ and 120 min, respectively. The pseudo first-order equation describes the kinetic data (pH, 7.0 ± 0.1; temperature, 303 ± 1.6 K) well. The equilibrium data (pH, 7.0 ± 0.2; temperatures (± 1.6 K), 288, 303, 318 and 333) fit very well with the Langmuir isotherm model. The Langmuir monolayer adsorption capacity is 66.96 ± 9.16 (mg of As per g. of HCC) at 303 K, and that increases with increasing temperature. The adsorption reaction is spontaneous ($\Delta G° =$ negative) and exothermic ($\Delta H° =$ negative), and that takes place with increasing entropy ($\Delta S° = 0.007$). The energy ($kJ\,mol^{-1}$) of As(III) adsorption and the FTIR analysis have suggested that the As(III) adsorption on HCC is of both physisorption as well as chemisorption in nature. The As(III) adsorption by HCC is negatively influenced by phosphate and sulfate ions. The regeneration of As(III) adsorbed material is possible maximum up to 75% with 1 M NaOH solution.

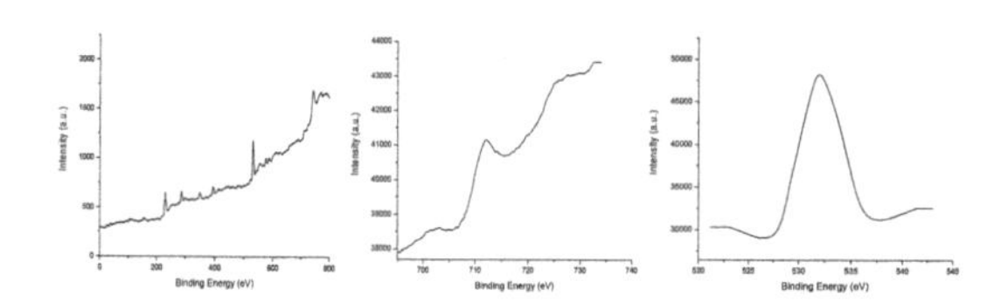

Figure 1. XPS of (1A) HCC, (1B) Fe 2p spectrum and (1C) O 1s spectrum.

ACKNOWLEDGEMENTS

The authors are thankful to Sripat Singh College for providing laboratory facilities.

REFERENCES

Biswas, B.K., Dhar R.K., Samanta, G., Mandal, B.K., Faruk, I., Islam, K.S., Chowdhury, M.M., Islam, A., Roy, S. & Chakraborti, D. 1998. Detailed study report of Samata, one of the arsenic affected villages of Jessore district, Bangladesh. *Curr. Sci.* 74(2): 134–145.

Mazumder, D.N.G., Ghosh, A., Majumdar, K.K., Ghosh, N., Saha, C. & Mazumder, R.N.G. 2010. Arsenic contamination of ground water and its health impact on population of district of Nadia, West Bengal. India. *Indian J. Community Med.* 35(2): 331–338.

Ritchie, J.A. 1961. Arsenic and antimony in some New Zealand thermal waters. *NZJ Sci.* 4: 218–229.

Smedley, P.L. & Kinniburgh, D.G. 2002. A review of the source, behaviour and distribution of arsenic in natural waters. *Appl. Geochem.* 17(5): 517–568.

Vaughan, D.J. 2006. Arsenic, *Elements* 2: 71–75.

Wang, L., Chen, A.S.C., Tong, N. & Coonfare, C.T. 2007. Arsenic removal from drinking water by ion exchange, U.S. EPA demonstration project at Fruitland, ID, six month evaluation report. EPA/600/R-07/017. United States Environmental Protection Agency, Water Supply and Water Resources Division, National Risk Management Research Laboratory, Cincinnati, OH.

Environmental Arsenic in a Changing World –
Zhu, Guo, Bhattacharya, Ahmad, Bundschuh & Naidu (Eds)
ISBN 978-1-138-48609-6

Iron-based subsurface arsenic removal by aeration (SAR) – results of a pilot-scale plant in Vietnam

V.T. Luong[1], E.E. Cañas Kurz[2,3], U. Hellriegel[2], L.L. Tran[3], J. Hoinkis[2] & J. Bundschuh[1]
[1]*Faculty of Health, Engineering and Sciences, University of Southern Queensland, Toowoomba, QLD, Australia*
[2]*Department of Mechatronics and Sensor Systems Technology, Vietnamese-German University, Binh Duong Province, Vietnam*
[3]*Center of Applied Research, Karlsruhe University of Applied Sciences, Karlsruhe, Germany*

ABSTRACT: Arsenic contamination in groundwater is a critical issue and one that causes great concern around the world with many negative health impacts on the human body. *In-situ* subsurface arsenic immobilization by aeration has shown to be a promising, convenient technology with high treatment efficiency. In contrast to most of other As-remediation technologies, in-situ subsurface immobilization offers the advantage of negligible waste production and hence has the potential of being a long-term, sustainable treatment option. A pilot scale plant (capacity = 2 m^3 d^{-1}) for the subsurface arsenic removal (SAR) was tested in the Mekong Delta region in South Vietnam. Within the first two weeks of operation the initial concentrations of 81 ± 14 μg As L^{-1} were successfully lowered to below the guideline value limit for drinking water recommended by the WHO of 10 μg L^{-1}. Results indicated adsorption and co-precipitation with iron oxides as the principal mechanism responsible for the arsenic removal. Evaluation of the results demonstrates the feasibility of *in-situ* technology for arsenic mitigation. However, difficulties in manganese removal arose due to existing high ammonia concentrations and natural occurring geochemical reducing conditions.

1 INTRODUCTION

Arsenic (As) is considered to be one of the most serious inorganic contaminants in groundwater and it is recognized as a significant environmental cause of cancer mortality globally (Martinez *et al.*, 2011).

There are many established methods to treat arsenic laden water including precipitation, membrane technologies, adsorption, ion exchange and capacitive deionization which show many different advantages and disadvantages (Nicomel *et al.*, 2015). The main disadvantage of – *ex-situ* – technologies is the generation of arsenic-laden waste which has to be either regenerated or disposed safely (Clancy *et al.*, 2013).

As questions regarding the efforts, safety and costs for the disposal of arsenic waste arise, alternative techniques with a sustainable approach are needed. The *in-situ* arsenic remediation by aeration is a technique based on the subsurface iron removal (SIR), which has proven for many years as a technically feasible technique in a variety of hydrogeochemical settings (Rott & Kauffmann, 2008). However, subsurface arsenic removal (SAR) is not yet an established solution since it shows vulnerability to diverse geochemical conditions such as pH, Fe:As ratio, and the presence of co-ions making it sometimes difficult to comply with the stringent guideline value for drinking water of 10 μg L^{-1} (WHO, 2011). However, the salient advantage relies on its low operating and maintenance costs, and its negligible waste production, as arsenic is bound to the subsurface matrix (Rott & Kauffmann, 2008).

This study shows the results of a SAR pilot plant installed for the *in-situ* treatment of As contaminated groundwater in the An Giang Province, Vietnam.

2 METHODS

2.1 *Subsurface arsenic removal plant*

The subsurface arsenic removal bases on the adsorption and co-precipitation of arsenic onto iron-(hydr)oxides. The operation of the SAR plant took place in the three following steps: *1. Aeration*: groundwater is extracted and aerated; *2. Infiltration*: oxygen-rich water is re-injected into the aquifer through the infiltration well inducing the formation of oxidation and adsorption zones around the tubewell; *3. Adsorption*: arsenic adsorbs onto the formed iron-oxides and co-precipitates in the oxide-matrix. Treated water can be extracted from the well.

The SAR pilot plant consists of an air-injection nozzle for aeration, a storage tank (390 L) and the delivery pump with pressure vessel. This study shows the results of the first five months of operation, including commissioning and installation of an extra tank (1000 L) after three months operation for increased

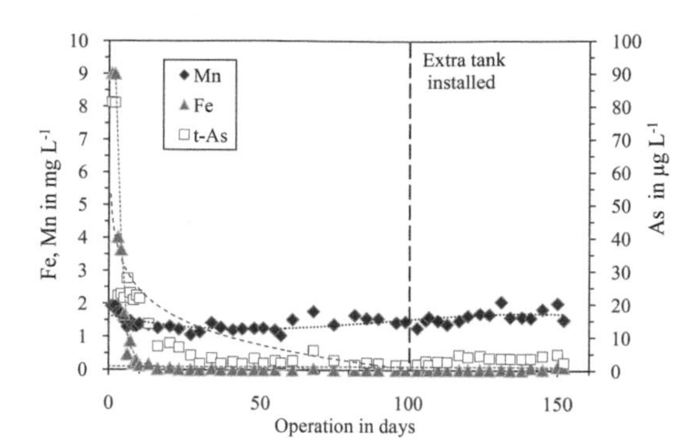

Figure 1. Iron, manganese (left axis) and total arsenic concentration (right axis) of treated groundwater.

infiltration volume (V_{total} = 1390 L). The *aeration* and *infiltration* cycles (max. 2 cycles per day) were regulated by a consumption-based controller. With an infiltration carried out twice a day, a total max. Volume $V_{treated}$ = 1,900 L could be extracted daily.

3 RESULTS AND DISCUSSION

3.1 *Iron and arsenic removal*

The initial concentrations of iron 8.2 ± 1 mg Fe L^{-1} were lowered to below Vietnamese drinking water standard value of 0.3 mg Fe L^{-1} within the first week of operation. Compliance for arsenic standard of 10 µg As L^{-1} was achieved after 10 days.

The results demonstrate that the oxidation of dissolved Fe^{2+} to particulate Fe^{3+} occurs rapidly after the first infiltration of oxygen into the aquifer, as shown in Figure 1. The mitigation of arsenic is directly proportional to the oxidation of iron, thus validating the proposed adsorption mechanisms of arsenic onto iron(hydr)oxides.

3.2 *Manganese and ammonium*

Achieving manganese drinking standard (0.3 mg Mn L^{-1}) was not accomplished within the first 5 months of operation. The small changes on manganese concentrations shown in Fig. 1 (average 1.5 ± 0.2 mg Mn L^{-1}) can be associated to natural hydrological fluctuations. The limitations on Mn removal might be related to the rather absent microbial activity and unfavorable natural low redox potential which makes the – already slow Mn oxidation – challenging. However, long-time experience with SIR shows that a complete Mn removal can take up to 6 months, so that mitigation of Mn with SAR is not to be excluded (Grischek *et al.*, 2015).

High initial ammonium concentrations (1.1 ± 0.1 mg $NH_4^+ L^{-1}$) can also be associated to the poor Mn mitigation, as the oxidation of NH_4^+ is highly oxygen consuming and the oxidation of Mn^{2+} can take place only after NH_4^+ has been fully oxidized. Further monitoring of Mn concentrations is required to assess feasibility for Mn mitigation under the different conditions in Vietnam.

In order to achieve Mn standard for drinking purpose, a downstream treatment such as greensand filters (MnO_2) might be a viable low-cost option.

4 CONCLUSIONS

Results show that subsurface arsenic removal (SAR) is a feasible technique for achieving drinking water standard of 10 µg As L^{-1}, also for low-income countries such as Vietnam. The salient advantage of this technique is the negligible waste production. Hydrogeochemical conditions may affect the efficiency for achieving drinking water standard when high NH_4^+ and Mn concentrations are present.

ACKNOWLEDGEMENTS

The project was funded by the German Federal Ministry of Education and Research (BMBF) under the grant number 02WAV1413A.

REFERENCES

Clancy, T.M., Hayes, K.F. & Raskin, L. 2013. Arsenic waste management. a critical review of testing and disposal of arsenic-bearing solid wastes generated during arsenic removal from drinking water. *Environ. Sci. and Technol.* 47(19): 10799–10812.

Grischek, T., Winkelnkemper, T., Ebermann, J. & Herlitzius, J. 2015. Small scale subsurface iron removal in Germany: *5th GEOINDO 2015*, Khon Kaen, Thailand.

Martinez, V.D., Vucic, E.A., Becker-Santos, D.D., Gil, L. & Lam, W.L. 2011. Arsenic exposure and the induction of human cancers. *J. Toxicol.* 2011: 1–13.

Nicomel, N.R., Leus, K., Folens, K., van der Voort, P. & Du Laing, G. 2015. Technologies for arsenic removal from water: current status and future perspectives. *Int. J. Environ. Res. Public Health* 13(1): 62.

Rott, U. & Kauffmann, H. 2008. A contribution to solve the arsenic problem in groundwater of Ganges Delta by in-situ treatment. *Wat. Sci. Tech.* 58(10): 2009–2015.

WHO, 2011. Guidelines for drinking-water quality. World Health Organization, Geneva.

Environmental Arsenic in a Changing World –
Zhu, Guo, Bhattacharya, Ahmad, Bundschuh & Naidu (Eds)
ISBN 978-1-138-48609-6

A new material could effectively treatment of arsenic-contaminated water

Y.F. Li[1], X. Li[1], D. Wang[1], B. Li[1], Q.M. Zheng[2], L.J. Dong[2] & G.F. Sun[3]
[1]*Research Center of Environment and Non-Communicable Disease, School of Public Health, China Medical University, Shenyang, P.R. China*
[2]*Sanitation & Environment Technology Institute, Soochow University, Jiangsu, P.R. China*
[3]*Hawaii Institute of Interdisciplinary Research, Hawaii, USA*

ABSTRACT: Arsenic-contaminated water is one of the seriously environmental problems around the world. Developing arsenic removal materials have become the major focuses. Mesopaper, which is synthesized by nano-scale porous powder and paper in a sandwich structure, have been developed recently. In this study, we selected three arsenic-contaminated water samples, including two samples collected from different Chinese arsenic-endemic areas and one sample collected from Chinese Shimen realgar mine, to test its effectiveness in arsenic removal. Our findings have shown that arsenic concentrations of the three samples were significantly decreased after filter through one layer and two layers of Mesopaper. The arsenic removal efficiency for the two layers of Mesopaper was significantly higher than that shown in one layer. Additionally, the adsorbed arsenic leaching rate of nano-scale powder, which the active media for arsenic removal in Mesopaper, in different pH value solutions were all relatively low. Together, all above findings have suggested that the Mesopaper has high effectiveness in treatment of arsenic-contaminated water and without secondary arsenic pollution.

1 INTRODUCTION

Arsenic is a naturally occurring metalloid and ubiquitously distributed in the environment. People could exposure to arsenic via water, air and soil. On the global scale, drinking arsenic contaminated water is the most common way for people exposure to arsenic (Naujokas *et al.*, 2013; Tang *et al.*, 2016). It has been well known that long-term arsenic exposure could induce enormous health hazards. And therefore, developing technologies and materials that could treatment of arsenic-contaminated water could be the only effective option to minimize the health hazards. Up to now, technologies including oxidation, phytoremediation, coagulation-flocculation, adsorption and membrane have been used to remove arsenic from contaminated drinking water (Singh *et al.*, 2015). A number of materials developed based on these technologies have also been reported. However, most of these materials could not achieve satisfactory arsenic removal results in field testing and their by-products could be further potential secondary arsenic pollution. The two drawbacks have become major limitations for these materials applied in practice. In this report, we describe a new material named as Mesopaper that can effectively remove arsenic from natural water and we hope it will give a new insight on the arsenic removal.

2 METHODS/EXPERIMENTAL

2.1 *Synthesis of Mesopaper*

The Mesopaper is synthesized by nano-scale powder and paper in sandwich structure. The nano-scale powder is composed of porous ceramic coated with nano zero-valent iron (NZVI), clay and carbon. The detailed processes for producing the Mesopaper has been listed in patent PCT/US17/53452.

2.2 *Water samples collection*

In this study, we selected three natural arsenic-contaminated water samples, including two groundwater samples and one mine wastewater sample, to test the removal efficacy of the Mesopaper. The groundwater samples were collected from Sasuoma Village located in Gansu Province and Gangfangying Village located in Inner Mongolia Autonomous Region, respectively. The two regions are both naturally serious arsenic pollution areas in China. The mine wastewater sample was collected from Shimen realgar mine, located in Hunan Province in central south China. The Shimen realgar mine area is the largest realgar deposit in Asia and had been halted production in 2011. However, a large amount of arsenic residue waste and arsenic ash were dumped on the soil surface and subjected to erosion and weathering processes, causing long-term and severe impact on local environment. Therefore, the arsenic content in the water in the mine area was extensively high. All of the three water samples were collected and then transported to our laboratory.

2.3 *Arsenic filtration experiment*

A square box with basal area of 88 cm^2 was used as filter container. The Mesopaper was folded well and putted in the filter and to ensure no water leakage.

Table 1. The arsenic concentrations of water before and after filtered through Mesopaper.

	Gansu ($\mu g\ L^{-1}$)		Innner Mongolia ($\mu g\ L^{-1}$)		Hunan ($\mu g\ L^{-1}$)	
	One layer	Two layer	One layer	Two layer	One layer	Two layer
Startup	729	729	1658	1658	86580	86580
0.2 L	3	0	26	0	47849	1125
0.4 L	5	0	82	5	79454	27277
0.6 L	13	0	557	38	80745	53796
0.8 L	21	0	685	80	77566	65696
1.0 L	33	0	843	129	82520	68370

And then, 1 liter (L) water sample was prepared and poured through the Mesopaper, with 0.2 L every time. The arsenic concentrations were determined at accumulated filtration volume of 0.2 L, 0.4 L, 0.6 L, 0.8 L and 1 L. In this study, we tested the arsenic removal efficiency of one layer and two layers of Mesopaper, respectively.

2.4 *Arsenic leaching experiment*

The major activated media for Mesopaper to remove arsenic is the nano-scale powder. And then, we directly tested the arsenic leaching rate of nano-scale powder equipped with enough arsenic. 2 g nano-scale powder that had been adsorbed nearly 3.56 mg arsenic were weighted and putted into 100 mL solutions with different pH values of 3, 5, 7 and 9, respectively. After stirring for 5 minutes, the supernatant liquid was extracted to determination of arsenic concentrations.

2.5 *Arsenic determination*

The arsenic concentrations of all water samples in this report were measured using hydride generation-atomic fluorescence spectroscopy (HF-AFS).

3 RESULTS AND DISCUSSION

3.1 *Arsenic removal efficiency*

Table 1 displays the removal effect of As by Mesopaper. It was obvious that the arsenic concentrations of the three samples were all significantly decreased after filter through one layer and two layers of Mesopaper. In Gansu samples, for example, more than 90% arsenic was adsorbed by one-layer Mesopaper and the final arsenic concentrations was below 10 $\mu g\ L^{-1}$ at the sampling point of 0.2 L and 0.4 L. In addition, our findings have also suggested that the arsenic removal efficiency for two layers of Mesopaper was significantly higher than that shown in one layer. We supposed that the better efficiency might be associated with the relatively much more nano-scale powder and large surface area presented in two layers of Mesopaper. Together, all above findings have suggested that the Mesopaper has highly effective in treatment of field arsenic-contaminated water.

Table 2. The arsenic leaching rate of nano-scale powder.

pH values	Arsenic concentration ($\mu g\ L^{-1}$)	Arsenic leaching rate (%)
3	150	0.42
5	10	0.03
7	7.5	0.02
9	10	0.03

Note: The arsenic leaching rate were calculated as [(arsenic concentration * 100 mL)/3.56 mg].

3.2 *Arsenic leaching rate*

As it listed in Table 2, the arsenic leaching rate in solutions with pH values of 3, 5, 7 and 9 were all relatively low. The findings give evidence on that the nano-scale powder used in Mesopaper has no secondary arsenic pollution.

4 CONCLUSIONS

The Mesopaper has high effectiveness in treatment of field arsenic-contaminated water and without secondary arsenic pollution.

ACKNOWLEDGEMENTS

This work was sponsored by National Natural Science Foundation of China (Code number: 81703172).

REFERENCES

Naujokas, M.F., Anderson, B., Ahsan, H., Aposhian, H.V., Graziano, J.H., Thompson, C. & Suk, W.A. 2013. The broad scope of health effects from chronic arsenic exposure: update on a worldwide public health problem. *Environ. Health Perspect.* 121(3): 295.

Singh, R., Singh, S., Parihar, P., Singh, V.P. & Prasad, S.M. 2015. Arsenic contamination, consequences and remediation techniques: a review. *Ecotoxicol. Environ. Saf.* 112: 247–270.

Tang, J., Liao, Y., Yang, Z., Chai, L. & Yang, W. 2016. Characterization of arsenic serious-contaminated soils from Shimen realgar mine area, the Asian largest realgar deposit in China. *J. Soils Sediments* 16(5): 1519–1528.

Environmental Arsenic in a Changing World –
Zhu, Guo, Bhattacharya, Ahmad, Bundschuh & Naidu (Eds)
ISBN 978-1-138-48609-6

Decrease of arsenic in water from the Pampean plain (Argentina) by calcium salts addition

L. Cacciabue[1], S. Dietrich[1], P.A. Weinzettel[1], S. Bea[1], L. Sierra[1] & C. Ayora[2]
[1]*Instituto de Hidrología de Llanuras "Dr. Eduardo J. Usunoff" (IHLLA), Azul, Buenos Aires, Argentina*
[2]*Instituto de Diagnóstico Ambiental y Estudios del Agua (IDAEA-CSIC), Barcelona, Spain*

ABSTRACT: Arsenic (As) presence in groundwater from Argentinean Pampean plain is favored by high pH and Na-HCO_3^- type waters. Our hypothesis was that by adding Ca to the aquifer system, pH could decrease due to calcite precipitation and As may be retained in solid phase (sediments). Batch experiments were performed on natural sediments in contact with a Na-HCO_3^- solution, where $CaSO_4$ or $CaCl_2$ were added. Ca salts addition decreased down to 0.3 points of pH solution and as a result, As concentration in solution decreased. Two mechanisms are proposed. One is that a small decrease in pH enhanced As adsorption on Fe oxides. The second is that calcite precipitated and could trap As inside its structure.

1 INTRODUCTION

High arsenic (As) groundwater concentrations remain a critical concern in the Pampean plain of Argentina (Nicolli *et al.*, 2012). Groundwater from the Pampean aquifer presents a Na-HCO_3 hydrochemical signature and pH values higher than 7.5. Stratigraphic sequence in the area is composed of loess sediments which are interbedded with numerous petrocalcic layers (known as calcrete or tosca) indicating semi-arid paleoclimatic conditions (Zárate & Tripaldi, 2012). Natural occurrence of high As-bearing carbonates was reported in the literatures (Alexandratos *et al.*, 2007; Winkel *et al.*, 2013), suggesting that As retention by calcite precipitation could be a natural process in the groundwater systems of the Pampean plain. The aim of this work is two-fold: (1) to assess a potential As removal treatment based on the addition of Ca salts that induce the calcite precipitation, pH decrease, and As retention; (2) to propose the aforementioned mechanisms as the ones responsible for the high As concentrations that were measured in some tosca layers.

2 METHODS/EXPERIMENTAL

In order to verify the effectiveness of the proposed geochemical mechanisms, the experiment was simulated with the model PHREEQC (Parkhurst, 1995), by adding Ca salts to a Na-HCO_3^- initial solution. This simulation allowed calculating the Ca mass needed to produce a decreasing of pH from 9 to 7.

Prior to batch experiments, a synthetic Na-HCO_3^- solution was prepared with miliQ water and NaCl (EMSURE), $MgCl_2$ (Sigma-Aldrich) $NaHCO_3$ (Merck) and $CaSO_4$ (Merck) salts. Solution pH was adjusted with KOH.

Table 1. Textural and mineralogical characteristics of selected samples.

Sample	Minerals	Texture	Type
M12	Calcite, quartz, Na rich anorthite, Na-montmorillonite	Silt 54%, Sand 26%, Clay 20%	Tosca
M22	Quartz, calcite, Ca rich albite, Na-montmorillonite	Sand 55%, Silt 35%, Clay 10%	Sandy loess
M41	Quartz, Na rich anorthite, Na-montmorillonite	Silt 58%, Sand 27%, Clay 16%	Silty loess

Sediment samples were obtained from a 70 m deep borehole located in Tres Arroyos city, Buenos Aires province, Argentina (Sierra *et al.*, 2016). Selected samples correspond to 12, 22 and 41 m depth (M12, M22 and M41, respectively). Main features of selected samples are shown in Table 1. Sediment samples showed similar composition, although M41 is the only sample where calcite was not observed (Table 1). Previous studies showed that samples M12 and M41 are of silt grain size whereas M22 is a fine sand (Sierra *et al.*, 2016).

Five suspensions were prepared with the following solids: M12, M22, M41, M41Fe (1% of ferrihydrite added) and MQz (pure quartz sand, considered as a blank). Six 50 mL-tubes were filled with 2 g of each sample, which were washed with MiliQ-water, centrifuged and dried.

Then, 10 mL of $NaHCO_3$ synthetic solution with 500 $\mu g L^{-1}$ of As were added to each tube. After 24 h shaking, 3 and 7 mmol of $CaSO_4$ and $CaCl_2$ salts were added, respectively. After 48 h shaking, final pH was measured. Samples were centrifuged at 4500 rpm

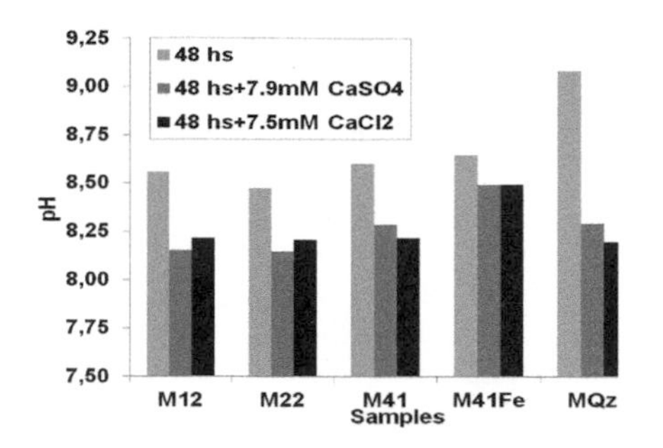

Figure 1. pH variations on samples through the experiments.

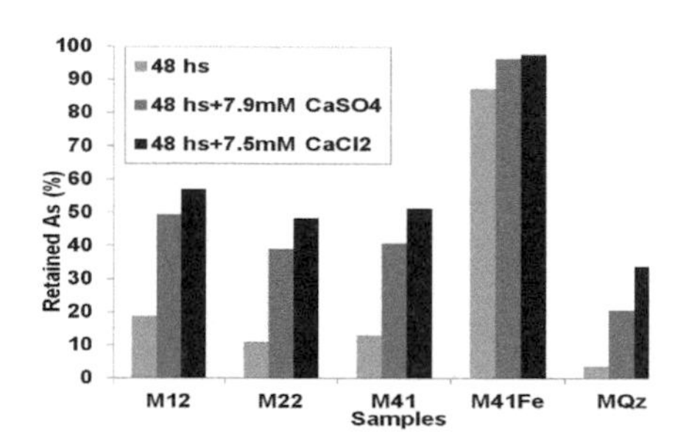

Figure 2. Retained As percentages on samples through the experiments.

during 15 min and filtered with nylon membrane of 0.2 μm pore. Aliquots of 5 mL were collected from each tube and acidified with HNO_3 10%. Arsenic concentrations were measured by inductively coupled plasma mass spectrometry (ICP-MS). Chloride interference was verified with an As blank solution.

3 RESULTS AND DISCUSSION

PHREEQC calculations yielded a decrease from 9.1 to 7.1 pH units in one liter of water by adding 7.9 mM of $CaSO_4$ or 7.5 mM of $CaCl_2$.

After 48 h, samples equilibrated with Na-HCO_3 solution at an approximate pH of 8.5 without salts addition (Fig. 1). This is because carbonates, which are present in almost all samples, have buffer properties. In this sense, quartz sample (MQz) had no initial change in pH due to its low reactivity with solution. After salts addition, pH showed a slight decrease, with a maximum of 0.3 units, compared with that obtained from PHREEQC.

Regarding As retention in absence of Ca salts, samples M12, M22 and M41 retained 13, 11 and 19% of total initial As, respectively (Fig. 2). Ca addition enhanced the As retention in all samples, which increased to 51, 48 and 57%, respectively.

The addition of $CaCl_2$ showed slightly more efficiency than $CaSO_4$ in the As retention. Sample M41Fe achieved 87% of As retention in the absence of Ca salts due to the presence of ferrihydrite, which increases the specific surface area practically adsorbing the initial mass of As (Dixit & Hering, 2003). The Ca salts addition in this sample raised the As retention in a 10% only. Sample MQz denoted an As retention of 30% by adding Ca salts, suggesting that the precipitated calcite could retain As into its structure or may adsorbed on its surface.

4 CONCLUSIONS

Ca salts addition to a synthetic water sample, similar to Pampean groundwaters, generated only a slight pH decrease. However, As retention on solid samples increased up to the double of its retaining capacity.

Although a decrease in pH favored As sorption on Fe oxides, calcite precipitation retained almost 30% of total As.

Arsenic uptake by calcite may be explained by entering calcite structure or by adsorbing on calcite surface.

ACKNOWLEDGEMENTS

This work was funded by the research project PID-075/2012 titled "Spatial-Temporal evolution of arsenic in groundwater of Argentina Republic". Authors gratefully acknowledge the cooperation of Institut de Diagnosi Ambiental i Estudis de l'Aigua (Barcelona) to provide working place and chemical analyses. Special thanks to Dr. Ester Torres who had suggested optimization procedures for the experiments.

REFERENCES

Alexandratos, V.G., Elzinga, E.J. & Reeder, R.J. 2007. Arsenate uptake by calcite: macroscopic and spectroscopic characterization of adsorption and incorporation mechanisms. *Geochim. Cosmochim. Acta* 71(17): 4172–4187.

Dixit, S. & Hering, J.G.2003. Comparison of arsenic (V) and arsenic (III) sorption onto iron oxide minerals: implications for arsenic mobility. *Environ Sci. Technol.* 37(18): 4182–4189.

Nicolli, H.B., Bundschuh, J., Blanco, M.D.C., Tujchneider, O.C., Panarello, H.O., Dapena, C. & Rusansky, J.E. 2012. Arsenic and associated trace-elements in groundwater from the Chaco-Pampean plain, Argentina: results from 100 years of research. *Sci. Total Environ.* 429: 36–56.

Parkhurst, D.L. 1995. User's guide to PHREEQC–a computer program for speciation, reaction-path, advective transport and inverse geochemical calculations. U.S.G.S. Water Resources Investigations Report. 95–4227.

Sierra, L., Cacciabue, L., Dietrich, S., Weinzettel, P.A. & Bea, S.A. 2016. Arsenic in groundwater and sediments in a loessic aquifer, Argentina. In: Arsenic Research and Global Sustainability: *Proceedings of the 6° International Congress on Arsenic in the Environment.* Stockholm, Sweden.

Winkel, L.H.E., Casentini, B., Bardelli, F., Voegelin, A., Nikolaidis, N.P. & Charlet, L. 2013. Speciation of arsenic in Greek travertines: co-precipitation of arsenate with calcite. *Geochim. Cosmochim. Acta* 106: 99–110.

Zárate, M. & Tripaldi, A. 2012. The aeolian system of central Argentina. *Aeolian Res.* 3(4): 401–417.

Environmental Arsenic in a Changing World –
Zhu, Guo, Bhattacharya, Ahmad, Bundschuh & Naidu (Eds)
ISBN 978-1-138-48609-6

Adsorption and photocatalytic study of calcium titanate ($CaTiO_3$) for the arsenic removal from water

R.M. Tamayo Calderón, R. Espinoza Gonzales & F. Gracia Caroca
Department of Chemical Engineering Biotechnology and Materials, Faculty of Physical and Mathematical Sciences, University of Chile, Santiago de Chile, Chile

ABSTRACT: We report the arsenite adsorption in dark conditions and different pH by $CaTiO_3$ perovskite nanoparticles. Its photocatalytic properties for oxidizing the arsenite As(III), to arsenate As(V), under UV light source have been studied in this research. $CaTiO_3$ (CTO), nanoparticles have a high As(III) removal capacity. Nearly 70% As(III) was oxidized in few minutes, and achieves full oxidizing of As(III) to As(V) in around 90 min. Our results suggest the prepared CTO is a promising candidate for arsenic removal.

1 INTRODUCTION

Arsenic is one of the most dangerous contaminants in drinking water, which is present in different parts of the world, like in northern Chile and southern Peru. Some studies have proposed the chronic exposure to the arsenic contaminated water as the cause of diverse diseases (Mandal, 2002).

There are different methods to remove the arsenic from the water, being one of the most promising method due to its low-cost, high efficiency and its facile operation. The water purification by adsorption using nanoparticles (NPTs) and, in particular using perovskites, has a new potential as environmental remedial. The CTO is a perovskite known for its electrical and optical properties (Singh *et al.*, 2014), and recently it has also been studied for the As(III) removal using its photocatalytical properties (Zhuang *et al.*, 2014).

In this study, CTO nanoparticles were synthesized, characterized and applied as a photoactive material to oxidize As(III) to As(V) by photocatalysis and remove the arsenic from the water by adsorption process.

2 EXPERIMENTAL

2.1 *Materials*

All reagents were of analytical grade and were used as received without further purification.

The reagents calcium nitrate tetrahydrate (80%), titanium (IV) isopropoxide (90%), citric acid (65%), 2-propanol (90%) and sodium arsenite ($NaAsO_2$, 95%) were purchased from Sigma Aldrich. The stock solution of $15\,mg\,L^{-1}$ arsenite was obtained by dissolving $NaAsO_2$ in 1 L deionized water. The pH was adjusted to 1, 3, 5 and 7 using hydrochloric acid.

2.2 *Synthesis of $CaTiO_3$*

The CTO was prepared by sol-gel method, calcium nitrate tetrahydrate, titanium (IV) isopropoxide, and citric acid, were simultaneously stirred in 2-propanol respectively for 30 min, then calcium nitrate solution and citric acid solution were added dropwise to titanium solution, and stirred for 30 min, after the deionized water were added forming a gel. The gel was dried at 50°C for 4 days, the resulting powder was calcined at 600°C for 1 h, obtaining the CTO perovskite.

2.3 *Characterization of $CaTiO_3$*

The crystal structure of CTO was analyzed by X-ray diffraction (Bruker, D8), the size and shape were measured by Transmission electron microscopy technique (FEI, Tecnai ST F20), the specific surface area was measured by BET sorptometer (Micromeritics ASAP 2010), while the zeta potential of CTO nanoparticles was measured by (ZEN 3600, Malvern). The arsenic species concentration was determined using a coupled high-performance liquid chromatography (HPLC) hydride generation (HG) atomic fluorescence spectrometry (AFS).

2.4 *Arsenic removal test*

The adsorption tests were developed at room temperature and in dark conditions and different pH. The oxidation of As(III) to As(V) was made at room temperature and under UV light radiation (OSRAM, 254 nm) and pH = 7. All the tests began with 300 mL of water with a $13\,mg\,L^{-1}$ of As(III) concentration. The dosage of CTO in all the solutions was $1\,g\,L^{-1}$, the test was made under constant stirring with the rate of 300 rpm.

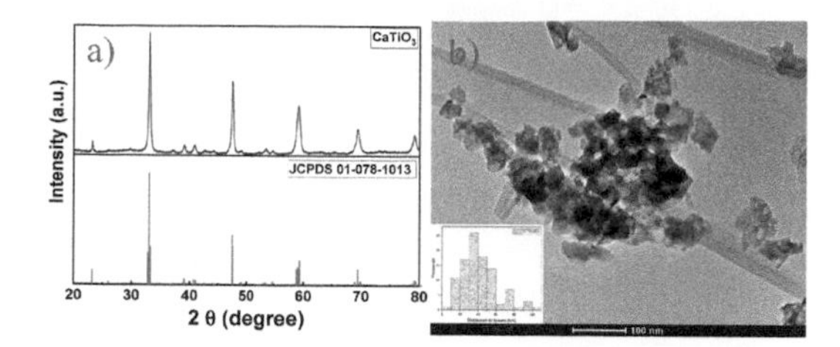

Figure 1. CTO a) X-ray diffraction pattern, b) TEM micrograph.

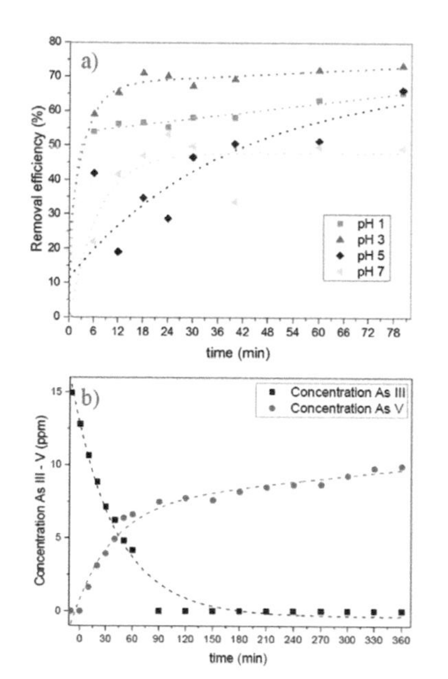

Figure 2. Arsenic removal in a) dark and b) under UV radiation.

3 RESULTS

3.1 *Characterization results*

The X-ray diffraction result has good agreement with JCPDS 01-078-1013 (Fig. 1a) corresponding to calcium titanate perovskite with orthorhombic structure. The crystallite size was calculated by Scherrer equation giving a 20.9 nm. As shown in Figure 1b. The nanoparticles have an irregular shape and the most frequent size is between 10 nm to 60 nm. The specific surface area of CTO nanoparticles reaches 43.85 m^2 g^{-1}. Zeta potential results indicate to pH = 3.5 as zero charge point, which indicates a positively charged outer surface of CTO nanoparticles.

3.2 *Arsenic removal results*

The As(III) adsorption was studied at pH 1, 3, 5 and 7. The best performance was obtained at pH 3, where CTO achieves ~70% of As removal in 18 min (Fig. 2a) and it increased with the pH increase. At pH less than

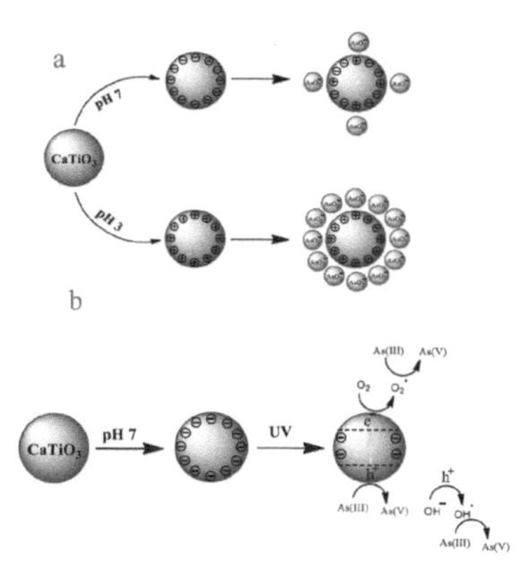

Figure 3. Proposed mechanism for a) adsorption in dark conditions and different pH, b) oxidation of arsenic (III) in arsenic (V) by photocatalysis.

3.5 the mechanism is governed by electrostatic adsorption, while at high pH, the adsorption is related to oxygen vacancies and Ca^+ and Ti^{4+} cations exposure on the surface of CTO nanoparticles (Fig. 3a).

The results showed in Fig. 2b indicate that the oxidation process of As(III) to As(V) is completed in 90 min, having a high oxidation rate during the first 30 min. The oxidation process is described by the following reactions:

$$CaTiO_3 + h\nu \rightarrow CaTiO_3(e^-_{cb}) + CaTiO_3(h^+_{vb})$$
$$CaTiO_3(h^+_{vb}) + As(III) \rightarrow CaTiO_3 + As(V)$$
$$CaTiO_3(e^-_{cb}) + As(III) \rightarrow CaTiO_3 + \cdot O_2^-$$
$$O_2^- + H^+ + As(III) \rightarrow As(V) + HO_2^-$$

The proposed mechanism is shown in Figure 3.

4 CONCLUSIONS

CTO perovskite nanoparticles was successfully synthesized by sol-gel method. The results of adsorption and oxidation of As(III) suggest that the prepared $CaTiO_3$ is a promising candidate for being used in arsenic removal applications.

ACKNOWLEDGEMENTS

Authors would like to thank the projects: FONDECYT PROJECT N° 1150652, FINCyT N° 078-2014.

REFERENCES

Mandal, B. 2002. Arsenic round the world: a review. *Talanta* 58(1): 201–235.

Singh, L., Rai, U.S., Mandal, K.D. & Singh, N.B. 2014. Progress in the growth of $CaCu_3Ti_4O_{12}$ and related functional dielectric perovskites. *Prog. Cryst. Growth Ch.* 60(2): 15–62.

Zhuang, J., Tian, Q., Lin, S., Yang, W., Chen, L. & Liu, P. 2014. Precursor morphology-controlled formation of perovskites $CaTiO_3$ and their photo-activity for As(III) removal. *Appl. Catal. B* 156: 108–115.

Environmental Arsenic in a Changing World –
Zhu, Guo, Bhattacharya, Ahmad, Bundschuh & Naidu (Eds)
ISBN 978-1-138-48609-6

TiO_2 facet-dependent arsenic adsorption and photooxidation: Spectroscopic and DFT study

L. Yan & C.Y. Jing
Research Center for Eco-Environmental Sciences, Chinese Academy of Sciences, Beijing, P.R. China

ABSTRACT: Anatase TiO_2 nanomaterials have been widely used in arsenic (As) remediation, although reports on their adsorption and photocatalytic capacity have been controversial. The motivation for our study is to explore the As adsorption and photooxidation processes on different TiO_2 facets at the molecular level. Our results from multiple complementary characterization techniques suggest that anatase {001} facets have stronger Lewis acid sites than those on {101} facets, resulting in a higher As adsorption affinity. Density functional theory (DFT) calculations confirmed that the As surface complex is more energetically favorable on {001} than on {101} facets. In addition, the strong interaction of {001} facets with molecular O_2 facilitates the transfer of photo-excited electrons to the adsorbed O_2 to generate superoxide radical ($O_2^{\bullet-}$), which is the primary As(III) oxidant as evidenced by our radical-trapping experiments. The insights gained from this study provide a firm basis for the proposition that As adsorption and photoactivity can be mediated by tailoring the exposed TiO_2 facets, which is of essence in the design and application of TiO_2-based environmental technologies.

1 INTRODUCTION

Adsorption and photooxidation on TiO_2 nanomaterials provide a promising technique for arsenic (As) removal. Though the As adsorption mechanism is well known to involve the formation of a bidentate binuclear surface complex (Hu *et al.*, 2015), even TiO_2 of the same anatase phase from different sources exhibit distinct adsorption and photo-catalytic capacities. The lack of conformity of many experimental observations with the general belief that particle size or surface area regulates TiO_2 adsorption and photocatalysis motivates our study.

Recent intriguing studies have shown that the adsorption and photocatalytic activity of TiO_2 largely depend on its surface atomic structure and the exposed crystal facets (Liu *et al.*, 2014). The majority of exposed facets for most anatase are less-reactive and thermodynamically stable {101} facets (about 94%) with a minor proportion of {001} facets. Conversely, water and dye molecules prefer to bind to anatase {001} rather than {101} facets. However, the molecular level knowledge on TiO_2 facets determining As adsorption is not fully understood.

The high photocatalytic performance for anatase {001} facets has been explained by hydroxyl radical (•OH), which is proposed to be the main oxidant for As(III) photooxidation in the TiO_2/UV system. In contrast, other studies suggested that superoxide radical ($O_2^{\bullet-}$) should play a dominant role. The controversy in the As(III) photooxidation mechanism may stem from the knowledge gap concerning the TiO_2 facets effect on the generation of reactive oxygen species (ROS). Therefore, we hypothesize that anatase TiO_2 facets may affect ROS generation and ultimately determine photocatalytic performance.

The objective of this study was to identify the primary factor influencing As adsorption and photocatalysis on anatase TiO_2, as well as its mechanism. Multiple complementary characterization methods and density functional theory (DFT) calculations were employed to explore the molecular-level processes on the {001} and {101} facets. The intrinsic facet-dependent mechanism obtained from this study would be fundamental in developing TiO_2-based environmental technologies with high adsorption capacity and photocatalytic activity.

2 METHODS/EXPERIMENTAL

2.1 *As(III) and As(V) adsorption on TiO_2*

Three different kinds of TiO_2 nanoparticles (NPs), including one homemade (HM) and two commercial samples (JR05 and TG01), were used in this study. Adsorption isotherm experiments were performed to determine As(III) and As(V) adsorption capacity on the three types of TiO_2 in 0.04 M NaCl solution at pH 7. Suspension samples containing 0.31–3450 mg L^{-1} As(III) or 0.32–4484 mg L^{-1} As(V) and 5 g L^{-1} TiO_2 were adjusted to pH 7 with NaOH and HCl. The samples were covered with aluminum foil to prevent light exposure and mixed on a rotator for 72 h. Then, the final pH was measured and the suspensions were filtered through a 0.22 μm membrane filter for measurement of soluble As using a furnace atomic absorption spectrometer (FAAS, Perkin-Elmer AAS-800).

2.2 *As(III) photooxidation*

As(III) photooxidation were carried out in an Erlenmeyer flask with 14 mg-As(III) L^{-1} in a suspension containing 0.3 g L^{-1} TiO_2 and 0.04 M NaCl at pH 7. The sample was stirred in the dark for 2 h to achieve equilibrium before being illuminated by a tubular mercury UV lamp (CEL-WLPM10-254, wavelength 254 nm) with an incident light intensity of about 4000 μW cm^{-2}. The photocatalysis process was monitored as a function of illumination time by taking 1 mL aliquots from the suspension and filtering through a 0.22-μm membrane. The As concentration and speciation were determined using high performance liquid chromatography (HPLC) coupled with a hydride generation atomic fluorescence spectrometer (HG-AFS, Jitian, P.R. China).

3 RESULTS AND DISCUSSION

3.1 *Facet dependence of As adsorption*

The adsorption isotherms of As(III) and As(V) on the three TiO_2 NPs conformed to the Langmuir model. Unexpectedly, no linear dependence was observed for the maximum adsorption capacity (q_m) on S_{BET}. For example, the S_{BET} of JR05 (167 m^2 g^{-1}) was about half that of HM (336 m^2 g^{-1}); however, the q_m of JR05 for As(III) (138 mg g^{-1}) and As(V) (144 mg g^{-1}) were comparable to those of HM (136 and 140 mg g^{-1}, respectively, for As(III) and As(V)). The results suggest that S_{BET} may not be the key factor influencing the As adsorption capacity on TiO_2 NPs. Notably, the acid-site-normalized loadings of As(III) and As(V) on JR05, 51.1 and 53.3 mol per mol acid, respectively, were approximately twice as high as for the other two TiO_2 NPs, indicating that the acid sites on JR05 have a greater strength. The NH_3-TPD characterization indicated that 46% strong acid sites existed on JR05 sample.

The distribution of acid sites is closely related to surface atomic coordination and arrangement, indicating that different crystal facets may endow the same Lewis acid sites with different strengths. This assumption was justified by our DFT calculations. The NH_3 adsorption energies were estimated to be −2.43 and −1.14 eV for {001} and {101} facets, respectively, indicating that {001} facets exhibit stronger acid sites than those on {101} facets. We then proposed that the high As adsorption capacity on JR05 should be related to its exposed facets and their relative proportions.

DFT calculations were performed to compare the As adsorption energies on TiO_2 {001} and {101} facets. The results indicated that the surface complexes are more energetically favorable on {001} than on {101} facets due to the 100% five coordinated (Ti_{5c}) sites on {001} facet.

3.2 *As(III) oxidation on three types of TiO_2*

As(III) oxidation on the surfaces of the three TiO_2 NPs follows first-order kinetics. The first-order rate constant, k, varied in the order TG01 (0.0225 min^{-1}) > JR05 (0.0213 min^{-1}) > HM (0.006 min^{-1}). The different As(III) photooxidation rate inspired our further study to characterize ROS effects. Our ESR and fluorescence experimental results show that both •OH and $O_2^{\bullet -}$ existed for all TiO_2 samples. The radical trapping experiments suggested that $O_2^{\bullet -}$ dominated As(III) photooxidation on TiO_2. The $O_2^{\bullet -}$ formation is the result of the trapping photogenerated electrons by molecular O_2 adsorbed on the TiO_2 surface. Because different TiO_2 facets exhibit distinct adsorption affinities for NH_3 and As, a reasonable postulation is that the exposed TiO_2 facets would influence the adsorption of molecular O_2, and as a result, affect the formation of $O_2^{\bullet -}$.

To understand the facet effect on $O_2^{\bullet -}$ generation, DFT calculations were performed. The results suggested a more negative adsorption energy (−2.48 eV) and shorter Ti-O bond distance (2.012 Å) on {001} compared to those on {101} facets (−1.79 eV and 2.033 Å). The strong interactions between O_2 and TiO_2 {001} facet may facilitate the electron transfer from the surface to O_2, forming $O_2^{\bullet -}$.

4 CONCLUSIONS

This work exploited structural effects on the adsorption and photocatalytic performance of TiO_2. We found the distribution of exposed crystal facets is the key factor. TiO_2 {001} facets, with stronger Lewis acid Ti^{4+} atoms, exhibit higher adsorption affinity than {101} facets. Furthermore, the strong interactions of {001} facets with oxygen molecules lead to more photoelectrons being transferred to the adsorbed O_2 to generate superoxide radical, which is the primary oxidant in the As(III) photooxidation. Our experimental and theoretical results provide new insights into the intrinsic origin of the facet-activity relationship. The mechanism of the facet dependence of adsorption and photocatalysis should be applicable to other pollutant molecules beyond arsenic.

ACKNOWLEDGEMENTS

We acknowledge the financial support of the Strategic Priority Research Program of the Chinese Academy of Sciences (XDB14020201).

REFERENCES

Hu, S., Yan, L., Chan, T.S. & Jing, C.Y. 2015. Molecular insights into ternary surface complexation of arsenite and cadmium on TiO_2. *Environ. Sci. Technol.* 49(10): 5973–5979.

Liu, G., Yang, H.G., Pan, J., Yang, Y.Q., Lu, G.Q. & Cheng, H.M. 2014. Titanium dioxide crystals with tailored facets. *Chem. Rev.* 114(19): 9559–9612.

Environmental Arsenic in a Changing World –
Zhu, Guo, Bhattacharya, Ahmad, Bundschuh & Naidu (Eds)
ISBN 978-1-138-48609-6

Sorption studies and characterization of developed biochar composites for As(III) adsorption from water

P. Singh & D. Mohan
School of Environmental Sciences, Jawaharlal Nehru University, New Delhi, India

ABSTRACT: Arsenic in groundwater is a serious problem. This paper investigates the role of biochar derived from agricultural waste byproducts to remediate arsenic from water. Composites were synthesized by thermal and chemical modifications in biomass. The biochar composites were characterized for their BET surface area and porosity. XPS, SEM, TEM, XRD, FT-IR, PPMS, ICP-MS were studied to analyze their magnetic moment, surface chemistry, mineralogy, crystallinity, elemental composition and functional group identification. Batch sorption studies, sorption equilibrium and kinetic studies were conducted using various mathematical equations and isotherm models to find sorption efficiency. As(III) removal occurred in pH 7.5–9.0. High adsorption capacities were reported for rice husk and wheat husk composites. Therefore, it can be considered to replace commercial adsorbents. Plausible mechanism and chemistry of As(III) adsorption was thus established.

1 INTRODUCTION

Arsenic is one of the most toxic contaminants in water. WHO has reported arsenic contamination in drinking water to have caused greatest number of deaths as compared to other heavy metals. WHO's permissible limit for arsenic in drinking water is 0.01 mg L^{-1}. In India, states of West Bengal, Jharkhand, Bihar, Uttar Pradesh, Assam, Manipur and Chhattisgarh are reported to be most affected (Bhattacharya *et al.*, 2011; Singh *et al.*, 2015). Long term exposure is fatal to humans. Thus, removal of arsenic has become subject of great importance and attention. Out of many existing forms of As, As(III) is more toxic than As(V). Thus, low cost, simple performing techniques and methods have been used to obtain materials with high As removal efficiency (Hu *et al.*, 2015; Ming *et al.*, 2013; Mohan, 2007; Sun *et al.*, 2015).

2 METHODS/EXPERIMENTAL

2.1 *Material*

Adsorption technique because of its simplicity and cost effectiveness has been used to remove arsenic. All chemicals, iron salt, ascorbic acid, potassium iodide, sodium borohydride were obtained from Merck. pHs were maintained using 0.1 N HNO_3 and 0.1 N NaOH. All studies were conducted using double distilled water. Analysis was carried out using AAS over mercury hydride system. The biochar was prepared using muffle furnace. The resulting biochar was used for arsenic removal.

2.2 *Method*

Biochar composite is developed using slow pyrolysis method in a muffle furnace. Biomasses were soaked in solution prepared from a known iron compound. The soaked biomasses in iron compound solution were then subjected to slow pyrolysis in muffle furnace under suitable thermal and temporal conditions. Iron biochar composite so obtained were washed, dried and used for conducting studies. pH optimization studies have been carried out in range 2–10. As(III) solution concentrations taken for conducting kinetic and isotherm studies ranged from 50–200 ug L^{-1}. Characterization of material was carried over various instruments before and after arsenic adsorption to find the functional group, mineralogy, crystallinity (Table 1). Plausible mechanism for As(III) adsorption over biochar surface was thus established.

3 RESULTS AND DISCUSSION

3.1 *Characterization of adsorbents*

Both biochar composites were characterized for their surface area and porosity. High surface area and porosity was reported. Qualitative and quantitative analysis was done to find mineral composition, % C, H, N and S content for both the biochar composites. Magnetic moment, oxidation states of various elemental species, mineralogy, crystallinity, elemental composition and functional groups were identified using PPMS, XPS, SEM, EDXRF, SEM-EDX, XRD, FTIR and ICP-MS (Fig. 1). Surface chemistry and hence the functional group identification made it possible to establish the mechanism involved in the adsorption of As(III) from water.

3.2 *Effect of initial pH*

Sorption studies were performed in batch mode to obtain the equilibrium and kinetic constants and pH impact over arsenic adsorption from water on bio-char

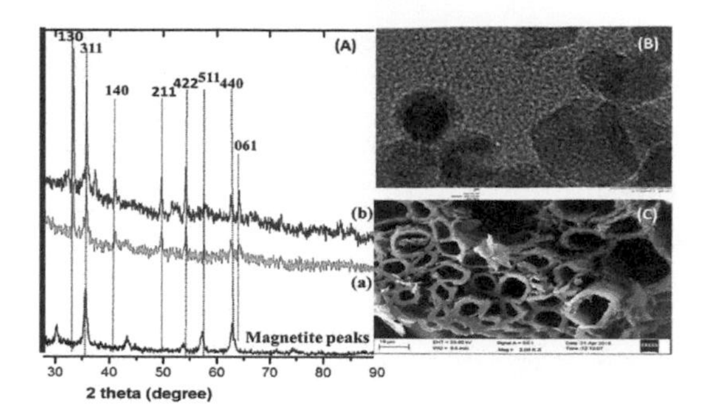

Figure 1. (A) XRD peaks (a) rice husk and (b) wheat husk biochar composite, (B) TEM image and (C) SEM image of biochar composites.

Table 1. Instruments used for adsorbent characterization.

Characterization techniques	Purpose
XPS	Oxidation states of elemental species
Quantachrome surface area analyzer	Specific surface area
SEM/EDX	Surface morphology
TEM	Crystallinity, magnetic domains & stress
FT-IR	Functional group identification
FT-Raman	Compositional studies
XRD	Mineral composition
EDXRF	Elemental composition
PPMS	Magnetic moment

composites. Batch studies are considered because of their simplicity.

Preliminary results have shown good removal and efficiency so far. The maximum arsenic occurred in a pH range of 7.5–9.0 which is near to neutral pH. This is advantageous as this will not require the adjustment of pH of water after arsenic removal.

3.3 *Adsorption kinetics and isotherm*

Kinetic studies were performed to examine the effect of adsorbent dose, contact time and temperature on arsenic adsorption rate. Pseudo second and first order rate equations were applied to establish adsorption kinetics and rate. Effect of adsorbent dose ($2\,g\,L^{-1}$, $1\,g\,L^{-1}$ and $0.5\,g\,L^{-1}$) on arsenic removal was studied. Kinetic experiments have showed good arsenic removal at low adsorbent dose. To establish mechanism and find the affinity between sorbate and sorbent, isotherm studies were conducted. Experimental values of adsorption capacities for rice husk and wheat husk biochar composites obtained were higher.

Adsorption data was modeled using Freundlich, Langmuir, Redlich-Peterson, Sips, Koble Corrigan, Radke-Prausnitz and Toth models to find various parameters necessary to design fixed bed reactors and find monolayer adsorption capacity. Information obtained from characterization has been applied to establish the probable mechanism for arsenic adsorption.

3.4 *Thermodynamic studies*

Thermodynamic studies were conducted to examine the adsorption behavior of arsenic onto the biochar surface. The parameters give the information about the feasibility and the spontaneity of the adsorption in terms whether adsorption process is endothermic or exothermic in nature. In following studies, the process was found to be endothermic in nature.

Desorption studies recover the toxic heavy metals from adsorbent for safe disposal as well as in keeping the process cost down.

4 CONCLUSIONS

Results from sorption studies have showed maximum removal of As(III). Kinetic have shown good adsorption capacity at lower optimum dose. Characterization studies provided information from which probable mechanisms, conditions, properties etc. of As(III) adsorption was established. This will now be utilized to conduct experiments with real groundwater samples to be collected from study area.

Here biochar is obtained from agricultural waste product that is very cost effective. These biochars can easily be used by developing modified biochar candles or columns to obtain arsenic free drinking water. Thus, composite biochars is considered to be a novel method for arsenic remediation. The biochar composites exhibited comparable sorption efficiency for arsenic removal. Therefore, this can easily replace commercial adsorbents.

ACKNOWLEDGEMENTS

One of the authors (PS) thanks to DST-INSPIRE for providing financial support to this work.

REFERENCES

Bhattacharya, P., Mukherjee, A. & Mukherjee, A.B. 2011. Arsenic in groundwater of India. In: J.O. Nriagu (ed) *Encyclopedia of Environmental Health*, vol. 1. Burlington: Elsevier, pp. 150–164.

Hu, X., Ding, Z., Zimmerman, A.R., Wang, S. & Gao, B. 2015. Batch and column sorption of arsenic onto iron impregnated biochar synthesized through hydrolysis. *Water Res.* 68: 206–216.

Ming, Z., Gao, B., Varnoosfaderani, S., Hebard, A., Yao, Y. & Inyang, M. 2013. Preparation and characterization of a novel magnetic biochar for arsenic removal. *Bioresour. Technol.* 130: 4457–462.

Mohan, D. 2007. Arsenic removal from water/waste using adsorbents – a critical review. *J. Hazard. Mater.* 142(1–2): 1–53.

Singh, S.K. 2015. Groundwater arsenic contamination in the Middle-Gangetic plain, Bihar (India): the danger arrived. *Int. Res. J. Environ. Sc.* 4(2): 70–76.

Sun, L., Chen, D., Shungang, W. & Zebin. Y. 2015. Performance, kinetics and equilibrium of methylene blue adsorption on biochar derived from eucalyptus saw dust modified with citric, tartaric, and acetic acids. *Bioresour. Technol.* 198: 300–308.

Environmental Arsenic in a Changing World –
Zhu, Guo, Bhattacharya, Ahmad, Bundschuh & Naidu (Eds)
ISBN 978-1-138-48609-6

Evaluating the arsenic removal potential of Japanese oak wood biochar in aqueous solutions and groundwater

N.K. Niazi[1], I. Bibi[1], M. Shahid[2] & Y.S. Ok[3]
[1]*Institute of Soil and Environmental Sciences, University of Agriculture Faisalabad, Faisalabad, Pakistan*
[2]*COMSATS Institute of Information and Technology, Vehari, Pakistan*
[3]*Korea Biochar Research Center, Korea University, Republic of Korea*

ABSTRACT: This study evaluated arsenic (As) removal efficiency of Japanese oak wood biochar (JOW-BC) produced at 500°C in aqueous environments. Langmuir model was the best to fit arsenite (As(III)) and arsenate (As(V)) sorption on JOW-BC, with slightly higher sorption affinity, Q_L, for As(V) than As(III). The maximum As removal was 81% and 84% for As(III)- and As(V)-JOW-BC systems at pH 7 and 6, respectively, which decreased above these pH values. Surface functional moieties contributed to As sequestration by the biochar examined here. Arsenic K-edge XANES spectroscopy demonstrated complex redox transformation of As(III)/As(V) with JOW-BC. The JOW-BC led to remove As from As-contaminated drinking well water with As depletion ranging from 92–100% of total As.

1 INTRODUCTION

Arsenic contamination of groundwater represents a global, health and environmental issue due to its toxic nature. Although many types of sorbents (e.g., activated carbon, Fe-coated granular AC, Fe-containing fly ash, nano-adsorbents), Japanese oak wood biochar (JOW-BC) has not received attention to remove As(III/V) from aqueous solutions, as well as As-laced groundwater. In this study, we examined removal efficiency of JOW-BC, produced at low 500°C temperature, for As(III) and As(V) in aqueous solutions, and explored its use for removal of As in As-contaminated drinking well water.

2 METHODS/EXPERIMENTAL

2.1 *Biochar production and analysis*

The Japanese oak wood chipping waste was collected, washed with tab water followed by rinsing with deionized water to remove impurities. After washing, wood chippings were air-dried and ground (<2 mm) to yield uniform size particles prior to pyrolysis process. Various physicochemical properties of JOW-BC were determined.

2.2 *Sorption experiments*

Batch sorption experiments were performed in 50 mL polyethylene (reaction) vials using the JOW-BC in a background electrolyte solution of 0.01 M NaCl. All experiments were performed at a constant temperature of 20 ± 1°C with an equilibration time of 2 h. Effect of pH (pH 3–10) was evaluated at an initial As(V) and As(III) concentrations of 4 mg L^{-1} and constant biochar suspension density of 1 g L^{-1} Suspension pH was maintained by adding 0.01 M HCl or NaOH solutions.

Sorption isotherm experiments were performed at initial As(V) and As(III) concentrations of 0.05–7.0 mg L^{-1}. Arsenic was analyzed using a hydride generation-atomic absorption spectrometer (HG-AAS; Agilent AA240 with VGA 77). Fourier transform infrared spectroscopy, SEM-EDX and X-ray absorption near edge structure (XANES) spectroscopy were employed to determine surface functional groups, morphology and speciation of As in As(III)- and As(V)-loaded biochar samples.

3 RESULTS AND DISCUSSION

Arsenic sorption isotherm experimental data indicated that As removal by JOW-BC increased the increasing initial As(V) and As(III) concentrations. Relatively greater sorption was obtained for As(V) (0.046–1.97 mg g^{-1}) than that of As(III) (0.038–1.76 mg g^{-1}). The isotherm modeling of As(III) and As(V) sorption data showed that, among the four models, Langmuir model achieved the best fit for As(III) and As(V) sorption to JOW-BC (Fig. 1), with a slightly better fit for As(V) than that of As(III) ($R^2 = 0.91$ and 0.85, respectively). This indicates that JOW-BC led to a monolayer (homogeneous) As sorption (Abid *et al.*, 2016).

The effect of solution pH on As(III) and As(V) removal percentage by JOW-BC indicated an increase in As(III) removal (74–80%) at pH from spanning 3–7, with the highest As(III) removal at pH ~7. Whilst,

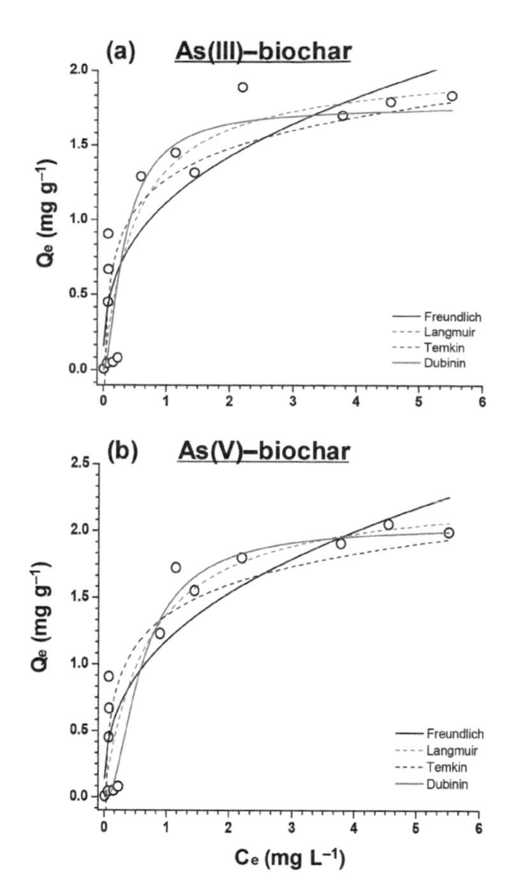

Figure 1. Sorption isotherm models of (a) As(III); and (b) As(V), sorption to JOW-BC, at pH 7.0, initial As(III) and As(V) concentrations of 0.05 to 7 mg L^{-1} and temperature 20 ± 1°C in aqueous solutions.

above pH 7 a decline was observed in the As sorption (76–69%) in As(III)-JOW-BC sorption systems. Conversely, As(V) removal percentage was the highest at pH 6 (85%) in As(V)-JOW-BC experiments (Bibi *et al.*, 2017).

Different surface functional groups, mainly –OH, –COOH, -C-O, $-CH_3$, are thought to be responsible in As removal by JOW-BC, suggesting the surface complexation, precipitation and/or electrostatic interaction of As on JOW-BC surface (Abid *et al.*, 2016; Bibi *et al.*, 2017).

The scanning electro microscopy combined with energy dispersive X-ray spectroscopy (SEM-EDX) indicated relatively more rough, irregular and porous micro-structure of As-unloaded JOW-BC, which is due to its high specific surface area (476 $m^2 g^{-1}$). The elemental dot maps from SEM-EDX analysis indicated, although partially, that As was mainly sorbed via interaction with carbon and oxygen bearing groups and/or calcium-bearing mineral species in JOW-BC. Arsenic association, mainly as As(V), could occur with calcium (present as $CaCO_3$-like species in biochar), thus making relatively insoluble Ca-As precipitates to immobilize As in biomass.

Arsenic K-edge X-ray absorption near edge structure (XANES) spectroscopy revealed that added As(III) was partially oxidized to As(V) (38%) in As(III)-JOW-BC system, and in As(V)-JOW-BC system, 46% of added As(V) was reduced to As(III).

We observed that 91 to 100% of As was removed in As-contaminated drinking well water (total As: 25–146 μg L^{-1}; total samples = 10), showing a great potential for remediation of As-rich well water.

4 CONCLUSIONS

The JOW-BC efficiently removed both As(III) and As(V) in aqueous solutions, as well as As from As-contaminated drinking well water. Langmuir model demonstrated the best fit for As(III) and As(V) sorption to JOW-BC, which was also supported by our FTIR spectroscopy and SEM-EDX elemental dot mapping data.

Our XANES data revealed complex redox transformation of As(III)/As(V) in the As(III)- and As(V)-JOW-BC systems. Application of JOW-BC to natural As-contaminated drinking well water showed up to 100% As removal nevertheless in the presence of a broad suite of anions and cations co-occurring in well water. These results are crucial for understanding As fate in water and/or in soil and sediment pore water, as well as for developing an efficient biochar technology in environmental remediation programs of As-contaminated systems.

ACKNOWLEDGEMENTS

We are thankful to the IFS (Sweden; W/5698-1) for financial support. This work was supported by the National Research Foundation of Korea (NRF) (NRF-2015R1A2A2A11001432). Thanks are extended by Dr Bibi (Ref 3.5 – PAK – 1164117 – GFHERMES-P) to the Alexander von Humboldt Foundation for a Postdoctoral Research Fellowship. The authors thank the members of the Pohang Accelerator Laboratory (PAL), Korca for providing synchrotron user facilities for XAFS spectroscopy experimentation.

REFERENCES

Abid, M., Niazi, N.K., Bibi, I., Farooqi, A., Ok, Y.S., Kunhikrishnan, A., Ali, F., Ali, S., Igalavithana, A.D. & Arshad, M. 2016. Arsenic(V) biosorption by charred orange peel in aqueous environments. *Int. J. Phytoremediation* 18(5): 442–449.

Bibi, S., Farooqi, A., Yasmin, A., Kamran, M.A. & Niazi, N.K. 2017. Arsenic and fluoride removal by potato peel and rice husk (PPRH) ash in aqueous environments. *Int. J. Phytoremediation* 19(11): 1029–1036.

Environmental Arsenic in a Changing World –
Zhu, Guo, Bhattacharya, Ahmad, Bundschuh & Naidu (Eds)
ISBN 978-1-138-48609-6

Modeling arsenic removal by co-precipitation under variable redox conditions

M.W. Korevaar[1], D. Vries[1] & A. Ahmad[1,2,3]
[1]*KWR Watercycle Research Institute, Nieuwegein, The Netherlands*
[2]*KTH-International Groundwater Arsenic Research Group, Department of Sustainable Development, Environmental Science and Engineering, KTH Royal Institute of Technology, Stockholm, Sweden*
[3]*Department of Environmental Technology, Wageningen University and Research (WUR), Wageningen, The Netherlands*

ABSTRACT: Drinking water companies in the Netherlands are actively investigating routes to reduce arsenic (As) to $<1\,\mu g\,L^{-1}$ in drinking water. Co-precipitation of As with iron during groundwater treatment is a promising method. When As(III) is present in raw water, permanganate (MnO_4) can be dosed to oxidize As(III) to As(V) in order to improve As removal efficiency. The dosages of MnO_4 and Fe(III) to achieve $<1\,\mu g\,L^{-1}$ As in the treatment effluents depend on the composition of raw water. The coprecipitation of As(III) and As(V) with ferrihydrite under variable raw water composition and redox environments, controlled by oxygen (O_2) or MnO_4 is modeled in this study by the generalized double layer model, and redox equilibrium reactions. Results show that the pH of the treatment process is critical to determine the As removal efficiency. At pH = 8 the highest As removal is obtained, followed by pH = 6 while pH = 7 gives the least removal. HCO_3, PO_4 and H_4SO_4 hamper the adsorption of As(V). In future work, the model outcome will be assessed by experiments. Furthermore, the model will be extended with oxidation kinetics in case oxidation (by e.g. oxygen) occurs at a slower rate than the (mean) residence time of the water in the process.

1 INTRODUCTION

Drinking water companies in the Netherlands are actively investigating methods to reduce As to $<1\,\mu g\,L^{-1}$ in drinking water (Ahmad *et al.*, 2014). At groundwater treatment plants (WTPs), indigenous Fe(II) in raw water is oxidized to form insoluble ferrihydrite. Also, Fe(II) or Fe(III) may be dosed in the filter influent to introduce or increase As removal. After Fe(II) gets oxidized to Fe(III) by O_2, Fe(III) undergoes hydrolysis which results in the formation of ferrihydrite which in turn can adsorb As (coprecipitation). These As-laced ferrihydrite precipitates are then removed in a filter placed downstream. The raw water at most WTPs in the Netherlands contain As(III) in variable proportions. In order to achieve higher As removal efficiency, complete oxidation of As(III) to As(V) is necessary through dosing a strong oxidant, such as permanganate (MnO_4) (Ahmad *et al.*, 2017). The dosages of MnO_4 and Fe(III) to achieve $<1\,\mu g\,L^{-1}$ As in the treatment effluent depend on the composition of raw water. The objective of this study is to model the coprecipitation of As(III) and As(V) with ferrihydrite under variable raw water composition and redox environments, controlled by oxygen (O_2) or MnO_4.

2 METHODS

2.1 *Model description*

The treatment process is simplified to batch mode and the raw water composition represents typical groundwater quality in the Netherlands (see Table 1). It is assumed that As only coprecipitates with ferrihydrite. Adsorption of ferrihydrite is modeled with the Generalized Double Layer Model (DLM) based on the work of (Dzombak & Morel, 1990). For this work, a modified PHREEQC database wateq4f is used. Arsenic adsorption constants and reactions are modified according to the average values presented in (Gustafsson & Bhattacharya, 2007). The log(k) values of (de)protonation of As species is also set equal to what Gustafsson & Bhattacharya, 2007 used to maintain consistency of the database. The adsorption of H_4SiO_4 is modeled according to (Stollenwerk *et al.*, 2007), adsorption of carbonate is described by reactions and constants as reported by (Appelo *et al.*, 2002). Furthermore, the system is assumed to be in thermodynamic equilibrium as most kinetics are much faster than the modeled residence time of the water. This means that MnO_4 is assumed to be fully reduced to Mn(II) (1 mole MnO_4 is assumed to oxidize 5 moles

Table 1. Overview of the concentrations used in the simulations.

Compound	μmol/L	mg L^{-1}
Fe(II)	0–70	0–3.9
Mn(II)	3.6	0.2
As(III)	0.3	0.02
Ca	1497	60
Mg	329.1	8
Cl	3840.9	136.2
Na	3229.1	74.2
HCO_3	2458.2	150
Si	356	10
PO_4	5.3	0.5

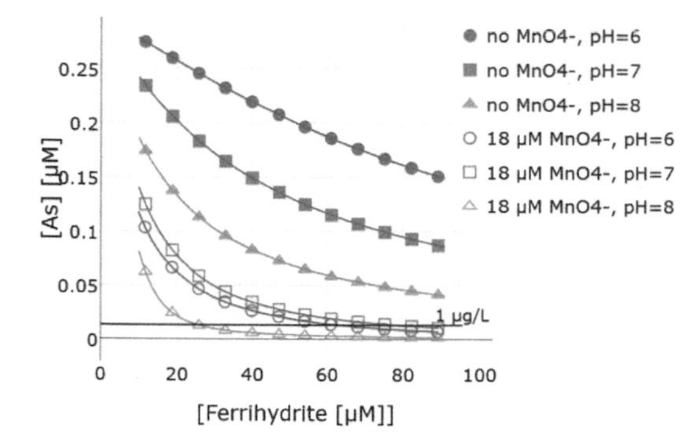

Figure 1. Removal of As(V) as function of ferrihydrite formed at different pH and MnO_4-dosages. The horizontal line indicates the target As in drinking water in the Netherlands.

of Fe(II) to produce ferrihydrite). The only exception to this thermodynamic equilibrium is the slow oxidation of As(III) and Mn(II) by oxygen.

3 RESULTS AND DISCUSSION

Figure 1 presents residual As(V) as a function of the precipitated ferrihydrite at different pH and MnO_4 dosages. The simulation shows that the amount of ferrihydrite is limited by the amount of available Fe(II) in raw water and MnO_4 dose. Interestingly, As(V) removal is higher for pH = 6 compared to pH = 7, while pH = 8 yields the highest As(V) removal. This observed trend may be due to the presence of H_4SiO_4, PO_4 and HCO_3 in water. Figure 2 shows the removal of As(III) as function of the amount of formed ferrihydrite at different pH and MnO_4 dosages. It confirms that the oxidation of As(III) by MnO_4 significantly improves As(III) removal. Figure 3 and 4 show the effect of HCO_3, PO_4 and H_4SiO_4 on the removal of As(V) and As(III) respectively. Both figures simulate the same system but the difference is the dosage of MnO_4. In Figure 3 MnO_4 is dosed and thus As(III) is oxidized to As(V) while in Figure 4 MnO_4 is not dosed. Moreover, in Figure 3 Fe(II) is oxidized by MnO_4 while

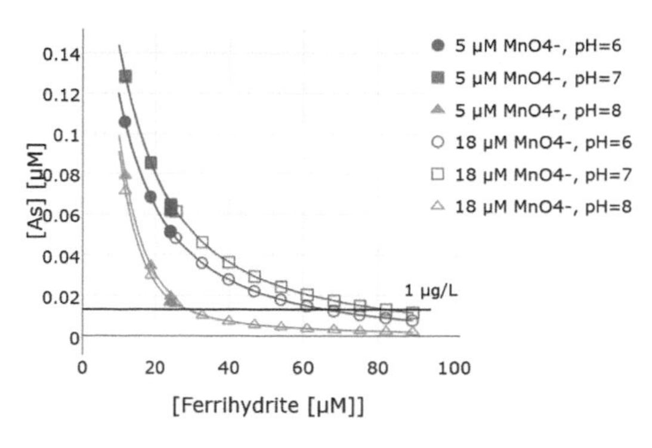

Figure 2. Removal of As(III) for different pH and MnO_4 dosages. The horizontal line indicates the target As in drinking water in the Netherlands.

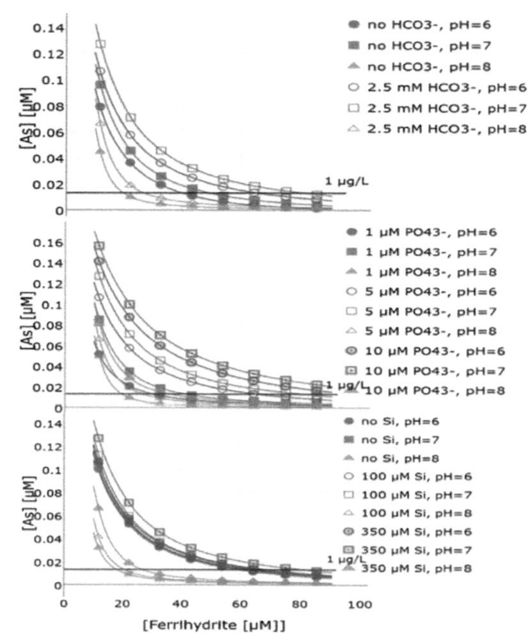

Figure 3. Removal of As(III) with MnO_4 dosing for different pH as function of the amount of ferrihydrite formed.

in Figure 4 Fe(II) is oxidized by O_2. For all the cases in Figure 3, the highest removal is achieved at pH = 8, the lowest at pH = 7. The presence of both HCO_3, PO_4 and H_4SiO_4 decreases coprecipitation of As. This is in agreement with literature (Antelo *et al.*, 2010; Appelo *et al.*, 2002; Stollenwerk *et al.*, 2007). Indeed, to such an extent that dosage of iron and MnO_4^- needs to be increased to meet the goal of 1 μg L^{-1}. E.g. due to 2.5 mM HCO_3 at a pH of 8.0, 27 μM instead 20 μM of ferrihydrite is required. For lower pH this difference is larger. It also holds for PO_4 that the influence of PO_4 is larger at lower pH and to an even stronger extent this holds for H_4SiO_4 as well.

This simulation shows that in order to achieve high process efficiency (i.e. low doses and higher removal of As) operating the treatment process at pH 8.0 is desirable. At pH = 8 the sensitivity of the required

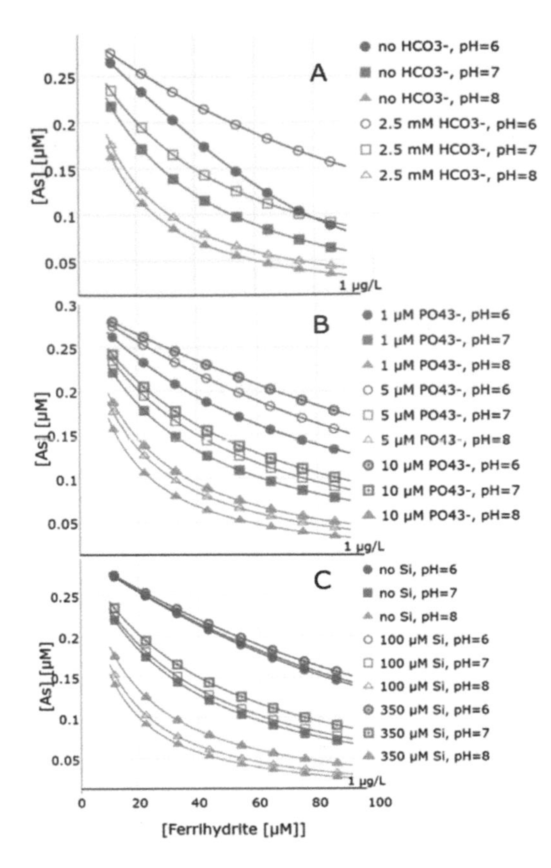

Figure 4. Removal of As(III) with O_2 dosing for different pH as function of the amount of ferrihydrite formed. Figure A, B and C illustrate the competition between the adsorption of As(V) and HCO_3, H_4SiO_4 and PO_4 respectively.

dosages to HCO_3, PO_4 and H_4SiO_4 concentrations is relatively low. The results, however, also show that the removal efficiency rapidly decreases when pH decreases to 7.0. This makes pH a critical parameter in designing and operating the As removal at groundwater treatment plants by Fe based coprecipitation under variable redox conditions.

4 CONCLUSIONS AND OUTLOOK

The co-precipitation of As(III) and As(V) with ferrihydrite under variable raw water composition and redox environments, controlled by oxygen (O_2) or MnO_4 is modeled by the generalized double layer model, and redox equilibrium reactions. Results show that As(III) and As(V) can be removed by Fe based coprecipitation. The coprecipitation efficiency is much higher for As(V) than As(III) and this supports that As(III) should be oxidized to improve removal rates. The pH of the treatment process is critical to determine the As removal efficiency. At pH = 8 the highest As removal is obtained, followed by pH = 6 while pH = 7 gives the least removal. HCO_3, PO_4 and H_4SO_4 hamper the adsorption of As(V). This is true for all tested pH conditions, but predominantly for pH = 6 and 7.In future work, the model outcome will be assessed by experiments. Furthermore, the model will be extended with oxidation kinetics in case oxidation (by e.g. oxygen) occurs at a slower rate than the (mean) residence time of the water in the process.

ACKNOWLEDGEMENT

The authors acknowledge support from RVO (TKI Watertechnologie) and Evides, Dunea, Brabant Water, Aqua Minerals, Carus and Maltha group.

REFERENCES

Ahmad, A., Van De Wetering, S., Groenendijk, M. & Bhattacharya, P. 2014. Advanced oxidation-coagulation-filtration (AOCF) – an innovative treatment technology for targeting drinking water with <1 µg/L of arsenic. Sustainable Development Environmental Science & Engineering, pp. 817–819.

Ahmad, A., van Dijk, T.G., J., Van de Wetering, S., Groenendijk, M. & Bhattacharya, P. 2017. Remediation case study: drinking water treatment by AOCF to target <1 µg L^{-1} effluent arsenic concentration. In: Bhattacharya, P., Polya, D.A. and Jovanovic, D. (eds) *Best Practice Guide on the Control of Arsenic in Drinking Water.* -IWA Publishing, UK, pp. 219–225.

Antelo, J., Fiol, S., Pérez, C., Mariño, S., Arce, F., Gondar, D. & López, R. 2010. Analysis of phosphate adsorption onto ferrihydrite using the CD-MUSIC model. *J. Colloid Interface. Sci.* 347(1): 112–119.

Appelo, C.A.J., Weiden, M.J.J.V, Tournassat, C. & Charlet, L. 2002. Surface complexation of ferrous iron and carbonate on ferrihydrite and the mobilization of arsenic. *Environ. Sci. Technol.* 36(14): 3096–3103.

Dzombak, D. & Morel, F. 1990. Surface *Complexation Modeling: Hydrous Ferric Oxide.* John Wiley & Sons. New York.

Gustafsson, J.P. & Bhattacharya, P. 2007. Geochemical modelling of arsenic adsorption to oxide surfaces. In: P. Bhattacharya, A.B. Mukherjee, J. Bundschuh, R. Zevenhoven, & R.H. Loeppert (eds) *Trace Metals and Other Contaminants in the Environment.* Elsevier.

Stollenwerk, K.G., Breit, G.N., Welch, A.H., Yount, J.C., Whitney, J.W., Foster, A. L., Uddin, M.N., Majumder, R.K. & Ahmed, N. 2007. Arsenic attenuation by oxidized aquifer sediments in Bangladesh. *Sci. Total Environ.* 379(2–3): 133–150.

Environmental Arsenic in a Changing World –
Zhu, Guo, Bhattacharya, Ahmad, Bundschuh & Naidu (Eds)
ISBN 978-1-138-48609-6

Visual MINTEQ simulation for prediction of the adsorption of arsenic on ferrihydrite

R. Irunde[1,2], P. Bhattacharya[2], J. Ijumulana[1,2], F.J. Ligate[1,2], A. Ahmad[2,3,4], F. Mtalo[1] & J. Mtamba[1]
[1]*DAFWAT Research Group, Department of Water Resources Engineering, College of Engineering and Technology, University of Dar es Salaam, Dar es Salaam, Tanzania*
[2]*KTH-International Groundwater Arsenic Research Group, Department of Sustainable Development, Environmental Science and Engineering, KTH Royal Institute of Technology, Stockholm, Sweden*
[3]*KWR Watercycle Research Institute, Nieuwegein, The Netherlands*
[4]*Department of Environmental Technology, Wageningen University and Research (WUR), Wageningen, The Netherlands*

ABSTRACT: The surface of ferrihydrite adsorbs arsenic (As) effectively. In this investigation, the As laced water samples collected from Geita and Mara regions within the Lake Victoria Basin (LVB) under DAFWAT project were simulated on Visual MINTEQ 3.1 software to determine the amount of ferrihydrite required to adsorb a given amount of As from water. Model simulations show that As concentration of $\leq$1 mM can be completely adsorbed by 4 g L^{-1} ferrihydrite. Previous studies show that the lower pH 4 to 4.5 influences adsorption, while it decreases as pH increases as well as when As concentration increases. The increase of adsorbent dose to 4 g L^{-1} has shown to improve As(V) adsorption on pH 5 to 8 at 100%. The amount of adsorbent can now be used for laboratory adsorption experiments by using iron-based materials or commercial ferrihydrite.

1 INTRODUCTION

High levels of arsenic (As) have been reported both in surface water and groundwater in several African countries (Ahoulé *et al.*, 2015). Elevated concentrations of As has recently reported in the northern part of Tanzania, particularly around the Lake Victoria Basin (LVB) and the Lake Victoria Goldfields (LVGF) (Ijumulana *et al.*, 2016a, 2016b, 2017; Kassenga & Mato, 2008; Lucca, 2017; Mnali, 2001).

Geochemical modeling tools such as PHREEQC (Parkhurst, 1995; Parkhurst & Appelo, 1999) and Visual MINTEQ (Gustafsson, 2009) are useful to model the mobility and fate of arsenic in water sources. The present study aims to use Visual MINTEQ for prediction of amount of iron oxide that can be applied on arsenic laced water samples for adsorptive removal. The simulation results of groundwater samples collected from Geita and Mara are presented.

2 MATERIALS AND METHODS

2.1 *Study area and water sampling*

Lake Victoria Basin, Tanzanian part, is one of the 9 river basins in Tanzania mainland covering area of 119,442 sq. km. The region has little seasonal variation but the eastern section where the study area lies (Geita and Mara), average only 750–1000 mm of rain. Favorable climatic conditions for agriculture and livestock and the abundance of natural resources have supported the livelihood of the rural population of over 35 million people (Lucca, 2017). The geology of the Tanzanian LVB consists of Archean granitoids-greenstone belts hosted in the Tanzanian craton. More than 80% of rural population depends on groundwater resources for various uses.

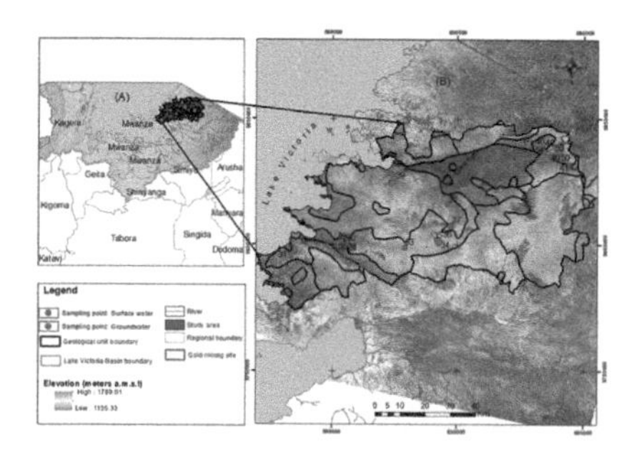

Figure 1. Map of the study area around the LVB region (A) and water sampling area locations (B).

Water sampling was carried out at the end of dry season during October 2016 from Geita and Mara (Fig. 1). A total of 29 water samples collected from Mara and 18 samples from Geita were simulated using the Visual MINTEQ.

2.2 *Visual MINTEQ parameter setting*

Groundwater quality data for Geita and Mara such as pH and the concentration of As(V) and As(III) (Lucca, 2017) were imported in the Visual MINTEQ. The following parameters were fixed in the model

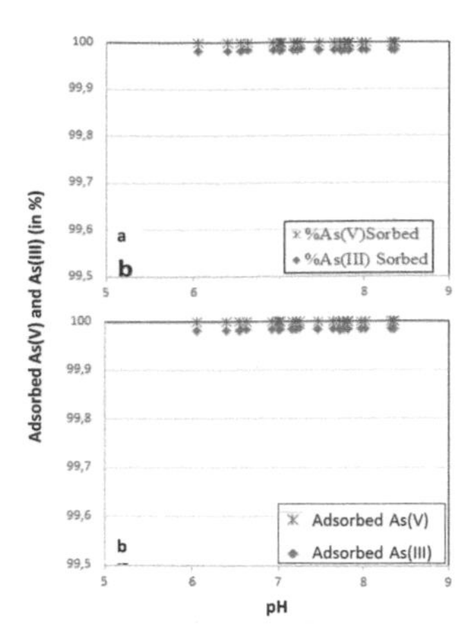

Figure 2. Adsorption of arsenic onto ferrihydrite from water sample from a) Geita and b) Mara sites.

for Geita sample; NaCl = 0.01M, temperature = 25°C, pH = 6.3 and ferrihydrite = 4 g L^{-1}; and for Mara sample; NaCl = 0.01M, temperature = 26°C, pH = 7.3 and ferrihydrite = 4g L^{-1}.

3 RESULTS AND DISCUSSION

The results of the geochemical modelling were concentrations of As adsorbed onto ferrihydrite. The percentage of As sorbed was calculated using the following equation:

$$As_{sorbed} = (\text{Final } As_{conc}/\text{Initial } As_{conc}) \times 100\% \quad \text{(eq. 1)}$$

The percentage of adsorbed As(V) and As(III) on ferrihydrite were plotted against pH to visualize the strength of Visual MINTEQ 3.1 on arsenic adsorption (Fig. 2).

The As(V) concentrations at Geita and Mara ranged from 10 μg L^{-1} to 300 μg L^{-1} which is equivalent to $\leq$1 mM and hence only 4 g L^{-1} ferrihydrite added in the Visual MINTEQ was adequate to remove As from water samples in the pH range of 5 to 8.5.

The model results from Geita (Fig. 2a) and Mara (Fig. 2b) show maximum adsorption of As(V) leading to 100% removal, while As(III) is removed at 99%. This is because As(V) a negatively charged species, is strongly bound to surface of the ferrihydrite and is thus easily removable from water up to a pH close to 8.5 (Ahmad *et al.*, 2014). The groundwater of Geita and Mara have higher concentrations of As(V) because of an oxidizing environment where As(III) can be oxidized to As(V) as well as Fe^{2+} being oxidized to Fe^{3+}.

4 CONCLUSION

The Visual MINTEQ software is a useful tool to gain insight on the adsorption processes. The software is recommended for prediction of adsorbent amount that is required for As removal from water samples. Simulation results predict 100% of As(V) removal from Geita and Mara in the pH range of 5 to 8.5 when 4 g L^{-1} ferrihydrite was dosed.

ACKNOWLEDGEMENTS

We acknowledge the Swedish International Development Cooperation Agency (Sida) for supporting the DAFWAT program (Contribution: 51170072).

REFERENCES

Ahmad, A., van de Wetering, S., Groenendijk, M. & Bhattacharya, P. 2014. Advanced Oxidation-Coagulation-Filtration (AOCF) – an innovative treatment technology for targeting drinking water with <1 µg/L of arsenic. In: M.I. Litter, H.B. Nicolli, M. Meichtry, N. Quici, J. Bundschuh, P. Bhattacharya & R. Naidu (eds. *One Century of the Discovery of Arsenicosis in Latin America (1914–2014) As 2014*. CRC Press, Boca Raton, FL, pp. 817–819.

Ahoulé, D.G., Lalanne, F., Mendret, J., Brosillon, S. & Maïga, A.H. 2015. Arsenic in African waters: a review. *Water Air Soil Poll.* 226(9): 302.

Gustafsson, J.P. 2009. Visual-MINTEQ, version 2.61: a Windows version of MINTEQA 2 (version 4.0) http://hem.bredband.net/b108693/.

Ijumulana, J., Mtalo, F. & Bhattacharya, P. 2016a. Spatial modelling of arsenic distribution and human health effects in LakeVictoria basin, Tanzania. *Geophysical Research Abstracts*, 18, EGU2016-15869-1,

Ijumulana, J., Bhattacharya, P. & Mtalo, F. 2016b. Arsenic occurrence in groundwater sources of Lake Victoria basin in Tanzania. In: P. Bhattacharya, M. Vahter, J. Jarsjö, J. Kumpiene, A. Ahmad, C. Sparrenbom, G. Jacks, M.E. Donselaar, J. Bundschuh, & R. Naidu (eds) *Arsenic Research and Global Sustainability As2016*. CRC Press, Boca Raton, FL, pp. 86–87.

Ijumulana, J., Lucca, E., Bhattacharya, P. & Mtalo, F. 2017 Mineral solubility controls on drinking water quality in the areas of gold mining in Geita and Mara regions of northern Tanzania. *Geol. Soc. Amer. Abstracts with Programs* 49(6): doi: 10.1130/abs/2017AM-308005.

Kassenga, G.R. & Mato, R.R. 2008. Arsenic contamination levels in drinking water sources in mining areas in Lake Victoria Basin, Tanzania, and its removal using stabilized ferralsols. *Int. J. Biol. Chem. Sci.* 2(4): 389–400.

Lucca, E. 2017. *Geochemical Investigation of Arsenic in Drinking Water Sources in Proximity of Gold Mining Areas in the Lake Victoria Basin, in Tanzania*. KTH Royal Institute of Technology, MSc Thesis.

Mnali, S. 2001. Assessment of heavy metal pollution in the Lupa gold field, SW Tanzania. *Tanzania J. Science* 27(2): 15–22.

Parkhurst, D.L. 1995. Users Guide to PHREEQC - a computer program for speciation, reaction-path, advective-transport, and inverse geochemical calculations. U.S. Geol. Surv. Water Resour. Invest. Rep. 95-4227.

Parkhurst, D.L. & Appelo, C.A.J. 1999. User's guide to PHREEQC (version 2), U.S. Geol. Surv. Water Resour. Invest., 99-4529.

Environmental Arsenic in a Changing World –
Zhu, Guo, Bhattacharya, Ahmad, Bundschuh & Naidu (Eds)
ISBN 978-1-138-48609-6

Clay-biochar composite for arsenic removal from aqueous media

M. Vithanage[1], L. Sandaruwan[2], G. Samarasinghe[2] & Y. Jayawardhana[3]
[1]*Faculty of Applied Science, University of Sri Jayewardenepura, Nugegoda, Sri Lanka*
[2]*S. de S. Jayasinghe Central College, Dehiwala, Sri Lanka*
[3]*Environmental Chemodynamics Project, National Institute of Fundamental Studies, Kandy, Sri Lanka*

ABSTRACT: Arsenic is present mainly in groundwater in the forms of highly toxic arsenate and arsenite. Biochar has been a promising cost effective and carbon negative material for many different contaminants such as trace metals, antibiotics, pesticides etc., however, not for anionic metalloids as arsenic. This study intended to evaluate the potential of a novel composite prepared using biochar, derives from fibrous fraction of municipal solid waste in Gohagoda landfill site, Kandy, Sri Lanka and red earth clay for remediate As(III) in aqueous media. Pyrolyzed biochar+red earth composite was characterized for its physicochemical properties. Furthermore, As(III) pH dependency (pH 3–9), kinetics behavior and sorbate (50–1000 $\mu g L^{-1}$) concentrations were investigated using a batch sorption technique. The concentrations of As(III) in aqueous media were measured by inductively coupled plasma optical emission spectrometry. For arsenic (III), pH 6–7 range was favorable for adsorption process. Whereas, highest adsorption at 24 hours reaction time at pH 6–7 recorded as around 25% (12.5 $\mu g g^{-1}$) for arsenic (III). Moreover, well fitted pseudo second order ($R^2 = 0.928$) could suggest chemical adsorption mechanism rather than a physical adsorption mechanism in to the adsorbent. Hence, process involved with chemisorption can be suggested as the As(III) removal mechanism. However, further isotherm experiments are needed with expanded concentration range to mechanism identification.

1 INTRODUCTION

Presence of arsenic in water sources has been a serious concern worldwide. Arsenic is present mainly in groundwater in the forms of highly toxic inorganic forms of arsenate and arsenite. Literature reveals that many techniques have been examined for arsenic removal from water. The most commonly used methods include coagulation with iron and aluminum salts; ion-exchange, reverse osmosis and electro-dialysis; adsorption onto activated alumina/carbon, activated bauxite, clay minerals and iron oxides (Nicomel *et al.*, 2016). Most of the methods are expensive and labor intense, which hinder the application in community scale whereas the chemical methods yield large quantities of solid sludge, which demands further treatment. Since adsorption process is simple, easy to handle and being regenerable, it has been widely used to remove arsenic (Duan *et al.*, 2017).

Biochar (BC) has been widely examined as adsorbent for removal of containments such as heavy metals and organic compounds in water and soil has also been reported (Ahmad *et al.*, 2014). As many waste materials can be used to produce biochar, there has been a great interest in using municipal solid waste as a feedstock thus supportive in reducing waste quantities (Jayawardhana *et al.*, 2017). However, its capacities for anionic metalloid removal have been low rendering it suitable for arsenic removal application (Vithanage *et al.*, 2017).

Clay has been widely used in arsenic removal whereas Natural Red Earth (NRE) exhibited excellent performances for both As(III) and As(V) removal at the same instance without any environmental change (Vithanage *et al.*, 2006, 2007). Hence, this study intended to evaluate the potential of a novel clay-BC composite prepared using biochar, derives from fibrous fraction of municipal solid waste in Gohagoda landfill site, Kandy, Sri Lanka and red earth clay for remediate As in aqueous media.

2 METHODS

2.1 *Feedstock for biochar and natural red earth*

The segregated organic fraction of the municipal solid waste (MSW) obtained from dump site Gohagoda, Kandy, Sri Lanka. Natural red earth (NRE) was obtained from the limestone quarry of the Aruwakkalu, Sri Lanka.

2.2 *Composite preparation*

Stable clay suspensions of MT and/or RE were prepared separately (50 g clay in 2 L deionized (DI) water)

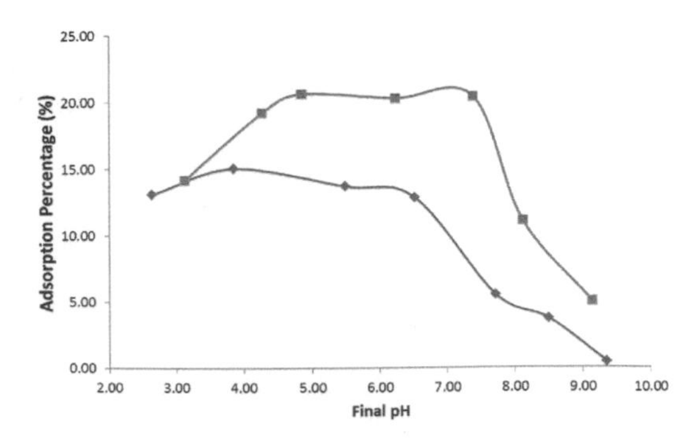

Figure 1. Arsenic (III) and (V) sorption to clay-BC composite at 2 g L^{-1}. Arsenic (III) in square shape symbols whereas arsenic (V) in diamond symbols.

followed by sonication of the mixture for 30 minutes in an ultrasonicator. Then 250 g of MSW biochars were added to clay suspensions and the mixtures were shaken for 2 hours. The clay-biochar suspensions were filtered and BC-clay feedstocks were oven dried at 80°C in an oven overnight. The clay treated biomass feedstocks were packed tightly into ceramic crucibles and slow pyrolysis was done at a rate of 7°C min^{-1} under oxygen limited conditions in a muffle furnace (MTI, Richmond, CA). The pyrolysis temperature was increased to 500°C and held constant at 500°C for 30 minutes. Untreated MSW feedstock was also pyrolyzed under same pyrolysis conditions. All the biochar samples were washed with DI water several times to remove impurities, dried in an oven at 80°C and sealed in a container for further testing.

2.3 *Batch experiments*

The effect of pH on the sorption was studied by adjusting the pH in the range of 3–9. The adsorbent concentration was kept at 2.00 g L^{-1} of solution containing 125 $\mu g L^{-1}$ As at 25°C for overnight. The kinetics experiments were conducted for 24 h at the same concentrations of sorbents and sorbate. The samples were then filtered using 0.45 μm PVDF disposable filters and As concentrations in the solutions were measured by the inductive coupled plasma optical emission spectroscopy (ICP-OES, Thermo).

3 RESULTS AND DISCUSSION

For As(III), pH 6–7 range was favorable for adsorption process (Fig. 1). The highest adsorption at 24 hours reaction time at pH 6–7 recorded as around 25% (12.5 $\mu g g^{-1}$) for As(III).

Arsenic (III) kinetic data exhibited no equilibrium at 24 h at 150 $\mu g L^{-1}$. However, the data were well fitted to pseudo second order ($R^2 = 0.928$) could suggest chemical adsorption mechanism rather than a physical adsorption mechanism in to the adsorbent.

4 CONCLUSIONS

The Clay-BC composite indicated high removal of arsenic from water at environmental pH conditions with strong chemisorption interactions. Further, arsenic removal to be investigated for sorbent concentrations and column experiments.

ACKNOWLEDGEMENTS

Chairperson, Director and the Science Popularization Division of the National Science Foundation, Sri Lanka and Principal, S. de S. Jayasinghe College are acknowledged for the support.

REFERENCES

Ahmad, M., Rajapaksha, A.U., Lim, J.E., Zhang, M., Bolan, N., Mohan, D., Vithanage, M., Lee, S.S. & Ok, Y.S. 2014. Biochar as a sorbent for contaminant management in soil and water: a review. *Chemosphere* 99: 19–33.

Duan, X., Zhang, C., Srinivasakannan, C. & Wang, X. 2017. Waste walnut shell valorization to iron loaded biochar and its application to arsenic removal. *Res-Efficient Technol.* 3(1): 29–36.

Jayawardhana, Y., Mayakaduwa, S., Kumarathilaka, P., Gamage, S. & Vithanage, M. 2017. Municipal solid waste-derived biochar for the removal of benzene from landfill leachate. *Environ. Geochem. Hlth.*: 1–15.

Nicomel, N.R., Leus, K., Folens, K., Van Der Voort, P. & Du Laing, G. 2015. Technologies for arsenic removal from water: current status and future perspectives. *Int. J. Environ. Res. Public Health* 13(1): 62.

Vithanage, M., Chandrajith, R., Bandara, A. & Weerasooriya, R. 2006. Mechanistic modeling of arsenic retention on natural red earth in simulated environmental systems. *J. Colloid Interface Sci.* 294(2): 265–272.

Vithanage, M., Herath, I., Joseph, S., Bundschuh, J., Bolan, N., Ok, Y.S., Kirkham, M. & Rinklebe, J. 2017. Interaction of arsenic with biochar in soil and water: a critical review. *Carbon* 113: 219–230.

Vithanage, M., Senevirathna, W., Chandrajith, R. & Weerasooriya., R. 2007. Arsenic binding mechanisms on natural red earth: a potential substrate for pollution control. *Sci. Total Environ.* 379(2–3): 244–248.

Environmental Arsenic in a Changing World –
Zhu, Guo, Bhattacharya, Ahmad, Bundschuh & Naidu (Eds)
ISBN 978-1-138-48609-6

Iron coated peat as a sorbent for the simultaneous removal of arsenic and metals from contaminated water

A. Kasiuliene[1], I. Carabante[1], J. Kumpiene[1] & P. Bhattacharya[2]
[1]*Waste Science and Technology Research Group, Department of Civil, Environmental and Natural Resources Engineering, Lulea University of Technology, Lulea, Sweden*
[2]*KTH-International Groundwater Arsenic Research Group, Department of Sustainable Development, Environmental Science and Engineering, KTH Royal Institute of Technology, Teknikringen, Stockholm, Sweden*

ABSTRACT: This study aimed at combining peat, an industrial residue, with Fe(II)-Fe(III) compound to produce a sorbent suitable for a simultaneous removal of arsenic (As) and metals (cadmium, copper, nickel, lead, zinc) from a contaminated water. Using a newly produced sorbent – iron-peat – the removal of As from contaminated water was almost 17 times higher than using an uncoated peat. On the other hand, the removal of metals by the iron-peat was slightly less efficient in comparison to the uncoated peat. Simultaneous removal of As and metals could be seen as an advantage over multiple-step treatment of contaminated groundwater.

1 INTRODUCTION

In Sweden, due to the legacies of past industrial activities, such as glass works and wood impregnation, multiple point sources of arsenic (As) contamination in soil and water bodies are scattered over the country. In several of these sites, other metals, such as cadmium (Cd), copper (Cu), nickel (Ni), lead (Pb) or zinc (Zn) occur as co-contaminants at varying concentrations depending on the type of industrial activities.

A common method to remove As from contaminated water is by adsorbing them onto reactive media and separating from solution. Iron (Fe) oxides have an affinity for arsenates and are used as an effective and potentially inexpensive adsorbent to treat As contaminated water (Carabante, *et al.*, 2014). To avoid clogging of filters (Mohan & Pittman, 2007) and overcome low hydraulic permeability (Theis *et al.*, 1992), Fe-containing filters commonly are produced by coating a bulk material, e.g. sand (Devi *et al.*, 2014) or activated carbon (Yürüm *et al.*, 2014) with Fe oxides. Activated carbon has an advantage over sand, because it can target contaminants as well. Replacing activated carbon with cheaper and readily available materials, e.g. peat could lead to a production of a more cost-efficient sorbent. Furthermore, peat by itself can target divalent metal ions.

The aim of this study was to develop a residue-based sorbent for the simultaneous removal of As, Cd, Cu, Ni, Pb, and Zn from contaminated groundwater, which after the use could be combusted for energy recovery and mass reduction.

2 MATERIALS AND METHODS

2.1 *Materials*

Heat-treated peat powder was obtained from Geogen Produktion AB, Sweden. This company produces heat-treated peat granulate as environmentally compatible oil-absorption agent. During its production, powder particles smaller than 2 mm are discarded and considered as a waste.

Ferriferous hydrosol (FFH) was used for coating the peat. The hydrosol is a product of Fe(II) and Fe(III) hydrated compounds. FFH is a commercially available product produced from the Fe waste; it was obtained from Rekin, Lithuania.

Solution containing As and metals used for adsorption test was prepared from chemicals of analytical grade provided by Fluka or Merck: NaH_2AsO_4, $CdCl_2$, $CuCl_2*2H_2O$, $NiCl_2$, $Pb(NO_3)_2$, and $ZnCl_2$. Concentrations of As and metals in the initial solution before adsorption experiment were detected with ICP-OES (Perkin Elmer Optima 2000V). Solution's pH was set to 5.0 using NaOH solutions.

2.2 *Methods*

Batches of iron-peat sorbent were prepared mixing FFH and peat powder at three different ratios: 1:1, 1:10, and 1:20, respectively. Each sorbent (in triplicate) was mixed with the prepared As and metal solution at a liquid to solid ratio (L/S) = 1000. Uncoated peat was used as reference. Mixtures were shaken for 24 h, then samples were filtered through 0.45 μm

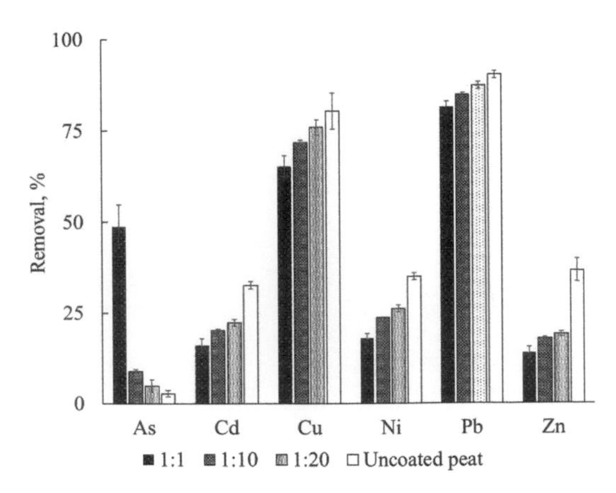

Figure 1. Removal of As and metals from contaminated solution using iron-peat at three different ratios and uncoated peat as a reference.

nitrocellulose filters and leachates were analyzed with ICP-OES.

The data were processed by analysis of variance (ANOVA) using the software Minitab 18. A two-sample t-test ($p < 0.05$) was applied to differentiate between sample means.

3 RESULTS AND DISCUSSION

3.1 *Arsenic removal efficiency*

Concentration of As in the initial solution was $0.54\,mg\,L^{-1}$. Iron-peat, where FFH was coated onto peat at a 1:1 ratio, was the most effective to adsorb As from contaminated water. The removal efficiency was about 50% (Fig. 1). Other treatments (1:10, 1:20, uncoated peat) were significantly weaker and the removal of As remained below 10%. Obtained results well coincide with other studies (e.g. Kumpiene *et al.*, 2009; Vitkova *et al.*, 2017) where high affinity of arsenate to be adsorbed on Fe oxides has been reported.

3.2 *Metal removal efficiency*

Metal concentrations in the initial solution were: $0.80\,mg\,L^{-1}$ of Cd, $4.3\,mg\,L^{-1}$ of Cu, $0.80\,mg\,L^{-1}$ of Ni, $2.2\,mg\,L^{-1}$ of Pb, and $4.0\,mg\,L^{-1}$ of Zn. Contrary to the removal of As, all investigated metals were better adsorbed onto uncoated peat. A clear metal removal efficiency pattern of the four sorbents can be observed from Figure 1. The results indicate that peat can efficiently bind to divalent metal ions due to its polar characteristics. It can be assumed that Fe compounds, occurring due to the coating, were blocking organic groups in the peat. Therefore, the removal rate of metals using coated peat was weaker in comparison to the adsorption by uncoated peat.

3.3 *Overall performance of the iron-peat*

Simultaneous sorption of As and metals from contaminated aqueous media was achieved using a newl produced iron-peat sorbent. Process optimization for specific groundwater conditions could further increase the sorption capacity of the sorbent. Adjustment of pH and reduction of contaminated water loads could substantially enhance the removal efficiency. Iron-peat sorbent could be used in filters, packed-bed reactors, or in passive permeable barriers to treat contaminated water. Number of treatment steps could be reduced due to the sorbent's ability simultaneously adsorb As and metals.

Results obtained from this study has not only environmental impact on cleaning contaminated water, but also emphasizes utilization of by-products, which also contribute to the volume reduction of waste materials to be treated.

4 CONCLUSIONS

Iron coating on peat (1:1) significantly increased As removal from contaminated water in comparison to other treatments. Metal removal was slightly higher onto uncoated peat in comparison to iron-peat. The results indicate that iron-peat simultaneously targets both As and divalent metals and could be used as a single-step treatment for contaminated water.

ACKNOWLEDGEMENTS

The authors would like to thanks the Swedish Research Council for Sustainable Development (Formas) for the research grant for this study.

REFERENCES

Carabante, I., Mouzon, J., Kumpiene, J., Gran, M., Fredriksson, A. & Hedlund, J. 2014. Reutilization of porous sintered hematite bodies as effective adsorbents for arsenic(V) removal from water. *Ind. Eng. Chem. Res.* 53(32): 12689–12696.

Devi, R.R., Umlong, I.R., Das, B., Borah, K., Thakur, A.J., Raul, P.K., Banerjee, S. & Singh, L. 2014. Removal of iron and arsenic(III) from drinking water using iron oxide-coated sand and limestone. *Appl. Water Sci.* 4(2): 175–182.

Kumpiene, J., Ragnvaldsson, D., Lovgren, L., Tesfalidet, S., Gustavsson, B., Lattstrom, A., Leffler, P. & Maurice, C. 2009. Impact of water saturation level on arsenic and metal mobility in the Fe-amended soil. *Chemosphere* 74(2): 206–215.

Mohan D. & Pittman, C.U. 2007. Arsenic removal from water/wastewater using adsorbents – a critical review. *J. Hazard. Mater.* 142(1–2): 1–53.

Theis, T.L., Iyer, R. & Ellis, S.K. 1992. Evaluating a new granular iron oxide for removing lead from drinking water. *J. Am. Water Works Assoc.* 84(7): 101–105.

Vitkova, M., Rakosova, S., Michalkova, Z. & Komarek, M. 2017. Metal(oid)s behaviour in soils amended with nano zero-valent iron as a function of pH and time. *J. Environ. Manage.* 186: 268–276.

Yürüm, A., Kocabaş-Ataklı, Z.Ö., Sezen, M., Semiat, R. & Yürüm, Y. 2014. Fast deposition of porous iron oxide on activated carbon by microwave heating and arsenic (V) removal from water. *Chem. Eng. J.* 15: 321–332.

4.2 Ion exchange and membrane technologies

Environmental Arsenic in a Changing World –
Zhu, Guo, Bhattacharya, Ahmad, Bundschuh & Naidu (Eds)
ISBN 978-1-138-48609-6

Development of membrane by using iron-containing synthetic materials for arsenic removal from water

L.R. Velázquez[1], R.V. Moreno[1], R.R. Mendoza[1], C. Velázquez[1] & S.E. Garrido Hoyos[2]
[1]*Facultad de Química, Universidad Autónoma de Querétaro, Santiago de Querétaro, Qro, Mexico*
[2]*Subcoordinacción de Posgrado, Instituto Mexicano de Tecnología del Agua, Jiutepec, Mor., Mexico*

ABSTRACT: The presence of arsenic in the environment has been one of the main causes of contamination in surface and groundwater. Arsenic can be found in four oxidation states, such as trivalent arsenic (As(III)), pentavalent arsenic (As(V)), arsine (As(-III)) and elemental arsenic (As^0). It is mainly found as As(III) and As(V) in groundwater. In the present work, MDP's membranes were developed by an adsorbent material and two additives. The material was synthesized and characterized, obtaining a surface area of 55.005 $m^2\ g^{-1}$ and a pore diameter of 34.11 Å, the material has a crystalline structure and tends to change its chemical composition from 281.14°C. A 2^3 factorial design was performed for to obtain the optimal conditions for the preparation of the membranes for the factors of adsorbent material goethite (FeOOH), binder (Al_2O_3) and lubricant ($C_{18}H_{36}O_2$). The membranes were pressed to 1.5 tons and sintered at 600°C, converting the goethite material (FeOOH) to hematite (Fe_2O_3). The membrane was analyzed by physisorption of nitrogen and was obtained a surface area of 39.4249 $m^2\ g^{-1}$ and a pore diameter of 83.01 Å. Finally the adsorption kinetics was performed for five membranes with the same characteristics of the MDP6 membrane. According to the kinetics, it was observed that the MDP2 membrane showed an elimination of 41% of arsenic As(V) at an initial concentration of 0.49 mg L^{-1}.

1 INTRODUCTION

Arsenic in groundwater is due to natural activities such as the dissolution of minerals, geogenic activities (Tuesca *et al.*, 2015). In groundwater is common to find arsenic as arsenate (As(V)) and arsenite (As(III)) (Sklari *et al.*, 2015). In Mexico, several cases of groundwater contamination have been reported in the states of Durango, Coahuila, Hidalgo, Sonora and Querétaro. The consumption of water contaminated with arsenic can produce skin lesions with a hyper-pigmentation of the chest, neck, and trunk, with concomitant or later appearance of hyperkeratosis (Karagas *et al.*, 2015). Consequently, Mexican Standard NOM-127-22A1-1994 set a limit of 0.025 mg L^{-1} of arsenic in drinking water. Finally, the use of porous membranes has been proposals as technology for the removal of contaminants in water. The aim of this work is the development of porous membranes by synthetic materials with iron oxyhydroxides for the removal of arsenic in water.

2 EXPERIMENTAL SECTION

2.1 *Preparation and characterization of adsorbent media*

The goethite was synthesized for the preparation of porous membranes following the methods by Garrido & Romero (2015). The synthesized material was characterized for the phase identification by X-ray diffraction (XRD) Diffractometer D8 Advance Bruker radiation Cu Kα ($\lambda = 0.15406$) range between 10–80° to 2θ. The textural properties (surface area, pore volume and size distribution) of the synthesized goethite were analyzed by Autosorb IQ2. The thermal stability of the synthetic goethite was also studied for TA instruments model Q500 temperature ranges from 10–1000°C and temperature ramp 5°C min^{-1}.

Table 1. Experimental design.

	Intervals	
Factors	1	−1
Binder (g)	0.45	0.375
Lubricant (g)	0.075	0.025
Synthetic goethite (g)	2.1	1.975

2.2 *Development membranes MDP's*

The development of eight membranes was carried out following the proposed methodology by Pérez *et al.* (2004). The experiments were carried out using a 2^3 factorial design Statgraphics Centurion XV (Table 1).

Subsequently the material mixture was compacted in a hydraulic press at 1.5 ton and sintered at 600°C for three hours. The kinetics was performed in a continuous process for 8 h. The As(V) concentrations were prepared with sodium arsenate heptahydrate (Na_2HasO_4 $7H_2O$, 98% purity, Sigma-Aldrich).

Samples were taken at time intervals of 30 min. The sample was filtered using a 0.45 μm membrane. Arsenic was determined using the Wagtech photometer method.

3 RESULTS AND DISCUSSION

3.1 *Characterization of synthetic goethite FeOOH*

According to the results of XRD the PDF database used was 29-0713. Spectra of synthetic goethite shows the characteristic reflections of goethite in positions 21.223, 33.242, 36.650 and 53.238 2θ. The structure corresponds to a type of orthorhombic family and its crystal structure can be indexed between the crystallographic plans 110, 130, 111, 221. The results were compared with those reported by Jaiswal *et al.* (2013). Nitrogen adsorption, showed a type IV isotherm which corresponds to mesoporous solid and a type IV hysteresis curve. The characterization of synthetic goethite shows 55.005 $m^2 g^{-1}$ surface area, pore volume of 0.137 $cm^3 g^{-1}$ and pore diameter (dV/d) 34.11 Å. Nevertheless Garrido & Romero (2015) reported a pore diameter of 119.05 Å is 3.49 times greater than the reported in this work. The thermogravimetric analysis indicates that the main mass loss occurs around 281.14°C. The material presents a mass decrease which is attributed to the loss of iron hydroxides converting the material into hematite.

3.2 *Development membranes MDP's*

The particle size was 0.5 mm. A statistically significant difference ($p < 0.005$) was found for development membranes show lubricant. The optimal conditions for development membranes were: synthetic goethite of 2.1 g, binder 0.450 g and lubricant 0.025 g. The use of lubricant is essential to give resistance to membrane fragmentation. The membranes were characterized by XRD and physisorption of nitrogen. The selected membrane was analyzed from the experimental design and after a thermal sintering process of 600°C according to PDF 03-6664. The material is hematite since it is characterized by the main reflections in positions 24.138, 33.153, 35.612, 49.480 and 54.091 in 2θ with the crystallographic planes of 012, 104, 110, 024 and 116 according to reported by Zhang *et al.* (2017). The surface area obtained in the membrane was 39.4249 $m^2 g^{-1}$.

3.3 *Kinetics sorption for membranes MDP's*

Figure 1 shows the five membranes selected to perform the adsorption kinetics MDP1, MDP2, MDP3, MDP4 and MDP5.

The values fitted to pseudo-second order equation, and the best results were for MDP2 membrane with kinetics constants: Adsorption capacity,

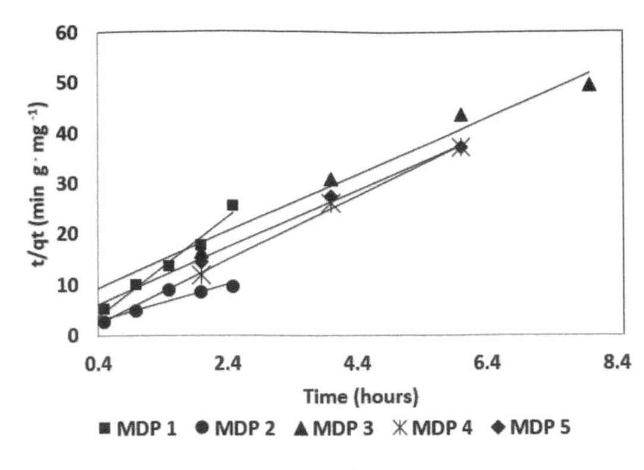

Figure 1. Kinetics sorption of As(V).

qe: 0.287 $mg\,g^{-1}$; Adsorption rate kad: 3.67 $mg\,min^{-1}$; and initial velocity rate, h: 0.303 $mg\,g^{-1}$ with R^2: 0.8503.

4 CONCLUSIONS

In this experiment the addition of lubricant was of great importance, without this additive the membranes were easily broken. However, it is recommended to use moderate amounts of lubricant, to obtain a better removal performance of membranes conforming by Hematite. The membrane MDP2 had an arsenic removal of 41% with an initial concentration of 0.49 $mg\,L^{-1}$ of As(V).

REFERENCES

Garrido, S. & Romero, L. 2015. Synthesis of minerals with iron oxide and hydroxide contents as a sorption medium to remove arsenic from water for human consumption. *Int. J. Environ. Res. Public Health* 13(1): 69.

Jaiswal, A., Banerjee, S., Mani, R. & Chattopadhyaya, M.C. 2013. Synthesis, characterization and application of goethite mineral as an adsorbent. *J. Environ. Chem. Eng.* 1(3): 281–289.

Karagas, M.R., Gossai, A., Pierce, B. & Ahsan, H. 2015. Drinking water arsenic contamination, skin lesions and malignancies: a systematic review of the global evidence. *Curr. Environ. Health Rep.* 2(1): 52–68.

Pérez, M.V., Castro, A.J. & Balmori, R.H. 2004. Characterization and preparation of porous membranes with a natural Mexican zeolite. *J. Phys. Condens. Matter* 16(22): S2345.

Sklari, S., Pagana, A., Nalbandian, L. & Zaspalis, V. 2015. Ceramic membrane materials and process for the removal of As (III)/As (V) ions from water. *J. Water Process Eng.* 5: 42–47.

Tuesca, R., Ávila, H., Sisa, A. & Pardo, D. 2015. Fuentes de abastecimiento de agua para consumo humano: análisis de tendencia de variables para consolidar mapas de riesgo. Barranquilla, Colombia: Universidad del Norte.

Zhang, Y., Dong, K., Liu, Z., Wang, H., Ma, S., Zhang, A. & Li, Y. 2017. Sulfurized hematite for photo-Fenton catalysis. *Pro. Nat. Sci-Mater.* 27(4): 443–45.

Environmental Arsenic in a Changing World –
Zhu, Guo, Bhattacharya, Ahmad, Bundschuh & Naidu (Eds)
ISBN 978-1-138-48609-6

As(V) rejection by NF membranes for drinking water from high temperature sources

B.J. Gonzalez, S.G.J. Heijman, A.H. Haidari, L.C. Rietveld & D. van Halem
Faculty of Civil Engineering and Geosciences, Delft University of Technology, Delft, The Netherlands

ABSTRACT: Geothermally influenced waters are frequently found to be contaminated with arsenic (as As(V)). This study investigated the effect of high-temperatures (50°C), as found in geothermal source waters, on the rejection of monovalent $H_2AsO_4^-$ and divalent $HAsO_4^{2-}$ species (at pH 6 and 8) during NF membrane filtration of a multi-component solution containing Cl^- and HCO_3^-. In this multi-component solution As(V) rejection was found to be enhanced at higher temperatures, whereas so far it was assumed that temperature rise had a negative effect on As(V) rejection. Previous studies were conducted with ultrapure waters, where pore size expansion dominated As(V) rejection, however, in the presence of other anions – like in natural water – As(V) rejection is promoted at higher temperatures. The enhancement of As(V) rejection at high temperature was associated with the presence of HCO_3^- and Cl^-, which are considerably more permeable than both As(V) species. Additional advantage of the higher temperature was the lower feed pressure (down to 3 bar) needed to operate these NF membranes, compared to colder waters. Temperatures shows the potential application of this technology for efficient treatment of As(V) contaminated, geothermally influenced waters.

1 INTRODUCTION

Geothermal systems are known to be a source of arsenic (As) contamination since geothermal fluids are responsible for transporting As and heavy metals, contaminating both surface water and groundwater. These geothermally influenced waters may thus have high As concentrations in combination with high temperatures. The contamination of water sources by geothermal As has been reported around the world in locations such as Waikato River in New Zealand, Eastern Sierra Nevada in USA, and Telica in Nicaragua (Johnson, 1965; Polhill, 1982).

The As speciation in natural groundwater (pH 6.5–8.5) is mainly determined by redox potential (Eh) and pH. Depending on the conditions, As is present as arsenate (As(V)) or arsenite (As(III)). As(V) is a charged ion and is commonly found under oxidizing conditions. For a pH lower than 6.9, As(V) exists mainly as the monovalent $H_2AsO_4^-$ species, and for a higher pH, the divalent $HAsO_4^{2-}$ species is dominant. On the other hand, As(III) is predominantly found under reducing conditions and as an uncharged ion when the pH is below 9.2. In geothermal fluids, the dominant species of As primarily consists of the reduced form As(III). However, as geothermal water ascends to the surface and mixes with oxygenated water from shallow aquifers, it is oxidized to As(V). Therefore, arsenic-contaminated drinking water sources in geothermally influenced waters are dominated by As(V).

The objective of this paper is to demonstrate the effect of high-temperature (50°C) waters, such as geothermally influenced waters, on the rejection of monovalent $H_2AsO_4^-$ and divalent $HAsO_4^{2-}$ species during NF membrane filtration of a multi-component solution containing Cl^- and HCO_3^-.

2 METHODS/EXPERIMENTAL

Dow NF270 and Alfa Laval were used to assess the As(V) species rejection. Rejection of $H_2AsO_4^-$ and $HAsO_4^{2-}$ was evaluated by maintaining the temperature constant (25°C or 50°C) and varying the pH (pH 8 and pH 6) of the feed water. The rejection experiments were conducted using feed flow of $100\ L\ h^{-1}$ (velocity $= 0.15\ m\ s^{-1}$), a flux of $40\ L\ m^{-2} h^{-1}$ and a recovery of 10%. All experiments were executed either at 25°C or 50°C. The concentration of arsenic in the samples were analyzed using an inductively coupled plasma mass spectrometry (ICP-MS MS Thermo – XSERIES ll, Thermo Fisher Scientific, USA).

3 RESULTS AND DISCUSSION

An improvement in rejection for $H_2AsO_4^-$ and $HAsO_4^{2-}$ was achieved as the effect of increasing temperature (Fig. 1). The rise of rejection of As(V) species could attribute to the presence of other anions (Cl^- and HCO_3^-) in the water matrix composition, used for pH and ionic strength buffer. Hodgson (1970) defined the terms less and more permeable ions, to describe the difference in behavior between co-ions (ions with same charge of the membrane). Anions such as Cl^- and

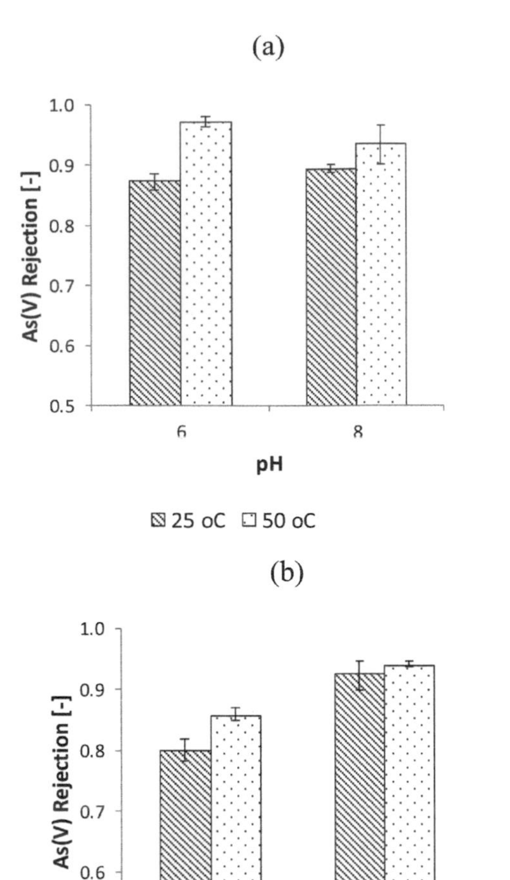

Figure 1. Removal efficiency of As(V) (300 μg L^{-1}) for (a) Dow NF270 and (b) Alfa Laval NF as a function of a constant Temperature (25°C or 50°C) and variable pH.

HCO_3^- with an equal or lower valence and higher diffusivity are more permeable than $H_2AsO_4^-$ and $HAsO_4^{2-}$. Therefore Cl^- and HCO_3^- will have a beneficial effect on the rejection of As(V) species (less permeable). The enhanced rejection of As(V) at 25°C, due to the presence of other -more permeable- co-ions has been reported by other researchers (Nguyen et al., 2009; Vrijenhoek & Waypa, 2000).

Additionally, to explain the improvement in As rejection, Vrijenhoek & Waypa (2000) pointed out the importance of the relatively small concentration of As compared with the concentration of the others co-ions present in the solution. The so-called more permeable ions can penetrate through the membrane more easily. Because of the electroneutrality condition need to keep in the bulk solution, the rejection of less permeable ions increases. The deterioration of the size exclusion mechanism and the increase in diffusion transport across the membrane due to temperature effect will enhance the permeation of the more permeable ions and as a consequence will improve the rejection of $H_2AsO_4^-$ and $HAsO_4^{2-}$ which are less permeable. The results in Figure 1 indicate that this phenomenon is temperature dependent and aiding in the rejection of As(V) at higher temperatures.

4 CONCLUSIONS

The presented study assessed the effect of high-temperature (50°C) waters, like geothermally influenced waters, on the rejection of monovalent $H_2AsO_4^-$ and divalent $HAsO_4^{2-}$ species during NF membrane filtration of a multi-component solution where Na^+, Cl^-, and HCO_3^- were present. It can be concluded that the higher the temperature, the better the rejection achieved for both As(V) species, with the added advantage that less transmembrane pressure was required to operate these NF membranes at higher temperatures.

Our findings differ from previous work where the As(V) rejection showed a decline as the consequence of temperature increase. This difference can be explained by the presence of HCO_3^- and Cl^- in our study. The experiments indicate that the enhancement of As(V) rejection was directly related to the presence of HCO_3^- and Cl^-, which are considerably more permeable than both As(V) species. The enlargement of the pore size and the increase of the mass transfer across the membrane due to temperature effect enhanced the permeation of these more permeable ions and, as a consequence, improved the rejection of $H_2AsO_4^-$ and $HAsO_4^{2-}$, which are less permeable. In particular, HCO_3^- played a vital role in As(V) retention during NF membrane filtration at higher temperatures.

ACKNOWLEDGEMENTS

This research study was supported by Nuffic. The authors want to thank to Ir. Abel Heinsbroek for the design and construction of the flat-sheet cross-flow module used during this research.

REFERENCES

Hodgson, T.D. 1970. Desalination. In: C. Viggiani (ed) *Geotechnical Engineering for the preservation of monuments and historical sites; Proc. Intern. Symp.*, Napoli, 3–4 October 1996. Rotterdam: Balkema.

Johnson, H.L. 1965. Artistic development in autistic children. *Child Dev.* 65(1): 13–16.

Nguyen, C.M., Bang, S., Cho, J. & Kim, K.W. 2009. Performance and mechanism of arsenic removal from water by a nanofiltration membrane. *Desalination* 245(1–3): 8294.

Polhill, R.M. 1982. Crotalaria in Africa and Madagascar. Rotterdam: Balkema.

Vrijenhoek, E.M. & Waypa, J.J. 2000. Arsenic removal from drinking water by a loose nanofiltration membrane. *Desalination* 130(3): 265–277.

Environmental Arsenic in a Changing World –
Zhu, Guo, Bhattacharya, Ahmad, Bundschuh & Naidu (Eds)
ISBN 978-1-138-48609-6

As(III) oxidation during full-scale aeration and rapid filtration

J.C.J. Gude, L.C. Rietveld & D. van Halem
Department Water Management, Section Sanitary Engineering, Delft University of Technology, The Netherlands

ABSTRACT: Arsenic (As) oxidation during full-scale aeration and rapid filtration was investigated. In the supernatant water, As(III) showed no apparent oxidation and meagre adsorption to HFO was observed. However, in the top of the filter bed rapid As(III) oxidation occurred and the subsequently formed As(V) was efficiently adsorbed. Although MnO_2 could potentially oxidise As(III) it was found that As(III) oxidation was inhibited in the presence of Mn(II) and Fe(II). During ripening of virgin sand filters, As(III) oxidation was complete after 21 days of operation, before Mn(II) and NH_4^+ oxidation. Therefore, it may be concluded that formation of MnO_2 could not have been responsible for As(III) oxidation, strongly suggesting that this is a microbial process. This process could develop amongst common groundwater bacteria and mineral precipitates, directly leading to an increased As removal in the filter bed.

1 INTRODUCTION

In the Netherlands, groundwater treatment commonly consists of aeration, with subsequent sand filtration without using chemical oxidants like chlorine. Oxidation of As(III) is imperative for efficient removal in this process (Bissen & Frimmel, 2003; Gude *et al.*, 2017), but the oxidation reaction between As(III) and O_2 is sluggish and takes days (Kim & Nriagu, 2000). However, As(III) oxidation was observed in the top layer of rapid sand filters without the use of chemicals (Gude *et al.*, 2016). It was hypothesised that As(III) oxidation was caused by bacteria and/or MnO_2. Both are present in the filter bed and could be capable of (catalysing) the oxidation reaction (Huang, 2014; Manning *et al.*, 2002). However conclusive evidence whether bacteria and MnO_2 are actually oxidising As(III) in rapid sand filters is scarce. For the purpose of identifying contributors to natural As(III) oxidation, jar tests with MnO_2 grains and pilot plant research using sand filter columns were conducted.

2 MATERIALS AND METHODS

2.1 *MnO_2 jar tests*

Jar tests experiments were executed in 1.8 L, buffered, demineralised water. $20\,\mu g\,L^{-1}$ As(III), $2\,mg\,L^{-1}$ Fe(II), $2\,mg\,L^{-1}$ Fe(III), $2\,mg\,L^{-1}$ Mn(II) and $100\,mg\,L^{-1}$ pure MnO_2 power were used to investigate As(III) redox and As sorption behaviour. The jars were constantly kept at pH 7 and magnetically stirred. Samples were taken after 2, 5, 10, 20, 30, 60 and 120 minutes contact time. Sample preparation consisted of immediately filtering through $0.45\,\mu m$ and acidifying to pH 1 until analysis by ICP-MS.

2.2 *Biological ripening experiment*

For this experiment 3 sand columns were loaded/ripened with aerated, natural groundwater containing As(III) for 50 days. The remaining As in the filtrate was speciated to follow As(III) oxidation in the filter bed. Filtration rate was set to $1\,m\,h^{-1}$ and supernatant level between 5 and 20 cm. Principal water quality parameters are depicted in Table 1.

Table 1. Water quality natural groundwater

Parameter	Unit	Value
As	$\mu g\,L^{-1}$	13
Fe	$mg\,L^{-1}$	1.4
Mn	$mg\,L^{-1}$	0.04
PO_4	$mg\,L^{-1}$	0.45
NH_4	$mg\,L^{-1}$	0.62
pH		7.54
HCO_3	$mg\,L^{-1}$	246
Temperature	°C	12.4

3 RESULTS AND DISCUSSION

3.1 *MnO_2 jar tests*

In order to investigate whether MnO_2 could be responsible for As(III) oxidation in the top of the filter bed, As(III) was exposed to MnO_2 grains with possible competing substances: Mn(II) and Fe(II)/Fe(III). These results are depicted in Figure 1. Fig. 1 left shows As(III) in combination with Fe(II) only, Fe(II) and MnO_2 and Fe(II)-MnO_2-Mn(II). The same experiments were executed for Fe(III) and are depicted in Fig. 1 right.

Fig. 1 left shows that As(III) removal is similar for Fe(II) with and without MnO_2 grains with As(III) removal of 60% after 120 min. On the other hand Fe(III) and MnO_2 removed 98% of As(III) in 120 min (Fig. 1B). It was observed that the presence of Fe(II) and Mn(II) inhibited As(III) oxidation on MnO_2. At pH 7, Fe(II) and Mn(II) adsorption and oxidation were preferred over As(III) on the MnO_2 surface.

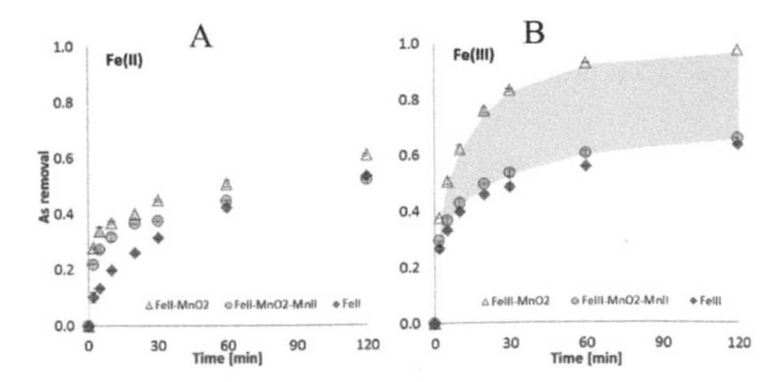

Figure 1. As(III) removal by Fe(II) and Fe(III) in presence and absence of MnO_2 and Mn(II) at pH 7 over 120 min. Concentrations were 20 µg L^{-1} As(III), 2 mg L^{-1} Fe, 2 mg L^{-1} Mn and 100 mg L^{-1} MnO_2 powder. (A) is As(III) removal by Fe(II) and (B) Fe(III) (Figure taken from Gude *et al.*, 2017).

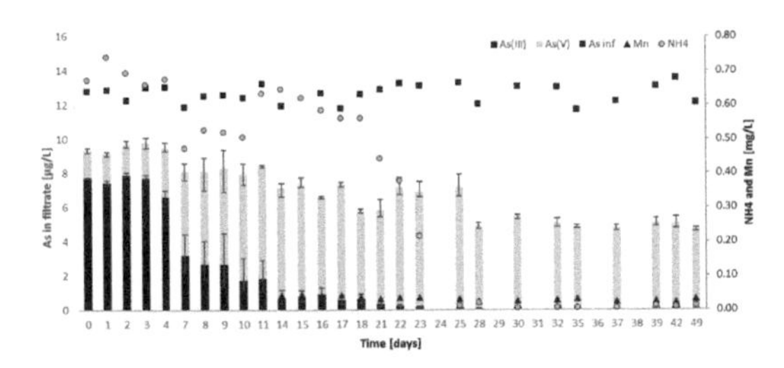

Figure 2. As(III) and As(V) in the filtrate of As(III) preloaded (top) and virgin sand columns (bottom) over time. Filtration velocity 1 m h^{-1} and bed height 0.5 m. Initial concentration of the groundwater: Fe 1.4 mg L^{-1}; As 13 µg L^{-1} (as As(III)) also depicted<■>; PO_4 0.45 mg L^{-1}; Mn 0.04 mg L^{-1}; NH_4 0.62 mg L^{-1}. As concentration in the filtrate is depicted as the sum of the As(III) and As(V) species.

However, unlike Fe(II), the addition of Fe(III) did not hinder As(III) oxidation on the MnO_2 surface; resulting in subsequent effective As(V) removal by the flocculating hydrous ferric oxides. Mn(II) and Fe(II) are typically present in the pore volume of the top of the filter bed, therefore it can be concluded that just because MnO_2 is present in a filter bed, it does not necessarily mean that MnO_2 in the top of a sand filter is available to oxidise As(III).

3.2 *Biological As(III) oxidation*

Biological As(III) oxidation is a possible alternative for MnO_2 to catalyse the aerobic oxidation reaction of As(III) and O_2 in the top of the filter bed. This was investigated by ripening a sand filter with aerated natural groundwater water containing As(III) for 50 days. During rapid filtration of natural groundwater many biological processes take place, in this experiment it was tested whether As(III) present in low concentrations would be able to oxidise. Results shown in Figure 2 are averaged from triplicates and As in the filtrate is depicted as the sum of the As(III) and As(V) species.

As(III) was oxidized to As(V) gradually in the first 5 days (10%) and showed a steep increase in the next 5 days (>80%). This pattern follows a typical biological process where after a slow lag phase, a rapid log phase follows. Mn(II) removal was not observed within the 50 day experiment and NH_4^+ removal started after 15 days. Therefore it is concluded that independently of biological NH_4^+ and Mn oxidation processes biological As(III) occurred in the sand filter.

The oxidation of As(III) was accompanied by a slight increase in As removal. During the ripening of the virgin filter sand As(III) oxidation was complete after 21 days of operation, before Mn(II) and NH_4^+ oxidation. Therefore, it may be concluded that formation of MnO_2 could not have been responsible for As(III) oxidation, strongly suggesting that this is a microbial process.

4 CONCLUSIONS

As(III) oxidation in the top of the filter bed is crucial for efficient As removal in rapid sand filters. It was found with jar tests that the presence of Fe(II) and Mn(II) inhibited As(III) oxidation by MnO_2. Therefore it is concluded that just because MnO_2 is present in a filter bed, it does not necessarily mean that MnO_2 will be available to oxidise As(III).

In addition it was found that the ripening of a virgin sand filter As(III) oxidation was complete before Mn(II) and NH_4^+ oxidation, strongly suggesting that As(III) oxidation is a microbial process. This process could develop amongst common groundwater bacteria and mineral precipitates, directly leading to an increased As removal in the filter bed.

ACKNOWLEDGEMENTS

This research is supported by the Dutch Technology Foundation STW, which is part of the Netherlands Organisation for Scientific Research (NWO), and which is partly funded by the Ministry of Economic Affairs.

REFERENCES

Bissen, M. & Frimmel, F.H. 2003. Arsenic—a review. Part II: Oxidation of arsenic and its removal in water treatment. Acta Hydrochim. *Hydrobiol*. 31: 97–107.

Gude, J.C.J., Rietveld, L.C. & van Halem, D. 2017. As(III) oxidation by MnO_2 during groundwater treatment. *Water Res*. 11:41–51.

Gude, J.C.J., Rietveld, L.C. & van Halem, D. 2016. Fate of low arsenic concentrations during full-scale aeration and rapid filtration. *Water Res*. 88: 566–574.

Huang, J.H. 2014. Impact of microorganisms on arsenic biogeochemistry: A review. Water. Air. *Soil Pollut*. 225(2): 1848.

Kim, M.J. & Nriagu, J. 2000. Oxidation of arsenite in groundwater using ozone and oxygen. *Sci. Total Environ*. 247, 71–79.

Manning, B.A., Fendorf, S.E., Bostick, B. & Suarez, D.L., 2002. Arsenic(III) oxidation and arsenic(V) adsorption reactions on synthetic birnessite. *Environ. Sci. Technol*. 36: 976–981.

4.3 Nanotechnological applications in arsenic treatment

Environmental Arsenic in a Changing World –
Zhu, Guo, Bhattacharya, Ahmad, Bundschuh & Naidu (Eds)
ISBN 978-1-138-48609-6

Combined effect of weak magnetic fields and anions on arsenite sequestration by zerovalent iron

Y. Sun & X. Guan
State Key Laboratory of Pollution Control and Resources Reuse, Tongji University, Shanghai, P.R. China

ABSTRACT: In this study, the effects of major anions (e.g., ClO_4^-, NO_3^-, Cl^-, and SO_4^{2-}) in water on the reactivity of zerovalent iron (ZVI) toward As(III) sequestration were evaluated with and without a weak magnetic field (WMF). Without WMF, ClO_4^- and NO_3^- had negligible influence on As(III) removal by ZVI but Cl^- and SO_4^{2-} could improve As(III) sequestration by ZVI. A synergetic effect of WMF and individual anion on improving As(III) removal by ZVI was observed for each of the investigated anion, which became more pronounced as the concentration of anion increased. The coupled influence of anions and WMF was associated with the simultaneous movement of anions with paramagnetic Fe^{2+} to keep local electroneutrality.

1 INTRODUCTION

There are many factors determining the reactivity of ZVI under realistic conditions, one of the most widely recognized and thoroughly studied being solution chemistry, especially the co-existing anions (Guan *et al.*, 2015; Sun *et al.*, 2016). In our recent studies, we have demonstrated that the application of a weak magnetic field (WMF) can accelerate the ZVI corrosion and thus enhance the removal of many toxic ions by ZVI. Taken together, the major anion effects on ZVI reactivity with the presence of a WMF may be quite different from that without WMF. However, this hypothesis has not been validated, to the best of our knowledge. So, the goals of this work are 1) to compare the effects of major anions on the kinetics of As(III) sequestration by ZVI with and without WMF; 2) investigate the potential mechanisms of coupled effects of anions and WMF on enhancing the ZVI reactivity.

2 METHODS/EXPERIMENTAL

2.1 *Materials*

All chemicals employed in this study were of analytical grade and used as received. The sodium salts of the anions were used. The iron particles were obtained from the Shanghai Jinshan reduced iron powder factory (China), which had a mean diameter of 40 μm and a BET surface area of 0.76 $m^2 g^{-1}$. All experiments were conducted using the ultrapure water (UP water) produced by a Milli-Q Reference water purification system.

2.2 *Bach experiments*

The experimental setup described in our previous study was also used in this study (Sun *et al.*, 2014). Briefly, a permanent magnet with a diameter of 20 mm and thickness of 1 mm was placed under the reactor, which could provide a maximum magnetic field flux intensity of ~15 mT at the bottom of the reactor. The working solutions (1000 $\mu g L^{-1}$ As(III), Se(IV), Sb(V), or Cr(VI) solutions with different co-existing anions of different concentrations) were freshly prepared for each batch test. If it was not otherwise specified, the pH_{ini} value of the working solution was adjusted to 7.0 by dropwise addition of NaOH after the dosing of ClO_4^-, NO_3^-, Cl^-, SO_4^{2-}, and/or $H_2PO_4^-$ to avoid introducing other anions.

3 RESULTS AND DISCUSSION

3.1 *Effect of major anions on As(III) removal with and without WMF*

Without WMF, ClO_4^- and NO_3^- had negligible influence on As(III) removal by ZVI but Cl^- and SO_4^{2-} could improve As(III) sequestration by ZVI. Moreover, the WMF-enhancing effect on As(III) removal by ZVI was minor in ultrapure water. A synergetic effect of WMF and individual anion on improving As(III) removal by ZVI was observed for each of the investigated anion, which became more pronounced as the concentration of anion increased, as shown in Figure 1.

Based on the extent of enhancing effects, these anions were ranked in the order of $SO_4^{2-} > Cl^- > NO_3^- \approx ClO_4^-$ (from most to least enhanced). Compared to ClO_4^- and NO_3^-, Cl^- is believed to be much more aggressive towards passivating oxides because they diffuse readily into the film and form strong complexes with iron centers. These complexes can enhance the dissolution of iron oxide and often cause pitting corrosion. Accordingly, it was not unusual to

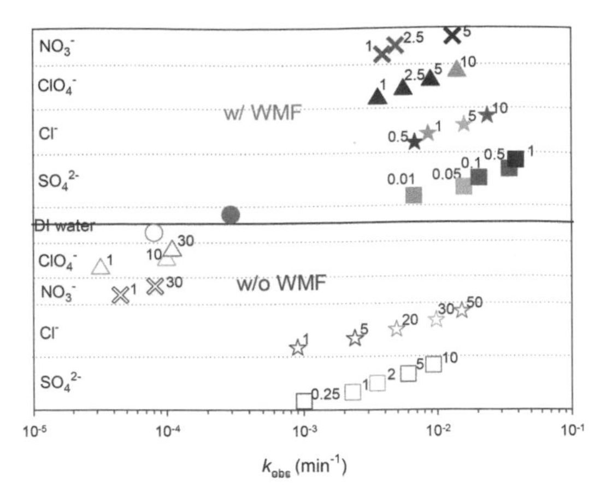

Figure 1. Rate constants of As(III) removal by ZVI with different anions in the absence or presence of WMF ($ZVI = 0.20\,g\,L^{-1}$, $As(III) = 1000\,\mu g\,L^{-1}$, $T = 25°C$). The annotating numbers represent the initial concentrations of corresponding anions (mM).

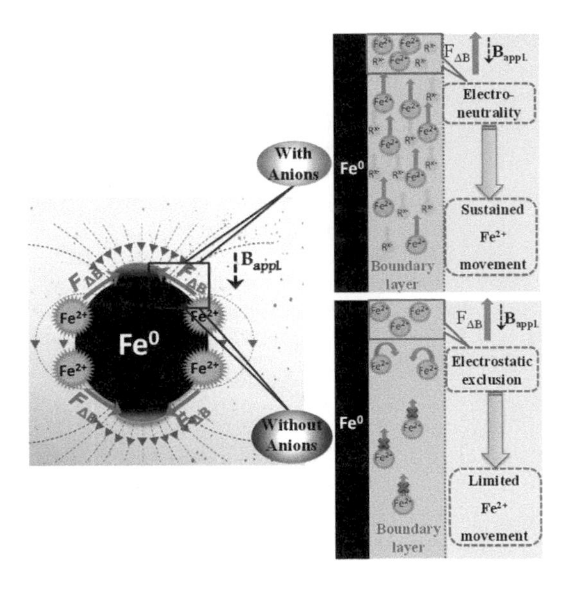

Figure 2. Proposed schematic illustration of the combined effects of WMF and anions on ZVI corrosion.

observe a better ZVI performance with Cl^- irrespective of the presence of WMF. Although many studies have reported that SO_4^{2-} could improve the ZVI performance, it was surprised to find that the performance with the presence of 0.01–1 mM SO_4^{2-} in the presence of WMF was much better than Cl^-.

3.2 *Role of co-existing anions on the reactivity of ZVI with WMF*

Recently, it has been verified that magnetic field gradient force ($F_{\Delta B}$) was the major driving force for the WMF-induced improvement of ZVI reactivity. Being pulled by $F_{\Delta B}$, the paramagnetic Fe^{2+} generated in ZVI corrosion tends to move along the magnetic lines to the place with higher magnetic field flux intensity, thereby resulting in an uneven distribution of Fe^{2+} and eventually localized distribution of oxide films on ZVI surface (Li *et al.*, 2017).

With regard to the role of anions, the principle of local electroneutrality should be emphasized. Given Fe^{2+} is positively charged, its movement will lead to an accumulation of positive charges in the high $|B\Delta B|$ regions. Thus, for the anion-free cases, due to electrostatic exclusion, the accumulation of Fe^{2+} would suppress the further movement of Fe^{2+}. However, for anion-containing scenario, since the co-existing anions can balance the positive charges, the transport of Fe^{2+} can be maintained and thereby the ZVI corrosion process can be sustained (Fig. 2).

4 CONCLUSIONS

These findings improved our understanding of the effects of anions on ZVI performance and thus could benefit the design and operation of a ZVI based technology. Another implication of this work should be a new method by taking advantage of the combined effect of WMF (or pre-magnetization) and anions (e.g., SO_4^{2-}) can be developed either to enhance iron reactivity or to overcome the adverse effect of co-solutes. It should be noted that, the introduction of WMF/anions could enhance the ZVI reactivity towards many other oxyanions besides As(III), for example, Cr(VI) and Se(IV).

ACKNOWLEDGEMENTS

This work was supported by the National Natural Science Foundation of China (Grants 21522704, 51478329, 51608431, and U1532120).

REFERENCES

Guan, X.H., Sun, Y.K., Qin, H.J., Li, J.X., Lo, I.M., He, D. & Dong, H.R. 2015. The limitations of applying zero-valent iron technology in contaminants sequestration and the corresponding countermeasures: the development in zero-valent iron technology in the last two decades (1994–2014). *Water Res.* 75: 224–248.

Li, J., Qin, H., Zhang, W., Shi, Z., Zhao, D. & Guan, X. 2017.Enhanced Cr(VI) removal by zero-valent iron coupled with weak magnetic field: role of magnetic gradient force. *Sep. Purif. Technol.* 176: 40–47.

Sun, Y., Li, J., Huang, T. & Guan, X. 2016. The influences of iron characteristics, operating conditions and solution chemistry on contaminants removal by zero-valent iron: a review. *Water Res.* 100: 277–295.

Sun, Y.K., Guan, X.H., Wang, J.M., Meng, X.G., Xu, C.H. & Zhou, G.M. 2014. Effect of weak magnetic field on arsenate and arsenite removal from water by zerovalent iron: an XAFS investigation. *Environ. Sci. Technol.* 48(12): 6850–6858.

Environmental Arsenic in a Changing World –
Zhu, Guo, Bhattacharya, Ahmad, Bundschuh & Naidu (Eds)
ISBN 978-1-138-48609-6

Remediation of arsenic contaminated groundwater with magnetite (Fe_3O_4) and chitosan coated Fe_3O_4 nanoparticles

S. Ahuja, C. Mahanta, S. Sathe, L.C. Menan & M. Vipasha
Department of Civil Engineering, Indian Institute of Technology, Guwahati, India

ABSTRACT: Nanoparticles, due to their high surface area, regenerative properties and possibility of in situ treatment, exhibited their application potential in capturing arsenic (As). The objective of the current study is to explore the potential of As adsorption by Fe_3O_4 nanoparticles (IONPs) and Chitosan coated Fe_3O_4 nanoparticles (Ch-IONPs). Batch experiments performed with an initial As(III) and As(V) concentration of 0.125 mg L^{-1}, revealed a maximum As removal of 99.9% in 0.4 g L^{-1} IONPs and 0.6 g L^{-1} Ch-IONPs, at a pH equal to 7.0 and room temperature. The maximum adsorption capacity of IONPs and Ch-IONPs was 213.25 and 118.67 $\mu g\,g^{-1}$ for As(III) and 239.50 and 116.33 $\mu g\,g^{-1}$ for As(V), respectively. The Pearson Correlation value showed a better correlation with Freundlich isotherm ($R^2 = 0.96$) than Langmuir isotherm ($R^2 = 0.87$). The results of this work suggest that the incorporation of IONPs and Ch-IONPs as adsorbents, offer a promising option for As removal in water treatment.

1 INTRODUCTION

Arsenic (As) contamination in groundwater is a major global health concern. Arsenite [As(III)] is supposed to be 60 times more poisonous and mobile than arsenate [As(V)] (De *et al.*, 2009). Among various technologies already in use for removal of As from contaminated groundwater, namely coagulation, filtration, membrane separation and ion exchange; adsorption offers many advantages like ease of operation, simple waste handling and low operation cost. Magnetite (Fe_3O_4) nanoparticles exhibit excellent magnetic properties, which are being investigated for their application in water purification for the removal of heavy metals, such as lead and As (Feng *et al.*, 2012; Mayo *et al.*, 2007). The surface of these nanoparticles can be revamped according to desired functionalities, by methods like coating with surfactants or polymers. The aim of the current study is to find out the optimum conditions for the removal of As(III) and As(V) species by laboratory-synthesized Fe_3O_4 nanoparticles (IONPs) and chitosan coated Fe_3O_4 nanoparticles (Ch-IONPs). The removal capacity was studied by applying batch experiments and inferred by isotherm study.

2 MATERIALS AND METHODS

2.1 *Synthesis of IONPs and Ch-IONPs*

For the synthesis of IONPs by co-precipitation, 10 mg $FeCl_3 \cdot 6H_2O$ was dissolved in 200 mL distilled water. To this, 4 mg $FeSO_4 \cdot 7H_2O$ was added, and the mixture heated to 55°C for 10 min. NH_4OH was rapidly added to the heated mixture for maintaining pH at 10. The final solution was centrifuged, washed with distilled water and ethanol, finally oven dried for 12 hours. For the Ch-IONPs' synthesis, 5 mg $FeCl_3 \cdot 6H_2O$ was mixed with 2 mg $FeCl_2 \cdot 4H_2O$ in 100 mL distilled water. Chitosan, 0.2 mg, was added in a 20 mL mixture of 1 mL acetic acid (2N) and 19 mL distilled water. Both the above mixtures were then mixed and pH was adjusted to 6.9. NH_4OH was then added after passing dry N_2 gas at 80°C. The final solution was centrifuged, washed with distilled water and ethanol, finally oven dried for 12 h.

2.2 *Effect of pH and adsorbent dose on As(III) and As(V) removal*

To study the effect of pH on As adsorption behavior of IONPs and Ch-IONPs, 50 mg of each of IONPs and Ch-IONPs were added in 50 mL $NaAsO_2$ solution (0.125 mg L^{-1}) and 50 mL $NaHAsO_4$ solution (0.125 mg L^{-1}), respectively. Batch study for IONPs and Ch-IONPs was performed by adjusting the pH at 4.0, 5.0, 6.0, 7.0, 8.0 and 9.0 in different batches (shaken for 60 minutes), adding 0.01 M HCl or 0.1 M NaOH. The study suggested an optimum pH of 7.0, which was kept constant and the dosage of IONPs and Ch-IONPs was varied, i.e. 5, 10, 15, 20, 30, 40 and 50 mg in each batch of 50 mL $NaAsO_2$ solution and 50 mL $NaHAsO_4$ solution, respectively, to optimize the adsorbent dose for As(III) and As(V) adsorption and shaken for 60 min.

3 RESULTS AND DISCUSSION

3.1 *Characterization of IONPs*

The morphology of IONPs was studied using Field Emission Scanning Electron Microscope (FESEM), which confirms the formation of peanut shaped IONPs with an average size of 30 to 80 nm. The FESEM images of IONPs and Ch-IONPs are shown in

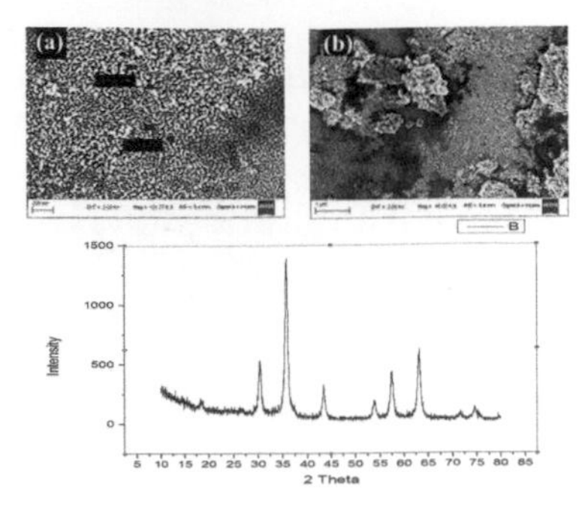

Figure 1. (a) FESEM image of IONPs with particle size (b) FESEM image of Ch-IONPs (c) XRD analysis image of IONPs.

Table 1. Effect of pH on As(III) and As(V) adsorption onto IONPs and Ch-IONPs, % removal values.

	As(III)		As(V)	
pH	IONP	Ch-IONP	IONP	Ch-IONP
4.0	66.92	82.40	86.72	45.92
5.0	99.90	87.84	89.28	88.36
6.0	99.90	93.92	99.80	95.48
7.0	99.90	99.90	99.90	97.84
8.0	99.90	87.80	99.90	85.40
9.0	99.90	76.96	94.68	79.64

Figure 1(a) and (b), respectively. The X-ray diffraction (XRD) analysis confirmed the existence of IONPs and Ch-IONPs by showing clear peaks at diffraction angles of 35.246, 42.831, 53.128 and 74.530 (Okudera *et al.*, 1996). The XRD image of the formed IONPs is shown in Figure 1(c).

3.2 *Batch studies for As(III) and As(V) removal*

A maximum As(III) and As(V) removal capacity of 99.9% was achieved at an initial concentration of $0.125\,mg\,L^{-1}$ As(III) and As(V), at pH equal to 7.0, which can be attributed to the fact that metal ions get easily adsorbed due to the protonated surface amine groups. Table 1 summarizes the results of pH study. The critical dose for As(III) and As(V) adsorption was observed as $0.4\,g\,L^{-1}$ for IONPs and $0.6\,g\,L^{-1}$ Ch-IONPs. A rapid increase in % As removal with an increase in the adsorbent dose of IONPs and Ch-IONPs (5.0 mg to 50.0 mg, in 50 ml solution) was observed. This is due to the increase in available adsorption sites on the surface of the adsorbent. Table 2 summarizes the results of adsorbent dose study.

3.3 *As(III) and As(V) adsorption isotherm*

The isotherm studies reveal the maximum adsorption capacity of IONPs equal to $213.25\,\mu g\,g^{-1}$ for As(III) and $239.50\,\mu g\,g^{-1}$ for As(V). A decrease of adsorption capacity in case of Ch-IONPs to $118.67\,\mu g\,g^{-1}$ for As(III) and $116.33\,\mu g\,g^{-1}$ for As(V) was observed. This can be explained by Chitosan's coating on the adsorption sites available on virgin IONPs, which

Table 2. Effect of adsorbent dose on As(III) and As(V) adsorption onto IONPs and Ch-IONPs, % removal values.

Adsorbent	As(III)		As(V)	
Dose (mg)	IONP	Ch-IONP	IONP	Ch-IONP
5.0	65.00	47.64	79	60.32
10.0	75.40	59.16	87.90	72.16
15.0	77.20	86.24	88.40	75.12
20.0	99.90	99.90	88.08	99.90
30.0	99.90	99.90	99.90	99.90
40.0	99.90	99.90	99.90	99.90
50.0	99.90	99.90	99.90	99.90

reduces the removal capacity of Ch-IONPs. The Pearson Correlation value showed a better correlation with Freundlich isotherm ($R^2 = 0.96$) than Langmuir isotherm ($R^2 = 0.87$).

4 CONCLUSIONS

Considering the average As concentration ($\sim 100\,\mu g\,L^{-1}$) in the Brahmaputra flood plain (as per literature reviewed), maximum As removal was achieved at a dose of $0.4\,g\,L^{-1}$ for IONPs and $0.6\,g\,L^{-1}$ for Ch-IONPs. As(V) removal was observed to be 20% (approx.) higher than As(III) removal, both for IONPs and Ch-IONPs. Further, the results also showed that the removal efficiency of IONPs was 1.5 times more than Ch-IONPs'. Both these nanomaterials can be locally (in house) used for a low As concentration as well as for a higher As concentration. Also, IONPs and Ch-IONPs used in this adsorption process can be separated easily with a magnet, washed and their surface can be regenerated for reuse.

ACKNOWLEDGEMENTS

Authors sincerely thank the Department of Civil Engineering, IIT Guwahati for providing the lab facilities and chemicals. Authors also acknowledge the Central Instruments Facility, IIT Guwahati for FESEM and XRD analysis.

REFERENCES

De, D., Mandal, S.M., Bhattacharya, J., Ram, S. & Roy, S.K. 2009. Iron oxide nanoparticle-assisted arsenic removal from aqueous system. *J. Environ. Sci. Health A*. 44(2): 155–162.

Feng, L., Cao, M., Ma, X., Zhu, Y. & Hu, C. 2012. Superparamagnetic high-surface-area Fe_3O_4 nanoparticles as adsorbents for arsenic removal. *J. Hazard. Mater.*, 217: 439–446.

Mayo, J.T., Yavuz, C., Yean, S., Cong, L., Shipley, H., Yu, W., Falkner, J., Kan, A., Tomson, M. & Colvin, V.L. 2007. The effect of nanocrystalline magnetite size on arsenic removal. *Sci. Technol. Adv. Mater*. 8(1–2): 71–75.

Okudera, H., Kihara, K. & Matsumoto, T. 1996. Temperature dependence of structure parameters in natural magnetite: single crystal X-ray studies from 126 to 773 K. *Acta Crystallogr. B* 52(3): 450–457.

Environmental Arsenic in a Changing World –
Zhu, Guo, Bhattacharya, Ahmad, Bundschuh & Naidu (Eds)
ISBN 978-1-138-48609-6

Removal of arsenic from wastewater treated by means of nanoparticles and magnetic separation

M.F. Isela[1], R.C. Mercedes E.[1] & G.M. Rocío[2]
[1]*Posgrado de Ingeniería, UNAM, Instituto Mexicano de Tecnología del Agua, Jiutepec, Mor., Mexico*
[2]*Centro de Ciencias de la Atmósfera, UNAM, Ciudad Universitaria, Coyoacán, Mexico*

ABSTRACT: This work is carried out based on the importance of the enormous consumption of water, which is obliged to comply with current regulations for the use to which it is intended. Because the arsenic is a toxic element and exposure to it, even in small amounts, can cause serious health problems, the water supplied to the services must comply with the maximum permissible limits. Therefore, it will use two iron oxide nanoparticles that allow the elimination of said pollutant and, in turn, the nanoparticles will be recovered through the application of a magnetic field for its regeneration and subsequent reuse, determining its life cycle. By establishing the optimal adsorption conditions for each of the nanoparticles, the removal of arsenic in a contact device water – nanoparticles – magnetic field using only the most efficient nanoparticles and the use wastewater treated real will be carried out.

1 INTRODUCTION

Arsenic (As) is a natural element from earth's crust; widely distributed through all the environment. Ingestion of small amounts of As can cause chronic effects due to its bioaccumulation in the organism. Prolonged exposure to inorganic As, through the consumption of contaminated water or food prepared with it and food crops watered with As-rich water can cause chronic intoxication.

The treatment train that municipal wastewater in Mexico receives includes physical, chemical and biological processes generally, under the operating conditions in which these processes are realized, certain pollutants, such as heavy metals, are not removed (Abejón & Garea, 2015).

Nanoparticles (NPs) are materials with a nanometric size on its three dimensions; as a result of its small size, they possess structural and morphological characteristics which make them ideal for some kind of applications. Specifically, magnetic NPs are efficient for As polluted water treatment up to the grade of obtaining water suitable for human consumption, sanitizing and irrigation. Due to the big use of superficies in proportion to its volume and they link easily to chemical substances, they can be eliminated, applying a magnetic field (Liyun *et al.*, 2012; Yavuz *et al.*, 2006).

Magnetic NPs base on iron, both magnetic and oxidized iron offer stronger magnetic susceptibility, high chemical stability, and low toxicity, so its use on water treatment results efficient, obtaining water with As concentrations under maximum level permissions established by Mexican Official Norms (NOM-127-SSA1-1994).

Table 1. NPs to use.

NPs	Description	Physical appearance
Iron oxide (Fe_2O_3)	Magnetic and semiconductor NPs	Fine reddish powder without characteristid aroma
Graphene functionalized with Fe_2O_3	Graphene microflake reduced and functionalized with Fe_2O_3 (2% a 5%)	Black pasta

2 METHODOLOGY

2.1 *Selection of parameters*

This first stage will be carried out using commercial iron oxide NPs (Table 1), model solutions of wastewater treated with known concentrations of As was prepared.

The optimal work parameters will be established: concentration As-dose NPs, pH, temperature, agitation and contact time. In order to identify which of the NPs has a greater capacity for removing As kinetics and isotherms of adsorption will be carried out with each of them.

2.2 *Design of contact device*

Once the working parameters have been optimized and NPs have been identified that allow for greater elimination of As, a prototype device at laboratory scale

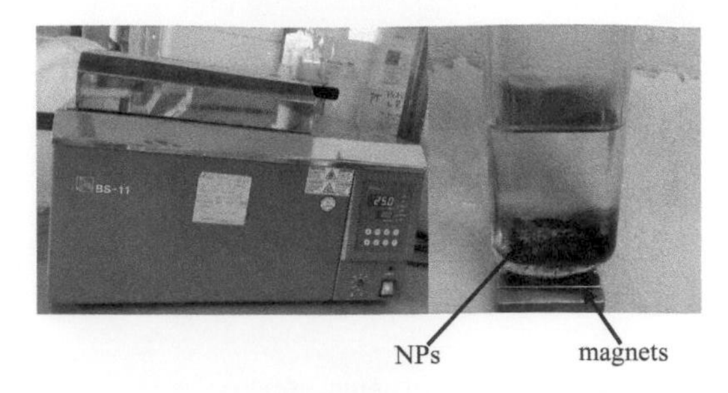

Figure 1. Experimental system for adsorption tests.

will be designed and assembled to evaluate the best contact between these NPs and the water. In this stage, real samples of treated wastewater will be used to evaluate the effects due to interferences in the elimination process.

A magnetic device will be assembled together to allow separation of the NP from the water in process. After this separation, the analysis of the effluent will be carried out by Atomic Absorption Spectrophotometry.

2.3 *Regeneration of NPs*

For the regeneration study of the NPs, an acid wash will be carried out, eliminating the As, centrifuged and reused the NPs, determining its adsorption capacity after each use, thus obtaining its life cycle.

3 RESULTS AND DISCUSSION

The experimental system for the adsorption tests consists of 150 mL glass bottles. A continuous oscillatory agitation device at 100 rpm is used, with a sample volume of 100 mL, this adsorption process will be carried out with different doses of the NPs under study, at room temperature (25°C). Samples are taken at different contact time intervals (1, 5, 8, 15, 24, 36, 48, 72, 96 h). Once this time has elapsed, a magnetic field is applied with the objective of separating the NPs from the water and proceeding to a filtration of the sample, using Whatman® filter paper (45 μm), eliminating the color of the treated water. All tests will be done in triplicate (Fig. 1).

The residual concentration of As in the effluent will be quantified, in this way both the kinetics and the adsorption isotherms will be obtained.

So far, a removal percentage of 94.67% has been obtained in 48 h, using an As concentration of 1.5 mg L^{-1} in synthetic water and a dose of NPs of 1 g L^{-1}. A greater removal is expected based on what is reported in the literature. It is worth mentioning that an average pH value of 8.02 was obtained for the effluent, which is why it is not an additional parameter to be modified in the effluent.

4 CONCLUSIONS

With the concentration of As that has been used for the synthetic water (1.5 mg L^{-1}) and the dose of NPs of Fe_2O_3 (1 g L^{-1}) the feasibility of the removal of said metal is verified, nevertheless, it is necessary to continue with experimentation to find the equilibrium conditions.

The agitation imposed on the system (100 rpm) makes it possible to keep the NPs in suspension, which improves the contact surface, although it is worth mentioning that after certain hours the nanoparticles sedimented, so this parameter must also be optimized.

REFERENCES

Abejón, R. & Garea, A. 2015. A bibliometric analysis of research on arsenic in drinking water during the 1992–2012 period: an outlook to treatment alternatives for arsenic removal. *J. Water Process Eng.* 6: 105–119.

Liyun, F., Minhua, C., Xiaoyu, M., Zhu, Y. & Hu C. 2012. Superparamagnetic high-surface-area Fe_3O_4 nanoparticles as adsorbents for arsenic removal. *J. Hazard. Mater.* 217–218: 439–446.

Official Mexican Standard. NOM-127-SSA1-1994 (as amended in 2000). Environmental health, water for human use and consumption – permissible quality limits and treatments to which water must be submitted for drinking water. Mexico.

Yavuz, C.T.J., Mayo, T., Yu, W.W., Prakash, A., Falkner, J.C., Yean, S., Cong, L., Shipley, H.J., Kan, A., Tomson, M., Natelson, D. & Colvin, V.L. 2006. Low-field magnetic sep-aration of monodisperse Fe_3O_4 nanocrystals. *Science* 314(5801): 964–967.

Environmental Arsenic in a Changing World –
Zhu, Guo, Bhattacharya, Ahmad, Bundschuh & Naidu (Eds)
ISBN 978-1-138-48609-6

A synergistic Cu-Al-Fe nano adsorbent for significant arsenic remediation and As(0) supported mitigation in aqueous systems

Y.K. Penke[1], G. Anantharaman[2], J. Ramkumar[3] & K.K. Kar[3]
[1]*Materials Science Programme, IIT Kanpur, India*
[2]*Department of Chemistry, IIT Kanpur, India*
[3]*Department of Mechanical Engineering, IIT Kanpur, India*

ABSTRACT: In this work arsenic adsorption in aqueous systems is studied onto copper based ternary metal oxide (Cu-Al-Fe) nano adsorbents. In vibrational spectroscopy analysis, various As-O and As-OH related stretching vibrations were observed in 800–850 cm^{-1} band. Adsorption kinetics study is observed with Pseudo Second Order (PSO) model and Freundlich model is observed for adsorption isotherms. Quantitative studies infer better As(III) adsorption in basic conditions and better As(V) adsorption in acidic conditions. XPS study of individual As(3d) spectra observed with multiplet peak behavior attributed various arsenic signals. In adsorbed systems significant proportions of As(0) signals are observed around 10 at.% for both As(III) and As(V) systems in pH 7 condition. Active redox behavior of Cu-Al-Fe resulted in better As(III) mitigation effect ability.

1 INTRODUCTION

Arsenic contamination in the South-Asian nations (India, Bangladesh, Pakistan etc.) is considered to be one of the largest mass poising incidents of human in history. This contamination scenario is affecting around 150–200 million population by means of different health disorders (Yan, 2012). Different types of arsenic remediation processes are in use for various purposes to reduce the final arsenic to less than 10 $\mu g L^{-1}$ (WHO guidelines). Adsorption based arsenic remediation method is considered to be one of the efficient and economical processes. Zero valent iron (ZVI) and pyrites (FeS_2) adsorbent systems were known for arsenic remediation by means of redox (i.e., simultaneous reduction and oxidation), surface complexation behavior and zero valent arsenic (i.e., As(0)) formation (Penke, 2016; Yan, 2012). The formation of As(0) onto adsorbent is useful in the prevention of post adsorption leaching effect of adsorbed arsenic back to the open environment. In this study we analyzed copper-based ternary metal oxides for arsenic mitigation by As(0) formation.

2 EXPERIMENTAL

$CuCl_2 \cdot 2H_2O$ (Loba Chemie, India), $Al(NO_3)_3 \cdot 9H_2O$ (Merck, India), $Fe(NO_3)_3 \cdot 9H_2O$ and Liq. Ammonia (A.R. grade, Qualigens Chemicals, India) of A.R grade were used for synthesizing aluminum substituted copper ferrite as per the reported procedure. In preliminary qualitative studies, 100 mL aliquots of arsenic solutions (57.6 $\mu g L^{-1}$ for As(III) and 24 $\mu g L^{-1}$ for As(V)) were dispersed with 0.1 g of Cu-Al-Fe adsorbent and agitated for 24 h using an orbital shaker at pH 7. The supernatant solutions were vacuum filtered using membrane filters. The separated adsorbent powders were dried and further preceded for IR, Raman and XPS studies. For XPS analysis of aliquot systems, filtered adsorption isotherms aliquots ($C_i \sim 100\, mg L^{-1}$) were evaporated under natural conditions at R.T in open atmosphere and scratched samples were proceeded for characterization. The supernatant aqueous solutions were advanced to ICP-MS based quantitative analysis. In quantitative analysis adsorption isotherm (24 h), absorption kinetics (6 h) and pH variation studies were performed. In all adsorption systems pH adjustment was done using standard HCl (1 M) and NaOH (1 M) solution.

3 RESULTS AND DISCUSSION

3.1 *FESEM and vibrational spectroscopy results*

After the arsenic adsorption agglomeration, and flocculation behavior was observed in adsorbent powders. This agglomeration and flocculation behavior which may be due to the formation of different immobilized arsenic species (As-O-M kind). This kind of studies were earlier reported for various metal-based (e.g., Fe, and Al) adsorbent systems. The weight percentage (i.e., EDS spectra) of arsenic onto these adsorbents is observed around 0.9 and 0.4 wt.% for As(III) and As(V) systems (Fig. 1). Vibrational spectra at pH 7 systems were observed with signals around 796, and 834 cm^{-1} for arsenic (As(III), and As(V)) adsorbed systems (Fig. 1). These vibrational spectra signals observed in the 800–850 cm^{-1} band inform the formation of various arsenic surface complex structures (e.g., As-O, As-OH) of ν and ν_{as} kind.

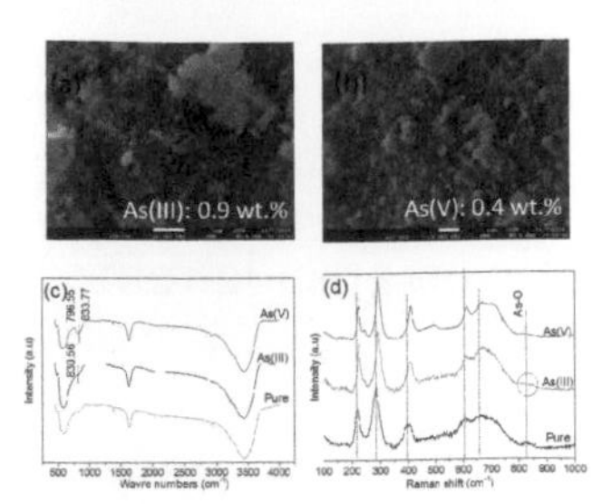

Figure 1. (a,b) FESEM, (c) IR and (d) Raman spectra of As(III), and As(V) adsorbed Cu-Al-Fe particles in pH 7.

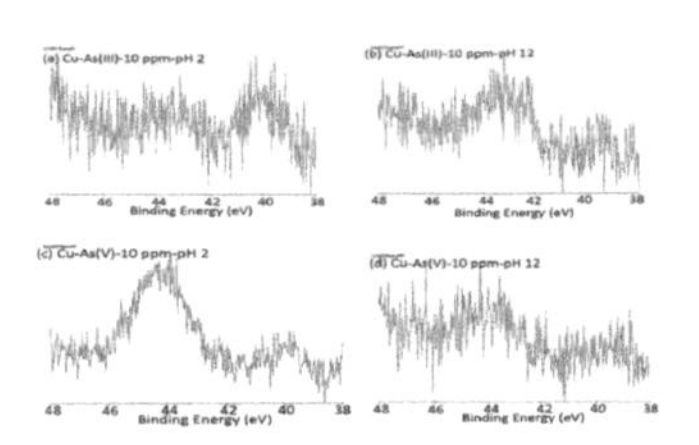

Figure 2. As(3d) plots of Cu-Al-Fe in pH 2, and pH 12 ($C_i \sim 10\,\mathrm{mg\,L^{-1}}$, m: $0.4\,\mathrm{g\,L^{-1}}$) (a,b) As(III)-pH 2, 12 (c,d) As(V)-pH 2, 12.

3.2 *XPS results*

The multiple peak behavior was observed in As(III) and As(V) adsorbed systems 43–45 eV in pH 7 confirms the occurrence of redox reactions.

In As(III) adsorbed systems, surface species corresponding to As(III) state were observed around 31.04 at.% whereas the remaining arsenic species were observed with less toxic As(V) state (~60.01 at.%). In As(V) adsorbed systems (pH 7) nearly 90 at.% of adsorbed arsenic was observed with As(V) state. Apart from these peaks an additional signals were observed in 39.5–40.5 eV band attributed to As(0) onto the adsorbent (Fig. 2). This behavior may be due to the reduction of adsorbed As(III) and As(V) species. In the total adsorbed arsenic around 9 at.% (As(III)), and 10 at.% (As(V)) of As(0) species were observed onto the adsorbent.

3.3 *Mitigation results*

The evaporated As(III) aliquot samples were observed with redox supported multiplet peak behavior corresponding to As(0) (50.5 at.%), and As(V) (49.5 at.%) (Fig. 3).

3.4 *Adsorption mechanism*

Our previous studies with Ni-Al-Fe and Co-Al-Fe were observed with a redox active behavior in arsenic remediation where cobalt systems were observed with As(0) (2–10 at.%) (Penke, 2016, 2017). The present study with copper-based Cu-Al-Fe adsorbent is observed with higher As(0) proportion (10–60 at.%) which is first of its kind for a copper-based metal oxide adsorbent (Fig. 4). The better arsenic remediation ability, significant As(0) formation (~60 at.%), and better

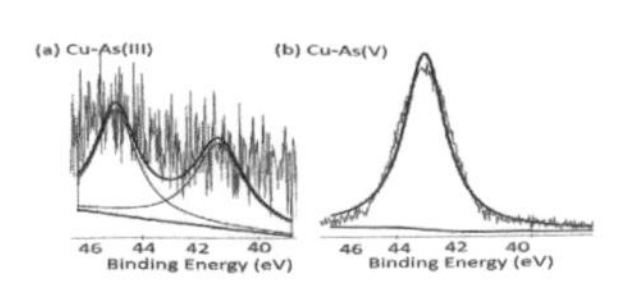

Figure 3. As 3d plots of evaporated As(III) aliquot samples in pH 7 ($C_i \sim 100\,\mathrm{mg\,L^{-1}}$, m: $0.4\,\mathrm{g\,L^{-1}}$)) (a) As(III) and (b) As(V).

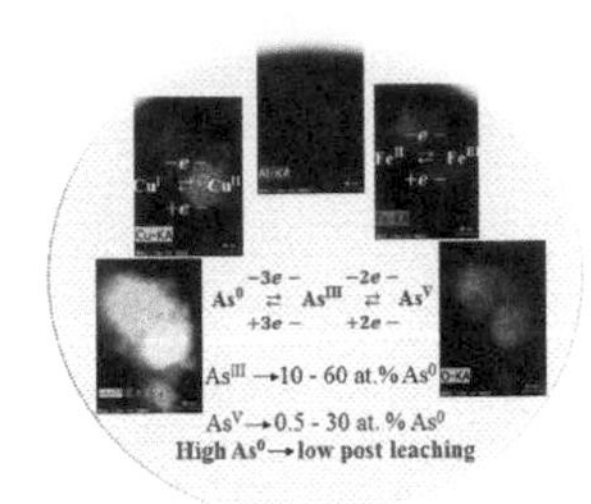

Figure 4. Scheme of arsenic mitigation on to Cu-Al-Fe.

mitigation ability (90–100%) are due to the active redox potential behavior (E°) of Cu-Fe systems. In addition, aluminum (Al) substitution in Cu-Fe system also emphasized the arsenic adsorption behavior by enhancing the surface related properties.

4 CONCLUSIONS

Copper-based ternary metal oxides (Cu-Al-Fe) are verified for probable arsenic remediation systems from aqueous systems. The adsorption phenomena of As(III)/As(V) were evaluated using various spectroscopy tools. The outcome of the results suggest that better agglomeration behavior was observed in morphology study which is due to the formation immobile arsenic surface complexes. Redox active nature of Cu-Al-Fe synergistically enhanced the As(III) and As(V) mitigation through As(0) formation. A significant As(0) formation in XPS (at%) resembles the environmentally responsible adsorbent.

ACKNOWLEDGEMENTS

The authors would like to thank Indian Institute of Technology Kanpur (MHRD, Government of India).

REFERENCES

Penke, Y. K. 2016. Aluminum substituted nickel ferrite (Ni–Al–Fe): a ternary metal oxide adsorbent for arsenic adsorption in aqueous medium. *RSC Adv.* 6(60), 55608–55617.

Penke, Y.K. 2017. Aluminum substituted cobalt ferrite (Co-Al-Fe) nano adsorbent for arsenic adsorption in aqueous systems and detailed redox behavior study with XPS. *ACS Appl. Mater. Interfaces* 9: 11587–11598.

Yan, W. 2012. As(III) sequestration by iron nanoparticles: study of solid-phase redox transformations with X-Ray photoelectron spectroscopy. *J. Phys. Chem. C* 116(9): 303–5311.

Environmental Arsenic in a Changing World –
Zhu, Guo, Bhattacharya, Ahmad, Bundschuh & Naidu (Eds)
ISBN 978-1-138-48609-6

Exploring the scope of nanoparticles for arsenic removal in groundwater

A. Kumar[1], H. Joshi[1] & A. Kumar[2]
[1]*Department of Hydrology, Indian Institute of Technology, Roorkee, India*
[2]*Department of Chemistry, Indian Institute of Technology, Roorkee, India*

ABSTRACT: Nanoadsorbents have gained considerable attention for arsenic removal in groundwater among scientific communities from last decade. Nanotechnology based water treatment systems are logical choice in respect to resource and energy efficiency. The literature is widely replete in cases of development of different nanoadsorbents which are explored for arsenic removal. Among these, nanoscale zero-valent iron has been observed extensively employed both at laboratory and pilot scale studies due to its strong affinity, easy availability and environmental friendly nature. Its unstable nature limits the application at large scale efficiently. In present study, γ-Fe_2O_3 nanoparticles were evaluated for arsenic removal using Taguchi's design of experimental methodology for real world water conditions by formulating artificial groundwater.

1 INTRODUCTION

Arsenic removal from groundwater is a great concern among the scientific communities due to its potential health impacts. Both developed and developing countries are under a potential threat of its groundwater contamination. Among the several treatment technologies, adsorption has been reported to be viable due to its easy operational processes. It is therefore, different types of conventional adsorbents have been widely employed in the removal of arsenic from groundwater.

In past two decades, several metallic nanoadsorbents including oxides of Fe, Al, Ce, Cu, Zr, Ti have been developed for arsenic removal. The metallic iron based nanoadsorbents have been widely explored due to their strong affinity towards arsenic and ecological-friendly nature. Among various polymorphs, nZVI (zero-valent iron) has been extensively explored in the literature both at laboratory and pilot scale studies. These NPs are unstable in natural environmental conditions and prone to oxidize into iron oxides/hydroxides after oxidation which limit its application for pilot scale studies significantly. However, these nanoparticles have gained notable attention for pilot scale studies in developed countries.

The aim of present study is to synthesize the stable polymorph of $Fe^{(III)}$ oxide (γ-Fe_2O_3) NPs and their evaluation for the possible scope of *in situ* sequestration of arsenic. These nanoparticles were explored for the real-world water conditions. The samples were collected from two locations of district Ballia, Uttar-Pradesh, India. The laboratory scale experiments are not feasible using actual groundwater due to requirement of large samples volume and fluctuating quality issues. Therefore, a computational analysis using mathematical matrix was exercised in calculating the required quantity of available salt compounds for synthesis of artificial groundwater (Adams 1998). Taguchi's methodology (Design Expert 7.0.0) was adopted for experimental design which investigate effect of different parameters on the mean and variance of performance characteristics.

2 METHODS/EXPERIMENTAL

2.1 *Synthesis of γ-Fe_2O_3 nanoparticles and characterization techniques*

The γ-Fe_2O_3 NPs used in this study were synthesized following a procedure adapted from (Kaloti & Kumar 2016) with few notable deviations. Phase identification was examined using XRD pattern which further confirmed through RAMAN spectrum. The surface morphology, charge and functionalities were explored through SEM/HRTEM, zeta potential and FTIR analysis.

2.2 *Taguchi's methodology and artificial groundwater formulation*

Taguchi's methodology is an effective statistical tool in developing a suitable approach for laboratory investigations at optimized parametric levels. It depends on modeling of experiments classifying the engineering aspects for the cost-effectiveness of process along with to overcome drawbacks related to conventional removal techniques. The orthogonal arrays (OAs), S: N analysis and variance are expressed as significant tools to analyze the outcomes of parameter design.

In this study, the calculated degree of freedom is 26 [= no. of parameters (7) × {no. of levels (3) − 1} + {no. of two-PI (3) × no. of PI (2) × no. of

Table 1. Process parameters for experimental design using Taguchi's orthogonal array (OAs) of methodology.

	Parameters	level 1	level 2	level 3
A	Arsenic conc. ($\mu g\ L^{-1}$)	55	127.5	200
B	TDS ($mg\ L^{-1}$)	350	900	1450
C	Shaking speed (rpm)	100	170	240
D	Temp. (°C)	10	20	30
E	pH	7	8	9
F	Dose ($g\ L^{-1}$)	0.05	0.10	0.15
G	Contact time (min)	2	53	104

columns assigned for each PI (2)}]. Hence, a standard three level OA of $L_{27}(3^{13})$ was selected for investigation. The details of experimental parameters are shown in Table 1. In groundwater formulation, a multiplication of inverse matrix with the targeted values of concentration was done to calculate the number of compounds required to achieve the desired element concentration in formulation of water.

The formulation array was customized by varying the compounds containing common ions and arrangement of array to goal the appropriate results.

3 RESULTS AND DISCUSSION

3.1 *Physio-chemical characteristics of NPs*

The nanoparticles were found to be polycrystalline in nature. The diffraction patterns match with the reflections which correspond to cubic structure of γ-Fe_2O_3 (JCPDS file no. 39-1346). Using Debye-Scherrer's equation, the average crystallite size of iron oxide NPs was calculated to be 14.61 ± 2.43 nm In FTIR spectra, presence of large surface hydroxyl moieties was observed. The average particle size (nm) distribution for these analyses were calculated to be 57.92 ± 15.51 and 15.72 ± 8.15, respectively. The large value of calculated average particle size during FESEM analysis indicates formation of agglomerates which occur due to the magnetic properties these NPs. However, a significant agreement between the calculated average particle size using XRD and HRTEM techniques was observed which validate the analysis of results obtained. The ζ-potential for these nanoparticles was observed in acidic range ($pH_{pzc} \sim 2.3$).

3.2 *Effect of process parameters*

Low temperature favors removal capacity indicates process of physisorption. Maximum shaking speed is affecting the adsorption capacity of the nanoparticles which might be due to the desorption of adsorbed As(V) into the aqueous solution (Fig. 1). Further, the maximum value of q_e is observed at the larger value of contact time (L_3) showing that the As(V) oxyanions and hydroxyl groups of nanoparticles are interacting with the weak electrostatic forces. A shift for the parameter (C) (one place) towards the higher value

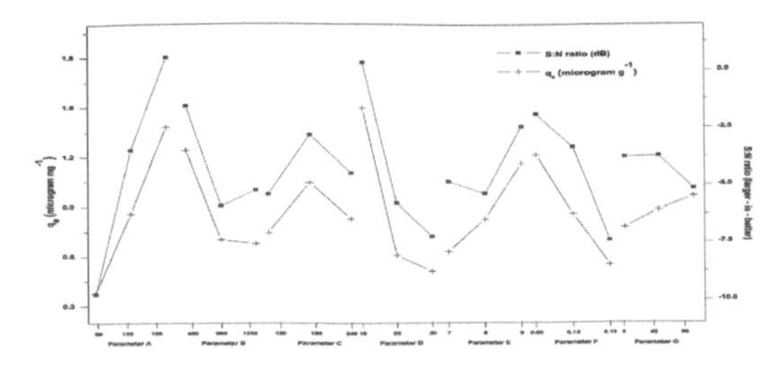

Figure 1. Response curve showing interaction between the process parameters in multi-ionic system for As(V) removal onto γ-Fe_2O_3 nanoparticles.

showing that the adsorption is occurring through weak electrostatic interaction which needs to care of in field scale application of these nanoparticles.

$$Fe-O-H+H^+ \xrightarrow{pH < pH_{PZC}} Fe-O^+H_2 \ (pH < 6.1)$$
$$Fe-O-H+OH^- \xrightarrow{pH > pH_{PZC}} Fe-O^- + H_2O \ (pH\ 7-9)$$
$$Fe-O-H+H^+ + 2(H-OH) \xrightarrow{pH > pH_{PZC}} Fe-O^- + 2H_3O^+ \ (pH\ 6.1-7.0)$$

The q_e value is increased as contaminant concentration increases which is due to a decrease in the hindrance for the uptake of As(V) as the mass-transfer operating force increased. This small variation explains that ions in solution are providing extra sites for adsorption by acting as counter ion providing which led to increase in the q_e value. As few authors have been reported the increase in removal capacity of nanoparticles in the presence of ions such as Nitrate and bicarbonates.

4 CONCLUSIONS

The environmental stable γ-Fe_2O_3 NPs were found efficient in removal of arsenic. Investigations using PHREEQ in exploring the formation of chemical process, surface complexation and mineral precipitation reactions are desirable for futuristic studies.

ACKNOWLEDGEMENT

This work is supported by University Grant Commission (UGC), New Delhi, India by providing financial assistance in the form of research fellowship under Grant Number 7411-29-061-429.

REFERENCES

Adams, G. & Bubucis, P.M. 1998. Calculating an artificial sea water formulation using spreadsheet matrices. *Aquarium Sci. Conserv.* 2(1): 35–41.

Kaloti, M. & Kumar, A. 2016. Synthesis of chitosan-mediated silver coated γ-Fe_2O_3 (Ag–γ-Fe_2O_3@ Cs) superparamagnetic binary nanohybrids for multifunctional applications. *J. Phys. Chem. C* 120(31): 17627–17644.

Environmental Arsenic in a Changing World –
Zhu, Guo, Bhattacharya, Ahmad, Bundschuh & Naidu (Eds)
ISBN 978-1-138-48609-6

Immobilization of magnetite nanoparticles for the removal of arsenic and antimony from contaminated water

G. Sun & M. Khiadani
School of Engineering, Edith Cowan University, Joondalup, WA, Australia

ABSTRACT: Magnetite (Fe_3O_4) nanoparticles were synthesized and immobilized in a synthetic resin polymethyl methacrylate (PMMA). The Fe_3O_4 nanoparticle-PMMA composites were studied for their efficiencies of removing dissolved arsenic (As) and antimony (Sb). The effects of major environmental and operating parameters on the removal of As and Sb were investigated in batch experiments. Singular and competitive adsorption of As and Sb onto the composites were studied. The results demonstrated the capability of the Fe_3O_4-PMMA composites for removing dissolved metalloids.

1 INTRODUCTION

Dissolved metalloids, including As and Sb, can be present in contaminated surface waters or groundwater. The health risk posed by arsenic, in particular when it is in inorganic dissolved form, has been widely recognized. Antimony is considered an emerging contaminant that is linked to skin, lung, and eye diseases (Cooper & Harrison, 2009). The removal of these metalloids from contaminated waters is essential for protecting human health and the environment.

A variety of methods have been studied for the removal of the two metalloids, including adsorption, membrane, coagulation/flocculation, oxidation/precipitation, etc. Among these methods, adsorption is considered a low cost conventional technique. Various high-tech or low-tech adsorbents, such as synthetic resins, activated carbon, agro-wastes, and mineral clays, etc., have been studied in lab experiments (Ungureanu *et al.*, 2015). Currently, many researchers are focusing their efforts on discovering innovative adsorbents, to further enhance adsorption efficiency and reduce cost.

Magnetite (Fe_3O_4) nanoparticles have the potentially to be used as an adsorbent or electron donor for the removal of water contaminants, due to their high surface areas, reactivity and well-established synthesis methods. Another advantage of the magnetite-based adsorbents is that they can be recovered through magnetic field. Their disadvantages include mobility in water, agglomeration, and oxidation by non-target compounds in water; these have limited the application of iron-based nanoparticles to treat contaminated waters. Appropriate methods of applying the particles, such as entrapment in porous media or carriers, need to be investigated. In this study, comparative experiments were carried out to study the efficiency of magnetite nanoparticles immobilized in a synthetic PMMA resin for the adsorption of As and Sb.

2 MATERIALS AND METHODS

2.1 *Synthesis of magnetite nanoparticles*

AR grade ferric chloride hexahydrate ($FeCl_3.6H_2O$), ferrous chloride tetrahydrate ($FeCl_2.4H_2O$), 2-(4-chlorosulfonylphenyl) ethyl-trichlorosilane (CTCS), methyl methacrylate (MMA), and 30% ammonium hydroxide (NH_4OH) were purchased from Sigma Aldrich. Milli-q water was produced by a Merck Millipore instrument. Magnetic iron oxide nanoparticles were synthesized by co-precipitation of Fe^{2+} and Fe^{3+} ions to form Fe_3O_4 precipitates via the following reaction.

$$Fe^{2+} + 2Fe^{3+} + 8OH^{-} \rightarrow Fe_3O_4\downarrow + 4H_2O \quad (1)$$

4.31 g of ferrous chloride tetrahydrate and 11.68 g of ferric chloride hexahydrate were added into a reaction chamber (500 mL triple neck round bottomed flask) pre-filled with 200 ml water (deoxygenated by purging with N_2 gas). The chamber was connected to a condenser and a mechanical mixer and placed in an oil bath. Ammonium hydroxide solution was gradually (dropwise) added to the solution. The color of the solution gradually changed from orange to dark brown and finally black. This was continued until the pH reached 8.0. A magnet was used to separate solid precipitates of Fe_3O_4 from the solution. The Fe_3O_4 precipitates were rinsed repeatedly with milli-q water and saline (0.1 M NaCl) water, vacuum-dried and stored.

2.2 *Synthesis and characterization of Fe_3O_4-PMMA composite*

Three grams of the Fe_3O_4 precipitates were mixed in 40 mL of dehydrated toluene. 1.3 mmol CTCS was added to the solution, and kept in room temperature for 24 h, to produce Fe_3O_4 particles immobilized with CTCS. The particles were separated from the solution, washed, dried and weighted. For polymerization, 1.6 g of these particles were added to 80 mL toluene and mixed with 50 g MMA in the reaction chamber in N_2 environment. The reactor was submerged in the oil bath at 80°C, and the solution was stirred by a mechanical impeller for 4 h. By the end of the polymerization period, magnetic composites in the liquid were separated by the magnet, collected, washed, centrifuged, and dried. Surface characteristics of the Fe_3O_4 nanoparticles and Fe_3O_4-PMMA composites were analyzed by a scanning electron microscope (JCM-6000, JOEL). An X-ray powder diffraction apparatus (PANAnalytical) was used to analyze the chemical compositions of the Fe_3O_4 nanoparticles.

2.3 *Batch testing of As and Sb removal by the nanoparticles*

An arsenic stock solution was prepared by diluting 1000 mg L^{-1} As(III) standard solution (Agilent) to 100 mg L^{-1}. An 274 mg L^{-1} antimony (III) stock solution by dissolving measured amount of antimony potassium tartrate $K_2Sb_2(C_4H_2O_6)2.3H_2O$ in milli-q water. A series of batch adsorption tests were then carried out to study As and Pb removal by the Fe_3O_4 nanoparticles, in mobile (bare particles) and PMMA immobilized forms. The batch adsorption studies were carried out using 100 mL flasks, each filled with 50 mL of As, Sb, or mixed As and Sb solutions, which were placed on an orbital shaker rotated at 180 rpm for 90 minutes. As and Pb concentrations of the solutions, at the beginning and end each batch test, were analyzed by a MP-AES instrument (Agilent).

3 RESULTS AND DISCUSSION

3.1 *Characteristics of the synthesized nanoparticles and Fe_3O_4-PMMA Composites*

Figure 1 shows the surface of the synthesized nanoparticles and Fe_3O_4-PMMA composites. It appears that the sizes of the particles were indeed in nm range. The Fe_3O_4 impregnated PMMA composites were observed to have more rugged surfaces, compared with the surfaces of blank (without particle impregnation) MMA polymer.

XRD analyses of the synthesized nanoparticles and a commercial Fe_3O_4 sample produced similar profiles. Both synthesized and commercial samples had six significant diffraction peaks observed in both samples (at 2θ of 511, 442, 440, 400, 311 and 220), confirming the presence of hematite in the synthesized nanoparticles.

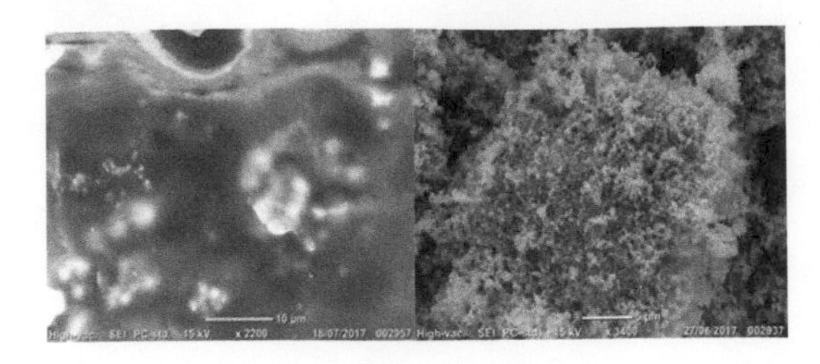

Figure 1. The SEM images of the Fe_3O_4 nanoparticles (left) and Fe_3O_4-PMMA composites (right).

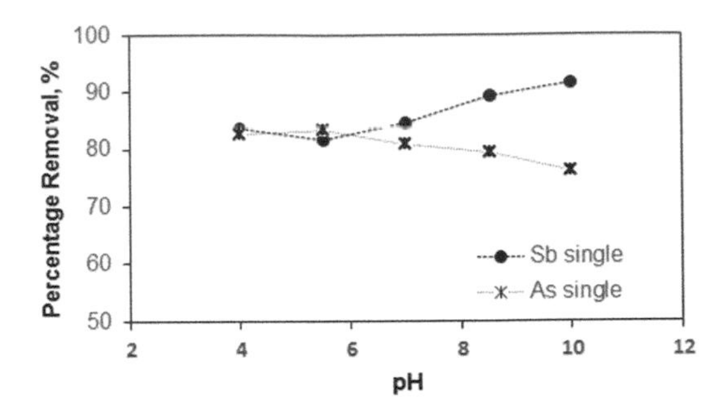

Figure 2. Percentage removal of As and Sb by the Fe_3O_4-PMMA composites at different pH.

3.2 *Batch testing results of As and Sb removal*

Figure 2 shows the percentage removal of As and Sb at different pH, when As and Sb were present as single contaminants in separate solutions. Efficiencies of commercial and synthesized nanoparticles, dosage, and co-existence (competitive removal) of these metalloids were also investigated.

4 CONCLUSION

Magnetite nanoparticle-PMMA composites were synthesized and tested for the removal of As and Sb. The results demonstrated their potential as an effective, recoverable adsorbent for *in situ* remediation of metalloid contaminated water.

ACKNOWLEDGEMENTS

The authors acknowledge Mr. Seyed Seyedi and Mr. Willie Tan, who contributed to the experiments.

REFERENCES

Cooper, R.G. & Harrison, A.P. 2009. The exposure to and health effects of antimony. *Indian J. Occup. Environ. Med.* 13(1): 3–10.

Ungureanu, G., Santos, S., Boaventura, R. & Botelho, C. 2015. Arsenic and antimony in water and wastewater: overview of removal techniques with special reference to latest ad-vances in adsorption. *J. Environ. Manag.* 151: 326–342.

Environmental Arsenic in a Changing World –
Zhu, Guo, Bhattacharya, Ahmad, Bundschuh & Naidu (Eds)
ISBN 978-1-138-48609-6

Arsenic retention on technosols prepared with nanoparticles for treatment of mine drainage water

D. Bolaños[1], V. Sánchez[1], J. Paz[2], M. Balseiro[3] & L. Cumbal[2,4]
[1]*Department of Earth and Construction Science, Universidad de las Fuerzas Armadas ESPE, Sangolquí, Ecuador*
[2]*Department of Life Science and Agriculture, Universidad de las Fuerzas Armadas ESPE, Sangolquí, Ecuador*
[3]*Department of Chemical Engineering, Centre for Research in Environmental Technologies (CRETUS), Universidade de Santiago de Compostela, Spain*
[4]*Center of Nanoscience and Nanotechnology (CENCINAT), Universidad de las Fuerzas Armadas ESPE, Sangolquí, Ecuador*

ABSTRACT: Arsenic (As) concentrations of 4.8 to 27.5 $\mu g\,L^{-1}$ have been detected in wastewater of gold mines in southern Ecuador (Portovelo) thus there is a need of applying remediation techniques to avoid superficial and groundwater contamination. In this study we have prepared a technosol composed by a ferric soil collected in the mining area and multicomponent nanoparticles synthetized using sodium borohydride and orange peel extract as co-reductants. The sorption capacity of the technosol was experimentally characterized using sorption isotherms. Langmuir model fits the experimental results and its parameters such as maximum sorption capacity and adsorption bond energy are 71185 $mg\,kg^{-1}$ 7.50 $kg\,mg^{-1}$, respectively. Based on these preliminary results, it seems that the as-prepared technosol will be a cheap cost solution for capturing arsenic from the mine drainage water.

1 INTRODUCTION

Mine drainage water usually contains high concentrations of heavy metals and other compounds. Due to poor handling, this contaminated water may leach and migrate to ground and spring water, becoming a potential contaminant of drinking water sources (Bolaños, 2015; Polizzotto *et al.*, 2006). There are many available technologies to cleanup As-contaminated water, whose efficiency and applicability will mainly depend on the chemical form of As in water (Cumbal, 2004). Adsorption of As in specific sorbents has been widely used due to its high versatility, easy operation and handling, and low costs. The application of specifically "tailored" technosols, composed of soil components and non-harmful residues with sorption capacities for specific contaminants, has shown successful results on the restoration of degraded mining areas contaminated with a wide variety of trace metals and metalloids (Macías, 2012). On the other hand, nanosorbents has emerged as a promising alternative for the cleanup of As-contaminated water (zero valent iron and iron oxide nanoparticles), because of their distinctive and advantageous properties, such as the small size, high surface area, and high reactivity due to the large number of sorption active sites (Pérez-Esteban *et al.*, 2016).

The objective of the present study was to find out the arsenic sorption capacity of a tailored technosol composed of a ferralsol, with 25,531 $mg\,kg^{-1}$ of Fe (2% *w/w*, associated to Fe and Mn oxides) and multicomponent nanoparticles.

2 METHODS

Water samples were collected at three discharges of artesian gold-processing plants located in the border of the Amarillo River (Portovelo, Ecuador), the main collector of liquid wastes from the gold-processing artesian plants. pH, Eh, electric conductivity were measured on site and heavy metals and As were analyzed with an atomic absorption spectrometer using standardized methods. For the fabrication of multicomponent nanoparticles (MCNPs) $FeCl_3.6H_2O$ and Na_2SO_4 solutions were mixed and purged with nitrogen. Then, $NaBH_4$ and orange peel extract were added as co-reductants. The technosol was prepared mixing 12.5 mg of MCPNPs and 4,987.5 mg of soil. Arsenic sorption was performed in batch tests mixing 5 g of technosol with 100 mL of arsenic contaminated water at neutral pH.

Mine drainage water was doped with six different concentrations of arsenic (25, 50, 100, 200, 400, 600 $\mu g\,L^{-1}$). The amount of arsenic sorbed on the technosol was determined by mass balance using the equation (Vanderborght & Van Grieken, 1977):

$$q = V/m\,(C_i - C_f) \quad \text{......................} \quad (1)$$

where q = concentration of As in the soil, V = volume of water contaminated with As, m = mass of soil and C_i and C_f initial and final As concentrations.

Table 1. Metal concentration in soil, used for preparing of technosol.

Soil Fraction	Cu mg kg^{-1}	Cd	Zn	Cr	Pb	Fe	As
Exchangeable	0.37	0.56	0.78	1.65	2.26	0.8	0.00
Sorbed-carbonate	1.23	0.50	2.11	0.96	5.70	3.43	0.00
Oxidizable (Fe/Mn oxides)	1.89	0.00	154.32	1.12	6.42	432.88	0.09
Reducible	3.54	0.15	5.27	2.78	1.67	190.09	0.00
Residual	4.5	0.00	6.94	0.75	3.21	24904	0.02
Total	11.53	1.21	169.42	7.26	19.26	25521.24	0.11

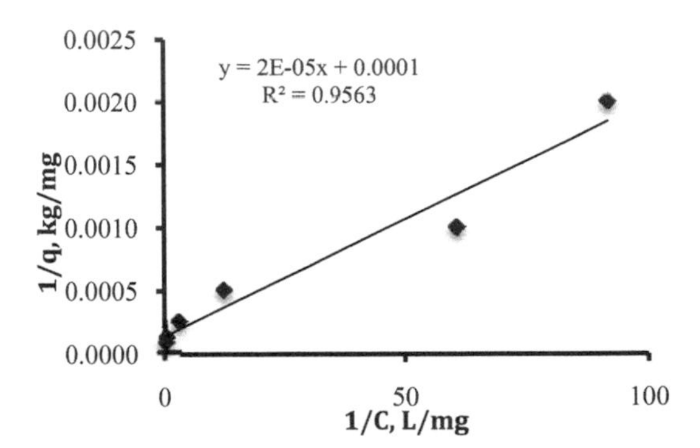

Figure 1. Sorption of arsenic on technosol.

3 RESULTS AND DISCUSSION

Water samples collected from the mine drainage had pH of 8.56 and Eh −107.50 mV. Under these conditions, arsenic is probably precipitating into the sediments basin. However, the variation of these parameters can easily lead to the release and mobilization of arsenic into the water (Polizzoto *et al.*, 2006) what is evidenced by the variation of arsenic concentrations from 4.8, 8.6 and 27.5 μg L^{-1}. Concentrations of metals in soil are given in Table 1. Soil with concentration of 25,531 mg of total iron kg^{-1} (432.9 mg kg^{-1} of Fe associated to oxides) was selected for preparing the technosol. The As-prepared MCNPs showed sizes in the range of 5 to 20 nm containing zero valent iron in the core and iron sulfide in the coverage similar to those reported by Cumbal *et al.* (2015). With the As-prepared technosol, we run adsorption tests for arsenic. In Figure 1, it is observed that experimental data fit very well the Langmuir isotherm model:

$$q = (Q_{max}\, b\, C)/(1+bC). \qquad (2)$$

The maximum adsorption capacity and adsorption bond energy (Langmuir, 1918) (Q_{max}) and b are 71,185 mg kg^{-1} and 7.50 kg mg^{-1}, respectively. The high sorption capacity of the technosol for arsenic could be associated to iron oxides (Cumbal, 2004) contained in soil and the zero valent iron and the iron sulfide of the nanoparticles (Cumbal *et al.*, 2015).

4 CONCLUSIONS

Technosol prepared with iron-rich soil mixed with multicomponent nanoparticles shows a good sorption capacity for arsenic dissolved in the mine drainage water. Adsorption results fit very well a Langmuir isotherm. Approximately 95% of arsenic is captured from the aqueous phase contaminated with 4.5 mg L^{-1}.

REFERENCES

Bolaños, D. 2015. Aplicación de Tecnosoles para la recuperación de suelos y aguas afectados por actividades de obras civiles, urbanas y minería. PhD Tesis, Universidad de Santiago de Compostela, Santiago de Compostela, Spain.

Cumbal, L. 2004. Polymer-supported hydrated iron oxide (HFO) nanoparticles: characterization and environmental applications. PhD Dissertation, Lehigh University, Bethlehem, PA, USA.

Cumbal, L., Debut, A., Delgado, D., Bastidas, C. & Stael, C. 2015. Synthesis of multicomponent nanoparticles for immobilization of heavy metals in aqueous phase. *NanoWorld J.* 1(2): 105–111.

Langmuir, I. 1918. The adsorption of gases on plane surfaces of glass, mica and platinum. *J. Am. Chem. Soc.* 40(9): 1362–1403.

Macías, F., Caraballo, M.A., Rötting, T.S., Pérez-López, R., Nieto, J.M. & Ayora, C. 2012. From highly polluted Zn-rich acid mine drainage to non-metallic waters: implementation of a multi-step alkaline passive treatment system to remediate metal pollution. *Sci. Total Environ.* 433(1): 323–330.

Pérez-Esteban, J., Caballero-Mejía, B., Masaguer, A. & Moliner, A. 2016. Effect of magnetite nanoparticles on heavy metals behavior in contaminated soils. *VII Congresso Ibérico das Ciências do Solo (CICS 2016), VI Congresso Nacional de Rega e Drenagem*, pp. 419–422.

Polizzotto, M.L., Harvey, C.F., Li, G., Badruzzman, B., Ali, A., Newville, M., Sutton, S. & Fendorf, S. 2006. Solid-phases and desorption processes of arsenic within Bangladesh sediments. *Chem. Geol.* 228(1–3): 97–111.

Vanderborght, M. & Van Grieken, E. 1977. Enrichment of trace metals in water by adsorption on activated carbon. *Anal. Chem.* 49(2): 311–316.

4.4 Arsenic solidification and immobilization for contaminated soils

Environmental Arsenic in a Changing World –
Zhu, Guo, Bhattacharya, Ahmad, Bundschuh & Naidu (Eds)
ISBN 978-1-138-48609-6

Reciprocal influence of arsenic and iron on the long-term immobilization of arsenic in contaminated soils

I. Carabante[1,2], J. Antelo[3], J. Lezama-Pacheco[2], S. Fiol[3], S. Fendorf[2] & J. Kumpiene[1]
[1]*Waste Science and Technology, Luleå University of Technology, Luleå, Sweden*
[2]*Department of Earth System Science, Stanford University, Stanford, CA, USA*
[3]*Technological Research Institute, University of Santiago de Compostela, Santiago de Compostela, Spain*

ABSTRACT: The main aim of this work was to evaluate the fate of arsenic associated with iron minerals in contaminated soils. The ageing behavior of synthetic arsenic-bearing poorly crystalline minerals – ferrihydrite and schwertmannite – was studied. Arsenic showed a passivation effect on poorly crystalline minerals, delaying their transformation towards more crystalline iron oxides. These results agreed well with studies performed on contaminated soils and sediments. These results are relevant in order to understand the long-term mobility of arsenic in contaminated soils and sediments. Iron oxides sequesters arsenic efficiently and, reciprocally, arsenic stabilize the mineral, delaying its transformation towards more crystalline phases.

1 INTRODUCTION

The main aim of this work was to evaluate the fate of arsenic (As) associated with iron (Fe) minerals in contaminated soils. In particular two different systems were studied: i) remediation of arsenic contaminated soils by addition of zero-valent iron (ZVI) and ii) immobilization of arsenic by secondary iron precipitates on acid mine drainage (AMD) systems.

A cost-effective method to reduce the risk, i.e. mobility and bioavailability, of As in a contaminated site is to apply in situ stabilization techniques, keeping the soil on site. This technology is based on changing the As geochemistry of the soil by adding a chemical amendment which promotes As immobilization by for example precipitation or adsorption. Addition of ZVI to the soil, which will immobilize arsenate after ZVI oxidation, is usually proposed as an effective in-situ remedation technique reducing As leaching up to 98% (Kumpiene *et al.*, 2008; 2006; Mench *et al.*, 2002). The applicability of these techniques is becoming possible as more European countries are accepting a risk-based approach when defining whether the site is contaminated or not. In this approach the main risk-defining factor of a site is the contaminant bioavailability and mobility and not on its total concentration in the soil. Although the promising results obtained using this technique, questions regarding the long-term stability of the immobilized contaminant in the soil must be answered before establishing in-situ chemical stabilization as a widely accepted technique. The particle size of the Fe oxide particles in the soil will ultimately define the arsenate adsorption (immobilization), e.g. the smaller the Fe oxide particles are the higher the arsenate adsorption capacity of the Fe oxide (Waychunas *et al.*, 2005). Since very small Fe oxide nanoparticles are formed upon the oxidation product of metallic ZVI, the immobilization of As in soils by Fe amendments can reach a high efficiency. The oxidation products of ZVI would be at first small Fe oxy-hydroxides such as ferrihydrite or lepidocrocite (γ-FeOOH) (Cornell & Schwertmann, 1996). The ageing of these nanoparticles in the soil might induce crystal growth and/or phase transformation, which would plausibly decrease the effectiveness of the remediation method with time.

Oxidation of iron sulfide minerals present in mining areas lead to AMD. AMD causes acidification of surface waters in combination with the release of trace elements (TE), such as As, Cr, U, Mo, Zn and Cu. AMD has been identified as the main cause of surface water contamination on the mid-Atlantic region of the USA (U.S. EPA, 2018). Worldwide, about 20,000 km of water streams and about 72,000 ha of surface waters were estimated to be seriously damaged as a consequence of AMD (Johnson & Hallberg, 2005).

Iron secondary minerals initially sequester large amounts of trace elements (TE), such as arsenic. However, the long-term immobilization of TE in these mineral phases will be defined by the stability of these minerals. Schwertmannite is a metastable mineral, transforming eventually into more stable phases such as hematite and goethite (Bigham & Nordstrom, 2000). Jarosite is stable under acidic conditions while it dissolves at natural pH (Smith et al., 2006). Interestingly, certain cations have been observed to increase the stability of secondary minerals related to AMD.

For instance, the presence of Cu significantly enhances the stability of schwertmannite (Antelo *et al.*, 2013) and some results show evidences pointing towards arsenate stabilizing schwertmannite (Regenspurg & Peiffer, 2005). Yet, the number of studies assessing the effect of TE on the stability and reactivity of iron oxy-hydroxysulfates is limited. Summarizing, the bonding mechanism of sequestration and the effect of TE on the reactivity and stability of iron oxyhydroxysulfates are understudied and can be considered key knowledge to better understand the mobility of TE in AMD systems.

2 METHODS/EXPERIMENTAL

2.1 *Ageing experiments*

As-bearing ferrihydrite: Synthetic ferrihydrite was used for this experiment. Three different As loadings were studied, arsenic to iron molar ratio of 0, 0.01 and 0.1. The transformation of ferrihydrite into hematite was studied by means of differential scanning calorimetry (DSC). Dried-powder samples of the three materials were heated up in the DSC furnace from room temperature up to 800°C at a heating rate of $10°C\,min^{-1}$ under a synthetic air atmosphere. In order to better understand the processes taking place along the transformation curves, samples were heated up to different temperatures in a muffle oven at the same heating rate than in the DSC analysis, $10°C\,min^{-1}$. This different samples were spare for As K-edge EXAFS, Fe K-edge EXAFS, sequential extraction and FTIR measurements.

ZVI amended soils: Different sites in Europe that had been previously in-situ treated by addition of 2% ZVI were selected in this study. The time from which the soils were treated changed from 2 to 15 years for the different sites. We had access to both soil that had been treated with ZVI but also control samples to which ZVI was not added. The iron mineralogy of the different soils was extensively evaluated by means of Fe-K edge EXAFS as well as sequential extraction.

As-bearing AMD systems: the adsorption of arsenic on synthetic ferrihydrite was characterized by batch adsorption experiments as a function of pH, from pH 3 to 9 and from an arsenic to iron molar ratio from 0 to 0.7. Subsequently schwertmannite bearing arsenic to iron molar ratio of 0.1 and 0 was aged in an electrolyte background suspension at pH 3, 5 and 7. The Portapego stream at Mina Touro, Spain, shows evidences of acid mine drainage. Sediments from this stream was collected. The iron mineralogy of the sediments were mostly schwertmannite. The same aging experiments, at pH 3, were carried out with these sediments.

2.2 *X-ray absorption spectroscopy*

Fe K-edge XAS and As K-edge XAS spectra were collected, at ambient temperature and pressure, on beam line 7.3, 4.3 and 11.2 at the Stanford Synchrotron Radiation Lightsource at SLAC National Laboratory Accelerator, beam line I811 at Max Lab Synchrotron facility in Sweden, beam line 22 CLAESS at ALBA Synchrotron light facility and at beam line BM25 at ERSF. Fe K-edge data of soil samples were measured on fluorescence mode using a PIPPS detector whereas synthetic iron oxides were measured on transmission mode using ionization chambers as I0 and I1. The data was self-calibrated with a Fe foil placed between I1 and I2. As K-edge spectra were obtained on fluorescence mode using a solid state multiple germanium elements detector.

3 RESULTS AND DISCUSSION

3.1 *As-bearing ferrihydrite*

The DSC curve of pristine ferrihydrite showed a sharp transformation event at 435°C. This event corresponds to the transformation of ferrihydrite into hematite. The DSC curves of the As-bearing ferrihydrite showed a more complex transformation pattern. More importantly, the transformation events were delayed: at 490°C for an arsenic to iron molar ratio of 0.01 and at 647°C for an As/Fe molar ratio of 0.1. The transformation processes assumed by the DCS curves were confirmed by Fe-EXAFS. The adsorption of arsenic thus delayed the thermal transformation of ferrihydrite.

Arsenic fractionation changed upon the thermal transformation of ferrihydrite. A higher fraction of arsenic was associated with the ferrihydrite fraction instead of with the hematite fraction. Interestingly, only 5% of arsenic became labile in the exchangeable fractions. Although this is a low percentage, this could explain fast arsenic leaching from soils after thermal treatment. The arsenic atomic environment changed also upon the thermal transformation of iron as shown by As K-edge EXAFS.

3.2 *ZVI amended soils*

The iron mineralogy of all the sites analyzed in this studied had strong dependence on the initial mineralogy of the site and conditions such as humidity, temperature and organic content. Our results indicated that the fraction of amorphous iron oxides, i.e. the most reactive fraction regarding the immobilization of arsenic, increased upon addition of ZVI. This trend was observed in all sites independently on the time after which the amendment was applied to the soils, 2 years, 6 years, 8 years, 12 years and 15 years. These results showed that the immobilization of arsenic and other TE was still a success after almost a few decades from the addition of ZVI.

3.3 *AMD systems*

The transformation of synthetic schwertmannite as well as schwertmannite sediments from Mina Touro had a strong dependence on the presence of arsenic in the system during the 300 days in which the transformation was monitored. Higher transformation rate were observed in the absence of arsenic. In particular schwertmannite did not transformed into goethite, or if it did it was to a minor extend, at pH 3 at an arsenic to iron molar ratio of 0.1. The transformation of schwertmannite was also pH dependent, obtaining higher degree of transformations at higher pH values.

4 CONCLUSIONS

One of the main concerns regarding the long-term immobilization of arsenic is that the main iron oxides involved in such immobilization are metastable. This study shows that a metastable iron oxide such as ferrihydrite delays its transformation towards more crystalline phases in the presence of arsenic. Interestingly, the immobilization of arsenic contaminated soils by addition of ZVI was a success for about two decades. These results are relevant in order to understand the long-term mobility of arsenic in ZVI-amended soils and from AMD systems. Iron oxides sequesters arsenic efficiently and, reciprocally, arsenic stabilize the mineral, delaying its transformation towards more crystalline phases.

ACKNOWLEDGEMENTS

The Wallenberg Foundation and Å Forsk Foundation are acknowledged for financial support. SSRL, ALBA, Max-Lab and ESRF are acknowledged for granting our beam time proposals under which X-Ray absorption measurements presented in this work were performed.

REFERENCES

Antelo, J., Fiol, S., Gondar, D., Pérez, C., López, R. & Arce, F. 2013. Cu(II) incorporation to schwertmannite: effect on stability and reactivity under AMD conditions. *Geochim. Cosmochim. Acta* 119: 149–163.

Bigham & Nordstrom. 2000. Iron and aluminum hydroxysulfates from acid sulfate waters. *Rev. Mineral. Geochem.* 40(1): 351–403.

Cornell, R.M. & Schwertmann, U. 2004. The Iron Oxides: Structure, Properties, Reactions, Occurrences and Uses. Wiley & Sons.

Johnson, D.B. & Hallberg, K.B. 2005. Acid mine drainage remediation options: a review. *Sci. Total Environ.* 338(1–2): 3–14.

Kumpiene, J., Lagerkvist, A. & Maurice, C. 2008. Stabilization of As, Cr, Cu, Pb and Zn in soils using amendments – a review. *Waste Manag.* 28(1): 215–225.

Kumpiene, J., Ore, S., Renella, G., Mench, M., Lagerkvist, A. & Maurice, C. 2006. Assessment of zerovalent iron for stabilization of chromium, copper and arsenic in soil. *Environ. Pollut.* 144(1): 62–69.

Mench, M., Vangronsveld, J., Clijsters, H., Lepp, N.W. & Edwards, R. 2000. *In situ* immobilization and phytostabilization of contaminated soils. In: N. Terry, G. Bañuelos (eds) *Phytoremediation of Contaminated Soils and Waters.* CRC Press, Boca Raton, FL, pp. 323–358.

Regenspurg, S. & Peiffer, S. 2005. Arsenate and chromate incorporation in schwertmannite. *Appl. Geochem.* 20(6): 1226–1239.

Smith, A.M.L., Hudson-Edwards, K.A., Dubbin, W.E. & Wright, K. 2006. Dissolution of jarosite $[KFe_3(SO_4)_2(OH)_2]$ at pH 2 and 8: insights from batch experiments and computational modelling. *Geochim. Cosmochim. Acta* 70(3): 608–621.

U.S. Environmental Protection Agency (2018) www.sosbluewaters.org/epa-what-is-acid-mine-drainage%5B1%5D.pdf. (Accessed on April 3, 2018).

Waychunas, G.A., Kim, C.S. & Banfield, J.F. 2005. Nanoparticulate iron oxide minerals in soil and sediments: unique properties and contaminant scavenging mechanisms. *J. Nanopart. Res.* 7(4–5): 409–433.

Environmental Arsenic in a Changing World – Zhu, Guo, Bhattacharya, Ahmad, Bundschuh & Naidu (Eds)
ISBN 978-1-138-48609-6

Evaluation of chemical stabilizers for the retention in a mining tailing contaminated soil

C.G. Sáenz-Uribe[1], M.A. Olmos-Márquez[1], M.T. Alarcón-Herrera[2] & J.M. Ochoa-Rivero[3]
[1] *Universidad Autónoma de Chihuahua, Chihuahua, Chih., Mexico*
[2] *Centro de Investigación en Materiales Avanzados, SC, Chihuahua, Chih., Mexico*
[3] *Instituto Nacional de Investigaciones Forestales, Agrícolasy Pecuarias, Aldama, Chih., Mexico*

ABSTRACT: This study focused on studying the chemical stabilization of arsenic (As) in mining tail from a contaminated site in the north of Mexico. The main objective was to evaluate the behavior of As solubility in the contaminated soil after its amendment with manganese oxide (cryptomelane) prepared by two different procedures (McKenzie and Gangas). The initial concentration of total As was 2797 mg kg^{-1} during 10 days of experimentation under controlled conditions; it was found that the reduction of soluble arsenic occurred since the second day using cryptomelane obtained by Gangas procedure.

1 INTRODUCTION

Contamination of the environment by arsenic (As) from both natural and anthropogenic sources has occurred in many parts of the world and it is nowadays recognized as a global problem (Vodyanitskii, 2009). One of the principal anthropogenic sources for soil contamination by As include the mining of lead and zinc (Stafilov *et al.*, 2010). Mining and metallurgical activities in Chihuahua City have a long story; one of the most recent case is the Avalos smelter, which is located in the southeast of the city and produced lead and zinc from the 1908s until 1993. These wastes contain high contents of heavy metals and metalloids such As, which have caused environmental pollution in the area (Puga *et al.*, 2006). The high cost of traditional soil remediation techniques (excavation and landfilling) and limited resources allocated to remediate contaminated sites prompted the development of alternative techniques that are cost-effective and less disruptive to the environment such as soil stabilization (Mulligan *et al.*, 2001). The objective of this study was to evaluate the effects of As bioavailability in a contaminated soil, after amended with a manganese oxide (cryptomelane), prepared by two different techniques.

2 METHODS AND EXPERIMENT

2.1 *Soil selection and As extraction*

The soil collected for this study was taken from the surroundings of Avalos smelter (28°37′17.32″N, 106°00′12.73″W). Soil samples were taken at 0.30 m of depth from six points near to the tailing dam. These soil samples were joined and sieved through a 0.83-mm opening sieve (#20) to remove large particles and provide a homogeneous soil size. For initial concentration of total and soluble As determination, were taken two simple samples (S1 and S2) of 2 kg in duplicate. For the arsenic determination, the soil samples were acid digested with HNO_3 in Teflon™ containers using a microwave digester CEM model MARS-X®. The extractions of soluble As were performed through two extraction techniques: a) PEC method using glacial acetic acid (PEC), according NOM-053-SEMARNAT-1993 and b) ABA modified method, using carbonic acid (ABA), according NOM-141-SEMARNAT-2003. The As extracted material was vacuum filtered with TLPC filters. After that, the samples were analyzed by atomic absorption (AA) according NMX-AA-132-SCFI-2016, using HACH® spectrometer model DR2000™.

2.2 *Preparation of chemical stabilizers*

For the chemical stabilization of polluted soil samples, 2 cryptomelane (KMn_8O_{16}) stabilizers were prepared. The first was obtained according to the Ganga (G) method (Hettiarachchi & Pierzynski, 2000). The second method considered was the McKenzie (M) method (McKenzie, 1971); both stabilizers were analyzed by X-ray diffraction in Xpert Phillips MPD® instrument. The surface area was determinate by gas adsorption using nitrogen in analyzer Autosorb® 6B™ with software for Windows® version 1.16. Stabilization tests with the polluted soil were carried out with 0.75 g of stabilizing agent in two subsamples with 150 g of soil (G-S1, G-S2, M-S1 and M-S2).

3 RESULTS AND DISCUSSION

3.1 *Soil As concentrations*

The initial concentrations of total and soluble As were similar to the composite samples of contaminated soil

Table 1. Comparison of initial total As concentrations vs. maximal levels references.

Sample	S1	S2	Avg	Reference for agricultural/ residential land use[1]	Reference for industrial land use[1]
Total As ($mg\,kg^{-1}$)	2842.5	2750.8	2796.6	22	260
Soluble As ($\mu g\,L^{-1}$)	10.4	10.7	10.5	–	–

[1]NOM-147-SEMARNAT/SSA1-2003.

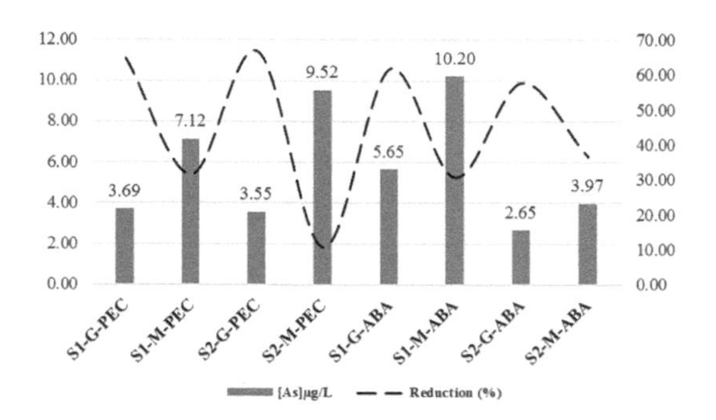

Figure 1. As concentration in soil samples, after the chemical stabilization.

(S1 and S2). The results showed that total As concentrations in soil was above the maximal levels allowed by Mexican regulations for industrial land use. Table 1 shows the average concentrations of total and soluble As and its normative references.

3.2 *Chemical stabilizers performance*

About the properties of the chemical stabilizers, X-ray diffraction indicated that the peaks obtained by the process recommended by Ganga stabilizer (G) tied perfectly with the graphs from X'pert Highscore software™, with a surface area of 130.37 $m^2\,g^{-1}$. On the other hand, McKenzie stabilizer (M) has a surface area of 302.22 $m^2\,g^{-1}$, but the peaks shown had not defined in different sections on the graph, which indicate that the compound did not crystallize. In order to compare the efficiency of the stabilizers G and M, they were treated with ABA modified and PEC methods of extraction. For soluble As extractions, the method PEC and Ganga stabilizer were significant more effective. Figure 1 shows the values and percentage reduction in the concentration of soluble As.

3.3 *Stabilization time*

It was obtained the velocity of stabilization through G stabilizer during 10 days.

The greatest total As reduction was detected after 4 days, with a reduction of 38.86%. The soluble As

Table 2. Concentrations of total and soluble As and its reduce percentage after 10 days of stabilization time.

	Total As		Soluble As	
Time days	($mg\,kg^{-1}$)	% reduction	($\mu g\,L^{-1}$)	% reduction
2	2109.66	24.56	N.D.	100
4	1709.67	38.86	N.D.	100
6	2005.92	28.27	N.D.	100
8	2055.03	26.51	N.D.	100
10	2342.06	16.25	N.D.	100
Avg	2044.47	26.89	N.D.	100

N.D.: Not detected.

was reduced to no detected values (ND) since the 2nd day; that indicates that the contaminated soil can be stabilized in less than 2 days. Table 2 shows the values of total and soluble As. Additional analyzes will be carried out to determine the average stabilization time.

4 CONCLUSIONS

The stabilization that gave the highest percentage of retention of arsenic and conversion of soluble arsenic to insoluble was using the Ganga stabilizer (G), with the technique of extraction PEC.

ACKNOWLEDGEMENTS

National Council Science and Technology (CONACYT) financially supported this work, within the program Attention to National Problems.

REFERENCES

Hettiarachchi, G.M. & Pierzynski G.M. 2000. The use of phosphorus and other soil amendments for *in situ* stabilization of soil lead. *Proc. of the 2000 Conference on Hazardous Waste Research*, pp. 125–133.

McKenzie, R.M. 1971. The synthesis of birnessite, cryptomelane, and some other oxides and hydroxides of manganese. *Mineralogical Mag.* 38(296):493–502.

Mulligan, C.N., Yong, R.N. & Gibbs, B.F. 2001. Remediation technologies for metal-contaminated soils and groundwater: an evaluation. *Eng. Geol.* 60(1–4):193–207.

Puga, S., Sosa, M., Lebgue, T., Quintana, C. & Campos, A. 2006. Contaminación por metales pesados en suelo provocada por la industria minera. *Ecol. Apl.* 5(1–2): 149–155.

Stafilov, T., Šajn, R., Pančevski, Z., Boev, B., Frontasyeva, M.V. & Strelkova, L.P. 2010. Heavy metal contamination of surface soils around a lead and zinc smelter in the Republic of Macedonia. *J. Hazard. Mater.* 175(1–3): 896–914.

Vodyanitskii, Y.N. 2009. Chromium and arsenic in contaminated soils (review of publications). *Eurasian Soil Sci.* 42(5): 507–515.

Environmental Arsenic in a Changing World –
Zhu, Guo, Bhattacharya, Ahmad, Bundschuh & Naidu (Eds)
ISBN 978-1-138-48609-6

Treatment of low-level As contaminated excavated soils using ZVI amendment followed by magnetic retrieval

J.N. Li[1,2], S. Riya[1], A. Terada[1] & M. Hosomi[1]
[1]*Department of Chemical Engineering, Tokyo University of Agriculture and Technology, Tokyo, Japan*
[2]*Research Fellow of Japan Society for the Promotion of Science, Tokyo, Japan*

ABSTRACT: This study investigates the remediation of low-level As contaminated excavated soils from construction projects by ZVI amendment followed by magnetic retrieval. A case study was conducted with an alkaline excavated soil (total As only 7.5 mg kg^{-1}). The remediation efficiency was assessed using sequential leaching tests. The results showed ZVI amendment (1% ZVI and 0.5 mL kg^{-1} H_2SO_4 of soil) followed by magnetic retrieval could significantly decrease water-leachable As (from 33 μg L^{-1} to <3 μg L^{-1}), also the potential leachability of As in this excavated soil. These initial results indicate that this treatment could be developed for the efficient remediation of low-level As contaminated excavated soils.

1 INTRODUCTION

In Japan, excavated soils with low-level As obtained from construction projects during city development have been of great concern because in many cases the water-leachable As is higher than the environment standard (Soil Leachate Standard in Soil Contamination Countermeasures Law of Japan, 10 μg L^{-1}) (Li *et al.*, 2016, 2017). Now water washing is usually used for the remediation of As-contaminated soils, after which the large soil grains can be reused immediately while the fine soil grains contain high levels of As need to be further treated. After washing, the fine soil grains are more like "sludge" and they are usually treated in disposal sites, which will greatly increase the transportation cost and the load of disposal sites. Hence, the development a cost-effective disposal system to treat low-level As-contaminated excavated soils is urgently needed. The objective of this study was to probe the effectiveness of the zero valent iron (ZVI) amendment followed by magnetic retrieval for remediation of low-level As contaminated excavated soils.

2 MATERIALS AND METHODS

2.1 *Soil sample and ZVI product*

The excavated soil was obtained from one construction project in Tokyo, Japan. The sample was air-dried, crushed and sieved through a 2-mm opening mesh. Soil moisture content was 32.5%; soil pH was 9.0 (soil to water ratio; 1:10); total As was 7.5 ± 0.13 mg kg^{-1}. The ZVI product was obtained from DOWA ECO-SYSTEM, Japan. The Fe content was 940 ± 7.4 g kg^{-1}; the As content was 13.4 ± 0.5 mg kg^{-1}.

2.2 *Soil treatment*

Before amendment experiment, water content of the soil was adjusted to 15% using deionized water. This water content is very low and avoids producing sludge during treatment process. The detailed information about different soil treatments was shown in Table 1. In all treatments, the samples were agitated intensely for 15 min and then air-dried at room temperature. For magnetic retrieval, two steps were conducted: (1) a rectangular magnet was held in contact with the treated soils' surfaces and moved horizontally back and forth until no solids could be pulled out of the soils. The remaining soils were considered as non-magnetic (NMS); (2) the mixtures obtained at previous step were further magnetically-separated to retrieve ZVI powders only (post-test ZVI); and the remaining soils after this step were considered as weakly-magnetic (WMS). The As and other elements in different parts were digested using a strong acid ($HClO_4$-HNO_3-HF) digestion method.

Table 1. Different soil treatments.

Test	Description
T1	Original soil (Control 1)
T2	No ZVI; No H_2SO_4 (Control 2)
T3	No ZVI; Add H_2SO_4 (0.5 mL kg^{-1} of soil)
T4	Add ZVI (10 g kg^{-1}); No H_2SO_4
T5	Add ZVI (10 g kg^{-1}); Add H_2SO_4 (0.5 mL kg^{-1})

ZVI and H_2SO_4 were added on the basis of dry mass of soil.

2.3 *Sequential leaching tests (SLT)*

The parameters of each step of SLT were generally in accord with the standard leaching test in Japan. 3 g of sample was mixed with 30 mL deionized water in a 50 mL polypropylene centrifuge tube. The tubes were shaken using a lateral-reciprocating shaker at 200 rpm for 6 h at room temperature, after which the pH values of the suspensions were measured immediately. The suspensions were then centrifuged at about 6000 rpm

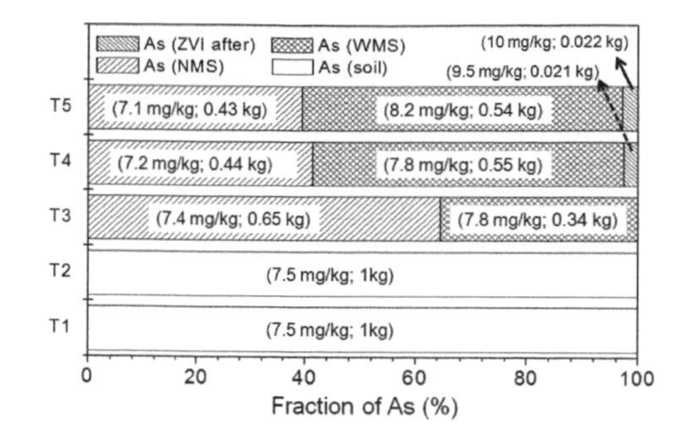

Figure 1. Fractions of As in different separated soils and ZVI.

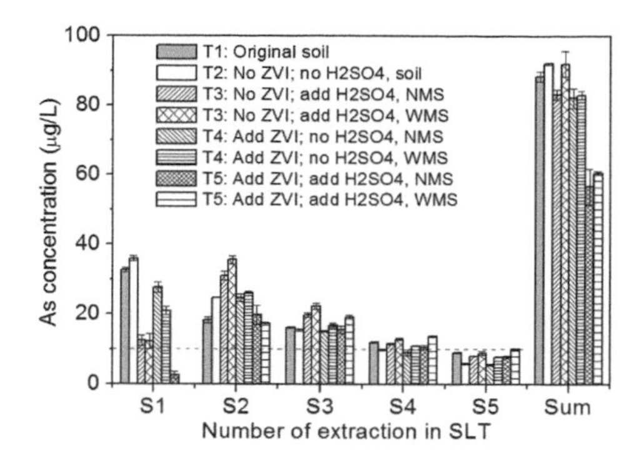

Figure 2. Leaching of As in sequential leaching tests.

(4700 g) for 15 min, followed by filtration through 0.45 μm membrane filters. The residual soils remaining in the tubes were stored at 4°C overnight and were re-suspended in fresh deionized water the next day for the next leaching step. This procedure was repeated 5 times in total.

3 RESULTS AND DISCUSSION

3.1 *Distributions of As before and after treatment*

The distributions of As in soils and ZVI are shown in Figure 1. The As concentrations in WMS of T3, T4 and T5 were higher than those in NWS. Similarly, the Fe, Mn, Cr, Cu and Zn concentrations in WMS were higher than those in NWS (data not shown). This can be attributed to the separation of magnetic minerals in WMS that are often associated with heavy metal(loid)s. As shown in Figure 1, the As concentrations in post-test ZVI (T4: 10 mg kg^{-1}; T5: 9.5 mg kg^{-1}) were lower than that in pre-test ZVI; but the removal of As from soils was also confirmed due to the mass of post-test ZVI increased more than two times. Nevertheless, the removal efficiency of As by ZVI was not high (lower than 0.6%).

3.2 *Arsenic leaching concentrations*

The As concentrations at the first step of T1 (33 μg L^{-1}) and T2 (36 μg L^{-1}) were all higher than the standard 10 μg L^{-1}, indicating that this excavated soil was considered as slightly contaminated and needed to be remediated. When only the ZVI was used (T4), the As leaching concentrations decreased but still more than 20 μg L^{-1} in the leachate. When only acid was added (T3), the As concentrations decreased more than those in T3 but still slightly higher than the standard. Only when both the ZVI and acid was used, the As concentrations at the first step of T5 were decreased a lot and lower than the standard, indicating this treatment can efficiently treat this excavated soil. As described in previous section, it seems that the removal of As by ZVI was not the key factor decreasing the As leaching at the first step of SLT. The slight pH decrease due to the addition of acid in T3 (mean: 8.1) and T5 (mean: 8.0) contributed the As leaching decrease because the As mobility was quite limited in the circumneutral pH (6–8) (Tabelin *et al.*, 2014). The comparison between T3 and T5 also confirmed the ZVI addition contributed to the As leaching decrease after treatment.

Interestingly, the As release during SLT could last for a long period. For soils in T3 and T5, the As concentrations increased at the second step and then decreased at later steps. It required about 5 times until the As concentrations in the leachates lower than the standard 10 μg L^{-1}. Figure 2 also shows that the T5 treatment significantly reduced the potential leachibility of As in soils (sum of released As in five-step SLT).

4 CONCLUSIONS

This study suggests that ZVI amendment followed by magnetic retrieval is technologically feasible to treat low-level As contaminated excavated soils and reveals promising perspective for its practical application.

ACKNOWLEDGEMENTS

This research was financially supported by an Environmental Research and Technology Development Fund (5-1606) from the Ministry of the Environment, Japan and a Grant-in-Aid for Scientific Research (No. 17J07673) from the JSPS.

REFERENCES

Li, J.N., Kosugi, T., Riya, S., Hashimoto, Y., Hou, H., Terada, A. & Hosomi, M. 2017. Use of batch leaching tests to quantify arsenic release from excavated urban soils with relatively low levels of arsenic. *J. Soil. Sediment.* 17(8): 2136–2143.

Li, J.N., Kosugi, T., Riya, S., Hashimoto, Y., Hou, H., Terada, A. & Hosomi, M. 2018. Pollution potential leaching index as a tool to assess water leaching risk of arsenic in excavated urban soils. *Ecotoxicol. Environ. Saf.* 147: 72–79.

Tabelin, C.B., Hashimoto, A., Igarashi, T. & Yoneda, T. 2014. Leaching of boron, arsenic and selenium from sedimentary rocks: II. pH dependence, speciation and mechanisms of release. *Sci. Total Environ.* 473: 244–253.

Environmental Arsenic in a Changing World –
Zhu, Guo, Bhattacharya, Ahmad, Bundschuh & Naidu (Eds)
ISBN 978-1-138-48609-6

Arsenic remediation through magnetite based *in situ* immobilization

J. Sun[1], B.C. Bostick[2], S.N. Chillrud[2], B.J. Mailloux[3] & H. Prommer[4]
[1] *University of Western Australia & CSIRO Land and Water, Perth, WA, Australia*
[2] *Barnard College and Lamont-Doherty Earth Observatory, Columbia University, New York, NY, USA*
[3] *Department of Environmental Sciences, Barnard College, New York, NY, USA*
[4] *University of Western Australia & CSIRO Land and Water, Perth, WA, Australia*

ABSTRACT: The remediation of arsenic-contaminated aquifers is a formidable challenge to achieve, in part because geochemical conditions often do not favor the stabilization of arsenic within the solid phase. A promising remedial approach involves stimulating iron mineral transformations that immobilize arsenic through sorption or precipitation. Despite intense research, the current immobilization methods are still often ineffective, in part because many iron minerals are susceptible to redox gradients common in subsurface environments. We have been conducting a series of studies to illustrate the potential of nanoparticulate magnetite (Fe_3O_4) to sequester arsenic. Magnetite is stable under most redox conditions in aquifers, and able to co-precipitate and adsorb arsenic. Here, we present results from microcosm and column experiments using sediments and groundwater from U.S. Superfund sites, and from reactive transport modelling. All these results demonstrate that *in situ* formation of nanoparticulate magnetite can be achieved by the combination of nitrate and ferrous iron, and that it should be feasible to produce an *in situ* reactive filter by such nitrate-iron(II) co-injection and immobilize arsenic in contaminated aquifers.

1 INTRODUCTION

Groundwater arsenic contamination is a global public health problem and also a concern at hundreds of U.S. Superfund sites (ATSDR, 2015). Mitigating groundwater arsenic contamination is urgently required, however, has proven difficult. One attractive remedial option is *in situ* immobilization, which stimulates changes in the chemical speciation of arsenic and/or other elements within the aquifer to adsorb and/or precipitate arsenic. Magnetite (Fe_3O_4) can potentially be an advantageous host-mineral for arsenic immobilization in that it is stable under a wide range of aquifer conditions including both oxic and iron(III)-reducing environments. We, therefore, have been investigating methods to stimulate *in situ* magnetite formation, and evaluating the effect of magnetite formation on immobilizing arsenic, through both laboratory experiments and numerical modelling.

2 MATERIALS AND METHODS

2.1 *Microcosm and column experiments*

A series of microcosms were performed by mixing sediments and groundwater from the Vineland Chemical Company Superfund site, and then amended as: (i) control, (ii) sulfate-lactate treatment, (iii) nitrate-only treatment and (iv) nitrate-iron(II) treatment. To test the stability of the neoformed minerals, lactate was added in nitrate-only and nitrate-iron(II) microcosms on day 10 to stimulate microbial reduction. The microcosms were incubated semi-anaerobically in sealed bottles. The duration was 38 days, during which solution composition and mineralogical transformations were determined.

To conduct column experiment, sediments from the Dover Municipal Landfill Superfund site were used. Four experimental stages were designed. Stage I used artificial groundwater (A-GW) with lactate and arsenic (III), to equilibrate the sediments and mimic landfill conditions. Stage II used A-GW with lactate, arsenic (III), iron (II) and nitrate, to produce minerals and test if they could immobilize arsenic. Stage III used A-GW with elevated concentration of lactate, to examine the stability of the neoformed minerals. Stage IV used the same composition with Stage I, to return to landfill condition and test if the neoformed minerals could immobilize additional transported arsenic. The influents were purged with $N_2(g)$ and injected at 2 pore volumes (PVs) day^{-1}. The duration was 4 months, during which solution composition and mineralogical transformations were determined.

2.2 *Reactive transport modeling*

The initial geochemical and flow condition of the Dover columns were reconstructed using PHREEQC and MODFLOW, which formed the basis for the subsequent simulations with the reactive transport code PHT3D (Prommer *et al.*, 2003). The reactive transport model contained two distinct sets of reactions: one

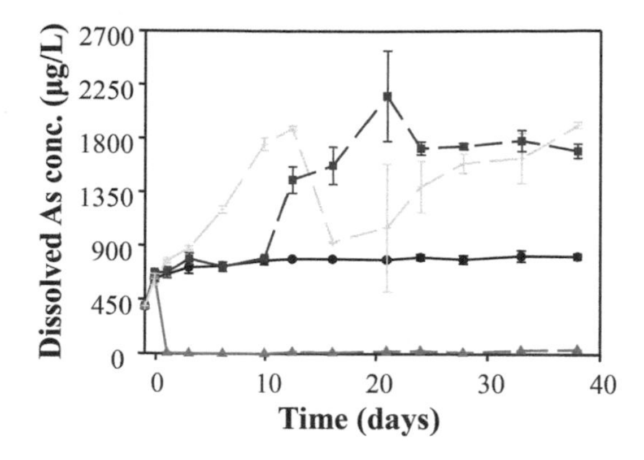

Figure 1. Dissolved arsenic concentrations in the microcosms. The dashed lines indicate the portion where lactate was present. Modified from Sun *et al.* (2016a and 2016b).

set focused on mineral (trans)formation, whereas the other focused on the effect of the mineralogical changes on surface sites and associated surface species. Except for the few reactions with no known rate constants, most initial parameters were from literatures. A calibration procedure was then employed to refine the parameters, to match up simulation with observation temporally and spatially. Based on the calibrated parameters, 2D simulations was then conducted, to predict the efficiency of the nitrate-iron(II) strategy on immobilizing arsenic at field-scale.

3 RESULTS AND DISCUSSION

In the nitrate-iron(II) microcosms, effective arsenic removal from solution was observed, even under sustained microbial reduction (Fig. 1). The enhanced arsenic retention was mainly attributed to co-precipitation within magnetite and adsorption on a mixture of magnetite and ferrihydrite. Ferrihydrite was unstable under reducing conditions, while magnetite and the associated arsenic were stable. Arsenic solubility increased in the other microcosms including those designed to stimulate sulfate reduction.

In the column experiment, prior to nitrate-iron(II) addition, arsenic was not effectively retained within the sediments (Fig. 2). Following nitrate-iron(II) addition, magnetite and ferrihydrite formed in the columns, which rapidly decreased effluent arsenic concentrations to $<10\,\mu g\,L^{-1}$. This magnetite persisted in the columns even as conditions became reducing, whereas ferrihydrite was transformed to stable iron oxides. This magnetite incorporated arsenic into its structure during precipitation and subsequently adsorbed arsenic. Adsorption to the minerals kept effluent arsenic concentrations $<10\,\mu g\,L^{-1}$ for more than 100 PVs despite considerable iron reduction.

Modeling is used to quantify the processes involved the magnetite-based strategy and to scale laboratory results to field environments. Such modelling suggests

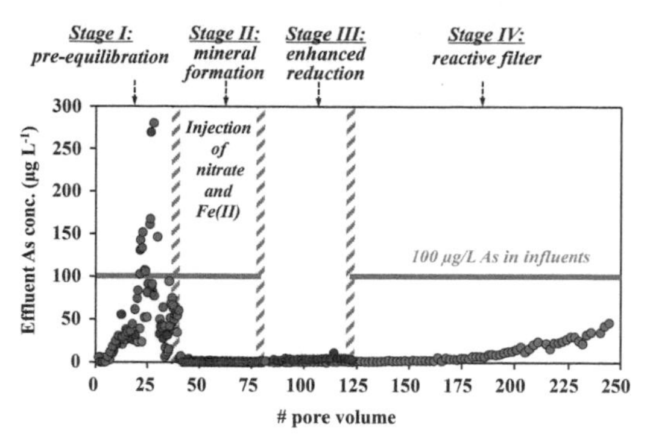

Figure 2. Effluent arsenic concentrations from the columns as a function of PVs. Modified from Sun *et al.* (2016c).

that the ratio between iron (II) and nitrate in the injectant regulates the formations of magnetite and ferrihydrite, and thus regulates the long-term effectiveness of the strategy. The results from field-scale models favor scenarios that rely on the chromatographic mixing of iron (II) and nitrate after injection.

4 CONCLUSIONS

In situ formation of magnetite can be achieved by the microbial oxidation of ferrous iron with nitrate. Magnetite can incorporate arsenic into its structure during formation, forming a stable arsenic sink. Magnetite, once formed, can also immobilize arsenic by surface adsorption, and thus serve as a reactive filter when contaminated groundwater migrates through the treatment zone. These results suggest that a magnetite-based strategy may be a long-term remedial option for arsenic contaminated aquifers.

ACKNOWLEDGEMENTS

U.S. National Institute of Environmental Health Sciences (grant ES010349 and ES009089).

REFERENCES

ATSDR, 2015. Summary data for 2015 priority list of hazardous substances.

Prommer, H., Barry, D.A. & Zheng, C. 2003. MOD-FLOW/MT3DMS-based reactive multicomponent transport modeling. *Groundwater* 41(2): 247–257.

Sun, J., Chillrud, S.N., Mailloux, B.J., Stute, M., Singh, R., Dong, H., Lepre, C. J. & Bostick, B.C. 2016a. Enhanced and stabilized arsenic retention in microcosms through the microbial oxidation of ferrous iron by nitrate. *Chemosphere* 144: 1106–1115.

Sun, J., Quicksall, A.N., Chillrud, S.N., Mailloux, B.J. & Bostick, B.C. 2016b. Arsenic mobilization from sediments in microcosms under sulfate reduction. *Chemosphere* 153: 254–261.

Sun, J., Chillrud, S.N., Mailloux, B.J. & Bostick, B.C. 2016c. In situ magnetite formation and long-term arsenic immobilization under advective flow conditions. *Environ. Sci. Technol.* 50(18): 10162–10171.

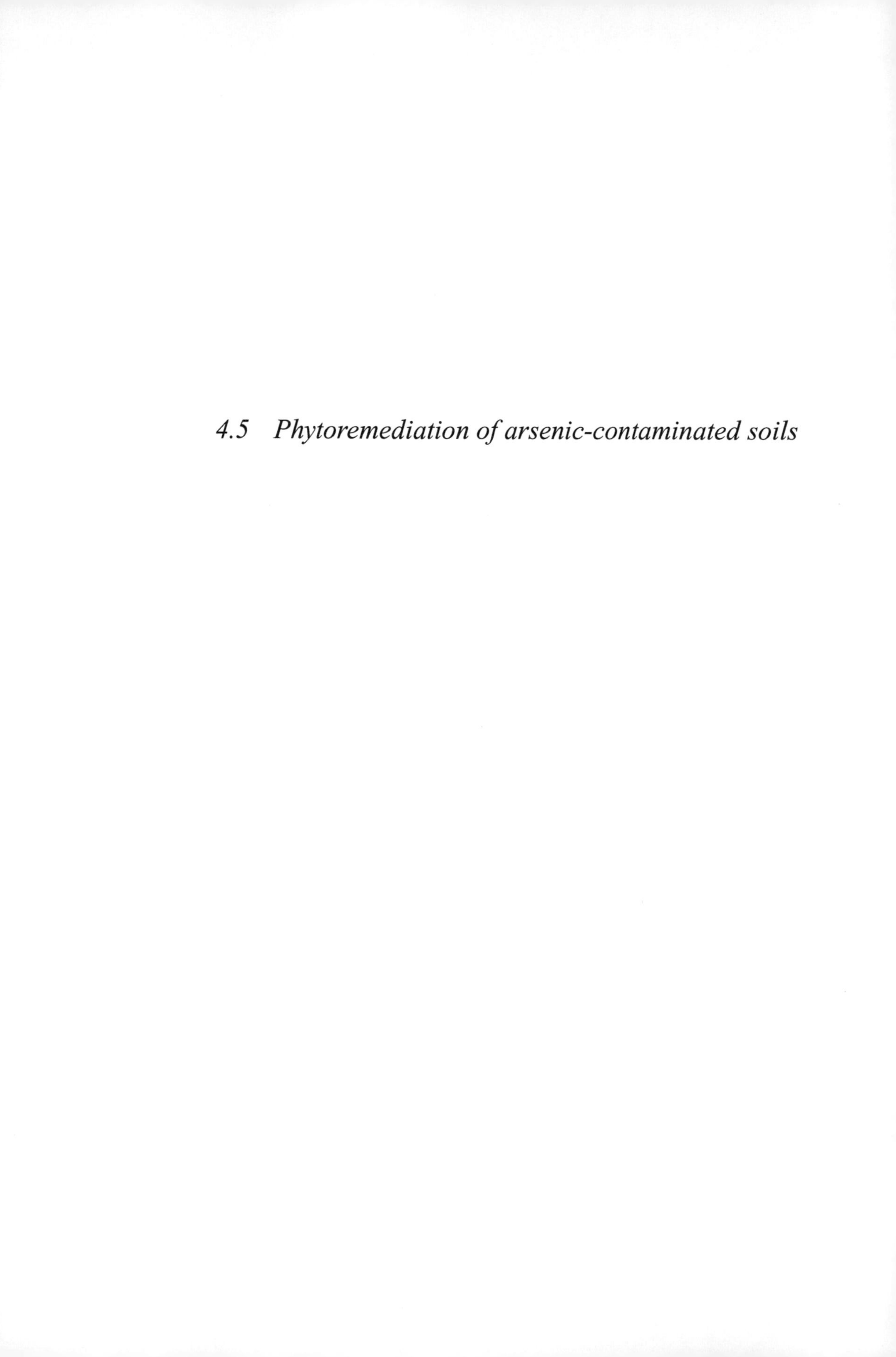

4.5 Phytoremediation of arsenic-contaminated soils

Environmental Arsenic in a Changing World –
Zhu, Guo, Bhattacharya, Ahmad, Bundschuh & Naidu (Eds)
ISBN 978-1-138-48609-6

Phytate enhanced dissolution of As-goethite and uptake by As-hyperaccumulator *Pteris vittata*

X. Liu[1], J.W. Fu[2], Y. Cao[2], Y. Chen[2] & L.Q. Ma[1,3]
[1]*Research Center for Soil Contamination and Environment Remediation, Southwest Forestry University, Kunming, P.R. China*
[2]*State Key Lab of Pollution Control and Resource Reuse, School of the Environment, Nanjing University, Jiangsu, P.R. China*
[3]*Soil and Water Science Department, University of Florida, Gainesville, FL, USA*

ABSTRACT: This work investigated dissolution of goethite with pre-adsorbed arsenate (As(V); As-goethite) by phytate and its effect on As uptake by As-hyperaccumulator *P. vittata*. The results showed that phytate promoted As-goethite dissolution, thus enhanced plant growth. In addition, while inorganic phosphorus (P_i) decreased As uptake, organic phytate increased As uptake by *P. vittata*.

1 INTRODUCTION

In soils, As is often present in its oxidized form arsenate (As(V)), which is a phosphate (P) analog (Meharg & Hartley-Whitaker, 2002). During plant uptake, As(V) and P compete for root P transporters (Ditusa *et al.*, 2016).

Phytate is the predominant form of organic P in soils (Turner *et al.*, 2006). As the dominant organic acid in *P. vittata* rhizosphere, phytate may affect As uptake by *P. vittata*, thus its phytoremediation efficiency.

Pteris vittata (PV), the first known As-hyperaccumulator (Ma *et al.*, 2001), prefers to grow in calcareous soils with low available Fe and As. Siderophores and root exudates are effective in dissolving Fe-As minerals (Liu *et al.*, 2015). However, limited information is available regarding how their co-presence affects goethite dissolution. Besides, it is important to understand how phytate influences As and P uptake and plant growth in *P. vittata*.

Therefore, the objective of this study was to examine the effect of phytate on As-goethite dissolution and the subsequent As and P uptake in *P. vittata*.

2 METHODS/EXPERIMENTAL

2.1 *Synthesis of goethite*

Goethite was synthesized following Schwertmann and Cornell (1991) and As(V)-adsorbed goethite was prepared following Wolff-Boenisch & Traina (2007). The goethite contained 0.25 mM g^{-1} As and is referred as As-goethite. To examine the effect of root exudates on As-goethite dissolution and the subsequent Fe and As plant uptake and plant growth, *P. vittata* was cultured in sterile Fe-free 0.2X HS (pH = 6.0) spiked with 0.25 g L^{-1} As-goethite with following treatments (four replicates): (1) control; (2) 50 μM oxalate; (3) 50 μM phytate; (4) 50 μM DFO-B; (5) 50 μM PG12-siderophore (PG12-S); (6) 50 μM oxalate + 50 μM PG12-S; and (7) 50 μM phytate + 50 μM PG12-S.

2.2 *Effect of phytate on plant As uptake*

To examine the effect of phytate on As uptake by *P. vittata*, uniform sporophytes were cultured in sterile media including following treatments of three replicates each: (1) 0.5-strength MS medium control; (2) 50 μM As (As_{50}); (3) 50 μM phytate ($phytate_{50}$); (4) As_{50} + $phytate_{50}$; (5) 500 μM phytate ($phytate_{500}$); and (6) As_{50} + $phytate_{500}$. The non-hyperaccumulator fern *P. ensiformis* (PE) was used for comparison.

After treatment, freeze-dried materials were digested with HNO_3/H_2O_2 using USEPA Method 3050B. Arsenic was analyzed with ICP–MS. Iron was measured by FAAS. Phosphorus was analyzed using a modified molybdenum blue method after removing As(V) interference via cysteine reduction (Singh & Ma, 2006).

3 RESULTS AND DISCUSSION

3.1 *Iron solubility of As-goethite*

Iron solubility of As-goethite was lower than goethite. In the suspensions containing As-goethite, Fe was undetectable. In the presence of phytate, soluble Fe was 0.08 mg L^{-1}, indicating 98.5% less Fe was dissolved from As-goethite than from goethite, thus pre-adsorbed As(V) decreased goethite dissolution. Phytate released As from As-goethite, increasing As uptake by *P. vittata* (435 and 78 mg kg^{-1} for fronds and roots) (Fig. 1).

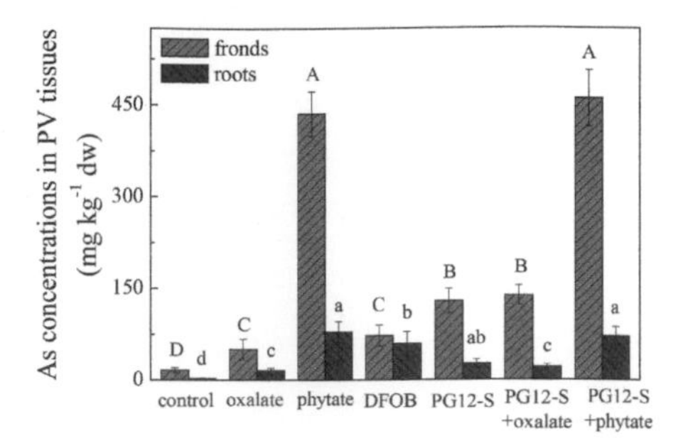

Figure 1. Arsenic concentration in *P. vittata* after growing for 14 d in Fe-free 0.2-strength Hoagland nutrient solution spiked with 0.25 g L^{-1} As-goethite.

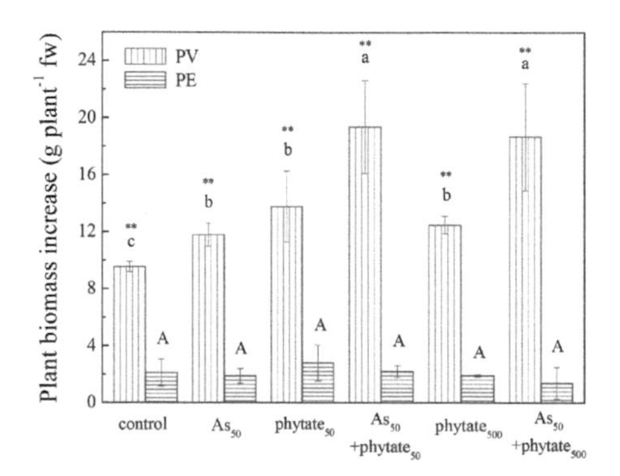

Figure 2. Increase in plant biomass of PV and PE after growing for 60 d in 0.5-strength MS agar amended with As and/or phytate.

3.2 *Growth of plant biomass*

After growing for 60 d, PV grew better than PE in all treatments. Enhanced growth of PV was observed in As and/or phytate treatments, with the most being in As + phytate. In contrast, the growth of PE showed no significant difference among treatments (Fig. 2).

3.3 *Arsenic in plant biomass*

The As concentrations in dry PV fronds and roots were 69.8 and 10.1 mg kg^{-1} in As_{50} treatment (Fig. 3), which increased by 2.2- and 3.1-fold to 157 and 31 mg kg^{-1} in the As_{50} + $phytate_{50}$ treatment, indicating phytate promoted As uptake in PV.

4 CONCLUSIONS

The results showed that As pre-adsorbed on goethite inhibited its dissolution. However, Fe uptake in *P. vittata* in bi-ligand systems was significantly higher than the sum of the corresponding one-ligand systems. This indicated a synergic effect between root exudates and microbial siderophore in promoting Fe dissolution.

Inorganic P suppressed As uptake by PV, which is unfavorable for phytoremediation of As contaminated soils. In contrary, phytate coupled with As promoted

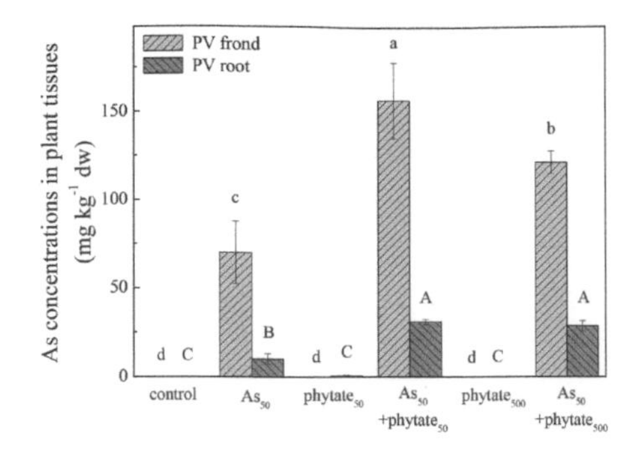

Figure 3. Arsenic concentration in PV tissues after growing for 60 d in 0.5-strength MS agar, which was served as control containing 63 μM P and was amended with 50 μM As (As_{50}) and/or 50 or 500 μM phytate ($phytate_{50}$ or $phytate_{500}$).

both P and As uptake by PV, indicating a beneficial role of phytate in enhancing As uptake.

ACKNOWLEDGEMENTS

This work was supported by the State Key Program of National Natural Science Foundation of China (Grant No. 21637002).

REFERENCES

Ditusa, S.F., Fontenot, E.B., Wallace, R.W., Silvers, M.A., Steele, T.N., Elnagar, A.H., Dearman, K.M. & Smith, A.P. 2016. A member of the phosphate transporter 1 (Pht1) family from the arsenic-hyperaccumulating fern *Pteris vittata* is a high-affinity arsenate transporter. *New Phytol.* 209(2): 762–772.

Liu, X., Yang, G.M., Guan, D.X., Ghosh, P. & Ma, L.Q. 2015. Catecholate-siderophore produced by As-resistant bacterium effectively dissolved $FeAsO_4$ and promoted *Pteris vittata* growth. *Environ. Pollut.* 206: 376–381.

Ma, L.Q., Komar, K.M., Tu, C., Zhang, W.H., Cai, Y. & Kennelley, E.D. 2001. A fern that hyperaccumulates arsenic: a hardy, versatile, fast-growing plant helps to remove arsenic from contaminated soils. *Nature* 409: 579–579.

Meharg, A.A. & Hartley-Whitaker, J. 2002. Arsenic uptake and metabolism in arsenic resistant and non-resistant plant species. New Phytol. 154: 29–43.

Schwertmann, U. & Cornell, R.M. 1991. Iron oxides in the laboratory. VCH: Weinheim.

Singh, N. & Ma, L.Q., 2006. Arsenic speciation, and arsenic and phosphate distribution in arsenic hyperaccumulator *Pteris vittata* L. and non-hyperaccumulator *Pteris ensiformis* L. *Environ. Pollut.* 141(2): 238–246.

Turner, B.L., Richardson, A.E. & Mullaney, E.J. 2006. Inositol phosphates in soil: amounts, forms and significance of the phosphorylated inositol stereoisomers (Chapter 12). In: B.L. Turner, A.E. Richardson, E.J. Mullaney (eds) *Inositol phosphates: Linking Agriculture and the Environment*, CAB International, pp. 186–206 (ISBN 9781845931520).

Wolff-Boenisch, D. & Traina, S.J., 2007. The effect of desferrioxamine B, enterobactin, oxalic acid, and Na-alginate on the dissolution of uranyl-treated goethite at pH 6 and 25°C. *Chemical Geology* 243(3–4): 357–368.

Environmental Arsenic in a Changing World –
Zhu, Guo, Bhattacharya, Ahmad, Bundschuh & Naidu (Eds)
ISBN 978-1-138-48609-6

Development of a phyto-stabilization strategy based on the optimization of endogenous vegetal species development on a former arsenic-bearing mine waste

H. Thouin[1], M.P. Norini[1], P. Gautret[1], M. Motelica[1], F. Battaglia-Brunet[2], L. Le Forestier[1], L. De Lary De Latour[2] & M. Beaulieu[2]
[1]*BRGM, ISTO, Orléans, France*
[2]*BRGM, Orléans, France*

ABSTRACT: In order to promote the development of a vegetal cover on a dump of mining residues, endogenous plants that started to colonize the waste material were collected and identified. The strategy of remediation included the selection of amendments that could optimally promote plant growth and decrease the mobility and bio-availability of the main toxic elements present in the residue, i.e., arsenic, lead and barium.

1 INTRODUCTION

Mine residues sometimes present risks associated with the dispersion of solid particles containing toxic elements, such as arsenic, and can also induce a pollution of groundwater and surface water through leaching. Phyto-stabilization strategies have been proposed as promising remediation options in order to decrease the risk of transfer of arsenic and heavy metals from the residues to the surrounding environment and to the biotope (Hattab *et al.*, 2015, Moreno-Jimenez *et al.*, 2017). However, the harmonious development of a vegetal cover that will efficiently stabilize the toxic elements for a long time is difficult to control: scientific and technical obstacles linked to the specific needs of the endogenous plants in terms of soil texture and nutrient needs must be linked to the minimization of toxic elements mobility. We present here the first steps of the development of the phytostabilization strategy for an arsenic-bearing mining waste.

2 METHODS

2.1 *Site description and residue sampling*

The materials were sampled on a former silver and lead mining site, exploited from the antiquity but with the highest activity in the second half of the XIXth century (Sabourault *et al.*, 2016). This mine is located in medium mountain (altitude 750 m). Mine residues ($3\,m^3$) were recovered from the 0–60 cm level of a mine waste dump, using a power shovel. Residues were taken in seven discrete zones in a $25\,m^2$ area. Measurements of arsenic and metals concentrations were performed on site using an X-ray fluorescence apparatus Niton Xlt999. The material was coarsely mixed with the shovel and transported to the laboratory for multi-scale experiments.

2.2 *Sampling of vegetal specimens*

Plants were sampled with their balled roots, in several types of substrates: (1) on the margin of the mining residues, where pedogenesis supported by the natural supply of organic litter already generated a soil-like structure; (2) directly in the sand-like yellow residues, and (3) in the intermediary substrate, presenting finer and darker phases than the sandy waste. The plants were immediately installed in pots, brought to the laboratory and maintained out-door in big tanks filled with mine residues, pure or mixed with organic litter from the site. The organic litter, composed of leaves from trees (oak and birch mainly) was sampled on the margin of the mining dump.

3 RESULTS AND DISCUSSION

3.1 *Concentrations in arsenic and metals*

The on-site quantification of arsenic and metals concentrations in the residues showed high concentrations in lead (from 12 to $38\,g\,kg^{-1}$, average value $17\,g\,kg^{-1}$), arsenic (from 0.5 to $2.2\,g\,kg^{-1}$, average value $1.6\,g\,kg^{-1}$) and zinc (from 0.13 to $2.7\,g\,kg^{-1}$, average value $0.47\,g\,kg^{-1}$).

3.2 *Sampling and identification of endogenous vegetal species*

The endogenous plants collected on site, resistant to arsenic and lead, were identified as belonging to

Figure 1. Endogenous plants (a) *Teucrium scorodonia, Achillea millefolium* growing on a soil-like substratum at the margin of the dump, (b) *Agrostis* and (c) *Festuca*) directly in the sandy waste, and (d) *Rumex*) in an intermediary substratum.

diverse genera that seemed to colonize three different types of substrates on the dumping areas (Fig. 1):

- The margin of the residue, where pedogenesis supported by the natural supply of organic litter already generated a soil-like structure was colonized by *Teucrium scorodonia* L., *Achillea millefolium* L.
- The mine residue, i.e. the sand-like yellow residues, was colonized by *Agrostis* sp. and *Festuca* sp., two genera previously found on metals-polluted sites (Ernst, 2006).
- The intermediary substrate, presenting finer and darker phases than the sandy waste, supported the growth of *Lotus corniculatus, Hypochaeris radicata* L., *Polygonum aviculare* L. *Persicaria* sp., *Scrofularia* sp., *Linaria repens, Silene vulgaris* (Moench) Garcke, *Cerastium* sp.

Rumex acetosella L. was found on different substrates. This species is otherwise known to be able As accumulation in this tissue contrary to excluder species which restricted As uptake (Otones *et al.*, 2011).

These vegetal specimens are thus adapted to both toxic elements (arsenic, lead), and to the site climate, with negative winter temperatures.

3.3 *Stabilization of arsenic in the mine residue*

The next steps of the research program will tend to optimize the stabilization of arsenic and other toxic elements (lead, zinc and barium) combining the stimulation of colonization by vegetal species and the geochemical and biogeochemical stabilization of pollutants. Different amendments (iron- and organic matter- bearing materials) will be tested, alone and in combination. The mobility of arsenic and other toxic elements in presence of plants will be assessed from the unsaturated to the water saturated zone in mesocosm experiments. The evolution of microbial communities, including As-transforming microbes, will be monitored along with the evolution of vegetal cover and water/soil geochemistry.

4 CONCLUSIONS

A range of vegetal species are specifically adapted to the presence of arsenic and to the physico-chemical conditions of each specific mine site, including climate. Here, a methodology has been developed in ordcr to optimizc thc phyto-stabilization proccss using a multi-scale experimental program closely linking site, laboratory and mesocosm experiments.

ACKNOWLEDGEMENTS

This research work is performed in the frame of Phytoselect project funded by the Region Centre Val de Loire, contract N°2016-00108485, and by the Labex Voltaire (ANR-10-LABX-100-01). We gratefully acknowledge the financial support provided to the PIVOTS project by the Région Centre – Val de Loire (ARD 2020 program and CPER 2015–2020) and the French Ministry of Higher Education and Research (CPER 2015–2020 and public service subsidy to BRGM). This operation is co-funded by European Union. Europe is committed to the Center-Val de Loire region with the European Regional Development Fund.

REFERENCES

Ernst, W.H.O. 2006. Evolution of metal tolerance in higher plants. *For. Snow & Landsc. Res.* 80(3): 251–274

Hattab, N., Motelica-Heino, M., Faure, O. & Bouchardon, J.-L. 2015. Effect of fresh and mature organic amendments on the phytoremediation of technosols contaminated with high concentrations of trace elements. *J. Environ. Manag.* 159: 37–47.

Moreno-Jiménez, E., Sepúlveda, R., Esteban, E. & Beesley, L. 2017. Efficiency of organic and mineral based amendments to reduce metal [loid] mobility and uptake (*Lolium perenne*) from a pyrite-waste contaminated soil. *J. Geochem. Explor.* 174: 46–52.

Otones, V., Álvarez-Ayuso, E., García-Sánchez, A., Santa Regina, I. & Murciego, A. 2011. Mobility and phytoavailability of arsenic in an abandoned mining area. *Geoderma* 166(1): 153–161.

Sabourault, P., Niemiec, N., Pidod, A. & Girardeau, I. 2016. Le district minier de plomb-argentifère de Pontgibaud (France): vers une résilience des anciens dépôts de résidus de traitement du minerai. GESRIM, *2nd International Congress on the Management of Mining Wastes and Post-Mining.*

Environmental Arsenic in a Changing World –
Zhu, Guo, Bhattacharya, Ahmad, Bundschuh & Naidu (Eds)
ISBN 978-1-138-48609-6

Phytoremediation of arsenic using a chemical stabilizer and *Eleocharis macrostachya* in a contaminated mining soil

J.M. Ochoa-Rivero[1], M.A. Olmos-Márquez[2], C.G. Sáenz-Uribe[2] & M.T. Alarcón-Herrera[3]
[1]*Instituto Nacional de Investigaciones Forestales, Agrícolas y Pecuarias (INIFAP)-CIRNOC-Experimental Station La Campana, Aldama, Chih., Mexico*
[2]*Facultad de Zootecnia y Ecología, Universidad Autónoma de Chihuahua, Chihuahua, Chih., Mexico*
[3]*Centro de Investigación en Materiales Avanzados (CIMAV, S.C.), Chihuahua, Chih., Mexico*

ABSTRACT: Arsenic (As) is a highly toxic metalloid that occurs naturally in the environment, however, anthropogenic activities have also caused this element occurrence in water and soil. The objective of this study was to evaluate phytoremediation capacity of *Eleocharis macrostachya* to remove As from the mining soils using cryptomelane (KMn_8O_{16}) as a chemical stabilizer. Experiment was established under greenhouse conditions during 70 days (10 weeks) using 20 pots with capacity of 3 kg of soils. 15 pots were applied 5 g of cryptomelane, while five pots were considered as a blank. The variables determined were soluble and total As in soil and plant (root and stem). 89% of the As was retained in the root and the rest in stem. The arsenic concentration, both in root and stem, was higher in the control soil than that with cryptomelane addition, which confirms that the chemical stabilizer retains or maintains less bioavailable arsenical compounds.

1 INTRODUCTION

Arsenic (As) is a highly toxic metalloid that occurs naturally in the environment, however, anthropogenic activities have also caused this element occurrence in water and soil (Han *et al.*, 2003; Hettiarachchi et al., 2000). Once in the ecosystem, As affects the quality of soils and waters used for various productive activities (Singh *et al.*, 2015). In contrast to conventional treatments (physicochemical), which require long periods of time, involve high costs and can generate residual compounds, phytoremediation that develops the stabilization, transformation and degradation of pollutants, has the characteristics of being an ecological alternative, implemented in situ and have social acceptability (Wani *et al.*, 2017; Singh et al., 2015). The objective of this study was to evaluate phytoremediation capacity of *Eleocharis macrostachya* to remove As in the mining soils using cryptomelane (KMn_8O_{16}) as a chemical stabilizer.

2 METHODS/EXPERIMENTAL

2.1 *Soil sampling and experimental*

The soil collected for this study was taken from the surroundings of Avalos smelter (28°37′17.32″ N, 106°00′12.73″ W). Soil samples were taken at 0.30 m of depth from six points near to the tailing dam. Experiment was under greenhouse conditions during 70 days (10 weeks) an experiment of 20 pots with capacity of 3 kg of soils, 15 pots were applied 5 g of cryptomelane (KMn_8O_{16}) chemical stabilizer while 5 pots were used as control. After ten days of the stabilizer addition in the soil, *Eleocharis macrostachya*, collected of natural wetland in Chihuahua, Mexico, was planted. The variables that were measured are soluble and total As in soil and plant (root and stem).

2.2 *Arsenic analysis*

Arsenic concentrations in soil and plant were analyzed after 70 days of the experiment. The samples were analyzed for total and soluble arsenic by atomic absorption spectrometer (AAS) according NMX-AA-132-SCFI-2016, using HACH® spectrometer model DR2000™.

3 RESULTS AND DISCUSSION

3.1 *Total and soluble arsenic in soil*

Table 1 shows the average concentrations of total and soluble As determined in soil at the end of the experiment.

3.2 *Roots and stem*

89% of the As stabilized by soil was retained in the root and the rest in stem, both for the samples treated with the stabilizer and for the control targets. When stabilizer was used, the average As concentration in

Table 1. Total and soluble As concentrations in soil after 70 days.

Treatment	Total As (mg kg^{-1})	Soluble As (μg L^{-1})	U.S. EPA Permissible limit (mg kg^{-1})
Initial As concentration in soil	2,796.62	8.07	24 mg kg^{-1}
As concentration in soil (after stab.)	2044.47	N/D	–
As concentration in soil (after phyt. + estab.)	2409.97	7.97	–

Table 2. Descriptive statistics of concentrations of As in roots (a) and stem (b) of *Eleocharis macrostachya*.

(a) Roots

Treatment	n	mean	SE	min	max
Phytoremediation (stabilizer)	15	779.26	135.61	361.69	1119.55
Phytoremediation (without stab.)	5	877.72	122.45	173.18	1869.01

*SE: Standard Error

(b) Stem

Treatment	n	mean	SE	min	max
Phytoremediation (stabilizer)	15	105.17	15.80	25.58	252.30
Phytoremediation (without stab.)	5	77.66	20.87	27.34	154.47

*SE: Standard Error

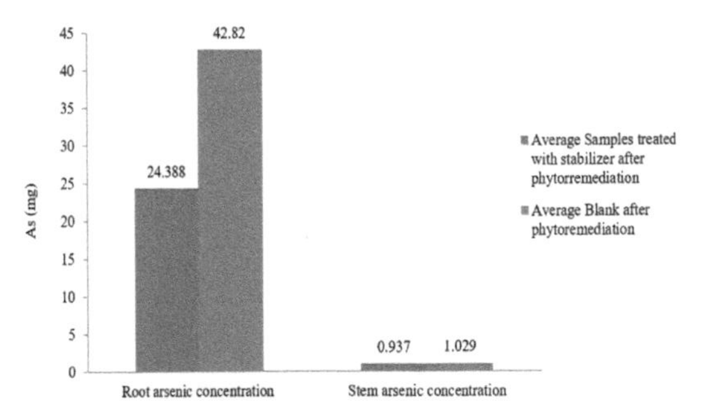

Figure 1. The average As concentrations in roots and stem after stabilizer apply.

the root was 779.26 mg kg^{-1}, while for the stem it was 105.17 mg kg^{-1} (Fig. 1). In the controls average As concentration in root was higher than plants with stabilizer, with an As concentration of 877.72 mg kg^{-1}.

It is important to mention that reproduction and development of plants in both treatments were similar (Fig. 2). Tendency was positive during experiment (70 days or 10 weeks).

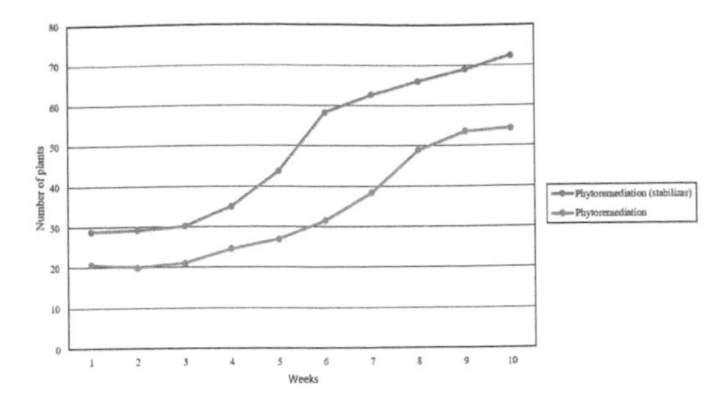

Figure 2. Reproduction of *Eleocharis macrostachya* in both treatments.

4 CONCLUSIONS

The maximum decrease in the soluble arsenic concentration was presented with the application of the chemical stabilizer. The concentration of arsenic after phytoremediation in the control targets and in the soil samples treated with the stabilizer was similar. On the other hand, the arsenic concentration, both in root and stem, were higher in the plants planted in the control soil, which confirms that the chemical stabilizer retains or maintains less bioavailable arsenical compounds.

ACKNOWLEDGEMENTS

This work was financially supported by National Council for Science and Technology (CONACYT), within the program Attention to National Problems.

REFERENCES

Han, F.X., Su, Y., Monts, D.L., Plodinec, M.J., Banin, A. & Triplett, G.E. 2003. Assessment of global industrial-age anthropogenic arsenic contamination. *Naturwissenschaften*, 90(9): 395–401.

Hettiarachchi, G.M., Pierzynski, G.M. & Ransom, M.D. 2000. In situ stabilization of soil lead using phosphorus and manganese oxide. *Environ. Sci. Technol.* 34(21): 4614–4619.

Razo, I., Carrizales, L., Castro, J., Díaz-Barriga, F., & Monroy, M. 2004. Arsenic and heavy metal pollution of soil, water and sediments in a semi-arid climate mining area in Mexico. *Water Air Soil Pollut.* 152(1–4): 129–152.

Singh, R., Singh, S., Parihar, P., Singh, V.P.Y. & Prasad, S.M. 2015. Arsenic contamination, consequences and remediation techniques: a review. *Ecotoxicol. Environ. Saf.* 112: 247–270.

Wani, R.A., Ganai, B.A.M. & Shah, A. 2017. Heavy metal uptake potential of aquatic plants through phytoremediation technique – a review. *J. Bioremediat. Biodegrad.* 8(404): 2.

4.6 Innovative technologies

Environmental Arsenic in a Changing World –
Zhu, Guo, Bhattacharya, Ahmad, Bundschuh & Naidu (Eds)
ISBN 978-1-138-48609-6

Removal of arsenic by fungal strains of *Fusarium*

E.E. Pellizzari, G.R. Bedogni, E.H. Monzon & M.C. Gimenez
Laboratorio de Microbiología General, Universidad Nacional del Chaco Austral, P.R. Sáenz Peña, Chaco, Argentina

ABSTRACT: Bioremediation of groundwater contaminates with arsenic using microorganisms has long been a research topic, has the advantage of being environmentally friendly and cost effective. Many fungi studied are resistant to arsenic such as *Trichoderma, Penicillium* and *Fusarium*. The genus *Fusarium* is very common contaminant in our region. Different strains of Fusarium were exposed to Arsenic. The strains were cultured for 2, large mycelia growth was observed. The level of arsenic decreased markedly. The reduction of the arsenic content in the water was very good, so it is feasible to carry out removal processes.

1 INTRODUCTION

In Argentina, the arsenic road begins in Jujuy and Salta, moves through Tucumán, La Rioja, Catamarca, San Juan, Chaco and Santiago del Estero; other provinces, until reaching the Atlantic coast, on that route, there is the dissolution of arsenical compounds, contaminating the aquifers that move through much of the territory of Argentina (Bhattacharya *et al.*, 2006; Bundschuh *et al.*, 2008, 2012; Nicolli *et al.*, 2012). The use of bacteria in ex situ groundwater bioreactor action processes, using equipment specially designed to carry out the process and reaching a complete biological treatment is possible (Pellizzari *et al.*, 2015).

The objective of our study is to observe the different mechanisms that can be performed by strains of the genus *Fusarium*, to develop bioassay systems of arsenic and other trace elements.

2 METHODS/EXPERIMENTAL

2.1 *Materials*

The experiments were carried out using *Fusarium* isolated in the province of Chaco. The colonies were obtained from the collection of microbiology, from cotton harvested in the region and from contaminated soil.

Fusarium culture was performed on potato dextrose agar (PDA) using 5, 10, 15, 20, 30 and 50 mgAs(V) L^{-1} stock solutions by dissolving sodium arsenate ($Na_2HAsO_4.7H_2O$) in deionizing water. The serial dilution and counting methods were used in extended plates. *Fusarium* populations were counted as colony forming units (cfu) in triplicate in medium plates and colonies were counted after 7 days of incubation at 30°C.

The strains were purified on Bengal pink agar, which was incubated at 30°C for 7 days, mycelia growth was measured radials 8 hours.

2.2 *Method*

To isolate metal resistant fungal strains, the cultivation medium PDA with As(V). This experiment was carried out in triplicate with completely randomized design from 20 day at 30°C old pure cultures of each fungal isolate were inoculated individually onto Petri plates containing different concentrations of arsenate. Colony diameter for radial growth was measured during the incubation period. The degree of tolerance was compared to a control without As(V).

The cultures were identified at the specie level on the basis of macroscopic and microscopic characteristics. Colony diameter for radial growth (cm) was measured during the incubation period. Slide extension with cotton blue lactophenol for the observation of hyphae, conidiophores and conidia, examined under a Leica-DM2500 microscope in a 100× magnification. Pure cultures of common *Fusarium* were identified by comparing the phenotypic characteristic features of each fungi described in the Fungi and Food Spoilage (Pitt & Hocking, 2009).

Pure fungal isolates were maintained on agar Rose Bengal, under laboratory conditions. Arsenic tolerance was detected by growth in the presence of arsenic. Comparisons of growth were made, considering high tolerance to the strains that presented the highest growth rate.

All *Fusarium* strains were purified on Bengal Rose agar, which was incubated at 30°C for 7 days, mycelia growth was measured radials every 8 hours.

Phenotypic changes of *Fusarium* in the presence of arsenic were determined by sowing the strains in Bengal Rose Agar, formulated with stock solution of

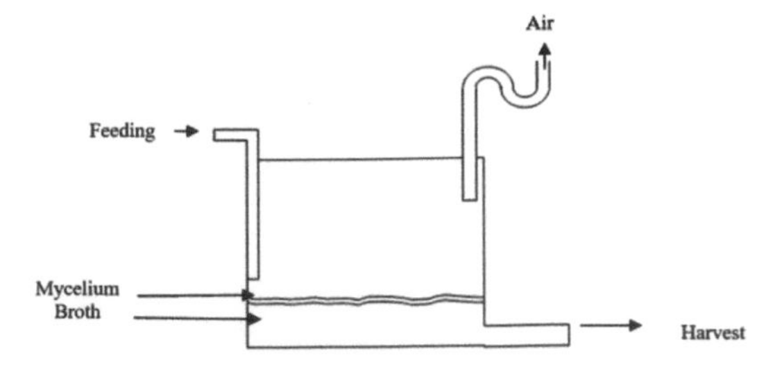

Figure 1. System designed with natural air circulation.

arsenic at 1, 5, 10, 20, 40, 50, 60 and 80 mg As(V) L^{-1}, respectively. Fungal growth was controlled by radial measurement of mycelium every 8 hours for 7 days. Seeds were grown in arsenic-free medium as a control.

For the study of the removal of arsenic, the *Fusarium* inoculate were prepared from a pure fungal culture of 15 days and inoculated in a system designed with natural air circulation (Fig. 1), with a depth of 3 cm broth formulated with 0.5% of NaCl, 1% glucose, and 10 mg As(V) L^{-1}, using sodium arsenate salt.

3 RESULTS AND DISCUSSION

3.1 *Results*

The purpose of this investigation was to obtain soil fungi from arsenic-contaminated agricultural soils for their possible exploitation in arsenic bioremediation studies. A total of 5 *Fusarium* strains were enumerated. The most frequently encountered isolates from the soil samples of all the sites were *Fusarium oryzae*.

It was observed that the growth rate of all strains was similar below 30 mg As(V) L^{-1}. At higher concentrations, the species that presented considerable growth was *Fusarium oryzae*, is significantly high and can accumulate important amounts of arsenic, which reached a tolerance of 60 mg As(V) L^{-1}, above this value, no strain had positive growth at pH 7.

Fusarium oryzae grew very well in the culture trays with 1 mg As(V) L^{-1}. The mycelium covered the surface of the tray in 24 h. The flotation of the mycelium allowed draining the liquid of removal easily. The percentage of removal of arsenic was 90% in 15 days. We can say that *Fusarium* would be very good organism to work with toxic metals.

In the growth of *Fusarium* with As(V), phenotypic differences of the fungus were observed in comparison with the control culture. Sexual sporulation is observed, produced by cellular stress, which generates the presence of arsenic in the culture medium.

The pH of medium can affect metal-fungal responses by its effects on metal speciation and cell physiology and metabolism, indirectly and the capacity of bio-volatilization of arsenic by *Trichoderma* sp., *Neocosmospora* sp., *Rhizopus* sp. and *Penicillium* sp., were considered as good strains for the removal of arsenic (Srivastava *et al.*, 2011). *Fusarium oryzae*, was the only one that reached 60 mg As(V) L^{-1}, being the most tolerant (Monzon & Pellizzari, 2016).

4 CONCLUSIONS

The objective of this work was to study the biological removal of arsenic using strains of *Fusarium*, which in previous studies demonstrated high tolerance. The behavior of different species of *Fusarium* with highly aggressive metabolism against arsenic has been observed. The results obtained allow us to deduce that *Fusarium* has a high capacity to remove arsenic. Our fundamental purpose is to improve the quality of water in the area in order to reduce or reduce exposure to arsenic of the resident population through the application of low cost methodologies to improve water quality in general.

ACKNOWLEDGEMENTS

This work was supported with funds Universidad Nacional del Chaco Austral.

REFERENCES

Bhattacharya, P., Claesson, M., Bundschuh, J., Sracek, O., Fagerberg, J., Jacks, G., Martin, R.A. & Storniolo, A.R. 2006. Distribution and mobility of arsenic in the Rio Dulce alluvial aquifers in Santiago del Estero Province, Argentina. *Sci. Total Environ.* 358(1–3):97–120.

Bundschuh, J., Litter, M.I. & Bhattacharya, P. 2012. Arsenic in Latin America, an unrevealed continent: occurrence, health effects and mitigation. *Sci. Total Environ.* 429: 1.

Bundschuh, J., Nicolli, H.B., Blanco, M.C, Blarasin, M., Farías S., Cumbal, L., Cornejo, L., Acarapi, J., Lienqueo, H., Arenas, M., Guérèquiz, R., Bhattacharya, P., García, M.E., Quintanilla, J., Deschamps, E., Viola, Z., Castro de Esparza, M.L., Rodríguez, J., Pérez-Carrera, A. & Fernández Cirelli, A. 2008. Distribución de arsénico en la región Sudamericana. In: J. Bundschuh, A. Perez Carrera, M.I. Litter M.I. (eds) *Distribución del Arsénico en las Regiones Ibérica e Iberoamericana.* CYTED, IBEROARSEN, pp. 137–185.

Monzón, E.H. & Pellizzari, E.E. 2016. Tolerancia al arsénicopor *Fusarium oryzae*. IV Simposio Argentino de Procesos Biotecnológicos. Universidad Nacional de Quilmes.

Nicolli, H.B., Bundschuh, J., Blanco, M.C, Ofelia C. Tujchneider, O.C., Panarello, H.O., Dapeña, C. & Rusansky, J.E. 2012. Arsenic and associated trace-elements in ground water from the Chaco-Pampean plain, Argentina: results from 100 years of research. *Sci. Total Environ.* 429: 36–56.

Pellizzari E.E., Marinich L.G., Flores Cabrera, S.A. & Gimenez, M.C. 2015. Application of processes for the removal of arsenic by bacteria from groundwaters. *International Conference on Food and Biosystems Engineering*, 28–31 May 2015, Mykonos Island, FaBE 2015-044.

Pitt, J.I. & Hocking, A.D. 2009. The ecology of fungal food spoilage. In: *Fungi and Food Spoilage*. Springer US.

Srivastava, P.K., Vaish, A., Dwivedi, S., Chakrabarty, D., Singh, N. & Tripathi, R.D. 2011. Biological removal of arsenic pollution by soil fungi. *Sci. Total Environ.* 409(12): 2430–2442.

Environmental Arsenic in a Changing World –
Zhu, Guo, Bhattacharya, Ahmad, Bundschuh & Naidu (Eds)
ISBN 978-1-138-48609-6

Permanent remediation of toxic arsenic trioxide in Canada's North

P.E. Brown
Giant Mine Oversight Board, Yellowknife, NT, Canada

ABSTRACT: The former Giant Mine is North America's largest and most technically challenging arsenic contaminated site. During the operational phase of the mine, some 237,000 metric tons of arsenic trioxide dust were stored in underground vaults within the mine. An interim mine site remediation plan has been approved for a maximum of 100 years, subject to a number of pre-conditions and regulatory approvals. Among these conditions is the requirement for a research program to identify methods to permanently eliminate the risks posed by the arsenic trioxide dust. As a first step, the independent Giant Mine Oversight Board initiated a comprehensive review of potentially promising methods.

1 BACKGROUND

Located in Canada's sub-arctic, the former Giant Mine is adjacent to the City of Yellowknife on the shores of Great Slave Lake, approximately 300 km south of the Arctic Circle. Gold ore at the Giant Mine is associated with an arsenic-bearing mineral known as arsenopyrite. The process used to release the gold from the arsenopyrite led to the production of arsenic-rich gas as a by-product. Operators of the mine captured this gas in the form of arsenic trioxide dust which was transferred to large underground rock chambers within the mine. Throughout the 50-year operational life of the mine, a total of 237,000 metric tons of arsenic trioxide dust was produced. The arsenic trioxide dust, which is approximately 60% arsenic, is hazardous to both people and the environment (Arcadis Canada Inc., 2017).

2 REMEDIAL STRATEGY

During mining operations, it was originally assumed that the naturally frozen rock surrounding the dust chambers would immobilize the arsenic waste. However, mining activities resulted in the thawing of the rock and the waste is no longer effectively contained. Of particular concern, arsenic in the dust is water soluble and has the potential to contaminate groundwater and downstream surface water bodies. Arsenic concentrations of up to 4,000 mg L^{-1} have been measured in the mine water.

2.1 *Interim remediation*

Following the bankruptcy of the mine operator, the Government of Canada became responsible for managing the risks associated with the mine. After conducting a comprehensive review of alternatives, it determined that the preferred method for managing the arsenic trioxide dust was to freeze it in place. Referred to as the "frozen block" method, the approach involves artificially cooling the surrounding rock using refrigeration and passive thermo-syphons. A pilot test of the technology was successfully performed and full-scale implementation of the frozen block method is projected to begin in 2020, subject to regulatory approvals.

2.2 *Requirement for a permanent solution*

While the Mackenzie Valley Environmental Impact Review Board approved the frozen block method, it was accepted only as an interim solution for a maximum of 100 years, primarily because the approach requires active long-term care and is not considered to be permanent. There is, therefore, a requirement to develop a remedial strategy that will *permanently* mitigate the risks associated with the arsenic trioxide dust.

Figure 1. Testing of the Frozen Block method.

2.3 *Independent oversight and research*

An independent body (the Giant Mine Oversight Board – GMOB) was established to oversee the remediation of the Giant Mine, including the implementation of the frozen block method. GMOB is also responsible for designing and managing a research program to identify and evaluate technologies that could lead to a permanent solution for the arsenic trioxide waste currently stored at the site.

3 TECHNOLOGY REVIEW

Prior to initiating the design of its arsenic trioxide management research program, GMOB commissioned a "State of Knowledge" (SOK) review of potentially promising technologies. The review was performed to establish a technology baseline that would inform the design of the research program.

3.1 *Research methods*

The SOK review assessed a wide range of management options and technologies that can be broadly grouped into the following categories: 1) *in situ* management; 2) arsenic trioxide dust extraction; 3) *ex situ* waste stabilization/processing; and physical isolation and disposal. Each technology was assessed against a set of weighted criteria that included but were not limited to the following: permanence (i.e., long-term waste stability); technical maturity; occupational risks; operation, maintenance and mon-itoring requirements; compatibility with future land use; contingencies; and cost

GMOB recognizes that individual technologies are unlikely to resolve the problem on their own; instead, it is likely that a combination of technologies may be required, e.g., extraction followed by *ex situ* treatment and disposal.

3.2 *Research findings*

In situ methods

To establish a baseline for comparison, the SOK review evaluated the approved 'frozen block' method (Fig. 1). The method performed well for both technical soundness and safety but scored poorly in the critically important criteria of permanence. An alternative *in situ* technology, nano-scale zero-valent iron which involves injection of very small iron particles to create a barrier to arsenic movement was also evaluated. The technique, which has been used effectively at other contaminated sites, was determined to be impractical as a primary mitigation at the Giant Mine.

Arsenic trioxide dust extraction

Dust extraction or mining would remove the arsenic trioxide dust from underground for processing. To be effective, a high degree of removal efficiency (i.e., >98%) would be necessary to minimize the risk of residual arsenic contamination. In an effort to limit occupational risks, remote mechanical mining methods were evaluated. While recent technology advances have increased the effectiveness and safety of remote mining, the occupational risks have not been eliminated. As a result, mining of the dust generally scored low in the safety category. However, hydraulic borehole mining, which uses high-pressure liquid or steam to remove the dust was assessed to be the safest and most effective of the mining methods.

Ex situ waste stabilization/processing

Multiple *ex situ* waste stabilization technologies were assessed including cement stabilization, vitrification, cement paste backfill, mineral precipitation and biological precipitation. The most promising of the *ex situ* techniques was vitrification which involves encasing the arsenic trioxide dust in a glass matrix. Key advantages of the technique include long-term stability of the resulting glass and moderate overall costs.

Physical isolation and disposal

Physical containment or disposal of untreated arsenic trioxide dust was not considered due to the ongoing risks associated with the dust. However, long-term storage of the treated arsenic would be required for all ex situ methods. Under the current review, only one potential method was reviewed. The technique would involve placing the treated product underground in the mine within concrete vaults surrounded by sand and/or gravel to provide protection from ground movement.

4 CONCLUSIONS

The SOK Review evaluated a wide range of technologies, some of which show potential for the effective and long-term management of arsenic trioxide dust. There is, however, a need for targeted research to further assess the viability of the approaches that were reviewed, other emerging technologies, and the integration of technologies. GMOB is using the findings of the SOK Review to assist with the design of this research program.

REFERENCE

Arcadis Canada Inc. 2017. *Giant Mine State of Knowledge Review: Arsenic Dust Management Strategies*.

Environmental Arsenic in a Changing World – Zhu, Guo, Bhattacharya, Ahmad, Bundschuh & Naidu (Eds)
ISBN 978-1-138-48609-6

Identification arsenic (V) by cyclic voltammetry and recovery of arsenic by electrodeposition

H.I. Navarro Solis[1], G. Rosano Ortega[1] & S.E. Garrido Hoyos[2]
[1]*Escuela de Ingeniería Ambiental, Universidad Popular Autónoma del Estado de Puebla A.C., Puebla, Pue., Mexico*
[2]*Instituto Mexicano de Tecnología del Agua, Subcoordinación de Posgrado, Jiutepec, Mor., Mexico*

ABSTRACT: This work is focuses in to develop an As(V) identification technique in an aqueous medium with KCl and KI as background electrolytes, in low and high concentrations from water purification technologies and water wells. A technique of electro recovery of arsenic was developed by the application of a potential in acid solution in order to reduce As(V) and As(III) present in drinking and rejection water coming to mining activities, and process like Capacitive deionization and inverse osmosis.

1 INTRODUCTION

Electrochemical methods are an alternative for the determination of arsenic in its reduced state, determinations have been reported in the range of 0.007 to 270.0 mg L^{-1} and have the advantage over the spectroscopic in sensitivity, selectivity and especially in speciation (Mays & Hussam, 2009). The redissolution voltammetry technique is characterized by its selectivity to As(III), high sensitivity and low detection limits that can be compared with other techniques. Considering the advantages of electrochemical methods such as Anodic Redissolution Voltammetry (AVS), the present work aims to develop a technique to identify the As(V) metalloid in an aqueous medium different process, like inverse osmosis and capacity deionization for low concentration and reject water from mines, for the investigation was using a platinum working electrode with a KCl bottom solution (Cavicchioli *et al.*, 2004; Gibbon-Walsh *et al.*, 2012; Litter *et al.*, 2009; Noskova *et al.*, 2012).

2 METHODS/EXPERIMENTAL

2.1 *Identification of arsenic (V)*

All the solutions used in the development of this research were standard solution at low concentrations of 0.01 to 1 mg L^{-1}, a medium concentrations 1 to 10 mg L^{-1}, and highs of 30, 50, 80 and 100 mg L^{-1}, As carrier electrolyte 0.001 mol L^{-1} KCl. All As(V) samples were added with IK and adjusted to pH 2 with 0.1 mol L^{-1} HCl.

Three electrode cell was used in the detection of arsenic (V). A double- jacketed Ag/AgCl (E = ECS) as the reference electrode (RE), a platinum electrode as the working electrode (WE), that was manufactured with a glass tube and epoxy resin and a graphite electrode as an auxiliary electrode (AE).

Readings of cyclic voltammetry were carried out using a potentiostat/galvanostat/ZRA 09096 – Gamry instruments, with an electrochemical window of initial potential of 0.0 to +0.800 V with a scanning potential of 100 mV s^{-1}.

2.2 *Arsenic (V) recovery by electrodeposition*

Arsenic recovery tests were carried out with solutions containing a mixture of 5 mg L^{-1} As(V) and As(III) with a pH adjusted to 2 by HCl, these solutions were prepared with analytical grade reagents. Potentiostatic test were performed in a range of 0.8 V at −0.6 V in a potentiostat/galvanostat/ZRA 09096, in order to observe at cathodic or reduction potential necessary to electrodeposit the arsenic in the metal electrodes, were used a working electrode, with 1 cm^2 area, encapsulated and mirror polished. A double jacket Ag/AgCl reference electrode (E = ECS) and the graphite auxiliary electrode, said process was carried out in a hermetically sealed glass cell. Finally, the electrodeposition of As(III) and As(V) was performed using 3 × 3 cm, two-sided copper electrodes 9 cm^2 of area, for 60 min at different voltages of 0.5, 1.0 and 1.5 V using an Agilent brand energy source, model E3642A. In order to eliminate the oxygen present and avoid undesirable reactions was maintained constant bubbling of N_2.

3 RESULTS AND DISCUSSION

Experimental tests were performed with high (10 to 100 mg L^{-1}), medium (1 to 10 mg L^{-1}) and low (0.001 to 1.0 mg L^{-1}) concentrations pH 2 adjusted with HCl, experiment was exposed to constant bubbling to N_2. Figure 1 shows the displacement of the anodic peak by identifying As(V) at 10 mg L^{-1} at an amperage of −3.84 μA and −4.39 μA at a concentration of 100 mg L^{-1} confirming that the magnitude of

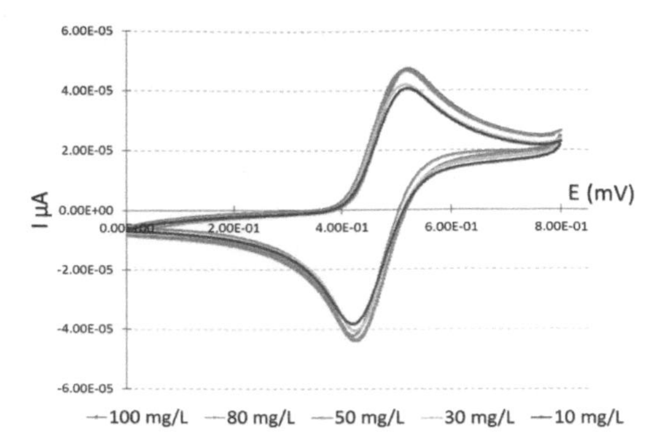

Figure 1. Voltamperogram high concentrations of As(V) at 1.33×10^{-4} M, 4.00×10^{-4} M, 6.67×10^{-4} M, 1.06×10^{-3} M, 1.33×10^{-3} M, plus 0.001 M KCl and a concentration of 0.01 M IK.

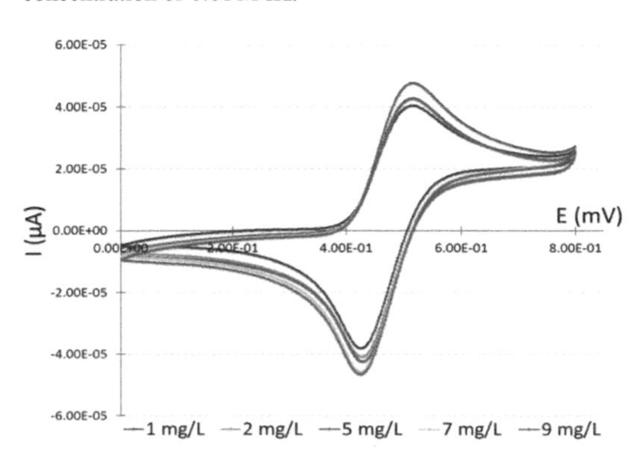

Figure 2. Voltamperogram medium concentrations of As(V) 1.33×10^{-5} M, 2.66×10^{-5} M, 4.00×10^{-5} M, 5.33×10^{-5} M, 6.67×10^{-5} M, 8.00×10^{-5} M, 9.34×10^{-5} M, 1.06×10^{-4} M, 1.20×10^{-4} M, 1.33×10^{-4} M, of 0.001 KCl and 0.01 M IK.

the anodic peak is proportional to the Amount of As present. Identifying the oxidation and reduction of the metalloid. The behavior of the oxidation reduction in medium concentrations is shown in Figure 2. Figure 3 shows the behavior of low concentration of As(V).

3.1 *Electrodeposition of arsenic*

The electrodeposition was developed with a concentration 5 mg L^{-1} of As(III) and As(V), with electrodes of carbon with a 9 cm^2 of area. Theoretically using the Faraday equation, it is note than it is possible remove up 45% with us conditions 1 V during 60 min. However in reality the experiments show that removed 62% but a percentage of this Arsenic is removed that arsine gas. Figure 4 shows the electrode with arsenic electrodeposited.

4 CONCLUSIONS

The identification of As(V) by cyclic voltammetry is a technologically viable option due to the reduction

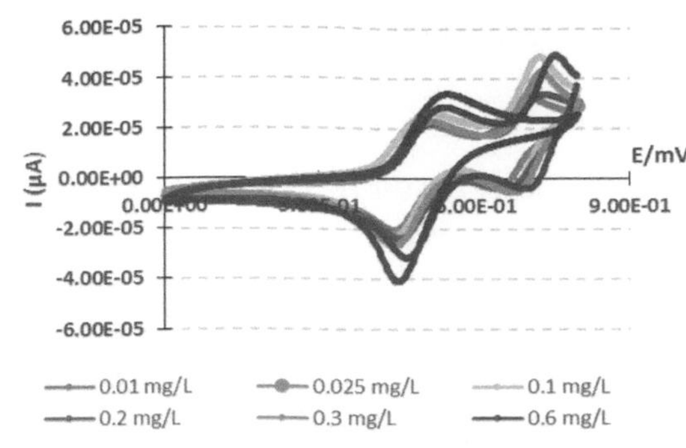

Figure 3. Voltamperogram low concentrations of As(V) 1.0×10^{-7} M a 1×10^{-5} M, of 0.001 KCl and 0.01 M IK.

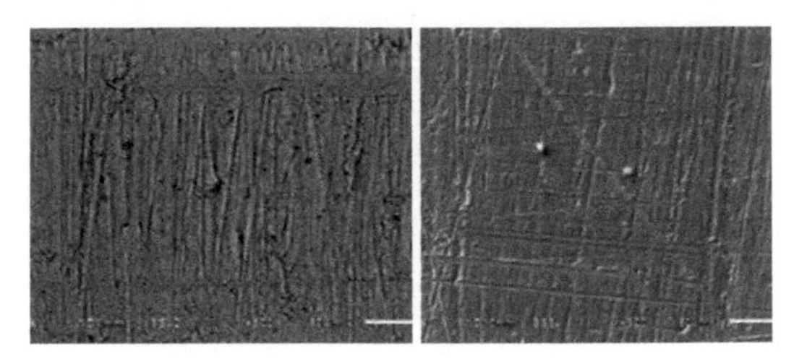

Figure 4. Images obtained in the MED, metal with arsenic.

of time and costs, with high sensitivity and friendly to the environment. It is possible to recover the arsenic by electrodeposition with efficiencies greater than 50%.

ACKNOWLEDGEMENTS

We thank with IMTA and ININ for their facilities and the support provided to the development of this work. We thank CONACYT for providing the scholarships of the students involved in this research work.

REFERENCES

Cavicchioli, A., La-Scalea, M.A. & Gutz, I.G.R. 2004. Analysis and speciation of traces of arsenic in environmental, food and industrial samples by voltammetry: a review. *Electroanalysis* 16(9): 697–711.

Gibbon-Walsh, K., Salaün, P. & Van den Berg, C.M.G. 2012. Determination of arsenate in natural pH seawater using a manganese coated gold microwire electrode. *Anal. Chim. Acta* 710: 50–57.

Litter, M.I., Farías, S.S. & Armienta, M.A. 2009. *Metodologías Analíticas para la Determinación y Especiación de Arsénico en Aguas y Suelos.* CYTED. Argentina (Adobe Digital Editions Versión: http://docplayer.es/6722881-Iberoarsen-metodologias-analiticas-para-la-determinacion-y-especiacion-de-arsenico-en-aguas-y-suelos.html).

Mays, D.E. & Hussam, A. 2009. Voltammetric methods for determination and speciation of inorganic arsenic in the environment – a review. *Anal. Chim. Acta* 646(1–2): 6–16.

Noskova, G.N., Zakharova, E.A., Kolpakova, N.A. & Kabakaev, A.S. 2012. Electrodepositión and stripping voltammetry of arsenic(III) and arsenic(V) on a carbon black-polyethylene composite electrode in the presence of iron ions. *J. Soil State Electrochem.* 16(7): 2459–2472.

Environmental Arsenic in a Changing World –
Zhu, Guo, Bhattacharya, Ahmad, Bundschuh & Naidu (Eds)
ISBN 978-1-138-48609-6

Constructed wetlands as an alternative for arsenic removal

C.E. Corroto[1], A. Iriel, E. Calderón[2], A. Fernádez-Cirelli[2] & A.L. Pérez Carrera[2]
[1]*Agua y Saneamientos Argentinos S.A. (AySA S.A.) and Centro de Estudios Transdisciplinarios del Agua, Universidad de Buenos Aires and Buenos Aires, Argentina*
[2]*Facultad de Ciencias Veterinarias, Centro de Estudios Transdisciplinarios del Agua (CETA – UBA), Universidad de Buenos Aires, Buenos Aires, Argentina*

ABSTRACT: In Argentina, approximately 10% of the population consumes water with arsenic concentrations higher than the WHO and Argentine Food Code recommendations ($10\,\mu g\,L^{-1}$). Because of this, in some places reverse osmosis represented an immediate and effective solution to solve this problem. However, this process has a secondary effluent known as "concentrate" where all the elements present in the feed flow to this system are more concentrated. Sometimes, this residue is discharged in natural water bodies. In this work, the phytoremediation with constructed wetlands was proposed to solve the problem. The study was made from the design and construction of 3 prototypes, one of them implanted with *J. effusus*, another with *C. haspan* and a control one with substrate only. The percentages of arsenic removal were close to 50% with *C. haspan* and 80% with *J. effusus*. The technology was considered an efficient and environmentally sustainable solution to reverse osmosis effluent treatment and disposal.

1 INTRODUCTION

Arsenic (As) is a pollutant of groundwater, present in different regions of Argentina. Particularly in district of La Matanza, province of Buenos Aires, the reverse osmosis is used for As removal. The disadvantage of this technology is the secondary effluent (concentrate) that contains high levels of salts. The As concentration in the concentrate may be between 2 and 4 times higher than in the feed flow. This effluent is discharged in the river and may cause an important environmental problem due to the As concentration. For this reason, the objective of this study is to analyze the possibility of using constructed wetlands (CW) as a sustainable, cost-effective, and environmentally friendly alternative for As removal. Wetlands are considered as a complex bioreactor due to the interaction between microbial communities, plants, soils and sediments (Kadlec & Wallace, 2009). Vegetal species used in CW should be tolerant to heavy metals, capable to immobilize the pollutant in roots, cheap, environmentally sustainable, easy to implement and should not have translocation of heavy metals to shoots and stems (Ciarkowska *et al.*, 2017; Yadav *et al.*, 2012). The most important factors that should be considered in the implementation and operation are: As speciation, vegetation, substrate, microorganism, environmental conditions and physicochemical parameters (pH, oxide-reduction potential and temperature) (Lizama *et al.*, 2011; Valles Aragón *et al.*, 2013).

2 MATERIALS AND EXPERIMENTAL METHODOLOGY

2.1 *Materials*

Working solutions were obtained from the secondary effluent of a reverse osmosis process plant in Buenos Aires province.

In each of the CW, 220 kg of a mixture of inert gravel (average diameter = 2 mm) and 10% of laterite were used as substrate. The pH, electrical conductivity and textural composition were determined. The vegetal species used in this work were: *Cyperus haspan* (*C. haspan*) and *Juncus effusus* (*J. effusus*).

2.2 *Experimental methodology*

The experiments were performed by using a continuous regime. Three prototypes were put in operation inside a greenhouse, two of them with different plants (A: *Cyperus haspan;* B: *Juncus effusus*) and one only with the substrate (C: *Control*). The initial concentration of As in the concentrate of reverse osmosis was, on average, $100\,\mu g\,L^{-1}$, the liquid level was 3 cm below the surface. The experiments were carried out for 419 days at room temperature. The monitored parameters were: flow, pH, conductivity, oxidation reduction potential, humidity and temperature. During this time, several samples were taken in the inflow and outflow in all the wetlands. Samples were filtered (0.45 µm) for solid removal and As concentrations were determined in

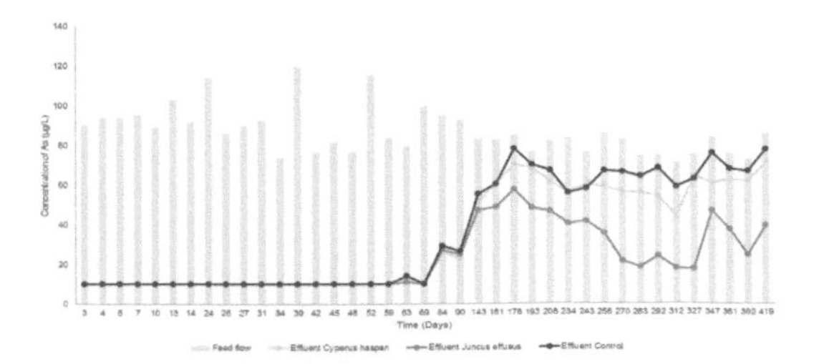

Figure 1. Profile of As concentration in the feed and out flow of each constructed wetlands during the period of operation.

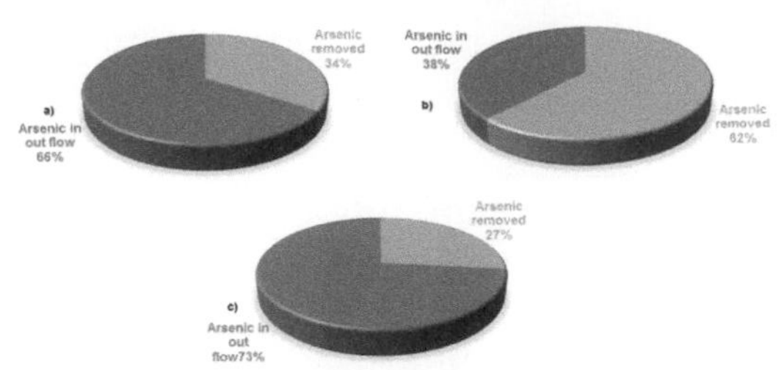

Figure 2. Percentage of As mass in the feed and out flow of constructed wetland during the operation period where these are, a) constructed wetland implanted with *Cyperus haspan* (PA), b) constructed wetland implanted with *Juncus effuses* (PB) and c) constructed wetland not implanted (Control or PC).

the supernatant, the As analysis was carried out using ICP-MS (Agilent, model 7500 CX with collision/reaction cell).

3 RESULTS AND DISCUSSION

3.1 *Concentrate physicochemical properties*

The results of the analysis of the physicochemical parameters in the reverse osmosis concentrate were: pH=8.2, conductivity=1394 μS cm^{-1}, Cl^-= 32.7 mg L^{-1}, SO_4^{-2}=31.6 mg L^{-1}, NO_3^-=76.3 mg L^{-1}, NO_2^-=0.02 mg L^{-1}, F^-=2.0 mg L^{-1}, alkalinity= 653 mg L^{-1}, Na^+=292 mg L^{-1}, K^+=14 mg L^{-1}, Fe (total) <0.05 mg L^{-1}, Mg^{2+}=17 mg L^{-1}, total hardness (CO_3Ca)=140 mg L^{-1}, NH_4^+ <0.05 mg L^{-1}, PO_4^{-3}=1.44 mg L^{-1}.

3.2 *Substrate characterization*

The analysis of textural characteristic of the laterite and gravel were made with Bouyoucos method. The results showed that laterite has 55% clay, 17.50% silt and 27.50% sand and its class is clayey. The pH of the laterite was 5.56 and the electrical conductivity, 38 μS cm^{-1}. The principal components were Al 37%, Si 27% and Fe 33%. On the other hand, the physicochemical characteristics of the gravel presented pH of 7.22, and electrical conductivity 21 μS cm^{-1}. The textural class was 1.25% clay and 98.75% sand.

3.3 *Arsenic removal*

The Figure 1 shows the As concentration profile, as well as the inflow and outflow in CW systems during the 419 days of operation. In the stabilization period (90 days), As concentration in the outflow was $\leq$10 μg L^{-1}. This effect was probably favored by laterite presence in the substrate. After this stage, in each of the CW the As concentration in outflow increases as a result of substrate saturation. In the case of the prototype planted with *Juncus effusus* (PB) the vegetal species were protagonists in the As removal. On the contrary, the trend of the curves of the prototype planted with *Cyperus haspan* (PA) and the prototype without plants (PC), in the last time, the values of the As concentration were similar in the feed and outflow.

However, in the curve for PB (Fig. 1) it is possible to see that the values of As concentration in output flow were lower than in the input, this indicates that there was removal of the contaminant.

During all the period of operation, the PB presented higher percentage of As removal and it was also possible to observe after performing an analysis of the percentage of mass in feed and out flow as shown in Figure 2. In this graph the As removal in PA was similar to PC, this shows the low efficiency of *C. haspan* to removal the pollutant in working conditions.

4 CONCLUSIONS

The maximum As removal was efficient, CW planted with *C. haspan* and *J. effusus* achieved removal values of 50% and 80% respectively, while the unplanted, only 35%. It should be noted that during the stabilization period (90 days) the maximum As removal was higher due to the laterite presence. However, the *J. effusus* species was found to be the most appropriate for the treatment (with a general percentage of removal above 50%) and it was easily adapted to the environmental conditions. Therefore, it is possible to consider this sustainable technology using vegetal species to remove As.

ACKNOWLEDGEMENTS

University of Buenos Aires, Consejo Nacional de Investigación Científica y Tecnológica and Agua y Saneamientos Argentinos S.A.

REFERENCES

Ciarkowska, K., Hanus-Fajerska, E. & Gambu, F. 2017. Phytostabilization of Zn-Pb ore flotation tailings with *Dianthus carthusianorum* and *Biscutella laevigata* after amending with mineral fertilizers or sewage sludge. *J. Environ. Manag.* 189: 75–83.

Kadlec, R.H. & Wallace, S.D. 2009. *Treatments Wetlands*. 2nd ed., CRC Press.

Lizama, K., Fletcher, T.D. & Sun, G. 2011. Removal processes for arsenic in constructed wetlands. *Chemosphere* 84(8): 1032–1043.

Valles-Aragon, M.C., Olmos-Marquez, M.A., Llorens, E. & Alarcón-Herrera, M.T. 2013. Redox potential and pH behavior effect on arsenic removal from water in a constructed wetland mesocosm. *Environ. Prog. Sustain. Energy* 33(4): 1332–1339.

Yadav, A.K., Abbassi, R., Kumar, N., Satya, S., Sreekrishnan, T.R. & Mishra, B.K. 2012. The removal of heavy metals in wetland microcosms: effects of bed depth, plant species, and metal mobility. *Chem. Eng. J.* 211–212: 501–507.

Environmental Arsenic in a Changing World –
Zhu, Guo, Bhattacharya, Ahmad, Bundschuh & Naidu (Eds)
ISBN 978-1-138-48609-6

Simultaneous electricity production and arsenic mitigation in paddy soils by using microbial fuel cells

Z. Chen[1], W. Gustave[1], Z.F. Yuan[1], R. Sekar[1], H.C. Chang[1], P. Salaun[2] & J.Y. Zhang[3]
[1]*Department of Environmental Science, Xi'an Jiaotong-Liverpool University, Suzhou, Jiangsu, P.R. China*
[2]*Department of Environmental Science, University of Liverpool, Brownlow Hill, Liverpool, UK*
[3]*College of Resources and Environmental Sciences, Nanjing Agricultural University, Nanjing, P.R. China*

ABSTRACT: Arsenic behavior in paddy soils is known to couple with the redox process of iron (Fe) minerals. When soil is flooded, Fe oxides are transformed to soluble ferrous ions by accepting the electrons from Fe reducers. In this study, we tried to manipulate the Fe redox processes in paddy soils by deploying sediment microbial full cells (sMFC). The results showed that the sMFC bioanode can modulate soil porewater Fe and arsenic (As) concentrations. At the end of the experiment, Fe and As contents around sMFC anode were 65.0% and 47.0% of the control respectively. A similar trend was observed in the sMFC bulk soil, where the Fe and As contents were 67.0% and 89.0% of the control respectively. This decrease in Fe and As concentrations could be attributed to the enhanced organic matter (OM) removal by sMFC. In the vicinity of bioanode, OM removal efficiencies were 10.3% and 14.0% higher than the control for lost on ignition carbon and organic carbon respectively. Furthermore, sequencing of the 16S rRNA genes suggested that the change in microbial community structure was minimal. Moreover, during the experiment a maximum current and power density of 0.31 mA and 12.0 mWm^{-2} were obtained, respectively. This study shows a novel way to make good use of As contaminated paddy soils, which is to simultaneously generate electricity and reduce the As mobility.

1 INTRODUCTION

Rice paddy soils are rich in iron (Fe) oxides, which serve as a potent repository for arsenic (As). When flooded, the reduction of Fe in turn increases As bioavailability and subsequent uptake of As by rice plants. Sediment MFCs are environmentally friendly bioelectrochemical systems that have the capacity to simultaneously produce electricity and remove organic pollutants. Previous studies have shown that the sMFC can immobilize and even extract metal ions by changing their redox state or via electrokinetic processes. Recent studies have provided evidence that the Fe minerals reduction in flooded soils might also be influenced by the buried bioanode (Touch et al., 2017; Yang *et al.*, 2016). However, it is still unknown whether the bioanode can affect the behaviors of Fe and sediment trace elements, especially As bound with Fe minerals, in paddy soils.

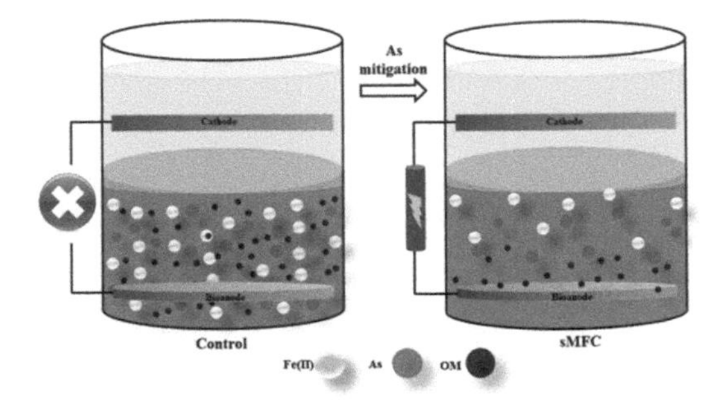

Figure 1. The SMFC assembly and the schematic diagram of the mechanism for As mitigation by the bioanode.

2 METHODS/EXPERIMENTAL

2.1 *Sediment microbial fuel cell assembly*

Subsurface paddy soil (~15 cm) was collected from an arsenic contaminated paddy field in Shangyu, Zhejiang China and transported directly to the lab. Eight sMFCs, were assembled as showed in Figure 1. Each sMFC contained 1 kg (dry weight) of paddy soil. A light-proof columnar polyethylene terephthalate container (10 cm diameter × 15 cm depth) was used to construct each sMFC. All of the cells were incubated in the dark for 60 days at 28°C.

2.2 *Chemical and microbial analysis*

Total As and Fe concentrations in soil porewater were determined by ICP-MS and AAS, respectively, from day 0 to 50 at 10 days intervals. The sediments from

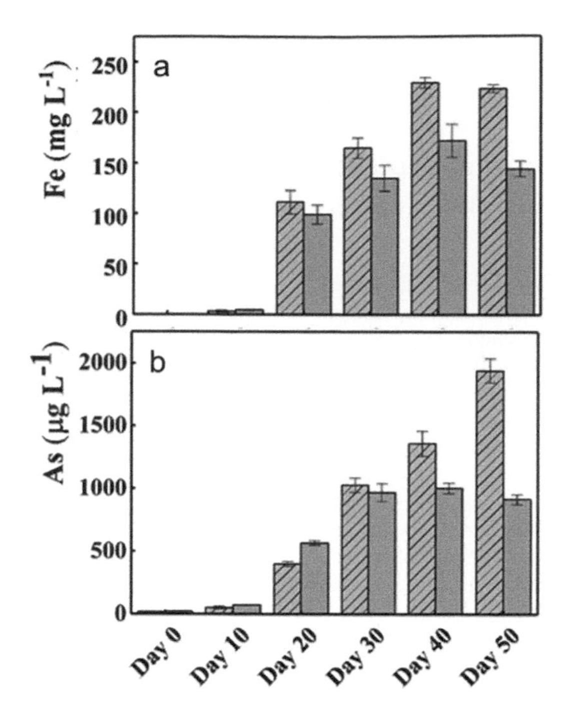

Figure 2. The SMFC assembly and the schematic diagram of the mechanism for As mitigation by the bioanode.

top, middle and bottom layers were collected on day 60 from both the sMFC and the control. The samples were homogenized and the content of total organic carbon (TOC) and loss on ignition (LOI) carbon was determined. TOC and DOC content were determined with a TOC analyzer.

Genomic DNA was extracted from bacterial cells using Powersoil DNA isolation Kit. Next generation sequencing library preparations and Illumina MiSeq2500 PE250 platforms sequencing were conducted at GENEWIZ, Inc.

3 RESULTS AND DISCUSSION

3.1 *Power output*

The produced current increased sharply after connecting the 500 ohms external resister without any lag time and reached the first peaked at around 0.2 mA on day 10. Afterwards, the current was quasi steady with slight fluctuation until the end of the experiment. The maximum power density (normalized to the anode geometric surface area) was found to be 12.0 mW m^{-2}, when the current density was 50.0 mA m^{-2} and the external resistance was 1000 ohms.

3.2 *Change in Fe and As concentration*

Average Fe concentration in the soil porewater of the controls was higher than that of the sMFC (Fig. 2a). These results revealed that the bioanode significantly reduced Fe reduction in its vicinity with time.

Similarly to Fe, the release of As in the control was significantly ($P < 0.05$) enhanced compared to that of the sMFC. (Fig. 2b) The As concentration increased sharply with increasing incubation time in both the control and the sMFC. However, the rate of As released in the control was higher than that of the sMFC on average from day 30–50. On day 0, 10 and 20, the As concentration (μg L^{-1}) was lower in the control (17.6 vs 19.5, 51.8 vs 71.6 and 400 vs 567, respectively), compared to the sMFC.

3.3 *The interaction between anodes and Fe oxides*

Our results indicate that significantly higher ($P < 0.05$) DOC concentration in the control than the sMFC was responsible for the enhanced Fe reduction in paddy soil. Although the sMFC altered the microbial community, the change is minimal.

4 CONCLUSIONS

In conclusion, the results presented here demonstrate that the sMFC can generate electricity in As-contaminated paddy soils and at the same time decrease the Fe and As in paddy soil porewater. Thus, sMFC can be used as a novel tool for As immobilization for soil remediation. The ability of the bioanode to immobilize As can be attributed to the rapid exhausting of organic substrate in the bioanode vicinity, furthermore, slowing down the reduction of Fe (oxy)hydroxide as it serves as reservoir for As in soil solid phase. The findings point out a novel way for the good use of As polluted paddy soils.

ACKNOWLEDGEMENTS

This work was supported by the National Science Foundation of China (41571305).

REFERENCES

Touch, N., Hibino, T., Morimoto, Y. & Kinjo, N. 2017. Relaxing the formation of hypoxic bottom water with sediment microbial fuel cells. *Environ. Technol.* 38(23): 3016–3025.

Yang, Q., Zhao, H., Zhao, N., Ni, J. & Gu, X. 2016. Enhanced phosphorus flux from overlying water to sediment in a bioelectrochemical system. *Bioresour. Technol.* 216, 182–187.

Environmental Arsenic in a Changing World –
Zhu, Guo, Bhattacharya, Ahmad, Bundschuh & Naidu (Eds)
ISBN 978-1-138-48609-6

Arsenic removal from groundwater by capacitive deionization (CDI): findings of laboratory studies with model water

E.E. Cañas Kurz[1], V.T. Luong[2], U. Hellriegel[1], J. Hoinkis[3] & J. Bundschuh[2]
[1]*Center of Applied Research, Karlsruhe University of Applied Sciences, Karlsruhe, Germany*
[2]*Faculty of Health, Engineering and Sciences, University of Southern Queensland, Toowoomba, QLD, Australia*
[3]*Department of Mechatronics and Sensor Systems Technology, Vietnamese-German University, Binh Duong Province, Vietnam*

ABSTRACT: The water supply of many countries in Southeast Asia has significant water problems and it is being affected, amongst others, by both high salinity and high arsenic concentrations in groundwater. This study presents an overview of a laboratory study using a lab-scale membrane capacitive deionization (MCDI) unit and arsenic-spiked model water. CDI is a novel technology for the electro-chemical desalination of brackish water with low energy consumption. In addition, removal performance of co-ions (NH_4^+; Mn^{2+}) typically found in groundwater in the region, and that have a negative influence on other conventional arsenic removal technologies, were also studied with increasing concentrations of NaCl (0–2.0 g L^{-1}). Results show a high dependency of the removal efficiency of selected ions on the TDS concentration, resulting in lower adsorption rates for substances in µg-range such as arsenic.

1 INTRODUCTION

1.1 *Project background*

Many countries in Southeast Asia such as Vietnam, Cambodia, Laos and Myanmar have significant water problems. The strong economic growth and social development, in connection with the growing population, increasingly exert pressure on existing water resources. The water supply of these countries is also being threatened by climate change. This includes in particular the rising sea levels and an increase in extreme weather conditions, what leads to an increased salinization of groundwater and surface water. An *et al.* (2014) have studied coastal ground- and surface water on Cu Lao Dung island in southern Vietnam and found a strong salinization of groundwater with electrical conductivities of up to 24,000 µS cm^{-1}.

Besides salinization problems many groundwater sources here have high concentration of geogenic iron (Fe) and arsenic (As). In particular, the Red River Delta has high levels of arsenic, with concentrations up to >600 µg L^{-1}, which are significantly higher than those recommended by the WHO drinking water standards (10 µg L^{-1}). A study of Agusa *et al.* (2014) has shown that a large part of the Vietnamese population is exposed to high levels of arsenic (300–600 µg L^{-1}), although sand filters are used in many households for As-removal.

1.2 *Objectives*

This study is part of the project, modular concept for sustainable desalination using capacitive deionization on the example of Vietnam (www.wakap.de). The overall objective is the development and piloting of a novel, more energy-efficient, modular combined process for the desalination of groundwater consisting of a capacitive deionization (CDI) unit and an upfront *in-situ* treatment by aeration to remove problematic compounds such as iron (Fe^{2+}) and arsenite (As(III)).

2 EXPERIMENTAL

2.1 *Capacitive Deionization (CDI)*

CDI is a novel desalination process in which saline water flows between two porous electrodes made of activated carbon. When a voltage is applied, salt ions are drawn to the electrodes and stored in an electrical double layers within the carbon electrodes (*purification* phase) (Biesheuvel *et al.*, 2015). When the electrodes become fully charged, the polarity of the electrodes is changed and stored ions can be discharged into a brine or concentrate stream (*regeneration* phase). In order to enhance the desalination efficiency with increased charge efficiency and avoiding co-ion transport during the *regeneration* phase, ion exchange membranes (IEM) were introduced to the CDI process, now so-called MCDI (Choi *et al.*, 2016).

2.2 *Materials and method*

A lab-scale MCDI unit has been used to pre-evaluate the removal efficiency of As, Mn and NH_4^+ in model water simulating the high natural concentrations seen

Table 1. Operational parameters MCDI.

Phase	Cycle	Flow $L\,min^{-1}$	Time sec	Current A
1	Regeneration	0.25	115	57.6
2	Pre-purify*	1.0	40	18.1
3	Purify	1.0	310	18.1

*Pre-phase to ensure quality in purification cycle.

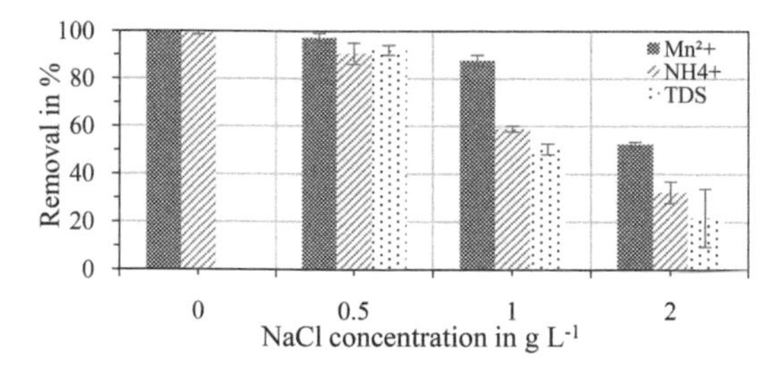

Figure 1. Removal of manganese ($C_{0,Mn} = 6\,mg\,L^{-1}$), ammonium ($C_{0,NH4+} = 25\,mg\,L^{-1}$) and total dissolved solids (TDS) with respects to their initial concentrations C_0 vs. increasing initial NaCl. Initial pH and temperature were 7.1 ± 0.6 and $25 \pm 3°C$.

in the groundwater in Vietnam. The initial TDS concentration was varied between 0; 0.5; 1.0 and 2.0 $g\,L^{-1}$ NaCl. *Purification* and *regeneration* cycles were kept equal in all single-pass experiments. Table 1 shows the operational settings of the MCDI chosen based on experimentally determined configurations for a high-energy-efficient operation.

3 RESULTS AND DISCUSSION

3.1 *Ammonium and manganese removal*

The results (Fig. 1) show that the removal of Mn is more favorable over NH_4^+ and the overall adsorption of TDS (NaCl), what can be explained by the higher valence state (2+) of Mn over NH_4^+, Cl^- and Na^+.

The higher removal of NH_4^+ over the NaCl might be due to its smaller hydrated ion radius, which allows a higher ion mobility and increased diffusion towards the electrodes and through the IEM. Moreover, with increasing NaCl concentrations, the removal efficiency for selected ions drops as more ions compete for adsorption places and the electrodes become saturated more rapidly. This behavior is in accordance with previous studies (Fan *et al.*, 2016).

3.2 *Arsenic removal*

Adsorption experiments for different As(V) concentrations (50–200 $\mu g\,L^{-1}$) are shown in Figure 2. Results show that for the given low concentrations of arsenic (μg) the selective arsenic removal with MCDI, even with NaCl concentrations as low as 0.5 $g\,L^{-1}$, was very

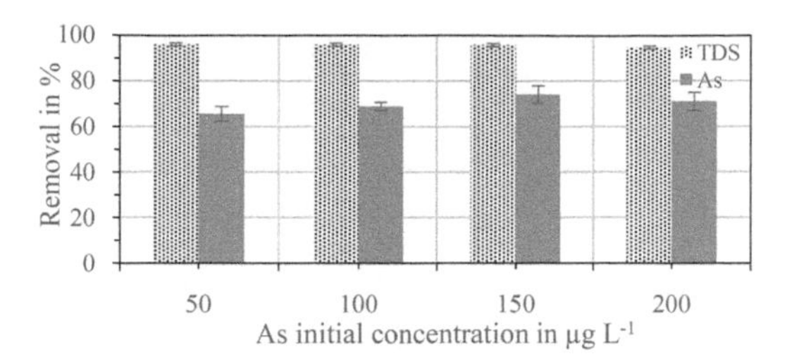

Figure 2. Removal of arsenic with MCDI (TDS = 0.5 $g\,L^{-1}$). Initial pH = 6.5 ± 0.5 and T = $18 \pm 0.2°C$.

limited. An average max. As-removal of only $70 \pm 3\%$ was achieved, so that complying with drinking water standard of 10 μg As L^{-1} was not possible under given conditions. Salt removal was stable at $96 \pm 1\%$.

Low As-removal can be mainly explained by the energy-efficient operational settings chosen as well as the extreme low As-concentrations, which makes the selective As-elimination with MCDI, when dissolved salts are present, challenging.

4 CONCLUSIONS

The decrease in the removal capacity with increasing TDS concentrations can be attributed to the electrosorbtion competition by Na^+ and Cl^--ions. In order to improve removal efficiency at higher TDS, higher currents (increased energy consumption) and shorter purify phases (lower recovery) should be considered.

ACKNOWLEDGEMENTS

The project was funded by the German Federal Ministry of Education and Research (BMBF) under the grant number 02WAV1413A.

REFERENCES

Agusa, T., Trang, P.T.K., Lan, V.M., Anh, D.H., Tanabe, S., Viet, P.H. & Berg, M. 2014. Human exposure to arsenic from drinking water in Vietnam. *Sci. Total Environ.* 488–489: 562–569.

An, T.D., Tsujimura, M., Le Phu, V., Kawachi, A. & Ha, D.T. 2014. Chemical characteristics of surface water and groundwater in coastal watershed, Mekong Delta, Vietnam. *Procedia Environ. Sci.* 20: 712–721.

Biesheuvel, P.M., Hamelers, H. & Suss, M.E. 2015. Theory of water desalination by porous electrodes with immobile chemical charge. *Colloid. Interface Sci. Commun.* 9: 1–5.

Choi, J., Lee, H. & Hong, S. 2016. Capacitive deionization (CDI) integrated with monovalent cation selective membrane for producing divalent cation-rich solution. *Desalination* 400: 38–46.

Fan, C.-S., Liou, S.Y.H. & Hou, C.-H. 2016. Capacitive deionization of arsenic-contaminated groundwater in a single-pass mode. *Chemosphere* 184: 924–931.

Environmental Arsenic in a Changing World –
Zhu, Guo, Bhattacharya, Ahmad, Bundschuh & Naidu (Eds)
ISBN 978-1-138-48609-6

As(III) removal in natural groundwaters: influence of HPO_4^{2-}, H_4SiO_4, Ca^{2+}, Mg^{2+} and humic acid in complex, realistic water matrices

D.J. de Ridder & D. van Halem
Sanitary Engineering Section, Faculty of Civil Engineering and Geoscience, Delft University of Technology, Delft, The Netherlands

ABSTRACT: The influence of phosphate, silicate, calcium, magnesium and humic acid on iron flocculation and arsenic removal has been shown in synthetic waters, but these findings may not necessarily be representative for actual groundwaters.In our study, three different groundwaters were selected which varied in composition, and with jar test, the removal of As and changes in iron floc size were monitored. The groundwaters were used as is, or after modification. The main findings are that the iron floc size can be related to the molar (Ca+Mg)/P ratio, and that As removal is closely related to Fe removal, which would indicate that adsorption competition is of lesser importance than hindered flocculation

1 INTRODUCTION

Arsenic removal is well known to be negatively affected by the presence of phosphate, silicate, and humic acid either (directly) due to competition for adsorption sites on iron oxides, or (indirectly) due to hindering the formation of iron oxides themselves (Guan *et al.*, 2009). Ca^{2+} has been shown to (partly) neutralize the latter (Kaegi *et al.*, 2010), and Mg^{2+} is expected to be similarly effective. Natural groundwaters are a mixture of all these ions, and as such, may not be represented well when laboratory experiments are carried out with synthetic water that only includes a selection.

Our aim was to systematically investigate the influence of phosphate, humic acid, calcium and magnesium on As(III) adsorption and Fe(II) flocculation in natural groundwaters. For this purpose, groundwaters were selected with different initial composition, and were used in their original state, and after manipulation of phosphate, calcium, magnesium and humic acid concentrations by either addition or removal by anion/cation exchange. Calcium and magnesium were dosed in equimolar concentrations, in an attempt to separate the effect of potential precipitation (Ca^{2+}) from sole charge neutralization (Mg^{2+}).

2 METHODS/EXPERIMENTAL

Three natural groundwaters with different water composition were selected. The initial concentrations of As and Fe were equalized to 20 $\mu g\ L^{-1}$ and 5 $mg\ L^{-1}$. These waters were used as is, or modified by dosing Ca^{2+}, Mg^{2+}, HPO_4^{2-}, humic acid, or precontacted (anaerobically) with anion exchange resin (Amberlite IRA400) or cation exchange resin (Amberlite IR120). After anion exchange, 150 $mg\ L^{-1}$ HCO_3^- was dosed back to compensate for its removal, and the pH was corrected to pH 7. Jar tests were carried out with 1 L solution in baffled jars at a mixing rate of 80 rpm and at 15–16°C. In the first 30 s of the jar test, the samples were aerated intensively and an O_2 concentration of 9.7 $mg\ L^{-1}$ was reached. Water samples were collected before aeration, and after 1, 5, 15, 30 and 60 min, filtered over 0.45 μm, and analyzed for As, Fe, Ca, Mg, Si, P, S by ICP/MS. Before aeration, HCO_3^- was determined by titration with 0.01M HCl to an endpoint of pH 4.3, and DOC was measured before aeration and after 60 min. The pH was measured continuously. After 60 min, the particle size and quantity distribution were measured with a particle counter (HIAC Royco model 9703, Pacific Scientific). All experiments were carried out in duplicate.

3 RESULTS AND DISCUSSION

3.1 *Baseline*

The composition of the three groundwaters after As and Fe equalization is shown in Table 1.

Iron removal was most rapid in KG water, followed by AMD water, followed by AP water. This is in accordance with Morgan and Lahav (2007), who found that iron oxidation rates increase at higher pH values. The total iron floc volume after 60 min was similar for KG and AMD water, but much less for AP water. Nevertheless, adsorption of As was hindered for AMD water, as compared to KG water. Most likely, this can be attributed to adsorption competition with Si.

3.2 *Modified groundwaters*

The presence of Ca^{2+} and Mg^{2+} was instrumental for iron flocculation. Their removal reduced iron flocculation in all groundwaters, and this effect was

Table 1. Groundwater composition, baseline.

	Fe (mg/l)	As (ug L^{-1})	Ca (mg L^{-1})	Mg (mg L^{-1})	Si (mg L^{-1})	P (mg L^{-1})	HCO_3^- (mg L^{-1})	pH
AMD	6.9	23.0	43.5	4.1	11.9	0.18	146	7.01
AP	4.9	21.8	20.9	2.5	7.5	0.34	103	6.83
KG	4.8	25.8	28.4	11.5	5.9	0.26	196	7.31

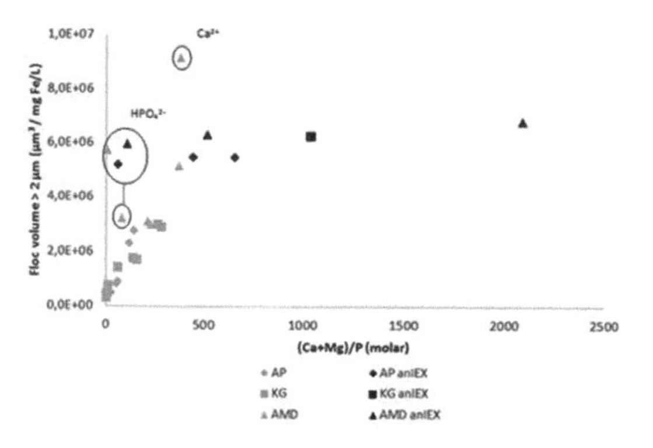

Figure 1. Floc volume VS (Ca+Mg)/P ratio.

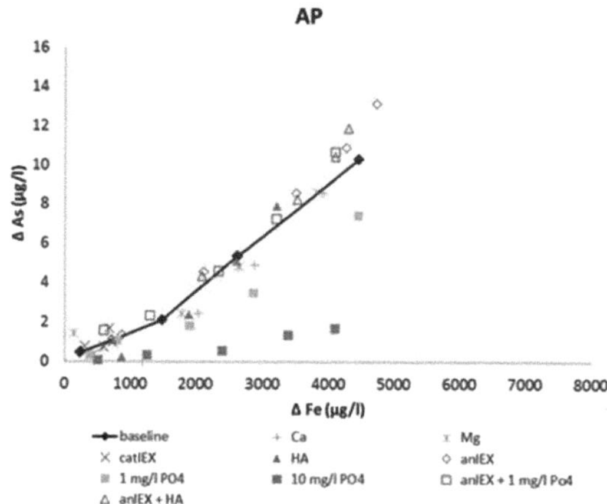

Figure 2. Relationship between iron and arsenic removal for AP groundwater.

more severe for groundwaters with higher P/Fe ratios. Equimolar addition of Ca^{2+} and Mg^{2+} increased the total floc volume to a similar quantity in KG and AP waters, but Ca^{2+} created significantly larger floc volumes in AMD water. Precipitation of hydroxyapatite could (partially) account for this, but is not strongly supported in literature. Possibly, more voluminous (and less dense) iron flocs are formed at high Si and Ca^{2+} concentrations.

Dosing phosphate slightly reduced iron flocculation in KG and AP water, and actually enhanced it in (Ca^{2+}-rich) AMD water. Dosing humic acid either did not influence iron flocculation (KG, AP water), or enhanced it (AMD water). Iron removal was enhanced for all groundwaters after anion exchange.

3.3 *(Ca+Mg)/P molar ratio*

Since Si is uncharged in our experiments, and humic acid carries a low molar charge density, we used the molar (Ca+Mg)/P ratio to represent the charge balance for iron flocs, and compared this to the total floc volume (Fig. 1). For KG and AP waters, a relatively linear relationship is found, which levels off at ratios higher than 500 where either all Fe has been flocculated, or where complete charge neutralization has occurred. In AMD water, larger iron flocs are created than expected based on this ratio.

3.4 *Arsenic removal*

Arsenic removal was largely depending on iron removal (Fig. 2). Dosing phosphate did show a reduction in As adsorption at an unrealistically high dose of 10 mg L^{-1}. Adsorption competition with Si was more evident.

4 CONCLUSIONS

Removal of As was closely related to removal of Fe, indicating that optimized Fe removal could optimize As removal simultaneously. The (Ca+Mg)/P ratio seems capable to predict iron floc growth and could be used as a tool to optimize iron removal in rapid sand filters. Our results seem to indicate that under environmental conditions, Si can hinder As due to adsorption competition to a larger extent than HPO_4^{2-}.

ACKNOWLEDGEMENTS

We thank Stephan van de Wetering and Tim van Dijk for their support during the experiments, and STW for their financial support.

REFERENCES

Guan, X., Dong, H., Ma, J. & Jiang, L. 2009. Removal of arsenic from water: effects of competing anions on As(III) removal in $KMnO_4$-Fe(II) process. *Wat. Res.* 43(15): 3892–3899.

Kaegi, R., Voegelin, A., Folini, D. & Hug, S.J. 2010. Effect of phosphate, silicate, and Ca on the morphology, structure and elemental composition of Fe(III)-precipitates formed in aerated Fe(II) and As(III) containing water. *Geochim. Cosmochim. Acta* 74(20): 5798–5816.

Environmental Arsenic in a Changing World –
Zhu, Guo, Bhattacharya, Ahmad, Bundschuh & Naidu (Eds)
ISBN 978-1-138-48609-6

Fecal contamination of drinking water in arsenic-affected area of rural Bihar: tube-well and storage container survey

M. Annaduzzaman, L.C. Rietveld & D. van Halem
Sanitary Engineering Section, Faculty of Civil Engineering and Geoscience, Delft University of Technology, Delft, The Netherlands

ABSTRACT: The study aimed to investigate the fecal contamination in shallow (depth <50 m) drinking water tube-wells (TWs) and household storage containers in the arsenic-affected area of rural Bihar. In this study, 365 TWs surveyed for As contamination, 164 TWs, and 68 storage containers for the presence/absence of fecal contamination. The results reveal that 32% (n = 111) of surveyed TWs (n = 365) exceeded WHO guideline value for As of $10\,\mu g\,L^{-1}$ and 11% (n = 40) of $50\,\mu g\,L^{-1}$. Within 164 TWs and 68 storage containers, 25% (n = 40) and 43% (n = 29) respectively showed presence of fecal contamination. Arsenic contamination did not show significant relation to TWs depth, but deeper (18–50 m) TWs confirm the presence of fecal contamination. Results also depict that fecal contamination both in source and storage container influenced by socio-economic, sanitation practices, the presence of latrine, livestock close to source or containers.

1 INTRODUCTION

In Bihar and neighboring areas of large Ganges-Brahmaputra-Meghna (GBM) Delta, drinking water interventions have much focused on arsenic (As) contamination in shallow groundwater tube-wells installed in millions of households. However, these shallow sandy aquifers are also vulnerable to fecal contamination (Embrey & Runkle, 2006). In addition, As mitigation technologies might lead exposure to harmful microorganisms ((Leber *et al.*, 2011), as well as improper water storage practices at the household level. Fecal (re)contamination of drinking water may result in transmittance of pathogenic microorganisms leading to an outbreak of diseases including cholera, dysentery, diarrhea, hepatitis, miscarriage and even cryptosporidiosis.

The aim of this research project is to develop a holistic approach towards arsenic mitigation, without increasing the exposure to fecal contamination. In this research, a first step is taken by investigating the state of exposure to fecal contamination via TWs and storage containers in the arsenic-affected areas in Bihar, India.

2 MATERIALS AND METHODS

Seven densely populated and geologically comparable rural villages of Barhara Block in Bhojpur district, Bihar were selected for this study. A total of 365 tube-wells water sampled in April 2017 for As and other trace elements analysis via ICP-MS (at TU Delft). Information of sanitation practices was collected through a systematic questionnaire amongst 164 households, using AKVO FLOW smartphone app (AKVO, 2017). For assessment of fecal contamination, the qualitative (presence/absence) Sumeet H_2S test for *E. coli* was used: 20 mL water sample was directly collected from 164 tube-wells and 68 household containers for analysis.

3 RESULTS

3.1 *Presence of arsenic in tube-well water*

Of the 365 surveyed tube-wells, the As level varied from <1 to $400\,\mu g\,L^{-1}$. According to WHO guideline value of $10\,\mu g\,L^{-1}$ As, 69.6% and 30.4% TWs was safe and unsafe respectively – 19.4% were in the As range of $10–50\,\mu g\,L^{-1}$. In all villages As contamination was observed to exceed WHO value, with one village with unusually high As concentrations (G). In this village 47 wells exceeded the WHO guideline of $10\,\mu g\,L^{-1}$ (Fig. 1).

3.2 *Fecal contamination in tube-well water*

Among the surveyed 164 tube-wells water for presence/absence of fecal contamination, 75% TWs were safe and 25% TWs were contaminated. Within

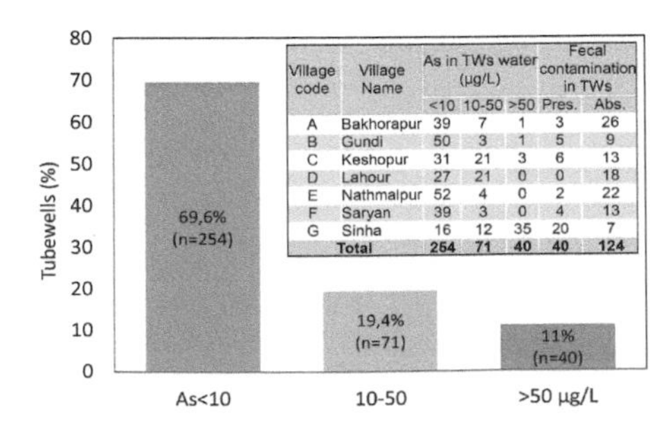

Village code	Village Name	As in TWs water (µg/L)			Fecal contamination in TWs	
		<10	10-50	>50	Pres.	Abs.
A	Bakhorapur	39	7	1	3	26
B	Gundi	50	3	1	5	9
C	Keshopur	31	21	3	6	13
D	Lahour	27	21	0	0	18
E	Nathmalpur	52	4	0	2	22
F	Saryan	39	3	0	4	13
G	Sinha	16	12	35	20	7
	Total	**254**	**71**	**40**	**40**	**124**

Figure 1. Arsenic distribution in surveyed tube-wells.

Table 1. Absence or presence of *E. coli* per sampled TW (n = 164) related to TW characteristics.

Categories	Fecal contamination	
	Absence	Presence
Tube-well depth		
<18 m	18	12
18–50 m	106	28
Toilet distance from TWs		
<10 m or NT	105	38
>10 m	19	2
TWs distance to livestock		
<10 m	50	17
>10 m or no LS	74	23
TWs Platform condition		
Good	59	18
Poor/none	65	22

Table 2. Absence or presence of *E. coli* per water storage practice in 68 sampled households.

Categories	Fecal contamination	
	Absence	Presence
Container height from ground		
<40 cm	8	21
>40 cm	31	8
Light condition to container		
Shadow and cool	15	20
Sunny	28	9
Container clean enough		
Yes	19	23
No	20	6
Container cover		
Yes	30	8
No	10	21
Presence of flies near container		
Yes	8	24
No	31	5
Toilet distance to container		
<10 m or NT	15	25
>10 m	24	4
Container distance to livestock		
<10 m	12	21
>10 m or no LS	27	8

the fecal contaminated TWs, it was found that above 70% TWs were from deep aquifer. In case of presence of toilet (or no toilet at all) within 10 m of the TWs, in 38 cases found fecal contamination.

The presence of livestock close to TWs (<10 m) or platform condition (e.g. damaged) did not show a difference in contamination (Table 1).

3.3 *Fecal contamination in storage container*

The presence or absence of fecal contamination in the storage container and the possible pathway are given in Table 2. It was found that, within 68 sampled storage container water, 57% and 43% cases were safe and contaminated respectively. Of the households that abstracted microbially safe water from their tube-well, 10% (n = 16) had contaminated water in the storage container outflow. Vice versa, 5% (n = 7) of the households abstracted contaminated water from their well, but this was not observed in their container.

Table 2 provides an overview of the water storage practices and their corresponding absence or presence of *E. coli*. From these results, it may be concluded that presence of a cover and height from ground (>40 cm) prevent re-contamination, but also the placement of the container in sunlight. Heating of the water, as well as UV-irradiation, are known to cause a die-off of pathogens (Sommer *et al.*, 1997). Indicators of hygiene practices, like distance to the toilet, livestock, and presence of flies, clearly show more presence of *E. coli* under poor hygienic conditions.

4 CONCLUSION

Besides As contamination in TWs water, fecal contamination in TWs and storage water is a considerable drinking water quality problem in the study area. Arsenic contamination did not show a clear relation to TWs depth, but both shallow and deeper TWs (18–50 m) were found to be fecally contaminated in 25% of the cases. After household water storage – at the point of consumption – an additional estimated 10% was contaminated; particularly due to poor container placement and sanitation/hygiene practices. Altogether it may be concluded that it is essential to include microbial contamination risks when developing arsenic mitigation strategies.

ACKNOWLEDGEMENTS

This study supported by DELTAP project funded by NWO-WOTRO research grant. We thank students from AN College, Patna, Prof. Ashok Ghosh and his team for their help during intensive fieldwork.

REFERENCES

AKVO 2017. https://akvo.org/ (accessed on 30th November 2017).

Embrey, S.S. & Runkle, D.L. 2006. Microbial quality of the nation's ground-water resources, 1993–2004. Scientific Investigations Report 2006-5290, U.S. Geological Survey, Reston, VA, USA.

Leber, J., Rahman, M.M., Ahmed, K.M., Mailloux, B. & van Geen, A. 2011. Contrasting influence of geology on E. coli and arsenic levels in sedimentary aquifers. *Groundwater* 49(1): 111–123.

Sommer B., Marino, A., Solarte, Y., Salas, M.L., Dierolf, C., Valiente, C., Mora, D., Rechsteiner, R., Setter, P., Wirojanagud, W., Ajarmeh, H., Al-Hasan, A. & Wegelin, M. 1997. SODIS: an emerging water treatment process. *J. Water SRT-Aqua* 46(3): 127–137.

Environmental Arsenic in a Changing World –
Zhu, Guo, Bhattacharya, Ahmad, Bundschuh & Naidu (Eds)
ISBN 978-1-138-48609-6

Biomineralization of charophytes and their application in arsenic removal from aquatic environment

S. Amirnia, T. Asaeda & C. Takeuchi
Department of Environmental Science, Saitama University, Saitama, Japan

ABSTRACT: As(III) is present in neutral form ($H_3AsO_3^0$) at the pH ranges of natural waters; therefore, sorption and biosorption are not much effective removal methods for dissolved arsenic. In this study, we investigated the biomineralization of charophyte *C. braunii* in Ca and Mn-containing media and its potential for remediation of As(III) in aquatic environments. Relative proportions of the arsenic adsorbed in the plant biomass and those incorporated into calcium and manganese deposits were discerned using a modified sequential extraction. Under favorable conditions, over 50% of the accumulated arsenic by the plant was Mn-bound.

1 INTRODUCTION

Arsenic is a notorious metalloid for its adverse effects on human and environmental health. From As(III) and As(V), arsenite (As(III)) is more toxic and more difficult to be removed from water due to its neutral charge in natural water pHs (Vaclavikova *et al.*, 2008).

Charophytes, common submerged plants in freshwaters, are known for their ability to form calcite encrustation and contribute to nutrient cycling, such as phosphorous, in water columns (Asaeda *et al.*, 2014; Kufel *et al.*, 2016). They are also useful for remediation of various pollutants, such as chromium and cadmium from contaminated waters (Gomes & Asaeda, 2013).

Calcite encrustation in *Chara*, genera of charophytes, is known to be inevitable due to bicarbonate assimilation and as a by-product of photosynthetic activity of the plants (McConnaughey, 1991). In this work, we reported formation of a secondary type of biogenic deposits on the cell walls of calcifying charophyte, *Chara braunii* (common to rice fields) in Ca/Mn-containing media, and our objective was to assess the capability of the plant to accumulate arsenic in the context of charophytes biomineralization.

2 METHODS/EXPERIMENTAL

We examined As(III) accumulation in calcifying charophyte *Chara braunii* in a laboratory experiment for 4 weeks. Experimental plants were placed in a temperature-controlled incubator at $20 \pm 2°C$ with 12 h: 12 h light: dark photoperiod.

The plant growth in the presence of 0.5 mg As L^{-1}, and varying concentrations of Mn(II) and Ca(II) was tested, and shoot elongation, as a parameter for the growth, was calculated relative to the initial length of the plants. Chlorophyll Florescence parameters (F_v/F_m) were analyzed by an auto imager (Olympus Tokyo) after a 20 min of dark adaptation. Element concentrations in water samples were measured by an ICP-OES (Optima 5300 DV).

The extent of arsenic storage in the plant biomass and of that trapped on biomineral deposits on the plant by a modified chemical fractionation method. Scanning electron microscopy (Hitachi S-3400N) was employed to study the plant surface morphology and encrustations.

3 RESULTS AND DISCUSSION

Temporal variation of shoot length of *C. braunii* showed an average decrease of ~48% in one week and ~16% in four weeks indicating adaptation of the plants to the new environment. Plants exposed to arsenic exhibited some reduction of chloroplast resulting in chlorosis. However, since the arsenic level was the same in all the studied experimental units, no significant differences was found between the treatments for F_v/F_m values ($P > 0.05$).

SEM analyses helped us to identify the formation of volcano shaped deposits with diameters of 5–10 μm on the plant internodal cells and branchlets (Fig. 1). EDX (energy dispersive X-ray) elemental mapping revealed that the main metal constituent of these unique deposits is manganese (MnO_x), which was formed in the presence of manganese in the nutrient culture.

Figure 2 shows the relative proportions of the arsenic adsorbed in the plant biomass and those incorporated into calcium and manganese deposits determined by a modified sequential extraction method. Following the exchangeable fraction, the manganese deposits retained a major fraction of the arsenic, while the Ca-bound together with the residual/organic-bound arsenic were the least dominant fractions.

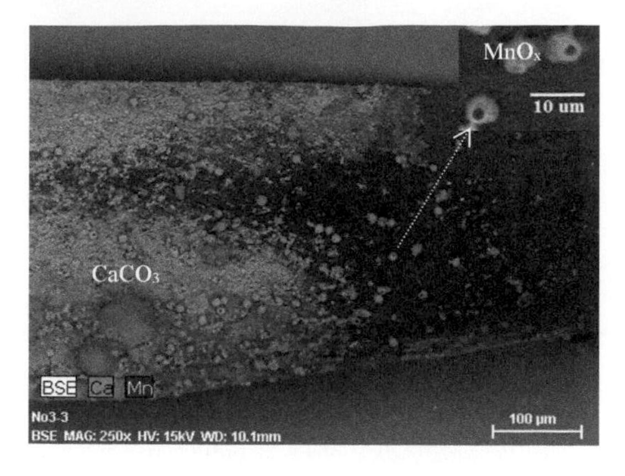

Figure 1. SEM-EDX micrograph of *C. barunii* internodal cell wall surface. $CaCO_3$ and volcano-shaped deposits of MnO_x were formed on the plant.

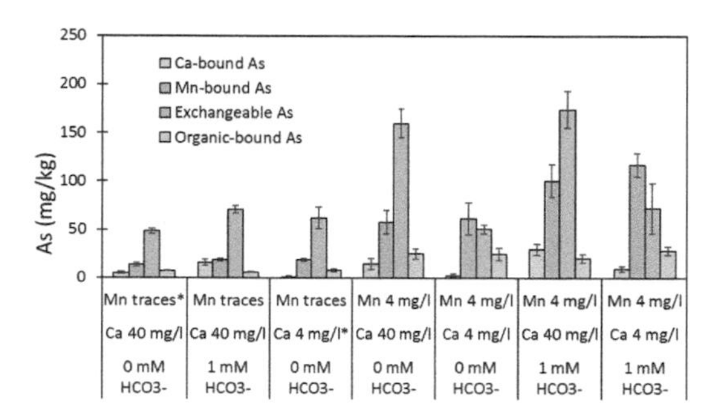

Figure 2. As concentration (mg kg^{-1} DW) and relative distribution within each chemical fraction of the plant. As concentration in all treatment were 0.5 mg L^{-1} ($* = 4$ mg L^{-1} Ca and also trace Mn levels in water were due to mixing with substrate and were not added).

The amount of arsenic associated with Ca and organic were found little (1.0–29.5 mg kg^{-1} for Ca-bound, and 7.3–28.3 mg kg^{-1} for residual/organic-bound, respectively). Arsenic percentages associated to organic fraction were the lowest among all the studied fractions.

As depicted in Figure 2, over 50% of the total As in *C. braunii* was present in Mn-oxides phase in the assay with low calcium and high level of Mn in water in the presence of bicarbonate. The manganese oxide bound arsenic had an increasing trend with increase in Mn concentration with bicarbonate ions in water and reached its highest value (~120 mg kg^{-1}) in the treatment with 4 mg Mn L^{-1}, 40 mg Ca L^{-1} with addition of bicarbonate compared to the other conditions.

Mechanism of arsenic bound to calcite and MnO_x deposits, and chemical nature of As-Mn and As-Ca complexes are still a matter of debate. As(III) might have reacted with solid-state Mn-oxide as electron acceptor, and might have oxidized As(III) in solution to As(V), with or without disproportionation. Oxidation of As(III) by manganese oxide followed by co-precipitation/adsorption of As(V) with manganese oxide on the cell walls of *C. braunii* was likely to be the mechanism for arsenic removal.

4 CONCLUSIONS

Appearance of two types of biominerals on the surface of charophyte *C. braunii*, which were mediated by simultaneous presence of Ca and Mn in the culture medium, promoted the arsenic accumulation capacity of the plant. Arsenic partitioned differently amongst the two deposits, and MnO_x retained a larger amount of the arsenic removed by the plant as compared to calcite deposits in all water conditions in this work. The portion of arsenic co-precipitated and trapped during the calcite encrustation and MnO_x formation processes may not leach back easily to water upon plants' senescence and decay.

ACKNOWLEDGEMENTS

This work was supported by JSPS KAKENHI Grant Number 16F16750.

REFERENCES

Gomes, P.I.A. & Asaeda, T. 2013. Phytoremediation of heavy metals by calcifying macro-algae (*Nitella pseudoflabellata*): implications of redox insensitive end products. *Chemosphere* 92(10): 1328–1334.

McConnaughey, T. 1991. Calcification in *Chara corallina*: CO_2 hydroxylation generates protons for bicarbonate assimilation. *Limnol. Oceanogr.* 36(4): 619–628.

Kufel, L., Strzałek, M. & Biardzka, E. 2016. Site- and species-specific contribution of charophytes to calcium and phosphorus cycling in lakes. *Hydrobiologia* 767(1): 185–195.

Vaclavikova, M., Gallios, G.P., Hredzak, S. & Jakabsky, S. 2008. Removal of arsenic from water streams: an overview of available techniques. *Clean Technol. Environ. Policy.* 10(1): 89–95.

Environmental Arsenic in a Changing World –
Zhu, Guo, Bhattacharya, Ahmad, Bundschuh & Naidu (Eds)
ISBN 978-1-138-48609-6

Efficient generation of aqueous Fe in electrocoagulation systems for low-cost arsenic removal

S. Müller, T. Behrends & C.M. van Genuchten
Department of Earth Sciences-Geochemistry, Faculty of Geosciences, Utrecht University, Utrecht, The Netherlands

ABSTRACT: Iron-electrocoagulation (Fe-EC), the electrochemical generation of Fe adsorbents in situ, is an effective treatment solution for arsenic (As) contaminated water in decentralized areas. However, extended field trials also revealed problems regarding the efficiency over long-term operation. To investigate how Fe generation is affected by repeated operation of Fe-EC systems, we performed laboratory experiments over a range of electrochemical and solution chemical conditions for 15–35 runs in batch mode. Our results show that Fe generation declines continuously during repeated operation under typical field conditions, resulting in a lower Fe dose than expected based on Faraday's law. In addition, we find that efficient Fe generation can be maintained in electrolytes free of oxyanions or by applying charge dosage rates $\geq 15\ \mathrm{C\,L^{-1}\,min^{-1}}$. Based on these results, we discuss potential strategies to maintain the efficiency of Fe-EC field systems under realistic conditions.

1 INTRODUCTION

Arsenic (As) contamination of drinking water is an ongoing public health disaster, especially for decentralized communities without existing water infrastructure. Iron-electrocoagulation (Fe-EC) is a promising alternative treatment method to achieve As $<10\ \mu\mathrm{g\,L^{-1}}$, which has been demonstrated in West Bengal, India.[1] In Fe-EC, a steel anode is electrochemically oxidized to generate Fe^{2+} in solution, which is further oxidized by dissolved oxygen to form iron oxide precipitates. Arsenic is adsorbed on the surface of the in-situ formed iron particles, which can be removed by gravitational settling or filtration.

Although Fe-EC performs well under ideal laboratory conditions, the long-term operation of Fe-EC under field conditions revealed a number of issues: The As removal efficiency (the mass of As removed per charge passed for final [As] $<10\ \mu\mathrm{g\,L^{-1}}$) is up to five times lower than the efficiency of systems tested in the laboratory. In addition, extensive surface layers formed on the electrodes after repeated operation in the field (Amrose *et al.*, 2013, van Genuchten *et al.*, 2016). In this study, we track Fe production and surface layer formation in laboratory Fe-EC systems over repeated operation to identify the key parameters that impact extended Fe-EC efficiency.

2 METHODS

2.1 *Experiments*

Experiments were conducted in downscaled EC cells that typically consisted of two steel electrodes spaced 1 cm apart with a submerged anode surface area of $18\ \mathrm{cm^2}$ in 200 mL of electrolyte, stirred continuously and open to the atmosphere. In all experiments, a constant current was applied to achieve a coulombic dose of $450\ \mathrm{C\,L^{-1}}$ and the electrodes were stored open to the atmosphere between experiments. The charge dosage rate (Amrose *et al.*, 2014) (*CDR*, between 4 and $54\ \mathrm{C\,L^{-1}\,min^{-1}}$), Fe anode purity (lab electrodes = 99% Fe, field electrodes = 92% Fe), and electrolyte composition (synthetic groundwater (SGW) with several modifications) were investigated over 15–35 runs.

2.2 *Analysis*

The anodic interface potential and cell voltage was measured during each experiment and the electrolyte conductivity, pH, and oxygen concentration were measured before and after each experiment. The total concentration of Fe in the bulk electrolyte was determined using a modified SMWW protocol (phenanthroline method) (APHA, 2015). Using the electrochemical operating parameters, we calculated the Faradaic efficiency *FE* for each experiment, which is the measured Fe concentration normalized by the Fe concentration based on Faraday's law.

$$[\mathrm{Fe}]_{\mathrm{faraday}} = I * t * (n * F * V)^{-1} \qquad (1)$$

n = 2 for Fe(II); Faraday's constant $F = 96485\ \mathrm{C\,mol^{-1}}$

Electrode surface layers were mechanically removed only after the last run and weighed.

3 RESULTS AND DISCUSSION

3.1 *Faradaic efficiency*

Figure 1 shows the *FE* over 20 runs for duplicate experiments with high and low purity anodes

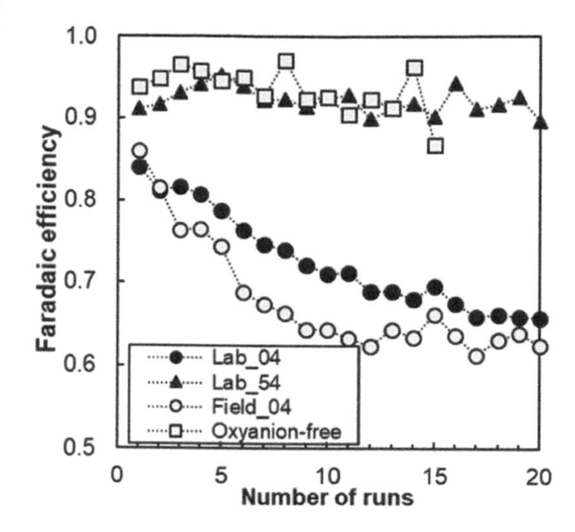

Figure 1. Faradaic efficiency (*FE*) as a function of the number of runs of the experiment with lab electrodes at *CDR* of $4°C\,L^{-1}\,min^{-1}$ in SGW (Lab_04). Compared to Lab_04, experiments were performed in a modified electrolyte (Oxyanion-free), at higher charge dosage rate (Lab_54), or with a low purity anode instead of high purity anodes (Field_04).

(Lab_04 & Field_04) at a *CDR* of $4\,C\,L^{-1}\,min^{-1}$ in SGW. The *FE* of Lab_04 and Field_04 started below 1 and decreased over time, i.e. less Fe was present in the bulk solution than expected based on Faraday's law. Starting at a *FE* of 0.85 (85% of the charge was converted to Fe in the bulk solution), the laboratory anodes showed a gradual reduction in *FE* to 0.66 over 20 runs, which includes electrolysis and storage open to the atmosphere overnight between runs. The *FE* of the field anode over time was similar to the laboratory anodes, decreasing to 0.62 after 20 runs, which is equal to a final Fe dose of approximately $80\,mg\,L^{-1}$. The similarity in *FE* between the field and laboratory electrodes suggests that a high purity Fe electrode does not necessarily translate to sustained high *FE* at the investigated *CDR* of $4\,C\,L^{-1}\,min^{-1}$ and that oxidation reactions of other steel components such as Mn are not responsible for the low *FE*. This result suggests that locally available steel of unknown purity can be implemented in large scale EC systems without a loss of *FE*.

Figure 1 also compares Lab_04 to two experiments where another parameter than anode purity was varied: a higher *CDR* (Lab_54), and a different electrolyte without bicarbonate, silicate, and phosphate (oxyanion-free). In contrast to the *FE* of Lab_04, the *FE* of Lab_54 with a *CDR* of $54\,C\,L^{-1}\,min^{-1}$ remained above 0.85 throughout the entire 20 runs. Remarkably, no differences in long-term *FE* were observed between the different experiments at elevated *CDR* ($15, 32, 54\,C\,L^{-1}\,min^{-1}$) and between the field and lab anode at high *CDR* (data not shown). Furthermore, the same trend of sustained $FE > 0.8$ was also observed at a *CDR* of $4\,C\,L^{-1}\,min^{-1}$ when oxyanions (P, Si, HCO_3) were not present in the electrolyte. In all experiments with a high *FE*, the mass of iron deposits on the anode was consistently lower compared to Lab_04 and Field_04, which suggests that the formation of anodic surface layers is correlated to the loss in *FE* over time.

For the operation of EC systems in the field, the most significant result of this work is the strong impact of the *CDR* on the *FE* because the *CDR* can be adjusted easily in Fe-EC using existing infrastructure. Because the performance of Fe-EC hinges on maintaining a high *FE*, operating EC systems at $CDR > 15\,C\,L^{-1}\,min^{-1}$ can ensure efficient and reliable treatment. Furthermore, increasing *CDR* also decreases total treatment time, which is beneficial in areas with intermittent electricity supply. Increasing the *CDR* beyond $15\,C\,L^{-1}\,min^{-1}$ had no observable effect on the *FE* and is only recommended if shorter treatment times are required.

4 CONCLUSIONS

The aim of this study was to determine if long-term operation leads to a decrease in *FE*, a potential reason for the lower As removal efficiency observed for field EC-systems relative to those in the lab. Our results showed that the *FE* continuously decreased over time under conditions similar to a system used for As removal in West Bengal, India. This drop in *FE* should be addressed in the extended operation of Fe-EC systems in the field, possibly by increasing the electrolysis time incrementally over several months.

Our results also showed that the decrease in *FE* was not caused by a low Fe anode purity. Instead, the electrolyte composition and the *CDR* were identified as key parameters governing the *FE*. Applying a $CDR \geq 15\,C\,L^{-1}\,min^{-1}$ or excluding oxyanions from the electrolyte improved the long-term *FE*. These modifications also correlated with surface layers of lower mass on the anode surface. Hence, a high and stable *FE* in field systems can be achieved by increasing the applied current (*CDR*).

ACKNOWLEDGEMENTS

CMvG acknowledges funding support by a NWO Veni grant (Project No. 14400).

REFERENCES

Amrose, S., Gadgil, A., Srinivasan, V., Kowolik, K., Muller, M., Huang, J. & Kostecki, R. 2013. Arsenic removal from groundwater using iron electrocoagulation: effect of charge dosage rate. *J. Environ. Sci. Health A* 48(9): 1019–1030.

Amrose, S.E., Bandaru, S.R., Delaire, C., van Genuchten, C.M., Dutta, A., DebSarkar, A., Orr, C., Roy, J., Das, A. & Gadgil, A.J. 2014. Electro-chemical arsenic remediation: field trials in West Bengal. *Sci. Total Environ.* 488: 539–546.

APHA. 2015. *Standard Methods for the Examination of Water & Wastewater*. American Public Health Association (APHA), Washington, DC, USA.

van Genuchten, C.M., Bandaru, S.R., Surorova, E., Amrose, S.E., Gadgil, A.J. & Pena, J. 2016. Formation of macroscopic surface layers on Fe(0) electrocoagulation electrodes during an extended field trial of arsenic treatment. *Chemosphere* 153: 270–279.

Environmental Arsenic in a Changing World –
Zhu, Guo, Bhattacharya, Ahmad, Bundschuh & Naidu (Eds)
ISBN 978-1-138-48609-6

Arsenic removal without thio-As formation in a sulfidogenic system driven by sulfur reducing bacteria under acidic condition

Y. Hong, J. Guo, J. Wang & F. Jiang
School of Chemistry and Environment, South China Normal University, Guangzhou, P.R. China

ABSTRACT: Sulfidogenic treatment using sulfate-reducing bacteria has been used to remediate Acid Mine Drainage (AMD). With sulfide, As(III) can be precipitated as As_2S_3. However, thio-arsenic formation with the pH increased by sulfate reduction, and results in the failure of As(III) removal. In this study, we proposed a novel sulfidogenic system driven by sulfur reducing bacteria (S^0RB), which worked under acidic condition to reduce elemental sulfur into sulfide without pH elevation. Then the acidic sulfide-rich effluent mixed with the arsenic-containing AMD to remove arsenic. In the long-term lab-scale test, S^0RB activities maintained for over 100 days under acidic condition (pH 4.3). Over 99% of the influent arsenic (10 mg L^{-1}) were removed by the acidic sulfide-rich solution. The results of batch tests show that, higher As removal rate can be obtained under lower pH conditions, under the same As-to-S molar ratio. The Eh-pH plot demonstrates that, sulfide production under acidic condition avoided the formation of thio-arsenic compounds (thio-As), ensuring the high removal efficiency of As(III) from AMD.

1 INTRODUCTION

Removal of trivalent arsenic is a major concern in the treatment of arsenic-contaminated acid mine drainage (AMD) due to its high toxicity. In recent years, as an attractive AMD treatment technology, sulfate-reducing bacteria (SRB) was used to remove As(III) due to the immobilization of arsenic through the reaction of H_2S with As(III) (Eq.1). However, sulfate reduction by SRB leads to a significant increase in pH (Eq.3), promoting the formation of dissolved thio-As (Eq.2) and resulting in inefficiency of arsenic removal (Johnston *et al.*, 2016). As an alternative, this study proposed a novel sulfidogenic process driven by sulfur-reducing bacteria (S^0RB) at acidic condition, to generated sulfide-rich solution with low pH to precipitate As(III) and avoid the formation of thio-As. In contrast to sulfate reduction, elemental sulfur reduction does not generate OH^- (Eq. 4), which ensures the acidic condition remains stable and control the formation of thio-As. Therefore, it is possible to efficiently remove As(III) from AMD by sulfidogenic process with S0RB under acidic conditions. In our previous studies, high-rate sulfur reduction process under neutral and alkaline condition has been achieved (Sun *et al.*, 2017). However, sulfur reduction under acidic condition has not been tested yet, and the performance of As removal shall be investigated.

$$H_3AsO_3 + 3/2\ H_2S = 1/2\ As_2S_3 + 3H_2O \quad (1)$$

$$As_2S_3 + HS^- + OH^- = 2AsS\ (OH)\ (HS)^- + H_2O \quad (2)$$

$$SO_4^{2-} + 2H_2O + 2C_{org} = H_2S + 2CO_2 + 2OH^- \quad (3)$$

$$2S^0 + C_{org} + 2\ H_2O = CO_2 + 2\ H_2S \quad (4)$$

2 METHODS AND MATERIALS

In this study, we established a lab-scale sulfidogenic arsenic removal system including a sulfur reduction reactor (SRR) and an arsenic-sulfide precipitation tank. The effluent of the sulfur reduction reactor is mixed with arsenic-contaminated acid mine drainage at a volumetric ratio of 1:4 in the arsenic-sulfide precipitation tank. The sulfur reduction reactor was packed with sulfur lumps and fed with 600 mg C/d sodium acetate. The arsenic removal system operated continuously for more than 140 days. The influent pH of the sulfur reduction reactor gradually decreased from 7.2 to 4.0 within 40 days, then remained stable for 100 days. In addition, batch tests were also carried out to study the effect of pH and As/S molar ratio on arsenic removal.

3 RESULTS AND DISCUSSION

Long-term performances of the elemental sulfur reduction and arsenic removal system are shown in Figure 1. The effluent pH gradually decreased from 7.5 to 4.5, and S^0RB activities kept stable to produce 50–150 mg L^{-1} H_2S for arsenic removal. Under neutral condition, the efficiency of the As(III) removal was 60–80%, and then increased to more than 99% when the effluent pH decreased to 4.5. The results of

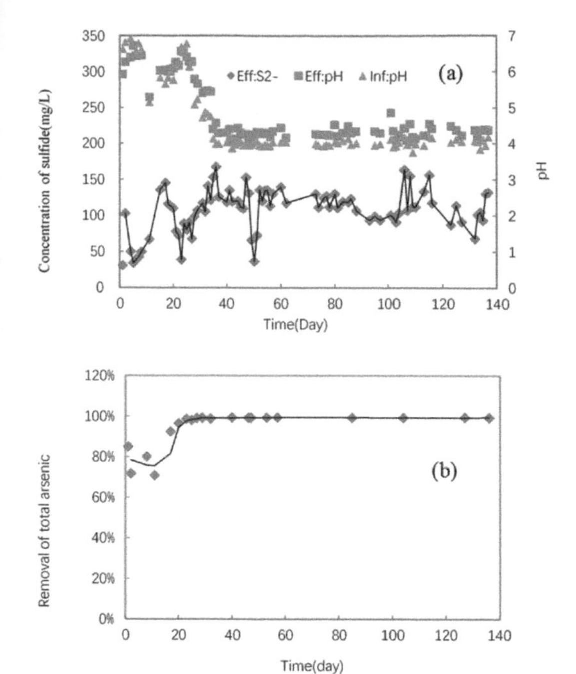

Figure 1. The performances of elemental sulfur reduction and arsenic removal system (a. the inlet and outlet pH and concentration of the effluent H2S; b. the removal efficiency of arsenic in the arsenic and sulfur reaction tank).

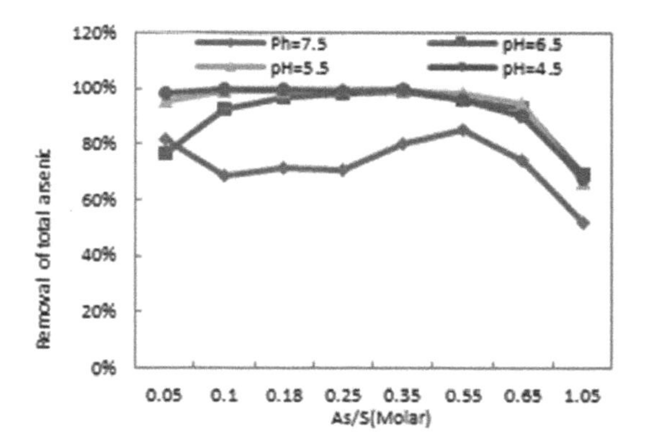

Figure 2. The removal efficiency of arsenic S with As/S, and pH changes.

batch tests showed that the arsenic removal efficiency decreased with the increase in pH from 4.5 to 7.5 (Fig. 2). When the pH was 4.5, the efficiency reached more than 99% when As/S molar ratios were less than 0.35. The results of these batch tests were consistent with the performance of the sulfidogenic arsenic removal system. It explained the high efficiency of

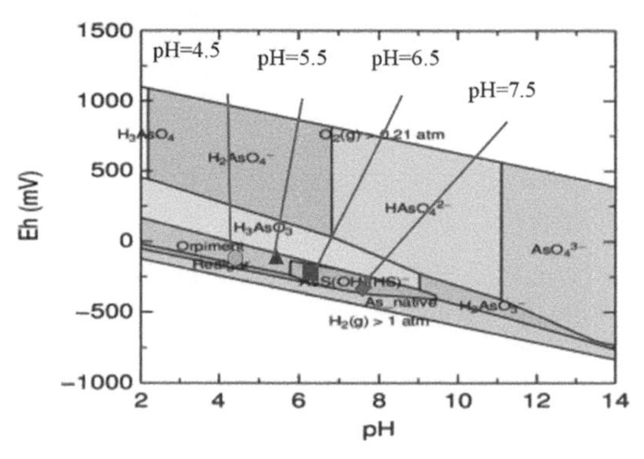

Figure 3. Distribution of arsenic with As/S molar ratio of 0.18.

arsenic removal in the sulfidogenic arsenic removal systems even though the sulfide concentrations of SRR effluent varied from 50 to 150 $mg\,L^{-1}$ (Fig. 1b). To discuss the mechanism of this phenomenon, the Eh-pH relationship was plotted in Figure 3. It shows that, when the As/S molar ratio is 0.18, thio-As only forms when the pH is between 6.0 to 8.5 (Fig. 3). When the pH dropped to 4.5, no thio-As will be formed. It suggested that, arsenic could be completely transformed into As_2S_3 when sulfide presented under acidic condition.

4 CONCLUSION

Maintaining the acidic condition of AMD during the sulfidogenic process, is the key factor to ensure the efficiently removal of As from of arsenic-contaminated wastewater without thio-As generation. In this study a sulfidogenic process driven by S^0RB was proposed and tested to generate sulfide under acidic condition without pH elevation, and its sulfide-rich effluent efficiently precipitated over 99% of the As(III) from wastewater. This sulfidogenic arsenic removal process is a promising technology in arsenic-contaminated AMD remediation

REFERENCES

Johnston, S.G., Burton, E.D. & Moon, E.M. 2016. Arsenic mobilization is enhanced by thermal transformation of schwertmannite. *Environ. Sci. Technol.* 50(15): 8010–8019.

Sun, R., Zhang, L., Zhang Z, Chen, G.H. & Jiang, F. 2017. Realizing high-rate sulfur reduction under sulfate-rich conditions in a biological sulfide production system to treat metal-laden wastewater deficient in organic matter. *Water Res.* 131: 239–245.

Environmental Arsenic in a Changing World –
Zhu, Guo, Bhattacharya, Ahmad, Bundschuh & Naidu (Eds)
ISBN 978-1-138-48609-6

Arsenic retention and distribution in a treatment wetland prototype

M.T. Alarcón Herrera[1], M.A. Olmos-Márquez[2], J. Ochoa[3] & I.R. Martin-Domínguez[1]
[1]*Centro de Investigación en Materiales Avanzados SC (CIMAV), Chihuahua, Chih., Mexico*
[2]*Universidad Autónoma de Chihuahua (UACH), Chihuahua, Chih., Mexico*
[3]*Instituto Nacional de Investigaciones Forestales, Agrícolas y Pecuarias, Campo Experimental La Laguna, Matamoros, Coah., Mexico*

ABSTRACT: The aim of this study was to investigate the plant's arsenic mass retention and the distribution of As along the wetland flow gradient and the soil in the wetland mesocosmos. Experiments were carried out in laboratory-scale wetland prototypes, two planted with *E. macrostachya* and one without plants. Samples of water were taken at the inlet and outlet of the wetlands during the 33-week test period. At the end of the experiment, plants and soil (silty-sand) from each prototype were divided in three equal segments (entrance, middle and exit) and analyzed for their arsenic content. Results revealed that the planted wetlands have a higher As-mass retention capacity (87–90% of the total As inflow) than prototypes without plants (27%). Results As mass balance in the planted wetlands revealed that 78% of the total inflowing As was retained in the soil bed. Nearly 2% was absorbed in the plant roots, 11% was flushed as outflow, and the fate of the remaining 9% is unknown. In the prototype without plants, the soil retained 16% of As mass, 72% of the arsenic was accounted for in the outflow, and 12% was considered unknown.

1 INTRODUCTION

Constructed wetlands (CWs) are an innovative technology in arsenic water treatment, which has proven to be both effective and affordable. They utilize the interaction of plants and microorganisms in the removal of different pollutants, among them metals and metalloids (Vymazal, 2005). To understand As-retention capacities, it is necessary to analyze the specific behavior of arsenic during uptake by plants, as well as the influence of media (soil) retention (Frohne et al., 2011). The mass balance of As in treatment wetlands, both in the presence and absence of wetland plants, could yield insights on the capacity of CWs to retain arsenic from water. Therefore, the objectives of this study were to investigate the distribution of total arsenic in main wetland compartments (plants and soil bed) and in three segments (entrance, middle and exit) of each unit along the flow path.

2 METHODS/EXPERIMENTAL

2.1 *Experimental design: laboratory-scale prototypes*

The three wetland prototypes used in this study were built with acrylic (length: 150 cm, width: 50 cm, height: 50 cm). The prototypes were uniformly filled with 200 kg of silt sand (density of 0.90 g cm^{-3}, porosity of 31%, and hydraulic conductivity of 6.89×10^{-4} cm s^{-1}). The support for the plants was a bed 30 cm in depth, composed of silty sand with a particle size between 0.007 and 0.47 cm. Rough gravel (2.5–4.0 cm) was used to create uniform water distribution at the entrance and exit of each unit. The water level was adjusted to 5 cm below the surface of the sand bed. Two prototype wetlands (W1and W2) were planted with *E. macrostachya*, and one prototype remained unplanted (W3) as a control.

2.2 *Arsenic mass balance determination*

After 33 weeks of operation, the experiment was ended. An arsenic mass balance was performed in each prototype unit and within the three segments by considering the total As-mass input, the total As-mass output, and the total As retained in the soil and plant biomass. The remaining (loss or gain) of arsenic from the mass balance was considered to be unknown. The total As-mass input and output in each unit was calculated from the cumulative total As-mass inflow and outflow during the whole operation period of 33 weeks.

3 RESULTS AND DISCUSSION

3.1 *Arsenic removal efficiencies of constructed wetland systems*

Figure 1 shows the total amount of arsenic mass inflow and the total arsenic retention in all three corresponding units during the whole study period of 33 weeks. A total inflow As-mass of 1,441 mg was fed to each prototype, and a total of 182, 140, and 1,034 mg As mass was flushed through the outlet of units W1, W2 and W3, respectively. Therefore, a cumulative total mass of 1,259, 1,301 and 407 mg As mass was retained in each unit. This means a retention of nearly 87%, 90% and 28% in units W1, W2 and W3, respectively (Fig. 1).

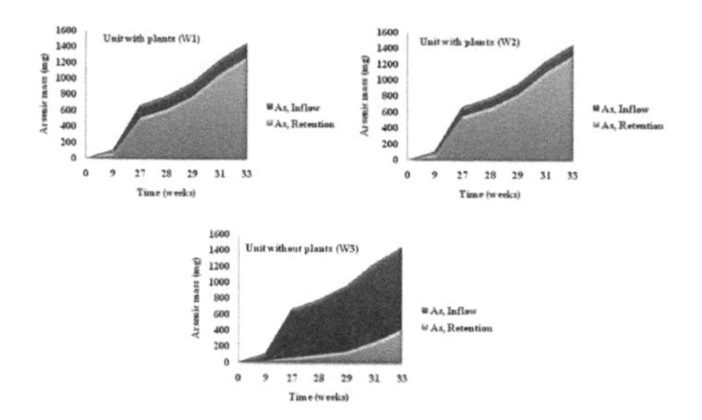

Figure 1. Cumulative As-mass inflow and retention in the three prototype units.

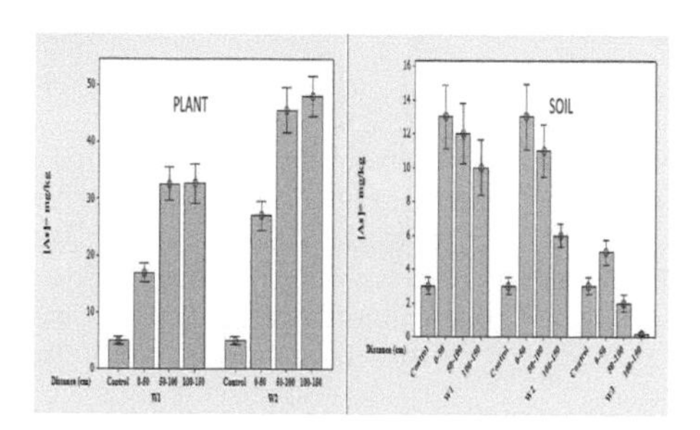

Figure 2. Concentration of total arsenic in plant roots and soil measured in each segment of the prototype (W1, W2 and W3).

3.2 *Arsenic retention by E. macrostachya*

We quantified the arsenic retained by plants in each of the three segments in each prototype. The prototypes with *E. macrostachya* showed greatest arsenic retention in the exit section, reaching average root arsenic concentrations of 33 ± 1.37 and $47 \pm 1.42\,\text{mg kg}^{-1}$. The entrance sections retained 17 ± 0.70 and $27 \pm 1.01\,\text{mg kg}^{-1}$ in W1 and W2, respectively (Fig. 2). The mean total arsenic concentration in the roots of the planted units (W1 and W2) was 27 ± 1.9 and $40 \pm 2.8\,\text{mg kg}^{-1}$ (dry weight), respectively. This concentration was also remarkably low in the shoots of the respective prototypes (0.30 ± 0.02 and $0.73 \pm 0.05\,\text{mg kg}^{-1}$).

3.3 *Arsenic retention in soil*

At the end of the experiment, soil was also collected from six different segments of each prototype. Results are presented in Figure 2, which shows how As concentrations decreased between the entrance and exit segments from 13.05 ± 0.75 to $5.86 \pm 0.26\,\text{mg kg}^{-1}$ (dry weight). In the unplanted prototype, W3, the arsenic concentration values presented the same trend but also significantly lower concentrations in all three segments (4.91 ± 0.30, 1.86 ± 0.20 and $0.14 \pm 0.05\,\text{mg kg}^{-1}$ (dw) for the entrance, middle and exit segments, respectively) (Fig. 3). Arsenic retention in the soil of the prototypes showed an inverse behavior to that of the plants. A better retention of Arsenic was obtained at the entrance section of each unit.

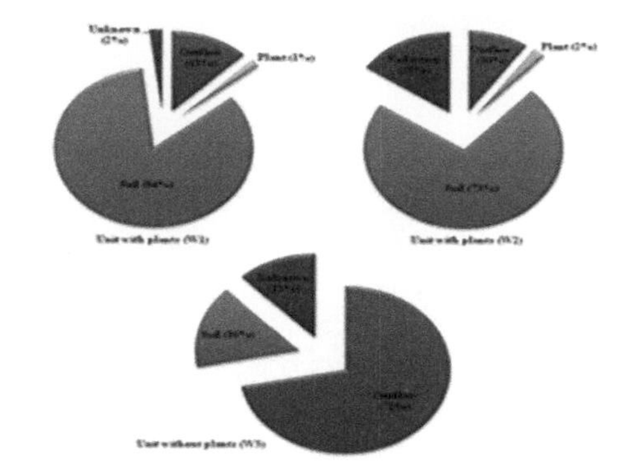

Figure 3. Mass balance of arsenic as percentage of inflowing total arsenic mass in the prototypes of subsurface flow constructed wetlands W1, W2 and W3.

3.4 *Arsenic mass balance and distribution*

Figure 3 shows the mass balance obtained for the different components of the system (plants, soil, outlet water, and unknown), calculated as a percentage of the As mass in the feed water in each unit.

The total mass balance showed that the soil with the plants is the main source of As retention system in units W1 and W2. The mass of arsenic that was retained in the units with plants was 1,210 and 1,044 mg for W1 and W2, respectively, while the unit without plants (W3) retained only 230 mg. The units with plants (W1, W2) presented a smaller quantity of arsenic in the exit water (182 and 140 mg, respectively) than the unit without plants W3 (1,034 mg). The plants' retention of arsenic was very small (2%) in comparison to that obtained from the soil-bed of units W1 and W2, which retained 84% and 72%, respectively.

4 CONCLUSIONS

The mass balance allows us to better analyze the arsenic distribution in the different parts of the system, as well as the retention of arsenic in the different segments of the CW prototypes. The plants play an important role in the As retention through different chemical, physical and microbiological process which do not occur in the CW without plants.

ACKNOWLEDGEMENTS

This work was financially supported by National Council for Science and Technology (CONACYT), within the Program Attention to National Problems.

REFERENCES

Frohne, T., Rinklebe, J., Diaz-Bone, R. & Du Laing, G. 2011. Controlled variation of redox conditions in a floodplain soil: impact on metal mobilization and biomethylation of arsenic and antimony. *Geoderma*. 160(3–4): 414–424.

Vymazal, J. 2005. Removal of heavy metals in a horizontal sub-surface flow constructed wetland. *J. Environ. Sci. Health Part A* 40(6–7): 1369–1379.

Environmental Arsenic in a Changing World –
Zhu, Guo, Bhattacharya, Ahmad, Bundschuh & Naidu (Eds)
ISBN 978-1-138-48609-6

Reactive transport modeling to understand attenuation of arsenic concentrations in anoxic groundwater during Fe(II) oxidation by nitrate

D.B. Kent[1], R.L. Smith[2], J. Jamieson[3], J.K. Böhlke[4], D.A. Repert[2] & H. Prommer[3]
[1]*U.S. Geological Survey, Menlo Park, CA, USA*
[2]*U.S. Geological Survey, Boulder, CO, USA*
[3]*University of Western Australia, Perth, WA, Australia*
[4]*U.S. Geological Survey, Reston, VA, USA*

ABSTRACT: A previously published field-experimental investigation showed that injection of nitrate in anoxic groundwater that contained aqueous and sediment-bound Fe(II) diminished concentrations of As(V) and As(III) to below drinking-water limits. In the current study, reactive transport modeling confirmed that the observed attenuation was consistent with oxidation of Fe(II) by nitrate, leading to precipitation of hydrous ferric oxide, which, in turn, sorbed both As(V) and As(III). After calibration with site-specific observations, reactive transport modeling could aid in designing effective treatment to remove arsenic using injection of nitrate to oxidize Fe(II).

1 INTRODUCTION

Hazardous levels of arsenic in groundwater are often associate with high concentrations of aqueous and sediment-bound Fe(II). Field-scale observations in a contaminant plume and field-scale reactive transport experiments showed that injection of nitrate into anoxic groundwater with elevated Fe(II) concentrations resulted in oxidation of Fe(II) to hydrous ferric oxide (HFO), with concomitant attenuation of As(V) and As(III) concentrations (Smith *et al.*, 2017). The objectives of this study were to examine controls on Fe(II) oxidation by nitrate using reactive transport modeling (RTM) and the utility of RTM to aid in designing effective treatment of arsenic-contaminated groundwater using Fe(II) oxidation by nitrate.

2 METHODS

Two sets of field-scale experiments were conducted in which nitrate (NO_3^-) was injected into anoxic groundwater with elevated concentrations of dissolved and sediment-bound Fe(II) (Smith *et al.*, 2017). Each set of experiments involved withdrawing the anoxic groundwater, adding NaBr and $NaNO_3$ anaerobically, and reinjecting. Concentrations of nitrogen-species, major and minor cations and anions, Fe(II), As(V), As(III), and other selected solutes were monitored with time at various transport distances downgradient from the injection.

One-dimensional reactive transport simulations of experimental observations were conducted with PHREEQC (Appelo and Parkhurst, 2013). Parameters for cation and anion sorption reactions on aquifer sediments were calibrated using experimental data from the first set of field experiments. Parameters for As(V) sorption on aquifer sediments were derived from laboratory experiments conducted with site-specific sediments. Parameters for sorption on HFO were taken from Dzombak & Morel (1990) and Gustafsson and Bhattacharya (2007). Transport parameters – pore-water velocity, dispersivity, and mass of injected tracers – were determined by fitting bromide (Br^-) time histories (e.g., Kent *et al.*, 2007).

3 RESULTS AND DISCUSSION

Time histories for NO_3^- were well-described using a rate expression that was first-order in NO_3^- concentration and biomass, causing N and O isotopic fractionation of the NO_3^- and producing transient accumulation of nitrous oxide (N_2O). Time histories for N_2O were also well described assuming a rate expression first-order in nitrite and biomass to produce N_2 (Fig. 1). The time-histories were also well-described using Monod expressions in NO_3^- and N_2O at the expense of an additional parameter that required calibration.

Aqueous Na^+, K^+, Mg^{2+}, Ca^{2+}, NH_4^+ concentrations could be described using equilibrium cation exchange reactions. In addition to equilibrium cation exchange, Fe(II) time-histories showed clear evidence for loss of Fe(II) mass, which could be described as a rate-limited oxidation of Fe(II) by NO_3^- and N_2O according to the reactions:

$$2NO_3^- + 8Fe^{2+} + 19H_2O = N_2O + 8Fe(OH)_{3,s} + 14H^+$$
$$N_2O + 2Fe^{2+} + 5H_2O = N_2 + 2Fe(OH)_{3,s} + 4H^+.$$

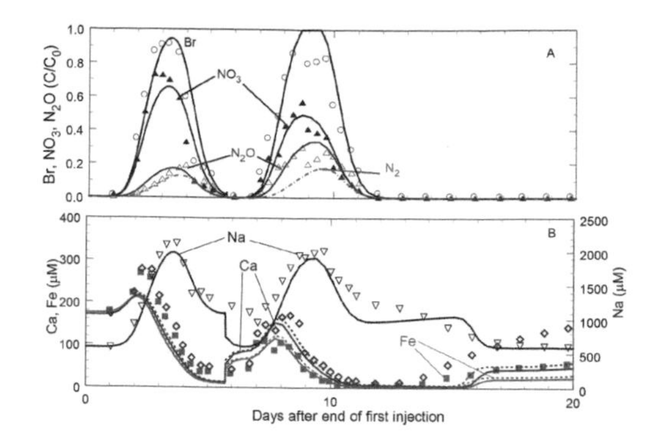

Figure 1. Experimentally measured and simulated time histories of Br^-, NO_3^-, N_2O (panel A) and Na^+, Ca^{2+}, and Fe(II) (panel B) 1.0 meter downgradient of the location where anoxic groundwater amended with approximately 1 mmol L^{-1} each of NaBr and NaNO3 was injected in two pulses separated by several days.

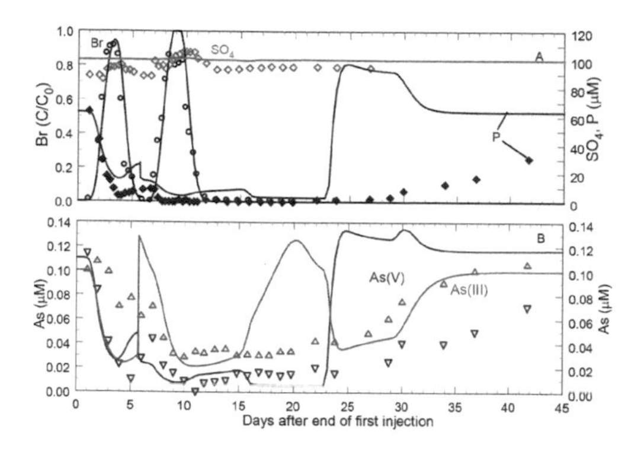

Figure 2. Experimentally measured and simulated time histories of SO_4^{-2} and total dissolved P (panel A) and As(V) and As(III) at the same location described in Figure 1. Note that the WHO drinking water limit for As is 0.13 μM.

Sorption of Ca^{2+} or Fe^{2+} on the freshly precipitated HFO had little impact on simulated time-histories (Fig. 1).

No attenuation of SO_4^{-2} concentrations was observed (Fig. 2). In contrast, significant attenuation of total dissolved phosphorus (P) (independently shown to equal phosphate concentrations), As(V), and As(III) concentrations was observed. Simulated time histories of all four oxyanions assuming equilibrium sorption on HFO generated by oxidation of aqueous and sorbed Fe(II) using published surface complexation parameters were in good agreement with the experimental data. Arsenic concentrations remained below the WHO drinking water limited for several weeks after the nitrate injections, after which they slowly returned to pre-treatment concentrations.

Simulations based on alternative hypotheses for reactions involving NO_3^- and Fe(II) yielded significant discrepancies compared to the experimental data. Reduction of NO_3^- and N_2O by organic carbon rather than Fe(II) yielded insufficient HFO generation to match the observed attenuation of P, As(V), and As(III) concentrations. Reduction of NO_3^- to NH_4^+ rather than N_2 resulted in much higher NH_4^+ concentrations than observed and excessive generation of HFO, resulting in much larger decreases in SO_4^{-2}, P, As(V), and As(III) concentrations than observed.

4 CONCLUSIONS

Experimental data show that injection of NO_3^- in anoxic groundwater with aqueous and sediment-bound Fe(II) can greatly diminish both As(V) and As(III) concentrations. Reactive transport model simulations show that the observations are consistent with oxidation of Fe(II) by NO_3^- to HFO, producing transient N_2O and N_2. Sorption of As(V) and As(III) on HFO can maintain arsenic concentrations below drinking water limits for several weeks. Once calibrated using site-specific observations, reactive transport models can provide guidance in designing treatments that minimize As(V/III) and nitrate concentrations.

ACKNOWLEDGEMENTS

Funding was provided by the U. S. Geological Survey through the Toxic Substances Hydrology Program and Water Mission Area programs. Critical review by Dr. Aria Amirbahman greatly improved the quality of the exposition.

REFERENCES

Gustafsson, J.P. & Bhattacharya, P. 2007. Geochemical modelling of arsenic adsorption to oxide surfaces; In: Bhattacharya, P., Mukherjee, A.B., Bundschuh, J., Zevenhoven, R. & Loeppert, R.H. (eds) *Arsenic in Soil and Groundwater Environment.* Elsevier, Amsterdam, pp. 153–200.

Dzombak, D. & Morel, F.M.M. 1990. *Surface Complexation Modeling: Hydrous Ferric Oxide.* John Wiley and Sons, New York NY, USA.

Kent, D.B., Wilkie, J.A. & Davis, J.A. 2007. Modeling the movement of a pH perturbation and its impact on adsorbed zinc and phosphate in a wastewater-contaminated aquifer. *Water Resour. Res.* 43(7): W07440.

Parkhurst, D. & Appelo, C.A.J. 2013. Description of input and examples for PHREEQC Version 3 – a computer program for speciation, batch-reaction, one-dimensional transport, and inverse geochemical calculations. U.S. Geol. Surv. *Tech. Methods, book 6.* Chap. A43.

Parkhurst, D.L., Stollenwerk, K.G. & Coleman, J.A. 2003. Reactive-transport simulation of phosphorus in the sewage plume at the Massachusetts Military Reservation, Cape Cod, Massachusetts. U.S. Geol. Surv., WRIR 03-4017.

Smith, R.L., Kent, D.B., Repert, D.A. & Böhlke, J.K. 2017. Anoxic nitrate reduction coupled with iron oxidation and attenuation of dissolved arsenic and phosphate in a sand and gravel aquifer. *Geochim. Cosmochim. Acta* (196): 102–120.

Environmental Arsenic in a Changing World –
Zhu, Guo, Bhattacharya, Ahmad, Bundschuh & Naidu (Eds)
ISBN 978-1-138-48609-6

Simultaneous oxidation of As(III) and reduction of Cr(VI) by *Alcaligenes* sp.

N. Rane, V. Nandre, S. Kshirsagar, S. Gaikwad & K. Kodam
Department of Chemistry, Savitribai Phule Pune University, Pune, India

ABSTRACT: Arsenic (III) and chromium (VI) are most common environmental contaminants due to its tremendous applications in dye, tannery and pulp industries. They are non-biodegradable as they are heavy metals, and therefore, of major concern. *Alcaligenes* sp. had the ability to efficiently oxidize As(III) to As(V) and reduce Cr(VI) to Cr(III) separately. It could also simultaneously oxidize As(III) in presence of Cr(VI) under shaking conditions. Simultaneous reduction of 5 mg L^{-1} Cr(VI) and oxidation of 5 mM As(III) was carried out where Cr(VI) was reduced up to 90% and oxidized 36% arsenic in 24 h. With respect to As(III) oxidation, there was no significant decrease in the rate of oxidation of As(III) due to presence of Cr(VI). Whereas, in case of only Cr(VI) exposure, only 40% of Cr(VI) reduction was observed in 24 h. In presence of arsenic the isolate did not show any effect on growth rate, while rate of chromium reduction was enhanced in arsenic presence. So, further scale up experiments can lead to potential use of Alcaligenes sp. to simultaneously biotransform toxic forms of both arsenic and chromium to less toxic forms.

1 INTRODUCTION

Chromium (VI) and arsenic (III) are non-biodegradable heavy metal and metalloids and potential carcinogens, and hence, of major concern. Therefore, it is important that the remediation method should be such that brings chromium and arsenic within the range of the permissible limits before the industrial effluent is discharged. This warrants the need for a potential isolate to biotransform more than one heavy metal (Bachate *et al.*, 2013). Several different strategies are adopted by microorganisms for heavy metals removal, mostly involving biosorption and biotransformation or both. With this background, we studied the effect of simultaneous arsenic and chromium detoxification by *Alcaligenes* sp. which was isolated from tannery industry waste. This bacterial strain showing a simultaneous reduction of Cr(VI) and oxidation of As(III) and is a potential candidate for bioremediation of environments contaminated with both of these toxic metal species.

2 METHODS

2.1 *Growth curve*

The growth of *Alcaligenus* sp. was studied in order to get sufficient biomass in less time interval with its high tolerance to As(III) and Cr(VI). The optimal media composition was found to contain Tris Minimal Media (Bachate *et al.*, 2013) with yeast extract (0.2%) and Na-acetate at 20 mM concentration. 1% of pre-inoculum was used for further studies.

2.2 *As(III) oxidation and Cr(VI) reduction*

Arsenite oxidation was checked by molybdenum blue method (Nandre *et al.*, 2017) under aerobic condition 37°C, 150 rpm. Chromium reduction at was checked using diphenyl carbazide method (Chaudhari *et al.*, 2013) for 24 h 37°C, static condition.

2.3 *Evaluation of siderophore production*

Siderophore production was evaluated by chrome azurole sulfonate (CAS) test at 630 nm (Louden *et al.*, 2011).

3 RESULT AND DISCUSSION

3.1 *Arsenic oxidation and simultaneous biotransformation of chromium (VI) and arsenic (III) under shaking condition*

Arsenic (III) at a concentration of 5 mM was oxidized to arsenic (V) in 24 h (Fig. 1) whereas when *Alcaligenes* sp. exposed to arsenic and chromium both there was simultaneous biotransformation. The rate of arsenic oxidation did not increase or decreased in presence of chromium. Arsenic and chromium are often abundant constituents of acid mine drainage and also of tannery industry. Hexavalent chromium Cr(VI) and arsenite As(III) are the most toxic forms of chromium and arsenic respectively, and reduction of Cr(VI) to Cr(III) and oxidation of As(III) to As(V) has great environmental implications as they affect toxicity and mobility of these toxic species. To simultaneously change their oxidation state from As(III) to As(V),

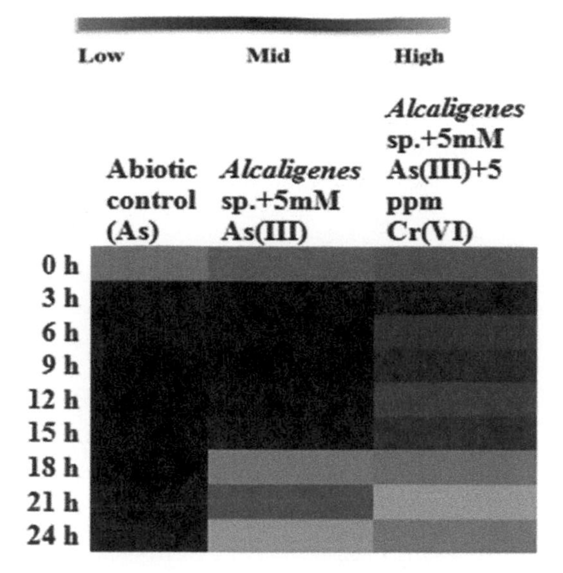

Figure 1. Percent of As(III) oxidized by *Alcaligenes* sp. i) Abiotic control had 5 mM As(III) ii) Isolate +5 mM As(III) iii) Isolate +5 mM As(III) +5 mg L^{-1} Cr(VI).

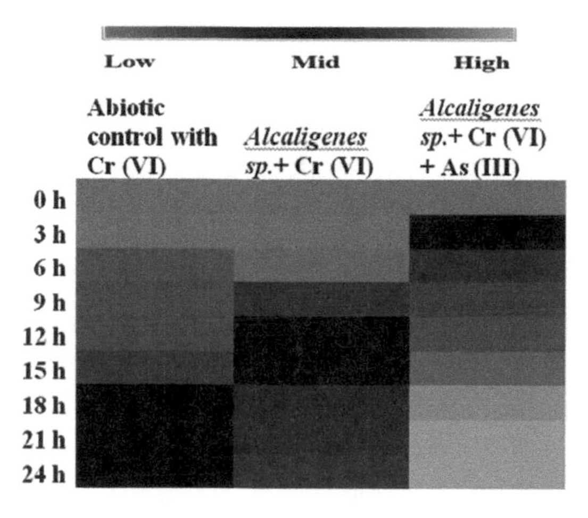

Figure 2. Percent of Cr(VI) reduced by *Alcaligenes* sp. i) Abiotic control having 5 mg L^{-1} Cr(VI) ii) Isolate +5 mg L^{-1} Cr(VI) iii) Isolate +5 mM As(III) +5 mg L^{-1} Cr (VI).

and Cr(VI) to Cr(III), is a potentially effective and attractive strategy for environmental remediation.

3.2 *Chromium reduction and simultaneous redox conversion of chromium (VI) and arsenic (III)*

Chromium (VI) at a concentration of 5 mg L^{-1} was reduced to chromium (III) in 24 h whereas when Alcaligenes sp. exposed to arsenic and chromium both the rate of chromium reduction was found to increase as compared with chromium alone. 5 mM of As(III) and 5 mg L^{-1} of Cr(VI) was simultaneously biotransformed (Fig. 2). Also, there is an increase in growth rate when arsenic was present in the minimal media.

3.3 *Siderophore production*

Siderophore production was screened at 5 mg L^{-1} of Cr(VI) and 5 mM of As(III) individually and in combination using CAS assay. Maximum siderophore production was observed at 48 h in case of both individual 5 mM of As(III) or 5 mg L^{-1} of Cr(VI). Whereas, in case of exposure of combination of 5 mM of As(III) and 5 mg L^{-1} of Cr(VI), maximum siderophore concentration was observed at 20 h. Maximum siderophore concentrations at 20 h indicated their co-involvement in biotransformation of arsenic and chromium.

4 CONCLUSIONS

The *Alcaligenes* sp. could not only individually oxidize As(III) to As(V) but also reduce Cr(VI) to Cr(III) separately. The potency of the organism extended to simultaneous arsenic oxidation and chromium reduction. Simultaneously it reduced 90% Cr(VI) and oxidized 36% arsenic in 24 h. Whereas, in case of only Cr(VI) exposure, only 40% of Cr(VI) reduction was observed in 24 h. With respect to As(III) oxidation, there was no significant increase in the rate of oxidation of As(III) due to presence of Cr(VI). *Alcaligenes* sp. exposed to arsenic and chromium both the rate of chromium reduction was found to increase as compared with chromium alone. The isolate has not shown any effect on growth rate in presence of arsenic. So, further scale up experiments can lead to potential use of *Alcaligenes* sp. to simultaneously biotransform toxic forms of both arsenic and chromium to less toxic forms.

ACKNOWLEDGEMENTS

Authors would like to thank UGC Dr. D. S. Kothari fellowship for funding.

REFERENCES

Bachate, S.P., Nandre, V., Ghatpande, N.S. & Kodam, K.M. 2013. Simultaneous reduction of Cr(VI) and oxidation of As(III) by *Bacillus firmus* TE7 isolated from tannery effluent. *Chemosphere* 90(8): 2273–2278.

Chaudhari, A.U., Tapase, S.R., Markad, V.L. & Kodam, K.M. 2013. Simultaneous decolorization of reactive Orange M2R dye and reduction of chromate by *Lysinibacillus* sp. KMK-A. *J. Hazard. Mater.* 262: 580–588.

Louden, B.C., Haarmann, D. & Lynne, A. 2011. Use of blue agar CAS assay for siderophore detection. *J. Microbiol. Biol. Educ.* 12(1): 51–53.

Nandre, V.S., Bachate, S.P., Salunkhe, R.C., Bagade, A.V., Shouche, Y.S. & Kodam, K.M. 2017. Enhanced detoxifica-tion of arsenic under carbon starvation: a new insight into microbial arsenic physiology. *Curr. Microbiol.* 74(5): 614–622.

Environmental Arsenic in a Changing World –
Zhu, Guo, Bhattacharya, Ahmad, Bundschuh & Naidu (Eds)
ISBN 978-1-138-48609-6

Nitrate respirers mediate anaerobic As(III) oxidation in filters

C.Y. Jing & J.L. Cui
Research Center for Eco-Environmental Sciences, Chinese Academy of Sciences, Beijing, P.R. China

ABSTRACT: Microorganisms play a key role in the redox transformation of arsenic in aquifers. In this study, the impact of indigenous bacteria, especially the prevailing nitrate respirers, on arsenite oxidation was explored during groundwater filtration using granular TiO_2 and subsequent spent TiO_2 anaerobic landfill. X-ray absorption near edge structure spectroscopy analysis showed As(III) oxidation (46% in 10 days) in the presence of nitrate in the simulated anaerobic landfills. The Fe phase showed no change during the anaerobic incubation. Incubation experiments implied that the indigenous bacteria completely oxidized arsenite to arsenate in 10 days using nitrate as the terminal electron acceptor under anaerobic conditions. Phylogenetic tree analysis revealed that Proteobacteria was the dominant phylum, with Hydrogenophaga (34%), Limnohabitans (16%), and Simplicispira (7%) as the major bacterial genera. The nitrate respirers especially from the Hydrogenophaga genus anaerobically oxidized As(III) using nitrate as an electron acceptor instead of oxygen. Our study implied that microbes can facilitate the groundwater As oxidation using nitrate on the adsorptive media.

1 INTRODUCTION

Various treatment technologies have been extensively studied during last decades to mitigate the severe problem of arsenic (As) tainted groundwater (Hu *et al.*, 2015). As a consequence of enhanced As removal, a significant amount of As-bearing solid residuals is generated and disposed of in landfills, where As speciation and fate are generally determined by the ubiquitous microbes in the aquifer (Clancy *et al.*, 2013).

A series of arsenite (As(III)) oxidizing bacteria have been successfully isolated and employed to transform As(III) to arsenate (As(V)) for As remediation, because As(V) is much more strongly adsorbed than As(III). Considering the ubiquitous existence of As(III) oxidizing bacteria in groundwater, we hypothesize that the indigenous As(III) oxidizing bacteria in groundwater should affect the speciation of adsorbed As in filters.

In the subsurface, As is often released into the groundwater coupled with a microbial-reductive process. Under such anoxic conditions, microorganisms critically influence As release using common oxyanions including sulfate and nitrate. High nitrate concentrations often restrain the release of As, which is attributed to the nitrate-dependent bacterial oxidation of As(III). A variety of bacterial isolates or communities can oxidize As(III) with nitrate as an electron acceptor under anaerobic conditions. However, the impact of indigenous nitrate respires in groundwater on As(III) oxidation in filters and subsequent landfill of the As-laden materials remains unclear. Considering high nitrate concentrations in groundwater (up to 25–198 mg L^{-1}) due to agricultural activities, the effect of nitrate and indigenous nitrate respirers on As speciation in filters and landfills motivated our study.

The objective of this study was to investigate the role of indigenous bacteria and nitrate on the transformation of As during groundwater filtration and in landfills. The change in As and Fe speciation on the spent $GTiO_2$ under anaerobic conditions was studied using X-ray absorption near edge structure (XANES) spectroscopy. The microbial diversity of the indigenous bacteria was characterized using 16S rRNA sequences. By combining XANES and 16S rRNA analysis, we intend to elucidate the effect of microbe and nitrate supplementation on As transformation. The result of our study will improve the understanding of the fate of As during filtration and anaerobic landfills.

2 METHODS/EXPERIMENTAL

2.1 *Filtration*

Field filtration experiments were conducted on site at Shanxi, China. Approximately 14.6 g $GTiO_2$ were loaded in a 1.2-cm diameter column, resulting in a bed volume (BV) of 14.5 mL. The empty bed contact time (EBCT) for the filter was 0.21 min.

Three treatments of $GTiO_2$ columns in duplicates were (1) GGW, (2) GGW amended with 10 mg L^{-1} nitrate, and (3) SGW. The influent and effluent were collected periodically, passed through a 0.22 μm membrane filter, and then preserved in the dark with 5% HCl at 4°C before analysis. At the end of a filtration cycle, the columns were stored on dry ice in a cooler in the field and transported to the lab for various analysis. The indigenous microorganisms were obtained by extracting the spent $GTiO_2$.

2.2 *Anaerobic incubation with indigenous microorganisms*

To study microbial As(III) oxidation, the indigenous microorganisms were mixed with SGW containing 1.8 mg L^{-1} As(III) and 10.0 mg L^{-1} nitrate under anaerobic incubations (glovebox with 100% N_2). For comparison, incubation was also performed with no addition of nitrate. To test the role of microbes, the particles extracted from the spent $GTiO_2$ were autoclaved and added into the incubation experiment as a control study.

Bacterial DNA was extracted from the indigenous particles from the spent $GTiO_2$ using a power soil DNA kit. Standard procedures were used for DNA amplification, ligation and cloning. DNA sequencing was performed by the TSINGKE Company (Beijing, China). Sequence alignments were performed using the Clustal X2 program with the sequences of the GenBank database. The sequence similarity was assessed using BLAST. Phylogenetic trees were generated with the neighbor-joining method using MEGA version 5.0.

3 RESULTS AND DISCUSSION

3.1 *Adsorption filtration*

The As(III) breakthrough point at 10 μg L^{-1} occurred earlier in GGW (~600 BVs) than in SGW (~2000 BVs). At this breakthrough point, the corresponding As adsorption was 0.40 ± 0.04 mg-As g^{-1} $GTiO_2$ in GGW and 1.34 ± 0.07 mg g^{-1} in SGW. The lower As adsorption in GGW than in SGW may be attributed to strong competition from coexisting anions, viz. silicate and carbonate, at high concentrations. The addition of 10 mg L^{-1} nitrate to GGW did not inhibit the uptake of As, suggesting the stronger affinity of As to TiO_2 as an inner-sphere surface complex compared with outer-sphere nitrate complex (Cui *et al.*, 2018).

3.2 *Speciation and transformation of As and Fe*

The speciation of the adsorbed As under anaerobic conditions was characterized using XANES. At the beginning of the anaerobic incubation after filtration experiment ($t = 0$ d), As(III) on spent $GTiO_2$ for SGW, GGW, and GGW + NO_3^- were 87%, 71%, and 71%, respectively.

The Fe K-edge XANES analysis showed that the retained Fe on $GTiO_2$ was dominated by amorphous ferric arsenate ($amFeAsO_4$) (67–69%), ferrihydrite (24–26%), and a small amount of goethite (5–8%). An insignificant difference ($p < 0.01$, $R = 0.932$) of Fe species on the spent $GTiO_2$ was found for GGW and GGW + NO_3^-. The Fe phase showed no change during the 10-d incubation under anaerobic conditions, suggesting that the oxidation of adsorbed As(III) was not related to Fe.

3.3 *As(III) oxidation by indigenous microorganism*

To examine the hypothesis that indigenous microorganisms involved in As(III) oxidation under anaerobic condition, the retained particles on the surface of $GTiO_2$ were collected and mixed with 1.8 mg L^{-1} As(III) in SGW. As(III) was not oxidized in SGW alone, sterilized SGW controls, or the control samples with nitrate amendment. In contrast, As(III) was completely oxidized to As(V) in the SGW amended with nitrate in 10 d. The comparison results indicate that the microbes in GGW should facilitate As(III) oxidation in the presence of nitrate under anaerobic conditions, while chemical oxidation of As(III) can be neglected.

Microbial-assisted As(III) oxidation should play a major role in As transformation on spent $GTiO_2$, particularly when nitrate is available. Phylogenetic tree analysis revealed that *Proteobacteria* was the dominant phylum, with *Hydrogenophaga* (34%), *Limnohabitans* (16%), and *Simplicispira* (7%) as the major bacterial genera. The nitrate respirers especially from the *Hydrogenophaga* genus anaerobically oxidized As(III) using nitrate as an electron acceptor instead of oxygen.

4 CONCLUSIONS

The adsorbed As(III) in filters for tainted groundwater can be oxidized to As(V) in the presence of nitrate without oxygen. Nitrate absence during microbial incubation brought no redox change in As species. Microbial community analysis of the groundwater indicated *Hydrogenophaga* genus contributed to 34% of the total cloned sequences. Some indigenous bacteria belonging to *Hydrogenophaga* genus may anaerobically oxidize As(III) using nitrate as the electron acceptor. Our study highlighted that the presence of nitrate improves microbial As(III) oxidation in groundwater and facilitates in situ immobilization of As on spent adsorptive media under anaerobic conditions. Further isolation of such kinds of bacterial species is helpful to strengthen their role on As(III) oxidation and nitrate reduction, remediating both pollutants.

ACKNOWLEDGEMENTS

We acknowledge the financial support of the Strategic Priority Research Program of the Chinese Academy of Sciences (XDB14020201).

REFERENCES

Clancy, T.M., Hayes, K.F. & Raskin, L. 2013. Arsenic waste management: A critical review of testing and disposal of arsenic-bearing solid wastes generated during arsenic removal from drinking water. *Environ. Sci. Technol.* 47: 10799–10812.

Cui, J., Du, J., Yu, S. & Jing, C. 2018. Groundwater arsenic removal using granular TiO_2: integrated laboratory and field study. *Environ. Sci. Pollut. Res.* 22(11): 8224–8234.

Hu, S., Shi, Q.T. & Jing, C.Y. 2015. Groundwater arsenic adsorption on granular TiO_2: integrating atomic structure, filtration, and health impact. *Environ. Sci. Technol.* 49(16): 9707–9713.

Section 5: Sustainable mitigation and management

5.1 Societal involvement for mitigations of long-term exposure

Environmental Arsenic in a Changing World –
Zhu, Guo, Bhattacharya, Ahmad, Bundschuh & Naidu (Eds)
ISBN 978-1-138-48609-6

Sustainable arsenic mitigation and management through community participation

A.K. Ghosh, A. Kumar, R. Kumar & M. Ali
Mahavir Cancer Sansthan & Research Centre, Patna, Bihar, India

ABSTRACT: In recent times water pollution has become a gigantic health issue globally. Inorganic arsenic is one of water contaminant naturally present in very high concentration in the groundwater of many countries, creating an important public health issue affecting a population of 200 million globally. Arsenic exposure appears linked to increase in cancer, heart disease, and developmental problems. In Bihar 18 districts are affected with arsenic poisoning in groundwater. The maximum arsenic contamination in ground reported in Bihar till date is 1908 $\mu g\,L^{-1}$ observed village Tilak Rai Ka Hatta (TRKH), Simri Block of Buxar District. The arsenic exposed population of study area has developed many visible symptoms of arsenic poisoning like hyperkeratosis of sole/palm along with melanosis and rain drop pigmentation. In the primary school, the highest arsenic concentration in drinking water observed was 857 $\mu g\,L^{-1}$ while in the children hair sample the maximum value was 12.609 $mg\,kg^{-1}$. Finally, a community based arsenic filter based on adsorbent Hybrid Anion Exchange Nano Resin has been installed though joint initiative of Mahavir Cancer Sansthan & Research Centre (MCSRC), Patna in collaboration with Lehigh University, Pennsylvania, USA, which changed the life of the population with symptomatic relief related to arsenic poisoning.

1 INTRODUCTION

The Gangetic flood plain region of Bihar is the most severely arsenic affected area with more than 5 million population. In Buxar district of Bihar, the Tilak Rai Ka Hatta (TRKH) village is the severely arsenic exposed area. The major chunk of population exhibits typical symptoms of arsenicosis. The highest arsenic concentration recorded in ground water was 1908 $\mu g\,L^{-1}$. Many persons residing in this village had health related issues but the major intervention was made through installation of a community based arsenic filter, which changed the life of the population. There was significant decrease in the arsenic caused health issues in persons residing in study area.

2 METHODS/EXPERIMENTAL

2.1 *Location*

The study was done at Tilak Rai Ka Hatta village (25°41′36″N, 84°07′51″E) of Buxar district of Bihar.

The population of the Tilak Rai Ka Hatta village is 5,348 with 340 households (Census, 2011).

2.2 *Sample collection & survey*

A total of 65 hair samples of children aged between 4 years to 15 years from the primary school of the village with 03 sources of drinking water along with 80 water samples were randomly collected in duplicates from hand pumps of each household situated at every 50 m. The depths of the handpumps were also recorded

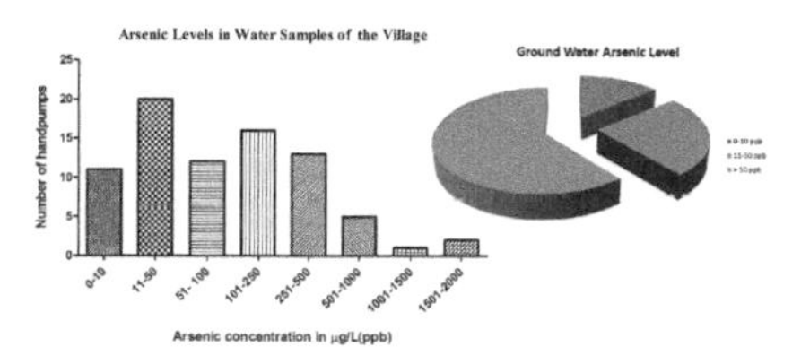

Figure 1. Arsenic concentration in water samples of Handpumps of Tilak Rai Ka Hatta analyses through GF-AAS with Pie chart (ANOVA-Dunnett's Test, $P < 0.05$).

for the correlation of arsenic concentration. After collection, all the hair samples were digested using concentrated HNO_3 on hot plate under fume hood and total arsenic concentration (inorganic form) was estimated as per the protocol of (NIOSH, 1994) through graphite furnace atomic absorption spectrophotometer (Pinnacle 900T, Perkin Elmer, Singapore). Simultaneously, health assessment of the population was also done through a health survey questionnaire proforma. For determining the exact location of the hand pump, hand held Global Positioning System (GPS) receivers (Garmin etrex10, of USA) with an estimated accuracy of $\approx$10 m were utilized.

3 RESULTS AND DISCUSSION

3.1 *Arsenic assessment*

The study shows novel findings ever explored in this area. In the village TRKH the maximum arsenic concentration in water sample observed was 1908 $\mu g\,L^{-1}$ (Fig. 1).

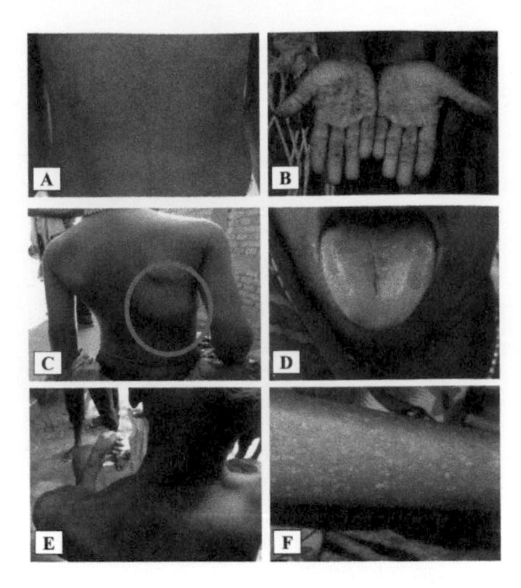

Figure 2. Showing arsenicosis symptoms in the population.

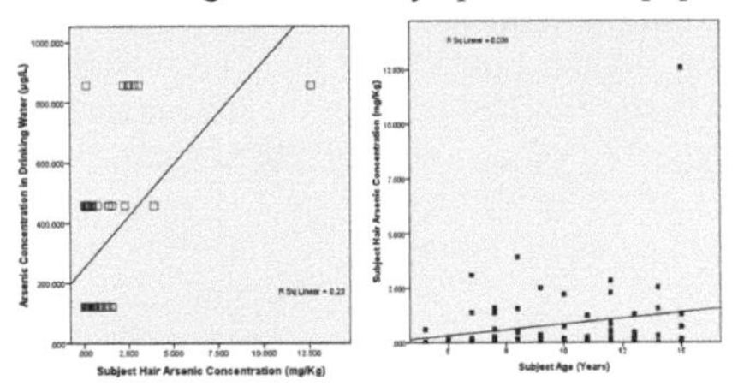

Figure 3. The correlation coefficient between arsenic concentration in drinking hand pump and arsenic concentration in the hair samples of the subjects ($r = 0.23$ & $P < 0.05$) and arsenic concentration in drinking hand pump and subject age (years) ($r = 0.039$ & $P < 0.05$).

3.2 *Health assessment*

The rural population exhibited the typical symptoms of arsenicosis like hyperkeratosis in palm and sole, melanosis in palm and sole, blackening of tongue, skin irritation, anemia, gastritis, constipation, loss of appetite, bronchitis and cough, etc. (Fig. 2).

Correlation coefficient of arsenic concentration in handpump water and arsenic concentration in the hair samples of the subjects shows the maximum arsenic concentration in the hand pump of the school as $857\,\mu g\,L^{-1}$ while the maximum arsenic concentration in the hair sample was $12.609\,mg\,kg^{-1}$ correlates the arsenic exposure ($r = 0.23$, $P < 0.05$; Fig. 3).

Arsenic determination in urine, hair and nail are considered as most reliable indicator of exposure. Hair samples are used as a biomarker for arsenic exposure because inorganic arsenic and Dimethylarsinic acid are stored in hair root and thus reflect past exposure (Yoshida *et al.*, 2004). An elevated arsenic level in hair indicates the past exposure of 6–12 months (NRC, 2001). In the present study, very high concentration of arsenic contamination was observed in the hair samples of the children. They were drinking water from three highly arsenic contaminated handpumps. The most fascinating result was that many cancer cases were also reported from these villages i.e. skin cancer, gall bladder cancer and breast cancer cases.

The other health related assessments showed hormonal imbalance in the population. The rural population exhibited elevated levels of serum estrogen while decreased levels of serum testosterone levels denotes that the arsenic contamination in groundwater and its consumption by the rural population has caused severe health problems to them (Kumar *et al.*, 2015, 2016). The entire study thus reveals that high arsenic concentration in groundwater in these two villages and drinking of this contaminated water has led to severe health related problems in the population.

3.3 *Mitigation*

A community based arsenic filter was installed in the study area with the help of community participation through collaborative effort of Mahavir Cancer Institute and Research Centre, Patna and Lehigh University, Pennsylvania, USA, which is based on adsorbent Hybrid Anion Exchange Nano Resin. This filter is providing arsenic safe water giving good health outcomes.

4 CONCLUSIONS

The survey of the village TRKH of Buxar district, demonstrated very high arsenic concentration in drinking water as well as in the hair samples of the school children. Many arsenic induced health related problems were observed in the population like keratosis, melanosis, leuco-melanosis, hormonal imbalance and few cases of cancer. The arsenic filter installed through community participation was successful and sustainable initiative. Installation of arsenic filter has changed the health status of the arsenic exposed village population.

ACKNOWLEDGEMENTS

The authors acknowledge Tagore-Sengupta Foundation for providing fund for installation of Arsenic filter.

REFERENCES

Census 2011. Interim Report of Population Census of India. http://www.censusindia.gov.in/2011.

Kumar, A., Ali, M., Rahman, S.M., Iqubal A.M., Anand, G. & Niraj, P.K. 2015. Ground water arsenic poisoning in "Tilak Rai Ka Hatta" village of Buxar District, Bihar, India causing severe health hazards and hormonal imbalance. *J. Environ. Anal. Toxicol.* 5(4): 1–7.

Kumar, A., Kumar, R., Rahman, M.S., Iqubal, M., Ali, M., Niraj, P.K., Anand, G., Prabhat, K., Abhinav & Ghosh, A. K. 2016. Ground water arsenic contamination: a local survey in India. *Int. J. Prev. Med.* 7: 100.

NRC 2001. Arsenic in drinking water. 2001 update. The National Academies Press, Washington, DC.

Yoshida, T., Yamauchi, H. & Fan, S.G. 2004. Chronic health effects in people exposed to arsenic via the drinking water: dose-response relationships in review. *Toxicol. Appl. Pharmacol.* 198(3): 243–252.

Environmental Arsenic in a Changing World –
Zhu, Guo, Bhattacharya, Ahmad, Bundschuh & Naidu (Eds)
ISBN 978-1-138-48609-6

Community effects on safe water selection – the case of West Bengal

M. Sakamoto[1], S. Mukhopadhyay[2], K. Bakshi[2] & S. Roy[2]
[1]*The University of Tokyo, Tokyo, Japan*
[2]*Kalyani Institute for Study, Planning and Action for Rural Change (KINSPARC), Kalyani West Bengal, India*

ABSTRACT: Since an arsenic patient was firstly reported in 1983 in West Bengal, 35 years has been passed. Recently, pipelines using the Ganges river water have been intensively constructed in the state to provide arsenic-free drinking water in rural areas. Although development has been going on and safe water sources are more available to local people, there still exists not a small number of people who do not choose safe water sources. This study aims to investigate factors influencing their drinking water source selection. The questionnaire survey was conducted for the whole households in three villages which included one of the villages surveyed during 1983–87. Individual variables, as well as community variables, were considered in a statistical model by applying Multi-level analysis. As a result, communities had a similarity in water source selection rather than selection is made purely with individual rationality. Average community educational level had an influence rather than personal educational level.

1 INTRODUCTION

After an arsenic-caused patient was firstly reported in West Bengal in 1983, the following survey was conducted in 61 villages in the state from 1983 to 1987 (Saha, 1995). Recently, pipelines have been intensively constructed in West Bengal to carry water from the Ganges to rural areas for mitigating the arsenic problem. It seems that the pipeline development solves the arsenic problem in the region, however, from the recent field observations, we could say that not a small number of local people do not choose the pipeline water as their drinking water source and stick to the present water sources, namely deep tubewells or shallow tubewells that most probably contain arsenic. In 2015, we conducted a questionnaire survey at three villages in West Bengal to know the situation and people's perceptions, including one of those villages surveyed during 1983–1987. Although our survey was conducted before the pipeline construction in the villages, it would be helpful to analyze the data to understand people's responses to arsenic risk. Hence, the aim of this study is to investigate factors influencing their decisions on choosing water sources from safe water sources or contaminated water sources, especially from viewpoints of social, cultural, and community effects on their water source selection.

2 METHODOLOGY

2.1 *Multi-level analysis*

The multi-level analysis is one kind of regression analysis that enables researchers to simultaneously estimate individual effects as well as group effects of an explanatory variable on a dependent variable (Hox, 2002). People are often influenced by people surrounding himself/herself when he/she makes a decision. People who belong to the same community may have similarities on their preferences, behaviors, and tendency in making decisions. In such a situation, it may be hard to do random sampling because samples from the same community may not be independent to each other on a certain attribute, and it may cause a problem in estimates of a statistical model. Therefore, we employed multi-level analysis to avoid such a bias due to community similarities. For statistical modeling, HAD ver. 16 was used.

2.2 *Community detection*

In an application of multi-level analysis, groups, where people belong, are necessary to define. Usually, administrative boundaries or affiliations, such as districts, clubs, and schools, are used. In our case, carrying water is women's role in the region, and they usually do not go far beyond their neighborhood. Moreover, water source selection is a daily activity so that the clustering to be used in the analysis should be rooted in their everyday life circles. In this study, community detection based on network analysis is employed. The procedure is as follows.

1) Geographical information of houses of interviewees is recorded and read on the software R.
2) Neighborhood contiguity is defined by a distance between houses and determined with R package spdep). The distance parameter is adjusted for each village by considering physical and social clustering based on the field observation.
3) Communities are extracted with R package igraph.

Table 1. Descriptive statistics.

Village	Sample (households)	Contaminated water users	Arsenicosis patients
A	121	56 (53.72%)	20
B	135	3 (2.22%)	1
C	244	25 (10.25%)	4
Total	**500**	**93 (18.60%)**	**25**

3 RESULTS

3.1 *Status of drinking water*

Table 1 shows the descriptive statistics from the survey at three villages. The questions were asked to all the household wives who were present durmg the survey. All these villages are located in Nadia district and one of these villages was surveyed during 1983–87 survey. All three villages had been identified at risk of arsenic contamination, although number of arsenic removal plants, deep tubewells, and community-size rainwater tank were installed by governmental institutes and/or NGOs. Tragically, one of these villages where arsenic was firstly reported in the region, most of installed water facilities were abandoned, and significant number of people still took water from contaminated water sources, and as a consequence there exists more arsenicosis patients than other two villages.

3.2 *Logistic regression on waters source selection*

Communities were detected by physical distance of each house. For A village, 14 communities were identified by clustering houses located within 40 m to each other; for B village, 7 communities were identified with 40 m threshold; for C village, 16 communities were identified with 50 m threshold. Logistic re-gression was applied to analyze factors influencing households' water source selection. A categorical variable is used as a dependent variable, which represents if a household uses a safe water source (deep tubewell, filtered shallow tubewell, or bottle water) or an arsenic contaminated water source (non-filtered shallow tubewell). Table 2 shows the results.

Model A includes four categorical explanatory variables. Availability means if a safe water source is available to a household or not. The sign is positive and significant. Therefore, if a water source is available to a household, the household is more likely to choose the safe water source. Hindu means if a household is Hindu or Muslim. The sign is negative and it is significant. Therefore, if a household is Hindu, the family is less likely to choose a safe water source than Muslim family. The variables regarding education were not significant. Therefore, educational experiences may not influence risk perception on arsenic. Random intercept was significant. This means that there is a similarity within communities in water source selection.

Model B includes three explanatory group variables in addition to the categorical variable Availability. Ratio of Hindu is negatively significant. It means that if the ratio of Hindu in the community where a household belongs increases, the household is less likely to choose a safe water source. Ratio of upper secondary school is positively significant. Therefore, if there exist more people who have higher education in a community where a household belongs, the household is more likely to choose a safe water source.

Table 2. Results of multi-level logistic regression.

Model A	Estimate	SE
Fixed parameters		
Intercept	0.710	0.040**
Availability	0.318	0.063**
Hindu[a]	−0.092	0.04*
Under secondary school[b]	0.008	0.031
Upper secondary school[b]	0.052	0.034
Random parameter		
Intercept	0.275	0.075**

Reference of categorical variables: [a]Muslim [b]No education.
**$p < 0.01$, *$p < 0.05$ Adjust R2: 0.054

Model B	Estimate	SE
Fixed parameters		
Intercept	0.801	0.027**
Availability	0.319	0.062**
Ratio of Hindu	−0.225	0.041**
Ratio of upper secondary school	0.178	0.086*
Ratio of under secondary school	−0.081	0.097
Random parameter		
Intercept	0.268	0.072**

**$p < 0.01$, *$p < 0.05$ Adjust R2: 0.108

4 DISCUSSION AND CONCLUSIONS

It has been recognized that higher education enhances better risk perception. However, the result shows that community-level education had a significant influence on water source selection rather than individual-level education. Highly educated people are often wealthy, and therefore they tend to have more accessibility to safe water sources than less educated people. The variable availability may have controlled this kind of effect of education on water source selection. Therefore, if people have more chances to talk with highly educated people beyond communities, it would encourage people to choose safe water sources. The influence of the religion is not certain at the moment. More detailed filed work would be necessary.

REFERENCES

Hox, J.J., Moerbeek, M. & Schoot, R. 2002. *Multilevel Analysis: Techniques and Applications*. Routledge Academic.
Saha, K. 1995. Chronic arsenical dermatoses from tube-well water in West Bengal during 1983–87. *Indian J. Dermatol.* 40(1): 1–12.

Environmental Arsenic in a Changing World –
Zhu, Guo, Bhattacharya, Ahmad, Bundschuh & Naidu (Eds)
ISBN 978-1-138-48609-6

Likelihood of adoption of arsenic-mitigation technologies under perceived risks to health, income, and social discrimination to arsenic contamination

S.K. Singh & R.W. Taylor
Department of Earth and Environmental Studies, Montclair State University, Montclair, NJ, USA

ABSTRACT: This study aims to assess the likelihood of adoption of arsenic-mitigation technologies under perceived risks in an Indian rural region that confronts severe and chronic groundwater arsenic contamination. A total of 340 households were surveyed in three arsenic-affected villages of Bihar, India. The average population perceives greater health risks and economic risks to arsenic-contaminated groundwater than social discrimination risk, therefore, are willing to adopt arsenic-mitigation technologies. Caste, education, income, sanitation practices, people's prioritization of socio-environmental problems, arsenic awareness, and social capital, were the strongest predictors of perceived economic risk. The same variables as above (with the exception of income, sanitation practices, and social capital) with additional factors like agricultural landholdings, and social trust, were the strongest predictors of perceived health risks. However, in the case of perceived social discrimination risk, the respondents' agricultural landholdings, people's prioritization of social problems, arsenic awareness, and social capital, were the strongest predictors.

1 INTRODUCTION

Risk perceptions play a vital role in the development of government policies and their subsequent adoption, as they significantly influence individuals' proclivities for hazards management that could affect human safety and health, or ecological conditions (Gerber and Neeley 2005). Until now the risk perception framework has been mostly applied in research related to climate change adaptations and natural hazards, and its application in decision-making for areas such as flood insurance, coastal environment protection, landslides, heat waves, chemical industries, air pollution and hurricanes is a recent phenomenon (Bickerstaff, 2004; Carlton and Jacobson, 2013; Damm *et al.*, 2013; Huang *et al.*, 2013; Jones *et al.*, 2013; Peacock, *et al.*, 2005; Tam and McDaniels, 2013). On the issue of groundwater contamination, there are only a handful of studies that have addressed the relationship between perceived risks of groundwater contamination and a significant increase in expenditures on activities aimed at mitigation (Abdalla, 1992; Janmaat, 2007). There are some studies that have applied the risk perception framework to water quality issues (Dupont and Krupnick, 2010; Jakus *et al.*, 2009; McSpirit and Reid, 2011; Nguyen *et al.*, 2010; Onjala *et al.*, 2014; Walker *et al.*, 2006). Studies of decision-making under perceived health risk are still in the rudimentary stage. We did not encounter any existing studies that have evaluated decision-making for better environmental services under perceived risks to income (economic risk) or risk of social discrimination.

Also, none of the studies has attempted to investigate the causal relationships between socioeconomic and demographic factors, knowledge and awareness, institutional trust, social capital, and social trust to risk perceptions in the context of arsenic-contamination. This study aims to address two primary goals related to the aforementioned gaps: a) to investigate people's decision-making to adopt arsenic-mitigation under perceived risks to health, income (economic), and social discrimination to arsenic; and b) to analyze the socioeconomic, demographic, and other factors that construct people's risk perception to groundwater arsenic-contamination.

2 METHODS

2.1 *Administration of the survey*

Based on the current levels of arsenic severity and the arsenic-mitigation initiatives, three villages, namely Suarmarwa, Rampur Diara, and Bhawani Tola of the Maner block of Patna district in Bihar, India were selected for this study (Singh, 2015; Singh *et al.*, 2018) (Fig. 1). Detailed description of the survey administration, arsenic contamination, and socioeconomic models of arsenic mitigation are provided in other studies (Singh, 2015; Singh *et al.*, 2016, 2017), in brief, the surveyed populations were randomly selected and stratified by the variable of caste, as the castes (forward, backward, and scheduled caste) are the sub-groups of the people in the project area, which are internally homogeneous, but externally heterogeneous (UNSD, 2008).

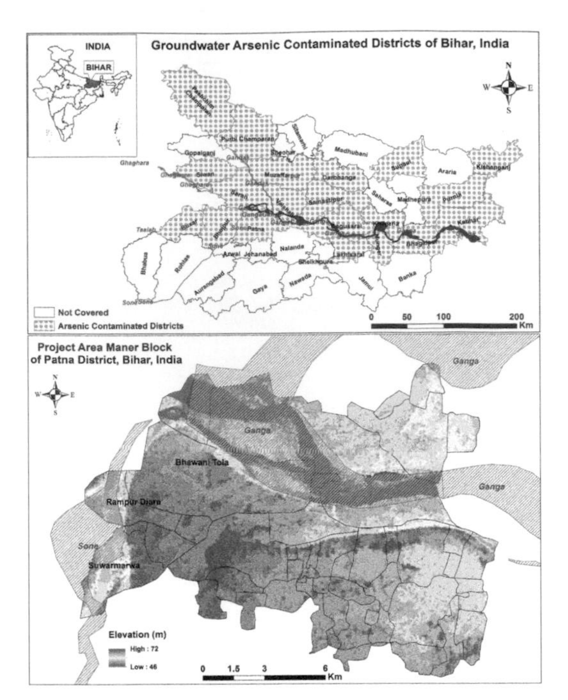

Figure 1. A map showing the arsenic affected districts of Bihar and the three villages selected for this study with their elevations (Singh *et al.*, 2018).

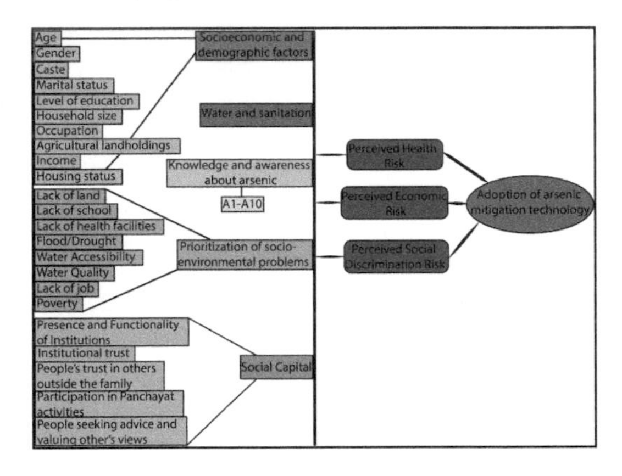

Figure 2. The conceptual framework of risk perception and decision-making of arsenic-mitigation technology.

2.2 *Conceptual framework of risk perception and decision-making of arsenic mitigation technology*

The arsenic risk perception model has been conceptualized based on the hypothesis that; a) the arsenic-exposed communities will adopt arsenic-mitigation technologies or programs under perceived health risk; perceived economic risk; and perceived risk of social discrimination; and b): the existing socioeconomic and demographic factors, people's prioritization of socio-environmental problems, and social capital, are significant drivers in constructing people's perceived risks (Fig. 2).

Table 1. Socioeconomic and demographic characteristics of the studied populations.

	Variables	Suamarwa	Rampur Diara	Bhawani Tola	p-value
Gender	Female	10	2	5	0.038
	Male	101	117	105	
Age group	18-28	9	15	8	0.049
	29-39	16	27	21	
	40-50	36	28	36	
	51-61	23	18	12	
	62 and above	16	12	23	
Caste	Scheduled caste	23	19	20	<0.001
	Backward caste	77	18	37	
	Forward caste	0	62	43	
Marital status	Single	5	9	1	0.015
	Married	95	91	99	
Education level	Illiterate	49	11	14	<0.001
	Primary education	41	40	41	
	Secondary education	8	31	31	
	College	1	18	14	
Household size	≤5	43	36	34	0.569
	>5-10>	51	56	61	
	>10	5	8	5	
Occupation	Unemployed	22	3	2	<0.001
	Labor	51	30	35	
	Agriculture	20	41	39	
	Job + Business	7	25	24	
Income group	BPL(Rs.500<per month)	27	4	2	<0.001
	Lower APL(>Rs.500-Rs.10,000 per month)	60	76	78	
	Upper APL(>Rs.10,000 per month)	13	20	20	
Agricultural landholdings	No landholdings	48	65	55	0.024
	Landholdings	52	35	45	
Housing status	Straw made	19	7	3	0.001
	Thatched roof	29	21	24	
	Kachcha house	30	39	44	
	Pucca house	22	34	29	

2.3 *Data analysis*

Statistical Package for Social Sciences (SPSS), IBM version 21 in Windows environment, was used for statistical analysis (IBM, 2012; SPSS, 2012). A bivariate analysis was performed between the risk perceptions (health, economic, and social discrimination) and socioeconomic, demographic and other variables (IBM, 2012; SPSS, 2012; Warner, 2012).

3 RESULTS AND DISCUSSION

3.1 *Socioeconomic and demographic characteristics of the surveyed population*

The detailed socioeconomic and demographic survey results are explained in (Singh, 2015), here we present a village-wise comparison of socioeconomic and demographic survey results that help to understand the socioeconomic and demographic characteristics of the surveyed population (Table 1).

3.2 *Perceived risks and decision-making for arsenic-mitigation technologies*

The vast majority of individuals (91%) perceived health risk due to drinking arsenic-contaminated water

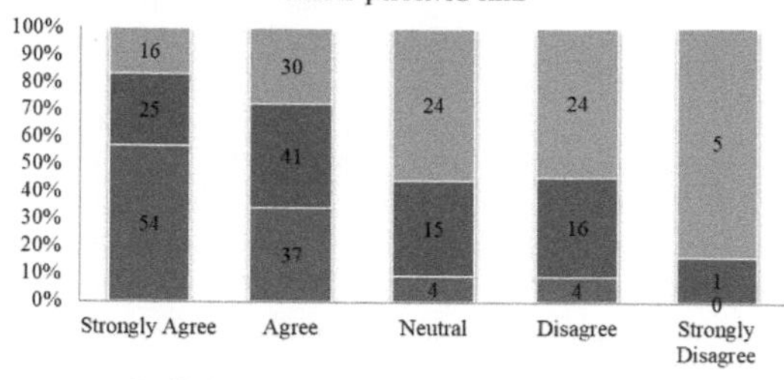

Figure 3. Participants' responses to statements on perceptions in the context of the adoption of arsenic mitigation options.

and was willing to adopt arsenic-mitigation technologies (Fig. 3).

About 66% of the respondents perceive economic risk from arsenic contamination. Only 46% of the surveyed population connects risk of social discrimination with arsenic contamination. The respondents showed a neutral response to perceived social discrimination risk, with the numbers almost equally split between those who perceive social discrimination risk and others who disagree or are neutral. The respondents of Rampur Diara village were slightly more likely (93%) to adopt arsenic-mitigation under perceived health risk than the respondents of Suarmarwa (90%) and Bhawani Tola (89%) villages.

3.3 *Factors influencing likelihood of adoption of arsenic-mitigation technologies under perceived risks*

Variables such as caste, education level, agricultural landholdings, housing status, presence and functionality of *Anganwadi*, and social capital (people who participate in panchayat activities), were found to have a contributing effect on communities' perceived health risk. Moreover, people's perceived economic risk was found to be strongly positively correlated with the respondents' health risk perception. In sum, educated higher caste communities who own land, live in better housing structures, and had trust in *Anganwadi*[1] in the area, tend to perceive more health risk. Therefore, under perceived health risk, communities that possess the above mentioned characteristics will adopt arsenic mitigation technology in the arsenic-contaminated areas (Table 2). The results of bivariate analyses showed that there was a very strong positive correlation between respondents' perceived health risk, social discrimination risk, and their perceived economic risk. It further shows that the communities that had high perceived health risk and social discrimination risk were likely to perceive more economic risk (Table 2).

Table 2. Predictors of risk perceptions to health risk, economic risk, and social discrimination risk.

	Health risk perception	Economic risk perception	Social discrimination risk perception
Caste	.174**	.242**	-.062
Education level	.111*	.184**	.102
Agricultural landholdings	.156**	.047	-.127*
Housing status	.112*	.124*	.036
Time spent for water collection	.059	.093	-.171**
Distance from water source	.057	.065	-.120*
Place for defecation	.077	.170**	.082
Materials used to wash hands after defecation	.062	.162**	.082
Lack of land	.081	.184**	.017
Lack of school	.047	.265**	.072
Lack of health facilities	-.099	-.295**	-.093
Flood	-.035	-.122*	-.016
Water accessibility	.046	-.158**	.071
Water quality	-.059	-.228**	.000
Poverty	.032	.325**	-.018
Arsenic awareness	.032	.234**	.110*
Opinion about the presence and functionality of *Anganwadi*	.139*	-.158**	-.058
Opinion about the presence and functionality of *Mahila Samakhya*	-.076	-.042	-.197**
Opinion about the presence and functionality of *Self Help Groups*	.026	.057	-.226**
Trust in *NGOs*	.106	.161**	.232**
Trust in *Panchayat Raj Institutions*	.039	.141**	.251**
Trust in private agencies	-.018	.169**	.077
Participation in panchayat activities	.107*	.046	.008
Neighbors and others seek advice and value views	.091	.397**	.191**
Health risk perception	1.000	.268**	.068
Economic risk perception	.268**	1.000	.233**
Social discrimination risk perception	.068	.233**	1.000

***. Correlation is significant at the 0.01 level (2-tailed).*
**. Correlation is significant at the 0.05 level (2-tailed).*

Likewise, a strong positive correlation was derived between caste, education level, housing status, sanitation behavior (place for defecation and material used to wash hands after defecation), people's prioritization of social problems (lack of land, lack of school, and poverty), arsenic awareness, trust in agencies (NGOs, *Panchayati Raj Institutions*, private agencies), and individual social capital, as neighbors and others sought and valued their advice, with the residents' perceived economic risk (Table 2). However, a strong negative correlation was observed between people's prioritization of socio-environmental problems (lack of health facilities, flood/drought, water accessibility, and water quality), and presence and functionality of institutions (*Anganwadi*), with the respondents' perceived economic risk.

In sum, socio-demographic factors (caste, education, and housing status), sanitation practice, people's prioritization of socio-environmental problems, arsenic awareness, presence and functionality of institution (*Anganwadi*), trust in agencies, and social capital, were found to be strong predictors of the perceived economic risk of the communities (Table 2). In comparison to the other two perceived risks (economic and health risks), we found fewer predictors of perceived risk of social discrimination. None of the socioeconomic and demographic factors, except agricultural landholdings, was found to be a statistically significant predictor of the communities' social risk perception. People with larger agricultural landholdings, who spent more time collecting water (and traveled a distance of >50 m), and the communities

[1] *Anganwadi* (courtyard shelter) is a government of India's child-care and mother-care unit at panchayat levels, comprised of mostly female health workers.

that had higher trust in *Mahila Samakhya*[2] and *Self-Help Groups*[3] were strongly negatively correlated with their social discrimination risk perception. In contrast, those who were aware of arsenic contamination, who had trust in *NGOs* and *Panchayati Raj Institutions*, and the respondents with higher social capital were found to be strongly positively correlated with their perceived social discrimination risk. Furthermore, communities' social discrimination risk perception was found to be strongly positively correlated with their perceived economic risk (Table 2).

4 CONCLUSIONS

This study provides insights into communities' decision-making to adopt arsenic-mitigation technologies under their perceived risks to health, income, and social discrimination to groundwater arsenic-contamination. The findings are potentially advantageous to policymakers, as this help identify the mechanisms of decision-making to adopt better environmental services for the communities' wellbeing. Risk perception in general, but health, economic and social discrimination risk perceptions in particular, cause individuals to take precautionary measures and support policies aimed at arsenic-mitigation (Botzen *et al.*, 2009). A "one-size-fits-all" approach to promoting arsenic-mitigation is unlikely to succeed owing to the heterogeneous nature of the communities affected. As is evident from this paper, different castes and economic classes, constituting the affected communities, have differing perceptions of health, economic and social risk, because of their varied perceptions and priorities. Upper caste communities should be targeted for immediate arsenic-mitigation, as they are more likely to adopt. This might also help to enhance technology adoption among lower castes due to the fact that the latter seek to emulate the former in many situations. However, it appears that, unfortunately, health concerns take a backseat to pressing issues of economic survival for lower-caste and impoverished households. In the short-term, education and awareness campaigns aimed at arsenic contamination and, specifically, its adverse impacts on livelihoods should be implemented among the lower caste communities, including backward and scheduled caste. Communities with minimal trust in the institutions and low social capital, both owing perhaps to their social marginalization, were most concerned with economic risks and social discrimination risks. However, underprivileged communities for whom social problems were the highest priority had low or no arsenic awareness. Therefore, to raise arsenic awareness in these communities, campaigns should focus, in the medium to long-term, on strengthening institutions and enhancing social capital activities. Most importantly, these institutions should be involved in village level arsenic mitigation-programs. That groundwater arsenic may not be a priority among poor communities should not be seen as a discouraging finding. On the contrary, it should be seen as an affirmation of the fact that environmental and health perceptions are related to socio-economic problems like lack of jobs, lack of land, lack of schools, or lack of health facilities. Understanding these linkages could be very useful for understanding how people perceive risks, what the predictors are, and how decisions are made for ultimately delivering better social or environmental services. The identified predictors of the three risk perceptions to groundwater arsenic-contamination will help us model the risk perceptions to arsenic-contamination. The results of this study suggest that severity of groundwater, soil, or food arsenic contamination, while important, should not be the only control on mitigation policies. People's risk perception plays a vital role in the adoption of arsenic-mitigation interventions. Caste, broad measures of economic well-being and markers of social status within the community are important drivers of the different risk perceptions. However, it appears that social marginality, which is a relative social position (as determined by the compounding effect of education, landholding, arsenic awareness, among other factors), is a crucial factor, perhaps more so than absolute measures of income and education. Therefore, socioeconomic and demographic factors, awareness, institutional trust, social trust, and social capital, etc., should necessarily be evaluated before implementing any arsenic-mitigation policy. This approach would help increase adoption of arsenic-mitigation interventions and achieve their sustainable use.

[2] The *Mahila Samakhya* is a women's empowerment program of the government of India, designed to aid socioeconomically impoverished women through increased literacy.

[3] *Self Help Groups* are comprised of socioeconomically disadvantaged women in villages in India and other South Asian countries primarily established for social empowerment through micro-finance.

REFERENCES

Abdalla, C.W., Brian, A.R. & Donald, J.E. 1992. Valuing environmental quality changes using averting expenditures: an application to groundwater contamination. *Land Econ.* 68(2): 163–169.

Bickerstaff, K. 2004. Risk perception research: socio-cultural perspectives on the public experience of air pollution. *Environ. Int.* 30(6): 827–840.

Botzen, W.J.W., Aerts, J.C.J.H. & van den Bergh, J.C.J.M. 2009. Dependence of flood risk perceptions on socio-economic and objective risk factors. *Water Resour. Res.* 45(10): 455–464.

Carlton, S.J. & Jacobson, S.K. 2013. Climate change and coastal environmental risk perceptions in Florida. *J. Environ. Manage.* 130: 32–39.

Damm, A., Eberhard, K., Sendzimir, J. & Patt, A. 2013. Perception of landslides risk and responsibility: a case study in eastern Styria, Austria. *Nat. Hazards* 69(1): 165–183.

Dupont, D. & Krupnick, A. 2010. Differences in water consumption choices in Canada: the role of socio-demographics, experiences, and perceptions of health risks. *J. Water Health* 8(4): 671–686.

Gerber, B.J. & Neeley, G.W. 2005. Perceived risk and citizen preferences for governmental management of routine hazards. *Policy Stud. J.* 33(3): 395–418.

Huang, L., Ban, J., Sun, K., Han, Y.T., Yuan, Z.W. & Bi, J. 2013. The influence of public perception on risk acceptance of the chemical industry and the assistance for risk communication. *Safety Sci.* 5 (1): 232–240.

IBM 2012. IBM SPSS statistics for Windows, version 21.0. IBM Corp Armonk, New York.

Jakus, P.M., Shaw, W.D., Nguyen, T.N. & Walker, M. 2009. Risk perceptions of arsenic in tap water and consumption of bottled water. *Water Resour. Res.* 45(5).

Janmaat, J. 2007. Divergent drinking water perceptions in the Annapolis Valley. *Can. Water Resour. J.* 32(2): 99–110.

Jones, E.C., Faas, A.J., Murphy, A.D., Tobin, G.A., Whiteford, L.M. & McCarty, C. 2013. Cross-cultural and site-based influences on demographic, well-being, and social network predictors of risk perception in hazard and disaster settings in Ecuador and Mexico. *Human Nature* 24(1): 5–32.

McSpirit, S. & Reid, C. 2011. Residents' perceptions of tap water and decisions to purchase bottled water: a survey analysis from the Appalachian, Big Sandy Coal Mining Region of West Virginia. *Soc. Natur. Resour.* 24(5): 511–520.

Nguyen, T.N., Jakus, P.M., Riddel, M. & Shaw, W.D. 2010. An empirical model of perceived mortality risks for selected US arsenic hot spots. *Risk Anal.* 30(10): 1550–1562.

Onjala, J., Ndiritu, S.W. & Stage, J. 2014. Risk perception, choice of drinking water and water treatment: evidence from Kenyan towns. *J. Water Sanit. Hyg. Dev.* 4(2): 268–280.

Peacock, W.G., Brody, S.D. & Highfield, W. 2005. Hurricane risk perceptions among Florida's single family homeowners. *Landsc. Urban Plan.* 73(2): 120–135.

Singh, S.K. 2015. *Assessing and Mapping Vulnerability and Risk Perceptions to Groundwater Arsenic Contamination: Towards Developing Sustainable Arsenic Mitigation Models*. Available from ProQuest Dissertations & Theses Full Text. (1681668682). PhD Dissertation/Thesis, Earth and Environmental Studies, Montclair State University.

Singh, S.K., Brachfeld, S.A. & Taylor, R.W. 2016. Evaluating hydrogeological and topographic controls on groundwater arsenic contamination in the Middle-Ganga plain in India: towards developing sustainable arsenic mitigation models. In: A. Fares (ed) *Emerging Issues in Groundwater Resources*. pp. 263–287.

Singh, S.K., Taylor, R.W. & Su, H. 2017. Developing sustainable models of arsenic-mitigation technologies in the Middle-Ganga Plain in India. *Curr. Sci.* 113(1): 80–93.

Singh, S.K., Taylor, R.W., Rahman, M.M. & Pradhan, B. 2018. Developing robust arsenic awareness prediction models using machine learning algorithms. *J. Environ. Manage.* 211: 125–137.

SPSS, IBM 2012. IBM SPSS Statistics Vcrsion 21. Boston, Mass. International Business Machines Corp.

Tam, J. & McDaniels, T. L. 2013. Understanding individual risk perceptions and preferences for climate change adaptations in biological conservation. *Environ Sci Policy* 27: 114–123.

UNSD 2008. Designing household survey samples: practical guidelines. Vol. 98. USA: United Nations Statistical Division (UNSD).

Walker, M., Shaw, W.D. & Benson, M. 2006. Arsenic consumption and health risk perceptions in a rural western US area. *J. Am. Water Resour. Assoc.* 42(5): 1363–1370.

Warner, R.M. 2012. *Applied Statistics: From Bivariate Through multivariate Techniques*. Sage Publications.

Environmental Arsenic in a Changing World –
Zhu, Guo, Bhattacharya, Ahmad, Bundschuh & Naidu (Eds)
ISBN 978-1-138-48609-6

Social approach to arsenic mitigation in Gangetic belt, India

S. Singh & Chandrbhushan
Innervoice Foundation, Varanasi, India

ABSTRACT: Arsenic contamination of groundwater from two Indian states (Bihar and Uttar Pradesh) has been reported during 2003–2004. Although several small scale studies documented the arsenic concentrations in groundwater, source of arsenic in groundwater, effect of arsenic-contaminated in food crops and the health effects due to the consumption of polluted water. Very little is known about the social approach on arsenic mitigation, which is crucial. This paper highlights the importance of social approach to arsenic mitigation.

1 INTRODUCTION

Arsenic contamination in groundwater of Bihar and Uttar Pradesh (UP) was documented during 2003–2004 (Chakraborti *et al.*, 2003, 2004; MDWS, 2011; Mukherjee *et al.*, 2006; Ramanathan *et al.*, 2009). During last 15 years, various experts have been working on this topic. It has been estimated that about 2 million people are exposed and currently at risk from groundwater arsenic contamination from both Indian states. Reduction of arsenic in public water is important to minimize the exposure to human. Therefore, mitigation approach is crucial to provide arsenic-safe water to the people living in these areas (Hossain *et al.*, 2015; SASMIT, 2014). However, immediate mitigation intervention is not possible due to the negligence of the problem, lack of acceptance of the existing problem by the government and other agencies who are responsible for the safe water supply.

In developing nations, there is lack of adequate water legislation, and periodic screening program of the drinking water is absent. Although water legislations are present, no implementations are likely to occur due to lack of awareness of the village communities and adequate capacity of the government officials in arsenic mitigation programs. Thus, arsenic mitigation efforts are often missing in the affected villages. As a result, people are suffering from various arsenical diseases due to the consumption of the arsenic contaminated water. On the other end, it can be realized that without community's active participation, it is challenging to fight against the arsenic toxicity. Therefore, we need first to educate the villagers as well as we need to train them about the effective arsenic mitigation approaches for the operative mitigation program.

2 METHODS

Based on the available village-wise baseline data from the study area, mapping of arsenic distribution in groundwater were conducted. The information gathered was critically analyzed to understand the current knowledge and gaps in the existing mitigation approaches followed by implementation of the effective mitigation approaches. Community-based meetings were organized with the affected villagers and local authorities to determine the problems and the existing health concerns. Their responses were evaluated and included in the viable mitigation approach to the impacted areas. This will lead to developing a holistic strategy to combat arsenic mitigation in rural parts of the Gangetic belt.

3 RESULTS AND DISCUSSION

Proper mapping of arsenic-contaminated sites is required involving technical experts, scientists, experts in water law, policymakers and members from affected communities. Primary and secondary level data from affected areas are required for this purpose. Reports from local newspapers could be used to form such database.

Discussion with the community will be carried out, and the results will be recorded in a questionnaire format. Assessing community's need often requires special insight at a preliminary level. Questionnaire format developed in consultation with the affected communities provides a significant understand of the real problems faced by the communities in the affected regions. From previous experience, we came to know that villagers were entirely unaware about the issue and therefore not in a position to speak on the subject at this stage as they did not know the basic facts of arsenic contamination, which was affecting their health and livelihood beyond measures. It proved to be helpful if the community is sensitized and educated on the subject at this stage. Making community speak out on the subject marked significant information/suggestion inflow on the subject (Rajya Sabha TV, 2015). Various reports in the media have proved to have considerable impact, as exemplified by the report on arsenic contamination in Ballia based on the complaint addressed by the affected people of Ballia. A National Level Monitor was appointed to investigate the complaint on the water quality by the Ministry of Rural Development, Government of India (India Water Portal, 2011; MDWS, 2011).

These exercises have a very positive impact on society, which also helped in ensuring their active participation in subsequent arsenic mitigation programmes. Opening up a dialogue with Central, State, and Local authorities are also crucial.

4 CONCLUSIONS

There is a lack of legislative structure in developing countries including India, which is necessary to address such sensitive issue. To solve this issue, we need concerted efforts apart from technological efforts. Media can play a very significant role. At this very crucial stage of 'Social Arsenicosis' program, utmost care and patience are required in choosing appropriate reports, facts and platform to gather support for the cause. Care should be taken not to offend sensibilities of any group or set of persons. Also, it can have the very negative impact on campaign build over the period.

REFERENCES

Chakraborti, D., Mukherjee, S.C., Pati, S., Sengupta, M.K., Rahman, M.M., Chowdhury, U.K., Lodh, D., Chanda, C.R., Chakraborti, A.K. & Basu, G.K. 2003. Arsenic groundwater contamination in Middle Ganga Plain, Bihar, India: a future danger? *Environ. Health Perspect.* 2111(9):1194–1201.

Chakraborti, D., Sengupta, M.K., Rahman, M.M., Ahamad, S., Chowdhury, U.K., Hossain, M.A., Mukherjee, S.C., Pati, S., Saha, K.C., Dutta, R.N. & Quamruzzaman, Q. 2004. Groundwater arsenic contamination and its health effects in the Ganga-Meghna-Brahmaputra plain. *J. Environ. Monit.* 6(6): 74N–83N.

India Water Portal 2011. Investigation and assessment report: Arsenic in drinking water sources and related problems of Ballia district of Uttar Pradesh. http://www.indiawaterportal.org/articles/investigation-and-assessment-report-arsenic-drinking-water-sources-and-related-problems (accessed on 30 April, 2018)

MDWS 2011. Report of the Central Team on Arsenic mitigation in rural drinking water sources in Ballia district, Uttar Pradesh State. Ministry of Drinking Water and Sanitation, Government of India, New Delhi, India. 25p. (Downloadable from: http://www.mdws.gov.in/sites/default/files/ballia-finalreport_0.pdf (accessed on 14 April, 2018).

Mukherjee, A.B., Bhattacharya, P., Jacks, G. Banerjee, D.M., Ramanathan, A.L., Mahanta, C. Chandrashekharam, D., Chatterjee, D. & Naidu, R. 2006. Groundwater arsenic contamination in India: extent and severity. In: R. Naidu, E. Smith, G. Owens, P. Bhattacharya & P. Nadebaum. (eds) *Managing Arsenic in the Environment: From soil to human health.* CSIRO Publishing, Melbourne, Australia, pp. 533–594.

Ramanathan, AL., Tripathi, P., Kumar, M., Bhattacharya, P., Thunvik, R. & Bundschuh, J. 2009. Arsenic distribution in the groundwater in Central Gangetic Plains of Uttar Pradesh, India. In: J. Bundschuh, M.A. Armienta, P. Birkle, P. Bhattacharya, J. Matschullat & A.B. Mukherjee (eds): *Natural Arsenic in Groundwater of Latin America – Occurrence, health impact and remediation.* CRC Press/Balkema, Leiden, The Netherlands, pp. 215–224.

Rajya Sabha TV 2015. Special Report – Arsenic Poisoning of Water. (Pani mein zahar). (Viewable link: https://www.youtube.com/watch?v=GvsbS4zzosw&t=106s (accessed on 30 April, 2018).

5.2 Policy instruments to regulate arsenic exposure

Environmental Arsenic in a Changing World –
Zhu, Guo, Bhattacharya, Ahmad, Bundschuh & Naidu (Eds)
ISBN 978-1-138-48609-6

Integrating policy, system strengthening, research and harmonized services delivery for scaling up drinking water safety in Bangladesh

B. Onabolu[1], E. Khan[2], J. Chowdhury[2], N. Akter[1], S.K. Ghosh[2], M.S. Rahman[2], P. Bhattacharya[3], D. Johnston[1], K. Alam[1], F. Ahmed[1], R. Amin[1], S. Uddin[4], S. Khanam[5], M. Hassan[6], M. von Brömssen[7] & K.M. Ahmed[8]

[1]*UNICEF Bangladesh, Dhaka, Bangladesh*
[2]*Department of Public Health Engineering, Dhaka, Bangladesh*
[3]*KTH-International Groundwater Arsenic Research Group, Department of Sustainable Development, Environmental Science and Engineering, KTH Royal Institute of Technology, Stockholm, Sweden*
[4]*Asia Arsenic Network, Dhaka, Bangladesh*
[5]*Environment and Population Research Center, Dhaka, Bangladesh*
[6]*Village Education and Resource Center, Dhaka, Bangladesh*
[7]*Ramböll Sweden AB, Stockholm, Sweden*
[8]*Department of Geology, University of Dhaka, Dhaka, Bangladesh*

ABSTRACT: The WASH Sector Development Plan (2011–2025) recognizes the absence of harmonized approaches as the root challenge to scaling up drinking water safety in Bangladesh. UNICEF is supporting the Government of Bangladesh (GoB) through the Department of Public Health Engineering (DPHE) and Policy Support Branch to re-engineer its approach to drinking water safety by integrating policy, systems strengthening & sustainable services delivery at national and sub-national levels. Some of the key contributions to the sector include the implementation of the UNICEF-DPHE Arsenic Safe Union model with the declaration of 106 Arsenic Safe Villages, reduction in arsenic contamination rates of new tubewells, a system for preventing elite capture of water points, and the adoption of the ASU model in a $240 million arsenic mitigation drinking water project using domestic resources. Relatedly, a Policy Support Branch has been established, the sector coordination mechanisms revised, WASH bottleneck analysis is ongoing & the National Implementation Plan on Arsenic has been reviewed: Next steps include the professionalisation of drilling by local drillers through a partnership with KTH, Sweden, the GoB and UNICEF.

1 INTRODUCTION

Bangladesh has made laudable progress towards achieving its goal of ensuring access to safe drinking water for all its citizens and achieved the Millennium Development Target for drinking water (GED, 2016). However in the context of the SDG 6.1 there is a gap between coverage and safety of the improved water sources. Although 97.8% of the population have access to improved water sources nationally, about 65% of the population lack access to drinking water that is arsenic safe and free from microbial contamination and 19.7 million people are exposed to arsenic from improved water sources that exceed the BDWS (50 μg/L) (BBS and UNICEF, 2015).

The Sector Development Plan (2011-2025) identifies the challenges to scaling up drinking water safety in Bangladesh as geogenic arsenic, iron, manganese & salinity; environmental vulnerability; inadequate prioritization of arsenic prone areas; private sector capacity; and absence of harmonized sector wide approaches. This paper focuses on the sustainable services delivery component.

2 PROCESS

2.1 *Sustainable services delivery -scalable DPHE-UNICEF Arsenic Safe Union Model*

The Arsenic Safe Union (ASU) model uses the village as the unit of intervention and not the individual water source. It emphasizes vulnerability risk assessment for selecting most arsenic prone areas, local government ownership, provision of sanitation and safe water supply, pro-poor selection criteria, geocoded water point installations, real time reporting and information management, evidence based decision making and advocacy. The final step is the declaration, verification, and certification of the union as arsenic safe.

The ASU Model is built on the UNICEF-DPHE arsenic mitigation protocol (Fig. 1) and arsenic safe

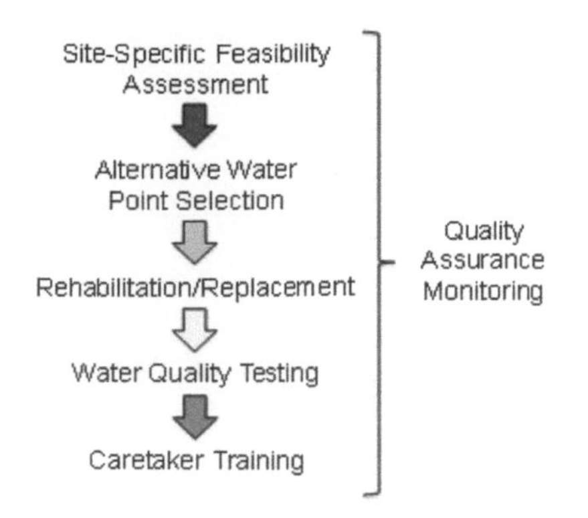

Figure 1. Arsenic Mitigation Protocol.

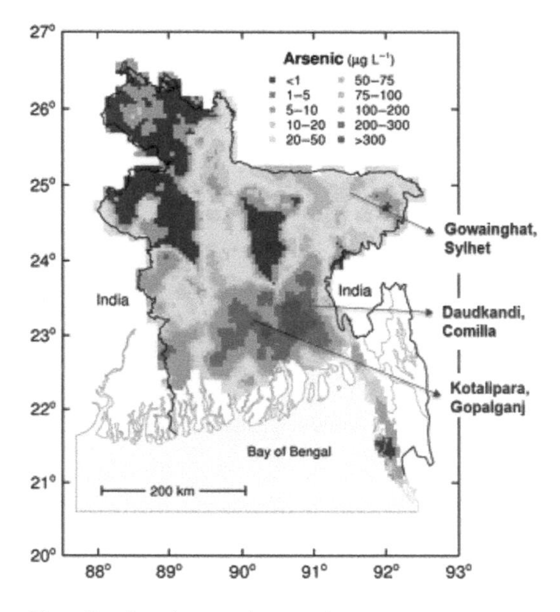

Figure 2. Arsenic prone intervention areas.

village concept with the inclusion of vulnerability risk assessments to select most vulnerable administrative units for intervention.

2.2 *Scalable DPHE-UNICEF Arsenic Safe Union model*

UNICEF in collaboration with DPHE designed the ASU concept in 2015. The concept was implemented between January 2016 and December 2017 in a total of 15 unions in three Sub-districts of three districts in Bangladesh (Fig. 2) with funding from Swedish International Development Agency and UNICEF.

The project was implemented in partnership with Asia Arsenic Network, Environment and Population Research Center and Village Education and Research Center. The Project aimed to reach 100,000 people with safe drinking water and 300,000 people with safe sanitation and increased awareness about arsenic and hygiene.

2.3 *Integration with policy, planning, systems strengthening and resource mobilization*

In addition to advocating for a common approach to services delivery, UNICEF Bangladesh engaged with government, development partners and the sector to improve coordination, domestic and foreign resource mobilization for scaling up drinking water safety, improved information management and mechanisms to close the gap between policies and implementation of the policies.

2.4 *Adoption of vulnerability risk assessment by national and sub-national government for selection of most vulnerable areas*

Secondary data was used to select the As prone districts and sub-districts (Upazilas) (BBS & UNICEF, 2015; Flanagan *et al.*, 2012). Then vulnerability risk assessments were used to select 15 most As contaminated unions in the intervention upazilas. The vulnerability indices were percentage of As contamination, number of arsenicosis patients, safe water supply to population ratio. The acceptance of this by government is a significant achievement and change from the past when allocation was not based on need or degree of contamination.

3 RESULTS: KEY CONTRIBUTIONS TO THE SECTOR

3.1 *Blanket arsenic screening*

All drinking water sources ($n = 21{,}540$) in the intervention sites were screened for As and tubewell depth Some of the findings of the screening results are presented in Table 1.

The survey indicated that nearly 88% of the wells were shallow (<500 ft) and the remaining 12% were deep wells, and nearly 97% of the wells were functional. Combining both the shallow and the deep wells about 54% were found to be safe in terms of the Bangladesh drinking water standard ($50\,\mu g\,L^{-1}$).

Arsenic contamination was mostly prevalent in the shallow tube-wells (52%) and mostly confined to those provide privately.

3.2 *Quality assurance (QA) monitoring*

About 80% of samples were declared either safe or unsafe ($0.05\,mg\,L^{-1}$) by both field kit (Econo-quick) and laboratory (Atomic Absorption Spectrophotometer) while some false positives (9%) and false negatives (11%) indicated discrepancy between the two methods.

3.3 *Baseline survey alignment of indicators with SDG 6.1*

A water and sanitation KAP baseline survey was conducted in 750 households using SDG 6.1 and 6.2 indicators. The findings indicated a significant decrease from coverage figures when safety and access on premises was considered (Fig. 3).

Table 1. Summary of the results of the screening of the drinking water sources.

#	Criteria	Percentage
1.	Functional status	97%
2.	Depth of the wells	
	>500 ft	12%
	<500 ft	88%
3.	Arsenic contamination status	
	Shallow and deep wells (combined)	
	Safe	54%
	Unsafe	46%
	Arsenic contamination in deep tube-wells	1%
	Arsenic contamination in shallow tube-wells	45%
4.	Provider (Shallow and deep wells, combined)	
	Private	83%
	Government	12%
	NGO	5%
5.	Provider of shallow wells	
	Private	83%
	Government	5%
	NGO	12%
6.	Arsenic contamination of TW by provider	
	Private	59%
	NGO	41%

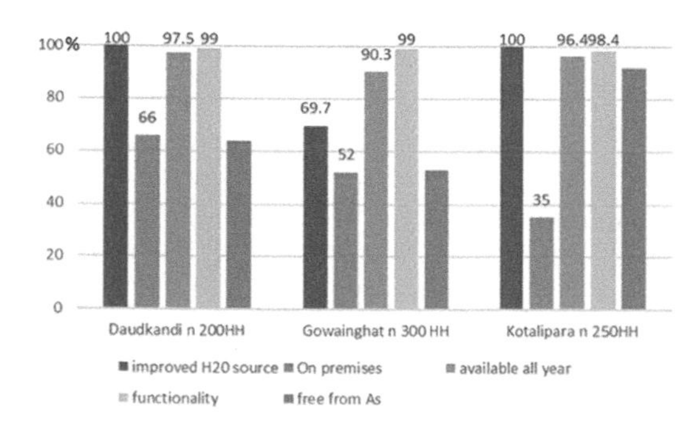

Figure 3. SDG 6.1 – Baseline survey results

3.4 *The use of GIS mapping for evidence based advocacy and decision making*

The data from site-specific feasibility assessment (before intervention) and new water points was converted into GIS maps and used for advocacy to government to prioritize vulnerable villages and change from the practice of equal allocations and as evidence of the successful interventions.

3.5 *Three tier monitoring – reduction in proportion of arsenic contaminated wells*

The proportion of new wells that were As contaminated ($0.05\,mg\,L^{-1}$) reduced from 9% to 3% and As, Fe, Mn and salinity levels were determined for mitigation before handing over to users.

3.6 *Three tier monitoring – reduction in elite capture of project tubewells*

Site selection criteria integrated with GPS coordinates of agreed and actual sites were monitored to ensure water sources were allocated to the poor as planned.

3.7 *Scaling up – adoption of Arsenic Safe Union model by Government of Bangladesh*

DPHE has adopted the ASU model for use in a $240 million arsenic mitigation project in 29 arsenic prone districts.

4 CONCLUSION AND NEXT STEPS

The Arsenic Safe Union model if integrated with policy and systems strengthening interventions is scalable for sector wide As mitigation. UNICEF is working with KTH Royal Institute of Technology in Sweden to integrate science with indigenous knowledge using the Sediment Colour Tool to improve private sector effectiveness in safe water drilling. UNICEF is supporting the development of low cost As test kit locally, arsenic safe union operational guidelines, integration of arsenicosis in the health information system and the Implementation of national plan for arsenic mitigation by all stakeholders.

ACKNOWLEDGEMENTS

The financial contribution of the Swedish International Development Agency (Sida Contribution 52170040) is appreciated.

REFERENCES

Bangladesh Bureau of Statistics (BBS) and UNICEF Bangladesh, 2015. *Bangladesh Multiple Indicator Cluster Survey 2012-2013, Progotir Pathey: Final Report*. Dhaka, Bangladesh: Bangladesh Bureau of Statistics (BBS) and UNICEF Bangladesh.

General Economics Division (GED), 2016. *Millennium Development Goals (MDGS): End-Period Stocktaking and Final Evaluation (2000-2015)*. Dhaka, Bangladesh: General Economics Division (GED), Bangladesh Planning Commission, Government of the People's Republic of Bangladesh.

Flanagan, S.V., Johnston, R.B., & Zheng, Y., 2012. Arsenic in tube well water in Bangladesh: health and economic impacts and implications for arsenic mitigation. *Bulletin of the World Health Organization, Volume 90, Number 11, November 2012, 839-846*. Available at: https://www.who.int/bulletin/volumes/90/11/11-101253/en/ [Last accessed: 31 May 2018]

5.3 Risk assessments and remediation of contaminated land and water environments – Case studies

Environmental Arsenic in a Changing World –
Zhu, Guo, Bhattacharya, Ahmad, Bundschuh & Naidu (Eds)
ISBN 978-1-138-48609-6

Effect of arsenic risk assessment in Pakistan on mitigation action

J.E. Podgorski[1], S.A.M.A.S. Eqani[2] & M. Berg[1]
[1]*Eawag, Swiss Federal Institute of Aquatic Science and Technology, Dübendorf, Switzerland*
[2]*COMSATS Institute of Information Technology, Islamabad, Pakistan*

ABSTRACT: A recently published study on the potential extent of arsenic contamination in groundwater in Pakistan led to a large media response that spurred widespread interest by the public, health agencies and the government. An effect was the commitment of the WHO and the Pakistani government to enact immediate mitigation measures. This highlights the significant effect that contaminant prediction mapping can have on raising awareness and initiating mitigation action.

1 INTRODUCTION

The heterogeneous nature of aquifers and arsenic contamination ultimately mean that it is necessary to test all wells in an area in order to determine if unsafe levels of arsenic are present. Contamination prediction maps help to efficiently use limited testing resources in a targeted manner by identifying areas that are more likely to exhibit high arsenic levels. A recent study on the previously unmapped extent of arsenic contamination in groundwater throughout Pakistan (Podgorski *et al.*, 2017) used a statistically based approach to map areas of high probability of arsenic concentration exceeding the 1.5 $\mu g\ L^{-1}$ World Health Organization (WHO) guideline of arsenic in drinking water. This resulted in a prediction that 50–60 million people are at risk to exposure to unacceptably high levels of arsenic in their drinking water. The ensuing media response has incited a nationwide debate about the seriousness of the problem and how to best address it. In this paper, we look at how this debate has unfolded and what concrete effect the originally published study is having on the mitigation of the arsenic problem in Pakistan.

2 METHODS

2.1 *Statistical modeling*

The arsenic study of Podgorski *et al.* (2017) made nearly 1200 arsenic measurements around Pakistan, which were then modelled to create a prediction map of arsenic concentrations in groundwater exceeding the WHO guideline of 1.5 $\mu g\ L^{-1}$. Logistic regression was used, which finds correlations between known concentrations and predictor variables that act as proxies for the processes of arsenic accumulation in groundwater. As opposed to interpolation, logistic regression enables model prediction based on actual physical relationships, rather than merely the geographical distribution of sampled groundwater wells. The probability of the model selected as the cutoff to distinguish between areas of high and low risk was determined by the best tradeoff between the prediction accuracy of low and high values (below or above 1.5 $\mu g\ L^{-1}$). The high-hazard area so determined was then compared with a population map to establish the number of people at risk in this area, also taking into account the proportion of people consuming groundwater.

2.2 *Dissemination*

The findings of this study were disseminated via three channels: the publication itself in the journal (Science Advances), a press release issued by the lead author's home institution (Eawag, Switzerland) and a press package released by the publisher to EurekAlert and other media dissemination outlets that then made the information about the study visible to media outlets worldwide.

3 RESULTS AND DISCUSSION

3.1 *Statistical modeling*

The main finding of the Podgorski *et al.* (2017) study was a prediction model showing a high probability of elevated arsenic concentrations in groundwater throughout the Indus Valley and its tributaries (Fig. 1A). After trying about 10 different predictor variables, the best model was found using five variables: fluvisols, Holocene fluvial sediments, soil organic carbon, soil pH and slope. The cutoff used for the probability at which the model is accurate at predicting low and high values was 0.60. The high-hazard area thus delineated encompasses the plains adjacent to the Indus River and its tributaries in the states of Punjab and Sindh. Taking into account that approximately 60–70% of the Pakistani population relies on

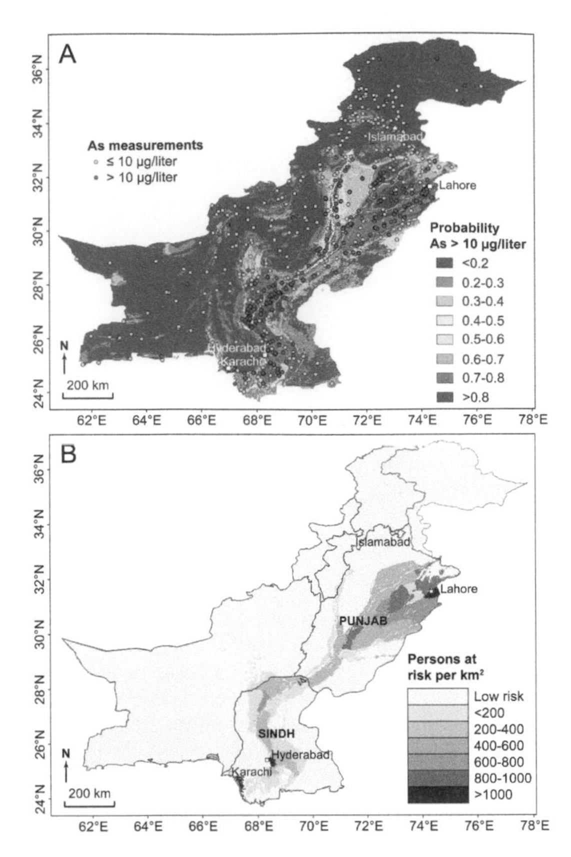

Figure 1. (A) Probability map of arsenic in groundwater in Pakistan exceeding the WHO guideline of $1.5\,\mu g\,L^{-1}$. Points indicate measured concentrations meeting or exceeding this guideline. (B) Population consuming groundwater at risk to arsenic concentrations in excess of $1.5\,\mu g\,L^{-1}$.

groundwater for drinking, approximately 50–60 million have been estimated to be at risk to elevated arsenic levels (Fig. 1B). This represented the first time that the full extent of arsenic contamination had been mapped in detail across Pakistan.

3.2 *Media response*

Following the press package released by the publisher prior to the actual publication of the study, many interviews were given on the study's findings. Interviews were made with journalists from internationally recognized media outlets such as the Associated Press, BBC and CNN who prepared their articles about the study for immediate release coinciding with the study's publication. The high level of interest was apparently due to the great number of people potentially affected. The article by the Associated Press, in particular, resulted in broad dissemination of the arsenic study's findings due to many other media outlets carrying this article, e.g. New York Times, Washington Post. In terms of the quality and quantity of online attention received by the study, it ranks in the top 0.07% of all published research ever evaluated by Altmetric (Altmetric score 608).

3.3 *Reaction in Pakistan*

In Pakistan, the release of the study was a big story that resulted in a very high level of interest over the following weeks. Debates were carried out in the internet, newspapers, television and the government about the ramifications of the arsenic risk assessment and how best to go about addressing the problem. As a direct consequence of the finding's study, the WHO sent a special envoy to Pakistan to conduct a week-long mission to visit arsenic hotspots and investigate the initiation of mitigation measures. Despite the senate initially rejecting the study, the Pakistani government then committed to joining the WHO in launching a nationwide survey of groundwater quality, which is the first action that needs to be conducted in addressing the problem.

4 CONCLUSIONS

The widespread media response received by the Podgorski *et al.* (2017) Pakistan arsenic study and the ensuing public outcry led to the WHO and the Pakistani government enacting an immediate mitigation plan. Aside from the original intent of predicting areas likely containing high contaminant concentrations, this case shows that statistically produced prediction maps, given an adequate level of dissemination, can have an immediate direct effect in promoting awareness and prompting mitigation measures.

ACKNOWLEDGEMENTS

We thank the Swiss Agency for Development and Cooperation (7F-09010.01), the Swiss Federal Institute of Aquatic Science and Technology (Eawag), the Higher Education Commission of Pakistan (project no. PM-IPFP/HRD/HEC/2012/3599), the National Natural Science Foundation of China (Research Fellowship for International Young Scientists, No. 21450110419), and the Institute of Urban Environment of the Chinese Academy of Sciences (CAS President's International Fellowship for Postdoctoral Researchers, No. 2016PB009) for their support in carrying out this project.

REFERENCE

Podgorski, J.E., Eqani, S.A.M.A.S., Khanam, T., Ullah, R., Shen, H. & Berg, M. 2017. Extensive arsenic contamination in high-pH unconfined aquifers in the Indus Valley. *Sci. Adv.* 3. e1700935.

Environmental Arsenic in a Changing World –
Zhu, Guo, Bhattacharya, Ahmad, Bundschuh & Naidu (Eds)
ISBN 978-1-138-48609-6

Alleviation of altered ultrastructure in arsenic stressed rice cultivars under proposed irrigation practice in Bengal Delta Basin

A. Majumdar[1], A. Barla[1], S. Bose[1], M.K. Upadhyay[2] & S. Srivastava[2]
[1]*Earth and Environmental Science Research Laboratory, Department of Earth Sciences, Indian Institute of Science Education and Research Kolkata (IISER-K), Mohanpur, West Bengal, India*
[2]*Institute of Environment & Sustainable Development, Banaras Hindu University, Varanasi, India*

ABSTRACT: Application of a periodical intermittent water cycle during rice cultivation proved its potential for minimizing considerable bioavailability of arsenic (As) in soil. We also propose two parametric equations, determining arsenic release from seasonal field conditions. Soil samples were determined through Raman spectroscopy and ICP-MS. The practice also alleviate stress from ultra-physiological xylem-phloem integrity of plant parts compared to conventional flooded cultivation, as observed under FE-SEM. Rice grains were analyzed for As concentrations using Synchrotron-XRF for two rice cultivar with subsequent internal and external surface topography determination under AFM showing greater amplitude of roughness in As stressed grains from conventional cultivation. Fresh plants were analyzed for biomass with pigmentation and stress regulator enzymes viz. malondialdehyde (MDA), catalase (CAT), superoxide dismutase (SOD), guaiacol peroxidase (GPX) and total protein from both conditions and found to be better in proposed cultivation method with better sustainable productivity. Isolated rhizospheric bacteria also plays crucially in arsenic uptake and metabolism.

1 INTRODUCTION

Arsenic hazards in South-East Asia compromises socio-economic and health perspective of millions of peoples (Meharg, 2004). Arsenic contamination to the groundwater and thereafter to the rice grain needs a sustainable restriction approach and scientific communities around the globe are trying to minimise this natural poison to get entered in the food chain. Rice plant can take up As more efficiently compared to other cereals (Williams *et al.*, 2007) which is why agronomic practices and soil physico-chemical parameters plays crucially in arsenic mobility (Barla *et al.*, 2017). Accumulation of arsenic can severely damage internal plant cell organelles and altered vesicles (Li *et al.*, 2006) with releasing high amount of reactive oxygenic species (ROS) resulting in elevation of stress regulating enzymes (Shri *et al.*, 2009). Rhizospheric microbes on the other hand ensures some partial bio-rendering of As changing redox status and being transported to the rice root system covered with iron plaques (Zecchin *et al.*, 2017).

2 EXPERIMENTAL

2.1 *Soil-groundwater sampling and assessment*

Samples were collected from two different experimental fields, intermittent flooded (IF) and conventional flooded (CF), a part of Nadia district, West Bengal, India. Both the samples were analysed by both ICP-MS (acid digested) and Raman Spectroscopy (untreated) for arsenic and other element determination.

2.2 *Plant arsenic determination*

Rice plants were sampled, segmented and finely chopped to make pallets for assessment under synchrotron X-ray fluorescence (Indus-2 beamline, India) for elemental analysis with subsequent rice grain internal and external surface topography determination by atomic force microscope (AFM) to check the physiological effect of As on rice grain. Leaf and nodal xylem-phloem integrity were also observed under field emission scanning electron microscope (FE-SEM).

2.3 *ROS scavenging enzymes*

Fresh plant tissues were measured for the production of anti-stress regulator enzymes. Comparison between IF field plant's biomass, total protein, pigment content and CF field plants were made.

2.4 *Microbial amendments*

Arsenic resistant Fe oxidizing-reducing bacteria were isolated and screened for the highest As tolerance level subsequently applied in soil As restriction.

3 RESULTS AND DISCUSSIONS

3.1 *Arsenic contentment*

Both soil and rice grains contains arsenic beyond permissible limit by WHO-FAO (2016) due to continuous application of As rich groundwater in CF field whereas

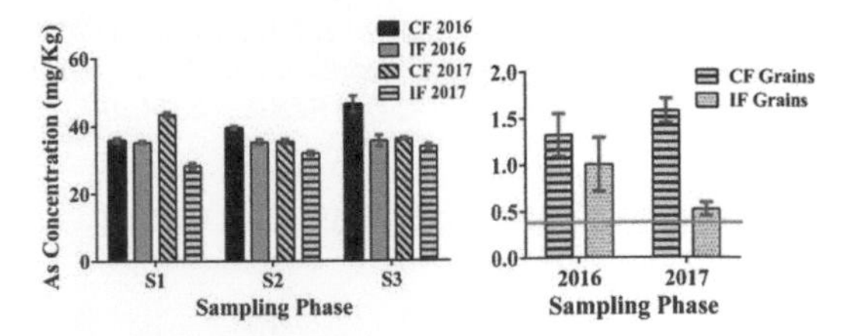

Figure 1. Arsenic in samples from two different irrigation practices.

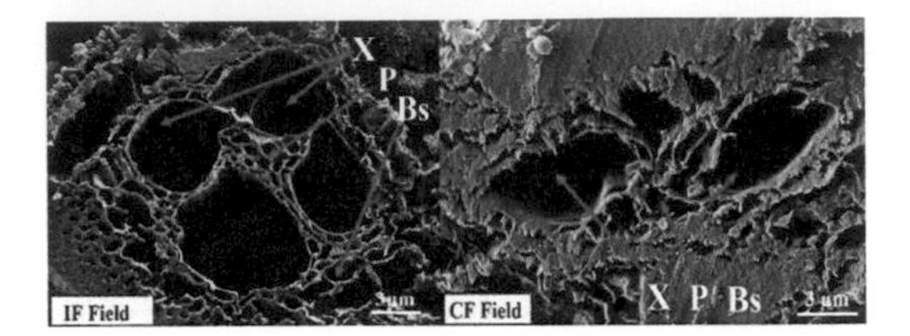

Figure 2. Ultrastructure comparison in rice.

in IF field, the As load in rice grain is much less and ensures to be lesser in subsequent practice (Figure 1). The average concentration of As in selected area lies within 75–90 $\mu g\ L^{-1}$.

Graphical representations are showing As concentrations in groundwater, soil and rice grains in consecutive 2016 and 2017 samples as a mean of triplicated data. Green line indicates permissible limit of As in rice grains. Application of IF system involves redox change and decreased bioavailability (BA) of As resulting in lesser As in rice grains.

3.2 *Alleviated ultrastructure*

Translocation of As proceeds through vesicular network that gets structurally distorted in the presence of high ROS content. Reduced BA affects less to the internal vesicular organelles like xylem (X), phloem (P) and bundle sheath (Bs) in IF compared to CF as shown in Figure 2.

3.3 *Variation in ROS scavenging enzymes*

Fresh plants from the CF field showed much higher production of MDA, CAT, SOD and GPX compared to IF field plants, indicating a greater extent of As phytoaccumulation in CF plants (Figure 2). This causes an internal degenerative stress molecules triggering ROS scavenger enzymes in CF plants.

3.4 *Microbial enhancement*

Isolated bacterial species can tolerate up to 400 $mg\ kg^{-1}$ of arsenite and 1000 $mg\ kg^{-1}$ of lead concentrations with 30% of that are being bioaccumulated within the cells itself. This proved to be a fascinating bioremediation agent with some potential nutritional enrichments.

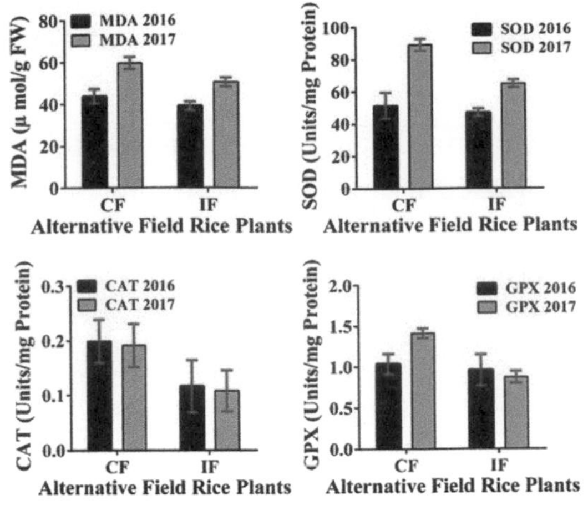

Figure 3. Stress responsive enzymes in rice plant. Data are the mean of triplicated values.

4 CONCLUSION

Implementation of intermittent dry-wet irrigation over the conventional flooded practice amended with biorestrictors makes internally healthier and As-less rice with low risk of health hazards.

ACKNOWLEDGEMENT

Authors are thankful to MoES, Govt. of India for providing funds, and IISER Kolkata, BHU and RRCAT (Indus-2 BL-16) central institutional facilities.

REFERENCES

Barla, A., Shrivastava, A., Majumdar, A., Upadhyay, M.K. & Bose, S. 2017. Heavy metal dispersion in water saturated and water unsaturated soil of Bengal delta region, India. *Chemosphere* 168: 807–816.

Li, W.X., Chen, T.B., Huang, Z.C., Lei, M. & Liao, X.Y. 2006. Effect of arsenic on chloroplast ultrastructure and calcium distribution in arsenic hyperaccumulator *Pteris vittata* L. *Chemosphere* 62(5): 803–809.

Meharg, A.A., 2004. Arsenic in rice–understanding a new disaster for South-East Asia. *Trends Plant Sci.* 9(9): 415–417.

Shri, M., Kumar, S., Chakrabarty, D., Trivedi, P.K., Mallick, S., Misra, P., Shukla, D., Mishra, S., Srivastava, S., Tripathi, R.D. & Tuli, R. 2009. Effect of arsenic on growth, oxidative stress, and antioxidant system in rice seedlings. *Ecotoxicol Environ Saf.* 72(4): 1102–1110.

Williams, P.N., Villada, A., Deacon, C., Raab, A., Figuerola, J., Green, A.J., Feldmann, J. & Meharg, A.A. 2007. Greatly enhanced arsenic shoot assimilation in rice leads to elevated grain levels compared to wheat and barley. *Environ. Sci. Technol.* 41(19): 6854–6859.

Zecchin, S., Corsini, A., Martin, M., Romani, M., Beone, G.M., Zanchi, R., Zanzo, E., Tenni, D., Fontanella, M.C. & Cavalca, L. 2017. Rhizospheric iron and arsenic bacteria affected by water regime: Implications for metalloid uptake by rice. *Soil Biol. Biochem.* 106: 129–137.

Environmental Arsenic in a Changing World –
Zhu, Guo, Bhattacharya, Ahmad, Bundschuh & Naidu (Eds)
ISBN 978-1-138-48609-6

Arsenic linked to a former mining activity in the Hunan province: distribution at the local scale and bacterial As(III) oxidation

F. Battaglia-Brunet[1], S. Touzé[1], C. Joulian[1], C. Grosbois[2], M. Desmet[2], Q. Peng[3], F. Zhang[4], C. Meng[4], J.Y. Zhang[4], L. Luo[4], X. Li[5], Q. Li[5] & J. Paing[6]
[1]*BRGM, Orléans, France*
[2]*University of Tours, Tours, France*
[3]*ASEM Water Resources Research and Development Center, Changsha, Hunan, P.R. China*
[4]*Hunan Agricultural University, Changsha, Hunan, P.R. China*
[5]*Central South University, Changsha, Hunan, P.R. China*
[6]*O-pure, Beaumont-la-Ronce, France*

ABSTRACT: The impact of a former small Pb-Zn mine in Hunan (China) was studied in terms of arsenic (As) and metals concentrations in water streams and sediments, and from the perspective of biogeochemistry of As. Three water streams were characterized, two from galleries and one seeping at the base of the mine waste rocks dump. Contrary to the water from galleries, this last stream was acidic and contained relatively high As concentration. High As levels were also found in sediments downstream of the mine drainage flowing from the first gallery. Diverse As(III)-oxidizing bacteria were enriched and isolated from the mine sediments, some presenting ability to grow at low pH or in autotrophic conditions.

1 INTRODUCTION

Mining activities have caused major environmental problems on a global scale. Huge amounts of waste are generated when these geological materials are treated to extract the mining resources. Mine residues contain large fractions of sulfide materials that are vulnerable to oxidation when coming into contact with air and rainwater. These processes generate mine water containing a range of toxic metals and metalloids often including As (Johnson and Hallberg, 2005). Bioaccumulations and bioamplifications of this toxic element can lead to human intoxication, especially in fishing populations. Here, the distribution of the toxic element in different compartments of a small Pb-Zn former mine site was considered. Then, As(III)-oxidizing bacterial strains that may be involved in remediation processes were enriched and isolated.

2 METHODS

2.1 *Site description and sampling*

The site is a small disused Pb-Zn mine located in the region of Aotoushan, # 200 km southwest from Changsha, Hunan. The mine works include two galleries and one waste dump composed of coarse materials presenting visible sulfide crystals. Water flowed from two galleries and seeped at the base of the dump. The first gallery was located above the mine waste dump. The second gallery was in a discrete place, 300 m from the first gallery and the dump. The water flowing from the first gallery was discharged, at less partly, into the dump. Samples of water, bed sediments from the three water streams were collected in 200 mL sterile bottles. A sediment core was collected a few km downstream in a dam reservoir.

2.2 *Chemical analyzes*

At the laboratory, part of the samples was used to inoculate enrichment media, and the other part was devoted to chemical analyzes. Water phase was filtrated at 0.45 μm and acidified with concentrated HNO_3, then analyzed by ICP-MS. The solid phases were dried at 50°C, crushed and analyzed using X-ray fluorescence apparatus (NITON®).

2.3 *Enrichment and isolation of strains*

As(III)-oxidizing cultures were obtained by inoculating and sub-culturing the site samples in minimal medium CAsO1 (Battaglia-Brunet *et al.*, 2002) at pH 3, 5 and 6, in autotrophic and heterotrophic (with 0.2 g L^{-1} yeast extract), in aerobic conditions. Three sub-cultures were performed, each being incubated for 3 weeks. Bacterial counting was performed by microscopy using a Thoma cell and by MPN technique for As(III)-oxidizing microbes. Isolation was performed using the same medium but solidified with

Table 1. Main characteristics of water streams.

	pH	As $\mu g\ L^{-1}$	Pb $\mu g\ L^{-1}$	Pb $\mu g\ L^{-1}$
Gallery 1	5.3	7	19	22.8
Gallery 2	5.6	19	19	1.9
Dump seepage	3.1	373	865	2018

agar. Strains recovered as colonies on the plates were screened for their ability to oxidize As(III). Purification of As(III)-oxidizing strains is performed through five successive cultures on solid then liquid medium alternatively. Identification of the purified strains will be performed by sequencing their 16S rDNA genes.

3 RESULTS AND DISCUSSION

3.1 *Arsenic and metals distribution in waters and sediments*

The water streams from galleries presented slightly acidic pH values, low As and Pb concentrations, but Zn concentrations higher than 1 mg L^{-1} (Table 1). In contrast, the dump seepage was more acidic a contained significant concentrations in As and heavy metals.

The surface sediment sampled in the mine drainage a few meters from the outlet of gallery 1 contained 2% As, 30% lead, 6% zinc and 20% total S. These solid materials are polluted by As- and metals-containing dust of sulfidic ores from the mine exploitation. They could contribute to As and metal concentrations in water and sediment downstream. As a fact, river sediments, a few km downstream the mine, were enriched in As, Pb and Zn with a maximum peak at 8–10 cm deep (550 mg kg^{-1} of As).

3.2 *Enrichment and isolation of bacterial strains*

The bacterial enrichment procedures focused on As(III)-oxidizing organisms produced five microbial communities, two at pH 3 from the dump seepage (one autotrophic and one heterotrophic), one at pH 5 from the first gallery (heterotrophic), and two at pH 6 from the second gallery (one autotrophic and one heterotrophic). The highest proportion of As(III)-oxidizing microbes was found in the heterotrophic enrichment from gallery 2, where more than 10% of the cells were As(III)-oxidizing microbes. The isolation step resulted in a collection of strains from these enrichments. Isolation of strains from the first gallery

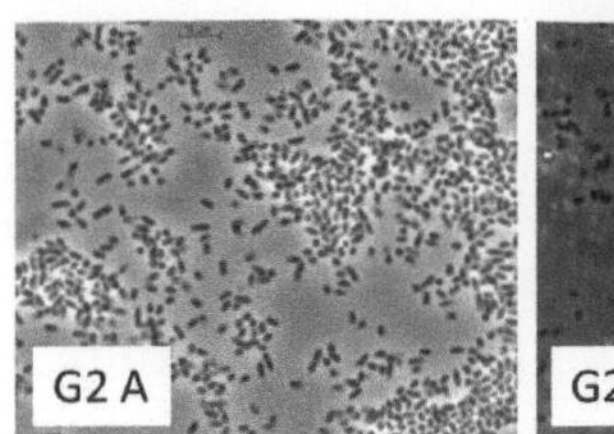

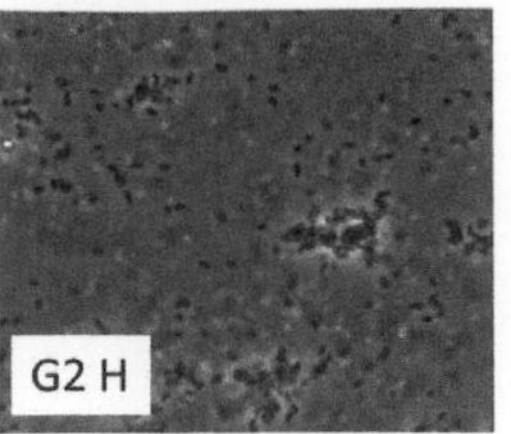

Figure 1. Strains from gallery 2 observed under an optical microscope (×1000), G2 A: autotrophic; G2 H: heterotrophic.

was not efficient, with only one heterotrophic strain obtained. The waste seepage yielded five strains oxidizing As(III) at pH 3, including one autotrophic. The enrichments from gallery 2 allowed the isolation of the highest number of As(III)-oxidizing strains, 65 colonies from the autotrophic medium, and 84 colonies from the heterotrophic medium (Fig. 1).

4 CONCLUSIONS

A range of As(III)-oxidizing strains were isolated from a Pb-Zn disused mine in Hunan. The availability of several strains from the dump seepage, able to grow and oxidize As(III) at pH 3, present a particular interest for the development of bioremediation of acid mine drainages. The identification of the isolated strains will complete the characterization of the biodiversity of cultivable As(III)-oxidizing bacteria in the different water streams of the mine.

ACKNOWLEDGEMENTS

This research work has been performed in the frame of ATIM-Hunan project funded by the Region Centre Val de Loire, contract N 2015-00099715.

REFERENCES

Battaglia-Brunet, F., Dictor, M.C., Garrido, F., Crouzet, C., Morin, D., Dekeyser, K., Clarens, M. & Baranger, P. 2002. An arsenic (III) – oxidizing bacterial population: selection, characterization, and performance in reactors. *J. Appl. Microbial.* 93(4): 656–667.

Johnson, D.B. & Hallberg, K.B. 2005. Acid mine drainage remediation options: a review. *Sci. Total Environ.* 338(1–2): 3–14.

Environmental Arsenic in a Changing World –
Zhu, Guo, Bhattacharya, Ahmad, Bundschuh & Naidu (Eds)
ISBN 978-1-138-48609-6

Assessing the phytoremediation potential of a flowering plant *Zinnia angustifolia* for arsenic contaminated soil

Poonam & S. Srivastava
Institute of Environment and Sustainable Development, Banaras Hindu University, Varanasi, India

ABSTRACT: Arsenic (As)-based chemicals, insecticides, herbicides, rodenticides and irrigation with As containing water have resulted in As contamination in farmland, lawns and residential areas. This study focused on phytoremediation potential of a seasonal flowering plant *Zinnia angustifolia* for As removal from contaminated soil. Plants were subjected to As stress (100 mg kg^{-1} soil) in field conditions for 60 days and their tolerance was evaluated through growth and antioxidant assays. Root tissue was found to accumulate more As (1085.42 $\mu g\ g^{-1}$ dry weight (DW)) in comparison to shoot (279.28 $\mu g\ g^{-1}$ DW) and flowers (228.67 $\mu g\ g^{-1}$ DW) at 60 d.

1 INTRODUCTION

About 100 million people, living in 23 countries are exposed to As toxicity. In Bengal delta (comprising of West Bengal, India and Bangladesh), source of As pollution is geogenic. The extraction of toxic metalloid pollutants from contaminated soil and water is needed. There are various physical, chemical and biological methods available. However, the expensive high-tech remedial measures are not easy for developing countries like India and Bangladesh; hence emphasis has to be on remediation by cost-effective methods. Phytoremediation fulfills these requirements as it is cost-effective solar driven technology for metal removal (Chen *et al.*, 2015; Reed *et al.*, 2015). Ornamental and flowering plants have not been much investigated for phytoremediation potential of As contaminated soils. Flowering plants can help in reclamation of contaminated land and through production of cut flowers and other marketable commodities (Srivastava *et al.*, 2007, 2011). This can manage the cost involved in the phytoremediation for farmers. These plants can also provide an aesthetic value to people and surrounding locality on contaminated sites.

2 METHODS

2.1 *Soil preparation and nursery development*

The plant used in the study is zinnia (*Zinnia angustifolia*). To achieve the set objective, pot experiments were performed. Soil preparation was done a month before. Soil was crushed and filtered to get fine particles so that As (100 mg kg^{-1}) can be equally mixed. All the experiments were carried out in the poly house under control conditions and duration of experiment was 60 d from transplanting to harvest at 20 d intervals (20, 40 and 60 d). Plant nursery was developed and healthy plants of equal heights (30 d old) from the nursery were selected for the experiment. In due course of time, the soil was under incubation and was mixed regularly so that As got homogenously distributed in soil. Triplicate set of pots were used for the control and As treatments for the experiment.

2.2 *Growth, enzyme and metabolite assays and arsenic analysis*

Plant growth parameters like fresh biomass and shoot-root growth in term of length were measured. Plants samples (root, shoot and flowers) were oven dried and acid digested for As estimation. Arsenic concentration was determined by inductively coupled plasma mass spectrometer (ICP-MS; Agilent7500 cx). The standard references materials were used for calibration and quality assurance for each analytical batch. For accuracy, repeated analysis (n = 10) of quality control samples was done, and the results were found within the certified values. Fresh plant sample were crushed in liquid N_2 and used for chlorophyll analysis and enzyme extraction by following standard protocols and readings were taken using spectrophotometer.

3 RESULTS AND DISCUSSION

3.1 *Plant growth and physiological response*

Zinnia plants grown in As treated soil were observed to have shorter root (2.16 cm) and shoot (3.07 cm) length as compared with plant grown in control soil (shoot = 15.00–29.40 cm, root = 8.46–14.00 cm). Flower bud appeared between 20 to 40 d and control plants showed maximum number of flower buds. Total number of fresh leaves also varied at different

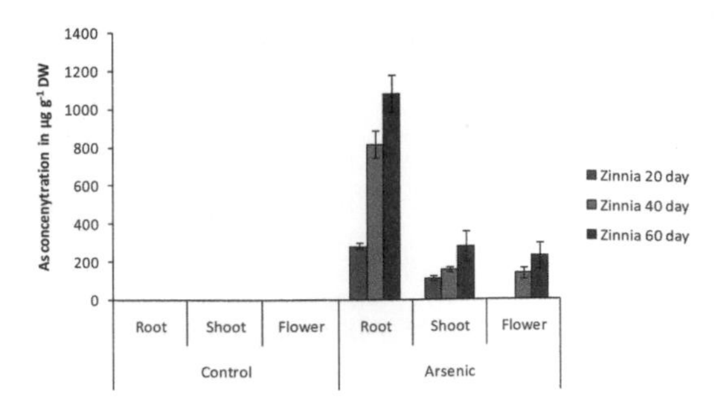

Figure 1. Arsenic removal by zinnia during 1–60 day of growing period.

Table 1. Cysteine content (ηmol g^{-1} FW) observed in different parts of zinnia.

Cysteine content				
		Root	Shoot	Flower
20 d	Control	33.80	22.41	–
	Arsenic	22.72	38.52	–
40 d	Control	32.83	22.23	–
	Arsenic	20.44	30.92	–
60 d	Control	30.15	21.34	17.47
	Arsenic	18.39	27.68	16.97

harvesting time between control and As treated plants. Roots from the treatments were prone to damage and break as compared with control plants. Roots were hard and easily breakable in As treated plants. Roots showed high As accumulation in comparison to shoots (Fig. 1).

3.2 *Biochemical parameters*

To analyze the effect of As on biochemical parameters of plants; chlorophylls, enzymatic and non-enzymatic parameters were assayed. Chlorophyll content of zinnia in As treated soil was higher as compared to control plants at 20 and 40 d and decreased at 60 d. Ascorbate peroxidase (APX) and guaiacol peroxidase (GPX) activities were reported to increase with each harvesting time in As treated plants as compared to control plants. Cysteine content was decreased in roots while increased in shoot in As treated plants as compared to that in control plants on all durations (Table 1). In case of proline content, it was reported less in treated plants than the control in both root and shoot of plants during 60d of exposure. But with increasing toxicity, proline was found to increase with highest value at 60 d (root = 9.39 and shoot = 3.61 mg g^{-1} fresh weight (FW)) in As treated plants.

4 CONCLUSIONS

From our study based on arsenic tolerance analysis of zinnia, it was found that plant was able to grow under 100 mg kg^{-1} As in soil. The stress caused by As was evident in terms of growth reduction. However, it could be tackled effectively by the increase in antioxidant enzyme like APX and GPX activity in root and shoot and important metabolites like cysteine. Overall, the use of flowering plants could offer a viable approach of not only making contaminated site aesthetic and reclaimed but also decontaminating that with time. The major benefit with flowering plants is their non-edible nature and hence this would pose no risk of any contamination to humans.

ACKNOWLEDGEMENTS

Authors gratefully acknowledge the support from funding received from DST-SERB (GOI) under Young Scientist Award (project code: P07/610).

REFERENCES

Chen, G., Liu, X., Brookes, P.C. & Xu, J. 2015. Opportunities for phytoremediation and bioindication of arsenic contaminated water using a submerged aquatic plant: *Vallisneria natans* (Lour.) Hara. *Int. J. Phytoremediation* 17(3): 249–255.

Reed, S.T., Ayala-Silva, T., Dunn, C.B. & Gordon, G.G. 2015. Effects of arsenic on nutrient accumulation and distribution in selected ornamental plants. *Agric. Sci.* 6(12): 1513–1531.

Srivastava, S., Mishra, S., Tripathi, R.D., Dwivedi, S., Trivedi, P.K. & Tandon, P.K. 2007. Phytochelatins and antioxidant systems respond differentially during arsenite and arsenate stress in *Hydrilla verticillata* (L.f.) Royle. *Environ. Sci. Technol.* 41(8): 2930–2936.

Srivastava, S., Shrivastava, M., Suprasanna, P. & D'souza, S.F. 2011. Phytofiltration of arsenic from simulated contaminated water using *Hydrilla verticillatain* field conditions. *Ecol. Eng.* 37(11): 1937–1941.

Environmental Arsenic in a Changing World –
Zhu, Guo, Bhattacharya, Ahmad, Bundschuh & Naidu (Eds)
ISBN 978-1-138-48609-6

Evaluation of leaf proteomics responses during selenium mediated tolerance of arsenic toxicity in rice (*Oryza sativa* L.)

R. Chauhan[1], S. Awasthi[1], S. Mallick[1], V. Pande[2], S. Srivastava[3] & R.D. Tripathi[1]
[1] *CSIR - National Botanical Research Institute, Lucknow, Uttar Pradesh, India*
[2] *Department of Biotechnology, Kumaun University, Bhimtal, Nainital, Uttarakhand, India*
[3] *Institute of Environment & Sustainable Development, Banaras Hindu University, Varanasi, Uttar Pradesh, India*

ABSTRACT: Arsenic (As) is a toxic element, which poses significant threat to human health due to frequency of occurrence, toxicity and various routes for human exposure. Arsenic contamination of water and food, especially rice, is a serious issue. Selenium (Se) is an essential element for human and other animals and is beneficial for plants as well. Selenium has been found to antagonize the toxicity of As in plants. The proteomics study was performed by using 2-dimensional gel electrophoresis and MALDI-TOF/TOF. The present study provides new insights into As-Se interaction in rice leaves at the proteome level and identifies crucial proteins involved in Se mediated amelioration of As toxicity in rice.

1 INTRODUCTION

Arsenic (As) is an environmental toxin and class I human carcinogen. Arsenic contamination is a major environmental hazard of recent times which poses a health risk for human through food chain contamination. Nearly 200 million people are at risk of As toxicity. Rice is the dietary staple for half of the world's population and a good accumulator of As. Thus, rice consumers are at great risk. Arsenic exerts toxicity to plants in various ways and hampers plants growth. Increased production of reactive oxygen species (ROS) and oxidative damage to plants are apparent symptoms of As toxicity. Selenium (Se) is an essential nutrient element for human and is also beneficial element for plants which enables them to counter against various abiotic stresses at low dosages. Selenium acts as an antioxidant and is involved in the detoxification of heavy metals including As by alleviating the oxidative stress and antagonizing the uptake of heavy metals (Chauhan *et al.*, 2017). Thus, supplementation of Se in As contaminated environment might be an important strategy to reduce As uptake and associated phytotoxicity in rice plant. The present study analyzes As-Se interaction in rice leaves at the proteome level.

2 METHODS

2.1 *Plant material and experimental conditions*

Rice variety, Triguna was used for the experiments. Rice seedlings were exposed to As and Se for 15 days using sodium arsenite (As(III), $NaAsO_2$) and sodium selenite (Se(IV), Na_2SeO_3). Plants were exposed to different treatments; control, 25 μM As(III), 25 μM Se(IV) and 25 μM (As + Se) for 15 days. After 15 d exposure, the plants were harvested in triplicate for proteome analysis and As accumulation.

2.2 *Protein extraction, 2DE analysis, image analysis, protein identification and MALDI-TOF-TOF analysis*

Rice leaf proteins from three biological replicates were extracted using ammonium acetate/acetone precipitation method followed by phenol extraction as per method described by Agrawal *et al.* (2008). Total shoot protein was determined by Bradford assay (Bio-Rad, USA). The 13 cm IEF strips (pH 4–7) were rehydrated with protein (250 μg) for 16 h and electrofocused using an IPGphor system (Bio-Rad, USA) at 20°C up to 25,000 Vh. After reduction with DTT and alkylation by using IAA, the strips were then loaded on top of 12.5% poly acrylamide gels for SDS-PAGE. The gels were fixed and stained with silver stain (silver stain plus kit; Bio-Rad, USA). The silver stained gel obtained after 2DE were scanned for image acquisition using the Bio-Rad FluorS system equipped with a 12-bit camera. Images from three biological replicate 2-DE gels were taken for the data analysis in PD Quest version 8.0.1 (Bio-Rad). Protein spots showing significant changes in abundance between different treatments were selected and excised manually and processed for peptide identification by MALDI-TOF-TOF (Model 4800, Applied Biosystems, USA). In all the protein identifications, probability scores were greater than the score fixed by Mascot as significant with a $p < 0.05$.

2.3 *Arsenic and selenium quantification*

For estimation of total As and Se, dried plant samples (0.2 g) were powdered and digested in 3 mL HNO_3 at 120°C for 6 h (Dwivedi *et al.*, 2010). Total As and Se were quantified with the help of inductively coupled plasma mass spectrometer (ICP-MS, Agilent 7500 cx). The multielement standard solution (Agilent, Part # 8500–6940) was used for the calibration and quality assurance for each analytical batch. Known concentrations of spiked samples were prepared to check the As and Se recovery. Selenium and As were recovered about 93.5% (±2.3; n = 5) and 92.5% (±3.1; n = 5), respectively.

3 RESULTS AND DISCUSSION

3.1 *Expression profile of rice leaf proteins*

In the present study, the protein expression profile of rice leaf proteins under As stress and Se supplementation has been investigated that provides a fine picture of protein networks and metabolic pathways involved in cellular detoxification and tolerance mechanism. Results showed that Se supplementation (25 μM Se(IV)) to As (25 μM As(III)) exposed rice plants improved the plant growth in comparison to As alone treatment, exhibiting antagonistic interaction between As and Se. Photosynthetic pigments were also significantly increased upon Se supplementation during As stress. The improved growth of rice in As+Se treatment was thus attributable to the improved levels of photosynthetic pigments which contribute towards biomass gain and in addition to lesser As uptake. By comparison of gel images of different treatments viz., control, As and Se with the help of PDQuest software, 145 differentially regulated spots were subjected for MALDI-TOF MS-MS analysis out of which 110 protein spots identified which further led to identification of 85 proteins with a significant score (Fig. 1). Functional categorization and cluster analysis (SOTA analysis) of As and Se responsive differentially expressed proteins revealed proteins of defense, photosynthesis, ROS pathway, energy metabolism, antioxidant system, amino acid metabolism and transport to be the responsive proteins. Protein ubiquitination and phosphorylation also appeared to play important roles in Se mediated amelioration of As toxicity in rice. In conclusion, the present study provides new insights into As-Se interaction in rice plant at the proteome level and identifies crucial proteins involved in Se mediated As toxicity amelioration in rice.

3.2 *Effect of selenium on arsenic accumulation*

When Se and As were supplied together the accumulation of As was reduced significantly ($p < 0.05$) in both roots and shoots. The maximum reduction in As accumulation was also observed at equimolar concentration of As(III) and Se(IV) (25 μM) with about 68% and 45% reduction in roots and shoots respectively in comparison to As(III) alone treated plants. On the contrary, the accumulation of Se increased in both the tissues of rice plant. Thus, Se helps in reducing As uptake and subsequently increase uptake of other divalent cations required for antioxidant defense mechanism.

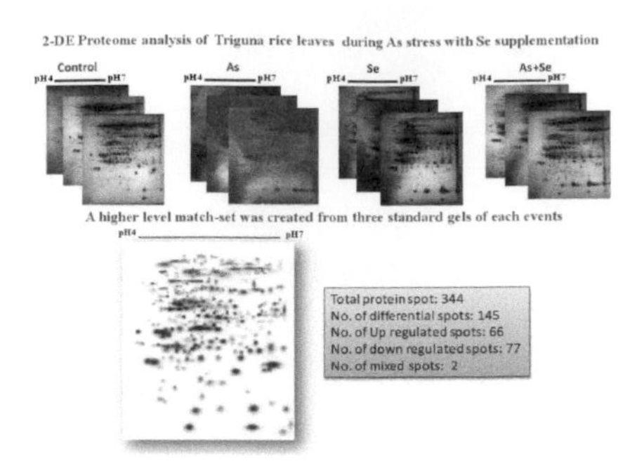

Figure 1. Representation of 2DE maps of rice leaf proteins isolated from control, As, Se and As + Se treatments. Three replicate silver stained gels for each stage were computationally combined using PDQuest software and one representative master standard gel images were generated.

4 CONCLUSIONS

Present study with the help of comparative proteomic study provides a systemic molecular overview of Se mediated modulation of As toxicity in rice plant. Functional classification revealed that maximum number of proteins fall in the category of bioenergy and metabolism followed by those involved in cell defense. Photosynthesis is the most affected biological process of rice shoot by As treatment. Selenium supplementation repaired the photosynthetic system and the proteins involved in photosynthesis were found upregulated in Se treatment during As stress. Stress responses and cell redox homeostasis were also found enhanced on Se supplementation. Selenium treatment also activated antioxidant system to enhance As resistance in rice plant. Present proteomics study could represent only a small part of the rice proteome during As and Se interaction, further investigation to assign their putative biological functions may be useful for a better understanding of complex biological traits.

ACKNOWLEDGEMENT

Reshu Chauhan is thankful to Department of Science and Technology (DST) New Delhi, India for the award of DST-INSPIRE Junior/Senior Research Fellowship.

REFERENCES

Agrawal, L., Chakraborty, S., Jaiswal, D.K., Gupta, S., Datta, A. & Chakraborty, N. 2008. Comparative proteomics of tuber induction, development and maturation reveal the complexity of tuberization process in potato (*Solanum tuberosum* L.). *J. Proteome Res.* 7(9): 3803–3817.

Chauhan, R., Awasthi, S., Tripathi, P., Mishra, S., Dwivedi, S., Niranjan, A., Mallick, S., Tripathi, P., Pande, V. & Tripathi, R.D. 2017. Selenite modulates the level of phenolics and nutrient element to alleviate the toxicity of arsenite in rice (*Oryza sativa* L.). *Ecotox. Environ. Saf.* 138: 47–55.

Dwivedi, S., Tripathi, R., Tripathi, P., Kumar, A., Dave, R., Mishra, S., Singh, R., Sharma, D., Rai, U. & Chakrabarty, D. 2010. Arsenate exposure affects amino acids, mineral nutrient status and antioxidant in rice (*Oryza sativa* L.) genotypes. *Environ. Sci. Technol.* 44: 9542–9549.

Environmental Arsenic in a Changing World –
Zhu, Guo, Bhattacharya, Ahmad, Bundschuh & Naidu (Eds)
ISBN 978-1-138-48609-6

Evaluation of proteomics responses of rice (*Oryza sativa*) during arsenic toxicity amelioration by a potential microbial consortium

S. Awasthi[1], R. Chauhan[1], S. Srivastava[2] & R.D. Tripathi[1]
[1]*CSIR – National Botanical Research Institute, Varanasi, Uttar Pradesh, India*
[2]*Institute of Environment & Sustainable Development, Banaras Hindu University, Varanasi, Uttar Pradesh, India*

ABSTRACT: Environmental pollution of arsenic (As) is a serious problem that is not only affecting the local people but also population throughout the world through rice and rice-based food products. Rice accumulates As in significantly greater amounts in comparison to other cereal crops. Microbiological interventions can serve as protective agents to ameliorate As stress in rice. The experiments were conducted in hydroponically grown in rice seedlings subjected to arsenate [As(V)] stress for 15 d. Two-dimensional gel electrophoresis analysis of 340 spots revealed that 145 spots were differentially expressed; out of these, 87 spots were identified using MALDI-TOF-TOF. The differentially expressed proteins during supplementation of microbial consortium under As stress belonged to defense, growth promotion, energy metabolism, cellular and metabolic processes.

1 INTRODUCTION

Arsenic (As) exposure from rice consumption has now become a global health issue. Rice (*Oryza sativa* L.), a staple food in many Asian countries is a crop with high accumulation of As, which are highly toxic and non-essential elements for human body. Arsenic is initially absorbed from the soil through the roots and enters the shoot via the xylem and from the xylem of stem it is transported to the phloem and finally into the grain. Microbes that are able to grow in As contaminated environment are equipped with genetic architecture, metabolic pathways and enzymatic systems that enable them either to tolerate As toxicity via transformation of toxic inorganic forms into less toxic organic ones or to use As as a part of metabolism. Proteins are functionally versatile macromolecules that play major role in living cells. In the present study, a comparative proteomics analysis coupled with photosynthesis and antioxidant enzymes analysis on rice plants subjected to As stress has been adopted, to identify the protein profile associated with consortium of *Pseudomonas putida* MTCC5279 (P. putida) and Chlorella vulgaris induced resistance to As stress and to explore the possible molecular mechanisms involved.

2 METHODS

2.1 *Plant material and experimental conditions*

Oryza sativa L. var. Triguna was used for the experiments. Plants were exposed to different treatments [50 μM As(V)], *P. putida* and *C. vulgaris* and co-inoculation of *P. putida* and *C. vulgaris* for 15 d. After 15 days exposure, 3 replicates were harvested for proteome analysis, expression of the selected genes and As accumulation.

2.2 *Protein extraction, 2DE analysis, image analysis, protein identification and MALDI-TOF-TOF analysis*

Shoot tissue (5 g) was pulverized to a fine powder in liquid nitrogen and suspended in 10 mL lysis buffer. Suspension was mixed with Tris saturated phenol (pH 8.0) for 30 min with gentle shaking and centrifuge at 9000 × g for 10 min. After that tissue was precipitated in 0.1 M ammonium acetate for overnight at −200°C. Again centrifuged and the dried pellet was dissolved into rehydration buffer. Total shoot protein was determined by Bradford assay (Bio-Rad, USA). Isoelectric focusing was carried out with 250 μg of protein as given by Agrawal *et al.* (2008). The strips were then loaded on top of 12.5% poly acrylamide gels for SDS-PAGE. The gels were fixed and stained with a silver stain plus kit as per protocol (Bio-Rad, USA). The silver stained gel obtained after 2DE were scanned for image acquisition using the Bio-Rad Fluor S system equipped with a 12-bit camera. Images from three biological replicate 2-DE gels were taken for the data analysis in PD Quest version 8.0.1 (Bio-Rad). The silver stained protein spots were digested for MALDI MS/MS were digested with trypsin (Promega Corporation, MA, USA) as per manufacturer instruction. The stained protein spots were excised manually, washed thoroughly with water and crushed in small pieces and digested with trypsin. Peptide solution mixed with 5 mg mL^{-1} CHCA (a-Cyano-4-hydroxy

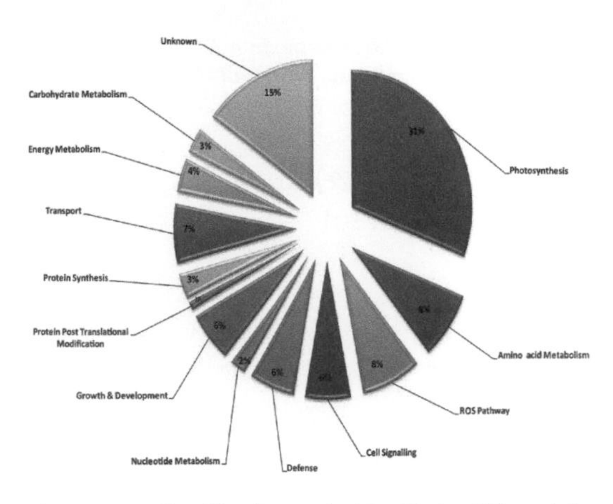

Figure 1. Classification of identified differentially expressed rice shoot proteins in various functional classes based on their putative functions assigned them using protein database.

cinnamic acid) matrix and spotted on to MALDI plate. A 4800 Proteomics Analyzer (Applied Biosystems, USA) with TOF/TOF optics was used for all MALDI-MS and MS/MS applications. In all the protein identifications, probability scores were greater than the score fixed by Mascot as significant with a $p < 0.05$.

3 RESULTS AND DISCUSSION

3.1 *Identification of differentially abundant proteins in rice*

The present study indicates a practical potential for microbial application as an intervention strategy in mitigating As toxicity and reducing As uptake and translocation in rice plants. Supplementation of *P. putida* and *C. vulgaris* with 50 µM As(V) significantly alleviated As induced growth inhibition in rice seedlings (Fig. 1), improved photosynthetic and antioxidant machinery and significantly decreased rice As concentration. Many proteins were differentially induced in rice seedlings by *P. putida* and *C. vulgaris* under As stress. The differentially expressed proteins were mainly involved to regulate defense system, photosynthesis, detoxification, plant hormones and energy metabolism. Heat shock protein are also found to be accumulated, which were a class of functionally related proteins involved in repair and aid in the renaturation of stress-damaged proteins, and reestablish cellular homeostasis, protect cells against the damage of stress (Ahsan *et al.*, 2009). This study also revealed the expression of IAP100 (Inhibitor of Apoptosis) in the presence of consortium under As stress; IAP100 has ability to regulate caspases and also influence ubiquitin (Ub)-dependent pathways that modulate innate immune signaling. In conclusion, results provide new insights about the proteins and mechanisms involved in microbial consortium mediated As toxicity amelioration in rice.

4 CONCLUSIONS

The comparative proteomic analyses of rice leaf revealed that there were several complex metabolic interactions leading to consortium mediated resistance against As stress. The present study provides important protein profile on the basis of consortium mediated improved resistance against As stress in rice. The identified proteins would be helpful for better understanding the complex response network under As stress and may open new avenues for seeking new strategies for As detoxification in rice plants.

ACKNOWLEDGEMENT

Surabhi Awasthi is thankful to Council of Scientific & Industrial Research for the award of Senior Research Fellow.

REFERENCES

Agrawal, L. Chakraborty, S. Jaiswal, D. K. Gupta, S. Datta, A. & Chakraborty, N. 2008. Comparative proteomics of tuber induction, development and maturation reveal the complexity of tuberization process in potato (*Solanum tuberosum* L.). *J. Proteome Res.* 7(9): 3803–3817.

Ahsan, N. Renaut, J. & Komatsu, S. 2009. Recent developments in the application of proteomics to the analysis of plant responses to heavy metals. *Proteomics* 9(10): 2602–2621.

Environmental Arsenic in a Changing World –
Zhu, Guo, Bhattacharya, Ahmad, Bundschuh & Naidu (Eds)
ISBN 978-1-138-48609-6

Evaluation of nitrogen supply on arsenic stress responses of rice (*Oryza sativa* L.) seedlings

S. Srivastava
Institute of Environment and Sustainable Development, Banaras Hindu University, Varanasi, India

ABSTRACT: In the present study, the effects of nitrate supply (low nitrogen: LN and high nitrogen: HN) on arsenate [As(V)] stress (25 M) responses in rice seedlings were monitored for 7 d. The LN + As treatment resulted in significant decline in As concentration (848 $g\,g^{-1}$ DW) than that in As alone treatment (1434 $g\,g^{-1}$ DW) in roots but no significant effect was seen in shoot. In contrast, HN + As treatment showed significant increase in shoot As (6.86 $g\,g^{-1}$ DW) as compared to As alone treatment (3.43 $g\,g^{-1}$ DW). The level of nitrate was increased in roots but declined in shoots in As alone treatment. Surprisingly, no improvement in nitrate level was seen in HN + As as compared to that in As treatment in both root and shoot. The expression analysis of transporters of As (*Pht1;1, Lsi2* and *NIP1;1*) and nitrate (*NRT2;1, NRT2;3a, NRT2;4*) showed significant differences in expression patterns in As, LN+As and HN+As treatments.

1 INTRODUCTION

Arsenic (As) is a nonessential carcinogenic metalloid whose entry into grains of rice plants has raised health concerns for millions of people dependent on rice as staple food. Further, several studies have been conducted till date to see the effects of elements like silica (Si), phosphate, selenium (Se) and sulfur (S) on As stress responses and As concentrations in rice. Nitrogen (N) is a macronutrient required for the biosynthesis of nucleic acids, proteins, chlorophyll, and phytohormones and its availability regulates plant growth. The conventional methods of crop cultivation rely on the application of N fertilizers in high amounts to obtain high yields from high yielding short duration varieties. The impact of nitrogen (N) supply on rice plants has not been studied with respect to As stress. It was considered worthwhile since N not only may impact As responses in rice but also can have positive impact in terms of reducing N fertilizer cost. Hence, the present experiments were, hence, planned to evaluate the impact of variable N supply on rice plants subjected to As(V) stress for 3 and 7 d.

2 METHODS

2.1 *Plant material, growth and treatments*

The study was conducted on Oryza sativa var. IR64. The 4-d old seedlings were placed in 500 mL beaker having 300 mL of 1/2 Kimura solution (pH 5.5, with 0.457 mM N). At 9th d, seedlings were divided into three sets. One set continued to grow in 1/2 Kimura medium with 0.457 mM N (control set). Of the other two sets, one set received 0.0457 mM N (Low N) and other received 0.914 mM S (High N). The Kimura medium contained three sources of N: KNO_3, $Ca(NO_3)_2$ and $(NH_4)_2SO_4$. For Low N treatment, the amount of each source was reduced in equal proportion and salts were replaced with KCl, $CaCl_2$ and $MgSO_4$ in order to avoid deficiency of K, Ca and S. For High N treatment, an additional 0.457 mM of $(NH_4)_2NO_3$ was added. Then on 16th d, each of the three set was divided into two separate sets constituting: one control for each condition and one subjected to 25 μM arsenate [As(V)]. The harvesting was done at 3 d and 7 d.

2.2 *Biochemical, molecular and arsenic analysis*

The root and shoot were then separated and oven-dried at 80–85°C till constant dry weight. The acid digested sample was analyzed with ICP-MS. Cysteine level was measured following the protocol of Gaitonde (1967). The activities of superoxide dismutase and guaiacol peroxidase were assayed as described in Srivastava *et al.* (2006). For proline estimation, the method of Bates *et al.* (1973) was followed. The level of nitrate was estimated following the protocols of Kamphake *et al.* (1967). All the primers used for real-time PCR were from exon-intron boundary and designed using web-based Quant-prime tool. The protocols for RNA extraction, cDNA synthesis, real-time PCR analyses and log2fold-calculation were same as described in our earlier work (Srivastava *et al.*, 2014).

3 RESULTS AND DISCUSSION

3.1 *Arsenic concentrations*

Arsenic was found to be predominantly concentrated in roots than in shoots. The concentrations of As ($\mu g\,g^{-1}$

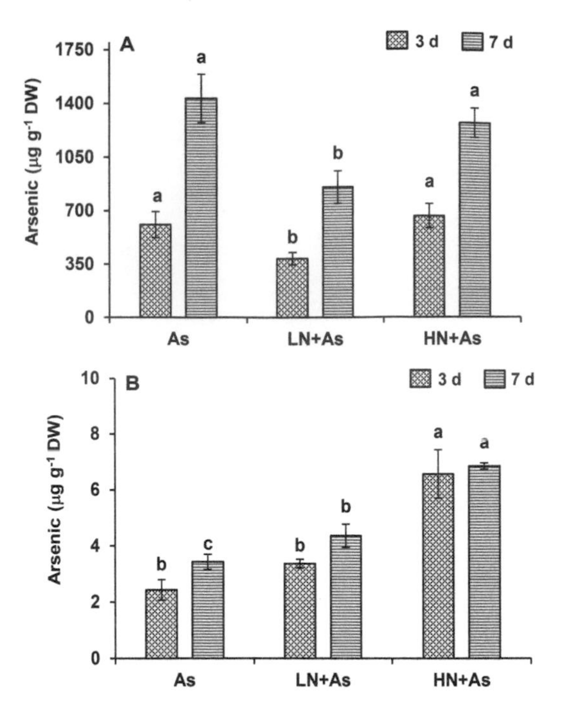

Figure 1. Accumulation of arsenic by rice roost (A) and shoot (B) exposed to arsenate for 3 and 7 d. All values are means of triplicates ± S.D. ANOVA significant at $p \leq 0.01$. Different letters indicate significantly different values at a particular duration (DMRT, $p \leq 0.05$).

DW) were 608 and 1434 in roots and 2.44 and 3.45 in shoot at 3 d and 7 d, respectively. In shoot, As concentrations increased further 3.3 and 4.4 in LN + As and to 6.6 and 6.9 in HN + As at 3 d and 7 d, respectively. In roots, LN + As treatment lead to reduction in As concentrations to 383 (37%) and 848 (41%) $\mu g g^{-1}$ DW as compared to As alone treatment. However, HN + As treatment caused further increase in As concentrations to 655 (8%) at 3 d but induced a decline to 1264 (11%) at 7 d (Fig. 1).

3.2 *Biochemical parameters*

The level of cysteine showed a significant increase in response to As stress and N treatments. However, HN and HN + As treatment induced higher increases in root cysteine as compared to LN and LN + As treatment that caused greater increases in shoot cysteine levels (Fig. 2). There were changes in activities of various antioxidant enzymes like superoxide dismutase and guaiacol peroxidase. Nitrate, proline and photosynthetic pigment level were also altered. The expression analysis of important transporters of nitrate and As was performed at 3 d and 7 d in both roots

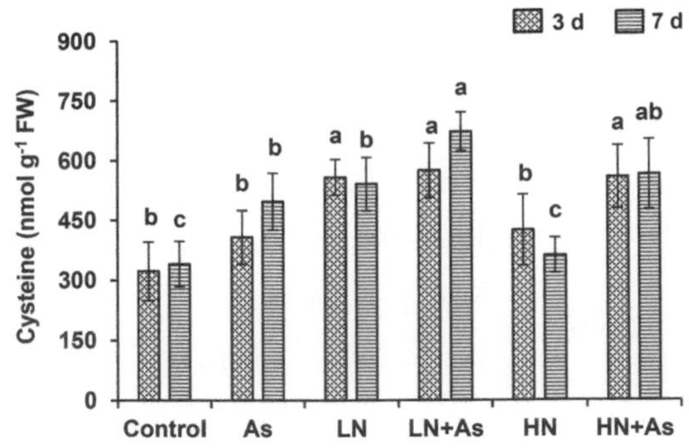

Figure 2. Effect of As and N treatments on cysteine level in shoot of rice. All values are means of triplicates ± S.D. ANOVA significant at $p \leq 0.01$. Different letters indicate significantly different values at a particular duration (DMRT, $p \leq 0.05$).

and shoot. Variable expression patterns in different treatments were observed for *Pht1;1, Lsi2, NIP1;1, NRT2;1, NRT2.3a,* and *NRT2.4.*

4 CONCLUSIONS

In conclusion, nitrate supply has profound influence on responses and As concentration of rice plants exposed to As. The strategy needs to be analyzed further at field scale in future.

ACKNOWLEDGEMENTS

Authors gratefully acknowledge the support from funding received from UGC Startup Grant (F.30-112/2015(BSR).

REFERENCES

Bates, L.S. 1973. Rapid determination of free proline for water-stress studies. *Plant Soil* 39(1): 205–207.

Gaitonde, M. K. 1967. A spectrophotometric method for the direct determination of cysteine in the presence of other naturally occurring amino acids. *Biochem. J.* 104(2): 627–633.

Kamphake, L.T., Hannah, S.A. & Cohen, T.M. 1967. Automated analysis for nitrate by hydrazine reduction. *Water Res.* 1(3): 2015–216.

Srivastava, A.K., Srivastava, S., Mishra, S. & D'Souza, S.F. 2014. Identification of redox-regulated components of arsenate (As(V)) tolerance through thiourea supplementation in rice. *Metallomicms* 6(9): 1718–1730.

Srivastava, S., Mishra, S., Tripathi, R.D., Dwivedi, S. & Gupta, D.K. 2006. Copper-induced oxidative stress and responses of antioxidants and phytochelatins in *Hydrilla verticillata* (L.f.) Royle. *Aquat. Toxicol.* 80(4): 405-415.

Environmental Arsenic in a Changing World –
Zhu, Guo, Bhattacharya, Ahmad, Bundschuh & Naidu (Eds)
ISBN 978-1-138-48609-6

Remediation of a heavy metals contaminated site in urban area: a case study from southern China

H. Huang, Y.F. Wang, S.J. Wang, Y.C. Liu, B. Zhang & Y. Yang
Research and Development Department, China State Science Dingshi Environmental Engineering Co., Ltd, Beijing, P.R. China

ABSTRACT: The reuse of former industrial sites is recognized as one solution to increase land use in major cities in China. As many former industrial sites are located near the heart of a city, the restoration of these sites would increase revenues and reacquire what was once prime real estate. A 76 years old paper mill has been closed in 2012 and left a site contaminated with As, Pb Cu and Zn. The successful remediation of the site has been achieved by the CSSDS using stabilization/solidification with the independently developed in situ deep injection-mixing equipment.

1 INTRODUCTION

A former paper mill which was established in 1938 and was one of the largest news-printing paper providers in China, the mill was closed in 2012. The abandoned site was planned to reuse as local government reserve land for future development. Considering the site investigation results (Table 1), top soils were excavated and disposed. Heavy metals in deep soil layers were stabilized/solidified by selected amendments. The appropriate amendments were injected into soils with an in situ deep injection-mixing equipment. The S/S project lasted for over four months and treated soils on the contaminated site have been assessed by the certified organization.

2 METHODS

2.1 *Site description*

The site locates in a former paper mill in a southern city in China. The contaminated site was used as the pulp workshop, thermal power station and wastewater treatment plants. The field investigation showed that the contaminated area was 11,735 m^2 and the main contaminants were As, Pb Cu and Zn. Total concentrations of each heavy metal are: As 56–1652 mg kg^{-1}, Cu 310–673 mg kg^{-1}, Pb 181–10600 mg kg^{-1}, Zn 324–26500 mg kg^{-1}.

Table 1. Site characterization and remediation targets.

	Depths (m)	Project quantity (m^3)	Remediation technologies	HM con. in soil leachate (mg L^{-1})	Remediation targets* (mg L^{-1})
1	0.2~0.0	2347	excavation and disposal	–	–
2	0.0~–0.3	3521			
3	–0.3~–6.0	66888	In situ stabilization/ solidification	As:0.1–0.3	0.05
				Cu:0.2–0.6	1.5
				Pb:0.2–1.0	0.1
				Zn:0.1–1.5	5

*: Treated soils are leached and the HM concentration in the leachate qualifies the IV criteria of 'Chinese Quality Standard for Groundwater' (GB/T14848-1993).

2.2 *Remediation technologies*

Top soils (0.2–0.3 m) were excavated and disposed (Table 1).

For deeper soils, stabilization/solidification was used to immobilize As, Pb Cu and Zn. Amendments used in this project were mixtures of lime, iron-based material and clay (lime: iron-based material: clay = 2:1:1). The application rate of amendments to soil was between 1% and 4%.

Depending on the site condition, cement was also applied with rate of 5%–10%.

2.3 *In situ deep injection-mixing equipment*

The independently developed in situ deep injection-mixing equipment was used to inject amendments and mix amendments with soils. An injection-mixing system consists of two identical agitating shafts and one injection pipe. The JBS-II biaxial mixing drill (Fig. 1) was used in this project. Parameters of the equipment are shown in Table 2. Total drilling points on the site was 19,558, with depth of 5.7 m for each borehole.

The engineering process is shown in Figure 2.

2.4 *Assessment of remediation*

After the completion of remediation, 187 soil samples were collected on site. All soil samples were analyzed according to the standard extraction toxicity methods.

Figure 1. JBS-II biaxial mixing drilling system.

Table 2. Parameters of the in situ deep injection-mixing equipment.

Components	Parameters
Borehole (mm)	1200 × 700
Drilling depth (m)	18~20
Max. torque (kN · m)	29.2
Power head drilling speed (r min^{-1})	24.2
Step length (m)	1.1

Figure 2. Engineering flow diagram of stabilization/solidification.

3 RESULTS AND DISCUSSION

3.1 *Stabilization/solidification of HMs*

Solidification/stabilization reduced the mobility of contaminants in the environment through both physical and chemical mechanism (Kumpiene *et al.*, 2008). The concentrations of As, Cu, Pb and Cu in the leachate of treated soil were below the IV criteria of 'Chinese Quality Standard for Groundwater' (GB/T14848-1993). Over 50% to 80% of contaminants on site were not presented in soil leachate, which indicated a relatively high stabilization performance. The selected amendments consist of iron-based material which has been demonstrated to effectively reduce As mobility and phytoavailability (Warren *et al.*, 2003). Lime is a commonly used alkaline amendment for mobilizing HMs by raising pH and co-precipitation. It is effective for Cu and Pb but increase of Cr and As (Seoane and Leiros, 2001). Research also indicates that Cu, Zn and Pb could be retained in soil with iron or clay materials (Wang *et al.*, 2001).

3.2 *Deep injection-mixing equipment*

The equipment has the advantage of combining injection and mixing together during the drilling, this could ensure the complete reaction of amendments with HMs presented in soils.

Treatment process is not affected by the soil texture, especially for the soils with low permeability and for medium/hard formations. Soils in south

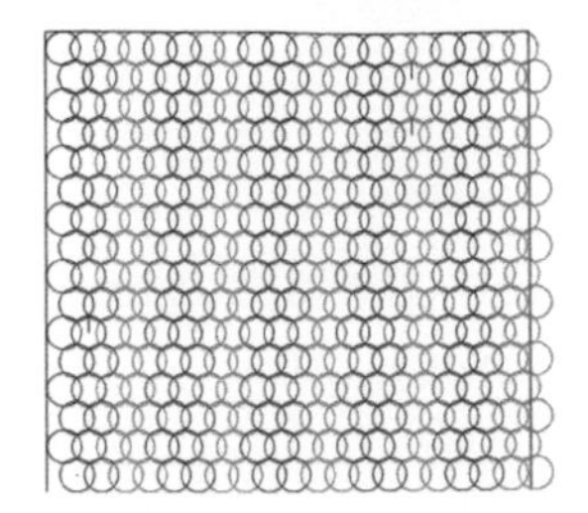

Figure 3. Schematics of borehole distributions.

China are loam and with high water content. When amendments injected into soils by common injection system, amendments diffusion is constrained by the soil physiochemical characteristics. Therefore, mixing soils with amendments during the injection process is necessary for this site remediation. Soils are under an efficient mixing could maintain a turbulence status, therefore amendments and soils are completely mixed, providing an optimal condition for stabilization reaction.

The equipment has a larger injection range, which is 1200 mm × 700 mm (Table 2). Proper drilling point distribution design guarantees no gaps between adjacent drilling points (Fig. 3).

3.3 *Project cost*

Total cost of the project is 18.69 million CNY, and the cost for treating a cubic meter of soil by using this technology is approximate 300 CNY.

4 CONCLUSIONS

A former paper mill site was contaminated by HMs and remediated by stabilization/solidification. Selected amendments were applied by using the independently developed in situ deep injection-mixing equipment. The contaminants in soils were successfully immobilized.

ACKNOWLEDGEMENTS

We appreciate our colleagues from R&D Department and Engineering Technology Department for their technical supports during the project.

REFERENCES

Kumpiene, J., Lagerkvist, A. & Maurice, C. 2008. Stabilization of As, Cr, Cu, Pb and Zn in soil using amendments-A review. *Waste Manage.* 28(1): 215–225.

Seoane, S. & Leiros, M.C. 2001. Acidifiction-neutralisation in a linite mine spoil amended with fly ash or limestone. *J. Environ. Qual.* 30(4):1420–1431.

Wang, Y.M., Chen, T.C., Yeh, K.J. & Shue, M.F. 2001. Stabilization of an elevated heavy metal contaminated site. *J. Hazard. Mater.* 88(1): 63–74.

Warren, G.P., Alloway, B.J., Leep, N.W., Singh, B., Bochereau, F.J.M. & Penny, C. 2003. Field trails to assess the uptake of arsenic by vegetables from contaminated soils and soil remediation with iron oxides. *Sci. Total Environ.* 311(1–3): 19–33.

Environmental Arsenic in a Changing World –
Zhu, Guo, Bhattacharya, Ahmad, Bundschuh & Naidu (Eds)
ISBN 978-1-138-48609-6

Influence of pH in the conditioning and dehydration processes of arsenic-containing sludge

S.E. Garrido Hoyos[1], K. García[1], J. Briseño[2] & B. Lopez[2]
[1] *Instituto Mexicano de Tecnología del Agua, Subcoordinación de Posgrado, Jiutepec, Mor., Mexico*
[2] *Universidad Politécnica del Estado de Morelos. Jiutepec, Mor., Mexico*

ABSTRACT: The arsenic that is removed from the water, for human use and consumption generates a residue that is toxic. The objective of this study was to analyze the influence of pH in the conditioning process of sludge produced in the removal of As(V) by the coagulation-flocculation process. The properties that influence the chemical conditioning and kinetics of drainage in the dewatering process are, pH, polymer dosage, time and speed of agitation. It was determined that the pH in the conditioning process if it influences the dewatering of sludge with As, being the pH 7, which shows significant differences in the conditioning of the mud, improving the dehydration stage.

1 INTRODUCTION

Arsenic is considered to be a metalloid that produces toxicity to humans at acute exposures: $>2\,mg\,kg^{-1}\,d^{-1}$ and chronic exposures: $0.03–0.1\,mg\,kg^{-1}\,d^{-1}$. The regulations for water consumption are: WHO-EPA $0.010\,mg\,L^{-1}$ and NOM-127-SSA1-1994 $0.025\,mg\,L^{-1}$. In the removal of arsenic from water for human consumption through different treatment systems, waste is generated, which must be minimized (Hoyos *et al.*, 2013). So, it is important to treat sludge with As generated in the water treatment plants, as well as to eliminate the greater amount of water and thus reduce its volume to facilitate transport and final disposal operations. Among the methods of treatment to reduce the volume of sludge are the following: thickening, chemical conditioning and dewatering.

2 METHODS/EXPERIMENTAL

Optimization of the conditions to obtain mud As(V) Firstly for the production of water and mud, we used a central composite design, $N = 2^3$, where the factors were: A: pH 6.5 to 7.5; B: Dosage of $FeCl_3$ 20 to 40 mg L^{-1}; C: Poacrylamine cationic polymer dosed; 0.5 to 1.5 mg L^{-1}. The initial concentration of As(V) was 0.050–0.100–0.150 mg L^{-1}, a rapid mixture: 400 s^{-1}, 10 s, slow mixture: 60, 24, 12, 10 s^{-1} for 5 min and sedimentation 20 min, obtaining the values of R^2 and Raj^2.

2.1 *Chemical conditioning at different pH's*

Secondly, for the chemical conditioning of the sludge previously obtained and thickened by gravity during 12 h, the adjustment was made to different pH's, (3, 4, 5, 5.5, 6, 6.5, and 7), 800 mg L^{-1}, of polymer AN 913 VHM, a speed in the rapid mixture was dosed: 100 rpm / 9 s; Slow mixing: 30 rpm/14 s and a sedimentation of 1 min. Subsequently, the sludge obtained at different pH's was analyzed, the characteristics and physical and chemical properties were analyzed according to the Mexican standards, (Garrido *et al.*, 2016).

2.2 *Dehydration of conditioned sludge*

Third, the dewatering of the sludge, chemically conditioned at different pH's, was performed in a band filter equipment (Bootest-IFTL). The test was carried out with a volume of 500 mL, in which the drainage kinetics was determined by the amount of total solids retained in an NFMM 302-style polypropylene mesh with a permeability of 80–100 $m^3\,s^{-1}\,m^{-2}$. On the other hand, the total suspended solids in the drained volume were determined, to obtain the drainage index (Eg).

3 RESULTS AND DISCUSSION

3.1 *Optimal conditions in obtaining mud*

The optimal values of the variables in the removal of arsenic for an initial concentration of As of 0.150 mg L^{-1} and the polymer C-496 HMW which was the one that performed best in the tests: (A) pH: 7.20; (B) Dosage of $FeCl_3$: 34.33 mg L^{-1}; (C) Polymer dosage: 0.89 mg L^{-1}. The variability of the concentration response of As in the supernatant was <0.003 mg L^{-1} and SSED of 7.59 mL L^{-1}. With $R^2 = 85.09\%$ and R^2 (adjusted by g.l.) = 68.31%.

Table 1. Quality physico-chemical of the $FeCl_3$ sludge of the conditioning

Parameter	pH 3	4	5	5.5	6	6.5	7
Quality physico							
r (cm g^{-1})	$3.0E^{+11}$	$1.3E^{+11}$	$7.8E^{+10}$	$9.8E^{+10}$	$1.4E^{+11}$	$7.1E^{+10}$	$5.7E^{+10}$
tF (s)	698	289	203	246	214	150	100
Eg	4.8	−2.2	−3.1	−2.5	−3.2	−3.7	−3.5
Willcomb index	6	6	8	8	8	8	*10
Total solids (g L^{-1})	10.6	12.9	11.9	12	7.2	9.8	10
Total solids in wet filter cake (g L^{-1})	6.2	67	6.0	5.9	5.8	8.1	5.3
Initial sludge moisture content (Ci) (%)	92	90	91	91	92	90	89
Final sludge moisture content (cf) (%)	73	75	70	68	73	81	68
Potential Z (mV)	−13	−28	−32	−38	−39	−48	−50
Quality chemical							
pH (final)	2.9	4.3	6.0	6.6	6.9	7.2	7.4

*Sized flocs precipitate relatively quickly.

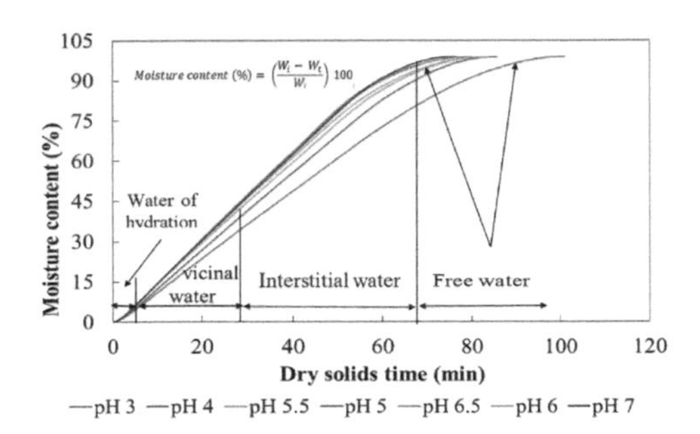

Figure 1. Drying characteristic curve, where Wi is the initial weight and Wf is the weight at any time after drying at 105°C.

3.2 *Chemical conditioning at different pH's*

The F-Q characteristics of the conditioned sludge show that at a higher pH lower PZ, with pH 7 being the lowest with a value of −50 mV, confirming its stability in the colloids in the mud.

Regarding the final pH, it indicates that pH 4–7 is an increase, due to the characteristics of the proportioned polymer. And for SSED, 500 mL L^{-1} was obtained in the different pH's (Table 1).

The moisture content in the sludge of the chemical conditioning process is related to the distribution of the water content, for the pH 7 The interstitial humidity contains 13.07%, the vicinal water of 69.28%, free humidity 9.15% (without associated with solid particles) and finally the chemically bound moisture contains 8.49%, obviously having a shorter evaporation time in the mud (76.5 min) (Fig. 1).

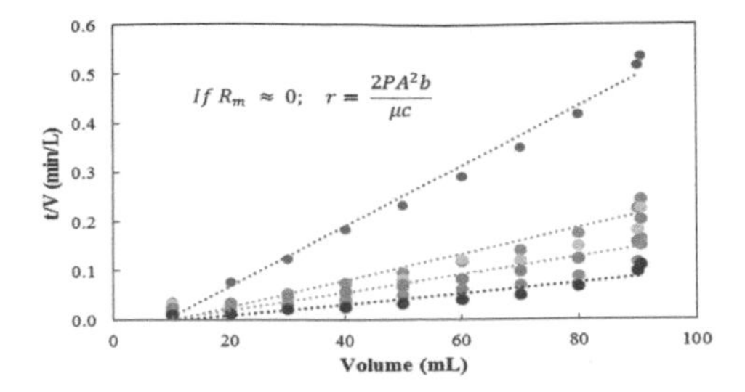

Figure 2. Comparison of time and volume interactions in the REF technique, of sludge conditioned at different pHs.

When comparing the time to filtration (tF) at pH 3 (698 s) and 7 (100 s) and as soon as the Specific Resistance to Filtration (REF) it was obtained that the sludge at pH 3 (3.0 + 11) unlike the pH 7 (5.7 E+10) they present a higher REF, this due to the composition of the mud and the distribution that the particles present, as well as the size and resistance of the floc.

3.3 *Dewatering of conditioned sludge*

For a pH of 3 the speed in the drainage kinetics of the mud is greater than for the pH of 7, because it has the smallest size of flocs (0.5–0.75 mm), Willcomb index, stability of the cake moisture on the mesh and higher concentration of suspended solids in the filtered volume (Fig. 2).

4 CONCLUSIONS

It was determined that the pH does influence the dewatering of sludge. The pH 7, shows a greater% of superficial humidity (69.28%) in the conditioning of the mud, improving the dehydration stage. The sludge with As obtained contributes to reducing disposal costs for potabilization plants and damages to public health.

ACKNOWLEDGEMENTS

Consejo Nacional de Ciencia y Tecnología (CONACyT), FIDEICOMISO Instituto Mexicano de Tecnología del Agua (IMTA), Universidad Politécnica del Estado de Morelos (UPEMOR).

REFERENCES

Garrido, S. & Garcia, K. 2016. Evaluation of dewatering performance and physicalchemical characteristics of iron chloride sludge. *Proceedings Arsenic in the Environment. Arsenic Research and Global Sustainability As2016*. CRC Press, Boca Raton, FL, pp. 490–492.

Hoyos, S.E.G., Flores, M.A., Gonzalez, A.R., Fajardo, C.G., Zoloeta, S.C. & Orozco, H.V. 2013. Comparation two operating configurations in a fullscale arsenic removal plant: Case study: Guatemala. *Water*. 5(2): 834–851.

Environmental Arsenic in a Changing World –
Zhu, Guo, Bhattacharya, Ahmad, Bundschuh & Naidu (Eds)
ISBN 978-1-138-48609-6

Assessment of environmental and health risks of arsenic in agricultural soils

C. Jones-Johansson[1], M. Elert[1], S. Vijayakumar[1,2], P. Bhattacharya[2], I. Jordan[3], I. Mueller[4], F. Battaglia-Brunet[5] & M.L. Guédard[6]
[1]*Kemakta Konsult AB, Stockholm, Sweden*
[2]*KTH-International Groundwater Arsenic Research Group, Department of Sustainable Development, Environmental Science and Engineering, KTH Royal Institute of Technology, Stockholm, Sweden*
[3]*G.E.O.S. Ingenieurgesellschaft mbH, Halsbrücke, Sachsen, Germany*
[4]*Saxon State Office for Environment, Agriculture and Geology, Freiberg, Germany*
[5]*BRGM, Orléans, France*
[6]*LEB Aquitaine Transfert-ADERA, Villenave d'Ornon, France*

ABSTRACT: Within the framework of the AgriAs project, assessment of risks to health and the environment from arsenic in the agricultural soils and recipient water bodies is being carried out. Criteria for risk assessment from different organizations and countries have been collated, and an exposure model for arsenic in agricultural soils is being developed. The exposure model will be applied to assess the present-day situation at two test sites where elevated arsenic concentrations are found in agricultural soil. The model will also be used to assess the risk reduction achieved by proposed methods of amelioration and treatment.

1 INTRODUCTION

Arsenic (As) in agricultural soils, and its subsequent transport in the food chain and recipient waters potentially causes risks to human health (Murcott, 2012). Arsenic can accumulate in crop plants and high arsenic concentrations may also reduce crop yields. The present study, performed within the framework of the Water JPI AgriAs project, aims to develop tools for the risk assessment of arsenic in agricultural soils.

The risk assessment tools will be used to assess the risks at two test sites; one site is the former chemical ammunition destruction facility near Verdun, France, where arsenic-containing ammunition was destroyed in the inter-war period. The other site is located in Saxony, Germany, where mining and associated industries over a period of 800 years have led to elevated concentrations of arsenic in soils, groundwater and surface waters. Data from investigations at these sites about the speciation of arsenic and its bioavailability and ecotoxicity are being collected as part of the AgriAs project. The present-day situation will be studied and the effects of different methods of amelioration and treatments at these sites will be studied. The results of the work will be used in the development of guidelines for the sustainable management of the risks identified.

2 METHODS

2.1 *Collation of risk assessment criteria and site data*

Relevant criteria for use in health and environmental risk assessments of As in agricultural soils have been identified. The compiled criteria include toxicological reference values, limiting values for the concentrations of arsenic in different media through which exposure occurs (food, fodder and drinking water) and limit or guideline concentrations for arsenic in soils.

Data from the two test sites are also being compiled regarding concentrations of arsenic in soils, groundwater, surface water and plants.

Risk assessment has then been carried out by comparing measured concentrations in environmental media with the limit or guideline values.

2.2 *Development of an exposure model*

An exposure model for As in agricultural soils and water has been developed taking into account the main exposure pathways (Fig. 1).

The model can be adapted to different sites by using site-specific parameter values, especially in relation to the speciation of As and its effect on transport to and in groundwater, bioavailability and toxicity. The model than can be used in site-specific assessments of the risks from arsenic.

3 RESULTS AND DISCUSSION

Criteria for arsenic are not available for a comprehensive range of foodstuffs; at present criteria are available only for arsenic and one or two other groups of foodstuffs. In addition, criteria for arsenic in soil may lead to an overestimation of the risks from arsenic in agricultural soils as they do not take into account

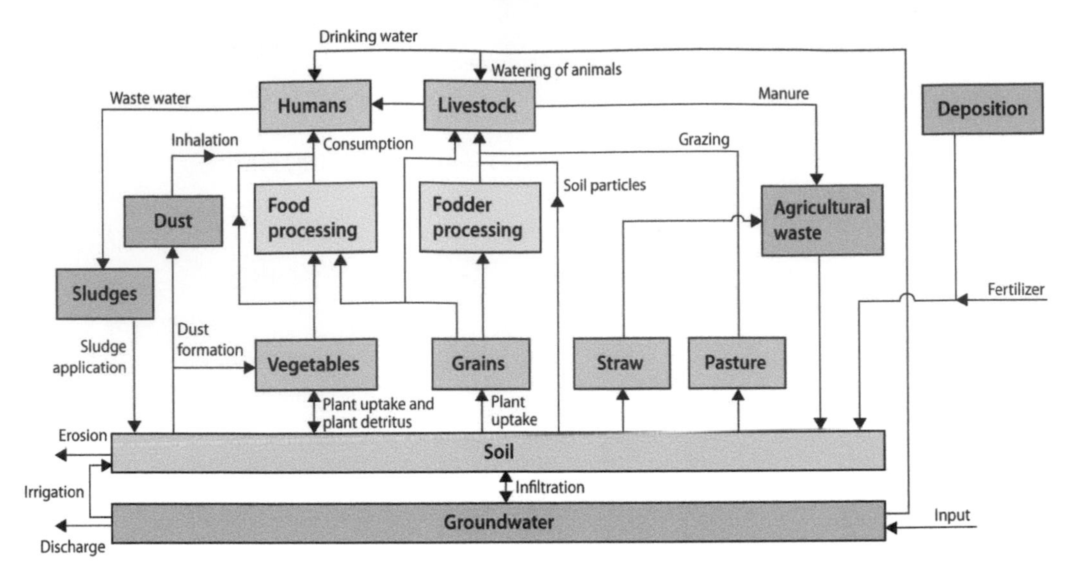

Figure 1. Conceptual exposure model for arsenic in agricultural soils.

site specific variations in bioavailability, mobility and uptake into the food chain.

The exposure model will be used for assessment of the present-day situation at the two field sites. The model will also be used to identify the major uncertainties regarding exposure to arsenic in agricultural soils and to identify areas where site specific data can improve exposure estimates. After calibration with site specific data, the model will be used to evaluate the risk reduction achieved by the proposed amelioration methods tested within the AgriAs project and the treatment methods developed and tested within the project. This information will be an important input for the development of recommendations for the sustainable management of risks from arsenic in agricultural soils.

ACKNOWLEDGEMENTS

AgriAs is co-funded by the EU and the Academy of Finland, L'Agence Nationale de la Recherche, Bundesministerium für Ernährung und Landwirtschaft and Forskningsrådet FORMAS under the ERA-NET Cofund WaterWorks 2015 Call. This ERA-NET is an integral part of the 2016 Joint Activities developed by the Water Challenges for a Changing World Joint Programme Initiative (Water JPI).

REFERENCES

AgriAs. 2108. Collation of evaluation criteria for assessment of the risks from arsenic in agricultural soils (to be published).

Murcott S. 2012. *Arsenic Contamination in the World: An International Sourcebook*. IWA Publishing. London.

Environmental Arsenic in a Changing World –
Zhu, Guo, Bhattacharya, Ahmad, Bundschuh & Naidu (Eds)
ISBN 978-1-138-48609-6

Acid induced arsenic removal from soil amended with clay-biochar composite

M. Vithanage[1], L. Weerasundara[2] & A.K. Ghosh[3]
[1]*Ecosphere Resilience Research Center, Faculty of Applied Science, University of Sri Jayewardenepura, Nugegoda, Sri Lanka*
[2]*Environmental Chemodynamics Project, National Institute of Fundamental Studies, Kandy, Sri Lanka*
[3]*Environmental Mahavir Cancer Institute and Research Centre, Patna, India*

ABSTRACT: Phytoremediation itself is not efficient enough to remove As from soil as all As bound fractions are not available for plants. Therefore, the use of extracting agent has been common. However, in general high potential for extraction have been seen by extracting agents like EDTA which is recognized as somewhat toxic. Not many studies showed the potential of organic acids with composites enhancing the extracting capacity. Therefore, three organic acids; acetic, citric and oxalic, and three inorganic acids; sulfuric, nitric, hydrochloric, were used to evaluate the desorption capacity of As into the soil solution from As contaminated soil together with two types of modified BC composites; Biochar-montmorillonite (BC-MT) and Biochar-Natural Red-Earth (BC-RE). Citric acid showed the better performance with As release compared to all other organic and inorganic acids which is 60% of the aqua regia digested arsenic in soil at highest acid concentration of 10 mM. The use of BC composites further enhanced the capability of As desorption with citric acid. The biochar-montmorillonite (BC-MT) composite showed 90% of desorption capacity with citric acid and for biochar-red earth (BC-RE) composite it was a 95% of escalation. Therefore, the citric acid with the presence of BC-MT and BC-RE composites have higher capacity to enhance the As desorption from As contaminated soils.

1 INTRODUCTION

Phytoextraction has many limitations, since minute concentrations are taken up by plants whereas chemically induced phytoextraction has been proposed as an alternative to clean up metal-contaminated soils (Alkorta *et al.*, 2004). Various types of chelating agent, including ethylene diamine tetraaceticacid (EDTA) is used as efficient synthetic chelating agent for increasing the concentration of various metals in the aboveground biomass of plants however, EDTA leaching results surface and groundwater pollution (Evangelou *et al.*, 2007). In comparison with EDTA, low molecular weight (LMW)-organic and inorganic acids, are alternatives to assist with metal phytoextraction. At the same time, Biochar (BC) is a soil amendment which is mostly used to immobilize the heavy metals and metalloids in soil (Kumarathilaka *et al.*, 2018). Hence, three organic and inorganic acids were assessed in order to evaluate the possibility of cleaning As contaminated soil through enhancing its mobility and two BC composites; biochar-montmorillonite (BC-MT) and biochar-red earth (BC-RE) were used to assess the capability to trigger the clean-up process in As contaminated soils.

2 METHODS

Arsenic contaminated soil was obtained from As contaminated site in Naya Tola, Maner, India, was used. The soil was air-dried and mechanically sieved to <1 mm fraction. The pH and electrical conductivity (EC) of the soil were measured in 1:10 suspension of soil-to-water whereas the total As concentration in soil was measured with aqua regia. Cation exchange capacity (CEC), Total organic carbon (TOC), available phosphate (PO_4^{3-}) and nitrate (NO_3^-). As per the methods in Vithanage *et al.*, (2014).

2.1 *Preparation of biochar composites*

Montmorillonite (MT) and red earth (RE) clay were dissolved in distilled water to prepare 25 g L^{-1} clay suspensions separately. 250 g of BC was added into 2 L of solution and then the mixtures were shaken for two hours in a mechanical shaker and oven dried. Then the slow pyrolysis process was used at the rate of 7°C min^{-1} under limited oxygen condition at 500°C and held constant for 30 min.

2.2 *Acid experiments*

Three organic (acetic, citric and oxalic) and three inorganic (sulfuric, nitric and hydrochloric) acids of different concentrations (0.05, 0.1, 0.5, 1.0, 5.0 and 10 mM) were used to evaluate the As release in the As contaminated soil. Approximately, 1 mL of chloroform was added per liter of all organic acid solutions in order to prevent microbial breakdown of the organic acids. As contaminated soil (20 g L^{-1}) was placed in polyethylene tubes and 40 mL of each acid was added with different concentrations to assess the As release

Table 1. Physico-chemical properties of As contaminated soil and biochar composites.

Soil		BC-MT composite		BC-RE composite	
pH	7.47	pH	9.51	pH	8.99
EC ($\mu S\,cm^{-1}$)	7.21				
Total As ($mg\,kg^{-1}$)	4.61				
Available NO_3^- ($mg\,kg^{-1}$)	2.19	EC ($\mu S\,cm^{-1}$)	1799	EC ($\mu S\,cm^{-1}$)	1049
Available PO_4^{-3} ($mg\,kg^{-1}$)	2.79				
TOC %	2.59				
CEC ($cmol\,kg^{-1}$)	7.16				

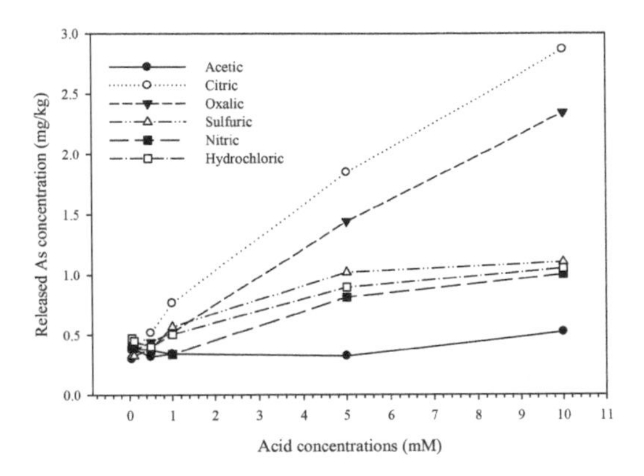

Figure 1. As release with organic and inorganic acids.

rate in bare soil. To assess the influence of BC composites 2.5 wt% BC were added to the same weight of soil and 20 mL of each acid was added separately. These tubes were equilibrated for 24 h at room temperature (~26°C) and agitated on a mechanical shaker at 100 rpm. The supernatant was filtered after centrifugation. The filtered solutions were analyzed for pH and As concentrations were measured with ICP-OES.

3 RESULTS AND DISCUSSION

The soil and BC characterization data are depicted in Table 1. The total and DTPA extractable metal concentration are almost same for As contaminated soil and therefore it can be suggested that all the As in soil is in its bioavailable form.

The As release rates with different organic acids and inorganic acids and inorganic acids are presented in Figure 1.

The inorganic acids showed the weakest As desorption rates compared to organic acids except acetic acid. However, acetic acid is at the lowest place in the means of As release. Citric acid shows the most effective with As desorption. Addition of citric acid could decrease the soil pH by approximately 3 units at the highest concentration (10 mM). Therefore, use of higher concentrations of citric acid may not efficient

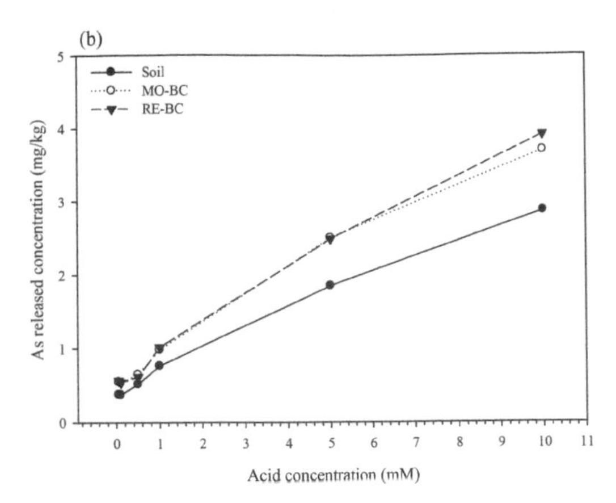

Figure 2. Arsenic release with citric acid with the presence of biochar composites.

for the As desorption in the means of effects on soil quality and plant growth.

Figure 2 shows the way in which different inorganic and organic acids are responded to the BC-MT and BC-RE in the means of As desorption. Citric acid could release almost all the bio-available As fraction into the soil solution with the percentage of 95% with BC-RE composite. However, the two BC composite types did not show significant different as the BC-MO showed 90% release of As into the soil solution.

4 CONCLUSIONS

Citric acid showed the highest performance for As desorption than any other organic and/or inorganic acid type with approximately 70% of desorption capacity. Both BC types were showed significant triggering capability for As desorption with citric acids as 90% for BC-MT and 95% for BC-RE. Therefore, the use of BC-MT and BC-RE with citric acid pose significant potential to desorption of As from As contaminated soils.

REFERENCES

Alkorta, I., Hernández-Allica, J., Becerril, J.M., Amezaga, I., Albizu, I., Onaindia, M. & Garbisu, C. 2004. Chelate-enhanced phytoremediation of soils polluted with heavy metals. *Rev. Environ. Sci. Bio.* 3(1): 55–70.

Evangelou, M.W.H., Ebel, M. & Schaeffer, A. 2007. Chelate assisted phytoextraction of heavy metals from soil. Effect, mechanism, toxicity, and fate of chelating agents. *Chemosphere* 68(6): 989–1003.

Kumarathilaka, P., Ahmad, M., Herath, I., Mahatantila, K., Athapattu, B., Rinklebe, J., Ok, Y.S., Usman, A., Al-Wabel, M.I. & Abduljabbar, A.. 2018. Influence of bioenergy waste biochar on proton-and ligand-promoted release of Pb and Cu in a shooting range soil. *Sci. Total Environ.* 625: 547–554.

Vithanage, M., Rajapaksha, A.U., Oze, C., Rajakaruna, N. & Dissanayake, C. 2014. Metal release from serpentine soils in Sri Lanka. *Environ. Monit. Assess.* 186(6): 3415–3429.

5.4 Mitigation and management of arsenic in a sustainable way

Environmental Arsenic in a Changing World –
Zhu, Guo, Bhattacharya, Ahmad, Bundschuh & Naidu (Eds)
ISBN 978-1-138-48609-6

Implementation of arsenic mitigation: insights from Araihazar and Matlab – two extensively studied areas in Bangladesh

K.M. Ahmed[1], A. van Geen[2] & P. Bhattacharya[3]
[1]*Department of Geology, University of Dhaka, Dhaka, Bangladesh*
[2]*Lamont Doherty Earth Observatory, Columbia University, New York, USA*
[3]*KTH-International Groundwater Arsenic Research Group, Department of Sustainable Development, Environmental Science and Engineering, KTH Royal Institute of Technology, Stockholm, Sweden*

ABSTRACT: Despite many efforts over the last 25 years, millions of people are still drinking groundwater having arsenic above Bangladesh and World Health Organization limits. This can be linked, to some extent, to a lack of pragmatic policies and implementation strategies. Important lessons have been learned from extensive studies in Araihazar and Matlab in Bangladesh. Also better mitigation strategies for reducing arsenic exposure have been demonstrated at both locations. New policy and mitigation strategies should focus on mapping using mobile technology for village scale mapping. Dissemination of test result is extremely important in increasing awareness of the users and a new three-color scheme (Blue, Green, Red corresponding to <10, 10–50, and $>50\,\mu g\,L^{-1}$) instead of existing Red-Green color scheme should be adopted to lower arsenic intake. Safe wells can be installed at appropriate locations and depth by combining hydrogeological and social criteria aided by applications-based score ranking.

1 INTRODUCTION

Arsenic contamination of groundwater remains a major problem in access to safe water in Bangladesh although various arsenic mitigation plans have been tested and implemented in the country over the last 25 years. About 20 million people are still drinking water with more than $50\,\mu g\,L^{-1}$ arsenic while the number increases to 40 million is $10\,\mu g\,L^{-1}$ is considered as the safe limit. Inability to provide safe water to a large number of people is linked with the issue of pragmatic implementable policy of arsenic mitigation (Pfaff *et al.*, 2017). Various national and local level studies have focused on scientific and societal aspects in prioritizing arsenic mitigation in Bangladesh. Araihazar in Narayangaj and Matlab in Chandpur are the two extensively studied areas located in two different types arsenic exposure scenarios (Fig. 1) where mitigation has been designed based on hydrogeological and socio-economic aspects. Lessons learned from the two areas can help in upgrading policies and implementation plans for arsenic in areas where people are still drinking arsenic above safe limits (Hossain *et al.*, 2014; van Geen *et al.*, 2016).

2 METHODS

Occurrences and distribution of arsenic has been thoroughly investigated in Araihazar in relation to a health impact study by a multidisciplinary team of scientists from Columbia University, USA and University of Dhaka Bangladesh since 2001. A detailed study has also been carried out in Matlab area by a consortium of the Royal Institute of Technology, Sweden; University of Dhaka and NGO Forum for Public Health, Bangladesh. Both the studies included blanket testing of arsenic; detailed subsurface mapping; analysis of arsenic and other water quality parameters; investigations on spatial, vertical and temporal variations of groundwater arsenic; association between sediment color and groundwater chemistry; village scale mapping of arsenic; installation of safe wells based on sediment color; exploration of intermediate aquifer as a source of safe water. Detailed and accurate mapping at village scales of locations of unsafe wells along with geology were studied for installing safe well in Araihazar; whereas, socio-economic status of the villagers was also recorded for placing safe wells in Matlab.

3 RESULTS AND DISCUSSION

Blanket testing and mapping using Mobile Technology: Studies in Araihazar showed that there are many untested wells with arsenic concentrations above allowable limit and those wells are being used by the local communities. There is a need for national blanket testing using smart technologies and apps so that data can be stored and used for mapping down to village scale.

Not just Red and Green, we need a Blue: It is very important to display test results on wells. Red and Green color of well spouts were used in earlier testing program for marking safe and unsafe wells according to Bangladesh standards. As now there is more

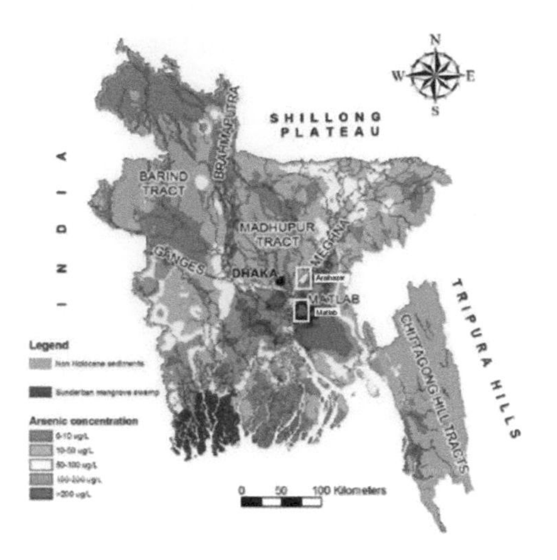

Figure 1. Location of Araihazar and Matlab on arsenic distribution map of Bangladesh (modified from Hossain *et al.*, 2014).

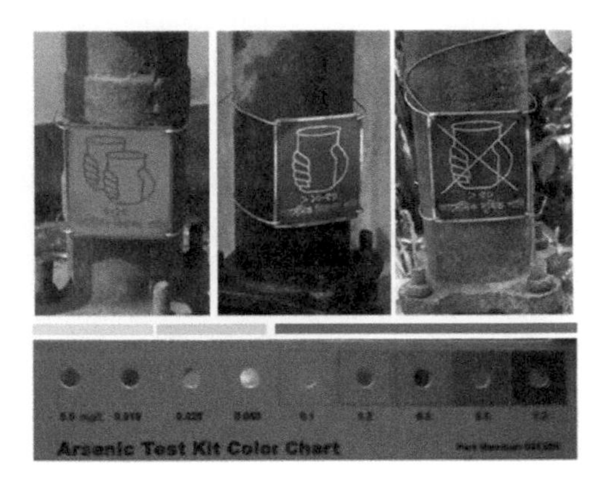

Figure 2. Three color placards used in Araihazar for disseminating test results (from van Geen *et al.*, 2014).

information regarding the relative health risk of 50 and $10\,\mu g\,L^{-1}$, users must be informed about it. This can be achieved by adopting a three-color system instead of two by including Blue for wells up to $10\,\mu g\,L^{-1}$ (Fig. 2).

Village scale implementation plan (VSIP): Studies both in Araihazar and Matlab demonstrated that the best way to plan for arsenic mitigation is through village scale mapping of spatial and vertical variations. Social criteria can be added to the arsenic map to ensure safe water to the poorer communities.

Score-based safe well location: Relative scoring for various locations can be introduced by using mobile phone apps. This can ensure better location of safe wells for providing maximum coverage.

Not only STW and HTW, there can be IDTW: It has been found at both the study area that there is an intermediate aquifer between 100 to 200 m at most villages. If intermediate deep wells (IDTW) are installed targeting this depth, lot of resource can be compared

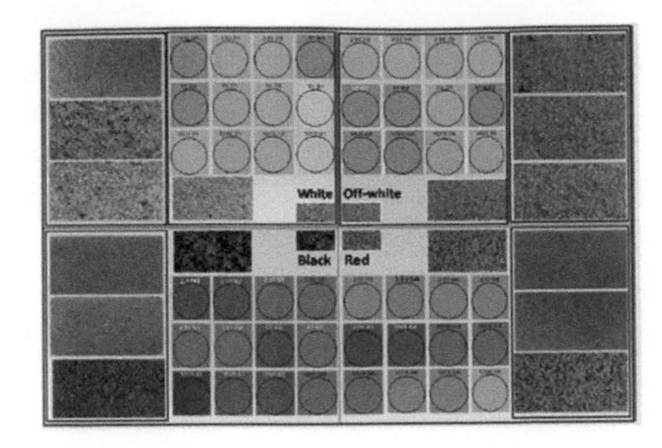

Figure 3. Sediment color tool for use by the local drillers in targeting arsenic safe aquifers (from Hossain *et al.*, 2014).

to installation of deep wells only at depth more than 200 m.

Capacity building of drillers, the Sediment Color Tool: Sediment color has links with arsenic, manganese and iron concentrations in groundwater. Local drillers are aware of this but their capacity can be further enhanced by providing simple tools like the sediment color tool developed from the Matlab study (Fig. 3).

4 CONCLUSIONS

Key findings from two extensive studies should be used to update and harmonize arsenic mitigation policies. Issues that need to be included in any new policy of arsenic mitigation include promotion and regulation of commercial tube-well testing; launching national campaign to update households on risks of arsenic exposure; dissemination of arsenic-safe depth at the village level and capacity building at local levels to encourage drillers and households to target low-arsenic aquifers. Future safe tube-well installations should be based on geochemical as well as hydrogeological suitability and should take into account social factors in order to maximize impact.

REFERENCES

Hossain, M., Bhattacharya, P., Frape, S.K., Jacks, G., Islam, M.M., Rahman, M.M., Hasan, M.A. & Ahmed, K.M. 2014. Sediment color tool for targeting arsenic-safe aquifers for the installation of shallow drinking water tubewells. *Sci. Total Environ.* 493: 615–625.

Pfaff, A.A. Walker, S., Ahmed, K.M. & A. van Geen 2017. Reduction in exposure to arsenic from drinking well-water in Bangladesh limited by insufficient testing and awareness. *J. Wat. San. Hyg. Dev.* 7(2): 331–339.

van Geen, A., Sumon, E.B.A., Pitcher, L., Mey, J.L., Ahsan, H., Graziano, J.H. & Ahmed, K.M. 2014. Comparison of two blanket surveys of arsenic in tubewells conducted 12 years apart in a $25\,km^2$ area of Bangladesh. *Sci. Total Environ.* 488–489: 484–492.

van Geen, A., Ahmed, K.M., Ahmed, E.B., Choudhury, I., Mozumder, M.R., Bostick, B.C. & Mailloux, B.J. 2016. Inequitable allocation of deep community wells for reducing arsenic exposure in Bangladesh. *J. Wat. San. Hyg. Dev.* 6(1): 142–150.

Environmental Arsenic in a Changing World –
Zhu, Guo, Bhattacharya, Ahmad, Bundschuh & Naidu (Eds)
ISBN 978-1-138-48609-6

Mitigation actions performed to the remediation of groundwater contamination by arsenic in drinking water sources in Chihuahua, Mexico

M.A. Olmos-Marquez[1,2], C.G. Sáenz-Uribe[1], J.M. Ochoa-Riveros[4], A. Pinedo-Alvarez[1] & M.T. Alarcón-Herrera[3]
[1] *Universidad Autónoma de Chihuahua, Chihuahua, Chih., Mexico*
[2] *Junta Municipal de Agua y Saneamiento, Chihuahua, Chih., Mexico*
[3] *Centro de Investigación Materiales Avanzados (Sede Chihuahua), Chihuahua, Chih., Mexico*
[4] *Sitio Experimental la Campana, CIRNOC, Instituto Nacional de Investigaciones Forestales, Agrícolas y Pecuarias (INIFAP), Aldama, Chih., Mexico*

ABSTRACT: In this work, current information about the contamination of groundwater by arsenic (As) from geogenic sources in Chihuahua state, Mexico is presented together with a possible emerging mitigation solution. The problem in Chihuahua is of the same order of magnitude as other world regions, such as Argentina. Despite the studies undertaken by numerous local researchers, and the identification of proven treatment methods for the specific water conditions encountered, no technologies have been commercialized to treat reject water from the Reverse Osmosis System (ROS). The use of constructed wetlands like an emerging, low-cost technologies to mitigate the problem of As in reject water from ROS that are installed for rural and urban areas have been evaluated. This technology generally use simple and low-cost equipment that can easily be handled and maintained by the local population and the water obtained can be used to irrigation of some crops, resulting in a benefit to community.

1 INTRODUCTION

Arsenic contaminations of groundwater in several parts of the world are the results of natural and/or anthropogenic sources and have a large impact on human health (Geschwind *et al.*, 1992). Millions of people from different countries rely on groundwater containing As for drinking purposes (Mondal *et al.*, 2013; Olmos–Márquez et al., 2011). A great amount of communities in Chihuahua extract groundwater for domestic water uses, and many communities are entirely dependent on groundwater. Chihuahua groundwater generally contains As in concentrations above the Mexican regulation (25 $\mu g\,L^{-1}$). Therefore, the institution in charge of drinking water supply in Chihuahua, Junta Central de Agua y Saneamiento, (JCAS for its Spanish acronym), has exigent implications for the provision of safe drinking water in Chihuahua. To put the arsenic impact on Chihuahua State into perspective, the total number of municipalities affected by As presence in groundwater represents roughly 75%. Since the beginning of the year 2000 the JCAS, has been working to reduce exposure to As by drinking water and foods consumption, in the communities affected by As occurrence, installing a large number of reverse osmosis system and development innovative projects to treat the reject water.

2 SITUATION OF GROUNDWATER CONTAMINATION BY ARSENIC IN CHIHUAHUA STATE

2.1 *Description of the study area*

Chihuahua state is located in the north of Mexico, its conformed by 67 municipalities, with a surface of 247,460 km^2 and a population of 3,556,574 inhabitants, which is the biggest state of Mexico. A 60% from the water used in the state comes from groundwater sources.

2.2 *Arsenic distribution in groundwater*

50 of the 67 municipalities from Chihuahua State have at least one of its groundwater supply with arsenic concentrations above the Mexican drinking water regulation. 24 municipalities have more than 50% of its sources of drinking water out of normative. The municipalities with the highest arsenic concentration are: Julimes, Camargo, Jimenes, Delicias, Valle de Zaragoza, Meoqui, La Cruz, Saucillo, Rosales and San Francisco de Conchos. The average As concentrations in these areas are between 25 and 910 $\mu g\,L^{-1}$ (Fig. 1).

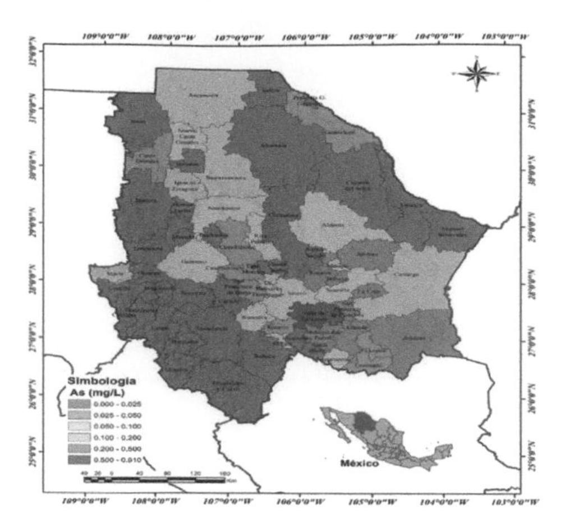

Figure 1. Arsenic groundwater concentrations in Chihuahua state.

3 CURRENT STATUS OF REMEDIATION EFFORTS MAKE BY CHIHHUAHUA STATE GOVERNMENT FOR SOLVE THE PROBLEMS WITH ARSENIC IN DRINKING WATER

3.1 *Mitigation actions to provide arsenic free drinking water*

The government of the State of Chihuahua, Mexico through the JCAS has undertaken several actions to respond to the high levels of As in the water sources. Since year 2000 there have been installed 370 systems of reverse osmosis in 50 of the affected municipalities that have As concentration above the safe standards marked in the Mexican Norm (NOM-127-SSA1-2000). The population benefited for the program is of 720,176 inhabitants scattered in 309 communities. The investments was of 93,612,385 MeX$ (5,200,688 USD). However, there is a large percentage (50%) on the reject stream that is not been disposed accordingly. To reduce this percentage the Universidad Autónoma de Chihuahua in cooperation with the Board Commission for Water are developing a technology that allows to reduce the As in the reject stream that was a volume 1,480,000 L day^{-1}, and thus make that water available for other uses such agricultural irrigation, family and municipal gardens, among others.

3.2 *Constructed wetlands pilot plant*

A pilot-level system was designed and built in Julimes municipality (28°25′25″N; 105°25′32″W) which has the one of the highest As concentrations in the state (488 $\mu g\,L^{-1}$). The system is located inside a greenhouse of 10 m × 5 m × 3 m and consists of two fiberglass constructed wetlands (CWs) with dimensions of 4.5 m × 1.5 m × 0.70 m. The CWs were filled with silty sand up to 0.50 m and planted with *Schoenoplectus americanus* and *Elleocharys macrsotachya*. The average As concentration of the ROS reject water (feed water of CWs) is in the range of 300 $\mu g\,L^{-1}$, the average water flow treated is 1400 L d^{-1} with a residence

Figure 2. Pilot system implemented to remove arsenic from reject water of reverse osmosis system through constructed wetlands in Julimes, Chihuahua Mexico.

time of 2 d. Water treated by the wetland system is collected and pumped to an 800 L water storage tank, for use in the irrigation of a small family orchard (Fig. 2). The system has removal As efficiencies of 85–99% and the operation and maintenance cost is very low, because it only uses a pump of 0.5 hp to pump the water and the rest of the hydraulic process is by gravity, the system should be supervised only once week. It is important to mention that if this water is not treated it implies some risks to health and environment.

4 CONCLUSIONS

The state of Chihuahua has a serious impact by As in drinking water, the state government has taken actions to provide safe water to the inhabitants of the affected communities, installing inverse osmosis plants. However, it is necessary to solve the problem of adequate disposition of the reject water generated by reverse osmosis systems. Constructed wetlands have proven to be a viable and economical option to solve this problem, as well as to provide a reuse capacity of treated water in activities that economically benefit the communities where these systems are installed.

ACKNOWLEDGEMENTS

This work was financially supported by National Council for Science and Technology (CONACYT), within the Program Attention to National Problems.

REFERENCES

Geschwind, S.A., Stolwijk, J.A., Bracken, M., Fitzgerald, E., Stark, A., Olsen, C. & Melius, J. 1992. Risk of congenital malformations associated with proximity to hazardous waste site. *Am. J. Epidemiol.* 135(11), 1197–1207.

Mondal, P., Bhowmick, S., Chatterjee, D., Figoli, A. & Van der Bruggen, B. 2013. Remediation of inorganic arsenic in groundwater for safe water supply: a critical assessment of technological solution. *Chemosphere* 92(2): 157–170.

Olmos-Márquez, M.A., Alarcón-Herrera, M.T. & Martin-Domínguez, I.R. 2011. Performance of *Eleocharys macrostachya* and its importance for arsenic retention in constructed wetlands. *Environ. Sci. Pollut. R.* 19(3): 763–771.

Environmental Arsenic in a Changing World –
Zhu, Guo, Bhattacharya, Ahmad, Bundschuh & Naidu (Eds)
ISBN 978-1-138-48609-6

Microorganism: natural sweepers of arsenic in industrial wastewater

A.B. Kashif Hayat[1,2], C.D. Saiqa Menhas[2], J. Bundschuh[3], P. Zhou[1] & H.J. Chaudhary[2]
[1]*School of Agriculture and Biology, Key Laboratory of Urban Agriculture, Ministry of Agriculture, Bor S. Luh Food Safety Research Center, Shanghai Jiao Tong University, Shanghai, China*
[2]*Department of Plant Sciences, Faculty of Biological Sciences, Quaid-i-Azam University, Islamabad, Pakistan*
[3]*UNESCO Chair on Groundwater Arsenic within the 2030 Agenda for Sustainable Development, University of Southern Queensland, Toowoomba, QLD, Australia*

ABSTRACT: The presence of arsenic (As), a potent carcinogen, in industrial wastewater poses a serious threat to plant, animal, and public health, especially in the transition and developing countries. Recent developments in the field of bioremediation have enhanced our understanding about the microbial biotechnology processes governing As uptake and detoxification on laboratory and pilot scale. In particular, the biofilm and consortia remediation will prove invaluable in the developments of new strategies to mitigate this threat at industrial scale also. The role of microbial remediation of As in industrial wastewater cannot be neglected due to its proven significance as a natural sweeper that can mitigate As in wastewater without production of solid/liquid waste and reduce it to low toxic level.

1 INTRODUCTION

One of the most burning issues confronting the world is the administration and sustainable management of a silent killer and carcinogen, i.e. arsenic (As) in industrial wastewater. In fact, many conventional techniques being applied to treat As in industrial wastewater face limitations in the form of missing technical expertise, low effectiveness, high costs, and production of toxic waste (Bundschuh and Maity, 2015). In this regard, microbial biotechnology emerged as an effective process for As mitigation in industrial wastewater (Hayat *et al.*, 2017).

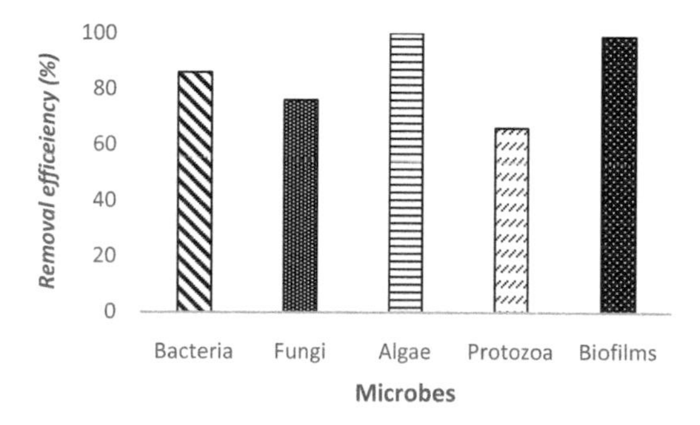

Figure 1. Removal efficiency (%) of arsenic by microbes in industrial wastewater (Hayat *et al.*, 2017).

2 MICROBES: AN INTENGIBLE SWEEPERS OF ARSENIC IN INDUSTRIAL WASTEWATER

Microorganisms may be termed as an intangible and invisible natural sweeper of As in industrial wastewater. A diverse group of microorganisms plays a key role in the fate of As the biochemical cycle in industrial wastewater (see Fig. 1). Microorganisms have developed a lot of mechanisms to decrease the adverse effects of As in industrial wastewater, thus leading to positive effects on environment and human health. The key benefit of these microbes is that it may not produce any waste and cope with the As-rich environment effectively (Hayat *et al.*, 2017). These microbes may be used individually or as a contingent termed as "biofilm or consortia". The application of algae and biofilm biotechnology is turning to be the most effective approach to treat As (98.8 to 100%) in industrial wastewater. Beside all other microbial technologies, biofilm (600 times more resilient than individual microbes) has been successfully used on a pilot scale due to its long-lasting effect (Battaglia-Brunet *et al.*, 2006; Dastidar and Wang, 2009; Hayat *et al.*, 2017).

3 CONCLUDING REMARKS AND FUTURE PROSPECTS

In the last decade, the scientific community focused on the microbial removal of As from industrial wastewater and it was revealed that all the microbes follow a common ideal mechanism of As mitigation by which they can remove As releasing it into the atmosphere without producing solid/liquid waste. Thus, the treated wastewater can be regarded as within the safe limit by the environmentalist and other global scientific community. Moreover, following the recent successful trend of using natural as well as genetically engineered (GE) microbes (without avoiding the GE conflict) for As removal from industrial wastewater, there are also some other options such as nanotechnology and DNA editing technology (CRISPR) and other new molecular mechanism that should be applied and tested to reduce As concentration in industrial wastewater. However, microbial biotechnology has not yet been applied on an industrial scale (and only biofilms on pilot scale). Therefore, more insights into the microbial arsenic mitigation in industrial wastewater are urgently required.

REFERENCES

Battaglia-Brunet, F., Joulian, C., Garrido, F., Dictor, M.-C., Morin, D., Coupland, K., Johnson, D.B., Hallberg, K.B. & Baranger, P. 2006. Oxidation of arsenite by *Thiomonas* strains and characterization of *Thiomonas arsenivorans* sp. nov. *Antonie van Leeuwenhoek*. 89(1): 99–108.

Bundschuh, J. & Maity, J.P. 2015. Geothermal arsenic: occurrence, mobility and environmental implications. *Renew. Sust. Energ. Rev* 42, 1214–1222.

Dastidar, A. & Wang, Y.T. 2009. Arsenite oxidation by batch cultures of *Thiomonas arsenivorans strain* b6. *J. Environ. Eng.* 135(8): 708–715.

Hayat, K., Menhas, S., Bundschuh, J. & Chaudhary, H.J. 2017. Microbial biotechnology as an emerging industrial wastewater treatment process for arsenic mitigation: a critical review. *J. Clean. Prod*, 151: 427–438.

Environmental Arsenic in a Changing World –
Zhu, Guo, Bhattacharya, Ahmad, Bundschuh & Naidu (Eds)
ISBN 978-1-138-48609-6

Smart phone Fe test kit as quick screening tool for identification of high risk areas for arsenic exposure

D. Halem & A. Mink
Watermanagement Department, Faculty of Civil Engineering and Geosciences, Delft University of Technology, Delft, The Netherlands

ABSTRACT: Arsenic contamination of reduced shallow aquifers, as in deltas around the world, including Bangladesh and India has been found to frequently co-occur with dissolved ferrous iron. In this study the deployment of smart phone Fe test kit for identification of arsenic-affected villages was found to be a powerful, quick screening instrument for problem mapping in areas dependent for their water supply on tubewell groundwater from arsenic-contaminated, reduced aquifers. The research consisted of laboratory investigation of smart phone test kit accuracy under different conditions (light, exposure time, smart phone brand) and a proof-of-concept field study by screening 365 tube-wells in a potentially arsenic-affected area in Bhojpur, Bihar (India).

1 INTRODUCTION

Mobile crowd participation is gaining interest in the field of water research as a novel tool for inclusion of end-users, a critical ingredient of sustainable water supply (Prahalad, 2012; Wilkinson *et al.*, 2014). Smart phone water quality test kits (Akvo.org) could provide end-users an active role in water quality monitoring programs and raise awareness amongst rural communities. Smart phone coverage has rapidly increased the past years, particularly amongst the youth (GSMA, 2017, Mink *et al.*, 2017).

Arsenic contamination of reduced shallow aquifers, as found in deltas around the world, including Bangladesh and India has been found to frequently co-occur with dissolved ferrous iron. Given the fact that arsenic analysis in the field is time-consuming and relatively expensive, it was the aim of this study to investigate the use of inexpensive Fe smartphone test kits as a quick screening tool for identification of high risk areas for arsenic occurrence. The research consisted of laboratory investigation of smart phone test kit accuracy under different conditions (light, exposure time, smart phone brand) and a proof-of-concept field study by screening 365 tubewells in a potentially arsenic-affected area in Bhojpur, Bihar (India).

2 METHODS

2.1 *Laboratory tests: smart phone test accuracy*

The Akvo Caddisfly and Akvo Flow app were both installed on different smart phones (including Samsung Galaxy J7) and used for the reading of HACH Fe test strip discoloration with a reference card (Fig. 1). $Fe(II)SO_4$ solutions were prepared, ranging from

Figure 1. Smart phone and reference card (Akvo.org).

0.1–4 mg L^{-1}. Pictures were taken directly upon finishing the test in natural sunlight and indoors; additionally, repetitive pictures were taken after 5 and 10 min exposure time.

2.2 *Proof-of-concept field study*

In Bhojpur, Bihar, 365 shallow tubewells were sampled for immediate analyses with smartphone Fe test kit. Subsequently 15 mL sample vials were transported to the laboratory for ICP-MS analyses (As and Fe) after 0.45 μm filtration and acidification with nitric acid.

3 RESULTS AND DISCUSSION

3.1 *Smart phone test kit accuracy*

Figure 2 depicts the Fe concentrations detected by the smart phone test kit in direct sunlight against the

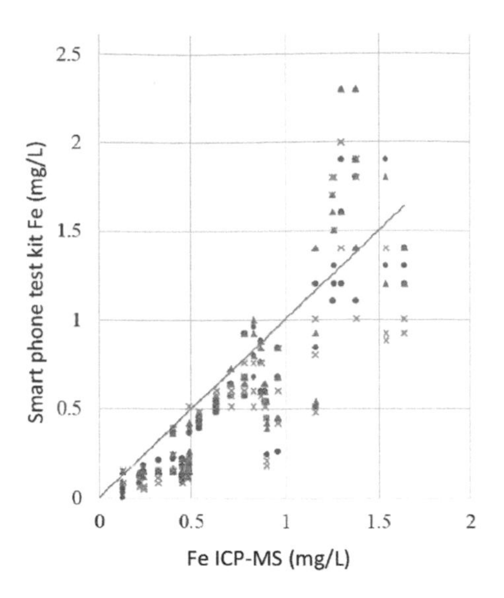

Figure 2. Smart phone test kit accuracy in direct sunlight.

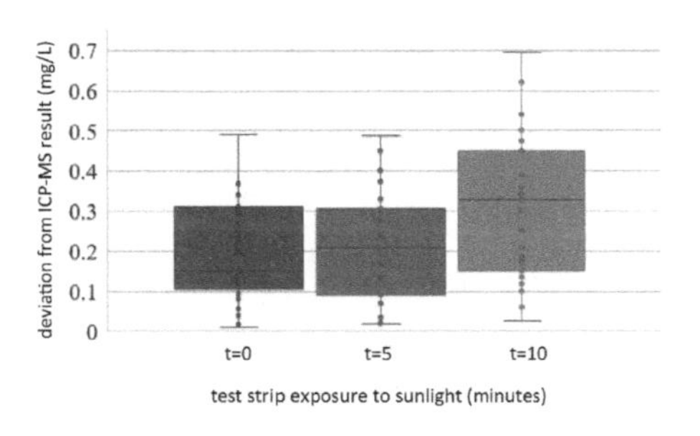

Figure 3. Effect of test strip sunlight exposure on documented Fe result.

values measured with ICP-MS. The test kit systematically underestimated the actual Fe concentrations, with increasing inaccuracy at higher concentrations ($>1\ \mathrm{mg\,L^{-1}}$). The results were consistent, independent of mobile phone brand and lighting (results not shown). Only when test strips were not directly tested, the color of the strip faded with a consequently lower reading by the smart phone after 10 min (Fig. 3).

3.2 *Fe-As co-occurrence in Bhojpur*

The ICP-MS results for Fe and As from 365 tubewell samples from Bhojpur, showed that all tubewells with As concentrations $>50\ \mu\mathrm{g\,L^{-1}}$ contained Fe concentrations $>1\ \mathrm{mg\,L^{-1}}$ (results not shown). The test kit Fe results showed some outliers, but altogether the results are alike – except for the test kit detection limit beyond $5\ \mathrm{mg\,L^{-1}}$. Based on this As-Fe relationship, it may be concluded that >50 As $\mu\mathrm{g\,L^{-1}}$ tubewells can be identified with the smart phone Fe test kit. Identification of $<10\ \mu\mathrm{g\,L^{-1}}$ tubewells (WHO guideline) was not achieved for all wells, as 21 tubewells had Fe concentrations $<1\ \mathrm{mg\,L^{-1}}$, but As concentrations between 10 and $50\ \mu\mathrm{g\,L^{-1}}$. It should be noted, though, that the smart phone test kit is only considered a powerful tool when "crowd" data are collected – providing a high density of data points and with that a higher accuracy towards detection of high-risk areas.

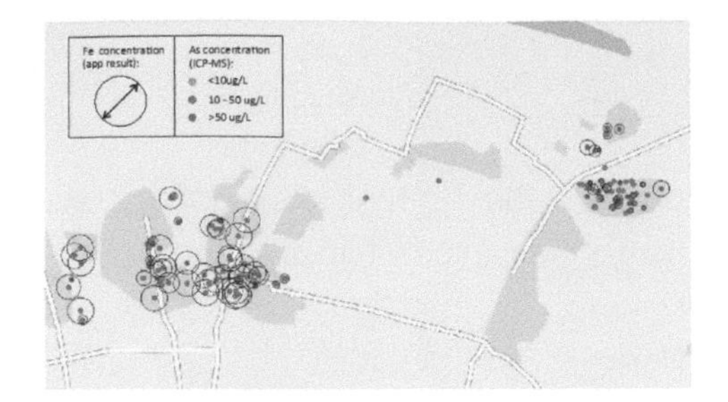

Figure 4. As (ICP-MS) and Fe (smart phone test kit) results for two neighboring villages in Bihar, India.

3.3 *Smart phone results as screening tool*

Figure 4 depicts the GPS-stamped data collected with the smart phone Fe test kit (black circles) together with the ICP-MS results for As (colored dots). The smart phone Fe test kit results clearly show the elevated Fe concentrations in the western village, compared to the eastern village. This finding is in agreement with the ICP-MS measurements for As, showing that that the smart phone Fe test kit functions as a quick screening tool for high risk As villages in Bhojpur. This finding is supported by measurements in four surrounding villages; and will be further investigated in arsenic-affected areas with similar geochemical settings.

4 CONCLUSIONS

In this study the deployment of smart phone Fe test kit for identification of arsenic-affected villages was found to be a powerful, quick screening instrument for problem mapping in areas dependent for their water supply on tubewell groundwater from arsenic-contaminated, reduced aquifers.

ACKNOWLEDGEMENTS

This research was funded through the "Urbanising Deltas of the World" programme of NWO WOTRO – Science for Global Development.

REFERENCES

Akvo Foundation, www.akvo.org.

Gsma 2017 Economic Impact: Bangladesh Mobile Industry. Groupe Speciale Mobile Association, 7.

Mink, A., Hoque, B., & van Halem, D. 2017. Mobile Crowd Participation, *IWA Water and Development Congress, Argentina.*

Prahalad, C. 2012 Bottom of the pyramid as a source of breakthrough innovations. *J. Product. Innov. Man.* 29: 6–12.

Wilkinson, C.R., & De Angeli, A., 2014. Applying user centred and participatory design approaches to commercial product development. *Design Studies* 35: 614–631.

Environmental Arsenic in a Changing World –
Zhu, Guo, Bhattacharya, Ahmad, Bundschuh & Naidu (Eds)
ISBN 978-1-138-48609-6

Thiourea supplementation reduces arsenic accumulation in two selected rice (*Oryza sativa* L.) cultivars in a field study in Bengal Delta Basin, India

M.K. Upadhyay[1], S. Srivastava[1], A. Majumdar[2], A. Barla[2] & S. Bose[2]
[1]*Institute of Environment and Sustainable Development, Banaras Hindu University, Varanasi, U.P., India*
[2]*Earth and Environmental Science Research Laboratory, Department of Earth Sciences, Indian Institute of Science Education and Research Kolkata (IISER-K), Mohanpur, West Bengal, India*

ABSTRACT: Arsenic (As) accumulation in rice is a severe problem in Southeast Asia including West Bengal, India. Thiourea (TU) is a redox active thiol (-SH) based molecule, which has been found to regulate redox equilibrium and As level in lab study. In this work, we, for the first time, evaluated the effect of TU supplementation on growth and yield of rice plants, biochemical changes and As concentrations in rice grains. Both rice cultivars (Satabdi-*IET4786* and Gosai) were grown in three fields (Control, Low As and High As) in Nadia, West Bengal. The result indicated positive impact of TU supplementation on rice growth and yield and most importantly, showed reduction in As concentrations in rice grains.

1 INTRODUCTION

Arsenic (As) is well known class 1 carcinogen classified by IARC. Presently, 150 million people around the world are affected by As hazard. The condition is severe in Bengal delta basin region. There are many districts in West Bengal, where groundwater As concentration is $>50\,\mu g\,L^{-1}$ including Nadia. This groundwater is used for irrigation in rice fields throughout the concerned area. Therefore, rice accumulates As in significant concentrations due to its favorable flooded cultivation, suitable transporters (channel proteins) and reducing environment. Being a staple crop worldwide, rice is the major source of As exposure to humans. Sulphur deficiency, redox imbalance, induced generation of reactive oxygen species (ROS) are key symptoms of As toxicity. Sulfur is important for plants as it affects As uptake, translocation and accumulation in rice plants (Srivastava *et al.*, 2007). TU is sulfur containing redox regulator and ROS scavenger, which has already applied in variety of stresses in plants (Kelner *et al.*, 1990). We applied TU and assessed its impacts on As accumulation in rice grains, yield, growth and antioxidant defense system.

2 METHODS/EXPERIMENTAL

2.1 *Site description*

The study was conducted in Sarapur, Dewali Gram panchayat 23°01′07.8″N88°39′43.1″E-23° 01′14.3″N 88°38′24.7″E, Chakdaha block in Nadia district of West Bengal, India. This area is known for prolonged use of As contaminated groundwater for irrigation purposes. Here agriculture is the primary profession of the native people and so, this region may act as representative site for agriculture based As study.

2.2 *Growth of plants, collection and processing the samples*

We choose two rice genotypes (Satabdi-IET4786 and Gosai), which are the local and predominant variety being used in the study area. The cultivation of rice was performed in Monsoon (Aman, July–October 2016) in three selected fields viz- Low As ($38.74 \pm 0.13\,mg\,kg^{-1}$), High As ($84.77 \pm 0.51\,mg\,kg^{-1}$) and Control ($22.15 \pm 0.2\,mg\,kg^{-1}$) with/without thiourea (TU) application ($500\,\mu g\,L^{-1}$ for seed soaking and foliar application at the flowering stage). The area of per selected field was 400 square feet. The plant samples were taken to lab preserved in ice filled box. Plants were analyzed for pigment and antioxidant enzyme analyses after crushing in liquid N_2. The remaining plants were washed thrice, air dried after evaluating length, tiller no. and other growth parameter. Further, they were separated into different parts i.e. root, straw, husk and grains etc. and after digestion it has been used for ICP-MS analysis. The crushed sample of shoot, root and grain were also used for synchrotron based XRF analysis after making its pellet. Apart from this, leave's xylem-phloem internal integrity was also assessed under field emission scanning electron microscope (FE-SEM).

3 RESULTS AND DISCUSSIONS

3.1 *Arsenic (As) concentration in rice grains*

In Aman (monsoon), 2016, the concentration of As (all in $mg\,kg^{-1}$) in rice grains in Gosai, Satabdi, Gosai

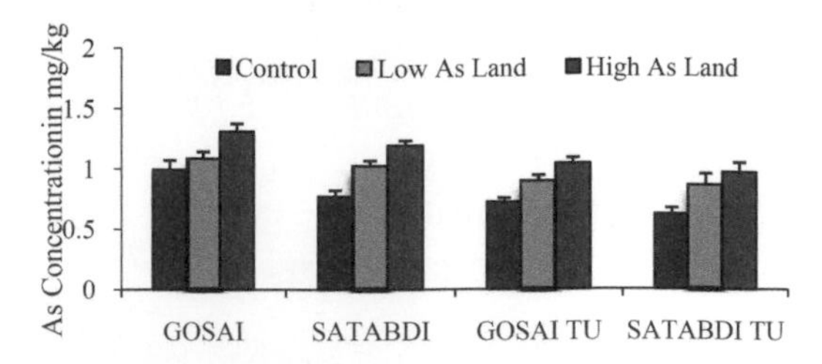

Figure 1. Arsenic (As) concentration in mg kg^{-1} in two cultivars of rice with and without thiourea (TU). All values are means of triplicates $\pm$SD.

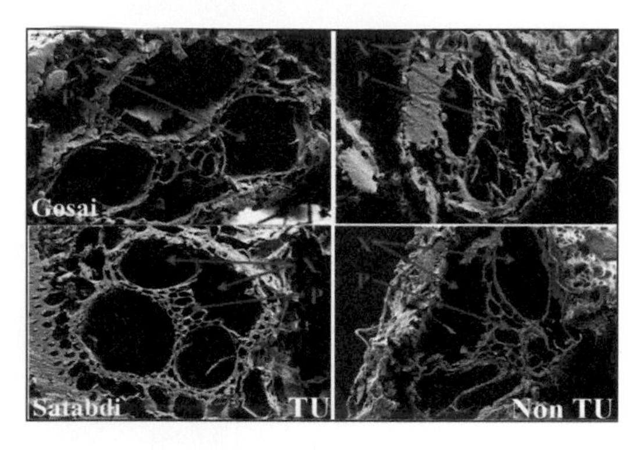

Figure 2. FE-SEM images of differentiated xylem (X) and phloem (P) integrity in TU amended and non-TU rice plant leaf of both cultivars.

(TU) and Satabdi (TU) were found 1.086, 1.030, and 0.900, 0.866, respectively in low As, 1.31, 1.196, 1.043 and 0.97, respectively in high As and 0.994, 0.77, 0.723, 0.62, respectively in control (Fig. 1). The accumulation of As was thus significantly reduced in grains under TU supplementation.

3.2 *Growth and yield parameters*

The growth and yield were also improved in two cultivars under TU supplementation as compared to without TU plants. The biomass, tiller no., leaf area were increased in rice plants. The yield of rice grains was also increased in TU ameliorated cultivars in all the fields (Table 1).

Table 1. This table shows the yield of rice in kg.

	Satabdi (IET-4786)		Gosai	
Setting	TU	NON TU	TU	NON TU
Control field	14.4	12.9	14.0	13.8
Low As field	16.1	15.4	15.3	13.5
High As field	12.9	12.6	12.1	11.4

3.3 *Alteration in integrity of xylem/phloem*

Internal organs (xylem phloem, bundle sheath) OF rice leaves integrity were distorted due to As toxicity. TU actually improved the vascular organ's integrity (Fig. 2). So, this could be a possible reason behind better biomass growth and yield in rice.

4 CONCLUSIONS

This study indicates that TU can reduce the As load in rice grain, which is a significant result. Further, TU supplementation also improved growth of plants along with yield. Hence, TU supplementation through seed soaking and spraying at the time of flowering can be a feasible strategy to mitigate As toxicity through reduce its translocation to grains in contaminated areas and grow rice with low arsenic in rice grains.

ACKNOWLEDGEMENTS

The authors are thankful to IESD-BHU, Varanasi, India for the lab facility. We are thankful to core instrumentation facilities of IISER-Kolkata, Mohanpur. MKU is thankful to UGC-BHU for Ph. D fellowship.

REFERENCES

Kelner, M.J., Bagnell, R. & Welch, K.J. 1980. Thioureas react with superoxide radicals to yield a sulfhydryl compound. Explanation for protective effect against paraquat. *J. Biological Chemistry* 265(3): 1306–1311.

Srivastava, S. & D'souza. S.F. 2009. Increasing sulfur supply enhances tolerance to arsenic and its accumulation in *Hydrilla verticillata* (L.f.) Royle. *Environmental Science & Technology* 43(16): 6308–6313.

Environmental Arsenic in a Changing World –
Zhu, Guo, Bhattacharya, Ahmad, Bundschuh & Naidu (Eds)
ISBN 978-1-138-48609-6

Effects of boron nutrition on arsenic uptake and efflux by rice seedlings

R.L. Zheng[1] & G.X. Sun[2]
[1]*Beijing Research & Development Center for Grasses and Environment, Beijing Academy of Agriculture and Forestry Sciences, Beijing, P.R. China*
[2]*State Key Laboratory of Urban and Regional Ecology, Research Center for Eco-Environment Sciences, Chinese Academy of Sciences, Beijing, P.R. China*

ABSTRACT: The impacts of boron (B) through root application and foliar spray on arsenic (As) uptake and efflux by/in rice seedlings (*Oryza sativa* L.) were investigated in three hydroponic experiments. Addition of B in culture medium didn't alter concentrations of arsenite [As(III)], arsenate [As(V)] and total As in rice seedlings under either As(III) or As(V) exposure. Foliar B supply decreased root As concentrations by 20.9% under As(V) treatment and by 12.6% under As(III) treatment, respectively, yet didn't significantly decrease shoot As concentrations. Concentrations of As(V) were positively related to B concentration in rice root under As(V) treatment following foliar B supply ($P < 0.05$). Rice seedlings extruded 105.2% more As after As(III)-pretreatment than after As(V)-pretreatment. Foliar B supply increased the amount of As excreted by As(III)-pretreated rice root by 14.0%–16.9% ($P > 0.05$) and made no effect on As efflux for As(V)-pretreatment seedlings. These results indicate that root application of B at four times the concentration of As can hardly decrease As accumulation by rice, whereas foliar B supply is conducive to a decline in As acquisition by rice root. It is likely that B channel is at least not the main pathway for As(III) entering into rice root, and As(V) distribution mechanism in rice plant may share with B.

1 INTRODUCTION

Boron (B) mainly occurs in soils or organisms as $B(OH)_3$ which is an uncharged molecule structurally similar to $As(OH)_3$. Many studies have shown that NIP channel proteins such as OsNIP3;1, AtNIP5;1, AtNIP6;1 which facilitate As uptake or transport are also $B(OH)_3$ transporters essential for B uptake by rice or thale cress roots (Hanaoka *et al.*, 2014; Miwa and Fujiwara, 2010). PIP channel proteins (OsPIP2;4 and OsPIP2;7) from rice mediated not only B but also As transport (Kumar *et al.*, 2014). Furthermore, the expression of these aquaporins can be regulated by B level in the environment (Kumar *et al.*, 2014). The tissue specificity of OsNIP3;2 expression is similar to that of B transporter OsNIP3;1 (Chen *et al.*, 2017). It remains unknown whether As uptake and efflux are influenced by B addition.

2 METHODS/EXPERIMENTAL

2.1 *Effect of culture solution B-level on As uptake*

After seedlings of rice were pre-cultured in a full strength nutrient solution for one week, 6 treatments including three B levels (0, 10 and 40 $\mu mol\,L^{-1}$) with 10 $\mu mol\,L^{-1}$ As(III) or As(V) in solution were imposed. Plants were harvested after As exposure for 2 d. Shoot and root were quickly separated, washed with deionized water, blotted dry and ground in a mortar and pestle with liquid N for further measurements after the fresh weight was recorded.

2.2 *Effect of foliar B supply on As uptake*

Rice seedlings subjected to preculture in a nutrient solution without B for one week were transferred to nutrient solutions containing 10 $\mu mol\,L^{-1}$ As(III) or As(V) whilst no B addition. Boron was applied through foliage including foliar sprays of deionized water (control), and 9, 36 $mmol\,L^{-1}$ H_3BO_3 solution. Foliar sprays (3 mL per pot) were applied once a day. Plants were harvested as described above after As exposure and foliar B supply for one week. Specially, shoot was rinsed with deionized water containing Triton X-100 three times first to remove B adhering on the surface of leaves.

2.3 *Effect of foliar B supply on As uptake*

Rice seedlings were pre-cultured in a nutrient solution without B for one week, and 10 $\mu mol\,L^{-1}$ As(III) or As(V) was added into culture solution afterwards. After As exposure for one week, plant roots were rinsed in an ice-cold desorption solution, and then soaked in the same ice-cold desorption solution for 10 min to remove apoplastic As. After roots were washed with deionized water, plants were transferred to 50 mL lightproof centrifuge tubes containing 40 mL nutrient solution without As and B, and foliar B supply

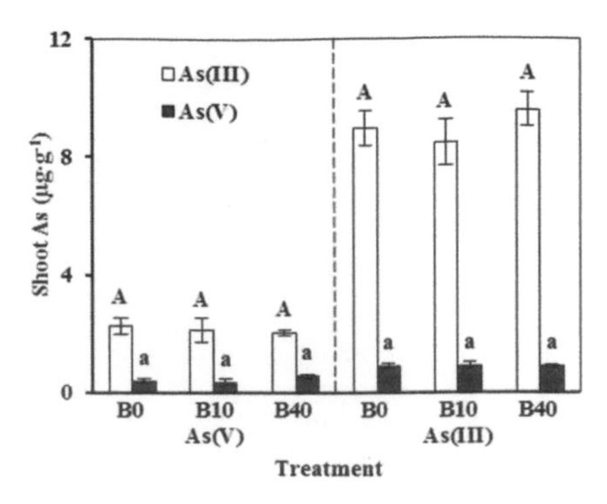

Figure 1. Effects of B level of culture solution on As(III) and As(V) concentrations in rice shoot.

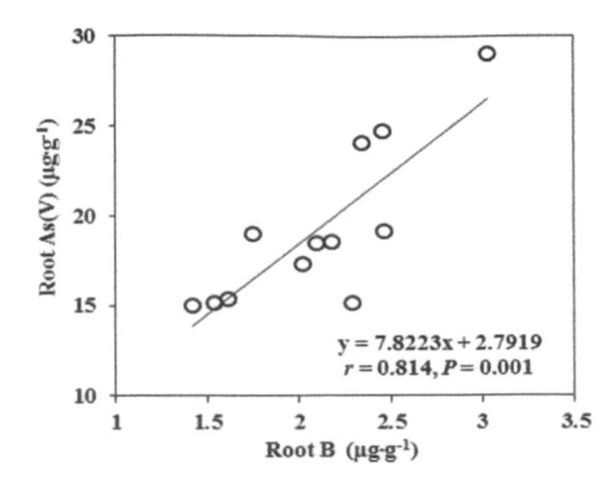

Figure 2. Relationship between root B concentration vs root As(V) concentration of rice seedlings under foliar sprays of different B levels.

Table 1. Total As and B concentrations in shoot and root of rice seedlings grown in culture solutions with As (III) or As(V) under foliar sprays of different B levels.

As species	Foliar B supply	Shoot As ($\mu g\, g^{-1}$)	Root As ($\mu g\, g^{-1}$)	Shoot B ($\mu g\, g^{-1}$)	Root B ($\mu g\, g^{-1}$)
	B0	7.4	27.7	4.2	2.5
As(V)	B9	7.0	23.1	21.7	2.1
	B36	6.5	21.9	70.1	1.7
	B0	19.2	57.2	5.8	1.1
As(III)	B9	19.6	60.0	24.9	1.2
	B36	18.2	50.0	70.4	1.1
Significance of					
As species		***	***	ns	***
Foliar B supply		ns	*	***	*
As × B		ns	ns	ns	*

commenced as described above once plants were transferred. One week later, plants were harvested.

3 RESULTS AND DISCUSSION

3.1 *Effect of B level in culture solution on As uptake*

With increasing B concentration in culture solution, up to four times that of As, concentrations of both As(III) and As(V) in shoots and roots of rice under either As(III) or As(V) exposure were not changed statistically (Fig. 1). It may be attributed to the highly strict control of the permeability to $B(OH)_3$ by ar/R selectivity filter in comparison to As(III).

3.2 *Effects of foliar B supply on As uptake*

Root As concentrations declined significantly by 20.9% under As(V) treatment and by 12.6% under As(III) treatment, respectively ($P < 0.05$), with increasing foliar B level supply (Table 1). However, foliar B supply didn't decrease shoot As concentrations ($P < 0.05$) under either As(V) or As(III) exposure. Root B concentrations decreased by up to 47.1% ($P < 0.05$) with increasing foliar B levels in As(V) treatment but not in As(III) treatment. Root As(V) concentration decreased by up to 29.8% following foliar B supply in As(V) treatment ($P < 0.05$). Further analysis showed significant positive relationships between root B concentration versus root As(V) concentration (Fig. 2) under As(V) exposure ($P < 0.05$).

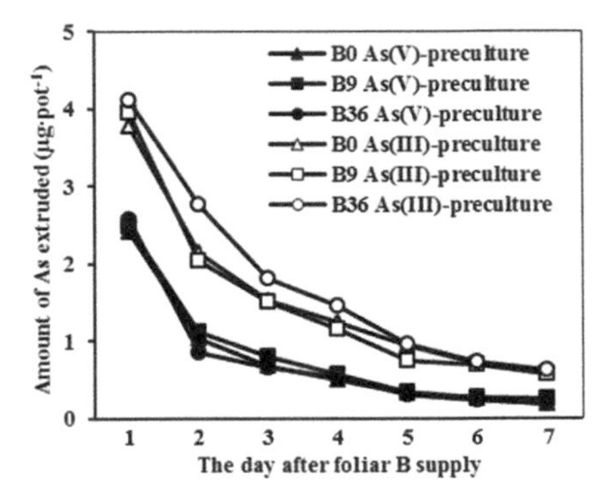

Figure 3. Everyday amount of As excreted into culture solutions by rice seedlings under foliar sprays of B.

3.3 *Effects of foliar B supply on As efflux*

Rice seedlings extruded 105.2% more As after As(III)-pretreatment than after As(V)-pretreatment. Foliar B supply increased the amount of As excreted by As(III)-pretreated rice root by 14.0%–16.9% ($P > 0.05$) and made no effect on As efflux for As(V)-pretreatment seedlings (Fig. 3).

4 CONCLUSIONS

Boron level in culture solution made no effect on the uptake of As(III) and As(V) by rice roots. Foliar B supply decreased concentrations of total As and As(V) in root significantly but didn't markedly decrease shoot As concentrations. There was a significant relationship between B and As(V) concentrations in rice root under As(V) exposure following foliar B supply. Foliar B supply didn't increase As efflux under As(III)- or As(V)-preculture treatment. It was likely that B and As(V) share identical regulatory mechanism inside rice plant, and B channels contributed little to As entering into rice roots.

REFERENCES

Chen, Y., Sun, S.K., Tang, Z., Liu, G.D., Moore, K.L., Maathuis, F.J.M., Miller, A.J., McGrath, S.P. & Zhao, F.J. 2017. The Nodulin 26-like intrinsic membrane protein OsNIP3;2 is involved in arsenite uptake by lateral roots in rice. *J. Exp. Bot.* 68(11): 3007–3016.

Hanaoka, H., Uraguchi, S., Takano, J. Tanaka, M. & Fujiwara T. 2014. OsNIP3;1, a rice boric acid channel, regulates boron distribution and is essential for growth under boron-deficient conditions. *Plant J.* 78(5): 890–902.

Kumar, K., Mosa, K.A., Chhikara, S., Musante, C., White, J.C. & Dhankher, O.P. 2014. Two rice plasma membrane intrinsic proteins, OsPIP2;4 and OsPIP2;7, are involved in transport and providing tolerance to boron toxicity. *Planta* 239(1): 187–198.

Miwa, K. & Fujiwara, T. 2010. Boron transport in plants: coordinated regulation of transporters. *Ann. Bot.* 105(7): 1103–1108.

Environmental Arsenic in a Changing World –
Zhu, Guo, Bhattacharya, Ahmad, Bundschuh & Naidu (Eds)
ISBN 978-1-138-48609-6

Sustainable management of groundwater resources in China: the impact of anthropogenic and natural occurring arsenic pollution

X.Y. Jia & D.Y. Hou
School of Environment, Tsinghua University, Beijing, P.R. China

ABSTRACT: As the official data published by Ministry of Health PRC in 2012 indicated, 1.85 million people were still exposed to drinking water with arsenic level above 0.05 mg L^{-1}. Xinjiang, Inner Mongolia, Shanxi, Shandong, Jilin Province are identified as "hotpots", which are seriously affected by arsenic-contaminated groundwater. In the present study, the authors intend to explore the relationship between arsenic contamination and sustainable management of groundwater resources. The temporal trend and spatial pattern of groundwater resources will be examined, and the impact of anthropogenic arsenic pollution and elevated natural background will be studied. A framework for evaluating the sustainability of groundwater resource management systems will be proposed and used to assess the impact of arsenic pollution. The results of the study may be used to provide a basis for government to conduct effective water quality management in the future.

1 INTRODUCTION

Chronic exposure to arsenic through drinking water will cause various diseases, such as cancers. However, due to lack of water, numerous places in China have been using arsenic-contaminated groundwater for drinking since 1960s. Arsenic can enter groundwater mainly from two ways, natural processes and human activities (He and Laurent Charlet, 2013). Geogenic factors, such as the processes including reductive dissolution and mineral desorption (Smedley and Kinniburgh, 2002; Welch *et al.*, 1988), are responsible for high concentration of arsenic in groundwater. Additionally, aquifers with alkaline pH can also cause elevated concentration of arsenic. As for anthropogenic sources, mining and smelting activities, pesticides use and chemical industries are known as important threats for freshwater aquifers (Breit and Guo, 2012).

2 METHODS

Recently, more and more researches have concentrated on arsenic contamination in groundwater, and there is a great deal of meaningful results. Unfortunately, these excellent results have dispersed in papers, which will impede us to get instructions from them efficiently when we want to sustainably manage groundwater with arsenic contamination. So, in this study, the results of these significant researches will be integrated. And, we will collect data of groundwater published by the government to learn the situation of groundwater. All the data we get will be incorporated in GIS as spatial patterns. The policies about groundwater with arsenic contamination also will be gathered to help us learn the relationship between arsenic contamination and sustainable management of groundwater.

3 RESULTS

The temporal trend and spatial pattern of groundwater resources have been examined by data published by the government. The impact of human activities has significantly affected arsenic concentration in groundwater. Also, the natural background of arsenic has been elevated in this study. Finally, the framework for the groundwater management sustainability evaluating and arsenic impact assessing has been proposed.

4 CONCLUSIONS

The study attempts to explore the relationship between arsenic contamination and sustainable management of groundwater resources. As the results show, arsenic contamination has become a major threaten for groundwater in China. In the future, to cope with the problem, the government needs to work out plans or establish policies to manage groundwater sustainably.

ACKNOWLEDGEMENTS

The authors would like to acknowledge the institutions and individuals who have contributed to this study.

REFERENCES

Breit, G. & Guo, H.M. 2012. Geochemistry of arsenic during low-temperature water-rock interaction – preface. *Appl. Geochem.* 27(11): 2157–2159.

He, J. & Charlet, L. 2013. A review of arsenic presence in China drinking water. *J. Hydrol.* 492: 79–88.

Ministry of Health PRC, National Development, Reform Commission, Ministry of Finance PRC, 2012. "12th five-year plan" on Control of National Endemic Diseases. Ministry of Health PRC.

Welch, A.H., Lico, M.S. & Hughes, J.L. 1988. Arsenic in ground water of the western United States. *Ground Water*. 26(3): 333–347.

Smedley, P.L. & Kinniburgh, D.G. 2002. A review of the source, behaviour and distribution of arsenic in natural waters. *Appl. Geochem* 17(5): 517–556.

Environmental Arsenic in a Changing World –
Zhu, Guo, Bhattacharya, Ahmad, Bundschuh & Naidu (Eds)
ISBN 978-1-138-48609-6

Simultaneous removal of arsenic and fluoride from water using iron and steel slags

J. García-Chirino[1], B.M. Mercado-Borrayo[1], R. Schouwenaars[2],
J.L. González-Chávez[3] & R.M. Ramírez-Zamora[1]
[1]*Instituto de Ingeniería, Universidad Nacional Autónoma de México (UNAM), Mexico City, Mexico*
[2]*Facultad de Ingeniería, Universidad Nacional Autónoma de México (UNAM), Mexico City, Mexico*
[3]*Facultad de Química, Universidad Nacional Autónoma de México (UNAM), Mexico City, Mexico*

ABSTRACT: Arsenic and fluoride are recognized as the most toxic inorganic contaminants in water resources that require the implementation of cost-effective treatment processes and materials for their removal. In this context, slags have been evaluated to remove arsenic individually, but to the best of our knowledge the simultaneous removal of both pollutants has not been performed. The aim of this work is to evaluate the performance of four types of iron and steel slags for the simultaneous removal of arsenic and fluoride present at high concentrations in water. Batch adsorption tests were carried out at room constant temperature and natural pH to determine the capacity of these materials to remove these pollutants. Two treatment stages with steel slag produced an effluent with residual concentrations of As and F lower than the limits established for drinking water.

1 INTRODUCTION

Chronic natural and anthropogenic contamination of water resources by arsenic (As) and fluoride (F) is frequently observed around the world, which has severely affected millions of people (Ali, 2014; Jadhav, 2015). According to the World Health Organization (WHO), the maximum limit of As in drinking water is 0.010 mg L^{-1} and 1.5 mg L^{-1} for F. Various technologies have been evaluated for the simultaneous removal of As and F to reach these limits. Adsorption is regarded as one of the most promising techniques, especially if non-conventional economical materials can be applied, such as slags, with a great potential to remove simultaneously both pollutants at these levels. Slag is an industrial by-product issued from the pyro metallurgical processing of various ores. Mexico is the 2nd largest steel producer in Latin America and number 13 worldwide, and in 2016 generated 5.6 million tons of slag. Slags have been evaluated to remove several inorganic pollutants from water, but the simultaneous removal of As and F at high concentrations has not been investigated.

2 METHODS/EXPERIMENTAL

2.1 *Conditioning and characterization of slags*

The slags used in the experiments were generated by different processes, as seen in Table 1. These by-products were pretreated by grinding followed by sieving using meshes 100 and 200. The quantification of the main elements of all slag samples was carried out using X-ray fluorescence, the mineral phases were determined by X-ray diffraction analysis. The slags were washed several times with deionized water and dried in an oven at 100°C for 5 h to remove moisture before using. The lixiviated contents of Ca, Fe and Mg were measured using standardized methods.

Table 1. Generation processes of the evaluated slags.

	Type	Process
Slag 1	Steel scrap	Electric arc furnace
Slag 2	Iron	Electric arc furnace
Slag 3	Iron	Basic oxygen furnace
Slag 4	Iron	Blast furnace

2.2 *Removal experiments of As and F*

Batch experiments were performed adding separately 8 g L^{-1} of the four slags to an aqueous solution prepared using a mixture of sodium arsenate ($Na_2HAsO_4 \cdot 7H_2O$) and sodium fluoride (NaF) at concentrations of 1.4 mg L^{-1} of As and 2.6 mg L^{-1} of F. The concentrations of arsenic and fluoride were established according to one of the highest values detected in Mexican groundwater. The slag dose was an average of the best values reported in previous studies (Mercado, 2013) carried out for the removal of arsenic. After the addition of the slags, the suspensions were placed in an oscillating shaker (Barnstead/Lab-lion) and put under stirring at 190 rpm at constant room temperature (23°C) and initial pH of 10 for different periods of time (15, 30, 60, 120 and 180 min). Arsenic concentrations in water were determined using the molybdenum blue colorimetric method (Lenoble, 2013), and fluoride concentration was determined with the SPANDS reagent (Standard Methods Committee, 2010).

Table 2. Chemical composition of slag samples.

%	Fe_2O_3	CaO	MgO	Al_2O_3	SiO_2	MnO	TiO_2
S1	45.9	26.4	2.8	5.6	7.0	5.4	0.8
S2	36.5	36.8	4.4	5.7	12.6	0.7	0.7
S3	23.7	43.3	11.2	3.7	11.9	2.9	0.6
S4	0.4	56.4	7.0	6.4	25.5	0.2	0.6

Table 3. Removal efficiencies of As and F using four steel and iron slag samples (%).

	Slag 1		*Slag 2*		*Slag 3*		*Slag 4*	
$t_{(\text{min})}$	As	F	As	F	As	F	As	F
15	87.8	23.0	23.4	28.0	6.0	36.4	14.7	36.6
30	87.5	23.0	12.0	23.2	17.0	28.8	28.2	23.0
60	86.7	23.0	2.6	23.3	34.1	30.6	31.8	23.0
120	85.3	38.0	33.4	25.9	50.9	39.6	23.5	23.0
180	64.4	50.7	28.7	32.3	35.2	27.8	32.4	23.0

3 RESULTS AND DISCUSSION

3.1 *Characterization of slag samples*

Table 2 shows the chemical composition of the steel and iron slag samples. Iron, calcium and magnesium were identified as the main constituents. The X-ray diffraction analysis were focused on the identification of the crystalline phases of these elements in the slag samples. Magnetite Fe_3O_4 (52%) and fayalite $FeSiO_4$ (48%) were identified for slag 1; magnetite (19%) and ilvaite $CaFe_3^2(SiO_4)_2OH$ (81%) for slag 2; magnetite (74%), brucite $Mg(OH)_2$ (9%) and calcite $CaCO_3$ (17%) for slag 3; slag 4 mainly contained amorphous material.

Some hydroxides and oxides of iron, aluminum, calcium and magnesium are efficient to remove arsenic and fluoride (Jadhav, 2015), thus considering the content and type of the minerals identified in most of slag samples, these materials may have a great potential for the efficient simultaneous removal of both pollutants present in high concentrations in water.

3.2 *Simultaneous removal of arsenic and fluoride*

The simultaneous removal efficiencies of As and F for every slag are shown in Table 3.

These results show that slag 1 achieved the best removals, at least for arsenic at the end of 180 min. Based on this, this slag was selected to perform kinetics experiments at longer contact times (360 min) to improve efficiency and also to evaluate the influence of a washing on the removal efficiency of this material.

3.3 *Removal tests of As and F using the most efficient slag*

Figure 1 presents the removal efficiencies of As and F using slag 1, washed and unwashed, with contact times ranging from 15 to 360 min. On one hand, fluoride showed a slower removal kinetics with respect As, even

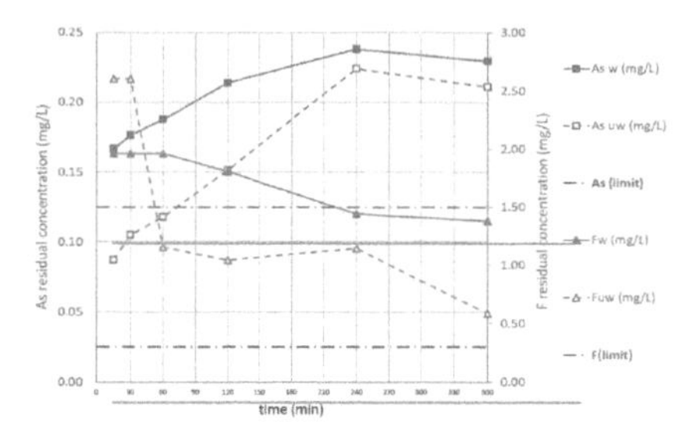

Figure 1. Residual contents of As and F using Slag 1 ($8\,g\,L^{-1}$).

though, the residual concentration met the WHO and Mexican limits ($1.5\,mg\,F\,L^{-1}$) after 360 min. On the other hand, in the firsts 15 min the removal of As was greater than the value of F, the residual concentration of arsenic did not accomplish the WHO guidelines or the Mexican drinking water standards ($0.025\,mg\,L^{-1}$). An additional treatment stage with fresh slag reduced the As concentration in the final effluent to meet this level. The unwashed slag showed a better performance than the washed one. This is probable due to the lixiviated high quantities of calcium ($120\,mg\,L^{-1}$) after washing.

4 CONCLUSIONS

This work showed that slag with high contents of iron, calcium and magnesium oxides and hydroxides was efficient for the simultaneous efficient removal of As and F present at high concentrations. Thus, the final produced effluent met the WHO guidelines and the Mexican drinking water standards for F and As, as far as a second treatment stage with fresh slag is added. Calcium was probably the main responsible of the removal of both pollutants due to the high lixiviation levels from steel slag.

ACKNOWLEDGEMENTS

The project was supported by DGAPA grant IV100616. J. García-Chirino gratefully acknowledges National Council of Science and Technology of Mexico (CONACYT) for the Master grant.

REFERENCES

Ali, I. 2014. Removal of arsenic species from water by batch and column operations on bagasse fly ash. *Environ. Sci. Pollut. Res.* 21(5):3218–3229.

Jadhav, S.V., Bringas, E., Yadav, G., Rathod, V., Ortiz, I. & Marathe, K. 2015. Arsenic and fluorine contaminated groundwaters: a review of current technologies for contaminants removal. *J. Environ. Manage.* 162: 306–325.

Lenoble, V., Deluchat, V., Serpaud, B. & Bollinger, J.C. 2003. Arsenite oxidation and arsenate determination by the molybdene blue method. *Talanta* 61(3):267–276.

Mercado, B., Schouwenaars, R., González-Chávez, J.L. & Ramírez-Zamora, R.M. 2013. Multi-analytical assessment of iron and steel slag characteristics to estimate the removal of metalloids from contaminated water. *Environ. Sci. Health* 48(8): 887–895.

Environmental Arsenic in a Changing World –
Zhu, Guo, Bhattacharya, Ahmad, Bundschuh & Naidu (Eds)
ISBN 978-1-138-48609-6

Mitigation of As accumulation in paddy rice (*Oryza sativa* L.) by amendments containing iron and manganese

M. Lei[1], L. Tang[1], H.H. Du[1], L. Peng[1], B.Q. Tie[1] & G.X. Sun[2]
[1]*College of Resource and Environment, Hunan Agricultural University, Changsha, P.R. China*
[2]*State Key Laboratory of Urban and Regional Ecology, Research Center for Eco-Environmental Sciences, Chinese Academy of Sciences, Beijing, P.R. China*

ABSTRACT: In order to mitigate As accumulation in rice, amendments such as iron ore tailing (FeT) and manganese ore tailing (MnT) were employed in As contaminated soil. Before experiment, the toxicities of both amendments were assessed with Toxicity Characteristic Leaching Procedure (TCLP) and horizontal vibration method (HVM), respectively. The ranges of contents of As were 0.13–1.54 mg kg^{-1} in husk samples, and 0.06–0.42 mg kg^{-1} in brown rice, respectively. Among both amendments treatment, the lowest concentrations of As in rice grain were found by MnT treatment due to its structure and composition. There was significantly negative relation ($p < 0.05$) between Mn and As, while significantly positive relationship between Fe and As, and ratio of Fe/Mn and As in rice plant ($p < 0.05$).

1 INTRODUCTION

Iron (Fe) and manganese (Mn) are the first (5.6×10^4 mg kg^{-1}) and third (9.5×10^2 mg kg^{-1}) most abundant transition metals in the earth's crust (Cox, 1995). Due to the redox chemistries of Fe (II/III) and Mn (II/III/IV) having important roles and impacts in the environment (Stone and Morgan, 1987), Fe and Mn oxides and hydroxides have been incorporated into metal-contaminated soil, especially for controlling As adsorption-desorption reactions in soil. Moreover, arsenic is a toxic and ubiquitous element in the environment, especially in rice plants because rice easily uptakes As from soil, even in uncontaminated soil compared to other cereals under reduced conditions. The aim of this study was to study the effects of Fe and Mn from FeT and MnT on As uptake by rice plant and transfer from soil to rice and assess the toxicity of FeT and MnT with the toxicity characteristic leaching procedure (TCLP) of USA and horizontal vibration method (HVM) of China before pot experiment.

2 METHODS/EXPERIMENTAL

2.1 *Materials*

The paddy soils (0–20 cm) were collected from one smelting-impacted area in Chenzhou city, Hunan province. The amendments such as iron mine tailing (FeT) and manganese mine tailing (MnT) were collected from Xiangxiang iron work and Xiangxi manganese ore factory, respectively. Both amendments were grounded and passed through a 2-mm sieve. The dry samples of FeT and MnT were characterized by SEM-EDAX under KV 20.0, Mag 1000, TILT 0.0, Micronsperpixy 0.119 conditions. Moreover, the toxicities of both amendments were assessed with toxicity characteristic leaching procedure (TCLP) and horizontal vibration method (HVM), respectively.

2.2 *Experiment design and sample collection*

Before rice was planted, each pot was filled with 5.0 kg soil, mixed with either FeT (0.0%, 1.0%, and 2.0%) and MnT (0.0%, 1.0%, and 2.0%), respectively. The aging time was 30 days. At the start of the pot experiment, the solid fertilizers (0.36 g kg^{-1} soil of $CaH_2PO_4 \cdot 2H_2O$, 0.43 g kg^{-1} soil of $CO(NH_2)_2$, and 0.41 g kg^{-1} soil of KCl) were added to ensure adequate nutrition for the growth of rice seedlings. About 3–4 cm water above the soil level in the pot was maintained 2 days before seedling transplantation. Growth of rice plants continued for 4 months. Plants were watered with distilled water (18 mΩ, Millipore) without heavy metals and fertilizer, one to two times per week. Pots were designed with four replications, randomized and alternated weekly in greenhouse. At harvest, Rice grains were separated from the plant by handpicking, straws were cut at 4 cm above the soil, and roots were pulled out from the paddy soil. Iron plaque on fresh root surfaces was extracted using dithionite–citrate–bicarbonate (DCB) solution according to the method of Taylor and Crowder (1983). All rice plant samples were tagged, stored in polyethylene bags, and transported to the laboratory within 4 h for sample processing.

3 RESULTS AND DISCUSSION

3.1 *As contents in paddy soil and both amendments*

The soil As content with 62.54 mg kg^{-1} was 2.5 times higher than the national standard value (25.0 mg kg^{-1}) for paddy soil in China. The total As concentrations in FeT (33.43 mg kg^{-1}) and MnT (20.54 mg kg^{-1}) were lower than that in agricultural use (GB4284-1984), while much higher than the world As mean value, which suggested that FeT and MnT should be pre-treated before utilization.

Figure 1 demonstrated that the particles of MnT were not dense and seems to be constituted of an agglomeration of plate or lamellar particles. The SEM showed that the MnT (a) adsorbed As better than that of FeT(b). There were many elements such as Fe, Mg, Al, Si, Ca, O and C in FeT (c), while only three elements like Mn, O and C were found in MnT (d).

The extraction of As concentrations from FeT and MnT with TCLP were much higher than those with HVM. It was observed that the As contents in extract liquids with both regents from FeT and MnT in TCLP exceed that allowable regulatory limit 0.1 mg L^{-1}, and were much higher than those leached with HVM due to the low pH (2.9) of the leachate solution 2 in TCLP. Judged by Chinese identification standards for hazardous wastes-identification for extraction toxicity (GB5085.3-2007), the FeT and MnT employed in present study should be pretreated with distilled water until As and other metals weren't determined in extract solution.

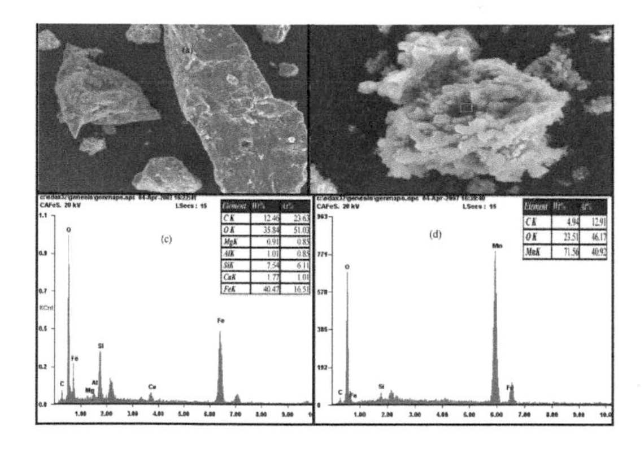

Figure 1. Images of surface FeT (a) and MnT (b) with SEM, and quantitative compositions of FeT (c) and MnT (d) with EDAX.

3.2 *Relationship between Mn and Fe in rice plant*

The significantly negative relationship between Mn and Fe was found in rice plant ($p < 0.05$), indicating that increasing Mn concentrations caused Fe concentration to decrease in rice plant. Among all parts of rice plant, the highest concentrations of Fe (27077.62–47804.89 mg kg^{-1}) were found in DCB corresponding addition dosages, followed by Mn (4761.70–5492.93 mg kg^{-1}). The rice root, especially root surface accumulated Fe concentrations with 1279.68–47804.89 mg kg^{-1} were almost 10-time higher than those of Mn (116.58–5492.93 mg kg^{-1}). While, the concentrations of Mn in straw and grain were higher than those of Fe. The transfer factors (TFs) of Mn ranged 0.74–1.93 from root to rice plant above ground, and ranged from 0.35–0.43 from straw to grain, while the TFs of Fe were 0.01–0.02 from root to rice plant above ground, and ranged from 0.10–0.25 from straw to grain, indicating the TFs of Mn were higher those of Fe in rice plant.

3.3 *Concentrations of As, Fe, and Mn in rice grain*

The ranges of As concentrations in husk samples were 0.13–1.54 mg kg^{-1}, and in brown rice were 0.06–0.42 mg kg^{-1}, respectively (Table 1). Comparison of CK, the contents of As in husk and brown rice were not significantly reduced with increasing amounts of FeT and MnT ($p = 0.39 > 0.05$). The distribution of As in the soil–rice plant system with unamend (CK) and FeT treatment followed: DCB > root > soil > straw > husk > brown rice, indicating As contents in DCB (225.57–286.78 mg kg^{-1}) were much higher than that of soil (62.54 mg kg^{-1}). However, due to the high Mn concentrations in DCB solution, the order of As distribution was: root > soil > DCB > straw > husk > brown rice with MnT treatment. Among both amendments, the concentrations of As in husk and brown rice were the lowest, especially treatment with 2.0% of MnT, with 0.13 mg kg^{-1} and 0.06 mg kg^{-1} in husk and brown rice, indicating the MnT was the best material for controlling As transfer from soil to rice plant.

4 CONCLUSIONS

There was significantly negative relationship between Fe and Mn, and Mn and As in rice plant, while, there

Table 1. Contents of Mn and Fe in husk and brown rice.

	Husk (mg kg^{-1})			Brown rice (mg kg^{-1})		
Treatment	As	Fe	Mn	As	Fe	Mn
CK	1.54 ± 0.24	25.46 ± 1.69	275.11 ± 8.95	0.42 ± 0.11	6.46 ± 0.12	13.92 ± 1.42
1.0% FeT	1.23 ± 0.21	41.58 ± 2.28	254.93 ± 54.00	0.34 ± 0.16	8.91 ± 1.32	13.58 ± 0.59
2.0% FeT	1.11 ± 0.41	51.89 ± 10.45	278.71 ± 63.12	0.28 ± 0.10	17.76 ± 1.45	15.09 ± 4.64
1.0% MnT	0.49 ± 0.04	33.07 ± 4.17	676.71 ± 179.23	0.09 ± 0.01	2.67 ± 0.75	39.12 ± 4.06
2.0% MnT	0.13 ± 0.01	21.57 ± 2.63	605.85 ± 142.16	0.06 ± 0.02	5.18 ± 1.72	37.88 ± 0.42

was positive correlation between Fe and As, and the ratio of Fe/Mn and As in rice plant. After application of both amendments, the soil pH increased, and the As contents in rice grain decreased, especially Mn mine tailing is the best material for controlling As transfer from soil to rice plant due to transfers factor of Mn higher than that of Fe.

ACKNOWLEDGEMENTS

This work was supported by the National Science Foundation of China (41671475/21007014).

REFERENCES

Cox P.A. 1995. The Elements on Earth: Inorganic Chemistry in the Environment. New York, Oxford University Press.

Stone A.T. & Morgan J.J. 1987. Reductive dissolution of metal oxides, In: W. Stumm (ed) Aquatic Surface Chemistry. New York: Wiley, pp. 221–254.

Taylor G.J. & Crowder A.A. 1983. Use of DCB technique for extraction of hydrous iron oxides from roots of wetland plants. *Am. J. Bot.* 70: 1254–1257.

Environmental Arsenic in a Changing World –
Zhu, Guo, Bhattacharya, Ahmad, Bundschuh & Naidu (Eds)
ISBN 978-1-138-48609-6

Arsenic removal from water of the Peruvian rivers using low cost carbon base adsorbents produced from agro industry waste

G. Cruz[1], M.M. Gomez[2] & J.L. Solis[2]
[1] *Universidad Nacional de Tumbes, Campus Universitario, Tumbes, Peru*
[2] *Universidad Nacional de Ingeniería, Lima, Peru*

ABSTRACT: Microporous-mesoporous activated carbons or biochar were produced from sawdust from Peruvian forestry species and agro-industrial wastes as potential materials for application in water and wastewater treatment. The used sawdust samples were from *Cedrelinga catenaeformis* Ducke and *Colicodendron scabrida* and the agro-industrial wastes were cocoa pod husk, corncob, red mombin seed, internal and external part of the mango pit, and coffee husk. Hybrid materials based on activated carbons with chitosan or ZnO nanoparticles were produced in order to add an antibacterial capacity or photocatalytic properties. The impregnated samples showed adsorption capacity of Arsenic from Tumbes River water with removal capacities between 80–100% at first 5 min of experiments. Besides, the antibacterial properties against *E. coli* and *Shigella flexneri* were evident.

1 INTRODUCTION

Arsenic contamination of water can occur naturally or as a consequence of human activities such as mining. Natural sources of arsenic in Peru are mainly enargite-bearing copper, zinc and lead deposits. In the South of Peru (Arequipa, Tacna and Moquegua) the groundwater has dangerous levels of arsenic. Peru is a mining country and represents around 60% of its exports, primarily gold, copper, and zinc. In South America the most water-stressed country is Peru and every year mining and metallurgy release over 13 billion m^3 of effluents into Peru's bodies of water (Bebbington and Williams, 2008).

Connections between mining and water risks are many. Méndez performed a study in the Rímac river basin and showed that mine tailings resulted in elevated arsenic, copper and lead concentrations in the river and its tributaries. C. de Meyer *et al.* (2017) have detected health-threatening concentrations of arsenic, manganese, and aluminum in almost half of the groundwater resources investigated from the Peruvian Western Amazonia.

In Peru, the current national regulatory standards for arsenic in drinking water are based on WHO's recommended limit; however, little is known about the extent of arsenic contamination of the drinking water and about its health implications in the country. George *et al.* (2014) have found that in several districts of Peru, drinking water shows widespread arsenic contamination, exceeding the WHO arsenic guideline. This poses a public health threat requiring further investigation and action especially in the rural areas where there is not water treatment.

Peru is a mainly agricultural country with a high level of production of fruits, cereals, and different vegetables for internal and external markets. The environmental problems related to the management of the residual agricultural biomass have become alarming to social and governmental entities. The inadequate disposal and open field burning of these materials produce different pollutants, affecting mainly the air and soil quality and the surrounding population.

The search for new renewable precursors for production of activated carbons (AC) with highly microporous-mesoporous structure and adsorption capacities comparable with those of commercial adsorbents remains an important issue for industrial application in water treatment.

2 EXPERIMENTAL

To produce activated carbons, raw materials were collected in the agricultural areas in Piura and Tumbes in Peru. The raw materials were washed with potable water and then dried at 80°C until constant weight was reached. The dried raw materials were ground and sieved to obtain 0.5–1 mm particle-size.

Raw materials were mixed with $ZnCl_2$ in a proportion 1/1 (raw material/chemical reagent) and then put in a ceramic reactor in a horizontal oven. The activation and carbonization were carried out in a single step in presence of N_2 flow (150 mL min^{-1}), heated up in the oven until 600°C was reached. The carbonization time was 2 h and then the material was cooled by keeping in the inert atmosphere. Finally, the carbonized materials were washed with a 0.15 M hydrochloride acid solution then with distilled water (Cruz *et al.*, 2015).

Prepared activated carbons were codified as follows: RMS-AC, red mombin seed activated carbon; GS-AC, ice-cream bean activated carbon; CPH-AC, cocoa pod husk activated carbon; CH-AC, coffee husk activated carbon; MSEP-AC, mango seed external part activated carbon; MSIP-AC mango seed internal part activated carbon; and CC-AC, corncob activated carbon (Cruz *et al.*, 2017).

Chitosan was impregnated onto the bare activated carbons using the wet impregnation method. A mixture of 50 mL of chitosan solution and 1 g of activated carbon was stirred and heated at 100°C during 1 h in a closed steel reactor. The pressure was increased until 239.2 kPa and kept during the impregnation. The samples code for activated carbons from corncob and red mombim seed impregnated with chitosan were CC-Ch and RMS-Ch, respectively.

For the production of activated carbons impregnated with ZnO nanoparticles, both precursors were mixed with zinc chloride ($ZnCl_2$) as the chemical activator and zinc acetate ($Zn(C_2H_3O_2)_2$) dissolved in distilled water as the ZnO nanoparticles precursor in a proportion 1/1/0.5, respectively. The mixture was dried overnight at 80°C and then put in a ceramic tube to be carbonized into a metallic reactor to be introduced in a horizontal oven and heated to 600°C. This temperature was kept during 2 h under nitrogen atmosphere. The material was cooled and then washed repeatedly. These samples were codified as: CCmod and RMSmod, respectively.

Kinetic experiments were performed with real polluted water to test the As adsorption capacity of the activated carbons with and without impregnation.

A sample of polluted river water was taken from Tumbes River in Peru and preserved with concentrated nitric acid. At the laboratory, the water sample was filtered and the pH of the samples was re-adjusted to pH around 6 before the adsorption experiments. Different aliquots were taken at the beginning and at different times between 5 and 300 min with a 10 mL syringe. Aliquots were filtered to retain activated carbon particles with a 0.2 μm filter, removing the first 1 mL to avoid the effect of the filter in the concentration of heavy metals. Then, 10 mL sample was acidified with nitric acid p.a. to be sure that the pH of the sample was reduced at level less than 3 and then storage at 4°C until the analysis.

3 RESULTS AND DISCUSSION

Since the adsorption of As and Pb was very fast by the samples and the adsorption equilibrium was reached in the earlier stages of the experiments, only the adsorption data of the first 90 min was used in the evaluation. The initial concentration of heavy metals was 1750 μg L^{-1} of Al, 56.7 μg L^{-1} of As, 2920 μg L^{-1} of Fe, 224 μg L^{-1} of Pb and 272 μg L^{-1} of Zn.

The Figure 1 shows that both bare activated carbons exhibit high removal capacity of As from the polluted river water, reaching removal levels close to 100% within the first five min. Almost similar behavior presents the samples impregnated with chitosan and the samples impregnated with ZnO. All

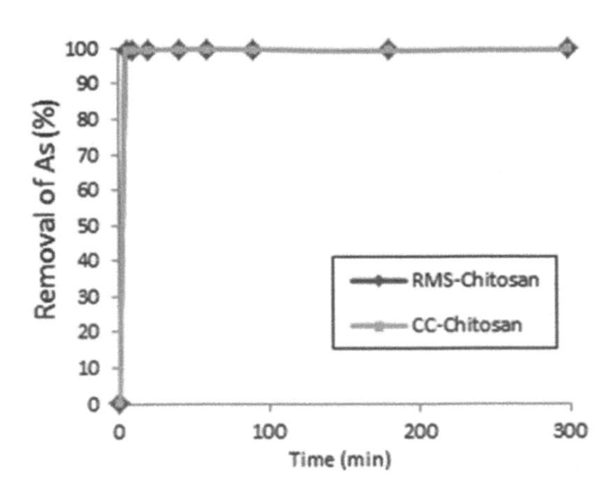

Figure 1. Arsenic removal kinetics of RMS-Ch and CC-Ch.

adsorbents impregnated or not had the ability to eliminate *Escherichia coli* and *Shigella flexneri.*

4 CONCLUSIONS

The produced activated carbons impregnated with chitosan or ZnO are promising materials for application in removal of arsenic from contaminated water.

ACKNOWLEDGEMENTS

This work was financially partially supported by CONCYTEC. The Universidad Nacional de Tumbes provided important financial support for activated carbon production (CANON funding).

REFERENCES

Bebbington, A. & Williams, M. 2008. Water and mining conflicts in Peru, *Mountain Res. Dev.* 28(3): 190–195.

Cruz, G., Matějová, L., Pirilä, M., Ainassaari, K., Canepa, C.A., Solis, J., Cruz, J.F., Šolcová, O. & Keiski, R. 2015. A comparative study on activated carbons derived from a broad range of agro-industrial wastes in removal of large-molecular-size organic pollutants in aqueous phase. *Water Air & Soil Poll.* 226(7): 214.

Cruz, G., Kubonova, L., Aguirre, D., Matejova, L., Peikertova, P., Troppova, I., Cegmed, E., Wach, A., Kustrowski, P., Gomez M. & Obalova L. 2017. Activated carbons prepared from a broad range of residual agricultural biomasses tested for xylene abatement in the gas phase. *ACS Sust. Chem. Eng.* 5: 2368–2374.

de Meyer, C., Rodríguez, J., Carpio, E., García, P., Stengel, C. & Berg, M. 2017. Arsenic, manganese and aluminum contamination in groundwater resources of Western Amazonia (Peru). *Sci. Total Environ.* 607–608:1437–1450.

George, C., Sima, L., Arias, M., Mihalic, J., Cabrera, L., Danz, D., Checkley, W. & Gilman, R. 2014. Arsenic exposure in drinking water: an unrecognized health threat in Peru. *Bull. World. Health. Organ.* 92:565–572.

Méndez, W. 2005. *Contamination of Rímac River basin Peru, due to Mining Tailings.* MSc Thesis. Royal Institute of Technology, Stockholm, Sweden.

Environmental Arsenic in a Changing World –
Zhu, Guo, Bhattacharya, Ahmad, Bundschuh & Naidu (Eds)
ISBN 978-1-138-48609-6

Arsenic oxidation by hypertolerant *Bacillus* sp. L-148 in artificial groundwater microcosm

A.V. Bagade[1], D. Paul[2], A. Giri[3], D. Dhotre[2], S. Pawar[2] & K. Kodam[1]
[1]*Department of Chemistry, Savitribai Phule Pune University, Pune, India*
[2]*Division of Biochemical Sciences, CSIR-National Chemical Laboratory, Pune, India*
[3]*Microbial Culture Collection, National Centre for Cell Science, Pune, India*

ABSTRACT: A microcosm mimicking artificial groundwater harboring high (75 mM) AS(III) concentration was developed to study the rate of arsenic oxidation by *Bacillus* sp. L-148 under different conditions. Na. acetate and glucose as carbon sources were used along with and/or absence of Fe ions. Morphological studies revealed elongation of 75 mM As(III) exposed bacteria along with terminal spore formation. Decrease in pH with time and As(V) formation was seen. Maximum 77% oxidation in presence of Na. acetate and Fe ions on 15th day was seen. Growth studies were in analogy with As oxidation studies. Microcosm for As oxidation in presence of 0.4 mM mixture of heavy metals Cd, Co, Pb, Li, As(III), As(V), Sb was studied. Effect of heavy metals on growth and As(III) oxidation was assessed. Contrastingly, here glucose served as a better carbon source but completely oxidized As(III). Thus, this microcosm study concluded that *Bacillus* sp. L-148 can be used for bioremediation of heavy arsenic contaminated site along with presence of other heavy metals.

1 INTRODUCTION

Arsenic (As) is ranked first on the" Superfund Priority List" of hazardous materials by Environmental Protection Agency and World Health Organization because of its potent carcinogenic properties. Leaching of As from the minerals into groundwater creates the contamination issues and enhances population risk of deleterious cancerous effects. In spite of the toxic nature of arsenic, microorganisms have evolved different tolerance mechanisms to reduce its harmful effects.

The bioremediation by these bacteria is usually checked by setting up microcosm at times at pilot level. As contaminated sites having 215–12500 mg kg^{-1} As in Italy (Lucia *et al.*, 2015) and 100–2500 mg kg^{-1} As in Chile soil (Oskar *et al.*, 2016) warrants a need for a hypertolerant bacterial isolate for bioremediation. Considering this background, a microcosm of artificial groundwater with 75 mM As(III) was set up and assessed for As oxidation, growth, pH variation and morphological studies. Other heavy metal contamination in groundwater is also reported, hence mixture of heavy metal in microcosm was studied for their effect on As oxidation.

2 METHODS

2.1 *Microcosm set up*

A 15-day experiment with 30 mL artificial groundwater along with 75 mM As(III) in presence of 2% inoculum of *Bacillus* sp. L-148 was set up. Two different carbon sources (0.164%) Na. acetate and glucose were used separately with and without presence of 20 mM Fe. Respective abiotic control, carbon starvation control, no Fe control, carbon starvation with Fe control were also taken. A 2nd set of microcosms to study the effect of 0.4 mM heavy metals (Cd, Co, Pb, Li, As(III), As(V), Sb) each on As oxidation and growth was also studied. Abiotic control was used for this set.

2.2 *Morphology and pH assessment*

Samples were withdrawn every 2 days and checked for pH change. morphological studies were done by SEM.

2.3 *Growth and rate of As oxidation*

Growth of the isolate was checked by serial diluting the sample and spread plating on 0.1 X tryptic soya agar to get the colony forming unit mL^{-1}. As(III) oxidation of the aliquoted samples were estimated by using molybdenum blue method (Bachate *et al.*, 2009).

3 RESULTS AND DISCUSSION

3.1 *Morphological studies*

Effect of high arsenic concentration on morphology of *Bacillus* sp. L-148 was studied. Exposure of 75 mM arsenic increased the length of the isolate almost by $1 \pm 0.4\,\mu m$. Exposing the isolate to 0.4 mM each of Cd, Co, Pb, Li, As(III), As(V), Sb also exhibited increase in cell length around $0.26 \pm 0.09\,\mu m$. This increase in cell length is attributed to limited nutrient access which affects the cell size. A large surface to

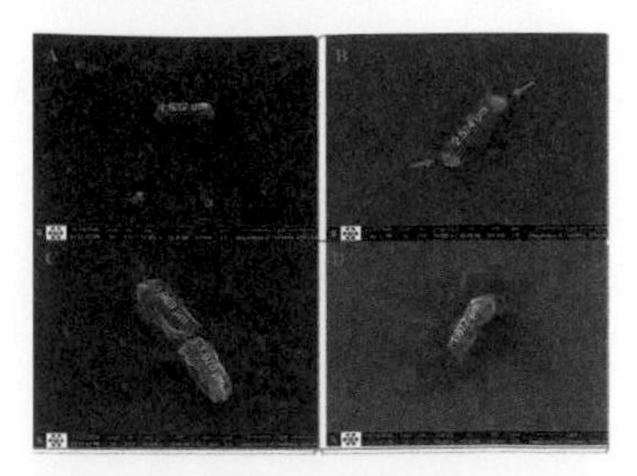

Figure 1. S.E.M of *Bacillus* sp. L-148: A) Control cells, B) Na. acetate + Fe (endosporulation shown by red arrows); C) Na. acetate + Fe + multi metals; D) Glucose + Fe + multi metals.

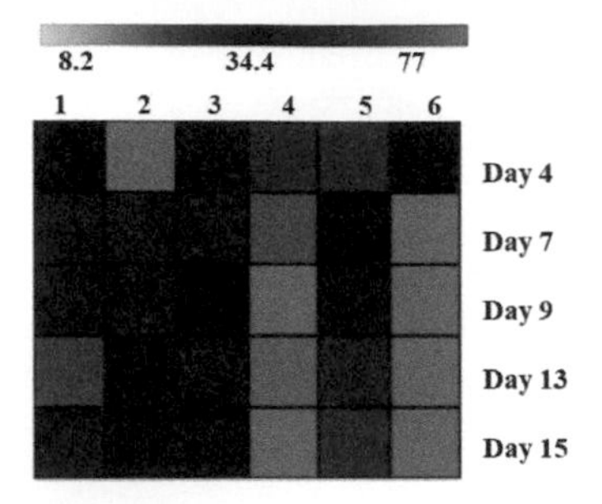

Figure 2. Heatmap showing percent As(III) oxidation 1) No carbon 2) Glucose 3) Na. acetate 4) No carbon + Fe 5) Glucose + Fe 6) Na. acetate + Fe.

volume ratio is achieved to support the internal biochemistry (Young *et al.*, 2007). Spore formation was visible (Fig. 1B red arrows) in 75 mM exposed cells with Na. acetate + Fe. Spore formation was visible on the 15th day which indicated the nutrient insufficiency, stationary phase, presence of extracellular deleterious metabolites in the artificial groundwater microcosm under study.

Influence of pH changes the solubility of arsenic in arsenic contaminated soil or water (Gersztyn *et al.*, 2013). Change in pH of artificial groundwater microcosm was assessed throughout 15 days. In the samples containing different carbon sources and Fe there was gradual decrease in pH values as arsenic was oxidized with days. Contrastingly the metal mixture showed no decrease in pH. As(III) oxidation decreases the pH but as both As(III) and As(V) were added in small concentrations along with other metals, the pH change was not visible in metal mixture.

3.2 *Growth and arsenic oxidation*

The growth pattern in microcosm in presence of different carbon sources, iron and their respective controls was evaluated to get a brief idea about the amount of cells involved in high arsenic oxidation and effect of the different conditions exposed on growth of the isolate. This would be important while performing on field bioremediation of arsenic contaminated sites. Growth rate was faster where Na. acetate + Fe and only Na. acetate was provided where the CFU mL^{-1} decreased after 5 days as against 9 days seen in other conditions. We speculate this to the loss of limited nutrients provided in the microcosm (Riess *et al.*, 2008). Sodium acetate is a metabolic byproduct of glucose catabolism hence might be preferred over glucose at times (Thomas *et al.*, 2014).

Maximum percent As(III) oxidation was seen in sodium acetate + Fe sample (Fig. 2) which was attributed to synergistic effect of bacterial oxidation and chemical oxidation of arsenic by Fe (Hug, 2003). The decreasing order of rate of oxidation was as follows: sodium acetate + Fe $>$ carbon starvation + Fe $>$ glucose + Fe $>$ sodium acetate $\geq$ glucose $\geq$ carbon starvation (Fig. 2). Growth and As(III) oxidation data were in correlation with each other. On the other hand, the multimetal with 0.4 mM of As(III) and As(V) showed more growth (Fig. 1C and 1D) in glucose as carbon source. The excess stress of all heavy metals together and minimal nutrients might have rendered utilization of glucose as carbon source. As(III) 0.4 mM in presence of heavy metals was fully oxidized on 9th day in both glucose and sodium acetate. This indicated that the isolate slowly oxidized As(III) because of the toxicity and presence of other heavy metals.

4 CONCLUSIONS

Bacillus sp. L-148 could oxidize 77% of 75 mM As(III) present in artificial groundwater microcosm using Na. acetate as carbon source and in presence of Fe. The length of the organism increased to suffice the toxicity. It could also oxidize As(III) in presence of other heavy metals. Hence *Bacillus* sp. L-148 holds potential to be used for bioremediation of high arsenic contaminated sites.

ACKNOWLEDGEMENTS

IIT-B,SAIF for ICP-AES analysis. University Grant Commission, India for funding.

REFERENCES

Gersztyn, L. 2013. Influence of pH on the solubility of arsenic in heavily contaminated soils. *Environ. Prot. Nat. Resour.* 3(57): 7–11.

Hug, J. 2003. Iron-catalyzed oxidation of arsenic(III) by oxygen and by hydrogen peroxide: pH-dependent formation of oxidants in the Fenton reaction. *Environ. Sci. Technol.* 37(12): 2734–2742.

Riess, T. 2008. Analysis of a novel insect cell culture medium-based growth medium for *Bartonella* species. *Appl. Environ. Microbiol.* 74(16): 5224–5227.

Thomas, V.C. 2014. A central role for carbon-overflow pathways in the modulation of bacterial cell death. *PLoS Pathog.* 10(6): 1–13.

Environmental Arsenic in a Changing World –
Zhu, Guo, Bhattacharya, Ahmad, Bundschuh & Naidu (Eds)
ISBN 978-1-138-48609-6

Relation between As(III) oxidation potential and siderophore production: a study of tannery As(III) oxidizers

V. Nandre & K. Kodam
Department of Chemistry, Savitribai Phule Pune University, Pune, India

ABSTRACT: Herein we report the relationship between bacterial ability in arsenic (As) transformation, As resistance and siderophore production. It was investigated using different 40 As-resistant bacteria isolated from the tannery environment. Out of the bacterial genera isolated the 7 As(III) oxidizing genera and 4 As resistant genera produced siderophore. The bacterial genera producing more siderophores showed greater As(III) oxidation and As(III) resistance than those producing less siderophore.

1 INTRODUCTION

Arsenic (As) is a potent poison present in the environment in different oxidation states depending on the existing environment. The As(III) form is more toxic which is present in reducing environment than As(V) present in oxidizing environment (Oremland and Stolz, 2003). The microbes evolved number of strategies to survive against the toxic environment. One of these is the production of iron chelator i.e., siderophore in the environment. The siderophores are low molecular weight organic compound secreted by microorganisms in the environment (Miethke and Marahiel, 2007). Nearly, all living organisms require iron for metabolic activities and growth. It is not readily available to microorganisms and plants under normal conditions. The basic important role of siderophores is involved in the high affinity acquisition and receptor dependent transport of ferric ion. Regulation of the siderophore production is based on the concentration of iron in the environment (Miethke, 2013). The function of these compounds is to scavenge iron from the environment and to make it available to the microbial cell. Iron is soluble in ferrous form and can be easily acquired by microbial cell (Andrews *et al.*, 2003). The production of iron binding compounds, siderophores is the most studied mechanism amongst the various means used by microorganisms (Grove 1980; Jappelli & Marconi, 1997). Herein we studied the relation between different genera for As(III) oxidation and siderophore production isolated from tannery environment.

2 METHODS

2.1 *Isolation and identification of arsenite oxidizing bacteria*

Tannery soil and effluent samples were taken from waste disposal site Miraj, India. In enrichment, 3 g soil was inoculated in Tris-mineral medium, supplemented with 0.04% yeast extract (TMM) and 5 mM of As(III). Flasks were incubated on shaker for As(III) at 30°C for 5 days and 3 mL of this enrichment culture was inoculated in fresh medium. This procedure was repeated two times. The isolates yielded were initially screened by using silver nitrate plate assay. Sequencing of 16S rRNA gene was performed using genomic DNA isolated by alkaline lysis method (Fig. 1).

2.2 *Screening of arsenite oxidizers and siderophore producers*

Arsenic oxidation ability isolated strains were estimated by using molybdenum blue method (Bagade *et al.*, 2016). Siderophore production ability was tested on the universal CAS plate assay (Schwyn *et al.*, 1987).

3 RESULTS AND DISCUSSION

3.1 *Isolation and identification of arsenite oxidizing bacterial strains*

Out of the total 40 heterotrophic cultures isolated from enrichment of tannery effluent and soil, 18 different genera were identified. The isolates were identified by partial (800 to 1300 bp) sequencing of 16S rDNA and checked for its As transformation ability. The isolates after sequencing showed bacteria belonging to different total 11 genera namely *Achromobacter*, *Ochrobactrum*, *Sinorhizobium*, *Alcaligenes*, *Agrobacterium*, *Bordetella*, *Rhizobium*, *Stenotrophomonas*, *Bacillus*, *Microbacerium* and *Pseudomonas*.

3.2 *Screening of arsenic redox isolates and siderophore producers*

Out of the total cultures screened, seven cultures showed As(III) oxidizing ability which were grouped according to the genera they belonged. The seven genera showed As(III) oxidizing ability

Table 1. Different bacterial genera showing arsenic oxidizing genera and/or siderophore producing ability.

Isolate	As(III) oxidation	CAS positive
Achromobacter sp. T-14	+	+
Ochrobactrum sp.T-5	+	+
Sinorhizobium sp. T-20	+	+
Alcaligenes sp.R-3-2	+	+
Agrobacterium sp. R-7	+	+
Bordetella sp. R-8	+	+
Rhizhobium sp. R-6-2	+	+
Stenotrophomonas sp. R-4-1	−	+
Bacillus cereus sp. R-4-2	−	+
Pseudomonas sp. T-1-1	−	+
Microbaterium sp. T-2	−	+

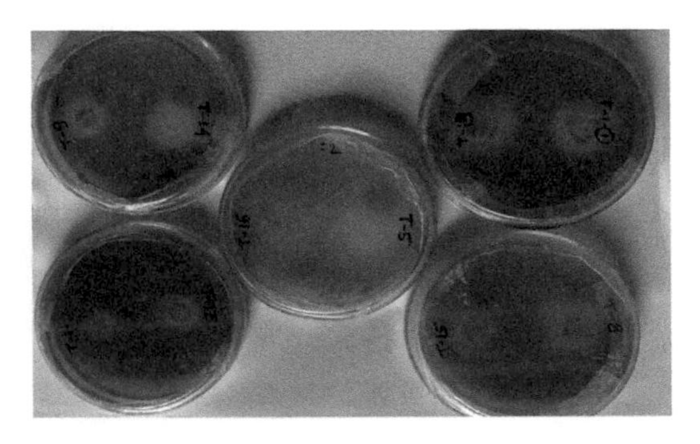

Figure 1. Representative figure of few bacterial genera showing yellow zone around the cultures indicating CAS test positive.

are namely *Achromobacter*, *Ochrobactrum*, *Sinorhizobium*, *Alcaligenes*, *Agrobacterium*, *Bordetella*, *Rhizobium*. While the four genera *Stenotrophomonas, Bacillus, Microbacerium* and *Pseudomonas* only showed siderophore producing ability (Table 1).

There was a relation between biotransformation of arsenicals *viz.*, As(III), As(V) and siderophore production. Out of the bacterial genera isolated the As(III) oxidizer and As(V) reducers the bacterial strains producing more siderophores showed greater As(III) oxidation and As(III) resistance than those producing less siderophore.

4 CONCLUSIONS

We found in our study that arsenite oxidizer showed siderophore production ability that could be an exception for previously reported results. There are total 29 isolates obtained from this study distributed in 11 different genera having the ability of siderophore production.

There is a strong correlation we found from our study that pathogenic bacteria specifically synthesize the siderophore. Due to huge diversity and growth condition we have not performed screening of chemical type of siderophore by different isolates. However, it is quite interesting to observe that those isolates responsible for As speciation can also synthesis the siderophore.

ACKNOWLEDGEMENTS

We would like to thank University Grants Commission (UGC), New Delhi, India and University with Potential for Excellence (UPE-II).

REFERENCES

Andrews, S.C., Robinson, A.K. & Rodríguez-Quiñones, F. 2003. Bacterial iron homeostasis. *FEMS Microbiol. Rev.* 27(2–3): 215–237.

Bagade, A.V., Bachate, S.P., Dholakia, B.B., Giri, A.P. & Kodam, K.M. 2016. Characterization of *Roseomonas* and *Nocardioides* sp. for arsenic transformation. *J. Hazard. Mater.* 318: 742–750.

Grove, A.T. 1980. Geomorphic evolution of the Sahara and the Nile. In: M.A.J. Williams & H. Faure (eds) *The Sahara and the Nile*. Balkema, Rotterdam, pp. 21–35.

Jappelli, R. & Marconi, N. 1997. Recommendations and prejudices in the realm of foundation engineering in Italy: a historical review. In: C. Viggiani (ed) *Geotechnical Engineering for the Preservation of Monuments and Historical Sites; Proceedings International Symposium, Naples, 3–4 October 1996*.

Miethke, M. 2013. Molecular strategies of microbial iron assimilation: from high-affinity complexes to cofactor assembly systems. *Metallomics* 5(1): 15–28.

Miethke, M. & Marahiel, M.A. 2007. Siderophore-based iron acquisition and pathogen control. *Microbiol. Mol. Biol. Rev.* 71(3): 413–451.

Oremland, R.S. & Stolz, J.F., 2003. The ecology of arsenic. *Science* 300(5621), 939–944.

Schwyn, B. & Neilands, J.B. 1987. Universal chemical assay for the detection and determination of siderophores. *Anal. Biochem.* 160: 47–56.

Environmental Arsenic in a Changing World –
Zhu, Guo, Bhattacharya, Ahmad, Bundschuh & Naidu (Eds)
ISBN 978-1-138-48609-6

Arsenic removal to $<1\,\mu g\,L^{-1}$ by coprecipitation with in-situ generated Fe(III) precipitates with and without advanced pre-oxidation

A. Ahmad[1,3,4], J. van Mook[2], B. Schaaf[2] & A. van der Wal[2,3]
[1]*KWR Water Cycle Research Institute, Nieuwegein, The Netherlands*
[2]*Evides Water Company N.V. Rotterdam, The Netherlands*
[3]*Department of Environmental Technology, Wageningen University and Research (WUR), Wageningen, The Netherlands*
[4]*KTH-International Groundwater Arsenic Research Group, Department of Sustainable Development, Environmental Science and Engineering, KTH Royal Institute of Technology, Stockholm, Sweden*

ABSTRACT: The aim of this study is to investigate removal of low As concentrations from groundwater at a Drinking Water Treatment Plant (DWTP) in The Netherlands in order to achieve $<1\,\mu g\,L^{-1}$ As in produced drinking water. Two iron based emerging technologies, both relying on in-situ generated Fe(III) precipitates for As adsorption, were investigated. These include: 1) Advanced Oxidation-Coprecipitation-Filtration (AOCF) and 2) Coprecipitation prior to ultrafiltration (C-UF). We show that most of the As removal occurs in the top half of a Rapid Sand Filter (RSF) bed. In this part we also observe the conversion of As(III) into As(V). The mechanism of As(III) oxidation to As(V) in the RSF is still not understood, however we hypothesize that either the manganese oxides or the biological activity in the filter bed may be responsible for this conversion. In agreement with this observation, we also notice that drinking water only contains As(V) and that the levels of As(III) are negligible. The experiments have shown that both AOCF and C-UF are promising emerging technologies to reduce arsenic levels to below $1\,\mu g\,L^{-1}$ which is the agreed target in The Netherlands between the Dutch water companies.

1 INTRODUCTION

Arsenic (As) is a highly toxic metalloid which occurs naturally in the earth's crust. It is frequently encountered in natural (ground) waters around the world. According to the World Health Organization (WHO), long term exposure to As from drinking water and food can cause cancer and skin lesions. The WHO guideline for As in drinking water is provisionally set at $10\,\mu g\,L^{-1}$, however lower levels are recommended technically and economically feasible (WHO, 2011). In 2016, the drinking water sector in The Netherlands agreed to aim for drinking water production whereby As $<1\,\mu g\,L^{-1}$. The aim of this study was to investigate removal of low As concentrations from groundwater at a drinking water treatment plant (DWTP) in The Netherlands to achieve $<1\,\mu g\,L^{-1}$ As in produced drinking water. Two emerging technologies, both relying on in-situ generated Fe(III) precipitates for As adsorption, were investigated. These include: 1) Advanced Oxidation-Coprecipitation-Filtration (AOCF) and 2) Coprecipitation prior to ultrafiltration (C-UF). The study includes sampling campaigns at the full-scale treatment trains of the DWTP, as well as batch and pilot experiments using actual process water of the treatment facility.

2 MATERIALS AND METHODS

The study was carried out at DWTP Ouddorp which uses infiltrated dune water as its source. The treatment consists of the following steps: 1) rapid sand filtration (RSF), 2) granular activated carbon filtration (GAC) and 3) ultrafiltration (UF). Firstly, multiple sampling campaigns were carried out at the full-scale treatment trains to understand the mobility and fate of As and its species. Subsequently, batch experiments were carried out using the actual process water samples collected at different points along the treatment train. The water samples that contained As(III) were pre-treated with $KMnO_4$ to oxidize As(III) to As(V). This may be called as the advanced oxidation process. The samples that contained As(V) only were directly dosed with $FeCl_3$ to produce Fe(III) precipitates for As(V) adsorption. Eventually, the As laced Fe(III) precipitates were removed from water using a $0.45\,\mu m$ syringe filter. After gaining preliminary indications of As removal efficiency by different coprecipitation approaches, pilot experiments were carried out to investigate the issues of practical concern and long-term treatment efficiency. Samples were preserved by acidifying with concentrated nitric acid. As(III) and As(V) were measured using ion chromatography coupled to ICP-MS detection, whereby the samples were stabilized with EDTA.

3 RESULTS AND DISCUSSION

The levels of total dissolved As as well as the As(III) and As(V) species in the raw (ground) water are given in Table 1. Most (>65%) of the dissolved As in the raw water is available as As(III), which is more difficult to

Table 1. Average dissolved arsenic levels ($\mu g\ L^{-1}$) in the raw water for the DWTP (>38 samples, 2014–2016).

Total As	As(III)	As(V)
9.10 ± 1.67	6.00 ± 1.61	2.99 ± 0.99

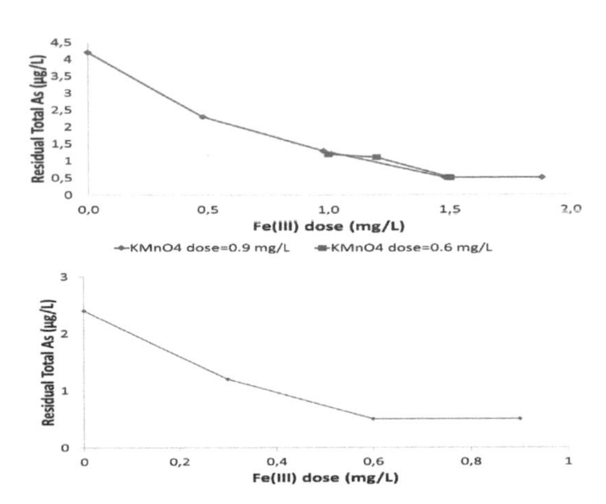

Figure 1. Arsenic profile, including its species in the rapid sand filter (top) as well as arsenic speciation in produced drinking water (bottom).

remove than As(V), due to its charge neutrality below pH 9.2. Figure 1 (top) shows the profile of As and its species over the height of the RSF bed. Most of the As removal occurs in the top half, i.e. in the top 800 mm of the filter bed. In the bottom half of the filter bed, only very low concentration of As, probably co-precipitated with Fe(III) precipitates, is being removed. Moreover, Figure 1 shows that most of the transformation of dissolved As(III) into As(V) takes place in the top half of the filter bed. Figure 1 (bottom) shows that the drinking water contains only As(V) and no As(III), which is consistent with the previous observation that As, while passing through the filter bed, is effectively removed and that the dissolved As(III) becomes oxidized into As(V) (Gude *et al.*, 2016). This has also been observed at other drinking water production locations in The Netherlands (Ahmad *et al.*, 2014). Nevertheless, this observation is remarkable, considering that oxidation of As(III) by atmospheric oxygen/air is known to be a slow process (Bissen and Frimmel, 2003). So far, the mechanism of oxidation of As(III) into As(V) in the RSF is not fully understood.

We hypothesized that by oxidizing As(III) into As(V) in the influent water of the RSF, more of the capacity of the RSF becomes available for As removal, which should lead to an increased As removal efficiency. In order to validate this, jar tests were performed with the raw (ground)water, whereby As(III) was converted to As(V) by dosing $KMnO_4$ as an oxidant, prior to filtration. Figure 2 (top) shows the

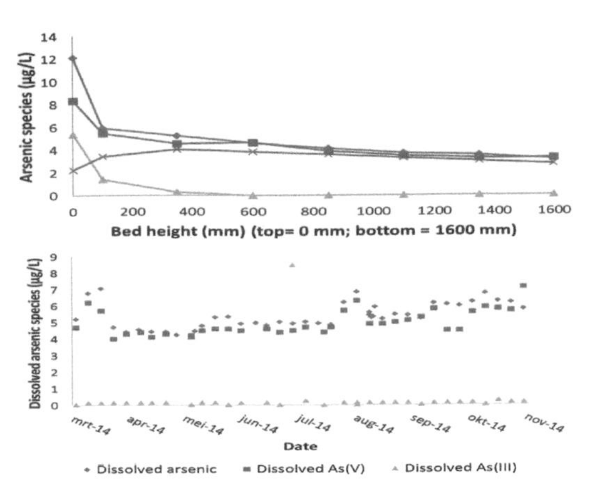

Figure 2. Residual arsenic levels in raw (ground) water in AOCF jar test (top) and C-UF lab test (bottom).

reduction in As levels to below $1\ \mu g\ L^{-1}$ by dosing $FeCl_3$ at levels higher than $1.3\ mg\ L^{-1}$ in the RSF influent, in which $KMnO_4$ (0.6 and $0.9\ mg\ L^{-1}$) is also present. The same result could also be obtained by dosing $FeCl_3$ into UF influent, which is shown in Figure 2 (bottom).

4 CONCLUSIONS AND OUTLOOK

It has been shown that most of As removal from groundwater occurs in the top half of a Rapid Sand Filter bed. In this part we also observed the conversion of As(III) into As(V). The mechanism of oxidation of As(III) into As(V) in the RSF is still not understood, however we hypothesize that either the manganese oxides or the biological activity in the filter bed may be responsible for this conversion. In agreement with this observation, we have also observed that drinking water only contains As(V) whereas the levels of As(III) are negligible. The experiments have shown that both AOCF and C-UF are promising emerging technologies to reduce arsenic levels to below $1\ \mu g\ L^{-1}$.

REFERENCES

Ahmad, A., Van De Wetering, S., Groenendijk, M. & Bhattacharya, P. 2014. Advanced oOxidation-Coagulation-Filtration (AOCF) – an innovative treatment technology for targeting drinking water with $<1\ \mu g/L$ of arsenic. *5th International Congress on Arsenic in the Environment, As 2014, Buenos Aires, Argentina, 11–16 May 2014*. CRC Press, pp. 817–819.

Bissen, M. & Frimmel, F.H. 2003. Arsenic – a review. Part II: Oxidation of arsenic and its removal in water treatment. *Acta Hydrochim. Hydrobiol.* 31(2): 97–107.

Gude, J.C.J., Rietveld, L.C. & van Halem, D. 2016. Fate of low arsenic concentrations during full-scale aeration and rapid filtration. *Water Res.* 88: 566–574.

WHO (2011) Arsenic in drinking-water: background document for development of WHO guidelines for drinking-water quality. WHO Press, Geneva, Switzerland.

Environmental Arsenic in a Changing World –
Zhu, Guo, Bhattacharya, Ahmad, Bundschuh & Naidu (Eds)
ISBN 978-1-138-48609-6

Reduction of low arsenic concentrations in drinking water to below 1 μg L^{-1} by adsorption onto granular iron (hydr)oxides

A. Jeworrek[1], A. Ahmad[1,2,4], B. Hofs[3], J. van Mook[3] & A. van der Wal[1,3]
[1]*Department of Environmental Technology, Wageningen University and Research (WUR), Wageningen, The Netherlands*
[2]*KWR Water Cycle Research Institute, Nieuwegein, The Netherlands*
[3]*Department of Watertechnology, Evides Waterbedrijf, Rotterdam, The Netherlands*
[4]*KTH-International Groundwater Arsenic Research Group, Department of Sustainable Development, Environmental Science and Engineering, KTH Royal Institute of Technology, Stockholm, Sweden*

ABSTRACT: Arsenic in drinking water should be reduced as much as possible, because it is amongst carcinogenic substances. FerroSorp® Plus, Huijbergen and Spannenburg GIH could reduce As(V) levels in Ouddorp water below the desired 1 μg L^{-1}. Particularly Huijbergen GIH had according to non-linear Freundlich modeling the higher K_F value, translating into a potentially higher effective adsorption capacity. One of the characteristics that probably determines the success of a GIH is its physical properties. Since the effective adsorption capacity increases with higher initial As(V) concentrations, Huijbergen GIH is a promising adsorbent to reduce a wide range of arsenic concentrations from drinking water. Therefore, this affordable technology is not only reducing ultra-low arsenic concentrations to even lower concentrations in Ouddorp water but is also promising for developing countries that are seriously affected by high concentrations of arsenic in their drinking water.

1 INTRODUCTION

More than 100 million people worldwide are drinking As contaminated water (Mandal and Suzuki, 2002). Arsenic (As), in concentrations lower than the WHO guideline (10 μg L^{-1}), may be toxic to humans (Ahmad *et al.*, 2014; Banerjee *et al.*, 2008). At water treatment plant (WTP) Ouddorp in The Netherlands the produced drinking water contains 3.3 μg L^{-1} arsenate (As(V)). According to Vewin (2016), As concentrations in the Dutch drinking water need to be below 1 μg L^{-1} by 2020. One of the low-cost options to reduce As concentrations below 1 μg L^{-1}, is to make use of the adsorption potential of the iron (hydr)oxide-based groundwater treatment residuals (Ahmad *et al.*, 2017; Hofman-Caris *et al.*, 2017; Ociński *et al.*, 2016). In this study we have investigated the As(V) adsorption potential of GIHs originating from different sources.

2 METHODS

2.1 *Adsorption isotherms & Freundlich model fitting*

In order to determine the adsorption capacities of the GIHs, batch adsorption isotherm experiments were carried out with equilibration time of 114 h. GIHs including FerroSorp® Plus, Huijbergen and Spannenburg were compared. As(V) loadings of 1.1, 11.1, 111.1, 1111.1, 11111.1 and 50001.1 μg As(V) per g GIH were tested at 15°C using Ouddorp water containing only As(V). The experimental results were fitted with the linear and non-linear Freundlich model. K_F and n from the linear plotting ($\log q_e = n \log C_e + \log K_F$) where used for non-linear fitting where the effective adsorption capacity was calculated with the following non-linear Freundlich equation: $q_e = K_F C_e^n$.

3 RESULTS AND DISCUSSION

FerroSorp® Plus Spannenburg and Huijbergen GIH succeeded in reducing the As(V) levels below 1 μg L^{-1}, with residual As concentrations reaching down to 0.66, 0.68 and 0.86 μg L^{-1} respectively (Fig. 1). At an initial As(V) concentration of 3.3 μg L^{-1} the effective adsorption capacities per 3 g of GIH were around 2 μg As(V) per g GIH for all three adsorbents. All three GIHs can reduce initial As(V) concentrations of 300 μg L^{-1} to below the WHO maximum permitted concentration of 10 μg As L^{-1} (WHO, 2011). The effective adsorption capacities at 300 μg As(V) L^{-1} initial As(V) concentrations are around 98 to 100 μg As(V) g^{-1} GIH, making this adsorption technique applicable for a wide range of As(V) concentrations.

Due to a change in chemical equilibrium at higher initial As(V) concentrations in solution, the

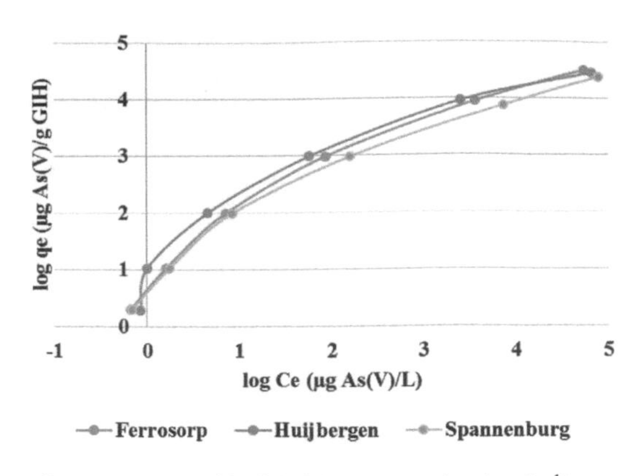

Figure 1. Log residual As(V) concentration (µg L^{-1}) versus log effective adsorption capacity (µg As(V) g^{-1} GIH) of FerroSorp® Plus, Huijbergen GIH and Spannenburg GIH in Ouddorp water with a pH of 7.8; experimental data.

Table 1. Values for n, K_F, log SSE and R^2 of the GIH in non-linear model fitting.

GIH	n	K_F^*	log SSE	R^2
Ferrosorp plus	0.79	13.3	0.14	0.96
Huijbergen	0.75	20.8	0.18	0.94
Spannenburg	0.73	12.4	0.12	0.96

* $(\mu g/g)/(\mu g/L)^n$.

experimental effective adsorption capacity of a GIH increases. Huijbergen GIH showed the highest effective adsorption capacity at the different initial As(V) concentrations (Fig. 1).

Huijbergen GIH featured, according to non-linear model fitting the highest K_F value (Table 1), translating into the highest effective adsorption capacity, making it the most suitable GIH to adsorb As(V) in Ouddorp water.

The better effective performance of Huijbergen GIH can probably be explained by its physical parameters: a comparably higher surface area and a more suitable pore size distribution for As(V) adsorption. At pH of 7.8, potential adsorption competition of same charged species (silicate, phosphate) as As(V) was seen, which indicates that the pH and the water matrix collude in determining the effectiveness of an adsorbent.

4 CONCLUSIONS

All tested GIH achieved the desired residual As(V) concentrations below 1 µg L^{-1}. Nevertheless, Huijbergen GIH showed the higher effective adsorption capacity according to non-linear Freundlich fitting. GIHs, especially HUY, are therefore not only applicable to reduce low arsenic concentrations below 1 µg L^{-1}, but also to reduce arsenic concentrations of at least 300 µg L^{-1}, which makes this affordable adsorption technology interesting for countries such as India or Bangladesh which suffer from high arsenic concentrations in their drinking water sources.

ACKNOWLEDGEMENTS

The authors thank the TKI program of the Dutch Government project for partially financing this research. This work was performed in collaboration with Evides Water Company and KWR Water Cycle Research Institute. The authors like to thank the Arsenic reduction team of Evides for the shared knowledge and input.

REFERENCES

Ahmad, A., Van De Wetering, S., Groenendijk, M. & Bhattacharya, P. 2014. Advanced Oxidation-Coagulation-Filtration (AOCF) – an innovative treatment technology for targeting drinking water with <1 µg/L of arsenic. *5th International Congress on Arsenic in the Environment, As 2014, Buenos Aires, Argentina, 11–16 May 2014*. CRC Press, pp. 817–819.

Ahmad, A., Richards, L.A. & Bhattacharya, P. 2017. Arsenic remediation of drinking water: an overview. In: P. Bhattacharya, D.A. Polya & D. Jovanovic (eds) *Best Practice Guide on the Control of Arsenic in Drinking Water*. IWA Publishing UK, pp. 79–98.

Banerjee, K., Amy, G., Prevost, M., Nour, S., Jekel, M., Gallagher, P. & Blumenschein, C. 2008. Kinetic and thermodynamic aspects of adsorption of arsenic onto granular ferric hydroxide (GFH). *Water Res.* 42(13): 3371–3378.

Hofman-Caris, R., Ahmad, A., Siegers, W., Rahn, S., Voort, J. & van der Kolk, O. 2017. *Pelletized drinking water treament residuals for the removal of arsenic and phosphate from water.* KWR Water Research Institute.

IARC. 2012. Arsenic and arsenic compounds. *IARC*, 100 C 6.

Mandal, B. & Suzuki, K. 2002. Arsenic round the world: a review. *Talanta* 58(1): 201–235.

Ociński, D., Jacukowicz-Sobala, I., Mazur, P., Raczyk, J. & Kociołek-Balawejder, E. 2016. Water treatment residuals containing iron and manganese oxides for arsenic removal from water – characterization of physicochemical properties and adsorption studies. *Chem. Eng. J.* 294: 210–221.

Vewin. 2016. Kerngegevens Drinkwater.

WHO. 2011. Guidelines for drinking-water quality, 4th Edition. World Health Organization, Geneva.

Environmental Arsenic in a Changing World –
Zhu, Guo, Bhattacharya, Ahmad, Bundschuh & Naidu (Eds)
ISBN 978-1-138-48609-6

Mitigating of arsenic accumulation in rice (*Oryza sativa* L.) from typical arsenic contaminated paddy soil of southern China using α-MnO_2 nano-flowers: pot experiment and field application

B. Li, S. Zhou & M. Lei
College of Resource & Environment, Hunan Agricultural University, Changsha, P.R. China

ABSTRACT: Hunan province of China is famous in the world for its plenty of non-ferrous metal ore (tungsten, bismuth, realgar) deposit. In the past several decades, intensive mineral exploitation, ore extraction, and refining activities has caused a large amount of waste such as mining waste water, together with the dumping of mining and smelting waste (arsenic residue, arsenic ash) on the soil surface and subjected to erosion and weathering processes, greatly affected the surrounding area and causing long-term and severe impact on soil and water environment. Pollution of paddy fields by Arsenic caused a severe threat to human health in Hunan province, because rice is a staple food and could accumulates higher levels of As compared to other crops. In the present study, a novel Nano manganese oxide with higher reactivity and oxidation ability has been synthesized. Pot experiment combined with field application has been conducted to explorer the potential of α-MnO_2 Nano manganese oxides for the remediation of As polluted paddy soil. Results turns out that the application of α-MnO_2 Nano-flowers can effectively control the transportation and accumulation of arsenic from soil to rice.

1 INTRODUCTION

The biogeochemical cycle of As in paddy soils at varies conditions (pH, Eh, periodically flooded) was rather complexity. As mobilization in flooded paddy environment was subjected to two main processes. First, the reductive dissolution of iron (oxyhydr) oxides triggered by soil flooding which caused the sorbed solid phase As releasing into the liquid phase. Second, the sorbed arsenates were reduced to arsenite under the reductive circumstances and the latter has a greater tendency to partition into the liquid phase than arsenate. Thus, effective measures must be taken to reduce the bio-availability and mobility of As in paddy soils (Suda and Makino, 2016; Xu *et al.*, 2017; Yu *et al.*, 2017).

The major goal of this present study was to investigate the potential of synthetic α-MnO_2 nano-flowers for the mitigation of As accumulated in rice. A series of experiments, including soil incubation, green house pot experiments combined with filed applications, were designed and conducted using a typical As contaminated paddy soil from southern China in Hunan Province.

2 METHODS/EXPERIMENTAL

2.1 *Pot experiment design*

The pot experiment was carried out in a greenhouse of Hunan agricultural university. The soil was amended with α-MnO_2 Nano-flower at rates of 0.0%, 0.2%, 0.5%, 1.0% and 2.0% of soil weight. Each pot was then saturated with distilled water and drained down to an equilibrium state for 7 days under natural conditions. All the treatments were triplicated and randomly arrangement. Rice seedlings were transplanted to each pot after 30 days. During the whole growth period, all pots were irrigated with distilled water daily to maintain water level of 3 cm above the soil surface. Rice was harvested at day 105. At each growth stage (tillering, heading and maturing stages) rice plants will collected, digested, and the concentrations of Mn and As in the organs (roots, stems, leaves and grains) of rice plants were analyzed. The pore water was extracted during the same time and the concentrations of Fe, Mn, As will analyzed.

2.2 *Field application experiment design*

The field experiment was carried out in 2013 at a paddy field in Dengjiatang, Chenzhou City, Hunan Province, (25°36′N, 113°00′E). NanO-MnO_2 were added into the soil at a rate of 0.0%, 0.2%, 0.5%, 1.0% and 2.0% of soil weight, respectively, and mixed thoroughly with top soil. All treatment was conducted in triplicate with a completely randomized factorial design, and three blank control is provided. Rice plants were harvested at each growth stage (tillering, heading and maturing stages). the concentrations of Mn and As in the organs (roots, stems, leaves and grains) of rice plants were analyzed. The pore water was extracted during the same time and the concentrations of Fe, Mn, As will analyzed.

3 RESULTS AND DISCUSSION

3.1 *Characterization of synthesized α-MnO_2 nano-flowers*

High-magnification SEM images show that the α-MnO_2 is a flower-like spherical with diameter of 1–1.5 μm, which consists of several straight and

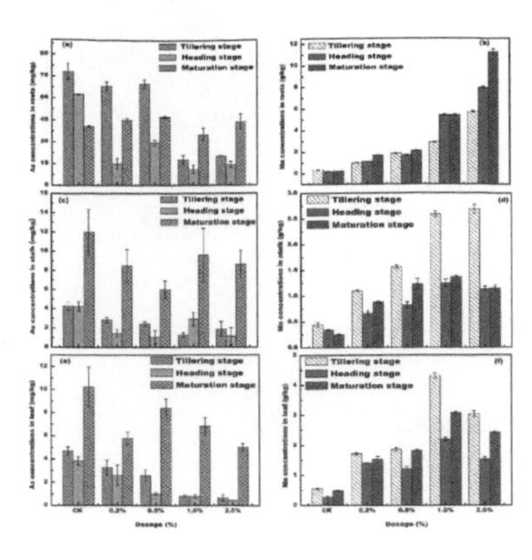

Figure 1. Concentrations of As (a) and Mn (b) in rice stalk under different dosage of α-MnO_2.

radially grown nano-rods with uniform diameter around 30–40 nm. The synthesized α-MnO_2 nanoparticles has an obvious characteristic diffraction peak in the XRD pattern, which indicates that the high crystallization degree of nanometer manganese dioxide, and the characteristic peak was in agreement with the standard data given in its JCPDS card (24-0072).

3.2 *Pot experiment*

In greenhouse pot experiments, whether or not to add α-MnO_2 Nano-flowers material, the content of arsenic in rice was in the order of root > stem > leaf > husk > brown rice. The As concentrations in soil solution decreased with the increasing dosage of 0.5%, 1.0%, and 2.0% in soils and thus greatly reduced the accumulation of As in rice. The relationship between manganese and arsenic in rice plants showed a significant negative correlation ($p < 0.05$), indicated that the concentration of arsenic in rice decreased with the increasing amounts of manganese in rice. The total As concentration in brown rice was reduced by 65.4% with the addition of 1.0% α-MnO_2 nano-flowers, and the corresponding inorganic As content was decreased to 0.19 mg kg^{-1} which lower than the limited content of inorganic As in food.

3.3 *Field application*

In field experiment, the concentrations of As in roots were 2–7 times greater than the background value of the soil, and significantly higher than those in the stems and leaves, which demonstrated that the content of As in rice followed the order of root > stem > leaf > husk > brown rice (Table 1). There was a significant positive correlation between amount of iron plaque and As concentration in root at mature stage.

Interestingly, the amounts of iron plaque decreased with increasing of α-MnO_2 nano-flowers, indicated that there was more Mn plague deposited on the rice roots. Meanwhile, a significant correlation between trans-location factors of As from iron plaque, root, stem and leaf to seed in mature stage has been observed, which also indicated that there was a significant difference in As concentrations between same part of rice with different α-MnO_2 nano-flowers treatments ($p < 0.05$). The content of arsenic in rice were below the limit of arsenic in food with the dosage of 0.2%, 0.5% and 2.0%.

Table 1. Concentrations of As and Mn in rice husk and brown rice (field application).

		Total As/mg kg^{-1} (n = 3)		
	Treatments	Tillering stage	Heading stage	Maturation stage
Root	CK	476.20 ± 35.69	370.99 ± 42.36	552.57 ± 58.20
	0.2%	478.37 ± 42.36	380.30 ± 18.96	433.95 ± 32.15
	0.5%	297.53 ± 18.54	266.91 ± 32.58	368.46 ± 18.94
	1.0%	382.23 ± 21.95	126.28 ± 21.25	446.39 ± 45.96
	2.0%	153.45 ± 10.23	110.19 ± 10.21	321.93 ± 35.68
Stalk	CK	4.46 ± 2.32	7.12 ± 0.84	18.81 ± 2.65
	0.2%	0.89 ± 0.54	1.00 ± 0.08	8.64 ± 4.51
	0.5%	0.74 ± 0.36	1.21 ± 0.54	4.92 ± 1.20
	1.0%	2.43 ± 1.03	0.64 ± 0.24	6.10 ± 1.36
	2.0%	0.97 ± 0.38	1.24 ± 0.21	5.32 ± 0.98
Leaf	CK	4.73 ± 1.55	3.69 ± 1.27	12.11 ± 4.03
	0.2%	3.37 ± 0.23	1.92 ± 0.89	3.17 ± 0.25
	0.5%	1.56 ± 0.24	2.59 ± 1.23	1.77 ± 0.15
	1.0%	2.06 ± 1.03	0.66 ± 0.32	8.96 ± 1.08
	2.0%	1.17 ± 0.12	1.14 ± 0.11	6.24 ± 0.94

4 CONCLUSIONS

In this study, the potential of synthetic α-MnO_2 Nano-flowers for the remediation of arsenic polluted paddy soil has been investigated. Results turns out that the application of α-MnO_2 Nano-flowers can effectively reduce the bio-availability of As in paddy soil and control the transportation and accumulation of As in rice. These results highlight the potential of α-MnO_2 Nano-flowers to mitigate the phyto-accumulation of arsenic by rice and to thereby reduce As exposure risk associated with rice consumption.

ACKNOWLEDGEMENTS

The authors are grateful for financial support from the National Science Foundation of China (41671475/21007014).

REFERENCES

Suda, A. & Makino, T. 2016. Functional effects of manganese and iron oxides on the dynamics of trace elements in soils with a special focus on arsenic and cadmium: a review. *Geoderma* 270: 68–75.

Xu, X., Chen, C., Wang, P., Kretzschmar, R. & Zhao, F.J. 2017. Control of arsenic mobilization in paddy soils by manganese and iron oxides. *Environ. Pollut.* 231(Pt 1): 37–47.

Yu, Z., Qiu, W., Wang, F., Lei, M., Wang, D. & Song, Z. 2017. Effects of manganese oxide-modified biochar composites on arsenic speciation and accumulation in an indica rice (*Oryza sativa* L.) cultivar. *Chemosphere* 168: 341–349.

Environmental Arsenic in a Changing World –
Zhu, Guo, Bhattacharya, Ahmad, Bundschuh & Naidu (Eds)
ISBN 978-1-138-48609-6

Phycoremediation of arsenic by *Chlorella* sp. CB4

M.O. Alam, S. Chakraborty & T. Bhattacharya
Department of Civil and Environmental Engineering, Birla Institute of Technology, Mesra, India

ABSTRACT: At the dawn of 21st century, phycoremediation has received significant attention as a feasible technique for arsenic removal from contaminated water. Here a study was attempted to assess the potential of microalgae *Chlorella* sp. CB4 to remove As(III) and As(V) from aqueous phase in optimized conditions. Removal efficiency achieved for As(V) was 87% and for As(III) was 86% in optimized condition. In groundwater the removal efficiency was 64% for As(V) and 75% for As(III). *Chlorella* sp. CB4 provided a viable option for treatment of As laden water.

1 INTRODUCTION

Interests seemed to rise in applying biological approaches for arsenic removal with plants (terrestrial and aquatic) and a plethora of microorganisms. It offered advantage over other methods being efficient, environmentally sustainable, and socio-economically acceptable. In recent days attempts were made to use microalgae, living cells or dead biomass for removing heavy metals from contaminated water (Kumar and Oommen, 2012; Podder and Majumder, 2015). Microalgae living at the very bottom of the food chain, are widely dwelling in terrestrial and aquatic environments and play an important role in arsenic cycling (Duncan *et al.*, 2015; Wang *et al.*, 2014). They reduce arsenic toxicity by transforming pollutants into non-hazardous materials by certain mechanism like cell surface binding (adsorption), As(III) oxidation, As(V) reduction, methylation, transformation into arsenosugars or arsenolipids, chelation of As(III) with glutathione and phytochelatins, as well as excretion from cells, etc. (Jiang *et al.*, 2011) Several flocculating microalgae have also been reported, such as *Chlorella vulgaris, Ettliatexensis*, and *Scenedesmus obliquus* to remove arsenic from aqueous phase. Present study was attempted to assess the potential of microalgae *Chlorella* sp. CB4 to remove As(III) and As(V) from aqueous phase in optimized conditions.

2 EXPERIMENTAL

Chlorella sp. CB4 (Gene Bank ID No. JQ710683; NCBI databank), a freshwater green microalga was collected from the Bioengineering repository, BIT, Mesra, Jharkhand, India and grown in sterile Tris Acetate buffer (TAP) media. The culture was incubated at 25°C under light (8000 lux) with photoperiod of 16:8 h (light: dark). In addition to simple microscopic examination samples were examined under SEM to observe morphological changes in cell structure and/or after adsorption. FTIR in infrared spectra region 4000–500 cm^{-1} was used to determine the responsible functional group for adsorption.

Batch experiments were performed for arsenic removal from the artificial TAP media to investigate their removal potentiality of As(V) and As(III) and optimize initial concentration of As(V) and As(III), pH of the media, temperature and inoculum dose for best removal efficiency. The samples were collected from the culture broth at 3 days interval initially for As(III) but later it was observed that removal for As(V) was a rapid process and hence it was harvested every 8 hours up to 3 days to see maximum removal. The residual As(III) and As(V) concentrations in the culture broth were analyzed in ICP-OES (Perkin Elmer Optima 2100DV). Algal removal of arsenic was finally conducted with two experimental set up in glass tank (30 cm × 30 cm × 20 cm dimensions' volume 18000 cm^3). 5 L of As free groundwater (tubewell) was dosed with 1 mg L^{-1} of $NaAsO_2$ and Na_3AsO_4 for As(III) and As(V), respectively. Reaction conditions, optimized in the batch experiments were maintained throughout the removal process. Samples were collected at an interval of 8 h from As(III) set up (up to 72 h) and 3 days from As(V) set up (up to 24 days).

The metal ion removal efficiency was calculated using equation (1) as follows (Ji *et al.*, 2011).

$$E= (C_0-C_f / C_0) x100 \quad (1)$$

where E represents: removal efficiency of metal ions (%); C_0 and C_f: initial and final concentrations of As (mg L^{-1}) in the medium respectively.

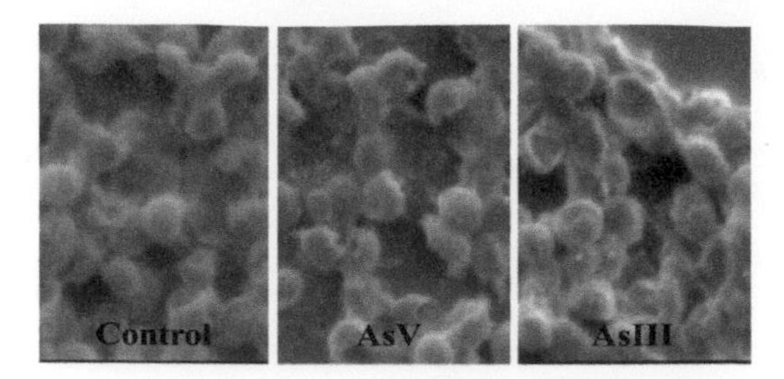

Figure 1. SEM image of microalgae *Chlorella* sp. CB4 before adsorption (control), after adsorption of As(V) and As(III).

Table 1. Optimized physical environmental condition.

	Optimized condition	
Parameters	As(V)	As(III)
Initial concentration	$1\ mg\,L^{-1}$	$1\ mg\,L^{-1}$
pH	pH 7	pH 7
Temperature	25°C	25°C
Inoculum dose	1:50	1:10
Final removal	87%	86%

3 RESULTS AND DISCUSSION

3.1 *Characterization*

In SEM images, the smooth and homogenous surface of Chlorella sp. CB4 turned out to be rough and heterogeneous on adsorption of As(III) and As(V) (Fig. 1). Presence of As did not affect the density of the nodules, suggesting that the growth of the nodular consortium remains unaffected by the presence of As in the growth medium. The nodules appeared to be glued which might be due to the excessive extra-cellular polysaccharides (EPS) production, a well-known response against stress (Podder and Majumder, 2016).

EDX confirms the adsorption of As(III) and As(V) in the surface of the algae. FTIR spectrum of Chlorella sp. CB4 suggested that C-H bond of CH_2 groups, hydroxyl and carboxyl groups are chiefly responsible for arsenic blinding.

3.2 *Optimization of adsorption parameters*

Removal efficiency achieved for As(V) was 87% and for As(III) was 86%. The optimized environmental conditions are presented in Table 1.

3.3 *Kinetics study*

The kinetics study confirmed that the current biosorption was best fitted to pseudo-first order kinetic model for both arsenic species (As(III) – $R^2 = 0.8761$) and (As(V) – $R^2 = 0.0961$).

3.4 *Glass tank experiment*

In true groundwater, when the removal experiment was conducted with the optimized parameters from the batch experiments, a maximum As(V) removal of 64% was obtained on 64th hour of experiment. Similarly, the maximum As(III) removal of 76% was obtained on 20th day of experiment.

4 CONCLUSION

The experimental results of phycoremediation of As by living microalgae *Chlorella* sp. CB4 from aqueous phase proved the technique to be feasible, which could remove up to 64% of As(V) and 75% of As(III) from groundwater in optimum conditions. However, for making the water fit for consumption, it has to undergo the removal process twice, so that, the residual concentration of arsenic is within the permissible limit. Further, it can be concluded that, Chlorella sp. CB4 can be a viable option for treatment of As laden water due to their rapid growth rate and good removal capacity.

ACKNOWLEDGEMENTS

Department of Science and Technology, New Delhi is highly acknowledged for funding the SERB Young Scientist Project SR/FTP/ES-2/2013 for this study.

REFERENCES

Duncan, E., Maher, W. & Foster, S. 2015. Contribution of arsenic species in unicellular algae to the cycling of arsenic in marine ecosystems. *Environ. Sci. Technol.* 49(1): 33–50.

Ji, L., Xie, S., Feng, J., Li, Y. & Chen, L. 2011. Heavy metal uptake capacities by the common fresh water green alga *Cladophora fracta*. *J. Appl. Phycol.* 24(4): 979–983.

Jiang, Y., Purchase, D., Jones, H. & Garelick, H. 2011. Effects of arsenate (As^{5+}) on growth and production of glutathione (GSH) and phytochelatins (PCS) in Chlorella vulgaris. *Int. J. Phytoremed.* 13(8): 834–844.

Kumar, J.I. & Oommen, C. 2012. Removal of heavy metals by biosorption using freshwater alga Spirogyra hyalina. *J. Environ. Biol.* 33(1): 27–31.

Podder, M.S. & Majumder, C.B. 2015. Phycoremediation of arsenic from wastewaters by Chlorella pyrenoidosa. *Groundwater Sust. Develop.* 1(1–2): 78–91.

Podder, M.S. & Majumder, C.B. 2016. Arsenic toxicity to *Chlorella pyrenoidosa* and its phycoremediation. *Acta Ecol. Sin.* 36(4): 256–268.

Wang, N.X., Huang, B., Xu, S., Wei, Z.B., Miao, A.J., Ji, R. & Yang, L.Y. 2014. Effects of nitrogen and phosphorus on arsenite accumulation, oxidation, and toxicity in *Chlamydomonas reinhardtii*. *Aquat. Toxicol.* 157: 167–174.

Environmental Arsenic in a Changing World –
Zhu, Guo, Bhattacharya, Ahmad, Bundschuh & Naidu (Eds)
ISBN 978-1-138-48609-6

Arsenic removal from natural contaminated groundwaters in Calabria Region (Italy) by nanofiltration

A. Figoli[1], I. Fuoco[2], C. Apollaro[2], R. Mancuso[3], G. Desiderio[4], R. De Rosa[2], B. Gabriele[3] & A. Criscuoli[1]

[1]*Institute on Membrane Technology (ITM-CNR), Rende (CS), Italy*
[2]*Department of Biology, Ecology and Earth Sciences (DIBEST), University of Calabria, Rende (CS), Italy*
[3]*LISOC Group, Department of Chemistry and Chemical Technologies, University of Calabria, Rende (CS), Italy*
[4]*DeltaE, University of Calabria, Rende (CS), Italy*

ABSTRACT: Nanofiltration (NF) has been applied for treating arsenic contamination of groundwater from an area located in the Sila Massif in Calabria (Italy). High values of As in different environmental matrices, such as rocks, soils and waters, were observed. In fact, the geochemical analysis of 3 groundwaters (GWs) samples collected show As values which exceeded the permitted levels of $10\,\mu g\,L^{-1}$. The Arsenic is present mainly in the pentavalent form (As(V)). Different types of NF membranes, operating conditions and the competing effects of different ions with Arsenic in the GWs samples have been evaluated for analyzing their effect on the arsenic removal.

1 INTRODUCTION

Arsenic is a toxic inorganic pollutant for both the environment and human health. The removal of arsenic is one of most problematic targets of hydrogeochemical research (Ahmad *et al.*, 2017; Bhattacharya *et al.*, 2002; Mohan and Pittman, 2007; Smedley and Kinniburgh, 2002). The WHO drinking water guideline for As has been set to $10\,\mu g\,L^{-1}$ and it has been adopted by many countries as a drinking water standard (WHO, 2001). For reaching this new value, several separation techniques have been studied. Membrane technology is one of these and pressure driven operation as nanofiltration (NF) and reverse osmosis (RO), often applied on synthetic solutions, have been already proved to accomplish the arsenic value below the drinking water standard (Ahmad *et al.*, 2017; Figoli *et al.*, 2016). In this study, NF has been applied for treating As contaminated groundwaters, coming from an area located in the Sila Massif (Calabria, Italy). Sila Mas-sif represents the major morphostructural high of the Ionian margin of northeastern Calabria and fall in the northern sector of the Calabrian Peloritan Arc (CPA). The water samples collected, labeled GW1, GW2, GW3, have interacted with the Calabride Complex formed by Hercynian and pre-Hercynian gneiss, granite and phyllite, which underwent intense weathering processes (Van Dijk *et al.*, 2002). The performance (water flux and arsenic rejection) of the membranes has been evaluated changing the operating pressures.

2 METHODS

The three water samples (GW1, GW2, GW3) differ by the arsenic concentration, which is about 60, 120 and $430\,\mu g\,L^{-1}$, respectively. The arsenic is present mainly in the pentavalent form (As(V)).

NF experiments were performed by using a laboratory pilot unit (SEPA CF), consisting of three major components: a cell body, cell holder, and feed pump. Four types of membrane modules commercialized by GE Osmonics, named HL (polyamide), DK (proprietary thin-film) and CK (cellulose acetate) and by Microdyn Nadir, named NP030P (polyethersulfone), were used.

The samples, before and after membrane treatment, have been analyzed, determining the major elements HPLC (Dionex ICS 1100). The total arsenic was evaluated by ICP-MS, Perkin Elmer/SCIEX, Elan DRCe.

3 RESULTS AND DISCUSSION

In Figure 1, it is reported the water flux plotted versus the trans-membrane pressure (TMP) variation, for the GW2 sample (feed stream).

All the investigated membranes show a linear increase of water flux at higher TMP. In particular, only the NF membrane Type HL shows much higher water flux compared to the other ones, which could be

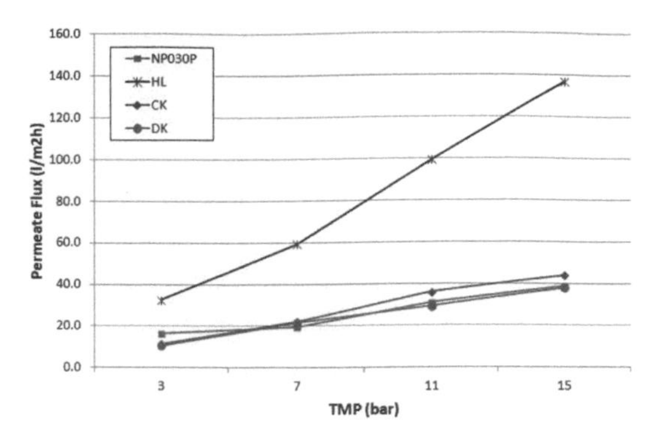

Figure 1. Effect of TMP on permeate flux for GW2 contaminated As(V) sample.

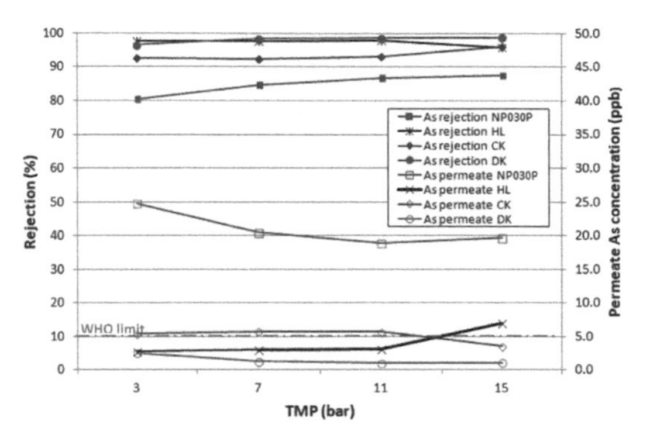

Figure 2. Effect of TMP on rejection (%) and permeate arsenic concentration for GW2 contaminated As(V) sample.

explained by the polymeric nature of the material as well as by its higher pore-size. The same results have obtained also for the other GW samples containing a lower and higher concentration of As(V).

In Figure 2, it is reported the As(V) rejection and the arsenic concentration detected in the permeate (purified water) after the membrane treatment of GW2 contaminated As (V) sample.

The results show that all the investigated NF membranes, except the NP030P, reject the As(V) and that the As(V) concentration in the purified permeate water is below the WHO drinking water limit value of $10\,\mu g\,L^{-1}$.

4 CONCLUSIONS

The results clearly report that NF can be considered a valid technique for arsenic removal in the natural water, producing a permeate (purified water) with the As(V) concentrations within the allowed WHO limits for most of the membrane investigated.

ACKNOWLEDGMENT

The work has been supported by the project "AsSe" n. CUP: J28I17000030006, cofunded by Fondo FESR POR Calabria FESR FSE 2014-2020-Azione 1.2.2.

REFERENCES

Ahmad, A., Richards, L.A. & Bhattacharya, P. 2017. Arsenic remediation of drinking water: an overview. In: P. Bhattacharya, D.A. Polya & D. Jovanovic (eds) *Best Practice Guide on the Control of Arsenic in Drinking Water. Metals and Related Substances in Drinking Water Series*. IWA Publishing, UK, pp. 79–98.

Bhattacharya, P., Frisbie, S.H., Smith, E., Naidu, R., Jacks, G. & Sarkar, B. 2002. Arsenic in the environment: a Global perspective. In: B. Sarkar (ed) *Handbook of Heavy Metals in the Environment*, Marcell Dekker Inc., New York, pp. 147–215.

Figoli, A., Hoinkis, J. & Bundschuh, J. 2016. *Membrane Technology for Water Treatment, Removal of Toxic Trace Elements with Emphasis on Arsenic, Fluoride and Uranium*. CRC Press, Boca Raton, FL.

Mohan, D. & Pittman, C.U. Jr. 2007. Arsenic removal from water/wastewater using adsorbents – a critical review. *J. Hazard. Mater.* 142(1–2): 1–53.

Smedley, P.L. & Kinniburgh, D.G. 2002. A review of the source, behaviour and distribution of arsenic in natural. waters. *Appl. Geochem.* 17(5): 517–568.

Van Dijk, J.P., Bello, M., Brancaleoni, G.P., Cantarella, G., Costa, V., Frixa, A., Golfetto, F., Merlini, S., Riva, M., Torricelli, S., Toscano, C. & Zerilli, A. 2000. A regional structural model for the northern sector of the Calabrian Arc (southern Italy). *Tectonophysics* 324: 267–320.

WHO 2001. Guideline for drinking-water quality: arsenic in drinking water. Fact Sheet No. 210. World Health Organization, Geneva. Switzerland.

Environmental Arsenic in a Changing World –
Zhu, Guo, Bhattacharya, Ahmad, Bundschuh & Naidu (Eds)
ISBN 978-1-138-48609-6

Towards harmonizing approaches for scaling up access to arsenic safe water in Bangladesh: The Arsenic Safe Village Concept

N. Akter[1], B. Onabolu[1], M. Bolton[2] & H. Sargsyan[1]
[1] *UNICEF Bangladesh, Dhaka, Bangladesh*
[2] *Golder Associates Ltd., Canada*

ABSTRACT: Although 97.8% of the population in Bangladesh haves access to improved water sources, 19.7 million people use water sources that are contaminated above the Government of Bangladesh (GOB) standards for arsenic (As, 50 μg L^{-1}), posing a significant health risk. There were many initiatives taken by Government and donors on As mitigation, however the situation did not improve significantly. UNICEF Bangladesh with partners implemented a new approach of As mitigation by addressing entire villages to provide As-safe water and with improved sanitation and hygiene behavior. Rather than thinking about individual water points, consider the entire population in a village. The core objectives of this approach are a quantitative, measurable, reduction in As exposure, and the aim is to building up areas of contiguous As safety in the sequence of village to union, union to upazila and then to district. Using this approach 126 villages have been declared 'Arsenic-safe village with improved sanitation and hygiene'.

1 INTRODUCTION

According to the MICS 2012–2013 report, nationwide, 19.7 million people use water sources that are contaminated above the Government of Bangladesh (GOB) standards for arsenic (As, 50 μg L^{-1}) and twice that number drink water with levels above the WHO guidelines (10 μg L^{-1}).

Arsenic mitigation in Bangladesh has proceeded piecemeal, through installation of devices that serve 100–200 persons or less, in villages of typically 2,000–5,000 people, resulting in a patchy distribution of safe water supplies. Various alternative water supplies and As removal technologies have been endorsed in the National Policy for Arsenic Mitigation (GOB, 2004) for use in As affected areas. Deep tube-wells have generally been the preferred option, as there have been operational challenges with the As removal technologies and sustainability is uncertain. Based on the technical challenges and scale of the As contamination, As mitigation has been neglected in some areas. In the absence of specific targets, for the country or individual agencies, progress since the mid-2000s stagnated. In the meantime, several million new wells have been installed; however, many have not been tested for As.

Despite the money and effort that has been committed, attempts to resolve the As issue have lost momentum and worse, As poisoning is becoming accepted as the inevitable fate of the Bangladeshi people. A comprehensive approach to As mitigation is required to address this crisis and to provide equitable access to all members in a community, including the most vulnerable.

Figure 1. UNICEF DPHE Protocol for scaling arsenic safety in Bangladesh.

2 CONCEPT

The UNICEF DPHE Protocol was used to implement an innovative approach to As-mitigation by intervening in entire villages and promoting sanitation and hygiene. Rather than thinking about individual water points, consider the entire population in a village and construct appropriate technologies to provide effective coverage based on the site-specific information and options. The aim is to build up areas of contiguous As safety in the sequence (Fig. 1).

3 APPLICATION/ACTION

From August 2012 to December 2014, with technical, financial and monitoring support from UNICEF, a project with the specific objective of "*creating arsenic-safe model villages with sustainable improvement in water, sanitation and hygiene*" was implemented in Comilla, Brahmanbaria and Narail districts. With technical assistance from Department of Public Health Engineering (DPHE) and Local Government Institutions (LGIs), UNICEF implemented this project in partnership of two national NGOs, Village Education Resource Centre (VERC) and Environment and Population Research Centre (EPRC).

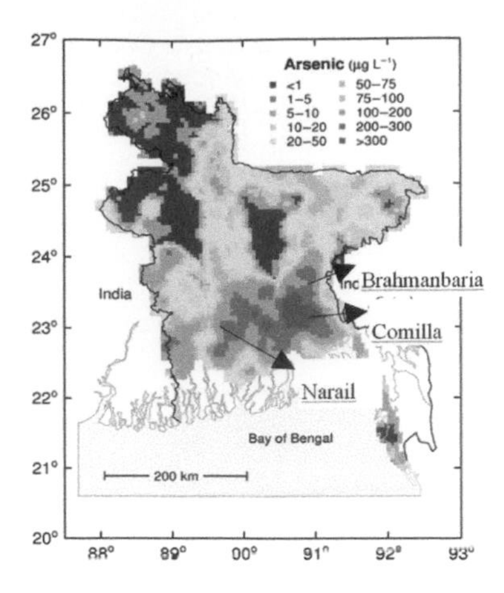

Figure 2. Targeting the high priorityintervention areas in Comilla based on vulnerability risk assessment.

4 MAJOR COMPONENTS

4.1 *Vulnerability risk assessment*

The project areas were selected on the basis of vulnerability risk assessment using ratio of As contamination, existing safe water coverage, number of arsenicosis patients and extreme poor. Selected 26 very high priority/high priority unions (DPHE and JICA, 2010) of Manoharganj upazila under Comilla, Banchharampur and Sarail upazilas under Brahmanbaria and Lohagora and Kalia upazilas under Narail District (Fig. 2).

4.2 *Community planning and leadership*

The Community Action Plan (CAP) is a key element of the project. The CAP process aims to assess the WASH situation in a community and to identify an appropriate approach to make improvements across the community, with special consideration to the needs of the poorest.

4.3 *Arsenic screening and monitoring*

To facilitate planning and target the most severely As affected people, all the tube-wells in the project areas were screened and a periodic monitoring system was established.

4.4 *Selection of appropriate technologies*

Based on the hydrogeological conditions and water quality considerations in the project areas, through community consultation different technological options are provided, including deep tube-wells, multiple connections to As-safe wells, rainwater harvesting systems, and Piped Water Supply Systems (Fig. 3).

4.5 *Water quality testing*

For each new water point, water quality testing is conducted with field test kits. 10% of the samples cross checked in a laboratory. Following confirmation of acceptable water quality, water point is handed over to the community.

Figure 3. Collecting water from arsenic safe DTW.

4.6 *Caretaker training and water safety plans*

Caretakers are trained and provided with tools for system operation and maintenance (O&M) and water safety.

4.7 *Social mobilization for sanitation and hygiene education and School programmes*

Communities are informed about the ill-effects of drinking As-contaminated water, open defecation and unhygienic behavior through different social mobilization tools and techniques and sanitation promoted.

Arsenic safe water facilities installed and sanitation blocks are constructed (with running water, separate latrines for boys and girls).

4.8 *Arsenic-safe village declaration*

When all the households of a village receive access to As-safe water and ensure use of hygienic latrines, the village is declared as an 'As-safe village with improved sanitation and hygiene'.

5 MAJOR OUTCOMES

126 villages in three districts have been declared 'As-safe with improved sanitation and hygiene' and become models for replication in other affected areas of Bangladesh. The Department of Public Health Engineering and UNICEF Bangladesh are scaling up this concept in 15 unions (Sub-districts in Bangladesh).

6 CONCLUSION

The use of a community led demand responsive approach to provide As-safe water is a critical step towards harmonizing approaches for scaling up As mitigation in drinking water in Bangladesh.

ACKNOWLEDGEMENTS

We acknowledge German National Committee (UNICEF), the Embassy of the Kingdom of The Netherlands and the Swedish International Development Cooperation Agency (Sida).

REFERENCES

DPHE & JICA 2010. Situation analysis of arsenic mitigation 2009. Local Government Division, Government of Bangladesh.

MICS 2012–2013. Bangladesh Multiple Indicator Cluster Survey 2012–2013, Progotir Pathey: Final Report. Bangladesh Bureau of Statistics (BBS) and UNICEF Bangladesh, 2014, Dhaka, Bangladesh.

Environmental Arsenic in a Changing World –
Zhu, Guo, Bhattacharya, Ahmad, Bundschuh & Naidu (Eds)
ISBN 978-1-138-48609-6

Harmonizing sector approaches for scaling up access to arsenic safe water in Bangladesh: The DPHE-UNICEF Arsenic Mitigation Protocol

B. Onabolu[1], S.K. Ghosh[2], N. Akter[1], S. Rahman[2] & M. Bolton[3]
[1] *UNICEF Bangladesh, Dhaka, Bangladesh*
[2] *Department of Public Health Engineering, Ministry of Local Government, Rural Development and Cooperatives, Dhaka, Bangladesh*
[3] *Golder Associates Ltd., Canada*

ABSTRACT: Progress in improved water coverage has not been matched by progress in access to arsenic safe water in Bangladesh for social, institutional and technical reasons, one of which is the use of different approaches by the numerous stakeholders involved in drinking water provision in arsenic contaminated areas. In response to this challenge, the Department of Public Health Engineering and UNICEF Bangladesh implemented an arsenic mitigation project in selected sub-districts in 10 severely arsenic-contaminated districts in Bangladesh using the DPHE-UNICEF arsenic mitigation protocol. the successful provision of 1733 arsenic-safe drinking water devices using this protocol highlights the crucial importance of adopting a sector wide harmonized approach and protocol to scaling up arsenic mitigation in Bangladesh.

1 INTRODUCTION

Bangladesh has made significant progress towards achieving its goal of universal access to improved water supply and improved sanitation for all its citizens. Access to improved water sources increased from 68% to 97.9% between 1998 and 2012–2013 (MICS 2012–2013), however there are challenges related to the drinking water quality. Although 97.9% of the population has access to improved water sources, about 19.7 million people are exposed to arsenic concentrations in drinking water above the Government of Bangladesh (GoB) standard (0.05 mg L^{-1}) and twice the number of people are exposed to concentrations above the WHO recommended guidelines of 0.01 mg L^{-1} (MICS 2012–2013). The government has clearly articulated its commitment to ensuring universal access to safe drinking water in various strategic documents and policies including its 7th 5-year strategic plan (FY 2016–2020). The concerted efforts of the GoB and its development partners, resulted in the reduction of the number of exposed people from 22 to 19.7 million between 2009 and 2012 (MICS 2009; 2012–2013), however the arsenic in drinking water still persists.

Some of the challenges to scaling up arsenic mitigation include the geogenic nature of arsenic contamination, the magnitude of the problem and poor private sector capacity in locating and constructing arsenic safe water options. The absence of a sectoral adopted and harmonized protocol or approach for arsenic-safe water provision also contributes significantly to arsenic contamination of water points provided by the sector. In response to this challenge, UNICEF Bangladesh developed a Protocol for Mitigating Arsenic in Drinking Water. The protocol was used by The Department of Public Health and Engineering (DPHE), and selected NGOs to rehabilitate 1733 arsenic-contaminated water points in 42 sub-districts in 10 of the most arsenic contaminated districts of Bangladesh (DPHE and JICA, 2010) between 2014 and 2015.

2 METHODS

2.1 *Overview of the protocol*

The protocol consists of five major steps integrated by a 3-tier quality assurance monitoring (Fig. 1). The steps include:

Site-specific feasibility assessment
The depth, functionality and water quality (arsenic, iron and manganese) of the contaminated well and 5–10 surrounding water points were recorded and data uploaded on the national DPHE MIS Database and GIS maps produced.

Alternative water point selection
The results of feasibility assessment were reviewed in conjunction with the use of technology selection maps, borehole logs after which recommendations for safe water options and alternative water point sites

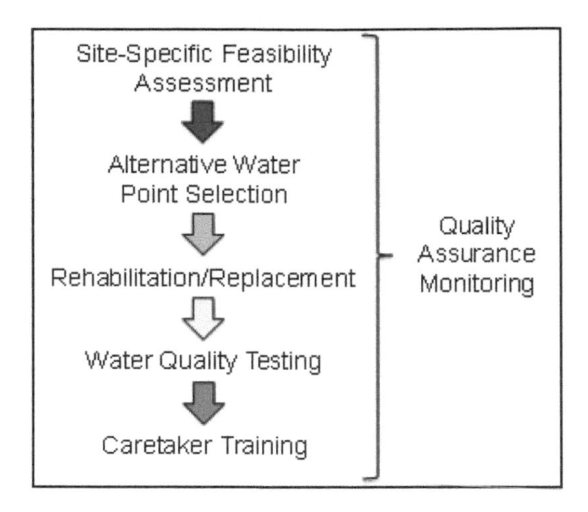

Figure 1. Arsenic Mitigation Protocol.

were proposed and reviewed by a technical review committee made up of UNICEF and DPHE.

1. Arsenic-Safe Water Options: The community was consulted, safe water options agreed and caretaker from community selected. Water points were constructed according to agreed specifications.
2. Water quality testing: Each new water point was tested by an independent 3rd party laboratory.
3. Caretaker training: Caretaker training for operation and maintenance and water safety planning was conducted. Water point was handed over to the community if water point was certified safe by third party.
4. Quality Assurance (QA) monitoring: In addition to Quality Assurance (QA) to confirm quality of the materials and workmanship. A three-tier water quality testing was conducted:
   - All samples were tested with field kits (Econo quick test kit).
   - All samples were analyzed at an independent laboratory (International Centre for Diarrhoeal Disease Research, Bangladesh; ICDDRb).
   - 20% of the samples also tested in DPHE's central laboratory.

2.2 *Mitigation options*

The approved mitigation measures comprise of switching to neighboring wells that are arsenic-safe, re-drilling wells to arsenic-safe depths; installing treatment plants/filters; and developing alternative water sources (Fig. 3). Information from the site-specific feasibility assessment underpinned the decision making.

2.3 *The use of GIS mapping for decision making*

Due to the spatial variability of arsenic contamination in Bangladesh, the area wise technology maps

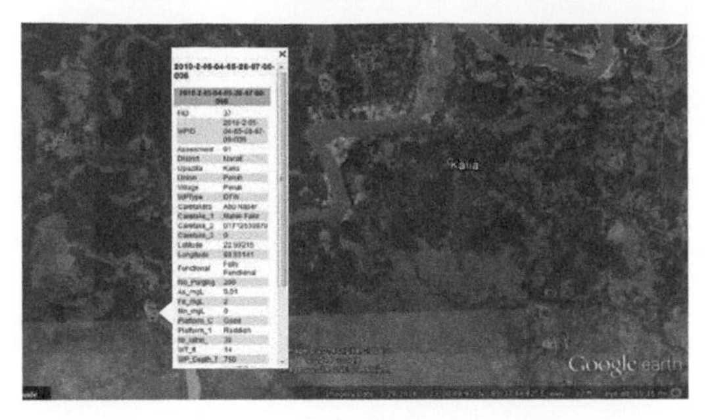

Figure 2. KMZ file for Arsenic Rehabilitation Programme water points in Kalia Upazila of Narail District viewed on Google Earth software. Feasibility assessment data for each water point can be viewed with the software to support geospatial analysis and selection of alternative waterpoints.

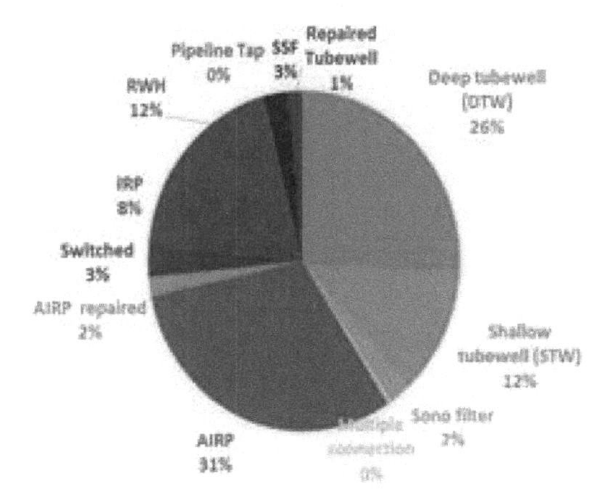

Figure 3. Distribution of arsenic mitigation options based on the DPHE-UNICEF Arsenic Mitigation Protocol.

are not sufficient for decision making. The data from site-specific feasibility assessment was converted by DPHE MIS unit to KMZ files to improve the precision of arsenic safe water provision (Fig. 2).

3 RESULTS

3.1 *Immediate outputs*

Using the protocol, the DPHE and selected NGOS were able to provide 1733 drinking water options which were certified by third party monitors (ICDDRB) as meeting the Bangladesh Standard for arsenic in drinking water. The type of options provided varied from deep tube wells (26%) to arsenic and iron removal plants (31%).

3.2 *Contributions to sector*

- Development of a 20-digit unique coding system for water points currently being used by the World Bank and JICA.

- Support to DPHE to establish/strengthen its MIS/GIS unit.
- Nationwide water point mapping of 150,000 water points installed by DPHE between 2006 and 2012.
- Strengthening of the DPHE capacity to monitor quality of water points provided.
- Production of sub-district wide technology selection maps.
- Revised DPHE guidelines on drilling contract conditions.

4 CONCLUSION

The DPHE-UNICEF arsenic mitigation protocol was successfully used by DPHE and four national NGOS to provide arsenic safe water in severely arsenic contaminated districts. This suggests that the adoption of a harmonized protocol is a critical step towards improving the sector's precision in providing arsenic safe water and scaling up arsenic mitigation in drinking water in Bangladesh.

REFERENCES

BBS/UNICEF 2014. Bangladesh Multiple Indicator Cluster Survey 2012–2013, Progotir Pathey: Final Report. Bangladesh Bureau of Statistics (BBS) and UNICEF Bangladesh, 2014, Dhaka, Bangladesh.

DPHE & JICA 2010. *Situation analysis of arsenic mitigation 2009*. Local Government Division, Government of Bangladesh.

Environmental Arsenic in a Changing World –
Zhu, Guo, Bhattacharya, Ahmad, Bundschuh & Naidu (Eds)
ISBN 978-1-138-48609-6

Iran's first waterworks with granular ferric hydroxide-based dearsenification – a look back over the first two years of operation

C. Bahr[1], F. Tarah[2] & M. Mahdyarfar[2]
[1]*GEH Wasserchemie GmbH & Co. KG, Osnabrück, Germany*
[2]*Delta Niroo Gameron Co., Tehran, Iran*

ABSTRACT: Like many other countries, Iran has numerous regions with high levels of geogenic arsenic (As) in groundwater which jeopardize safe drinking water supply in these areas. In an effort to address this problem in Iran's Kerman Province, a large-scale drinking water dearsenification plant was erected at a waterworks there. The plant, which utilizes ferric hydroxide adsorption beds making it the first of its kind in the country, has now been in operation for over two years' time. Due to the relatively high As content of the groundwater treated, the plant had to be designed with a combined process approach to realize optimum treatment performance. The process configuration selected, comprising (a) pH reduction of the inlet raw water, (b) multiple adsorber lines operated in parallel with time-staggered adsorbent exchange and (c) lead/lag configured beds and a standby bed making up each line, permits realization of long bed life of the adsorbent media used.

1 INTRODUCTION

Geogenic arsenic (As) contamination of drinking water is a health threat encountered in many regions world-wide. This is well documented, e.g. by a recently published international sourcebook (Murcott, 2012) providing updated As contamination data – which can vary substantially among regions within a given country – for a total of 105 countries around the world. The precarious situation in India and Bangladesh in this regard has been a much-discussed topic for many years. High As contamination levels have been determined in many other countries as well, calling attention to the need for urgent action in this respect.

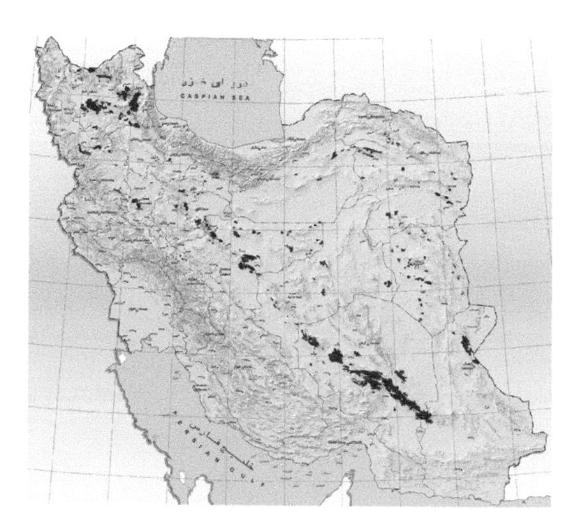

Figure 1. Regions with potentially high arsenic contamination. Geological Survey of Iran (MIM, 2015).

Iran is another example, where As concentrations ranging up to 1.48 mg L^{-1} have been recently been established in the Kurdistan Province. Earliest reports of As-caused diseases in the area date from the 1980's (Mosaferi *et al.*, 2003). Iran's groundwater regions potentially subject to high As contamination can be thought of roughly as a belt extending from the country's northwest corner to its southeast corner (MIM, 2015) (see Fig. 1).

As compared with various other methods used for dearsenification, such as flocculation, ion exchange and reverse osmosis, adsorber beds have proven to be the most reliable and easiest method to implement from a technical point of view. Ferric hydroxide-based granular materials have proven to be the best adsorbents for As removal due to their very high selectivity and adsorption capacity (Amy *et al.*, 2005).

Accordingly, decision makers in Iran opted to move ahead with construction of a large-scale drinking water dearsenification plant using adsorber bed technology.

2 METHODS

2.1 *Bardsir drinking water treatment plant*

Erected in 2013, Bardsir waterworks is located in Kerman Province in Southeastern Iran. The Iranian company Delta Niroo Gameron Co. designed, financed and constructed this plant as a BOO project (Build Own Operate) and now operates it on behalf of municipal authorities to supply drinking water to local consumers. The Bardsir waterworks provides a volume treatment capacity of up to 4 million m^3 $year^{-1}$ and is therefore able to meet the needs of roughly 85,000 consumers for As-free drinking water supply. The plant comprises 5 treatment lines operated in parallel, each of which can be switched on or off-line as required to achieve the desired water production rate. Each treatment line consists of 3 adsorber beds, two of which are operated in a lead/lag configuration while the other is held in standby mode. Each bed contains 5.3 metric tons of granular adsorbent.

2.2 *Granular ferric hydroxide adsorbent*

The adsorbent media used for As removal is GEH® 102 granular ferric hydroxide. This media, developed in Germany in the 1990s at Technische Universität Berlin, was the first ferric hydroxide-based granular adsorbent available for drinking water treatment (Driehaus *et al.*, 1998).

GEH® 102 is a synthetically manufactured ferric hydroxide composed of akaganeite (β-FeOOH) with a ferrihydrite ($Fe(OH)_3$) component. It has a particle size distribution ranging from 0.2 to 2.0 mm and a high specific surface area for adsorption, approx. 300 m^2 g^{-1} as determined by the BET method. The adsorbent is certified in accordance with NSF/ANSI Standard 61 and other internationally recognized specifications for use in drinking water treatment and has been utilized successfully for al-most 2 decades by users worldwide.

3 RESULTS AND DISCUSSION

3.1 *Raw water*

The raw water fed into the plant is groundwater extracted from a depth of 170 m below ground level. It has relatively high As content, i.e. concentrations ranging up to 250 $\mu g\,L^{-1}$. The As content averaged 145 $\mu g\,L^{-1}$ over the year 2015. The typical composition of the water is given in Table 1.

3.2 *Operation, optimization and performance*

The plant is designed for around-the-clock (24/7) operation. Each treatment line can be operated at a volumetric flow rate of up to 94 $m^3\,h^{-1}$. This flow rate corresponds to a flow speed of 16 $m\,h^{-1}$ and an empty bed contact time (EBCT) of 3 min through each adsorber bed. The plant's internal As limit, set voluntarily at a lower level than the WHO guidelines to ensure additional consumer safety, in the blended clean water is $<5\,\mu g\,L^{-1}$. The As concentrations in the raw water, the individual bed effluents and the blended clean water are determined at regular intervals using a portable rapid tester.

Table 1. Raw water composition.

Ca^{2+} $mg\,L^{-1}$	Mg^{2+} $mg\,L^{-1}$	Na^{+} $mg\,L^{-1}$	HCO_3^- $mg\,L^{-1}$	SO_4^{2-} $mg\,L^{-1}$	Cl^- $mg\,L^{-1}$	NO_3^- $mg\,L^{-1}$
53	13	48	176	78	66	12

pH –	Conductivity $\mu S\,cm^{-1}$	TDS $mg\,L^{-1}$	As $\mu g\,L^{-1}$	Hardness $mmol\,L^{-1}$
68	598	407	250 (max)	19

Three technological approaches were combined to realize optimum dearsenification performance:

a) The 2 active adsorber beds in each line are operated in a lead/lag configuration, therefore permitting better utilisation of each bed's adsorption capacity.
b) Exchange of exhausted adsorbent media is done in a time-staggered sequence, i.e. for only one treatment line at one time and only when the internal As limit has been reached in the blended clean water. The averaging effect provided by blending permits operation of individual lines with higher effluent concentrations, therefore further improving utilisation of adsorption capacity.
c) Sulfuric acid is added to the raw water as required to reduce its pH level from 6.8 to 6.0. This increases the adsorption capacity of the media by roughly 20%. This is later counteracted by addition of caustic soda to the clean water to bring its pH up to neutral.

Evaluation of plant operating data shows that the adsorbents used under these conditions attain bed lives in the range of 50,000 to 60,000 BV. In practical terms this means that one of the 15 adsorber beds in the plant must be recharged with fresh adsorbent on an average of every 3–4 months. The As-free water is sold for a price of roughly €0.60 per m^3 water supplied.

4 CONCLUSIONS

Experience gathered to date in the Bardsir waterworks shows that its granular ferric hydroxide-based adsorber process provides clean drinking water which consistently complies with the WHO As guideline value ($As < 10\,\mu g\,L^{-1}$) in spite of the challenging nature, i.e. relatively high As content, of the raw water treated. The process has proven to be a reliable and cost-efficient water treatment technology requiring only minimum maintenance. The granular ferric hydroxide-based dearsenification plant in Bardsir has achieved recognition as a beacon project for future waterworks of this type in Iran.

REFERENCES

Amy, G., Chen, H.-W., Drizo, A. & Brandhuber, P. 2005. Adsorbent treatment technologies for arsenic removal.

AWWA Research Foundation & AWWA. ISBN 1-58321-399-6.
Driehaus, W., Jekel, M., Hildebrandt, U. 1998. Granular ferric hydroxide – a new adsorbent for the removal of arsenic from natural water. *J. Water Supply Res. T.* 47(1): 30–35.
Ministry of Industries and Mines of Iran (MIM) 2015. Favora-ble Environments for As. Geological Survey of Iran http://www.gsi.ir/.
Mosaferi, M., Yunesian, M., Mesdaghinia, A., Nadim, A., Nasseri, S. & Mahvi, A.H. 2003. Arsenic occurrence in drinking water of I.R. of Iran: the case of Kurdistan Province. *Proceedings of BUET-UNU International Symposium "Fate of Arsenic in the Environment*, Dhaka, Bangladesh.
Murcott, S. 2012. *Arsenic Contamination in the World. An International Sourcebook.* IWA Publishing, London, UK.

5.5 Drinking water regulations of water safety plan

Environmental Arsenic in a Changing World –
Zhu, Guo, Bhattacharya, Ahmad, Bundschuh & Naidu (Eds)
ISBN 978-1-138-48609-6

Arsenic pollution in shallow drinking wells in Yuncheng Basin, China: occurrence and mechanisms

C.C. Li
School of Environmental Studies, China University of Geosciences, Wuhan, P.R. China

ABSTRACT: High concentrations of arsenic (As) in groundwater has led to adverse effects on human health and ecosystem due to the risk of As toxicity from drinking water sources. Diverse mechanisms govern arsenic mobilization in groundwater. In this study, scatter plot, cluster analysis and factor analysis were employed to investigate the occurrence and mechanisms controlling As concentrations in shallow groundwater at Yuncheng Basin, China. The research shows: (1) As contents in shallow groundwater is closely related to human activity. (2) It is favorable for the release of As in aquifer under reducing environment. Our findings suggest that cautions should be made for the management of shallow As contaminated groundwater.

1 INTRODUCTION

As a highly toxic element, arsenic (As) widely exists in natural environment (Ayotte *et al.*, 2015; Mahanta *et al.*, 2015; Smedley and Kinniburgh, 2002). Contamination of groundwater, either from natural or anthropogenic processes, has now caused worldwide attention (Currell *et al.*, 2011; Nordstrom, 2002). Millions of people in many countries including China are exposed to elevated levels of As through the uptake of As-rich groundwater. Recently, waterborne As poisoning (arsenicosis) has been found in aquifers in Yanhu District, Yongji City and Linyi County at Yuncheng Basin, northern China.

2 METHODS AND MATERIALS

Forty-six shallow groundwater samples were collected across the basin from July to October, 2016. Before sampling, the wells were pumped for over 1 h. At the time of sampling, all groundwater samples were filtered through 0.45 um membranes on site and stored in three new 300 mL polyethylene bottles which had been rinsed three times with deionized water. For cation analysis, reagent-quality nitric acid (HNO_3) was added to one of the polyethylene bottles until the pH of the water sample was less than 2. Unstable parameters like temperature (T), pH and electrical conductivity (EC) were measured in situ using portable Hanna EC and pH meters that had calibrated before use.

Alkalinity was measured on the sampling day using the Gran titration method. Concentrations of anions were determined using IC (Dionex 120). An Intrepid II XSP ICP-AES was used for cation measurements.

3 RESULTS AND DISCUSSION

3.1 *Scatter plot*

The scatter plot shows As contents in shallow groundwater versus alkalinity, major ions (Na^+, Ca^{2+}, Mg^{2+}, Cl^-, SO_4^{2-}, HCO_3^- and NO_3^-) concentrations.

There was a well-defined positive correlation between alkalinity, SO_4^{2-}, Cl^- and NO_3^- with As, indicating the positive contribution of these components to shallow groundwater As (only As vs. NO_3^- show in Fig. 1). Due to the lack of natural nitrate in most geologic formations, Figure 1 shows that the arsenic

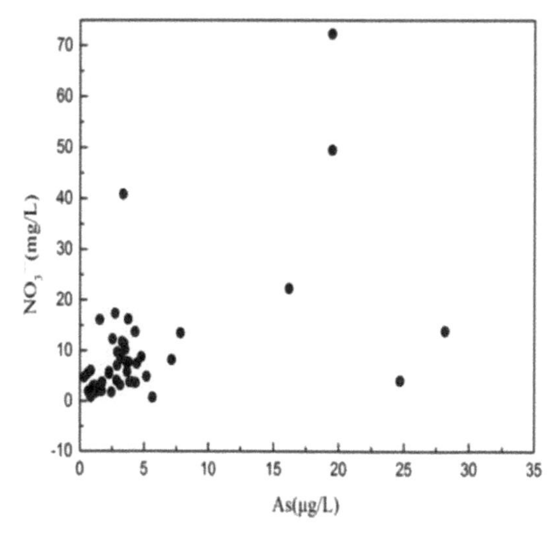

Figure 1. Scatter plot of As and NO_3^- contents in shallow groundwater.

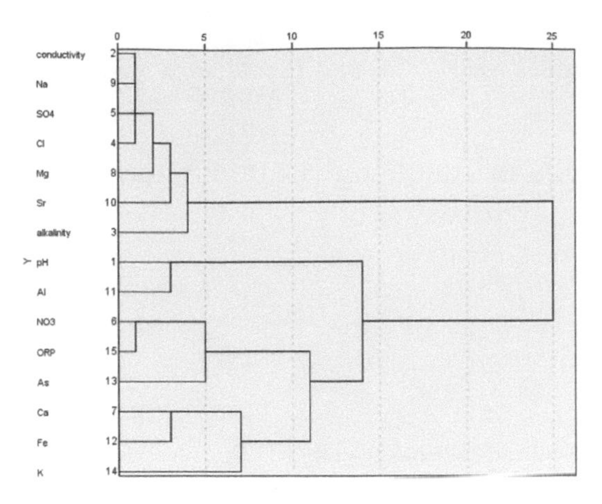

Figure 2. The results of cluster analysis in shallow groundwater.

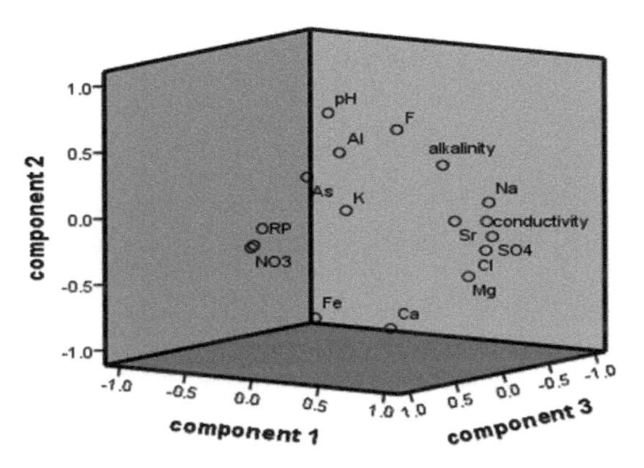

Figure 3. Factor analysis screen plot for shallow groundwater.

in shallow groundwater is closely related to human activities.

3.2 *Cluster analysis*

The cluster analysis (Fig. 2) in the shallow groundwater has shown that the NO_3^-, ORP and As fall into the same cluster. The ORP has a great influence on the enrichment of arsenic in the shallow aquifer and it is favorable for the release of arsenic in aquifer under reducing conditions. Human activity is another important factor affecting the concentration of arsenic in shallow groundwater.

3.3 *Factor analysis*

Factor 1 accounts for most of the total variance (36.18%) and is characterized by the associations of conductivity, Cl^-, SO_4^{2-} and Na^+, representing anthropogenic pollution such as farmland irrigation and fertilization (Fig. 3).

4 CONCLUSIONS

The results of this research showed that:

(1) The concentrations of As in shallow groundwater are affected by the NO_3^- contents, suggesting the influence of human activities on As in the shallow aquifer.
(2) The ORP has a great influence on the enrichment of arsenic in the shallow aquifer and it is favorable for the release of arsenic in aquifer under reducing conditions.

REFERENCES

Ayotte, J.D., Belaval, M., Olson, S.A., Burow, K.R., Flanagan, S.M., Hinkle, S.R. & Lindsey, B.D. 2015. Factors affecting temporal variability of arsenic in groundwater used for drinking water supply in the United States. *Sci. Total Environ.* 505: 1370–1379.

Currell, M., Cartwright, I., Raveggi, M. & Han, D.M. 2011. Controls on elevated fluoride and arsenic concentrations in groundwater from the Yuncheng Basin, China. *Appl. Geochem.* 26(4):540–552.

Mahanta, C., Enmark, G., Nordborg, D., Sracek, O., Nath, B., Nickson, R.T., Herbert, R.B., Jack, G., Mukherjee, A., Ramanathan, A., Choudhury, R. & Bhattacharya, P. 2015. Hydrogeochemical controls on mobilization of arsenic in groundwater of a part of Brahmaputra river floodplain, India. *J. Hydrol. – Regional Studies* 4: 154–171.

Nordstrom, D.K. 2002. Worldwide occurrences of arsenic in ground water. *Science* 296(5576): 2143–2145.

Smedley, P. & Kinniburgh, D. 2002. A review of the source, behaviour and distribution of arsenic in natural waters. *Appl. Geochem.* 17(5): 517–568.

Environmental Arsenic in a Changing World –
Zhu, Guo, Bhattacharya, Ahmad, Bundschuh & Naidu (Eds)
ISBN 978-1-138-48609-6

Analysis of tubewell arsenic concentration test results using an updated arsenic information management system in Nepal

R. Ogata[1,2] & M. Sakamoto[2]
[1]*Japan International Cooperation Agency (JICA), Japan*
[2]*The University of Tokyo, Tokyo, Japan*

ABSTRACT: Testing of more than one million tubewells for arsenic contamination in 20 districts in Nepal was carried out between 2003 and 2008 by the Government of Nepal with development partners and NGOs. The results showed that 1.69% of tubewells were contaminated with arsenic at levels of more than $50\,\mu g\,L^{-1}$. A computerized system to manage and analyze the test results, including GPS location data for individual tube wells, was created in 2009 and updated in 2016. It was found that locations of contaminated areas are quite concentrated and a large number of people are exposure to high levels of arsenic. There were 8,660 people living in the wards where more than 80% of tube-wells are arsenic contaminated, and 55,176 people and 116,081 people living in wards where the tube-wells were 50–80% and 30–50% contaminated, respectively. It is recommended that immediate and permanent safe water options to be installed in the high arsenic contaminated wards with localized plan of ward level arsenic contamination analysis because lager area contamination status may dilute significance of high arsenic contamination in hotspots.

1 INTRODUCTION

Arsenic contamination in the groundwater was found in West Bengal, India in 1983 and it has been estimated that 10 million people may be at risk of consumption of water contaminated by arsenic at $>50\,\mu g\,L^{-1}$. (Nickson *et al.*, 2007). Arsenic contamination in Bangladesh was confirmed by the Department of Public Health Engineering in 1993; the approximate number of people exposed to arsenic contaminated drinking water was between 21 and 28 million in 2003 (Huq *et al.*, 2010). In Nepal, the Department of Water Supply and Sewerage (DWSS) of the Government of Nepal conducted a research study in 1999 with the assistance of the World Health Organization (WHO) and found arsenic present at levels higher than $50\,\mu g\,L^{-1}$. Consequently, DWSS, with the assistance of UNICEF and various other organizations, carried out blanket testing for arsenic in 20 districts of the Terai area (Bhattacharya *et al.*, 2003; Tandukar *et al.*, 2006). This testing was completed in 2008. The aim of this paper is to identify the actual magnitude of arsenic contamination in Nepal using newly developed software to analyze the test results from the 20 districts.

2 DATA MANAGEMENT SYSTEM

For the purpose of data entry, data management, GIS mapping, analysis, and reporting of the measured water arsenic concentrations, the arsenic information management system (AIMS) was developed as an integrated software system by DWSS. This was not possible with the previous version of the software which was not fully functional on recent Windows operating systems (Windows 7, 8, and 10) and was unable to analyze precise localized contamination status. In 2016, it was updated as the new arsenic information management system (NAIMS), with financial and technical assistance from the Japan International Cooperation Agency (JICA). The main features of NAIMS are as follows.

- Compatible with all recent versions of operating systems including Windows 7, 8 and 10.
- Can use GIS maps while offline.
- Arsenic contamination map can show up to six levels of contamination in different colors, as compared to the previous version which could show only two levels in red or green.
- Map of percentage contamination is available up to the ward level.
- List of areas (wards and VDCs) highly contaminated by arsenic, in order of percentage of wells contaminated, is also available.

3 RESULTS

3.1 *Overall tube-well test result analysis*

The revised database of tube-well test results by NAIMS revealed the following facts. The total number of tube-well tests in the 20 districts was 1,097,015. Of these, 7.01% (76,952) were found to have arsenic concentrations of more than $10\,\mu g\,L^{-1}$, and 1.74% (19,112) had arsenic concentrations above $50\,\mu g\,L^{-1}$. The breakdown of arsenic concentrations of more than $50\,\mu g\,L^{-1}$ was 16,583 wells (1.51%) with concentrations of 51–100 $\mu g\,L^{-1}$, 2,147 wells (0.2%) with 101–300 $\mu g\,L^{-1}$, 341 wells (0.03%) with 301–500 $\mu g\,L^{-1}$, and 41 wells (0.004%) with arsenic concentrations above $500\,\mu g\,L^{-1}$.

3.2 *Hotspot area*

Highly arsenic contaminated areas called hotspots, are shown in Table 1. Although the nationwide proportion of wells in Nepal with arsenic contamination is only 1.69%, 8,660 people live in wards where more than 80% of tube-wells have arsenic concentrations above $50\,\mu g\,L^{-1}$. Likewise, 63,836 and 179,917 people live in the wards where more than 50% and 30% of tube-wells have arsenic concentrations above $50\,\mu g\,L^{-1}$, respectively. This result describes that a large number of people are at high risk of arsenic exposure in Nepal.

3.3 *Distribution of hotspot wards*

The GIS-based contamination map produced by NAIMS shows that highly arsenic contaminated areas were concentrated in particular areas in most districts. Figure 1 shows the ward-level arsenic contamination

Table 1. Hotspot areas and population.

Percentage of wells contaminated by arsenic	Number of wards	Total population living in wards
>80	14	8,660
>50, ≤80	72	55,176
>30, ≤50	127	116,081
Total	213	179,917

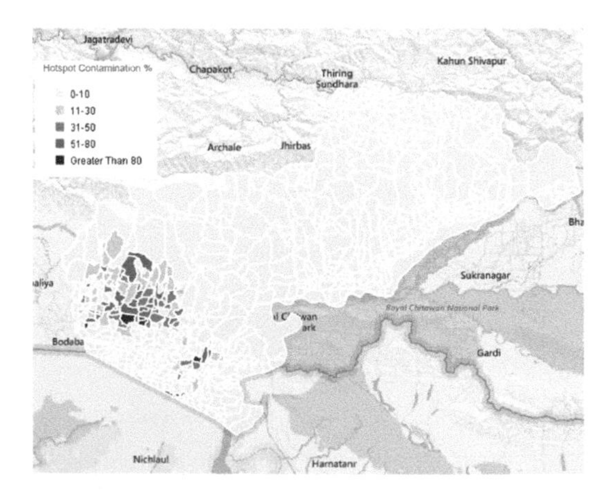

Figure 1. Map of ward-level arsenic contamination based on tube-well test results in Nawalparasi district.

proportions in Nawalparasi district, where the proportion of arsenic contaminated tube-wells is the highest in Nepal at 11.69%. All the highly arsenic contaminated wards are located in the western part of the district; in six wards, more than 80% of the wells are contaminated with arsenic at levels above $50\,\mu g\,L^{-1}$.

Figure 2 shows the ward-level arsenic contamination proportions in Siraha district, where the overall proportion of arsenic-contaminated wells is only 2.57%. As in Figure 1, the highly arsenic contaminated wards are located in the western part of the district and in five wards more than 80% of the wells have concentrations of arsenic of more than $50\,\mu g\,L^{-1}$. Most of the districts showed patterns of arsenic contamination similar to those of Nawalparasi and Siraha.

4 DISCUSSION AND CONCLUSIONS

The overall arsenic contamination level in Nepal is less than in India and Bangladesh. However, hotspot areas are found in concentrated locations and a large number of residents in these areas are exposed to high levels of arsenic contamination. UNICEF and Nepal Red Cross Society conducted a health survey in four districts in Nepal in 2003 and identified over 4,000 suspected cases of arsenicosis (Pahari, 2009; Neupane *et al.*, 2014). This highlights the potential danger of arsenic contamination in Nepal. The National Arsenic Steering Committee (2011) reported arsenic contamination status by district and VDC. However, a more localized mitigation plan, based on information at the ward or village level, may be more appropriate for establishing safe water facilities because status of arsenic contamination in larger area may dilute the significance of village level high arsenic contamination. The hotspot wards where many people have developed serious arsenic-caused disease need immediate and permanent safe water facilities (e.g., yellow, red and brown areas shown in Figs. 1 and 2) even though district wise contamination is low.

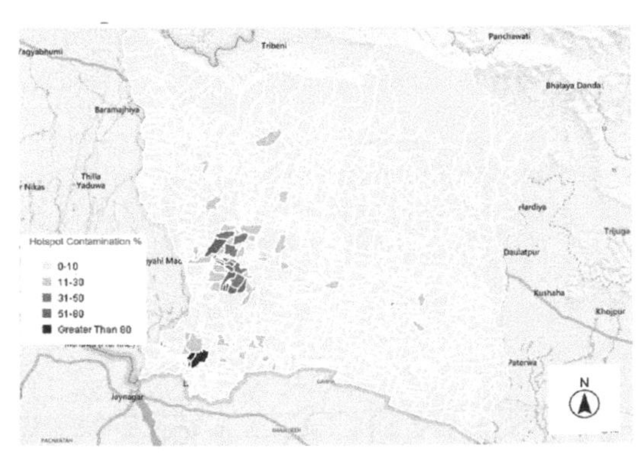

Figure 2. Map of ward-level arsenic contamination based on tube-well test results in Siraha district.

ACKNOWLEDGEMENTS

The authors express their gratitude to Sunil Kumar Das, Narayan Prasad Khanal, and Manina Baidya of DWSS for their continuous support on NAIMS development, and to Anish Jyoshi and Ajay Mishra for technical contributions to NAIMS.

REFERENCES

Bhattacharya, P., Tandukar, N., Neku, A., Valero, A.A., Mukherjee, A.B. & Jacks, G., (2003) Geogenic arsenic in groundwaters from Terai alluvial plain of Nepal. *Jour. de Physique IV France* 107: 173–176.

Department of Public Health Engineering (DPHE) and Japan International Cooperation Agency (JICA). 2010. Situation analysis of arsenic mitigation 2009. Dhaka, Bangladesh: Government of Bangladesh, DPHE, JICA Bangladesh.

National Arsenic Steering Committee (Nepal). 2011. The state of arsenic in Nepal—2011. Draft December, 2011.

Neupane, M., Thakur, J.K., Gautam, A., Dhakal, A. & Pahari, M. 2014. Arsenic aquifer sealing technology in wells: a sustainable mitigation option. Water Air Soil Pollut. 225: 2087.

Nickson, R., Sengupta, C., Mitra, P., Dave, S.N., Banerjee, A.K., Bhattacharya, A. & Deverill, P. 2007. Current knowledge on the distribution of arsenic in groundwater in five states of India. *J. Environ. Sci. Health A* 42(12): 1707–1718.

Pahari, M. 2009. Arsenic Testing, Mitigation and Information Management in Nepal. UNICEF Report.

Tandukar, N., Bhattacharya, P., Neku, A. & Mukherjee, A.B. (2006) Extent and severity of arsenic poisoning in Nepal. In: R. Naidu, E. Smith, G. Owens, P. Bhattacharya & P. Nadebaum. (Eds.) *Managing Arsenic in the Environment: From soil to human health*. CSIRO Publishing, Melbourne, Australia, pp. 595–604.

Environmental Arsenic in a Changing World –
Zhu, Guo, Bhattacharya, Ahmad, Bundschuh & Naidu (Eds)
ISBN 978-1-138-48609-6

Integrating arsenic in water safety planning in The Netherlands

P. van der Wens[1] & A. Ahmad[2,3,4]
[1]*Brabant Water, Department of Drinking Water Production, Breda, The Netherlands*
[2]*KTH-International Groundwater Arsenic Research Group, Department of Sustainable Development, Environmental Science and Engineering, KTH Royal Institute of Technology, Stockholm, Sweden*
[3]*KWR Watercycle Research Institute, Nieuwegein, The Netherlands*
[4]*Department of Environmental Technology, Wageningen University and Research (WUR), Wageningen, The Netherlands*

ABSTRACT: The Dutch drinking water sector complies with the most restrictive guidelines in the world and has a long-standing history of striving for excellence in water quality standards. The Dutch Association of Drinking Water Companies (Vewin) concluded in 2015 to lower the standard on arsenic in drinking water to $1\,\mu g\,L-1$. Following the new recommendation Brabant Water, a major water supply company in The Netherlands developed its masterplan on arsenic reduction by analysing the presence of arsenic in the systems from source to tap. Several measures were put in place to integrate arsenic into water safety planning in order to manage the risks. The underlying cost-benefit study and its implication on water safety planning at Brabant Water is discussed in this presentation.

1 INTRODUCTION

Drinking water is a major source of inorganic arsenic (As) consumed in EU (EFSA, 2014). The WHO guideline for As in drinking water ($10\,\mu g\,L-1$) is fully met in The Netherlands. The WHO guideline is provisional and therefore every effort should be made to keep the As concentrations as low as reasonably possible in drinking water (WHO, 2006, 2011). In The Netherlands a cost benefit analysis was carried out which revealed that it is advantageous to remove As to $<1\,\mu g\,L^{-1}$ (Van der Wens *et al.*, 2016). Brabant Water, the water supply company in the North Brabant Province of The Netherlands integrated the findings in its water safety planning. In this presentation the underlying cost-benefit study and its implication on water safety planning at Brabant Water is discussed.

2 COST-BENEFIT ANALYSIS

Figure 1 shows the linearly modelled dose-response relationship which is based on the epidemiological study in Chile and the conclusions of the National Research Council (NRC, 2001) on the increased risk of lung cancer by 41% due to lifelong exposure to $50\,\mu g\,L^{-1}$ As. The relationship led to a calculated health benefit of 10.7 M€ per year using the Disability Adjusted Life Years (DALY) method of the National Health Council (NHC, 2007) which is more than twice the calculated annual costs for As removal when the target is set on $1\,\mu g\,L^{-1}$ (Van der Wens *et al.*, 2016).

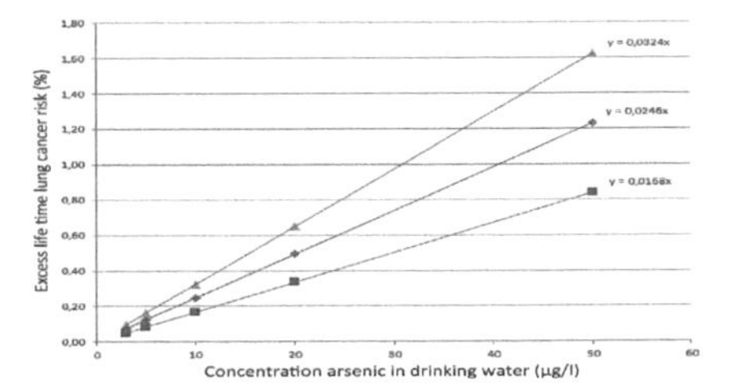

Figure 1. Estimated excess life time lung cancer risk (95% c.i.) in the Dutch population related to As concentrations in drinking water.

3 WATER SAFETY PLANNING

The water safety planning process led to a revision of the water quality policy of Brabant Water (Brabant Water, 2015). Depending on long term investments and relatively slow raw groundwater quality changes an internal set of water quality targets were set. In this way investments were expected to be stable leading to robust treatment works reflecting high quality at low costs. An extensive literature survey was carried out in which three issues were under focus: 1) the high-level health protection of the WHO guideline for borium, 2) the low-level health protection of the WHO guideline for As and 3) the high societal cost benefit ratio of lowering the internal guideline for total hardness.

The second area of results from the water safety planning is insight into the fate of As from source to tap. Figure 2 indicates four issues that will be discussed. The main process for As removal is removal

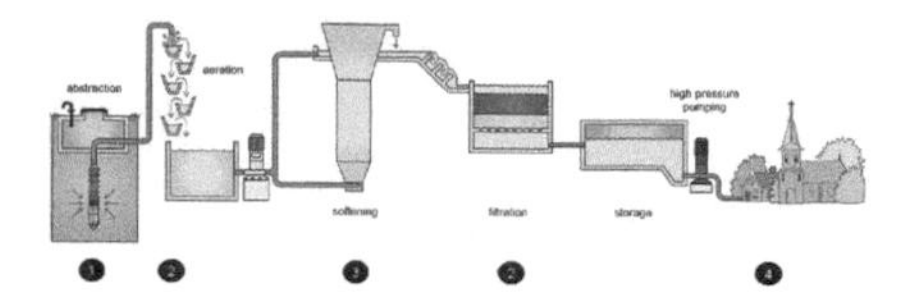

Figure 2. Water safety planning in water treatment.

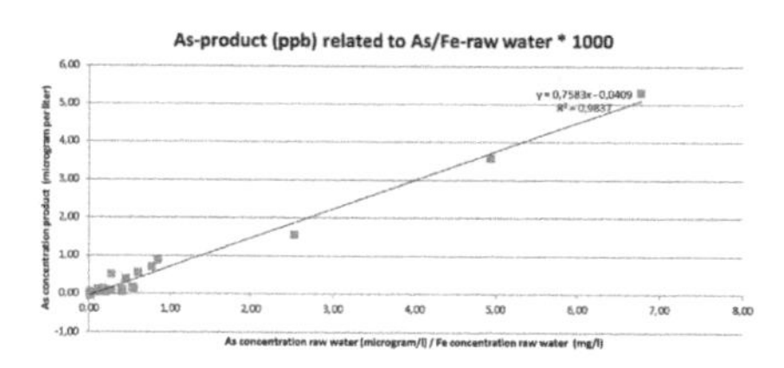

Figure 3. As-product vs As/Fe-ratio for well fields of 30 water treatment plants of Brabant Water.

of iron (Fe) flocs in the filtration process which are laced with sorbed As. The Fe/As ratio and the time for the As(III) to oxidize to As(V) are important factors that determine removal efficiency (Groenendijk, 2016). Implementation of the advanced oxidation-coagulation filtration (AOCF) was done with absence of the softening process. In 2017 the next observations and conclusions were part of the WSP:

1) In Brabant, in the southern part ofThe Netherlands, severe differences between individual wells occur. These differences are not easy to avoid. Variations in arscnic concentrations from less than $0.1\,\mu g\,L^{-1}$ to more than $30\,\mu g\,L^{-1}$ may vary within well fields at the same depth (approx. 250 m deep) at less than hundred meters distance. It seems unavoidable to mix water from wells with high and low concentrations because of the large spatial variation. If the As/Fe ratio in the mixed raw water is low enough then As levels will be $<1.0\,\mu g\,L^{-1}$. The presence of enough iron appears to be a very important factor. Figure 3 illustrates this linear relationship. For the 30 spatially divided well fields, the average As/Fe ratio is plotted against the average As concentration in the drinking water;
2) In the aeration process As(III) is not oxidized to As(V). This process can be completed by adding a strong oxidant like permanganate to ensure As(V) occurs as main As species which sorbs better to Fe flocs. This however creates slightly different sludge composition. Although the sludge has good settling conditions, it has poor dewatering characteristics. This may add to financial and environmental costs in the backwash water treatment and sludge removal. Although this appears to be a large factor the cost-benefit ratio remains positive and current research leads to adjustments in the AOCF-process that creates sludge with better dewatering characteristics;
3) At two water treatment locations introducing softening to the process scheme caused elevation of produced drinking water As concentrations. This is caused by the removal of Fe in the softening reactors. The iron is caught in the $CaCO_3$ pellets. The Fe concentration in the bypass is too low to sorb the residual As. Adding a small amount of Fe is expected to improve As removal;
4) After adjusting the treatment process, it is necessary to clean the distribution network. Although concentrations of dissolved As will not be elevated, the amount of accumulated As in the sediments can cause surge of As. Total As up to 40 times higher than normal dissolved concentrations have been registered during hydraulic disturbances (Blokker *et al.*, 2016).

4 CONCLUSIONS

Using the precautionary principle and the modelled dose-effect model of arsenic in drinking water the Dutch water sector aims at reducing As to levels lower than $1.0\,\mu g\,L^{-1}$ which are currently ‘as low as reasonably possible’. Treatment measures have already been taken and proven to be effective.

The water safety planning process helps to clearly understand of the fate of arsenic though the process from source to tap. Issues form source to tap need to be attended.

ACKNOWLEDGEMENTS

The authors thank the Dutch Water Companies for their contribution to the policy process and KWR Water research and Brabant Water for the research projects.

REFERENCES

Blokker, E.J.M., van Vossen, J., Schriks, M., Bosch, J. & van der Haar, M. 2016. Arsenic accumulation in drinking water distribution networks. In: P. Bhattacharya, M. Vahter, J. Jarsjö, J. Kumpiene, A. Ahmad, C. Sparrenbom, G. Jacks, M.E. Donselaar, J. Bundschuh, & R. Naidu (Eds.) *"Arsenic Research and Global Sustainability As 2016"*.CRC Press/Taylor and Francis, pp. 644–645.

Brabant Water 2015. Corporate Water Quality Standards. Internal document. Brabant Water, The Netherlands.

Groenendijk, M. 2016. Removing arsenic to $<1\,\mu g\,L^{-1}$ in conventional groundwater treatment plants: practical tips and tricks. In: P. Bhattacharya, M. Vahter, J. Jarsjö, J. Kumpiene, A. Ahmad, C. Sparrenbom, G. Jacks, M.E. Donselaar, J. Bundschuh, & R. Naidu (Eds.) *"Arsenic Research and Global Sustainability As 2016"*. CRC Press/Taylor and Francis, pp. 519–520.

Van der Wens, P., Baken, K. & Schriks, M. 2016. Arsenic at low concentrations in Dutch drinking water: assessment of removal costs and health benefits. In: P. Bhattacharya, M. Vahter, J. Jarsjö, J. Kumpiene, A. Ahmad, C. Sparrenbom, G. Jacks, M.E. Donselaar, J. Bundschuh, & R. Naidu (Eds.) *"Arsenic Research and Global Sustainability As 2016"*. CRC Press/Taylor and Francis, pp. 563–564.

WHO (World Health Organization) 2006. Water Safety Plan Manual. WHO, Geneva, Switzerland.

WHO (World Health Organization) 2011. Guidelines for Drinking Water Quality, Fourth Edition. WHO, Geneve, Switzerland.

5.6 Arsenic in drinking water and implementation plan for safe drinking water supply from sustainable development perspectives

Environmental Arsenic in a Changing World –
Zhu, Guo, Bhattacharya, Ahmad, Bundschuh & Naidu (Eds)
ISBN 978-1-138-48609-6

Sustainable small-scale, membrane based arsenic remediation for developing countries

J. Hoinkis[1], E.E. Cañas Kurz[1], U. Hellriegel[1], T.V. Luong[2,3] & J. Bundschuh[3]
[1]*Center of Applied Research, Karlsruhe University of Applied Sciences, Karlsruhe, Germany*
[2]*Vietnamese-German University, Ho Chi Minh City, Vietnam*
[3]*UNESCO Chair on Groundwater Arsenic within the 2030 Agenda for Sustainable Development & Faculty of Health, Engineering and Sciences, University of Southern Queensland, Toowoomba, Australia*

ABSTRACT: Membrane technologies for small-scale arsenic removal are currently mainly involving nanofiltration, reverse osmosis, membrane distillation and electrodialysis. Their advantages are the simplicity of the technology and the scalability down to very small modules. Forward osmosis is a relatively new technology for arsenic removal which was mainly applied so far on laboratory scale. Basically, all membrane technologies, with the exception of membrane distillation, show significantly lower rejection for As(III) than for As(V). In order to treat anoxic groundwater with high As(III), which is usually associated with high Fe, a pre-oxidation step and the removal of Fe(oxy)hydroxide is mandatory to avoid membrane clogging. Generally, safe disposal of the occurring arsenic rich sludges and liquid waste streams pose a difficult issue.

1 INTRODUCTION

Membrane processes are becoming more and more attractive in drinking water treatment due to their salient features such as i) no addition of chemicals needed, ii) separation carried out in mild environmental conditions, iii) easy to up-scale. Therefore, membrane processes have been also extensively studied for arsenic removal mainly from groundwater (Figoli, *et al.*, 2010a, 2016; Shih, 2005). However, the use of membrane technologies still suffers from main disadvantages such as i) membrane fouling and ii) safe disposal of concentrates particularly for arsenic-laden waters.

Recently, using small-scale membrane units powered by renewable energies (solar, wind) is gaining significant interest particularly for seawater desalination in remote areas and in developing countries (Ali *et al.*, 2018). Especially solar energy is abundant in many of these countries and costs have dropped significantly in the last years worldwide. The most growth among the renewable energy technologies has been observed in solar energy. However, the real contribution of solar energy in desalination is less than 0.02% (Ali *et al.*, 2018).

To date, small-scale renewable energy powered membrane units are mainly applied for desalination of seawater e.g. on ships, however, for arsenic removal from groundwater these have only been studied by numerous pilot trials that have however, not entered the market and remain still as field of research (Schmidt *et al.*, 2016). The objective of this paper is to give a short review of the current research in this field.

2 METHODS

2.1 *Membrane processes*

Different membrane technologies are employed in arsenic removal. Among them pressure-driven processes such as microfiltration (MF), ultrafiltration (UF), nanofiltration (NF) and reverse osmosis (RO) were studied most commonly (Figoli *et al.*, 2010a). Hereby a water solution (feed) is passed along the membrane as crossflow and a pressure gradient forces the treated water through the membrane while retaining particulates down to solutes. However, by use of MF and UF with pore sizes between 0.01 and 0.2 μm dissolved species cannot be removed (Shih, 2005).

Besides pressure-driven membrane processes, also thermally and electrically driven membrane processes are used on small-scale. Membrane distillation (MD) is a thermally driven process based on vapor pressure difference maintained across a microporous hydrophobic membrane. The diffusing vapor then condenses at the permeate membrane interface. The main advantage of MD is the fact that solar heat can be directly used without employing high pressure pumps (Manna and Pal, 2016).

In the electrodialysis (ED) process, an electrical potential difference is used as the driving force to transport the dissolved ions through ion exchange membranes. Hereby, cation and anion exchange membranes are placed alternatingly between the anode and cathode. Finally, when a salt containing feed solution is fed to the electrodialysis unit one cell of the pair becomes depleted of ions while the adjacent cell becomes enriched in ions. The salient feature of ED is that only low-pressure pumps and no high-pressure pumps as for RO are needed.

A relatively new membrane process for water treatment that has been studied on laboratory scale for arsenic removal is forward osmosis (FO). The principle of FO process relies on using the natural osmotic process to draw water across a semipermeable membrane from a saline feed water to a higher concentrated solution, namely the draw solution (DS). The driving force is therefore created naturally by the difference in osmotic pressure between the DS and the feed solution (FS). This offers several advantages over conventional hydraulic pressure-driven membrane processes (e.g. RO) such as lower energy requirements and reduced membrane fouling potential (McCutcheon and Elimelech, 2006). However, regarding drinking water production it should be pointed out that due to the fact that water is always drawn into a saline DS an additional separation step to recover the water from the DS is needed.

2.2 *Arsenic speciation*

The most dominant forms of arsenic in the environment are the pentavalent arsenate As(V) and trivalent arsenite As(III). Redox and pH conditions of the water control the speciation of As (Sharma and Sohn, 2009). Typical As(V) species being present under oxidizing aerobic conditions are the mono- and divalent oxyanionic forms $H_2AsO_4^-$ and $HAsO_4^{2-}$, respectively, while the neutral As(III) species H_3AsO_3 exists in anaerobic groundwater of pH lower than 9 (Camacho *et al.*, 2011). Arsenic desorption under oxidizing conditions and high pH explains the high As(V) concentrations in the groundwater of extended areas of the Chaco-Pampean plain in Argentine and its continuation into the adjacent plains of Uruguay (Bundschuh *et al.*, 2010). In West Bengal (India), Bangladesh down to South East Asia (Cambodia, Vietnam) the groundwater is typically anoxic and arsenite As(III) predominates (Kar *et al.*, 2010). In this region high arsenic concentration is typically associated with high dissolved Fe and Mn (Kar *et al.*, 2010).

As a matter of fact, the toxicity and the removability of arsenic differ between As(III) and As(V). As(III) is considered to be more toxic and generally more difficult to remove from water than As(V) (USEPA, 2001). The WHO recommended maximum contaminant level (MCL) of total As is set at $10\,\mu g\,L^{-1}$.

3 RESULTS AND DISCUSSION

3.1 *Nanofiltration*

Experimental studies have shown that NF membranes cannot significantly reject arsenite at neutral and lower pH (Uddin *et al.*, 2007). This can be explained by the fact that NF process is strongly dependent on charge exclusion effects caused by interaction of the charged membrane and the charged species such as arsenate whereas arsenite is neutral in charge (see 2.2). Uddin *et al.* (2007) compared the performance of the *"loose"* Dow NF200 with the *"dense"* Dow NF90 in the range of 100–1000 $\mu g\,L^{-1}$ by use of synthetic water. Arsenic rejection of the NF90 was always significantly higher than for the NF200. The NF90 achieved highest rejection for arsenate (>98%) and was able to comply with the MCL of $10\,\mu g\,L^{-1}$ for up to $1000\,\mu g\,L^{-1}$ feed concentrations (10 bar, pH 7.3, 25°C). However, the removal of arsenite under the same conditions was only 53–59% and hence the MCL could not be achieved in the entire feed range.

Figoli *et al.* (2010b) studied the influence of operating parameters on the removal of pentavalent As(V) from synthetic water by use of the Dow NF90 and the Microdyn-Nadır N30F. The As rejection of the *"dense"* NF90 was significantly higher (>91%) than for the *"loose"* N30F.

3.2 *Reverse osmosis*

Many studies were conducted regarding arsenic removal by RO (e.g. Geucke *et al.*, 2009; Schmidt *et al.*, 2016). It has been shown that rejection of arsenate is at least as efficient as for the *dense* NF, however, As(III) rejection is only slightly better. Geucke *et al.* (2009) studied the performance of the Dow XLE, Dow TW30 and Dow SW30 in a small-scale RO desalinator. They showed that for trivalent arsenic the arsenic values in permeate could be kept below the MCL of $10\,\mu g\,L^{-1}$ only up to a feed concentration of approximately $350\,\mu g\,L^{-1}$. As (V) was rejected efficiently up to a feed concentration higher than $2000\,\mu g\,L^{-1}$, without crossing the MCL level in permeate. As for the NF process, this fact can be explained by a charge exclusion effect.

Schmidt *et al.* (2016) conducted pilot trials at two sites namely Bind Toli and Ramnagar near Patna, Bihar, India with a small technical RO desalinator. The arsenic in feed was 480 and $67\,\mu g\,L^{-1}$, respectively. The total arsenic reduction on both grounds was around 99% and in most of the cases the arsenic concentration in permeate was in compliance with the National Indian Standard of $10\,\mu g\,L^{-1}$. In order to avoid clogging of the membrane they conducted an upfront oxidation by air and granular media filtration (sand, anthracite) step. Hereby the majority of dissolved iron was removed as iron(oxy)hydroxide (Fe(O)OH). Thus, a great part of arsenic could be also removed by co-precipitation and filtration, so that the RO unit served only as polisher to comply with the stringent MCL of $10\,\mu g\,L^{-1}$ (Schmidt *et al.*, 2016). The RO unit made use of an energy recovery system in order to keep the energy consumption low (~3–4 $Wh\,L^{-1}$).

3.3 *Membrane distillation*

The main advantages of MD are the direct use of thermal energy (e.g. solar energy) and the high As(III) rejection that makes it attractive for application in remote areas particularly in India and Southeast Asia (Table 1). Criscuoli *et al.* (2013) studied

Table 1. Overview of advantages and disadvantages of membrane technologies on arsenic removal.

Type of membrane process	Removal efficiency		Advantages	Disadvantages
	As(V)	As(III)		
NF	+	–	• Capable of removal of other groundwater contaminants (salinity, micropollutants) • Can be operated at low pressure (<6 bar)	Low efficiency for As(III) removal
RO	++	+/o	• High removal rate of other groundwater contaminants (salinity, micropollutants)	Relatively high energy consumption
MD	++	++	• High efficiency for As(III) removal • Direct use of heat for driving the process	Relatively low water flow per unit area (flux)
ED	+	–/o	• No need for pressure pumps • Direct use of electrical energy	Low efficiency for As(III) removal
FO	++	–/o	• Low fouling propensity • Double barrier due to necessity of additional process for regeneration of draw solution	Has to be operated as combined process due to regeneration of draw solution

arsenic removal by use of vacuum membrane distillation (VMD) with synthetic aqueous solutions up to $0.5\,\mu g\,L^{-1}$ As(III) and As(V), respectively. Arsenic in the produced permeate was always below the detection limit. However, permeate water flux was typically lower compared to NF or RO (Criscuoli *et al.*, 2013). Mann and Pal (2016) used a pilot-scale flash vaporization membrane distillation (FVMD) to treat an arsenic groundwater with $396\,\mu g\,L^{-1}$ As from a well in West Bengal, India. After 40 h long operation, flux remained almost constant at around $36\,kg\,m^2\,h^{-1}$ while arsenic was below detection limit in the distillate sample.

3.4 *Electrodialysis*

The strength of ED is its direct use of electric energy and the necessity of only small low-pressure pumps. Gonzales *et al.* (2017) studied purification of brackish groundwater in northern Chile (TDS $= 5482\,mg\,L^{-1}$) with high concentration of arsenic ($2.04\,mg\,L^{-1}$) using a solar powered ED unit. The purification process partially combined the ED system with upfront ion exchange and adsorption technologies (column filtration system). The ion exchanger was operated to remove hardness and the adsorption unit was used for arsenic removal. For the untreated feed with high arsenic the removal efficiency achieved 93–94% and for the combined process 99.9% (being in compliance with MCL). The lowest specific electricity consumption was $2.16\,kWh\,m^{-3}$ for the ED and $5.46\,kWh\,m^{-3}$ for the combined system (Gonzales *et al.* 2017).

3.5 *Forward osmosis*

Mondal *et al.* (2014) studied arsenic removal at laboratory scale with a commercial FO membrane (provided by HTI) by use of artificial water solutions and two draw solutions (glucose, $MgSO_4$). The removal efficiency was more than 98% when the initial As(V) concentration was $500\,\mu g\,L^{-1}$ in a neutral pH range (Mondal *et al.*, 2014). The rejection increased with increasing pH of the feed solution and the As concentration was below the MCL under optimized conditions. However, for As(III), a significant lower rejection was observed (<20% at pH 7) and only oxidation of As(III) at neutral pH increased the rejection to 95.7%.

Hoinkis *et al.* (2016) conducted laboratory trials with a commercial membrane provided by Porifera and As(III) and As(V) model solutions (As feed concentration up to $400\,\mu g\,L^{-1}$). The experiments showed that rejection of As(V) is significantly better than of As(III). For As(III) from $50\,\mu g\,L^{-1}$ in feed solution arsenic level in draw solution exceeded the MCL of $10\,\mu g\,L^{-1}$. However, it should be noted that FO has to be run as combined process with e.g. RO or MD to regenerate the draw solution what serves as a second barrier and further reduces arsenic.

3.6 *Concept for As(III) removal*

As mentioned above, most of the membrane processes show low removal efficiency for As(III) and pre-oxidation is needed (e.g. by H_2O_2, O_3, MnO_2). Moreover, when treating groundwater in the developing regions like India and Southeast Asia in many cases oxidation is required as pre-treatment since As(III) in the typically anoxic water is commonly associated with high Fe and Mn. When Fe is oxidized it forms sparingly soluble Fe(O)OH which needs to be separated by filtration otherwise severe membrane clogging may occur also for MD which is basically able to remove trivalent arsenite. Due to co-precipitation this Fe(oxy)hydroxide sludge also contains high arsenic level and it is consequently difficult to dispose it safely. An interesting option of waste-free pre-treatment is offered by subsurface arsenic remediation by aeration (van Halem, 2010).

In order to comply with the stringent MCL subsequently to the oxidation step NF or MD can be employed. NF is a simple and robust treatment, which can be operated at lower pressure compared to RO, and MD can make direct use of solar heat.

4 CONCLUSIONS

Existing membrane technologies such as NF, RO, ED and MD offer promising opportunities to treat arsenic-laden groundwater by renewable energy in developing countries. Their advantages are the simplicity of the technology and the scalability down to very-small modules. However, in order to treat anoxic groundwater with high As(III), which is usually associated with high Fe, pre-oxidation and removal of Fe(oxy)hydroxide is mandatory. The safe disposal of the occurring arsenic rich sludges pose still a difficult issue.

REFERENCES

Ali, A., Tufa, R.A., Macedonio, F., Curcio, E. & Drioli, E. 2018. Membrane technology in renewable energy driven desalination. *Renew. Sust. Energy. Rev.* 81: 1–21.

Bundschuh, J., Litter, M., Ciminelli, V.S.T., Morgada, M.E., Cornejo, L., Hoyos, S.G., Hoinkis, J., Alarcon-Herrera, M.T., Armienta, M.A. & Bhattacharya, P. 2010. Emerging mitigation needs and sustainable options for solving the arsenic problems of rural and isolated urban areas in Latin America – a critical analysis. *Water Res.* 44(19): 5828–5845.

Camacho, L.M., Gutierrez, M., Alarcon-Herrera, M.T., de Lourdes Villaba, M. & Deng, S. 2011. Occurrence and treatment of arsenic in groundwater and soil in northern Mexico and southwestern USA. *Chemosphere* 83(3): 211–225.

Criscuoli, A., Bafaro, P. & Drioli, E. 2013. Vacuum membrane distillation for purifying waters containing arsenic. *Desalination* 323: 17–21.

Figoli, A., Criscuoli, A. & Hoinkis, J. 2010a. Review of membrane processes for arsenic removal from drinking water. In: N. Kabay, J. Bundschuh, B. Hendry, M. Bryjak, K. Yoshizuka, P. Bhattacharya & S. Anac (eds) *The Global Arsenic Problem: Challenges for safe water production.* CRC Press, Boca Raton, FL, pp. 131–145.

Figoli, A., Cassano, A., Criscuoli, A., Mozumder, M.S.I., Uddin, M.T., Islam, M.A. & Drioli, E. 2010b. Influence of operating parameters on the arsenic removal by nanofiltration. *Water Res.* 44(1): 97–104.

Figoli, A., Hoinkis, J. & Bundschuh, J. (eds) 2016. *Membrane Technologies for Water Treatment: Removal of Toxic Trace Elements with Emphasis on Arsenic, Fluoride and Uranium.* CRC Press, Boca Raton, FL.

Gonzales, A., Grageda, M. & Ushak, S. 2017. Assessment of pilot-scale water purification module with electrodialysis technology and solar energy. *Appl. Energy* 206: 1643–1652.

Hoinkis, J., Jentner, J., Buccheri, S., Deowan, S.A., Figoli, A. & Bundschuh, J. 2016. Forward osmosis – challenges and opportunities of a novel technology for arsenic removal from groundwater. *Arsenic Research and Global Sustainability – As 2016: Proceedings of the 6th International Congress on Arsenic in the Environment, 19–23 June 2016, Stockholm, Sweden*, pp. 523–524.

Kar, S., Maity, J.P., Jean, J.S., Liu, C.C., Nath, B., Yang, H.J. & Bundschuh, J. 2010. Arsenic-enriched aquifers: occurrence and mobilization of arsenic in groundwater of Ganges Delta Plain, Barasat, West Bengal, India. *Appl. Geochem.* 25(12): 1805–1814.

Manna, A.K. & Pal, P. 2016. Solar-driven flash vaporization membrane distillation for arsenic removal from groundwater: Experimental investigation and analysis of performance parameters. *Chem. Eng. Prog.* 99: 51–57.

McCutcheon, J.R. & Elimelech, M. 2006. Influence of concentrative and dilutive internal concentration polarization on flux behavior in forward osmosis, *J. Membr. Sci* 284(1–2): 237–247.

Mondal, P., Hermans, N., Tran, A.T.K., Zhang, Y., Fang, Y., Wang, X. & Van der Bruggen, B. 2014. Effect of physico-chemical parameters on inorganic arsenic removal from aqueous solution using a forward osmosis membrane. *J. Environ. Chem. Eng.* 2(3): 1309–1316.

Schmidt, S.A., Gukelberger, E., Hermann, M., Fiedler, F., Großmann, B., Hoinkis, J., Ghosh, A., Chatterjee, D. & Bundschuh, J. 2016. Pilot study on arsenic removal from groundwater using a small-scale reverse osmosis system – Towards sustainable drinking water production. *J. Hazard. Mater.* 318: 671–678.

Sharma, V.K. & Sohn, M. 2009. Aquatic arsenic: toxicity, speciation, transformation and remediation. *Environ. Int.* 35(4): 743–759.

Shih, M.C. 2005. An overview of arsenic removal by pressure-driven membrane processes. *Desalination* 172. 85–97.

Uddin, M.T. Mozumder, M.S.I. Islam, M.A., Deowan, S.A. & Hoinkis, J. 2007. Nanofiltration membrane process for the removal of arsenic from drinking water. *Chem. Eng. Technol.* 30(9): 1248–1254.

USEPA 2001. National Primary Drinking Water Regulations: Arsenic and Clarifications to Compliance and New Source Contaminants Monitoring.

Van Halem, D., Heijman, S.G.J., Johnston, R., Huq, I.M., Ghosh, S.K., Verberk, J.Q.J.C., Amy, G.L. & Van Dijk, J.C. 2010. Subsurface iron and arsenic removal: low-cost technology for community-based water supply in Bangladesh. *Water Sci. Technol.* 62(11): 2702–2709.

Environmental Arsenic in a Changing World –
Zhu, Guo, Bhattacharya, Ahmad, Bundschuh & Naidu (Eds)
ISBN 978-1-138-48609-6

Identifying the arsenic-safe aquifers of the Ganges Delta: some insights into sustainable aquifer management

M. Chakraborty[1], A. Mukherjee[1,2,3], K.M. Ahmed[4], P. Bhattacharya[5] & A.E. Fryar[6]
[1]*Department of Geology and Geophysics, Indian IIT Kharagpur, Kharagpur, West Bengal, India*
[2]*School of Environmental Science and Engineering, IIT Kharagpur, Kharagpur, West Bengal, India*
[3]*Applied Policy Advisory to Hydrogeosciences Group, IIT Kharagpur, Kharagpur, West Bengal, India*
[4]*Department of Geology, University of Dhaka, Curzon Hall Campus, Dhaka, Bangladesh*
[5]*KTH-International Groundwater Arsenic Research Group, Department of Sustainable Development, Environmental Science and Engineering, KTH Royal Institute of Technology, Stockholm, Sweden*
[6]*Department of Earth and Environmental Sciences, University of Kentucky, Lexington, KY, USA*

ABSTRACT: The widespread health impacts on millions of people from consumption of arsenic (As) contaminated groundwater of the Ganges delta necessitate an effort to locate As-safe aquifer zones and to develop a sustainable aquifer management policy. We look into the aquifer geometries, sediment characteristics and groundwater As concentrations to develop an understanding of the factors that inhibit As invasion/mobilization within the As-safe aquifers. However, the sustainability of the present-day safe aquifer is at risk due to the heavy groundwater pumping for irrigation and thus, promoting water conserving agricultural practices is extremely vital. We suggest that, formulating a sustainable As mitigation plan should also involve the non-scientific communities such as the local tubewell drillers and farmers for widespread implementation of the plan.

1 INTRODUCTION

Groundwater has been the main source of drinking and irrigation water for the inhabitants of the Ganges delta plains, since early 1970s. However, over the years, widespread occurrence of toxic levels of arsenic (As) in groundwater has been detected in the Ganges delta aquifers (Ahmed *et al.*, 2004). Presently, about 60 million people living in the delta are at risk from consumption of high levels of As in drinking water (Chakraborti *et al.*, 2009; Hossain *et al.*, 2015).

In order to protect the fundamental human need to access safe drinking water, several mitigation efforts have been undertaken in the past, but with limited success (Hossain *et al.*, 2015). Such efforts have revealed that installing hand-pump tubewells in As-safe aquifers is the most successful mitigation option, both in terms of community acceptance and costs involved. Hence, targeting As-safe aquifers is key to providing a sustainable mitigation plan for yielding low-As, potable drinking water.

Delineating As-safe aquifers involves a high-resolution investigation of the aquifer architecture and sediment characteristics as well as the groundwater chemistry, flow dynamics and abstraction patterns. However, a large-scale implementation of this knowledge base in providing As-safe groundwater also requires formulation of simple tools and techniques for easy identification of such As-safe aquifers by the non-scientific community, e.g., the tube well construction planners and the drillers (Hossain *et al.*, 2015).

The sustainability of the current safe aquifers may also be endangered by the heavy groundwater abstraction rates within the delta plains. Deep irrigational pumping may contaminate the low-As deeper aquifers by promoting enhanced inflow of As-rich groundwater from the shallow contaminated aquifers (Mukherjee *et al.*, 2011). Hence, the long-term success of As mitigation programs depends on planning and implementation of sustainable groundwater abstraction policies to protect the currently safe aquifers against As invasion/mobilization in future.

The transboundary nature of the Ganges delta aquifer demands participation from both India and Bangladesh to uphold the policies of such As-mitigation programs as an effort to provide safe drinking water to the inhabitants of the delta.

1.1 *The study area*

The Ganges delta forms a triangular lobe, being bounded by the Indian craton to its west, the main Ganges River to its north and east and the Bay of Bengal to the south (Fig. 1).

1.2 *Materials and methods*

This study involves development of a hydrostratigraphic model in RockWorks 15 (RockWare, USA) using ~2500 lithologs from the study area (collected from West Bengal Public Health Engineering Department (WBPHED) and Directorate of Public Health

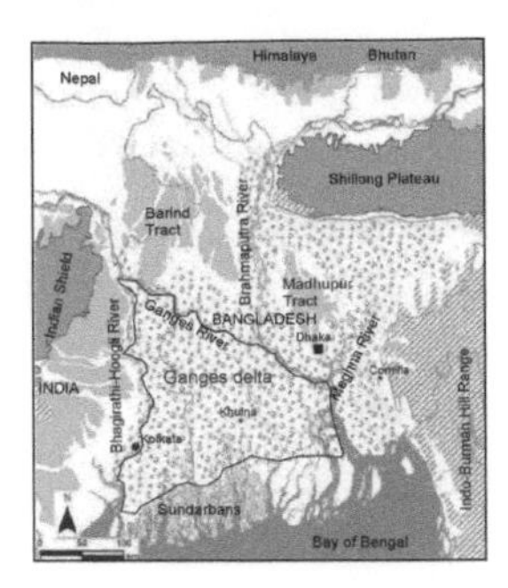

Figure 1. Map showing the study area (bordered by blue). Area with red dots marks the As-affected zone (modified from Burgess *et al.* (2010)).

Engineering (DPHE), Bangladesh) along with the analysis of groundwater As data.

2 RESULTS AND DISCUSSION

The hydrostratigraphic model shows the presence of multiple semi-connected aquifers separated by aquitards of varying thickness and extent in most parts of the delta; although, such interlayered geometry is especially well-developed to the south. These clay inter-layers hinder connectivity between aquifer units and act as a natural barrier against inflow of As or DOC from shallower depths. However, mostly towards the north, there are thick sequences of aquifer sands without sizeable clay inter-layers, resulting in free mixing of shallow As-rich waters with the relatively uncontaminated deeper groundwater (Ahmed *et al.*, 2004). Hence, delineation of site-specific aquifer geometry is crucial to the identification of potential As-safe aquifers in an area.

The sediment color of the aquifers varies from blackish and greyish, to whitish to yellowish, reddish and brownish, being indicative of the redox status of the aquifers (Hossain *et al.*, 2014). While the reduced grey/black sand aquifers usually have high As content, mainly due to reductive dissolution of iron hydroxides, the oxidized brown/yellow/red sand aquifers show lower As levels. Thus, sediment color may be used as one of the proxies to groundwater As content within the aquifers. However, such inferences must be integrated with an understanding of the aquifer geometry and recharge pattern.

The sustainability of the currently safe aquifers against invasion of As or DOC under future pumping conditions depends upon the pumping rates, depth of abstraction, the aquifer geometry and adsorption capacity of sediments. Although the discontinuous clay layers hinder the drawdown of DOC and As-rich water from above, the gaps in these impermeable layers allow such infiltration, under the enhanced vertical hydraulic gradient caused by irrigational pumping. Thus, deep irrigational pumping might eventually lead to large-scale contamination of the safe-water resources at depths, and hence must not be practiced.

3 CONCLUSION

Understanding the aquifer behaviors is key to the formulation of an effective As mitigation plan. A holistic understanding of the both the regional and site-specific aquifer geometry and characteristics (e.g., redox states), groundwater chemistry, flow dynamics and abstraction patterns is needed to identify the As-safe and sustainable aquifer zones.

Developing simple techniques to identify As-safe aquifers (i.e., sediment color, presence/absence of clay caps), followed by involving and training the local tubewell construction planners and drillers is crucial for large-scale implementation of the As-mitigation plan. Efficient use of groundwater in irrigation and water conserving agricultural practices should also be widely promoted among the farmers.

ACKNOWLEDGEMENTS

We acknowledge West Bengal Public Health Engineering Department (WBPHED) and Directorate of Public Health Engineering (DPHE), Bangladesh for providing some of the data for this study. We also acknowledge the Swedish International Development Cooperation Agency (Sida) for the financial assistance (Contribution 75000854).

REFERENCES

Ahmed, K.M., Bhattacharya, P., Hasan, M.A., Akhter, S.H., Alam, S.M.M., Bhuyian, M.A.H., Imam, M.B., Khan, A.A. & Sracek, O. 2004. Arsenic contamination in groundwater of alluvial aquifers in Bangladesh: An overview. *Appl. Geochem.* 19(2): 181–200.

Burgess, W.G., Hoque, M.A., Michael, H.A., Voss, C.J., Breit, G.N. & Ahmed, K.M. 2010. Vulnerability of deep groundwater in the Bengal aquifer system to contamination by arsenic. *Nat. Geosci.* 3 (2): 83–87.

Chakraborti, D., Das, B., Rahman, M.M., Chowdhury, U.K., Biswas, B., Goswami, A.B., Nayak, B., Pal, A., Sengupta, M.K., Ahamed, S., Hossain, A., Basu, G., Roychowdhury, T. & Das, D. 2009. Status of groundwater arsenic contamination in the state of West Bengal, India: a 20-year study report. *Mol. Nutr. Food Res.* 53 (5): 542–541.

Hossain, M., Bhattacharya, P., Frape, S.K., Jacks, G., Islam, M.M., Rahman, M.M., Hasan, M.A. & Ahmed, K.M. 2014. Sediment color tool for targeting arsenic-safe aquifers for the installation of shallow drinking water tubewells. *Sci. Total Environ.* 493: 615–625.

Hossain, M., Rahman, S.N, Bhattacharya, P., Jacks, G., Saha, R. & Rahman M. 2015. Sustainability of arsenic mitigation interventions – an evaluation of different alternative safe drinking water options provided in Matlab, an arsenic hot spot in Bangladesh. *Front. Environ. Sci.* 3 (30): 1–15.

Mukherjee, A., Fryar, A., Scanlon, B., Bhattacharya, P. & Bhattacharya, A. 2011. Elevated arsenic in deeper groundwater of the western Bengal basin, India: extent and controls from regional to local scale. *Appl. Geochem.* 26(4): 600–613.

Environmental Arsenic in a Changing World –
Zhu, Guo, Bhattacharya, Ahmad, Bundschuh & Naidu (Eds)
ISBN 978-1-138-48609-6

ASMITAS – a novel application for digitalizing the SASMIT Sediment Color Tool to identify arsenic safe aquifers for drinking water supplies

S. Sharma[1], P. Bhattacharya[2], D. Kumar[1], P. Perugupalli[3], M. von Brömssen[4], M.T. Islam[2] & M. Jakariya[5]
[1]*Exceldot AB, Bromma, Sweden*
[2]*KTH-International Groundwater Arsenic Research Group, Department of Sustainable Development, Environmental Sciences and Engineering, KTH Royal Institute of Technology, Stockholm, Sweden*
[3]*Spectralinslights Pvt Ltd., Bangalore, India*
[4]*Ramböll Sweden AB, Stockholm, Sweden*
[5]*Department of Environmental Science and Management, North South University, Dhaka, Bangladesh*

ABSTRACT: Arsenic (As) contamination in groundwater is an environmental health hazard in several part of the world and a large number of population has affected due to its toxic nature. In Bangladesh, a comprehensive research was accomplished which established a very strong correlation between the aquifer sediments and As concentration in the groundwater. The local drillers practice this knowledge over a time. Based on an action research conducted by the SASMIT project team from KTH Royal Institute of Technology, Sweden in collaboration with the local drillers in Bangladesh and a novel handheld sediment color tool was produced to facilitate the local drillers to target safe aquifers for safe tube-well installation. This study was made to advance this tool to be developed as a fully artificial intelligence (AI) based digital ASMITAS (Arsenic Mitigation at Source) tool based on spectral scanning of the sediment color. An HSI camera enables us to capture continuous spectral channels from a given object ranging from IR bands, through VR to UV ranges and hence scores comprehensive imaging output over traditional RGB data based color cameras, thus increase the perception of the color with more accuracy and precision. This new tool can be scalable to cover large geographical area and hence can be very useful tool for local drillers and other stakeholders for smart decision making for installation of safe tube-wells for mitigating As at source level.

1 INTRODUCTION

Safe drinking water is defined as a basic human right and a specific goal for sustainable development (SDG 6) (Hossain *et al.*, 2017; Johnston, 2016) that serves as a vital component to protect public health. The safe water access is severely impaired due to the widespread occurrence of natural arsenic (As) in groundwater in Bangladesh. Since the discovery of As in the country in 1993; several million population are still exposed to As at levels above the WHO guideline ($10\,\mu g\,L^{-1}$) and the Bangladesh drinking water standard (BDWS; $50\,\mu g\,L^{-1}$). Keeping in view the magnitude of the human health impacts and the outcomes of the mitigation programs, the main challenge is to develop a sustainable mitigation strategy to scale up safe water access (Ahmad *et al.*, 2017; Hossain *et al.*, 2017). Tube-wells are most widely accepted drinking water option (Jakariya *et al.*, 2007a, 2007b; Hossain *et al.*, 2015), and ~90% of these tube-wells are installed by local tube-well drillers, which emphasizes the important role of the tube-well drillers.

The mechanism behind the naturally occurring excess As concentrations in groundwater in the Bengal delta is the reducing condition in some sediment sections where Fe-oxyhydroxides are reduced releasing As to the groundwater (Bhattacharya *et al.*, 1997; Ahmed *et al.*, 2004). This implies that there is a covariation of As and ferrous iron in the groundwater. The local drillers have learnt to search for low iron groundwater and this coincides with low As (Jonsson and Lundell, 2004). The low iron and As groundwater occurs in oxidized sediments which can be characterized by the red color of the sediment (Hossain *et al.*, 2014; von Brömssen *et al.*, 2007) generally occurring at depths within 100 m, accessible for the local drillers through hand percussion drilling. Since the cost of drilling with hand-percussion technology to a depth of 100 m is affordable for most households, it is important to enhance the capacity of the local drillers to perceive and classify sediments and identify safe aquifers. Identifying the local drillers as the main driver for the provision of safe drinking water, a novel handheld Sediment Color Tool (SCT) was developed based on the correlation of the sediment color with the As concentration in groundwater. This study advances the SCT using artificial intelligence (AI)-based digital ASMITAS (Arsenic Mitigation at Source) tool based on spectral scanning of the sediment color. Using a HSI camera, continuous spectral channels from a given sediment ranging from IR bands, through VR to UV ranges is captured and hence provides comprehensive imaging output over traditional color cameras using RGB data base and have a clear advantage to increase the perception of the color with more accuracy and precision.

2 METHODS/EXPERIMENTAL

2.1 *Understanding the scientific logic for intervention*

Having SASMIT as a concept for practical and low-cost solution to identify the aquifer sediments with low As groundwater was a major breakthrough in understanding the groundwater science. However, one of the major challenges of the SASMIT was inability to scale for other geographic location and often failed to produce reliable data on the presence of As in new areas where sediments are inadequately characterized. Digitalization of the SASMIT tool would therefore prove to be an important step to overcome some of the potential issues related to upscaling of the use of the color pallet matching tool in unexplored areas. Some of these technical challenges in SASMIT color tool are as follow:

a) The handheld tool developed in SASMIT project was based on the eye-ball matching of sediments colors and manually matching to the nearest Munsell colors, which were often challenging to define the exact reference color;
b) The correlation between the sediments colors and the As concentration in groundwater varies from place to place due to the intrinsic characteristics of the sediment layers;
c) The eye ball matching has a considerable scope for errors due to its perception under varying ambient light conditions, presence of moisture in sediments and human error in recognizing the sediment colors; and
d) The handheld sediment color tool can house the dry sediments that will have different color appearance as compared to wet color.

2.2 *Spectral imaging of the sediments*

Since over three decades, scientist and researchers are using Image Spectrometry (IS) technique to monitor environment resources, natural disasters, climate change and other natural media. The remote sensing satellite are equipped with high-resolution spectral camera to monitor earth round the clock. The new CMOS based sensor (Strle *et al.*, 2015; Wong, 1997) has helped to design compact HSI camera and hence its affordability has resulted in its wider adoptability in several applications. The HSI camera captures several hundred of continuous spectral channels from a given object, which ranges from IR bands, through VR to UV ranges and hence scores comprehensive imaging output over traditional RGB data based color cameras. The spectrum obtained from an HSI camera has a profile of reflection absorption ranging from 0 to 1 for a given object with respect to various spectral band.

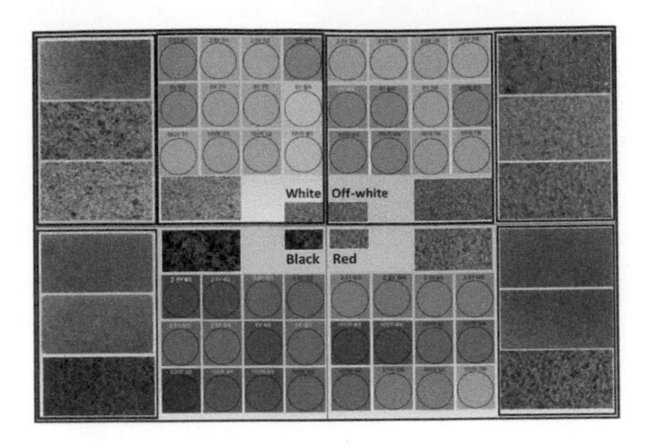

Figure 1. Color variations observed in aquifer sediment during the development of the Sediment Color Tool during the SASMIT project based on matching with Munsell color codes (Hossain *et al.*, 2014; SASMIT, 2014).

The spectral profiles using HSI camera could be compared with the traditional RGB sediment color and respective Munsell colors and hence can be calibrated with high level of precision.

ASMITAS constitutes two major steps, which include: i) digitalizing the reference sediments baselined with absolute As concentration in water with respect to the depth information of the tube-well based on the SASMIT study (Hossain *et al.*, 2014; SASMIT, 2014); and ii) analysing the new samples with the reference library built on basis of the baseline sediments colors.

2.3 *ASMITAS Deep Learning Tool*

ASMITAS consists of an advanced analytic tool available both in cloud based and node-based system. The node based analytic tool provide As profiling in time span of minutes. The reference tool is being enhanced by the reference sediments images and state of the art techniques are being used to capture image of new samples and a normalized image data that is compared with the reference library. Predictive analysis tools accomplish the regression and predict the aquifers with safe levels of As in groundwater at the given depth. This can be further refined on the basis of real time analysis of the water samples and through incorporation new hyperspectral image data of the sediment color from larger geographical areas with diverse color specificities.

3 RESULTS AND DISCUSSION

3.1 *Core scientific logic for intervention*

ASMITAS overcomes the above challenges by means of digitalizing the sediments colors (Fig. 1) and comparing with a base reference library with very advanced algorithm based on Artificial Intelligence to estimate the color of the sediments followed by determination of As concentration on real time basis.

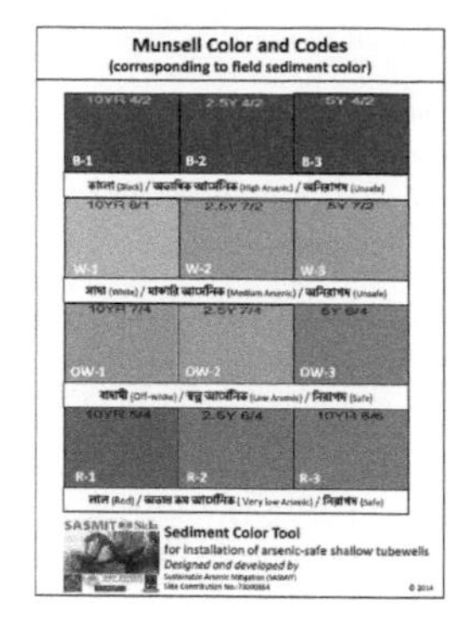

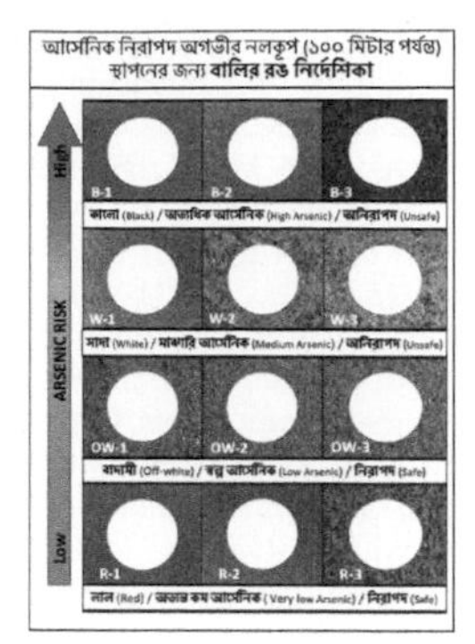

Figure 2. The SASMIT Sediment Color Tool showing the matching Munsell color codes and the risk of As in groundwater abstracted from the sediments (Hossain *et al.*, 2014; SASMIT, 2014).

A distinct relationship of sediment color and corresponding As concentrations in well water. has been documented by the SASMIT research (Hossain *et al.*, 2014, SASMIT, 2014; von Brömssen *et al.*, 2007). Arsenic concentration in more than >90% of the shallow wells installed in black sands is high with an average of 239 $\mu g\,L^{-1}$ and therefore the local drillers have evolved the practice of shallow tube-well installation in red sediments where groundwater As levels are below the WHO drinking water guideline (10 $\mu g\,L^{-1}$). Arsenic levels are also low in off-white sediments, however targeting off-white sands is limited due to uncertainties of proper identification of color, and dependent light is a an important factor must be avoided. Based on these findings a simple color based tool for targeting shallow aquifers for the installation of arsenic safe community tube-wells was developed for the local drillers (Fig. 2). The low As wells installed in red colored sediments comply with the drinking water standards for As, although concentrations of Mn in many of these wells are above national drinking water standards, As warrants highest attention due to its acute health effects.

3.2 *Spectral imaging of the sediments*

The spectral profile of Munsell colors were determined and compared to the sediments samples spectra profile for 0–350 intermediate bands of spectrum. The original samples were used in the study were snapshot by Hyper Spectral Saber Camera provided by Spectral Insight, Bangalore for extending the SASMIT study. Spectral data for all the samples and Munsell colors were obtained to develop a reference library. This library can be used for analyzing the blind samples and the analytic tool can determine the As level.

The spectrum obtained from an HSI camera (Fig. 3) has a profile of reflectance based on the degree of absorption ranging from 0 to 1 for a given sediment with respect to various spectral bands related to the specific attributes of each sediment. The degree of absorption 0 represents dark sediment where all light is

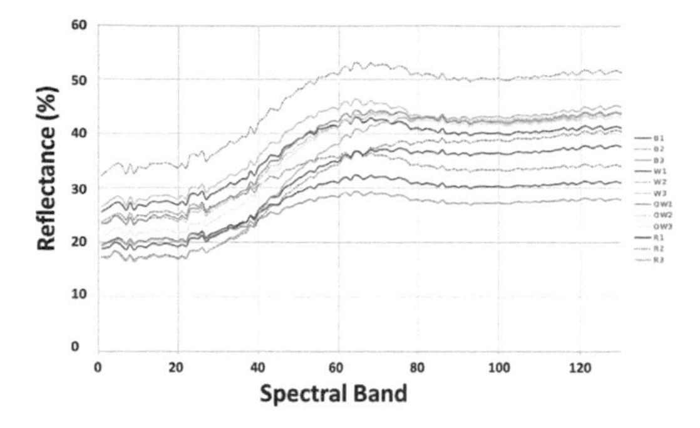

Figure 3. Spectral profiles of the reference aquifer sediments representing the four sediment color classes used for developing the Sediment Color Tool during the SASMIT project.

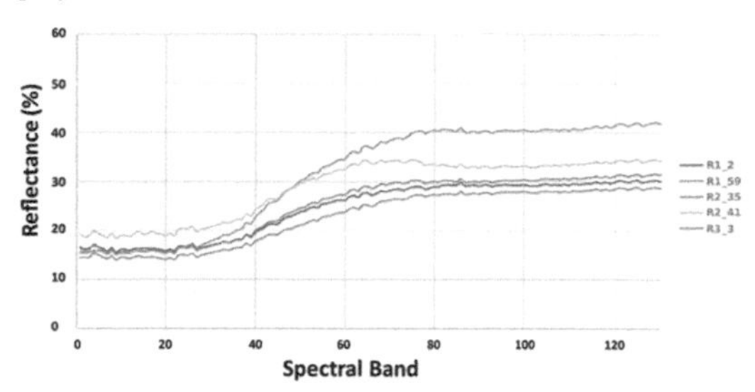

Figure 4. The spectral profiles of the blind aquifer sediments representing the five distinctive samples.

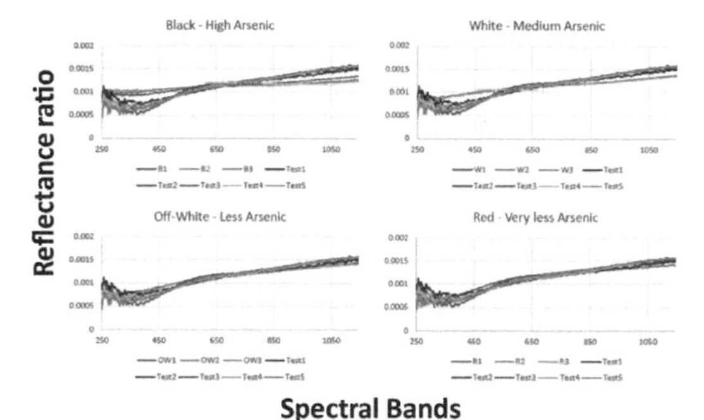

Figure 5. Comparison of the spectral profiles with the sediments color determined with respective to the Munsell colors.

absorbed and 1 represent complete white where 100% light is reflected.

In order to establish the hypothesis and its validation, spectral scans for a set of blind samples were taken by the HSI camera and the spectral profiles are presented in Figure 4.

These blind samples were further subjected to regression analysis using a statistical analysis tool. Figure 5 shows the comparison of the spectral profiles with the sediments color determined with respective to the Munsell color spectral data. Finally, all the blind samples were tested with the reference library built with digitalized aquifer sediments.

The tool suggests a very strong correlation between the blind samples with OW1, R1, R2 and R3 sets of

Table 1. Regression analysis (R^2 error) of blind samples and reference sediments.

Sediments	R1_2	R1_59	R2_35	R2_41	R3_3
B1	0.56	0.75	1.04	0.79	0.98
B2	0.86	1.07	1.35	1.11	1.3
B3	0.77	0.97	1.25	1.01	1.2
W1	0.53	0.68	0.95	0.71	0.88
W2	0.51	0.69	0.97	0.72	0.9
W3	0.55	0.7	0.96	0.74	0.9
OW1	0.35	0.41	0.64	0.42	0.57
OW2	0.37	0.41	0.63	0.42	0.57
OW3	0.42	0.38	0.55	0.41	0.49
R1	0.48	0.4	0.52	0.43	0.47
R2	0.61	0.36	0.23	0.36	0.205
R3	0.66	0.42	0.27	0.4	0.205

sediments (Table 1). The tool is also capable of offset the white correction, wet and dry sample reflectance difference and profiling of As flow in targeted area and hence offers an interesting opportunity for scaling this tool to different location.

4 CONCLUSION AND FUTURE OUTLOOK

The outcomes of the present study indicate that ASMITAS has the potential to overcome the challenges by means of digitalizing the sediments colors and comparing with a base reference library with very advanced algorithm based on Artificial Intelligence to estimate the color of the sediments followed by determination of As concentration on real time basis.

The ASMITAS tool can be used by the local drillers to identify the color of the sediments in a more accurate manner for targeting the safe aquifers for installation of safe drinking water wells. The tool can also be upgraded with the input of the concentration of extractable As in the sediments for characterization of the toxicity in different sediments.

ACKNOWLEDGEMENTS

The authors thank Sida-SASMIT (Contribution number 73000854) and the Sida-UNICEF project – Systems strengthening and scaling up Drinking Water Safety in Bangladesh (Contribution number 52170040) for the baseline study in the process of upgrading the visual Sediment Color tool into a digitalized sediment color tool. We also thank Spectral Insights, Bangalore India for providing the Hyper Spectral Camera and the interpretative support during the realization of this study.

REFERENCES

Ahmad, A., Richards, L.A. & Bhattacharya, P. 2017. Arsenic remediation of drinking water: an overview. In: P. Bhattacharya, D.A. Polya & D. Jovanovic (eds.) *Best Practice Guide on the Control of Arsenic in Drinking Water.* Metals and Related Substances in Drinking Water Series, IWA Publishing, UK, pp. 79–98.

Ahmed, K.M., Bhattacharya, P., Hasan, M.A., Akhter, S.H., Alam, S.M.M., Bhuyian, M.A.H., Imam, M.B., Khan, A.A. & Sracek, O. 2004. Arsenic contamination in groundwater of alluvial aquifers in Bangladesh: an overview. *Appl. Geochem.* 19(2): 181–200.

Bhattacharya, P., Chatterjee, D. & Jacks, G. 1997. Occurrence of arsenic contaminated groundwater in alluvial aquifers from Delta Plains, Eastern India: options for safe drinking water supply. *Int. J. Wat. Res. Manag.* 13(1): 79–92.

Hossain, M., Bhattacharya, P., Frape, S.K., Jacks, G., Islam, M.M., Rahman, M.M., Hasan, M.A. & Ahmed, K.M. 2014. Sediment color tool for targeting arsenic-safe aquifers for the installation of shallow drinking water tubewells. *Sci. Total Environ.* 493: 615–625.

Hossain, M., Rahman, S.N., Bhattacharya, P., Jacks, G., Saha, R. & Rahman, M. 2015. Sustainability of arsenic mitigation interventions – an evaluation of different alternative safe drinking water options provided in Matlab, an arsenic hot spot in Bangladesh. *Front. Env. Sci.* 3(30): 1–15.

Hossain, M., Bhattacharya, P., Jacks, G., von Brömssen, M., Ahmed, K.M., Hasan, M.A. & Frape, S.K. 2017 Sustainable arsenic mitigation – from field trials to implementation for control of arsenic in drinking water supplies in Bangladesh. In: P. Bhattacharya, D.A. Polya & D. Jovanovic (eds) *Best Practice Guide on the Control of Arsenic in Drinking Water.* Metals and Related Substances in Drinking Water Series, IWA Publishing, UK, pp. 99–116.

Jakariya, M., Vahter, M., Rahman, M., Wahed, M.A., Hore, S.K., Bhattacharya, P., Jacks, G., Persson, L.-Å. 2007a. Screening of arsenic in tubewell water with field test kits: evaluation of the method from public health perspective. *Sci. Total Environ.* 379(2–3): 167–175.

Jakariya, M., von Brömssen, M., Jacks, G., Chowdhury, A.M.R., Ahmed, K.M. & Bhattacharya, P. 2007b. Searching for sustainable arsenic mitigation strategy in Bangladesh: experience from two upazilas. *Int. J. Environ. Pollut.* 31(3–4): 415–430.

Johnston, R.B. 2016. Arsenic and the 2030 Agenda for Sustainable Development. In: P. Bhattacharya, M. Vahter, J. Jarsjö, J. Kumpiene, A. Ahmad, C. Sparrenbom, G. Jacks, M.E. Donselaar, J. Bundschuh, & R. Naidu (eds) *Arsenic Research and Global Sustainability As 2016.* CRC Press, Boca Raton, FL, pp. 12–14.

Jonsson, L. & Lundell, L. 2004. Targeting safe aquifers in regions with arsenic-rich groundwater in Bangladesh. Case study in Matlab Upazila. Minor Field Studies No 277. Swedish University of Agricultural Sciences, SLU External Relations, Uppsala. ISSN 1402–3237.

Strle, D., Nahtigal, U., Batistell, G., Zhang, V., Ofner, E., Fant, A. & Sturm, J. 2015. Integrated high resolution digital color light sensor in 130 nm CMOS technology. *Sensors* 15(7): 17786–17807.

von Brömssen, M., Jakariya, Md., Bhattacharya, P., Ahmed, K. M., Hasan, M.A., Sracek, O., Jonsson, L., Lundell, L. & Jacks, G. 2007. Targeting low-arsenic aquifers in groundwater of Matlab Upazila, Southeastern Bangladesh. *Sci. Total Environ.* 379(2–3): 121–132.

Wong, H.S. 1997. CMOS image sensors – recent advances and device scaling considerations, *IEEE IEDM Technical Digest*, pp. 8.5.1–8.5.4.

Environmental Arsenic in a Changing World –
Zhu, Guo, Bhattacharya, Ahmad, Bundschuh & Naidu (Eds)
ISBN 978-1-138-48609-6

Deep hand tube-well water for achieving sustainable development goal in arsenic affected villages: Bangladesh experience

B.A. Hoque[1], S. Khanam[1], M.A. Zahid[1], M.M. Hoque[1], N. Akter[2],
M.N. Mahmud[2], S. Huque[1] & S. Ahmed[1]
[1]*Environment and Population Research Centre (EPRC), Dhaka, Bangladesh*
[2]*UNICEF Bangladesh, Dhaka, Bangladesh*

ABSTRACT: Arsenic mitigation rural water supply is a continued challenge in developing countries. Here we present data in water supply through highly demanded community-level deep hand tube-well (DTW) and discuss perspectives for achieving Sustainable Development Goal (SDG-6.1) in rural Bangladesh. Equity in access to the DTWs was addressed by applying GPS for distance between DTWs and houses of about 34,500 beneficiaries. About 95% and 50% of the water point water samples respectively conformed 1) arsenic and bacteriological, and ii) arsenic, bacteriological, manganese and iron quality standards of Bangladesh. Improved technology/hand pump at premises reduced to about 5% from 50%. Safely managed drinking water service after the SDG-6.1 remained questionable. Research on comprehensive program for safely managed pipe water, appropriate technologies and other issues towards SDG recommended.

1 INTRODUCTION

Consumption of arsenic contaminated water has been a public health problem among hundreds of millions of population in Asia and other countries (Shankar *et al.*, 2014; Jack *et al.*, 2003). Smith *et al.* (2000) claimed more than 1.5 decades ago, that Bangladesh had 'the largest mass poisoning of a population in history because groundwater used for drinking has been contaminated with naturally occurring inorganic arsenic'. Although the country has reached MDG drinking water target (UNICEF and WHO, 2015); in 2013 more than 19 million populations (12.4%) were exposed to the risks of drinking arsenic contaminated water (BBS and UNICEF, 2015). The population exposed to arsenic contaminated water reduced 1.0 percentage point from 2009 to 2013. Community-level hand deep tubewell (DTW) has been the predominant and people's most in-demand arsenic mitigation technology in rural Bangladesh (Bilqis *et al.*, 2006; Kundu *et al.*, 2016; Ravenscroft *et al.*, 2014). Most of the DTWs provided with arsenic and bacteriological/fecal contamination safe water (BBS and UNICEF, 2015; Bilqis *et al.*, 2006). Inequitable distribution of DTWs has been a common problem (Geen *et al.*, 2016).

Here we have presented our experiences gained in scopes of DTW based arsenic mitigation water supply in SDG 6.1 perspectives, particularly under safely managed safe drinking water and updated 'JMP ladders for drinking water'.

2 METHODS/EXPERIMENTAL

An action research on community-based arsenic safe drinking water supply was conducted among approximately 0.2 million rural populations of 10 Unions in Narail district by EPRC in collaboration with UNICEF-Bangladesh and DPHE. The project was funded by Embassy of The Netherlands in Bangladesh. One of the main objectives of the project was to create access to safe drinking water among identified worst arsenic exposed approximately 55,000 populations. The screening was done from November'13 to January 2014. One group of the volunteers (01 male & 01 female) visited every house with a tubewell in a village and conducted the arsenic test on tubewell water sample. They painted arsenic safe ($\leq$0.05 mg L^{-1} arsenic) tubewells with green color and arsenic unsafe ($>$0.05 mg L^{-1} arsenic) tubewells red based on the test results. Project officers measured GPS values of every tested tubewell and randomly checked about 5% of the results. Out of the 144 project villages in the 10 Unions approximately 75 worst arsenic affected villages were selected initially for priority water services based on arsenic screening and baseline results. Then the households, with local women, were demarked into a preliminary cluster based on all/majority households which did not meet the UNICEF suggested 150 m access to safe drinking water device. A cluster included about 20 households per arsenic DTW.

According to the Bangladesh national strategies, Chairmen of Union Councils take final decision on sites proposed for improved water technologies based on needs. The arsenic screening results, maps and lists of the proposed DTW sites were sent to every Union Council for their necessary actions and approval. The lists of the proposed sites were prepared by project officers, with the local women and men, based on the following project criteria for site and technology selection: families in the cluster do not have access to a safe drinking water technology within 150 m, community (women-led) chosen DTW sites, about 20 families per DTW/cluster, poor and vulnerable families and technical feasibility. Maps on proposed and installed DTW sites were prepared for every Union/village as shown in Figure 1.

Arsenic in water point samples were measured by ECONO quick arsenic test kit supplied by UNICEF. Membrane filter technique by Potatest was used to measure fecal coliform bacteria (cfu 100 mL) in the sites.

Here we present a selected part of the data related to arsenic safe 343 DTWs (>150 m depth) installed by the project and construction completed about one month before the water quality measurement.

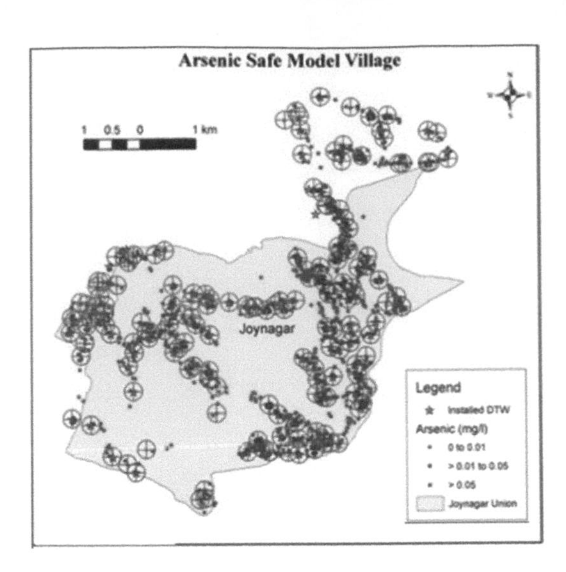

Figure 1. The documents on proposed sites with arsenic values and GPS map.

3 RESULTS AND DISCUSSION

3.1 *Equity and conflict resolution*

Overall, about 61% of the screened TWs were found contaminated with >0.05 mg L^{-1} arsenic (Bangladesh standard) and 73.8% when the reference value is >0.01 mg L^{-1} (WHO) standard. The demand for deep tubewells was very high and led to serious conflicts on the proposed potential sites among local leaders, influential families and potential beneficiaries. Most of the members and chairmen of the Union Councils and other leaders were not willing to accept/approve the proposed sites submitted to them by the project and local communities. Also they were not happy that women were given the leading roles in site selection. As a consequence, they threatened to stop the project activities. The project team met with the authorities, chairmen and leaders to discuss/show papers on project objectives, government approved criteria for equitable access to water technology, contamination distribution, the documents on proposed sites with arsenic values and GPS maps. The participants to the meeting being convinced with the used criteria and GPS maps, agreed to the proposed sites and the project was smoothly completed.

3.2 *Collection time*

Here 150 m criteria were adopted for maximum distance between households and DTWs in a cluster as suggested by UNICEF. Collection time between households and project installed DTWs were found within 30 minutes, when measured in about randomly selected 30% of the clusters.

3.3 *Location on premises*

The rate of households with a shallow or deep tubewell at premises was about 50% before the intervention. 85% of the owners of those tube-wells claimed that they could collect water any time and felt safe; though water from about half of those tube-wells had arsenic and/or bacteriological contamination. The rate with the DTWs at premises reduced 5% after the intervention. One-third of the DTW users reported that they felt safe to collect water any time.

3.4 *Water quality*

Out of the 344 arsenic safe DTWs by Bangladesh standard (0.05 mg L^{-1}), water samples from 94.5% conformed Bangladesh standard on fecal contamination (0 cfu per 100 mL). WHO guideline values (combined arsenic and fecal coliform) were met in water samples of 90.1% DTWs.

Bangladesh has water quality standards for manganese and iron. Water samples from 51% of DTWs conformed to Bangladesh standards against arsenic, fecal coliform, iron and manganese. WHO guideline does not show health risk values for Manganese and Iron. Here the significant differences in values between water quality (for the four parameters) by Bangladesh standards and WHO guideline values may create confusions on the achievements.

Approximately 90% of the DTWs beneficiaries (households) reported that they stored the drinking water overnight or more hours. 57% of the stored water samples from the DTWs with 0 cfu per 100 mL count at water discharge point were found contaminated. Bacteriological contamination of water between technology/supply point and household storage points have been reported earlier. Chlorination and other disinfection methods can control risks of bacteriological contamination at supply and consumption points. An earlier study on regular chlorination at household level in a part of the same area (kalia) was not appreciated by the same community. They preferred chlorination with pipe water supply (Bilqis *et al.*, 2005).

4 CONCLUSIONS AND RECOMMENDATIONS

The objective about creating equitable access to improved drinking water technology through deep tube-wells in terms of safe arsenic and free from fecal contamination conditions was achieved (95%). The results met the 'basic' level of SDG. The equity in the service and undue conflict resolutions were realized at community level by developing and implementing the innovative technique incorporating GPS coordinates with arsenic values and common/poor beneficiaries, leaders, politicians other locations. The results showed significant questions and policy gaps in regard to 'safely managed' level. Also the risk for exposure to the drinking of bacteriological contaminated water at consumption point (from storage container) remained very high.

We recommend research on comprehensive program for safely managed pipe water, community-based appropriate technologies, gaps between national and international issues, women in decision making, policy and other related aspects towards SDG.

ACKNOWLEDGEMENTS

We acknowledge Narail people and the administration for participation and support. We sincerely thank P. Ravenscroft, M. Bolton, Hrachya Sargsyan, M.S. Rahman, M. Slotema, Khaleduzzaman, M. Siddik, A. Rahman and other colleagues from UNICEF, DPHE, Dutch Embassy and EPRC for their contributions to the project.

REFERENCES

BBS & UNICEF 2015. Multiple indicator cluster survey. Progotir Pathey, 2012–2013, Government of the People's Republic of Bangladesh.

Bilqis, A.H., Khanam, S. & Motin, A. 2005. Assessment and management of risks in supply of safe drinking water through alternative water supply options in arsenic affected area. Report submitted to UNICEF & DFID by EPRC.

Bilqis, A.H., Yamaura, S., Sakai, A., Khanam, S., Mahbooba, K., Yamen, H., Sanowar, H., Islam, S. & Obaid, H. 2006. Arsenic mitigation for water supply in Bangladesh: appropriate technological and policy perspectives. *Water Qual. Res. J. Can.* 41(2): 226–234.

Geen, A.V., Ahmed, K.M., Ahmed, E.B., Choudhury, I., Mozumder, M.R., Bostick, B.C. & Mailloux, B.J. 2016. Inequitable allocation of deep community wells for reducing arsenic exposure in Bangladesh. *J. Water Sanit. Hyg. Dev.* 6(1): 142–150.

Jack, C.N., Wang, J. & Shraim, A. 2003. A global health problem caused by arsenic from natural sources. *Chemosphere* 52(9): 1353–1359.

Kundu, D.K., Bas, J.M., Vliet, V. & Aarti, G. 2016. The consolidation of deep tube well technology in safe drinking water provision: the case of arsenic mitigation in rural Bangladesh. *Asian J. Technol. Innov.* 24(2): 254–273.

Ravenscroft, P., Kabir, A., Hakim, S.A.I., Ibrahim, A.K.M., Ghosh, S.K., Rahman, M.S., Akter, F. & Sattar, M.A. 2014. Effectiveness of public rural water points in Bangladesh with special reference to arsenic mitigation. *J. Water Sanit. Hyg. Dev.* 4(4): 545–562.

Shankar, S., Shanker, U. & Shikha. 2014. Arsenic contamination of groundwater: a review of sources, prevalence, health risks, and strategies for mitigation. *The Scientific World Journal*, 304524.

Smith, A.H., Lingas, E.O. & Rahman, M. 2000. Contamination of drinking-water by arsenic in Bangladesh: a public health emergency. *Bull. World Health Organ.* 78(9): 1093–1103.

UNICEF and World Health Organization (WHO) 2015. Progress on sanitation and drinking water – 2015 update and MDG assessment. WHO Press, World Health Organization, Geneva, Switzerland, 80p.

Environmental Arsenic in a Changing World –
Zhu, Guo, Bhattacharya, Ahmad, Bundschuh & Naidu (Eds)
ISBN 978-1-138-48609-6

Small-scale piped water supply: end-user inclusive water research in arsenic affected areas in India and Bangladesh (DELTAP)

D. van Halem[1], B.A. Hoque[2], A.K. Ghosh[3,4], M.E. Donselaar[1,5], E. Karana[6], P. Bhattacharya[7] & K.M. Ahmed[8]

[1]*Faculty of Civil Engineering and Geosciences, Delft University of Technology, Delft, The Netherlands*
[2]*Environment and Population Research Centre (EPRC), Dhaka, Bangladesh*
[3]*Mahavir Cancer Institute and Research Centre, Patna, India*
[4]*Bihar State Pollution Control Board, Patna, India*
[5]*KU Leuven, Department of Earth and Environmental Sciences, Leuven-Heverlee, Belgium*
[6]*Faculty of Industrial Design Engineering, Delft University of Technology, Delft, The Netherlands*
[7]*KTH-International Groundwater Arsenic Research Group, Department of Sustainable Development, Environmental Science and Engineering, KTH Royal Institute of Technology, Stockholm, Sweden*
[8]*Department of Geology, University of Dhaka, Dhaka, Bangladesh*

ABSTRACT: DELTAP is a multi-disciplinary research project, where geologists, water treatment scientists and industrial design engineers join forces to develop an integrated approach towards small-scale piped water supply (SPWS) systems in the arsenic-affected Ganges-Brahmaputra-Meghna Delta. The project has started in 2016 with a field study in Bihar (India) with a focus on water quality mapping with mobile crowd participation. The coming years the research will continue with 3 PhD candidates, both in India and Bangladesh, with the ultimate aim to develop blueprints for end-user inclusive SPWS systems.

1 INTRODUCTION

Arsenic-contaminated groundwater causes serious health risks affecting millions of people living in the Ganges-Brahmaputra-Meghna Delta. Mitigation attempts for this unresolved problem have taught critical success factors to be: true local embedding (social, technical and economical), adaptability and dynamics. DELTAP's integrative approach (Fig. 1) aims for sustainable innovation to reduce contaminant exposure with Small-scale Piped Water Supply (SPWS) and active end-user participation through mobile crowd participation (MCP) and Do It Yourself (DIY) Laboratory. SPWS combines the attractiveness of socially/economically feasible tap connections with smart targeting of safest sources, centralized treatment and end-user inclusion.

In urban laboratories in Bangladesh and India, DELTAP integrates MCP with:

- Dynamic geological models for prediction of arsenic patches in deltaic systems (Donselaar *et al.*, 2017),
- Blueprints for source-to mouth safe water supply, and,
- Development of DIY materials and products which take human practice and local sources as the main drive in the design process (Rognoli *et al.*, 2015).

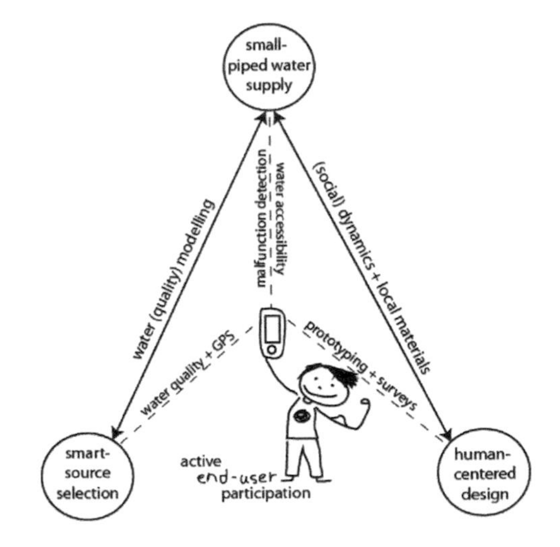

Figure 1. DELTAP's integrative approach for small-scale piped water supply with active end-user participation through mobile crowd participation and DIY practices.

2 METHODS

2.1 *Mobile crowd participation*

The first focus of the DELTAP research team – in close collaboration with A.N. College Patna and UNICEF

Bihar – was to investigate arsenic distribution in groundwater wells in Bihar (India) by combining traditional laboratory analyses with new approaches for mobile crowd monitoring of water quality (Mink, 2017). For this purpose, the Akvo.org Caddisfly app was used, in combination with HACH test kits for iron and phosphate.

2.2 *Geological survey*

Arsenic concentrations in the aquifer are highly variable, and directly linked to the geomorphological setting of the affected areas. Satellite imaging surveys, in combination with ground-truth data from sediment cores and electric logs from shallow wells, and ground-penetrating radar surveys will yield detailed maps and predictive 3D models of As-polluted and As-free areas.

2.3 *DIY materials and water*

As an initial exploratory study with DELTAP, it was investigated whether plastic waste could be used to create new materials to be potentially used in making water and/or hygiene products in Dhaka, Bangladesh. For this research, through the 'material-driven design' approach, the mechanical performance of the plastics as well as their experiential qualities (i.e. how they are appreciated by people) have been explored. The machines developed by www.preciousplastic.com were used for DIY manufacturing of the end products.

2.4 *Co-production: engaging end-users*

Stakeholder co-production has a dominant role in the research methodology, at all levels: End-users produce MCP data through interactive apps; SMEs co-develop apps and DIY products; NGOs build SPWS for pilot research; joint science/NGO policy briefings. This co-production results in strong embedding of results, optimizing feasibility of highly needed impact for sustainable and inclusive development: creating pathways for safe drinking water for millions.

3 RESULTS

3.1 *Mobile crowd participation*

A surprising high number of households in rural Bihar owned a 4G-connected smartphone and was willing and capable of executing the smartphone water quality test independently. With this survey a first step is taken towards development of mobile crowd participation as an innovative methodology for water research and monitoring. The results of 346 household tubewells will feed into the development of a predictive geological model to visualize arsenic-free aquifers in deltaic systems.

3.2 *Geological survey*

Sand-prone fluvial point-bar deposits surrounded by clay-filled former meander bends in a Holocene fluvial and deltaic setting are the most common geomorphological type in all As-polluted areas. The sediment type differences have inherent porosity and permeability heterogeneity, which is the cause of the large spatial variability in As concentration in the aquifer space. Generic models taking into account the geomorphology-pollution relation have the potential to be applied world-wide in similar settings.

3.3 *DIY materials: water container lid*

The new approach of linking human-centered design with local materials and DIY practices to water products; has resulted in a spin-off product that prevents Kolshi containers from microbial contamination. This innovative process of local material reuse is an example of end-user inclusive water supply, although the potential water quality risks of plastic reuse need further investigation.

4 CONCLUSIONS

The DELTAP project is focusing on demonstrating the opportunities of end-user inclusive water research in arsenic affected areas. Initial results have shown the potential of mobile crowd participation Bihar (India) and DIY reuse of local materials in Dhaka (Bangladesh) as new methods to support sustainable safe water supply.

ACKNOWLEDGEMENTS

This research was funded through the "Urbanising Deltas of the World" programme of NWO WOTRO – Science for Global Development.

REFERENCES

Donselaar, M.E., Bhatt, A.G. & Ghosh A.K. 2017. On the relation between fluvio-deltaic flood basin geomorphology and the wide-spread occurrence of arsenic pollution in shallow aquifers. *Sci. Total Environ.* 574: 901–913.

Mink, A. 2017. Mobile crowd participation to root water supply systems in developing regions. *IWA–Water and Development Congress*, Buenos Aires, Argentina.

Rognoli, V., Bianchini, M., Maffei, S., Karana, E. 2015. DIY Materials. *Materials and Design* 86: 692–702.

Author index